MW01152988

Management, Body Systems, and Case Studies in COVID-19

Management, Body Systems, and Case Studies in COVID-19

Edited by

Rajkumar Rajendram
Department of Medicine, King Abdulaziz Medical City, King Abdullah International Medical Research Center, Ministry of National Guard-Health Affairs, Riyadh, Saudi Arabia; Department of Medical Education, College of Medicine, King Saud bin Abdulaziz University for Health Sciences, Riyadh, Saudi Arabia

Victor R. Preedy
Faculty of Life Sciences and Medicine, King's College London, London, United Kingdom; Department of Clinical Biochemistry, King's College Hospital, London, United Kingdom

Vinood B. Patel
Centre for Nutraceuticals, School of Life Sciences, University of Westminster, London, United Kingdom

Colin R. Martin
Clinical Psychobiology and Applied Psychoneuroimmunology and Institute for Health and Wellbeing, University of Suffolk, Ipswich, United Kingdom

ELSEVIER

ACADEMIC PRESS
An imprint of Elsevier

Academic Press is an imprint of Elsevier
125 London Wall, London EC2Y 5AS, United Kingdom
525 B Street, Suite 1650, San Diego, CA 92101, United States
50 Hampshire Street, 5th Floor, Cambridge, MA 02139, United States

Copyright © 2024 Elsevier Inc. All rights are reserved, including those for text and data mining, AI training, and similar technologies.

Publisher's note: Elsevier takes a neutral position with respect to territorial disputes or jurisdictional claims in its published content, including in maps and institutional affiliations.

No part of this publication may be reproduced or transmitted in any form or by any means, electronic or mechanical, including photocopying, recording, or any information storage and retrieval system, without permission in writing from the publisher. Details on how to seek permission, further information about the Publisher's permissions policies and our arrangements with organizations such as the Copyright Clearance Center and the Copyright Licensing Agency, can be found at our website: www.elsevier.com/permissions.

This book and the individual contributions contained in it are protected under copyright by the Publisher (other than as may be noted herein).

Notices
Knowledge and best practice in this field are constantly changing. As new research and experience broaden our understanding, changes in research methods, professional practices, or medical treatment may become necessary.

Practitioners and researchers must always rely on their own experience and knowledge in evaluating and using any information, methods, compounds, or experiments described herein. In using such information or methods they should be mindful of their own safety and the safety of others, including parties for whom they have a professional responsibility.

To the fullest extent of the law, neither the Publisher nor the authors, contributors, or editors, assume any liability for any injury and/or damage to persons or property as a matter of products liability, negligence or otherwise, or from any use or operation of any methods, products, instructions, web links, or ideas contained in the material herein.

ISBN: 978-0-443-18703-2

For information on all Academic Press publications
visit our website at https://www.elsevier.com/books-and-journals

Publisher: Stacy Masucci
Editorial Project Manager: Timothy Bennett
Production Project Manager: Swapna Srinivasan
Cover Designer: Christian Bilbow

Typeset by STRAIVE, India

Working together
to grow libraries in
developing countries
www.elsevier.com • www.bookaid.org

Dedication

This book is dedicated to my fabulous daughter Dr. Caragh Brien, a caring and compassionate junior doctor of whom I am very proud.

–Colin R. Martin

Contents

Section B
Management

40. Patients with autoimmune liver disease and the impact of Sars-COV-2 infection

Annarosa Floreani, Sara De Martin, and Nora Cazzagon

41. COVID-19 severity and nonalcoholic fatty liver disease

Nina Vrsaljko, Branimir Gjurašin, and Neven Papić

42. Changes in obesity and diabetes severity during the COVID-19 pandemic at Virginia Commonwealth University health system

Asmaa M. Namoos, Vanessa Sheppard, NourEldin Abosamak, Martin Lavallee, Rana Ramadan, Estelle Eyob, Chen Wang, and Tamas S. Gal

43. COVID-19 associated rhino-orbital-cerebral mucormycosis in patients with diabetes and comorbid conditions

Caglar Eker

44. Tissue location of SARS-CoV-2 RNA: A focus on bone and implications for skeletal health

Edoardo Guazzoni, Luigi di Filippo, Alberto Castelli, Andrea Giustina, and Federico Grassi

Section H
Resources

Contributors

Numbers in parenthesis indicate the pages on which the authors' contributions begin.

NourEldin Abosamak (465), Virginia Commonwealth University, Richmond, VA, United States

Ademola Aiyenuro (431), Research4Knowledge, Lagos, Nigeria; University of Cambridge, Cambridge, United Kingdom

Mirza H. Ali (311), Department of Internal Medicine, Lehigh Valley Health Network, Allentown, PA, United States

Hamdan Al-Jahdali (43), Department of Medical Education, College of Medicine, King Saud bin Abdulaziz University for Health Sciences; Division of Pulmonology, Department of Medicine, King Abdulaziz Medical City, King Abdullah International Medical Research Center, Ministry of National Guard-Health Affairs, Riyadh, Saudi Arabia

Sara Allievi (325), Division of Vascular Surgery, Department of Cardiovascular Surgery, Fondazione Poliambulanza, Brescia, Italy

Juliana Amorim dos Santos (557), Laboratory of Oral Histopathology, Health Science Faculty, University of Brasília, Campus Darcy Ribeiro—Asa Norte, Brasília, Brazil

Maria Margarida Andrade (175), Department of Internal Medicine, Hospital de Cascais, Dr. José de Almeida, Alcabideche, Portugal

Saša Anžej Doma (397), University Medical Centre Ljubljana, Hematology Department; Faculty of Medicine, University of Ljubljana, Ljubljana, Slovenia

Salvador Arias-Santiago (3), Dermatology Department, Virgen de las Nieves University Hospital; Instituto de Investigación Biosanitaria ibs; Dermatology Department, Faculty of Medicine, University of Granada, Granada, Spain

Wilbert S. Aronow (365), Cardiology Division, Department of Medicine, Westchester Medical Center and New York Medical College, Valhalla, NY, United States

Şebnem Ataman (111), Division of Rheumatology, Medical School, Ankara University, Ankara, Turkey

Luca Attisani (325), Division of Vascular Surgery, Department of Cardiovascular Surgery, Fondazione Poliambulanza, Brescia, Italy

Eleonora Avenatti (365), Department of Cardiology, Houston Methodist DeBakey Heart & Vascular Center, Houston, TX, United States

Akshat Banga (507), Department of Internal Medicine, Sawai Man Singh Medical College, Jaipur, India

Vikas Bansal (507,523), Division of Pulmonary and Critical Care Medicine, Department of Internal Medicine; Department of Critical Care Medicine, Mayo Clinic, Rochester, MN, United States

Ferran Barbé (285), Translational Research in Respiratory Medicine, University Hospital Arnau de Vilanova and Santa Maria, IRBLleida, Lleida; CIBER of Respiratory Diseases (CIBERES), Institute of Health Carlos III, Madrid, Spain

Giacomo Barberini (77), Department of Biomolecular Sciences, School of Pharmacy, University of Urbino Carlo Bo, Urbino, Italy

Sarah R. Barenbaum (169), Division of Endocrinology, Diabetes and Metabolism, New York-Presbyterian Hospital/Weill Cornell Medical College, Comprehensive Weight Control Center, New York, NY, United States

Erzsébet Bartolák-Suki (269), Department of Biomedical Engineering, Boston University, Boston, MA, United States

Raffaello Bellosta (325), Division of Vascular Surgery, Department of Cardiovascular Surgery, Fondazione Poliambulanza, Brescia, Italy

Bahadır M. Berktaş (561), Department of Pulmonology, University of Health Sciences, Atatürk Sanatorium Training and Research Hospital, Ankara, Turkey

Luca Bertolaccini (211), Department of Thoracic Surgery, IEO, European Institute of Oncology IRCCS, Milan, Italy

Girish Chandra Bhatt (499), Division of Pediatrics Nephrology, Department of Pediatrics, AIIMS Bhopal, Bhopal, India

Christian Booz (245), Division of Experimental Imaging, Department of Diagnostic and Interventional Radiology, University Hospital Frankfurt, Frankfurt am Main, Germany

José Jesús Broseta (423), Department of Nephrology and Renal Transplantation, Hospital Clínic of Barcelona, Barcelona, Spain

Adele S. Budiansky (133), Department of Anesthesiology and Pain Medicine, University of Ottawa, Ottawa, ON, Canada

Luigi Cante (349), Cardiology Unit, Department of Medical Translational Sciences, University of Campania "Luigi Vanvitelli", Naples, Italy

Adriano Caputo (349), Cardiology Unit, Department of Medical Translational Sciences, University of Campania "Luigi Vanvitelli", Naples, Italy

M. Ludovica Carerj (245), Radiology Unit, BIOMORF Department, University Hospital Messina, Messina; Centro Cardiologico Monzino IRCCS, Milan, Italy

Rainier Luiz Carvalho da Silva (557), CMF Odontologia Hospitalar, Rede Santa, Brasília, Brazil

Antonio Cascio (303), Infectious Disease Unit, P. Giaccone University Hospital, Palermo, Italy

Monica Casiraghi (211), Department of Thoracic Surgery, IEO, European Institute of Oncology IRCCS; Department of Oncology and Hemato-Oncology, University of Milan, Milan, Italy

Mario Castano (383), Cardiac Surgery Department, University Hospital of León, León, Spain

Alberto Castelli (485), Department of Orthopaedics and Traumatology, IRCCS Fondazione Policlinico San Matteo, University of Pavia, Pavia, Italy

Laura Castillo (383), Cardiac Surgery Department, University Hospital of León, León, Spain

Nora Cazzagon (443), Department of Surgery, Oncology and Gastroenterology, University of Padova, Padova, Italy

Arthur J. Chang (221), Department of Pediatrics, University of Nebraska Medical Center, Omaha, NE, United States

Carlos Chiesa-Estomba (89), Department of Otorhinolaryngology—Head and Neck Surgery, Hospital Universitario de Donostia, Donosti, Spain

Tumul Chowdhury (133), Department of Anesthesia and Pain Management, Toronto Western Hospital, University of Toronto, Toronto, ON, Canada

Emmanuel Gomes Ciolac (539), Exercise and Chronic Disease Research Laboratory, Department of Physical Education, School of Sciences, São Paulo State University, São Paulo, Brazil

Giovanni Domenico Ciriello (359), Electrophysiology and Cardiac Pacing Unit, Pellegrini Hospital, Naples, Italy

Diogo Cruz (175), Department of Internal Medicine, Hospital de Cascais, Dr. José de Almeida, Alcabideche; Faculty of Medicine, Lisbon University, Lisbon, Portugal

Elena Cuadrado-Payán (423), Department of Nephrology and Renal Transplantation, Hospital Clínic of Barcelona, Barcelona, Spain

Antonello D'Andrea (349), Cardiology Unit, Umberto I Hospital, University of Campania "Luigi Vanvitelli", Nocera Inferiore, Italy

Tommaso D'Angelo (245), Radiology Unit, BIOMORF Department, University Hospital Messina, Messina, Italy; Department of Radiology and Nuclear Medicine, Erasmus MC, Rotterdam, The Netherlands

David de Gonzalo-Calvo (285), Translational Research in Respiratory Medicine, University Hospital Arnau de Vilanova and Santa Maria, IRBLleida, Lleida; CIBER of Respiratory Diseases (CIBERES), Institute of Health Carlos III, Madrid, Spain

Sara De Martin (443), Department of Pharmaceutical and Pharmacological Sciences, University of Padova, Padova, Italy

Miriam De Michele (303), Provincial Health Authority of Palermo, Palermo, Italy

Carmen Del Giudice (349), Cardiology Unit, Department of Medical Translational Sciences, University of Campania "Luigi Vanvitelli", Naples, Italy

Luigi di Filippo (485), Institute of Endocrine and Metabolic Sciences, Università Vita-Salute San Raffaele, IRCCS Ospedale San Raffaele, Milan, Italy

Erica Diani (259), Department of Diagnostics and Public Health, Division of Microbiology and Virology, University of Verona, Verona, Italy

Precious Chisom Dimo (431), Department of Medical Laboratory Sciences, University of Nigeria, Nsukka, Nigeria

Vanessa Teixeira do Amaral (539), Exercise and Chronic Disease Research Laboratory, Department of Physical Education, School of Sciences, São Paulo State University, São Paulo, Brazil

Valentino Ducceschi (359), Electrophysiology and Cardiac Pacing Unit, Pellegrini Hospital, Naples, Italy

Caglar Eker (473), Department of Otolaryngology, Faculty of Medicine, Balcali Hospital, University of Cukurova, Saricam, Adana, Turkey

Emmanuel Ebuka Elebesunu (431), Department of Medical Laboratory Sciences, University of Nigeria, Nsukka, Nigeria

Ángel Estella (285), Department of Medicine, Intensive Care Unit University Hospital of Jerez, University of Cádiz, INIBiCA, Cádiz, Spain

Estelle Eyob (465), Department of Biology & Public Health, William & Mary, Sadler Center, Williamsburg, VA, United States

Malachy Ekene Ezema (431), Faculty of Pharmaceutical Sciences, University of Nigeria, Nsukka, Nigeria

Dario Fabiani (349), Cardiology Unit, Department of Medical Translational Sciences, University of Campania "Luigi Vanvitelli", Naples, Italy

Marco Fabiani (27), Department of Human Genetics, Altamedica, Rome, Italy

Emmanuel J. Favaloro (201), Haematology, Institute of Clinical Pathology and Medical Research (ICPMR), Sydney Centres for Thrombosis and Haemostasis, NSW Health Pathology, Westmead Hospital, Westmead; School of Dentistry and Medical Sciences, Faculty of Science and Health, Charles Sturt University, Wagga; School of Medical Sciences, Faculty of Medicine and Health, University of Sydney, Westmead Hospital, Westmead, NSW, Australia

Annarosa Floreani (443), Scientific Institute for Research, Hospitalization and Healthcare, Verona; Department of Surgery, Oncology and Gastroenterology, University of Padova, Padova, Italy

Larissa Moraes dos Santos Fonseca (15), SENAI Institute for Advanced Health Systems, SENAI CIMATEC University Center; Gonçalo Moniz Institute (IGM) Oswaldo Cruz Foundation (Fiocruz), Rua Waldemar Falcão, Salvador, BA, Brazil

Tamas S. Gal (465), Department of Research Informatics, Wright Center for Clinical and Translational Research and Massey Cancer Center, Virginia Commonwealth University, Richmond, VA, United States

Francesco Giallauria (349), Department of Translational Medical Sciences, University of Naples "Federico II", Naples, Italy

Davide Gibellini (259), Department of Diagnostics and Public Health, Division of Microbiology and Virology, University of Verona, Verona, Italy

Claudio Giorlandino (27), Department of Human Genetics, Altamedica, Rome, Italy

Andrea Giustina (485), Institute of Endocrine and Metabolic Sciences, Università Vita-Salute San Raffaele, IRCCS Ospedale San Raffaele, Milan, Italy

Branimir Gjurašin (457), Department for Intensive Care and Neuroinfections, University Hospital for Infectious Diseases Zagreb, Zagreb, Croatia

Silvia Gómez (285), Translational Research in Respiratory Medicine, University Hospital Arnau de Vilanova and Santa Maria, IRBLleida, Lleida; CIBER of Respiratory Diseases (CIBERES), Institute of Health Carlos III, Madrid, Spain

Federico Grassi (485), Department of Orthopaedics and Traumatology, IRCCS Fondazione Policlinico San Matteo, University of Pavia, Pavia, Italy

Licia Grazzi (123), Headache Center, Neuroalgology Department, IRCCS Foundation "Carlo Besta" Neurological Institute, Milan, Italy

Javier Gualis (383), Cardiac Surgery Department, University Hospital of León, León, Spain

Edoardo Guazzoni (485), IRCCS Humanitas Research Hospital, Rozzano, Milan; Fondazione Livio Sciutto Onlus, Campus Savona, Università Degli Studi di Genova, Savona, Italy

Eliete Neves Silva Guerra (557), Laboratory of Oral Histopathology, Health Science Faculty, University of Brasília, Campus Darcy Ribeiro—Asa Norte, Brasília, Brazil

Rahul Gupta (311), Lehigh Valley Heart Institute, Lehigh Valley Health Network, Allentown, PA, United States

Daniel Gyamfi (573), The Doctors Laboratory Ltd, London, United Kingdom

Jesús Herranz-Larrañeta (89), Department of Otorhinolaryngology—Head and Neck Surgery, Complexo Hospitalario Universitario A Coruña, A Coruña, Spain

Nicholas Herzer (411), Department of Biology, Western Kentucky University and Western Kentucky Heart and Lung/Med Center Health Research Foundation, Bowling Green, KY, United States

Alessandro Domingues Heubel (539), Cardiopulmonary Physiotherapy Laboratory, Physiotherapy Department, Federal University of São Carlos, São Paulo, Brazil

Mark D. Hicar (221), Department of Pediatrics, Jacobs School of Medicine and Biomedical Sciences, University at Buffalo, State University of New York; John R. Oishei Children's Hospital, Buffalo, NY, United States

Katharine Valéria Saraiva Hodel (15), SENAI Institute for Advanced Health Systems, SENAI CIMATEC University Center, Salvador, BA, Brazil

Chiara Iaria (303), Infectious Disease Unit, ARNAS Civico-Di Cristina Hospital, Palermo, Italy

Claudia Imburgia (303), Infectious Disease Unit, ARNAS Civico-Di Cristina Hospital, Palermo, Italy

Gowthami Sai Kogilathota Jagirdhar (523), Department of Gastroenterology, Saint Michael's Medical Center, Newark, NJ, United States

Riecha Joshi (507), Department of Pediatrics, Government Medical College, Kota, India

Abdulvahap Kahveci (111), Division of Rheumatology, Medical School, Ankara University, Ankara; Rheumatology Clinic, Kastamonu Training and Research Hospital, Kastamonu, Turkey

Rahul Kashyap (507,523), Department of Research, WellSpan Health, York, PA; Department of Critical Care Medicine, Mayo Clinic, Rochester, MN, United States

Mohammad Ayaz Khan (43), Department of Medical Education, College of Medicine, King Saud bin Abdulaziz University for Health Sciences; Division of Pulmonology, Department of Medicine, King Abdulaziz Medical City, King Abdullah International Medical Research Center, Ministry of National Guard-Health Affairs, Riyadh, Saudi Arabia

Kazuo Kobayashi (147), Committee of Hypertension and Kidney Disease, Kanagawa Physicians Association; Department of Medical Science and Cardiorenal Medicine, Yokohama City University Graduate School of Medicine, Yokohama, Japan

Ramesh Kordi (221), Department of Pediatrics, Jacobs School of Medicine and Biomedical Sciences, University at Buffalo, State University of New York; John R. Oishei Children's Hospital, Buffalo, NY, United States

Anna Lagni (259), Department of Diagnostics and Public Health, Division of Microbiology and Virology, University of Verona, Verona, Italy

Gregorio Laguna (383), Cardiac Surgery Department, University Hospital of León, León, Spain

Ludovica R.M. Lanzafame (245), Radiology Unit, BIOMORF Department, University Hospital Messina, Messina, Italy

Martin Lavallee (465), Department of Biostatistics, Virginia Commonwealth University, Richmond, VA, United States

Alejandro Lazo-Langner (337), Division of Hematology, Department of Medicine and Department of Epidemiology and Biostatistics, Western University, London, ON, Canada

Naiara Tais Leonardi (539), Cardiopulmonary Physiotherapy Laboratory, Physiotherapy Department, Federal University of São Carlos, São Paulo, Brazil

Stephanie Nogueira Linares (539), Cardiopulmonary Physiotherapy Laboratory, Physiotherapy Department, Federal University of São Carlos, São Paulo, Brazil

Giuseppe Lippi (201), Section of Clinical Biochemistry, University of Verona, Verona, Italy

Olga Neva López-García (63), Department of Gastroenterology, Hospital Universitario Clínico San Carlos, Madrid, Spain

András Lorx (269), Department of Anesthesiology and Intensive Therapy, Semmelweis University, Budapest, Hungary

Virginia Lotti (259), Department of Diagnostics and Public Health, Division of Microbiology and Virology, University of Verona, Verona, Italy

Luca Luzzani (325), Division of Vascular Surgery, Department of Cardiovascular Surgery, Fondazione Poliambulanza, Brescia, Italy

Bruna Aparecida Souza Machado (15), SENAI Institute for Advanced Health Systems, SENAI CIMATEC University Center, Salvador, BA, Brazil

Naveed Mahmood (43), Department of Medicine, King Abdulaziz Medical City, King Abdullah International Medical Research Center, Ministry of National Guard-Health Affairs; Department of Medical Education, College of Medicine, King Saud bin Abdulaziz University for Health Sciences, Riyadh, Saudi Arabia

Pasquale Maiorano (383), Cardiac Surgery Department, University Hospital of León, León, Spain

Katia Margiotti (27), Department of Human Genetics, Altamedica, Rome, Italy

Elio Martín-Gutiérrez (383), Cardiac Surgery Department, University Hospital of León, León, Spain

Miguel Mayo-Yáñez (89), Department of Otorhinolaryngology—Head and Neck Surgery, Complexo Hospitalario Universitario A Coruña, A Coruña, Spain

Silvio Mazziotti (245), Radiology Unit, BIOMORF Department, University Hospital Messina, Messina, Italy

Doug McElroy (411), Department of Biology, Western Kentucky University and Western Kentucky Heart and Lung/Med Center Health Research Foundation, Bowling Green, KY, United States

Priyal Mehta (523), Department of Medicine, M.W. Desai Hospital, Mumbai, Maharashtra, India

Renata Gonçalves Mendes (539), Cardiopulmonary Physiotherapy Laboratory, Physiotherapy Department, Federal University of São Carlos, São Paulo, Brazil

Tazim Merchant (99), Feinberg School of Medicine, Northwestern University, Chicago, IL, United States

Alvaro Mesoraca (27), Department of Human Genetics, Altamedica, Rome, Italy

Antonino Micari (245), Radiology Unit, BIOMORF Department, University Hospital Messina, Messina, Italy

Shehab Mohamed (211), Department of Thoracic Surgery, IEO, European Institute of Oncology IRCCS, Milan, Italy

Alejandro Molina-Leyva (3), Dermatology Department, Virgen de las Nieves University Hospital; Instituto de Investigación Biosanitaria ibs, Granada, Spain

Marta Molinero (285), Translational Research in Respiratory Medicine, University Hospital Arnau de Vilanova and Santa Maria, IRBLleida, Lleida; CIBER of Respiratory Diseases (CIBERES), Institute of Health Carlos III, Madrid, Spain

Francesca Monaco (27), Department of Human Genetics, Altamedica, Rome, Italy

Trinidad Montero-Vilchez (3), Dermatology Department, Virgen de las Nieves University Hospital; Instituto de Investigación Biosanitaria ibs, Granada, Spain

Danilo Antonio Montisano (123), Headache Center, Neuroalgology Department, IRCCS Foundation "Carlo Besta" Neurological Institute, Milan, Italy

Guillermo Muniz-Albaiceta (383), Intensive Care Medicine Department, University Central Hospital of Asturias, Oviedo, Spain

Hiroshi Nakase (493), Department of Gastroenterology and Hepatology, School of Medicine, Sapporo Medical University, Chuo-ku Sapporo, Hokkaido, Japan

Asmaa M. Namoos (465), Department of Social and Behavioral Sciences, School of Medicine, Virginia Commonwealth University, Richmond, VA, United States

Bharat Narasimhan (365), Department of Cardiology, Houston Methodist DeBakey Heart & Vascular Center, Houston, TX, United States

Chrystie Nguyen (411), Lake Cumberland Regional Hospital, Somerset, KY, United States

Marta Afonso Nogueira (175), Department of Cardiology, Hospital de Cascais, Dr. José de Almeida, Alcabideche, Portugal

Samuel Ogunsola (431), Department of Physiology and Pathophysiology, Max Rady College of Medicine, Rady Faculty of Health Science, University of Manitoba, Winnipeg, MB, Canada

Francesco Onorato (303), Infectious Disease Unit, ARNAS Civico-Di Cristina Hospital, Palermo, Italy

Chijioke Obiwe Onyeani (431), Department of Medical Laboratory Sciences, University of Nigeria, Nsukka, Nigeria; Department of Microbiology and Physiological Systems, University of Massachusetts Chan Medical School, Worcester, MA, United States

Stefano Palermi (349), Public Health Department, University of Naples "Federico II", Naples, Italy

Neven Papić (457), Department of Infectious Diseases, School of Medicine, University of Zagreb, Zagreb, Croatia

Pablo Parente-Arias (89), Department of Otorhinolaryngology—Head and Neck Surgery, Complexo Hospitalario Universitario A Coruña, A Coruña, Spain

Leonardo Pasalic (201), Haematology, Institute of Clinical Pathology and Medical Research (ICPMR), Sydney Centres for Thrombosis and Haemostasis, NSW Health Pathology, Westmead Hospital, Westmead; School of Dentistry and Medical Sciences, Faculty of Science and Health, Charles Sturt University, Wagga; Westmead Clinical School, Faculty of Medicine and Health, University of Sydney, Westmead, NSW, Australia

Vinood B. Patel (573), Centre for Nutraceuticals, School of Life Sciences, University of Westminster, London, United Kingdom

Matteo Alberto Pegorer (325), Division of Vascular Surgery, Department of Cardiovascular Surgery, Fondazione Poliambulanza, Brescia, Italy

Manola Peverini (77), Hospital Pharmacy of Urbino, ASUR Marche AV1, Urbino, Italy

Giuseppe Pipitone (303), Infectious Disease Unit, ARNAS Civico-Di Cristina Hospital, Palermo, Italy

Mateo Porres-Aguilar (337), Department of Internal Medicine, Divisions of Hospital and Adult Clinical Thrombosis Medicine, Texas Tech University Health Sciences Center and Paul L. Foster School of Medicine, El Paso, TX, United States

Victor R. Preedy (573), Faculty of Life Sciences and Medicine, King's College London; Department of Clinical Biochemistry, King's College Hospital, London, United Kingdom

Md Moshiur Rahman (567), Neurosurgery Department, Holy Family Red Crescent Medical College, Dhaka, Bangladesh

Naveed Rahman (311), Department of Internal Medicine, Lehigh Valley Health Network, Allentown, PA, United States

Wesley Rajaleelan (133), Department of Anesthesiology and Pain Medicine, University of Ottawa, Ottawa, ON, Canada

Rajkumar Rajendram (43,53,573), Department of Medicine, King Abdulaziz Medical City, King Abdullah International Medical Research Center, Ministry of National Guard-Health Affairs; Department of Medical Education, College of Medicine, King Saud bin Abdulaziz University for Health Sciences, Riyadh, Saudi Arabia

Rana Ramadan (465), Virginia Commonwealth University, Richmond, VA, United States

Gianluca Rigatelli (295), Department of Cardiology, Schiavonia General Hospital, Padova, Italy

Paul Rizzoli (123), John Graham Headache Center, Faulkner & Brigham Hospital, Harvard Medical School, Boston, MA, United States

Vinícius Pinto Costa Rocha (15), SENAI Institute for Advanced Health Systems, SENAI CIMATEC University Center, Salvador, BA, Brazil

Diana Rodríguez-Espinosa (423), Department of Nephrology and Renal Transplantation, Hospital Clínic of Barcelona, Barcelona, Spain

Carlos Rodríguez-Muñoz (285), Translational Research in Respiratory Medicine, University Hospital Arnau de Vilanova and Santa Maria, IRBLleida, Lleida; CIBER of Respiratory Diseases (CIBERES), Institute of Health Carlos III, Madrid, Spain

Noel Roig-Marín (237), Independent Researcher, Alicante, Spain

Vincenzo Russo (349), Cardiology Unit, Department of Medical Translational Sciences, University of Campania "Luigi Vanvitelli", Naples, Italy

Francesco Sabatella (349), Cardiology Unit, Department of Medical Translational Sciences, University of Campania "Luigi Vanvitelli", Naples, Italy

Victor Sagredo (383), Intensive Care Medicine Department, University Hospital of Salamanca, Salamanca, Spain

Massimo Sartelli (303), Department of Surgery, Macerata Hospital, Macerata, Italy

Aanchal Sawhney (311), Department of Internal Medicine, Crozer Chester Medical Center, Upland, PA, United States

Giuseppe Seghieri (377), Epidemiology Unit, Regional Health Agency of Tuscany; Department of Experimental and Clinical Biomedical Sciences, University of Florence, Florence, Italy

Aayush Shah (365), Department of Cardiology, Houston Methodist DeBakey Heart & Vascular Center, Houston, TX, United States

Tanya Sharma (499), Department of Pathology & Laboratory Medicine, AIIMS Bhopal, Bhopal, India

Mark Sheldon (99), Department of Philosophy, Weinberg College of Arts and Sciences, Medical Humanities and Bioethics Program, Feinberg School of Medicine, Northwestern University, Chicago, IL, United States

Vanessa Sheppard (465), Department of Health Behavior and Policy, Massey Cancer Center, School of Medicine, Virginia Commonwealth University, Richmond, VA, United States

Alpana P. Shukla (169), Division of Endocrinology, Diabetes and Metabolism, New York-Presbyterian Hospital/Weill Cornell Medical College, Comprehensive Weight Control Center, New York, NY, United States

Aniruddha Singh (411), Western Kentucky Heart and Lung/Med Center Health Research Foundation and Tower Health Medical Group Cardiology-West Reading, West Reading, PA, United States

Harminder Singh (157), Department of Neurosurgery, Stanford University School of Medicine, Stanford, CA, United States

Milena Botelho Pereira Soares (15), SENAI Institute for Advanced Health Systems, SENAI CIMATEC University Center; Gonçalo Moniz Institute (IGM) Oswaldo Cruz Foundation (Fiocruz), Rua Waldemar Falcão, Salvador, BA, Brazil

Alberto Soto-Moreno (3), Dermatology Department, Virgen de las Nieves University Hospital, Granada, Spain

Lorenzo Spaggiari (211), Department of Thoracic Surgery, IEO, European Institute of Oncology IRCCS; Department of Oncology and Hemato-Oncology, University of Milan, Milan, Italy

Aarushi Sudan (507), Department of Internal Medicine, Jacobi Medical Center, The Bronx, NY, United States

Béla Suki (269), Department of Biomedical Engineering, Boston University, Boston, MA, United States

Mohamad B. Taha (365), Department of Cardiology, Houston Methodist DeBakey Heart & Vascular Center, Houston, TX, United States

Kouichi Tamura (147), Department of Medical Science and Cardiorenal Medicine, Yokohama City University Graduate School of Medicine, Yokohama, Japan

Daniele Ugo Tari (187), Department of Breast Imaging, District 12, Caserta Local Health Authority, Caserta, Italy

Carlos Taxonera (63), Head Inflammatory Bowel Disease Unit, Department of Gastroenterology, Hospital Universitario Clínico San Carlos, Madrid, Spain

Clara-Amanda Ureña-Paniego (3), Dermatology Department, Virgen de las Nieves University Hospital, Granada, Spain

Nina Vrsaljko (457), Department of Infectious Diseases, University Hospital for Infectious Diseases Zagreb, Zagreb, Croatia

Apurva Vyas (311), Lehigh Valley Heart Institute, Lehigh Valley Health Network, Allentown, PA, United States

Chen Wang (465), Department of Biostatistics, Virginia Commonwealth University, Richmond, VA, United States

Brittany Woods (523), Department of Hospital Medicine, Advent Health Redmond, Rome, GA, United States

Yogendra Singh Yadav (499), Department of Pediatrics, AIIMS Bhopal, Bhopal, India

Yoshihiro Yokoyama (493), Department of Gastroenterology and Hepatology, School of Medicine, Sapporo Medical University, Chuo-ku Sapporo, Hokkaido, Japan

Fletcher G. Young (411), Lake Cumberland Regional Hospital, Somerset, KY, United States

Michael Zhang (157), Department of Neurosurgery, Stanford University School of Medicine, Stanford, CA, United States

James Zhou (157), California Northstate University College of Medicine, Elk Grove, CA, United States

Marco Zuin (295), Department of Translational Medicine, University of Ferrara, Ferrara, Italy

Preface

In late 2019, a virulent strain of coronavirus emerged. This was given the name Severe Acute Respiratory Coronavirus 2 (SARS-CoV-2). The disease resulting from symptomatic infection was termed coronavirus disease 2019 (COVID-19) shortly thereafter. In March 2020, the World Health Organization (WHO) declared COVID-19 a pandemic.

The trajectory was terrifying. Only 3 months later, there were 15 million confirmed cases, and more than half a million deaths associated with this disease worldwide. Millions more will have had unconfirmed infection with SARS-CoV-2. By the end of the pandemic and a possible second wave, it was predicted that there would be as many as half a million deaths in the United States alone.

Most individuals infected with the virus have mild, if any, symptoms, and eventually seem to recover to full health. A small proportion, however, die mainly as a consequence of respiratory failure. However, the long-term consequences of COVID-19, whether mild or severe, remain unknown.

The initial focus was on the respiratory effects of COVID-19. Yet as time progressed, it became clear that COVID-19 is a multisystem disease that affects a variety of organs, tissues, and pathways. These include the neurological, haematological, hepatobiliary, gastrointestinal, and cardiovascular systems. It is now being argued that virtually every tissue system is affected either directly or indirectly. To ensure the best outcomes, all medical specialities from diagnostics to nutrition must be involved in the management of patients with COVID-19.

The impact on the wider society and resulting culture shock have been monumental. All elective healthcare services were initially stopped. When these were restarted, healthcare providers had to adapt to the highly contagious nature of SARS-CoV-2. There are regular updates on procedural guidelines for general practitioners, medical doctors, and healthcare specialists. Advances in telemedicine have accompanied the progression and severity of the COVID-19 pandemic.

The impact of enforced changes in behavior on the mental health of the populace has been eye opening. This has evolved our understanding of how people respond to instructions, containment procedures, and governmental protocols. In recent months, there have also been developments in animal models investigating COVID-19 and the use of *in vitro* systems has advanced the understanding of molecular mechanisms.

The drive to discover or repurpose pharmacological and nonpharmacological treatments is all consuming. However, the media has predominantly focused on vaccine development. It is clear that a variety of disciplines are engaged in understanding and documenting the nature of the COVID-19 pandemic.

Much of the information on COVID-19 is scattered throughout the literature with little regard for the intellectual or educational expertise of the readership. To address this, we have compiled **Thematic Approaches to COVID-19.**

Currently, there are four volumes:

- **Management, Body Systems, and Case Studies in COVID-19**
- *Features, Transmission, Detection, and Case Studies in COVID-19*
- *International and Life Course Aspects of COVID-19*
- *Linking Neuroscience and Behavior in COVID-19*

In **Management, Body Systems, and Case Studies in COVID-19,** we have more than 50 chapters in the following subsections:

- **Section A: Introductory chapters and setting the scene**
- **Section B: Management**
- **Section C: Guidelines for different treatment clinics**
- **Section D: Impact on the respiratory system**
- **Section E: Effects on cardiovascular and hematological systems**

- **Section F: Effects on body systems**
- **Section G: Case studies with mini review**
- **Section H: Resources**

To foster understanding and engagement across different disciplines, we include the following features in each chapter:

An Abstract *that is published online.*
A Mini-dictionary of terms *to aid the novice and to bridge the transdisciplinary and intellectual divides.*
Policy and procedure *recommendations or suggestions for research, strategies, guidelines, prevention, and treatment.*
Applications to other areas *highlight the translational natures of the chapter. These could include points that are relevant to additional fields of study related to COVID-19 or other coronaviruses and their impact on the body and community.*
Summary points *encapsulating the whole chapter in a series of single sentences.*

Management, Body Systems, and Case Studies in COVID-19 is designed for research and teaching purposes. It is suitable for virologists, health scientists, public health workers, doctors, nurses, pharmacologists, and research scientists. It is valuable as a personal reference book and also for academic libraries that cover the domains of virology, health, and medical sciences. Contributions are from leading national and international experts, including those from world-renowned institutions. It is suitable for undergraduates, postgraduates, lecturers, and academic professors.

The Editors

Section A

Introductory chapters and setting the scene

Chapter 1

Hand hygiene strategies

Trinidad Montero-Vilchez[a,b], Clara-Amanda Ureña-Paniego[a], Alberto Soto-Moreno[a], Alejandro Molina-Leyva[a,b], and Salvador Arias-Santiago[a,b,c]
[a]*Dermatology Department, Virgen de las Nieves University Hospital, Granada, Spain,* [b]*Instituto de Investigación Biosanitaria ibs, Granada, Spain,* [c]*Dermatology Department, Faculty of Medicine, University of Granada, Granada, Spain*

Abbreviations

ABHS	alcohol-based hand sanitizer
ACD	allergic contact dermatitis
CDC	Centers for Disease Control and Prevention
HIV	human immunodeficiency virus
HSV	herpes simplex virus
ICD	irritant contact dermatitis
TEWL	transepidermal water loss
WHO	World Health Organization

Introduction

Hand hygiene is the most important strategy to prevent infectious diseases (Moore et al., 2021; World Health, 2009). There are several hand hygiene products, but they can be easily classified iton two different groups: soaps and hand sanitizers. Many studies have been conducted to assess their effectiveness at eliminating bacteria, viruses, and fungi; their impact on skin function; and their tolerability by users with controversial results (Montero-Vilchez et al., 2021). Moreover, after the COVID-19 outbreak, the interest in hand hygiene strategies and the frequency of hand hygiene have greatly increased (Moore et al., 2021). So, the objective of this chapter is to review the different preparations used for hand hygiene; compare them in terms of effectiveness, skin impairment, and tolerability; and evaluate the impact of COVID-19 on hand hygiene.

Soaps

Types of soaps

Plain (nonantimicrobial) soap or classic soaps are detergent-based products containing esterified fatty acids and oils (of animal or vegetable origin) and sodium, potassium, or magnesium hydroxide. These fatty acid salts have the ability to form mycelia in aqueous media, removing the lipid content of the stratum corneum, and causing dirt and various organic substances from the hands to adhere to them thanks to their surfactant properties (Castanedo-Cázares et al., 2020). The antimicrobial activity of soaps is based on their ability to dissolve the lipid membrane of bacteria, thereby inactivating them. In any case, the antimicrobial capacity of plain soaps is low and most of their mechanism of action is based on the mechanical dragging of dirt and the superficial lipid component of the skin (Jing et al., 2020). Soap classification is shown in Table 1.

Classification by chemical synthesis: Soaps and syndets

Soaps were the only cleansing agent for years, but a new generation of cleansers, the so-called synthetic detergents or syndets, has been developed during the last decades (Mijaljica et al., 2022). Chemically, soap is obtained by the saponification of a fatty acid of animal or vegetable origin with a salt such as sodium hydroxide (caustic soda/leach), potassium hydroxide (potash), or magnesium hydroxide. Syndets are chemically synthesized from fats, petroleum/petrochemicals or oil-based products (oleochemicals), and alkali by a combination of chemical processes other than saponification. A syndet generally comprises a mixture of synthetic compounds such as fatty acid isothionates or esters of sulfosuccinic acid (Mijaljica et al., 2022). Syndets are often the superior choice for all skin types, as the processes used to create them allow

Management, Body Systems, and Case Studies in COVID-19. https://doi.org/10.1016/B978-0-443-18703-2.00017-7
Copyright © 2024 Elsevier Inc. All rights are reserved, including those for text and data mining, AI training, and similar technologies.

TABLE 1 Soap classification.

Type of soap	
Classification by chemical synthesis	Soaps and syndets
Classification by type of surfactant	Nonionic, anionic, cationic, and amphoteric
Classification by commercial presentation	Soap, tissue, leaf, and liquid

their pH to be adjusted to be similar to that of the skin, resulting in minimal disruption to the skin barrier and native microflora (Draelos, 2018). Soaps easily extract endogenous skin components, such as proteins and lipids, during cleansing and remain on the skin surface after rinsing. The development of soaps with new surfactants could improve the safety of liquid soaps for hand washing, reducing the safety difference with respect to other hand sanitizers (Klimaszewska et al., 2022).

Classification by type of surfactant

Surfactants are the main constituents of most soap and syndet formulations, as they are responsible for their cleansing action and antimicrobial activity. The different types of surfactants are used individually or, more commonly, in combination with each other to integrate suitable properties. Surfactants are classified into four main types according to the charge present on the hydrophilic head group: nonionic, anionic, cationic, and amphoteric (Mijaljica et al., 2022).

- Nonionic surfactants: They have no electrical charge on their hydrophilic head. Their scrubbing ability and foam characteristics are rather weak. Therefore, nonionic surfactants usually have the lowest skin irritation potential among the different types of surfactants. Examples of nonionic surfactants are fatty alcohols and alcohol ethoxylates, and the polyoxyethylene family (Corazza et al., 2010).
- Anionic surfactants: They are the most common type of surfactant used in both soaps and syndets. Members of this group have the highest cleaning ability while they are generally considered to be potent skin irritants. Anionic surfactants can be subdivided according to the chemistry of their polar group into carboxylates, sulfates, sulfonates, phosphate esters, and isethionates (Mijaljica et al., 2022).
- Cationic surfactants: This group of surfactants has higher antimicrobial activity than anionic surfactants. Cationic surfactants have a similar or higher irritation potential than anionic surfactants, and lower cleaning capacity. The most common cationic surfactants are amine salts and quaternary ammonium salts such as benzalkonium chloride. Benzalkonium chloride has demonstrated antimicrobial activity against fungi as well as gram-positive and lipophilic viruses (Corazza et al., 2010).
- Amphoteric/zwitterionic surfactants: They are less aggressive to the skin than anionic surfactants. Amphoteric surfactants are often used in combination with anionic or cationic surfactants to improve mildness. They demonstrate good cleansing ability and foam characteristics, moderate antimicrobial activity, and good mild to moderate antimicrobial activity. Commonly used amphoteric surfactants include cocamidopropyl betaine, lauryl betaine, sodium cocoamphoacetate, and disodium cocoamphodiacetate (Mijaljica et al., 2022).

The surfactants used in the manufacture of soaps and syndets can be classified according to their cleaning capacity, from highest to lowest capacity, as follows: anionic > cationic > amphoteric > nonionic. Similarly, they can be classified according to their antimicrobial capacity as follows, from highest to lowest capacity: cationic > anionic = amphoteric > nonionic (Corazza et al., 2010).

Classification by commercial presentation

There are several forms available on the market including bar soap as well as tissue, leaf, and liquid preparations. Apparently, the commercial presentation is oriented to consumer preferences, but the truth is that there are differences in the chemical composition of the product according to its presentation (Draelos, 2018).

Liquid soaps are numerous on the market and are most often used in hand hygiene. From a chemical point of view, they are not soaps (salts of higher fatty acids formed in the saponification reaction with hydroxides), but are usually liquid solutions of surfactants with various additives. Surfactants of the anionic, nonionic, and amphoteric groups are present in the composition of soaps (Klimaszewska et al., 2022).

The presentation can also provide guidance on classical soap production: the choice between caustic soda and potash will influence the nature of the resulting soap at room temperature. Caustic soda soaps are solids (known as "white soaps") while potash soaps are liquids (known as "black soaps") (Mijaljica et al., 2022).

Effectiveness of soap in hand hygiene

The Centers for Disease Control and Prevention (CDC) recommend hand washing with soap and water whenever possible, as it significantly reduces the amount of bacteria and dirt on the surface of the skin (Jing et al., 2020). This efficacy depends on the time of action, for example, washing hands with soap and water for 15 seconds reduces bacterial counts on the skin by 0.6–1.1 log10 while washing for 30 seconds reduces counts by 1.8–2.8 log10 (World Health, 2009).

In other studies, hand washing with ordinary soap failed to eliminate pathogens, and was even associated with a paradoxical increase in the number of bacteria on the skin (Bottone et al., 2004). However, there is sufficient evidence to consider that the actual risk of transmission of microorganisms through hand washing with soap is negligible (World Health, 2009).

Hand washing with water and nonantibacterial soap is more effective in removing bacteria than hand washing with water alone (Burton et al., 2011). In a trial with healthy volunteers, hand washing with water alone reduced the presence of fecal bacteria to 23% ($P < .001$) while hand washing with plain soap and water reduced the presence of bacteria to 8% ($P < .001$) (Burton et al., 2011). Increasing the number of hand washes with soap and water has shown contradictory results, with one randomized clinical trial finding no significant differences in reducing the number of colony-forming units (Trinidad Montero-Vilchez et al., 2022). The progressive accumulation of organisms on hands after repeated episodes of contamination could explain these findings (Sickbert-Bennett et al., 2005).

Furthermore, soaps are widely used in decontamination protocols for the wide variety of chemical contaminants; however, their usefulness is a matter of debate. In a systematic review involving three in vivo animal models and 11 different contaminants, soap and water decontamination solutions resulted in complete decontamination in 6.9% ($n = 8/116$) of protocols, and incomplete decontamination in 92.2% ($n = 107/116$) of protocols (Burli et al., 2021). The ineffectiveness as a decontaminant of water and soap had already been suggested in several in vitro animal models (Green et al., 2022). The decontamination efficacy of soap is also questioned in human models. A systematic review that included seven studies in in vivo human models in which soap and water were applied showed a partial decontamination of 92.9%, and a minimal or no effect in 7.1% (Kashetsky et al., 2022). Although water only, or soap and water, is considered a gold standard for skin decontamination, most studies investigated suggest that water only, and soap and water, provided incomplete decontamination (Kashetsky et al., 2022).

Adverse skin manifestations related to soaps

Soaps have the ability to alter and potentially damage bilayer lipids by extracting endogenous cholesterol, fatty acids, and ceramides or by intercalating them into the lipid bilayer. Dissolution of the lipid barrier has numerous clinical consequences for the skin, including dryness (Navare et al., 2019), increased transepidermal water loss (TEWL) (Khosrowpour et al., 2019), and irritant and allergic contact dermatitis (Boyce, 2019).

In a prospective randomized trial where irritation and dryness were evaluated by self-assessment and visual assessment scores, skin irritation and dryness increased significantly when participants washed their hands with the plain soap (Boyce, 2019). This alteration of the skin barrier has been shown to correspond with increased cytotoxicity in vitro; higher TEWL and erythema numbers were observed when applying the soaps that showed the highest toxicity in keratinocyte cultures on the skin of healthy participants (Castanedo-Cázares et al., 2020).

An alkalinization of the skin surface has been described in several studies. This alkalinization can be explained by the presence of anionic surfactants in classic soaps (Khosrowpour et al., 2022). The increase in pH maintained over time can compromise the control of microflora by altering the function of the acid mantle of the skin and can cause direct clinical irritation (Mijaljica et al., 2022).

The higher number of adverse effects described for soaps may compromise adherence to hand hygiene practices. Syndets are a less harmful alternative to classic soaps as they have a pH closer to the physiological pH of the skin and use mild surfactants in their production. Unfortunately, even syndets with favorable mildness can potentially remove essential skin constituents, compromise the integrity and functionality of the skin, and inevitably lead to some weakening of the skin barrier, sensitization, and irritation. On the other hand, alcohol-based hand sanitizer (ABHSs) are tolerated better than soap, may be more secure and accepted, and may lead to improved hand-hygiene practices (Khosrowpour et al., 2019, 2022).

Hand sanitizers

Types of hand sanitizers

Hand sanitizer can generally be classified into two groups: ABHS or alcohol-free hand sanitizer (Jing et al., 2020); see Table 2. The most frequent alcohols used in ABHS are ethanol, isopropyl alcohol, n-propanol, or a combination of these. Alcohol-free preparations usually include other antiseptics different from alcohol such as chlorhexidine, chloroxylenol, iodine/iodophors, quaternary ammonium compounds, or triclosan (Jing et al., 2020). Hand sanitizers also contains humectants that prevent skin dehydration, and excipients that help stabilize the product and increase its biocidal activity by prolonging the time needed for alcohol evaporation (Golin et al., 2020). In this chapter, we will focus on ABHSs, which are more frequently used for hand hygiene.

Effectiveness of hand sanitizers in hand hygiene

ABHSs reduce microbes effectively and quickly, covering a broad germicidal spectrum without the need for water drying with towels. Their antimicrobial activity is due to lipid membrane dissolution and protein denaturation (Haft et al., 2014). They have broad bactericidal activity against most vegetative bacteria–those undergoing metabolism and binary fission, fungi, and enveloped viruses–including human immunodeficiency virus (HIV) and herpes simplex virus (HSV). However, they have a short-lived antimicrobial effect and weak activity against protozoa, some nonenveloped (nonlipophilic) viruses, and bacterial spores. It has been proposed that the addition of hydrogen peroxide (3%) may be a solution to the lack of antibacterial spore effect (Golin et al., 2020).

Bacteria on hands can be categorized as resident, which colonize deep layers of the skin that are resistant to mechanical removal, and transient floras, which colonize the superficial layers of skin. Resident floras include Staphylococcus aureus, Staphylococcus epidermidis, and Enterococcus faecalis; transient floras comprise S. aureus, Escherichia coli, and Pseudomonas aeruginosa (Di Muzio et al., 2015). Moreover, there are other bacterial strains that can be transmitted to the host from other sources and can develop into several bacterial infections. Alcohol is quickly effective at destroying many pathogens, mainly against gram-positive and gram-negative bacteria. So, ABHSs have excellent in vitro antimicrobial activity, including against multidrug-resistant pathogens such as methicillin-resistant S. aureus or vancomycin-resistant Enterococcus (Jain et al., 2016). Furthermore, ABHSs also have antimicrobial activity in eliminating bacteria from contaminated hands, which has been demonstrated in several studies in vivo (Golin et al., 2020). Comparing different types of alcohol, N-propanol is the most frequently used alcohol compound and the most effective one, followed by isopropanolol (Golin et al., 2020). It is important to mention that some bacteria could develop tolerance mechanisms to alcohol.

Several studies have evaluated the effectiveness of ABHSs on viruses, although they are more difficult to directly study in vivo compared to bacteria. They are effective against bovine viral diarrhea virus, hepatitis C virus, Zika virus, murine norovirus, and coronaviruses as shown with effective inactivation in quantitative suspension tests (Castanedo-Cázares et al., 2020). Moreover, ABHSs could decrease the likelihood of nosocomial and community transmission of SARS CoV-239; in vitro studies confirmed that ABHSs decrease the viral load and inactivate the virus below the limit of detection. They mainly act against the viral envelope, derived from host lipid envelopes, that contains and protects the genetic material and the genetic material itself. Sterillium, a formulation containing isopropanol as the main ingredient, also completely inactivated enveloped enteric and respiratory viruses, such as H1N1 influenza A virus, but failed to inactivate nonenveloped viruses, except rotavirus (Tuladhar et al., 2015). Ethanol has a greater virucidal activity than propanol. In fact, a high concentration of ethanol is highly effective against enveloped viruses and the majority of clinically relevant viruses. Higher alcohol concentrations and extended contact times could even reach satisfactory activity against nonenveloped viruses, and adding acids to ethanol solutions can increase their efficacy against resistant viruses (Kampf, 2018). Despite all these facts, most ABHSs are still ineffective against nonenveloped viruses.

TABLE 2 Hand sanitizer classification.

Type of hand sanitizer	Type of alcohol
Alcohol-based hand sanitizer	Ethanol, isopropyl alcohol, n-propanol
Alcohol-free	Chlorhexidine, chloroxylenol, iodine/iodophors, quaternary ammonium compounds, or triclosan

Several formulations have been used to prepare ABHSs, including gel, foam, cream, spray, and wipes (Jing et al., 2020). Gels and foams are more widely accepted compared to liquids due to their fast absorption, soft hand feel, and decreased smell (Greenaway et al., 2018). Nevertheless, there are few studies that compare the efficacy of various sanitizer delivery systems on virucidal efficacy. One study compared the reductions of H1N1 viral counts on finger pads and did not find any difference among foams, gels, and wipes (Larson et al., 2012). Another study comparing complete hand coverage upon the use of equal volumes between gel and foam showed similar results with both formulations (Kampf et al., 2013). Another study comparing gel to wipes showed a similar effectiveness in removing fungi and bacteria, but greater rates of skin impairment and less tolerability when using wipes (Trinidad Montero-Vilchez et al., 2022).

Adverse skin manifestations related to hand sanitizers

Almost half of the users of hand hygiene products could develop skin reactions (Montero-Vilchez et al., 2021). The following conditions were related to a higher rate of adverse skin events: the female sex, working in units with COVID-19 patients, hand washing more than 10 times a day, an alcohol concentration >60%, and using gloves (Montero-Vilchez et al., 2021). ABHSs can damage the skin by denaturation of the stratum corneum proteins, decrease the corneocyte cohesion, alter the intercellular lipids, and reduce the stratum corneum water-binding capacity.

Irritant contact dermatitis (ICD) and allergic contact dermatitis (ACD) are the most frequent skin reactions related to ABHSs (Ale & Maibach, 2014). The symptoms of ICD can range from mild to severe dryness, pruritus, erythema, and bleeding. Similar to ACD, the symptoms can either be mild and localized or severe and generalized, with respiratory distress or anaphylactic symptoms in the most severe cases. It is frequently difficult to differentiate between ICD and ACD, with the patch test as the gold standard diagnosis tool, but even a negative test can't completely rule out ACD (Ale & Maibach, 2014). Several factors contribute to an increased rate of ICDs, including lack of use of supplementary emollients, friction due to wearing, and low relative humidity. Comparing the types of alcohol, ethanol has a lower skin-irritant property compared to n-propanol and isopropanol (Erasmus et al., 2010).

The adverse skin events related to ABHSs can be prevented by identifying the trigger and following appropriate measures such as selecting products with a less irritating agent, moisturizing skin after hand sanitation, and avoiding habits that may cause or aggravate skin irritation.

Comparison between soaps and hand sanitizers

There are controversial results comparing effectiveness between soaps and ABHSs. Clinical trials observed high rates of microorganism decontamination using soaps compared to ABHSs (Girou et al., 2002; Zaragoza et al., 1999); see Table 3. As compared to soap, ABHSs do not eliminate all types of germs, including norovirus and Clostridium difficile, the common pathogens that can cause diarrhea. Also, sanitizer may not work well when the hands are grossly dirty or contaminated with harmful chemicals. Nevertheless, high rates of microbiological decontamination were found when using ABHSs compared to soaps in everyday use (Khairnar et al., 2020; Trinidad Montero-Vilchez et al., 2022; Winnefeld et al., 2000), in agreement with in vitro studies (Jain et al., 2016). Soaps decrease the microbial load by removing debris while ABHSs kill microorganism by penetrating though their membrane and inducing cellular lysis (Jing et al., 2020).

As viruses are more difficult to directly study in vivo, there are few studies that compare different types of hand hygiene products regarding virus load reduction. In vitro, both soaps and ABHS are effective at inactivating enveloped viruses (Golin et al., 2020). In an vitro quantitative suspension test comparing three different ethanol-based hand sanitizers and three different antimicrobial soaps, all demonstrated a 4 log10 (>99.99%) reduction in the test enveloped viruses (Steinmann et al., 2012).44. It has been suggested that ABHSs are more effective at eliminating enveloped viruses but less effective for nonenveloped viruses than soaps (Siddharta et al., 2017).

TABLE 3 Comparison between soaps and hand sanitizers.

In favor of soap	In favor of hand sanitizer
ABHS do not eliminate all types of germs, including norovirus and Clostridium difficile	High rates of microbiological decontamination were found when using ABHS compared to soaps in everyday use
The sanitizer may not work well when the hands are grossly dirty or contaminated with harmful chemicals	Lower rates of skin impairment with ABHS, reflected in lower TEWL

Regarding the combination of soap and ABHS, a single-blind randomized controlled trial conducted in 90 participants found no significant differences between the group that used both strategies and an ABHS and the groups that used only one of the two strategies (Khairnar et al., 2020). In addition, it should be noted that excessive duration of hand hygiene may compromise adherence (Sickbert-Bennett et al., 2005). Furthermore, not all hand sanitizers can be combined with soap because chlorhexidine is cationic, so it is advisable to avoid using products containing chlorhexidine with natural soaps and hand creams containing anionic surfactants as they may cause inactivation or precipitation of chlorhexidine, thus reducing its efficacy (Jing et al., 2020). On the other hand, water immersion such as hand washing with soap may increase the disruption of the skin barrier function of other hand sanitizers, such as ABHSs (Trinidad Montero-Vilchez et al., 2022).

Regarding skin barrier impairment, it has been observed that both soaps and alcohol increase TEWL (Khosrowpour et al., 2019; Loden, 2020; Rudolph & Kownatzki, 2004). High TEWL values are related to skin barrier dysfunction. Hand hygiene products remove lipids and damage protein in the stratum corneum attenuating the skin barrier, which is reflected in increased TEWL values (Rundle et al., 2020). Skin impairment increases the epidermal penetration of irritants, microorganisms, and allergens (Rundle et al., 2020). Two studies compared the impact on skin barrier function between soaps and ABHSs. One did not show any differences in TEWL values between soap and ABHSs (Winnefeld et al., 2000). However, the other study reflected high TEWL using a soap compared to ABHSs (Trinidad Montero-Vilchez et al., 2022) and even higher when using wipes instead of a gel (Trinidad Montero-Vilchez et al., 2022). The lower rates of skin impairment when using ABHSs could be because ABHSs contain additional skin care substances, such as glycerin, which could help to replace some of the water that is stripped by the alcohol (Jing et al., 2020). It has also been observed that pH increases using soaps but not ABHSs (Trinidad Montero-Vilchez et al., 2022). The epidermis maintains an acidic pH to provide structural integrity, and a pH increase due to soaps is related to stratum corneum swelling and lipid rigidity and may be a contributing factor for increasing irritation (Ali & Yosipovitch, 2013).

Concerning tolerability and acceptability, both ABHSs and soap are well-rated by the user. Previous studies showed that ABHSs are well accepted and tolerated between healthcare workers (Tarka et al., 2019), and during working hours they could be even more time-saving than soaps (Gon et al., 2020).

Impact of COVID-19 on hand hygiene

Since the 2020 COVID-19 pandemic, hand hygiene has received unprecedented attention. Hand hygiene is one of the pillars of COVID-19 prevention, along with wearing face masks and physical distancing. SARS-COV-2 is transmitted person to person through aerosols, droplets, and contact transmission (Meyerowitz et al., 2021). Even though the main route of transmission is through respiratory particles, contact contamination still constitutes a relevant source of SARS-COV-2 infection (Warnes et al., 2015). The infectious virus may stay on environmental surfaces even when air samples are found to be negative for as long as 6 days (Otter et al., 2013). Furthermore, it has been shown that SARS-COV-2 remains active for longer periods of time on skin surfaces than influenza A virus, augmenting its potential for infection (Hirose et al., 2021).

The role of hand hygiene in preventing COVID-19 spread is crucial. Hands can easily be contaminated from environmental manipulation, initiating SARS-COV-2 into the mucous membranes of the nose, eyes, or mouth (Otter et al., 2013). Fortunately, the ethanol in hand sanitizers inactivates SARS-COV-2 within 15 seconds (Hirose et al., 2021). In consequence, the CDC and the World Health Organization (WHO) have encouraged quality and consistent hand hygiene since the start of the pandemic with soap and water or at least 60% alcohol-based hand sanitizers (World Health, 2009).

Changes in hand hygiene products during COVID-19

Along with other hygiene items such as face masks, commercial hand disinfectants were rapidly depleted worldwide during the pandemic, even affecting healthcare centers. In a survey conducted in March 2020 among infection preventionists in the United States, 71.03% of the respondents reported their stock of hand sanitizer was completely depleted or running low. Hand soap was depleted in only 32.92% of the facilities. The impact of the pandemic on hand hygiene product consumption and trends for the public is more difficult to quantify. In the United States, soap sales increased up to 60% while Malaysia recorded an increase of 800% in sales of hand sanitizer in the first half of 2020 compared to the previous year (Lufkin, 2020).

With an online questionnaire, the authors of a Korean study assessed the use patterns and perceptions of bar and liquid soap as well as hand sanitizer before and after COVID-19. There was a 24.4% decrease in the use of bar soap along with a reduction in the perception of its preventive efficacy against COVID-19. Hand sanitizer consumption increased from a baseline of 14.6%–89.8%, and 35.2% more of the respondents carried hand sanitizers with them after COVID-19. Also, the frequency and duration of handwashing increased by 1.43 times and 1.57 min, respectively (Choi et al., 2021). In another European survey, 95% and 67.8% of participants declared that they washed their hands more frequently with soap and hand sanitizer, respectively, while 49.6% reported carrying hand sanitizer when going out. Interestingly, when stratified by countries, a compliance gradient could be noted in which countries with higher SARS-COV-2 incidence and more rigid lockdowns would report higher rates of implementation of preventive measures (Meier et al., 2020).

Social awareness

Social awareness on hand hygiene would be expected to massively increase during and after COVID-19, but the results on this are contradictory. In several self-reported questionnaires conducted during the pandemic, the general public reported a high degree of hand hygiene compliance and knowledge. Respondents who were female and older with a higher education and income level were more sensitized toward this issue. A high level of concern about COVID-19 among participants and personal experience with the disease (having acquaintances who had tested positive or been hospitalized with SARS-COV-2) would boost hand washing adherence (Bazaid et al., 2020). It also seems that these behaviors are associated with the COVID-19 pandemic trajectory. Szczuka et al. (2021) described in a study conducted in 14 countries that hand washing adherence was related to punctual spikes in the morbidity and mortality of COVID-19, rather than the accumulated number of deaths (Szczuka et al., 2021).

The awareness of populations that are less compliant regarding hand hygiene should prompt targeted interventions in an attempt to improve real COVID-19 risk perception and the role hand hygiene (or lack thereof) can have on transmission. It is important to note that low-income respondents were less likely than their counterparts to use hand sanitizer. Thus, an increase in hand hygiene in this population could be done through the arrangement of contactless dispensers with adequate signage in public spaces such as gas stations, banks, libraries, or markets (Lunn et al., 2020).

The attitudes and behaviors of healthcare workers toward public health measures have been in the spotlight during the pandemic due to their role in the transmissibility and prevention of SARS-COV-2. Hand hygiene compliance of 100% in healthcare settings is possible under adequate circumstances (Wong et al., 2020). Despite this, the fact that hand hygiene compliance does not happen in a vacuum should not be dismissed. A significant decrease in hand washing has been reported during high workload periods, especially with the management of complex patients. Physicians were described as having less hand hygiene compliance than other healthcare workers such as nurses. Interventions directed toward reasonable staffing during higher workload periods and monitoring hand hygiene performance with direct observation could be of use (Chang et al., 2022). Unfortunately, there is evidence suggesting that the long-term effect of hand washing programs fade after time even during a pandemic, which implies that continuous improvement initiatives are pivotal for consistent change (Stangerup et al., 2021).

For both the general population and healthcare workers, misinformation seems to be a strong predictor of low adherence to hand washing and other preventive measures (Meier et al., 2020). Thus, educational interventions directed at combatting misinformation are key to improving compliance on hand hygiene.

COVID-19 has resulted in a paradigm shift in hand hygiene, in which hand washing has become a means of saving lives and reducing contagions. Such momentum should be taken advantage of as an invaluable opportunity to increase hand hygiene awareness and adherence.

Skin manifestations

Cutaneous manifestations of COVID-19 were described relatively late during the pandemic and may have been overlooked for this reason. Aside from the skin presentations directly related to SARS-COV-2 infection, behavioral changes such as frequent hand washing and continuous use of gloves have taken a toll on skin health among healthcare workers and the general population (Darlenski & Tsankov, 2020). The risk factors for greater skin damage are atopic diathesis, frequency of hand washing, wet work, longer work shifts, and glove usage (Christopher et al., 2022).

Friction, water, surfactants, and disinfectants involved in the hand hygiene process can injure the epidermal barrier (Robinson et al., 2010). As a byproduct that is exacerbated by glove use, maceration and erosions can develop on the hands, resulting in many cases of contact dermatitis (Darlenski & Tsankov, 2020). In a survey conducted among Chinese healthcare workers, skin damage on the hands was reported by 74.5% of the participants, who had symptoms of dryness, tenderness, itching, and burning (Lan et al., 2020). Besides, cutaneous adverse effects can be aggravated by chronic psychosocial stressors, as could be the case for the COVID-19 epidemic (Garcovich et al., 2020).

Contact hand dermatitis during hand hygiene intensification in the pandemic can lead to irritant contact dermatitis (ICD) and allergic contact dermatitis (ACD). ICD represents the most frequent occupational dermatosis, especially among healthcare workers, with a prevalence of 30%, which has probably increased after COVID-19 (Jakasa et al., 2018). ICD is induced by the disruption of the skin barrier caused by irritants such as hot water, iodophors, soaps, detergents, alcohols, and other additives involved in hand washing. Detergents are the most noxious agents for skin integrity due to the severe damage they inflict on the lipid layer of the stratum corneum. On the other side, alcohol-based sanitizers seem to be safer for the skin (Trinidad Montero-Vilchez et al., 2022). ACD ensues when an individual gets sensitized to a specific allergen, eliciting a local cutaneous inflammatory response after successive exposures. Preservatives, surfactants, and fragrances contained in soaps, detergents, antiseptics, and hand sanitizers have the potential of producing ACD. ACD presents in an acute (with erythema oedema and vesicle formation), subacute (crusts and scaling), or chronic (lichenification) pattern. Patch testing should be performed in all individuals suspected of having ACD (Rundle et al., 2020).

Preventive measures to minimize hand dermatitis include reducing risk factors and prioritizing alcohol-based sanitizers over soap. When this isn't possible, people should use of pH neutral soap with lukewarm water as well as apply skin emollients and barrier creams frequently, especially after hand washing. Avoiding allergenic surfactants, preservatives, fragrances, or dyes should also be part of this strategy. Curative treatment includes short courses of topical corticosteroid or calcineurin inhibitors, especially for the most severe cases. Additionally, increasing awareness and providing education on proper hand hygiene are paramount for reducing the perils of inadequate hand washing.

The importance of the prevention of skin disorders within this context lies in hand hygiene compliance. The pain and discomfort of using disinfectants and detergents might prevent users from correct hand hygiene. Despite how effortless the preceding measures should be, compliance to skin hydration seems low. Reducing the incidences of cutaneous adverse effects will likely improve hand washing compliance. In a study carried out in Indonesia in healthcare workers, despite the significant impact of hand hygiene measures on the quality of life of the respondents, only 38.1% made use of moisturizers to prevent cutaneous adverse effects (Christopher et al., 2022).

Dermatologists should lead along other healthcare professionals in giving informed advice to patients on hand hygiene, its means, optimal product use, and possible cutaneous skin effects.

Policies and procedures

In this chapter, we reviewed different hand hygiene strategies. Proper hand hygiene is one of the most important measures to prevent the transmission of infectious diseases. We mainly compared two different hand strategies: hand sanitizers and soaps. The WHO recommends washing hands with soap and water whenever possible because handwashing reduces the amounts of all types of germs and chemicals on the hands. If soap and water are not available, using a hand sanitizer is recommended. We reviewed their effectiveness at eliminating different types of microorganisms from the skin, including SARS-COv-2. We also evaluated the influence of these products on skin properties and the frequency of side events related to their use. Moreover, after the COVID-19 outbreak, the interest in hand hygiene strategies and the frequency of hand hygiene have greatly increased, so we also evaluated in this chapter how COVID-19 has changed hand hygiene habits.

Applications to other areas

This chapter has a translational nature. Hand hygiene is important in the entire hospital environment, as secondary infections are the leading cause of death in hospital admissions. Moreover, performing proper hand washing has become particularly relevant during the outbreak of the COVID-19 pandemic. Hand hygiene thus spans medical and social disciplines. Inculcating in the population the different hand hygiene products available, when to use each one, which ones combat which microorganisms, the possible side effects associated with hand hygiene, and what measures can be taken to improve hand care are of particular relevance in our society.

Mini-dictionary of terms

- Alcohol-based hand sanitizer. An alcohol-containing preparation (gel, liquid foam) is applied on hands to inactivate microorganisms and/or temporarily suppress their growth. It should contain one or more types of alcohol, other active ingredients with excipients, and humectants.
- Hand cleansing. Action of performing hand hygiene to physically or mechanically remove dirt, organic material, and/or microorganisms.
- Hand disinfection. This term can refer to antiseptic handwash, antiseptic hand rubbing, hand antisepsis/decontamination/degerming, handwashing with an antimicrobial soap and water, hygienic hand antisepsis, or a hygienic hand rub.

- Handwashing. Washing hands with plain or antimicrobial soap and water.
- Soap, detergent, or surfactant. A compound that has a cleaning action. It is composed of a hydrophilic and a lipophilic part and can be divided into four groups: anionic, cationic, amphoteric, and nonionic.

Summary points

- Proper hand hygiene is one of the most important measures to prevent the transmission of infectious diseases, including COVID-19.
- After COVID-19, interest in hand hygiene strategies and the frequency of hand hygiene have greatly increased
- There are two main type of hand hygiene products: soaps and alcohol-based hand sanitizers.
- It is recommended to wash hands with soap and water whenever possible because handwashing reduces the amounts of all types of germs and chemicals on hands.
- If soap and water are not available, using a hand sanitizer with a final concentration of at least 60% ethanol or 70% isopropyl alcohol is recommended.

References

Ale, I. S., & Maibach, H. I. (2014). Irritant contact dermatitis. *Reviews on Environmental Health*, *29*(3), 195–206. https://doi.org/10.1515/reveh-2014-0060.

Ali, S. M., & Yosipovitch, G. (2013). Skin pH: From basic science to basic skin care. *Acta Dermato-Venereologica*, *93*(3), 261–267. https://doi.org/10.2340/00015555-1531.

Bazaid, A. S., Aldarhami, A., Binsaleh, N. K., Sherwani, S., & Althomali, O. W. (2020). Knowledge and practice of personal protective measures during the COVID-19 pandemic: A cross-sectional study in Saudi Arabia. *PLoS One*, *15*(12), e0243695. https://doi.org/10.1371/journal.pone.0243695.

Bottone, E. J., Cheng, M., & Hymes, S. (2004). Ineffectiveness of handwashing with lotion soap to remove nosocomial bacterial pathogens persisting on fingertips: A major link in their intrahospital spread. *Infection Control and Hospital Epidemiology*, *25*(3), 262–264. https://doi.org/10.1086/502388.

Boyce, J. M. (2019). Best products for skin antisepsis. *American Journal of Infection Control*, *47S*, A17–A22. https://doi.org/10.1016/j.ajic.2019.03.012.

Burli, A., Kashetsky, N., Feschuk, A., Law, R. M., & Maibach, H. I. (2021). Efficacy of soap and water based skin decontamination using in vivo animal models: A systematic review. *Journal of Toxicology and Environmental Health. Part B, Critical Reviews*, *24*(7), 325–336. https://doi.org/10.1080/10937404.2021.1943087.

Burton, M., Cobb, E., Donachie, P., Judah, G., Curtis, V., & Schmidt, W.-P. (2011). The effect of handwashing with water or soap on bacterial contamination of hands. *International Journal of Environmental Research and Public Health*, *8*(1), 97–104. https://doi.org/10.3390/ijerph8010097.

Castanedo-Cázares, J. P., Cortés-García, J. D., Cornejo-Guerrero, M. F., Torres-Álvarez, B., & Hernández-Blanco, D. (2020). Study of the cytotoxic and irritant effects of skin cleansing soaps. *Gaceta Médica de México*, *156*(5), 418–423. https://doi.org/10.24875/GMM.M20000430.

Chang, N.-C. N., Schweizer, M. L., Reisinger, H. S., Jones, M., Chrischilles, E., Chorazy, M., & Herwaldt, L. (2022). The impact of workload on hand hygiene compliance: Is 100% compliance achievable? *Infection Control and Hospital Epidemiology*, *43*(9), 1259–1261. https://doi.org/10.1017/ice.2021.179.

Choi, K., Sim, S., Choi, J., Park, C., Uhm, Y., Lim, E., & Lee, Y. (2021). Changes in handwashing and hygiene product usage patterns in Korea before and after the outbreak of COVID-19. *Environmental Sciences Europe*, *33*(1), 79. https://doi.org/10.1186/s12302-021-00517-8.

Christopher, P. M., Roren, R. S., Tania, C., Jayadi, N. N., & Cucunawangsih, C. (2022). Skin damage and quality of life among healthcare workers providing care during the COVID-19 pandemic: A multicenter survey in Banten Province, Indonesia. *Indian Journal of Dermatology*, *67*(3), 313. https://doi.org/10.4103/ijd.ijd_645_21.

Corazza, M., Lauriola, M. M., Zappaterra, M., Bianchi, A., & Virgili, A. (2010). Surfactants, skin cleansing protagonists. *Journal of the European Academy of Dermatology and Venereology: JEADV*, *24*(1), 1–6. https://doi.org/10.1111/j.1468-3083.2009.03349.x.

Darlenski, R., & Tsankov, N. (2020). COVID-19 pandemic and the skin: What should dermatologists know? *Clinics in Dermatology*, *38*(6), 785–787. https://doi.org/10.1016/j.clindermatol.2020.03.012.

Di Muzio, M., Cammilletti, V., Petrelli, E., & Di Simone, E. (2015). Hand hygiene in preventing nosocomial infections: A nursing research. *Annali di Igiene*, *27*(2), 485–491. https://doi.org/10.7416/ai.2015.2035.

Draelos, Z. D. (2018). The science behind skin care: Cleansers. *Journal of Cosmetic Dermatology*, *17*(1), 8–14. https://doi.org/10.1111/jocd.12469.

Erasmus, V., Daha, T. J., Brug, H., Richardus, J. H., Behrendt, M. D., Vos, M. C., & van Beeck, E. F. (2010). Systematic review of studies on compliance with hand hygiene guidelines in hospital care. *Infection Control and Hospital Epidemiology*, *31*(3), 283–294. https://doi.org/10.1086/650451.

Garcovich, S., Bersani, F. S., Chiricozzi, A., & De Simone, C. (2020). Mass quarantine measures in the time of COVID-19 pandemic: Psychosocial implications for chronic skin conditions and a call for qualitative studies. *Journal of the European Academy of Dermatology and Venereology: JEADV*, *34*(7), e293–e294. https://doi.org/10.1111/jdv.16535.

Girou, E., Loyeau, S., Legrand, P., Oppein, F., & Brun-Buisson, C. (2002). Efficacy of handrubbing with alcohol based solution versus standard handwashing with antiseptic soap: Randomised clinical trial. *BMJ*, *325*(7360), 362. https://doi.org/10.1136/bmj.325.7360.362.

Golin, A. P., Choi, D., & Ghahary, A. (2020). Hand sanitizers: A review of ingredients, mechanisms of action, modes of delivery, and efficacy against coronaviruses. *American Journal of Infection Control*, *48*(9), 1062–1067. https://doi.org/10.1016/j.ajic.2020.06.182.

Gon, G., Virgo, S., de Barra, M., Ali, S. M., Campbell, O. M., Graham, W. J., & de Bruin, M. (2020). Behavioural determinants of hand washing and glove recontamination before aseptic procedures at birth: A time-and-motion study and survey in Zanzibar Labour Wards. *International Journal of Environmental Research and Public Health*, *17*(4). https://doi.org/10.3390/ijerph17041438.

Green, M., Kashetsky, N., Feschuk, A. M., & Maibach, H. I. (2022). Efficacy of soap and water-based skin decontamination using in vitro animal models: A systematic review. *Journal of Applied Toxicology: JAT*, *42*(6), 942–949. https://doi.org/10.1002/jat.4274.

Greenaway, R. E., Ormandy, K., Fellows, C., & Hollowood, T. (2018). Impact of hand sanitizer format (gel/foam/liquid) and dose amount on its sensory properties and acceptability for improving hand hygiene compliance. *The Journal of Hospital Infection*, *100*(2), 195–201. https://doi.org/10.1016/j.jhin.2018.07.011.

Haft, R. J., Keating, D. H., Schwaegler, T., Schwalbach, M. S., Vinokur, J., Tremaine, M., & Landick, R. (2014). Correcting direct effects of ethanol on translation and transcription machinery confers ethanol tolerance in bacteria. *Proceedings of the National Academy of Sciences of the United States of America*, *111*(25), E2576–E2585. https://doi.org/10.1073/pnas.1401853111.

Hirose, R., Ikegaya, H., Naito, Y., Watanabe, N., Yoshida, T., Bandou, R., … Nakaya, T. (2021). Survival of severe acute respiratory syndrome coronavirus 2 (SARS-CoV-2) and influenza virus on human skin: Importance of hand hygiene in Coronavirus Disease 2019 (COVID-19). *Clinical Infectious Diseases: An Official Publication of the Infectious Diseases Society of America*, *73*(11), e4329–e4335. https://doi.org/10.1093/cid/ciaa1517.

Jain, V. M., Karibasappa, G. N., Dodamani, A. S., Prashanth, V. K., & Mali, G. V. (2016). Comparative assessment of antimicrobial efficacy of different hand sanitizers: An in vitro study. *Dental Research Journal (Isfahan)*, *13*(5), 424–431. https://doi.org/10.4103/1735-3327.192283.

Jakasa, I., Thyssen, J. P., & Kezic, S. (2018). The role of skin barrier in occupational contact dermatitis. *Experimental Dermatology*, *27*(8), 909–914. https://doi.org/10.1111/exd.13704.

Jing, J. L. J., Pei Yi, T., Bose, R. J. C., McCarthy, J. R., Tharmalingam, N., & Madheswaran, T. (2020). Hand sanitizers: A review on formulation aspects, adverse effects, and regulations. *International Journal of Environmental Research and Public Health*, *17*(9). https://doi.org/10.3390/ijerph17093326.

Kampf, G. (2018). Efficacy of ethanol against viruses in hand disinfection. *The Journal of Hospital Infection*, *98*(4), 331–338. https://doi.org/10.1016/j.jhin.2017.08.025.

Kampf, G., Ruselack, S., Eggerstedt, S., Nowak, N., & Bashir, M. (2013). Less and less-influence of volume on hand coverage and bactericidal efficacy in hand disinfection. *BMC Infectious Diseases*, *13*, 472. https://doi.org/10.1186/1471-2334-13-472.

Kashetsky, N., Law, R. M., & Maibach, H. I. (2022). Efficacy of water skin decontamination in vivo in humans: A systematic review. *Journal of Applied Toxicology: JAT*, *42*(3), 346–359. https://doi.org/10.1002/jat.4230.

Khairnar, M. R., Anitha, G., Dalvi, T. M., Kalghatgi, S., Datar, U. V., Wadgave, U., … Preet, L. (2020). Comparative efficacy of hand disinfection potential of hand sanitizer and liquid soap among dental students: A randomized controlled trial. *Indian Journal of Critical Care Medicine*, *24*(5), 336–339. https://doi.org/10.5005/jp-journals-10071-23420.

Khosrowpour, Z., Ahmad Nasrollahi, S., Ayatollahi, A., Samadi, A., & Firooz, A. (2019). Effects of four soaps on skin trans-epidermal water loss and erythema index. *Journal of Cosmetic Dermatology*, *18*(3), 857–861. https://doi.org/10.1111/jocd.12758.

Khosrowpour, Z., Ahmad Nasrollahi, S., Samadi, A., Ayatollahi, A., Shamsipour, M., Rajabi-Esterabadi, A., … Firooz, A. (2022). Skin biophysical assessments of four types of soaps by forearm in-use test. *Journal of Cosmetic Dermatology*, *21*(7), 3127–3132. https://doi.org/10.1111/jocd.14589.

Klimaszewska, E., Wieczorek, D., Lewicki, S., Stelmasiak, M., Ogorzałek, M., Szymański, Ł., … Markuszewski, L. (2022). Effect of new surfactants on biological properties of liquid soaps. *Molecules (Basel, Switzerland)*, *27*(17). https://doi.org/10.3390/molecules27175425.

Lan, J., Song, Z., Miao, X., Li, H., Li, Y., Dong, L., … Tao, J. (2020). Skin damage among health care workers managing coronavirus disease-2019. *Journal of the American Academy of Dermatology*, *82*(5), 1215–1216. https://doi.org/10.1016/j.jaad.2020.03.014.

Larson, E. L., Cohen, B., & Baxter, K. A. (2012). Analysis of alcohol-based hand sanitizer delivery systems: Efficacy of foam, gel, and wipes against influenza A (H1N1) virus on hands. *American Journal of Infection Control*, *40*(9), 806–809. https://doi.org/10.1016/j.ajic.2011.10.016.

Loden, M. (2020). Ethanol-based disinfectants containing urea may reduce soap sensitivity. *Dermatitis*, *31*(5), 328–332. https://doi.org/10.1097/DER.0000000000000612.

Lufkin, B. (2020). *Coronavirus: The psychology of panic buying*. Retrieved from https://www.bbc.com/worklife/article/20200304-coronavirus-covid-19-update-why-people-are-stockpiling.

Lunn, P. D., Belton, C. A., Lavin, C., McGowan, F. P., Timmons, S., & Robertson, D. A. (2020). Using behavioral science to help fight the coronavirus. *Journal of Public Administration*, *3*(1). https://doi.org/10.30636/jbpa.31.147.

Meier, K., Glatz, T., Guijt, M. C., Piccininni, M., van der Meulen, M., Atmar, K., … on behalf of the COVID-19 Survey Study group. (2020). Public perspectives on protective measures during the COVID-19 pandemic in the Netherlands, Germany and Italy: A survey study. *PLoS One*, *15*(8), e0236917. https://doi.org/10.1371/journal.pone.0236917.

Meyerowitz, E. A., Richterman, A., Gandhi, R. T., & Sax, P. E. (2021). Transmission of SARS-CoV-2: A review of viral, host, and environmental factors. *Annals of Internal Medicine*, *174*(1), 69–79. https://doi.org/10.7326/M20-5008.

Mijaljica, D., Spada, F., & Harrison, I. P. (2022). Skin cleansing without or with compromise: Soaps and syndets. *Molecules (Basel, Switzerland)*, *27* (6). https://doi.org/10.3390/molecules27062010.

Montero-Vilchez, T., Cuenca-Barrales, C., Martinez-Lopez, A., Molina-Leyva, A., & Arias-Santiago, S. (2021). Skin adverse events related to personal protective equipment: A systematic review and meta-analysis. *Journal of the European Academy of Dermatology and Venereology*, *35*(10), 1994–2006. https://doi.org/10.1111/jdv.17436.

Montero-Vilchez, T., Martinez-Lopez, A., Cuenca-Barrales, C., Quiñones-Vico, M. I., Sierra-Sanchez, A., Molina-Leyva, A., … Arias-Santiago, S. (2022). Assessment of hand hygiene strategies on skin barrier function during COVID-19 pandemic: A randomized clinical trial. *Contact Dermatitis*, *86*(4), 276–285. https://doi.org/10.1111/cod.14034.

Moore, L. D., Robbins, G., Quinn, J., & Arbogast, J. W. (2021). The impact of COVID-19 pandemic on hand hygiene performance in hospitals. *American Journal of Infection Control*, *49*(1), 30–33. https://doi.org/10.1016/j.ajic.2020.08.021.

Navare, B., Thakur, S., & Nakhe, S. (2019). A review on surfactants: Role in skin irritation, SC damage, and effect of mild cleansing over damaged skin. *International Journal of Advance Research, Ideas and Innovations in Technology*, *5*, 1077–1081.

Otter, J. A., Yezli, S., Salkeld, J. A. G., & French, G. L. (2013). Evidence that contaminated surfaces contribute to the transmission of hospital pathogens and an overview of strategies to address contaminated surfaces in hospital settings. *American Journal of Infection Control*, *41*(5), S6–S11. https://doi.org/10.1016/j.ajic.2012.12.004.

Robinson, M., Visscher, M., Laruffa, A., & Wickett, R. (2010). Natural moisturizing factors (NMF) in the stratum corneum (SC). II. Regeneration of NMF over time after soaking. *Journal of Cosmetic Science*, *61*(1), 23–29. Retrieved from http://www.ncbi.nlm.nih.gov/pubmed/20211114.

Rudolph, R., & Kownatzki, E. (2004). Corneometric, sebumetric and TEWL measurements following the cleaning of atopic skin with a urea emulsion versus a detergent cleanser. *Contact Dermatitis*, *50*(6), 354–358. https://doi.org/10.1111/j.0105-1873.2004.00368.x.

Rundle, C. W., Presley, C. L., Militello, M., Barber, C., Powell, D. L., Jacob, S. E., … Dunnick, C. A. (2020). Hand hygiene during COVID-19: Recommendations from the American Contact Dermatitis Society. *Journal of the American Academy of Dermatology*. https://doi.org/10.1016/j.jaad.2020.07.057.

Sickbert-Bennett, E. E., Weber, D. J., Gergen-Teague, M. F., Sobsey, M. D., Samsa, G. P., & Rutala, W. A. (2005). Comparative efficacy of hand hygiene agents in the reduction of bacteria and viruses. *American Journal of Infection Control*, *33*(2), 67–77. https://doi.org/10.1016/j.ajic.2004.08.005.

Siddharta, A., Pfaender, S., Vielle, N. J., Dijkman, R., Friesland, M., Becker, B., & Steinmann, E. (2017). Virucidal activity of World Health Organization—Recommended formulations against enveloped viruses, including zika, ebola, and emerging coronaviruses. *The Journal of Infectious Diseases*, *215*(6), 902–906. https://doi.org/10.1093/infdis/jix046.

Stangerup, M., Hansen, M. B., Hansen, R., Sode, L. P., Hesselbo, B., Kostadinov, K., & Calum, H. (2021). Hand hygiene compliance of healthcare workers before and during the COVID-19 pandemic: A long-term follow-up study. *American Journal of Infection Control*, *49*(9), 1118–1122. https://doi.org/10.1016/j.ajic.2021.06.014.

Steinmann, J., Paulmann, D., Becker, B., Bischoff, B., Steinmann, E., & Steinmann, J. (2012). Comparison of virucidal activity of alcohol-based hand sanitizers versus antimicrobial hand soaps in vitro and in vivo. *The Journal of Hospital Infection*, *82*(4), 277–280. https://doi.org/10.1016/j.jhin.2012.08.005.

Szczuka, Z., Abraham, C., Baban, A., Brooks, S., Cipolletta, S., Danso, E., … Luszczynska, A. (2021). The trajectory of COVID-19 pandemic and handwashing adherence: Findings from 14 countries. *BMC Public Health*, *21*(1), 1791. https://doi.org/10.1186/s12889-021-11822-5.

Tarka, P., Gutkowska, K., & Nitsch-Osuch, A. (2019). Assessment of tolerability and acceptability of an alcohol-based hand rub according to a WHO protocol and using apparatus tests. *Antimicrobial Resistance and Infection Control*, *8*, 191. https://doi.org/10.1186/s13756-019-0646-8.

Tuladhar, E., Hazeleger, W. C., Koopmans, M., Zwietering, M. H., Duizer, E., & Beumer, R. R. (2015). Reducing viral contamination from finger pads: Handwashing is more effective than alcohol-based hand disinfectants. *The Journal of Hospital Infection*, *90*(3), 226–234. https://doi.org/10.1016/j.jhin.2015.02.019.

Warnes, S. L., Little, Z. R., & Keevil, C. W. (2015). Human Coronavirus 229E remains infectious on common touch surface materials. *MBio*, *6*(6), e01697-01615. https://doi.org/10.1128/mBio.01697-15.

Winnefeld, M., Richard, M. A., Drancourt, M., & Grob, J. J. (2000). Skin tolerance and effectiveness of two hand decontamination procedures in everyday hospital use. *The British Journal of Dermatology*, *143*(3), 546–550. https://doi.org/10.1111/j.1365-2133.2000.03708.x.

Wong, S. C., AuYeung, C. H. Y., Lam, G. K. M., Leung, E. Y. L., Chan, V. W. M., Yuen, K. Y., & Cheng, V. C. C. (2020). Is it possible to achieve 100 percent hand hygiene compliance during the coronavirus disease 2019 (COVID-19) pandemic? *The Journal of Hospital Infection*, *105*(4), 779–781. https://doi.org/10.1016/j.jhin.2020.05.016.

World Health, O. (2009). *WHO Guidelines on Hand Hygiene in Health Care: First Global Patient Safety Challenge Clean Care is Safer Care*.

Zaragoza, M., Salles, M., Gomez, J., Bayas, J. M., & Trilla, A. (1999). Handwashing with soap or alcoholic solutions? A randomized clinical trial of its effectiveness. *American Journal of Infection Control*, *27*(3), 258–261. https://doi.org/10.1053/ic.1999.v27.a97622.

Chapter 2

Approved vaccines for the COVID-19 pandemic: Linking in future perspectives

Larissa Moraes dos Santos Fonseca[a,b], Katharine Valéria Saraiva Hodel[a], Vinícius Pinto Costa Rocha[a], Milena Botelho Pereira Soares[a,b], and Bruna Aparecida Souza Machado[a]

[a]*SENAI Institute for Advanced Health Systems, SENAI CIMATEC University Center, Salvador, BA, Brazil,* [b]*Gonçalo Moniz Institute (IGM) Oswaldo Cruz Foundation (Fiocruz), Rua Waldemar Falcão, Salvador, BA, Brazil*

Abbreviations

ACE-2	angiotensin converting enzyme-2
Ad	adenovirus
ADCC	antibody-dependent cellular cytotoxicity
cGAMP	cyclic guanosine monophosphate adenosine monophosphate
ChAdOx1	chimpanzee adenovirus-vectored vaccine from University of Oxford
DAMPs	damage-associated molecular patterns
EUL	emergency use listing
IM	intramuscular route
IRF	interferon regulatory factor
LAV	live attenuated vaccines
MDA5	melanoma differentiation-associated protein 5
MERS	Middle East respiratory syndrome
NALT	nasopharynx-associated lymphoid tissue
NF-kB	nuclear factor-Kb
NLRP	nucleotide-binding oligomerization domain, leucine rich repeat and pyrin domain containing
PEI	polyethylenimine
PAMPs	pathogen-associated molecular pattern
RBD	receptor-binding domain
RIG-I	retinoic-acid-inducible gene I
RNA	ribonucleic acid
S	spike glycoprotein
saRNA	self-amplifying RNA
SARS	severe acute respiratory syndrome
SARS-CoV-2	severe acute respiratory syndrome coronavirus 2
STAT	signal transducer and activator of transcription
STING	stimulator of interferon genes
TRL	toll-like receptors
VLP	virus-like particles
VoCs	variants of concern

Introduction

The coronavirus disease 2019 (COVID-19) pandemic enabled an unprecedented breakthrough in vaccine study and development. The humanitarian and economic impact spurred the evaluation of innovative technological platforms through new paradigms established to accelerate development. This allowed a revolution in the pharmaceutical industry, where vaccine production was affected in record time. Just one year after the discovery of severe acute respiratory syndrome coronavirus-2 (SARS-CoV-2), the causative agent of COVID-19, the Pfizer-BioNTech vaccine (Comirnaty) became the first to receive emergency use authorization from the US Food and Drug Administration (FDA) on Dec. 11, 2020 (FDA, 2021). By

Management, Body Systems, and Case Studies in COVID-19. https://doi.org/10.1016/B978-0-443-18703-2.00014-1
Copyright © 2024 Elsevier Inc. All rights are reserved, including those for text and data mining, AI training, and similar technologies.

November 2022, according to the World Health Organization (WHO), there were nine vaccines in ongoing clinical trials or already approved to prevent COVID-19. These were classified into five categories including inactivated, live attenuated, viral vector, protein subunit, and nucleic acid (DNA and RNA) vaccines (World Health Organization (WHO), 2022b).

Vaccination has shown its efficiency by reducing the number of cases and the infection rate in the population as well as preventing hospitalizations and especially reducing the number of associated deaths worldwide (Chen et al., 2022; Lyngse et al., 2022). Despite their high efficacy and safety, the currently available vaccines against SARS-CoV-2 infection have not yet been able to completely control the pandemic. This is due to the ongoing need to extend vaccination efforts, especially to reach low-income countries (UNSG Office, WHO, UNICEF, 2022). Overall, vaccines classified as RNA, protein subunit, and viral vector vaccines have shown high protection against COVID-19. However, the expectation of continued transmission of the virus cannot be ignored. The high mutation rates of the spike (S) glycoprotein affect the long-lasting protection and effectiveness of the vaccines. Thus, people who have already been immunized can become infected again with new variants of the virus (Harvey et al., 2021). For this reason, the development of alternative strategies capable of inducing long-lasting immune responses, including protection against new variants, and favoring rapid development and large-scale production remain necessary.

In this chapter, we will cover the main vaccines developed for COVID-19, discussing their rapid development and limits associated with new possible approaches and prospects.

Rapid COVID-19 vaccine development

Vaccination is the most effective way to contain the COVID-19 pandemic because it is through vaccines that it is possible not only to prevent SARS-COV-2 infection, or at least severe cases of the disease, but also to reduce human-to-human transmission rates (Watson et al., 2022). Furthermore, it is already well established that the capability of a virus to achieve spread to pandemic levels is decreased by achieving high levels of community (herd) immunity. This can happen through mass vaccination of the population or by recurrent waves of infection until approximately 70% of the people develop immunity (Graham, 2020). In light of the unprecedented health, economic, social, and environmental impacts caused by the pandemic, there was an urgent need for vaccine development. According to the WHO, by November 2022, about 172 vaccines were under clinical development, and 199 were under preclinical development (World Health Organization WHO, 2022a). This has become a priority for political decisions by governments, as well as for the scientific community, the pharmaceutical industry, and drug regulatory agencies. Consequently, this sense of social and political urgency resulted in the rapid availability of COVID-19.

Many steps are involved in making a vaccine available to the population, including not only the development stages, but also clinical trials, regulatory agency authorization, production, and distribution. In general, the development of traditional vaccines takes up to 15 years, starting with a prolonged discovery phase during which vaccines are designed and preclinical exploratory studies are performed. With COVID-19 vaccines, agreements were reached worldwide among these important actors, such as drug regulatory agencies with the criteria for approval of clinical trials in mind. In such a manner, necessary and appropriate requirements were defined to achieve a balance between regulatory considerations and the essential requirements for safe and effective pharmaceuticals. The rapid vaccine development happened due to several factors (Barrett et al., 2022; Li et al., 2020):

- **Technology advances make new vaccines possible.** The concepts elucidated in the fields of immunology, virology, and molecular biology obtained from decades of past research allowed the rapid development of COVID-19 vaccines. Researchers have been studying for years the coronaviruses that cause severe acute respiratory syndrome (SARS) and Middle East respiratory syndrome (MERS), and, with this prior knowledge combined with the rapid availability of the genome sequence of the SARS-CoV-2 (in January 2020), it was possible to accelerate vaccine development for COVID-19 (Li et al., 2020; Lu et al., 2020).
- **Worldwide collaboration in the race to develop safe and effective vaccines for COVID-19.** All over the globe, it was possible to observe an unprecedented collaboration among academia, ethical approval boards, manufacturers, and regulatory agencies in making safe and effective vaccines available to the population (Angelis et al., 2022). Encouraging initiatives such as worldwide data sharing, uncommon partnerships between scientists and industry, and the sharing of knowledge and intellectual property made it possible to accelerate the clinical development of safe and effective COVID-19 vaccines (Druedahl et al., 2021).
- **Funding for COVID-19 vaccine research.** Enormous funding ranging from governments to the private sector allowed the industry and academy to go through the rapid development, manufacturing, and acquisition of COVID-19 vaccines and anti-COVID-19 therapeutics (Ball, 2021). As a result, companies were able to conduct preclinical and clinical development phases in parallel, rather than sequentially, and take on high financial risks. This allowed these companies

to focus on large-scale testing as well as manufacturing vaccine candidates even before knowing if they were going to work (Ball, 2021).

- **Large-scale manufacturing of COVID-19 vaccines.** To be effective in containing the pandemic, the manufacturing of novel vaccine products should occur at an enormous scale and in parallel with clinical development. To this end, companies established production scale-up processes even without having obtained regulatory approval (Barrett et al., 2022). Low-cost processes, large-scale production (including global distribution), and good quality could be established through multisite technology transfer strategies. This accelerated plant installation as process development, validation, and regulatory activities occurred in parallel (Nicholson Price et al., 2020). In addition, the sharing of documents describing manufacturing processes incorporated a strong quality assurance strand and measures to ensure analytical comparability across sites.
- **Rapid enrollment of volunteers in COVID-19 vaccine candidate clinical trials.** In addition to these factors, it is also important to highlight the role of the large number of volunteers who applied to participate in the clinical trials. The pandemic has strengthened the connection between scientists and society, highlighting the significance of clinical trials and elevating scientists to key contributors in the rapid development of vaccines.

Through an extraordinary effort involving multiple areas of research and development in basic, preclinical, and clinical science that had already been underway over the past few years in association with high investment, it has been possible to develop highly effective vaccines at unprecedented speed. It is important to note that while the vaccines for COVID-19 were created quickly, every precaution was taken to ensure their safety and efficacy, with risk and benefit assessment always being considered (Haque & Pant, 2020).

Current and most common COVID-19 Vaccines

Four different classes of COVID-19 vaccines, distinct in origin, composition, and immunogenicity, have received the WHO's emergency use listing (EUL): protein subunit, messenger RNA (mRNA), nonreplicating viral vector, and inactivated vaccines (World Health Organization (WHO), 2022b). The active substance (antigen) in most of these vaccines is the viral spike glycoprotein, either alone or together with other viral proteins or inactivated virus vaccines. Each type of vaccine has a specific structure and mechanism of action linked to its advantages and disadvantages that are related to immunogenicity, safety, efficacy, and usability (Fathizadeh et al., 2021). In this context, the S glycoprotein from vaccines can be presented to the immune system in various ways, and is perceived in two main categories:

- **Protein-based vaccines:** vaccines that comprise protein-based approaches containing the S glycoprotein in various forms and combinations with adjuvants to generate a safe immune response. These are the subunit-based vaccines and inactivated vaccines.
- **Genetic vaccines:** vaccines that contain genetic information for the biosynthesis of the SARS-CoV-2 S glycoprotein in the cells of the immunized organism, with precise performance and in its proper conformation to B cells. These are vaccines based on mRNA technology and nonreplicating adenoviral vector (viral vector) vaccines.

Fig. 1 provides an overview of the various types of COVID-19 vaccines that have received approval. In general, the mechanism of action of most COVID-19 vaccines is to block the virus from entering the host cell by stimulating antibody production to the viral S proteins. Such proteins have several epitopes capable of stimulating CD4 and CD8-T cell responses. In Table 1, we present COVID-19 vaccines granted EUL by the WHO and each primary developer.

1. **Inactivated vaccines**
   Inactivated vaccines are those that use killed whole pathogens, as their virulence factors have been inactivated after exposure to some external agent. For viral vaccines, these molecules are grown in cell culture, especially Vero cells, and then exposed to chemicals (e.g., β-propyl lactone or formaldehyde) or physical agents (e.g., heat) to consolidate their inactivation (Sharpe et al., 2020). Vaccines based on this technology are known to stimulate the humoral and cellular systems by different mechanisms of action. It is noteworthy that within the context of COVID-19 vaccines, the presence of the whole virus induces an immune response against envelope, matrix, and nuclear proteins, going beyond the spike glycoprotein (the target commonly found in vaccines) (Li et al., 2022). Nevertheless, these vaccines have been commonly associated with low immunogenicity, making it necessary to associate them with adjuvant molecules (e.g., CpG oligodeoxynucleotide aluminum hydroxide) or more than one dose in a primary vaccination scheme. Because they are relatively simple to produce and easy to scale up, generating lower production costs, these vaccines were one of the big issues in the fight against the pandemic of COVID-19 (Iversen & Bavari, 2021). This is the case of the inactivated vaccines CoronaVac (formerly called PiCoVacc), Covaxin (codenamed BBV152), and Covilo (also known as BBIBP-CorV), which had Chinese and Indian companies as the main developers.

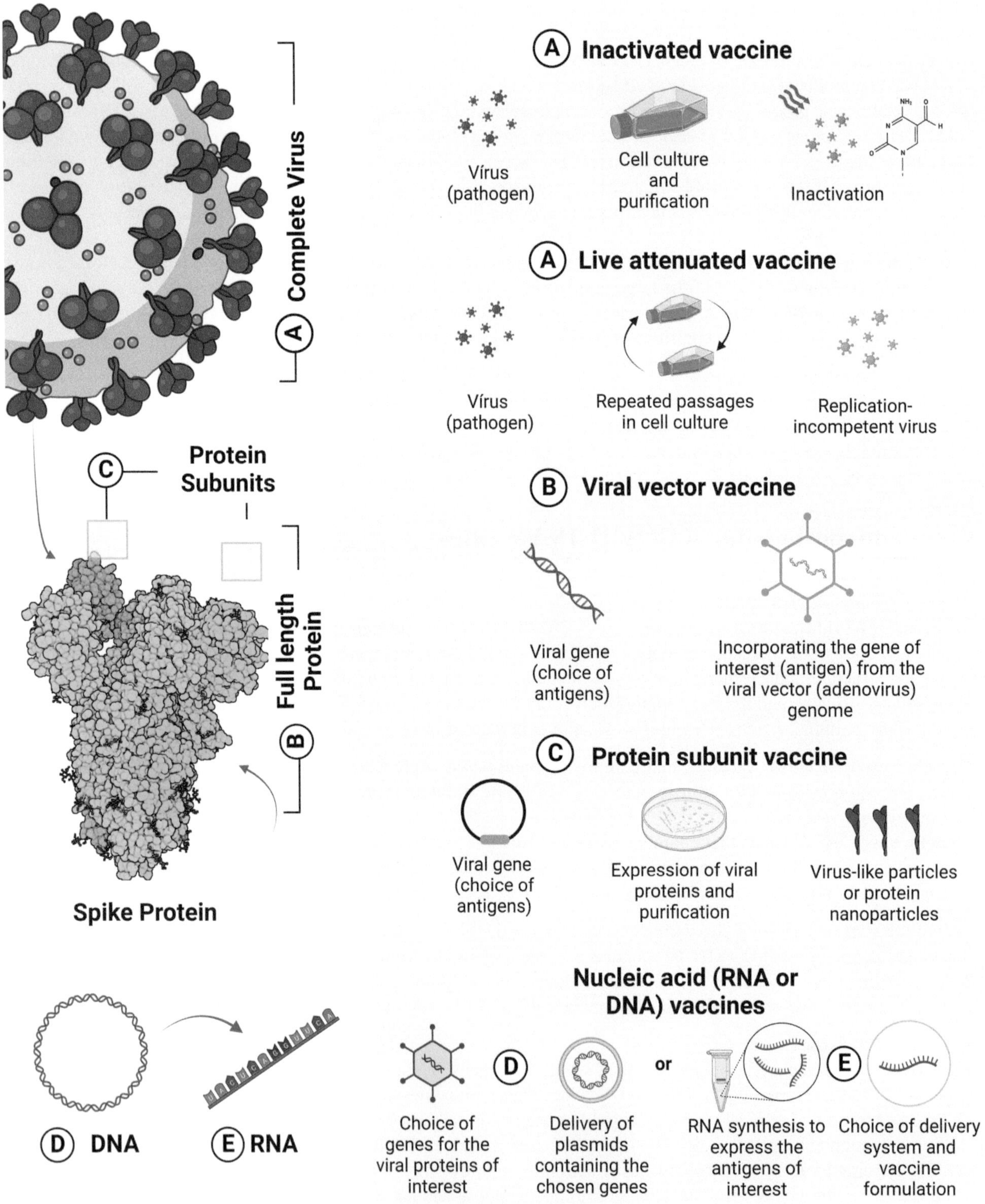

FIG. 1 Technological platforms of the main COVID-19 vaccines, considering the main steps in their production and the respective strategies for antigen selection. Each technological platform is associated with a specific part of the virus: (a) complete virus; (b) full length spike glycoprotein; (c) protein subunit; (d) nucleic acid technology using DNA, and (e) nucleic acid technology using RNA.

TABLE 1 Main approved COVID-19 vaccines.

Vaccine name	Primary developers	Technology	Approval/clinical trials
Comirnaty (also known as Tozinameran or BNT162b2)	Pfizer, BioNTech; Fosun pharma	RNA	Approved in 149 countries 100 Trials in 31 countries
Comirnaty Original/Omicron BA.1	BioNTech, Pfizer	RNA	Approved in 35 countries
Comirnaty Original/Omicron BA.4-5			Approved in 33 countries
Spikevax (also known as mRNA-1273 or Elasomeran or Moderna)	Moderna, NIAID, BARDA	RNA	Approved in 88 countries 70 Trials in 24 countries
Vaxzevria (also known as AZD1222, ChAdOx1 nCoV-19)	The University of Oxford; AstraZeneca; IQVIA; Serum Institute of India; BARDA, OWS	Viral vector	Approved in 149 countries 73 Trials in 34 countries
Covishield (Oxford/ AstraZeneca formulation)	Serum Institute of India	Viral vector	Approved in 49 countries 6 Trials in 1 country
Convidecia (also known as Ad5-nCoV)	CanSino Biologics	Viral vector	Approved in 10 countries 14 Trials in 6 countries
Jcovden (also known as Ad26. COV2.S, Ad26COVS1, JNJ-78436735)	Janessen Vaccines (Johnson & Johnson), BARDA, NIAID, and OWS	Viral vector	Approved in 113 countries 26 Trials in 25 countries
Covovax (also known as Novavax formulation)	Serum Institute of India	Protein subunit	Approved in 6 countries 7 Trials in 3 countries
Novavax (also known as NVX-CoV2373 or Nuvaxovid)	Novavax, CEPI, Serum Institute of India	Protein subunit	Approved in 40 countries 22 trials in 14 countries
CoronaVac (formerly PiCoVacc)	Sinovac	Inactivated	Approved in 56 countries 42 trials in 10 countries
Covilo (also known as BBIBP-CorV (VeroCells))	Beijing Institute of Biological Products; China National Pharmaceutical Group (Sinopharm)	Inactivated	Approved in 93 countries 39 Trials in 18 countries
Covaxin (also known as BBV152)	Bharat Biotech	Inactivated	Approved in 14 countries 16 Trials in 2 countries

Main COVID-19 vaccines granted emergency use listing by the World Health Organization.

2. **Live attenuated vaccines (LAV)**
   The technology of live attenuated vaccines (LAV) can be considered mature, as it has been applied in the development of influenza and polio vaccines. LAVs are characterized by the presence of live viruses that are capable of replication but lack virulence (Sharpe et al., 2020). This lack of virulence is acquired after repeated passages in cell cultures or chicken eggs. When compared to inactivated vaccines, LAVs appear with a more robust immunogenic capacity, generating greater cellular and humoral protection. This property offers the prospect of using LAVs in a single dose because they are able to simulate a viral infection without the onset of illness (Batty et al., 2021). It is important to highlight that this technological platform also has the advantage of possible intranasal administration, stimulating the immune

response system and in that specific region. It is noted that the first generation of COVID-19 vaccines did not exploit the LAV platform as much as other technologies, such as inactivated virus (Khoshnood et al., 2022). However, clinical trials with LAV-based candidates are being conducted; moreover, this technology may be an interesting alternative when it comes to vaccine development for new viral variants and long-lasting immune responses (Tang et al., 2022).

3. **Viral vector vaccines**
   In general, vaccines based on viral vector technology are designed to express one or more specific antigens from the virus' ability to carry relatively large genomes. For this, a modified virus is used that is not the one in which the antigens are associated, becoming a recombinant viral vector, which gives this platform a range of application possibilities. Within this perspective, it is recommended that the antigens chosen for the planning of a vaccine based on a viral vector can satisfactorily induce the humoral and cellular immune response without the need for adjuvants, mimicking natural infections (Chung et al., 2021). In addition, it is important to consider whether the vector is able to replicate (make copies of itself). Similar to vaccines for other pathologies, COVID-19 vaccines based on viral vectors primarily use adenoviruses (Ad) to deliver and express SARS-CoV-2 antigens, particularly the spike glycoprotein. This prominence can be attributed to the adenoviruses' low pathogenicity, high immunogenicity, and the absence of integration with the host genome, ensuring genetic safety (Travieso et al., 2022). Commercial COVID-19 vaccines Convidecia (utilizing replication-incompetent Ad type 5), Jcovden (utilizing replication-incompetent Ad type 26), and Vaxzevria (nonreplicating adenovirus chimpanzee) are based on viral vector technology that has already received WHO emergency use approval, opening up prospects for new vaccine candidates.
4. **Protein subunit vaccines**
   Subunit protein vaccines are made up of small parts of proteins intrinsically capable of inducing a protective immunogenic response. The production of these vaccines is known to consist of in vitro steps, where the direct manipulation of highly pathogenic agents is not required. Subunit vaccines usually elicit a stronger humoral response than a cellular response, necessitating the inclusion of adjuvant molecules in the formulations for an effective protective response (Batty et al., 2021). The Novavax subunit vaccine, also known as NVX-CoV2373 or Nuvaxovid, is included on the WHO's list of recommended COVID-19 vaccines. It consists of a prefusion spike glycoprotein administered with an adjuvant. Of note among the subunit vaccines are those that are virus-like particles (VLP) and those based on protein nanoparticles. Among them, VLPs are highlighted, and can be defined as a subunit vaccine that contains many copies of the antigen, thus mitigating the need for adjuvants (Shin et al., 2020). Due to its versatility in containing different SARS-CoV-2 protein subunits, including envelope, nucleocapsid, and membrane proteins, this platform is expected to be extensively explored for the development of new COVID-19 vaccines.
5. **Nucleic acid (RNA or DNA) vaccines**
   Vaccines based on nucleic acids (DNA and RNA) were undoubtedly a key protagonist within the development of new vaccines for COVID-19. The vaccines Comirnaty (also known as Tozinameran or BNT162b2) and Spikevax (also known as mRNA-1273, Elasomeran, or Moderna) were the first vaccines approved for emergency use for COVID-19 at the end of December 2020 and were also approved by the WHO (Rosa et al., 2021). Both vaccines are based on mRNA technology, which, when delivered to the cell's cytoplasm, triggers the translation and expression of the SARS-CoV-2 spike glycoprotein. The COVID-19 pandemic marked a significant a milestone for another RNA-derived technology, self-amplifying RNA (saRNA), which contains a specific ribonucleotide structure capable of inducing amplification through the presence of replicons. Thus, saRNA vaccines are expected to have a lower concentration than mRNA vaccines, which may be an interesting factor in terms of production costs and safety profile (Machado et al., 2021). Planning vaccines using these platforms must consider both antigen expression, and antigen delivery systems, and the inclusion of adjuvants to ensure a strong immune response against the target pathogen.

DNA vaccines use circular DNA strands to prime the immune system against the SARS-CoV-2 virus, administered through the skin (Mallapaty, 2021). In general, for DNA vaccines to elicit an immune response, the plasmid must be delivered to the cell nucleus, leading to the expression of the target antigen. While not fully understood, the use of this technology has raised questions about the genetic safety of these vaccines due to the possibility of plasmid DNA incorporation into the host genome for the mechanism to take place (Chung et al., 2021). Despite the challenges, the search for new COVID-19 vaccines for has led to the use of new generation vaccine technologies, including DNA and RNA vaccines, which represents a new era for obtaining more effective technologies.

Vaccination strategies to fight against variants of concern

The versatility of mRNA platform for vaccine development and the emergence of variants of concerns (VoCs)

Several SARS-CoV-2 variants have emerged since the COVID-19 pandemic. The RNA-based viral genome contributes to the mutation acquisition, which will result in viral transmissibility gain of function. The most transmissible viral strain will superpose the previous by infecting people, whether they are vaccinated or not. The world's population has experienced this fact with the emergence of the first mutation, D614G, which contributed to increasing the affinity between the viral spike glycoprotein and human ACE-2 (Plante et al., 2021). D614G is kept in all variants of concern, from alpha (B.1.1.7) to omicron (B.1.1.529) and its subvariants (BA.2, BA.2.12.1, BA.4, BA.5). Omicron has the highest number of accumulated mutations, which gives it the highest transmission capability, contributing to its status as the currently dominant circulating variant. However, new subvariants derived from omicron BA5, the BQ.1, BQ.1.1, BA.4.6, BF.7, and BA.2.75.2, have emerged and the number of infected people and hospitalizations has been increasing. Besides the increased transmissibility, the neutralization capability of the neutralizing antibodies elicited by the earlier SARS-CoV-2 variant (B.1)-based vaccine tends to decrease. The newest omicron subvariants share potent antibody resistance even after vaccination (Qu et al., 2022). Based on that, new approaches are pivotal to improve the neutralization of new variants and avoid transmissibility among the population.

The mRNA platform has all requisites to produce vaccines on demand to fight against new SARS-CoV-2 variants of concern as well as other potential pandemics caused by any pathogen. The target mRNA is easily produced by in vitro transcription (IVT), compared to the first- and second-generation vaccine platforms. It is completely cell-free, reducing time and costs for production. Because mRNA can be manipulated, it is possible to produce chimeric sequences harboring the most immunogenic peptides, important for B and T cell activation. Moreover, the safety of the mRNA platform is improved compared to DNA-based vaccines because mRNA translation occurs in the cytoplasm and does not produce sequence integration in the genome (Maruggi et al., 2019). Based on the versatility of the mRNA platform, researchers can respond quickly against new outbreaks or even pandemics.

Strategies for new vaccine development

The use of different mRNA sequences encoding spike glycoproteins from the most relevant variants of SARS-Cov-2 has been expected as new vaccines emerge. It can be a polyvalent or a pan-coronavirus vaccine that will provide high levels of systemic neutralizing antibodies and T-cell response to the current circulating viral variant. The mRNA platform is technically suitable to respond quickly against the emergence of any viral variant and provide a new version of a polyvalent or pan-coronavirus vaccine, based on which variants are circulating. The strategies for new vaccine development may include a seasonally adjusted variant spike or quimeric sequences containing relevant epitopes for B and T cells from spike variants or from the whole viral proteome, including different viruses, not only SARS-CoV-2, as a pan-vaccine (Altmann & Boyton, 2022). Recent clinical trial results clearly demonstrate that the combination of mRNA from spike glycoproteins is a strategy to quickly stimulate the immune response against a circulating variant and that this combination must be changed as variants are emerging to overcome the neutralizing escape caused by the mutations (Chalkias et al., 2022).

The spike glycoproteins have been used for most vaccines in use because this antigen generates immunoprotection. Other SARS-CoV-2 antigens exhibit this profile in addition to being immunoreactive. This is the case of the nucleoprotein, which is highly immunogenic and stimulates B and T cell responses. In animal models, the nucleoprotein stimulates virus clearance from the upper respiratory tract (Jiang et al., 2021). It is important to note that for vaccine development, immunogens capable of eliciting not only neutralizing antibodies but also nonneutralizing Fc-dependent effector functions, including antibody-dependent cellular cytotoxicity (ADCC), play a crucial role in contributing to protection (Zohar et al., 2020). Then other structural (membrane and envelope proteins) and nonstructural proteins may be explored for new vaccine development.

The technologies applied for vaccine development are in constant evolution. An important strategy to find promising sequences for in vivo proof of concept is reverse vaccinology. By applying this strategy, it is possible to have the most important polypeptide sequences for T and B lymphocyte activation.

Alternative routes of vaccine administration

Intranasal vaccine

In the beginning of the SARS-CoV-2 infection, the viruses were restricted to the upper airways until reaching the lungs and alveoli, causing SARS. The nose and upper airways represent the first site of infection by SARS-COV-2 and viral-host interaction. The main virus transmission is through aerosols into the nose or mouth, or via the conjunctiva of the eyes and drainage into the nasal passages through the lacrimal duct (Russell & Mestecky, 2022). Then, immunological elements able to block virus dissemination can control the community spread of the disease and, consequently, the pandemic. To achieve this pattern of blockage, it is necessary to have effective neutralizing antibodies in the upper airways. Despite the success of COVID-19 vaccination worldwide, based on the systemic generation of antibodies, mainly IgG, as well as T-cell mediated cytotoxicity, the intramuscular vaccines are not able to block virus transmission because they do not induce mucosal immunity in the upper airway (Acharya et al., 2022; Riemersma et al., 2022). Circulating antibodies do not reach the upper respiratory airways in biologically relevant amounts, so they are not effective at neutralizing viruses in the mucosa.

All clinically used COVID-19 vaccines are for intramuscular administration and are poor inducers of mucosal immunity. This highlights the importance of intranasal vaccine development to avoid the rapid spread of new SARS-COV-2 variants of concern. Intranasal vaccines present many advantages compared to intramuscular vaccines. Besides the induction of mucosal immunity, vaccination by the intranasal route is easier than the intramuscular route, does not require specialized people for administration, is friendlier to people who don't like needles such as children, and can be used in countries with fewer resources, improving vaccination coverage around the world (Alu et al., 2022). Despite the urgent need to stop viral transmission and the technical efforts, the development of vaccines able to induce mucosal immunity is still low.

The development of intranasal vaccines is quite challenging. Most of the preclinical developments use protein or a protein subunit and vector-based platforms to deliver the immunogen. However, the poor immunogenicity and stability require the use of adjuvants and stabilizing formulations as well as delivery systems to overcome the mucosal barrier and stimulate the antigen-specific response. Moreover, most antigens have low affinity to the epithelium and tend to be cleared easily by mucociliary removal (Alu et al., 2022). In this context, there are several strategies that can improve the immunogenicity of intranasal vaccines. The adjuvants can include bioproducts such as toxoid adjuvants and agonists of pattern recognition receptors as well as chemical and natural adjuvants (Bernasconi et al., 2021). Agonists of pattern recognition receptors can stimulate the innate immune response by targeting TRLs (TRL agonists), which constitute the major class of nasal adjuvants. TRL3, TRL9, and TRL5 can be activated by double-stranded viral RNA, synthetic bacterial DNA, and flagellin, respectively.

In summary, the development of intranasal vaccines is important to fight the COVID-19 pandemic in association with the intramuscular vaccines in use. Even the neutralizing antibody titre produced by intramuscular vaccines decreases after around 4–5 months. Intranasal and intramuscular routes of administration can act in synergy to promote high levels of neutralizing antibodies and T cell response systematically and at the mucosal site. The emerging omicron subvariants reinforce the importance in blocking the viral spread from upper airways, and this can be only achieved by intranasal vaccination.

Other mucosal routes of vaccination

The oral cavity is the most widely used and patient-friendly route of administration. However, the ingestion of some medicines and vaccines must be avoided because of the acidic and high proteolytical stomach microenvironment. The oral cavity can still be used to administer a medicine or vaccine exploring its mucosal region. In this case, sublingual and buccal routes can be applied to vaccination. These systemic routes can efficiently induce mucosal and systemic immunization due to their nonkeratinized and stratified epithelium. Because of a more elastic and permeable tissue, drug absorption is improved at this site, bypassing the enterohepatic circulation and thus the resulting first-pass effect of the hepatic metabolism (Trincado et al., 2021). Sublingual and buccal may also be promising routes to administer COVID-19 vaccines because of their ability to induce systemic and mucosal immunity. However, even though these routes seem to be promising for vaccine development, there are few studies regarding such COVID-19 vaccine development and no vaccines are in clinical trials.

Policies

In this chapter, we reviewed the impact of the rapid development and availability of new technology platforms in combating the COVID-19 pandemic caused by SARS-CoV-2. To this end, a detailed review of vaccine technologies and their mechanism of action was conducted, including the advantages and limitations of each. It is noted that the COVID-19 pandemic presented a particularity in the line of new vaccine development, as regulatory agencies in different countries allowed clinical phases to be conducted concurrently, resulting in recommendation for emergency use in less than 12 months when COVID-19 was first reported. This race was also characterized by the investment of different developers in conventional and new generation technological platforms, where vaccines based on mRNA technology were the major protagonists for their innovative character and the induction of a robust immune response against SARS-CoV-2. Despite this seemingly controlled context, the emergence of new viral variants with greater transmissibility and lethality when compared to the original strain has caused concern among the scientific and civil communities. Given the uncertainty about the efficacy of the vaccines already approved against new variants of SARS-CoV-2, it is essential that discussions be encouraged about updating the viral targets used. Additionally, the development of COVID-19 vaccines that can be administered by new routes is considered a prospect that can strongly contribute to the adherence of the population to mass immunization campaigns in the face of the expectation of less-invasive interventions, such as inhaled or oral vaccines. Thus, it is essential that institutions responsible for promoting advances in public health become aware of the need to foster scientific advances in this sector, especially in the use of new-generation platforms that can directly contribute to the epidemiological control of the disease.

Applications to other areas

SARS-CoV-2 is a coronavirus that emerged and spread in a very rapid and unprecedented way, leading to millions of deaths worldwide. For this reason, the clinical success of the vaccines was crucial in tackling the COVID-19 pandemic. This was accomplished thanks to their rapid development with the subsequent development steps done in parallel. Nevertheless, we see a wide range of applications for the technological platforms developed, especially RNA vaccines, not only as an immunizer against infectious agents, but also for use in immunotherapies and genetic reprogramming, among others. Although we have significant gaps in the pandemic to fill, many lessons have been learned in terms of health systems, drug availability, laboratory testing, research and development capacity, and global governance. This will need to be incorporated into various fields of knowledge, including psychology, economics, and environmental issues. The pandemic has highlighted the feasibility of accelerated vaccine development and the benefits of a swift, globally coordinated response. In addition, nonconventional technologies, such as intranasal vaccines, have garnered attention and investment, potentially serving as viable alternatives for other diseases. Notably, many of the COVID-19 vaccines developed or in development are adaptable technologies that can form the foundation for vaccines against SARS-CoV-2 and its variants, as well as other public health threats. Despite the challenges, these broad vaccine technology approaches will be best employed in conjunction with active surveillance for emerging variants or novel pathogens.

Mini-dictionary of terms

- RNA vaccine: Technology platform that uses RNA molecules to produce an immune response.
- Humoral immune response: Immune response based on the production and action of specific antibodies against an antigen.
- Cellular immune response: Immune response based on the production of specific cells (such as T cells) against a given antigen. Cellular immune responses play an important role against intracellular agents.
- Variant of concern: Variants in which increased transmissibility, disease severity, ability to neutralize antibodies generated after infection or previous vaccination, or failure of detection by diagnostic tests are evidenced.
- Spike glycoprotein: A glycoprotein that consists of two subunits and is present in a transmembrane form in SARS-CoV-2. It is also the main antigen used by COVID-19 vaccines.

Summary points

- Health crisis associated with technological knowledge, funding to private sectors, and large-scale production capacity allowed the rapid development of safe and effective COVID-19 vaccines.
- Common COVID-19 vaccines include those containing genetic information and protein-based vaccines, each eliciting distinct immune responses.
- Pfizer and Moderna's vaccines were the first RNA-based vaccines to receive approval from the WHO and globally.
- Despite the rising in SARS-CoV-2 variants and cases, RNA-based vaccines platforms offer the versatility needed to address emerging variants of concern.
- The intranasal vaccine for COVID-19 offer an efficient mechanism of action by stimulating a mucosal immune response in the upper airways.

References

Acharya, C. B., Schrom, J., Mitchell, A. M., Coil, D. A., Marquez, C., Rojas, S., Wang, C. Y., Liu, J., Pilarowski, G., Solis, L., Georgian, E., Belafsky, S., Petersen, M., Derisi, J., Michelmore, R., & Havlir, D. (2022). Viral load among vaccinated and unvaccinated, asymptomatic and symptomatic persons infected with the SARS-CoV-2 delta variant. *Open Forum Infectious Diseases*, *9*. https://doi.org/10.1093/OFID/OFAC135.

Altmann, D. M., & Boyton, R. J. (2022). COVID-19 vaccination: The road ahead. *Science*, *375*, 1127–1132. https://doi.org/10.1126/SCIENCE.ABN1755.

Alu, A., Chen, L., Lei, H., Wei, Y., Tian, X., & Wei, X. (2022). Intranasal COVID-19 vaccines: From bench to bed. *eBioMedicine*, *76*. https://doi.org/10.1016/J.EBIOM.2022.103841.

Angelis, A., Suarez Alonso, C., Kyriopoulos, I., & Mossialos, E. (2022). Funding sources of therapeutic and vaccine clinical trials for COVID-19 vs non–COVID-19 indications, 2020-2021. *JAMA Network Open*, *5*. https://doi.org/10.1001/JAMANETWORKOPEN.2022.26892. e2226892.

Ball, P. (2021). The lightning-fast quest for COVID vaccines—and what it means for other diseases. *Nature*, *589*, 16–18. https://doi.org/10.1038/D41586-020-03626-1.

Barrett, A. D. T., Titball, R. W., MacAry, P. A., Rupp, R. E., von Messling, V., Walker, D. H., & Fanget, N. V. J. (2022). The rapid progress in COVID vaccine development and implementation. *npj Vaccines*, *7*, 1–2. https://doi.org/10.1038/s41541-022-00442-8.

Batty, C. J., Heise, M. T., Bachelder, E. M., & Ainslie, K. M. (2021). Vaccine formulations in clinical development for the prevention of severe acute respiratory syndrome coronavirus 2 infection. *Advanced Drug Delivery Reviews*, *169*, 168–189. https://doi.org/10.1016/j.addr.2020.12.006.

Bernasconi, V., Norling, K., Gribonika, I., Ong, L. C., Burazerovic, S., Parveen, N., Schön, K., Stensson, A., Bally, M., Larson, G., Höök, F., & Lycke, N. (2021). A vaccine combination of lipid nanoparticles and a cholera toxin adjuvant derivative greatly improves lung protection against influenza virus infection. *Mucosal Immunology*, *14*, 523–536. https://doi.org/10.1038/S41385-020-0334-2.

Chalkias, S., Harper, C., Vrbicky, K., Walsh, S. R., Essink, B., Brosz, A., McGhee, N., Tomassini, J. E., Chen, X., Chang, Y., Sutherland, A., Montefiori, D. C., Girard, B., Edwards, D. K., Feng, J., Zhou, H., Baden, L. R., Miller, J. M., & Das, R. (2022). A bivalent omicron-containing booster vaccine against Covid-19. *The New England Journal of Medicine*, *387*, 1279–1291. https://doi.org/10.1056/NEJMOA2208343.

Chen, X., Huang, H., Ju, J., Sun, R., & Zhang, J. (2022). Impact of vaccination on the COVID-19 pandemic in U.S. states. *Scientific Reports*, *12*, 1–10. https://doi.org/10.1038/s41598-022-05498-z.

Chung, J. Y., Thone, M. N., & Kwon, Y. J. (2021). COVID-19 vaccines: The status and perspectives in delivery points of view. *Advanced Drug Delivery Reviews*, *170*, 1–25. https://doi.org/10.1016/j.addr.2020.12.011.

Druedahl, L. C., Minssen, T., & Price, W. N. (2021). Collaboration in times of crisis: A study on COVID-19 vaccine R&D partnerships. *Vaccine*, *39*, 6291. https://doi.org/10.1016/J.VACCINE.2021.08.101.

Fathizadeh, H., Afshar, S., Masoudi, M. R., Gholizadeh, P., Asgharzadeh, M., Ganbarov, K., Köse, Ş., Yousefi, M., & Kafil, H. S. (2021). SARS-CoV-2 (Covid-19) vaccines structure, mechanisms and effectiveness: A review. *International Journal of Biological Macromolecules*, *188*, 740–750. https://doi.org/10.1016/J.IJBIOMAC.2021.08.076.

Food and Drug Administration (FDA). (2021). *FDA approves first COVID-19 vaccine*. https://www.fda.gov/news-events/press-announcements/fda-approves-first-covid-19-vaccine.

Graham, B. S. (2020). Rapid COVID-19 vaccine development. *Science*, *1979*(368), 945–946. https://doi.org/10.1126/science.abb8923.

Haque, A., & Pant, A. B. (2020). Efforts at COVID-19 vaccine development: Challenges and successes. *Vaccines (Basel)*, *8*, 739. https://doi.org/10.3390/vaccines8040739.

Harvey, W. T., Carabelli, A. M., Jackson, B., Gupta, R. K., Thomson, E. C., Harrison, E. M., Ludden, C., Reeve, R., Rambaut, A., Peacock, S. J., & Robertson, D. L. (2021). SARS-CoV-2 variants, spike mutations and immune escape. *Nature Reviews. Microbiology*, *19*, 409–424. https://doi.org/10.1038/s41579-021-00573-0.

Iversen, P. L., & Bavari, S. (2021). Inactivated COVID-19 vaccines to make a global impact. *The Lancet Infectious Diseases*, *21*, 746–748. https://doi.org/10.1016/S1473-3099(21)00020-7.

Jiang, W., Shi, L., Cai, L., Wang, X., Li, J., Li, H., Liang, J., Gu, Q., Ji, G., Li, J., Liu, L., & Sun, M. (2021). A two-adjuvant multiantigen candidate vaccine induces superior protective immune responses against SARS-CoV-2 challenge. *Cell Reports*, *37*. https://doi.org/10.1016/J.CELREP.2021.110112.

Khoshnood, S., Arshadi, M., Akrami, S., Koupaei, M., Ghahramanpour, H., Shariati, A., Sadeghifard, N., & Heidary, M. (2022). An overview on inactivated and live-attenuated SARS-CoV-2 vaccines. *Journal of Clinical Laboratory Analysis*, *36*, e24418. https://doi.org/10.1002/jcla.24418.

Li, Y.-D., Chi, W.-Y., Su, J.-H., Ferrall, L., Hung, C.-F., & Wu, T.-C. (2020). Coronavirus vaccine development: from SARS and MERS to COVID-19. *Journal of Biomedical Science*, *27*, 104. https://doi.org/10.1186/s12929-020-00695-2.

Li, L., Wei, Y., Yang, H., Yan, J., Li, X., Li, Z., Zhao, Y., Liang, H., & Wang, H. (2022). Advances in next-generation coronavirus vaccines in response to future virus evolution. *Vaccines (Basel)*, *10*, 2035. https://doi.org/10.3390/vaccines10122035.

Lu, R., Zhao, X., Li, J., Niu, P., Yang, B., Wu, H., Wang, W., Song, H., Huang, B., Zhu, N., Bi, Y., Ma, X., Zhan, F., Wang, L., Hu, T., Zhou, H., Hu, Z., Zhou, W., Zhao, L., … Tan, W. (2020). Genomic characterisation and epidemiology of 2019 novel coronavirus: implications for virus origins and receptor binding. *The Lancet*, *395*, 565–574. https://doi.org/10.1016/S0140-6736(20)30251-8.

Lyngse, F. P., Mølbak, K., Denwood, M., Christiansen, L. E., Møller, C. H., Rasmussen, M., Cohen, A. S., Stegger, M., Fonager, J., Sieber, R. N., Ellegaard, K. M., Nielsen, C., & Kirkeby, C. T. (2022). Effect of vaccination on household transmission of SARS-CoV-2 Delta variant of concern. *Nature Communications*, *13*, 1–8. https://doi.org/10.1038/s41467-022-31494-y.

Machado, B. A. S., Hodel, K. V. S., dos Fonseca, L. M. S., Mascarenhas, L. A. B., da Andrade, L. P. C. S., Rocha, V. P. C., Soares, M. B. P., Berglund, P., Duthie, M. S., Reed, S. G., & Badaró, R. (2021). The importance of RNA-based vaccines in the fight against COVID-19: An overview. *Vaccines (Basel)*, *9*, 1345. https://doi.org/10.3390/vaccines9111345.

Mallapaty, S. (2021). India's DNA COVID vaccine is a world first—more are coming. *Nature*, *597*, 161–162. https://doi.org/10.1038/d41586-021-02385-x.

Maruggi, G., Zhang, C., Li, J., Ulmer, J. B., & Yu, D. (2019). mRNA as a transformative technology for vaccine development to control infectious diseases. *Molecular Therapy*, *27*, 757–772. https://doi.org/10.1016/j.ymthe.2019.01.020.

Nicholson Price, W., Rai, A. K., & Minssen, T. (2020). Knowledge transfer for large-scale vaccine manufacturing. *Science*, *369*, 912–914. https://doi.org/10.1126/science.abc9588.

Plante, J. A., Liu, Y., Liu, J., Xia, H., Johnson, B. A., Lokugamage, K. G., Zhang, X., Muruato, A. E., Zou, J., Fontes-Garfias, C. R., Mirchandani, D., Scharton, D., Bilello, J. P., Ku, Z., An, Z., Kalveram, B., Freiberg, A. N., Menachery, V. D., Xie, X., … Shi, P. Y. (2021). Spike mutation D614G alters SARS-CoV-2 fitness. *Nature*, *592*, 116–121. https://doi.org/10.1038/S41586-020-2895-3.

Qu, P., Evans, J. P., Faraone, J., Zheng, Y.-M., Carlin, C., Anghelina, M., Stevens, P., Fernandez, S., Jones, D., Lozanski, G., Panchal, A., Saif, L. J., Oltz, E. M., Xu, K., Gumina, R. J., & Liu, S.-L. (2022). Distinct neutralizing antibody escape of SARS-CoV-2 omicron subvariants BQ.1, BQ.1.1, BA.4.6, BF.7 and BA.2.75.2. *bioRxiv*. https://doi.org/10.1101/2022.10.19.512891.

Riemersma, K. K., Haddock, L. A., Wilson, N. A., Minor, N., Eickhoff, J., Grogan, B. E., Kita-Yarbro, A., Halfmann, P. J., Segaloff, H. E., Kocharian, A., Florek, K. R., Westergaard, R., Bateman, A., Jeppson, G. E., Kawaoka, Y., O'Connor, D. H., Friedrich, T. C., & Grande, K. M. (2022). Shedding of infectious SARS-CoV-2 despite vaccination. *PLoS Pathogens*, *18*. https://doi.org/10.1371/JOURNAL.PPAT.1010876.

Rosa, S. S., Prazeres, D. M. F., Azevedo, A. M., & Marques, M. P. C. (2021). mRNA vaccines manufacturing: Challenges and bottlenecks. *Vaccine*, *39*, 2190–2200. https://doi.org/10.1016/j.vaccine.2021.03.038.

Russell, M. W., & Mestecky, J. (2022). Mucosal immunity: The missing link in comprehending SARS-CoV-2 infection and transmission. *Frontiers in Immunology*, *13*. https://doi.org/10.3389/FIMMU.2022.957107.

Sharpe, H. R., Gilbride, C., Allen, E., Belij-Rammerstorfer, S., Bissett, C., Ewer, K., & Lambe, T. (2020). The early landscape of coronavirus disease 2019 vaccine development in the UK and rest of the world. *Immunology*, *160*, 223–232. https://doi.org/10.1111/imm.13222.

Shin, M. D., Shukla, S., Chung, Y. H., Beiss, V., Chan, S. K., Ortega-Rivera, O. A., Wirth, D. M., Chen, A., Sack, M., Pokorski, J. K., & Steinmetz, N. F. (2020). COVID-19 vaccine development and a potential nanomaterial path forward. *Nature Nanotechnology*, *15*, 646–655. https://doi.org/10.1038/s41565-020-0737-y.

Tang, P. C. H., Ng, W. H., King, N. J. C., & Mahalingam, S. (2022). Can live-attenuated SARS-CoV-2 vaccine contribute to stopping the pandemic? *PLoS Pathogens*, *18*, e1010821. https://doi.org/10.1371/journal.ppat.1010821.

Travieso, T., Li, J., Mahesh, S., Mello, J. D. F. R. E., & Blasi, M. (2022). The use of viral vectors in vaccine development. *npj Vaccines*, *7*, 75. https://doi.org/10.1038/s41541-022-00503-y.

Trincado, V., Gala, R. P., & Morales, J. O. (2021). Buccal and sublingual vaccines: A review on oral mucosal immunization and delivery systems. *Vaccines (Basel)*, *9*. https://doi.org/10.3390/VACCINES9101177.

UNSG Office, WHO, UNICEF, ACT-A. (2022). *WHO Director-General's opening remarks at the UNGA UNSG-hosted event: "Ending the pandemic through equitable access to COVID-19 vaccines, tests and treatments" – 23 September 2022*. https://www.who.int/director-general/speeches/detail/who-director-general-s-opening-remarks-at-the-unga-unsg-hosted-event---ending-the-pandemic-through-equitable-access-to-covid-19-vaccines--tests-and-treatments----23-september-2022. (Accessed 23 September 2022).

Watson, O. J., Barnsley, G., Toor, J., Hogan, A. B., Winskill, P., & Ghani, A. C. (2022). Global impact of the first year of COVID-19 vaccination: a mathematical modelling study. *The Lancet Infectious Diseases*, *0*. https://doi.org/10.1016/S1473-3099(22)00320-6.

World Health Organization (WHO). (2022a). *COVID-19 vaccine tracker and landscape*. https://www.who.int/publications/m/item/draft-landscape-of-covid-19-candidate-vaccines. (Accessed 22 September 2022).

World Health Organization (WHO). (2022b). *COVID19 vaccine tracker*. https://covid19.trackvaccines.org/agency/who/. (Accessed 25 October 2022).

Zohar, T., Loos, C., Fischinger, S., Atyeo, C., Wang, C., Slein, M. D., Burke, J., Yu, J., Feldman, J., Hauser, B. M., Caradonna, T., Schmidt, A. G., Cai, Y., Streeck, H., Ryan, E. T., Barouch, D. H., Charles, R. C., Lauffenburger, D. A., & Alter, G. (2020). Compromised humoral functional evolution tracks with SARS-CoV-2 mortality. *Cell*, *183*. https://doi.org/10.1016/J.CELL.2020.10.052. 1508–1519.e12.

Chapter 3

Molecular methods for SARS-CoV-2 variant detection

Marco Fabiani*, Katia Margiotti*, Francesca Monaco, Alvaro Mesoraca, and Claudio Giorlandino
Department of Human Genetics, Altamedica, Rome, Italy

Abbreviations

VOC	variants of concern
WGS	whole genome sequencing
ONT	Oxford Nanopore Technology
NGS	next-generation sequencing
HCE	hybrid capture-enrichment
RT-qPCR	reverse transcription quantitative PCR
RT-ddPCR	reverse transcriptase droplet digital PCR
FDA	Food and Drug Administration
RBD	receptor-binding domain
SARS-CoV-2	severe acute respiratory syndrome coronavirus 2
VOI	variants of interest
WHO	World Health Organization

Introduction

As of Dec. 23, 2022, more than 651 million positive cases of COVID-19 have been reported worldwide as a result of the infection caused by the novel coronavirus severe acute respiratory syndrome coronavirus-2 (SARS-CoV-2) https://covid19.who.int/, which is the pathogen that caused COVID-19 (Zhu et al., 2020). It is closely related to SARS-CoV, sharing a 79% genome sequence identity (Wu et al., 2020). The SARS-CoV-2 RNA genome encodes 16 nonstructural proteins (NSP) and at least 10 structural proteins including spike (S), ORF3a, envelope (E), membrane (M), open reading frame 6 (ORF6), ORF7a, ORF7b, ORF8, nucleocapsid (N), and ORF10 (Cagliani et al., 2020). It's important to note that SARS-CoV-2 and other RNA viruses have rapid mutation rates, and their genomes change considerably more quickly than those of larger organisms, as reported by several studies (Mohammadi et al., 2021). During the pandemic, several SARS-CoV-2 lineages developed with frequent transmission passes. The emergence of these hypermutated viruses resulted in changes in the pathogen's infectious properties, along with an increase in transmission capabilities (Bansal & Kumar, 2022; Scrima et al., 2022). In a recent study on the molecular evolutionary traits of SARS-CoV-2, researchers demonstrated that the nucleotide mutation rate of the whole genome was 6.7×10^{-4} substitutions per site per year, and the nucleotide mutation rate of the S gene was 8.1×10^{-4} substitutions per site per year, which was at a medium level compared with other RNA viruses (Wang et al., 2022). These findings confirmed the scientific hypothesis that the rate of evolution of the virus gradually decreases over time. The first SARS-CoV-2 whole genome sequences were made accessible on GISAID on Jan. 10, 2020 (www.gisaid.org). Since then, about 10 million sequences have been uploaded to GISAID, which has grown to be the biggest archive of SARS-CoV-2 sequences in the world (Alkhatib et al., 2022). GISAID provides genomic epidemiology and real-time monitoring to track the introduction of novel COVID-19 viral strains throughout the world (Shu & McCauley, 2017). It became important to develop a nomenclature to identify the varieties as they emerged.

*Contributed equally.

Management, Body Systems, and Case Studies in COVID-19. https://doi.org/10.1016/B978-0-443-18703-2.00038-4
Copyright © 2024 Elsevier Inc. All rights are reserved, including those for text and data mining, AI training, and similar technologies.

Because unscientific audiences frequently connected the variant names with the countries where they were first detected, the important variants have been assigned with letters of the Greek alphabet to avoid instances of stigmatization and discrimination toward those countries. Europe saw four distinct waves as a result of the spread of three major variants of concern (VOC). In late 2020, a variant first found in the United Kingdom became the most widespread in Europe; in December 2020, it was conferred VOC status by European Centre for Disease Prevention and Control (ECDC) and was called the alpha variant (B.1.1.7) based on the newly recommended World Health Organization (WHO) nomenclature [https://www.ecdc.europa.eu/en/covid-19/variants-concern]. In the United States, the frequency of this variant ranged from 30% to 47%. Due to a significant decrease in its distribution in Europe, the alpha variant was replaced by the delta strain. The delta variant (B.1.617.2) passed from a variant of interest (VOI) change to VOC through June 7, 2022. Months later, a new variant named omicron (B.1.1.529) was established and instantly categorized as a VOC in November 2021. As of January 2023, the only VOCs taken into consideration by the WHO are the BA.2, BA.4, and BA.5 omicron sublineages. As a result of increased concern to public health, in recent years there has been a rapidly growing demand for whole genome sequencing (WGS), which is the reference standard for detecting SARS-CoV-2 variants. However, WGS does not allow for the rapid identification of VOCs (at least 5 working days), delaying their spread control. These limitations prompted research into improving methods for an early and cost-effective strategy for detecting SARS-CoV-2 variants. It is of primary importance for patient care and public health surveillance and response that diagnostic platforms detect all SARS-CoV-2 variants with high levels of sensitivity and specificity. In this chapter, we will provide a summary of the various laboratory techniques available for sequencing and quick detection of SARS-CoV-2 variants in clinical samples (Fig. 1).

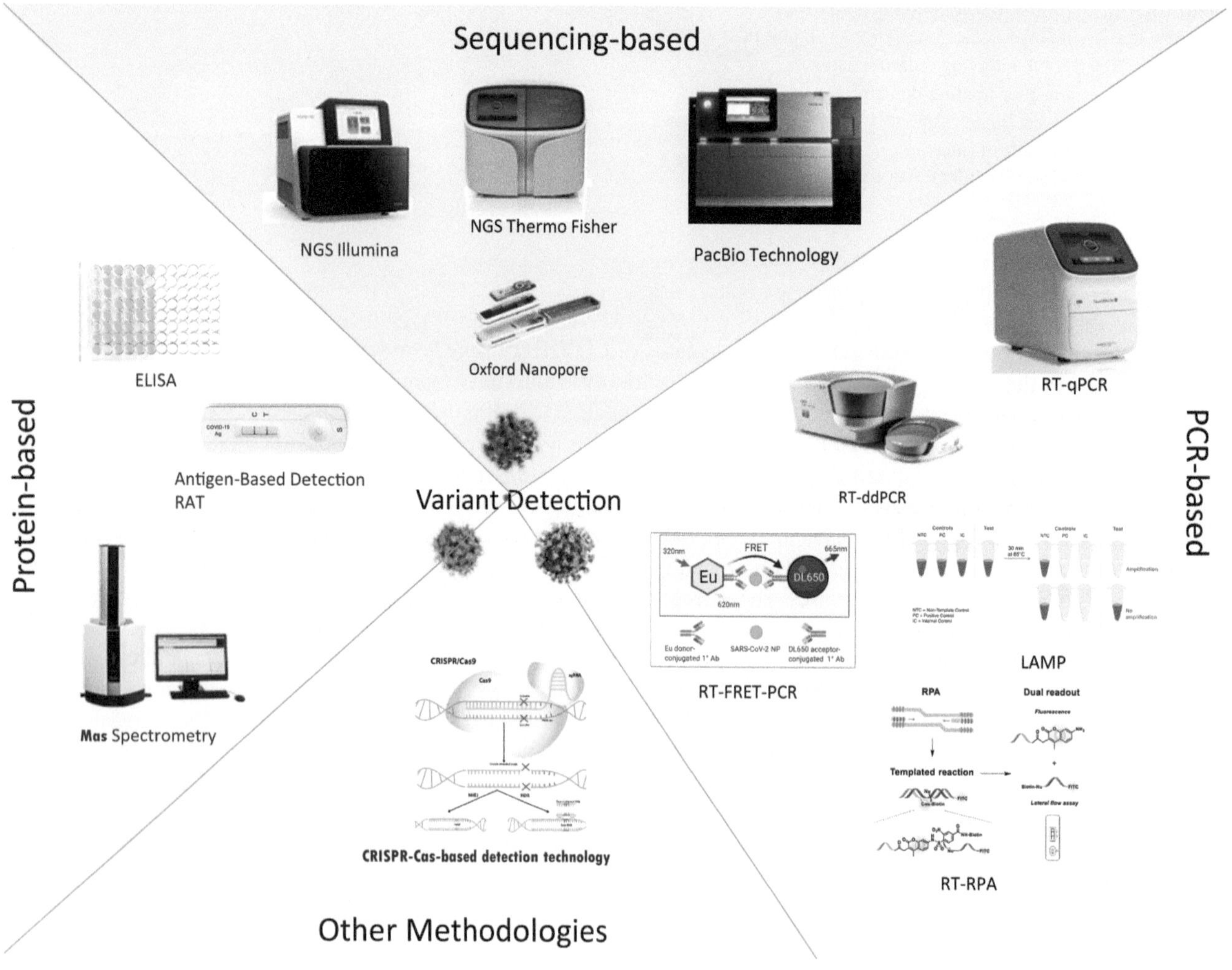

FIG. 1 Developed methods for detection of SARS-CoV-2.

Sequencing methods for detection of SARS-CoV-2 variants

Next-generation sequencing (NGS) technology

The gold standard in monitoring and identifying novel variants of SARS-CoV-2 is next-generation sequencing (NGS) technology (Berno et al., 2022). NGS is a massively parallel sequencing technology that offers ultrahigh throughput, scalability, and speed. The technology is used to determine the order of nucleotides in entire genomes or targeted regions of DNA or RNA. NGS has been extensively and successfully applied to the sequencing of all SARS-CoV-2 genes, including those encoding nonstructural proteins and intragenic regions (Campos et al., 2020; Pillay et al., 2020). Moreover, NGS is employed to conduct phylogenetic research, which produces more accurate findings than the sequencing of partial gene regions. These types of analyses are essential for tracking mutational trends among circulating strains and identifying the potential appearance of novel variants. NGS is the best method for identifying minority variants that may be beneficial for tracking escape variants in immunocompromised patients or in patients receiving monoclonal antibody therapy. Although NGS technologies provide high throughput and reliable approaches for detecting SARS-CoV-2 variants, identifying the variants from raw data is challenging and needs specialist bioinformatics tools. Bioinformatics experts created methods to analyze raw data and identify variants quickly using data from various sequencing strategies (Lo et al., 2022; Tilloy et al., 2022). NGS approaches are mainly divided into two different methods, target enrichment and shotgun metagenomic sequencing. Both approaches can be applied to sequence the genome of SARS-CoV-2.

NGS target enrichment

This method uses oligonucleotide probes designed to target regions of interest. Different NGS protocols have been developed based on the amplicons approach or based on capture enrichment on different sequencing platforms, especially Ion Torrent, Illumina, Oxford Nanopore Sequencers, and the PacBio Platform.

Even though amplified-based sequencing is very precise, it necessitates extensive a priori data on the target sequence. Amplicon-based methods for sequencing SARS-CoV-2 use a procedure for enrichment that starts first with strand cDNA synthesis and ends with multiplex polymerase chain reactions (PCRs) to amplify the genome. The goal is to create pools of amplicons that either span the full viral genome or specific subsets of it. The difference between amplicon-based and non-targeted techniques is mainly the use of specific primers designed for specific genome sequence investigations. Amplicon sequencing requires less time than the other techniques because nonviral reads are more rare with respect to other DNA present in the sample. It is also highly selective, and it is robust to low RNA levels or nonoptimal starting samples. Recently, a novel workflow using amplicon deep sequencing of the S gene was implemented for Illumina sequencing to provide a high throughput and unbiased identification workflow for VOC/VOI from clinical SARS-CoV-2 samples. This NGS approach allows the detection of rare clonal types, cells, or microbes comprising as little as 1% of the original sample. Sequencing only the S gene is a practical and cost-effective detection tool that can reduce the number of PCR cycles (Castañeda-Mogollón et al., 2022). Amplicon deep sequencing, which focuses on a specific gene or genomic region, provides greater sequencing depth and coverage while reducing costs and data burden. Recently, the use of Oxford Nanopore Technology (ONT) in combination with mutation-specific PCR has been shown to enable the identification of samples with extremely low virus loads as well as the quick and highly sensitive detection of VOCs (Dächert et al., 2022). More crucially, the approach can detect novel SARS-CoV-2 variants that may appear in the next few years as a consequence of gene sequencing. Using the ONT platform seems to be more effective for smaller quantities of samples and has quicker turnaround times than short-read sequencing technologies such as Illumina sequencing. The accuracy of ONT sequencing for SARS-CoV-2 variants, however, remains limited, showing poorer genome coverage and a larger number of indels than the Illumina sequences (Tshiabuila et al., 2022). The main drawback for amplicon-based NGS is that the primers were created using as a reference the Wuhan SARS-CoV-2 genome sequence, and the method may not be able to detect significant structural variations. Despite the fact that NGS is extensively utilized, there remains a considerable degree of uncertainty in the computational errors and bias in NGS (Abnizova et al., 2017). Another disadvantage is that specialized and costly sequencing equipment is required, and the readout procedure of such devices is quite slow and expensive. However, NGS-based testing can meet the need for orders of magnitude amplification and offer ubiquitous genotyping data, especially when amortizing numerous samples with manageable sequencing reagent costs. As a result, creating and upgrading NGS-based approaches is a promising way for future study. While the amplicon-based method is highly reliable for reconstructing the most widespread genome variant in a viral population, a recent study suggests that it provides a highly biased representation of minor allele frequencies when compared to metatranscriptomics experiments performed on the same samples (Xiao et al., 2020).

Another nontargeted sequencing approach is based on hybrid capture enrichment (HCE), which, like amplicon-based sequencing, allows targeting specific relevant regions. Target-enrichment technologies based on HCE were initially developed for human genome studies to allow for the quick and cost-effective sequencing of protein-coding gene exons (Albert et al., 2007).

In general, HCE approaches offer a more thorough profile of the target sequences than amplicon-based methods because they are based on a larger number of fragments/probes. Furthermore, capture using hybridization is often more resilient to genomic heterogeneity than PCR-amplicon production because the capture of target areas depends less on precise alignment. Even though Xiao et al. revealed that amplicon-based methods are more susceptible than hybrid capture sequencing for the genotyping of SARS-CoV-2 genomes and did not advise its application for challenging samples with low viral loads, enrichment by hybridization has been successful, even for samples with very low viral loads in other studies (Maurano et al., 2020; Xiao et al., 2020).

Shotgun metagenomic sequencing

This nontargeted sequencing technique exploits a strategy that can investigate all the DNA in a sample, allowing the characterization of complex communities of microorganisms without any prior knowledge of their genome sequences. Shotgun metagenomics acquires a particular value in clinical microbiology, where it may be used to guide treatment approaches by informing and giving quantitative information on the composition of microbial communities. From a variety of clinical sample types, shotgun metatranscriptomics has been successfully used to obtain the complete SARS-CoV-2 genome during the pandemic (Handelsman, 2004; H. Zhang et al., 2021). This methodology requires no prior knowledge of the viral sequence and avoids potential effects of divergent regions on capture and amplicon approaches. When adequate levels of genome coverage are obtained, in addition to some insight into host gene expression, this approach can provide an accurate evaluation of intrasample virus variants from quasispecies or coinfections, allowing insights into host gene expression patterns during infection. The major limitation of shotgun metatranscriptomics is the requirement for a high viral load to obtain complete virus assemblies. The viral load shows enormous variation in clinical specimens due to variations in sampling technique as well as from inherent differences in loads between patients. The proportion of reads derived from SARS-CoV-2 can vary greatly between samples, even where viral loads are similar (Butler et al., 2021; Xiao et al., 2020; H. Zhang et al., 2021).

Sanger sequencing

Because the time and money required to produce NGS might create a delay in getting findings, Sanger sequencing of the S gene may be timelier and more feasible than NGS in some situations. Several laboratories devised techniques for RT-qPCR followed by sequencing to amplify particular sections of the S-gene. With its greater than 99% accuracy and long-read capabilities, Sanger sequencing is the gold-standard sequencing technology. Sanger sequencing utilizes DNA polymerase selective incorporation of fluorescently labeled dideoxynucleotides, enabling the prolonged oligonucleotides to be selectively terminated at the G, A, T, or C bases. This technique, despite being developed several years ago, was also widely used for determining the type of SARS-CoV-2 VOCs (Daniels et al., 2021). It has the advantages of being easy to use, fast, and low cost; facilities for Sanger sequencing infrastructure are available in many laboratories. It is considered that the SARS-CoV-2 genomic RNA is too large for Sanger sequencing to analyze the entire genome; for this reason, researchers usually consider only the S gene. Primers were designed to target the S gene region because nucleotide changes in this gene are used for VOC definitions (Bloemen et al., 2022; Dorlass et al., 2021).Therefore, one study performed long-range RT-PCR amplification of the entire SARS-CoV-2 S gene, including a 4 kb region, and then the entire S gene sequence was analyzed by Sanger sequencing. In the study, the authors compared the analysis results with the reference SARS-CoV-2 genome, identified amino acid mutations in the sequence, and determined the type of VOC according to the nature of mutations, which could be used to track and monitor SARS-CoV-2 variants with conventional Sanger sequencing instruments (Matsubara et al., 2022).

PCR-based methods for detection of SARS-CoV-2 variants

Reverse transcription quantitative PCR (RT-qPCR)

Alternatively to sequencing, several other methods have been developed for the faster identification of SARS-CoV-2 variants. The vast majority of these methods for diagnostic screening mainly consist of nucleic acid amplification techniques based on PCR able to generate results in a few hours. RT-qPCR is a particular assay that is able not only to detect but also to quantify specific DNA or RNA sequences in a sample. The technique involves the use of fluorescent probes or dyes that bind to specific regions of the target DNA or RNA and emit a fluorescent signal when amplified by the PCR process. The fluorescence is detected in real time as the PCR reaction progresses, allowing for the quantification of the target DNA or

RNA present in the sample. RT-qPCR is a widely used method for the detection and quantification of SARS-CoV-2 as well as for genotyping different SARS-CoV-2 variants that are spreading. The technique involves the amplification of specific regions of the SARS-CoV-2 genome using primers and probes that are specific to the virus. The primers bind to the ends of the target region and serve as a starting point for PCR amplification while the probe is a fluorescent-labeled oligonucleotide that binds to a specific location within the target region. The probe emits a fluorescent signal when it is bound to the amplified target DNA, which is detected in real time during the PCR reaction. The real-time PCR test is very sensitive and can detect even small amounts of viral RNA in a sample. It is also highly specific, meaning that it can accurately distinguish between SARS-CoV-2 and other similar viruses. The test is typically performed on a sample of respiratory tract secretions, such as a nasal or throat swab, and results are typically available within a few hours. The test is considered to be highly accurate and is widely used for the diagnosis of COVID-19 in clinical settings as well as for monitoring the spread of the virus in communities. Mainly, commercially available methods based on RT-qPCR for identifying variants of SARS-CoV-2 have been developed (see Table 1).

Moreover, many laboratories developed their own in-house protocols for identifying SARS-CoV-2 variants. Despite the considerable advantage in terms of time and cost, RT-qPCR can help identify a variant only when the variants that circulate in a specific area are already known. Fabiani et al. developed an in-house protocol to distinguish the alpha, beta, gamma, and delta SARS-CoV-2 variants by detecting S gene mutations using a real-time RT-PCR. In this study, 132 positive

TABLE 1 Commercial methods for SARS-CoV-2 variant identification.

Company	Kit	Description	Website
Clonit S.r.l.	COVID-19 Variant Catcher	RT-PCR, which allows the identification of the S gene mutations 69–70del, E484K, N501Y, L452R, E484Q.	https://www.clonit.it/en/company/news/covid-19-ultra-variant-catcher-kit/
Seegene Inc. company	Novaplex SARS-CoV-2 Variants Assay	Multiplex real-time RT-PCR that detects different panels of spike protein mutations for each specific SARS-CoV-2 variant.	https://www.seegene.com/advantages/complete_solution_for_the_covid_19_response
Diasorin Molecular	Simplexa SARS-CoV-2 Variants Direct assay	Method capable of identifying the tip mutations of SARS-CoV-2 N501Y, E484K, E484Q, and L452R. The advantage of this assay is avoiding the extraction of RNA by operating directly on nasopharyngeal or nasal swabs.	https://molecular.diasorin.com/international/kit/simplexa-sars-cov-2-variants-direct-ruo/
ABL company	UltraGene Assay SARS-CoV-2 Multi Variants Deletions V1	Method using RT-PCR assay for the qualitative detection of SARS-CoV-2 69/70 (Δ69), Y144 (Δ144), and 242–244 (Δ242) deletions on the spike gene and 3675 deletion–3677 (Δ3675) on the ORF1ab gene.	https://www.ablsa.com/laboratory-applications/ultragene-combo2screen-3/
ELITech	SARS-CoV-2 Extended ELITe MGB	Molecular multitarget kit for the detection and discrimination of the mutations L452R, E484K, E484Q, and N501Y of the S gene of SARS-CoV-2.	https://www.elitechgroup.com/news/the-first-pcr-kit-that-differentiates-the-sars-cov-2-omicron-variant/
TIB Molbiol	VirSNip SARS Spike	Test for the detection of mutations in the new B.1.1.529 omicron SARS-CoV-2 variant capable of discriminating different SARS-CoV-2 mutations (ins214EPE, S371L, S373P, or E484A).	https://www.tib-molbiol.de/it/covid-19
Roche	Cobas SARS-CoV-2 Variant Set 1	RT-PCR assay for the rapid in vitro qualitative detection and discrimination of selected SARS-CoV-2 mutations E484K, N501Y, and deletion Δ69 in, for example, nasal and nasopharyngeal swab specimens.	https://diagnostics.roche.com/global/en/products/params/cobas-sars-cov-2-variant-set-1.html
ThermoFisher Scientific	TaqPath COVID19 CE-IVD RT-PCR Kit	This test uses a multitarget design that includes an S gene target for potential early identification of B.1.1.529.	https://www.thermofisher.com/it/en/home/clinical/clinical-genomics/pathogen-detection-solutions/covid-19-sars-cov-2/multiplex.html

patients were analyzed with RT-qPCR for genotyping validation (Fabiani et al., 2022). Among the 132 COVID-19-positive samples, the method was able to discriminate all the investigated SARS-CoV-2 variants with 100% concordance when compared with the NGS method (Fabiani et al., 2022).

Reverse transcription fluorescence resonance energy transfer polymerase chain reaction (RT-FRET-PCR)

Barua et al. developed a reverse transcription fluorescence resonance energy transfer polymerase chain reaction (RT-FRET-PCR) designed to identify the T478K mutation (present in 99.73% of the delta variant) that can be used both to diagnose COVID-19 patients and simultaneously identify whether they are infected with the delta variant (Barua et al., 2022). Many other groups have also developed other in-house techniques for the identification of single position mutations to quickly identify a single variant circulating on the territory (Chan et al., 2022), but the major limitation of these kinds of assays is the inability to detect all the other major variants because each variant is characterized by a multitude of mutations (Erster et al., 2021).

Allele-specific polymerase chain reaction (ASP)

Some laboratories describe in-house methods based on the allele-specific polymerase chain reaction (ASP) variant assay to detect SARS-CoV-2 VOCs. Brito-Mutunayagam et al. described an ASP based on three mutation targets–E484 K, L452R, and P681R–showing a sensitivity ranging from 67% to 100% in relation to the identified variant (Brito-Mutunayagam et al., 2022), whereas Borillo et al. described the same technology for N501Y, E484K, and S982A with a concordance of results of 100% (Borillo et al., 2022). Other groups describe a multiplex RT-qPCR capable of detecting nine mutations with specific primers and probes in the same mix; these PCR typing strategies allowed the detection of the major variants and also provided an open-source PCR assay that could rapidly be deployed in laboratories around the world (Chung et al., 2022). Given that, RT-qPCR-based genotyping approaches are rapid methods for monitoring SARS-CoV-2 variants, but they require continuous adaptation. Clark et al. described a multiplex fragment analysis approach using targeting mutations by size and fluorescent color. Eight SARS-CoV-2 mutational hot spots in VOCs were targeted. This kind of method could classify a variant with similar accuracy as sequencing without frequent target modification (Clark et al., 2022). Moreover, the use of several joint detection technologies has grown to compensate for the lack of single technology. Hernandez et al. described a method that combines multiplex PCR and mass spectrometry minisequencing (Hernandez et al., 2022). In addition, the prevalence of COVID-19 has led to increasing demand for high-throughput, multiplexed, and sensitive detection methods. This is what was made by Welch et al. that selected up to 26 mutations to distinguish between or detect mutations shared among the alpha, beta, gamma, delta, epsilon, and omicron variant lineages in a cost-effective virus and variant detection platform, which combines CRISPR-based diagnostics and microfluidics with a streamlined workflow for clinical use (Welch et al., 2022).

High-resolution melting (HRM) assay

HRM is a real-time PCR-based method that allows for the detection of small changes in the melting temperature (Tm) of the amplified product. This is achieved by monitoring the fluorescence of the amplified product while it melts (increases in temperature). Different variants of the virus can have slightly different Tm values in relation to purine and pyrimidine balance, which can be used to distinguish them. The HRM method was used to detect SARS-CoV-2 variants. It can be applied in conjunction with PCR-based assays, whole genome sequencing, or targeted sequencing. To verify its effectiveness in the detection of variants, different groups independently investigated different strains according to GISAID clades and different mutation sites, especially in the receptor-binding domain (RBD) both by NGS and HRM. For HRM, primers were designed and nested PCR amplification was performed according to the first RT-PCR amplification products; finally, the amplicons were analyzed by HRM (Miyoshi et al., 2022) (Aoki et al., 2021). The results showed that the HRM curve and Tm values were consistent with those of the NGS data and confirmed that HRM can be easily used for the detection of all SARS-CoV-2 variants only by designing oligonucleotides with a drastically reduced cost.

LAMP assay

The loop-mediated isothermal amplification (LAMP) LAMP assay is a method for detecting specific nucleic acid sequences using PCR at a constant temperature. It can be used to detect SARS-CoV-2 by amplifying a specific region of the virus's genetic material. LAMP is a cost-effective, simple, and rapid alternative to traditional PCR methods, and it does not require specialized equipment (Alves et al., 2021). Due to its advantages in instrument requirements, detection efficiency, operating methods, and result interpretation, LAMP is well suited for use in resource-limited environments and is an alternative technique to RT-PCR (Wong et al., 2018). However, LAMP assays are not as sensitive as RT-qPCR and may not be able to detect low levels of viral RNA. Moreover, the fluorescent readout-based PCR technique exploits the nonspecific binding of SYBR fluorescent dye to dsDNA, leading to false-positive results. To eliminate this nonspecific amplification and improve the sensitivity and specificity of the assay, fluorescence-quenched reverse transcription loop-mediated isothermal amplification (FQLAMP) was developed instead. The method was amplified using a fluorescently labeled reporter gene primer, a LoopB primer (FLB), and a short complementary oligonucleotide quenching agent (QLB) in a real-time thermal cycler for 30min at 65oC. Next, the reaction was cooled to room temperature to read the fluorescence signal. Because the assay results can be interpreted by the naked eye or by a mobile phone, FQLAMP is more suitable for POC or home testing environments than the standard RTLAMP (Jones et al., 2022). Santos et al. tested the sensitivity and specificity of LAMP by applying this method to 60 SARS-CoV-2 RNA samples extracted from swab samples; the results showed a sensitivity of 93% and a specificity of 89% for samples with a Ct $\leq$ 23 (dos Santos et al., 2022). Song et al. exploited both LAMP and CRISPR techniques, referring to them as DNAzyme reaction triggered by LAMP with CRISPR (DAMPR) assay, for detecting SARS-CoV-2 variant genes with high sensitivity within an hour. The CRISPR-associated protein 9 (Cas9) system eliminated false-positive signals of LAMP products, improving the accuracy of the DAMPR assay. Further, they fabricated a portable DAMPR assay system using a three-dimensional printing technique and developed a machine learning (ML)-based smartphone application to routinely check the diagnostic results of SARS-CoV-2 and variant detection. Among blind tests of 136 clinical samples, the proposed system successfully diagnosed COVID-19 patients with a clinical sensitivity and specificity of 100% each. More importantly, the D614G (variant-common), T478K (delta-specific), and A67V (omicron-specific) mutations of the SARS-CoV-2S gene were detected selectively, enabling the diagnosis of 70 SARS-CoV-2 delta or omicron variant patients. The DAMPR assay system is expected to be employed for onsite, rapid, accurate SARS-CoV-2 detection and genotyping of its variants (Song et al., 2022). In conclusion, because LAMP can use internal primers, external primers, and loop primers simultaneously, making it highly efficient and sensitive, as well as having the benefits of simplicity, affordability, speed, and readability of experimental results, it may be used for the detection of SARS-CoV-2 and its variants. The LAMP test showed potential for discriminating against VOCs and may be particularly helpful as a screening method for variants, particularly in countries with laboratories without sequencing technologies.

Reverse transcription-recombinase polymerase amplification (RT-RPA)

RT-RPA is a real-time, isothermal amplification method that can be used for both the diagnosis and surveillance of SARS-CoV-2 variants. The RT-RPA method uses a combination of reverse transcription and recombinase polymerase amplification (RPA) to amplify the viral RNA. The method is relatively simple and does not require specialized equipment or thermal cycling. RT-RPA has several advantages for the detection of SARS-CoV-2 variants. In contrast with HRM and LAMP, RT-RPA has high sensitivity as it is able to detect low levels of viral RNA, which makes it useful for detecting early-stage infections. Another advantage is that RT-RPA is a relatively fast method, with results available in as little as 30min. The main difference between LAMP and RT-RPA is that LAMP requires the use of a heating device to keep the temperature at 62oC, whereas RT-RPA ia amplified at 37–39oC, which can be achieved by using methods such as water baths, hand warmers, or low-cost equipment. It is a simple and inexpensive method that is ideal for use in resource-constrained environments (Cherkaoui et al., 2021). The comparison between RT-RPA and RT-PCR has shown that the sensitivity and specificity of SARS-CoV-2 positive strains were 96% and 97%, respectively, and researchers validated RT-RPA for the four main VOCs (alpha, beta, delta, and omicron). It is important to note that RT-RPA is a relatively new method, so further research is needed to determine its sensitivity and specificity compared to other methods such as NGS techniques.

RT-ddPCR (reverse transcriptase droplet digital PCR)

RT-ddPCR is a method for detecting and quantifying specific sequences, including those of SARS-CoV-2 variants. It combines the reverse transcription step with droplet digital PCR technology to provide a highly sensitive and specific method for detecting and quantifying viral RNA. The basic principle of RT-ddPCR is that it partitions the sample into thousands of tiny droplets, each containing a small amount of cDNA. Each droplet acts as an individual PCR reaction chamber, and the number of droplets that contain amplified target cDNA is used to calculate the starting number of target molecules in the original sample (Pekin et al., 2011). The main advantage of RT-ddPCR is its high sensitivity and specificity, which allow the detection of low levels of viral RNA, even in the presence of high levels of nonspecific background noise. Furthermore, it has a wide dynamic range, allowing the quantification of large differences in the amount of viral RNA present in a sample. RT-ddPCR was used to specifically detect mutational signatures in variants and in wastewater, and could be used to monitor the prevalence trend of future SARS-CoV-2 variants (Heijnen et al., 2021). Another study by Mills et al. investigated 419 clinical samples that were positive for SARS-CoV-2 and found 99.7% consistency between the sample mutations identified by RT-ddPCR and NGS data (Mills et al., 2022). Moreover, the authors were able to clearly distinguish omicron from delta variants (Mills et al., 2022). Another study found that 42.6% of the positive SARS-CoV-2 samples detected by RT-ddPCR were missed during sequencing, and 26.7% of the negative samples tested by sequencing were positive by RT-ddPCR. This shows that the sensitivity of RT-ddPCR to detect mutations is higher than that of sequencing (Lou et al., 2022). In summary, RT-ddPCR is a highly sensitive and specific method for detecting and quantifying SARS-CoV-2 and its variants, but it requires specialized equipment and trained personnel. It is an effective tool in detecting low levels of viral RNA and quantifying large differences in the amount of viral RNA present in a sample. From some points of view, it is even better than NGS strategies.

CRISPR-Cas-based detection

In the past decades, a novel gene-editing tool called CRISPR (clustered regularly interspaced short palindromic repeats) and CRISPR-associated (Cas) proteins have led to unprecedented advances in molecular biology (Komor et al., 2017). CRISPR is a genome editing tool that can be used to detect the presence of specific genetic sequences, including those of SARS-CoV-2 variants. CRISPR-based detection methods involve the use of specific guide RNAs that bind to target sequences within the SARS-CoV-2 genome and activate Cas enzymes, such as Cas12a or Cas13, to cleave or degrade the target sequence, generating a specific signal that can be detected by various methods such as fluorescent or electrochemical.

There are different CRISPR-based detection platforms for SARS-CoV-2 that have been developed, such as CRISPR-Cas12a, CRISPR-Cas12b, and CRISPR-Cas13, each with its own advantages and limitations. These platforms are highly sensitive and specific and can detect a wide variety of SARS-CoV-2 variants, including those with mutations in the spike protein, which is the target of most vaccines. Additionally, CRISPR-based detection can be multiplexed, meaning that it can detect multiple targets at once, making it a powerful tool for monitoring the emergence and spread of SARS-CoV-2 variants.

CRISPR-Cas12a diagnostic techniques designed for the synthesis of mismatched crRNAs or gRNA mutation sites for SARS-CoV-2 variants have been applied to the detection and identification of SARS-CoV-2 variants, such as the alpha, beta, gamma, delta, kappa, lambda, and apsilon variants (He et al., 2022; Huang et al., 2021; Liang et al., 2021; Ning et al., 2022). Furthermore, these assays used a two-pot strategy, a preamplification step prior to CRISPRCas detection that increased the reaction time and carryover contamination. Thus, in another investigation, CRISPR-BrCas12b and RT-LAMP were combined to create a one-pot detection reaction system known as CRISPR-SPADE. This was done by utilizing the high specificity of BrCas12b and its potent trans-cleavage activity at RT-LAMP reaction temperature. It could identify SARS-CoV-2 VOCs in 10–30 min, including alpha, beta, gamma, delta, and omicron, with high values of sensitivity (92%), specificity (99%), and accuracy (96%) . Arizti-Sanz et al. designed crRNA and RPA primers for spike protein-specific mutations and developed a Cas13-based nucleic acid diagnostic procedure for SINEv2 distinguishing between alpha, beta, gamma, and delta in 90 min (Arizti-Sanz et al., 2021). Using similar principles, Lin et al. combined isothermal recombinase-aided amplification for single-nucleotide mutations and CRISPR-Cas12a to achieve detection of alpha, beta, and delta variants and omicron sublines BA.1 and BA.2. The sensitivity was 100.0%, and the specificity was 94.9%–100.0% (Lin et al., 2022). A highly sensitive portable field assay for Δ69–70 mutations based on the CRISPR/Cas13a system was also investigated that could detect 1 copy/μL of the alpha and omicron variants (Niu et al., 2022). This CRISPR-Cas-based genotyping technology is suitable for POC diagnosis because equipment, reagents, and facilities can be shared, greatly reducing costs and detection time. Despite many research groups having studied and developed different strategies to apply CRISPR-Cas technology to SARS-CoV-2 variant detection, the CRISPR-based method is still in the development stage and it is not yet available for clinical or community use. However, it is considered a promising tool for the rapid identification of SARS-CoV-2 variants in the future.

Protein-based variant SARS-CoV-2 detection

Host infection with the SARS-CoV-2 virus produces structural and nonstructural proteins, triggering an immune response that leads to the generation of viral protein specific antibodies. Thus, viral proteins and/or specific antibodies against viral proteins can be used as detection targets to identify the presence of a host infection with SARS-CoV-2 strains (X. Li et al.,

2022). At present, there are many detection technologies and kits for the S protein and N protein of SARS-CoV-2 (Domenico et al., 2021; Le Hingrat et al., 2021), but due to the continuous mutation of SARS-CoV-2, the detection performance of these technologies needs to be tested, and methods need to be developed for these new mutations to identify variants and control the pandemic quickly.

Antigen-based detection

Rapid antigen tests (RATs) play an important role in detecting and controlling COVID-19 transmission; while they are less sensitive than PCR methodology, the results are faster, cheaper, and more suitable for point-of-care detection (Dinnes et al., 2021; Iglói et al., 2021). RATs using a colloidal gold immunochromatographic system can detect specific mutations in SARS-CoV-2 variants with high sensitivity and specificity. Research has also produced a multiantibody combination modified graphene transistor sensor for rapid and accurate detection of the virus. More testing is needed before these methods can be widely used in clinical settings. To verify the sensitivity and specificity of RATs for genotyping SARS-CoV-2 variants, Medoro et al. enrolled 584 SARS-CoV-2 positive patients undergoing RAT and sequencing (Medoro et al., 2022). The sensitivity to the delta/kappa variants at the L452R and E484Q S gene mutations was 97%, while the sensitivity to mutations in the N501Y S gene of the omicron variant was 91%. This RAT has some sensitivity for the detection of the widely dispersed omicron variant. (Medoro et al., 2022). New research has produced a multiantibody combination modified graphene transistor sensor, a detection device that targets different regions of the SARS-Cov-2 S protein by modifying S1 protein monoclonal, CR3022 monoclonal, and N3021 nano antibodies on the surface of graphene transistors (Dai et al., 2021). The results showed a 100% overlap with RT-PCR, demonstrating high accuracy and a detection time of less than 1 min. This assay technique solves the problem of low precision of antigen detection, and its rapid and high-throughput advantages make it a valuable tool for the rapid diagnosis and screening of COVID-19. However, more testing with much higher numbers of patients will be needed before clinical application.

Antibody-based detection

For virus serological antibody detection, methods commonly used include the enzyme-linked immunoassay (ELISA), the chemiluminescence immunoassay, the immunofluorescence assay, and the colloidal gold immunochromatographic assay. Among them, ELISA is the gold standard in serology, providing quantitative and extremely sensitive assays (Amanat et al., 2020). Antibody testing has been a complementary method to PCR for the detection of SARS-CoV-2 variants and helps to screen suspected and asymptomatic patients (Yong et al., 2020). Zhang et al. developed a rapid and sensitive magneto-fluidic immuno-PCR technique (P. Zhang et al., 2022), which preloads a programmable magnetic arm in the analytical reagent to attract and transport magnetically captured specific antibodies, followed by the use of a microthermal immuno-PCR performed with a circulator and a fluorescence detector to detect specific antibodies. To validate the sensitivity of the method, the team analyzed 108 clinical serum samples and obtained a sensitivity of 93.8% and a specificity of 98.3%. Magnetic fluid immuno-PCR overcomes lengthy workflows and bulky instrumentation and has great potential for rapid and sensitive serological testing. Variant detection has been studied using specific monoclonal antibodies that bind to specific antigens of SARS-CoV-2. For example, based on the change of N501Y in the SARS-CoV-2 variants, the recombinant monoclonal antibody design was carried out using 2E8 obtained from patients diagnosed with COVID-19, and the N501 and Y501 in the spike protein RBD were distinguished by ELISA. That is, 2E8 bound to variants without N501Y mutations, such as delta, but did not bind to alpha, beta, and gamma, and then combined with recombinant CB6 mAb to distinguish alpha and gamma from beta. Second, mutations in key regions of primer or antibody binding can affect the effectiveness of in vitro detection. Therefore, studies reported the development of specific monoclonal antibodies to detect the conservative epitope of the SARS-CoV-2 N protein by lateral flow immunoassay (LFIA). Moreover, 75E12 as a capture probe and 54G6/54G10 combined as a detection probe could confirm SARS-CoV-2 infection in the context of the continuous evolution of SARS-CoV-2 mutations and avoid false-negative elevation due to mutations (Lee, Jung, et al., 2022a). In addition, alpha and beta variants were successfully detected using anti-RBD antibodies (MD29 and MD65) as reporter antibodies and an anti-RBD antibody (BL6) and anti-NTD antibody (BL11) as capture antibodies (Barlev-Gross et al., 2021). It is difficult for an immunoassay to detect antigenic variants of the emerging SARS-CoV-2 spike protein because spike protein mutations can alter viral antigenicity. Angiotensin-converting enzyme 2 (ACE2) is a receptor necessary for SARS-CoV-2 entry into host cells, and S protein mutations enhance the affinity for binding to the ACE2 receptor (Q. Li et al., 2021). In using this mechanism, an ACE2 biosensor based on LFIA was designed that was based on the binding target antigen principle of the S1-mAb of red-labeled CNB and the 180E11 antibody of blue-labeled CNB. Then SARS-CoV-2 wild type, the S1 alpha mutant type, the S1 beta mutant type, and neutralizing antibodies were detected according to the line intensity and color change after 20 min (Lee, Lee, et al., 2022b).

Mass spectrometry

Matrix-assisted laser desorption/ionization (MALDI) is a key technique in mass spectrometry (MS)-based proteomics. It has been demonstrated that MALDI-MS approaches, in particular, offer advantages in terms of the speed and sensitivity of analysis where viral proteins are first isolated and then digested. Mass spectrometry was used for the analysis of SARS-CoV-2 variants and their peptide segments and offers a viable and complementary alternative (Griffin & Downard, 2021; Mann et al., 2022) to conventional gene-based sequencing strategies. However, because the sensitivity of protein-based assays needs to be improved, it is necessary to develop and select more specific monoclonal antibodies and tests against SARS-CoV-2 variants for detection in future studies.

Conclusions and perspectives

The SARS-Cov-2 pandemic has brought more virus detection technologies to the forefront. This chapter summarized and analyzed the latest advances in the detection methods of VOCs. As alternatives to sequencing methods, several molecular approaches have been developed for variant determination that are based on the detection of specific mutations associated with a certain variant. Overall, the multiplex real-time qualitative RT-PCR-based assays for SARS-CoV-2 variant identification target some of the mutations in the spike protein associated with the most widespread VOCs and represent simple, accurate, and fast methods, enabling the high-speed detection of known key viral variants. At present, NGS is still the gold standard for determining new emerging SARS-CoV-2 variants. Serum antibody and protein detection, on the other hand, are alternative and necessary complementary methods to nucleic acid detection that can supplement epidemiological information that is not available by other methods. Novel methods, such as isothermal amplification technology, CRISPR technology, biosensors, and microfluidic chip technology, are still under development, although these diagnostic techniques have the advantages of simplicity of operation, short detection time, inexpensive equipment, portability, availability of equipment, and high sensitivity and specificity. However, most of the above techniques have not been tested and validated on a large scale, or with different types of clinical samples. Quickly applying the best detection methods could be helpful for controlling the epidemic of COVID-19 and then taking corresponding health surveillance, isolation, and treatment measures for SARS-CoV-2 variants and their possible new variants in the future. In addition, future research on SARS-CoV-2 variants should focus more on developing and improving methods of detection that are portable, self-testing, mutation site specific, accurate, harmless, rapid, easy to observe, and interpretable results. It can reduce the working hours and labor intensity of healthcare workers and also meet the needs of family self-testing.

Mini-dictionary of terms

- **Spike:** Surface projection of viral glycoproteins of varying lengths spaced at regular intervals on the viral envelope, also called peplomers.
- **Mutation:** A mutation refers to a single change in a virus's genome. Mutations happen frequently, but only sometimes change the characteristics of the virus.
- **Lineage:** A lineage is a group of closely related viruses with a common ancestor. SARS-CoV-2 has many lineages.
- **Variant:** A variant is a viral genome that may contain one or more mutations. In some cases, a group of variants with similar genetic changes, such as a lineage or group of lineages, may be designated by public health organizations as a VOC or a VOI due to shared attributes and characteristics that may require public health action.
- **Amino acid substitution:** A change in a specific amino acid of a protein. This is caused by nonsynonymous mutations. By convention, an amino acid substitution is written in the form N501Y to denote the wild-type amino acid (N (asparagine)) and the substituted amino acid (Y (tyrosine)) at site 501 in the amino acid sequence.

Summary points

- The application of NGS and molecular methodologies to SARS-CoV-2 has been fundamental in epidemiological and other aspects of the fight against COVID-19.
- Different NGS approaches, with different advantages and limitations, can be applied to the sequencing of SARS-CoV-2 variants. Various considerations should influence the choice of approach in different clinical and research contexts.
- As alternatives to sequencing, other methods mainly based on PCR have been developed for the faster identification of SARS-CoV-2 variants.

- Different CRISPR-based detection platforms are also widely used for SARS-CoV-2 variant identification, such as CRISPR-Cas12a, CRISPR-Cas12b, and CRISPR-Cas13, each with its own advantages and limitations.
- Characterization of SARS-CoV-2 direct vs virus protein or serum antibody detection are alternative and necessary complementary approaches to nucleic acid detection that may improve epidemiological data that other methods do not provide.

References

Abnizova, I., te Boekhorst, R., & Orlov, Y. L. (2017). Computational errors and biases in short read next generation sequencing. *Journal of Proteomics & Bioinformatics*, *10*(1). https://doi.org/10.4172/jpb.1000420.

Albert, T. J., Molla, M. N., Muzny, D. M., Nazareth, L., Wheeler, D., Song, X., Richmond, T. A., Middle, C. M., Rodesch, M. J., Packard, C. J., Weinstock, G. M., & Gibbs, R. A. (2007). Direct selection of human genomic loci by microarray hybridization. *Nature Methods*, *4*(11), 903–905.

Alkhatib, M., Salpini, R., Carioti, L., Ambrosio, F. A., D'Anna, S., Duca, L., Costa, G., Bellocchi, M. C., Piermatteo, L., Artese, A., Santoro, M. M., Alcaro, S., Svicher, V., & Ceccherini-Silberstein, F. (2022). Update on SARS-CoV-2 omicron variant of concern and its peculiar mutational profile. *Microbiology Spectrum*, *10*(2), e0273221.

Alves, P. A., de Oliveira, E. G., Franco-Luiz, A. P. M., Almeida, L. T., Gonçalves, A. B., Borges, I. A., de Rocha, F. S., Rocha, R. P., Bezerra, M. F., Miranda, P., Capanema, F. D., Martins, H. R., Weber, G., Teixeira, S. M. R., Wallau, G. L., & do Monte-Neto, R. L. (2021). Optimization and clinical validation of colorimetric reverse transcription loop-mediated isothermal amplification, a fast, highly sensitive and specific COVID-19 molecular diagnostic tool that is robust to detect SARS-CoV-2 variants of concern. *Frontiers in Microbiology*, *12*, 713713.

Amanat, F., Stadlbauer, D., Strohmeier, S., Nguyen, T. H. O., Chromikova, V., McMahon, M., Jiang, K., Arunkumar, G. A., Jurczyszak, D., Polanco, J., Bermudez-Gonzalez, M., Kleiner, G., Aydillo, T., Miorin, L., Fierer, D., Lugo, L. A., Kojic, E. M., Stoever, J., Liu, S. T. H., … Krammer, F. (2020). *A serological assay to detect SARS-CoV-2 seroconversion in humans*. https://doi.org/10.1101/2020.03.17.20037713.

Aoki, A., Mori, Y., Okamoto, Y., & Jinno, H. (2021). Development of a genotyping platform for SARS-CoV-2 variants using high-resolution melting analysis. *Journal of Infection and Chemotherapy: Official Journal of the Japan Society of Chemotherapy*, *27*(9), 1336–1341.

Arizti-Sanz, J., Bradley, A., Zhang, Y. B., Boehm, C. K., Freije, C. A., Grunberg, M. E., Kosoko-Thoroddsen, T.-S. F., Welch, N. L., Pillai, P. P., Mantena, S., Kim, G., Uwanibe, J. N., John, O. G., Eromon, P. E., Kocher, G., Gross, R., Lee, J. S., Hensley, L. E., Happi, C. T., … Myhrvold, C. (2021). Equipment-free detection of SARS-CoV-2 and Variants of Concern using Cas13. *medRxiv: The Preprint Server for Health Sciences*. https://doi.org/10.1101/2021.11.01.21265764.

Bansal, K., & Kumar, S. (2022). Mutational cascade of SARS-CoV-2 leading to evolution and emergence of omicron variant. *Virus Research*, *315*, 198765.

Barlev-Gross, M., Weiss, S., Paran, N., Yahalom-Ronen, Y., Israeli, O., Nemet, I., Kliker, L., Zuckerman, N., Glinert, I., Noy-Porat, T., Alcalay, R., Rosenfeld, R., Levy, H., Mazor, O., Mandelboim, M., Mendelson, E., Beth-Din, A., Israely, T., & Mechaly, A. (2021). Sensitive Immunodetection of severe acute respiratory syndrome coronavirus 2 variants of concern 501Y.V2 and 501Y.V1. *The Journal of Infectious Diseases*, *224*(4), 616–619.

Barua, S., Bai, J., Kelly, P. J., Hanzlicek, G., Noll, L., Johnson, C., Yin, J.-H., & Wang, C. (2022). Identification of the SARS-CoV-2 Delta variant C22995A using a high-resolution melting curve RT-FRET-PCR. *Emerging Microbes & Infections*, *11*(1), 14–17.

Berno, G., Fabeni, L., Matusali, G., Gruber, C. E. M., Rueca, M., Giombini, E., & Garbuglia, A. R. (2022). SARS-CoV-2 variants identification: Overview of molecular existing methods. *Pathogens*, *11*(9), 1058. https://doi.org/10.3390/pathogens11091058.

Bloemen, M., Rector, A., Swinnen, J., Van Ranst, M., Maes, P., Vanmechelen, B., & Wollants, E. (2022). Fast detection of SARS-CoV-2 variants including omicron using one-step RT-PCR and sanger sequencing. *Journal of Virological Methods*, *304*, 114512.

Borillo, G. A., Kagan, R. M., & Marlowe, E. M. (2022). Rapid and accurate identification of SARS-CoV-2 variants using real time PCR assays. *Frontiers in Cellular and Infection Microbiology*, *12*, 894613.

Brito-Mutunayagam, S., Maloney, D., McAllister, G., Dewar, R., McHugh, M., & Templeton, K. (2022). Rapid detection of SARS-CoV-2 variants using allele-specific PCR. *Journal of Virological Methods*, *303*, 114497.

Butler, D., Mozsary, C., Meydan, C., Foox, J., Rosiene, J., Shaiber, A., Danko, D., Afshinnekoo, E., MacKay, M., Sedlazeck, F. J., Ivanov, N. A., Sierra, M., Pohle, D., Zietz, M., Gisladottir, U., Ramlall, V., Sholle, E. T., Schenck, E. J., Westover, C. D., … Mason, C. E. (2021). Shotgun transcriptome, spatial omics, and isothermal profiling of SARS-CoV-2 infection reveals unique host responses, viral diversification, and drug interactions. *Nature Communications*, *12*(1), 1660.

Cagliani, R., Forni, D., Clerici, M., & Sironi, M. (2020). Coding potential and sequence conservation of SARS-CoV-2 and related animal viruses. *Infection, Genetics and Evolution*, *83*, 104353. https://doi.org/10.1016/j.meegid.2020.104353.

Campos, G. S., Sardi, S. I., Falcao, M. B., Belitardo, E. M. M. A., Rocha, D. J. P. G., Rolo, C. A., Menezes, A. D., Pinheiro, C. S., Carvalho, R. H., Almeida, J. P. P., Aguiar, E. R. G. R., & Pacheco, L. G. C. (2020). Ion torrent-based nasopharyngeal swab metatranscriptomics in COVID-19. *Journal of Virological Methods*, *282*, 113888.

Castañeda-Mogollón, D., Kamaliddin, C., Fine, L., Oberding, L. K., & Pillai, D. R. (2022). SARS-CoV-2 variant detection with ADSSpike. *Diagnostic Microbiology and Infectious Disease*, *102*(3), 115606.

Chan, C. T.-M., Leung, J. S.-L., Lee, L.-K., Lo, H. W.-H., Wong, E. Y.-K., Wong, D. S.-H., Ng, T. T.-L., Lao, H.-Y., Lu, K. K., Jim, S. H.-C., Yau, M. C.-Y., Lam, J. Y.-W., Ho, A. Y.-M., Luk, K. S., Yip, K.-T., Que, T.-L., To, K. K.-W, & Siu, G. K.-H. (2022). A low-cost TaqMan minor groove binder probe-based one-step RT-qPCR assay for rapid identification of N501Y variants of SARS-CoV-2. *Journal of Virological Methods*, *299*, 114333.

Cherkaoui, D., Huang, D., Miller, B. S., Turbé, V., & McKendry, R. A. (2021). Harnessing recombinase polymerase amplification for rapid multi-gene detection of SARS-CoV-2 in resource-limited settings. *Biosensors & Bioelectronics*, *189*, 113328.

Chung, H.-Y., Jian, M.-J., Chang, C.-K., Lin, J.-C., Yeh, K.-M., Chen, C.-W., Hsieh, S.-S., Hung, K.-S., Tang, S.-H., Perng, C.-L., Chang, F.-Y., Wang, C.-H., & Shang, H.-S. (2022). Emergency SARS-CoV-2 variants of concern: Novel multiplex real-time RT-PCR assay for rapid detection and surveillance. *Microbiology Spectrum*, *10*(1), e0251321.

Clark, A. E., Wang, Z., Ostman, E., Zheng, H., Yao, H., Cantarel, B., Kanchwala, M., Xing, C., Chen, L., Irwin, P., Xu, Y., Oliver, D., Lee, F. M., Gagan, J. R., Filkins, L., Muthukumar, A., Park, J. Y., Sarode, R., & SoRelle, J. A. (2022). Multiplex fragment analysis for flexible detection of all SARS-CoV-2 variants of concern. *Clinical Chemistry*, *68*(8), 1042–1052. https://doi.org/10.1093/clinchem/hvac081.

Dächert, C., Muenchhoff, M., Graf, A., Autenrieth, H., Bender, S., Mairhofer, H., Wratil, P. R., Thieme, S., Krebs, S., Grzimek-Koschewa, N., Blum, H., & Keppler, O. T. (2022). Rapid and sensitive identification of omicron by variant-specific PCR and nanopore sequencing: Paradigm for diagnostics of emerging SARS-CoV-2 variants. *Medical Microbiology and Immunology*, *211*(1), 71–77.

Dai, C., Guo, M., Wu, Y., Cao, B.-P., Wang, X., Wu, Y., Kang, H., Kong, D., Zhu, Z., Ying, T., Liu, Y., & Wei, D. (2021). Ultraprecise antigen 10-in-1 Pool testing by multiantibodies transistor assay. *Journal of the American Chemical Society*, *143*(47), 19794–19801.

Daniels, R. S., Harvey, R., Ermetal, B., Xiang, Z., Galiano, M., Adams, L., & McCauley, J. W. (2021). A sanger sequencing protocol for SARS-CoV-2 S-gene. *Influenza and Other Respiratory Viruses*, *15*(6), 707–710.

Dinnes, J., Deeks, J. J., Berhane, S., Taylor, M., Adriano, A., Davenport, C., Dittrich, S., Emperador, D., Takwoingi, Y., Cunningham, J., Beese, S., Domen, J., Dretzke, J., Ferrante di Ruffano, L., Harris, I. M., Price, M. J., Taylor-Phillips, S., Hooft, L., Leeflang, M. M., … Cochrane COVID-19 Diagnostic Test Accuracy Group. (2021). Rapid, point-of-care antigen and molecular-based tests for diagnosis of SARS-CoV-2 infection. *Cochrane Database of Systematic Reviews*, *3*(3), CD013705.

Domenico, M. D., Di Domenico, M., De Rosa, A., & Boccellino, M. (2021). Detection of SARS-COV-2 proteins using an ELISA test. *Diagnostics*, *11*(4), 698. https://doi.org/10.3390/diagnostics11040698.

Dorlass, E. G., Lourenço, K. L., Magalhães, R. D. M., Sato, H., Fiorini, A., Peixoto, R., Coelho, H. P., Telezynski, B. L., Scagion, G. P., Ometto, T., Thomazelli, L. M., Oliveira, D. B. L., Fernandes, A. P., Durigon, E. L., Fonseca, F. G., & Teixeira, S. M. R. (2021). Survey of SARS-CoV-2 genetic diversity in two major Brazilian cities using a fast and affordable sanger sequencing strategy. *Genomics*, *113*(6), 4109–4115.

Erster, O., Mendelson, E., Levy, V., Kabat, A., Mannasse, B., Asraf, H., Azar, R., Ali, Y., Shirazi, R., Bucris, E., Bar-Ilan, D., Mor, O., Mandelboim, M., Sofer, D., Fleishon, S., & Zuckerman, N. S. (2021). Rapid and high-throughput reverse transcriptase quantitative PCR (RT-qPCR) assay for identification and differentiation between SARS-CoV-2 variants B.1.1.7 and B.1.351. *Microbiology Spectrum*, *9*(2), e0050621.

Fabiani, M., Margiotti, K., Sabatino, M., Viola, A., Mesoraca, A., & Giorlandino, C. (2022). A rapid and consistent method to identify four SARS-CoV-2 variants during the first half of 2021 by RT-PCR. *Vaccine*, *10*(3). https://doi.org/10.3390/vaccines10030483.

Griffin, J. H., & Downard, K. M. (2021). Mass spectrometry analytical responses to the SARS-CoV2 coronavirus in review. *TrAC Trends in Analytical Chemistry*, *142*, 116328. https://doi.org/10.1016/j.trac.2021.116328.

Handelsman, J. (2004). Metagenomics: Application of genomics to uncultured microorganisms. In. *Microbiology and Molecular Biology Reviews*, *68*(4), 669–685. https://doi.org/10.1128/mmbr.68.4.669-685.2004.

He, C., Lin, C., Mo, G., Xi, B., Li, A., Huang, D., Wan, Y., Chen, F., Liang, Y., Zuo, Q., Xu, W., Feng, D., Zhang, G., Han, L., Ke, C., Du, H., & Huang, L. (2022). Rapid and accurate detection of SARS-CoV-2 mutations using a Cas12a-based sensing platform. *Biosensors and Bioelectronics*, *198*, 113857. https://doi.org/10.1016/j.bios.2021.113857.

Heijnen, L., Elsinga, G., de Graaf, M., Molenkamp, R., Koopmans, M. P. G., & Medema, G. (2021). Droplet digital RT-PCR to detect SARS-CoV-2 signature mutations of variants of concern in wastewater. *The Science of the Total Environment*, *799*, 149456.

Hernandez, M. M., Banu, R., Gonzalez-Reiche, A. S., van de Guchte, A., Khan, Z., Shrestha, P., Cao, L., Chen, F., Shi, H., Hanna, A., Alshammary, H., Fabre, S., Amoako, A., Obla, A., Alburquerque, B., Patiño, L. H., Ramírez, J. D., Sebra, R., Gitman, M. R., … Paniz-Mondolfi, A. E. (2022). Robust clinical detection of SARS-CoV-2 variants by RT-PCR/MALDI-TOF multitarget approach. *Journal of Medical Virology*, *94*(4), 1606–1616.

Huang, X., Zhang, F., Zhu, K., Lin, W., & Ma, W. (2021). dsmCRISPR: Dual synthetic mismatches CRISPR/Cas12a-based detection of SARS-CoV-2 D614G mutation. *Virus Research*, *304*, 198530.

Iglói, Z., Velzing, J., van Beek, J., van de Vijver, D., Aron, G., Ensing, R., Benschop, K., Han, W., Boelsums, T., Koopmans, M., Geurtsvankessel, C., & Molenkamp, R. (2021). Clinical evaluation of Roche SD biosensor rapid antigen test for SARS-CoV-2 in municipal health service testing site, the Netherlands. *Emerging Infectious Diseases*, *27*(5), 1323–1329. https://doi.org/10.3201/eid2705.204688.

Jones, L., Naikare, H. K., Mosley, Y.-Y. C., & Tripp, R. A. (2022). Isothermal amplification using sequence-specific fluorescence detection of SARS coronavirus 2 and variants in nasal swabs. *BioTechniques*, *72*(6), 263–272.

Komor, A. C., Badran, A. H., & Liu, D. R. (2017). CRISPR-based Technologies for the Manipulation of eukaryotic genomes. *Cell*, *169*(3), 559. https://doi.org/10.1016/j.cell.2017.04.005.

Le Hingrat, Q., Visseaux, B., Laouenan, C., Tubiana, S., Bouadma, L., Yazdanpanah, Y., Duval, X., Burdet, C., Ichou, H., Damond, F., Bertine, M., Benmalek, N., Choquet, C., Timsit, J.-F., Ghosn, J., Charpentier, C., Descamps, D., Houhou-Fidouh, N., French, C. O. V. I. D., & Cohort Management Committee, & CoV-CONTACT Study Group. (2021). Corrigendum to "detection of SARS-CoV-2 N-antigen in blood during acute COVID-19 provides a sensitive new marker and new testing alternatives". *Clinical Microbiology and Infection: The Official Publication of the European Society of Clinical Microbiology and Infectious Diseases*, *27*(11), 1713.

Lee, J.-H., Jung, Y., Lee, S.-K., Kim, J., Lee, C.-S., Kim, S., Lee, J.-S., Kim, N.-H., & Kim, H.-G. (2022a). Rapid biosensor of SARS-CoV-2 using specific monoclonal antibodies recognizing conserved Nucleocapsid protein epitopes. *Viruses*, *14*(2). https://doi.org/10.3390/v14020255.

Lee, J.-H., Lee, Y., Lee, S. K., Kim, J., Lee, C.-S., Kim, N. H., & Kim, H. G. (2022b). Versatile role of ACE2-based biosensors for detection of SARS-CoV-2 variants and neutralizing antibodies. *Biosensors & Bioelectronics*, *203*, 114034.

Li, Q., Nie, J., Wu, J., Zhang, L., Ding, R., Wang, H., Zhang, Y., Li, T., Liu, S., Zhang, M., Zhao, C., Liu, H., Nie, L., Qin, H., Wang, M., Lu, Q., Li, X., Liu, J., Liang, H., … Wang, Y. (2021). SARS-CoV-2 501Y.V2 variants lack higher infectivity but do have immune escape. *Cell*, *184*(9). 2362–2371.e9.

Li, X., Wang, J., Geng, J., Xiao, L., & Wang, H. (2022). Emerging landscape of SARS-CoV-2 variants and detection technologies. *Molecular Diagnosis & Therapy*, 1–19.

Liang, Y., Lin, H., Zou, L., Zhao, J., Li, B., Wang, H., Lu, J., Sun, J., Yang, X., Deng, X., & Tang, S. (2021). CRISPR-Cas12a-based detection for the major SARS-CoV-2 variants of concern. *Microbiology Spectrum*, *9*(3), e0101721.

Lin, H., Liang, Y., Zou, L., Li, B., Zhao, J., Wang, H., Sun, J., Deng, X., & Tang, S. (2022). Combination of isothermal recombinase-aided amplification and CRISPR-Cas12a-mediated assay for rapid detection of major severe acute respiratory syndrome coronavirus 2 variants of concern. *Frontiers in Microbiology*, *13*, 945133.

Lo, C.-C., Shakya, M., Connor, R., Davenport, K., Flynn, M., Gutiérrez, A., Hu, B., Li, P.-E., Jackson, E. P., Xu, Y., & Chain, P. S. G. (2022). EDGE COVID-19: a web platform to generate submission-ready genomes from SARS-CoV-2 sequencing efforts. *Bioinformatics*, *38*(10), 2700–2704. https://doi.org/10.1093/bioinformatics/btac176.

Lou, E. G., Sapoval, N., McCall, C., Bauhs, L., Carlson-Stadler, R., Kalvapalle, P., Lai, Y., Palmer, K., Penn, R., Rich, W., Wolken, M., Brown, P., Ensor, K. B., Hopkins, L., Treangen, T. J., & Stadler, L. B. (2022). Direct comparison of RT-ddPCR and targeted amplicon sequencing for SARS-CoV-2 mutation monitoring in wastewater. *The Science of the Total Environment*, *833*, 155059.

Mann, C., Hoyle, J. S., & Downard, K. M. (2022). Detection of SARS CoV-2 coronavirus omicron variant with mass spectrometry. *The Analyst*, *147*(6), 1181–1190.

Matsubara, M., Imaizumi, Y., Fujikawa, T., Ishige, T., Nishimura, M., Miyabe, A., Murata, S., Kawasaki, K., Taniguchi, T., Igari, H., & Matsushita, K. (2022). Tracking SARS-CoV-2 variants by entire S-gene analysis using long-range RT-PCR and Sanger sequencing. *Clinica Chimica Acta; International Journal of Clinical Chemistry*, *530*, 94–98.

Maurano, M. T., Ramaswami, S., Zappile, P., Dimartino, D., Boytard, L., Ribeiro-Dos-Santos, A. M., Vulpescu, N. A., Westby, G., Shen, G., Feng, X., Hogan, M. S., Ragonnet-Cronin, M., Geidelberg, L., Marier, C., Meyn, P., Zhang, Y., Cadley, J., Ordoñez, R., Luther, R., … Heguy, A. (2020). Sequencing identifies multiple early introductions of SARS-CoV-2 to the new York City region. *Genome Research*, *30*(12), 1781–1788.

Medoro, A., Davinelli, S., Voccola, S., Cardinale, G., Passarella, D., Marziliano, N., & Intrieri, M. (2022). Assessment of the diagnostic performance of a novel SARS-CoV-2 antigen sealing tube test strip (colloidal gold) as point-of-care surveillance test. *Diagnostics (Basel, Switzerland)*, *12*(5). https://doi.org/10.3390/diagnostics12051279.

Mills, M. G., Hajian, P., Bakhash, S. M., Xie, H., Mantzke, D., Zhu, H., Perchetti, G. A., Huang, M.-L., Pepper, G., Jerome, K. R., Roychoudhury, P., & Greninger, A. L. (2022). Rapid and accurate identification of SARS-CoV-2 omicron variants using droplet digital PCR (RT-ddPCR). *Journal of Clinical Virology: The Official Publication of the Pan American Society for Clinical Virology*, *154*, 105218.

Miyoshi, H., Ichinohe, R., & Koshikawa, T. (2022). High-resolution melting analysis after nested PCR for the detection of SARS-CoV-2 spike protein G339D and D796Y variations. *Biochemical and Biophysical Research Communications*, *606*, 128–134.

Mohammadi, E., Shafiee, F., Shahzamani, K., Ranjbar, M. M., Alibakhshi, A., Ahangarzadeh, S., Beikmohammadi, L., Shariati, L., Hooshmandi, S., Ataei, B., & HaghjooyJavanmard, S. (2021). Corrigendum to: "Novel and emerging mutations of SARS-CoV-2: Biomedical implications" [Biomed. Pharmacother. 139 (2021) 111599]. *Biomedicine & Pharmacotherapy*, *140*, 111723.

Ning, B., Youngquist, B. M., Li, D. D., Lyon, C. J., Zelazny, A., Maness, N. J., Tian, D., & Hu, T. Y. (2022). Rapid detection of multiple SARS-CoV-2 variants of concern by PAM-targeting mutations. *Cell Reports Methods*, *2*(2), 100173.

Niu, M., Han, Y., Dong, X., Yang, L., Li, F., Zhang, Y., Hu, Q., Xia, X., Li, H., & Sun, Y. (2022). Highly sensitive detection method for HV69-70del in SARS-CoV-2 alpha and omicron variants based on CRISPR/Cas13a. *Frontiers in Bioengineering and Biotechnology*, *10*. https://doi.org/10.3389/fbioe.2022.831332.

Pekin, D., Skhiri, Y., Baret, J.-C., Le Corre, D., Mazutis, L., Salem, C. B., Millot, F., El Harrak, A., Hutchison, J. B., Larson, J. W., Link, D. R., Laurent-Puig, P., Griffiths, A. D., & Taly, V. (2011). Quantitative and sensitive detection of rare mutations using droplet-based microfluidics. *Lab on a Chip*, *11*(13), 2156–2166.

Pillay, S., Giandhari, J., Tegally, H., Wilkinson, E., Chimukangara, B., Lessells, R., Moosa, Y., Mattison, S., Gazy, I., Fish, M., Singh, L., Khanyile, K. S., San, J. E., Fonseca, V., Giovanetti, M., Alcantara, L. C., & de Oliveira, T. (2020). Whole genome sequencing of SARS-CoV-2: Adapting Illumina protocols for quick and accurate outbreak investigation during a pandemic. *Genes*, *11*(8), 949. https://doi.org/10.3390/genes11080949.

dos Santos, C. A., do Carmo Silva, L., de Souza Júnior, M. N., de Melo Mendes, G., Estrela, P. F. N., de Oliveira, K. G., de Curcio, J. S., Resende, P. C., Siqueira, M. M., Pauvolid-Corrêa, A., Duarte, G. R. M., & de Paula Silveira-Lacerda, E. (2022). Detecting lineage-defining mutations in SARS-CoV-2 using colorimetric RT-LAMP without probes or additional primers. *Scientific Reports*, *12*(1). https://doi.org/10.1038/s41598-022-15368-3.

Scrima, M., Cossu, A. M., D'Andrea, E. L., Bocchetti, M., Abruzzese, Y., Iannarone, C., Miarelli, C., Grisolia, P., Melisi, F., Genua, L., Di Perna, F., Maggi, P., Capasso, G., Noviello, T. M. R., Ceccarelli, M., Fucci, A., & Caraglia, M. (2022). Genomic characterization of the emerging SARS-CoV-2 lineage in two districts of Campania (Italy) using next-generation sequencing. *Frontiers in Virology*, *2*. https://doi.org/10.3389/fviro.2022.814114.

Shu, Y., & McCauley, J. (2017). GISAID: Global initiative on sharing all influenza data - from vision to reality. *European Communicable Disease Bulletin*, *22*(13). https://doi.org/10.2807/1560-7917.ES.2017.22.13.30494.

Song, J., Cha, B., Moon, J., Jang, H., Kim, S., Jang, J., Yong, D., Kwon, H.-J., Lee, I.-C., Lim, E.-K., Jung, J., Park, H. G., & Kang, T. (2022). Smartphone-based SARS-CoV-2 and variants detection system using colorimetric DNAzyme reaction triggered by loop-mediated isothermal amplification (LAMP) with clustered regularly interspaced short palindromic repeats (CRISPR). *ACS Nano*, *16*(7), 11300–11314.

Tilloy, V., Cuzin, P., Leroi, L., Guérin, E., Durand, P., & Alain, S. (2022). ASPICov: An automated pipeline for identification of SARS-Cov2 nucleotidic variants. *PLoS One*, *17*(1), e0262953.

Tshiabuila, D., Giandhari, J., Pillay, S., Ramphal, U., Ramphal, Y., Maharaj, A., Anyaneji, U. J., Naidoo, Y., Tegally, H., San, E. J., Wilkinson, E., Lessells, R. J., & de Oliveira, T. (2022). Comparison of SARS-CoV-2 sequencing using the ONT GridION and the Illumina MiSeq. *BMC Genomics*, *23*(1), 319.

Wang, S., Xu, X., Wei, C., Li, S., Zhao, J., Zheng, Y., Liu, X., Zeng, X., Yuan, W., & Peng, S. (2022). Molecular evolutionary characteristics of SARS-CoV-2 emerging in the United States. *Journal of Medical Virology*, *94*(1), 310–317.

Welch, N. L., Zhu, M., Hua, C., Weller, J., Mirhashemi, M. E., Nguyen, T. G., Mantena, S., Bauer, M. R., Shaw, B. M., Ackerman, C. M., Thakku, S. G., Tse, M. W., Kehe, J., Uwera, M.-M., Eversley, J. S., Bielwaski, D. A., McGrath, G., Braidt, J., Johnson, J., … Myhrvold, C. (2022). Multiplexed CRISPR-based microfluidic platform for clinical testing of respiratory viruses and identification of SARS-CoV-2 variants. *Nature Medicine*, *28*(5), 1083–1094.

Wong, Y.-P., Othman, S., Lau, Y.-L., Radu, S., & Chee, H.-Y. (2018). Loop-mediated isothermal amplification (LAMP): A versatile technique for detection of micro-organisms. *Journal of Applied Microbiology*, *124*(3), 626–643.

Wu, F., Zhao, S., Yu, B., Chen, Y.-M., Wang, W., Song, Z.-G., Hu, Y., Tao, Z.-W., Tian, J.-H., Pei, Y.-Y., Yuan, M.-L., Zhang, Y.-L., Dai, F.-H., Liu, Y., Wang, Q.-M., Zheng, J.-J., Xu, L., Holmes, E. C., & Zhang, Y.-Z. (2020). A new coronavirus associated with human respiratory disease in China. *Nature*, *579*(7798), 265–269.

Xiao, M., Liu, X., Ji, J., Li, M., Li, J., Yang, L., Sun, W., Ren, P., Yang, G., Zhao, J., Liang, T., Ren, H., Chen, T., Zhong, H., Song, W., Wang, Y., Deng, Z., Zhao, Y., Ou, Z., … Li, J. (2020). Multiple approaches for massively parallel sequencing of SARS-CoV-2 genomes directly from clinical samples. *Genome Medicine*, *12*(1), 57.

Yong, S. E. F., Anderson, D. E., Wei, W. E., Pang, J., Chia, W. N., Tan, C. W., Teoh, Y. L., Rajendram, P., Toh, M. P. H. S., Poh, C., Koh, V. T. J., Lum, J., Suhaimi, N.-A. M., Chia, P. Y., Chen, M. I.-C., Vasoo, S., Ong, B., Leo, Y. S., Wang, L., & Lee, V. J. M. (2020). Connecting clusters of COVID-19: An epidemiological and serological investigation. *The Lancet Infectious Diseases*, *20*(7), 809–815.

Zhang, H., Ai, J.-W., Yang, W., Zhou, X., He, F., Xie, S., Zeng, W., Li, Y., Yu, Y., Gou, X., Li, Y., Wang, X., Su, H., Zhu, Z., Xu, T., & Zhang, W. (2021). Metatranscriptomic characterization of coronavirus disease 2019 identified a host transcriptional classifier associated with immune signaling. *Clinical Infectious Diseases*, *73*(3), 376–385. https://doi.org/10.1093/cid/ciaa663.

Zhang, P., Chen, L., Hu, J., Trick, A. Y., Chen, F.-E., Hsieh, K., Zhao, Y., Coleman, B., Kruczynski, K., Pisanic, T. R., 2nd, Heaney, C. D., Clarke, W. A., & Wang, T.-H. (2022). Magnetofluidic immuno-PCR for point-of-care COVID-19 serological testing. *Biosensors & Bioelectronics*, *195*, 113656.

Zhu, N., Zhang, D., Wang, W., Li, X., Yang, B., Song, J., … Tan, W. (2020). A novel coronavirus from patients with pneumonia in China, 2019. *New England Journal of Medicine*, *382*(8), 727–733. https://doi.org/10.1056/NEJMoa2001017.

Section B

Management

Chapter 4

Systematic patient assessment for acute respiratory tract ailments (SPARTA): A simple tool to improve outcomes from COVID-19 and other pulmonary diseases

Rajkumar Rajendram[a,b], Naveed Mahmood[a,b], Mohammad Ayaz Khan[b,c], and Hamdan Al-Jahdali[b,c]

[a]*Department of Medicine, King Abdulaziz Medical City, King Abdullah International Medical Research Center, Ministry of National Guard-Health Affairs, Riyadh, Saudi Arabia,* [b]*Department of Medical Education, College of Medicine, King Saud bin Abdulaziz University for Health Sciences, Riyadh, Saudi Arabia,* [c]*Division of Pulmonology, Department of Medicine, King Abdulaziz Medical City, King Abdullah International Medical Research Center, Ministry of National Guard-Health Affairs, Riyadh, Saudi Arabia*

Abbreviations

A-a gradient	alveolar-arterial oxygen gradient
ADRs	adverse drug reactions
Alb	albumin
Alk P	alkaline phosphatase
ALT	alanine transaminase
AP	airway pressure
aPTT	activated partial thromboplastin time
AST	aspartate transaminase
BAL	bronchoalveolar lavage
BE	base excess
bili	bilirubin
BIOCHEM	biochemistry
BNP	brain natriuretic peptide
BP	blood pressure; ceiling of therapy, goals of care/treatment escalation plan/treatment limitations
CFS	clinical frailty scale
CoV	Coronavirus
CK	creatine kinase
CRP	C-reactive protein
CRT	capillary refill time
CURB 65	risk stratification for pneumonia [Confusion, Urea >7 mmoL/L (20 mg/dL), respiratory rate ≥ 30/min, systolic BP < 90 mmHg or diasystolic BP < 60 mmHg, age over 65 years]
CVP	central venous pressure
CXR	chest X-ray
DERM	dermatology
DVT	deep vein thrombosis
EWSS	early warning system score
ECG	electrocardiograph
Echo	echocardiogram
EPAP	expiratory positive airway pressure
ET CO2	end tidal carbon dioxide
fibrino	fibrinogen
FiO2	fraction of inspired oxygen

Management, Body Systems, and Case Studies in COVID-19. https://doi.org/10.1016/B978-0-443-18703-2.00032-3
Copyright © 2024 Elsevier Inc. All rights are reserved, including those for text and data mining, AI training, and similar technologies.

FTC	flow time corrected for heart rate
Fluid Bal	fluid balance
GAST RES	gastric residual
GCS	Glasgow Coma Score
GGT	gamma-glutamyl transferase
Gluc	glucose
HAEM	hematology
Hb	hemoglobin
HCO3	bicarbonate
HDU	high dependency unit
HFNO2	high flow nasal oxygen
HR	heart rate
INR	International Normalized Ratio
IPAP	inspiratory positive airway pressure
ITU	intensive care unit
JVP	jugular venous pressure
LL	lower limb
MAP	mean arterial pressure
(M)BAL	mini-bronchoalveolar lavage
MERS	Middle East respiratory syndrome
MERS-CoV	MERS Coronavirus
MICRO	microbiology
MODE	mode of ventilatory support
MSK	musculoskeletal
NBM	nil by mouth
NG	nasogastric tube
po	per os
NIV	noninvasive ventilation
NSAIDs	nonsteroidal antiinflammatory drugs
PaCO2	arterial partial pressure of carbon dioxide
PaO2	arterial partial pressure of oxygen
PCT	procalcitonin
plt	platelets
PPI	proton pump inhibitor
PT	prothrombin time
qSOFA	quick sequential organ function assessment; rehabilitation pathway, expected achievable functional status on recovery from acute illness
RR	respiratory rate
RRT	renal replacement therapy
ScvO2	central venous oxygen saturation
SOFA	sequential organ function assessment
SPARTA	Systematic Patient assessment for Acute Respiratory Ailments
SpO2	oxygen saturation
Temp	temperature
T Max °C	highest temperature in the preceding 24 h
TV	tidal volume
UL	upper limb
VTE	venous thromboembolism

Introduction

Coronavirus disease 2019 (COVID-19) is a major global public health threat (Ferguson et al., 2020). A similar coronaviral respiratory disease, the Middle East respiratory syndrome (MERS), was first reported in Saudi Arabia in 2012. Dromedary camels are a major reservoir for MERS coronavirus (MERS-CoV) and a source of human infection (World Health Organisation, 2020). From the MERS emergence in 2012 through Jan. 31, 2020, the World Health Organization (WHO) has confirmed 2519 cases of MERS with 866 associated deaths (mortality rate 34%) (World Health Organisation, 2020). About 80% of all reported cases have occurred in Saudi Arabia (World Health Organisation, 2020).

While MERS has spread to several other countries, its incidence outside the Middle East is significantly less than that of COVID-19 (World Health Organisation, 2020). Regardless, experience in the management of MERS was of great value in the fight against COVID-19. In Saudi Arabia, social distancing limited the spread of SARS-CoV-2. Elsewhere, measures to contain the virus have been less effective. Worldwide, by March 16, 2020, there were 164,837 cases and 6470 deaths due to COVID-19 (Ferguson et al., 2020).

A quote from Dr. Daniele Macchine of Italy on March 9, 2020, sums up the situation: "…and there are no more surgeons, urologists, orthopedists, we are only doctors who suddenly become part of a single team to face this tsunami that has overwhelmed us."

Excess deaths occur when hospital networks do not deliver healthcare (e.g., organ support) to patients who need it (Ferguson et al., 2020). The importance of ventilators and beds in intensive therapy units (ITUs) has been sensationalized. However, the capacity to deliver healthcare is, in fact, most limited by human resources. To improve patient outcomes from the COVID-19 pandemic, the capacity of healthcare systems to manage acute respiratory illnesses must be increased. This requires an increase in the number of staff who can do this.

Collaborative approach to quality improvement, training, and patient safety

Increasing healthcare capacity and maintaining standards despite the ongoing shortage of resources require a collaborative approach to quality improvement and training. The systematic patient assessment for acute respiratory tract ailments (SPARTA) proforma (Fig. 1) is the product of collaboration among frontline clinicians, academics, educationalists, and experts in quality and patient safety at King Abdulaziz Medical City (KAMC), Riyadh. It crystallizes our experience in the management of MERS and COVID-19. It is a quality improvement tool that we use to teach clinical skills for a single, defined syndrome (i.e., acute respiratory disease) in a way that can respond to changing needs, practices, and technology. The SPARTA approach to the assessment and treatment of MERS is currently being applied to patients presenting with symptoms suggestive of COVID-19 at KAMC, Riyadh.

The challenge of following guidelines during an evolving pandemic

Guidelines for the management of COVID-19 have been published (Alhazzani et al., 2020; Jin et al., 2020) and updated several times since the start of the pandemic (Alhazzani et al., 2021). Copious amounts of online educational material are available. The website of the Society of Critical Care Medicine contains links to many such resources (Society of Critical Care Medicine, 2020).

However, these resources do not help shattered clinicians, working outside their areas of expertise at the bedside of a deteriorating patient. Clear, practical, and specific guidance on how to assess and manage these patients is required. So, the

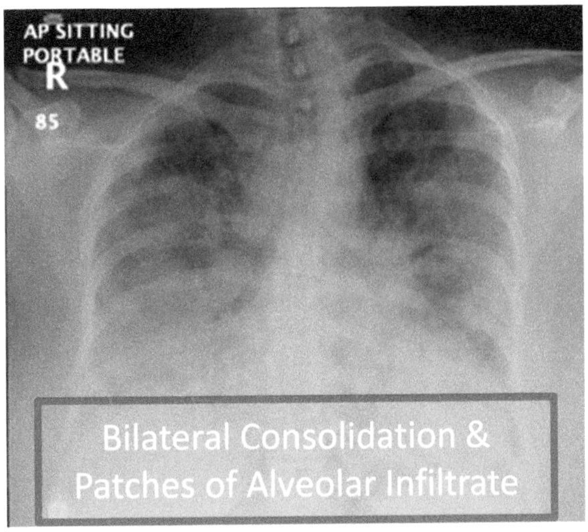

FIG. 1 A chest X-ray showing characteristic features of COVID-19. This chest x-ray shows bilateral alveolar shadowing and consolidation. These are typical features of COVID-19. *(Adapted from Rajendram, R, & Hussain, A. (2021) Severe COVID-19 pneumonia complicated by cardiomyopathy and a small anterior pneumothorax. BMJ Case Reports 14(9), e245900 with permission.)*

SPARTA proforma is far more useful than standard guidelines in an evolving pandemic. Armed with this comprehensive, didactic, cognitive aid, surgeons, urologists, and orthopods can deliver high-quality care to patients with acute respiratory disease alongside internists, pulmonologists, and intensivists.

Focusing on cognitive aids to reduce errors in patient assessment and diagnosis

Healthcare professionals under stress are more likely to overlook fundamental aspects of patient assessment and treatment. So, in emergency situations, important findings can be missed (Robinson et al., 1996; Solberg et al., 1995) and crucial information is routinely omitted from medical records (Osborn et al., 2005; Pullen & Loudon, 2006). This creates a perfect storm. Inadequate assessments greatly increase the risk of diagnostic error and suboptimal management while inaccessible, incomprehensible medical records delay treatment.

The use of structured checklist proformas for recording findings, rather than blank paper, reduces errors of omission and improves documentation (Osborn et al., 2005; Pullen & Loudon, 2006; Robinson et al., 1996; Solberg et al., 1995). Allowing timely access to relevant patient data enhances efficiency and patient safety (Wyatt & Wright, 1998). Indeed, surgeons will be very familiar with the use of the WHO safer surgery checklist, which improves perioperative outcomes (Abbott et al., 2018). The use of checklists in ITUs has also been shown to improve patient outcomes (Weiss et al., 2011). So, features of these checklists have been incorporated into the SPARTA proforma.

Development of the SPARTA proforma

The primary author (RR) has extensive experience in the development of proformas. Examples of proformas that he developed for use by ITUs, high dependency units (HDUs), and critical care outreach teams have been published previously (Rajendram, 2020).

The proformas have been modified extensively over the last 20 years based on his personal experience and extensive consultation with colleagues in several centers of excellence throughout the world. These centers are listed in Table 1. Various versions of these proformas are currently being used in many of the centers listed in Table 1 (Rajendram, 2020).

TABLE 1 Centers involved in the development of documents from which the SPARTA proforma evolved.
King Abdulaziz Medical City, Riyadh, Saudi Arabia
King Khalid University Hospital, King Saud University Medical City, Riyadh, Saudi Arabia
Oxford University Hospitals, United Kingdom (UK)
Queen Mary's Hospital, Sidcup, UK
Queen's Hospital, Romford, UK
Royal Brompton Hospital, London, UK
Royal Free London Hospitals, UK
Royal London Hospital, UK
Royal National Orthopedic Hospital, Stanmore, UK
St Helena General Hospital, St Helena
Stoke Mandeville Hospital, Aylesbury, UK
The Lister Hospital, Stevenage, UK
University College Hospital, London, UK
Victoria Hospital, Kirkaldy, UK

This table lists (in alphabetical order) the centers in which the documents that formed the basis of the SPARTA proforma have been reviewed by former or current colleagues of the primary author. The proforma remains in use in many of these centers.

Important features of the SPARTA proforma

The design of the SPARTA proforma outlines the systematic assessment of patients presenting to hospitals with acute respiratory illnesses. It contains a section for risk assessment for viral infection and a reminder to consider noninfective diagnoses.

The top of the form includes prompts to consider goals of care (i.e., ceilings of therapy and cardiopulmonary resuscitation status). Calculation of the clinical frailty score can help with these difficult decisions. Early warning scoring systems (EWSSs) improve recognition of patient deterioration and expedite intervention (Al Qahtani et al., 2019). They enhance communication and provide auditable standards for healthcare professional responses to deteriorating patients. Use of the SPARTA proforma also standardizes healthcare professional responses to deteriorating patients.

As in all patients admitted to hospitals, vital signs (i.e., mental state, oxygen saturation, respiratory rate, heart rate, blood pressure, temperature, urine output) and physical examinations are fundamental in assessing critically ill patients (Ferguson et al., 2020). Other biomarkers such as lactate can be used to detect organ dysfunction (Ferguson et al., 2020). These are equally relevant in the assessment of patients suspected to have COVID-19.

Imaging of the heart and lungs is extremely important. This has traditionally been dependent on a plain X-ray (Fig. 1) and computed tomography (CT; Fig. 2). However, point of care ultrasound (US; Fig. 3) has transformed the bedside assessment of this cohort (Rajendram et al., 2020). Echocardiography and lung US can reduce reliance on CT (Peng et al., 2020). A strategy for risk assessment of COVID-19 based on symptoms, oxygen requirement, and lung US findings has even been proposed (Stone, 2020). So, sections for imaging have been included on the SPARTA proforma.

Patients infected with SARS-CoV-2 may also be infected with other viruses and bacteria (Alhazzani et al., 2020; Jin et al., 2020). So, empirical antiviral and antibacterial therapy should be considered initially. Antimicrobial stewardship is critical to prevent the development of resistant organisms. So, it is equally important to consider deescalation of therapy when appropriate.

Many medications that are being recommended for the treatment of COVID-19 (e.g., hydroxychloroquine and azithromycin) increase the risk of torsade de pointes by prolonging the QT interval (Jin et al., 2020). So, the rate-corrected QT (QTc) interval should be measured before starting treatment and at least daily thereafter. Fig. 4 is an electrocardiograph showing prolongation of the QT interval in a patient who had recently had COVID-19 and illustrates how to measure the QT interval and calculate the QTc (Rajendram et al., 2022).

Patients with COVID-19 are also at high risk of developing renal and liver dysfunction. A reminder to check the QTc interval and adjust drug doses for liver and renal function has therefore been included on the SPARTA proforma.

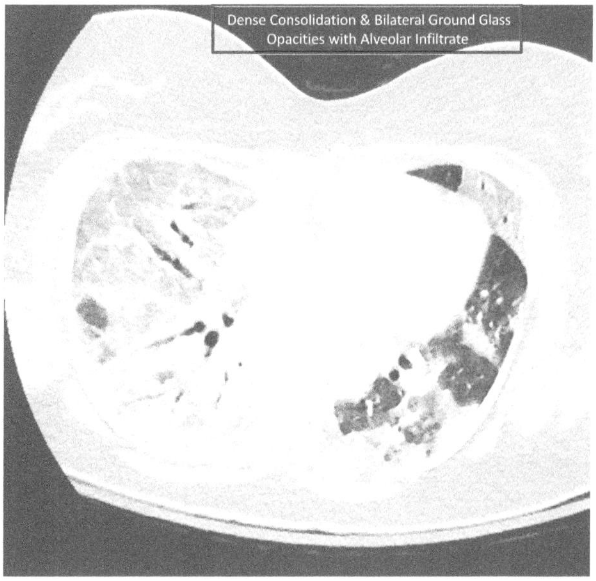

FIG. 2 A computed tomography scan of the chest showing characteristic features of COVID-19. This computed tomography scan shows bilateral ground glass shadowing and dense consolidation. These are typical features of COVID-19. This CT scan was also performed on the same day and on the same patient as the chest X-ray presented in Fig. 1. *(Adapted from Rajendram, R, & Hussain, A. (2021) Severe COVID-19 pneumonia complicated by cardiomyopathy and a small anterior pneumothorax. BMJ Case Reports 14(9), e245900 with permission.)*

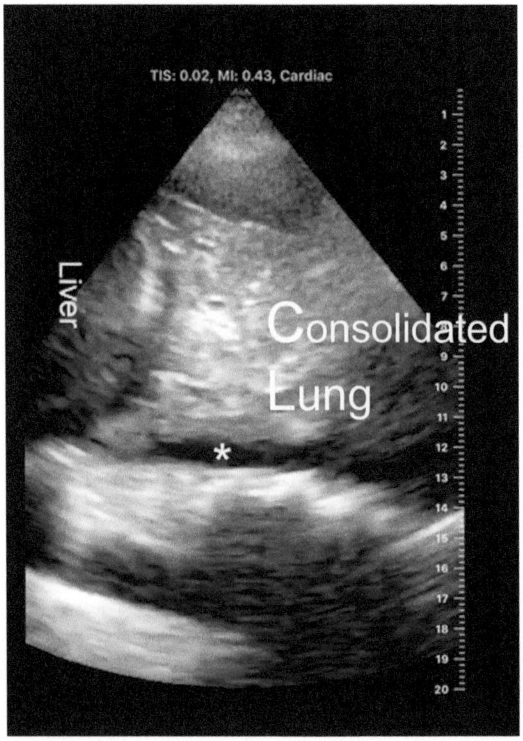

FIG. 3 This lung ultrasound scan shows typical features of COVID-19. This still image from a lung ultrasound scan demonstrates a large area of consolidated lung (C) surrounded by a moderate pleural effusion (*) in the right lung. These findings may be present in patients with COVID-19. *(Adapted from Rajendram, R, & Hussain, A. (2021) Severe COVID-19 pneumonia complicated by cardiomyopathy and a small anterior pneumothorax. BMJ Case Reports 14(9), e245900 with permission.)*

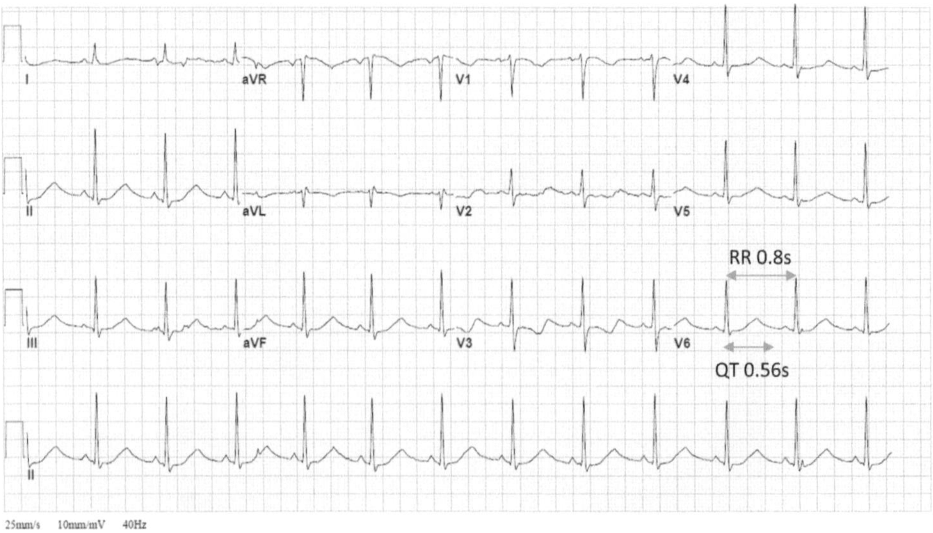

FIG. 4 An electrocardiograph showing acquired prolongation of the QT interval that illustrates how to measure the QT interval and calculate the QTc. The electrocardiograph shows sinus rhythm (heart rate 75 per minute). in the lead V5, the measured QT interval is 560ms. The RR interval is 0.8s. The QT interval corrected for the heart rate (QTc) is 622ms (using Bazett's formula, QT interval/$\sqrt{}$ RR interval$=0.56/\sqrt{0.8}$). *(Adapted from Rajendram, R., Alghamdi, A. A., & Alanazi, M. A. (2022). Acquired long QT syndrome due to antiemetics, COVID-19 and Blastocystis hominis induced exacerbation of congenital chloride losing diarrhoea. BMJ Case Reports 15, e246175 with permission.)*

After assessment of the patient, a clear management plan is required. This may involve isolation, admission to a ward, antimicrobial therapy, a request for HDU or ITU admission, follow up, deisolation, stop deescalation of antimicrobial therapy, step down from HDU to a medical ward, or discharge home. A checklist for this has therefore been included on the proforma.

Policies and procedures

Policies

When under sustained stress such as that induced by the COVID-19 pandemic, healthcare professionals can easily omit aspects of basic patient care. Indeed, documentation may be seen as an unnecessary chore. Yet medical records are important for many reasons besides direct patient care (Mann & Williams, 2003). These are described below.

However, it is particularly important to note that the data recorded by physicians are also used by professional coders to classify patient diagnoses and the treatment they received (Howard & Reddy, 2018). In some healthcare settings, this coding is used to claim reimbursement for the healthcare provided to individual patients. Physicians generally have a poor understanding of this process and the documentation required for accurate coding of clinical encounters (Howard & Reddy, 2018). The SPARTA proforma already incorporates features that facilitate the coding. However, further modifications can be made for specific institutional coding requirements. Thus, the use of the SPARTA proforma could have financial benefits.

Procedures: How to use the SPARTA proforma in clinical practice

Healthcare professionals with basic medical knowledge can use the SPARTA proforma appropriately without any specific training. Simply studying the SPARTA proforma for a few minutes can be enlightening. So, deployment of the SPARTA proforma virtually eliminates the need for formal training on the assessment, risk stratification, and initial treatment of unwell patients presenting with an acute respiratory illness.

Examples of the typical features and potential complications of COVID-19 are presented in Figs. 1–4. Providing these to trainees alongside the proforma further enhances their understanding of the disease process and facilitates their completion of the proforma with the requisite data.

The SPARTA proforma can obviously be downloaded onto handheld smart devices such as phones or tablets. However, armed with paper proformas, our trainees rapidly apply what they have learned without fear of contaminating handheld devices. At KAMC, it is currently used as a simple checklist. However, it is being integrated into the electronic healthcare record.

The importance of the SPARTA proforma as a cognitive aid should not be underestimated. Realizing that a crucial part of the assessment has been omitted after removing personal protective equipment (PPE) is soul destroying. Inexperienced clinicians have often either accepted an incomplete assessment or reflexively nipped back into isolation rooms without PPE. Both temptations must be resisted at all costs, but prevention with the assistance of a cognitive aid is much better than forcing a clinician to self-isolate following inadvertent exposure.

During a crisis such as the evolving COVID-19 pandemic, clinicians with limited experience can be asked to complete the entire SPARTA proforma for each patient that they assess. This can then be reviewed with a senior colleague who can rapidly use the data on patient comorbidities, baseline function, current physiology (i.e., vital signs and EWSS), and the systematic assessment to define treatment plans.

The SPARTA proforma evolved from proformas previously described for use in ITUs (Rajendram, 2020). The similarity of the forms standardizes patient assessment and the documentation of relevant data. Apart from the gain in efficiency from not attempting to reinvent the wheel, this strategy greatly facilitates interdisciplinary communication.

Applications to other areas

Guidance on the management of COVID-19 is being continuously updated. The SPARTA proforma can easily be modified in light of new developments. For example, remdesivir was added to the SPARTA proforma when this became available in the formulary at KAMC. The utility of the proforma is also greatly magnified by modification in response to feedback from junior and senior clinicians at the frontline. However, while local resources, treatment recommendations, and risk stratification may vary, the assessment of this cohort will not change significantly. So, very few if any changes will be required before the SPARTA proforma can be effectively deployed anywhere in the world.

The SPARTA proforma was primarily designed for use in emergency departments, acute medical units, COVID-19 isolation units, and pulmonology wards. However, the SPARTA proforma was developed from proformas used within the ICU and HDU settings (Rajendram, 2020). The SPARTA proforma can be used for patients admitted to critical care areas with any acute respiratory illness.

The primary aim of the SPARTA proforma is to improve patient care. However, it also facilitates documentation. In the 21st century, medical records have many functions besides the direct support of patient care (Mann & Williams, 2003). Contemporaneous documentation of clinical information serves as a medicolegal record of events as well as a source of data for research, audit, service evaluation, performance monitoring, and resource allocation (Mann & Williams, 2003). Thus, the use of the SPARTA proforma can facilitate all these activities.

It is at least possible, if not likely, that other contagious infectious diseases may become pandemics in the coming years. Monkeypox is but one example of this (Bunge et al., 2022). Monkeypox was first diagnosed in humans in the Democratic Republic of the Congo in 1970 (Bunge et al., 2022). Since then, it has spread to the rest of Africa (Bunge et al., 2022). Outbreaks of monkeypox outside the African continent confirm its relevance to the international community (Bunge et al., 2022).

The SPARTA concept and proforma can be applied to other acute illnesses besides COVID-19. The proforma can easily be adapted to include specific details relevant to other acute infectious and noninfectious diseases, including cardiac disease, renal disease, and liver disease.

Conclusion

Well-designed assessment and documentation aids such as the SPARTA proforma reduce the workload while improving data collection and facilitating data retrieval. Using such proformas trains junior physicians and other clinicians working outside their areas of expertise in how to approach patients with acute respiratory disease. So, the SPARTA proforma (Fig. 5) or similar tools could significantly increase the capacity of healthcare systems to deliver healthcare during a pandemic in the future. Such collaborative approaches to patient safety, quality improvement, and training are crucial to minimize casualties in the war against COVID-19.

Mini-dictionary of terms

- **Acute respiratory illness:** Any disease that suddenly affects the proper function of the airways and or lungs (i.e., pulmonary system). They typically manifest with symptoms such as cough and breathlessness, which can be distressing.
- **Early warning scoring systems:** A rapid indicator of the severity of a patient's illness. It is used by healthcare professionals and institutions that provide healthcare to facilitate communication and patient monitoring. These scoring systems are usually based on vital signs.
- **Medical proforma:** A document that satisfies the minimum requirements for the assessment of a patient presenting with a defined condition.
- **Middle East respiratory syndrome:** The Middle East respiratory syndrome is caused by the MERS coronavirus. This virus comes from the same family of viruses that causes COVID-19.
- **Systematic patient assessment for acute respiratory tract ailments:** An approach to the assessment of patients with acute respiratory disease that has been developed by the authors.

Summary points

- Well-designed assessment and documentation aids reduce the healthcare professional's workload while improving data collection and facilitating data retrieval.
- Using proformas can train clinicians working outside their areas of expertise in the approach to patients with acute respiratory disease.
- The use of a proforma could significantly increase the capacity of healthcare systems to deliver healthcare during the COVID-19 pandemic.
- Collaborative approaches to patient safety, quality improvement, and training are crucial to improve the outcomes of COVID-19.

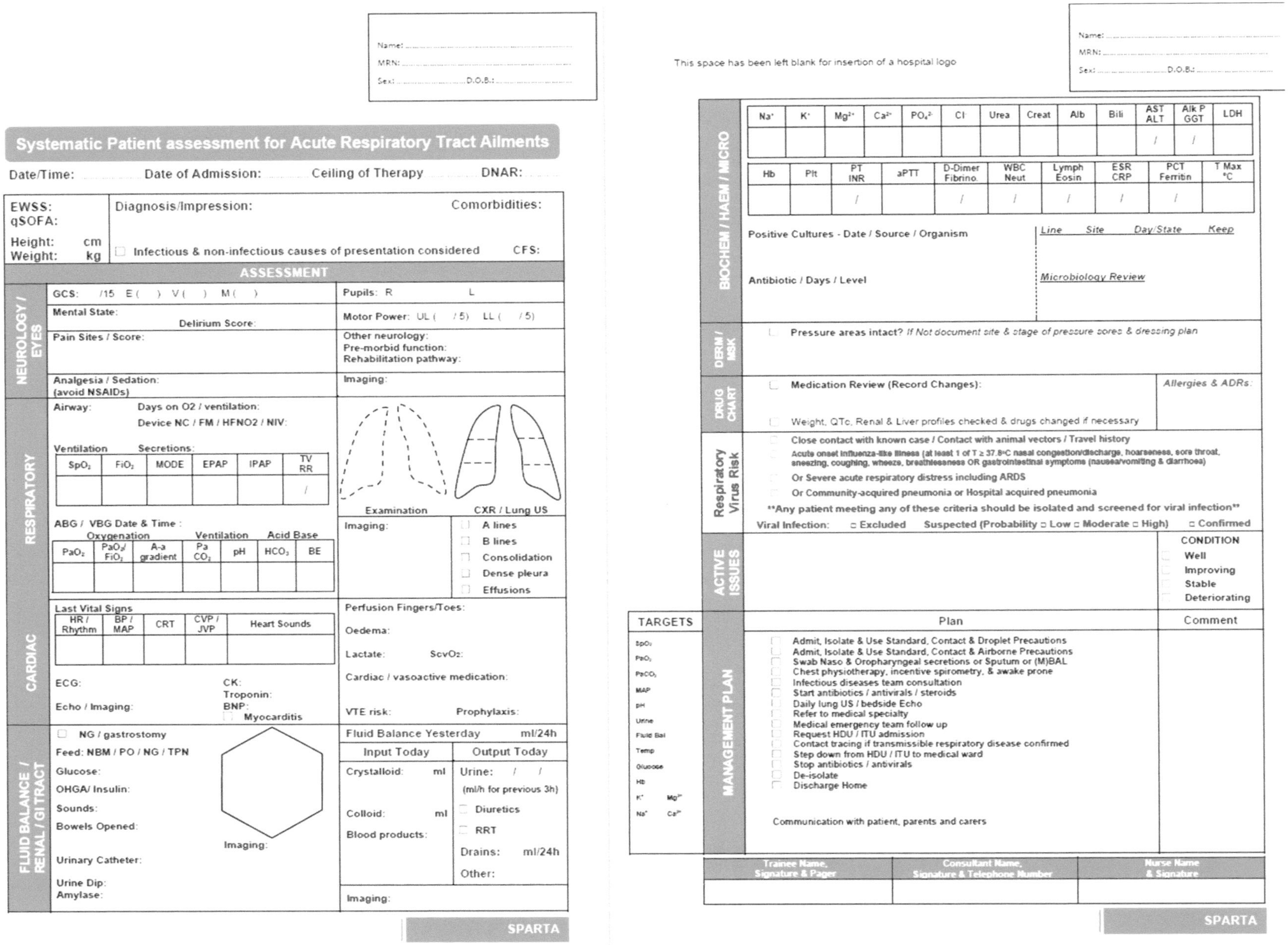
Name:
MRN:
Sex: D.O.B.:

Systematic Patient assessment for Acute Respiratory Tract Ailments

Date/Time: Date of Admission: Ceiling of Therapy DNAR:

EWSS:
qSOFA:
Height: cm
Weight: kg
Diagnosis/Impression:
Comorbidities:
☐ Infectious & non-infectious causes of presentation considered
CFS:

ASSESSMENT

NEUROLOGY / EYES
GCS: /15 E( ) V( ) M( )
Pupils: R L
Mental State:
Delirium Score:
Motor Power: UL (/5) LL (/5)
Pain Sites / Score:
Other neurology:
Pre-morbid function:
Rehabilitation pathway:
Analgesia / Sedation:
(avoid NSAIDs)
Imaging:

RESPIRATORY
Airway:
Days on O2 / ventilation:
Device NC / FM / HFNO2 / NIV:
Ventilation
Secretions:

SpO_2	FiO_2	MODE	EPAP	IPAP	TV RR
					/

Examination
CXR / Lung US
ABG / VBG Date & Time:

Oxygenation			Ventilation	Acid Base		
PaO_2	PaO_2/ FiO_2	A-a gradient	Pa CO_2	pH	HCO_3	BE

Imaging:
☐ A lines
☐ B lines
☐ Consolidation
☐ Dense pleura
☐ Effusions

CARDIAC
Last Vital Signs

HR / Rhythm	BP / MAP	CRT	CVP / JVP	Heart Sounds

Perfusion Fingers/Toes:
Oedema:
Lactate: ScvO2:
Cardiac / vasoactive medication:
ECG:
CK:
Troponin:
BNP:
☐ Myocarditis
Echo / Imaging:
VTE risk: Prophylaxis:

FLUID BALANCE / RENAL / GI TRACT
☐ NG / gastrostomy
Feed: NBM / PO / NG / TPN
Glucose:
OHGA/ Insulin:
Sounds:
Bowels Opened:
Imaging:
Urinary Catheter:
Urine Dip:
Amylase:
Fluid Balance Yesterday ml/24h
Input Today
Crystalloid: ml
Colloid: ml
Blood products:
Output Today
Urine: / /
(ml/h for previous 3h)
☐ Diuretics
☐ RRT
Drains: ml/24h
Other:
Imaging:

SPARTA

This space has been left blank for insertion of a hospital logo

Name:
MRN:
Sex: D.O.B.:

BIOCHEM / HAEM / MICRO

Na^+	K^+	Mg^{2+}	Ca^{2+}	PO_4^{2-}	Cl^-	Urea	Creat	Alb	Bili	AST ALT	Alk P GGT	LDH
										/	/	

Hb	Plt	PT INR	aPTT	D-Dimer Fibrino.	WBC Neut	Lymph Eosin	ESR CRP	PCT Ferritin	T Max °C
		/		/	/	/	/	/	

Positive Cultures - Date / Source / Organism
Line Site Day/State Keep
Antibiotic / Days / Level
Microbiology Review

DERM / MSK
☐ Pressure areas intact? *If Not document site & stage of pressure sores & dressing plan*

DRUG CHART
☐ Medication Review (Record Changes):
Allergies & ADRs:
☐ Weight, QTc, Renal & Liver profiles checked & drugs changed if necessary

Respiratory Virus Risk
☐ Close contact with known case / Contact with animal vectors / Travel history
☐ Acute onset influenza-like illness (at least 1 of T ≥ 37.8°C nasal congestion/discharge, hoarseness, sore throat, sneezing, coughing, wheeze, breathlessness OR gastrointestinal symptoms (nausea/vomiting & diarrhoea)
☐ Or Severe acute respiratory distress including ARDS
☐ Or Community-acquired pneumonia or Hospital acquired pneumonia
Any patient meeting any of these criteria should be isolated and screened for viral infection
Viral Infection: ☐ Excluded Suspected (Probability ☐ Low ☐ Moderate ☐ High) ☐ Confirmed

ACTIVE ISSUES
CONDITION
☐ Well
☐ Improving
☐ Stable
☐ Deteriorating

TARGETS
SpO_2
PaO_2
$PaCO_2$
MAP
pH
Urine
Fluid Bal
Temp
Glucose
Hb
K^+ Mg^{2+}
Na^+ Ca^{2+}

MANAGEMENT PLAN
Plan
☐ Admit, Isolate & Use Standard, Contact & Droplet Precautions
☐ Admit, Isolate & Use Standard, Contact & Airborne Precautions
☐ Swab Naso & Oropharyngeal secretions or Sputum or (M)BAL
☐ Chest physiotherapy, incentive spirometry, & awake prone
☐ Infectious diseases team consultation
☐ Start antibiotics / antivirals / steroids
☐ Daily lung US / bedside Echo
☐ Refer to medical specialty
☐ Medical emergency team follow up
☐ Request HDU / ITU admission
☐ Contact tracing if transmissible respiratory disease confirmed
☐ Step down from HDU / ITU to medical ward
☐ Stop antibiotics / antivirals
☐ De-isolate
☐ Discharge Home
Communication with patient, parents and carers
Comment

Trainee Name, Signature & Pager	Consultant Name, Signature & Telephone Number	Nurse Name & Signature

SPARTA

FIG. 5 Systematic patient assessment for acute respiratory tract ailments proforma. The abbreviations used in the SPARTA proforma are common in clinical practice in many countries. However, those wishing to use the forms must ensure that their teams are familiar with these abbreviations or should modify them if necessary.

References

Abbott, T., Ahmad, T., Phull, M. K., Fowler, A. J., Hewson, R., Biccard, B. M., Chew, M. S., Gillies, M., Pearse, R. M., & International Surgical Outcomes Study (ISOS) Group. (2018). The surgical safety checklist and patient outcomes after surgery: A prospective observational cohort study, systematic review and meta-analysis. *British Journal of Anaesthesia, 120*, 146–155.

Al Qahtani, M. A., Rajendram, R., Binsalih, S., et al. (2019). Using an early warning score system in the acute medicine unit of a medical city in Saudi Arabia is feasible and reduces admissions to intensive care. *Journal of Health Informatics in Developing Countries, 13*, 239.

Alhazzani, W., Evans, L., Alshamsi, F., Møller, M. H., Ostermann, M., Prescott, H. C., Arabi, Y. M., Loeb, M., Ng Gong, M., Fan, E., Oczkowski, S., Levy, M. M., Derde, L., Dzierba, A., Du, B., Machado, F., Wunsch, H., Crowther, M., Cecconi, M., … Rhodes, A. (2021). Surviving sepsis campaign guidelines on the management of adults with coronavirus disease 2019 (COVID-19) in the ICU: First update. *Critical Care Medicine, 49*, e219–e234.

Alhazzani, W., Møller, M. H., Arabi, Y. M., Loeb, M., Gong, M. N., Fan, E., Oczkowski, S., Levy, M. M., Derde, L., Dzierba, A., Du, B., Aboodi, M., Wunsch, H., Cecconi, M., Koh, Y., Chertow, D. S., Maitland, K., Alshamsi, F., Belley-Cote, E., … Rhodes, A. (2020). Surviving Sepsis campaign: Guidelines on the Management of Critically ill Adults with coronavirus disease 2019 (COVID-19). *Critical Care Medicine, 48*, e440–e469.

Bunge, E. M., Hoet, B., Chen, L., Lienert, F., Weidenthaler, H., Baer, L. R., & Steffen, R. (2022). The changing epidemiology of human monkeypox-a potential threat? A systematic review. *PLoS Neglected Tropical Diseases, 16*, e0010141.

Ferguson, N. M., Laydon, D., Nedjati-Gilani, G., et al. (2020). *Impact of non-pharmaceutical interventions (NPIs) to reduce COVID19 mortality and healthcare demand.* Imperial College London. https://www.imperial.ac.uk/media/imperial-college/medicine/sph/ide/gida-fellowships/Imperial-College-COVID19-NPI-modelling-16-03-2020.pdf (Accessed 01/08/22).

Howard, R., & Reddy, R. M. (2018). Coding discrepancies between medical student and physician documentation. *Journal of Surgical Education, 75*, 1230–1235.

Jin, Y. H., Cai, L., Cheng, Z. S., Cheng, H., Deng, T., Fan, Y. P., Fang, C., Huang, D., Huang, L. Q., Huang, Q., Han, Y., Hu, B., Hu, F., Li, B. H., Li, Y. R., Liang, K., Lin, L. K., Luo, L. S., Ma, J., … Evidence-Based Medicine Chapter of China International Exchange and Promotive Association for Medical and Health Care (CPAM). (2020). A rapid advice guideline for the diagnosis and treatment of 2019 novel coronavirus (2019-nCoV) infected pneumonia (standard version). *Military Medical Research, 7*, 4.

Mann, R., & Williams, J. (2003). Standards in medical record keeping. *Clinical Medicine (London, England), 3*, 329–332.

Osborn, G. D., Pike, H., Smith, M., Winter, R., & Vaughan-Williams, E. (2005). Quality of clinical case note entries: How good are we at achieving set standards? *Annals of the Royal College of Surgeons of England, 87*, 458–460.

Peng, Q. Y., Wang, X. T., Zhang, L. N., & Chinese Critical Care Ultrasound Study Group (CCUSG). (2020). Findings of lung ultrasonography of novel corona virus pneumonia during the 2019-2020 epidemic. *Intensive Care Medicine, 46*, 849–850.

Pullen, I., & Loudon, J. (2006). Improving standards in clinical record-keeping. *Advances in Psychiatric Treatment, 12*, 280–286.

Rajendram, R. (2020). Checklist Proformas to guide and document the assessment of critically Ill patients: A tool to standardize assessment and minimise diagnostic error. *American Journal of Anesthesia & Clinical Research, 5*, 27–36.

Rajendram, R., Alghamdi, A. A., & Alanazi, M. A. (2022). Acquired long QT syndrome due to antiemetics, COVID-19 and *Blastocystis hominis* induced exacerbation of congenital chloride losing diarrhoea. *BMJ Case Reports, 15*, e246175.

Rajendram, R., Hussain, A., Mahmood, N., & Kharal, M. (2020). Feasibility of using a handheld ultrasound device to detect and characterize shunt and deep vein thrombosis in patients with COVID-19: An observational study. *The Ultrasound Journal, 12*, 49.

Robinson, S. M., Harrison, B. D., & Lambert, M. A. (1996). Effect of a preprinted form on the management of acute asthma in an accident and emergency department. *Journal of Accident & Emergency Medicine, 13*(2), 93–97.

Society of Critical Care Medicine. (2020). *Critical Care for Non-ICU Clinicians.* https://www.sccm.org/Disaster/COVID19-ResourceResponseCenter. (Accessed 01 August 2022).

Solberg, E. E., Aabakken, L., Sandstad, O., Bach-Gansmo, E., Nordby, G., Enger, E., & Stene-Larsen, G. (1995). Journalen-innhold, forventning og kvalitet. En studie av 100 indremedisinske innkomstjournaler [The medical record- -content, interpretation and quality. Study of 100 medical records from a department of internal medicine]. *Tidsskrift for den Norske laegeforening: tidsskrift for praktisk medicin, ny raekke, 115*(4), 488–489.

Stone, M. (2020). *COVID-19 lung ultrasound triage.* Butterfly Inc. Available at https://assets.website-files.com/5a0cbe08f1138d000147a9d4/5e7263c656e99f4b26c9a75b_2020-03_COVID_treatment.pdf. (Accessed 1 August 2022).

Weiss, C. H., Moazed, F., McEvoy, C. A., Singer, B. D., Szleifer, I., Amaral, L. A., Kwasny, M., Watts, C. M., Persell, S. D., Baker, D. W., Sznajder, J. I., & Wunderink, R. G. (2011). Prompting physicians to address a daily checklist and process of care and clinical outcomes: A single-site study. *American Journal of Respiratory and Critical Care Medicine, 184*(6), 680–686.

World Health Organisation. (2020). *Middle East respiratory syndrome coronavirus (MERS-CoV)—The Kingdom of Saudi Arabia.* Disease Outbreak News: Update 24 February 2020. https://www.who.int/csr/don/24-february-2020-mers-saudi-arabia/en/. (Accessed 1 August 2022).

Wyatt, J. C., & Wright, P. (1998). Design should help use of patients' data. *Lancet (London, England), 352*(9137), 1375–1378.

Chapter 5

Point of care ultrasound for coronavirus disease 2019: The multiorgan approach to COVID-19

Rajkumar Rajendram[a,b]

[a]*Department of Medicine, King Abdulaziz Medical City, King Abdullah International Medical Research Center, Ministry of National Guard-Health Affairs, Riyadh, Saudi Arabia,* [b]*Department of Medical Education, College of Medicine, King Saud bin Abdulaziz University for Health Sciences, Riyadh, Saudi Arabia*

Abbreviations

BLUE	bedside lung ultrasound in emergency
COVID-19	coronavirus disease 2019
CT	computed tomography
PoCUS	point of care ultrasound
SARS-CoV-2	severe acute respiratory syndrome coronavirus 2
WINFOCUS	World Interactive Network Focused On Critical UltraSound

Introduction

The spread of severe acute respiratory syndrome coronavirus 2 (SARS-CoV-2) initiated a critical global health emergency at the end of 2019. The resulting coronavirus disease 2019 (COVID-19) pandemic resulted in millions of deaths worldwide.

Many healthcare systems worldwide were overwhelmed. Thus, the SARS-CoV-2 pandemic invoked major changes to the preexisting, well-established approaches to diagnosis and treatment. Vast numbers of patients had to be evaluated rapidly while the risk of viral transmission to healthcare professionals, other patients, and their relatives was minimized.

This necessity brought point-of-care ultrasound (PoCUS) into the limelight. PoCUS is performed by trained medical professionals who are not specialists in radiology. PoCUS is used for diagnosis, monitoring, or guiding procedures wherever a patient is being treated (i.e., at the bedside).

A "standard" ultrasound study performed by a radiologist or a sonographer is a comprehensive diagnostic imaging examination. PoCUS is a more simplified form of imaging that is integrated into the bedside assessment by the physician(s) attending the patient. Although both imaging modalities provide valuable information for clinicians treating patients, the roles of PoCUS and standard ultrasound differ.

When available and deployed effectively, PoCUS can streamline the approach to the diagnosis and treatment of COVID-19 and its complications (Hussain et al., 2020). The use of PoCUS reduces patients' exposure to the harmful ionizing radiation required for imaging techniques such as x-rays (Fig. 1) and computed tomography (CT; Fig. 2) (Ultrasound Guidelines, 2017).

This chapter reviews the role of PoCUS in the management of COVID-19. Several literature databases were searched to identify relevant articles for this narrative review. These included PubMed, Scopus, Embase, and Google Scholar. PubMed, for example, was searched from inception until Jan. 3, 2023, using the search query: (("Point-of-Care Systems"[Mesh]) AND ("Ultrasonography"[Mesh] OR "Ultrasonography, Doppler"[Mesh] OR "Ultrasonography, Interventional"[Mesh])) AND ("COVID-19"[Mesh] OR "SARS-CoV-2"[Mesh]) to identify all articles that were relevant to PoCUS and COVID-19. Surprisingly, there were very few data on the use of PoCUS in COVID-19 (Fig. 3). This search identified only 199 articles. For comparison, searching PubMed from inception until Jan. 3, 2023 using the search query: ("COVID-19"[Mesh] OR "SARS-CoV-2"[Mesh]) identified 224,675 articles.

Management, Body Systems, and Case Studies in COVID-19. https://doi.org/10.1016/B978-0-443-18703-2.00046-3
Copyright © 2024 Elsevier Inc. All rights are reserved, including those for text and data mining, AI training, and similar technologies.

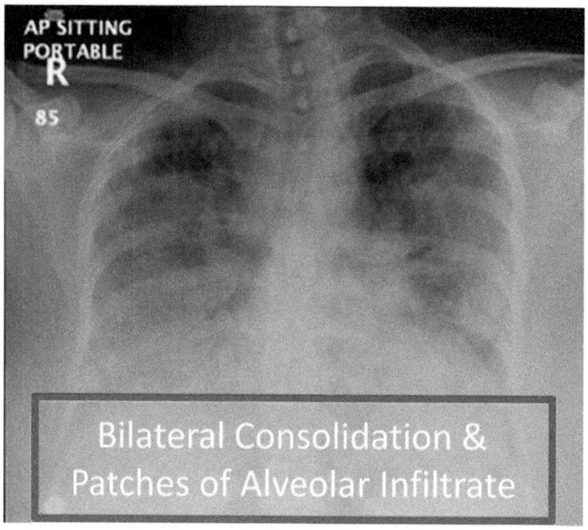

FIG. 1 A chest X-ray showing characteristic features of COVID-19. This chest X-ray shows typical features of COVID-19 pneumonia. There is bilateral alveolar shadowing and consolidation. *(Adapted from Rajendram, R., & Hussain, A. (2021). Severe COVID-19 pneumonia complicated by cardiomyopathy and a small anterior pneumothorax.* BMJ Case Reports, 14(9), *e245900, with permission.)*

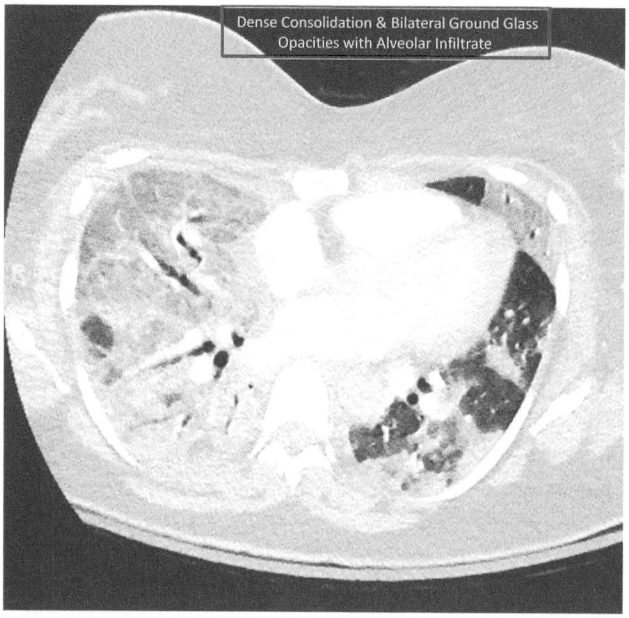

FIG. 2 A computed tomography scan of the chest showing characteristic features of COVID-19. This computed tomography scan shows bilateral ground glass shadowing and dense consolidation. These are typical features of COVID-19. However, the involvement of the right lung is much worse than the left. This CT scan was also performed on the same day and on the same patient as the chest X-ray presented in Fig. 1. *(Adapted from Rajendram, R., & Hussain, A. (2021). Severe COVID-19 pneumonia complicated by cardiomyopathy and a small anterior pneumothorax.* BMJ Case Reports, 14(9), *e245900, with permission.)*

Pulmonary pathology in COVID-19

Severe COVID-19 most commonly affects the lung. Thus, the imaging of patients with COVID-19 initially targeted pulmonary pathologies. In patients with COVID-19 pneumonia, a CT typically shows a bilateral interstitial pneumonia with irregular asymmetrical areas of consolidation or ground glass change that predominantly affects the peripheries (Zieleskiewicz et al., 2020; Fig. 2).

Ultrasound cannot penetrate deep into the lung parenchyma. The predominantly peripheral distribution of the lung involvement explains why PoCUS is so useful in the assessment of COVID-19 pneumonia.

The typical lung ultrasound findings in COVID-19 pneumonia include the presence of bilateral areas with confluent B-lines (Fig. 4) that alternate with spared areas (Volpicelli et al., 2020). This finding is not specific to COVID-19 and can be

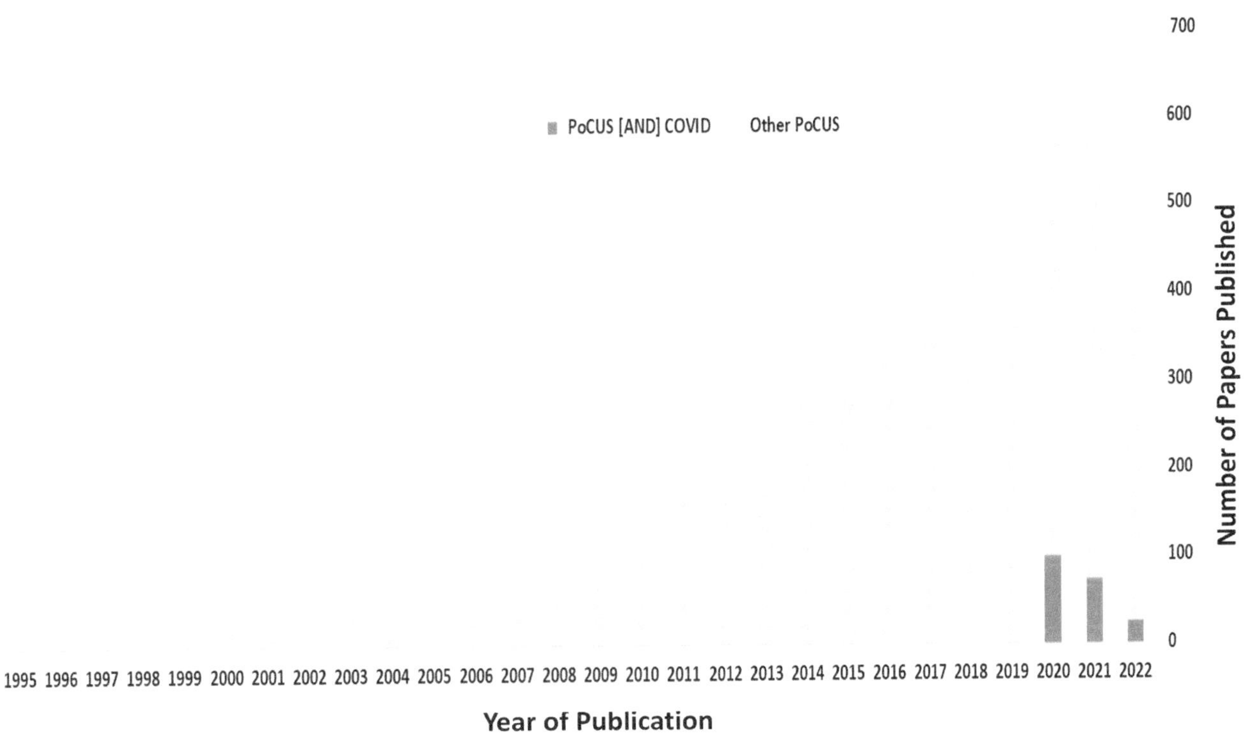

FIG. 3 The number of papers published on PoCUS and COVID-19. This figure illustrates the number of papers on PoCUS published and listed in PubMed since its inception. The subset of papers on PoCUS and COVID-19 are highlighted (orange bars). As SARS-CoV-2 was only identified in 2019, the first papers on COVID-19 and PoCUS were published in 2020. The figure was created using the search query in PubMed from inception to Dec. 31, 2022: (("Point-of-Care Systems"[Mesh]) AND ("Ultrasonography"[Mesh] OR "Ultrasonography, Doppler"[Mesh] OR "Ultrasonography, Interventional"[Mesh])) AND ("COVID-19"[Mesh] OR "SARS-CoV-2"[Mesh]) to identify all articles in PubMed that were relevant to PoCUS and COVID-19. For comparison, the articles identified using that search were subtracted from the articles identified using the search query in PubMed from inception to Dec. 31, 2022: (("Point-of-Care Systems"[Mesh]) AND ("Ultrasonography"[Mesh] OR "Ultrasonography, Doppler"[Mesh] OR "Ultrasonography, Interventional"[Mesh])). Thus, this figure presents the articles relevant to COVID-19 and PoCUS as a proportion of all articles on PoCUS. To contextualize this further, the search query "SARS-CoV-2"[Mesh] OR "COVID-19"[Mesh] in PubMed from inception to Dec. 31, 2022 identified 213,985 articles.

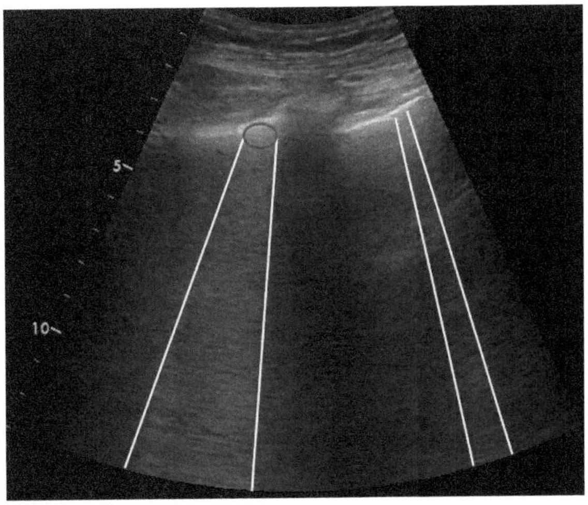

FIG. 4 Lung ultrasound scan showing B-lines in a patient with COVID-19. This still image from a lung ultrasound scan of the right lung of a patient with COVID-19 shows multiple B-lines (slightly hyperechoic vertical ultrasound reverberation artifacts outlined in yellow). They originate from the pleura, extend the full depth of the image, and move with lung sliding. The confluent area B-lines to the left of the image arise from a small peripheral consolidation (brown outline). A single B-line arises from the hyperechoic pleural line on the left of the image. These findings may be present in patients with COVID-19.

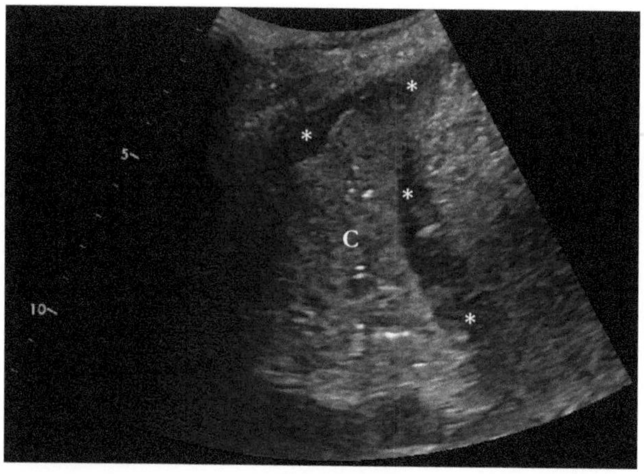

FIG. 5 Lung ultrasound scan showing consolidated lung in a patient with COVID-19. This still image from an ultrasound scan of a patient with COVID-19 demonstrates a large area of consolidated lung (C) surrounded by a moderate pleural effusion (*) in the right lung. Consolidated lung appears as focal hypoechoic areas at the periphery of the lung, which is known as the C-pattern of a lung ultrasound. Pleural effusion appears as an anechoic area on a lung ultrasound. These findings may be present in patients with COVID-19.

seen in patients with other diseases (Soldati et al., 2009; Vetrugno, Baciarello, et al., 2020). However, in patients with symptoms suggestive of COVID-19, this lung ultraound (LUS) pattern can help physicians detect lung involvement. Lung ultrasound may also identify consolidations (C-pattern; Fig. 5), signs suggestive of bacterial superinfection, or the presence of complications such as pneumothorax or pleural effusion (Rajendram et al., 2020; Rajendram & Hussain, 2021).

The progression, severity, and complications of COVID-19 can be detected and monitored with lung ultrasound. This information can be used to guide decisions about admission, discharge, or the clinical setting in which the patient should be treated (i.e., discharge home, admit to regular ward, or transfer to critical care area; Hussain et al., 2020; Peixoto et al., 2021). However, the patchy nature of pulmonary involvement by COVID-19 mandated the use of PoCUS protocols to standardize image acquisition.

Diagnosis of SARS-CoV-2 infection and pneumonia

Infection with SARS-CoV-2 is routinely identified by detection of the virus using reverse-transcriptase polymerase-chain-reaction (RT-PCR) on samples usually obtained via nasopharyngeal swab. In the hospital, COVID-19 pneumonia is often diagnosed on a chest x-ray or CT. These imaging modalities may reveal consolidation, ground-glass opacities, reticular shadowing, and "crazy paving" (Ye et al., 2020). Nonspecific signs suggestive of pneumonia or other complications of COVID-19 can be detected by PoCUS (Rajendram & Hussain, 2021; Volpicelli et al., 2020). A lung ultrasound can rule out COVID-19 pneumonia in patients presenting to the emergency department (ED; sensitivity 86%, specificity 92%, diagnostic accuracy 89%; Di Gioia et al., 2022). Patients without respiratory failure who are asymptomatic or only have mild symptoms benefit from screening for pulmonary involvement with PoCUS (Peixoto et al., 2021).

When x-ray and CT are not available (e.g., in prehospital and primary care settings or in patients' homes), PoCUS can identify those patients at highest risk of deterioration. This application of PoCUS improves decisions about who can safely be managed at home (Peixoto et al., 2021). Unnecessary hospital admissions can be prevented when PoCUS is deployed effectively (Peixoto et al., 2021).

Triage

At the peak of the COVID-19 pandemic, many healthcare systems organized well-defined pathways to manage the flow of patients with COVID-19 among prehospital services, primary care, and dedicated COVID-19 hospitals. The COVID-19 hospitals often set up pretriage units outside their ED. The structures of these services took various forms. However, most pretriage units had dedicated physicians or paramedics and the ability to perform rapid diagnostic tests for SARS-CoV-2 (i.e., RT-PCR, serology).

Pneumonia and many other complications of COVID-19 can be identified with PoCUS (Rajendram & Hussain, 2021; Volpicelli et al., 2020). So, many pretriage units used POCUS for the early diagnosis of COVID-19 and follow-up of patients (Brenner et al., 2021). A scanning protocol was often used to standardize the use of PoCUS.

PoCUS protocols and scoring systems for pulmonary pathology in COVID-19

Before the pandemic, the bedside lung ultrasound in emergency (BLUE) protocol was one of the most widely used pulmonary PoCUS scanning protocols worldwide. The BLUE protocol facilitates the differential diagnosis of patients presenting to the ED with acute respiratory failure (Lichtenstein & Mezière, 2008). Several new protocols were created during the pandemic. For example, an extended version of the BLUE protocol was used to detect high-risk patients with COVID-19 and stratify the level of observation the patients required (Morin et al., 2021). Another group described a 12-point protocol for the detection of COVID-19 (sensitivity 86%, specificity 71%, negative predictive value 89%; Bianchi et al., 2022).

During the pandemic, the pretest probability of SARS-CoV-2 infection was high in every patient. In that context, the identification of the LUS abnormalities described below could then be considered "diagnostic" of COVID-19 pneumonia. However, as the incidences of COVID-19 decreased, such nonspecific LUS findings could be associated with any cause of pneumonia (Vetrugno, Baciarello, et al., 2020).

Scoring systems such as the lung ultrasound score (LUS) have been derived to define the extent of lung diseases (Lichter et al., 2020). The LUS score is based on the presence of changes such as B-lines, pleural abnormalities, and consolidations (Lichter et al., 2020). A LUS score over 18 was associated with severe damage, deterioration, and the need for noninvasive ventilation (Lichter et al., 2020). However, the sensitivity of this cut-off (i.e., LUS score >18) is low (62%; Lichter et al., 2020). A cut-off of 20 more accurately predicts mortality from COVID-19 pneumonia (Tana et al., 2022).

Although the initial use of PoCUS during the COVID-19 pandemic targeted pulmonary pathologies, SARS-CoV-2 infection can involve the heart (Dweck et al., 2020; Hussain et al., 2020) and is also associated with a high risk of thromboembolism (Chi et al., 2020; Hussain et al., 2020). So, other PoCUS protocols were used to evaluate these complications in patients with COVID-19.

PoCUS for the cardiovascular complications of COVID-19

The use of POCUS to assess the heart is also known as focused cardiac ultrasound (FOCUS). The role, scope,and limitations of FOCUS were defined by the consensus of an international panel of experts in 2014 (Via et al., 2014). Albeit much more limited than echocardiography, FOCUS can detect preexisting heart disease and identify cardiac complications of COVID-19 such as cardiomyopathy (Hussain et al., 2020).

Dweck et al. (2020) conducted a prospective study that identified abnormalities in 667 (55%) of 1216 patients with COVID-19 on echocardiography. Abnormalities of the left and right ventricles were found in 479 (39%) and 397 (33%), respectively (Dweck et al., 2020). Echocardiography changed disease-specific therapy in 42% of cases (171/405; Dweck et al., 2020). These changes included the initiation of treatment for heart failure, acute coronary syndrome, pulmonary embolism, tamponade, and endocarditis (Dweck et al., 2020).

While Dweck et al. (2020) used formal echocardiography, many of the abnormalities they described fall within the scope of FOCUS as defined by Via et al. (2014). Indeed, FOCUS performed using a handheld ultrasound device was able to detect right-to-left shunt via a patent foramen ovale in a patient with severe COVID-19 (Rajendram et al., 2020; Rajendram, Hussain, et al., 2021). Thus, cardiac PoCUS (i.e., FOCUS) is a valuable tool for patients with COVID-19 in whom cardiac impairment is suspected.

PoCUS for the guidance of fluid management in patients with COVID-19

PoCUS can also be used to guide the administration of intravenous fluids in critically ill patients (Via et al., 2014). Thus, assessment of the collapsibility and/or distensibility of the inferior vena cava could be used to guide fluid management in patients with COVID-19 (Hussain et al., 2020).

PoCUS for the thromboembolic complications of COVID-19

The risk of venous thromboembolism (i.e., pulmonary embolism and deep vein thrombosis) is increased in patients with COVID-19 (Chi et al., 2020; Hussain et al., 2020). During the pandemic, PoCUS was used extensively to screen patients for deep vein thrombosis (Hussain et al., 2020).

Multiorgan PoCUS protocols

Recognizing the broad range of indications for PoCUS in COVID-19, an international panel of experts recommended a multiorgan approach when using PoCUS in patients infected with SARS-CoV-2 (Hussain et al., 2020). Multiorgan PoCUS is a well-tolerated, low-cost imaging modality that can be deployed rapidly for a quick bedside evaluation of the heart, lungs, and major blood vessels (Hussain et al., 2020). As a result, this tool was widely utilized on the frontlines of the battle against COVID-19.

The use of handheld ultrasound devices during the COVID-19 pandemic

The application of PoCUS during the pandemic was facilitated by ultraportable ultrasound devices (Khanji et al., 2020; Rajendram et al., 2020). These affordable, handheld devices only became available a few years before the start of the pandemic. Though small, these devices are able to provide remarkably good, diagnostic-quality images. Indeed, a handheld ultrasound device was able to detect right-to-left shunt via a patent foramen ovale (Rajendram et al., 2020).

These ultraportable devices greatly simplified the use of ultrasound in the diagnosis and management of patients with COVID-19 (Khanji et al., 2020). When available, these versatile devices reduced the risk of cross infection in comparison to laptop-sized and cart-based devices (Khanji et al., 2020). The use of ultraportable ultrasound devices was therefore better for patients and safer for staff (Khanji et al., 2020).

Policies and procedures

The World Interactive Network Focused On Critical UltraSound (WINFOCUS) is a leading scientific organization committed to the development of PoCUS. During the COVID-19 pandemic, the members of WINFOCUS promoted the practice of PoCUS and conducted research and training relevant to PoCUS. Hussain et al. (2020) organized a consensus conference to shape and define the role of PoCUS in COVID-19.

Thus, Hussain et al. (2020) highlighted the important role of PoCUS in the management of COVID-19. The multiorgan point-of-care ultrasound for COVID-19 (PoCUS4COVID) international expert consensus outlined the potential applications and limitations of PoCUS in patients with COVID-19 (Hussain et al., 2020). The nine domains covered by the expert consensus included the use of PoCUS in the diagnosis of Sars-Cov-2 infection, triage/disposition, diagnosis of COVID-19 pneumonia, diagnosis of cardiovascular complications, diagnosis of thromboembolism, guidance of respiratory support, management of fluid administration, monitoring of COVID-19, and infection control (Hussain et al., 2020).

In June 2021, WINFOCUS conducted an online conference that was free of charge to all participants worldwide. This conference delivered a huge amount of education about PoCUS to healthcare providers around the world. Another WINFOCUS annual conference held in December 2022 provided its PoCUS one world (POW) concept of in-person training courses in several countries in different time zones on the same day. Much of the content of these conferences focused on the role of PoCUS in COVID-19. However, the role of PoCUS for many other conditions was discussed.

While encouraging the use of PoCUS and providing education are urgently required, it is critical to ensure that PoCUS is applied correctly. Thus, clear pathways for training in PoCUS and obtaining accreditation and certification are necessary (Mahmood et al., 2020; Rajendram et al., 2022; Rajendram, Souleymane, et al., 2021). Examples of accreditation pathways for PoCUS include those developed by the Royal College of Radiologists for the training of nonradiologists in ultrasound (The Royal College of Radiologists, 2012) and the pathway for obtaining accreditation in FoCUS (also known as level 1 echocardiography) described by Colebourn et al. (2017). These training and accreditation pathways existed long before the COVID-19 pandemic. However, many healthcare professionals perform PoCUS without obtaining formal accreditation. This wide variation in practice may result in inappropriate use of PoCUS by poorly trained practitioners and may limit the applicability of PoCUS in some settings (Mahmood et al., 2020; Rajendram et al., 2022; Rajendram, Souleymane, et al., 2021).

Applications to other areas

The diagnostic role of POCUS was established well before SARS-CoV-2 was isolated in 2019. However, COVID-19 resulted in an exponential increase in the use of PoCUS. The pandemic catalyzed the bedside use of ultrasound, breaking down preexisting barriers to its uptake in many healthcare settings (Vetrugno, Bove, et al., 2020).

Most of the abnormalities that PoCUS can detect are not specific for COVID-19. As a result, PoCUS is applicable to many other conditions. Indeed, there are many potential applications of PoCUS. These include screening for abdominal aortic aneurysms, intrauterine pregnancy, hydronephrosis gallstones, severe left ventricular dysfunction, and musculoskeletal injuries (Steinmetz et al. [9]).

The widespread availability of portable ultrasound equipment has opened new horizons, in particular in terms of primary care health services at patients' homes and managing special populations, such as pregnant women (Vetrugno et al., 2022) However, after performing PoCUS it is important to correctly disinfect the ultrasound transducer to prevent cross-infection (Hussain et al., 2020).

Yet the data on diagnostic accuracy and outcomes remain limited. Furthermore, guidelines on PoCUS mainly focus on hospitalized patients. As such, not all applications of PoCUS are appropriate in primary care settings (Andersen et al., 2019).

Before the pandemic, the diagnostic importance and usefulness of ultrasound in general medicine had been demonstrated, showing that POCUS scans were reported to have a higher diagnostic accuracy and indicating that POCUS could reduce healthcare costs, even if there were still some uncertainties about the quality of ultrasound portable scans and about training (Andersen et al., 2019).

It is also important to underline the limits of diagnostic ultrasound. While sensitive for COVID-19 pneumonia in the setting of the pandemic, as the incidence of SARS-CoV-2 infection declines the specificity of a thoracic ultrasound decreases (Vetrugno, Baciarello, et al., 2020).

However, PoCUS is not universally available worldwide. Many countries lack sufficiently skilled PoCUS practitioners to fully realize the benefits of PoCUS (Mahmood et al., 2020; Rajendram et al., 2022; Rajendram, Souleymane, et al., 2021).

Conclusions

The use of POCUS through portable devices has been shown to have a role in monitoring the possible clinical evolution of patients with infection from SARS-CoV-2. The ultrasonographic features are helpful to manage SARS-CoV-2 patients in the primary care setting and can help physicians evaluate the presence of lung involvement and diagnose complications from SARS-CoV-2 infection involving the cardiovascular system. Driven by its proven benefits and emerging technologies, ultrasound is rapidly establishing itself as a transformative tool for primary care. Point-of-care ultrasound is opening doors to a future of real-time diagnostics that can improve patient care.

Mini-dictionary of terms

- **Focused cardiac ultrasound:** This is another name for cardiac POCUS. It is a simplified form of cardiac ultrasound that is also known as level 1 echocardiography.
- **Lung ultrasound B-lines:** B-lines are vertical ultrasound reverberation artifacts. They originate from the pleura, extend the full depth of the image, and move with lung sliding. The presence of more than two B-lines per ultrasound scan sector is pathological. This indicates increased fluid within the lung parenchyma (i.e., pulmonary oedema, pus, intraparenchymal hemorrhage, etc.)
- **Lung ultrasound C-pattern:** Consolidations (i.e., condensed lung tissue) cause the C pattern on a lung ultrasound. This appears as focal hypoechoic areas at the periphery of the lung.
- **Point-of-Care Ultrasound:** This imaging modality is performed at the patient's bedside and is interpreted by the physician attending the patient. The physician performing the scan will have been trained to perform PoCUS but will not be a specialist in radiology.
- **Radiologist:** A radiologist is a doctor who specializes in making diagnoses by looking at images of a patient's body. They usually report the findings of detailed x-rays, computed tomography, or ultrasound. Radiologists rarely interact with patients directly.
- **Ultrasound machine:** An ultrasound machine uses high-frequency sound waves to produce images of the body for diagnostic purposes. This allows organs to be assessed without ionizing radiation.
- **Ultraportable ultrasound device:** Ultraportable battery-powered ultrasound devices are small enough to be handheld. They are less powerful and generate lower-quality images than cart-based ultrasound devices.

Summary points

1. PoCUS was an important diagnostic tool before the start of the SARS-CoV-2 pandemic.
2. The COVID-19 pandemic greatly increased the use of PoCUS worldwide.
3. In patients infected with SARS-CoV-2, ultrasonography can help physicians detect the involvement of the lung and other organs.
4. Portable ultrasound devices can be used to monitor the progression of COVID-19 in many settings (e.g., hospital, primary care, prehospital care, and patient's homes).
5. Significant training is required to effectively use PoCUS for the bedside assessment of patients.

References

Andersen, C. A., Holden, S., Vela, J., Rathleff, M. S., & Jensen, M. B. (2019). Point-of-care ultrasound in general practice: A systematic review. *Annals of Family Medicine*, *17*(1), 61–69.

Bianchi, S., Savinelli, C., Paolucci, E., Pelagatti, L., Sibona, E., Fersini, N., Buggea, M., Tozzi, C., Allescia, G., Paolini, D., & Lanigra, M. (2022). Point-of-care ultrasound (PoCUS) in the early diagnosis of novel coronavirus 2019 disease (COVID-19) in a first-level emergency department during a SARS-CoV-2 outbreak in Italy: A real-life analysis. *Internal and Emergency Medicine*, *17*(1), 193–204.

Brenner, D. S., Liu, G. Y., Omron, R., Tang, O., Garibaldi, B. T., & Fong, T. C. (2021). Diagnostic accuracy of lung ultrasound for SARS-CoV-2: A retrospective cohort study. *The Ultrasound Journal*, *13*(1), 12.

Chi, G., Lee, J. J., Jamil, A., Gunnam, V., Najafi, H., Memar Montazerin, S., Shojaei, F., & Marszalek, J. (2020). Venous thromboembolism among hospitalized patients with COVID-19 undergoing Thromboprophylaxis: A systematic review and meta-analysis. *Journal of Clinical Medicine*, *9*(8), 2489.

Colebourn, C., Pierce, K., Sharma, V., & Steeds, R. (2017). *Delivering 7-day Services in Echocardiography–a Proposal for level 1 emergency echocardiography from the British Society of Echocardiography*. London, UK: British Society of Echocardiography.

Di Gioia, C. C., Artusi, N., Xotta, G., Bonsano, M., Sisto, U. G., Tecchiolli, M., Orso, D., Cominotto, F., Amore, G., Meduri, S., & Copetti, R. (2022). Lung ultrasound in ruling out COVID-19 pneumonia in the ED: A multicentre prospective sensitivity study. *Emergency Medicine Journal: EMJ*, *39*(3), 199–205.

Dweck, M. R., Bularga, A., Hahn, R. T., Bing, R., Lee, K. K., Chapman, A. R., White, A., Salvo, G. D., Sade, L. E., Pearce, K., Newby, D. E., Popescu, B. A., Donal, E., Cosyns, B., Edvardsen, T., Mills, N. L., & Haugaa, K. (2020). Global evaluation of echocardiography in patients with COVID-19. *European Heart Journal Cardiovascular Imaging*, *21*(9), 949–958.

Hussain, A., Via, G., Melniker, L., Goffi, A., Tavazzi, G., Neri, L., Villen, T., Hoppmann, R., Mojoli, F., Noble, V., Zieleskiewicz, L., Blanco, P., Ma, I. W. Y., Wahab, M. A., Alsaawi, A., Al Salamah, M., Balik, M., Barca, D., Bendjelid, K., … Arabi, Y. (2020). Multi-organ point-of-care ultrasound for COVID-19 (PoCUS4COVID): International expert consensus. *Critical Care (London, England)*, *24*(1), 702.

Khanji, M. Y., Ricci, F., Patel, R. S., Chahal, A. A., Bhattacharyya, S., Galusko, V., Narula, J., & Ionescu, A. (2020). Special article - the role of hand-held ultrasound for cardiopulmonary assessment during a pandemic. *Progress in Cardiovascular Diseases*, *63*(5), 690–695.

Lichtenstein, D. A., & Mezière, G. A. (2008). Relevance of lung ultrasound in the diagnosis of acute respiratory failure: The BLUE protocol. *Chest*, *134*(1), 117–125.

Lichter, Y., Topilsky, Y., Taieb, P., Banai, A., Hochstadt, A., Merdler, I., Gal Oz, A., Vine, J., Goren, O., Cohen, B., Sapir, O., Granot, Y., Mann, T., Friedman, S., Angel, Y., Adi, N., Laufer-Perl, M., Ingbir, M., Arbel, Y., … Szekely, Y. (2020). Lung ultrasound predicts clinical course and outcomes in COVID-19 patients. *Intensive Care Medicine*, *46*(10), 1873–1883.

Mahmood, N., Souleymane, M., Rajendram, R., Ghazi, A. M. T., Kharal, M., & AlQahtani, M. (2020). Focused cardiac ultrasound is applicable to internal medicine and critical care but skill gaps currently limit use. *Journal of the Saudi Heart Association*, *32*(4), 464–471.

Morin, F., Douillet, D., Hamel, J. F., Rakotonjanahary, J., Dupriez, F., Savary, D., Aubé, C., Riou, J., Dubée, V., & Roy, P. M. (2021). Point-of-care ultrasonography for risk stratification of non-critical COVID-19 patients on admission (POCUSCO): A study protocol of an international study. *BMJ Open*, *11*(2), e041118.

Peixoto, A. O., Costa, R. M., Uzun, R., Fraga, A. M. A., Ribeiro, J. D., & Marson, F. A. L. (2021). Applicability of lung ultrasound in COVID-19 diagnosis and evaluation of the disease progression: A systematic review. *Pulmonology*, *27*(6), 529–562.

Rajendram, R., Alrasheed, A. O., Boqaeid, A. A., Alkharashi, F. K., Qasim, S. S., & Hussain, A. (2022). Training medical students in physical examination and point-of-care ultrasound: An assessment of the needs and barriers to acquiring skills in point-of-care ultrasound. *Journal of Family & Community Medicine*, *29*(1), 62–70.

Rajendram, R., & Hussain, A. (2021). Severe COVID-19 pneumonia complicated by cardiomyopathy and a small anterior pneumothorax. *BMJ Case Reports*, *14*(9), e245900.

Rajendram, R., Hussain, A., Mahmood, N., & Kharal, M. (2020). Feasibility of using a handheld ultrasound device to detect and characterize shunt and deep vein thrombosis in patients with COVID-19: An observational study. *The Ultrasound Journal*, *12*(1), 49.

Rajendram, R., Hussain, A., Mahmood, N., & Via, G. (2021). Dynamic right-to-left interatrial shunt may complicate severe COVID-19. *BMJ Case Reports*, *14*(10), e245301.

Rajendram, R., Souleymane, M., Mahmood, N., Kharal, M., & AlQahtani, M. (2021). Point-of-care diagnostic lung ultrasound is highly applicable to the practice of medicine in Saudi Arabia but the current skills gap limits its use. *Annals of Thoracic Medicine*, *16*(3), 266–273.

Soldati, G., Copetti, R., & Sher, S. (2009). Sonographic interstitial syndrome: The sound of lung water. *Journal of Ultrasound in Medicine: Official Journal of the American Institute of Ultrasound in Medicine*, *28*(2), 163–174.

Tana, C., Ricci, F., Coppola, M. G., Mantini, C., Lauretani, F., Campanozzi, D., Renda, G., Gallina, S., Lugará, M., Cipollone, F., Giamberardino, M. A., & Mucci, L. (2022). Prognostic significance of chest imaging by LUS and CT in COVID-19 inpatients: The ECOVID multicenter study. *Respiration; International Review of Thoracic Diseases*, *101*(2), 122–131.

The Royal College of Radiologists. (2012). *Focused ultrasound training standards*. London: The Royal College of Radiologists. Available online at https://www.rcr.ac.uk/sites/default/files/publication/BFCR(12)18_focused_training.pdf. (Accessed 1 July 2023).

Ultrasound Guinelines. (2017). Ultrasound Guidelines: Emergency, point-of-care and clinical ultrasound Guidelines in medicine. *Annals of Emergency Medicine*, *69*(5), e27–e54.

Vetrugno, L., Baciarello, M., Bignami, E., Bonetti, A., Saturno, F., Orso, D., Girometti, R., Cereser, L., & Bove, T. (2020). The "pandemic" increase in lung ultrasound use in response to Covid-19: Can we complement computed tomography findings? A narrative review. *The Ultrasound Journal*, *12*(1), 39.

Vetrugno, L., Bove, T., Orso, D., Bassi, F., Boero, E., & Ferrari, G. (2020). Lung ultrasound and the COVID-19 "pattern": Not all that glitters today is gold tomorrow. *Journal of Ultrasound in Medicine: Official Journal of the American Institute of Ultrasound in Medicine*, *39*(11), 2281–2282.

Vetrugno, L., Sala, A., Orso, D., Meroi, F., Fabbro, S., Boero, E., Valent, F., Cammarota, G., Restaino, S., Vizzielli, G., Girometti, R., Merelli, M., Tascini, C., Bove, T., Driul, L., & PINK-CO Study Investigators. (2022). Lung ultrasound signs and their correlation with clinical symptoms in COVID-19 pregnant women: The "PINK-CO" observational study. *Frontiers in Medicine*, *8*, 768261.

Via, G., Hussain, A., Wells, M., Reardon, R., ElBarbary, M., Noble, V. E., Tsung, J. W., Neskovic, A. N., Price, S., Oren-Grinberg, A., Liteplo, A., Cordioli, R., Naqvi, N., Rola, P., Poelaert, J., Gulič, T. G., Sloth, E., Labovitz, A., Kimura, B., … International Conference on Focused Cardiac UltraSound (IC-FoCUS). (2014). International evidence-based recommendations for focused cardiac ultrasound. *Journal of the American Society of Echocardiography: Official Publication of the American Society of Echocardiography*, *27*(7), 683.e1–683.e33.

Volpicelli, G., Lamorte, A., & Villén, T. (2020). What's new in lung ultrasound during the COVID-19 pandemic. *Intensive Care Medicine*, *46*(7), 1445–1448.

Ye, Z., Zhang, Y., Wang, Y., Huang, Z., & Song, B. (2020). Chest CT manifestations of new coronavirus disease 2019 (COVID-19): A pictorial review. *European Radiology*, *30*(8), 4381–4389.

Zieleskiewicz, L., Markarian, T., Lopez, A., Taguet, C., Mohammedi, N., Boucekine, M., Baumstarck, K., Besch, G., Mathon, G., Duclos, G., Bouvet, L., Michelet, P., Allaouchiche, B., Chaumoître, K., Di Bisceglie, M., Leone, M., & Network, A. Z. U. R. E. A. (2020). Comparative study of lung ultrasound and chest computed tomography scan in the assessment of severity of confirmed COVID-19 pneumonia. *Intensive Care Medicine*, *46*(9), 1707–1713.

Chapter 6

COVID-19 in patients with inflammatory bowel diseases: Characteristics and guidelines

Carlos Taxonera[a] and Olga Neva López-García[b]
[a]*Head Inflammatory Bowel Disease Unit, Department of Gastroenterology, Hospital Universitario Clínico San Carlos, Madrid, Spain,* [b]*Department of Gastroenterology, Hospital Universitario Clínico San Carlos, Madrid, Spain*

Abbreviations

ACE2	Angiotensin-converting enzyme 2
CD	Crohn's disease
CDC	Centers for Disease Control and Prevention
COVID-19	Coronavirus disease 2019
GI	gastrointestinal
HR	hazard ratio
IBD	inflammatory bowel disease
IMID	immune-mediated inflammatory disease
JAK	Janus kinase
MERS	Middle East respiratory syndrome
mRNA	messenger RNA
OR	odds ratio
RR	risk ratio
SARS-CoV-2	severe acute respiratory syndrome coronavirus-2
SECURE-IBD	International Database Surveillance Epidemiology of Coronavirus Under Research Exclusion for IBD
TMPRSS2	transmembrane serine protease 2
TNFα	tumor necrosis factor α
UC	ulcerative colitis

Introduction

Coronavirus disease 2019 (COVID-19), caused by severe acute respiratory syndrome coronavirus-2 (SARS-CoV-2), was first reported in December 2019, and it rapidly spread throughout the world leading to a pandemic of international concern. Early in the pandemic, older age and comorbidities were recognized as risk factors associated with severe COVID-19 and mortality (Zhou et al., 2020). Inflammatory bowel diseases (IBD), including Crohn's disease (CD) and ulcerative colitis (UC), are chronic inflammatory disorders of the gastrointestinal (GI) tract caused by immune dysregulation that may affect people of all ages. Many IBD patients require immunosuppressive therapies for inducing and maintaining remission. Therefore, patients with IBD and other immune-mediated inflammatory diseases (IMIDs) have been of special interest during the pandemic because the inflammatory disease itself and/or associated use of immunosuppressive therapies may increase the risk of contagion or progression of serious viral infections, including COVID-19. In addition, patients with IBD might express increased concentrations of angiotensin-converting enzyme 2 (ACE2) receptor in the GI tract, potentially predisposing them to GI symptoms such as diarrhea, nausea, and vomiting. Importantly, GI manifestations may be the only initial symptoms of COVID-19 in some IBD patients (Taxonera et al., 2020).

Management, Body Systems, and Case Studies in COVID-19. https://doi.org/10.1016/B978-0-443-18703-2.00022-0
Copyright © 2024 Elsevier Inc. All rights are reserved, including those for text and data mining, AI training, and similar technologies.

In the first waves of the pandemic, health facilities in many countries were overwhelmed, leading to extensive reallocation of hospital resources to help manage COVID-19 patients. To harmonize the response during the pandemic, scientific societies, including the European Crohn's and Colitis Organization, the International Organization for the Study of IBD, and the American Gastroenterological Association, published recommendations for managing IBD during COVID-19 (Magro et al., 2020; Rubin et al., 2020). In most countries, the reorganization of the IBD clinic by switching from face-to-face visits to remote consultations was able to guarantee a minimum standard quality of care for the IBD population during the pandemic (Taxonera, Alba, Olivares, Martin, et al., 2021). It was considered crucial that IBD patients maintain their treatments during the pandemic, and to this end new strategies were implemented in day hospitals for intravenous (IV) biologics and in the form of delivery for subcutaneous (SC) biologics. The impact of medications on COVID-19 vaccines has raised concern regarding vaccine responses in IBD patients (Wong et al., 2021). The available evidence has provided insight into this and other clinical differences in IBD patients, and how best to manage this patient population.

SARS-CoV-2 and the gastrointestinal tract

Like the prior SARS-CoV and Middle East respiratory syndrome (MERS)-CoV coronaviruses, SARS-CoV-2 binds to its target cells via the mucosal membrane ACE2 receptor expressed in multiple tissues, and this enzyme is thought to enable viral invasion of human cells. The spike (S) protein of SARS-CoV-2 has a high affinity for ACE2, abundantly expressed by epithelial cells of the lung, intestine, kidney, and blood vessels. The S protein is then cleaved by the host transmembrane serine protease 2 (TMPRSS2), which facilitates viral entry into the cytoplasm of the host cell. ACE2 and TMPRSS2 concentrations in the small intestine and colon are among the highest in the body (Burgueno et al., 2020). This raised the possibility of interaction among COVID-19-associated inflammation, IBD, and immunosuppressive therapies. However, neither active inflammation nor IBD medications increase the expression of the viral entry molecules ACE2 and TMPRSS2 in the GI tract, supporting the notion that patients with IBD are not at increased risk of GI infection by SARS-CoV-2 (Burgueno et al., 2020). A recent study reported reduced ACE2 expression in the inflamed ileum, although the ACE2 expression remains significant in the small intestine. In contrast, IBD-associated inflammation enhances ACE2 and TMPRSS2 expression in the rectum (Suárez-Fariñas et al., 2021). Commonly used IBD medications, whether biologics or nonbiologics, do not significantly impact ACE2 and TMPRSS2 receptor expression.

GI tissue samples, obtained from a COVID-19 patient using endoscopy, tested positive for SARS-CoV-2 RNA with intracellular staining of the viral nucleocapsid protein in the gastric, duodenal, and rectal epithelia (Xiao et al., 2020). Importantly, SARS-CoV-2 RNA has been identified by reverse transcription-polymerase chain reaction (RT-PCR) in stool samples in over half the patients in the general population, suggesting possible transmission by a fecal-oral route (Wang et al., 2020). Moreover, Chen et al. (2020) reported that two-thirds of patients remained positive for viral RNA in feces after the pharyngeal swabs turned negative. A metaanalysis reported a pooled prevalence for stool samples positive for SARS-CoV-2 RNA of 48.1%; of these samples, 70.3% were collected after loss of the virus from respiratory specimens (Cheung et al., 2020). This suggested that the replication of the virus in the GI tract may not be consistent with that in the respiratory tract, and fecal-oral transmission may occur even after viral clearance in respiratory tissue, highlighting the importance of fecal tests to control spread. Although some authors have recommended routine testing for SARS-CoV-2 RNA in stools prior to discharge of hospitalized patients with COVID-19 (Xiao et al., 2020), this advice has not been adopted in clinical practice.

COVID-19 gastrointestinal symptoms

GI symptoms are common in patients with SARS, MERS, and COVID-19. Although the specific mechanisms involved in COVID-19-associated GI symptoms are not entirely known, diarrhea could be caused by productive viral infection in the enterocytes, leading to the disruption of tight and adherent junctions of the endothelium and intestinal epithelium. This may lead to leaky gut syndrome, local and systemic invasion of normal microbiota, and immune activation that evokes an inflammatory response in the intestines, which in turn is characterized by the production of various proinflammatory cytokines and chemokines, leading to progressive bowel dysfunction and diarrhea (Megyeri et al., 2021). A cohort study outside Wuhan, China, where the pandemic originated, reported that 11.4% of patients presented with at least one GI tract symptom (nausea, vomiting, and diarrhea) (Jin et al., 2020). Of particular interest were COVID-19 cases with only digestive symptoms and no fever on presentation. A case series reported that 16% of COVID-19 patients presented with GI symptoms only, with a paucity of other manifestations (Luo et al., 2020). Sometimes diarrhea might be the first symptom before diagnosis (Taxonera et al., 2020). Such patients could be overlooked, leading to potentially serious consequences to them and their contacts.

In a systematic review and metaanalysis of 60 studies, the pooled prevalence of all COVID-19 GI symptoms was 17.6% (95% CI 12.3–24.5) (Cheung et al., 2020). Although the impact of GI symptoms on COVID-19 prognosis is controversial, the metaanalysis showed that severe disease was more common in patients with GI symptoms than those without (17.1% vs. 11.8%). However, another comprehensive metaanalysis reported no significant difference in mortality among COVID-19 patients with GI symptoms (0.4%, 95% CI 0–1.1) compared with mortality in the overall population (2.1%, 95% CI 0.2–4.7; $P = 0.15$), although the authors recognized that this must be interpreted with caution due to variable follow-up and lack of adjustment for confounders, such as age and comorbidities (Tariq et al., 2020). Of note, the incidences of diarrhea and nausea/vomiting were higher in the subgroup of studies conducted outside China compared with studies from China, likely due to increasing awareness and reporting of GI manifestations as the pandemic progressed (Tariq et al., 2020). It has also been suggested that the increase in GI symptoms in the later phase of this pandemic could be motivated by the possible mutation of the virus toward greater transmissibility and decreased virulence (Jin et al., 2020).

Risk of COVID-19 in inflammatory bowel disease

Data on the incidence of COVID-19 in IBD have been contradictory and available evidence sometimes has methodological limitations. Initially, it was assumed that IBD patients, due to an altered immune system and the immunosuppressive effect of drug treatment, had a higher risk of SARS-CoV-2 infection. Conversely, the first reports from China suggested that IBD patients had a decreased risk of COVID-19 compared with the general population (Mao et al., 2020). Subsequently, a study reported that among 522 IBD patients followed in a tertiary center at Italy, no case of COVID-19 was diagnosed (Norsa et al., 2020). Later, a study reported that IBD patients had a lower adjusted incidence ratio of COVID-19 (OR 0.74, $P < 0.001$) compared to the general population, although the authors recognized that due to the small sample size and the high impact that missed cases of COVID-19 could have, this finding should be treated with caution (Taxonera et al., 2020). Additional data coming from Europe and the United States indicated that COVID-19 was not more prevalent in patients with IBD than in the general population (Taxonera, Alba, & Olivares, 2021). A systematic review and meta-analysis reported a pooled incidence rate of COVID-19 per 1000 patients with IBD and the general population of 4.02 (CI 1.44–11.17) and 6.59 (CI 3.25–13.35), respectively. The relative risk of the acquisition of COVID-19 in IBD patients did not significantly differ from that in the general population (0.47, 0.18–1.26) (Singh et al., 2021). The relative risk of the acquisition of COVID-19 was not different between patients with UC or CD, although a difference was apparent for age, with a pooled incidence of 2.06% in studies included in the metaanalysis with a mean/median age of 45 years compared with 4.44% in studies with an age over 45 years (Singh et al., 2021). As a limitation, in most studies incidence rates were not adjusted for confounding factors, including age, and this could lead to an overestimation of the risk of COVID-19 in IBD. One study reported that crude incidence rates of COVID-19 in IBD decreased by 20% after adjusting for age, and this reduction had a clear impact when compared to the general population (Taxonera et al., 2020).

Recent reports from large cohorts have provided additional evidence that the risk of COVID-19 in IBD patients is comparable with that of the general population. A large case–control study reported that patients with IBD in the US Veterans Affairs healthcare system had a similar incidence of confirmed COVID-19 compared with the general veterans affairs population (matched by age, sex, race, location, and comorbidities) (0.23% vs. 0.20%; $P = 0.29$) (Khan, Patel, et al., 2021). A nationwide cohort study from the Netherlands found an incidence of COVID-19 of 2.87 per 1000 IBD patients, comparable to the incidence in the general Dutch population (3.3 per 1000; $P = 0.15$) (Derikx et al., 2021). A population-based prospective cohort study from Denmark reported that IBD patients may have a lower prevalence of COVID-19 than the general population (2.5% vs. 3.7%, $P = 0.01$) (Attauabi et al., 2021).

Taken together, these data suggest that patients with IBD do not appear to be more susceptible to SARS-CoV-2 infection. This finding is noteworthy because a considerable proportion of IBD patients were receiving immunosuppressants and/or biologics and a smaller proportion systemic steroids. A review assessing the interaction between viral immunopathology and immunosuppressive and biologic drugs concluded that immunosuppressive therapy does not appear to have a major impact on infection with SARS CoV-1, MERS-CoV, or SARS-CoV-2 (Sebastian et al., 2020). A key question arising from this finding is whether immunological factors associated with immunosuppressive and biologic therapy could play a compensatory role in reducing the risk of COVID-19 among patients with IBD. Another possible explanation for this observation may be that IBD patients adhere more strictly to protection measures during the pandemic owing to a perceived higher risk.

Outcomes of COVID-19 in IBD

At the beginning of the pandemic, there was concern about whether patients with IBD acquiring SARS-CoV-2 infection may have a different clinical presentation or be more prone to severe COVID-19 than the general population. This question was first addressed by an Italian observational cohort study that enrolled 79 IBD patients with COVID-19. Forty-six percent of patients had COVID-19-related pneumonia, 9% required nonmechanical ventilation, 9 (11%) required continuous positive airway pressure therapy, 2 (3%) had endotracheal intubation and 6 (8%) died (Bezzio et al., 2020). At the same time, an observational case-series study reported a similar COVID-19-associated mortality ratio in IBD patients (OR 0.95, 95% CI 0.84–1.06; $P = 0.36$), compared with the general population (Taxonera et al., 2020).

COVID-19 gastrointestinal symptoms in IBD

The common presenting symptoms of COVID-19 in IBD patients are like those of the general population, with a predominance of extraintestinal symptoms, of which fever is the most common with a pooled prevalence of 67.5%, followed by cough (59.6%). Diarrhea could nevertheless occur in a quarter of patients (Singh et al., 2022). The high affinity of SARS-CoV-2 to the gut and differences in the expression of intestinal membrane-bound ACE2 or of plasma ACE2 observed in IBD patients could be associated with the high rate of GI symptoms in IBD patients with COVID-19 observed in several studies (Singh et al., 2020; Taxonera et al., 2020). A population-based retrospective cohort study in the United States reported a higher proportion of patients in the IBD group presenting with nausea and vomiting (10.8% vs. 4.3%, $P < 0.01$), diarrhea (8.2% vs. 5.1%, $P < 0.01$), and abdominal pain (7.7% vs. 2.7%, $P < 0.01$) (Singh et al., 2020). Although there are no direct comparisons, a metaanalysis reported that the pooled prevalence of diarrhea in IBD patients with COVID-19 of 27.3% may be higher than in the general population (Singh et al., 2022). Another notable finding in IBD patients was the high frequency of abdominal pain (13%), which is uncommon in the general population with COVID-19. Sometimes, diarrhea was the only symptom caused by SARS-CoV-2 infection. This could lead to the misdiagnosis of an IBD flare-up, with inappropriate initiation of treatment that may include corticosteroids. Therefore, a study recommended testing all IBD patients with diarrhea for SARS-CoV-2 infection during the pandemic (Taxonera et al., 2020). Doing so enables discrimination between an IBD flare-up and diarrhea due to SARS-CoV-2 infection, thereby avoiding the inappropriate use of corticosteroids or other therapies that may favor COVID-19 progression. In addition, this strategy helps contain the spread of the SARS-CoV-2 infection by detecting and isolating cases without common symptoms of COVID-19.

Risk of severe COVID-19 and mortality in IBD

There are conflicting data regarding the risk of poor COVID-19 outcomes in patients with IBD. A population-based retrospective cohort study in the United States estimated the risk of severe COVID-19 in patients with IBD compared with the general population (Singh et al., 2020). After propensity score matching, there was no difference in the risk of severe COVID-19 between the IBD and non-IBD groups (risk ratio 0.93, 95% CI 0.68–1.27; $P = 0.66$). A prospective, population-based Danish study reported a high risk of severe COVID-19 among patients with UC and CD as compared with the general population, with 13.6% needing hospitalization and 2.1% requiring intensive care (Attauabi et al., 2021). A population-based cohort study in Sweden reported that patients with IBD were at increased risk of hospital admission, but there was no increased risk of severe COVID-19 (adjusted HR = 1.12; 0.85–1.47) (Ludvigsson et al., 2021). A recent pan-European cross-sectional study showed no increase in severe COVID-19 rates in patients with IBD compared with the general population (standardized incidence ratio 0.69, 95% CI 0.35–1.20, $P = 0.93$) (Amiot et al., 2022). The initial report from SECURE-IBD reported a mortality ratio for IBD patients with COVID-19 that ranged from 1.45 to 1.76, suggesting a 50% higher COVID-19-related mortality in patients with IBD; however, these findings were not statistically significant (Brenner et al., 2020). COVID-19-infected patients with IBD were at increased risk for requiring hospitalization compared with the non-IBD population (RR: 1.17, 95% CI 1.02–1.34) with no differences in the need for mechanical ventilation or mortality (Hadi et al., 2022). Based on the current literature, there is no evidence of worse COVID-19 outcomes in IBD patients.

Risk factors for severe COVID-19

Patients with IBD have key risk factors in common with the general population for severe COVID-19, including advanced age and presence of severe comorbidities. Advanced age has been shown to be an independent risk factor associated with severe COVID-19 in IBD, but no established age cut-off for increased risk has been determined. In the initial report from SECURE-IBD, the primary outcome (intensive care unit [ICU] admission/ventilator support/death) occurred in 7%. However, among patients >60 years of age, 20% experienced this outcome versus 0% in patients <20 years (Brenner et al., 2020). Having ≥ 2 comorbidities was associated with a threefold increased risk of the primary outcome (adjusted OR 2.9, 95% CI 1.1–7.8). An prospective cohort study reported that age ≥ 60 years was the only predictor for severe COVID-19 (OR 4.59, 95% CI 1.3–15.9) while CD with an inflammatory behavior was protective for this outcome

(OR 0.29, 95% CI 0.09–0.89). Some studies reported an increased risk of severe COVID-19 in UC as compared to CD (Attauabi et al., 2022; Singh et al., 2021) while other studies showed similar outcomes (Ludvigsson et al., 2021). The increased risk of adverse outcomes in UC could be confounded by age, sex, and comorbidities. In addition, treatment is likely to be different in UC and CD, as UC patients are more likely to receive 5-aminosalicylates while CD patients are more likely to receive biologics. A validated prognostic model following adjustment for known risk factors including age, male gender, comorbidity, and corticosteroid and biologic use can effectively predict which patients with IBD may be at higher risk for COVID-19-related morbidity, showing an excellent discrimination for hospitalization, ICU admission, and death (Sperger et al., 2021).

Persistent COVID-19 in IBD

Some patients may experience signs and symptoms for several months after the resolution of SARS-CoV-2 infection, a condition that has been termed persistent COVID-19. A prospective Danish study showed that after a median follow-up of 5 months, an equal proportion of patients with UC (42.3%) and CD (45.9%, $P = 0.60$) reported the persistence of COVID-19 for at least 12 weeks, with the most frequent symptoms including fatigue, anosmia, and ageusia (Attauabi et al., 2021). An observational study showed that 40% of IBD patients were classified as persistent COVID-19 carriers, with no significant difference between CD and UC (Salvatori et al., 2021). Asthenia was the most frequent symptom, as this occurred in nearly two-thirds of the patients. A recent retrospective study reported that 42.5% of patients with IBD were diagnosed with persistent COVID-19 during a median follow-up of 8.4 months (Taxonera Samso et al., 2022). Multivariable analysis identified UC ($P = 0.053$), comorbidities ($P = 0.090$), and being diagnosed during the first wave of COVID-19 ($P = 0.011$) as risk factors for persistent COVID-19. These data are consistent with the frequency of persistent COVID-19 in the general population, further suggesting that IBD by itself is not associated with worse outcomes after SARS-CoV-2 infection.

Impact of COVID-19 on IBD disease activity and incident IBD

Isolated cases of incident IBD have been reported following SARS-CoV-2 infection (Taxonera, Fisac, & Alba, 2021). A retrospective propensity score cohort study evaluated the risk of developing an IBD flare-up, and the risk of de novo IBD after SARS-CoV-2 infection. At the 3-month follow-up, COVID-19 patients were 1.3 times (95% CI 1.18 to 1.51) more likely to have an IBD-related disease flare-up compared with those without COVID-19 (Hadi et al., 2022). The risk for incident IBD post-COVID-19 was lower than that seen in the non-COVID-19 population (RR: 0.64, 95% CI 0.54–0.65). The pan-European study did not observe any changes in IBD clinical activity scores during the first months of the European COVID-19 outbreak, with no differences between patients with COVID-19 compared with patients who did not develop infection (Amiot et al., 2022). A study reported that having persistent COVID-19 was not associated with IBD relapses (Salvatori et al., 2021). Overall, there is no clear signal that COVID-19 may influence IBD activity or trigger de novo IBD.

Impact of pandemic waves

Several studies evaluated the clinical characteristics of COVID-19 cases in the general population during the first wave compared with successive waves, but data are scarce in IBD patients. Although the SECURE-IBD registry records severe complications of COVID-19, such as hospitalization, ICU admission, and death, these outcomes were too few to allow for temporal trend analyses between pandemic waves stratified by COVID-19 severity (Kaplan et al., 2022). An observational cohort study reported that IBD patients diagnosed in the first wave versus the second and third waves were older and more symptomatic and had more comorbidities (Algaba et al., 2021). Although the rate of severe COVID-19 was higher in the first wave, this was attributed to the limited availability of tests in that period, and there were no differences in mortality or in the percentage of ICU admissions between pandemic waves. A single-center prospective cohort study compared 62 COVID-19 cases diagnosed during the first wave of the pandemic, and 54 and 44 cases during the second and third waves, respectively. In the multivariate analysis, first-wave cases were associated with a higher risk of progression to severe COVID-19 (OR 4.76, 95% CI 1.83–12.37, $P = 0.001$), and with the development of persistent COVID-19 (OR 2.4, 95% CI 1.16–4.95, $P = 0.018$) (Taxonera Samso et al., 2022). In a highly dynamic pandemic such as COVID-19, the changes observed between successive waves may reflect differences in the diagnostic yield, the population's structure, viral changes to less aggressive variants, and the effect of vaccines from their administration.

Vulnerable IBD populations

Like adults, pediatric IB patients receiving immunosuppressive drugs may be considered vulnerable and may have a higher potential risk of developing adverse outcomes with COVID-19. However, a study integrating results from two international databases showed that pediatric IBD patients have a relatively low risk of severe COVID-19 or related morbidity and mortality compared to adults, even when receiving biologic and/or other immunosuppressive therapies for their IBD (Brenner et al., 2021).

The management of pregnant women with IBD during the COVID-19 pandemic has been challenging. Pregnant women are more vulnerable to respiratory illnesses such as influenza and other coronaviruses, including SARS. Therefore, there was concern that pregnancy may be an independent risk factor for both acquiring COVID-19 or having worse outcomes following infection. A study suggested that although pregnant women did not appear to be at a higher risk of contracting COVID-19, once infected, they were more likely to need admission to the ICU and require invasive ventilation compared with women who were not pregnant (Allotey et al., 2020). Data on pregnancy outcomes in patients with IBD and COVID-19 are scarce (Selinger et al., 2021). In general, to prevent IBD relapse, it is recommended to maintain medication in pediatric IBD patients and pregnant women with IBD during the COVID-19 pandemic. IBD flare-ups must be prevented to ensure a favorable outcome of the pregnancy and to minimize the risk of worse COVID-19 outcomes.

Effect of IBD medications on COVID-19 outcomes

Drugs used for IBD include corticosteroids, 5-aminosalicylates, biologics, small molecules, and immunosuppressants. The disease itself and some of these medications are known to increase vulnerability to some viral, bacterial, and fungal infections (Kirchgesner et al., 2018), and therefore IBD patients have been considered a vulnerable population for COVID-19. IBD patients, aware of these risks, may have self-isolated more rigorously than the general population and therefore been less exposed to the virus.

Biologics and small molecules

At time of COVID-19 pandemic, biologics approved for IBD patients include four TNF-α antibodies (infliximab, adalimumab, golimumab, and certolizumab); vedolizumab, belonging to the antiintegrin class; and ustekinumab, an antibody targeting the interleukins 12/23 shared p40 subunit. Additionally, the small molecules tofacitinib and filgotinib from the superfamily of JAK inhibitors have also been approved for IBD. Biologics in IBD patients were not associated with an increased risk of COVID-19 infection or severe COVID-19 (Burke et al., 2021; Taxonera et al., 2020; Ungaro et al., 2022; Wetwittayakhlang et al., 2021; Zabana et al., 2022). The effect of these medications on SARS-COV-2 infection is not completely understood. It has been speculated that anti-TNF therapy could inhibit the cytokine cascade triggered by the virus or just improve COVID-19 outcomes by controlling IBD activity (Alrashed et al., 2022). Some authors reported a protective effect of anti-TNF therapy on COVID-19 mortality (Brenner et al., 2020; Rizk et al., 2020; Stallmach et al., 2020), but this could not be confirmed in two randomized clinical trials (Fakharian et al., 2021; Fisher et al., 2022). Although there has been some controversy, it seems that vedolizumab and the JAK inhibitor tofacitinib did not increase the risk of severe COVID-19 (Agrawal et al., 2021; Alrashed et al., 2022; Zabana et al., 2022). Moreover, tofacitinib reduced the risk of death or respiratory failure among hospitalized patients with COVID-19 pneumonia (Guimarães et al., 2021), and the agent is being evaluated in a clinical trial (Kramer et al., 2022).

Immunomodulators and combinations

The use of conventional immunomodulators (azathioprine, 6-mercaptopurine, or methotrexate) was not associated with an increased risk of COVID-19 acquisition or with the severity of infection (Burke et al., 2021; Zabana et al., 2022). However, according to data from SECURE-IBD, the combination of anti-TNF and thiopurine (not so with methotrexate) may increase the risk of hospitalization and/or death but not of severe COVID-19, particularly in older patients (Brenner et al., 2020; Ungaro et al., 2021, 2022). This could be explained by a possible increased risk of opportunist infections other than COVID-19 itself.

5-Aminosalicylates

There is still debate about whether 5-aminosalicylates (mesalamine and sulfasalazine) may impact COVID-19 severity. 5-aminosalicylates are not immunosuppressive drugs and they do not increase the risk of viral or opportunistic infections. These drugs are commonly used in UC, especially in older patients, sometimes to avoid more aggressive drugs such as biologics or immunomodulators. During the first wave of the pandemic, 5-aminosalicylates were initially noted as a risk factor for severe COVID-19 when compared with anti-TNFs (Attauabi et al., 2021; Singh et al., 2021; Ungaro et al., 2021). Initial data from the SECURE-IBD registry described an association of 5-aminosalicylates with hospitalization and with severe COVID-19 (Brenner et al., 2020; Ungaro et al., 2021). This initial effect could be due to reporting bias: mild cases of COVID-19 in patients taking 5-aminosalicylates were underreported at that time, whereas patients treated with anti-TNF and with tighter clinical control were screened for SARS-CoV-2 infection even in the presence of mild symptoms. The most recent SECURE-IBD report, including more than 6000 patients, did not find an association between 5-aminosalicylate use and worse COVID-19 outcomes (Ungaro et al., 2022). Several other studies have not found any association between 5-aminosalicylate use, COVID-19 acquisition, and severe COVID-19 (Bezzio et al., 2020; Burke et al., 2021).

Corticosteroids

Corticosteroids, in particular dexamethasone, are currently used in the treatment of COVID-19 with respiratory failure, as they have demonstrated a mortality benefit in the general population (The RECOVERY Collaborative Group, 2021). In IBD patients, corticosteroids are frequently used to treat IBD flare-ups. Corticosteroids were tapered and used for the shortest time possible to avoid the deleterious side effects of prolonged treatments, including immunosuppression predisposing to infections. The use of corticosteroids (but not budesonide) prior or at the time of acquiring SARS-CoV-2 infection, before the onset of the cytokine cascade, may alter the clearance of the virus. Several publications of the SECURE-IBD registry reported systemic corticosteroids to be independently associated with severe COVID-19 (Brenner et al., 2020; Ungaro et al., 2021, 2022). Other studies showed that corticosteroids increased the risk of hospitalization and sequelae (Khan, Mahmud, et al., 2021; Wetwittayakhlang et al., 2021; Zabana et al., 2022). In addition, it should be noted that corticosteroids are commonly prescribed in active IBD, a factor that in itself contributes to an increased risk of severe COVID-19.

Management of IBD during the COVID-19 pandemic

Early in the pandemic, the Centers for Disease Control and Prevention (CDC) established several risk factors for severe COVID-19, including the presence of comorbidities or being treated with steroids or immunosuppressants. Therefore, patients with IBD were initially considered a high-risk population for adverse COVID-19 outcomes. This pandemic has provided a challenge to all healthcare providers. Our society demonstrated adaptability to the situation, cooperative working, and knowledge sharing to provide help to IBD patients. As a result, the rapidly evolving national and international declarations regarding COVID-19 recommended avoiding face-to-face care for the treatment of patients; this would not only ease the burden on hospital resources but also eliminate the risk of infection in the clinic (Aysha et al., 2020; Rubin et al., 2020). The reorganization of the IBD clinic following a strategy of switching from face-to-face visits to remote consultations when possible was able to guarantee a minimum standard quality of care to our patients during the COVID-19 pandemic (Al-Ani et al., 2020; Taxonera, Alba, Olivares, Martin, et al., 2021). Early in the pandemic, social distancing practices were strongly recommended for IBD patients, emphasizing the importance of maintaining hygienic measures and wearing a mask. Several gastroenterology societies created platforms to share research about SARS-CoV-2 in IBD patients with care providers. The analysis of data from SECURE-IBD (Brenner et al., 2020), other online databases such as Research Electronic Data Capture (REDCap), together with gastroenterology societies and IBD organizations generated international and national guidelines on disease management during the pandemic, sometimes based on consensus expert opinions (Aysha et al., 2020; Magro et al., 2020; Rubin et al., 2020).

Treatments for IBD during pandemic

IBD patients were recommended to continue with their IBD prepandemic maintenance medication to avoid a disease flare-up that may increase the risk of SARS-CoV-2 infection due to the need for steroids or hospitalization (Kennedy et al., 2020; Lees et al., 2022). Most guidelines discouraged switching from IV biologics to SC formulations to avoid loss of response. In IBD patients diagnosed with COVID-19, it has been suggested that treatment with 5-aminosalicylates, budesonide, and rectal therapies should be maintained. Thiopurines, methotrexate, and tofacitinib should be discontinued (Magro et al., 2020; Rubin et al., 2020). Systemic corticosteroids should be tapered to <20 mg/day or be changed to budesonide whenever possible. In case of COVID-19, temporary discontinuation of biologics is recommended. Such medication should be discontinued for 10 days to 2 weeks until resolution of symptoms (at least 3 days after clearance of symptoms), decrease in viral load, or a negative SARS-CoV-2 test (Kamath & Brenner, 2022; Siegel et al., 2020). In severe cases of COVID-19 with respiratory failure or pneumonia requiring hospitalization, it was recommended to focus on life support and COVID-19 treatment with systemic corticosteroids, antiviral therapies, or antiinflammatory drugs according to international or local protocols (Lin et al., 2022).

Endoscopy during the COVID-19 pandemic and postpandemic period

Early in the pandemic, an endoscopy was considered a potential source of infection. Therefore, other noninvasive diagnostic methods for IBD diagnosis and monitoring, such as fecal calprotectin, blood tests, and imaging (ultrasound, magnetic resonance imaging, or computed tomography) were recommended instead (Kennedy et al., 2020). For an endoscopy, a prioritization system was developed, and most gastroenterology societies strongly recommended postponing elective endoscopies to avoid spread. In IBD, high-risk patients were those with a new diagnosis of moderate-to-severe IBD, patients admitted with a disease flare-up, or those with complications such as severe hemorrhage (Ng et al., 2020). Before performing an endoscopy, a SARS-CoV-2 test should be done. In the event of a positive test, and only if the procedure cannot be delayed, endoscopies should be performed under strict isolation conditions, with FFP2 masks, gloves, and appropriate clothing and footwear for the endoscopist and assistants, with subsequent special cleaning and disinfection of the room (Gralnek et al., 2020).

According to these recommendations, endoscopy procedures decreased compared to prepandemic levels. The pan-European study reported that the COVID-19 pandemic resulted in a drastic decrease in endoscopic procedures and morphologic examination during the first lockdown period from March to May 2020, compared with the same period in 2019 and 2018 ($P < 0.05$) (Amiot et al., 2022). This negatively impacted the diagnosis of other important diseases, such as cancer, in particular colorectal cancer. As countries have recovered from the pandemic and severe cases of COVID-19 are much less frequent, endoscopic activity has picked up. However, waiting lists have increased dramatically, so reaching prepandemic targets is a real challenge that depends on local demand and resources.

SARS-CoV-2 vaccines in patients with IBD

Vaccination has become a cornerstone of the strategy to control the SARS-CoV-2 pandemic. Several vaccines have demonstrated in clinical trials that they could prevent severe disease and death in the general population (Baden et al., 2021; Francis et al., 2022; Polack et al., 2020; Voysey et al., 2021). However, IBD patients were initially excluded from the pivotal vaccine trials and have thus been understudied. Certain immunosuppressive drugs are known to attenuate the immunological response to some vaccines, as is the case with infliximab/adalimumab and inactivated influenza virus or hepatitis B virus in IBD patients (Macaluso et al., 2021; Siegel et al., 2021). Therefore, it was important to recognize population groups at risk of having an attenuated response to SARS-CoV-2 vaccination.

Immunological response to vaccines

Antibodies and T-cells are both important in SARS-CoV-2 response. T-cell response constitutes an important component of the protective response against COVID-19 and is particularly decisive for those with low antibody levels, offering more durable protection (Jena, James, et al., 2022). Antibody response mechanisms are better known than T-cell response and have been widely studied. Seroconversion rates were consistently lower after incomplete vaccination, which was usually considered to be just one dose for mRNA vaccines, compared to complete vaccination. For incomplete vaccination, serologic response was also lower in IBD patients as compared with the general population (Jena, James, et al., 2022; Sung et al., 2022). This could be attributed to an altered antibody response to vaccines due to the disease itself or to the impact of immunosuppressive drugs. Seroconversion after complete SARS-CoV-2 vaccination was slightly lower in IBD patients compared with healthy controls, except for mRNA vaccines (Jena, James, et al., 2022). In IBD patients, the pooled seroconversion rate after complete mRNA vaccination was 0.97 (95% CI 0.96–0.98) (Jena, James, et al., 2022).

Impact of treatments on response to vaccines

Complete vaccination in patients treated with vedolizumab and ustekinumab showed good seroconversion rates (Jena, James, et al., 2022). However, seroconversion in patients receiving JAK inhibitors (Alexander et al., 2022) and anti-TNF therapies was debated (Pratt et al., 2018), but it seems that vaccination efficacy is quite like that of patients treated with other immunosuppressants (Bhurwal et al., 2022; Jena, Mishra, et al., 2022; Sung et al., 2022). For patients treated with steroids or with a combination of anti-TNF and immunomodulators, seroconversion rates were lower, but still >90% (Jena,

James, et al., 2022). The CLARITY study showed lower rates of seroconversion after a single dose of BNT162b2 vaccine in IBD patients on infliximab compared with vedolizumab (Kennedy et al., 2021). However, an Israeli study that also addressed real-world SARS-CoV-2 vaccine effectiveness did not show an increased incidence of COVID-19 in vaccinated IBD patients on anti-TNF (Lev-Tzion et al., 2022).

Efficacy of SARS-CoV-2 vaccines in IBD

The SARS-CoV-2 infection rate in vaccinated people was comparable between IBD patients and the general population (OR, 1.28, 95% CI 0.96–1.71; $P = 0.09$). The risk of acquiring the infection was significantly greater in nonvaccinated IBD patients compared with vaccinated ones, and the SARS-CoV-2 mortality rate was very low in vaccinated IBD patients (Sung et al., 2022). International and national societies strongly recommended SARS-CoV-2 vaccination for IBD patients. Based on previous experience, inactivated vaccines are preferred for IBD patients (Geisen et al., 2021; Kennedy et al., 2020; Rubin et al., 2020; Sung et al., 2022). There has been a concern that emerging live attenuated and replication-competent viral vector vaccines may cause illness in immunocompromised hosts, and these vaccines were discouraged by several societies. If a live attenuated SARS-CoV-2 vaccine is inevitable, it is recommended to take it at least 8 weeks after cessation of the immunosuppressant (Siegel et al., 2021; Sung et al., 2022).

The risk of COVID-19 in patients with IBD who received only a single dose of mRNA vaccine was like unvaccinated patients with IBD. However, complete SARS-CoV-2 vaccination in IBD patients is associated with good seroconversion rates, decreasing the risk of infection. However, the duration of this protection was a concern. A third dose or booster doses may help achieve seroconversion in nonresponders or improve the response, particularly in patients on treatment with immunosuppressants (Jena, James, et al., 2022; Lin et al., 2022). Moreover, patients treated with anti-TNF receiving a third mRNA vaccine dose increased their serum neutralizing antibody levels by more than 16-fold (Chen et al., 2021), highlighting the need for boosters to prevent SARS-CoV-2 infection in this susceptible population.

Safety

Multiple studies have shown SARS-CoV-2 vaccines to be safe in patients with IBD, including those on immunosuppressive therapy. Adverse events after administrating the vaccine were very frequent (>50%), but most of them were mild and limited to the injection site. Severe side effects in IBD patients were infrequent (2%), and the safety profile was like that of a matched cohort of healthy controls (Hadi et al., 2021). As in the general population, isolated cases of immune-mediated thrombocytopenia, Guillain-Barre syndrome, transverse myelitis, stroke, deep vein thrombosis, myocarditis, and myocardial infarction have been reported in IBD patients (James et al., 2022). In general, COVID-19 vaccination was safe in IBD patients, and it has been strongly recommended by all gastroenterological societies (Bhurwal et al., 2022).

Policies and procedures in IBD

In this chapter, we reviewed the impact of the coronavirus pandemic on IBD patients. Patients with IBD often require immunosuppressive therapies for inducing and maintaining remission. Therefore, patients with IBD and other IMIDs are of special interest during this pandemic because the disease itself and the use of immunosuppressive therapies may increase the risk of acquisition or progression of COVID-19. The dynamic replication of SARS-CoV-2 in the GI tract may not be consistent with that in the respiratory tract, highlighting the importance of fecal tests to control spread. The risk of acquisition of the infection compared to the general population, including adjustment for confounding factors, has been reviewed in detail. The possibility of worse COVID-19 outcomes, including hospitalization, intensive care, and mortality, in IBD patients was assessed. Of special interest was the detailed analysis evaluating the impact of treatments used in IBD on COVID-19 outcomes. Based on these findings, international societies have published recommendations for managing IBD during COVID-19 (Magro et al., 2020; Rubin et al., 2020). The reorganization of the IBD clinic, with a switch from face-to-face visits to remote telemedicine consultations when possible, was able to guarantee a minimum standard quality of care to our patients during the COVID-19 pandemic (Al-Ani et al., 2020; Taxonera, Alba, Olivares, Martin, et al., 2021). Procedures were developed for endoscopic examination, including strict isolation conditions with FFP2 masks, gloves, and appropriate clothing and footwear, for the endoscopist and assistants (Gralnek et al., 2020).

Applications to other areas

During the pandemic, IBD units and clinics have reorganized their structure and function to give patients the best care. Infusion centers for administering IV biologics have been reorganized. Thus, to reduce the risk of transmission within the infusion center, patients scheduled for an infusion should be contacted 1 day before their appointment to determine their risk of COVID-19. This should include screening for COVID-19 symptoms and confirmation of no direct contact with a confirmed case of COVID-19. If they have any risk factor, the infusion should be delayed (Al-Ani et al., 2020). This strategy has worked well, avoiding the need to switch patients from an effective IV biologic to a SC biologic, with the uncertainty of outcome that this switch entails. What has been learned in IBD patients is valid for other patient populations who need to go to day hospitals both to receive biologics and for other reasons, and in particular for patients with rheumatological or dermatological IMIDs. Another lesson learned was the strategy of home delivery of medication, avoiding the need for patients to come to the hospital. This worked very well for the distribution of SC biologics and small molecules. Thus, with current knowledge, in the event of a future pandemic, this strategy should be applied from the outset. Although the strategy of shifting from face-to-face visits to remote telemedicine consultations made it possible to care for patients during the worst of the pandemic, it has become clear in many centers in several countries that advanced telemedicine capabilities are limited (in many centers only telephone calls were used) and need to be further developed to better cope with a future pandemic.

Mini-dictionary of terms

- **Biologics used in IBD:** Large antibody that is derived from the clone of a single B cell and that is produced in large quantities of identical cells possessing affinity for a specific antigen: cytokine or integrin.
- **Inflammatory bowel disease:** A term covering two diseases (Crohn's disease and ulcerative colitis) that are characterized by chronic relapsing inflammation of the gastrointestinal tract.
- **Immune-mediated inflammatory diseases:** A group of diseases that involves an immune response that is inappropriate or excessive, and is caused, signified, or accompanied by dysregulation of the body's normal cytokine milieu.
- **JAK inhibitors:** Any agent that targets and inhibits the activity of one or more of the Janus kinase family of enzymes.
- **Leaky gut syndrome:** The theory of leaky gut syndrome suggests that anything that injures your gut lining can lead to intestinal permeability if the injury is persistent enough.

Summary points

- Patients with IBD do not appear to be more susceptible to SARS-CoV-2 infection.
- There is no current evidence of worse COVID-19 outcomes in IBD patients.
- Patients with IBD have key risk factors in common with the general population for severe COVID-19, including advanced age and presence of severe comorbidities.
- Immunosuppressants, biologics, and small molecules do not appear to have a major impact on infection with SARS-CoV-2.
- IBD patients were recommended to continue with their IBD prepandemic maintenance medication to avoid a disease flare-up that may increase the risk of SARS-CoV-2 infection.
- GI manifestations may be the only initial symptoms of COVID-19 in some IBD patients.
- Diarrhea caused by SARS-CoV-2 infection could lead to misdiagnosis of an IBD flare-up, and inappropriate initiation of treatment that may include corticosteroids. Therefore, it is advisable to test all IBD patients presenting with diarrhea for SARS-CoV-2 infection during the pandemic.
- Transmission through feces makes the digestive system a potential route for human-to-human transmission of SARS-CoV-2.

References

Agrawal, M., Zhang, X., Brenner, E. J., Ungaro, R. C., Kappelman, M. D., & Colombel, J. F. (2021). The impact of Vedolizumab on COVID-19 outcomes among adult IBD patients in the SECURE-IBD registry. *Journal of Crohn's & Colitis*, *15*(11), 1877–1884.

Al-Ani, A. H., Prentice, R. E., Rentsch, C. A., Johnson, D., Ardalan, Z., Heerasing, N., et al. (2020). Review article: Prevention, diagnosis and management of COVID-19 in the IBD patient. In. *Alimentary Pharmacology and Therapeutics*, *52*(1), 54–72.

Alexander, J. L., Kennedy, N. A., Ibraheim, H., Anandabaskaran, S., Saifuddin, A., Castro Seoane, R., et al. (2022). COVID-19 vaccine-induced antibody responses in immunosuppressed patients with inflammatory bowel disease (VIP): A multicentre, prospective, case-control study. *The Lancet Gastroenterology and Hepatology*, 7(4), 342–352.

Algaba, A., Guerra, I., Castro, S., & Bermejo, F. (2021). Infección por SARS-CoV-2 en pacientes con enfermedad inflamatoria intestinal en la segunda y tercera ola y su comparación con los datos de la primera ola. *Gastroenterología y Hepatología*. https://doi.org/10.1016/j.gastrohep.2021.11.001.

Allotey, J., Stallings, E., Bonet, M., Yap, M., Chatterjee, S., Kew, T., et al. (2020). Clinical manifestations, risk factors, and maternal and perinatal outcomes of coronavirus disease 2019 in pregnancy: Living systematic review and meta-analysis. *The BMJ*, *370*. https://doi.org/10.1136/bmj.m3320.

Alrashed, F., Alasfour, H., & Shehab, M. (2022). Impact of biologics and small molecules for inflammatory bowel disease on COVID-19-related hospitalization and mortality: A systematic review and meta-analysis. *JGH Open*, *6*(4), 241–250.

Amiot, A., Rahier, J.-F., Baert, F., Nahon, S., Hart, A., Viazis, N., et al. (2022). The impact of COVID-19 on patients with IBD in a prospective European cohort study. *Journal of Crohn's & Colitis*. https://doi.org/10.1093/ecco-jcc/jjac091.

Attauabi, M., Dahlerup, J. F., Poulsen, A., Hansen, M. R., Vester-Andersen, M. K., Eraslan, S., et al. (2022). Outcomes and long-term effects of COVID-19 in patients with inflammatory bowel diseases—A Danish prospective population-based cohort study with individual-level data. *Journal of Crohn's & Colitis*, *16*(5), 757–767.

Attauabi, M., Poulsen, A., Theede, K., Pedersen, N., Larsen, L., Jess, T., et al. (2021). Prevalence and outcomes of COVID-19 among patients with inflammatory bowel disease-a Danish prospective population-based cohort study. *Journal of Crohn's & Colitis*, *15*(4), 540–550.

Aysha, A. A., Rentsch, C., Prentice, R., Johnson, D., Bryant, R. V., Ward, M. G., et al. (2020). Practical management of inflammatory bowel disease patients during the COVID-19 pandemic: Expert commentary from the Gastroenterological Society of Australia inflammatory bowel disease faculty. *Internal Medicine Journal*, *50*(7), 798–804.

Baden, L. R., el Sahly, H. M., Essink, B., Kotloff, K., Frey, S., Novak, R., et al. (2021). Efficacy and safety of the mRNA-1273 SARS-CoV-2 vaccine. *New England Journal of Medicine*, *384*(5), 403–416.

Bezzio, C., Saibeni, S., Variola, A., Allocca, M., Massari, A., Gerardi, V., Casini, V., et al. (2020). Outcomes of COVID-19 in 79 patients with IBD in Italy: An IG-IBD study. *Gut*, *69*(7), 1213–1217.

Bhurwal, A., Mutneja, H., Bansal, V., Goel, A., Arora, S., Attar, B., et al. (2022). Effectiveness and safety of SARS-CoV-2 vaccine in inflammatory bowel disease patients: A systematic review, meta-analysis and meta-regression. *Alimentary Pharmacology and Therapeutics*, *55*(10), 1244–1264.

Brenner, E. J., Pigneur, B., Focht, G., Zhang, X., Ungaro, R. C., Colombel, J. F., et al. (2021). Benign evolution of SARS-Cov2 infections in children with inflammatory bowel disease: Results from two international databases. In. *Clinical Gastroenterology and Hepatology*, *19*(2). 394–396.e5.

Brenner, E. J., Ungaro, R. C., Gearry, R. B., Kaplan, G. G., Kissous-Hunt, M., Lewis, J. D., et al. (2020). Corticosteroids, but not TNF antagonists, are associated with adverse COVID-19 outcomes in patients with inflammatory bowel diseases: Results from an international registry. *Gastroenterology*, *159*(2). 481–491.e3.

Burgueno, J. F., Reich, A., Hazime, H., Quintero, M. A., Fernandez, I., Fritsch, J., et al. (2020). Expression of SARS-CoV-2 entry molecules ACE2 and TMPRSS2 in the gut of patients with IBD. *Inflammatory Bowel Diseases*, *26*(6), 797–808.

Burke, K. E., Kochar, B., Allegretti, J. R., Winter, R. W., Lochhead, P., Khalili, H., et al. (2021). Immunosuppressive therapy and risk of COVID-19 infection in patients with inflammatory bowel diseases. *Inflammatory Bowel Diseases*, *27*(2), 155–161.

Chen, Y., Chen, L., Deng, Q., Zhang, G., Wu, K., Ni, L., et al. (2020). The presence of SARS-CoV-2 RNA in the feces of COVID-19 patients. *Journal of Medical Virology*, *92*(7), 833–840.

Chen, R. E., Gorman, M. J., Zhu, D. Y., Carreño, J. M., Yuan, D., VanBlargan, L. A., et al. (2021). Reduced antibody activity against SARS-CoV-2 B.1.617.2 delta virus in serum of mRNA-vaccinated individuals receiving tumor necrosis factor-α inhibitors. *Med*, *2*(12), 1327–1341.e4.

Cheung, K. S., Hung, I. F. N., Chan, P. P. Y., Lung, K. C., Tso, E., Liu, R., et al. (2020). Gastrointestinal manifestations of SARS-CoV-2 infection and virus load in fecal samples from a Hong Kong cohort: Systematic review and Meta-analysis. *Gastroenterology*, *159*(1), 81–95.

Derikx, L. A. A. P., Lantinga, M. A., de Jong, D. J., van Dop, W. A., Creemers, R. H., Römkens, T. E. H., et al. (2021). Clinical outcomes of Covid-19 in patients with inflammatory bowel disease: A nationwide cohort study. *Journal of Crohn's & Colitis*, *15*(4), 529–539.

Fakharian, A., Barati, S., Mirenayat, M., Rezaei, M., Haseli, S., Torkaman, P., et al. (2021). Evaluation of adalimumab effects in managing severe cases of COVID-19: A randomized controlled trial. *International Immunopharmacology*, *99*, 107961. https://doi.org/10.1016/j.intimp.2021.107961.

Fisher, B. A., Veenith, T., Slade, D., Gaskell, C., Rowland, M., Whitehouse, T., et al. (2022). Namilumab or infliximab compared with standard of care in hospitalised patients with COVID-19 (CATALYST): A randomised, multicentre, multi-arm, multistage, open-label, adaptive, phase 2, proof-of-concept trial. *The Lancet Respiratory Medicine*, *10*(3), 255–266.

Francis, A. I., Ghany, S., Gilkes, T., & Umakanthan, S. (2022). Review of COVID-19 vaccine subtypes, efficacy and geographical distributions. *Postgraduate Medical Journal*, *98*(1159), 389–394.

Geisen, U. M., Berner, D. K., Tran, F., Sümbül, M., Vullriede, L., Ciripoi, M., et al. (2021). Immunogenicity and safety of anti-SARS-CoV-2 mRNA vaccines in patients with chronic inflammatory conditions and immunosuppressive therapy in a monocentric cohort. *Annals of the Rheumatic Diseases*, *80*(10), 1306–1311.

Gralnek, I. M., Hassan, C., Beilenhoff, U., Antonelli, G., Ebigbo, A., Pellisè, M., et al. (2020). ESGE and ESGENA position statement on gastrointestinal endoscopy and the COVID-19 pandemic. *Endoscopy*, *52*(6), 483–490.

Guimarães, P. O., Quirk, D., Furtado, R. H., Maia, L. N., Saraiva, J. F., Antunes, M. O., et al. (2021). Tofacitinib in patients hospitalized with Covid-19 pneumonia. *New England Journal of Medicine*, *385*(5), 406–415.

Hadi, Y., Dulai, P. S., Kupec, J., Mohy-Ud-Din, N., Jairath, V., Farraye, F. A., et al. (2022). Incidence, outcomes, and impact of COVID-19 on inflammatory bowel disease: Propensity matched research network analysis. *Alimentary Pharmacology & Therapeutics*, *55*(2), 191–200.

Hadi, Y. B., Thakkar, S., Shah-Khan, S. M., Hutson, W., Sarwari, A., & Singh, S. (2021). COVID-19 vaccination is safe and effective in patients with inflammatory bowel disease: Analysis of a large multi-institutional research network in the United States. In. *Gastroenterology*, *161*(4), 1336–1339.e3.

James, D., Jena, A., Bharath, P. N., Choudhury, A., Singh, A. K., Sebastian, S., et al. (2022). Safety of SARS-CoV-2 vaccination in patients with inflammatory bowel disease: A systematic review and meta-analysis. In. *Digestive and Liver Disease*, *54*(6), 713–721.

Jena, A., James, D., Singh, A. K., Dutta, U., Sebastian, S., & Sharma, V. (2022). Effectiveness and durability of COVID-19 vaccination in 9447 patients with IBD: A systematic review and Meta-analysis. In. *Clinical Gastroenterology and Hepatology*, *20*(7), 1456–1479.e18.

Jena, A., Mishra, S., Deepak, P., Kumar, M. P., Sharma, A., Patel, Y. I., et al. (2022). Response to SARS-CoV-2 vaccination in immune mediated inflammatory diseases: Systematic review and meta-analysis. *Autoimmunity Reviews*, *21*(1), 102927. https://doi.org/10.1016/j.autrev.2021.102927.

Jin, X., Lian, J. S., Hu, J. H., Gao, J., Zheng, L., Zhang, Y. M., et al. (2020). Epidemiological, clinical and virological characteristics of 74 cases of coronavirus-infected disease 2019 (COVID-19) with gastrointestinal symptoms. *Gut*, *69*(6), 1002–1009.

Kamath, C., & Brenner, E. J. (2022). The safe use of inflammatory bowel disease therapies during the COVID-19 pandemic. In *Vol. 3. Current research in pharmacology and drug discovery* Elsevier B.V. https://doi.org/10.1016/j.crphar.2022.100101.

Kaplan, G. G., Underwood, F. E., Coward, S., Agrawal, M., Ungaro, R. C., Brenner, E. J., et al. (2022). The multiple waves of COVID-19 in patients with inflammatory bowel disease: A temporal trend analysis. *Inflammatory Bowel Diseases*. https://doi.org/10.1093/ibd/izab339.

Kennedy, N. A., Jones, G. R., Lamb, C. A., Appleby, R., Arnott, I., Beattie, R. M., et al. (2020). British Society of Gastroenterology guidance for management of inflammatory bowel disease during the COVID-19 pandemic. *Gut*, *69*(6), 984–990.

Kennedy, N. A., Lin, S., Goodhand, J. R., Chanchlani, N., Hamilton, B., Bewshea, C., et al. (2021). Infliximab is associated with attenuated immunogenicity to BNT162b2 and ChAdOx1 nCoV-19 SARS-CoV-2 vaccines in patients with IBD. *Gut*, *70*(10), 1884–1893.

Khan, N., Mahmud, N., Trivedi, C., Reinisch, W., & Lewis, J. D. (2021). Risk factors for SARS-CoV-2 infection and course of COVID-19 disease in patients with IBD in the veterans affair healthcare system. *Gut*, *70*(9), 1657–1664.

Khan, N., Patel, D., Xie, D., Pernes, T., Lewis, J., & Yang, Y. X. (2021). Are patients with inflammatory bowel disease at an increased risk of developing SARS-CoV-2 than patients without inflammatory bowel disease? Results from a Nationwide veterans' affairs cohort study. *American Journal of Gastroenterology*, *116*(4), 808–810.

Kirchgesner, J., Lemaitre, M., Carrat, F., Zureik, M., Carbonnel, F., & Dray-Spira, R. (2018). Risk of serious and opportunistic infections associated with treatment of inflammatory bowel diseases. *Gastroenterology*, *155*(2), 337–346.e10.

Kramer, A., Prinz, C., Fichtner, F., Fischer, A.-L., Thieme, V., Grundeis, F., et al. (2022). Janus kinase inhibitors for the treatment of COVID-19. *Cochrane Database of Systematic Reviews*, *2022*(6).

Lees, C. W., Ahmad, T., Lamb, C. A., Powell, N., Din, S., Cooney, R., et al. (2022). Withdrawal of the British Society of Gastroenterology IBD risk grid for COVID-19 severity. *Gut*. BMJ Publishing Group.

Lev-Tzion, R., Focht, G., Lujan, R., Mendelovici, A., Friss, C., Greenfeld, S., et al. (2022). COVID-19 vaccine is effective in inflammatory bowel disease patients and is not associated with disease exacerbation. *Clinical Gastroenterology and Hepatology*, *20*(6), e1263–e1282.

Lin, S., Lau, L. H., Chanchlani, N., Kennedy, N. A., & Ng, S. C. (2022). Recent advances in clinical practice: Management of inflammatory bowel disease during the COVID-19 pandemic. *Gut*, *71*(7), 1426–1439.

Ludvigsson, J. F., Axelrad, J., Halfvarson, J., Khalili, H., Larsson, E., Lochhead, P., et al. (2021). Inflammatory bowel disease and risk of severe COVID-19: A nationwide population-based cohort study in Sweden. *United European Gastroenterology Journal*, *9*(2), 177–192.

Luo, S., Zhang, X., & Xu, H. (2020). Don't overlook digestive symptoms in patients with 2019 novel coronavirus disease (COVID-19). *Clinical Gastroenterology and Hepatology*, *18*(7), 1636–1637.

Macaluso, F. S., Liguori, G., & Galli, M. (2021). Vaccinations in patients with inflammatory bowel disease. *Digestive and Liver Disease*, *53*(12), 1539–1545.

Magro, F., Rahier, J. F., Abreu, C., MacMahon, E., Hart, A., van der Woude, C. J., et al. (2020). Inflammatory bowel disease management during the covid-19 outbreak: The ten do's and don'ts from the ECCO-COVID task force. *Journal of Crohn's & Colitis*, *14*, S798–S806.

Mao, R., Liang, J., Shen, J., Ghosh, S., Zhu, L. R., Yang, H., et al. (2020). Implications of COVID-19 for patients with pre-existing digestive diseases. *The Lancet Gastroenterology and Hepatology*, *5*(5), 426–428.

Megyeri, K., Dernovics, Á., Al-Luhaibi, Z. I. I., & Rosztóczy, A. (2021). COVID-19-associated diarrhea. *World Journal of Gastroenterology*, *27*(23), 3208–3222.

Ng, S. C., Mak, J. W. Y., Hitz, L., Chowers, Y., Bernstein, C. N., & Silverberg, M. S. (2020). COVID-19 pandemic: Which IBD patients need to be scoped-who gets scoped now, who can wait, and how to resume to normal. *Journal of Crohn's & Colitis*, *14*, S791–S797.

Norsa, L., Indriolo, A., Sansotta, N., Cosimo, P., Greco, S., & D'Antiga, L. (2020). Uneventful course in patients with inflammatory bowel disease during the severe acute respiratory syndrome coronavirus 2 outbreak in northern Italy. *Gastroenterology*, *159*(1), 371–372.

Polack, F. P., Thomas, S. J., Kitchin, N., Absalon, J., Gurtman, A., Lockhart, S., et al. (2020). Safety and efficacy of the BNT162b2 mRNA Covid-19 vaccine. *New England Journal of Medicine*, *383*(27), 2603–2615.

Pratt, P. K., David, N., Weber, H. C., Little, F. F., Kourkoumpetis, T., Patts, G. J., et al. (2018). Antibody response to hepatitis B virus vaccine is impaired in patients with inflammatory bowel disease on infliximab therapy. *Inflammatory Bowel Diseases*, *24*(2), 380–386.

Rizk, J. G., Kalantar-Zadeh, K., Mehra, M. R., Lavie, C. J., Rizk, Y., & Forthal, D. N. (2020). Pharmaco-immunomodulatory therapy in COVID-19. *Drugs*, *80*(13), 1267–1292.

Rubin, D. T., Feuerstein, J. D., Wang, A. Y., & Cohen, R. D. (2020). AGA clinical practice update on Management of Inflammatory Bowel Disease during the COVID-19 pandemic: Expert commentary. *Gastroenterology*, *159*(1), 350–357.

Salvatori, S., Baldassarre, F., Mossa, M., & Monteleone, G. (2021). Long COVID in inflammatory bowel diseases. *Journal of Clinical Medicine*, *10*(23). https://doi.org/10.3390/jcm10235575.

Sebastian, S., Gonzalez, H. A., & Peyrin-Biroulet, L. (2020). Safety of drugs during previous and current coronavirus pandemics: Lessons for inflammatory bowel disease. *Journal of Crohn's & Colitis*, *14*(11), 1632–1643.

Selinger, C. P., Fraser, A., Collins, P., Gunn, M., Chew, T. S., Kerry, G., et al. (2021). Impact of the coronavirus infectious disease (COVID-19) pandemic on the provision of inflammatory bowel disease (IBD) antenatal care and outcomes of pregnancies in women with IBD. *BMJ Open Gastroenterology*, *8*(1). https://doi.org/10.1136/bmjgast-2021-000603.

Siegel, C. A., Christensen, B., Kornbluth, A., Rosh, J. R., Kappelman, M. D., Ungaro, R. C., et al. (2020). Guidance for restarting inflammatory bowel disease therapy in patients who withheld immunosuppressant medications during COVID-19. *Journal of Crohn's & Colitis*, *14*, S769–S773.

Siegel, C. A., Melmed, G. Y., McGovern, D. P., Rai, V., Krammer, F., Rubin, D. T., et al. (2021). SARS-CoV-2 vaccination for patients with inflammatory bowel diseases: Recommendations from an international consensus meeting. *Gut*, *70*(4), 635–640.

Singh, A. K., Jena, A., Kumar, M. P., Jha, D. K., & Sharma, V. (2022). Clinical presentation of COVID-19 in patients with inflammatory bowel disease: A systematic review and meta-analysis. *Intestinal Research*, *20*(1), 134–143.

Singh, A. K., Jena, A., Kumar, M. P., Sharma, V., & Sebastian, S. (2021). Risk and outcomes of coronavirus disease in patients with inflammatory bowel disease: A systematic review and meta-analysis. *United European Gastroenterology Journal*, *9*(2), 159–176.

Singh, S., Khan, A., Chowdhry, M., Bilal, M., Kochhar, G. S., & Clarke, K. (2020). Risk of severe coronavirus disease 2019 in patients with inflammatory bowel disease in the United States: A multicenter research network study. *Gastroenterology*, *159*(4), 1575–1578.e4.

Sperger, J., Shah, K. S., Lu, M., Zhang, X., Ungaro, R. C., Brenner, E. J., et al. (2021). Development and validation of multivariable prediction models for adverse COVID-19 outcomes in patients with IBD. *BMJ Open*, *11*(11). https://doi.org/10.1136/bmjopen-2021-049740.

Stallmach, A., Kortgen, A., Gonnert, F., Coldewey, S. M., Reuken, P., & Bauer, M. (2020). Infliximab against severe COVID-19-induced cytokine storm syndrome with organ failure—a cautionary case series. *Critical Care*, *24*(1), 444.

Suárez-Fariñas, M., Tokuyama, M., Wei, G., Huang, R., Livanos, A., Jha, D., et al. (2021). Intestinal inflammation modulates the expression of ACE2 and TMPRSS2 and potentially overlaps with the pathogenesis of SARS-CoV-2-related disease. *Gastroenterology*, *160*(1), 287–301.

Sung, K. Y., Chang, T. E., Wang, Y. P., Lin, C. C., Chang, C. Y., Hou, M. C., et al. (2022). SARS-CoV-2 vaccination in patients with inflammatory bowel disease: A systemic review and meta-analysis. *Journal of the Chinese Medical Association: JCMA*, *85*(4), 421–430.

Tariq, R., Saha, S., Furqan, F., Hassett, L., Pardi, D., & Khanna, S. (2020). Prevalence and mortality of COVID-19 patients with gastrointestinal symptoms: A systematic review and Meta-analysis. *Mayo Clinic Proceedings*, *95*(8), 1632–1648.

Taxonera, C., Alba, C., & Olivares, D. (2021). What is the incidence of COVID-19 in patients with IBD in Western countries? *Gastroenterology*, *160*(5), 1901–1902.

Taxonera, C., Alba, C., Olivares, D., Martin, M., Ventero, A., & Cañas, M. (2021). Innovation in IBD care during the COVID-19 pandemic: Results of a cross-sectional survey on patient-reported experience measures. *Inflammatory Bowel Diseases*, *27*(6), 864–869.

Taxonera, C., Fisac, J., & Alba, C. (2021). Can COVID-19 trigger De novo inflammatory bowel disease? *Gastroenterology*, *160*(4), 1029–1030.

Taxonera, C., Sagastagoitia, I., Alba, C., Mañas, N., Olivares, D., & Rey, E. (2020). 2019 Novel coronavirus disease (COVID-19) in patients with inflammatory bowel diseases. *Alimentary Pharmacology and Therapeutics*, *52*(2), 276–283.

Taxonera Samso, C., Olivares, D., Blanco, I., Velasco, E., Molina, S., López, G., et al. (2022). P114 SARS-CoV-2 infection in IBD: Pandemic waves and predictors of severe or persistent COVID-19. *Journal of Crohn's & Colitis*, *16*(Suppl._1), i204–i206.

The RECOVERY Collaborative Group. (2021). Dexamethasone in hospitalized patients with Covid-19. *New England Journal of Medicine*, *384*(8), 693–704.

Ungaro, R. C., Brenner, E. J., Agrawal, M., Zhang, X., Kappelman, M. D., Colombel, J. F., et al. (2022). Impact of medications on COVID-19 outcomes in inflammatory bowel disease: Analysis of more than 6000 patients from an international registry. *Gastroenterology*, *162*(1), 316–319.e5. W.B. Saunders https://doi.org/10.1053/j.gastro.2021.09.011.

Ungaro, R. C., Brenner, E. J., Gearry, R. B., Kaplan, G. G., Kissous-Hunt, M., Lewis, J. D., et al. (2021). Effect of IBD medications on COVID-19 outcomes: Results from an international registry. *Gut*, *70*(4), 725–732.

Voysey, M., Clemens, S. A. C., Madhi, S. A., Weckx, L. Y., Folegatti, P. M., Aley, P. K., et al. (2021). Safety and efficacy of the ChAdOx1 nCoV-19 vaccine (AZD1222) against SARS-CoV-2: An interim analysis of four randomised controlled trials in Brazil, South Africa, and the UK. *The Lancet*, *397*(10269), 99–111.

Wang, W., Xu, Y., Gao, R., Lu, R., Han, K., Wu, G., et al. (2020). Detection of SARS-CoV-2 in different types of clinical specimens. *JAMA—Journal of the American Medical Association*, *323*(18), 1843–1844.

Wetwittayakhlang, P., Albader, F., Golovics, P. A., Hahn, G. D., Bessissow, T., Bitton, A., et al. (2021). Clinical outcomes of COVID-19 and impact on disease course in patients with inflammatory bowel disease. *Canadian Journal of Gastroenterology and Hepatology*. https://doi.org/10.1155/2021/7591141. eCollection 2021.

Wong, S. Y., Dixon, R., Martinez Pazos, V., Gnjatic, S., Colombel, J. F., Cadwell, K., et al. (2021). Serologic response to messenger RNA coronavirus disease 2019 vaccines in inflammatory bowel disease patients receiving biologic therapies. *Gastroenterology*, *161*(2), 715–718.e4.

Xiao, F., Tang, M., Zheng, X., Liu, Y., Li, X., & Shan, H. (2020). Evidence for gastrointestinal infection of SARS-CoV-2. *Gastroenterology*, *158*(6), 1831–1833.e3.

Zabana, Y., Marín-Jiménez, I., Rodríguez-Lago, I., Vera, I., Martín-Arranz, M. D., Guerra, I., et al. (2022). Nationwide COVID-19-EII study: Incidence, environmental risk factors and long-term follow-up of patients with inflammatory bowel disease and COVID-19 of the ENEIDA registry. *Journal of Clinical Medicine*, *11*(2).

Zhou, F., Yu, T., Du, R., Fan, G., Liu, Y., Liu, Z., et al. (2020). Clinical course and risk factors for mortality of adult inpatients with COVID-19 in Wuhan, China: A retrospective cohort study. *The Lancet*, *395*(10229), 1054–1062.

Chapter 7

Management of COVID-19 and clinical nutrition

Manola Peverini[a] and Giacomo Barberini[b]

[a]*Hospital Pharmacy of Urbino, ASUR Marche AV1, Urbino, Italy,* [b]*Department of Biomolecular Sciences, School of Pharmacy, University of Urbino Carlo Bo, Urbino, Italy*

Abbreviations

ARDS	acute respiratory distress syndrome
COPD	chronic obstructive pulmonary disease
EN	enteral artificial nutrition
ECMO	extracorporeal membrane oxygenation
ICU	intensive care unit
IOT	ortracheal intubation
NIV	noninvasive ventilation
PN	parenteral nutrition
TPN	total parenteral nutrition

Introduction

The nutritional status of an individual plays a key role both in the maintenance of a physiological state of health and under conditions of stressful events, such as the presence of infection. During the pandemic, the malnutrition status of the COVID-19 patient was not fully understood, and it was only later concluded that the prevention, diagnosis, and treatment of malnutrition should be included in the management of SARS-CoV-2 patients for a better short- and long-term prognosis. In patients who are not mechanically ventilated or undergoing noninvasive ventilation (NIV), depending on the patient's masticatory capacity and ability to swallow, spontaneous oral nutrition should certainly be preferred, even taking into account any comorbidities; where this is not possible, artificial nutrition, either enteral (EN) or parenteral (PN), should be used (Villarini et al., 2022). When spontaneous oral nutrition is not possible and artificial nutrition must be practiced, EN is always preferred as it maintains active trophism of the gastrointestinal (GI) tract, has a lower risk of infectious complications, and is easier managed. Of course, the composition of enteral nutrition is able to influence the colonization and maintenance of opportunistic and beneficial microbial compositions. Confirmation of this comes from a meta-analysis conducted by Marik and colleagues, which showed that early enteral nutrition was associated with fewer infectious complications and a reduced length of hospital stay (Marik & Zaloga, 2001). Similarly, in a separate meta-analysis, Doig et al. also demonstrated a reduction in pneumonia and mortality when enteral nutrition was started within 24 h (Tian et al., 2018). In the specific case of COVID-19 patients, however, the choice of administration route for nutrition is closely linked to the respiratory autonomy of the COVID-19 patient, who can easily be on more or less severe artificial ventilation (NIV-noninvasive ventilation/high flow; ECMO-extracorporeal membrane oxygenation; IOT-ortracheal intubation) (Barazzoni et al., 2020).

In the first pandemic waves, the SARS-CoV-2 positive patient often arrived at the hospital already malnourished due to the advanced state of the disease, and after hospitalization, sarcopenia, reduced muscle mass, and even more pronounced malnutrition not infrequently occurred, affecting the outcome of the disease. In addition, the urgent, brutal, and massive admissions of patients requiring urgent respiratory care and artificial ventilation led to the need to reorganize hospital care, wards, and staff. In this context, screening and nutritional care may not have been considered a priority. It must also be

Management, Body Systems, and Case Studies in COVID-19. https://doi.org/10.1016/B978-0-443-18703-2.00050-5
Copyright © 2024 Elsevier Inc. All rights are reserved, including those for text and data mining, AI training, and similar technologies.

considered that since the beginning of the epidemic, due to the shortage of masks and other protective material, the risk of contamination of healthcare workers led to the nonuse of enteral nutrition, although indicated, because the insertion of the nasogastric tube is an aerosol-generating procedure (Thibault et al., 2021). For the reasons listed above, although EN would be preferred, PN is often used, suggesting that there is a strong need to develop a specific nutritional pathway to support the management of COVID-19 patients. The nutrition-COVID-19 relationship and related dietary changes induce a vicious cycle of malnutrition, obesity, and undernutrition with micronutrient deficiencies, which promotes infection, disease progression, and potential death (Antwi et al., 2021). It should be noted that antiviral therapy and the intestinal involvement of the infection tend to further aggravate the nutritional status of COVID-19 positive patients, due to various side effects including lack of appetite, nausea, and diarrhea (Italian Society of Artificial Nutrition and Metabolism (SINPE), 2023). Malnutrition does not always result from hospitalization or the infectious state, but may be related to a preexisting condition of altered nutritional status of the subject that may modify the response to the virus (Villarini et al., 2022):

- Undernourished patients have a deficiency of adipose tissue as a source of adipocytokines and a reduction of macrophages and T-cells. These conditions imply a decreased response to the virus by the immune system (Nutrimi, 2021).
- Obese patients may develop an impaired immune response and may suffer from chronic obstructive pulmonary disease (COPD), cardiovascular disease, hypertension, and diabetes, all risk factors that worsen the clinical condition of the SARS-CoV-2 patient (Fedele et al., 2020).

If the patient's nutrition is not adequately supported during hospitalization, which often lasts for long periods, the patient's health status may be compromised even after discharge (Zhang & Liu, 2020).

Nutrition of the COVID-19 nondysphagic patient

When oral nutrition is possible and effective, 70% of caloric requirements should be provided through low-calorie nutrition within 3–7 days. The daily calories administered should be between 25 and 30 Kcal per kilogram of body weight while the amount of protein should be between 1.2 and 2 g/kg/day. After the first week, the requirement can be increased up to 100% of the estimated caloric intake (Fig. 1). Patients should be encouraged to eat small meals. Attention must be paid to the risk of aspiration in dysphagic patients. If oral nutrition is possible but the expected requirements cannot be met, extra nutrients can be considered through nutritional supplements (approx. 400–600 Kcal) (Stachowska et al., 2020). The Espen guidelines may provide further details in the nutrition of the critical patient.

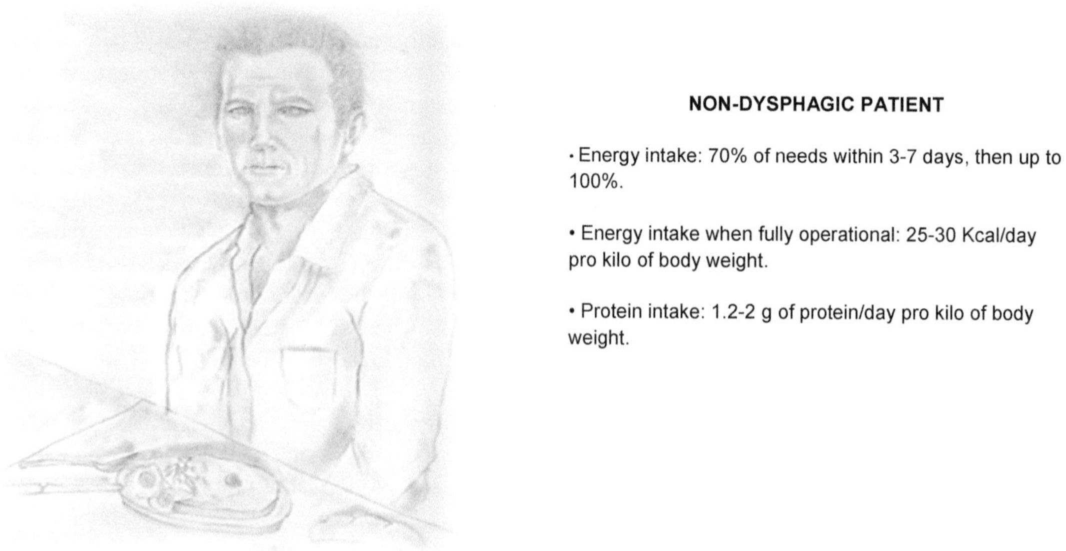

NON-DYSPHAGIC PATIENT
• Energy intake: 70% of needs within 3-7 days, then up to 100%.
• Energy intake when fully operational: 25-30 Kcal/day pro kilo of body weight.
• Protein intake: 1.2-2 g of protein/day pro kilo of body weight.

FIG. 1 Percentage of nutrients intake for nondysphagic person.

Artificial nutrition of the COVID-19 patient

Although enteral nutrition is the preferred artificial nutrition, it is contraindicated in uncontrolled shock, uncontrolled hypoxemia and acidosis, intestinal ischemia, intestinal obstruction, abdominal compartment syndrome, and high output fistula without distal feeding routes (Stachowska et al., 2020). It may be inadvisable in individuals with severe GI disorders (diarrhea, nausea, vomiting, abdominal discomfort, and, in some cases, gastrointestinal bleeding), which may sometimes occur before respiratory symptoms (Martindale et al., 2020). It is worth mentioning that 20% of COVID-19 patients present with GI symptoms (abdominal pain and diarrhea) either due to viral infection or as an adverse effect of antiviral drugs; these side effects may lead to an alteration of the intestinal microbiota. However, if the GI disorders listed above are not severe, maintaining EN together with adequate antiemetics, antidiarrheal drug therapy, and probiotic supplementation appears to be helpful in maintaining intestinal nutrient flow (Villarini et al., 2022). Enteral nutrition is contraindicated in patients treated with NIV. The latter modality is incompatible with enteral nutrition due to the risk of aspiration by gastric insufflation. When severe gastrointestinal symptoms occur and there is intolerance to enteral nutrition, supplemental or total parenteral nutrition is appropriate (Thibault et al., 2020). A return to enteral nutrition should be considered when gastrointestinal symptoms have subsided. The critical patient with rising lactate levels and hemodynamic instability (requiring increasing vasopressor support) should be given parenteral nutrition until stabilization. The reduced risk of ischemic bowel, the possible long stay in the intensive care unit (ICU), and the reduced possibility of aerosol development during positioning and maintenance maneuvers argue in favor of the early use of parenteral nutrition. In patients with total invasive mechanical ventilation (intubated in an induced coma), PN can be administered by continuous infusion; this reduces diarrhea and the number of contacts the healthcare team may have with the patient (Singer et al., 2019). PN can be total (TPN) and this must be taken into account to carry out proper weaning during reeducation to oral feeding (Barazzoni et al., 2020). Enteral nutrition should be started early (within 48 h after admission) (Thibault et al., 2020) at a low dosage, and then slowly increased over the first week to 25 Kcal/kg/day and a protein intake of 1.2–2 g protein per day. The recommended calorie intake is lower for a higher body mass index (BMI, overweight or obese) (Stachowska et al., 2020) (Fig. 2). Although adequate physical control of the patient and intestinal canalization are important practices during enteral nutrition, they must be carried out with particular care in the COVID-19 patient due to the high risk of contagion (Martindale et al., 2020). For the same reason, the indirect calorimetry method for estimating the patient's metabolism cannot be applied.

In the early acute phase of the disease, an isoosmotic polymeric mixture with a high protein content is chosen. When the patient's clinical situation improves, fiber may be added. In the case of diarrhea combined with polymeric enteral nutrition, the second choice is the semialimentary EN. Although the addition of fish oils may have theoretical utility in controlling the cytokine storm of acute respiratory distress syndrome, specific studies are still few but encouraging (Thibault et al., 2020). Several studies show the potential of probiotic benefits on patients in the ICU (Marik & Zaloga, 2001). Although some data

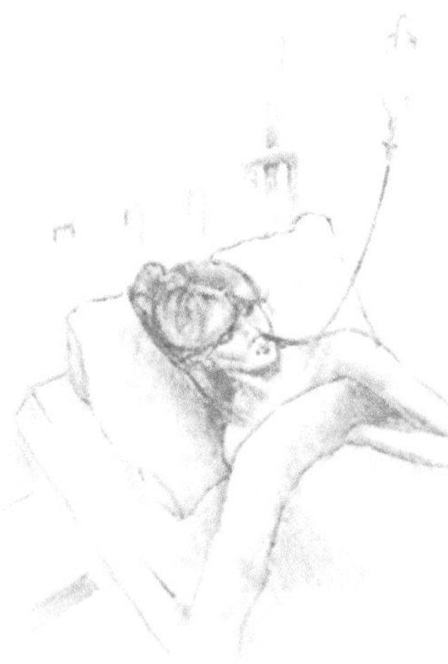
ENTERAL NUTRITION
Early, within 48 hours of admission.
Energy intake: gradually increased in the first week. Energy intake can be reduced in obese individuals.
Energy intake when fully operational: 25 Kcal/day pro kilogram of body weight.
Protein intake: 1.2-2 g/day pro kilogram of body weight.

FIG. 2 Constitution of the mixture to be used in enteral nutrition.

on the use of zinc in viral infections are interesting, supplementation of the same mineral in ICUs is not recommended (Duncan et al., 2012). Vitamin D seems to be particularly lacking in intensive care patients (Putzu et al., 2017). Although vitamin D supplementation in severely deficient patients has resulted in a significant decrease in in-hospital mortality, its supplementation to ICU patients with suboptimal blood levels does not seem to be recommended (Amrein et al., 2014).

Microbiota and COVID-19

General aspects

It is now known that microbial communities (bacteria, fungi, viruses, and protozoa) that inhabit the human gastrointestinal tract as well as the lungs, skin, and mouth have a commensal relationship with host cells, playing a positive role in health. It is also clear that the microbiota, normally residing in the mucosa of the upper and lower respiratory tracts and other body districts, is very important not only in health but also in disease. During the development of a respiratory tract disease (asthma, infections) or in lifestyle modification (smoking), the diversification as well as the microbial abundance of the two respiratory tracts undergoes an important change. Today, it is also clear that the health status of the upper and lower respiratory tracts is intimately connected with the intestinal microbiota. This is why we speak of the gut-lung axis; there is evidence that intestinal inflammation can lead to lung inflammation through so-called dysbiosis (Gasmi et al., 2021). It has been shown over time that SARS-Cov-2 can, in addition to the classical pulmonary symptoms, manifest itself at the gastrointestinal level with diarrhea or ulcerative colitis. More severe respiratory disorders have been correlated with the decline of certain microorganisms of the genera Lactobacillus and Bifidobacterium while the abundance of *Clostridium hathewayi*, *Clostridium ramosum*, and Coprobacillus is characteristic of individuals with more severe disease. The effects of SARS-Cov-2 on the microbiota can also be observed in other respiratory pathologies (asthma, bacterial infections, chronic obstructive bronchopathy, etc.) in addition to some probiotic benefits in terms of viral load and symptomatology, both on the upper respiratory tract (URTI) (Mullish et al., 2021) and in the avoidance of community-acquired lung infections (i.e., ventilator-associated pneumonia) (Siempos et al., 2010).

To understand the effect of probiotics against SARS-Cov-2 infection, it is necessary to illustrate the functioning of pattern recognition receptors (PRRs), the main sensors of innate immunity. The rapid immune response to viruses is based on the early recognition of pathogen-associated molecular patterns (PAMPs), that is, the essential components of viruses, by means of toll-like receptors (TLRs) found on dendritic and other inflammatory cells. Also very important as stimulating agents of TLRs are danger-associated molecular patterns (DAMPs), released from damaged or necrotic host tissues. TLRs are probably the most studied and characterized PRRs. This process leads to the activation of NKs via interferon. NK cells are of particular importance in defense against viruses (Mortaz et al., 2013; Stavropoulou & Bezirtzoglou, 2020). TLRs are not only a family of receptors related to the innate immune response but also act as a bridge between innate and adaptive immunity. TLR activation, through multistep signaling cascade processes, activates the NF-κB signal transduction pathway. Indeed, NF-κB signal transduction can initiate innate and adaptive immune responses directed at the pathogenic microorganism (e.g., the release of inflammation mediators such as IFN-β and TNF-α, neutrophil granulocyte chemotaxis, and lymphocyte activation) (Wu et al., 2013).

The action of probiotics is based on modulating TLRs, reducing or eliminating inflammation. A randomized study showed that the probiotic *Lactobacillus rhamnosus* GG was able to reduce the incidence of pneumonia associated with a pulmonary ventilator (Morrow et al., 2010). But the activation of certain TLRs (e.g., TLR3) could, conversely, activate the immune response by increasing respiratory tree clearance.

Several studies suggest that the microbiota is significantly involved in host responses to many viral infections through different mechanisms. Several mechanisms have been observed: binding and inactivation of the virus, production of certain antimicrobial products by the probiotics (organic acids, hydrogen peroxide, biotensioactives, and bacteriocins), and inhibition of viral replication. These probiotic mechanisms are often strain-specific (Lehtoranta et al., 2014).

In addition to these effects, the microbiota can exert antiviral effects through certain metabolites, in particular butyric acid (a short-chain fatty acid). The mechanism of action consists of enhancing the activity and energy consumption of regulatory T lymphocytes. Regulatory T cells are considered central in the suppression of allergic and inflammatory responses. The same short-chain fatty acid is also a source of energy for colon cells, where it suppresses NF-κB synthesis and exerts an antiinflammatory action. Butyric acid, like other short-chain fatty acids, plays an important role in promoting intestinal barrier function and reducing epithelial permeability (Hiippala et al., 2018). Butyric acid is produced by the intestinal fermentation of dietary fiber (Fig. 3) (Kim, 2021).

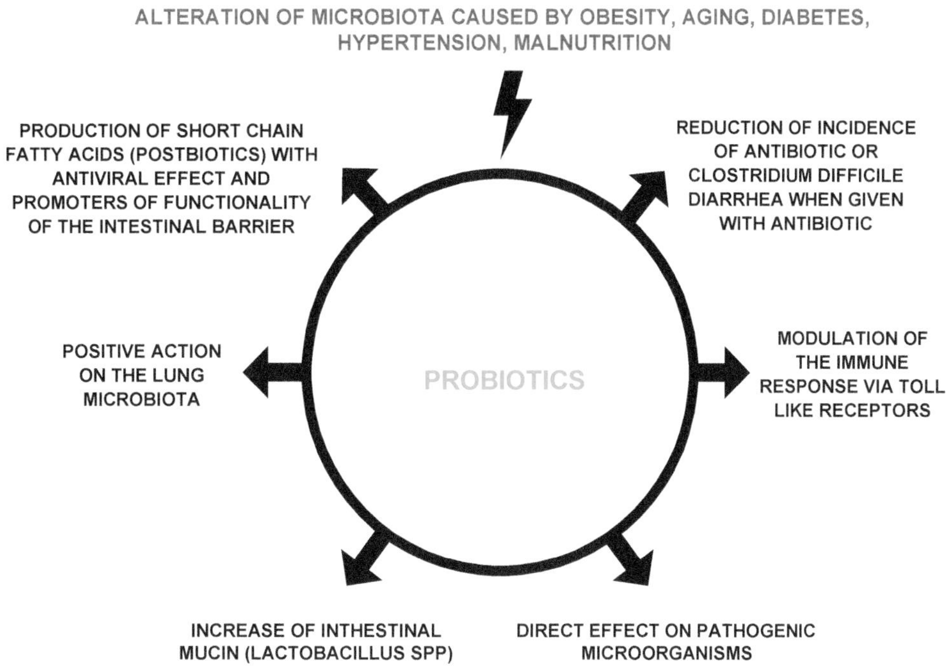

FIG. 3 Microbiota functions and disfunction.

Which subjects are at risk of dysbiosis

It has been observed that COVID-19 can occur, in its most severe form, in elderly, obese, diabetic, and hypertensive patients. In these cases, a profound alteration of the intestinal microbiota is often observed. Aging is associated with malnutrition and generalized atherosclerosis, resulting in decreased mucosal blood flow and atrophy of the intestinal villi. In obesity, the intestinal barrier is altered due to low-grade inflammation dependent on the higher basal production of inflammatory cytokines. The altered production of short-chain fatty acids may play a role in the development of hypertension. In addition, certain therapies, especially antibiotics, can profoundly alter the microbiota.

Use of probiotics and prebiotics in the nutrition of the COVID-19 patient

Probiotics are specific strains of the microbiota that can mediate beneficial effects in an organism and are frequently administered as dietary supplements. Although the efficacy and mechanism of action against the influenza virus of some of them have been defined (Park et al., 2013; Zhang et al., 2018), the use of probiotics to reduce the severity of SARS-Cov-2 infection seems promising, but the data are inconclusive (Din et al., 2021). The use of probiotics is in fact not supported by direct evidence but only indirectly (Mak et al., 2020). It is known that COVID-19 pathogenesis includes disruption of the intestinal barrier, inflammation, dysbiosis and increased production of inflammatory mediators (IL-6, TNF-a, C-reactive protein, IL-1b, IL2, IL7, IL10, GCSF, IP10, MCP1, MIP1A, and LDH). The use of probiotics could reduce inflammation by modulating the synthesis of inflammatory mediators by a mechanism involving TLRs. It is known that COVID-19 patients, whether hospitalized or not, are frequently treated with broad-spectrum antibiotics and that many of these patients develop diarrhea (as well as dysbiosis) as a side effect. The use of probiotics does not seem to modify the diarrhea symptom but plays a preventive role in some individuals with previous antibiotic diarrhea or those capable of developing *Clostridium difficile* diarrhea (Issa & Moucari, 2014). SARS-CoV-2 infection not only promotes intestinal inflammation, but also leads to a decrease in the production of antimicrobial peptides (AMP), leading to the development of secondary enteric infections. Mucus plays a very important role in antimicrobial defense; it is an extracellular secretion of the mucous membranes lining different body cavities (respiratory, digestive, and urogenital tracts). Mucin-glycoproteins are their main component and essentially perform a protective physical barrier function toward viruses, through the trapping and inhibition of viral replication. Moreover, mucin biopolymers have a broad-spectrum antiviral action. Lactobacillus spp. is a probiotic capable of increasing intestinal mucin synthesis, probably also having an antiviral action against SARS-Cov-2 (Din et al., 2021). The disruption of intestinal barrier function can increase bacterial translocation and trigger systemic inflammation and inflammatory response to other organs

(Battaglini et al., 2021). The action of probiotics would not in any case be limited to the gut; it is plausible that probiotics or their metabolites could be captured by dendritic cells or macrophages and transported from the blood to the lung.

Prebiotics, on the other hand, are nondigestible short-chain dietary fibers that nourish key strains of the microbiota and contribute to the production of short-chain fatty acids with an immunomodulatory function. The preferred sources of soluble fibers seem to be colored fruit and vegetables, as they contain antiinflammatory and immunomodulatory phytonutrients.

Use of probiotics in the critical patient

It is known that the critical patient shows intestinal and oral dysbiosis compared to the healthy subject, with a reduction of beneficial bacteria and a development of pathogens. The cause of this variation is not known. The use of probiotics, as already seen, has shown some efficacy in preventing ventilator-associated pneumonia in critically ill patients. The probiotic that seems to be most promising in this respect is *L. plantarum*, and the one to be avoided seems to be *S. boulardii* (Davison & Wischmeyer, 2019). The use of probiotics seems to have some efficacy in the prevention of antibiotic-associated diarrhea, in the critically ill patient and not, when the probiotic is administered during antibiotic therapy. On the other hand, there is no evidence on the prophylactic efficacy of probiotics against antibiotic-associated diarrhea, that is, administered in advance of antibiotics (Hempel et al., 2012). The use of probiotics appears to be effective in counteracting *C. difficile* diarrhea, especially in high-risk patients (Goldenberg et al., 2017). However, in the current state of knowledge, there are no specific recommendations on the type of strain, timing, daily dose, and duration of possible therapy (Manzanares et al., 2016).

The possible risks of probiotics in nutrition

Although most probiotics used are safe in the presence of many diseases, it may be useful to highlight some potential risks. The most frequent occurrence, following treatment with *Saccharomyces cerevisiae* or Saccharomyces boulardii, has been fungemia, evidenced in hemocultures (Doron & Snydman, 2015). Although very infrequently, cases of lactobacillus bacteremia and endocarditis have been reported in immunocompromised or diabetic subjects or patients recovering from recent surgery (Alexandre et al., 2014). In addition to being sustained by lactobacillus, sepsis has been associated with *S. boulardii* (and cerevisiae), *Bacillus subtilis*, Bifidobacterium brevis, or combined probiotics. Second, gene transfer via plasmids of antibiotic resistance from lactobacilli to various pathogenic microorganisms is theoretically possible, although it has never been observed. The third theoretical negative effect of probiotics would be excessive innate and adaptive immune stimulation, with the elevation of cytokine secretion. Such effects could lead to autoimmune or inflammatory responses, but they have never been observed (Doron & Snydman, 2015).

Conclusions

Probiotics seem very promising as an additional weapon against COVID-19 in noncritical patients, whereas great caution must be exercised in the critical patient due to the numerous overinfections found in COVID-19 clinical practice in ICUs. Their multiple actions observed against different microorganisms consist of a reduction of the infectious process, immunomodulation, and the reduction of some side effects of other treatments. These make probiotics among the most promising nonpharmacological treatments against SARS-Cov-2 infection. Evidence of the existence of a gut-lung axis reinforces the hypothesis of mechanisms of action not limited to the gut. Although very safe, the use of probiotics must be further investigated to confirm the action on COVID-19 as well as identify the correct probiotic type, the most active strain if any, and the relevant dosage and administration regimen.

Role of vitamins and minerals in the prevention and treatment of SARS-CoV-2 infection

Vitamin D

Vitamin D is a family of molecules in which, over the last decade, numerous immunomodulatory and antiinflammatory activities have been recognized. Although the sources of vitamin D can be diverse (egg yolk, fortified dairy products, ergocalciferol and cholecalciferol supplements), most of it is produced in the skin, where 7-dehydrocholesterol, involving sunlight or ultraviolet irradiation, is converted to previtamin D3 and then to vitamin D3. The most important compounds in this group of molecules are, in humans, vitamin D3 or cholecalciferol and vitamin D2 or ergocalciferol (from foods of plant origin). In the liver, cholecalciferol is converted to calcifediol (25-hydroxycholecalciferol) while ergocalciferol is converted to 25-hydroxyergocalciferol. These two vitamin D metabolites, called 25-hydroxyvitamin D or 25(OH)D, are the circulating form of vitamin D and are measured in serum to determine a subject's vitamin D status. In the body, 25-hydroxycholecalciferol is rehydroxylated in the kidney to generate the active form 1,25-dihydroxycholecalciferol or calcitriol (abbreviated as 1,25-(OH)2D3). Most of the effects of vitamin D are mediated by calcitriol (Moscatelli et al., 2021).

Vitamin D deficiency in the population

Plasma values of less than 50 nmol/L are indicative of 25-hydroxyvitamin D deficiency and may affect 40% of the European population and 50% of the world population. Interestingly, Italy and Spain—countries heavily affected by COVID-19—are among the European countries with the highest prevalence of hypovitaminosis D. Several studies have highlighted vitamin D deficiency in the population. In particular, vitamin D deficiency has been shown to affect healthy postmenopausal women (32% prevalence in the winter season) and institutionalized individuals (80%). Diabetes and obesity—known risk factors for the disease or its severity—are frequently characterized by vitamin D deficiency. Vitamin D deficiency has been correlated with inflammatory parameters and patients developing more severe forms of COVID-19, although further studies are needed (Bilezikian et al., 2020).

Proposed mechanisms

In rather general terms, vitamin D has a protective action against infection at three levels: it improves the physical biological barrier and stimulates innate and acquired immunity (Rondanelli et al., 2018). As stated earlier, SARS-CoV-2 entry involves viral interaction with the ACE2 enzyme and subsequent endocytosis of the complex. This is not a receptor in the strict sense, but an enzyme known to split angiotensin 2 into less active compounds and inhibit the renin-angiotensin-aldosterone (RAAS) system. Studies have shown that the interaction of the virus with the enzyme receptor results in a downregulation of the latter, leading to an accumulation of angiotensin II (a hypertensive and proliferative agent that can promote respiratory distress syndrome) and a reduction in angiotensin 1–7 (a vasodilating agent). ACE2 activity is also reduced in patients with cardiopulmonary diseases (Sahu et al., 2022). The role of vitamin D as a protective molecule against SARS CoV-2 is expressed in this context at various levels. Angiotensin 1–7 interacts with a G protein-coupled receptor called MAS. The angiotensin 1–7 interaction with the MAS receptor, in addition to mediating the vasodilation already mentioned, results in antiproliferative, antiinflammatory, and antifibrotic effects (Simões e Silva et al., 2013). Sufficient blood quantities of vitamin D are able to increase the expression of ACE2 and, as a negative endocrine regulator of the RAAS, induce the ACE2/Ang-(1–7)/MAS axis, which protects against acute lung damage and respiratory distress syndrome. However, the action of vitamin D could also be directed against SARS-CoV-2. Vitamin D3 via its receptor, the vitamin D receptor (VDR) induces in lung epithelia the expression of antimicrobial peptides such as cathelicidin 1 and defensin β4 as well as the macrophage production of cathelicidin 1 in large quantities. Stimulation of the VDR receptor would also be implicated in the modulation of immune responses (including IL-6, a cytokine) following infection with certain viruses (influenza, respiratory syncytial virus) (Ghelani et al., 2021). It seems clear that chronic activation of the innate immune response is not a useful means of suppressing SARS-CoV-2 infection but rather is responsible for the cytokine storm. 1,25(OH)2D is able to modulate the chronic innate immune response through a number of mechanisms, including the downregulation of TLRs and the inhibition of the direction of TNF/NFκB and IFNγ signaling pathways (Bilezikian et al., 2020).

Evidence supporting the importance of vitamin D in nutrition

Several studies have shown that overall mortality as well as susceptibility to respiratory infections are higher in the vitamin D-deficient population. Several studies (Grant et al., 2020; Sulli et al., 2021) have also shown that low serum vitamin D levels correlate with more severe lung involvement, prolonged disease, and increased risk of death. A systematic review and meta-analysis showed a significant relationship between vitamin D and severity/mortality from COVID-19 (Kazemi et al., 2021). In line with these studies are the conclusions of two meta-analyses (Chiodini et al., 2021; Wang et al., 2022) showing that individuals with vitamin D insufficiency or deficiency are more prone to an increased length of stay, various complications, respiratory distress, or death from respiratory insufficiency. An earlier meta-analysis (Liu et al., 2021) also showed that individuals with insufficient blood amounts of vitamin D are more prone to infectious risk and hospitalization. However, a more recent meta-analysis (Mishra et al., 2022), although showing a positive association between vitamin D deficiency and disease severity, does not suggest that subjects with suboptimal vitamin D concentrations are more exposed to infectious risk.

Vitamin C, zinc in COVID-19 patient

A balanced diet together with the presence of vitamins and minerals plays a key role in supporting and modulating the immune system and its response, thus preventing infection and modulating inflammation. Vitamin C (ascorbic acid) is probably the most studied vitamin in infectious diseases. Although most of the population does not suffer from severe forms of deficiency, 10% or more of individuals suffer from subclinical deficiency. Although these frequently happens in low- and middle-income countries, it is not uncommon to observe deficiencies of various levels in Western countries as well (Rowe & Carr, 2020). As is well known, large amounts of reactive oxygen species (ROS) are produced during an infectious process in various cells of the immune system. However, although ROS plays an important role in the destruction of various pathogenic microorganisms, they develop free radicals and lipid peroxidation and consequently immunosuppression. The role of vitamin C, which is contained in high concentrations within various cells of the immune system, is precisely to prevent the development and action of ROS, leading to an enhanced immune response. Ascorbic acid is involved in the production of important cytokines implicated in the immune response in the early stages of influenza pathology (Kim et al., 2013). Precisely because of the mechanisms highlighted, vitamin C seemed a promising product for the treatment of SARS-CoV-2 infection. In critical COVID-19 patients, vitamin C administration, although it does not improve 30-day in-hospital mortality, is able to reduce venous thrombosis (Al Sulaiman et al., 2021). Although some studies (Milani et al., 2021) show some activity in the treatment of COVID-19, they cannot be considered conclusive. In fact, a good part of the studies reviewed are observational or case reports, which referred to high intravenous dosages. These amounts are very different from those that would be achieved with a diet rich in ascorbic acid. In the meta-analysis "Vitamin C and COVID-19 treatment: A systematic review and meta-analysis of randomized controlled trials," different doses of the drug (low and high dose), different routes of administration (intravenous and oral), on different pictures of the disease (severe and nonsevere) were investigated, but no significant benefits were observed from taking ascorbic acid (mortality, ICU admission, hospital stay, mechanical ventilation) (Rawat et al., 2021). Vitamin C supplementation appears to have no preventive effect in the general population (adequately nourished, with physiological plasma concentrations). However, although it needs further confirmation, in some individuals (frail, CVD, metabolic diseases, athletes undergoing major workloads, or people undergoing major physical stress) vitamin C supplementation might reduce the incidence of infections, including COVID-19 (Cerullo et al., 2020).

Conclusions

Although different studies show different favorable effects of vitamin D in patients with plasma vitamin D deficiency, further investigations are needed to recommend vitamin D in special nutrition for the prevention or treatment of COVID-19 in nondeficient individuals (Bassatne et al., 2021; Mishra et al., 2022).

With regard to zinc, it can be stated that a low plasma concentration of zinc may pose a risk of increased infection and complications in the COVID-19 patient (Heller et al., 2021; Razeghi Jahromi et al., 2021). Several mechanisms have been proposed, but based on the available evidence, the use of zinc in the prevention and treatment of COVID-19 appears rather limited, needing further investigation. Zinc administration, even combined with ascorbic acid, did not reduce symptoms of SARS-CoV-2 infection (Suma et al., 2021).

Mini-dictionary of terms

- European countries with the highest prevalence of hypovitaminosis D.
- Vitamin D is a family of molecules in which, over the last decade, numerous immunomodulatory and antiinflammatory activities have been recognized.
- Vitamin D3 or cholecalciferol and vitamin D2 or ergocalciferol are the most important vitamin D types in humans.
- Vitamin D3 via its receptor VDR induces in lung epithelia the expression of antimicrobial peptides such as cathelicidin 1 and defensin β4.
- Vitamin D levels correlate with more severe lung involvement, prolonged disease, and increased risk of death.
- Danger-associated molecular patterns are important stimulating agents of TLRs.

- Main involved inflammatory mediators are IL-6, TNF-a, C-reactive protein, IL-1b, IL2, IL7, IL10, GCSF, IP10, MCP1, MIP1A, and LDH.
- Microbiota is significantly involved in host responses to many viral infections.
- Prebiotics are nondigestible short-chain dietary fibers. Probiotics are specific strains of the microbiota that can mediate beneficial effects in an organism and are frequently administered as dietary supplements.
- Respiratory tract disease: Asthma and infections are the most common.
- TLRs found on dendritic and other inflammatory cells are the most studied PRRs.

Summary points

- Nutrition represents, according to different available evidence, an important aspect of the patient hospitalized for COVID-19.
- The need for artificial ventilation largely conditions the type of nutrition of the patient who, if possible, should be fed by spontaneous or enteral feeding.
- Special attention has been paid to various vitamins, probiotics, and trace elements.
- Micronutrients that are important in counteracting the infectious process and modulating inflammation seem to be vitamin D, vitamin C, and zinc, especially when deficient in an organism. A special role is played by probiotics and prebiotics.
- The discovery of a gut-lung axis, modulation of TLRs and some clinical evidence of the usefulness of some probiotics in the prevention of diarrhea associated with antibiotic use and lower incidence of pneumonia in the patient undergoing pulmonary ventilation have aroused much interest in this class of products.
- Further studies are needed to define the type and strains of probiotics to be used in clinical settings.

References

Al Sulaiman, K., Aljuhani, O., Saleh, K. B., Badreldin, H. A., Al Harthi, A., Alenazi, M., Alharbi, A., Algarni, R., Al Harbi, S., Alhammad, A. M., Vishwakarma, R., & Aldekhyl, S. (2021). Ascorbic acid as an adjunctive therapy in critically ill patients with COVID-19: A propensity score matched study. *Scientific Reports*, *11*(1), 17648.

Alexandre, Y., Le Blay, G., Boisramé-Gastrin, S., Le Gall, F., Héry-Arnaud, G., Gouriou, S., Vallet, S., & Le Berread, R. (2014). Probiotics: A new way to fight bacterial pulmonary infections? *Médecine et Maladies Infectieuses*, *44*(1), 9–17.

Amrein, K., Schnedl, C., Holl, A., et al. (2014). Effect of high-dose vitamin D3 on hospital length of stay in critically ill patients with vitamin D deficiency: The VITdAL-ICU randomized clinical trial. *Journal of the American Medical Association*, *312*(15).

Antwi, J., Appiah, B., Oluwakuse, B., & Abu, B. A. Z. (2021). The nutrition-COVID-19 interplay: A review. *Current Nutrition Reports*, *10*(4), 364–374. Epub 2021 Nov 27 https://doi.org/10.1007/s13668-021-00380-2.

Barazzoni, R., et al. (2020). ESPEN expert statements and practical guidance for nutritional management of individuals with sars-cov-2 infection. *Clinical Nutrition*, *39*, 1631–1638.

Bassatne, A., Basbous, M., Chakhtoura, M., El Zein, O., Rahme, M., & El-Hajj, F. G. (2021). The link between COVID-19 and VItamin D (VIVID): A systematic review and meta-analysis. *Metabolism*, *119*, 154753. https://doi.org/10.1016/j.metabol.2021.154753.

Battaglini, D., Robba, C., Fedele, A., Trancă, S., Sukkar, S. G., Di Pilato, V., Bassetti, M., Giacobbe, D. R., Vena, A., Patroniti, N., Ball, L., Brunetti, I., Torres Martí, A., Rocco, P. R. M., & Pelosi, P. (2021). The role of dysbiosis in critically ill patients with COVID-19 and acute respiratory distress syndrome. *Frontiers in Medicine (Lausanne)*, *8*, 671714.

Bilezikian, J. P., Bikle, D., Hewison, M., Lazaretti-Castro, M., Formenti, A. M., Gupta, A., Madhavan, M. V., Nair, N., Babalyan, V., Hutchings, N., Napoli, N., Accili, D., Binkley, N., Landry, D. W., & Giustina, A. (2020). Mechanisms in endocrinology: Vitamin D and COVID-19. *European Journal of Endocrinology*, *183*(5), R133–R147.

Cerullo, G., Negro, M., Parimbelli, M., Pecoraro, M., Perna, S., Liguori, G., Rondanelli, M., Cena, H., & D'Antona, G. (2020). The long history of vitamin C: From prevention of the common cold to potential aid in the treatment of COVID-19. *Frontiers in Immunology*, *11*, 574029.

Chiodini, I., Gatti, D., Soranna, D., Merlotti, D., Mingiano, C., Fassio, A., Adami, G., Falchetti, A., Eller-Vainicher, C., Rossini, M., Persani, L., Zambon, A., & Gennari, L. (2021). Vitamin D status and SARS-CoV-2 infection and COVID-19 clinical outcomes. *Frontiers in Public Health*, *9*, 736665.

Davison, J. M., & Wischmeyer, P. E. (2019). Probiotic and synbiotic therapy in the critically ill: State of the art. *Nutrition*, *59*, 29–36.

Din, A. U., Mazhar, M., Waseem, M., Ahmad, W., Bibi, A., Hassan, A., Ali, N., Gang, W., Qian, G., Ullah, R., Shah, T., Ullah, M., Khan, I., Nisar, M. F., & Wu, J. (2021). SARS-CoV-2 microbiome dysbiosis linked disorders and possible probiotics role. *Biomedicine & Pharmacotherapy*, *133*, 110947.

Doron, S., & Snydman, D. R. (2015). Risk and safety of probiotics. *Clinical Infectious Diseases*, *60*(Suppl 2), S129–S134.

Duncan, A., Dean, P., Simm, M., O'Reilly, D. S., & Kinsella, J. (2012). Zinc supplementation in intensive care: Results of a UK survey. *Journal of Critical Care*, *27*(1), 102.e1–102.e6.

Fedele, D., et al. (2020). Obesity, malnutrition, and trace element deficiency in the coronavirus disease (COVID-19) pandemic: An overview. *Nutrition, 81*, 111016.

Gasmi, A., Tippairote, T., Mujawdiya, P. K., Peana, M., Menzel, A., Dadar, M., Benahmed, A. G., & Bjørklund, G. (2021). The microbiota-mediated dietary and nutritional interventions for COVID-19. *Clinical Immunology, 226*, 108725.

Ghelani, D., Alesi, S., & Mousa, A. (2021). Vitamin D and COVID-19: An overview of recent evidence. *International Journal of Molecular Sciences, 22*(19), 10559. https://doi.org/10.3390/ijms221910559.

Goldenberg, J. Z., Yap, C., Lytvyn, L., Lo, C. K., Beardsley, J., Mertz, D., & Johnston, B. C. (2017). Probiotics for the prevention of Clostridium difficile-associated diarrhea in adults and children. *Cochrane Database of Systematic Reviews, 12*(12), CD006095.

Grant, W. B., Lahore, H., McDonnell, S. L., Baggerly, C. A., French, C. B., Aliano, J. L., & Bhattoa, H. P. (2020). Evidence that vitamin D supplementation could reduce risk of influenza and COVID-19 infections and deaths. *Nutrients, 12*(4), 988.

Heller, R. A., Sun, Q., Hackler, J., Seelig, J., Seibert, L., Cherkezov, A., Minich, W. B., Seemann, P., Diegmann, J., Pilz, M., Bachmann, M., Ranjbar, A., Moghaddam, A., & Schomburg, L. (2021). Prediction of survival odds in COVID-19 by zinc, age and selenoprotein P as composite biomarker. *Redox Biology, 38*, 10176.

Hempel, S., Newberry, S. J., Maher, A. R., Wang, Z., Miles, J. N., Shanman, R., Johnsen, B., & Shekelle, P. G. (2012). Probiotics for the prevention and treatment of antibiotic-associated diarrhea: A systematic review and meta-analysis. *Journal of the American Medical Association, 307*(18), 1959–1969.

Hiippala, K., Jouhten, H., Ronkainen, A., Hartikainen, A., Kainulainen, V., Jalanka, J., & Satokari, R. (2018). The potential of gut commensals in reinforcing intestinal barrier function and alleviating inflammation. *Nutrients, 10*(8), 988.

Issa, I., & Moucari, R. (2014). Probiotics for antibiotic-associated diarrhea: Do we have a verdict? *World Journal of Gastroenterology, 20*(47), 17788–17795.

Italian Society of Artificial Nutrition and Metabolism (SINPE). (2023). *Practical recommendations for the nutritional treatment of patients with COVID-19*. p. 2–3–4-5-7-8-9.

Kazemi, A., Mohammadi, V., Aghababaee, S. K., Golzarand, M., Clark, C. C. T., & Babajafari, S. (2021). Association of Vitamin D Status with SARS-CoV-2 infection or COVID-19 severity: A systematic review and meta-analysis. *Advances in Nutrition, 12*(5), 1636–1658.

Kim, H. S. (2021). Do an altered gut microbiota and an associated leaky gut affect COVID-19 severity? *MBio, 12*(1). e03022–20.

Kim, Y., Kim, H., Bae, S., Choi, J., Lim, S. Y., Lee, N., Kong, J. M., Hwang, Y. I., Kang, J. S., & Lee, W. J. (2013). Vitamin C is an essential factor on the anti-viral immune responses through the production of interferon-α/β at the initial stage of influenza A virus (H3N2) infection. *Immune Network, 13*(2), 70–74.

Lehtoranta, L., Pitkäranta, A., & Korpela, R. (2014). Probiotics in respiratory virus infections. *European Journal of Clinical Microbiology & Infectious Diseases, 33*(8), 1289–1302.

Liu, N., Sun, J., Wang, X., Zhang, T., Zhao, M., & Li, H. (2021). Low vitamin D status is associated with coronavirus disease 2019 outcomes: A systematic review and meta-analysis. *International Journal of Infectious Diseases, 104*, 58–64.

Mak, J. W. Y., Chan, F. K. L., & Ng, S. C. (2020). Probiotics and COVID-19: One size does not fit all. *The Lancet Gastroenterology & Hepatology, 5*(7), 644–645.

Manzanares, W., Lemieux, M., Langlois, P. L., & Wischmeyer, P. E. (2016). Probiotic and synbiotic therapy in critical illness: A systematic review and meta-analysis. *Critical Care, 19*, 262.

Marik, P. E., & Zaloga, G. P. (2001). Early enteral nutrition in acutely ill patients: A systematic review. *Critical Care Medicine, 29*(12), 2264–2270.

Martindale, R., Patel, J. J., Taylor, B., Arabi, Y. M., Warren, M., & McClave, S. A. (2020). Nutrition therapy in critically ill patients with coronavirus disease 2019. *Journal of Parenteral and Enteral Nutrition, 44*, 1174–1184.

Milani, G. P., Macchi, M., & Guz-Mark, A. (2021). Vitamin C in the treatment of COVID-19. *Nutrients, 13*(4), 1172.

Mishra, P., Parveen, R., Bajpai, R., & Agarwal, N. (2022). Vitamin D deficiency and comorbidities as risk factors of COVID-19 infection: A systematic review and meta-analysis. *Journal of Preventive Medicine and Public Health, 55*(4), 321–333.

Morrow, L. E., Kollef, M. H., & Casale, T. B. (2010). Probiotic prophylaxis of ventilator-associated pneumonia: A blinded, randomized, controlled trial. *American Journal of Respiratory and Critical Care Medicine, 182*(8), 1058–1064.

Mortaz, E., Adcock, I. M., Folkerts, G., Barnes, P. J., Paul Vos, A., & Garssen, J. (2013). Probiotics in the management of lung diseases. *Mediators of Inflammation, 2013*, 751068.

Moscatelli, F., Sessa, F., Valenzano, A., Polito, R., Monda, V., Cibelli, G., Villano, I., Pisanelli, D., Perrella, M., Daniele, A., Monda, M., Messina, G., & Messina, A. (2021). COVID-19: Role of nutrition and supplementation. *Nutrients, 13*(3), 976.

Mullish, B. H., Marchesi, J. R., McDonald, J. A. K., Pass, D. A., Masetti, G., Michael, D. R., Plummer, S., Jack, A. A., Davies, T. S., Hughes, T. R., & Wang, D. (2021). Probiotics reduce self-reported symptoms of upper respiratory tract infection in overweight and obese adults: Should we be considering probiotics during viral pandemics? *Gut Microbes, 13*(1), 1–9.

Nutrimi, R. (2021). *COVID-19: squilibri dello stato nutrizionale influenzano il decorso della malattia*. [Internet] [updated 13 gennaio 2021; cited 23 febbraio 2021]. Available from: https://www.nutrimi.it/alimentazione-coronavirus-e-quarantena/.

Park, M. K., Ngo, V., Kwon, Y. M., Lee, Y. T., Yoo, S., Cho, Y. H., Hong, S. M., Hwang, H. S., Ko, E. J., Jung, Y. J., Moon, D. W., Jeong, E. J., Kim, M. C., Lee, Y. N., Jang, J. H., Oh, J. S., Kim, C. H., & Kang, S. M. (2013). Lactobacillus plantarum DK119 as a probiotic confers protection against influenza virus by modulating innate immunity. *PLoS One, 8*(10), e75368.

Putzu, A., Belletti, A., Cassina, T., Clivio, S., Monti, G., Zangrillo, A., & Landoni, G. (2017). Vitamin D and outcomes in adult critically ill patients. A systematic review and meta-analysis of randomized trials. *Journal of Critical Care, 38*, 109–114. https://doi.org/10.1016/j.jcrc.2016.10.029.

Rawat, D., Roy, A., Maitra, S., Gulati, A., Khanna, P., & Baidya, D. K. (2021). Vitamin C and COVID-19 treatment: A systematic review and meta-analysis of randomized controlled trials. *Diabetes and Metabolic Syndrome: Clinical Research and Reviews*, *15*(6), 102324.

Razeghi Jahromi, S., Moradi Tabriz, H., Togha, M., Ariyanfar, S., Ghorbani, Z., Naeeni, S., Haghighi, S., Jazayeri, A., Montazeri, M., Talebpour, M., Ashraf, H., Ebrahimi, M., Hekmatdoost, A., & Jafari, E. (2021). The correlation between serum selenium, zinc, and COVID-19 severity: An observational study. *BMC Infectious Diseases*, *21*(1), 899.

Rondanelli, M., Miccono, A., Lamburghini, S., Avanzato, I., Riva, A., Allegrini, P., Faliva, M. A., Peroni, G., Nichetti, M., & Perna, S. (2018). Self-care for common colds: The pivotal role of vitamin D, vitamin C, zinc, and echinacea in three Main immune interactive clusters (physical barriers, innate and adaptive immunity) involved during an episode of common colds-practical advice on dosages and on the time to take these nutrients/botanicals in order to prevent or treat common colds. *Evidence-based Complementary and Alternative Medicine*, *2018*, 5813095.

Rowe, S., & Carr, A. C. (2020). Global vitamin C status and prevalence of deficiency: A cause for concern? *Nutrients*, *12*(7), 2008.

Sahu, S., Patil, C. R., Kumar, S., Apparsundaram, S., & Goyal, R. K. (2022). Role of ACE2-Ang (1-7)-Mas axis in post-COVID-19 complications and its dietary modulation. *Molecular and Cellular Biochemistry*, *477*(1), 225–240.

Siempos, I. I., Ntaidou, T. K., & Falagas, M. E. (2010). Impact of the administration of probiotics on the incidence of ventilator-associated pneumonia: A meta-analysis of randomized controlled trials. *Critical Care Medicine*, *38*(3), 954–962.

Simões e Silva, A. C., Silveira, K. D., Ferreira, A. J., & Teixeira, M. M. (2013). ACE2, angiotensin-(1-7) and mas receptor axis in inflammation and fibrosis. *British Journal of Pharmacology*, *169*(3), 477–492.

Singer, P., Blaser, A. R., Berger, M. M., Alhazzani, W., Calder, P. C., Casaer, M. P., Hiesmayr, M., Mayer, K., Montejo, J. C., Pichard, C., Preiser, J. C., van Zanten, A. R. H., Oczkowski, S., Szczeklik, W., & Bischoff, S. C. (2019). ESPEN guideline on clinical nutrition in the intensive care unit. *Clinical Nutrition*, *38*(1), 48–79.

Stachowska, E., Folwarski, M., Jamioł-Milc, D., Maciejewska, D., & Skonieczna-Żydecka, K. (2020). Nutritional support in coronavirus 2019 disease. *Medicina (Kaunas, Lithuania)*, *56*(6), 289.

Stavropoulou, E., & Bezirtzoglou, E. (2020). Probiotics in medicine: A long debate. *Frontiers in Immunology*, *11*, 2192.

Sulli, A., Gotelli, E., Casabella, A., Paolino, S., Pizzorni, C., Alessandri, E., Grosso, M., Ferone, D., Smith, V., & Cutolo, M. (2021). Vitamin D and lung outcomes in elderly COVID-19 patients. *Nutrients*, *13*(3), 717.

Suma, T., et al. (2021). Effect of high-dose zinc and ascorbic acid supplementation vs usual care on symptom length and reduction among ambulatory patients with SARS-CoV-2 infection the COVID A to Z randomized clinical trial. *JAMA Network Open*, *4*(2), e210369.

Thibault, R., Coëffier, M., Joly, F., Bohé, J., Schneider, S. M., & Déchelotte, P. (2021). How the COVID-19 epidemic is challenging our practice in clinical nutrition-feedback from the field. *European Journal of Clinical Nutrition*, *75*(3), 407–416.

Thibault, R., Seguin, P., Tamion, F., Pichard, C., & Singer, P. (2020). Nutrition of the COVID-19 patient in the intensive care unit (ICU): A practical guidance. *Critical Care*, *24*(1), 447.

Tian, F., Heighes, P. T., Allingstrup, M. J., & Doig, G. S. (2018). Early enteral nutrition provided within 24 hours of ICU admission: A meta-analysis of randomized controlled trials. *Critical Care Medicine*, *46*(7), 1049–1056.

Villarini, A., Antonini, M., Teseo, G., Ricci, D., Cavaliere, A., & Peverini, M. (2022). Clinical nutrition and the role of hospital pharmacist in the management of covid patient. *Clinical Nutrition ESPEN*, *48*, 17–20.

Wang, Z., Joshi, A., Leopold, K., Jackson, S., Christensen, S., Nayfeh, T., Mohammed, K., Creo, A., Tebben, P., & Kumar, S. (2022). Association of vitamin D deficiency with COVID-19 infection severity: Systematic review and meta-analysis. *Clinical Endocrinology*, *96*(3), 281–287.

Wu, S., Jiang, Z. Y., Sun, Y. F., Yu, B., Chen, J., Dai, C. Q., Wu, X. L., Tang, X. L., & Chen, X. Y. (2013). Microbiota regulates the TLR7 signaling pathway against respiratory tract influenza A virus infection. *Current Microbiology*, *67*(4), 414–422.

Zhang, L., & Liu, Y. (2020). Potential interventions for novel coronavirus in China: A systematic review. *Journal of Medical Virology*, *92*(5), 479–490.

Zhang, H., Yeh, C., Jin, Z., Ding, L., Liu, B. Y., Zhang, L., & Dannelly, H. K. (2018). Prospective study of probiotic supplementation results in immune stimulation and improvement of upper respiratory infection rate. *Synthetic and Systems Biotechnology*, *3*(2), 113–120.

Chapter 8

The management of head and neck cancer in COVID-19

Jesús Herranz-Larrañeta[a], Pablo Parente-Arias[a], Carlos Chiesa-Estomba[b], and Miguel Mayo-Yáñez[a]

[a]*Department of Otorhinolaryngology—Head and Neck Surgery, Complexo Hospitalario Universitario A Coruña, A Coruña, Spain,* [b]*Department of Otorhinolaryngology—Head and Neck Surgery, Hospital Universitario de Donostia, Donosti, Spain*

Abbreviations

ED	emergency department
ENT	ear, nose, and throat
HNC	head and neck cancer
HNSCC	head and neck squamous cell carcinoma
HPV	herpes papilloma virus
PPE	personal protective equipment
PTC	papillary thyroid carcinoma
RT	radiotherapy

Introduction

Since its declaration as a public health emergency of international concern on Jan. 30, 2020 by the World Health Organization (World Health Organization, 2020), the severe acute respiratory syndrome coronavirus 2 (SARS-CoV-2) has represented a major challenge for healthcare systems worldwide, changing the way we understand medicine. The overload on the unprepared health system forced it to adapt its activity (CDC, 2020), changing primary care systems, restructuring hospital emergency departments, reorganizing medical and surgical activities, and redistributing health resources.

All these logistical challenges led to the reduction or to the temporary suspension of almost all elective surgical activity in many centers affected by the pandemic (Kowalski et al., 2020). This represents a serious problem in the case of head and neck cancer (HNC) in relation to its prognosis. These tumors, mostly squamous cell carcinomas, have a short tumor volume doubling time. Any delay in diagnosis or treatment can be fatal (Fakhry et al., 2020; Kowalski et al., 2020). In the initial phases of the COVID-19 pandemic, delays in oncologic care were reported in 40%–80% of specialized centers globally, with differences depending on the epidemiological situation. Likewise, it has been reported that >50% of the HNC surgeons were reassigned to other tasks outside their scope of practice (Mayo-Yáñez et al., 2021; Wu et al., 2020).

Multiple medical societies developed clinical practice guidelines regarding the diagnosis and treatment of HNC patients to offer the best care in an effective and safe manner, both for the patient and for healthcare workers (Di Martino et al., 2020; Fakhry et al., 2020; Finley et al., 2020; Givi et al., 2020; Kowalski et al., 2020; O'Connell et al., 2020; Zhao et al., 2020). Many of these recommendations were not based on scientific data established specifically for SARS-CoV-2, but on previous experience and knowledge of other respiratory viral infections as well as on basic precautionary principles (Alobid—Grupo de trabajo en COVID-19 de la SEORL-CCC, 2020; Mayo-Yánez, 2020). Due to the changing nature of the pandemic and the increasing knowledge of SARS-CoV-2, these recommendations evolved over time (COVIDSurg Collaborative, 2021; Gascon et al., 2022). This chapter covers four of the main aspects of care for the HNC patient: consultation care, diagnostic tests, surgical prioritization criteria, and treatment options.

Management, Body Systems, and Case Studies in COVID-19. https://doi.org/10.1016/B978-0-443-18703-2.00010-4
Copyright © 2024 Elsevier Inc. All rights are reserved, including those for text and data mining, AI training, and similar technologies.

Consultation care

As a result of organizational and material restructuring, mobility restrictions, recommendations to reduce social contact, fear of contagion, and closure of outpatient centers, the capacity for oncologic care was affected (Han et al., 2020; Lescanne et al., 2020). In the United Kingdom (UK), the impact on medical care resulting from COVID-19 and its consequences led to the underdiagnosis of up to 2300 new cancer cases per week (Solis et al., 2021). Suspected oncology cases referred to otorhinolaryngology consultations were reduced by 25%. The Piedmont region of Italy recorded a 28% reduction in head and neck cancer-related admissions in the first half of 2020 that was largely corrected by the end of 2020, although it had a more pronounced impact on older patients, those treated primarily surgically, and those with fewer comorbidities (Popovic et al., 2022). In Spain, 68.2% of the first consultations remained in person in addition to follow-up visits (90.9%) (Mayo-Yáñez et al., 2021).

To address this situation, it was recommended that telehealth be used to contact patients (Smith et al., 2020). This telematic assessment required ruling out the existence of COVID-19-type symptoms or the presence of risk contacts. Once these had been ruled out, triage was done to determine which patients required a face-to-face consultation (based on their baseline situation, their symptoms, or the poor evolution of a previous condition), which could be attended telematically with periodic follow-up, and which cases were susceptible to a longer delay in care based on their nonseverity. The presence of cervical masses with recent or rapidly progressive growth; the development of pain, dyspnea, stridor, or respiratory distress; the presence of recent dysphagia or dysphonia in patients with risk factors; or the suspicion of a deep cervical infection were signs and symptoms that warranted a face-to-face consultation. Another criterion was prioritizing the care of those patients with personal baseline conditions susceptible to complications in the event of a possible infection by COVID-19, seeking to reduce their presence in the hospital as much as possible (Hardman et al., 2021; Kaddour et al., 2022). Therefore, telehealth guided by a judicious and rational clinical attitude seems to be a good tool in situations where it is necessary to restructure and rethink patient care (Hardman et al., 2021; Kaddour et al., 2022; Paleri et al., 2020; Smith et al., 2020). In addition, studies show that the perception of care by patients attended telematically was good (Gómez González et al., 2021; Zhu et al., 2021).

Evaluation during the pandemic of HNC patients revealed highly variable data. During the first phases of the pandemic, there was a general increase in tumor burden in HNC patients, compared to prepandemic studies. In analyzing the studies separately, centers were detected that had greater delays in time to diagnosis (Curigliano et al., 2020; Riera et al., 2021); in others, no differences were found (Rygalski et al., 2021; Solis et al., 2021), or there was even a decrease in time (Zubair et al., 2022). Regarding time to treatment, there were no differences, but some centers detected more advanced clinical T and N stages at the initial consultation (Gazzini et al., 2022; Rygalski et al., 2021). No differences were detected with respect to presenting symptoms or their severity (Gazzini et al., 2022; Rygalski et al., 2021; Stevens et al., 2022), although some centers did report an increase in emergent consultations (Wilkie et al., 2021), with the consequent worse prognosis.

These variabilities detected among different studies may be related to a different epidemiological situation according to geographical location, the different pandemic phases in which the studies were carried out, the availability of out-of-hospital assessment, differences in the strictness of the different confinements, poor resolution capacity of the different telehealth triage programs, less strict remote follow-up schemes, or the fear of contagion from a highly immunocompromised population. All these reasons could have led to a delay in referral for HNC consultations.

Diagnostic tests

Minimally invasive examinations (endoscopic studies), imaging tests, and tissue sampling (biopsy) are the most common diagnostic tests required in HNC patients. In the pandemic context, these diagnostic tests were limited to cases in which their performance clearly conditioned subsequent management, cases in which the morbimortality associated with the underlying diagnostic suspicion was considerable, or situations in which the symptomatology demanded it (Vukkadala et al., 2020).

Flexible nasofibrolaryngoscopy is a simple and rapid procedure that provides a great deal of anatomical and functional information about the upper aerodigestive tract. It is, however, an aerosol-generating procedure. Given the high nasal viral load of SARS-CoV-2, it was recommended to reduce its use to the minimum necessary (MD Anderson Head and Neck Surgery Treatment Guidelines Consortium et al., 2020), and it was performed with extreme protective measures (good ventilation, use of negative pressure rooms and complete personal protective equipment (PPE), use of special FFP2 mask adapted to otolaryngological practice) (De Marrez et al., 2022). Prior to examination, the COVID-19 status should be known and an oral rinse with povidone iodine or hydrogen peroxide performed, aiming to reduce the viral load in saliva (Sánchez Barrueco et al., 2022). Recording the examination, thus avoiding subsequent repeated inspections that would increase the risk of contagion, is recommended (Solano et al., 2021). Unfortunately, there are reports confirming that these conditions could not be obtained in most cases. In Spain, the previsit SARS-CoV-2 symptom screening was not standardized, and was only done in 43.2% of the centers. Similarly, the use of nasofibrolaryngoscopy in consultations was maintained in 63.6% of the centers for 75% to 100% of the cases during initial visits and in 38.6% for follow-ups (Mayo-Yáñez et al., 2021).

Surgical prioritization criteria

Prioritization of treatments in a situation of asset shortage requires weighing two parameters: availability of resources and prognosis of the patient's underlying pathology. Resources availability evolved over time. The pandemic has evolved over the years, and with it the surgical criteria (Mehanna et al., 2020). Three scenarios can be discerned depending on hospital occupancy, the capacity to resolve the different services, and the availability of resources. These aspects will force us to decide which procedures are the most effective and efficient (Table 1).

The prognosis of a given disease requires an evaluation of each patient and his or her clinical condition. Delaying treatment in cancer patients increases the risk of requiring more aggressive procedures later on, or of not being able to develop certain treatments due to resectability criteria, thus affecting survival. Therefore, different categories are distinguished according to the urgency of the procedure, or the possibility of its deferral, based on the patient's clinical diagnosis and prognosis (Table 2).

The risk of SARS-CoV-2 infection requires knowing a patient's COVID-19 status before treatment, which may affect the patient's prognosis or turn the procedure into a dangerous technique for the medical staff (Ralli et al., 2022). Contacting the patient a few days before the procedure is recommended, as is advising him/her to minimize contacts as much as possible. It is advisable to perform at least one microbiological test in the 24–48h prior to the procedure, either medical or surgical. If this test is negative, it is recommended to proceed. If it is positive, and the patient's clinical situation allows it, it is recommended to postpone treatment for at least 14days. If the procedure is considered urgent, and the test is positive or indeterminate, it is recommended to proceed with all possible precautions (COVIDSurg Collaborative, 2020; Thomson et al., 2020; Topf et al., 2020).

TABLE 1 Surgical triage guidelines for the different phases of the pandemic.

	Triage 1 level Early in the Pandemic	Triage 2 level Worsening pandemic	Triage 3 level Worst-case scenario
Key signs of the stage	• Increased emergency department volumes	• Hospitals have surged to maximum bed capacity • EDs are overwhelmed • Insufficient ventilators • Hospital staff absenteeism >20%	• Hospitals have altered standards of care to accommodate expanded capacity • Hospital staff absenteeism >30%
Modifications to hospital care	• Preserve bed capacity by canceling elective surgeries requiring hospitalization • Increase patient care capacity • Implement enhanced infection control	• Preserve bed capacity by canceling all elective surgeries • Further increase patient care capacity • Preserve oxygen capacity	• Free bed capacity by facilitating early discharge • Preserve bed capacity by limiting urgent cases
Impact on cancer surgical care	• Only urgent cases (<30 d)	• Only urgent cases (<30 d) • Consider early postoperative discharge	• Only emergent cases ("serious threat to life") • Expedite early postoperative discharge

This table shows recommendations for triage and resource management during the different phases of the pandemic. Abbreviation: ED, emergency department. Adapted from Utah Hospital Pandemic Guidelines (2010). Utah Pandemic Influenza Hospital and ICU Triage Guidelines for ADULTS, January 28.

TABLE 2 Surgical priorities according to tumor etiology and location.

Urgent Proceed with surgery	• HPV-negative HNSCC • HPV-positive HNSCC with significant disease burden or delay in diagnosis • HNSCC patients with complications of cancer treatment • Recurrent HNSCC • Thyroid – Anaplastic thyroid carcinoma – Medullary thyroid carcinoma – Large (>4 cm) follicular lesions, neoplasms, or even indeterminate nodules – PTC with suspicion or identified metastatic disease, aggressive nature, or active progression • Parathyroidectomy with renal function declining • Skull base malignancy • Salivary cancer – Salivary duct carcinoma – High-grade mucoepidermoid carcinoma – Adenoid cystic carcinoma – Carcinoma ex pleomorphic adenoma – Acinic cell carcinoma – Adenocarcinoma – Other aggressive, high-grade salivary histology • Skin cancer – Melanoma >1 mm thickness – Merkel cell carcinoma – Advanced-stage, high-risk squamous cell carcinoma – Basal cell carcinoma in critical area (i.e., orbit)
Less urgent Consider postpone >30 days low risk	• Low-risk PTC without metastasis • Low-grade salivary carcinoma
Less urgent Consider postpone 30–90 days; reassess after pandemic appears to be resolving	• Thyroid – Goiter without airway/respiratory compromise – Routine benign thyroid nodules and thyroiditis – Revision PTC with stable or slow rate of progression • Parathyroidectomy with stable renal function • Benign salivary lesions • Skin cancer – Melanoma ≤1 mm thickness – Basal cell carcinoma where cosmetic impact/morbidity is likely low with further growth – Low-risk squamous cell carcinoma
Case-by-case basis	• Rare histology with uncertain rate of progression • Diagnostic procedures, such as direct laryngoscopy with biopsy

This table classifies surgical priorities according to tumor type. HPV, human papilloma virus; HNSCC, head and neck squamous cell carcinoma; PTC, papillary thyroid carcinoma.
Adapted from Topf, M. C., Shenson, J. A., Holsinger, F. C., Wald, S. H., Cianfichi, L. J., Rosenthal, E. L., & Sunwoo, J. B. (2020). Framework for prioritizing head and neck surgery during the COVID-19 pandemic. Head and Neck, 42(6), 1159–1167. https://doi.org/10.1002/hed.26184.

Treatment options

The selection of a particular treatment in oncological patients has been determined by a balance among the criteria of tumor resectability, patient status, health resource availability, and medical/surgical team experience. The presence of SARS-CoV-2 became a new parameter during the pandemic. It has been shown that perioperative infection by COVID-19 increases mortality in patients undergoing surgery due to pulmonary complications (COVIDSurg Collaborative, 2020). Given the high viral load of COVID-19 in otorhinolaryngology territory, the risk of generating aerosols during certain surgical techniques, or the limitation of the activity of some surgical services, the use of medical oncological treatment was recommended in all cases in which the prognosis and survival, whether undergoing surgery or chemotherapy or radiotherapy, were similar (Thomson et al., 2020). For example, half of the surveyed centers in Spain agreed that changes in therapeutic algorithms have been made during the pandemic. It is concerning that 29.5% reported that they had been "forced" to deviate from what they consider "standard of care" in the management of HNC cases due to the epidemiological situation. Despite these data, the reported proportion of HNC patients who had a shift from surgery to radiotherapy or chemotherapy-radiotherapy protocols was <25% in 75% of the centers (Mayo-Yáñez et al., 2021).

In a study conducted in 23 centers across the UK, significant changes in systemic and radiotherapeutic cancer treatment were found, although not in surgical treatment. Sixty-five percent of centers initiated a change in radical radiotherapy (RT), moving to a hypofractionation or acceleration schedule while for postoperative radiotherapy, 43.5% of the centers switched to a hypofractionation program. Regarding systemic treatment, 52.2% of centers discontinued neoadjuvant chemotherapy for all patients and 56.5% of centers followed the selective omission of chemotherapy in patients with concurrent chemoradiotherapy. Moreover, 73.9% of the centers switched the first-line chemotherapy treatment to pembrolizumab (following National Health Service England interim guidelines) and eight (34.8%) centers stopped treatment early or offered delays for patients already on systemic treatment (Vasiliadou et al., 2021).

For cases in which surgery was essential, recommendations were published that sought to make the surgical procedure as safe as possible. These included the creation of two independent surgical circuits (Glasbey et al., 2021), the use of negative pressure systems, minimization of the number of personnel in each surgery, the use of PPE, procedures done by the most experienced personnel, the use of cold material over electrical or robotic material, the use of local vasoconstriction whenever possible or favoring closures by primary intention before reconstruction, and, if this is not possible, the use of pedicled flaps rather than free flaps (Chiesa-Estomba et al., 2020; Mayo-Yánez et al., 2020.; Radulesco et al., 2020).

Another possible approach is to redirect head and neck cancer patients from overburdened hospitals to nearby hospitals with less burden. However, this strategy was severely limited by mobility restrictions and the lack of an overall organizational strategy (Shaw and COVIDSurg Collaborative, 2021).

Conclusions

The coronavirus disease has had an enormous impact on healthcare, which has had to focus on the emerging disease, leaving other diseases, including oncological ones, in the background. Thus, the diagnosis, treatment, and follow-up of patients with head and neck cancer were enormously affected in the first months of the pandemic, causing delays in diagnosis and treatment, modification of treatment protocols, and changes in follow-up that have varied according to the epidemiological situation of each territory.

These changes have been caused by the failure of health systems to cope with the burden of care demanded by COVID-19 and by changes in patient and professional safety protocols in routine diagnostic and therapeutic procedures. Unfortunately, these changes have resulted in a loss of opportunity, leading to increased mortality associated with oncological diseases without adequate control of the harm/benefit of such actions.

With the advent of vaccines and a better understanding of the disease, care times have improved, fast track systems have been reintroduced in head and neck cancer, and treatment protocols have returned to prepandemic standards. However, the use of new technologies in the doctor-patient relationship, such as telematic consultation or medical apps, will remain a tool to be used in the future.

Summary points

- During the pandemic, there was a 25%–30% reduction in first oncology consultations, resulting in delayed diagnosis.
- The variability detected between the different studies in relation to the care of head and neck cancer patients may be related to a different epidemiological situation according to geographical location.
- The performance of aerosol-generating diagnostic procedures was limited to cases in which their performance clearly conditioned subsequent management, in which the morbimortality associated with the underlying diagnostic suspicion was considerable or in which the symptomatology demanded it.
- The evaluation of each patient and his or her clinical condition is mandatory. A delay in treatment increases morbidity and affects survival. If the procedure is considered urgent, and the test is positive or indeterminate, it is recommended to proceed with all possible precautions
- More than 30% of centers were forced to make changes to their usual treatment regimen during the pandemic.

Policies and procedures

- HNC patients are a group especially susceptible to contagion, given the location of their pathology and their oncological situation. This forced the healthcare system to transform how it deals with this kind of patient, attending to their needs and balancing the risk of medical personnel–patient contagion (Prevention C for DC and. Strategies to Mitigate Healthcare Personnel Staffing Shortages. Coronavirus Dis 2019 [Internet]. Available from: https://www.cdc.gov/coronavirus/2019-ncov/hcp/mitigating-staff-shortages.html).
- The way in which the medical consultation was carried out was changed due to the pandemic.
- Telemedicine emerged as one of the most important elements, establishing itself as the first method of contact with the patient (Zhu et al., 2021.). It is essential to know the epidemiological situation as well as the COVID-19 status of the patient prior to care, classifying the patient as a suspected, probable, confirmed, or ruled out case (Lescanne et al., 2020). Patients who required face-to-face assessment, invasive examination, or surgery should be screened for active SARS-CoV-2 infection. The screening will consist of a clinical-epidemiological questionnaire and reverse transcription-polymerase chain reaction on a nasopharyngeal sample.
- In cases of aerosol-generating procedures, the door should remain closed, adequate air renewal should be ensured, and at the end of the consultation, the room should be kept aired for at least 15 min. Ventilation and air renewal become a priority. Also, the distance between the otolaryngologist and the patient must be maintained, giving some cameras and video devices even more prominence in this type of examination (Zhao et al., 2020). To minimize the risk of contagion, mandating the use of PPE has become the norm in a large number of medical centers Maza-Solano et al. (2021). PPE includes the use of a waterproof medical cap, an N95/FFP2 mask, an antipenetration isolation gown, latex gloves, tights, and goggles or a face shield if there is a risk of viral droplet production. In COVID-19 positive cases, an FFP3 mask is recommended.
- Prioritization of treatments in a situation of asset shortage requires weighing several parameters: availability of resources, prognosis of the patient's underlying pathology, and, in the SARS-CoV-2 pandemic, the inherent risk of contagion while undergoing hospital admission and a given treatment (Topf et al., 2020). The risk of SARS-CoV-2 infection requires knowing the patient's COVID-19 status prior to the procedure, which can affect their prognosis and make the procedure dangerous (Ralli et al., 2022). It is advisable to perform at least one microbiological test at least 48 h before the procedure, whether medical or surgical. The test result will determine the procedure to follow (Fakhry et al., 2020; Thomson et al., 2020).
- The selection of treatment in cancer patients will be determined by a balance among tumor resectability, patient operability, available healthcare resources, and the experience of the medical/surgical team. The presence or absence of SARS-CoV-2 became a new parameter. Perioperative COVID-19 infection has been shown to increase mortality in patients undergoing surgery due to pulmonary complications (COVIDSurg Collaborative, 2020; Crosetti et al., 2021; STARSurg Collaborative and COVIDSurg Collaborative, 2021). For cases with surgical indication, it is necessary to create circuits for positive and negative patients, negative pressure systems, minimization of the number of personnel in each of the surgeries, use of PPE, procedures done by the most experienced personnel, use of cold material over electrical or robotic material, use of local vasoconstriction whenever possible or favor closure by primary intention before reconstruction, and, if this is not possible, the use of pedicled flaps rather than free flaps (Glasbey et al., 2021; Mayo-Yánez et al., 2020).

Applications to other areas

- Lessons can be learned from the management of HNC patients during the SARS-CoV-2 pandemic.
- Telemedicine gained ground day by day throughout the most severe phases of the pandemic, emerging as a very useful tool. In addition, the creation of a COVID-19 free circuit and a "contaminated" circuit enhanced the safety of patients and staff in addition to the establishment of PPE as a basic tool to care for positive or suspected patients. The performance of diagnostic tests prior to any clinical assessment increased the safety of the examination procedures, many of them nasal endoscopies. All these variations in the usual mode of care in the ENT area can also be used by ophthalmologists, maxillofacial surgeons, dentists, and other specialists that have direct or indirect contact with the upper aerodigestive tract, the reservoir of COVID-19.
- The adaptation of oncological treatment to which otorhinolaryngologists have been subjected throughout the different stages of the pandemic made it necessary to carefully assess which patients were susceptible to surgical or medical treatment, based on the poor prognosis of a COVID-19 infection or a possible postoperative infection/complication. This situation is common to other surgical specialties whose therapeutic approaches were affected by the eruption of SARS-CoV-2 among their patients.

Minidictionary of terms

- **COVID-19 free circuit:** Medical pathway of consultation and complementary tests for confirmed COVID-19 negative patients.
- **Flexible nasofibrolaryngoscopy:** Diagnostic test that uses a flexible endoscope inserted through the nostril with the ability to evaluate the nasal cavity, paranasal sinuses, choanae and cavum, oropharynx, hypopharynx, larynx, and most of the cranial esophagus through direct vision.
- **Personal protective equipment:** Any equipment intended to be worn or held by the worker to protect from one or more risks that may threaten safety or health during professional exercise. During the SARS-CoV-2 pandemic, PPE consisted of goggles, a filtering mask (minimum FFP2 filter) and cap, gloves, and tights.
- **Telehealth:** Use of electronic information and telecommunication technology to receive and dispense medical care when face-to-face assessment is not possible or not advisable for geographic, epidemiological, etc., reasons.

References

Alobid—Grupo de trabajo en COVID-19 de la SEORL-CCC. (2020). *Sociedad Española de Otorriniolaringología—Cirugía de Cabeza y Cuello (SEORL-CCC), March 22*. https://seorl.net/wp-content/uploads/2020/03/Recomendaciones-de-la-SEORL-CCC-22-de-marzo-de-2020.pdf.

CDC. (2020). *Healthcare workers*. Centers for Disease Control and Prevention. , February 11 https://www.cdc.gov/coronavirus/2019-ncov/hcp/mitigating-staff-shortages.html.

Chiesa-Estomba, C. M., Lechien, J. R., Calvo-Henríquez, C., Fakhry, N., Karkos, P. D., Peer, M. D. S., Sistiaga-Suarez, J. A., Gónzalez-García, J. A., Cammaroto, G., Mayo-Yánez, M., Parente-Arias, P., Saussez, S., & Ayad, T. (2020). Systematic review of international guidelines for tracheostomy in COVID-19 patients. *Oral Oncology*. https://doi.org/10.1101/2020.04.26.20080242.

COVIDSurg Collaborative. (2020). Mortality and pulmonary complications in patients undergoing surgery with perioperative SARS-CoV-2 infection: An international cohort study. *Lancet (London, England)*, *396*(10243), 27–38. https://doi.org/10.1016/S0140-6736(20)31182-X. Epub 2020 May 29. Erratum in: Lancet. 2020 Jun 9. PMID: 32479829. PMCID: PMC7259900.

COVIDSurg Collaborative. (2021). Head and neck cancer surgery during the COVID-19 pandemic: An international, multicenter, observational cohort study. *Cancer*, *127*(14), 2476–2488. https://doi.org/10.1002/cncr.33320.

Crosetti, E., Tascone, M., Arrigoni, G., Fantini, M., & Succo, G. (2021). Impact of COVID-19 restrictions on hospitalisation and post-operative rehabilitation of head and neck cancer patients. *Acta Otorhinolaryngologica Italica*, *41*(6), 489–495. https://doi.org/10.14639/0392-100X-N1604. PMID: 34928262. PMCID: PMC8686804.

Curigliano, G., Banerjee, S., Cervantes, A., Garassino, M. C., Garrido, P., Girard, N., Haanen, J., Jordan, K., Lordick, F., Machiels, J. P., Michielin, O., Peters, S., Tabernero, J., Douillard, J. Y., Pentheroudakis, G., & Panel Members. (2020). Managing cancer patients during the COVID-19 pandemic: An ESMO multidisciplinary expert consensus. *Annals of Oncology: Official Journal of the European Society for Medical Oncology*, *31*(10), 1320–1335. https://doi.org/10.1016/j.annonc.2020.07.010.

De Marrez, L. G., Radulesco, T., Chiesa-Estomba, C. M., Hans, S., Baudouin, R., Remacle, M., Saussez, S., Michel, J., & Lechien, J. R. (2022). Proof-of-concept of a new FFP2 mask adapted to Otolaryngological practice in pandemic: A prospective study. *European archives of Oto-rhino-laryngology: Official journal of the European Federation of Oto-Rhino-Laryngological Societies (EUFOS): Affiliated with the German Society for Oto-Rhino-Laryngology - Head and Neck Surgery*, *279*(7), 3563–3567. https://doi.org/10.1007/s00405-022-07319-5.

Di Martino, M., García Septiem, J., Maqueda González, R., Muñoz de Nova, J. L., de la Hoz Rodríguez, Á., Correa Bonito, A., & Martín-Pérez, E. (2020). Cirugía electiva durante la pandemia por SARS-CoV-2 (COVID-19): Análisis de morbimortalidad y recomendaciones sobre priorización de los pacientes y medidas de seguridad. *Cirugía Española*, *98*(9), 525–532. https://doi.org/10.1016/j.ciresp.2020.04.029.

Fakhry, N., Schultz, P., Morinière, S., Breuskin, I., Bozec, A., Vergez, S., de Garbory, L., Hartl, D., Temam, S., Lescanne, E., Couloigner, V., Barry, B., French Society of Otorhinolaryngology, Head and Neck Surgery (SFORL), & French Society of Head and Neck Carcinology (SFCCF). (2020). French consensus on management of head and neck cancer surgery during COVID-19 pandemic. *European Annals of Otorhinolaryngology, Head and Neck Diseases*, *137*(3), 159–160. https://doi.org/10.1016/j.anorl.2020.04.008.

Finley, C., Prashad, A., Camuso, N., Daly, C., Aprikian, A., Ball, C. G., … Earle, C. C. (2020). Guidance for management of cancer surgery during the COVID-19 pandemic. *Canadian Journal of Surgery. Journal Canadien De Chirurgie*, *63*(22), S2–S4. https://doi.org/10.1503/cjs.005620.

Gascon, L., Fournier, I., Chiesa-Estomba, C., Russo, G., Fakhry, N., Lechien, J. R., Burnell, L., Vergez, S., Metwaly, O., Capasso, P., & Ayad, T. (2022). Systematic review of international guidelines for head and neck oncology management in COVID-19 patients. *European Archives of Oto-Rhino-Laryngology: Official Journal of the European Federation of Oto-Rhino-Laryngological Societies (EUFOS): Affiliated with the German Society for Oto-Rhino-Laryngology - Head and Neck Surgery*, *279*(2), 907–943. https://doi.org/10.1007/s00405-021-06823-4.

Gazzini, L., Fazio, E., Dallari, V., Accorona, R., Abousiam, M., Nebiaj, A., Giorgetti, G., Girolami, I., Vittadello, F., Magnato, R., Patscheider, M., Mazzoleni, G., & Calabrese, L. (2022). Impact of the COVID-19 pandemic on head and neck cancer diagnosis: Data from a single referral center, South Tyrol, northern Italy. *European Archives of Oto-Rhino-Laryngology: Official Journal of the European Federation of Oto-Rhino-Laryngological Societies (EUFOS): Affiliated with the German Society for Oto-Rhino-Laryngology - Head and Neck Surgery*, *279*(6), 3159–3166. https://doi.org/10.1007/s00405-021-07164-y.

Givi, B., Schiff, B. A., Chinn, S. B., Clayburgh, D., Iyer, N. G., Jalisi, S., … Davies, L. (2020). Safety recommendations for evaluation and surgery of the head and neck during the COVID-19 pandemic. *JAMA Otolaryngology. Head & Neck Surgery*, *146*(6), 579–584. https://doi.org/10.1001/jamaoto.2020.0780.

Glasbey, J. C., Nepogodiev, D., Simoes, J. F. F., Omar, O., Li, E., Venn, M. L., Abou Chaar, M. K., Capizzi, V., Chaudhry, D., Desai, A., Edwards, J. G., Evans, J. P., Fiore, M., Videria, J. F., Ford, S. J., Ganly, I., Griffiths, E. A., Gujjuri, R. R., … COVIDSurg Collaborative. (2021). Elective Cancer surgery in COVID-19-free surgical pathways during the SARS-CoV-2 pandemic: An international, multicenter, comparative cohort study. *Journal of Clinical Oncology: Official Journal of the American Society of Clinical Oncology*, *39*(1), 66–78. https://doi.org/10.1200/JCO.20.01933.

Gómez González, M. D. R., Piqueras Pérez, F. M., Guillamón Vivancos, L., Galindo Iñiguez, L., Jara Maquilón, A., & Martínez Alonso, J. A. (2021). Management of the ENT consultation during the COVID-19 pandemic alert. Are ENT telephone consultations useful? *Acta Otorrinolaringológica Española*, *72*(3), 190–194. https://doi.org/10.1016/j.otorri.2020.06.001.

Han, A. Y., Miller, J. E., Long, J. L., & St John, M. A. (2020). Time for a paradigm shift in head and neck Cancer management during the COVID-19 pandemic. *Otolaryngology—Head and Neck Surgery: Official Journal of American Academy of Otolaryngology-Head and Neck Surgery*, *163*(3), 447–454. https://doi.org/10.1177/0194599820931789.

Hardman, J. C., Tikka, T., Paleri, V., ENT UK, BAHNO, & INTEGRATE (The UK ENT Trainee Research Network). (2021). Remote triage incorporating symptom-based risk stratification for suspected head and neck cancer referrals: A prospective population-based study. *Cancer*, *127*(22), 4177–4189. https://doi.org/10.1002/cncr.33800.

Kaddour, H., Jama, G. M., Stagnell, S., Kaddour, S., Guner, K., & Kumar, G. (2022). Remote triaging of urgent suspected head and neck cancer referrals: Our experience during the first wave of the COVID-19 pandemic. *European Archives of Oto-Rhino-Laryngology: Official Journal of the European Federation of Oto-Rhino-Laryngological Societies (EUFOS): Affiliated with the German Society for Oto-Rhino-Laryngology - Head and Neck Surgery*, *279*(2), 1111–1115. https://doi.org/10.1007/s00405-021-07135-3.

Kowalski, L. P., Sanabria, A., Ridge, J. A., Ng, W. T., de Bree, R., Rinaldo, A., … Ferlito, A. (2020). COVID-19 pandemic: Effects and evidence-based recommendations for otolaryngology and head and neck surgery practice. *Head & Neck*, *42*(6), 1259–1267. https://doi.org/10.1002/hed.26164.

Lescanne, E., van der Mee-Marquet, N., Juvanon, J.-M., Abbas, A., Morel, N., Klein, J.-M., Hanau, M., & Couloigner, V. (2020). Best practice recommendations: ENT consultations during the COVID-19 pandemic. *European Annals of Otorhinolaryngology, Head and Neck Diseases*, *137*(4), 303–308. https://doi.org/10.1016/j.anorl.2020.05.007.

Mayo-Yánez, M. (2020). Research during SARS-CoV-2 pandemic: To "preprint" or not to "preprint", that is the question. *Medicina Clínica*, *155*(2), 86–87. https://doi.org/10.1016/j.medcli.2020.05.002.

Mayo-Yánez, M., Calvo-Henríquez, C., Lechien, J. R., Fakhry, N., Ayad, T., & Chiesa-Estomba, C. M. (2020). Is the ultrasonic scalpel recommended in head and neck surgery during the COVID-19 pandemic? State-of-the-art review. *Head & Neck*, *42*(7), 1657–1663. https://doi.org/10.1002/hed.26278.

Mayo-Yáñez, M., Palacios-García, J. M., Calvo-Henríquez, C., Ayad, T., Saydy, N., León, X., Parente, P., Chiesa-Estomba, C. M., & Lechien, J. R. (2021). COVID-19 pandemic and its impact on the Management of Head and Neck Cancer in the Spanish healthcare system. *International Archives of Otorhinolaryngology*, *25*(4), e610–e615. https://doi.org/10.1055/s-0041-1736425.

Maza-Solano, J. M., Plaza-Mayor, G., Jiménez-Luna, A., Parente-Arias, P., & Amor-Dorado, J. C. (2020). Strategies for the practice of otolaryngology and head and neck surgery during the monitoring phase of COVID-19. *Acta Otorrinolaringológica Española (English Edition)*, *71*(6), 367–378. https://doi.org/10.1016/j.otorri.2020.05.001. Epub 2020 May 25. PMID: 32600649. PMCID: PMC7834125.

MD Anderson Head and Neck Surgery Treatment Guidelines Consortium, Consortium members, Maniakas, A., Jozaghi, Y., Zafereo, M. E., Sturgis, E. M., Su, S. Y., Gillenwater, A. M., Gidley, P. W., Lewis, C. M., Diaz, E., Goepfert, R. P., Kupferman, M. E., Gross, N. D., Hessel, A. C., Pytynia, K. B., Nader, M.-E., Wang, J. R., Lango, M. N., … Lai, S. Y. (2020). Head and neck surgical oncology in the time of a pandemic: Subsite-specific triage guidelines during the COVID-19 pandemic. *Head & Neck*, *42*(6), 1194–1201. https://doi.org/10.1002/hed.26206.

Mehanna, H., Hardman, J. C., Shenson, J. A., Abou-Foul, A. K., Topf, M. C., AlFalasi, M., Chan, J. Y. K., Chaturvedi, P., Chow, V. L. Y., Dietz, A., Fagan, J. J., Godballe, C., Golusiński, W., Homma, A., Hosal, S., Iyer, N. G., Kerawala, C., Koh, Y. W., Konney, A., … Holsinger, F. C. (2020). Recommendations for head and neck surgical oncology practice in a setting of acute severe resource constraint during the COVID-19 pandemic: An international consensus. *The Lancet. Oncology*, *21*(7), e350–e359. https://doi.org/10.1016/S1470-2045(20)30334-X.

O'Connell, D. A., Seikaly, H., Isaac, A., Pyne, J., Hart, R. D., Goldstein, D., Yoo, J., & Canadian Association of Head and Neck Surgical Oncology. (2020). Recommendations from the Canadian Association of Head and Neck Surgical Oncology for the Management of Head and Neck Cancers during the COVID-19 pandemic. *Journal of Otolaryngology - Head & Neck Surgery = Le Journal D'oto-Rhino-Laryngologie Et De Chirurgie Cervico-Faciale*, *49*(1), 53. https://doi.org/10.1186/s40463-020-00448-z.

Paleri, V., Hardman, J., Tikka, T., Bradley, P., Pracy, P., & Kerawala, C. (2020). Rapid implementation of an evidence-based remote triaging system for assessment of suspected referrals and patients with head and neck cancer on follow-up after treatment during the COVID-19 pandemic: Model for international collaboration. *Head & Neck*, *42*(7), 1674–1680. https://doi.org/10.1002/hed.26219.

Popovic, M., Fiano, V., Moirano, G., Chiusa, L., Conway, D. I., Garzino Demo, P., Gilardetti, M., Iorio, G. C., Moccia, C., Ostellino, O., Pecorari, G., Ramieri, G., Ricardi, U., Riva, G., Virani, S., & Richiardi, L. (2022). The impact of the COVID-19 pandemic on head and neck Cancer diagnosis in the Piedmont region, Italy: Interrupted time-series analysis. *Frontiers in Public Health*, *10*, 809283. https://doi.org/10.3389/fpubh.2022.809283.

Radulesco, T., Lechien, J. R., Sowerby, L. J., Saussez, S., Chiesa-Estomba, C., Sargi, Z., … Michel, J. (2020). Sinus and anterior skull base surgery during the COVID-19 pandemic: Systematic review, synthesis and YO-IFOS position. *European Archives of Oto-Rhino-Laryngology*, *278*(6), 1733–1742. https://doi.org/10.1007/s00405-020-06236-9.

Ralli, M., Colizza, A., D'Aguanno, V., Scarpa, A., Russo, G., Petrone, P., Grassia, R., Guarino, P., & Capasso, P. (2022). Risk of SARS-CoV-2 contagion in otolaryngology specialists. *Acta Otorhinolaryngologica Italica: Organo Ufficiale Della Societa Italiana Di Otorinolaringologia E Chirurgia Cervico-Facciale*, *42*(Suppl. 1), S58–S67. https://doi.org/10.14639/0392-100X-suppl.1-42-2022-06.

Riera, R., Bagattini, Â. M., Pacheco, R. L., Pachito, D. V., Roitberg, F., & Ilbawi, A. (2021). Delays and disruptions in Cancer health care due to COVID-19 pandemic: Systematic review. *JCO Global Oncology*, *7*, 311–323. https://doi.org/10.1200/GO.20.00639.

Rygalski, C. J., Zhao, S., Eskander, A., Zhan, K. Y., Mroz, E. A., Brock, G., Silverman, D. A., Blakaj, D., Bonomi, M. R., Carrau, R. L., Old, M. O., Rocco, J. W., Seim, N. B., Puram, S. V., & Kang, S. Y. (2021). Time to surgery and survival in head and neck Cancer. *Annals of Surgical Oncology*, *28*(2), 877–885. https://doi.org/10.1245/s10434-020-09326-4.

Sánchez Barrueco, Á., Mateos-Moreno, M. V., Martínez-Beneyto, Y., García-Vázquez, E., Campos González, A., Zapardiel Ferrero, J., Bogoya Castaño, A., Alcalá Rueda, I., Villacampa Aubá, J. M., Cenjor Español, C., Moreno-Parrado, L., Ausina-Márquez, V., García-Esteban, S., Artacho, A., López-Labrador, F. X., Mira, A., & Ferrer, M. D. (2022). Effect of oral antiseptics in reducing SARS-CoV-2 infectivity: Evidence from a randomized double-blind clinical trial. *Emerging Microbes & Infections*, *11*(1), 1833–1842. https://doi.org/10.1080/22221751.2022.2098059.

Shaw, R., & COVIDSurg Collaborative. (2021). UK head and neck cancer surgical capacity during the second wave of the COVID-19 pandemic: Have we learned the lessons? COVIDSurg collaborative. *Clinical Otolaryngology: Official Journal of ENT-UK; Official Journal of Netherlands Society for Oto-Rhino-Laryngology & Cervico-Facial Surgery*, *46*(4), 729–735. https://doi.org/10.1111/coa.13749.

Smith, A. C., Thomas, E., Snoswell, C. L., Haydon, H., Mehrotra, A., Clemensen, J., & Caffery, L. J. (2020). Telehealth for global emergencies: Implications for coronavirus disease 2019 (COVID-19). *Journal of Telemedicine and Telecare*, *26*(5), 309–313. https://doi.org/10.1177/1357633X20916567.

Solano, J. M., et al. (2021). *Estrategias para el manejo del paciente orl durante la pandemia por la COVID19 (2021, February)*. https://seorl.net/wp-content/uploads/2021/03/ESTRATEGIAS-PARA-EL-MANEJO-DEL-PACIENTE-ORL-DURANTE-LA-PANDEMIA-POR-LA-COVID19.pdf.

Solis, R. N., Mehrzad, M., Faiq, S., Frusciante, R. P., Sekhon, H. K., Abouyared, M., Bewley, A. F., Farwell, D. G., & Birkeland, A. C. (2021). The impact of COVID-19 on head and neck Cancer treatment: Before and during the pandemic. *OTO Open*, *5*(4). https://doi.org/10.1177/2473974X211068075.

STARSurg Collaborative and COVIDSurg Collaborative. (2021). Death following pulmonary complications of surgery before and during the SARS-CoV-2 pandemic. *British Journal of Surgery*, *108*(12), 1448–1464. https://doi.org/10.1093/bjs/znab336. PMID: 34871379. PMCID: PMC10364875.

Stevens, M. N., Patro, A., Rahman, B., Gao, Y., Liu, D., Cmelak, A., Wiggleton, J., Kim, Y. J., Langerman, A., Mannion, K., Sinard, R. J., Netterville, J. L., Rohde, S. L., & Topf, M. C. (2022). Impact of COVID-19 on presentation, staging, and treatment of head and neck mucosal squamous cell carcinoma. *American Journal of Otolaryngology*, *43*(1), 103263. https://doi.org/10.1016/j.amjoto.2021.103263.

Thomson, D. J., Palma, D., Guckenberger, M., Balermpas, P., Beitler, J. J., Blanchard, P., Brizel, D., Budach, W., Caudell, J., Corry, J., Corvo, R., Evans, M., Garden, A. S., Giralt, J., Gregoire, V., Harari, P. M., Harrington, K., Hitchcock, Y. J., Johansen, J., … Yom, S. S. (2020). Practice recommendations for risk-adapted head and neck cancer radiotherapy during the COVID-19 pandemic: An ASTRO-ESTRO consensus statement. *Radiotherapy and Oncology: Journal of the European Society for Therapeutic Radiology and Oncology*, *151*, 314–321. https://doi.org/10.1016/j.radonc.2020.04.019.

Topf, M. C., Shenson, J. A., Holsinger, F. C., Wald, S. H., Cianfichi, L. J., Rosenthal, E. L., & Sunwoo, J. B. (2020). Framework for prioritizing head and neck surgery during the COVID-19 pandemic. *Head and Neck*, *42*(6), 1159–1167. https://doi.org/10.1002/hed.26184.

Vasiliadou, I., Noble, D., Hartley, A., Moleron, R., Sanghera, P., Urbano, T. G., Schipani, S., Gujral, D., Foran, B., Bhide, S., Haridass, A., Nathan, K., Michaelidou, A., Sen, M., Geropantas, K., Joseph, M., O'Toole, L., Griffin, M., Pettit, L., … Kong Conceptualisation, A. (2021). A multi-Centre survey reveals variations in the standard treatments and treatment modifications for head and neck cancer patients during Covid-19 pandemic. *Clinical and Translational Radiation Oncology*, *30*, 50–59. https://doi.org/10.1016/j.ctro.2021.06.002.

Vukkadala, N., Qian, Z. J., Holsinger, F. C., Patel, Z. M., & Rosenthal, E. (2020). COVID-19 and the otolaryngologist: Preliminary evidence-based review. *The Laryngoscope*, *130*(11), 2537–2543. https://doi.org/10.1002/lary.28672.

Wilkie, M. D., Gaskell, P., Hall, B., Jones, T. M., & Kinshuck, A. J. (2021). Emergency presentations of head and neck cancer: Our experience in the wake of the COVID-19 pandemic. *Clinical Otolaryngology: Official Journal of ENT-UK; Official Journal of Netherlands Society for Oto-Rhino-Laryngology & Cervico-Facial Surgery*, *46*(6), 1237–1241. https://doi.org/10.1111/coa.13821.

World Health Organization. (2020). *Coronavirus disease (COVID-19). Recuperado 3 de octubre de 2020, de*. https://www.who.int/emergencies/diseases/novel-coronavirus-2019.

Wu, V., Noel, C. W., Forner, D., Zhang, Z.-J., Higgins, K. M., Enepekides, D. J., Lee, J. M., Witterick, I. J., Kim, J. J., Waldron, J. N., Irish, J. C., Hua, Q.-Q., & Eskander, A. (2020). Considerations for head and neck oncology practices during the coronavirus disease 2019 (COVID-19) pandemic: Wuhan and Toronto experience. *Head & Neck*, *42*(6), 1202–1208. https://doi.org/10.1002/hed.26205.

Zhao, C., Viana, A., Wang, Y., Wei, H.-Q., Yan, A.-H., & Capasso, R. (2020). Otolaryngology during COVID-19: Preventive care and precautionary measures. *American Journal of Otolaryngology*, *41*(4), 102508. https://doi.org/10.1016/j.amjoto.2020.102508.

Zhu, Y., Chen, Z., Ding, A., Walter, H., Easto, R., & Wilde, A. (2021). Patient satisfaction with the head and neck Cancer telephone triage service during the COVID-19 pandemic. *Cureus*, *13*(9), e18375. https://doi.org/10.7759/cureus.18375.

Zubair, A., Jamshaid, S., Scholfield, D. W., Hariri, A. A., Ahmed, J., Ghufoor, K., & Ali, S. (2022). Impact of COVID-19 pandemic on head-neck cancer referral and treatment pathway in North East London. *Annals of the Royal College of Surgeons of England*. https://doi.org/10.1308/rcsann.2021.0360.

Chapter 9

The COVID-19 pandemic: Inventory management and allocation of personal protective equipment

Tazim Merchant[a] and Mark Sheldon[b]

[a]*Feinberg School of Medicine, Northwestern University, Chicago, IL, United States,* [b]*Department of Philosophy, Weinberg College of Arts and Sciences, Medical Humanities and Bioethics Program, Feinberg School of Medicine, Northwestern University, Chicago, IL, United States*

Abbreviations

NGO	nongovernmental organization
PPE	personal protective equipment

Introduction

An acquaintance relayed the following story. His son was in China right before COVID-19 swept through the country. He and his wife had had left temporarily to go to Sri Lanka for Chinese New Year, and with the rise in COVID-19 cases in Beijing, found themselves stuck there. They were forced to leave all their belongings—and their dog—in Beijing, and travel back to the United States. The son stepped onto the plane, a sea of masks before him. This posed a sharp contrast to what he experienced on landing in the United States: not a single person, save those on his flight, had masks on.

The stark difference in personal protective equipment (PPE) usage between each country was reflected in public health guidance expressed by officials, chiefly Anthony Fauci, the director of the National Institute of Allergy and Infectious Diseases (NIAID), early on in the pandemic. In a 60 Minutes interview in early March 2020, Fauci stated, "Right now, in the United States, people should not be walking around in masks … when you think of masks, you should think of healthcare providers needing them and people who are ill, people when you look at the films of foreign countries, and you see 85% of the people wearing masks, that's fine. I'm not against it, if you want to do it, that's fine … [but] it could lead to a shortage of masks for the people who really need it." (CBS, 2020).

Fauci's statement raised important ethical questions regarding scarce resource allocation and who is most deserving of limited PPE. Do healthcare workers deserve greater protection, which would be a question of justice? Or rather is it more a question of what the greatest *utility* of limited PPE is, and what does that mean (e.g., controlling viral spread, preventing deaths)?

This chapter will explore these questions, but most critically, delve into the question of what are the moral justifications, from a public health perspective, for how PPE should be managed in the event of scarcity.

An unprecedented global challenge

The degree of global supply chain collapse during the COVID-19 pandemic was unprecedented, causing an international PPE shortage. In the United States, between September 2020 and May 2021, nearly one-third of office-based physicians experienced a PPE shortage, with nearly 40% of them turning away COVID-19 patients or directing them elsewhere for care (Peters & Cairns, 2022). In Italy, medical students were brought into the hospital sooner, a product perhaps of the significant infection and death rates experienced particularly by healthcare professionals in the country (Ranney et al.,

Management, Body Systems, and Case Studies in COVID-19. https://doi.org/10.1016/B978-0-443-18703-2.00018-9
Copyright © 2024 Elsevier Inc. All rights are reserved, including those for text and data mining, AI training, and similar technologies.

2020). In South Korea, volunteer healthcare workers were hired and the army was brought in to combat COVID-19 (Tabari et al., 2020).

While there are general frameworks that exist for the ethical allocation of equipment (e.g., ventilators) in a hospital setting, there are fewer frameworks for the ethical allocation of scarce resources in pandemics, and even fewer for PPE specifically, with the majority focusing on PPE management within a hospital setting. Thus, nonprofits, governments, and others had to leverage existing frameworks on the allocation of other scarce resources to create an approach to PPE management. Such approaches are detailed below.

Ethical principles for scarce resource distribution in pandemics

Emmanuel et al.

Perhaps one of the most widely cited and well-known articles dealing with scarce resource allocation during COVID-19 is a piece written by Emanuel and colleagues titled, "Fair Allocation of Scarce Medical Resources in the time of COVID-19," and published in the *New England Journal of Medicine*. The authors discuss four guiding principles for scarce resource allocation: (1) maximizing benefits produced by scarce resources, (2) giving priority to the worst off, (3) promoting and rewarding instrumental value, and (4) treating people equally (Emanuel et al., 2020).

Consistent with a utilitarian framework, *maximizing benefits* is defined as saving the most lives. It also means provided that the survival odds are similar, prioritizing patients who can live longer after treatment (i.e., maximizing life-*years*). COVID-19 patients, notably, do not get special priority. The authors assert that one should consider what course of action would maximize benefits *regardless* of medical condition.

Giving priority to the worst off is defined as prioritizing patients who are the sickest or youngest (i.e., people who will have only lived for a short period of time) (Emanuel et al., 2020). The authors assert that this principle should only be used when it aligns with maximizing benefits (i.e., it would save the most lives/life-years in the former, or might for example be most conducive to controlling viral transmission in the latter) (Emanuel et al., 2020).

Instrumental value is considered important in resource allocation; that is, priority can be given to those who have the ability to treat others going forward (prospective), or *have* contributed in the past (retrospective). Healthcare workers, given their specialized training, have priority because of their important role in the pandemic. If many were to get sick, reduced healthcare system function would lead to increased mortality. Instrumental value can only be invoked if other factors (*prioritizing the worst off, maximization of benefits*) are equal (Emanuel et al., 2020). To illustrate, if two individuals had the same prognosis, likelihood of recovery, sickness, and age, Individual A, a physician, would be given priority for treatment over Individual B, a nonhealthcare worker.

Finally, Emanuel et al. (2020) note that *treating others equally* could theoretically be applied in two ways: (1) first-come, first-serve, and (2) random selection. However, first-come, first-serve, they assert, should not be used while random selection can only be used when all else is equal. To carry forward the previous example, if Individual A and B were both healthcare workers, and all other factors were equal, treatment would be randomly allocated to one person.

Dawson et al.

Dawson et al. (2020), based in Australia, provide values that somewhat overlap with those proposed in studies by Emanuel et al. (2020), but are applied differently. They discuss three key values: (1) getting the most out of resources, (2) giving priority to those in need, and (3) equal value to all (Dawson et al., 2020).

Dawson and colleagues argue that the primary obligation is extracting the most value from resource expenditure. *Getting the most out of resources* is something that can mean a variety of things–most efficient use of resources, most cost-effective use, etc. The authors therefore assert that this principle must be applied on a case-by-case basis (e.g., in case of intensive care unit (ICU) admission, prioritize individuals who are most likely to recover and can benefit the most from long-term ICU care (i.e., *specific* resources)) (Emanuel et al., 2020). In this work, similar to Emmanuel and colleagues, COVID-19 patients do not receive special priority (Emanuel et al., 2020).

Both Emmanuel and Dawson's papers assume accurate knowledge of prognoses. But early in the pandemic, the disease course was still being understood, and it varied for different people. Some individuals had mild illness, others life-threatening, and still others long COVID-19 symptoms with a devastating impact on the quality of life.

Giving priority to those in need was also considered. A few definitions were discussed–this principle could mean prioritizing the sickest or the most disadvantaged (i.e., those who have experienced societal injustice previously, or were more likely to suffer harm going forward) (Dawson et al., 2020). The authors purport that pregnant women, First Nation communities, and others could therefore be used as tie-breaking factors.

Finally, *equal value to all* is defined as ensuring factors such as gender, political views, wealth, etc., do not influence how scarce resources are distributed.

While less specific in their definitions than Emmanuel and colleagues about how to best define a given value (which could be viewed as both a detriment in one's ability to apply the principle, and a strength in terms of flexibility and specificity to circumstance), the authors bring up interesting considerations on more of a societal level as to what is owed to traditionally disadvantaged groups and how social determinants of health intersect with medical care.

Moodley et al.

Studies by Moodley et al. (2020) provide a direct commentary and expansion on Emmanuel and colleagues' work, examining the allocation of scarce resources in Africa. They focus on the principles of random selection and instrumental value, and additionally discuss equity.

They expand on the argument that a random, lottery-based allocation system should not be used, asserting that in practice such a system is difficult to regulate and more subject to corruption. For instance, they share that random selection in the Democratic Republic of the Congo would lead to more powerful members of Congolese society receiving scarce resources. They also note that a random selection system could appear to trivialize human life, so transparency would be key (Moodley et al., 2020).

Moodley and colleagues also discuss how healthcare equity in scarce resource distribution is more difficult in low-resource settings (Moodley et al., 2020). COVID-19 patients may use ventilators for longer than non-COVID-19 patients, an issue that can disrupt equity in low ICU bed settings. Prior to the pandemic, an individual perhaps would be able to access an ICU bed or a ventilator if they had pneumonia, but now are unable to do so.

Srinivas et al.

Srinivas and colleagues conducted a systematic review on healthcare resource rationing during emergency situations (e.g., pandemics/epidemics). The 26 included studies were not necessarily specific to the COVID-19 pandemic, but did reflect perspectives from different parts of the globe. A key goal, the authors assert, is to prevent people from becoming sick while also ensuring that society keeps functioning well (Srinivas et al., 2021). The authors distinguish between system-wide rationing and bedside rationing.

System-wide rationing: Utilitarianism (i.e., promoting the most societal benefit), egalitarianism (i.e., equality), and prioritizing the worst off, defined essentially as those experiencing the most societal injustice and fewest advantages, should be considered in a system/policy-level approach. Like Moodley, the authors emphasize transparency as well as inclusiveness, consistency, and accountability in ensuring allocation is truly fair.

Bedside/individual-level rationing: Interestingly, physicians should be encouraged to pursue random allocation/lottery because of the moral distress they might experience in allocation decisions. Principles such as *giving priority to the worst off* and *who will receive the most benefit* should also be considered. Additionally, the risk of the population should help determine who receives preventative treatments or a vaccine.

Though the authors don't specifically reference instrumental value, they do state that frontline healthcare workers "must be given top priority" because functionally, healthcare workers are necessary to treat as many patients as possible and are at greater risk of transmitting infection to others (Srinivas et al., 2021).

Summary

There are various studies on the ethical allocation of scarce medical resources during pandemics. Many of them consider similar values with different terminologies. The utilitarian principle of maximizing benefits was discussed where preserving life and life-years was a core principle (Dawson et al., 2020; Emanuel et al., 2020; Srinivas et al., 2021). Age, degree of sickness, and risk of population were brought up in relation to prioritizing vulnerable populations (Dawson et al., 2020; Emanuel et al., 2020; Srinivas et al., 2021). Correcting injustices and examining the application of values in the context of low-resource settings were equity-related approaches that were proposed (Dawson et al., 2020; Moodley et al., 2020). A lottery-based approach was generally not looked upon as favorably, as it would be difficult to implement fairly (Moodley et al., 2020).

Ethical allocation of personal protective equipment

Principles in considering PPE management strategies

Binkley and Kemp

Studies by Binkley and Kemp (2020) present viewpoints that are quite similar to previously discussed works. They discuss *utilitarianism*; *first-come, first-serve*; *social worth* (which is akin to instrumental value); *protecting vulnerable populations;* and the *principle of reciprocity*.

Binkley and Kemp apply *utilitarianism* in two ways: (1) concern for protecting the greatest number of clinicians, and (2) offering protection to clinicians who are able to do the most good for the greatest number of patients. The first approach would, practically speaking, result in recommending the minimum amount of protection necessary so that most individuals could receive PPE. It would also entail eliminating unnecessary elective procedures to conserve PPE. The second approach, which the authors favor less and suggest could cause moral distress, would entail giving more protection to providers who can save the most patients and recommending less PPE for others in ways that are not necessarily evidence-based.

First-come, first-serve is, according to Binkley and Kemp, untenable and ethically unjustifiable.

Instrumental value is repackaged under the term *social worth* by the authors. They assert that it can only be applied when "absolutely necessary," but don't elaborate significantly on the circumstances under which it can be used.

Similar to *prioritizing the worst off* described by Emanuel et al. (2020), studies by Binkley and Kemp (2020) state the importance of protecting vulnerable clinicians (i.e., immunosuppressed, pregnant). This goes hand-in-hand with a principle not detailed in the broader literature on scarce resource allocation during pandemics: the principle of reciprocity. The authors assert that because clinicians put themselves at a high level of risk, they should be prioritized and given the most protection. This is separate from the concept of "instrumental value" or "social worth." It is being given a reciprocal benefit based solely on the *risk* taken by the individual.

Sureka et al.

The value of the paper from Sureka et al. is primarily the way the authors engage in careful analysis and distinctions regarding the nature and kinds of PPE. This is extremely useful for getting an accurate picture of the many levels and contexts for the use of PPE, the needs at these various levels, and whether, for instance, it is for a "standard precaution" or "an expanded isolated precaution" (page 3 of 16). Also, the authors report that they follow the recommendations defined by the World Health Organization (WHO) and the Ministry of Health and Family Welfare. Included in the paper are graphs that make clear what type of PPE should be used in specific circumstances such as dental (N-95, goggles, face shield) or intubation (headgear, N95, gloves). A major concern that the authors have, and the reason for their very careful analysis of PPE types, is the concern that an institution not run out, touching on practical strategies for PPE and PPE sourcing.

Important to the focus of this paper is that the authors, toward the end, touch on what needs to be considered in regard to the ethical allocation of PPE. For the most part, this entails their setting out a list of values, many of which Binkley and Kemp identify, including *utilitarianism; prioritizing the worst off; social worth; the multiplier effect; first-come, first-serve*; and *the principle of reciprocity*. Sureka's application of *utilitarianism* is nearly identical to the second approach shared by Binkley and Kemp: protecting clinicians who can save the most number of patients. In practice, they assert that it is this principle that justifies avoiding elective surgeries and having a team-based care approach in operating.

Prioritizing the worst off in the context of PPE allocation means rationing PPE to healthcare workers taking care of the sickest patients (e.g., COVID-19 critical care, high dependent areas) who are most likely to get better (Sureka et al., 2021).

Social worth is also considered. Similar to Binkley and Kemp, the authors purport that it should not be used unless "absolutely necessary" (Sureka et al., 2021). Interestingly, the authors define social worth as distinct from instrumental value. They describe instead the *multiplier effect as instrumental value*. Rather than examining the overall worth of that person, the multiplier effect concept looks specifically at a person's ability to contribute in a way that prevents the disintegration of society, and save more lives than just themselves–hence the name, "multiplier effect." The authors do not discuss, however, when the multiplier effect should be considered.

The authors also mention the *principle of reciprocity*, but do not discuss the degree to which it should be weighed. Finally, consistent with other reports, *first-come, first-serve* and prioritization based on seniority or position should not be used in emergency situations.

However, ultimately what needs to be said is that while the authors list the considerations or values, they do not rank or give priority to these values. At the beginning of the paper, the authors say they value a "rational use of PPE equipment," but what this means, in terms of what they do in the paper, is that they identify in detail types of PPE and measure an institution's success by its not "running out" (Sureka et al., 2021). However, how ethical priorities in relation to this measure are implemented is not made clear.

Modeling fair allocation of PPE

Some authors tried to build a model to determine how to best allocate PPE. The utility in these articles lie in: (1) providing alternative principles to consider in allocation decisions at a more system level, and (2) better defining how to put those principles into practice.

Domnez et al.

Studies by Dönmez et al. (2022) are clearly committed to a utilitarian approach, and want to create the best model to achieve maximum efficiency as measured by minimizing disease transmission within hospitals. In the process of building their model, deprivation cost function is considered; it is defined as the "economic value of human suffering caused by lack of access to goods/services." The concepts of an opportunity cost in allocation and deprivation of goods from *lack* of access have not yet been explicitly mentioned by the authors thus far, but could be useful in calculating the course that would promote the most good.

Consistent with a utilitarian approach and with *prioritizing the worst off*, the authors also discuss how populations can be prioritized based on the risk of infection or death. Risk of infection is key for preventative therapies while the severity of symptoms (i.e., who needs the most help, or who could be the most helped) is more important when the resource is more therapeutic or remedial (e.g., medical staff, beds, drugs) (Dönmez et al., 2022).

The model also seeks to distribute limited PPE equitably between different healthcare institutions, which can be said to be grounded therefore in a principle of distributive justice and equity. This is further underscored in the text itself, and the authors bring up the concept of respecting ethical principles of humanitarian operations (Dönmez et al., 2022), and the importance of considering an equity-based approach that takes into account the impact on vulnerable populations.

Wang et al.

Studies by Wang et al. (2020) similarly operationalize *utilitarianism* in their model, considering the severity of infection as well as prioritizing the worst off by focusing on high-risk populations. They set high risk as being over 65 in their model given mortality and complication rates, and define the severity of infection in terms of the number of confirmed cases and the number of impacted individuals per case.

Summary

The few studies on the ethical allocation of PPE focused on similar values to previous authors, but the weights of those values were not as well characterized.

The utilitarian approach tended to be most well defined, and was a common theme among all papers, including modeling principles (even if not explicitly stated/referenced). In the context of PPE allocation, utilitarianism could mean protecting the most number of clinicians, or the clinicians who are able to do the most good for the greatest number of patients (Binkley & Kemp, 2020; Sureka et al., 2021). It also meant considering, per Dönmez et al. (2022), not only what could be done with PPE, but the opportunity cost and suffering caused by others *lacking* PPE. One approach to operationalizing utilitarianism is also in relation to minimizing suffering via lowering transmission, infection, and death (Dönmez et al., 2022; Wang et al., 2020). In practice, this entails quantifying the risk of infection in a population (with greater weight given to high-risk populations) and the severity of infection (how many individuals could be affected per case) (Wang et al., 2020). Prioritizing the worst off was to some extent touched upon, and meant considering the impact specifically on high-risk or vulnerable populations (Wang et al., 2020). Social worth and related concepts (e.g., the multiplier effect) should not be used in most cases, and first-come, first-serve is not ethically acceptable (Binkley & Kemp, 2020; Sureka et al., 2021). Justice and equity were also discussed briefly. Taking into account the impact of distribution strategies on vulnerable populations is key in addition to ensuring that allocation does not leave one institution disproportionately impacted and without PPE (Dönmez et al., 2022).

Public Health Guidelines on PPE allocation

Though modeling papers touch on how to best distribute PPE on a system level, more clarity on this subject is needed. County or state public health guidelines can, similar to modeling-focused work, provide more of an operational, practical framework for PPE distribution, and with further analysis, this can be tied to and grounded in specific ethical principles. In the United States, Washington's Department of Health is one such example (Public Health Seattle & King County, 2022).

Though the state does not provide explicit ethical principles to justify their strategies, they do provide factors to consider when prioritizing certain groups for distribution, namely the importance of—in no particular order—(1) protecting healthcare workers who are treating potential (and actual) COVID-19 patients, (2) controlling spread, especially in places with high-risk populations (e.g., nursing homes), (3) protecting those who are likely to be occupationally or otherwise exposed to COVID-19, and who play an important role in providing a service (i.e., essential workers), and (4) the disruption of the supply chain and limited ability to source PPE through traditional channels. These strategies have echoes of *utilitarianism* (1–3), *prioritizing the worst off* (e.g., high-risk populations in (3)), and *instrumental value* (1, 3) (Emanuel et al., 2020).

Prioritizing the worst off can also mean, on an institutional level, prioritizing institutions with the greatest need (Emanuel et al., 2020). Washington State Department of Health (2021) uses two approaches to quantify that need.

They first explicitly detail three states of need based on the CDC: (1) conventional, (2) contingency, and (3) crisis. These are defined by the supply of PPE and the ability to handle a surge (Table 1) (Centers for Disease Control and Prevention, 2020).

They go on to provide tiers of organizations, which align with the *utilitarian* concept of reducing viral transmission. The state prioritizes fulfilling orders of PPE based on the tier of the institution, and then the need at a given institution within a tier. Tiers are defined by: (i) the number of confirmed or potential COVID-19 cases, (ii) their ability to practice social distancing or utilize other methods of controlling spread, and (iii) the risk of contracting COVID-19 (Table 2) (Public Health Seattle & King County, 2022).

Though the Washington Department of Health doesn't provide significant justification for the tier system, a county-level document alludes to several principles that underlie the approach. Healthcare facilities fall under Tier I, given the *principle of reciprocity* (i.e., the high risk that healthcare workers will be exposed to COVID-19), and *instrumental value* (i.e., preventing healthcare workers from becoming sick will help maintain the healthcare infrastructure) (Binkley & Kemp, 2020; Emanuel et al., 2020). It also seems to suggest Tier I prioritization supports the worst off, given that healthcare workers at hospitals and long-term care facilities care for high-risk populations who are most at risk of complications or mortality (Emanuel et al., 2020).

Summary

Public health guidelines help operationalize ethical principles, and provide nuance on how to quantify institutional need. A tiered system helps one identify institutions that are most likely to be the *worst off*, that will *maximally reduce viral transmission*, and also prioritize those *instrumental* to healthcare infrastructure (Emanuel et al., 2020). Organizations can be stratified into tiers based on the number of potential and actual COVID-19 cases, their ability to use alternative methods to control spread, types of populations served (i.e., high risk or low risk in regard to COVID-19), and types of services provided (i.e., health, health-related, other essential functions). Within a single tier, organizations can be further prioritized based on their supply/ability to respond to a case surge, which is then reflected in the degree of conservation strategies employed.

TABLE 1 Urgency of need for PPE, Washington Department of Public Health.

State	Definition
Conventional	• Steady state • Enough PPE (30 day supply), can order enough additional PPE to handle surge
Contingency	• 7–14 day supply • Anticipate shortages, conserving PPE
Crisis	• <7–14 day supply

This table depicts the three levels of PPE need: conventional, contingency, and crisis.

TABLE 2 **Tier system, Washington Department of Public Health.**

Tier (highest = 1)	Examples of Facilities	Description
1	Hospitals, long-term care facilities	• Confirmed/potential COVID-19 cases • Hard to practice social distancing or control spread w/out PPE
2	COVID-19 test sites, outpatient sites w/out aerosolizing procedures	• Confirmed or potential cases • Can control spread via social distancing or other means
3	Pharmacies, funeral homes (providing social, behavioral services for vulnerable populations)	• Some possibility of encountering COVID-19 or actual cases
4	Volunteer-based organizations (providing essential services to public)	

This table depicts the three levels of PPE need: conventional, contingency, and crisis.
Based on content from the Washington State Department of Health. (2021, September 27). Prioritization Guidelines for Allocation of Personal Protective Equipment.

Grassroot initiative strategies

Though public health guidelines provide tiers that can be retrospectively connected to ethical principles such as social worth or supporting the worst off as done above, having a framework that marries an operational framework with moral justifications is still needed. Leaders of GetMePPE Chicago, a more regionally based organization in the United States that distributed close to a million units of PPE, provide the only known framework that explicitly operationalizes ethical values for PPE management (Clos et al., 2021). Clos et al. (2021) directly applied the values put forth by Emanuel et al. (2020) in the context of PPE allocation, in addition to detailing a separate algorithm for determining which institutions should receive PPE.

The authors' approach is grounded in utilitarianism, striving to determine where PPE can "do the most good" (Clos et al., 2021). In other writings, as we have seen, this has come to mean a variety of things such as protecting the most number of clinicians, protecting the healthcare workers who are most likely to do the most good for their patients, or protecting healthcare workers that take care of high-risk populations.

GetMePPE Chicago tends to apply utilitarianism to mean giving PPE to institutions to maximize control of viral transmission and prevent the most morbidity/mortality. The concept of giving priority to the worst off, the authors put forth, aligns exactly with utilitarianism when considering PPE allocation. By prioritizing institutions with a higher number of COVID-19 cases and who have an extremely low supply of PPE (which mirrors, in some way, Washington State's public health guidelines examined earlier), it necessarily follows that one is lowering the number of potential people who could get infected, and thereby morbidity/mortality.

To put this into practice, the authors took an approach that mirrors the local public health-based efforts detailed earlier. The authors' questionnaire/needs assessment examined PPE supply, determined for each type of PPE (e.g., gloves, N95s, etc.). Public health-based approaches determined <7–14 week supply as indicative of "crisis," whereas GetMePPE Chicago determined "dire" need to be <48 h supply of PPE ("high" need was <1 week supply) (Washington State Department of Health, 2021).

Determining who was the worst off and had the highest burden of COVID-19 was different and arguably more nuanced than previous approaches. Wang et al. (2020) considered the number of confirmed cases and the number of impacted individuals per case. GetMePPE Chicago looked instead at capacity (e.g., staff numbers, COVID-19 patient numbers, bed numbers) and resources.

Thus, an institution with a higher burden but fewer resources would receive PPE (e.g., a county hospital) over an institution with a higher burden but more resources (e.g., a private hospital). It was in this way that GetMePPEChicago tried to operationalize the principle of equity.

For two facilities with similar COVID-19 burdens and need, GetMePPE Chicago sought to meet the full need at one facility rather than the partial need at both. This approach, the authors asserted, would better control viral spread and maximize benefits (though that could be argued to depend on a case-by-case basis) (Clos et al., 2021).

For two facilities with the same case burden and resources, where their last donation was the same period of time previously, and neither facility would gain full coverage from the donation, random selection could be used to determine where PPE would be distributed (Clos et al., 2021).

GetMePPE Chicago, similar to other public health-based approaches, also provided tiers for types of facilities. This tiered system is where instrumental value came into play, as hospitals were the only organizations prioritized in Tier I while long-term care facilities (e.g., rehabilitation facilities, nursing homes, behavioral health hospitals, etc.), unlike governmental approaches, were put in Tier II. A tier's need would generally need to be exhausted before the next tier was considered, and decisions were made to distribute PPE between institutions within the highest tier (Clos et al., 2021; Washington State Department of Health, 2021).

The authors focused first on assessing the need of institutions within Cook County in all tiers (given proximity to the organization's base and the impact of COVID-19 in the county), and then institutions in other counties. Clos et al. (2021), similar to others, believe that first-come, first-serve is also not permissible, though they do not detail why it is not morally justifiable.

Recommendations for distribution and inventory management

Thus far, approaches to the distribution of scarce medical resources during pandemics as well as to PPE distribution in the literature and in public health contexts have been explored. The most effective arguments from each approach can be synthesized to inform principle-based recommendations for PPE distribution and inventory management. Table 3 depicts each principle, its definition, and application. They will be also summarized below:

TABLE 3 A synthesis of ethical principles to guide PPE management.

Principle	Definition	Application
Utilitarianism/Maximizing Benefits	(a) Save the most lives (preventing *mortality*) (b) Save the most life-years (preventing *morbidity*) (c) Save *quality* of life-years (preventing *morbidity)* (d) Reduce viral transmission (indirect method of achieving a-c) (Emanuel et al., 2020)	Most important; aligns directly with prioritizing the worst off; Entails providing full protection to one facility, rather than partial protection to two to better control transmission (Clos et al., 2021)

Prioritizing the worst off	Prioritize institutions with the greatest need
	Which types of institutions have the greatest need?
	
	• Tend to have a high number of potential or confirmed COVID-19 cases
	• Serve high-risk, vulnerable populations (Binkley & Kemp, 2020; Dawson et al., 2020; Wang et al., 2020)
	  ○ High risk of morbidity/mortality if infected
	  ○ Lower socioeconomic background
	  ○ Traditionally marginalized, disadvantaged populations
	• High risk of infection/exposure to COVID-19
	• Inability to social distance, or use other methods of controlling viral spread besides PPE
	
	(Public Health Seattle & King County, 2022; Washington State Department of Health, 2021)
	Within a given institutional type/tier, which institutions have the greatest need?
	Defined by:
	
	(a) **Urgency:**
	  a. Time until supply runs out (Clos et al., 2021; Washington State Department of Health, 2021)
	    i. <48 h supply: dire need
	    ii. <1 week supply: high need
		Most important; aligns directly with prioritizing the worst off; (Clos et al., 2021) consider opportunity cost of allocation decisions, and impact of *lack of PPE* (deprivation cost function) (Dönmez et al., 2022)

TABLE 3 A synthesis of ethical principles to guide PPE management—cont'd

Principle	Definition	Application
	b. Then number of units left (Clos et al., 2021) (b) **Highest COVID-19 burden relative to capacity** (measured by case number relative to number of staff in a 24 h period, ICU beds, active vs. overflow beds) (Clos et al., 2021)	
Instrumental value (related concepts include social worth and the multiplier effect)	An individual who: (a) provides a critical role to societal infrastructure (*social worth*) (b) as a result of the role, can save more lives, life-years, or substantially increase the quality of lives beyond their own (*multiplier effect*) (Binkley & Kemp, 2020; Emanuel et al., 2020; Sureka et al., 2021)	Somewhat important; justifies healthcare facilities in higher tiers
Equal value to all (its converse is first-come, first-serve)	Allocation decisions should not be based on demographic or other factors that would widen existing systemic disparities (e.g., wealth, position) (Binkley & Kemp, 2020; Dawson et al., 2020; Sureka et al., 2021).	Most important; always applied
Random selection	N/A	Should only be used when all other factors are equal Consider difficult to regulate, and is prone to corruption. Can also appear to undermine value of human life, so transparency would be critical (Moodley et al., 2020)
Principle of reciprocity	Those that put themselves in areas of high risk should be prioritized and given more protection; this is a reciprocal benefit based solely on the *risk* taken by the individual (Binkley & Kemp, 2020; Sureka et al., 2021).	Somewhat important; helps justify healthcare organizations in higher tiers
Distributive justice and equity	Support organizations that disproportionately experience societal injustice, serve marginalized populations (Dawson et al., 2020; Dönmez et al., 2022; Moodley et al., 2020; Srinivas et al., 2021)	Important; tends to align with prioritizing the worst off; justifies distribution to low resource settings
Duty obligations to donor(s) (where applicable)	Includes: (a) alignment with mission (b) alignment with donor preferences (Clos et al., 2021; Merchant et al., 2021)	Somewhat important; a loan can be given to an organization outside donor preferences/mission on case-by-case basis if there is reasonable certainty it can be returned, if it would maximize benefits/support those in the greatest need (i.e., in a tier of institution and level of need greater than those within the mission/donor preferences)

This table depicts a synthesis of ethical principles discussed in this chapter.

- *Utilitarianism/maximizing benefits:* This principle aims to maximally prevent mortality and morbidity, specifically by aiming to save the most lives, life-years, and *quality* of life-years as well as reduce viral transmission (Clos et al., 2021; Emanuel et al., 2020). Quality of life-years has not been discussed by authors thus far, but is proposed to take into account the variation of impact that acute and chronic courses of COVID-19 have on different people (e.g., life-threatening short-term course on one individual, and debilitating long COVID-19 in another). This principle is **most important.**
- *Prioritizing the worst off:* This principle aims to prioritize institutions with the greatest need. The greatest need institutions (i.e., those that have a high number of COVID-19 cases, serve high-risk vulnerable populations) can be separated into tiers, and then donations can be allocated based on urgency of need and the COVID-19 burden as a function of

resources (Clos et al., 2021; Public Health Seattle & King County, 2022; Washington State Department of Health, 2021). This is **most important**, and aligns directly with maximizing benefits (Clos et al., 2021).

- *Instrumental value (related to social worth, multiplier effect)*: This principle prioritizes those who provide a critical role in societal infrastructure, and can increase the number of lives and life-years beyond their own. This is **somewhat important**, and justifies healthcare institutions in higher tiers (Binkley & Kemp, 2020; Emanuel et al., 2020; Sureka et al., 2021).
- *Equal value to all:* This principle aims to ensure that allocation decisions aren't based on power, wealth, demographic factors, etc. This is **most important** and **always applied** (Dawson et al., 2020).
- *Random selection:* This should not be used unless all other factors are equal. This is difficult to regulate, and can appear to undermine the value of human life, so it should be used sparingly (Moodley et al., 2020).
- *Principle of reciprocity:* This principle is defined by providing a reciprocal benefit (i.e., PPE) based on the risk taken by an individual. This is **somewhat important**, and justifies healthcare institutions in higher tiers (Binkley & Kemp, 2020; Sureka et al., 2021).
- *Distributive justice and equity:* This principle aims to support organizations that disproportionately experience societal injustice. It is **important**, and tends to align with prioritizing the worst off (Dawson et al., 2020; Dönmez et al., 2022; Moodley et al., 2020; Srinivas et al., 2021).
- *Duty obligations to donor(s):* This principle includes aligning donations with the mission statement of an organization and with donor preferences. This is **somewhat important**. See Tables 3 and 4 for applications (Clos et al., 2021).

In addition to providing a distribution framework, GetMePPE Chicago leadership detailed approaches to specific inventory management and distribution issues: (1) donating PPE to other states (which can also be thought of as donating PPE to institutions outside mission or donor preferences more broadly), (2) receiving PPE donations from other states, and (3) building a PPE reserve. Table 4 reflects how to tackle these issues (Merchant et al., 2021).

TABLE 4 Inventory management and PPE distribution, adapted from "Lessons Learned" of Merchant et al. (2021).

Issue	Approach	Values	Additional considerations
When to build a PPE reserve	If Tier I facility need is exhausted, Tier II can be given PPE. A small reserve can be built in case a surge occurs and Tier I has increased need. Calculate the amount of PPE needed in a surge, estimated time to a potential surge, amount of donation expected during time to potential surge, and present amount of PPE to determine what PPE would be reserved.	Maximizing benefits (reducing transmission at present), and protecting the worst off (planning to protect Tier I facilities in the future and meet the needs of the worst off)	
Whether/when to donate PPE to another state/province/country	Provide a loan if the institution's tier in another state/province/country is higher than the types of institutions within the scope of the organization's reach. There should be fair certainty that the loan would be repaid.	Maximizing benefits (saving the most lives/life years) regardless of geographic location; prioritizing the worst off regardless of geographic location	The original framework was intended to assess NGO donations across state lines; governments choosing to donate or receive donations might have political implications
Whether/when to accept donations from another state/province/country	Assess case rate in other state/province/country. If lower than state of operation, accept donations. If not lower, direct to organization within state/province/country	Prioritizing the worst off (ensures PPE is directed toward the region of greatest need)	

This table depicts how to approach specific issues in PPE management.
Content of this entire table adapted from content in work of Merchant, T., Hormozian, S., Smith, R. S., Pendergrast, T., Siddiqui, A., Wen, Z., & Sheldon, M. (2021). Ethical principles in Personal Protective Equipment Inventory management decisions and partnerships across state lines during the COVID-19 pandemic. *Public Health Reports, 137(2)*, 208–212.

Policies and procedures

In this chapter, we reviewed approaches to the ethical distribution and management of PPE during the coronavirus pandemic. We reviewed in detail principles considered in scarce medical resource allocation during pandemics and in ethical PPE distribution strategies, including public health and grassroots/nongovernment organization approaches. These works were obtained through an extensive literature search and reflect the perspective of authors from across the globe. We summarized key aspects of each work, and drew parallels between the ethical values they mention. We also compared and contrasted arguments made regarding the weightage and importance of certain values. For works that did not explicitly name ethical values, we sought to identify the principles underpinning their approaches. At the end of the chapter, we compiled the values discussed, and shared recommendations on how to apply those values as well as additional ethical considerations in doing so. Where approaches may differ (e.g., how tiers are determined), we shared what we believe are the best guidelines for effective PPE management. We also shared recommendations on how to address specific issues in PPE management, drawn from one paper examining approaches to those challenges.

Given the limited literature on this topic, local, regional, and national governmental bodies should do further research using modeling to retrospectively analyze spread and determine what organizations are most prone to viral transmission (and thereby which should be prioritized). These bodies should can also retroactively assess community member perspectives on PPE distribution principles and how to allocate to save the most lives/life-years. This may identify areas of spread and need missed in this analysis, and help adjust tiering and other aspects of principle operationalization to promote the best utility of limited PPE in the future.

Applications to other areas

In this review, we discussed principles for ethical PPE management. However, many of the guidelines discussed could be extended to other preventative therapies (e.g., vaccines). Furthermore, although approaches detailed here focus on ethical allocation at a local or regional level, they can arguably apply across international lines (country to country). Thus, this work has broader implications for how we might think about sharing vaccine-related resources among countries, and how we might prioritize who gets vaccinated first even beyond COVID-19 (e.g., monkeypox). It is worth noting that there are additional political, economic, and social ramifications of allocation decisions made by governments on an international level that have not been discussed in this work. These would need to be explored as well, and may influence the calculus of how to maximize benefits to all. For instance, if a loan of COVID-19 vaccines is perceived well by the recipient nation, there could be economic benefits (e.g., sanctions lifted) to the country, which may need to be considered. Additionally, the authors of this chapter are based in the United States, so it is possible that the weightage of these principles and their associated application may need to be modified based on cultural context.

Finally, this review also briefly discusses the concepts of opportunity cost and lack of access to PPE. More broadly, from a public health perspective, it may be worthwhile to examine the opportunity cost (economic, health, or otherwise) and public health ramifications associated with diverting *general* resources away from other diseases (e.g., heart disease, mental health) to focus on developing COVID-19 boosters and COVID-19 specific treatments.

Mini-dictionary of terms

- **Instrumental value:** An individual has a critical role in societal infrastructure.
- **Prioritizing the worst off**: Prioritize institutions with the greatest need.
- **Principle of reciprocity**: People should receive a reciprocal benefit if they take on higher risk.
- **Personal protective equipment**: Items that protect the user from viral transmission; for instance, masks, N95s, etc.
- **Utilitarianism:** Seeks to promote the most good for the most people; can be defined differently based on the author and context.

Summary points

- Few studies characterize principles for PPE distribution and management.
- Approaches to scarce medical resources primarily discuss utilitarianism (saving the most lives/life years), but also equity; first-come, first-serve (not ethically justifiable); and prioritizing sicker/younger patients.

- Approaches to ethical PPE management in the literature discuss utilitarianism in a variety of ways: protecting the most number of clinicians, protecting clinicians who can save the most patients, determining suffering caused by lack of PPE, or minimizing viral trasmission, infection, and death.
- Approaches to ethical PPE management in the literature also consider risk of population, severity of infection, social worth, justice, and equity.
- Public health guidelines on PPE management help quantify need by assigning organizations to a tier based on COVID-19 cases and the ability to control viral transmission with alternative strategies, and prioritizing an organization within a tier based on supply levels and urgency of need.
- Grassroots initiatives can provide guidelines for PPE management that are explicitly grounded in core ethical values.

References

Binkley, C. E., & Kemp, D. S. (2020). Ethical rationing of personal protective equipment to minimize moral residue during the covid-19 pandemic. *Journal of the American College of Surgeons*, *230*(6), 1111–1113.

CBS. (2020). *Dr. Anthony Fauci Talks with Dr Jon LaPook about Covid-19*. YouTube. United States, Retrieved November 21, 2022, from https://www.youtube.com/watch?v=PRa6t_e7dgI&ab_channel=60Minutes.

Centers for Disease Control and Prevention. (2020, July 16). *Covid-19: Strategies for optimizing the supply of PPE*. Centers for Disease Control and Prevention. Retrieved November 21, 2022, from https://www.cdc.gov/coronavirus/2019-ncov/hcp/ppe-strategy/general-optimization-strategies.html.

Clos, A. L. T., Cohen, A. P., Edwards, L. E., Martin, A. H. H., Merchant, T. S., Pendergrast, T. R., Siegel, M. A., & Smith, R. S. (2021). Ethical considerations in PPE allocation during COVID-19: A case study. *Harvard Medical Student Review*. Retrieved November 21, 2022, from https://www.hmsreview.org/covid/ethical-considerations-in-ppe-allocation.

Dawson, A., Isaacs, D., Jansen, M., Jordens, C., Kerridge, I., Kihlbom, U., Kilham, H., Preisz, A., Sheahan, L., & Skowronski, G. (2020). An ethics framework for making resource allocation decisions within clinical care: Responding to COVID-19. *Journal of Bioethical Inquiry*, *17*(4), 749–755.

Dönmez, Z., Turhan, S., Karsu, Ö., Kara, B. Y., & Karaşan, O. (2022). Fair allocation of personal protective equipment to health centers during early phases of a pandemic. *Computers & Operations Research*, *141*, 105690. https://doi.org/10.1016/j.cor.2021.105690.

Emanuel, E. J., Persad, G., Upshur, R., Thome, B., Parker, M., Glickman, A., Zhang, C., Boyle, C., Smith, M., & Phillips, J. P. (2020). Fair allocation of scarce medical resources in the time of covid-19. *New England Journal of Medicine*, *382*(21), 2049–2055. https://doi.org/10.1056/nejmsb2005114.

Merchant, T., Hormozian, S., Smith, R. S., Pendergrast, T., Siddiqui, A., Wen, Z., & Sheldon, M. (2021). Ethical principles in Personal Protective Equipment Inventory management decisions and partnerships across state lines during the COVID-19 pandemic. *Public Health Reports*, *137*(2), 208–212.

Moodley, K., Rennie, S., Behets, F., Obasa, A. E., Yemesi, R., Ravez, L., Kayembe, P., Makindu, D., Mwinga, A., & Jaoko, W. (2020). Allocation of scarce resources in Africa during COVID-19: Utility and justice for the bottom of the pyramid? *Developing World Bioethics*, *21*(1), 36–43.

Peters, Z., & Cairns, C. (2022). Experiences related to the COVID-19 pandemic among U.S. physicians in office-based settings, 2020–2021. *National Health Statistics Reports*, *175*, 1–7.

Public Health Seattle & King County. (2022). *Personal Protective Equipment (PPE) Prioritization and Distribution for COVID-19 in King County*. June 13.

Ranney, M. L., Griffeth, V., & Jha, A. K. (2020). Critical supply shortages—the need for ventilators and personal protective equipment during the COVID-19 pandemic. *New England Journal of Medicine*, *382*(18), 41–43.

Srinivas, G., Maanasa, R., Meenakshi, M., Adaikalam, J. M., Seshayyan, S., & Muthuvel, T. (2021). Ethical rationing of healthcare resources during COVID-19 outbreak: Review. *Ethics, Medicine and Public Health*, 16.

Sureka, B., Nag, V. L., Garg, M. K., Tak, V., Banerjee, M., Bishnoi, A., & Misra, S. (2021). Rational use of PPE and preventing PPE related skin damage. *Journal of Family Medicine and Primary Care*, *10*(4), 1547.

Tabari, P., Amini, M., Moghadami, M., & Moosavi, M. (2020). International Public Health Responses to COVID-19 Outbreak: A Rapid Review. *Iranian Journal of Medical Sciences*, *45*(3), 157–169.

Wang, C., Deng, Y., Yuan, Z., Zhang, C., Zhang, F., Cai, Q., … Kurths, J. (2020). How to optimize the supply and allocation of medical emergency resources during Public Health Emergencies. *Frontiers in Physics*, *8*. Article 383.

Washington State Department of Health. (2021). *Prioritization Guidelines for Allocation of Personal Protective Equipment*. , September 27.

Chapter 10

Clinical management in the COVID-19 pandemic: Rheumatic disease

Abdulvahap Kahveci[a,b] and Şebnem Ataman[a]
[a]*Division of Rheumatology, Medical School, Ankara University, Ankara, Turkey,* [b]*Rheumatology Clinic, Kastamonu Training and Research Hospital, Kastamonu, Turkey*

Abbreviations

ACR	American College of Rheumatology
APLAR	The Asia Pacific League of Associations for Rheumatology
bDMARD	biological synthetic disease-modifying antirheumatic drug
COVID-19	novel coronavirus disease 2019.
csDMARD	conventional synthetic disease-modifying antirheumatic drug
EULAR	European League Against Rheumatism
GC	glucocorticoid
GSR	The German Society of Rheumatology
JAKinib	Janus kinase inhibitor
RD	rheumatic disease
SARS-CoV-2	severe acute respiratory syndrome coronavirus 2
TR	telerheumatology

Introduction

The novel coronavirus disease 2019 (COVID-19) is defined as an infectious disease caused by a new type of respiratory coronavirus (severe acute respiratory syndrome coronavirus 2; SARS-CoV-2). It first appeared in Wuhan, China, in December 2019 (Zhu et al., 2020). COVID-19 spread all around the world in a short time and was declared a pandemic by the World Health Organization on March 11, 2020 (https://www.who.int/emergencies/diseases/novel-coronavirus-2019). The acute inflammatory syndrome resulting from COVID-19 has led to millions of mortalities and morbidities, and still continues to be a problem.

Infectious diseases are serious conditions that negatively affect management in rheumatology practice. More explicitly, they have been a part of differential diagnosis of rheumatic diseases (RD), cause disease exacerbations, and may be complicated in immunocompromised patients who use antirheumatic drugs (Furer et al., 2019). COVID-19, which has been going on for almost three years, has created significant challenges for patients and rheumatologists (Kahveci et al., 2022; Mikuls et al., 2021; Specker et al., 2021; Varley et al., 2021; Zhu et al., 2021).

During the pandemic, the management of RD markedly differed from before the pandemic (Leipe et al., 2020). Also, along with the literature, rheumatology practices have changed and evolved to aim for ideal routines. However, consensus information about some of the domains of clinical practices such as treatment, prognosis, and vaccinations of RD in the pandemic has not yet been fulfilled. Accordingly, there are still many titles on the research agenda (Landewé et al., 2022; Mikuls et al., 2021; Specker et al., 2021).

At the beginning of the pandemic, early literature data based on mostly case reports showed the increased incidence, severity, and mortality of COVID-19 in patients with RD (Favalli, Ingegnoli, et al., 2020; Zhou et al., 2020). After that, global rheumatology registry reports indicated that COVID-19-related poor outcomes in patients with RD were associated with general factors (older age, male sex, and specific comorbidities) and disease-specific factors (high disease activity and specific medications) (Gianfrancesco et al., 2020; Strangfeld et al., 2021). They also reported that using moderate to high

Management, Body Systems, and Case Studies in COVID-19. https://doi.org/10.1016/B978-0-443-18703-2.00005-0
Copyright © 2024 Elsevier Inc. All rights are reserved, including those for text and data mining, AI training, and similar technologies.

doses of glucocorticoids (GCs) and some immunosuppressants (especially rituximab) were associated with a poor prognosis of COVID-19 (Gianfrancesco et al., 2020; Strangfeld et al., 2021).

So far, many recommendations, reviews, and original studies have been reported on RD management. Although the statements in the literature have changed over the pandemic, the current recommendations have become nearly parallel. Besides, important situations (new SARS-CoV-2 variants, results of a recent randomized controlled trial on antirheumatic drugs or vaccination, etc.) that might occur have the potential to change all these recommendations.

During the pandemic, it was indisputable that rheumatologists should update themselves with current and correct scientific information to adapt to the constant alternation of pandemic conditions. Therefore, as rheumatologists made critical decisions on managing RDs by using their experiences and new scientific data for each patient, we aim to review current and acceptable evidence and provide recommendations/statements for the clinical management of RD in the context of SARS-CoV-2 in this chapter.

Methods, policies, and procedures

In this chapter, we review the literature, including the clinical management of RD during COVID-19. We search the online databases (PubMed and Web of Science) for detailed case reports, cohort studies, survey studies, randomized controlled studies, reviews, metanalyses, and guidelines from the start of the pandemic to December 2022. The keywords of the search were "Management of rheumatic disease," "COVID-19 pandemic," "SARS-CoV-2," "Telerheumatology," and "Antirheumatic medication." Also, other related terms such as prevention, survey studies, glucocorticoids, conventional synthetic disease-modifying antirheumatic drugs (csDMARDs), biological disease-modifying antirheumatic drugs (bDMARDs), and immunosuppressants were included in the search.

The contents of this chapter are composed as answers to common questions we get in our rheumatology practice. The subjects of the chapters are as follows: (1) general recommendations for RD patients to prevent COVID-19, (2) telerheumatology-based management, (3) survey-based clinical management, and (4) evidence and guideline-based management of RD. Also, we summarize the management of antirheumatic drugs with COVID-19 vaccination in the applications to other areas section. We use the four society guidelines (the American College of Rheumatology (ACR), the European League Against Rheumatism (EULAR), the Asia Pacific League of Associations for Rheumatology (APLAR), and the German Society of Rheumatology (GSR)) as core references for this review (Landewé et al., 2022; Mikuls et al., 2021; Specker et al., 2021; Tam et al., 2021). In addition to evidence-based medicine, we report our own practices/experiences. The contents of this chapter could not cover all situations in clinical practice. Therefore, clinicians should make decisions on the management of their patients on a case-by-case basis.

General recommendation for RD patients to prevent COVID-19

During the pandemic, general preventive approaches were recommended by the World Health Organization, ministries of health, and various local and international health-related organizations (Kahveci & Ataman, 2021). Worldwide, governments have changed their strategies during the different stages of the COVID-19 pandemic. For example, in the first wave of the pandemic, the zero-case COVID-19 policy (lockdown, closing of schools, restriction on international flights, etc.) was followed by most countries. However, because of SARS-CoV-2 vaccination, a controlled social life and the protection of high-risk groups came to the fore later in the pandemic.

As with the general population, people with RD should be educated about prevention methods (social distancing, wearing protective mask, and hand hygiene) (Tam et al., 2021). Extra precautions should be taken to reduce the frequency of patient visits to the hospital to reduce exposure to SARS-CoV-2. Reducing the frequency of laboratory monitoring along with the use of telemedicine and widening the dose intervals of intravenous medications are examples of such precautions (Mikuls et al., 2021). Also, the ACR offered some recommendations for especially risky groups (geriatric, having comorbidities, etc.) such as preferring subcutaneous or oral forms instead of intravenous drugs, making necessary arrangements to maintain social distance in crowded places (house, office, etc.), and providing appropriate personal protective equipment (Mikuls et al., 2021).

To prevent sedentary behavior, patients should be trained in home exercises and given advice about staying healthy during a lockdown. In this direction, various exercise sets and videos have been published by EULAR to prevent pathologies that might occur in the musculoskeletal system due to sedentary behavior in patients staying home during the pandemic (https://www.eular.org/home_exercises_for_rmd_patients.cfm). Additionally, the British Society of Rheumatology also recommended that people who did not get enough sunlight should take 10mg per day of vitamin D (https://www.rheumatology.org.uk/practice-quality/covid-19-guidance).

Telerheumatology-based clinical management

Telemedicine, here referred to as telerheumatology (TR), had been rarely used in rheumatology before the pandemic (McDougall et al., 2017). However, during the pandemic, it was widely used at most rheumatology units worldwide (Kahveci et al., 2022; Tang et al., 2021) to reduce SARS-CoV-2 transmission. During the pandemic, it was used almost at the same rate as face-to-face examinations (Koonin et al., 2020). In addition, the international rheumatology societies such as the ACR, the EULAR, the APLAR, and the GSR recommended using TR to reduce SARS-CoV-2 transmission risk (Landewé et al., 2020; Mikuls et al., 2021; Specker et al., 2021; Tam et al., 2021).

TR has high satisfaction with patients and clinicians, and is not inferior to face-to-face visits regarding disease activity and function (Matsumoto & Barton, 2021). There were possible barriers tu using TR, such as concerns about being without a physical exam or lab tests, and insufficient technical tools (e.g., lack of internet, lack of communication device) (Matsumoto & Barton, 2021; Tang et al., 2021). A recent study identified two red flags in teleconsultation, which were patient-reported rheumatoid arthritis flares and increased C-reactive protein values. They suggested that these patients have rapid in-person examinations and modification in medications (Avouac et al., 2022).

Even though there were some barriers in using TR, it is an acceptable, feasible, and effective modality in managing RD, suggesting that it would continue to be an important instrument after the pandemic.

Survey based clinical management

Regarding managing RD during the pandemic, survey studies are a prominent directive topic. These clinician-based studies allow us to understand how rheumatologists in different countries made decisions and managed the disease process during the pandemic. In these surveys, at the beginning of the pandemic it was stated that the number of patients admitted to rheumatology clinics decreased significantly (Kahveci et al., 2022; Nune et al., 2021; Ziadé et al., 2020). The reasons were the risk of SARS-CoV-2 transmission, lockdown, supplying medicine from the pharmacy directly, and the use of TR (Akintayo et al., 2021; Kahveci et al., 2022; Nune et al., 2021; Ziadé et al., 2020).

In our survey study, clinicians reported that medical procedures (labial salivary gland biopsy, Schirmer's test, salivary flow rate test, nail bed video-capillaroscopy, musculoskeletal ultrasonography, and pathergy test) in which social distance cannot be maintained were less preferred (Kahveci et al., 2022).

Worldwide, in the early stages of the pandemic most surveys revealed that respondents had uncertainties about antirheumatic drug usage and preferred it less than before the pandemic (Kahveci et al., 2022). Some survey studies have reported that rheumatologists reduced the use of steroids, and tapered them early (Akintayo et al., 2021; Gupta et al., 2021; Kahveci et al., 2022). These studies and others also stated that clinicians hesitated to use and/or start bDMARDs (especially rituximab), immunosuppressants (especially cyclophosphamide), and Janus kinase (JAK) inhibitors (JAKinibs) (Akintayo et al., 2021; Batu et al., 2020; Gupta et al., 2021; Kahveci et al., 2022; Nune et al., 2021).

According to the population-based cohort studies and global rheumatology registry data, antirheumatic drugs (except rituximab, and >10 mg/day prednisolone equivalent) were not associated with a higher risk of COVID-19-related bad outcomes (Gianfrancesco et al., 2020; Santos et al., 2021). Eventually, we may predict that clinicians' constraints about medication preference would decrease as positive evidence increases in studies investigating the relationship between antirheumatic drugs and SARS-CoV-2.

Academic research and clinical learning activities were negatively affected in the early days of the pandemic (Ahmed et al., 2020). The studies reported that reducing in the scientific papers, preferring mainly online rheumatology platforms for education, research, and patients, and a decline in international collaboration between the colleagues (Abramo et al., 2022; Kasapçopur, 2020; Ziadé et al., 2020). After widespread SARS-CoV-2 vaccination, educational meetings and case sessions are being held face-to-face with controls, but virtual platforms are still widely used.

Evidence and guideline-based management of RD

Prevalence, severe disease, and mortality of COVID-19 in patients with RD

Based on the current literature, it is unclear whether the prevalence of COVID-19 in rheumatic patients is higher compared to the healthy (Wang et al., 2021; Zhu et al., 2021). Some retrospective cohort studies have reported that rheumatic patients have a significantly increased prevalence rate of COVID-19 (Ferri et al., 2020; Pablos et al., 2020). However, studies from different countries showed that the prevalence of COVID-19 in rheumatic patients was similar to the matched controls (Michelena et al., 2020; Zen et al., 2020).

Regarding the global rheumatology registry data, cohort studies, and reviews, COVID-19 as measured by hospitalization rate, intensive care unit admission, ventilation rate, and mortality did not present more in patients with RD compared with nonrheumatic controls (Batıbay et al., 2021; FAIR /SFR/SNFMI/SOFREMIP/CRI/IMIDIATE consortium and contributors, 2021; Gianfrancesco et al., 2020; Korolev et al., 2022; Strangfeld et al., 2021; Wang et al., 2021; Zhu et al., 2021). However, as in nonrheumatic patients, the risk of severe disease was higher in RD patients with general risk factors (older age, male gender, smoking) and comorbidities (obesity, diabetes mellitus, hypertension, premorbid lung, renal, or heart disease, etc.) (Gianfrancesco et al., 2020; Strangfeld et al., 2021; Wang et al., 2021).

The recommendations of the ACR, the EULAR, and the GSR guidelines indicated that the poor outcomes of COVID-19 were related to general risk factors (Landewé et al., 2022; Mikuls et al., 2021; Specker et al., 2021).

The impact of antirheumatic drugs on SARS-CoV-2

COVID-19 consists of different pathogenetic and clinical disease phases. Various clinical manifestations occur in each stage of viral pathogenesis. In the early phases of the disease, there are primarily similar clinical features as other viral respiratory infections (fever, cough, headache, etc.) (Tufan et al., 2020). In contrast, the clinical results of the host hyperinflammation (respiratory failure, macrophage activation syndrome, etc.) are dominant in the later stages (Tufan et al., 2020). To manage RD and treatments correctly, it is necessary to know the pathogenesis of COVID-19. Fig. 1 illustrates the disease pathogenesis, clinical manifestations, and potential therapeutic options for COVID-19.

Glucocorticoids

The chronic use of GCs causes the host to be susceptible to infections. The COVID-19 registries have also revealed a similar predisposition of SARS-CoV-2 regarding GC (Gianfrancesco et al., 2020; Santos et al., 2021; Strangfeld et al., 2021). Regarding the retrospective cohort studies, GC therapy was related to the increased risk of COVID-19 and a higher risk of severe disease (FAIR /SFR/SNFMI/SOFREMIP/CRI/IMIDIATE consortium and contributors, 2021; Favalli, Bugatti, et al., 2020). Also, in the global rheumatology registry, the mortality rate of COVID-19 patients was found to be 1.7-fold higher with using >10mg GC versus nonusers (Strangfeld et al., 2021).

On the other hand, the bad COVID-19 outcomes of the patients (such as systemic lupus erythematosus, myositis, and vasculitis) using high doses of steroids due to high disease activity suggest a confounding effect. In another global registry

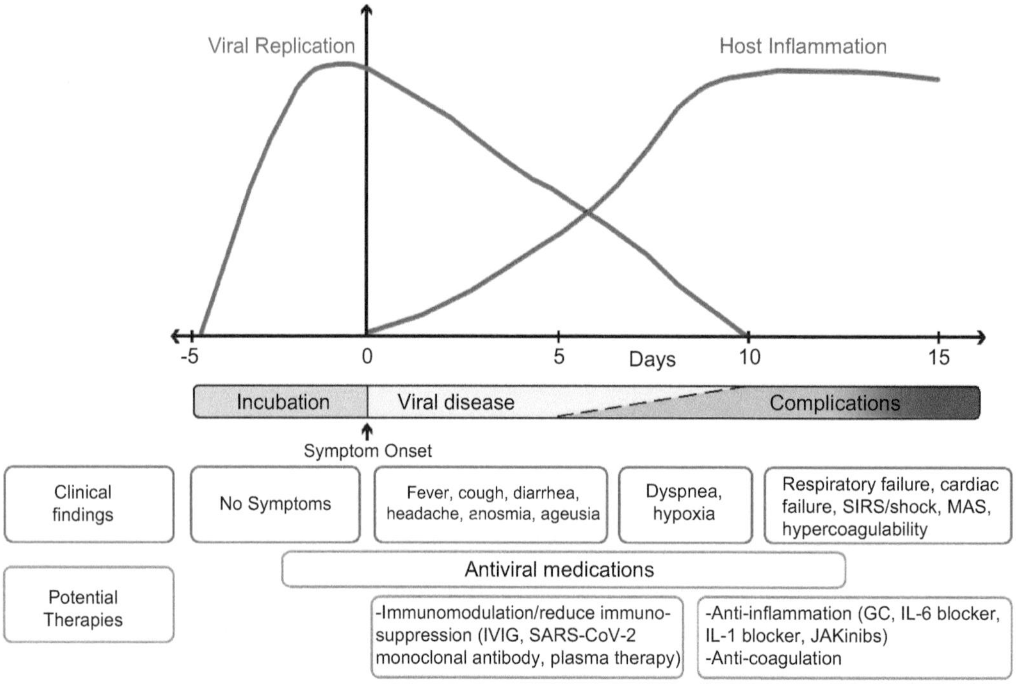

FIG. 1 The phases of COVID-19 pathogenesis, clinical findings, and potential therapies. IVIG, intravenous immunoglobulins; SARS-CoV-2, severe acute respiratory syndrome coronavirus 2; GC, glucocorticoid; IL-1, interleukin-1; IL-6, interleukin-6; JAKinibs, Janus kinase inhibitors; SIRS, systemic inflammatory response syndrome; MAS, macrophage activation syndrome.

report, steroid use above 10 mg was shown to be associated with severe disease and mortality independently of disease activity (Strangfeld et al., 2021). We know that GC therapy is crucial for some acute rheumatic conditions such as lupus nephritis, vasculitis, and sudden visual loss secondary for temporal arteritis. Therefore, the use of GC should be maintained on a case-by-case basis.

The suppression of the hypothalamus-pituitary–adrenal axis is a known result of chronic GC use. Therefore, adding a stress-dose GC may be required in these patients when they have a COVID-19 infection (Arlt et al., 2020). During the early days of the pandemic, a guideline was published by the European Society of Endocrinology to manage adrenal insufficiency (Arlt et al., 2020). This guideline recommended for patients with symptomatic COVID-19 infection that they should receive 10 mg prednisolone every 12 h if patients take 5–15 mg of prednisolone equivalent dose per day. Also, if they take over 15 mg prednisolone equivalent, it should be recommended to continue with own GC doses, but to get with at least two equal doses of 10 mg every 12 h. Also, hydration should be continued carefully to prevent dehydration (Arlt et al., 2020).

Considering these data, all society guidelines recommended that GC therapy ought to be used at the lowest possible dose to control RDs and should not be stopped abruptly in patients with RD during the pandemic. Additionally, high-dose GC therapy should not be avoided when it is necessary for end-organ involvement (Landewé et al., 2022; Mikuls et al., 2021; Specker et al., 2021; Tam et al., 2021).

Conventional synthetic disease-modifying antirheumatic drugs (csDMARDs)

Conventional synthetic disease-modifying antirheumatic drugs (csDMARDs) are major drug groups that have varied molecular mechanisms to treat RD. In the current literature, there were no data on the increased frequency or severity of COVID-19 in patients with RD who use hydroxychloroquine, sulfasalazine, methotrexate, and leflunomide versus nonusers (Favalli, Bugatti, et al., 2020; MacKenna et al., 2022). For this reason, it was generally recommended to continue csDMARDs in patients with RD who have no diagnosis or exposure to SARS-CoV-2 (Landewé et al., 2022; Mikuls et al., 2021; Specker et al., 2021; Tam et al., 2021).

When rheumatic patients have symptomatic and documented COVID-19, csDMARDs may be stopped on the basis of expert opinion and case conditions (Landewé et al., 2022; Mikuls et al., 2021; Tam et al., 2021). The GSR also recommended the use of cholestyramine to wash out leflunomide users (Specker et al., 2021). csDMARD therapy should be reinitiated after recovery from COVID-19. However, given the rheumatologist's expertise, the decision to withdraw or reinitiate the use of csDMARDs needs to be individualized regarding the prognosis of COVID-19 and RD (Landewé et al., 2022; Mikuls et al., 2021; Specker et al., 2021; Tam et al., 2021).

Early in the pandemic, hydroxychloroquine (HCQ) was used off-label for COVID-19 treatment regarding case reports and preclinic study (Wang et al., 2020). However, the positive effect of HCQ on COVID-19 could not be demonstrated in follow-up trials (Molina et al., 2020). Moreover, especially in hospitalized COVID-19 patients, the use of HCQ was associated with increased cardiac arrhythmias (QT prolongation etc.) (Desmarais et al., 2021).

Biological disease-modifying antirheumatic drugs (bDMARDs) and JAKinibs

According to the latest retrospective studies, biologic drugs (except rituximab) were disease-modifying antirheumatic agents with a safety profile for SARS-CoV-2 (Fernandez-Gutierrez et al., 2021; MacKenna et al., 2022; Santos et al., 2021). The IL-6 inhibitor-tocilizumab may also protect against severe COVID-19 (Santos et al., 2021). In the Randomized Evaluation of COVID-19 Therapy (RECOVERY) trials, tocilizumab may have therapeutic benefits (low intensive care duration, ventilations, low concomitant use of GC) during the hyperinflammatory phase of the COVID-19 course (Abani et al., 2021). However, these positive results could not be supported by other randomized controlled studies (Hermine et al., 2021; Rosas et al., 2021).

The global registry data indicated that JAK inhibitors were related to only a slightly higher risk of severe COVID-19 (Sparks et al., 2021). However, other studies did not show this risk (MacKenna et al., 2022). Also, in a randomized controlled trial, baricitinib, a JAK inhibitor drug, combined with remdesivir improved the outcome of hospitalized COVID-19 patients more than using only remdesivir (Kalil et al., 2021).

Based on society guidelines, when there is no active COVID-19, bDMARDs and JAKinibs should be continued during the pandemic (Mikuls et al., 2021; Specker et al., 2021; Tam et al., 2021). The only exception is the anti-CD20 monoclonal antibody, rituximab. Many cohort studies have revealed that rituximab was associated with increased COVID-19-related poor outcomes (Batıbay et al., 2021; Gianfrancesco et al., 2020; MacKenna et al., 2022; Santos et al., 2021). Therefore, the EULAR and the GSR guidelines recommended switching to alternative treatments or postponing the next cycle of rituximab on a case-by-case basis (Landewé et al., 2022; Specker et al., 2021). In case of vital-organ involvement in systemic inflammatory disease (ANCA associated vasculitis, myositis etc.), rituximab should not be delayed and ought to be administrated at the appropriate regimes (Mikuls et al., 2021).

Immunosuppressants

Immunosuppressants, including cyclophosphamide, mycophenolate mofetil, calcineurin inhibitors, and azathioprine, should be continued as per the standard of care during the pandemic without exposure or/and symptomatic SARS-CoV-2 (Mikuls et al., 2021; Specker et al., 2021; Tam et al., 2021). The APLAR guidelines recommend that alternative medications to cyclophosphamide may be considered according to the patients' medical conditions (Tam et al., 2021).

In patients with symptomatic or documented COVID-19, immunosuppressants should be discontinued until a certain period after the recovery (Landewé et al., 2022; Mikuls et al., 2021; Specker et al., 2021; Tam et al., 2021). The reinitiating time of immunosuppressants for all antirheumatic drugs should be individualized by case conditions such as the severity of symptoms, negative polymerase chain reaction testing, and complications.

Applications to other areas

The management of antirheumatic drugs after COVID-19 vaccination

This chapter reviewed how rheumatic patients were managed during the COVID-19 pandemic. The most important revolution in rheumatology and all medical fields is vaccination against COVID-19. Whereas the world is safer with immunization, we face numerous situations where we must make decisions in managing our patients. The issue of how to adjust the vaccination time in patients using antirheumatic drugs is one of these situations.

Antirheumatic drugs suppress the hyperinflammation of RD with vary degrees. As a result, they might lead to a decrease in vaccine response. The current data showed that most antirheumatic medications impaired COVID-19 vaccination immunogenicity (Arnold et al., 2021; Friedman et al., 2021). Treatment with rituximab (Auroux et al., 2022; Batıbay et al., 2022; Furer et al., 2021; Liao et al., 2022), abatacept (Auroux et al., 2022; Furer et al., 2021; Liao et al., 2022), methotrexate (Auroux et al., 2022; Sugihara et al., 2022), mycophenolate mofetil (Furer et al., 2021), leflunomide (Auroux et al., 2022), and GCs (Furer et al., 2021) were associated with significantly decreased COVID-19 vaccination response measured by mostly antibody titers.

Although there is not enough evidence for each drug, international societies have published recommendation sets, considering the existing evidence and the pharmacokinetics of the drugs (half-time, elimination etc.) (Curtis et al., 2022; Specker et al., 2021). Table 1 summarizes the ACR (Curtis et al., 2022) and GSR (Specker et al., 2021) recommendations about the timing/use of antirheumatic therapies in the context of vaccinations against COVID-19.

Conclusion

This review summarized the current literature regarding the clinical management of RD during the COVID-19 pandemic. Today, we have more evidence-based data and experiences about how to supervise our rheumatic patients than in the early days of the pandemic. However, the research agenda still includes many clinical issues. The recommendations are altered and diversified by the new results of research. Therefore, it is understandable that there were discrepancies between the recommendations due to the time and lack of high-level evidence-based data. The decision-making for each case together with evidence and clinician experience should be at the forefront of all guidelines.

Mini-dictionary of terms

- **Disease-modifying antirheumatic drugs:** A group of medications generally used in rheumatic patients. They work to suppress host inflammation, prevent organ damage, and conserve the function and structure of joints.
- **Nail bed video-capillaroscopy:** A diagnostic method to analyze microvascular conditions of nail bed capillaries in rheumatic disease.
- **Rituximab:** A monoclonal antibody that inhibits the CD20 antigen of B cells.
- **Pathergy test:** A nonspecific hypersensivity skin reaction induced by a needle prick that is performed to look for evidence of tissue reactivity. In rheumatology practices, it is used in the diagnosis of Behcet disease.

TABLE 1 Comparing the ACR and GSR recommendations about the timing/use of antirheumatic therapies in the context of vaccinations against COVID-19 (Curtis et al., 2022; Specker et al., 2021).

	ACR recommendations (Curtis et al., 2022)		GSR recommendations (Specker et al., 2021)	
Antirheumatic medications	**Statements**	**Level of consensus**	**Statements**	**LoA (± SD)**
Abatacept (IV)	Vaccination occurs 1 week prior to the next dose of abatacept (IV).	Moderate	Vaccination in the interval between two infusions; if possible 4 weeks after an infusion with delay of the next infusion by 1 week	8.53 (± 1.63)
Abatacept (SC)	Withhold for 1–2 weeks (as disease activity allows) after each COVID-19 vaccine dose.	Moderate	Pause for 1 week before and 1 week after each vaccination	8.47 (± 1.67)
NSAIDs and paracetamol	Assuming that disease is stable, withhold for 24 h prior to vaccination. No restrictions on use postvaccination if symptoms develop.	Moderate	Pause for 6–24 h (according to half-life of NSAID) before and 6 h after every vaccination	7.68 (± 2.77)
Belimumab (SC)	Withhold for 1–2 weeks (as disease activity allows) after each COVID-19 vaccine dose.	Moderate	No change	8.84 (± 1.63)
TNFi, IL-6R, IL-1Ra, IL-17, IL-12/23, IL-23, and other cytokine inhibitors	The task force failed to reach consensus on whether to temporarily interrupt these following each COVID-19 vaccine dose, including both primary vaccination and supplemental/booster dosing.	Moderate	No change	9.95 (± 0.22)
Rituximab	-Discuss the optimal timing of dosing and vaccination with rheumatology provider before proceeding. (Some practitioners measure CD19 B cells as a tool with which to time the booster and subsequent rituximab dosing. For those who elect to dose without such information, or for whom such measurement is not available or feasible, a supplemental vaccine dose 2–4 weeks should be provided before next anticipated rituximab dose (e.g., at month 5.0 or 5.5 in patients being administered rituximab every 6 months).	Moderate	– Consider alternative therapy and carry out vaccination – Postponement of the first or next RTX cycle to 2–4 weeks after completion of the vaccination series – If possible, vaccination at the earliest 4–6 months after the last RTX administration – For patients at risk, earlier vaccination if necessary	9.42 (± 1.27) 8.68 (± 1.92) 9.53 (± 0.75) 9.11 (± 1.17)
Hydroxychloroquine	No modifications to either immunomodulatory therapy or vaccination timing.	Strong	No change	10 (±0)

Continued

TABLE 1 Comparing the ACR and GSR recommendations about the timing/use of antirheumatic therapies in the context of vaccinations against COVID-19 (Curtis et al., 2022; Specker et al., 2021)—cont'd

	ACR recommendations (Curtis et al., 2022)		GSR recommendations (Specker et al., 2021)	
Antirheumatic medications	**Statements**	**Level of consensus**	**Statements**	**LoA (± SD)**
All other conventional and targeted immunomodulatory or immunosuppressive medications (e.g., JAK inhibitors, MMF) except for those listed above	Withhold for 1–2 weeks (as disease activity allows) after each COVID-19 vaccine dose.	Moderate	**Methotrexate**: Pause for 1(–2) weeks after each vaccination **Sulfasalazine, leflunomide, azathioprine, calcineurin inhibitors**: No change **JAK inhibitors:** Pause for 1–2 days before to 1 week after each vaccination **Mycophenolate**: Pause for 1 week after each vaccination	7.79 (±2.76) 9.63 (±0.67) 7.74 (±2.63) 7.53 (±2.11)
Cyclophosphamide (IV)	Time cyclophosphamide administration so that it will occur ~1 week after each vaccine dose, when feasible.	Moderate	–	
Prednisolone ≤ 10 mg per day	–		No change	9.74 (± 0.55)
Prednisolone > 10 mg per day	–		If possible, reduction to lower doses (≤10 mg daily)	8.63 (± 2.25)

RMD, rheumatic and musculoskeletal disease; IV, intravenous; SC, subcutaneous; NSAIDs, nonsteroidal antiinflammatory drugs; TNFi, tumor necrosis factor inhibitor; IL-6R, interleukin-6 receptor; IL-1Ra, interleukin-6 receptor antagonist; IL-17, interleukin-17; IL-12/IL-23, interleukin-12/23; MMF, mycophenolate mofetil; JAK, Janus kinase; RTX, rituximab.

- **Cyclophosphamide:** An alkylating immunosuppressive agent to treat severe manifestations of autoimmune and inflammatory syndromes.
- **Janus kinase inhibitors:** A small molecule used in immune-mediated diseases that inhibits Janus kinase and tyrosine kinase enzymes.

Summary points

- Rheumatic patients should also follow preventive methods such as social distancing, hand hygiene, and wearing masks.
- Telerheumatology is an acceptable and feasible modality to maintain patient care and educational activities in the pandemic.
- Clinician-based survey studies published by various countries represented the expert opinion/perspectives in the lack of the evidence.
- In rheumatic patients, general risk factors (age, male gender, smoking) and comorbidities were associated with poor outcomes of COVID-19 similarly with nonrheumatics.
- Within the antirheumatic drugs, rituximab and glucocorticoids (>10 mg/day prednisolone equivalent) have been related to severe COVID-19.
- The society guidelines are time-dependent and may conflict with each other.
- The vaccination time of SARS-CoV-2 ought to be adjusted with respect to antirheumatic medications to assure potent vaccine immunogenicity and efficiency.
- Rheumatologists should manage their patients through primarily high-level evidence.
- On a case-by-case basis, the expert opinion and clinician experience help in making critical decisions despite a lack of evidence.

References

Abani, O., Abbas, A., Abbas, F., Abbas, M., Abbasi, S., Abbass, H., … Abdelfattah, M. (2021). Tocilizumab in patients admitted to hospital with COVID-19 (RECOVERY): A randomised, controlled, open-label, platform trial. *The Lancet*, *397*(10285), 1637–1645.

Abramo, G., D'Angelo, C. A., & Di Costa, F. (2022). How the Covid-19 crisis shaped research collaboration behaviour. *Scientometrics*, *127*(8), 5053–5071.

Ahmed, S., Zimba, O., & Gasparyan, A. Y. (2020). Moving towards online rheumatology education in the era of COVID-19. *Clinical Rheumatology*, *39*, 3215–3222.

Akintayo, R. O., Akpabio, A. A., Kalla, A. A., Dey, D., Migowa, A. N., Olaosebikan, H., … Hamdi, W. (2021). The impact of COVID-19 on rheumatology practice across Africa. *Rheumatology*, *60*(1), 392–398.

Arlt, W., Baldeweg, S. E., Pearce, S. H., & Simpson, H. L. (2020). Endocrinology in the time of COVID-19: Management of adrenal insufficiency. *European Journal of Endocrinology*, *183*(1), G25–G32.

Arnold, J., Winthrop, K., & Emery, P. (2021). COVID-19 vaccination and antirheumatic therapy. *Rheumatology (Oxford)*, *60*(8), 3496–3502. https://doi.org/10.1093/rheumatology/keab223.

Auroux, M., Laurent, B., Coste, B., Massy, E., Mercier, A., Durieu, I., … Coury, F. (2022). Serological response to SARS-CoV-2 vaccination in patients with inflammatory rheumatic disease treated with disease modifying anti-rheumatic drugs: A cohort study and a meta-analysis. *Joint, Bone, Spine*, *89*(5), 105380. https://doi.org/10.1016/j.jbspin.2022.105380.

Avouac, J., Molto, A., Frantz, C., Wanono, S., Descamps, E., Fogel, O., … Allanore, Y. (2022). Evaluation of patients with rheumatoid arthritis in teleconsultation during the first wave of the COVID-19 pandemic. *The Journal of Rheumatology*, *49*(11), 1269–1275. https://doi.org/10.3899/jrheum.220073.

Batıbay, S., Koçak Ulucaköy, R., Özdemir, B., Günendi, Z., & Göğüş, F. N. (2021). Clinical outcomes of Covid-19 in patients with rheumatic diseases and the effects of the pandemic on rheumatology outpatient care: A single-Centre experience from Turkey. *International Journal of Clinical Practice*, *75*(9), e14442. https://doi.org/10.1111/ijcp.14442.

Batıbay, S., Ulucaköy, R. K., Günendi, Z., Fidan, I., Bozdayı, G., & Göğüş, F. N. (2022). Immunogenicity and safety of the CoronaVac and BNT162b2 Covid-19 vaccine in patients with inflammatory rheumatic diseases and healthy adults: Comparison of different vaccines. *Inflammopharmacology*, *30*(6), 2089–2096. https://doi.org/10.1007/s10787-022-01089-6.

Batu, E. D., Lamot, L., Sag, E., Ozen, S., & Uziel, Y. (2020). How the COVID-19 pandemic has influenced pediatric rheumatology practice: Results of a global, cross-sectional, online survey. *Seminars in Arthritis and Rheumatism*, *50*(6), 1262–1268. https://doi.org/10.1016/j.semarthrit.2020.09.008.

Curtis, J. R., Johnson, S. R., Anthony, D. D., Arasaratnam, R. J., Baden, L. R., Bass, A. R., … Mikuls, T. R. (2022). American College of Rheumatology Guidance for COVID-19 vaccination in patients with rheumatic and musculoskeletal diseases: Version 4. *Arthritis & Rhematology*, *74*(5), e21–e36. https://doi.org/10.1002/art.42109.

Desmarais, J., Rosenbaum, J. T., Costenbader, K. H., Ginzler, E. M., Fett, N., Goodman, S., … Werth, V. P. (2021). American College of Rheumatology white paper on antimalarial cardiac toxicity. *Arthritis & Rheumatology*, *73*(12), 2151–2160.

FAIR /SFR/SNFMI/SOFREMIP/CRI/IMIDIATE consortium and contributors. (2021). Severity of COVID-19 and survival in patients with rheumatic and inflammatory diseases: Data from the French RMD COVID-19 cohort of 694 patients. *Annals of the Rheumatic Diseases*, *80*(4), 527–538.

Favalli, E. G., Bugatti, S., Klersy, C., Biggioggero, M., Rossi, S., De Lucia, O., … Montecucco, C. (2020). Impact of corticosteroids and immunosuppressive therapies on symptomatic SARS-CoV-2 infection in a large cohort of patients with chronic inflammatory arthritis. *Arthritis Research & Therapy*, *22*(1), 290. https://doi.org/10.1186/s13075-020-02395-6.

Favalli, E. G., Ingegnoli, F., De Lucia, O., Cincinelli, G., Cimaz, R., & Caporali, R. (2020). COVID-19 infection and rheumatoid arthritis: Faraway, so close! *Autoimmunity Reviews*, *19*(5), 102523.

Fernandez-Gutierrez, B., Leon, L., Madrid, A., Rodriguez-Rodriguez, L., Freites, D., Font, J., … Abasolo, L. (2021). Hospital admissions in inflammatory rheumatic diseases during the peak of COVID-19 pandemic: Incidence and role of disease-modifying agents. *Therapeutic Advances in Musculoskeletal Disease*, *13*. https://doi.org/10.1177/1759720x20962692.

Ferri, C., Giuggioli, D., Raimondo, V., L'Andolina, M., Tavoni, A., Cecchetti, R., … Antonelli, A. (2020). COVID-19 and rheumatic autoimmune systemic diseases: Report of a large Italian patients series. *Clinical Rheumatology*, *39*(11), 3195–3204. https://doi.org/10.1007/s10067-020-05334-7.

Friedman, M. A., Curtis, J. R., & Winthrop, K. L. (2021). Impact of disease-modifying antirheumatic drugs on vaccine immunogenicity in patients with inflammatory rheumatic and musculoskeletal diseases. *Annals of the Rheumatic Diseases*, *80*(10), 1255–1265. https://doi.org/10.1136/annrheumdis-2021-221244.

Furer, V., Eviatar, T., Zisman, D., Peleg, H., Paran, D., Levartovsky, D., … Elkayam, O. (2021). Immunogenicity and safety of the BNT162b2 mRNA COVID-19 vaccine in adult patients with autoimmune inflammatory rheumatic diseases and in the general population: A multicentre study. *Annals of the Rheumatic Diseases*, *80*(10), 1330–1338. https://doi.org/10.1136/annrheumdis-2021-220647.

Furer, V., Rondaan, C., Heijstek, M., Van Assen, S., Bijl, M., Agmon-Levin, N., … Kapetanovic, M. C. (2019). Incidence and prevalence of vaccine preventable infections in adult patients with autoimmune inflammatory rheumatic diseases (AIIRD): A systemic literature review informing the 2019 update of the EULAR recommendations for vaccination in adult patients with AIIRD. *RMD Open*, *5*(2), e001041.

Gianfrancesco, M., Hyrich, K. L., Al-Adely, S., Carmona, L., Danila, M. I., Gossec, L., … Robinson, P. C. (2020). Characteristics associated with hospitalisation for COVID-19 in people with rheumatic disease: Data from the COVID-19 Global Rheumatology Alliance physician-reported registry. *Annals of the Rheumatic Diseases*, *79*(7), 859–866. https://doi.org/10.1136/annrheumdis-2020-217871.

Gupta, L., Misra, D. P., Agarwal, V., Balan, S., & Agarwal, V. (2021). Management of rheumatic diseases in the time of covid-19 pandemic: Perspectives of rheumatology practitioners from India. *Annals of the Rheumatic Diseases*, *80*(1), e1. https://doi.org/10.1136/annrheumdis-2020-217509.

Hermine, O., Mariette, X., Tharaux, P. L., Resche-Rigon, M., Porcher, R., & Ravaud, P. (2021). Effect of tocilizumab vs usual Care in Adults Hospitalized with COVID-19 and moderate or severe pneumonia: A randomized clinical trial. *JAMA Internal Medicine*, *181*(1), 32–40. https://doi.org/10.1001/jamainternmed.2020.6820.

Kahveci, A., & Ataman, Ş. (2021). Management of rheumatologic disease in COVID-19 pandemic. In O. Memikoğlu, & V. Genç (Eds.), *COVİD-19* (pp. 135–141). Ankara: E-publishing, Ankara Üniversitesi Basımevi, ISBN:978-605-136-516-9.

Kahveci, A., Gümüştepe, A., Güven, N., & Ataman, Ş. (2022). The impact of the ongoing COVID-19 pandemic on the management of rheumatic disease: A national clinician-based survey. *Rheumatology International*, *42*(4), 601–608.

Kalil, A. C., Patterson, T. F., Mehta, A. K., Tomashek, K. M., Wolfe, C. R., Ghazaryan, V., … Kline, S. (2021). Baricitinib plus remdesivir for hospitalized adults with Covid-19. *New England Journal of Medicine*, *384*(9), 795–807.

Kasapçopur, Ö. (2020). Scientific researches and academic publishing during the coronavirus pandemic. *Turkish Archives of Pediatrics/Türk Pediatri Arşivi*, *55*(3), 213.

Koonin, L. M., Hoots, B., Tsang, C. A., Leroy, Z., Farris, K., Jolly, B., … Tong, I. (2020). Trends in the use of telehealth during the emergence of the COVID-19 pandemic—United States, January–march 2020. *Morbidity and Mortality Weekly Report*, *69*(43), 1595.

Korolev, M. A., Letyagina, E. A., Sizikov, A. E., Bogoderova, L. A., Ubshaeva, Y. B., Omelchenko, V. O., … Kurochkina, Y. D. (2022). Immuno-inflammatory rheumatic diseases and COVID-19: Analysis of clinical outcomes according to the data of the register of patients of the Novosibirsk region receiving therapy with genetically engineered biological drugs. *Terapevticheskiĭ Arkhiv*, *94*(5), 636–641.

Landewé, R. B., Kroon, F. P., Alunno, A., Najm, A., Bijlsma, J. W., Burmester, G.-R. R., … Curtis, J. R. (2022). EULAR recommendations for the management and vaccination of people with rheumatic and musculoskeletal diseases in the context of SARS-CoV-2: The November 2021 update. *Annals of the Rheumatic Diseases*, *81*(12), 1628–1639.

Landewé, R. B., Machado, P. M., Kroon, F., Bijlsma, H. W., Burmester, G. R., Carmona, L., … Schulze-Koops, H. (2020). EULAR provisional recommendations for the management of rheumatic and musculoskeletal diseases in the context of SARS-CoV-2. *Annals of the Rheumatic Diseases*, *79*(7), 851–858. https://doi.org/10.1136/annrheumdis-2020-217877.

Leipe, J., Hoyer, B., Iking-Konert, C., Schulze-Koops, H., Specker, C., & Krüger, K. (2020). SARS-CoV-2 & rheumatic disease: Consequences of the SARS-CoV-2 pandemic for patients with inflammatory rheumatic diseases. A comparison of the recommendations for action of rheumatological societies and risk assessment of different antirheumatic treatments. *Zeitschrift für Rheumatologie*, *79*, 686–691.

Liao, H. T., Tung, H. Y., Chou, C. T., Tsai, H. C., Yen, Y. N., & Tsai, C. Y. (2022). Immunogenicity of the mRNA-1273 and ChAdOx1 nCoV-19 vaccines in Asian patients with autoimmune rheumatic diseases under biologic and/or conventional immunosuppressant treatments. *Scandinavian Journal of Rheumatology*, *51*(6), 500–505. https://doi.org/10.1080/03009742.2022.2062822.

MacKenna, B., Kennedy, N. A., Mehrkar, A., Rowan, A., Galloway, J., Matthewman, J., … Brown, J. (2022). Risk of severe COVID-19 outcomes associated with immune-mediated inflammatory diseases and immune-modifying therapies: A nationwide cohort study in the OpenSAFELY platform. *The Lancet Rheymatology*.

Matsumoto, R. A., & Barton, J. L. (2021). Telerheumatology: Before, during, and after a global pandemic. *Current Opinion in Rheumatology*, *33*(3), 262–269.

McDougall, J. A., Ferucci, E. D., Glover, J., & Fraenkel, L. (2017). Telerheumatology: A systematic review. *Arthritis Care & Research*, *69*(10), 1546–1557.

Michelena, X., Borrell, H., López-Corbeto, M., López-Lasanta, M., Moreno, E., Pascual-Pastor, M., … Antón, S. (2020). Incidence of COVID-19 in a cohort of adult and paediatric patients with rheumatic diseases treated with targeted biologic and synthetic disease-modifying anti-rheumatic drugs. *Seminars in Arthritis and Rheumatism*, *50*(4), 564–570.

Mikuls, T. R., Johnson, S. R., Fraenkel, L., Arasaratnam, R. J., Baden, L. R., Bermas, B. L., … Saag, K. G. (2021). American College of Rheumatology Guidance for the Management of Rheumatic Disease in adult patients during the COVID-19 pandemic: Version 3. *Arthritis & Rhematology*, *73*(2), e1–e12. https://doi.org/10.1002/art.41596.

Molina, J. M., Delaugerre, C., Le Goff, J., Mela-Lima, B., Ponscarme, D., Goldwirt, L., & de Castro, N. (2020). No evidence of rapid antiviral clearance or clinical benefit with the combination of hydroxychloroquine and azithromycin in patients with severe COVID-19 infection. *Médecine et Maladies Infectieuses*, *50*(4), 384.

Nune, A., Iyengar, K. P., Ahmed, A., Bilgrami, S., & Sapkota, H. R. (2021). Impact of COVID-19 on rheumatology practice in the UK—A pan-regional rheumatology survey. *Clinical Rheumatology*, *40*, 2499–2504.

Pablos, J. L., Abasolo, L., Alvaro-Gracia, J. M., Blanco, F. J., Blanco, R., Castrejón, I., … Gonzalez-Gay, M. A. (2020). Prevalence of hospital PCR-confirmed COVID-19 cases in patients with chronic inflammatory and autoimmune rheumatic diseases. *Annals of the Rheumatic Diseases*, *79*(9), 1170–1173.

Rosas, I. O., Bräu, N., Waters, M., Go, R. C., Hunter, B. D., Bhagani, S., … Malhotra, A. (2021). Tocilizumab in hospitalized patients with severe Covid-19 pneumonia. *The New England Journal of Medicine*, *384*(16), 1503–1516. https://doi.org/10.1056/NEJMoa2028700.

Santos, C. S., Férnandez, X. C., Moriano Morales, C., Álvarez, E. D., Álvarez Castro, C., López Robles, A., & Pérez Sandoval, T. (2021). Biological agents for rheumatic diseases in the outbreak of COVID-19: Friend or foe? *RMD Open*, *7*(1). https://doi.org/10.1136/rmdopen-2020-001439.

Sparks, J. A., Wallace, Z. S., Seet, A. M., Gianfrancesco, M. A., Izadi, Z., Hyrich, K. L., … Mateus, E. F. (2021). Associations of baseline use of biologic or targeted synthetic DMARDs with COVID-19 severity in rheumatoid arthritis: Results from the COVID-19 Global Rheumatology Alliance physician registry. *Annals of the Rheumatic Diseases*, *80*(9), 1137–1146.

Specker, C., Aries, P., Braun, J., Burmester, G., Fischer-Betz, R., Hasseli, R., … Schulze-Koops, H. (2021). Updated recommendations of the German Society for Rheumatology for the care of patients with inflammatory rheumatic diseases in the context of the SARS-CoV-2/COVID-19 pandemic, including recommendations for COVID-19 vaccination. *Zeitschrift für Rheumatologie*, *80*(Suppl 2), 33–48. https://doi.org/10.1007/s00393-021-01055-7.

Strangfeld, A., Schäfer, M., Gianfrancesco, M. A., Lawson-Tovey, S., Liew, J. W., Ljung, L., … Machado, P. M. (2021). Factors associated with COVID-19-related death in people with rheumatic diseases: Results from the COVID-19 Global Rheumatology Alliance physician-reported registry. *Annals of the Rheumatic Diseases*, *80*(7), 930–942. https://doi.org/10.1136/annrheumdis-2020-219498.

Sugihara, K., Wakiya, R., Kameda, T., Shimada, H., Nakashima, S., Kato, M., … Dobashi, H. (2022). Humoral immune response against BNT162b2 mRNA COVID-19 vaccine in patients with rheumatic disease undergoing immunosuppressive therapy: A Japanese monocentric study. *Medicine (Baltimore)*, *101*(42), e31288. https://doi.org/10.1097/md.0000000000031288.

Tam, L. S., Tanaka, Y., Handa, R., Li, Z., Lorenzo, J. P., Louthrenoo, W., … Haq, S. A. (2021). Updated APLAR consensus statements on care for patients with rheumatic diseases during the COVID-19 pandemic. *International Journal of Rheumatic Diseases*, *24*(6), 733–745. https://doi.org/10.1111/1756-185x.14124.

Tang, W., Khalili, L., & Askanase, A. (2021). Telerheumatology: A narrative review. *Rheumatology and Immunology Research*, *2*(3), 139–145.

Tufan, A., Güler, A. A., & Matucci-Cerinic, M. (2020). COVID-19, immune system response, hyperinflammation and repurposingantirheumatic drugs. *Turkish Journal of Medical Sciences*, *50*(9), 620–632.

Varley, C. D., Ku, J. H., & Winthrop, K. L. (2021). COVID-19 pandemic management and the rheumatology patient. *Best Practice & Research. Clinical Rheumatology*, *35*(1), 101663. https://doi.org/10.1016/j.berh.2021.101663.

Wang, M., Cao, R., Zhang, L., Yang, X., Liu, J., Xu, M., … Xiao, G. (2020). Remdesivir and chloroquine effectively inhibit the recently emerged novel coronavirus (2019-nCoV) in vitro. *Cell Research*, *30*(3), 269–271.

Wang, Q., Liu, J., Shao, R., Han, X., Su, C., & Lu, W. (2021). Risk and clinical outcomes of COVID-19 in patients with rheumatic diseases compared with the general population: A systematic review and meta-analysis. *Rheumatology International*, *41*(5), 851–861. https://doi.org/10.1007/s00296-021-04803-9.

Zen, M., Fuzzi, E., Astorri, D., Saccon, F., Padoan, R., Ienna, L., … Gasparotto, M. (2020). SARS-CoV-2 infection in patients with autoimmune rheumatic diseases in Northeast Italy: A cross-sectional study on 916 patients. *Journal of Autoimmunity*, *112*, 102502.

Zhou, F., Yu, T., Du, R., Fan, G., Liu, Y., Liu, Z., … Gu, X. (2020). Clinical course and risk factors for mortality of adult inpatients with COVID-19 in Wuhan, China: A retrospective cohort study. *The Lancet*, *395*(10229), 1054–1062.

Zhu, N., Zhang, D., Wang, W., Li, X., Yang, B., Song, J., … Lu, R. (2020). A novel coronavirus from patients with pneumonia in China, 2019. *New England Journal of Medicine*, *382*(8), 727–733.

Zhu, Y., Zhong, J., & Dong, L. (2021). Epidemiology and clinical management of rheumatic autoimmune diseases in the COVID-19 pandemic: A review. *Frontiers in Medicine*, *8*, 725226. https://doi.org/10.3389/fmed.2021.725226.

Ziadé, N., Hmamouchi, I., El Kibbi, L., Abdulateef, N., Halabi, H., Abutiban, F., … Masri, B. (2020). The impact of COVID-19 pandemic on rheumatology practice: A cross-sectional multinational study. *Clinical Rheumatology*, *39*, 3205–3213.

Chapter 11

Managing migraines during the COVID-19 pandemic: An Italian experience

Licia Grazzi[a], Danilo Antonio Montisano[a], and Paul Rizzoli[b]

[a]*Headache Center, Neuroalgology Department, IRCCS Foundation "Carlo Besta" Neurological Institute, Milan, Italy,* [b]*John Graham Headache Center, Faulkner & Brigham Hospital, Harvard Medical School, Boston, MA, United States*

Abbreviations

ACT	Acceptance commitment therapy
BDI	Beck's Depression Inventory
CM	chronic migraine
GSE	general self-efficacy scale
HFEM	high-frequency episodic migraine
ICHD3	International Classification of Headache Disorders 3rd Edition
ICU	intensive care unit
IHS	International Headache Society
IRCCS	Scientific Institute for Research, Hospitalization, and Healthcare
MOH	Medication overuse headache
PCS	Pain catastrophizing scale

Introduction

The COVID-19 emergency began in Italy on March 8, 2020, with a government issued "stay-at-home" order. Most Italian's lives changed significantly that day. Not only did the rhythm of school and work change, but the routine at our hospitals was also dramatically altered to manage the many COVID-19 patients requiring care, including those needing the intensive care unit (ICU). Many routine inpatient units were converted to emergency care units for COVID-19 patients. The changes were rapid and disruptive to all patients, even those not directly infected by the virus. The Lombardia region was the center of the SARS-CoV-2 epidemic in Italy and was hit the hardest.

The *IRCCS Neurological Institute "C. Besta" Foundation* is a neurological research institute located in Milan in the Lombardia Region that provides care to both adults and children with neurological and neurosurgical conditions from all regions of Italy (Pareyson et al., 2021). At the time of the lockdown, the Lombardia Health System suspended all nonurgent medical visits to concentrate efforts on fighting COVID-19 to avoid spreading the contagion to patients, caregivers, and health-system personnel. At the hospital, about 2000 planned neurological visits and consultations were canceled between March and May 2020. Also, many of the scheduled headache center therapies routinely performed at the hospital were suspended, and that impacted patient care. Our headache center treats patients from all regions of Italy, and we were tasked with rapidly finding solutions for how to continue to follow and care for our patients. The management of migraines, among other chronic diseases, that was already challenging under normal circumstances became much more difficult during the pandemic (Bobker & Robbins, 2020; Majersik & Reddy, 2020; Pareyson et al., 2021; Sellner et al., 2020).

A migraine is a complex and chronic, poorly understood neurological condition underdiagnosed and undertreated, in particular when the episodic form has evolved to a chronic form (Bigal et al., 2008; Headache Classification Committee of the International Headache Society (IHS), 2013). For multiple reasons, the pandemic caused an outsized impact on migraine patients and many patients did not do well. We were forced to quickly reassess these needs and form a plan to provide and to ensure access to optimal care for all our patients during this unusual and difficult experience (Bobker & Robbins, 2020; Chabra et al., 2022).

Management, Body Systems, and Case Studies in COVID-19. https://doi.org/10.1016/B978-0-443-18703-2.00021-9
Copyright © 2024 Elsevier Inc. All rights are reserved, including those for text and data mining, AI training, and similar technologies.

Migraine management became much more complex during the pandemic, mainly due to the widespread limitations on mobility and accessibility to a health system structure. As a result, providers implemented telemedicine and other web-based modalities for maintaining patient care, with the development of new, rigorous, and effective systems to remotely evaluate patients and to deliver therapeutic modalities, including psychotherapy and other behavioral therapy. The recent literature has demonstrated the utility and suitability of telemedicine services, not only for headache patients but also for those with other forms of pain (Adams et al., 2018; Alexander & Joshi, 2016; Friedmann et al., 2019; Muller et al., 2017; Rosser & Eccleston, 2011). Based on these encouraging results, telemedicine services were adopted during COVID-19 in different countries and medical centers (Chiang et al., 2021; Muller et al., 2017).

In recent decades at our center, we developed various behavioral treatment programs to educate and support both adolescents and adult patients suffering from migraines. Besides outpatient management, we also developed an in-hospital medication overuse withdrawal program for those with the diagnosis of chronic migraines and medication overuse headaches (CM-MOH). This program has been often combined with behavioral treatments, in particular mindfulness (Grazzi et al., 2017).

The behavioral programs were traditionally conducted at the hospital with weekly face-to-face sessions in small groups of adults and young patients (Andrasik et al., 2016; Grazzi et al., 2018; Raggi et al., 2018). The withdrawal program usually consisted of a 5–8 day program of daily hospital treatments. Of course, the pandemic forced us to abruptly change regular clinical practice, as patients could not attend the hospital. Therefore, we proposed a pilot study of the application of a home-based program for the withdrawal treatment procedure. We also designed a specific program to deliver behavioral treatment for young patients administered through the use of secure video smartphone or home computer applications that allowed for continuity of care (Grazzi & Rizzoli, 2020). Ultimately, though we learned as we went along and as forced by necessity, we were able to compile and report the results of these approaches.

Many of the services that we implemented during that time have now been shown to be effective for patient management, especially in migraine patients. This success could suggest that policymakers, decision makers, members of interdisciplinary migraine teams, and patient advocates could learn from this experience and further create innovative programs to improve the care of the migraine patient. As physicians caring for migraine patients, we learned a lot from this dramatic experience and, as a result, we hope to add some efficacious and innovative options for taking care of patients with headaches that can be applied in more normal times and for more routine clinical practice (Grazzi, Montisano, Raggi, & Rizzoli, 2022; Grazzi, Telesca, & Rizzoli, 2022; Muller et al., 2017).

Procedures

Telemedicine appears to be a safe and effective method to take care of patients, although it was not largely used in clinical practice until the COVID-19-pandemic.

Telemedicine allowed us to deliver care while preventing infections. The COVID-19 pandemic was a shocking experience for all of us and produced multiple challenges to routine medical practice; nevertheless, we learned a lot about the application of telemedicine and other technological modalities during these circumstances and this resource will be very helpful in the future too.

We developed programs to continue taking care of patients, not only for diagnostic assessment but also for therapy procedures (Table 1). In particular, the management of migraines and other painful conditions, problematic in a nonpandemic period, was crucial and behavioral treatments delivered by web platforms were an indispensable option.

The COVID-19 pandemic was a shocking experience for all of us and produced multiple challenges to routine medical practice; nevertheless, we learned a lot about the possibility of application of technology during those circumstances and this resource will be very helpful and they will support our clinical activities.

In our experience, the programs for delivering behavioral therapies with weekly sessions by web platforms seemed to be well accepted by patients. They were also useful, effective, and sustainable during the pandemic and should prove a useful adjunct in routine clinical practice to manage adults with chronic migraines–medication overuse headache (CM-MOH) and adolescents with high-frequency episodic migraines (HFEM) and CM. Although this analysis was conducted on a portion of the sample and at interim time points, it should be noted that none of the patients we followed were lost to follow-up during the observation period.

The chapter will review the different clinical experiences and modalities applied during the lockdown to continue the clinical activities at our headache center to guarantee patient follow-up.

TABLE 1 Programs developed during COVID-19 emergency.

Program	Patients Involved	Methods	Endpoint	Outcome
Be Home	25 patients involved diagnosed with CM-MO	Detox (steroids and benzodiazepine) 5 days Education on pain management Weekly phone calls Six weekly 1-h video mindfulness sessions (web platform) Mindfulness daily by a standard 12-minsession (smartphone) Follow-up visits at 3, 6, and 12 months	MO at 6 months Decrease of 50% in migraine days/month and medications/month Decrease of 50% in headache days/month	Compliance with pharmacological therapies was better after the program None of the patients recorded an MO condition at 6-month follow-up
Be Home Kids	25 adolescents involved (12–17 years old) diagnosed with CM and HFEM	Mindfulness daily by a standard 12-min session (smartphone) Six weekly 1-h video sessions (web platform) Follow-up every 3 months up to 1 year	Headache days/month Medications intake/month	Clinical improvement and good adherence to the program
ChroMigSmart	20 patients involved diagnosed with migraines	Daily sessions of 12 min and one weekly 1-h session of mindfulness Weekly video call to evaluate the clinical condition	Headache days/month Medications intake/month	Significant decrease in terms of clinical indexes, days of headache per month, and medication intake per month

Issues that needed to be addressed during the pandemic included:

1. How to handle patients on the waiting list for clinical evaluation.
2. How to manage patients who were waiting to start our behavioral program that was generally conducted in small face-to-face weekly group sessions.
3. How to handle patients who were waiting for a withdrawal program for chronic migraines with medication overuse.
4. How to handle the medication management of pediatric and adolescent patients.

How to handle patients on the waiting list for clinical evaluation

Due to the pandemic and restrictions, many patients who needed evaluation at our headache center could not be accommodated as the outpatient service was interrupted. To address this, we designed a telemedicine service at the *Besta Institute* that was approved by our hospital directors. This service included prescriptions, booking, consenting, privacy, and data protection. Also, a secure connection with patients was assured by using Microsoft Teams (Pareyson et al., 2021). There was excellent acceptance of this program by our patients. In fact, this modality allowed them to be screened and evaluated, assigned to appropriate therapy, and scheduled for follow-up without leaving their house. Results from a questionnaire revealed good satisfaction from patients or from their relatives; they confirmed that the service met their expectations and needs, and the technical problems were considered minor. Many patients expressed willingness to use this system in the future, and not exclusively related to the pandemic. Patients with cluster headaches who had a cluster onset and thus had even more acute needs also found the system helpful and effective. Moreover, telehealth was very well accepted by doctors and other healthcare personnel who participated (Majersik & Reddy, 2020; Pareyson et al., 2021).

Though telemedicine was born of necessity, it became apparent that there could be other benefits to the health system with this model, as patients living far from the hospital could still keep appointments without the need for long-distance travel or missing as much work. Also, these methods could help reduce health system costs.

It is possible that the COVID-19 pandemic helped create innovative methods for improved healthcare (Pareyson et al., 2021).

How to manage patients who were waiting to start our behavioral program that was generally conducted in small face-to-face weekly group sessions: *The CHROMIGSMART program*

One of our priorities was to find a way to continue behavioral treatments for our current patients. To accomplish this, we developed a specific "emergency" protocol through the use of smartphone and instant message applications (Grazzi, Montisano, et al., 2022; Grazzi, Telesca, et al., 2022). A group of patients with CM-MOH who regularly attended the in-hospital withdrawal program were on the waiting list for mindfulness sessions at the institute at the onset of the pandemic. In lieu of in-person sessions, patients were instructed to use their smartphone to follow the mindfulness program. Regular daily sessions of 12 min and one weekly 1-h session with the mindfulness trainer were implemented. The adherence of patients was high and patients attending these sessions regularly noted significant improvement over the following year (Grazzi, Montisano, et al., 2022; Grazzi, Telesca, et al., 2022).

Details of the program

Chronic migraines, a disabling condition that affects 2% of the migraine population (Bigal et al., 2008), is frequently associated with medication overuse (Chiang et al., 2016; Diener & Limmroth, 2004; Evers & Marziniak, 2010), which makes this condition more difficult to treat. Many prior investigations have confirmed the efficacy and importance of having patients withdraw from all medications contributing to this condition as well as the need to carefully follow patients after withdrawal to avoid relapses, to minimize the chances of resuming overuse, and to maintain the clinical benefits of whatever concurrent therapeutic approach was provided (Rossi et al., 2006). Also, clinical results are enhanced when traditional therapies are combined with behavioral approaches, in particular mindfulness, which helps patients become more conscious about their symptoms and learn techniques for managing pain without medications (Raggi et al., 2017, 2018). At the Besta Institute, we have been following a cohort of patients who have completed supervised medication withdrawal, followed by a specific prophylaxis for migraine but also a specially designed program for mindfulness practice. This program had been administered for six weekly 45-min face-to-face sessions in small groups.

Due to the lockdown, we investigated the feasibility and effectiveness of mindfulness delivered by smartphones combined with video calls for the clinical evaluation and follow-up of patients with CM-MOH after a withdrawal program and to confirm the effectiveness of this program after 1 year.

A group of patients recruited and treated in our neurology unit completed our standard medication withdrawal program. They were then provided the additional training needed for mindfulness on their smartphone, and practiced each day for 12 min. All remote sessions were recorded by the expert who generally manages the face-to-face sessions at the hospital. A separate weekly video call was made to evaluate the clinical condition and to encourage and reinforce the use of the pain management strategies being provided. Patients had to record their headache episodes on the daily headache diary, and they were followed with regular meetings every 3 months up to 1 year. Patients completed the treatment protocol to date, reporting a mean reduction of days of migraines per month of 50%, with a concurrent mean reduction of medication intake per month of the same amount. At the end of the program 1 year later, patients reported a diagnosis of migraines without aura at high frequency, and they did not report any relapse in medication overuse. Our preliminary findings allowed patients to follow therapeutic program remotely and to obtain significant clinical improvement (Grazzi & Rizzoli, 2020; Grazzi, Montisano, et al., 2022; Grazzi, Telesca, et al., 2022). Adherence to the treatment was judged as high based on recorded but nonsystematically elicited responses.

Our findings confirmed that this type of combined treatment, an in-clinic visit followed by smartphone delivery outside the clinic, can yield results such as those we obtained in previous studies relying on in-clinic treatment alone. This approach warrants further, more controlled investigations to document and replicate the clinical effectiveness, the savings in time by patients and staff, and the need for resources. The recent literature (Adams et al., 2018; Alexander & Joshi, 2016) promotes the use of smartphones or telemedicine for clinical and therapeutic applications with encouraging results. The modality we adopted with these programs conducted remotely allowed patients to be treated adequately.

How to handle patients who were waiting for a withdrawal program for CM-MOH: *The BeHome program*

Another experience concerned the management of patients with CM-MOH during the pandemic who could not attend the usual in-hospital withdrawal treatment due to mobility restrictions. For these patients, we developed the BeHome program, a home-based withdrawal program that also included a mindfulness-based treatment delivered through the web (the program was approved by the institute's ethical committee in May 2020). We assessed its feasibility and long-term effectiveness with a specific project and treatment protocol, designed during the COVID-19 pandemic. A group of patients with MOH were enrolled, and they went through an at-home withdrawal program that consisted of the oral administration of steroids and benzodiazepine for 5 days and education on pain management. Weekly phone calls were made to check the clinical course during the detoxification program. Six weekly 1-h video mindfulness sessions were delivered with a dedicated web-based platform and daily home practice was encouraged by 12-min specific mindfulness sessions provided to patients for use on a computer or smartphone. Follow-up visits were scheduled at 3, 6, and 12 months after withdrawal.

Details of the program

MOH is a disabling pain syndrome with a prevalence of 2% in the general population, and which affects patients suffering from primary headaches, mainly those with CM. The literature in recent decades confirmed the need to stop overuse through withdrawal or detoxification programs as an essential step in the treatment strategy. Remission rates range from 60% to 83% in studies with a 1-year follow-up (Diener & Limmroth, 2004; Evers & Marziniak, 2010).

Data from the literature showed that patients need to be carefully followed after withdrawal from medications overuse so that to avoid relapses in overuse and to enforce the clinical improvement. The migraine prophylaxis is mandatory after withdrawal to prevent migraine episodes (Diener & Limmroth, 2004). Furthermore, different clinical experiences confirmed the effectiveness of interventions based on outpatient withdrawal programs as well as education and support by standardized behavioral approaches. The feasibility of these treatments is not regularly reported in published studies, but it was estimated that a proportion of patients discontinue the treatment protocols, with a dropout rate around 15–20% at 1 year (Andrasik et al., 2016; Raggi et al., 2018).

According to our protocols, the withdrawal program is performed at our hospital, either in an inpatient setting or in a day hospital setting. This program includes abrupt interruption of overused drugs, intravenous supporting therapy, educational support, rescue treatments for severe headaches, and prescription of a specific pharmacological prophylaxis. Furthermore, we added a standardized behavioral program based on mindfulness practice (Grazzi et al., 2018), as this approach may be particularly useful in helping patients obtain a better outcome and avoid a relapse of overuse.

The aim of this pilot study was: to assess the feasibility and the long-term effectiveness of a specific protocol, designed during the COVID-19 pandemic, combining pharmacological treatment to behavioral approach like mindfulness delivered remotely. A group of patients with an MO diagnosis according to International Headache Society criteria was enrolled in the study. They performed the withdrawal program at home with the oral administration of steroids and benzodiazepine for 5 days, including education to manage pain. Weekly phone calls were done to check the clinical course during the detoxification program and six weekly 1-h video mindfulness sessions were organized, with home practice encouraged by 12-min mindfulness sessions on a smartphone. Follow-up visits were scheduled at 3, 6, and 12 months after withdrawal. Percentages of patients with an absence of MO at 6 months from withdrawal, assessed by the daily diary card, and percentages of patients with a decrease of at least 50% in the number of migraine days/month and medications/month were considered.

Other secondary outcomes were evaluated, such as the percentages of patients obtaining a decrease of at least 50% in the number of monthly headache days as well as changes in headache-related impact, decrease in quality-of-life scores, decrease in catastrophizing attitude, decrease in depression symptoms, and change in self-efficacy perception.

Patients enrolled achieved the 6-month follow-up and the results showed a significant decrease in medications/month (18 ± 8.3 at baseline vs. 6 ± 3.8 at 6 months), and headache days/month (15 ± 6.4 at baseline vs. 8 ± 4.1 at 6 months). None of the patients recorded an MO condition at the 6-month follow-up. Weekly video calls were scheduled to assess the frequency and symptomatic drug intake as well as to discuss strategies to manage pain and stressful situations that can induce pain episodes and to reinforce the mindfulness practice. Moreover, mindfulness sessions were administered by a therapist with expertise in this approach, combined with acceptance commitment therapy (ACT) elements. Sessions covered different topic, including: (1) creative helplessness: the problem of control; (2) introduction to mindfulness; (3) actions guided by values: working with thoughts; (4) working with acceptance and willingness; (5) committed actions: self as context; (6) integration: working with obstacles, and (7) wrap-up. In particular, mindfulness practice was emphasized and encouraged. Each session lasted approximately 75 min, with 15–20 min booster sessions, conducted at 2 and 4 weeks after completion of the six-session treatment. The sessions began with a mindfulness exercise, followed by a review of material previously covered. Patients were provided with written material during each session that addressed the topic covered in the specific session (Grazzi & Rizzoli, 2020).

Patients were instructed to practice mindfulness at home each day for 12 min; sessions were recorded by the same therapist who conducted the sessions online. We did not formally assess whether such home practice was conducted on a regular basis.

Clinical results are significant. The Be Home program seemed to be effective and sustainable during the COVID-19 pandemic, also at a medium-term follow-up, but also applicable in regular times for patients with migraine.

Another application of the online programs concerned patients with atypical facial pain and neuropathic pain evaluated at our neurology department. For them a specific program was organized and the Be Home program was tried with these patients during COVID-19.

Patients with atypical facial pain are disabled and often report an extremely low quality of life for their pain and difficulties to manage pain. In fact, pharmacological therapies are often inadequate, ineffective, and have several adverse effects for patients who do not completely adhere to the therapeutic program. Recently, studies in the literature have shown how educational support, rescue treatments for severe pain, and the use of behavioral approaches can be helpful for this category of patients when combined with specific pharmacological prophylaxis. Mindfulness in this case too may be considered an effective approach for helping patients manage their pain. It can help them be more conscious about their condition and use the medications for pain in the correct way.

Usually, the behavioral approach with mindfulness is delivered at the institute with face-to-face sessions featuring small groups of patients every week. Due to the COVID-19 emergency, this category of patients was involved. The aim of this study was to assess the feasibility and effectiveness on relevant outcomes of a specific protocol designed during the pandemic that has a specific behavioral program designed for patients with chronic neuropathic pain, in particular with atypical facial pain (PIFP-IHS), by the application of a web modality to administrate the mindfulness sessions.

Patients with a diagnosis of atypical facial pain were enrolled after a clinical evaluation. Pharmacological treatment was assigned to patients as well as a behavioral approach based on mindfulness. Mindfulness sessions were done online by a specific platform to enforce the ability of patients to cope with pain and to manage their pain by reducing the use of symptomatic medications. Specific instructions were given and all patients were involved in mindfulness practice through the web platform with 1-h weekly sessions in small groups. Also, daily standardized mindfulness sessions of 12 min were proposed. The sessions were recorded by the expert who guides their sessions at the hospital; face-to-face visits at follow-up every 3 months were scheduled up to 1 year out.

The outcomes evaluated were a decrease in pain intensity and medication intake; a decrease in catastrophizing attitude (revealed by the Pain Catastrophizing Scale), a decrease in depression symptoms (recorded by the Beck Depression Inventory), and a change in self-efficacy perception tested with standardized scale (the Global Self Efficacy questionnaire). Twenty-five patients were enrolled. We reported results at the 3-month follow-up. Although clinical indexes in terms of pain intensity and medication intake for pain did not change significantly, the questionnaire results demonstrated a slight decrease in catastrophizing attitude and depression level: BDI (17 ± 9 pretreatment vs. $12+8.1$ R 3 months), PCS (32 ± 10.8 pretreatment vs. 23 ± 10.5 R 3 months). Global self-efficacy did not show any change. Patients' adherence to the program was high; they did not miss any sessions and they practiced regularly at home. Compliance with the pharmacological therapies was better after the program and, although not specifically evaluated, patients improved their ability to cope with pain and to tolerate pharmacological approaches. Although we did not observe any clinical change up to now, the results confirm that the Be Home program with the use of a technological support protocol was effective and feasible, and allowed more patients to access this kind of behavioral approach.

How to handle the management of pediatric and adolescent patients: The BeHome kids program

A more challenging step was the extension of the Be Home program to adolescents, who are generally not exposed to a structured withdrawal. A total of 20 adolescents with migraines without aura HFEM and CM were on the waiting list to start the mindfulness sessions, so they were managed by a specific web-based program. They received six sessions of a mindfulness-based treatment and were followed up for 6 months as part of a larger study. Repeated measure analyses comprised 12 patients who completed the 6-month follow-up and who showed a significant improvement for headache frequency, symptoms of depression, and catastrophizing (Grazzi, Montisano, et al., 2022; Grazzi, Telesca, et al., 2022).

Details of the program

CM and HFEM are disabling conditions that affect 2% of the migraine population in adolescence. Common pharmacological treatments are often inadequate (Grazzi, Montisano, et al., 2022; Grazzi, Telesca, et al., 2022; Lopez-Bravo et al., 2020). It has been reported that clinical results can be improved when traditional therapies are combined with behavioral approaches; in particular, mindfulness seems to be effective at helping patients become more conscious about their symptoms and be able to manage pain without medication. Recent reviews have demonstrated that traditional preventive pharmacological therapies are not effective in young migraine patients. However, behavioral approaches have been successfully proposed and applied. According to our standard clinical practice, young migraine patients (12–17 years old) treated by mindfulness practice come to the hospital to practice mindfulness in small groups of patients for 6 weekly 45-min sessions (Grazzi et al., 2021). Due to the pandemic in Italy in spring 2020, young patients missed coming to the hospital for regular practice. Because of this, we developed a specific program consisting of sessions delivered by a web platform to allow our patients to follow the sessions so that they could continue to be monitored during their therapeutic process. This preliminary pilot study was conducted on 25 patients; now, 20 patients have been included age 12–17, and 12 of them achieved the 6-month follow-up. Patients were trained to practice mindfulness daily with a standard 12-min session on their smartphone. The session was recorded by the expert who manages the sessions at the hospital. Also, a weekly 1-h video session was done for 6 consecutive weeks through a specific web-platform to evaluate the clinical condition, to practice guided mindfulness sessions, and to encourage strategies for pain management. Follow-up sessions were planned every 3 months up to 1 year after treatment; these meetings at the hospital are face to face with every patient to check the clinical condition through the patient's diary. Clinical indexes in terms of days of headache per month and medication intake per month decreased significantly at 6 months (18.1 ± 2.7 vs. 7.7 ± 1.8; 6.2 ± 2.4 vs. 3.3 ± 1.4, respectively). This novel approach is suitable for young patients with migraines, as good clinical improvement and adherence were confirmed, although the results are very preliminary (Grazzi, Montisano, et al., 2022; Grazzi, Telesca, et al., 2022). This project has important implications for the future if we want to improve our opportunities to better treat young headache sufferers.

Conclusions

Given the need for social distancing during the pandemic, many of our usual methods for following and treating patients were converted to telemedicine.

Prepandemic randomized controlled trials showed the comparable efficacy of telemedicine compared with traditional face-to-face visits for headache patients. Moreover, a patient survey from the American Headache Foundation found that telemedicine was helpful in the care of patients suffering from migraines.

A Spanish survey (Lopez-Bravo et al., 2020) found that 86% of headache neurologists planned to continue and increase the use of telemedicine after the pandemic. This report reinforces the concept that telemedicine can be a simple and cost-effective choice of care for neurologists and their patients. It is important to note that because in-person therapies such as nerve blocks and botulinum toxin injections were delayed during the pandemic, patients reported worsening of their migraine attacks. Kristoffersen et al. (2020) reported that during the pandemic, only 36% of neurologists continued to provide botulinum therapy. Consequently, Ali (2020) stressed the importance of flexibility and creativity by headache neurologists, for example by finding alternative appropriate solutions such as noninvasive web-based neuromodulation techniques or web-based behavioral therapies. Not only was routine neurological care difficult during the pandemic, but there was a need for care related to new headache symptoms generated by COVID-19 infections.

Telemedicine appears to be a safe and effective method to deliver care while preventing infections, except in only very selected cases. The COVID-19 pandemic was a shocking experience for all of us and produced multiple challenges to routine medical practice; nevertheless, we learned a lot about the application of telemedicine during these circumstances and this resource will be very helpful in the future too.

Further, a British study of pain patients showed that digital sessions of psychological treatment, including cognitive behavioral therapy and ACT, were well accepted by most patients (Willcocks et al., 2023). Problems involved equity in the delivery of these services in that some populations were deprived of participation in virtual sessions for various reasons. Nevertheless, overall the report was positive, and the research should be confirmed and expanded to determine which populations can best benefit.

In our experience, the programs we developed seemed to be useful, effective, and sustainable during the pandemic and should prove a useful adjunct in routine clinical practice for the management of adults with CM-MOH and adolescents with HFEM and CM. Although this analysis was conducted on a portion of the sample and at interim time points, it also should be noted that none of the patients we followed were lost to follow-up during the observation period.

Limitations to the experiences herein reported include a small sample size, short-term follow-up, and the absence of a control group. Despite this, our pilot studies showed the feasibility and effectiveness of a web-based delivery of nonpharmacological therapies in both adults and adolescents. It is hoped that these small observations will initiate a larger application of these modalities in regular practice. These approaches can be expanded not only to patients with migraines but also patients with other forms of chronic pain.

The COVID-19 pandemic was an extraordinary opportunity for clinicians in this field to find new modalities to deliver our routine care, modalities that we would probably not have considered in the absence of the pandemic. Policymakers, members of interdisciplinary migraine teams, and patient advocates should learn from this lesson, and put together their expertise and knowledge to improve care of patients with migraines.

Applications to other areas

In March 2020, the hospital services were obliged to stop face-to-face consultations for patients due to lockdowns imposed to limit the spread of the COVID-19 pandemic.

Patients who did not suffer from COVID-19 were without any kind of support during the emergency.

The introduction of remote service consultations was a necessary measure to continue to care for patients with different kinds of clinical conditions.

Patients with chronic pain and headaches needed to be followed during the course of their therapy.

Telemedicine appeared to be a safe and effective method to deliver care while preventing infections. The pandemic was a shocking experience for all of us and produced multiple challenges to routine medical practice; nevertheless, we learned a lot about the application of telemedicine and this resource could be very helpful in the future too.

Policymakers, members of interdisciplinary migraine teams, and patient advocates should learn from this lesson, and put together their expertise and their knowledge to improve care of patients with migraine.

Remote digitalized interventions were acceptable to most patients, but attention should be paid to access and improving social aspects of delivery.

Telemedicine is an important social and cultural revolution that can make it easier to care for patients even in difficult situations when they are not able to come to the hospital.

We are perfectly aware that telemedicine cannot be a regular substitution for traditional health services, but it became an indispensable service that improved significantly the quality of the health system by delivering our expertise and support to an enormous population of patients.

Mini-dictionary of terms

- **Mindfulness.** A mental state achieved by focusing one's awareness on the present moment, without judgment, while calmly acknowledging and accepting one's feelings, thoughts, and bodily sensations, used in different clinical contests.
- **Behavioral treatment:** Ways to modify thoughts, feelings, and behaviors or behavioral triggers around medication taking for headaches and other clinical conditions; it is also the management of the physiological effects of stress and the prodromal symptoms of headache attacks through techniques.
- **Telemedicine web-based programs:** Programs for the evaluation and treatment of patients developed thanks to technology by specific web platforms.
- **Migraine:** Clinical condition determined by head pain, nausea, vomiting, photo phonophobia, and other symptoms that induce significant disability; the definition is according to the International Headache Classification Criteria 2018.
- **Pain:** Significant disability with physical and emotional components that can be treated by pharmacological therapies but that needs psychological support delivered by behavioral approaches such as mindfulness.

Summary points

COVID-19 was a life-threatening plague that affected the world and changed our clinical activity dramatically.

A migraine is usually a complex and challenging neurological condition. We developed various behavioral treatment programs to educate and support both adolescents and adults suffering from these conditions, usually provided face to- face.

During the pandemic, mainly due to the widespread limitations on mobility and less accessibility to health system structure, we were forced to quickly reassess clinical activity to provide and ensure access to optimal care for all our patients.

Telemedicine and other web-based modalities for maintaining patient care that were uncommon in past decades were implemented.

Different specific programs were developed *(Telemedicine, Be Home program, ChroMigSmart)* to manage patients on-demand from their home and to administer behavioral treatments through the web.

All the web programs were huge successes with clinicians and patients; they seemed to be useful, effective, and sustainable during the pandemic.

The web platforms offer a useful adjunct in routine clinical practice for the management of migraines and painful conditions even in a nonpandemic period.

References

Adams, C., Deeker-Van Weering, M. G. H., Van Etten-Jamaludin, F. S., & Stuiver, M. (2018). The effectiveness of exercise-based telemedicine on pain, physical activity and quality of life in the treatment of chronic pain: A systematic review. *Journal of Telemedicine and Telecare*, *24*(8), 511–526.

Alexander, J., & Joshi, G. P. (2016). Smartphone applications for chronic pain management: A critical appraisal. *Journal of Pain Research*, *9*, 731–7345.

Ali, A. (2020). Delay in OnabotulinumtoxinA treatment during the COVID-19 pandemic—Perspectives from a virus hotspot. *Headache*, *60*, 1183–1186.

Andrasik, F., Grazzi, L., D'Amico, D., Sansone, E., Leonardi, M., Raggi, A., & Salgado-García, F. (2016). Mindfulness and headache: A "new" old treatment, with new findings. *Cephalalgia*, *36*(12), 1192–1205.

Bigal, M. E., Serrano, D., Buse, D., Scher, A., Stewart, W. F., & Lipton, R. B. (2008). Acute migraine medications and evolution from episodic to chronic migraine: A longitudinal population-based study. *Headache*, *48*(8), 1157–1168. https://doi.org/10.1111/j.1526-4610.2008.01217.x. PMID:18808500.

Bobker, S. M., & Robbins, M. S. (2020). COVID-19 and headache: A primer for trainees. *Headache*, *60*, 1806–1811.

Chabra, N., Grill, M. F., & Halker Singh, R. B. (2022). Post COVID headache: A literature review. *Current Pain and Headache Reports*. https://doi.org/10.1007/s11916-022-01086-y.

Chiang, C. C., Halker Singh, R., Lalvani, N., Shubin Stein, K., Henscheid, L. D., Lay, C., Dodick, D. W., & Newman, L. C. (2021). Patient experience of telemedicine for headache care during the COVID-19 pandemic: An American migraine foundation survey study. *Headache*, *61*, 734–739.

Chiang, C. C., Schwedt, T. J., Wang, S. J., & Dodick, D. W. (2016). Treatment of medication-overuse headache: A systematic review. *Cephalalgia*, *36*, 371–386.

Diener, H. C., & Limmroth, V. (2004). Medication overuse headache: A worldwide problem. *Lancet Neurology*, *3*, 475–483.

Evers, S., & Marziniak, M. (2010). Clinical features, pathophysiology, and treatment of medication-overuse headache. *Lancet Neurology*, *9*, 391–401.

Friedmann, D. I., Rajan, B., & Seidmann, A. (2019). A randomized trial of telemedicine for migraine management. *Cephalalgia*, *39*, 1577–1585.

Grazzi, L., Grignani, E., Raggi, A., Rizzoli, P., & Guastafierro, E. (2021). Effect of a mindfulness-based intervention for chronic migraine and high frequency episodic migraine in adolescents: A pilot single-arm open-label study. *International Journal of Environmental Research and Public Health*, *18*(22), 11739. https://doi.org/10.3390/ijerph182211739. PMID: 34831494; PMCID: PMC8619568.

Grazzi, L., Montisano, D. A., Raggi, A., & Rizzoli, P. (2022). Feasibility and effect of mindfulness approach by web for chronic migraine and high-frequency episodic migraine without aura at in adolescents during and after COVID emergency: Preliminary findings. *Neurological Science*, *43*, 5741–5744. https://doi.org/10.1007/s10072-022-06225-2.

Grazzi, L., Raggi, A., D'Amico, D., Sansone, E., Leonardi, M., Andrasik, F., Gucciardi, A., & D'Andrea, G. (2018). A prospective pilot study of the effect on catecholamines of mindfulness training vs pharmacological prophylaxis in patients with chronic migraine and medication overuse headache. *Cephalalgia*, *13*. https://doi.org/10.1177/0333102418801584.

Grazzi, L., & Rizzoli, P. (2020). The adaptation of Management of Chronic Migraine Patients with Medication Overuse to the suspension of treatment protocols during the COVID-19 pandemic: Lessons from a tertiary headache Center in Milan, Italy. *Headache*, *60*, 1463–1464. https://doi.org/10.1111/head.13825.

Grazzi, L., Sansone, E., Raggi, A., D'Amico, D., De Giorgio, A., Leonardi, M., De Torres, L., Salgado-García, F., & Andrasik, F. (2017). Mindfulness and pharmacological prophylaxis after withdrawal from medication overuse in patients with chronic migraine: An effectiveness trial with a one-year follow-up. *The Journal of Headache and Pain*, *18*(1), 15.

Grazzi, L., Telesca, A., & Rizzoli, P. (2022). Management of chronic migraine with medication overuse by web-based behavioral program during the COVID-19 emergency: Results at 12 months. *Neurological Science*, *43*, 1583–1585. https://doi.org/10.1007/s10072-021-05836-5.

Headache Classification Committee of the International Headache Society (IHS). (2013). The international classification of headache disorder, 3rd edition (beta version). *Cephalalgia*, *33*(9), 629–808.

Kristoffersen, E. S., Faiz, K. W., Sandset, E. C., Storstein, M., et al. (2020). Hospital-based headache care during the Covid-19 pandemic in Denmark and Norway. *Journal of Headache and Pain*, *21*, 128.

Lopez-Bravo, A., Garcia-Azorin, D., Belvis, R., Gonzales-Oria, C., Latorre, G., Santos-lasosa, S., Guerrero-Peral, Á. L., et al. (2020). Impact of the COVID-19 pandemic on headache managemenet in Spain: An analysis of the current situation and future perspectives. *Neurología*, *35*, 372–380.

Majersik, J. J., & Reddy, V. K. (2020). Acute neurology during the COVID-19 pandemic: Supporting the front line. *Neurology, 94*, 1055–1057.

Muller, K. I., Alstadhaug, K. B., & Si, B. (2017). A randomized trial of telemedicine efficacy and safety for non-acute headaches. *Neurology, 89*, 153–162 (*Muller et all 2017*).

Pareyson, D., Pantaleoni, C., Eleopra, R., De Filippis, G., Moroni, I., Freri, E., Zibordi, F., Bulgheroni, S., Pagliano, E., Sarti, D., Silvani, A., Grazzi, L., Tiraboschi, P., Didato, G., Anghileri, E., Bersano, A., Valentini, L., Piacentini, S., Muscio, C., … Force, B.-T. T. (2021). Neuro-telehealth for fragile patients in a tertiary referral neurological institute during the COVID-19 pandemic in Milan, Lombardy. *Neurological Science, 42*(7), 2637–2644. https://doi.org/10.1007/s10072-021-05252-9. PMID:33929645. Epub 2021 Apr 30. PMCID: PMC8086222.

Raggi, A., Giovannetti, A. M., Leonardi, M., Sansone, E., Schiavolin, S., Curone, M., Grazzi, L., Usai, S., & D'Amico, D. (2017). Predictors of 12-months relapse after withdrawal treatment in hospitalized patients with chronic migraine associated with medication overuse: A longitudinal observational study. *Headache, 57*(1), 60–70.

Raggi, A., Grignani, E., Leonardi, M., Andrasik, F., Sansone, E., Grazzi, L., & D'Amico, D. (2018). Behavioral approaches for primary headaches: Recent advances. *Headache, 58*(6), 913–925.

Rosser, B., & Eccleston, C. (2011). Smartphone applications for pain management. *Journal of Telemedicine and Telecare, 17*, 308–312.

Rossi, P., Di Lorenzo, C., Faroni, J., Cesarino, F., & Nappi, G. (2006). Advice alone vs. structured detoxification programmes for medication overuse headache: A prospective, randomized, open-label trial in transformed migraine patients with low medical needs. *Cephalalgia, 26*(9), 1097–1105.

Sellner, J., Taba, P., Ozturk, S., & Helbok, R. (2020). The need for neurologists in the care of COVID-19 patients. *European Journal of Neurology, 27*, e31–e32 (*Sellner et all 2020*).

Willcocks, C., Loy, D., Seward, J., Mills, S., et al. (2023). Patient experiences of remote care in a pain service during a pandemic. *British Journal of Pain, 17*(1), 36–45.

Chapter 12

Managing acute ischemic stroke in the SARS-CoV-2 pandemic

Adele S. Budiansky[a], Wesley Rajaleelan[a], and Tumul Chowdhury[b]
[a]Department of Anesthesiology and Pain Medicine, University of Ottawa, Ottawa, ON, Canada, [b]Department of Anesthesia and Pain Management, Toronto Western Hospital, University of Toronto, Toronto, ON, Canada

Abbreviations

ACE	angiotensin converting enzyme
AIS	acute ischemic stroke
EVT	endovascular thrombectomy
GA	general anesthesia
LMIC	low- and middle-income countries
LVO	large vessel occlusion
MAC	monitored anesthesia care
rTPA	recombinant tissue plasminogen activator
TIA	transient ischemic attack

Introduction

A variety of neurological events have been reported in association with COVID-19 infection. These range from relatively mild symptoms such as anosmia and headache to severe and life-threatening ones such as encephalopathy (Chou et al., 2021). The risk of neurological complications with a COVID-19 infection increases for hospitalized patients and for patients with a higher severity of COVID-19 (Dangayach et al., 2022).

Evidence from the nascent stages of the pandemic suggested that acute ischemic stroke (AIS) was among the serious neurological events associated with COVID-19. The incidence of AIS was initially reported to be as high as 5% in those with COVID-19 (Li et al., 2020). Early case reports described the occurrence of AIS, and specifically large vessel occlusion (LVO), in young patients who did not have typical cardiovascular risk factors for stroke (Oxley et al., 2020). The median length from the first symptoms of COVID-19 to AIS onset was found to be 10 days, although some case studies reported an increased incidence of AIS even in the convalescent period of infection, with a median time of 54.5 days from the first positive COVID-19 serological test (Tu et al., 2021). Furthermore, in patients with a previous history of cerebrovascular disease, COVID-19 infection is associated with an increased risk of transient ischemic attack (TIA) compared to a matched cohort without COVID-19 (Nia et al., 2022). Of particular concern is the fact that patients with COVID-19 who develop AIS have a high mortality rate, particularly in the case of cryptogenic strokes (Chen et al., 2020; Ramos-Araque et al., 2021).

In addition to the challenge of COVID-19 infections being associated with AIS development, stroke care was significantly impacted during the pandemic for a multitude of other reasons. Particularly in the early stages of the pandemic, public fear and stay-at-home or isolation orders led to decreased presentations of patients to emergency departments (ED) for non-COVID-19 related conditions, including strokes (Kazakova et al., 2022).

AIS is a medical emergency that must be diagnosed and treated promptly, with thrombolytics, endovascular thrombectomy (EVT), or both, within certain time windows to reduce the risk of devastating disability or death. Adherence to performance metrics, such as the measured time from patient arrival to groin puncture for EVT, has been shown to be associated with improved stroke outcomes (Urimubenshi et al., 2017). To prevent COVID-19 transmission to other

Management, Body Systems, and Case Studies in COVID-19. https://doi.org/10.1016/B978-0-443-18703-2.00042-6
Copyright © 2024 Elsevier Inc. All rights are reserved, including those for text and data mining, AI training, and similar technologies.

patients and healthcare workers during the pandemic, major changes to stroke treatment protocols were implemented. Due to the emergent nature of AIS, most patients presenting for treatment did not have a COVID-19 test result prior to the initiation of care. Patients presenting with significant stroke symptoms may require urgent induction of general anesthesia (GA) for EVT, which carries a risk for aerosol generation and therefore COVID-19 spread. All these factors led to the development of "protected" stroke protocols (Khosravani et al., 2020), aimed at efficient yet safe AIS management.

COVID-19 infection and acute ischemic stroke

Pathophysiology of AIS in COVID-19

Several theories have been postulated for the development of AIS in COVID-19. Acute COVID-19 causes a dramatic inflammatory response, with a systemic release of cytokines. This inflammatory response may contribute to endothelial dysfunction and trigger coagulation pathways, thus resulting in thromboembolic events. This is supported by the fact that higher D-dimer and erythrocyte sedimentation rate (ESR), markers of abnormal coagulation and inflammation, are strong predictors of stroke in patients with COVID-19 infection (Goyal, Sodani, et al., 2021). Some of the damage to endothelial cells likely relates to the affinity of the SARS-CoV2 virus to endothelial angiotensin converting enzyme (ACE) II receptors, which are also present in brain tissue (Wang et al., 2020). The virus may affect brain tissue through a "Trojan horse" mechanism by infecting leukocytes that then enter the blood–brain barrier (Zubair et al., 2020). The transneuronal spread of the virus may also occur when the virus reaches the hippocampus and adjacent brain tissue from the olfactory receptor neurons via the olfactory nerve and olfactory bulb, as described in animal studies (Karimi-Galougahi et al., 2020). Once it has invaded, the virus affects brain tissue by directly damaging the neurons or the host immune response. Other proposed contributors to AIS occurrence and severity in COVID-19 include new onset arrythmias and hypoxemia (Wang et al., 2020).

Reviewing the evidence: What is the relationship between COVID-19 and AIS?

Despite early reports of a high incidence of AIS in COVID-19 infected patients, the exact relationship between COVID-19 and AIS remains uncertain. It is unclear whether COVID-19 itself has a significantly higher incidence of AIS events compared to other upper respiratory tract infections. For example, a large Danish study found that COVID-19 inpatients had 1.7 times more risk of ischemic stroke compared to inpatients with influenza, but not compared to inpatients with bacterial pneumonia under 80 years of age (Zarifkar et al., 2022). On the other hand, in another large study, the risk of arterial thromboembolism events, including AIS, was similar for patients with COVID-19 and patients with influenza (Lo Re et al., 2022).

As more data became available over time, it was determined that the incidence of AIS in COVID-19 was lower than initial estimates. Multiple large retrospective studies reported that the incidence of AIS in hospitalized patients with COVID-19 is closer to 1.3–3%, compared to 1% in patients without COVID-19 (Chou et al., 2021; Khandelwal et al., 2021; Qureshi et al., 2021; Zarifkar et al., 2022).

Characteristics of AIS in patients with and without COVID-19 infection

A large analysis of an American COVID-19 dataset (Qureshi et al., 2021) found the mean age of hospitalized patients with AIS and COVID-19 to be comparable to stroke patients without COVID-19 (Table 1). In addition, the investigators found

TABLE 1 Comparison of acute ischemic stroke patient characteristics with and without COVID-19 infection.

AIS patients with COVID-19	AIS patients without COVID-19
• Younger • Higher proportion Black (in United States) • More frequently males • LVO overrepresented • Fewer comorbidities • Worse morbidity and mortality	• Typically in >65 years old • Cardiovascular comorbidities usually present (hypertension, diabetes, CHF) • Higher discharge to home

This table compares the reported characteristics unique to acute ischemic stroke in COVID-19 patients with the typical characteristics of patients with non-COVID-19 stroke. AIS, acute ischemic stroke; CHF, congestive heart failure; LVO, large vessel occlusion.

that the proportion of patients with cardiovascular comorbidities, such as hypertension, diabetes, atrial fibrillation, and congestive heart failure, was the same in both groups. A higher proportion of Black patients with COVID-19 developed AIS. The in-hospital mortality was comparable for both stroke groups. Based on these findings, the investigators concluded that while COVID-19 may be associated with AIS, this is the case predominantly in patients with preexisting risk for stroke. It should be noted that the dataset in this study is based on diagnostic codes, which means that the study did not capture the severity of stroke. In addition, the non-COVID-19 patients in this study were those that presented with symptoms of respiratory tract infection but subsequently tested negative for COVID-19. The investigators acknowledge that this could mean that the non-COVID-19 cohort in this study could have been at an increased risk of AIS compared to patients without any respiratory tract symptoms (Qureshi et al., 2021).

Other large-scale studies have found a difference in stroke characteristics with COVID-19. An international metaanalysis of 129,491 patients found that patients presenting with stroke during the COVID-19 pandemic were significantly younger, more frequently male, and with a higher probability of LVO (Katsanos et al., 2021). Other large multicenter studies similarly reported patients with both stroke and COVID-19 to be younger and presenting with a significantly and disproportionately higher than expected rate of LVO (Dmytriw et al., 2022; Jabbour et al., 2022; Khandelwal et al., 2021). Patients with COVID-19 and LVO stroke had fewer cerebrovascular risk factors than would be expected. Importantly, patients with LVO and COVID-19 experienced higher morbidity and mortality compared to stroke patients without COVID-19, even following angiographically successful endovascular thrombectomy (EVT) (Al-Smadi et al., 2021; Douiri et al., 2021; Jabbour et al., 2022).

Therefore, while the incidence of AIS in COVID-19 may be lower than initially suspected, and most patients with COVID-19 who develop AIS have preexisting cardiovascular comorbidities, the evidence does suggest that some younger patients without usual stroke risk factors but with COVID-19 present more frequently and with more severe strokes than expected. A confounding factor, discussed later in this chapter, is the fact that patients with milder strokes were less likely to present to the hospital during the pandemic and therefore were likely underrepresented in most retrospective studies.

Characteristics of COVID-19 patients with and without AIS

In the same American dataset by Qureshi et al. (2021), COVID-19 patients that did have an AIS were older by 10–14 years compared to patients who did not have a stroke and had more cardiovascular comorbidities. Compared to patients who did not develop a stroke, patients with COVID-19 that had an AIS were more likely to develop other cerebral complications, such as cerebral edema or intracranial hemorrhage. They were also more likely to have other end-organ complications and higher mortality.

COVID-19 vaccination and AIS

Despite concerns of possible thromboembolic complications following vaccination against SARS-CoV-2, the prevalence of AIS in patients who received a COVID-19 vaccine was shown to be comparable to the prevalence of stroke in the general population, and lower than in patients with active COVID-19 infection (Stefanou et al., 2022). Vaccination against COVID-19 in patients considered to be at medium and high risk of stroke was similarly found to be safe (Wu et al., 2022).

Impact of the COVID-19 pandemic on AIS presentation and management

One of the major challenges for stroke care during the pandemic has been the significant decrease in overall stroke presentations to a hospital. Around the world, studies have consistently reported decreases in stroke admissions through the ED. The reported drop ranged from 10% to as much as 50% for stroke presentation (Hoyer et al., 2020; Peng et al., 2021). International studies found that admissions fell more for ischemic strokes than hemorrhagic strokes, and more so for older patients (Douiri et al., 2021; Jeong et al., 2022). Of note, much of the reduction occurred for milder strokes and transient ischemic attacks (TIAs) (Ishaque et al., 2022; Teo et al., 2020). Admissions for stroke mimics also decreased dramatically, particularly during the first wave of the pandemic (Ishaque et al., 2022).

It is unlikely that the incidence of milder strokes truly decreased during the pandemic; rather, patients with mild stroke symptoms were less likely to seek hospital care or delayed seeking care until the symptoms were more severe. This is supported by a large systematic review and metaanalysis of the effect of COVID-19 on stroke response times (Nawabi et al., 2022), which demonstrated that the last-known-well to arrival time increased by 24% during the pandemic. Many of the studies included in the analysis cited shelter-in-place advisories or patients' fear of seeking care in a hospital during the pandemic as perceived causes of delay. In a survey of community residents, 55% of respondents hesitated to seek medical attention after a suspected TIA because of fear of in-hospital infection (Yao et al., 2020).

This patient perspective is further confirmed by the fact that decreases in stroke presentation seem to be more prominent in areas where major outbreaks of the pandemic occurred (Jeong et al., 2022; Mitsuhashi et al., 2022). Furthermore, a study from Brazil (Cougo et al., 2022) determined that the severity of COVID-19 surges was independently associated with reduced stroke admissions, even during periods of relaxed social distancing policies. Other investigators have also reported a drop in primary care assessments and specialist neurology assessments for stroke (Myers et al., 2022). Because of these findings, multiple stroke care groups have released statements advocating for increased awareness and public education of the emergent nature of strokes as well as the need to seek care despite pandemic-related restrictions and hesitancy (Smith et al., 2020; Venketasubramanian et al., 2021).

Impact of the COVID-19 pandemic on endovascular thrombectomy of AIS

Timely access to endovascular thrombectomy for eligible patients with AIS directly impacts clinical outcomes (Giles et al., 2022). Rapid access to EVT and successful brain reperfusion require efficiency at multiple complex steps. The patient must arrive at a stroke treatment center within the window of EVT eligibility following the onset of stroke symptoms, be assessed and triaged quickly, and then undergo brain imaging. After arrival in the angiography suite, the patient receives sedation or a GA in an efficient manner so as not to delay brain reperfusion.

The COVID-19 pandemic has necessitated significant institutional changes in EVT care pathways, described in detail in the next section, to minimize the risk of COVID-19 transmission to patients and healthcare workers. These pandemic measures, such as the application of personal protective equipment (PPE), screening patients for COVID-19, and decontamination of angiography suites, would naturally be expected to negatively impact EVT performance. However, the impact of the pandemic and related precautions on EVT workflows was in fact heterogenous.

Perhaps surprisingly, many stroke centers internationally did not report major declines in EVT rates or performance. Despite a decrease in presentation of all stroke subtypes, the likelihood of being treated with thrombolysis or EVT did not decrease during as compared to before the pandemic (Ishaque et al., 2022). Contrary, the rate of EVT was higher for patients with AIS admitted during the pandemic, and this was possibly attributed to the higher rate of LVOs compared to milder stroke presentations (Katsanos et al., 2021; Rameez et al., 2020). Others reported stability in stroke treatment performance measures despite the challenges of the pandemic. A retrospective review of 14 comprehensive stroke centers in the United States did not find any difference in the primary outcome of door-to-groin-puncture time compared to pre-COVID-19 (Czap et al., 2022). In the Tokyo metropolitan area, stroke onset-to-door time was significantly longer in 2020 compared to prepandemic, yet door-to-puncture time was not affected (Katsumata et al., 2021). Similarly, a multicenter study from Germany found that intrahospital workflow time intervals did not change significantly (Tiedt et al., 2020).

However, others described substantive impacts on EVT care. A survey in China reported that nearly half of hospitals saw a decline of more than 50% in EVT cases, and that centers with larger declines were less likely to have PPE and more likely to have medical staff in quarantine (Peng et al., 2021). Data from France showed a significant increase in delays in time between imaging and groin puncture (Kerleroux et al., 2020) while in a United Kingdom study, 20 out of 24 neuroscience centers reported delays in patient pathways in the first few months of the pandemic (McConachie et al., 2020). It is difficult to know why some centers experienced significant delays in stroke care while others did not. In addition to differences in local healthcare resources and regional COVID-19 burden, it is possible that some of the stroke care delays can be explained by variations in pandemic-related precautionary protocols. Specifically, more substantive deviations from prepandemic EVT pathways, involving changes to patient transport, advanced PPE, etc., might have translated to more delays in care (Fig. 1).

It is possible that the implementation of modified protocols for stroke management, while improving safety in most circumstances, led to some unanticipated negative outcomes. For example, there have been instances of patients who developed oral angioedema after intravenous thrombolysis, which was not identified for a prolonged time because symptoms were obscured by a face mask in patients who were aphasic or dysarthric due to AIS (Éltes et al., 2022).

AIS presentation during subsequent pandemic waves

Multiple studies reported a lower decline in hospitalizations for AIS during the second and subsequent waves of the pandemic as compared to the first wave, despite the greater COVID-19 infection rate and severity (Dengler et al., 2022; Fuentes et al., 2021; Richter et al., 2022). Along these lines, there was also a decrease in stroke severity on presentation, including fewer patients presenting with concurrent COVID-19 illness (Fuentes et al., 2021). Stroke performance metrics also

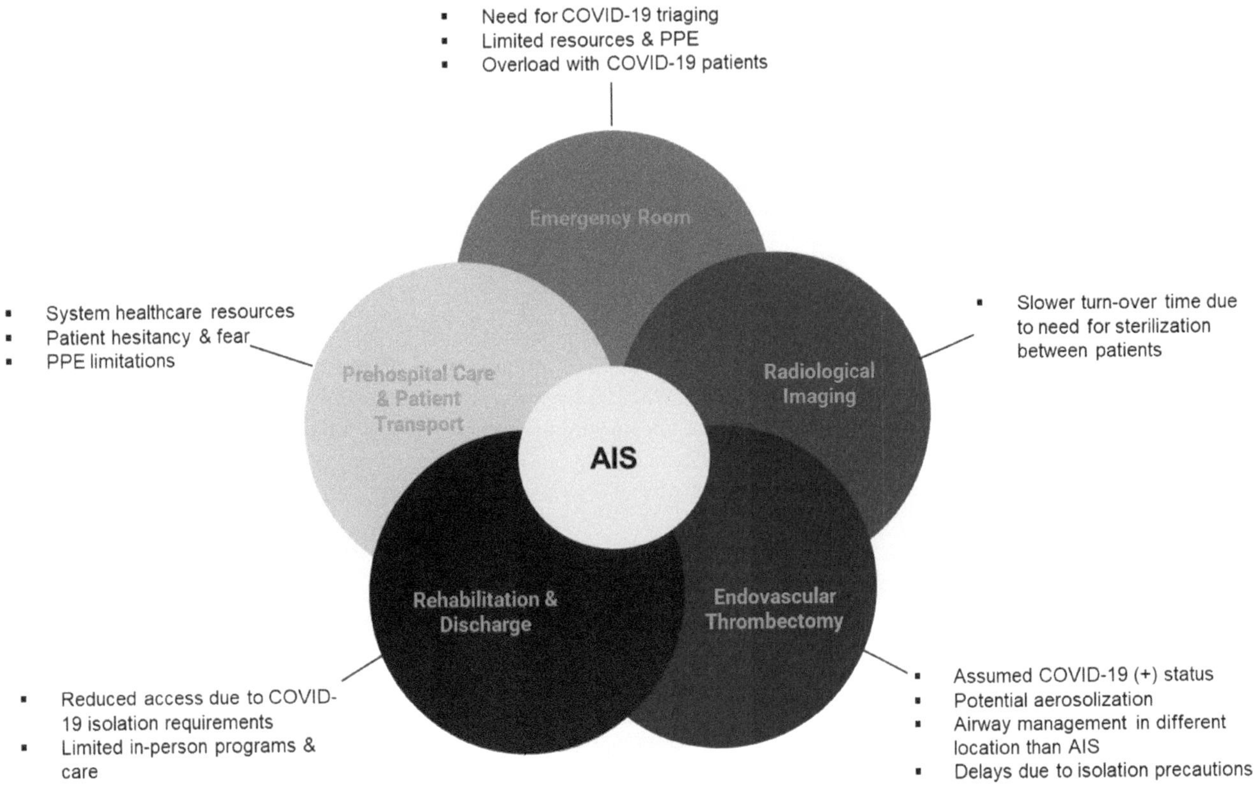

FIG. 1 Elements of acute ischemic stroke management and the COVID-19 pandemic. The figure describes primary elements of acute ischemic stroke management, and some of the barriers to delivering rapid stroke care during the COVID-19 pandemic. AIS, acute ischemic stroke; PPE, personal protective equipment.

improved during subsequent pandemic waves, with comparable or improved outcomes and mortality rates (Ramachandran et al., 2022) despite more COVID-19 infections among stroke care staff (Sobolewski et al., 2021). These findings likely reflect adaptation by both patients and hospital systems to the pandemic.

EVT protocols during the COVID-19 pandemic

During the pandemic, various AIS management protocols were developed and implemented both locally and internationally, with the aim of maintaining stroke care standards while adapting to the challenges of the pandemic. Maingard et al. (2021) provided a summary of guidelines from various neurointerventional organizations across the world for PPE. The unifying principle across all the recommendations is that PPE should protect against aerosolizing procedures. At the time of writing these guidelines, airway management maneuvers, which are not infrequently required during EVT for AIS, were considered to generate significant aerosols and would pose a high risk to stroke care team members.

In March 2020, the American Heart Association journal *Stroke* released a special report by Khosravani et al. (2020) of a "protected code stroke" protocol. In the report, they stated that appropriate PPE is the "cornerstone" of protected protocols. PPE recommended in this guideline includes contact and droplet precautions for most patient interactions, and a fit-tested N95 respirator for aerosolizing procedures. To reduce the risk of COVID-19 spread, the authors recommended that awake patients wear a surgical mask while patients who are obtunded or require significant oxygen supplementation be intubated according to local and regional COVID-19 protocols (Khosravani et al., 2020).

In April 2020, the Society of Neurointerventional Surgery released its own recommendations (Fraser et al., 2020). It emphasized that the pandemic should not alter the usual inclusion and exclusion criteria for EVT. To avoid exposing neurointerventional team members to COVID-19, the guidelines recommended a low threshold for intubation of COVID-19 positive patients prior to transfer to EVT. Specifically, the guidelines suggest that patients who present with high National Institutes of Health Stroke Scale (NIHSS) scores, decreased level of consciousness, dominant hemisphere LVOs, or posterior circulation strokes be considered for prophylactic intubation. Furthermore, the authors stated that patients should not be extubated in the interventional suite unless the suite is a negative pressure environment. Regarding preparation of the angiography suite, the guideline recommended that elective cerebrovascular cases be postponed until after the peak of the pandemic to minimize the burden of cleaning and room turnover in between many cases.

In 2021, an international panel of experts on stroke care released guidelines, based on the literature available through August 2020, for stroke care during the COVID-19 pandemic (Venketasubramanian et al., 2021). They too emphasized that intubation and extubation should be performed in controlled settings and preferably in negative pressure rooms. While still maintaining a low threshold for intubation, the panel recommended conscious sedation or monitored anesthesia care (MAC).

Anesthesia-specific guidelines for AIS management

The anesthetic management of patients presenting with AIS for EVT is challenging. Typically, the AIS patient is transferred to the EVT interventional suite in an emergency fashion, where the anesthesiologist has limited time to assess and optimize the patient prior to proceeding with the anesthetic. The anesthesiologist must quickly determine whether to proceed with MAC or GA based on the patient's stroke severity, cooperation, and cardiorespiratory status. The anesthetic is then rapidly initiated while stable hemodynamics are maintained so that the successful reperfusion of the ischemic brain can be attempted in as little time as possible.

A consensus statement from the Society for Neuroscience in Anesthesiology and Critical Care was published in July 2020 to address issues related to the anesthetic management of patients with AIS during the pandemic (Sharma et al., 2020). The main elements of the recommendations are summarized in Figs. 2–5. The consensus-based recommendations were created to minimize the risk of spreading COVID-19 among healthcare workers while facilitating the judicious allocation of available resources and seeking to achieve an optimal neurological outcome for the AIS patient. The recommendations aimed to be practical. For example, Sharma et al. (2020) acknowledged that it would be unrealistic to confirm a newly admitted patient's COVID-19 status prior to EVT because of the emergency nature of the procedure. Similarly, the authors stated that the choice of anesthesia technique should be individualized and is likely to be affected by the availability of equipment and personnel. In addition to AIS-related indications for intubation, they suggested including COVID-19-related criteria such as acute respiratory distress, high oxygen requirements, or active cough. Importantly, to reduce the risks to the entire stroke team, the recommendations stated that the decision to proceed with GA, and therefore airway manipulation, should be made well in advance. In keeping with other guidelines, the authors recommend that preplanned airway management occur in an airborne isolation area outside the interventional suite. At the same time, the authors acknowledged that individual institutions would likely tailor the recommendations to meet their practical logistical needs and limitations.

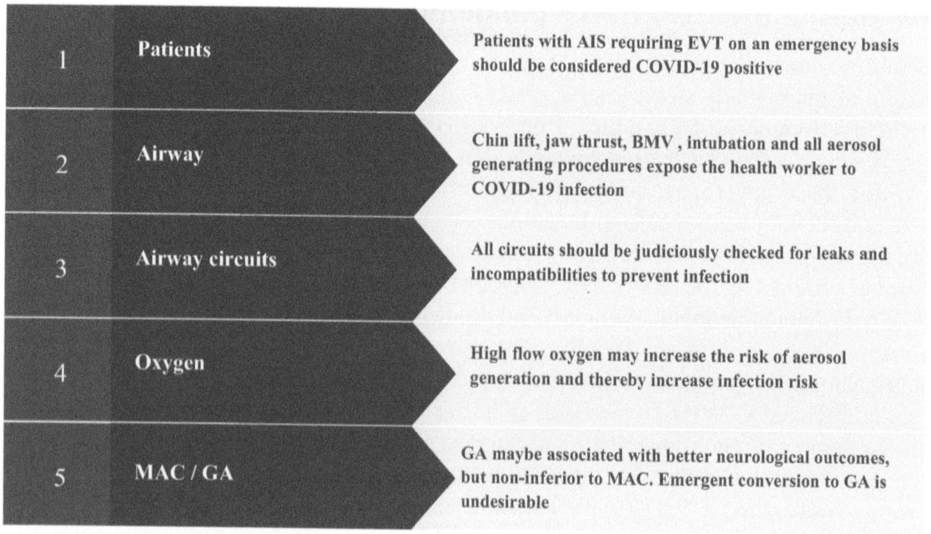

FIG. 2 General recommendations for anesthetic management of endovascular thrombectomy during the COVID-19 pandemic. These general principles highlight the importance of careful preparation and risk reduction through avoidance of aerosol generation. Summarized based on guidelines from the Society for Neuroscience in Anesthesiology and Critical Care by Sharma et al. (2020). AIS, acute ischemic stroke; BMV, bag-mask ventilation; EVT, endovascular thrombectomy; GA, general anesthesia; MAC, monitored anesthesia care.

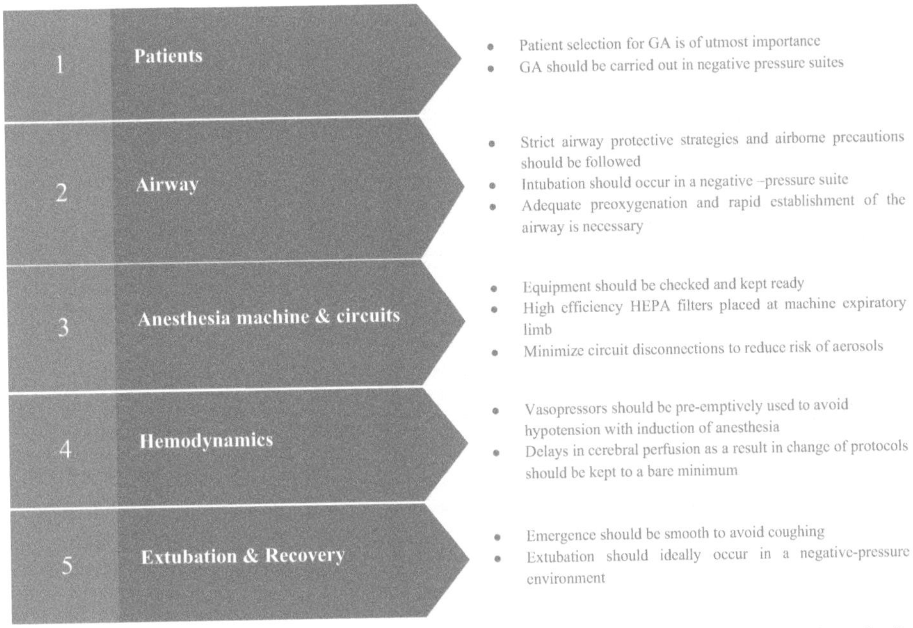

FIG. 3 Recommendations for general anesthesia management for endovascular thrombectomy during the COVID-19 pandemic. Summarized based on guidelines from the Society for Neuroscience in Anesthesiology and Critical Care by Sharma et al. (2020). GA, general anesthesia; HEPA filter, high efficiency particulate air filter.

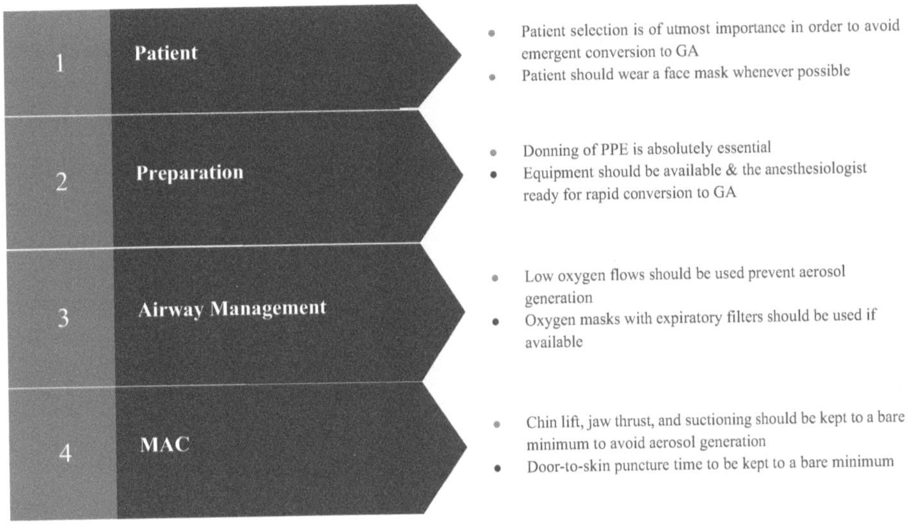

FIG. 4 Recommendations for monitored anesthesia care for endovascular thrombectomy during the COVID-19 pandemic. Emphasis is placed on minimizing aerosol-generating airway maneuvers and preparing for possible conversion to general anesthesia should the need arise. Summarized based on guidelines from the Society for Neuroscience in Anesthesiology and Critical Care by Sharma et al. (2020). GA, general anesthesia; PPE, personal protective equipment.

Variations in EVT practices during the COVID-19 pandemic

Even though recommendations for AIS management during the pandemic were relatively consistent among the various published guidelines, significant variation was seen in actual clinical practice. The variations in real-world EVT practices likely reflect local resources and practicalities that limit the application of some of the guideline recommendations.

Chowdhury et al. (2021) completed a global survey of EVT management practices during the pandemic, where 95% of respondents were anesthesiologists. The findings of the survey explored some of the variations in stroke care practices during the pandemic. Only 16% of responding centers used negative pressure rooms during EVT patient management. Full PPE was used by stroke team members regardless of COVID-19 status or anesthetic choice in only 50% of reported cases.

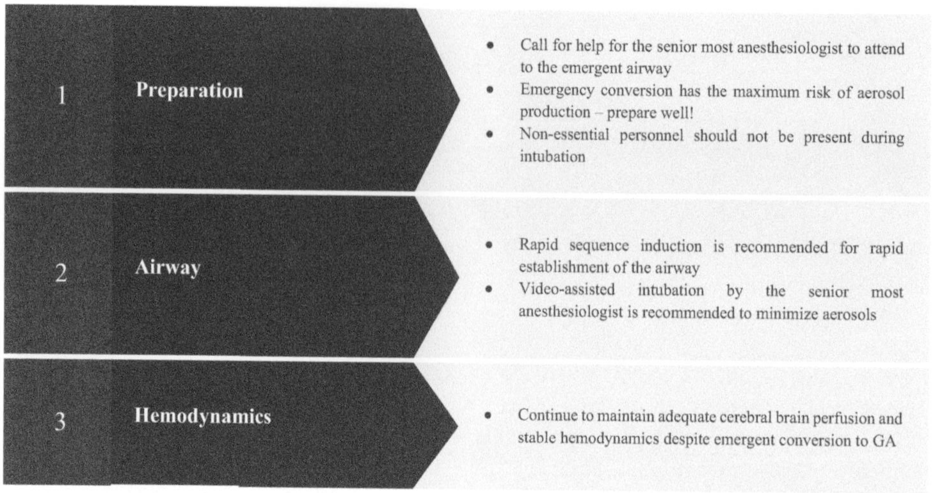

FIG. 5 Key management points of emergency conversion to general anesthesia from monitored anesthesia care during endovascular thrombectomy during the pandemic. This represents an emergency situation with high risk of aerosolization of COVID-19, requiring careful management while minimizing risk to the healthcare team. Summarized based on guidelines from the Society for Neuroscience in Anesthesiology and Critical Care by Sharma et al. (2020). GA, general anesthesia.

Despite recommendations for intubation to take place outside the interventional suite, 71% of respondents reported intubating patients for EVT in the interventional suite itself rather than a separate location. Furthermore, half of responding centers did not report any changes in their anesthetic management for EVT; a similar number of respondents shifted toward more GAs during the pandemic as others shifted toward more MACs.

There was also a wide variation in recommendations for obtaining chest imaging to assess for COVID-19 as part of EVT management. The international expert panel on stroke management during the pandemic (Venketasubramanian et al., 2021) suggested that it is reasonable to perform a low-resolution CT of the chest together with CT imaging of the brain to rule out COVID-19 infection. In contrast, the Canadian Stroke Best Practice Guidance (Smith et al., 2020) suggested not delaying EVT procedures to obtain chest imaging.

While there have not been any repeat or follow-up surveys of AIS practices since the first year of the pandemic, it is highly likely that practices have continued to deviate from the original guidelines as more has been learned about COVID-19. For example, while there were significant fears around aerosol generation during airway interventions, more recent evidence suggests that controlled airway maneuvers, such as intubation in a paralyzed patient, upper airway suctioning, and mask ventilation, generate fewer aerosols than coughing and exertional breathing (Pope et al., 2022; Shrimpton et al., 2022). It is possible that this new information has affected more recent preferences for GA or MAC for EVT care. Similarly, PPE practices may have become more relaxed over time, a change that may have impacted stroke performance measures.

Impact of the pandemic on the stroke clinician

Despite significant literature addressing the challenges faced by healthcare workers during the COVID-19 pandemic, there are very few studies focusing on the impact of the pandemic on AIS clinicians. In a survey of physicians at comprehensive stroke centers across the United States, many respondents felt that the outcome or care of AIS patients was negatively impacted by the pandemic (Kamdar et al., 2020) while in another survey, neurohospitalists reported worse personal well-being compared to before the pandemic (Goyal, Probasco, et al., 2021). Both residents and experienced stroke care providers felt less connected to patients during the pandemic and felt that they spent less time at the bedside (Kolikonda et al., 2022).

Impact of the COVID-19 pandemic on acute stroke care in low- and middle-income countries

From the start of the pandemic, low- and middle-income countries (LMICs) struggled to cope with the spread of the virus. Most LMICs were at the brink, with strained medical resources and severe challenges in the delivery of timely care to patients with AIS (Lele et al., 2022; Phuyal et al., 2021). For example, in an international survey of neurointerventional services during the COVID-19 pandemic, pooled respondents from Africa reported a greater impact on neurointerventional services, including suspension of services, compared to respondents from other continents (Fiehler et al., 2020).

In an observation study from Nepal, Phuyal et al. (2021) reported that the median time from onset of symptoms to hospital presentation was 8 h. LMICs are generally underrepresented in stroke care surveys, meaning that the magnitude of the proposed problem might be exponentially higher. Acute stroke management in LMICs is often restricted to a few urban hospitals, making access to care difficult for the people living in rural and remote areas. Phuyal et al. (2021) also reported that, in Nepal, local policies requiring a COVID-19 test for long-distance travel to enter cities to access healthcare caused significant delays, thereby increasing the mortality of patients presenting with AIS. They also reported that patients with AIS were managed with an ill-equipped infrastructure, inadequate resources, and limited staffing. The neurointerventional suites were being shared among infected and noninfected patients.

A survey of stroke care in Latin America (Pujol-Lereis et al., 2021) reported only a mild decrease in admissions of patients with AIS to the ER during the pandemic, except in Mexico, where the volume of AIS admissions was noticed to be significantly higher. They also observed that during the pandemic, there was a higher proportion of females admitted with AIS than males. Interestingly, they noticed that there was a substantial difference in stroke admissions among the surveyed countries, with the most significant reductions occurring in Chile and Argentina. Though they observed a substantial increase in the proportion of AIS admissions beyond 48 h from symptom onset (13.8% vs. 20.5%, $P < 0.001$), they found no significant differences in door-to-puncture time in both time periods. All-type mortality during hospitalization was almost doubled in 2020 compared to 2019 (Pujol-Lereis et al., 2021).

Prior to the COVID-19 pandemic, the mortality rate of stroke in sub-Saharan Africa was as much as five times greater than in Western countries; this has been shown to be related to socioeconomic status, particularly for young AIS patients (Leone et al., 2021). Adebayo et al. (2020) reported the use of telemedicine in stroke care in sub-Saharan Africa when neurological services came under severe duress due to the limited availability of specialists during the pandemic. Given the low median specialist-to-population ratio in Africa, it was postulated by the authors that teleneurology would find a niche in healthcare delivery models in Africa; however, substantial infrastructure development and financial support will be required.

The recommendations for AIS management in light of the pandemic advocate for comprehensive "protected" protocols for thrombolysis and EVT, and to designate stroke centers as COVID-19-ready centers with dedicated neuroradiology suites exclusively to treat patients with COVID-19. However, these guidelines are hardly applicable or practical in LMICs. While ensuring rapid and prompt response to AIS, ensuring the safety of healthcare workers also remains key. Clinicians and allied healthcare workers need to have adequate training in patient transport, donning and doffing PPE, and disinfection procedures.

The World Stroke Organization and the American Stroke and Heart Associations published pragmatic recommendations for the delivery of stroke care in LMICs during the COVID-19 pandemic, from prehospital care delivery through rehabilitation (Pandian et al., 2022). Some of the practical recommendations that are tailored to AIS management in LMICs are summarized as follows:

Prehospital Care

1. In patients with AIS, evaluation should not be delayed. Patients with minor symptoms can seek teleconsultation instead of admission to a hospital.
2. When an ambulance is not readily available, personal transport to the hospital should be used rather than delaying AIS care.

Emergency Room

1. Rapid triage and risk stratification in patients with AIS are the keys. Staff should don appropriate PPE during the triaging stages of COVID-19 patients.
2. High-flow nasal oxygen and the use of noninvasive positive pressure ventilation in patients with AIS with COVID-19 can aid in preserving beds with ventilator capacity for more critically ill patients when resources are limited in LMICs.

Radiological Imaging

1. All patients with AIS must undergo urgent imaging under institutional stroke protocols with dedicated transport pathways assigned to patients with COVID-19 or suspected of COVID-19.
2. The neuroradiology and interventional neuroradiology suites should be sanitized thoroughly between patients.

Thrombolysis

1. Intravenous thrombolysis should be initiated within 4 h of the onset of symptoms.
2. Institutions with limited capabilities and resources should have their stoke care protocols published.

Endovascular Thrombectomy

1. When resources are available, patients should be offered EVT for AIS irrespective of COVID-19 status.
2. To prevent aerosol generation, MAC is recommended over GA.
3. High-efficiency particulate absorbing (HEPA) filters and negative pressure suites are to be used for EVT in patients with COVID-19 or suspected COVID-19 if available, and the angiography suite sanitized after use.

Postprocedure Stroke Care

1. Exclusive stroke care units will have to be maintained to care for patients with AIS after EVT.
2. High-risk or COVID-19-positive patients should be cared for in designated ICUs and step-down ICUs with allocated staff and personnel with adequate critical care training.
3. Follow-up is crucial in the continued care of these patients and can be conducted using telemedicine or instant messaging applications such as WhatsApp.

Other challenges to AIS care include disparities in treatment availability and cost. For example, de Souza et al. (2022) conducted a global survey of 59 countries to determine the availability of the recombinant tissue plasminogen activator (rtPA), essential in thrombolysis for patients with AIS. They found significant variability in rtPA costs in LMICs when compared to higher-income countries, with the cost of rtPA in LMICs exceeding health expenditure resources. This proves that concerted efforts to improve rtPA availability as well as other treatment modalities for AIS are absolutely essential.

Lessons learned and future directions

The tremendous challenges of the COVID-19 pandemic led to significant changes in AIS presentation and management. Stroke centers had to rapidly adapt their care pathways in the face of many unknowns as well as limited protective equipment and depleted human resources. Because AIS is a medical emergency that must be rapidly diagnosed and treated, massive efforts had to be made by care teams around the world to maintain effective stroke care. Much can be learned from the evolution of AIS care during this period of crisis.

One of the major barriers to effective stroke care during the pandemic has been the reluctance of patients to seek hospital and ED care because of home-isolation policies or fear of contracting COVID-19. It is expected that future pandemics will result in the same issues. By studying the pattern of AIS admissions at the beginning of the pandemic and comparing it to the later COVID-19 waves, one can predict that a similar drop may occur in future pandemics, when there is significant uncertainty about the nature of the virus, its transmissibility, and which public containment measures are most appropriate. Therefore, public health campaigns should be launched very early in the course of a pandemic to educate patients on stroke signs and symptoms and to reinforce the importance of avoiding delays in care.

Because telestroke technology facilitates the assessment and AIS diagnosis of patients in remote locations who would otherwise have delays in treatment, telestroke has been shown to result in improved use of EVT and thrombolysis as well as a lower early mortality from stroke (Siegler et al., 2022). A study of AIS consultations in 171 hospitals in the United States using telestroke services found that telestroke assessments during the pandemic allowed for door-to-thrombolytic times similar to those prepandemic, despite reduced in-hospital staffing and resources (Sevilis et al., 2022). The dramatic increase in telestroke is thought to be due in part to the unique circumstances of the pandemic, during which legislative and regulatory barriers around telehealth technology were quickly relaxed (Harahsheh et al., 2022). This has led to a call by many stroke care organizations for telestroke services to continue to be supported and expanded internationally (Iodice et al., 2021; Shahid et al., 2021).

The growth of open-access scientific publications during the pandemic had a positive impact on timely and easy information sharing (Alemneh et al., 2020), which likely helped guide changes in AIS care pathways internationally. This facilitated sharing of evidence will hopefully be the standard for future stroke care, so that local and international teams can learn from each other and develop broadly applicable and transferrable recommendations.

Policies and procedures

Stroke is an important area of public health and imposes substantial comorbidities and mortality. In addition, resource utilization and economic burden on the healthcare system are quite significant in managing such a patient population. It involves care from many stakeholders, including multidisciplinary physicians, paramedics, allied health, the rehabilitation work force, and others. During the COVID-19 era, these resources and healthcare systems across the globe were under tremendous strain, which led to various policy/protocol modifications from various institutions/local governments.

A survey published by our group (Chowdhury et al., 2021) highlighted that the majority of institutions across more than 20 countries implemented the modified protocols for managing stroke patients during the COVID-19 pandemic. These included the adaptation of various safety features, patient flow schemes, anesthetic choices, a designated operating room (with a negative pressure room if available), emergence/extubation policies, limitation of manpower presence inside the operative room, provision of extra manpower outside the operating room, and others. Importantly, patients with AIS require immediate attention and intervention; therefore, these patients are considered as COVID-19 status unknown (or COVID-19 positive) due to testing limitations. It is important to know that local policies mainly depend upon the availability of resources, local practices, etc. In addition, public awareness remains the major key step for policy makers.

Applications to other areas

While stroke medicine is a complex field that has its unique challenges, many of the principles discussed here can be applied to other areas of emergency medical care. For example, there are many similarities between acute myocardial infarction (MI) and its management and AIS care. An MI is a medical emergency that requires, as in the case of AIS, rapid evaluation, triaging, and management. Early management of major MI events often involves thrombolysis or percutaneous coronary intervention (PCI). Just as neurological complications can occur from COVID-19, a range of cardiac complications can occur in infected patients, including cardiac dysrhythmias, heart failure, and thromboembolic events (Henning, 2022). Similarly, there have been reports of COVID-19-infected patients presenting with complications such as MI but without the usual predisposing cardiovascular risk factors (Filipiak-Strzecka et al., 2022). There are likely shared mechanisms between COVID-19-related MIs and AIS, including endothelial dysfunction and systemic inflammation.

As in the case of AIS, patients with COVID-19 who are admitted with an MI have a worse outcome than noninfected patients (Terlecki et al., 2022). The pandemic also resulted in treatment delays and less primary PCI (Huang et al., 2020). At the same time as guidelines were being developed for the "protected" management of EVT and AIS in general, cardiovascular societies were publishing similar protocols for the efficient yet safe management of acute MI and PCI (Chieffo et al., 2020). Because the two specialties faced similar challenges during the pandemic, there may in the future be a role for shared models of public health education, prehospital care, telemedicine assessment, and in-hospital triage to help reduce the strain on healthcare resources.

Mini-dictionary of terms

- **Acute ischemic stroke:** Abrupt interruption of blood supply to a part of the brain, typically caused by an arterial thromboembolic event.
- **Endovascular thrombectomy:** A time-sensitive interventional procedure used to try to reestablish blood flow to the ischemic brain through removal of the thrombus.
- **Large vessel occlusion:** An obstruction of a proximal and large cerebral artery. Such an occlusion would be expected to abruptly interrupt blood flow to large areas of the brain and therefore increase the risk for worse symptoms and outcome.
- **"Protected" code stroke:** A stroke care algorithm developed during the COVID-19 pandemic that involves steps such as COVID-19 screening and PPE to provide efficient AIS treatment while reducing the risk of infection.
- **Telestroke:** Video or digital platforms used for the remote assessment and triage of signs and symptoms of stroke, particularly when there is no access to a dedicated stroke neurologist.
- **Thrombolysis:** A time-sensitive treatment for AIS that involves the administration of medication to attempt to dissolve the thrombus and restore cerebral blood flow.

Summary points

- This chapter focused on the effect of the COVID-19 pandemic on acute ischemic stroke management and presentation.
- COVID-19 infection appears to be associated with a small but increased incidence of AIS. AIS has been reported in patients who are younger than expected and without preexisting cardiovascular comorbidities.
- At the same time, the early phase of the pandemic resulted in a significant drop in stroke presentations to hospitals, particularly for milder strokes. Public awareness campaigns should focus on patient education regarding the emergent nature of AIS.
- Multiple guidelines and recommendations were published for the management of AIS and specifically endovascular thrombectomy procedures while protecting staff and other patients from the risk of COVID-19 infection. There has been significant heterogeneity among centers in terms of impact on stroke performance measures.

- Low- and middle-income countries experienced similar but more pronounced strains on stroke care due to limitations of material and human resources.
- AIS management in future pandemics will benefit from the increased availability of telemedicine services and open-access research.

References

Adebayo, P. B., Oluwole, O. J., & Taiwo, F. T. (2020). COVID-19 and Teleneurology in sub-Saharan Africa: Leveraging the current exigency. *Frontiers in Public Health, 8*, 574505.

Alemneh, D. G., Hawamdeh, S., Chang, H. C., Rorissa, A., Assefa, S., & Helge, K. (2020). Open access in the age of a pandemic. *Proceedings of the Association for Information Science and Technology, 57*(1), e295.

Al-Smadi, A. S., Mach, J. C., Abrol, S., Luqman, A., Chamiraju, P., & Abujudeh, H. (2021). Endovascular thrombectomy of COVID-19-related large vessel occlusion: A systematic review and summary of the literature. *Current Radiology Reports, 9*(4), 4.

Chen, S., Pan, C., Zhang, P., Tang, Y., & Tang, Z. (2020). Clinical characteristics of inpatients with coronavirus disease 2019 and acute ischemic stroke: From epidemiology to outcomes. *Current Neurovascular Research, 17*(5), 760–764.

Chieffo, A., Stefanini, G. G., Price, S., Barbato, E., Tarantini, G., Karam, N., … Baumbach, A. (2020). EAPCI position statement on invasive Management of Acute Coronary Syndromes during the COVID-19 pandemic. *European Heart Journal, 41*(19), 1839–1851.

Chou, S. H., Beghi, E., Helbok, R., Moro, E., Sampson, J., Altamirano, V., … McNett, M. (2021). Global incidence of neurological manifestations among patients hospitalized with COVID-19-A report for the GCS-NeuroCOVID consortium and the ENERGY consortium. *JAMA Network Open, 4*(5), e2112131.

Chowdhury, T., Rizk, A. A., Daniels, A. H., Al Azazi, E., Sharma, D., & Venkatraghavan, L. (2021). Management of Acute Ischemic Stroke in the interventional neuroradiology suite during the COVID-19 pandemic: A global survey. *Journal of Neurosurgical Anesthesiology, 33*(1), 44–50.

Cougo, P., Besen, B., Bezerra, D., Moreira, R. C., Brandão, C. E., Salgueiro, E., … Cravo, V. (2022). Social distancing, stroke admissions and stroke mortality during the COVID-19 pandemic: A multicenter, longitudinal study. *Journal of Stroke and Cerebrovascular Diseases, 31*(5), 106405.

Czap, A. L., Zha, A. M., Sebaugh, J., Hassan, A. E., Shulman, J. G., Abdalkader, M., … Siegler, J. E. (2022). Endovascular thrombectomy time metrics in the era of COVID-19: Observations froPleam the Society of Vascular and Interventional Neurology Multicenter Collaboration. *Journal of NeuroInterventional Surgery, 14*(1). neurintsurg-2020-017205.

Dangayach, N. S., Newcombe, V., & Sonnenville, R. (2022). Acute neurologic complications of COVID-19 and postacute sequelae of COVID-19. *Critical Care Clinics, 38*(3), 553–570.

de Souza, A. C., Sebastian, I. A., Zaidi, W. A. W., Nasreldein, A., Bazadona, D., Amaya, P., … Lioutas, V. A. (2022). Regional and national differences in stroke thrombolysis use and disparities in pricing, treatment availability, and coverage. *International Journal of Stroke, 17*(9), 990–996.

Dengler, J., Prass, K., Palm, F., Hohenstein, S., Pellisier, V., Stoffel, M., … Rosahl, S. (2022). Changes in nationwide in-hospital stroke care during the first four waves of COVID-19 in Germany. *European Stroke Journal, 7*(2), 166–174.

Dmytriw, A. A., Ghozy, S., Sweid, A., Piotin, M., Bekelis, K., Sourour, N., … Jabbour, P. (2022). International controlled study of revascularization and outcomes following COVID-positive mechanical thrombectomy. *European Journal of Neurology, 29*(11), 3273–3287.

Douiri, A., Muruet, W., Bhalla, A., James, M., Paley, L., Stanley, K., … Bray, B. D. (2021). Stroke care in the United Kingdom during the COVID-19 pandemic. *Stroke, 52*(6), 2125–2133.

Éltes, T., Hajnal, B., & Kamondi, A. (2022). Concealment of allergic reactions to Alteplase by face masks in non-communicating acute stroke patients: A warning call to improve our physical examination practices during the COVID-19 pandemic. *The Tohoku Journal of Experimental Medicine, 257*(2), 157–161.

Fiehler, J., Brouwer, P., Díaz, C., Hirsch, J. A., Kulcsar, Z., Liebeskind, D., … Taylor, A. (2020). COVID-19 and neurointerventional service worldwide: A survey of the European Society of Minimally Invasive Neurological Therapy (ESMINT), the society of NeuroInterventional surgery (SNIS), the Sociedad Iberolatinoamericana de Neuroradiologia Diagnostica y Terapeutica (SILAN), the Society of Vascular and Interventional Neurology (SVIN), and the world Federation of Interventional and Therapeutic Neuroradiology (WFITN). *Journal of NeuroInterventional Surgery, 12*(8), 726–730.

Filipiak-Strzecka, D., Plewka, M., Szymczyk, E., Frynas-Jończyk, K., Szymczyk, K., Lipiec, P., & Kasprzak, J. D. (2022). Acute myocardial injury as a sole presentation of COVID-19 in patient without cardiovascular risk factors. *Cardiology Journal, 29*(5), 878–879.

Fraser, J. F., Arthur, A. S., Chen, M., Levitt, M., Mocco, J., Albuquerque, F. C., … Klucznik, R. P. (2020). Society of NeuroInterventional surgery recommendations for the care of emergent neurointerventional patients in the setting of COVID-19. *Journal of NeuroInterventional Surgery, 12*(6), 539–541.

Fuentes, B., Alonso de Leciñana, M., Rigual, R., García-Madrona, S., Díaz-Otero, F., Aguirre, C., … Díez Tejedor, S. (2021). Fewer COVID-19-associated strokes and reduced severity during the second COVID-19 wave: The Madrid stroke network. *European Journal of Neurology, 28*(12), 4078–4089.

Giles, J. A., Vellimana, A. K., & Adeoye, O. M. (2022). Endovascular treatment of acute stroke. *Current Neurology and Neuroscience Reports, 22*(1), 83–91.

Goyal, T., Probasco, J. C., Gold, C. A., Klein, J. P., Weathered, N. R., & Thakur, K. T. (2021). Neurohospitalist practice and well-being during the COVID-19 pandemic. *Neurohospitalist, 11*(4), 333–341.

Goyal, N., Sodani, A. K., Jain, R., & Ram, H. (2021). Do elevated levels of inflammatory biomarkers predict the risk of occurrence of ischemic stroke in SARS-CoV2 ?: An observational study. *Journal of Stroke and Cerebrovascular Diseases, 30*(11), 106063.

Harahsheh, E., English, S. W., Hrdlicka, C. M., & Demaerschalk, B. (2022). Telestroke's role through the COVID-19 pandemic and beyond. *Current Treatment Options in Neurology, 24*(11), 589–603.

Henning, R. J. (2022). Cardiovascular complications of COVID-19 severe acute respiratory syndrome. *American Journal of Cardiovascular Disease, 12*(4), 170–191.

Hoyer, C., Ebert, A., Huttner, H. B., Puetz, V., Kallmünzer, B., Barlinn, K., … Szabo, K. (2020). Acute stroke in times of the COVID-19 pandemic: A multicenter study. *Stroke*, *51*(7), 2224–2227.

Huang, B., Xu, C., Liu, H., Deng, W., Yang, Z., Wan, J., … Jiang, H. (2020). In-hospital management and outcomes of acute myocardial infarction before and during the coronavirus disease 2019 pandemic. *Journal of Cardiovascular Pharmacology*, *76*(5), 540–548.

Iodice, F., Romoli, M., Giometto, B., Clerico, M., Tedeschi, G., Bonavita, S., … Lavorgna, L. (2021). Stroke and digital technology: A wake-up call from COVID-19 pandemic. *Neurological Sciences*, *42*(3), 805–809.

Ishaque, N., Butt, A. J., Kamtchum-Tatuene, J., Nomani, A. Z., Razzaq, S., Fatima, N., … Shuaib, A. (2022). Trends in stroke presentations before and during the COVID-19 pandemic: A meta-analysis. *Journal of Stroke*, *24*(1), 65–78.

Jabbour, P., Dmytriw, A. A., Sweid, A., Piotin, M., Bekelis, K., Sourour, N., … Tiwari, A. (2022). Characteristics of a COVID-19 cohort with large vessel occlusion: A multicenter international study. *Neurosurgery*, *90*(6), 725–733.

Jeong, H. Y., Lee, E. J., Kang, M. K., Nam, K. W., Bae, J., Jeon, K., … Park, J. M. (2022). Changes in stroke Patients' health-seeking behavior by COVID-19 epidemic regions: Data from the Korean stroke registry. *Cerebrovascular Diseases*, *51*(2), 169–177.

Kamdar, H. A., Senay, B., Mainali, S., Lee, V., Gulati, D. K., Greene-Chandos, D., … Strohm, T. (2020). Clinician's perception of practice changes for stroke during the COVID-19 pandemic. *Journal of Stroke and Cerebrovascular Diseases*, *29*(10), 105179.

Karimi-Galougahi, M., Yousefi-Koma, A., Bakhshayeshkaram, M., Raad, N., & Haseli, S. (2020). (18)FDG PET/CT scan reveals hypoactive orbitofrontal cortex in anosmia of COVID-19. *Academic Radiology*, *27*(7), 1042–1043.

Katsanos, A. H., Palaiodimou, L., Zand, R., Yaghi, S., Kamel, H., Navi, B. B., … Tsivgoulis, G. (2021). Changes in stroke hospital care during the COVID-19 pandemic: A systematic review and meta-analysis. *Stroke*, *52*(11), 3651–3660.

Katsumata, M., Ota, T., Kaneko, J., Jimbo, H., Aoki, R., Fujitani, S., … Hirano, T. (2021). Impact of coronavirus disease 2019 on time delay and functional outcome of mechanical Thrombectomy in Tokyo, Japan. *Journal of Stroke and Cerebrovascular Diseases*, *30*(10), 106051.

Kazakova, S. V., Baggs, J., Parra, G., Yusuf, H., Romano, S. D., Ko, J. Y., … Jernigan, J. A. (2022). Declines in the utilization of hospital-based care during COVID-19 pandemic. *Journal of Hospital Medicine*.

Kerleroux, B., Fabacher, T., Bricout, N., Moïse, M., Testud, B., Vingadassalom, S., … Boulouis, G. (2020). Mechanical Thrombectomy for acute ischemic stroke amid the COVID-19 outbreak: Decreased activity, and increased care delays. *Stroke*, *51*(7), 2012–2017.

Khandelwal, P., Al-Mufti, F., Tiwari, A., Singla, A., Dmytriw, A. A., Piano, M., … Yavagal, D. R. (2021). Incidence, characteristics and outcomes of large vessel stroke in COVID-19 cohort: An international multicenter study. *Neurosurgery*, *89*(1), E35–e41.

Khosravani, H., Rajendram, P., Notario, L., Chapman, M. G., & Menon, B. K. (2020). Protected code stroke: Hyperacute stroke management during the coronavirus disease 2019 (COVID-19) pandemic. *Stroke*, *51*(6), 1891–1895.

Kolikonda, M. K., Blaginykh, E., Brown, P., Kovi, S., Zhang, L. Q., & Uchino, K. (2022). Virtual rounding in stroke care and neurology education during the COVID-19 pandemic—A residency program survey. *Journal of Stroke and Cerebrovascular Diseases*, *31*(1), 106177.

Lele, A. V., Wahlster, S., Alunpipachathai, B., Awraris Gebrewold, M., Chou, S. H., Crabtree, G., … Moheet, A. M. (2022). Perceptions regarding the SARS-CoV-2 Pandemic's impact on Neurocritical care delivery: Results from a global survey. *Journal of Neurosurgical Anesthesiology*, *34*(2), 209–220.

Leone, M., Ciccacci, F., Orlando, S., Petrolati, S., Guidotti, G., Majid, N. A., … Marazzi, M. C. (2021). Pandemics and burden of stroke and epilepsy in sub-Saharan Africa: Experience from a longstanding health programme. *International Journal of Environmental Research and Public Health*, *18*(5).

Li, Y., Li, M., Wang, M., Zhou, Y., Chang, J., Xian, Y., … Hu, B. (2020). Acute cerebrovascular disease following COVID-19: A single center, retrospective, observational study. *Stroke and Vascular Neurology*, *5*(3), 279–284.

Lo Re, V., 3rd, Dutcher, S. K., Connolly, J. G., Perez-Vilar, S., Carbonari, D. M., DeFor, T. A., … Cocoros, N. M. (2022). Association of COVID-19 vs influenza with risk of arterial and venous thrombotic events among hospitalized patients. *JAMA*, *328*(7), 637–651.

Maingard, J., Mont'Alverne, F. J., & Chandra, R. (2021). Staff and physician protection in neurointervention during the coronavirus disease-2019 pandemic: A summary review and recommendations. *Interventional Neuroradiology*, *27*(1_suppl), 24–29.

McConachie, D., McConachie, N., White, P., Crossley, R., & Izzath, W. (2020). Mechanical thrombectomy for acute ischaemic stroke during the COVID-19 pandemic: Changes to UK practice and lessons learned. *Clinical Radiology*, *75*(10). 795.e797–795.e713.

Mitsuhashi, T., Tokugawa, J., & Mitsuhashi, H. (2022). Long-term evaluation of the COVID-19 pandemic impact on acute stroke management: An analysis of the 21-month data from a medical facility in Tokyo. *Acta Neurologica Belgica*, 1–8.

Myers, L. J., Perkins, A. J., Kilkenny, M. F., & Bravata, D. M. (2022). Quality of care and outcomes for patients with acute ischemic stroke and transient ischemic attack during the COVID-19 pandemic. *Journal of Stroke and Cerebrovascular Diseases*, *31*(6), 106455.

Nawabi, N. L. A., Duey, A. H., Kilgallon, J. L., Jessurun, C., Doucette, J., Mekary, R. A., & Aziz-Sultan, M. A. (2022). Effects of the COVID-19 pandemic on stroke response times: A systematic review and meta-analysis. *Journal of NeuroInterventional Surgery*, *14*(7), 642–649.

Nia, A. M., Srinivasan, V. M., Lall, R. R., & Kan, P. (2022). COVID-19 and stroke recurrence by subtypes: A propensity-score matched analyses of stroke subtypes in 44,994 patients. *Journal of Stroke and Cerebrovascular Diseases*, *31*(8), 106591.

Oxley, T. J., Mocco, J., Majidi, S., Kellner, C. P., Shoirah, H., Singh, I. P., … Fifi, J. T. (2020). Large-vessel stroke as a presenting feature of Covid-19 in the young. *The New England Journal of Medicine*, *382*(20), e60.

Pandian, J. D., Panagos, P. D., Sebastian, I. A., Silva, G. S., Furie, K. L., Liu, L., … Alrukn, S. A. (2022). Maintaining stroke care during the COVID-19 pandemic in lower- and middle-income countries: World stroke organization position statement endorsed by American Stroke Association and American Heart Association. *Stroke*, *53*(3), 1043–1050.

Peng, G., Nie, X., Liebeskind, D., Liu, L., & Miao, Z. (2021). Management of endovascular therapy for acute ischemic stroke amid the COVID-2019 pandemic: A multicenter survey in China. *Neurological Research*, *43*(10), 823–830.

Phuyal, S., Lamsal, R., Shrestha, G. S., Paudel, R., & Thapa, L. (2021). Management of Stroke during COVID-19 pandemic: The challenges in Nepal. *Journal of Nepal Health Research Council*, *19*(1), 218–220.

Pope, C., Harrop-Griffiths, W., & Brown, J. (2022). Aerosol-generating procedures and the anaesthetist. *BJA Education, 22*(2), 52–59.

Pujol-Lereis, V. A., Flores, A., Barboza, M. A., Abanto-Argomedo, C., Amaya, P., Bayona, H., ... Ameriso, S. F. (2021). COVID-19 lockdown effects on acute stroke care in Latin America. *Journal of Stroke and Cerebrovascular Diseases, 30*(9), 105985.

Qureshi, A. I., Baskett, W. I., Huang, W., Shyu, D., Myers, D., Raju, M., ... Shyu, C. R. (2021). Acute ischemic stroke and COVID-19: An analysis of 27676 patients. *Stroke, 52*(3), 905–912.

Ramachandran, D., Panicker, P., Chithra, P., & Iype, T. (2022). Time metrics in acute ischemic stroke care during the second and first wave of COVID 19 pandemic: A tertiary care center experience from South India. *Journal of Stroke and Cerebrovascular Diseases, 31*(5), 106315.

Rameez, F., McCarthy, P., Cheng, Y., Packard, L. M., Davis, A. T., Wees, N., ... Min, J. (2020). Impact of a stay-at-home order on stroke admission, subtype, and metrics during the COVID-19 pandemic. *Cerebrovascular Diseases Extra, 10*(3), 159–165.

Ramos-Araque, M. E., Siegler, J. E., Ribo, M., Requena, M., López, C., de Lera, M., ... Vazquez, A. R. (2021). Stroke etiologies in patients with COVID-19: The SVIN COVID-19 multinational registry. *BMC Neurology, 21*(1), 43.

Richter, D., Eyding, J., Weber, R., Bartig, D., Grau, A., Hacke, W., & Krogias, C. (2022). A full year of the COVID-19 pandemic with two infection waves and its impact on ischemic stroke patient care in Germany. *European Journal of Neurology, 29*(1), 105–113.

Sevilis, T., McDonald, M., Avila, A., Heath, G., Gao, L., O'Brien, G., ... Devlin, T. (2022). Telestroke: Maintaining quality acute stroke care during the COVID-19 pandemic. *Telemedicine Journal and E-Health, 28*(4), 481–485.

Shahid, R., Al-Jehani, H. M., Zafar, A., & Saqqur, M. (2021). The emerging role of Telestroke in the Middle East and North Africa region in the era of COVID-19. *Primary Care Companion for CNS Disorders, 23*(5).

Sharma, D., Rasmussen, M., Han, R., Whalin, M. K., Davis, M., Kofke, W. A., ... Fraser, J. F. (2020). Anesthetic Management of Endovascular Treatment of acute ischemic stroke during COVID-19 pandemic: Consensus statement from Society for Neuroscience in Anesthesiology & Critical Care (SNACC): Endorsed by Society of Vascular & interventional neurology (SVIN), society of NeuroInterventional surgery (SNIS), Neurocritical care society (NCS), European Society of Minimally Invasive Neurological Therapy (ESMINT) and American Association of Neurological Surgeons (AANS) and Congress of Neurological Surgeons (CNS) cerebrovascular section. *Journal of Neurosurgical Anesthesiology, 32*(3), 193–201.

Shrimpton, A. J., Brown, J. M., Gregson, F. K. A., Cook, T. M., Scott, D. A., McGain, F., ... Pickering, A. E. (2022). Quantitative evaluation of aerosol generation during manual facemask ventilation. *Anaesthesia, 77*(1), 22–27.

Siegler, J. E., Abdalkader, M., Michel, P., & Nguyen, T. N. (2022). Therapeutic trends of cerebrovascular disease during the COVID-19 pandemic and future perspectives. *Journal of Stroke, 24*(2), 179–188.

Smith, E. E., Mountain, A., Hill, M. D., Wein, T. H., Blacquiere, D., Casaubon, L. K., ... Lindsay, M. P. (2020). Canadian stroke best practice guidance during the COVID-19 pandemic. *The Canadian Journal of Neurological Sciences, 47*(4), 474–478.

Sobolewski, P., Szczuchniak, W., Grzesiak-Witek, D., Wilczyński, J., Paciura, K., Antecki, M., ... Kozera, G. (2021). Stroke care during the first and the second waves of the COVID-19 pandemic in a community hospital. *Frontiers in Neurology, 12*, 655434.

Stefanou, M. I., Palaiodimou, L., Aguiar de Sousa, D., Theodorou, A., Bakola, E., Katsaros, D. E., ... Tsivgoulis, G. (2022). Acute arterial ischemic stroke following COVID-19 vaccination: A systematic review and Meta-analysis. *Neurology*.

Teo, K. C., Leung, W. C. Y., Wong, Y. K., Liu, R. K. C., Chan, A. H. Y., Choi, O. M. Y., ... Lau, K. K. (2020). Delays in stroke onset to hospital arrival time during COVID-19. *Stroke, 51*(7), 2228–2231.

Terlecki, M., Wojciechowska, W., Klocek, M., Olszanecka, A., Bednarski, A., Drożdż, T., ... Rajzer, M. (2022). Impact of concomitant COVID-19 on the outcome of patients with acute myocardial infarction undergoing coronary artery angiography. *Frontiers in Cardiovascular Medicine, 9*, 917250.

Tiedt, S., Bode, F. J., Uphaus, T., Alegiani, A., Gröschel, K., & Petzold, G. C. (2020). Impact of the COVID-19-pandemic on thrombectomy services in Germany. *Neurological Research and Practice, 2*, 44.

Tu, T. M., Seet, C. Y. H., Koh, J. S., Tham, C. H., Chiew, H. J., De Leon, J. A., ... Yeo, L. L. L. (2021). Acute ischemic stroke during the convalescent phase of asymptomatic COVID-2019 infection in men. *JAMA Network Open, 4*(4), e217498.

Urimubenshi, G., Langhorne, P., Cadilhac, D. A., Kagwiza, J. N., & Wu, O. (2017). Association between patient outcomes and key performance indicators of stroke care quality: A systematic review and meta-analysis. *European Stroke Journal, 2*(4), 287–307.

Venketasubramanian, N., Anderson, C., Ay, H., Aybek, S., Brinjikji, W., de Freitas, G. R., ... Hennerici, M. G. (2021). Stroke care during the COVID-19 pandemic: International expert panel review. *Cerebrovascular Diseases, 50*(3), 245–261.

Wang, Z., Yang, Y., Liang, X., Gao, B., Liu, M., Li, W., ... Wang, Z. (2020). COVID-19 associated ischemic stroke and hemorrhagic stroke: Incidence, potential pathological mechanism, and management. *Frontiers in Neurology, 11*, 571996.

Wu, G., Zhang, M., Xie, X., Zhu, Y., Tang, H., Zhu, X., ... Ke, S. (2022). A survey on the safety of the SARS-CoV-2 vaccine among a population with stroke risk in China. *Frontiers in Medicine, 9*, 859682.

Yao, S., Lin, B., Liu, Y., Luo, Y., Xu, Q., Huang, J., ... Liu, X. (2020). Impact of Covid-19 on the behavior of community residents with suspected transient ischemic attack. *Frontiers in Neurology, 11*, 590406.

Zarifkar, P., Peinkhofer, C., Benros, M. E., & Kondziella, D. (2022). Frequency of neurological diseases after COVID-19, influenza A/B and bacterial pneumonia. *Frontiers in Neurology, 13*, 904796.

Zubair, A. S., McAlpine, L. S., Gardin, T., Farhadian, S., Kuruvilla, D. E., & Spudich, S. (2020). Neuropathogenesis and neurologic manifestations of the coronaviruses in the age of coronavirus disease 2019: A review. *JAMA Neurology, 77*(8), 1018–1027.

Chapter 13

Impact of the coronavirus disease 2019 pandemic on patients with hypertension

Kazuo Kobayashi[a,b] and Kouichi Tamura[b]

[a]*Committee of Hypertension and Kidney Disease, Kanagawa Physicians Association, Yokohama, Japan,* [b]*Department of Medical Science and Cardiorenal Medicine, Yokohama City University Graduate School of Medicine, Yokohama, Japan*

Abbreviations

BP blood pressure
DCAP disaster cardiovascular prevention
RAS renin-angiotensin system

Introduction

Coronavirus disease 2019 (COVID-19), a respiratory disease caused by the novel coronavirus severe acute respiratory syndrome coronavirus 2 (SARS-CoV-2), was first reported in Wuhan, China, in December 2019 (Zhu et al., 2020). It was declared a pandemic by the World Health Organization on March 11, 2020. In Japan, the first case of COVID-19 was reported on Jan. 15, 2020, and the Japanese government recognized COVID-19 as a designated infectious disease on Feb. 1, 2020. Various measures were implemented worldwide to provide medical care for infected patients and prevent the spread of COVID-19. These measures placed significant restrictions and burdens on people's social lives. Consequently, there was a greater risk of worsening lifestyle-related diseases and increasing cerebrovascular and cardiovascular complications. Hypertension is an independent risk factor for cardiovascular disease, with incidence rates of 60% and 41% in men and women, respectively, aged 40–74 years; an estimated 43 million patients with hypertension live in Japan (Umemura et al., 2019). The COVID-19 pandemic had a considerable social impact and can be considered a major disaster. Therefore, the impact of the pandemic on blood pressure (BP) and relevant countermeasures against it should be considered.

Policies and procedures

In this chapter, we provide a narrative review for the COVID-19 pandemic and BP on patients with or without hypertension. We evaluated the change in BP in the office or at home as the primary outcome in this narrative review. Regarding the procedures, we initially searched literature databases (PubMed, Google Scholar, and Japan medical abstracts society) for reports or studies on clinical parameters relating to COVID-19 and BP (e.g., BP values in the office or at home, pulse rates, body weights, plasma glucose levels, and data for the lipid profile, kidney function, or liver function). Because our narrative review focused on the relationship between stress caused by the pandemic and BP management, data from patients with hypertension who were infected by SARS-CoV-2 were excluded. Finally, the reports or studies were included for this narrative review if they had relevant data pertinent to BP in people during the COVID-19 pandemic. In addition, studies on psychological stress in the COVID-19 pandemic, particularly anxiety and fear, were also reviewed. In those reports on stress, such as anxiety and fear, the results of questionnaires and stress scores are included. Various studies were reported on the COVID-19 pandemic in Japan or around the world. In addition to large-scale studies and surveys, there are many types of reports, (e.g., case reports, reports on conferences only, reports on websites, news, or official documents from governments). We tried to investigate as many reports as possible; however, it is undeniable that many more reports exist beyond the resources we have examined.

Management, Body Systems, and Case Studies in COVID-19. https://doi.org/10.1016/B978-0-443-18703-2.00029-3
Copyright © 2024 Elsevier Inc. All rights are reserved, including those for text and data mining, AI training, and similar technologies.

COVID-19 pandemic and social stress

The mortality rate in patients with COVID-19 in Wuhan at the beginning of 2020 was 2.3% (Wu & McGoogan, 2020). The US Centers for Disease Control and Prevention reported that the mortality rate in patients with COVID-19 was 106.5 per 100,000 people between 2019 and 2020, which was approximately six times higher than that caused by influenza (Centers For Disease Control And Prevention, 2022).

Although the fears of COVID-19 and death were prevalent, the primary cause of social stress during the pandemic was the highly developed information society, that is, the infodemic (Talevi et al., 2020; Vindegaard & Benros, 2020). In Japan, COVID-19 was confirmed in 10 passengers onboard the Diamond Princess while it was docked in the port of Yokohama on Feb. 4, 2020. During this time, the clinical characteristics of COVID-19 were not well-understood, no treatment strategy had been established, and only symptomatic treatment was provided, including isolation and respiratory management of patients. At the beginning of the pandemic, a limited number of medical facilities were able to respond adequately, and some medical facilities had to treat patients with severe COVID-19 or manage clusters of COVID-19 in the absence of a full-time respiratory specialist. In a medium-sized hospital in Sagamihara City, Kanagawa Prefecture, insufficient medical staff caused the spread of a cluster in the hospital, resulting in a medical collapse where the hospital's functions almost ceased and its reputation was damaged. As the number of community-acquired infections increased, the media reported on the situation with daily special features and realistic images. Furthermore, the controversial comments and speculations made by infection experts with limited evidence were disseminated rapidly on television and on various social networking services, and images of the worldwide spread of COVID-19 strengthened the public's anxiety regarding the disease. This anxiety resulting from news coverage was also referred to as the headline stress disorder (Bagus et al., 2021; Dong & Zheng, 2020). Governments and local authorities appealed daily to the public to maintain social distancing, avoid the so-called "three nectars" (closed spaces, crowded places, and close-contact settings), work from home if applicable, and avoid leaving the house unnecessarily to prevent the spread of COVID-19. Consequently, a society with extremely poor communication was created, contributing to the worsening of social stress. Nevertheless, the number of patients with COVID-19 increased, and the inability to provide necessary medical care (medical collapse) led many countries and cities to implement lockdowns as an intensive measure against the pandemic (BBC News Services, 2020). Lockdowns imposed strong restrictions on the rights of citizens, forbidding people from leaving their homes or having contact with each other, and penalties were sometimes imposed for noncompliance.

To prevent the further spread of COVID-19, a state of emergency was declared in Japan on April 7, 2020, initially in the metropolitan areas and then gradually throughout the country. According to a report by the Ministry of Health, Labor, and Welfare, the proportion of people who felt anxious about COVID-19 peaked immediately after the declaration of the state of emergency, exceeding 60% (Intage Research Co..Ltd, 2022). In a survey of 6000 people nationwide conducted by the Nissei Research Institute, the stronger the fear of acquiring COVID-19, the more an individual refrained from going out (83.6% of those with higher anxiety levels and 9.1% of those without anxiety refrained from going out) (Kuga, 2020), suggesting that the psychological impact also affected people's daily behaviors. In the United Kingdom (UK), 37% of citizens experienced high levels of anxiety, especially older people, when lockdown was imposed during the pandemic (Office for National Statistics, 2020). A total of 3097 adults participated in the COVID-19 Stress and Health Study conducted within the first 6 weeks of the implementation of social restriction measures in the UK; the study found that worries about acquiring COVID-19 as well as loneliness were associated with anxiety, depression, and stress (Jia et al., 2020). The psychosocial impact of the COVID-19 pandemic is summarized in Fig. 1 (created by the author with reference to Dubey et al. (Dubey et al., 2020)); the general public, who were not infected with COVID-19, were also part of the stress vortex.

COVID-19 as a special disaster for hypertension

In Japan, where natural disasters such as earthquakes, typhoons, torrential rains, and volcanic eruptions are common, social stress is common as a result. Kario et al. reported an increase of 14 mmHg in systolic BP and of 6 mmHg in diastolic BP at 1–2 weeks after the Hanshin-Awaji earthquake in 1995, with worsening of albuminuria and organ damage such as increased cardiovascular events (Kario et al., 2001). Miyakawa et al. reported an increase of 4 mmHg in systolic BP and 3 mmHg in diastolic BP at 1–3 weeks after the great East Japan earthquake in 2011, even in those living in areas outside the epicenter, and the elevated levels persisted for 5–7 weeks after the earthquake in patients with hypertension or advanced-stage chronic kidney disease (Miyakawa, 2012). Based on these findings, the concept of disaster-related cardiovascular disease has been proposed. Takotsubo cardiomyopathy, sudden death, or pulmonary embolism occurred immediately after the onset of a disaster, and hypertension-related diseases such as stroke, myocardial infarction, aortic dissection, or heart failure subsequently developed (Kario, 2012).

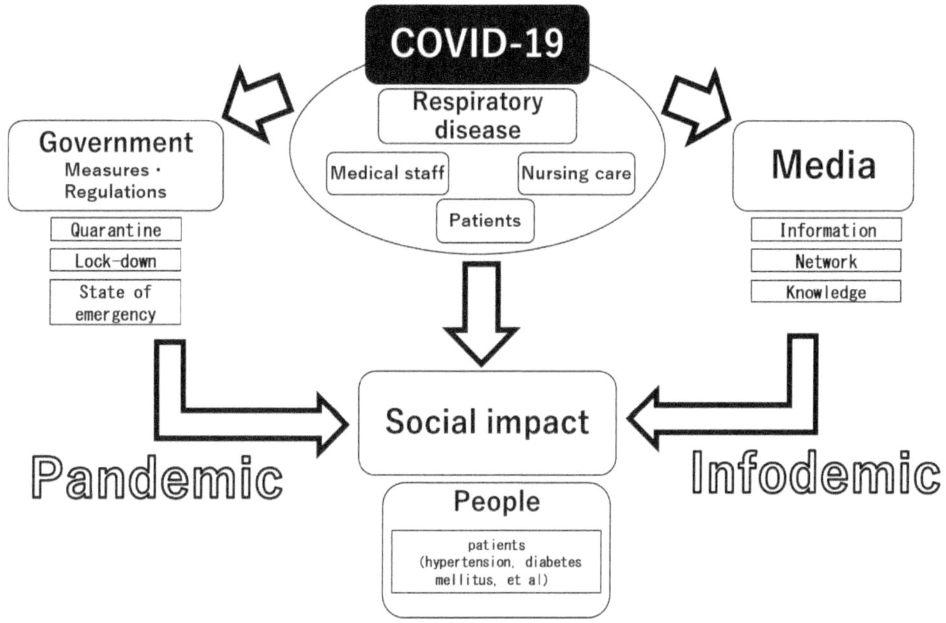

FIG. 1 Schema of COVID-19 and social stress.

In 2014, the Japanese Society of Cardiology, the Japanese Society of Hypertension, and the Japanese Society of Heart Disease published joint guidelines on the prevention and management of cardiovascular diseases during disasters, which indicated the importance of BP management from the onset of the disaster to recovery from it.

Other specific social stressors existed during the COVID-19 pandemic in patients with hypertension. The COVID-19-related mortality rates in Wuhan at the beginning of the pandemic were 0.9% in people without an underlying disease and 6% in patients with hypertension (Wu & McGoogan, 2020). The high mortality rate increased the fear of COVID-19 in many patients with hypertension. Furthermore, SARS-CoV-2 enters human cells via the angiotensin-converting enzyme 2 (ACE2) receptor, progressing to acute severe respiratory infection. The relationship between the renin-angiotensin system (RAS) and COVID-19 has been the focus of many studies. At the beginning of the pandemic, there were unfounded concerns that increased ACE2 levels in patients treated with RAS inhibitors might be associated with COVID-19 severity. In addition, approximately 60% of Japanese patients with hypertension were treated with RAS inhibitors such as ACE inhibitors or angiotensin receptor blockers (Ishida et al., 2019); thus, many patients with hypertension became interested in the association between RAS and COVID-19. Since March 2020, the European Society of Cardiology (18), the International Society of Hypertension (19), and the Japanese Society of Hypertension have responded to these concerns through their official websites or YouTube channels, stating that there is no evidence supporting the relationship between the use of RAS inhibitors and COVID-19 severity. Moreover, they advised that the treatment should not be easily interrupted and that patients should consult with doctors to continue their treatment.

Various studies have been conducted since the early stages of the COVID-19 pandemic, reporting that RAS inhibitors are not related to COVID-19 severity; however, they may be associated with the prevention of severe COVID-19. In February 2020, in Wuhan, no significant difference was observed in the risk of intensive care unit admission in individuals with or without hypertension (Huang et al., 2020). Zhang et al. reported a significant reduction in mortality rates in patients treated with RAS inhibitors (Zhang et al., 2020). In Japan, Matsuzawa et al. found no significant relationship between the severity of COVID-19 and the use of RAS inhibitors, contributing to a reduction in the incidence of complications of impaired consciousness (Matsuzawa et al., 2020).

In our survey of patients with hypertension who regularly visited general practitioners at clinics, 66% of patients had serious concerns regarding the association between COVID-19 severity and hypertension (Kobayashi et al., 2022). In a previous systematic review, hypertension was cited as one of the causes of worsening stress during the COVID-19 pandemic (Luo et al., 2020).

Changes in blood pressure during the COVID-19 pandemic

Table 1 shows the influence of the COVID-19 pandemic on BP. An internet survey of 30,000 patients with hypertension using antihypertensive medications was conducted by Omron Healthcare Corporation (Kyoto, Japan); approximately 90%

TABLE 1 Influence of the COVID-19 pandemic on blood pressure.

	Subjects		Before pandemic (mmHg)	After pandemic (mmHg)
Kobayashi et al. (2022)	Patients with hypertension (n=748)	In the office	136.5.17.5/78.2±12.0	138.6.18.6±79.0±12.2
		At home	128.2.10.3/75.8±8.8	126.9±10.2/75.2±9.0
Shimamoto (2020)	Patients with antihypertensive drugs (n=30,000)	At home	Increase; 7.6%, decrease; 9.6%	
Kobayashi, Kato, et al. (2021)	Medical health check-up (n=16,581)	Male	124.1±14.3/77.2±11.1	126.0±15.0/78.8±11.4
		Female	117.4±15.3/71.5±10.9	119.6±16.0/73.1±11.0
Satoh et al. (2022)	National database (n=157,510)	Without antihypertensive drugs	Male; 120.3±14.2/75.0±10.6 Female; 115.7±17.0/70.0±10.7	Increase by 1–2/0.5–1
		With antihypertensive drugs	Male; 130.9±15.5/81.4±10.2 Female; 133.7±16.7/78.4±10.0	
Endo et al. (2022)	Patients with diabetes (n=176)		130[122,138]/70[65, 76][a]	137[124,148]/74[67,85]

[a]*The values means the median [25th percentile, 75th percentile].*

of patients reported that the COVID-19 pandemic had an impact on their daily lives, whereas 17.2% reported a change in BP, of whom 7.6% reported an increase in BP (Shimamoto, 2020). In our survey of 748 patients with hypertension conducted in Sagamihara city, after the declaration of the state of emergency, "office BP" significantly increased from 136.5±17.5/78.2±12.0mmHg to 138.6±18.6/79.0±12.2mmHg (systolic BP, $P<0.001$; diastolic BP, $P=0.03$), whereas "home BP" in the morning decreased from 128.2±10.3/75.8±8.8mmHg to 126.9±10.2/75.2±9.0mmHg (systolic BP, $P<0.001$; diastolic BP, $P=0.01$)), resulting in the increased incidence of white-coat hypertension from 13% to 17% ($P<0.001$) (Kobayashi et al., 2022).

COVID-19 spread from the Tokyo metropolitan area, which includes Sagamihara, and the first death due to COVID-19 in Japan was reported in a hospital in Sagamihara. This might have strengthened the fear and stress regarding COVID-19 in patients with hypertension in Sagamihara, leading to an increase in office BP. By contrast, although COVID-19 cases were rarely reported in regional cities in Japan in 2020, a survey conducted as part of a routine health check-up in Niigata city showed an increase in BP from 124.1±14.3/77.2±11.1mmHg to 126.0±15.0/78.8±11.4mmHg in men and from 117.4±15.3/71.5±10.9mmHg to 119.6±16.0/73.1±11.0mmHg in women (Kobayashi, Kato, et al., 2021) This finding demonstrates how the fear of COVID-19 rapidly spread across the country. Furthermore, an analysis of large-scale health check-up data from 26 anonymous hospitals in Japan also reported an increase in systolic BP by from 1 to 2mmHg and diastolic BP from 0.5 to 1mmHg during the state of emergency (Satoh et al., 2022). In contrast, changes in BP in patients with diabetes have rarely been reported. Endo et al. reported that glycemic and BP control worsened in 176 patients with diabetes, and no improvement was observed during the state of emergency, even in patients who managed their diet and engaged in exercise (Endo et al., 2022).

Stress score and blood pressure

A checklist evaluating sleep, exercise, diet, weight, infection prevention measures, thromboprophylaxis treatment, adherence to medications, and BP control status (disaster cardiovascular prevention (DCAP) score) is available in the Guidelines for the Prevention and Management of Disaster-Related Cardiovascular Disease (Kario et al., 2005).

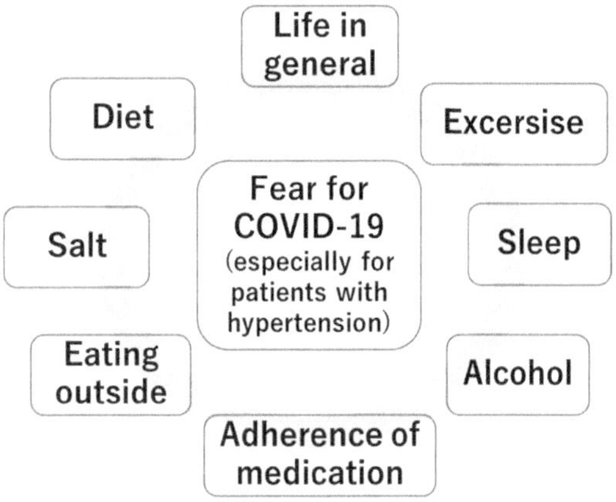

FIG. 2 Diagram of the nine stresses.

With reference to this checklist, we evaluated the levels of stress in the daily lives of patients in Sagamihara during the state of emergency. We surveyed nine factors; eight were related to the items mentioned in the checklist, and one was related to the fear of COVID-19, as shown in Fig. 2. Approximately 52% of the respondents reported worsening stress in their overall lives, and the proportion of respondents who reported a reduction in physical activity (39%) was higher than that of those who reported poor dietary intake (16%), which may be due to the "stay home" recommendation during the state of emergency. Multiple linear regression analysis revealed that poor dietary intake and poor adherence to medications were independent factors for elevated office BP, indicating a relationship between stress during the state of emergency and worsening of BP.

In addition, the scores of all nine factors were totaled (much improved, −2 points; little improved, −1 point; no change, 0 point; little worsened, +1 point; much worsened, +2 points), and the results of the two groups were compared. A total score of ≥3 points indicated higher levels of perceived stress while a total score of <3 points indicated lower levels of perceived stress. Office and early morning home BP were elevated in the group with a total stress score of ≥3 points, and albuminuria also worsened (Kobayashi et al., 2022). All nine stress-related factors evaluated in this study have the potential to accurately identify patients who experienced a chain of events such as "worsening of stress," "worsening of BP," and "organ damage" during the COVID-19 pandemic. However, further evaluation and investigation of their usefulness are required in the future.

COVID-19 pandemic and BP in the world

The COVID-19 pandemic had a public health impact in most parts of the world; however, the magnitude of transmission, public health measures, or political enforcement measures was not necessarily the same across countries. Therefore, the impact of the COVID-19 pandemic on BP was expected to vary across countries and regions. Zhang et al. surveyed a study of 283 patients with hypertension aged 60–80 years in Wuhan, China, and they reported that home BP levels temporarily worsened at the time of the COVID-19 outbreak compared to other regions; however, it subsequently returned to original levels (Zhang et al., 2021). In another report of 3724 elderly patients with hypertension in China, 7% of all patients felt anxiety for the COVID-19 pandemic, with a significant increase in systolic BP (1.2–1.7 mmHg, $P < 0.05$) and a high proportion without control (Zhang et al., 2022). Some surveys with a large sample size in the United States were reported regarding the COVID-19 pandemic and BP. Using data from a large national digital health BP disease management program with 72,706 participants, Shah et al. reported that higher systolic and diastolic BP at home were observed during the COVID-19 pandemic compared with prepandemic periods, with a higher proportion of uncontrolled and severely uncontrolled hypertension (Shah et al., 2022). In another large-scale survey with 464,585 participants in an annual employer-sponsored wellness program in the United States, significantly higher BP at home was seen during the COVID-19 pandemic compared to the previous year and larger increases were observed in females for both systolic and diastolic BP, in elder participants for systolic BP, and in younger participants for diastolic BP ($P < 0.0001$) (Laffin et al., 2022). In Spain, a prospective study with 6236 workers showed a significant increase in body weight, BP as well as low-density lipoprotein cholesterol and plasma glucose after the lockdown due to the COVID-19 pandemic; the relative risk was highest in BP with 1.28 [95% CI, 1.26, 1.31] (Ramerez Manent et al., 2022). Interestingly, in other parts of Europe such as Italy and France, significant home BP reductions were observed during the lockdown (Girerd et al., 2022;

Pengo et al., 2022). These results were consistent with our Japanese data (Kobayashi et al., 2022). Given this lack of uniform results, the ESH COVID-19 Task Force initiated the ESH ABPM COVID-19 study in Europe. The main goal of this study was to determine the impact of the COVID-19 lockdown on 24-h BP profiles and BP variability. Reports from Brazil, where the government took relatively passive pandemic measures despite a large number of infections, differed from other regions. In data from 24,227 untreated and 27,699 treated participants, a temporary mild increase in ambulatory BP was observed in the early stages of the COVID-19 pandemic; however, no significant change in BP was observed throughout the COVID-19 pandemic (Feitosa et al., 2022).

COVID-19 pandemic and access to medical care

The pandemic also had a significant impact on access to medical care. The pandemic reportedly caused disruptions in medical facilities, a reduction in regular medical services due to the expansion of the medical system for COVID-19, and the closure of hospitals associated with clusters or lockdowns (Skeete et al., 2020). In a survey conducted by Omron Healthcare Corporation (Kyoto), more than 70% of the respondents reported a change in attitude toward hospital visits. Many patients wanted to reduce or discontinue their hospital visits and felt anxiety over long waiting times in hospitals. These were natural measures undertaken to prevent infection (Shimamoto, 2020).

Yamaguchi et al. reported that the total number of hospital admissions and outpatient visits decreased by 27% and 22%, respectively, according to data obtained from 26 anonymous hospitals in Japan (Yamaguchi et al., 2022). In Singapore, owing to the risk of COVID-19 transmission by in-person interactions between patients and healthcare workers, exercise classes for cardiac rehabilitation were adjourned for up to 6 months during the pandemic; these delays may have resulted in suboptimal patient outcomes (Yeo et al., 2020) Moreover, substantial increases in the number of avoidable cancer deaths were expected as a result of diagnostic delays due to the COVID-19 pandemic in the UK (Maringe et al., 2020). Indeed, the total national healthcare expenditure amounted to 38.3 trillion yen from April 2020 to February 2021, which decreased by 4.4% annually (https://www.mhlw.go.jp/topics/medias/month/21/02.html).

Although the rates of hospital admissions for malignant neoplasms and maintenance hemodialysis and chemotherapy in outpatients were not significantly affected, the pandemic affected patients with chronic disease who showed few symptoms, suggesting that patients refrained from visiting general practitioners. With regard to clinical practice for lifestyle-related diseases, the use of hypoglycemic drugs and calcium channel blockers decreased by up to 15% in May 2020, and this decline continued until the end of 2020 (Yamaguchi et al., 2022). This indicates that the pandemic hindered the provision of necessary medical care to patients with hypertension or diabetes; this situation could be a major risk factor for the development of cardiovascular disease events. After Hurricane Katrina and the floods in New Orleans in August 2005, increased rates of unemployment, refraining from visiting medical centers, and smoking were associated with increased cardiovascular outcomes (Jiao et al., 2012).

Non-COVID-19-related deaths increased from July 2021 to October 2021 in England, compared to before the COVID-19 pandemic (Office for National Statistics, 2021), and this increase was speculated to be related to the presence or absence of diabetes. The cohort study showed that the completion of routine diabetes care processes, including glycated hemoglobin A1c, BP, cholesterol, serum creatinine, urine albumin, foot surveillance, BMI, and smoking status, largely decreased under the COVID-19 pandemic. This decrease was related to the increase in mortality in patients with diabetes in England (Valabhji et al., 2022). Although detailed BP values were not specified, a worsening of BP management was predicted.

In the United States, social disorganization caused problems in providing adequate healthcare in the early stages of the COVID-19 pandemic. Emergency department visits in the United States fell by up to 60% in some regions at the beginning of the pandemic; 40% were for nonurgent conditions that could be treated by primary care (European Society of Hypertension Corona-virus Disease 19 Task Force, 2021). Visits because of hypertension also decreased. It could be expected that patients with hypertension might refrain from visiting a doctor, which could lead to poor BP management. Other reports indicated that BP data were difficult to ascertain in the early stages of the COVID-19 pandemic (Meador et al., 2022).

Prevention of cerebrovascular and cardiovascular events during the COVID-19 pandemic

A trend to increase the length of the prescription validity for chronic disease treatments was observed during the COVID-19 pandemic (Yamaguchi et al., 2022). Although this is an effective approach to reduce the number of patient visits to general practitioners to reduce the risk COVID-19 infection, whether this is a fundamental solution remains unclear. Telemedicine has received the most attention for minimizing interruptions in medical care of chronic diseases during the COVID-19 pandemic (Wang et al., 2021). In the United States, telemedicine is gradually expanding, accounting for its advantages and disadvantages (Kichloo et al., 2020). In Japan, restrictions on online medical treatment have also been partially relaxed, albeit temporarily, making it possible to provide medical treatment online even on the first consultation with a general practitioner. However, future research and debates are warranted to determine whether these telemedicine systems are as effective as conventional face-to-face consultations in managing the risk factors for cardiovascular events, such as hypertension, diabetes, or dyslipidemia. While the use of telemedicine is effective, it may have had a negative impact on hypertension treatment in the early stages of the COVID-19 pandemic when society was in turmoil and isolation was significant. The proportion of patients with any in-person primary care visit dropped by almost 50% (33–17%), although telehealth visits grew by 9%, which may have led to poor BP control (Beckman et al., 2021). Telemedicine will be further developed in the future, and although some problems will need to be solved, recent international expert position papers provided evidence and recommendations on the use of telemedicine for the clinical management of hypertension (Kario et al., 2020; Omboni et al., 2020).

In fall 2022, many patients were still being infected with COVID-19, and the emergence of new variants of SARS-CoV-2 or the occurrence of different waves of COVID-19 infection required attention. With the persistence of COVID-19, practitioners needed to go beyond BP management for their patients and manage their stress. In addition to diet and exercise, dealing with various types of stress is necessary, which might not be feasible in telemedicine.

As the pandemic continued through 2022, reports regarding cardiovascular events also continued. Zhang et al. reported that patients with anxiety had an increased risk of cardiovascular events with a hazard ratio of 2.47 (95% CI, 1.10, 5.58; $P = 0.03$) (Zhang et al., 2022). An interesting report from Poland showed that atherosclerosis progression was reduced by unexpected reasons during the COVID-19 pandemic. Because aircraft traffic substantially decreased at the beginning of the pandemic, a reduction in noise damage by aircraft affected not only BP but also pulse wave velocity improvement (Wojciechowska et al., 2022). As the COVID-19 pandemic caused various stresses and the situation in each country was different, much further research is necessary to evaluate the relationship among COVID-19, BP, and cardiovascular prognosis.

Applications to other areas

Hypertension is an independent risk factor for cardiovascular and renal events. Despite the relative ease of diagnosis, the availability of effective and relatively inexpensive drugs, and the utilization of treatment guidelines in many countries around the world, many patients with hypertension do not reach their BP targets. This situation is known as the hypertension paradox, and it is an urgent problem that needs to be solved. Improved BP control is necessary to improve cardiovascular and renal events and extend healthy life expectancy. We have reviewed the association among COVID-19, BP, and social stress, and it is hoped that our review will be useful in improving BP control in the post-COVID-19 era as well as improving outcomes for patients with hypertension. Furthermore, the COVID-19 pandemic had as large an impact as natural disasters. As natural disasters are bound to occur in the future in any part of the globe, this review may also be useful for the management of patients with hypertension during disasters. The area where our review is most useful is telemedicine. With advances in the Internet of Things, telemedicine was in the spotlight even before the pandemic. The pandemic forced the world to push for the implementation of telemedicine in an undeniable way. Therefore, several problems became apparent in the early stages of the pandemic; however, the lessons learned from these problems can lead to further developments in telemedicine in the future.

Conclusion

We discussed the impact of the COVID-19 pandemic on patients with hypertension in clinical practice. The COVID-19 pandemic is considered a disaster. To our knowledge, no studies have accurately assessed the impact of a pandemic before its onset; currently, various studies are being conducted worldwide in a state of exploration. Making use of this study's findings to manage future pandemics (or disasters) is important.

Mini-dictionary of terms

- **BP at home.** Various practice guidelines recommend taking BP measurements at home (basically in the early morning) and monitoring them because they have better prognostic and predictive power than BP values measured in the office.
- **Guidelines for disaster medicine for patients with cardiovascular diseases:** After Kario et al. reported the increase in BP soon after the Hanshin-Awaji earthquake in 1995, disaster medicine got a lot of attention. The great East Japan

earthquake in 2011 caused many casualties and shocked Japan. These two major earthquakes mainly triggered a range of evidence and led to the publishing of these guidelines.

- **Hypertension paradox.** Although guidelines provide target BP values for each condition, the proportion of patients achieving target BP in practice is low, reported to be 25% in Japan.
- **Infodemic.** An infodemic is the phenomenon of the rapid spread of both uncertain and accurate information through social media. The term came into global use around the time of the 2003 SARS outbreak and spread quickly around the world after the COVID-19 pandemic.

Summary points

- The COVID-19 pandemic was a disaster.
- The COVID-19 pandemic caused increases in social stress.
- Medical care, especially by general practitioners, was influenced by the COVID-19 pandemic.
- Blood pressure in the office but not at home increased under the state of emergency.
- The social stress increased blood pressure in the office.
- Telemedicine was a useful method to provide care during the COVID-19 pandemic.

References

Bagus, P., Peña-Ramos, J. A., & Sánchez-Bayón, A. (2021). COVID-19 and the political economy of mass hysteria. *International Journal of Environmental Research and Public Health, 18*, 1376.

BBC News Services. (2020). *Coronavirus: The world in lockdown in maps and charts. https://www.bbc.com/news/world-52103747.*

Beckman, A. L., King, J., Streat, D. A., Bartz, N., Figueroa, J. F., & Mostashari, F. (2021). Decreasing primary care use and blood pressure monitoring during COVID-19. *The American Journal of Managed Care, 27*, 366–368.

Centers For Disease Control And Prevention, N. C. F. H. S. National Vital Statistics System, Mortality 1999–2020 on CDC Wonder Online Database, released in 2021. Data are from the Multiple Cause of Death Files, 1999–2020, as compiled from data provided by the 57 vital statistics jurisdictions through the Vital Statistics Cooperative Program. Accessed at http://wonder.cdc.gov/ucd-icd10.html on Jul 1, 2022 11:19:55 AM. *http://wonder.cdc.gov/ucd-icd10.html*, Accessed on July 1, 2022.

Dong, M., & Zheng, J. (2020). Letter to the editor: Headline stress disorder caused by Netnews during the outbreak of COVID-19. *Health Expectations, 23*, 259–260.

Dubey, S., Biswas, P., Ghosh, R., Chatterjee, S., Dubey, M. J., Chatterjee, S., Lahiri, D., & Lavie, C. J. (2020). Psychosocial impact of COVID-19. *Diabetes and Metabolic Syndrome: Clinical Research and Reviews, 14*, 779–788.

Endo, K., Miki, T., Itoh, T., Kubo, H., Ito, R., Ohno, K., Hotta, H., Kato, N., Matsumoto, T., Kitamura, A., Tamayama, M., Wataya, T., Yamaya, A., Ishikawa, R., & Ooiwa, H. (2022). Impact of the COVID-19 pandemic on glycemic control and blood pressure control in patients with diabetes in Japan. *Internal Medicine, 61*, 37–48.

European Society of Hypertension Corona-virus Disease 19 Task Force. (2021). The corona-virus disease 2019 pandemic compromised routine care for hypertension: A survey conducted among excellence centers of the European Society of Hypertension. *Journal of Hypertension, 39*, 190–195.

Feitosa, F., Feitosa, A. D. M., Paiva, A. M. G., Mota-Gomes, M. A., Barroso, W. S., Miranda, R. D., Barbosa, E. C. D., Brandão, A. A., Lima-Filho, J. L., Sposito, A. C., Coca, A., & Nadruz, W., Jr. (2022). Impact of the COVID-19 pandemic on blood pressure control: A nationwide home blood pressure monitoring study. *Hypertension Research, 45*, 364–368.

Girerd, N., Meune, C., Duarte, K., Vercamer, V., Lopez-Sublet, M., & Mourad, J. J. (2022). Evidence of a blood pressure reduction during the COVID-19 pandemic and associated lockdown period: Insights from e-health data. *Telemedicine Journal and E-Health, 28*, 266–270.

Huang, C., Wang, Y., Li, X., Ren, L., Zhao, J., Hu, Y., Zhang, L., Fan, G., Xu, J., Gu, X., Cheng, Z., Yu, T., Xia, J., Wei, Y., Wu, W., Xie, X., Yin, W., Li, H., Liu, M., … Cao, B. (2020). Clinical features of patients infected with 2019 novel coronavirus in Wuhan, China. *Lancet, 395*, 497–506.

Intage Research Co..Ltd. (2022). *Report on mental health and its effects related to COVID-19. https://www.mhlw.go.jp/content/12205000/syousai.pdf.*

Ishida, T., Oh, A., Hiroi, S., Shimasaki, Y., & Tsuchihashi, T. (2019). Current prescription status of antihypertensive drugs in Japanese patients with hypertension: Analysis by type of comorbidities. *Clinical and Experimental Hypertension, 41*, 203–210.

Jia, R., Ayling, K., Chalder, T., Massey, A., Broadbent, E., Coupland, C., & Vedhara, K. (2020). Mental health in the UK during the COVID-19 pandemic: Cross-sectional analyses from a community cohort study. *BMJ Open, 10*, e040620.

Jiao, Z., Kakoulides, S. V., Moscona, J., Whittier, J., Srivastav, S., Delafontaine, P., & Irimpen, A. (2012). Effect of Hurricane Katrina on incidence of acute myocardial infarction in New Orleans three years after the storm. *The American Journal of Cardiology, 109*, 502–505.

Kario, K. (2012). Disaster hypertension. *Circulation Journal, 76*, 553–562.

Kario, K., Matsuo, T., Shimada, K., & Pickering, T. G. (2001). Factors associated with the occurrence and magnitude of earthquake-induced increases in blood pressure. *The American Journal of Medicine, 111*, 379–384.

Kario, K., Morisawa, Y., Sukonthasarn, A., Turana, Y., Chia, Y. C., Park, S., Wang, T. D., Chen, C. H., Tay, J. C., Li, Y., & Wang, J. G. (2020). COVID-19 and hypertension-evidence and practical management: Guidance from the HOPE Asia network. *Journal of Clinical Hypertension (Greenwich, Conn.), 22*, 1109–1119.

Kario, K., Shimada, K., & Takaku, F. (2005). Management of cardiovascular risk in disaster: Jichi medical school (JMS) proposal 2004. *JMAL, 48*(7), 363–376.

Kichloo, A., Albosta, M., Dettloff, K., Wani, F., El-Amir, Z., Singh, J., Aljadah, M., Chakinala, R. C., Kanugula, A. K., Solanki, S., & Chugh, S. (2020). Telemedicine, the current COVID-19 pandemic and the future: A narrative review and perspectives moving forward in the USA. *Family Medicine and Community Health, 8*, e000530.

Kobayashi, A., Kato, K., Tanaka, K., Sotou, Y., Matsuda, K., & Sone, H. (2021). Changes in health status before and after the self-restraint life caused by a new type of coronavirus infection (COVID-19) as seen from the results of health checkups. *Official Journal of Japan Society of Ningen Dock, 36*(4), 582–589.

Kobayashi, K., Chin, K., Umezawa, S., Ito, S., Yamamoto, H., Nakano, S., … Tamura, K. (2022). Influence of stress induced by the first announced state of emergency due to coronavirus disease 2019 on outpatient blood pressure management in Japan. *Hypertension Research, 46*, 675–685.

Kuga, N. (2020). *Basic report: The anxiety by COVID-19; nationwide data consisted of 6,000 people* (pp. 1–7). *Nissei basic institute*.

Laffin, L. J., Kaufman, H. W., Chen, Z., Niles, J. K., Arellano, A. R., Bare, L. A., & Hazen, S. L. (2022). Rise in blood pressure observed among US adults during the COVID-19 pandemic. *Circulation, 145*, 235–237.

Luo, M., Guo, L., Yu, M., Jiang, W., & Wang, H. (2020). The psychological and mental impact of coronavirus disease 2019 (COVID-19) on medical staff and general public—A systematic review and meta-analysis. *Psychiatry Research, 291*, 113190.

Maringe, C., Spicer, J., Morris, M., Purushotham, A., Nolte, E., Sullivan, R., Rachet, B., & Aggarwal, A. (2020). The impact of the COVID-19 pandemic on cancer deaths due to delays in diagnosis in England, UK: A national, population-based, modelling study. *The Lancet Oncology, 21*, 1023–1034.

Matsuzawa, Y., Ogawa, H., Kimura, K., Konishi, M., Kirigaya, J., Fukui, K., Tsukahara, K., Shimizu, H., Iwabuchi, K., Yamada, Y., Saka, K., Takeuchi, I., Hirano, T., & Tamura, K. (2020). Renin-angiotensin system inhibitors and the severity of coronavirus disease 2019 in Kanagawa, Japan: A retrospective cohort study. *Hypertension Research, 43*, 1257–1266.

Meador, M., Coronado, F., Roy, D., Bay, R. C., Lewis, J. H., Chen, J., Cheung, R., Utman, C., & Hannan, J. A. (2022). Impact of COVID-19-related care disruptions on blood pressure management and control in community health centers. *BMC Public Health, 22*.

Miyakawa, M. (2012). The influence of the stress induced by East Japan Earthquake 2011 to blood pressure at home. *Ketsuatu, 19*, 935–938.

Office for National Statistics. (2020). *Coronavirus and anxiety, Great Britain: 3 April 2020 to 10 May 2020*. https://www.ons.gov.uk/peoplepopulationandcommunity/wellbeing/articles/coronavirusandanxietygreatbritain/3april2020to10may2020/previous/v1. Date accessed: Oct 1, 2022.

Office for National Statistics. (2021). *Deaths registered weekly in England and Wales, provisional*. https://www.ons.gov.uk/peoplepopulationandcommunity/birthsdeathsandmarriages/deaths/datasets/weeklyprovisionalfiguresondeathsregisteredinenglandandwales. Date accessed: Oct 1, 2022.

Omboni, S., McManus, R. J., Bosworth, H. B., Chappell, L. C., Green, B. B., Kario, K., Logan, A. G., Magid, D. J., McKinstry, B., Margolis, K. L., Parati, G., & Wakefield, B. J. (2020). Evidence and recommendations on the use of telemedicine for the Management of Arterial Hypertension: An international expert position paper. *Hypertension, 76*, 1368–1383.

Pengo, M. F., Albini, F., Guglielmi, G., Mollica, C., Soranna, D., Zambra, G., Zambon, A., Bilo, G., & Parati, G. (2022). Home blood pressure during COVID-19-related lockdown in patients with hypertension. *European Journal of Preventive Cardiology, 29*, e94–e96.

Ramerez Manent, J. I., Altisench Jané, B., Sanches Cortés, P., Busquets-Cortés, C., Arroyo Bote, S., Masmiquel Comas, L., … A. (2022). Impact of COVID-19 lockdown on anthropometric variables, blood pressure, and glucose and lipid profile in healthy adults: A before and after pandemic lockdown longitudinal study. *Nutrients, 14*(6), 1237.

Satoh, M., Murakami, T., Obara, T., & Metoki, H. (2022). Time-series analysis of blood pressure changes after the guideline update in 2019 and the coronavirus disease pandemic in 2020 using Japanese longitudinal data. *Hypertension Research, 45*, 1408–1417.

Shah, N. P., Clare, R. M., Chiswell, K., Navar, A. M., Shah, B. R., & Peterson, E. D. (2022). Trends of blood pressure control in the U.S. during the COVID-19 pandemic. *American Heart Journal, 247*, 15–23.

Shimamoto, K. (2020). *Changes in awareness and lifestyle in the pandemic of new corona virus infection*. https://www.healthcare.omron.co.jp/zeroevents/bloodpressure/topics/01.html.

Skeete, J., Connell, K., Ordunez, P., & Dipette, D. J. (2020). Approaches to the Management of Hypertension in resource-limited settings: Strategies to overcome the hypertension crisis in the post-COVID era. *Integrated Blood Pressure Control, 13*, 125–133.

Talevi, D., Socci, V., Carai, M., Carnaghi, G., Faleri, S., Trebbi, E., Di Bernardo, A., Capelli, F., & Pacitti, F. (2020). Mental health outcomes of the CoViD-19 pandemic. *Rivista di Psichiatria, 55*, 137–144.

Umemura, S., Arima, H., Arima, S., Asayama, K., Dohi, Y., Hirooka, Y., Horio, T., Hoshide, S., Ikeda, S., Ishimitsu, T., Ito, M., Ito, S., Iwashima, Y., Kai, H., Kamide, K., Kanno, Y., Kashihara, N., Kawano, Y., Kikuchi, T., … Hirawa, N. (2019). The Japanese Society of Hypertension Guidelines for the Management of Hypertension (JSH 2019). *Hypertension Research, 42*, 1235–1481.

Valabhji, J., Barron, E., Gorton, T., Bakhai, C., Kar, P., Young, B., Khunti, K., Holman, N., Sattar, N., & Wareham, N. J. (2022). Associations between reductions in routine care delivery and non-COVID-19-related mortality in people with diabetes in England during the COVID-19 pandemic: A population-based parallel cohort study. *The Lancet Diabetes & Endocrinology, 10*, 561–570.

Vindegaard, N., & Benros, M. E. (2020). COVID-19 pandemic and mental health consequences: Systematic review of the current evidence. *Brain, Behavior, and Immunity, 89*, 531–542.

Wang, H., Yuan, X., Wang, J., Sun, C., & Wang, G. (2021). Telemedicine maybe an effective solution for management of chronic disease during the COVID-19 epidemic. *Primary Health Care Research & Development, 22*, e48.

Wojciechowska, W., Januszewicz, A., DroŻDŻ, T., Rojek, M., Bączalska, J., Terlecki, M., Kurasz, K., Olszanecka, A., Smólski, M., Prejbisz, A., Dobrowolski, P., Grodzicki, T., Hryniewiecki, T., Kreutz, R., & Rajzer, M. (2022). Blood pressure and arterial stiffness in association with aircraft noise exposure: Long-term observation and potential effect of COVID-19 lockdown. *Hypertension, 79*, 325–334.

Wu, Z., & McGoogan, J. M. (2020). Characteristics of and important lessons from the coronavirus disease 2019 (COVID-19) outbreak in China. *JAMA*, *323*, 1239.

Yamaguchi, S., Okada, A., Sunaga, S., Ikeda Kurakawa, K., Yamauchi, T., Nangaku, M., & Kadowaki, T. (2022). Impact of COVID-19 pandemic on healthcare service use for non-COVID-19 patients in Japan: Retrospective cohort study. *BMJ Open*, *12*, e060390.

Yeo, T. J., Wang, Y.-T. L., & Low, T. T. (2020). Have a heart during the COVID-19 crisis: Making the case for cardiac rehabilitation in the face of an ongoing pandemic. *European Journal of Preventive Cardiology*, *27*, 903–905.

Zhang, S., Zhong, Y., Wang, L., Yin, X., Li, Y., Liu, Y., Dai, Q., Tong, A., Li, D., Zhang, L., Li, P., Zhang, G., Huang, R., Liu, J., Zhao, L., Yu, J., Zhang, X., Yang, L., Cai, J., & Zhang, W. (2022). Anxiety, home blood pressure monitoring, and cardiovascular events among older hypertension patients during the COVID-19 pandemic. *Hypertension Research*, *45*, 856–865.

Zhang, S., Zhou, X., Chen, Y., Wang, L., Zhu, B., Jiang, Y., Bu, P., Liu, W., Li, D., Li, Y., Tao, Y., Ren, J., Fu, L., Li, Y., Shen, X., Liu, H., Sun, G., Xu, X., Bai, J., … Cai, J. (2021). Changes in home blood pressure monitored among elderly patients with hypertension during the COVID-19 outbreak: A longitudinal study in China leveraging a smartphone-based application. *Circulation. Cardiovascular Quality and Outcomes*, *14*.

Zhang, P., Zhu, L., Cai, J., Lei, F., Qin, J. J., Xie, J., Liu, Y. M., Zhao, Y. C., Huang, X., Lin, L., Xia, M., Chen, M. M., Cheng, X., Zhang, X., Zhang, X., Guo, D., Peng, Y., Ji, Y. X., Chen, J., She, Z. G., … Li, H. (2020). Association of Inpatient use of angiotensin-converting enzyme inhibitors and angiotensin II receptor blockers with mortality among patients with hypertension hospitalized with COVID-19. *Circulation Research*, *126*, 1671–1681.

Zhu, N., Zhang, D., Wang, W., Li, X., Yang, B., Song, J., Zhao, X., Huang, B., Shi, W., Lu, R., Niu, P., Zhan, F., Ma, X., Wang, D., Xu, W., Wu, G., Gao, G. F., & Tan, W. (2020). A novel coronavirus from patients with pneumonia in China, 2019. *New England Journal of Medicine*, *382*, 727–733.

Chapter 14

Neurosurgical trauma management during COVID-19 restrictions

James Zhou[a], Michael Zhang[b], and Harminder Singh[b]
[a]*California Northstate University College of Medicine, Elk Grove, CA, United States*, [b]*Department of Neurosurgery, Stanford University School of Medicine, Stanford, CA, United States*

Abbreviations

COVID-19	coronavirus disease
ED	emergency department
LOS	length of stay
SCVMC	Santa Clara Valley Medical Center
SHC	Stanford Health Care
SIP	shelter-in-place
TBI	traumatic brain injury

Introduction

Starting in late 2019, the COVID-19 pandemic effected sweeping changes, not only to the types of neurosurgical injuries sustained but also the management of these patients. The rapid spread of the disease globally led to shelter-in-place (SIP) protocols, which placed harsh limits on person-to-person contact to limit transmission of the virus (Cucinotta & Vanelli, 2020). Subsequently, starting in spring 2020, there was a massive transition in how the general public interfaced with hospital systems. Neurosurgical trauma units are one component of a robust emergency medicine and trauma network found worldwide. The responsibilities of a neurotrauma service are focused on the identification, triage, and operation of procedures that affect the central nervous system and the surrounding bones. Throughout the pandemic, the changes effected by the COVID-19 pandemic shaped how neurosurgical trauma was managed. This chapter utilizes the experiences of many neurosurgical trauma units from across the globe to provide a summary of the adjustments required and offer insight into how this information can be used to model future responses.

Changes in neurosurgical trauma: An overview

On March 16, 2020, California Governor Gavin Newsom officially ordered a lockdown of the state via a shelter-in-place protocol, restricting activity outside the home as well as limiting interpersonal interaction in an effort to stop the spread of COVID-19. These restrictions, similar to those in place worldwide, drastically affected the patient population and the subsequent type of neurosurgical intervention necessitated in the following months (Bernucci et al., 2020; Pinggera et al., 2021; Zhang et al., 2021). Additionally, Santa Clara county, which encompasses Stanford University, also enacted additional hospital-based restrictions that limited the performance of elective surgeries during the lockdown in an effort to optimize neurosurgical utilization (Zhang et al., 2022). This chapter will highlight some of the experiences that neurosurgical services in Santa Clara County faced while dealing with shelter-in-place regulations and compare them to neurosurgical services under similar restrictions globally. Additionally, we will examine how the pandemic impacted the management of neurosurgical trauma globally by looking at guidelines for triage and case selection, with a discussion about how to

Management, Body Systems, and Case Studies in COVID-19. https://doi.org/10.1016/B978-0-443-18703-2.00049-9
Copyright © 2024 Elsevier Inc. All rights are reserved, including those for text and data mining, AI training, and similar technologies.

optimize neurosurgical utilization under future shelter-in-place restrictions. In summary, these experiences demonstrate how neurosurgical units reacted in the face of government-backed restrictions and provide insight into how to allocate resources in the event of future pandemics or public-health crises.

Neurosurgical trauma in Santa Clara County

Santa Clara County is home to two Level 1 trauma centers, Stanford Healthcare (SHC) and Santa Clara Valley Medical Center (SCVMC), which are a private academic center and a public hospital, respectively. The two centers each receive a different population and volume of neurosurgical trauma, and thus were likely to be impacted differently by the blanket restrictions enacted during the beginning of the SIP lockdown. To discern the impact of COVID-19 restrictions on neurosurgical trauma volume and patient population at these institutions, Zhang et al. (2021) did a retrospective review of patients with traumatic brain injury (TBI) and spinal fractures from 2018 to 2019 versus 2020, using previous years for historical comparison (Zhang et al., 2021). Their results showed a significant decrease in the total number of trauma admissions (376/937 (40%) to 106/399 (27%), $P<.0001$, Fig. 1), TBI and spine fracture admissions (238/585 (41%) to 58/238 (24%), $P<.0001$, Fig. 1), and neurosurgical consults (169/428 (39%) to 44/166 (27%), $P=.003$, Fig. 1) from pre- to post-lockdown periods in 2018 and 2019 versus 2020. Despite these decreases, additional analyses looking at the consult rate, emergency department (ED) to critical care service transfer rate, and hospital to rehab transfer rate showed no significant differences between pre- and postpandemic time periods. Comparing neurological trauma length of stay (LOS) showed a significant increase in pre- and postlockdown LOS during 2018 and 2019 but not 2020.

Comparing SHC to SCVMC showed no significant difference between hospitals in the number of total trauma admissions, TBI or spine fracture admissions, consults for TBI and spinal fracture, or neurosurgery procedures performed in the same time periods analyzed (Table 1). While the two hospitals cater to very different patient populations, it seemed that the restrictions enacted still had similar effects on each, with a global decrease in patient volume as well as TBI and spine fracture patients. In a similar fashion, analyses looking at consult rate, ED to Critical Care Service transfer rate, hospital to rehabilitation transfer rate or length of stay (LOS) showed no significant differences across institutions. These findings are similar to those reported by Johnson et al., who, in a multicenter review from 36 trauma centers in Michigan, found a significant decrease of TBI rate in the first 8 weeks of their lockdown (March 13–May 7, ERR 0.82, $P<.0001$) and a lower rate of patients presenting with an ED Glasgow Coma Scale Score of 15 (59% pre-COVID-19 2017–19 vs 56% COVID-19 2020, $P=.016$) (Johnson et al., 2022). In total, the lockdown period resulted in a significant decrease in total trauma, TBI, and neurosurgical consults, reflecting a patient population that was restricted from engaging in potentially high-risk activities as well as a pivot away from elective neurosurgical utilization across institutions, and coincided with a decrease in neurosurgical LOS.

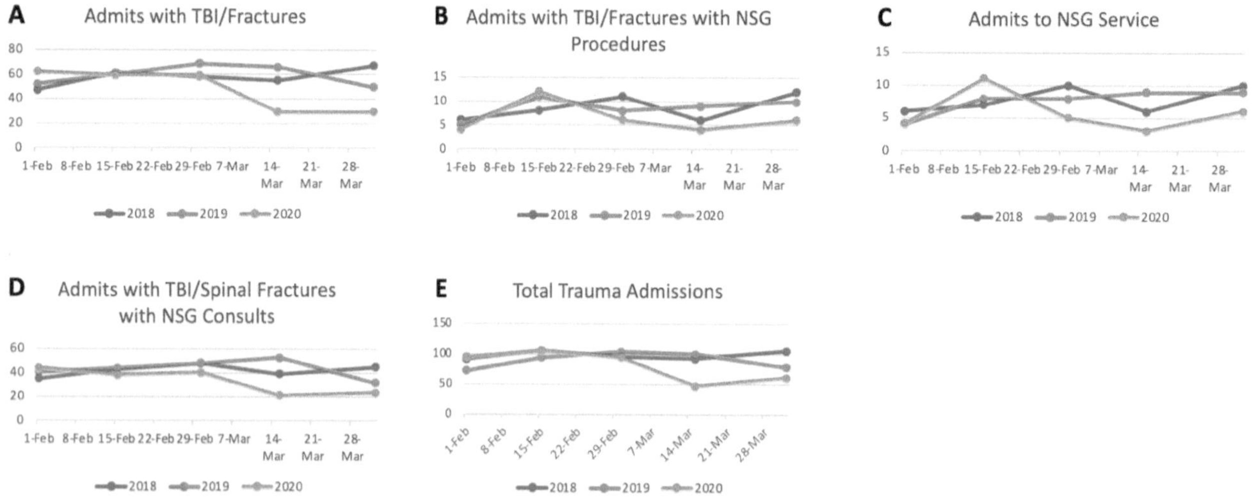

FIG. 1 Combined patient census for Stanford Hospital and Santa Clara Valley Medical Centers in the weeks before and after quarantine implementation.

Plots depict number of traumas with (A) TBI or spinal fractures, (B) neurosurgical procedures, (C) admission to the neurosurgery service, and (D) neurosurgery consults. Plot (E) depicting total traumas. *ED*, emergency department; *NSG*, neurosurgery. *From Zhang, M., Zhou, J., Dirlikov, B., Cage, T., Lee, M., & Singh, H. (2021). Impact on neurosurgical management in level 1 trauma centers during COVID-19 shelter-in-place restrictions: The Santa Clara County experience.* Journal of Clinical Neuroscience, 88, *128–134. https://doi.org/10.1016/j.jocn.2021.03.017. With permission from Elsevier Publishing Company.*

TABLE 1 Comparison of the neurosurgical trauma services between SCVMC and SHC, before and after the shelter-in-place policy of 2020.

	Pre-SIP	Post-SIP	Test statistic	*P*-value
TBI/fractures			0.184	.668
SHC	102	31		
SCVMC	78	27		
NSG procedure			–	1
SHC	15	7		
SCVMC	7	3		
NSG consults			0.372	.542
SHC	60	24		
SCVMC	62	20		
NSG admits			–	1
SHC	7	3		
SCVMC	0	0		
Trauma admits			1.09	.297
SHC	102	31		
SCVMC	191	75		

SIP, shelter-in-place; *SHC*, Stanford Hospital; *SCVMC*, Santa Clara Valley Medical Center; *TBI*, traumatic brain injury; –, Fisher's exact test.
From Zhang, M., Zhou, J., Dirlikov, B., Cage, T., Lee, M., & Singh, H. (2021). Impact on neurosurgical management in level 1 trauma centers during COVID-19 shelter-in-place restrictions: The Santa Clara County experience. *Journal of Clinical Neuroscience, 88*, 128–134. https://doi.org/10.1016/j.jocn.2021.03.017. With permission from Elsevier Publishing Company.

Following the SIP protocol enacted in March, there was a sharp decline in neurosurgical trauma at both SHC and SCVMC. Then, starting in May, the restrictions on elective surgeries were lifted and both institutions gradually started to return to normal operative capacity. To better understand the trends in types of surgeries performed after elective restrictions were lifted, an additional retrospective review looked at how both spinal and cranial surgeries performed at SCVMC were affected by SIP (Zhang et al., 2022). By delineating pre-SIP (February 15–March 15), SIP (March 15–April 15), post-SIP (June), and late (August) time periods, they were able to look at longitudinal trends both during 2020 and historically. They initially reported that post-SIP 2020, SCVMC saw a significant proportional increase in the total number of cases relative to 2018 and 2019, which served again as historical controls, without a significant difference between both pre-SIP and SIP or SIP and late. However, the categorization of surgeries as spinal or cranial showed a significant *decrease* in the proportion of total spinal cases in pre-SIP compared to SIP versus historical controls and then a subsequent significant *increase* in SIP compared to post-SIP in the same time periods analyzed. A similar pattern was not observed when looking at cranial surgeries during these time periods. They also looked at elective cases, both spinal and cranial, as compared to cranial cases during each time period, noting that the proportion of elective spinal cases significantly increased relative to previous years; however, they did not notice this trend in add-on spinal cases, elective, or add-on cranial cases (Fig. 2). These findings point to the idea that as hospitals cut elective surgeries to comply with government protocols, postponing elective *spinal* surgeries was the main driver of that decrease. A survey by the Lumbar Spine Research Society among its own spinal surgeon member base supports this idea, showing a decrease in elective spinal cases without a decrease in emergency spinal surgeries during this period in the pandemic (Arnold et al., 2021). Later, when elective restrictions were lifted, case volume started to return to normal values seen in previous years.

These experiences serve to paint a picture of the changes that occurred both before and after the COVID-19 SIP restrictions began at both SHC and SCVMC. Both trauma centers were significantly impacted by decreased patient volume, most notably affecting elective spinal surgeries. The changes here can best be utilized in planning for future pandemic restriction periods, where surgical indications are vastly different from daily operative norms and emergency surges are always possible.

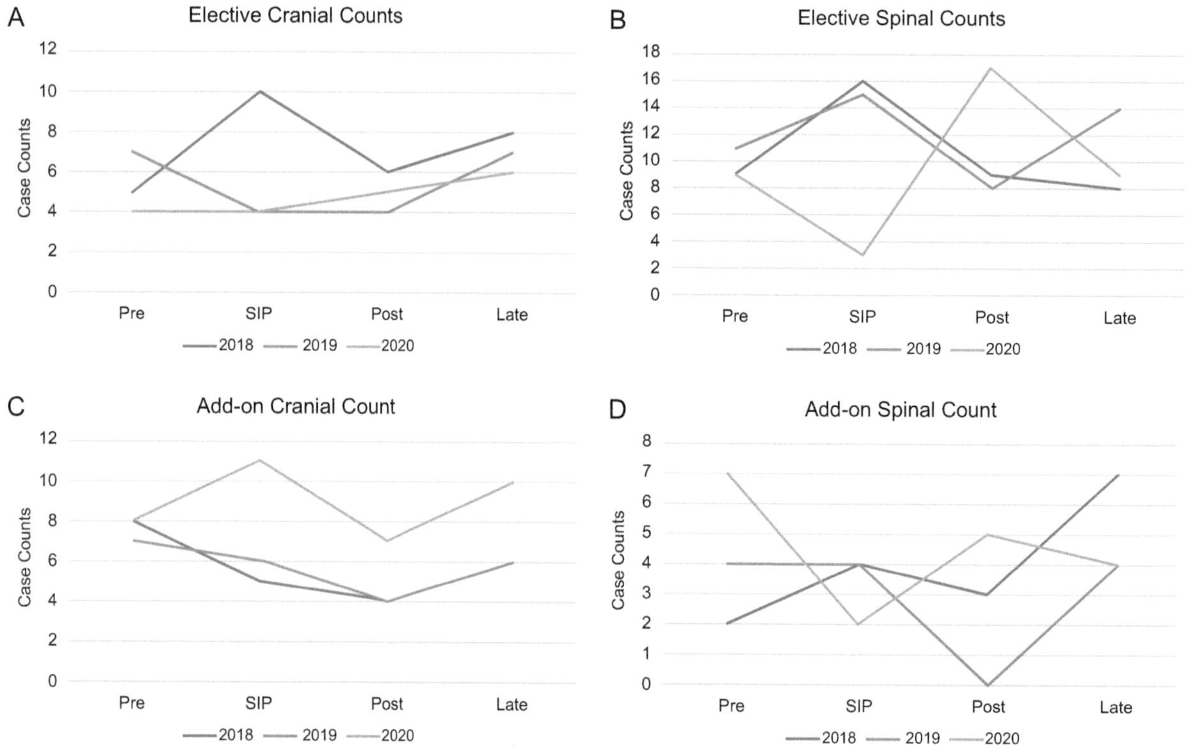

FIG. 2 Census of elective and add-on surgeries, either cranial or spinal during the pre-SIP, SIP, post-SIP, and late phases for 2018–20. (A) Total volume of elective cranial surgeries. (B) Total volume of elective spinal surgeries. (C) Total volume of add-on cranial surgeries. (D) Total volume of add-on spinal surgeries. *SIP*, shelter-in place. *Zhang, M., Zhou, J., Dirlikov, B., Cage, T., Lee, M., & Singh, H. (2022). Impact on neurosurgical management in a Level 1 trauma center post COVID-19 shelter-in-place restrictions. Journal of Clinical Neuroscience, 101, 131-136. https://doi.org/10.1016/j.jocn.2022.04*

Impact on neurosurgical trauma domestically

As mentioned before, the COVID-19 travel and interpersonal interaction restrictions were not unique to Santa Clara County and saw widespread adoption by many large municipalities across the United States. Neurosurgical trauma units working under these protocols saw many changes to both their total patient volume as well as the types of procedures performed (Figueroa et al., 2021; Kessler et al., 2020; Lara-Reyna et al., 2020; Saad et al., 2020). In this section, the efforts and experiences of some of these units will be detailed to examine similarities and differences among different neurosurgical practices within the United States.

Many hospital systems in the United States were placed under SIP restrictions similar to Santa Clara County, and several neurosurgical services reported their own experiences. Saad et al. in Atlanta, Georgia, reported how the entire neurosurgery department rescheduled cases from March 16, 2020, onward according to urgency. (Saad et al., 2020) Cases were separated by their need for completion within 1 h, 24 h, 4 weeks, or >1 month from presentation. What they observed was a significant decrease (80.0%, $P<.01$) in total neurosurgical cases in March and April 2020 compared to 2017–19, with decreases in five distinct subspecialties: functional (81%), spine (78%), vascular and endovascular (62%), tumor (55%), and trauma (51%). Notably, trauma was the least impacted by the rescheduling. Similar restrictions were put into place at Barrow Neurologic Institute in Phoenix, Arizona, whereby an indefinite SIP ordinance was put in place on March 12, 2020, with a concurrent cancellation of elective surgeries until May 1, 2020. Koester et al. examined neurosurgical call logs from pre-COVID-19 (January 2, 2020 to February 25, 2020) versus COVID-19 (March 20, 2020 to April 27, 2020), finding a similarly significant decrease in neurosurgical consultations (16.8 vs 5.6, $P<.001$) as well as an increase in the proportion of total trauma cases (18% vs 26%) (Koester et al., 2021). Moreover, they found that the percentage of total admissions for general pain (headache, back pain, or other general weakness) significantly decreased (41% vs 28%, $P<.001$) from pre-COVID-19 to COVID-19 time periods while the percentage of total admissions from the ED significantly increased (76% during the lockdown vs 85%, $P=.03$). These figures help paint a picture of a department that saw a mild uptick in neurosurgical trauma admissions while maintaining an overall decrease in total neurosurgical utilization.

Both studies detailed above show a decrease in the overall neurosurgical volume during SIP protocols, with the need for neurosurgical trauma services being preserved. This sentiment is expanded upon by Lara-Reyna et al., who showed that the mechanism for head injury pre-COVID-19 (November 1, 2019 to February 29, 2020) and post-COVID-19 (March 1, 2020 to April 26, 2020) at Mount Sinai Morningside in New York City, New York, significantly changed (Model $P=.03405$), with an increase in the total percentage of motor vehicle accidents (MVA) (5.0% vs 18.4%) and a decrease in falls (61.4% vs 40.8%) (Lara-Reyna et al., 2020). The differences in the change in etiology experienced by Lara-Reyna et al. showcase the diversity of trauma in different locales during this time. Interestingly, the rate of surgical management of neurotrauma did not see a significant change.

Global impact of COVID-19 on neurosurgical trauma

Neurosurgical units all over the globe felt the impact of COVID-19 both through the pandemic's effect on the general public as well as their country's SIP policies. As in the United States, many neurosurgery services saw a decrease in the number of patients presenting with head injury or other emergent cases (Ashkan et al., 2021; Bernucci et al., 2020; Goyal et al., 2020; Horan et al., 2021; Lester et al., 2021; Luostarinen et al., 2020; Pinggera et al., 2021; Raneri et al., 2020; Sinha et al., 2021). Looking at their experiences adds to our preparations for future pandemics.

Several European neurosurgical groups in England, Wales, Ireland, and Italy described their local responses to COVID-19 restrictions. Sinha et al. performed a retrospective review looking at the effects of lockdown in Britain from March 23 to April 22, 2020, finding that head trauma admissions dropped by roughly half (57 to 28) from 2019 to 2020 with a similar decrease in total operations on admitted patients during this period (Sinha et al., 2021). These findings follow a very similar trend to the experiences of other neurosurgical units in the United States. Ashkan et al., another group from England, found similar results in their own review, with COVID-19 restrictions bringing decreased total emergency (1846 vs 1227, $P<.01$) and trauma referrals but not a significant change in the rate of neurosurgical trauma operations (12.3% vs 17.9%, $P>.05$, Table 2) between pre-COVID-19 (January 18 to March 17, 2020) versus COVID-19 (March 18 to May 15, 2020) phases (Ashkan et al., 2021). Horan et al. in Ireland observed decreasing levels of TBI, traumatic subarachnoid hemorrhage, acute subdural hematoma, chronic subdural hematoma, contusion, intraventricular hemorrhage, cranial fracture, and spinal trauma between March 1 and May 31, 2019 versus 2020 (Horan et al., 2021). The primary driving factor for this decrease seemed to be the 14.3% decrease in falls between 2019 and 2020.

In Italy, Raneri et al. combined data from six neurosurgical departments during March each year from 2016 to 2020 in the Veneto region to better understand how the Italian SIP protocol affected neurosurgical care (Raneri et al., 2020). Each center experienced a decrease in total admissions (−17% to −51%, average −39%), emergency admissions (average −23%), operations (−33%), and 6-h surgical sessions (average −36%). These findings show a decrease not only in the total volume of patients, but also in the acuity of patient presentation in this area. Interestingly, in nearby Bergamo, Italy, Bernucci et al. experienced other causes of decreased neurosurgical utilization (Bernucci et al., 2020). As COVID-19 spread within this region, neurosurgical beds were converted to COVID-19 beds and neurosurgical staff were rerouted to serve on these COVID-19 units. In an areas with a population of more than a million people, as of the time of writing, the neurosurgical unit reported no new pediatric traumatic emergencies, a finding significantly below their own internal expectations.

Other trends in neurosurgical trauma during COVID-19

The COVID-19 pandemic had far-reaching impact globally and while some of its effects on patients who ended up in the hospital were quite profound, the effect COVID-19 had on persons residing at home also warrants examination. One area of interest was in pediatric neurotrauma, which saw an uptick during the pandemic due to a conflux of interpersonal factors including increased time at home due to social distancing restrictions, decreased childcare availability, and decreased access to pediatric medical professionals (Kovler et al., 2021; Sidpra et al., 2021).

Sidpra et al. in England found a 1493% increase in head trauma cases between March 23 and April 23, 2020 compared to the 3 years prior. Many of these patients demonstrated physical signs of child abuse, such as retinal hemorrhages (50%),

TABLE 2 Number and composition of adult neurosurgical operations performed by subspecialty.

Adult	Pre-COVID-19	COVID-19	*P* value
Number of operations in *N* (% of total)	408 (100.0)	173 (100.0)	
Functional	37 (9.1)	8 (4.6)	$P<.01$
DBSa for Parkinson's disease/tremor	8 (2.0)	0 (0.0)	
Intractable epilepsy/VNSb	18 (4.4)	1 (0.6)	
Peripheral nerve	3 (0.7)	1 (0.6)	
Baclofen pump/other	2 (0.4)	1 (0.6)	
DBS battery change	5 (2.7)	5 (2.9)	
Spinal	145 (35.5)	51 (29.5)	$P<.01$
Myelopathy	28 (6.9)	8 (4.6)	
Radiculopathy	73 (17.9)	13 (7.5)	
Cauda equina syndrome	26 (6.4)	14 (8.1)	
MSCCd and spinal tumor	14 (3.4)	15 (8.7)	
Spinal hematoma and other	4 (1.0)	1 (0.6)	
Trauma	50 (12.3)	31 (17.9)	$P>.05$
Acute subdural hematoma	9 (2.2)	3 (1.7)	
Chronic subdural hematoma	26 (6.4)	19 (11.0)	
Extradural hematoma	4 (1.0)	2 (1.2)	
Traumatic brain injury/other	0 (0.0)	1 (0.6)	
Traumatic vertebral fracture	11 (2.7)	5 (2.9)	
Vascular	22 (5.4)	9 (5.2)	$P>.05$
Aneurysm	9 (2.2)	1 (0.6)	
Intracranial hemorrhage	5 (1.2)	5 (2.9)	
Ischemic stroke	1 (0.2)	1 (0.6)	
Arteriovenous malformation	7 (1.7)	0 (0.0)	
Arteriovenous fistula	1 (0.2)	2 (1.2)	
Oncology	60 (14.7)	31 (17.9)	$P>.05$
Low-grade glioma	4 (1.0)	1 (0.6)	
High-grade glioma	31 (7.6)	12 (6.9)	
Cerebral metastasis	6 (1.5)	6 (3.5)	
Meningioma	13 (3.2)	7 (4.0)	
Other	6 (1.5)	5 (2.9)	
Skull base	36 (8.8)	7 (4.0)	$P>.05$
Pituitary adenoma/apoplexy	16 (3.9)	3 (1.7)	
Sphenoid wing meningioma	5 (1.2)	2 (1.2)	
Vestibular schwannoma	6 (1.2)	0 (0.0)	
Chiari malformation	5 (1.2)	0 (0.0)	
Chondrosarcoma	2 (0.5)	0 (0.0)	

TABLE 2 Number and composition of adult neurosurgical operations performed by subspecialty—cont'd

Adult	Pre-COVID-19	COVID-19	*P* value
Trigeminal neuralgia	1 (0.2)	0 (0.0)	
Craniopharyngioma	1 (0.2)	2 (1.2)	
Other	58 (14.2)	37 (21.4)	$P > .05$
Hydrocephalus	30 (7.4)	18 (10.4)	
Primary infections	7 (1.7)	4 (2.3)	
Secondary infections	15 (3.7)	11 (6.4)	
Postoperative hematoma	2 (0.5)	0 (0.0)	
CSF leak/pseudomeningocele	4 (1.0)	4 (2.3)	

(A) Deep brain stimulation. (B) Vagal nerve stimulator. (C) Metastatic spinal cord compression. (D) Cerebrospinal fluid.
From Ashkan, K., Jung, J., Velicu, A. M., Raslan, A., Faruque, M., Kulkarni, P., Bleil, C., Hasegawa, H., Kailaya-Vasan, A., Maratos, E., Grahovac, G., Vergani, F., Zebian, B., Barazi, S., Malik, I., Bell, D., Walsh, D., Bhangoo, R., Tolias, C., ... Gullan, R. (2021). Neurosurgery and coronavirus: Impact and challenges-lessons learnt from the first wave of a global pandemic. *Acta Neurochirurgica, 163*(2), 317–329. https://doi.org/10.1007/s00701-020-04652-8. With permission from Springer Publishing Company under Creative Commons license.

extensive bruising (50%), and scalp swelling (50%). In a similar study, Kovler et al. from Johns Hopkins University performed a retrospective review, finding that child abuse trauma totals from March 27 to April 27, 2020, exceeded both prior years combined (8 vs 3 in 2018 and 4 in 2019). Interestingly, the number of patients with head injuries accounted for 7 of 8 traumas reported, with 4 new reports of skull fracture compared to none in the 2 years prior.

One other aspect of neurosurgical trauma that was impacted by the COVID-19 pandemic is the training opportunities available to residents. Burks et al. from Miami reported on the challenges of obtaining adequate resident training volume across many subspecialties of neurosurgery, trauma included (Burks et al., 2020). A similar experience was shared in a multiinstitution review by Aljuboori et al., who saw a significant decrease in emergent procedures across eight residency programs during March 2020 compared to March 2019 ($P < .01$) (Aljuboori et al., 2021). Field et al. saw a similar trend of decreasing neurosurgical consults in Albany, New York, and attempted to use telehealth communication systems to maintain levels of patient interaction (Field et al., 2020). However, these visits were only used for outpatient communications, whereas most if not all traumas are seen inpatient. As communication channels evolve, it will be important to keep in mind where residency education will need to be directed to maximize their exposure for elective and emergent surgical decision making.

Critical to any neurosurgical program attempting to provide service during a pandemic setting is the availability of providers and staff who are capable of operating. With the high transmissibility of the virus in mind, Burke et al. from the University of California, San Francisco (UCSF) developed an algorithm for surgical scheduling and managing provider teams to maintain adequate operative status in a pandemic (Burke et al., 2020). They detailed a four-tier surge system with step-wise limitations in surgical volume as COVID-19 transmission increases (Fig. 3). Additionally, to provide adequate staffing, residents at UCSF were partitioned into teams, with two paired teams per site allowing constant coverage at the main, veteran's, trauma, and pediatric hospitals (Fig. 4). Virtual handoffs reduced the risk of virus transmission between teams. Programs looking to develop coverage systems for future pandemics can now model their scheduling in a similar manner.

Limitations of retrospective single-center reviews

Many of the examples used to highlight decreasing neurosurgical trauma volume, both domestically and worldwide, have been from single-center reviews. The advantage of these reviews is that they allow for very fast and impactful tracking of the effects of SIP protocols on neurosurgical utilization. However, it is important to note that many of these studies are subject to limitations, primarily that neurosurgical trauma within a region does not necessarily predict activity and behavior in another part of the globe. As catastrophe response reporting continues to occur, even more reliable patterns should continue to be revealed.

Surge level	Emergent and urgent cases	Elective and procedural cases	Transfers
Green: 1-9 community cases, or <6 COVID-19 + inpatients, and No staffing shortages	- Ensure ORs and surgeons available to rapidly operate and discharge ED cases (appy/chole/hip fx) - Maintain capacity for urgent cases	- proceed - Review and schedule all cases 7 days in advance for quick cancellation should surge level increase to yellow	Per service protocol
Yellow: 10-99 community cases, or 7-16 COVID-19 + inpatients, or < 20% staffing shortages	- Ensure ORs and surgeons available to rapidly operate and discharge ED cases (appy/chole/hip fx) - Maintain capacity for urgent cases	- Cap OR schedule for next 3 weeks - 25% reduction in procedural cases requiring overnight stay (ERCP/cath/IR) - Ramp up come and go procedures at outpatient facility - Reschedule OR cases so no more than 25 post-op admits over next 7 days	Time sensitive protocol
Red: >100 community cases, or >17 COVID-19 + inpatients, or > 21% staffing shortages	- Ensure ORs and surgeons available to rapidly operate and discharge ED cases (appy/chole/hip fx) - Maintain capacity for urgent cases	- Cap OR schedule for next 3 weeks - 50% reduction in procedural cases requiring overnight stay (ERCP/cath/IR) - no use of PACU for recovery - Cancel come and go procedures - Reschedule OR cases so no more than 12 post-op admits over next 7 days	Closed to all transfers
Black: Significant assistance needed from outside institutions	Emergent cases only	- Cancel all scheduled cases	Closed to all transfers

FIG. 3 Four-tier surge system with stepwise limitations on surgical volume.
The surge level criteria are shown in the left column, the emergent case and the elective case recommendations are shown in the middle two columns, and the transfer center recommendations are shown in the right column. *appy*, appendectomy; *cath*, cardiac catheterization; *chole*, cholecystectomy; *ED*, emergency department; *ERCP*, endoscopic retrograde cholangiopancreatography; *hip fx*, hip fracture; *IR*, interventional radiology; *OR*, operating room; *PACU*, postanesthesia recovery unit. Community cases refer to active cases in the local region. *Reproduced with permission from Wolters Kluwer Health, Inc.: Burke, J. F., Chan, A. K., Mummaneni, V., Chou, D., Lobo, E. P., Berger, M. S., Theodosopoulos, P. V., & Mummaneni, P. V. (2020). Letter: The coronavirus disease 2019 global pandemic: A neurosurgical treatment algorithm.* Neurosurgery, *87(1), E50–e56. https://doi.org/10.1093/neuros/nyaa116.*

Site 1: Main hospital			Site 2: VA hospital			Site 3: Trauma hospital			Site 4: Peds hospital			
Team 1	Team 2	Alternates	Team 1	Team 2	Alternates	Team 1	Team 2	Alternates	Team 1	Team 2	Alternates	Additional back-up coverage
PGY-2	PGY-2	PGY-2	PGY-4	PGY-7	PGY-7	PGY-4	PGY-7	fellow	PGY-3	PGY-4	fellow	fellow (spine)
PGY-3	PGY-3	PGY-5										fellow (spine)
PGY-6	PGY-6	PGY-5										fellow (peripheral nerve)
		PGY-5										fellow (tumor)
		PGY-6										fellow (vascular)

FIG. 4 Paired coverage model.
The following model of team-based paired coverage will go into effect during red levels of COVID-19 (see Fig. 1). In this model, each individual hospital (columns) will have three groups of providers: two teams that switch coverage on a 3-day cycle, and an alternate group that substitutes for any team member who shows signs of illness. In this way, if a team becomes contaminated, the other team will take over, and the alternates will fill the gap. Contact between teams and alternates is prohibited. *Peds*, pediatric; *PGY*, postgraduate year (resident level); *VA*, veterans affairs. *Reproduced with permission from Wolters Kluwer Health, Inc.: Burke, J. F., Chan, A. K., Mummaneni, V., Chou, D., Lobo, E. P., Berger, M. S., Theodosopoulos, P. V., & Mummaneni, P. V. (2020). Letter: The coronavirus disease 2019 global pandemic: A neurosurgical treatment algorithm.* Neurosurgery, *87(1), E50–e56. https://doi.org/10.1093/neuros/nyaa116.*

Conclusion

COVID-19 and the ensuing pandemic responses triggered large-scale healthcare change throughout the world. SIP protocols reduced public interfacing with healthcare systems and COVID-19 surgical protocols limited the performance of elective surgeries.

As a result, initially decreasing numbers of both emergent and elective cases were seen across the board. As a subspecialty, neurosurgical trauma was no exception, with most reports detailing a decrease in the total number of neurosurgical trauma cases. However, despite these decreases, in many places trauma was the least impacted neurosurgical subspecialty. The necessity of the ability to operate in and on emergent cases perhaps tethered the subspecialty closer to pre-COVID-19 norms. Operating in pandemic conditions also prompted a discussion on how to preserve neurosurgical trauma service and resident training during pandemic scenarios, leading to the development of new scheduling algorithms centered around surge planning. Adjacent to that is the recognition of telehealth as a means of managing neurosurgical trauma. In summation, this review documents the experiences of many neurosurgical trauma units around the world, synthesizing their perspectives to outline how to better prepare for future pandemics.

Policies and procedures

As the COVID-19 pandemic spread worldwide, many governments wisely implemented SIP policies that limited interpersonal interaction, and in turn, significantly decreased total patient volume through many hospital systems. Neurosurgery as a specialty was not immune to these changes and through worldwide reporting, there are now many accounts of the different challenges and changes neurosurgical units experienced throughout the pandemic. Many departments reported seeing a decrease in the total volume of both trauma and nontrauma admissions. Any sudden shift in neurosurgical utilization, especially trauma due to the emergent nature of cases, within a hospital system warrants further examination on how to allocate resources to mirror those changes. Based on the findings reported, it seems that a readjustment of the level of staffing to more impacted units, such as has been reported by Bernucci et al., can be sustainable (Bernucci et al., 2020). Burke et al. from UCSF detailed the need for more specialized resource allocation during the COVID-19 pandemic and outlined a three-tier surge algorithm for managing urgent surgeries that can serve as a template for future resource allocation discussions (Burke et al., 2020). Like UCSF, many other units, such as in Lara-Reyna et al., shifted to a tiering system that enabled emergent cases to be seen quickly while complying with operative restrictions (Lara-Reyna et al., 2020). However, a retrospective review of neurosurgical trauma management and operative triage by acuity across several hospital systems has yet to be performed. Such a study should be at the forefront of any policy discussions, both at the hospital and government levels.

Applications to other areas

While many of the studies examined focused on the topic of adult neurosurgical trauma and the experiences of operating neurosurgical units under SIP protocols, that is not the entirety of the scope of the impact of COVID-19, even within the specialty. One area that saw tremendous impact due to lockdown protocols was the development of telehealth communications as a means of managing neurotrauma patients. Koruga et al., from Osijek, Croatia, detailed using telehealth as a means of managing increased neurosurgical trauma consultation volume and total number of operations during 2020 versus 2019 (Koruga et al., 2022). The volume of management sustains their fundamental premise, which is that telemedicine can be an effective method of maintaining the same level of care, even in the absence of an onsite neurosurgeon, or avoiding needless transfers to tertiary care hospitals, both beneficial during impacted pandemic states (Klein et al., 2010; Olldashi et al., 2019).

Other areas of application include some of the topics that were covered in this chapter, namely pediatric neurosurgical trauma and resident training volume during pandemic conditions. Pediatric neurosurgical trauma shows its relevance in the context of greater levels of interpersonal conflict resulting from SIP protocols driving higher levels of pediatric abusive behavior (Kovler et al., 2021). Interestingly, in this study, Kovler et al. mentioned the use of a multidisciplinary team as part of their inclusion criteria, which can serve as a template for future management of child neurosurgical trauma during pandemic conditions. In anticipation of an increase in pediatric neurosurgical trauma, the creation of teams of surgeons, nurses, and social workers could serve as an important line of protection for these patients. Additionally, broadcasting the availability of the services will likely serve as effective community prophylaxis.

Lastly, the topic of neurosurgical trauma as part of the resident training process cannot be overlooked. Fortunately, many institutions reported similar levels of neurosurgical trauma during the pandemic and SIP protocols while others noted decreases in neurosurgical trauma utilization (Ashkan et al., 2021; Raneri et al., 2020; Zhang et al., 2021). Maintaining the ability to teach during these times should be an educational priority for programs, especially in the context of training to operate under similar conditions in the future.

Mini-dictionary of terms

- Shelter-in-place: Local, state, or federal government ordinances designed to reduce the spread of COVID-19 by restricting interpersonal movement and interaction.
- Neurosurgical trauma: Any emergent operation or consult that occurs because of damage to the head, skull, brain, or spinal cord, typically presenting through the emergency department.
- Elective surgery: Planned surgeries that then do not meet the criteria for medical emergency.
- Neurosurgical utilization: The total amount of neurosurgical consults seen and cases operated within a department or defined unit.
- Academic center: Typically, tertiary care hospitals that see a more specialized patient population.
- Community hospitals: Less-specialized local hospitals designed to provide care to larger catchments of patients.
- Single-center review: Retrospective review of patient data occurring at one hospital or site.
- Subspecialty: In the context of this review, the categorization of surgical procedures among trauma, functional, spine, vascular and endovascular, tumor, and others.
- Significance: A statistically meaningful relationship between two variables, determined by testing method and comparison cut-off.

Summary points

- COVID-19 and the ensuing SIP protocols saw a resultant decrease in the total volume of neurosurgical trauma at many institutions.
- While total volume decreased, many hospitals reported a stable proportion of trauma cases during SIP protocols relative to other subspecialties.
- These trends were seen both domestically at hospitals across the United States as well as abroad in England, Italy, and other parts of Europe.
- Neurosurgical trauma units in Italy and San Francisco faced issues of adequate staffing amid COVID-19 surges, and attempted to mitigate disruptions in unit operations with flexible planning.
- Resident training was impacted through reduced exposure to impacted case types and the threat of contracting COVID-19.
- Telehealth saw increased utilization in neurosurgical trauma during SIP protocols, with the ability to manage traumas even in the absence of onsite neurosurgeons.

References

Aljuboori, Z. S., Young, C. C., Srinivasan, V. M., Kellogg, R. T., Quon, J. L., Alshareef, M. A., Chen, S. H., Ivan, M., Grant, G. A., McEvoy, S. D., Davanzo, J. R., Majid, S., Durfy, S., Levitt, M. R., Sieg, E. P., Ellenbogen, R. G., & Nauta, H. J. (2021). Early effects of COVID-19 pandemic on neurosurgical training in the United States: A case volume analysis of 8 programs. *World Neurosurgery*, *145*, e202–e208. https://doi.org/10.1016/j.wneu.2020.10.016.

Arnold, P. M., Owens, L., Heary, R. F., Webb, A. G., Whiting, M. D., Vaccaro, A. R., Iyer, R. K., & Harrop, J. S. (2021). Lumbar spine surgery and what we lost in the era of the coronavirus pandemic: A survey of the lumbar spine research society. *Clinical Spine Surgery*, *34*(10), e575–e579. https://doi.org/10.1097/bsd.0000000000001235.

Ashkan, K., Jung, J., Velicu, A. M., Raslan, A., Faruque, M., Kulkarni, P., Bleil, C., Hasegawa, H., Kailaya-Vasan, A., Maratos, E., Grahovac, G., Vergani, F., Zebian, B., Barazi, S., Malik, I., Bell, D., Walsh, D., Bhangoo, R., Tolias, C., … Gullan, R. (2021). Neurosurgery and coronavirus: Impact and challenges-lessons learnt from the first wave of a global pandemic. *Acta Neurochirurgica*, *163*(2), 317–329. https://doi.org/10.1007/s00701-020-04652-8.

Bernucci, C., Brembilla, C., & Veiceschi, P. (2020). Effects of the COVID-19 Outbreak in Northern Italy: Perspectives from the Bergamo neurosurgery department. *World Neurosurgery*, *137*. https://doi.org/10.1016/j.wneu.2020.03.179. 465-468.e461.

Burke, J. F., Chan, A. K., Mummaneni, V., Chou, D., Lobo, E. P., Berger, M. S., Theodosopoulos, P. V., & Mummaneni, P. V. (2020). Letter: The coronavirus disease 2019 global pandemic: A neurosurgical treatment algorithm. *Neurosurgery*, *87*(1), E50–e56. https://doi.org/10.1093/neuros/nyaa116.

Burks, J. D., Luther, E. M., Govindarajan, V., Shah, A. H., Levi, A. D., & Komotar, R. J. (2020). Early changes to neurosurgery resident training during the COVID-19 pandemic at a large U.S. Academic Medical Center. *World Neurosurgery*, *144*, e926–e933. https://doi.org/10.1016/j.wneu.2020.09.125.

Cucinotta, D., & Vanelli, M. (2020). WHO declares COVID-19 a pandemic. *Acta Biomedica*, *91*(1), 157–160. https://doi.org/10.23750/abm.v91i1.9397.

Field, N. C., Platanitis, K., Paul, A. R., Dalfino, J. C., Adamo, M. A., & Boulos, A. S. (2020). Letter to the editor: Decrease in neurosurgical program volume during COVID-19: Residency programs must adapt. *World Neurosurgery*, *141*, 566–567. https://doi.org/10.1016/j.wneu.2020.06.141.

Figueroa, J. M., Boddu, J., Kader, M., Berry, K., Kumar, V., Ayala, V., Vanni, S., & Jagid, J. (2021). The effects of lockdown during the severe acute respiratory syndrome coronavirus 2 (SARS-CoV-2) pandemic on neurotrauma-related hospital admissions. *World Neurosurgery*, *146*, e1–e5. https://doi.org/10.1016/j.wneu.2020.08.083.

Goyal, N., Venkataram, T., Singh, V., & Chaturvedi, J. (2020). Collateral damage caused by COVID-19: Change in volume and spectrum of neurosurgery patients. *Journal of Clinical Neuroscience*, *80*, 156–161. https://doi.org/10.1016/j.jocn.2020.07.055.

Horan, J., Duddy, J. C., Gilmartin, B., Amoo, M., Nolan, D., Corr, P., Husien, M. B., & Bolger, C. (2021). The impact of COVID-19 on trauma referrals to a National Neurosurgical Centre. *Irish Journal of Medical Science (1971-)*, *190*(4), 1281–1293. https://doi.org/10.1007/s11845-021-02504-7.

Johnson, R. A., Eaton, A., Tignanelli, C. J., Carrabre, K. J., Gerges, C., Yang, G. L., Hemmila, M. R., Ngwenya, L. B., Wright, J. M., & Parr, A. M. (2022). Changes in patterns of traumatic brain injury in the Michigan trauma quality improvement program database early in the COVID-19 pandemic. *Journal of Neurosurgery*, 1–11. https://doi.org/10.3171/2022.5.Jns22244.

Kessler, R. A., Zimering, J., Gilligan, J., Rothrock, R., McNeill, I., Shrivastava, R. K., Caridi, J., Bederson, J., & Hadjipanayis, C. G. (2020). Neurosurgical management of brain and spine tumors in the COVID-19 era: An institutional experience from the epicenter of the pandemic. *Journal of Neuro-Oncology*, *148*(2), 211–219. https://doi.org/10.1007/s11060-020-03523-7.

Klein, Y., Donchik, V., Jaffe, D., Simon, D., Kessel, B., Levy, L., Kashtan, H., & Peleg, K. (2010). Management of patients with traumatic intracranial injury in hospitals without neurosurgical service. *Journal of Trauma and Acute Care Surgery*, *69*(3). https://journals.lww.com/jtrauma/Fulltext/2010/09000/Management_of_Patients_With_Traumatic_Intracranial.10.aspx.

Koester, S. W., Catapano, J. S., Ma, K. L., Kimata, A. R., Abbatematteo, J. M., Walker, C. T., Cole, T. S., Whiting, A. C., Ponce, F. A., & Lawton, M. T. (2021). COVID-19 and neurosurgery consultation call volume at a single large tertiary center with a propensity-adjusted analysis. *World Neurosurgery*, *146*, e768–e772. https://doi.org/10.1016/j.wneu.2020.11.017.

Koruga, N., Soldo Koruga, A., Rončević, R., Turk, T., Kopačin, V., Kretić, D., Rotim, T., & Rončević, A. (2022). Telemedicine in neurosurgical trauma during the COVID-19 pandemic: A single-center experience. *Diagnostics*, *12*(9), 2061. https://www.mdpi.com/2075-4418/12/9/2061.

Kovler, M. L., Ziegfeld, S., Ryan, L. M., Goldstein, M. A., Gardner, R., Garcia, A. V., & Nasr, I. W. (2021). Increased proportion of physical child abuse injuries at a level I pediatric trauma center during the Covid-19 pandemic. *Child Abuse & Neglect*, *116*(Pt 2), 104756. https://doi.org/10.1016/j.chiabu.2020.104756.

Lara-Reyna, J., Yaeger, K. A., Rossitto, C. P., Camara, D., Wedderburn, R., Ghatan, S., Bederson, J. B., & Margetis, K. (2020). "Staying home"-early changes in patterns of neurotrauma in New York City during the COVID-19 pandemic. *World Neurosurgery*, *143*, e344–e350. https://doi.org/10.1016/j.wneu.2020.07.155.

Lester, A., Leach, P., & Zaben, M. (2021). The impact of the COVID-19 pandemic on traumatic brain injury management: Lessons learned over the first year. *World Neurosurgery*, *156*, 28–32. https://doi.org/10.1016/j.wneu.2021.09.030.

Luostarinen, T., Virta, J., Satopää, J., Bäcklund, M., Kivisaari, R., Korja, M., & Raj, R. (2020). Intensive care of traumatic brain injury and aneurysmal subarachnoid hemorrhage in Helsinki during the Covid-19 pandemic. *Acta Neurochirurgica*, *162*(11), 2715–2724. https://doi.org/10.1007/s00701-020-04583-4.

Olldashi, F., Latifi, R., Parsikia, A., Boci, A., Qesteri, O., Dasho, E., & Bakiu, E. (2019). Telemedicine for neurotrauma prevents unnecessary transfers: An update from a nationwide program in Albania and analysis of 590 patients. *World Neurosurgery*, *128*, e340–e346. https://doi.org/10.1016/j.wneu.2019.04.150.

Pinggera, D., Klein, B., Thomé, C., & Grassner, L. (2021). The influence of the COVID-19 pandemic on traumatic brain injuries in Tyrol: Experiences from a state under lockdown. *European Journal of Trauma and Emergency Surgery*, *47*(3), 653–658. https://doi.org/10.1007/s00068-020-01445-7.

Raneri, F., Rustemi, O., Zambon, G., Del Moro, G., Magrini, S., Ceccaroni, Y., Basso, E., Volpin, F., Cappelletti, M., Lardani, J., Ferraresi, S., Guida, F., Chioffi, F., Pinna, G., Canova, G., d'Avella, D., Sala, F., & Volpin, L. (2020). Neurosurgery in times of a pandemic: A survey of neurosurgical services during the COVID-19 outbreak in the Veneto region in Italy. *Neurosurgical Focus*, *49*(6), E9. https://doi.org/10.3171/2020.9.Focus20691.

Saad, H., Alawieh, A., Oyesiku, N., Barrow, D. L., & Olson, J. (2020). Sheltered neurosurgery during COVID-19: The emory experience. *World Neurosurgery*, *144*, e204–e209. https://doi.org/10.1016/j.wneu.2020.08.082.

Sidpra, J., Abomeli, D., Hameed, B., Baker, J., & Mankad, K. (2021). Rise in the incidence of abusive head trauma during the COVID-19 pandemic. *Archives of Disease in Childhood*, *106*(3), e14. https://doi.org/10.1136/archdischild-2020-319872.

Sinha, S., Toe, K. K. Z., Wood, E., & George, K. J. (2021). The impact of COVID-19 on neurosurgical head trauma referrals and admission at a tertiary neurosurgical centre. *Journal of Clinical Neuroscience*, *87*, 50–54. https://doi.org/10.1016/j.jocn.2021.02.021.

Zhang, M., Zhou, J., Dirlikov, B., Cage, T., Lee, M., & Singh, H. (2021). Impact on neurosurgical management in level 1 trauma centers during COVID-19 shelter-in-place restrictions: The Santa Clara County experience. *Journal of Clinical Neuroscience*, *88*, 128–134. https://doi.org/10.1016/j.jocn.2021.03.017.

Zhang, M., Zhou, J., Dirlikov, B., Cage, T., Lee, M., & Singh, H. (2022). Impact on neurosurgical management in a Level 1 trauma center post COVID-19 shelter-in-place restrictions. *Journal of Clinical Neuroscience*, *101*, 131–136. https://doi.org/10.1016/j.jocn.2022.04.033.

Chapter 15

The COVID-19 pandemic and management of weight gain: Implications for obesity

Sarah R. Barenbaum and Alpana P. Shukla
Division of Endocrinology, Diabetes and Metabolism, New York-Presbyterian Hospital/Weill Cornell Medical College, Comprehensive Weight Control Center, New York, NY, United States

Abbreviations

ABOM	American Board of Obesity Medicine
ACE2	angiotensin-converting enzyme 2
BMI	body mass index
COVID-19	coronavirus disease 2019
SARS-CoV-2	severe acute respiratory syndrome coronavirus 2
WHO	World Health Organization

Introduction

The obesity pandemic is one of the most critical public health emergencies facing our world today. In the United States, the prevalence of obesity has been steadily increasing and in 2017–2018, 41.9% of American adults over the age of 20 had obesity (Stierman et al., 2021). According to the World Health Organization (WHO), in 2016, 39% of adults over 18 were overweight and 13% had obesity (World Health Organization Fact Sheet, 2021). Excess weight also impacts children, and in 2016 the WHO estimated that more than 340 million children and adolescents globally ages 5–19 had overweight or obesity. In 2020, the WHO further estimated that 39 million children under the age of 5 were overweight (World Health Organization Fact Sheet, 2021). Obesity is a chronic disease, and people with obesity are at increased risk of multiple comorbidities including cardiovascular disease, type 2 diabetes, obstructive sleep apnea, hypertension, hyperlipidemia, nonalcoholic fatty liver disease, and all-cause mortality (Must et al., 1999). The coronavirus disease 2019 (COVID-19) pandemic has further highlighted the crucial need to treat obesity, as it became clear early in the pandemic that those with obesity were at a significantly higher risk for COVID-19 related morbidity and mortality (Centers for Disease Control, 2022; Goyal et al., 2020).

Obesity as a risk factor for severe disease

Early in the COVID-19 pandemic, it was recognized that those with overweight and obesity were at increased risk for severe disease and death with severe acute respiratory syndrome coronavirus 2 (SARS-CoV-2) infection, which included those with lower levels of excess weight (Goyal et al., 2020; Hamer et al., 2020; Popkin et al., 2020; Rizzo et al., 2020; Simonnet et al., 2020). Obesity quickly became recognized as an independent risk factor for intensive care unit (ICU) admission, mechanical ventilation, and death (Gao et al., 2021; Klang et al., 2020). Several mechanisms have been suggested that may play a role in the pathogenesis of severe COVID-19 disease in those with obesity.

Those with obesity are at higher risk for severe illness with any infectious disease given existing cardiometabolic disease, kidney disease, pulmonary disease, and immune dysfunction (Sanchis-Gomar et al., 2020). In addition, obesity itself leads to a state of chronic, systemic low-grade inflammation (Minihane et al., 2015). Excess adipose tissue is

Management, Body Systems, and Case Studies in COVID-19. https://doi.org/10.1016/B978-0-443-18703-2.00035-9
Copyright © 2024 Elsevier Inc. All rights are reserved, including those for text and data mining, AI training, and similar technologies.

associated with an increased release of proinflammatory cytokines, including c-reactive protein, interleukin-6, tumor necrosis factor-alpha, and interleukin-1 beta, among others (Hotamisligil et al., 1995; Ouchi et al., 2011). Excess adipose tissue is further associated with the disproportionate recruitment of immune system cells, particularly macrophages (McNelis & Macrophages, 2014). This chronic state of inflammation combined with an impaired antiviral response may contribute to the increased risk of a cytokine storm in COVID-19 infection, which is a significant contributor to multi-organ failure (Yu et al., 2022).

Furthermore, comorbid pulmonary diseases such as obesity hypoventilation syndrome, obstructive sleep apnea, chronic obstructive pulmonary disease, pulmonary hypertension, asthma, etc., predispose to more severe respiratory disease with COVID-19 (Rychter et al., 2020). In addition, the mechanical restriction of pulmonary function due to adipose tissue mass can restrict diaphragmatic excursion and impair ventilation (Simonnet et al., 2020). Obesity also increases the risk of hypoventilation syndrome in patients admitted to the ICU, which can lead to respiratory failure in the setting of acute respiratory distress syndrome (Abou-Arab et al., 2020).

Moreover, SARS-CoV-2 penetrates human cells by binding to angiotensin-converting enzyme 2 (ACE2) receptors on cells throughout the body. Another possible contributor is that ACE2 expression is higher in adipose tissue, potentially leading to a susceptibility of adipose tissue to COVID-19 infection (Sanchis-Gomar et al., 2020). Finally, unrelated to the pathogenesis of severe illness itself, the stigma associated with overweight and obesity may delay those with the disease from seeking treatment, such that they present at a later and more severe state of COVID-19 infection (Townsend et al., 2020a).

Impact of the COVID-19 pandemic on obesity

Ironically, while it became clear early in the pandemic that those with overweight and obesity were at greater risk from COVID-19, the pandemic may have accelerated global rates of overweight and obesity. With the onset of the pandemic in the United States in early 2020, most Americans were asked to stay at home to prevent the spread of COVID-19 (FINRA, 2020). Similar directives were issued globally, which led to disruptions in supply chains. Global supply chain issues combined with local mandates led to the closure of businesses and restaurants and limited access to food and groceries. Stay-at-home orders forced nonessential workers to a work-from-home model and limited social engagements, which led to more sedentary behavior, changes in eating and exercise patterns, and changes in mental health (Robinson et al., 2020). All these factors may have contributed to weight gain.

Several studies found that many Americans experienced significant weight gain during the height of the COVID-19 pandemic (Bhutani et al., 2021; Lin et al., 2021; Zeigler, 2021). For example, a survey of 3013 American adults conducted by the American Psychological Association in February 2021 found that 42% of American adult respondents reported weight gain since the start of the pandemic. Of this group, the average reported weight gain was 29 pounds, with a median weight gain of 15 pounds (American Psychological Association, 2021). A smaller longitudinal cohort study of 269 participants weighing at home with Bluetooth scales from February–June 2020 reported a steady weight gain of 0.27 kg every 10 days (95% CI, 0.17 to 0.38 kg per 10 days; $P < 0.001$), which equaled roughly 1.5 pounds of weight gain per month during this period (Lin et al., 2021). A metaanalysis of 35 cross-sectional studies and one cohort study examining the impact on weight between March–May 2020 globally on those ages 16 and older found that 11.1–72.4% of individuals reported a significant increase in body weight (Bakaloudi et al., 2021).

Not all studies illustrated weight gain. A large retrospective study of an electronic health record database that included 4.25 million American adults ages 18–84 found that compared with the prepandemic weight trend, there was only a 0.1-kg increase in weight during the first year of the pandemic, suggesting that the overall mean change in weight during the pandemic may have been small (Freedman et al., 2022). However, even if weight gain during the pandemic itself was minimal, any increase in weight can be permanent. Studies have shown that even small incremental weight gain over vacations or during holidays can lead to substantial and permanent weight gain over time (Schoeller, 2014; Yanovski et al., 2000).

The management of patients with obesity during the pandemic was challenging, and published data endorse the utility of medical weight management during this critical period. A survey study of 970 patients at a large, multidisciplinary weight management center in the United States found that medical management, including behavioral modifications and antiobesity pharmacotherapy, protected against weight gain and also led to weight loss for certain patients (Barenbaum et al., 2022).

Importance of treating obesity

Although it will take more time and more studies to know the exact impact the COVID-19 pandemic has had on global rates of overweight and obesity, the pandemic has helped to expose the vulnerability of patients with obesity. It has become clear that patients with overweight and obesity must be offered treatment for their disease. Unfortunately, less than 2% of eligible patients receive medical treatment for obesity (Thomas et al., 2016). This may, in part, be due to the dearth of physicians trained in obesity medicine. In the United States, there are fewer than 7000 diplomates of the American Board of Obesity Medicine, providers who are trained to treat overweight and obesity medically (ABOM, 2023; Thomas et al., 2016). Given the scarcity of obesity medicine physicians, access to treatment is a real issue. Newer strategies developed during the pandemic, however, may help further the care of new and existing patients seeking treatment for excess weight. This includes the expansion of telehealth and telemedicine. Furthermore, while the obesity medicine field has seen dramatic growth in the development of highly effective antiobesity pharmacotherapies, coverage for such treatment remains very limited. Concerted efforts from policy makers and industry are needed to bridge this wide treatment gap in obesity.

The impact of the COVID-19 pandemic on current medical practice

In early 2020, many medical practices transitioned care from being completely in-person to either a hybrid of in-person and telemedicine or completely telemedicine. The value of telemedicine has become clear over time, particularly in its ability to help expand access to obesity treatment. One early cohort comparison study of all new and follow-up visits in 2019 and 2020 at two different multispecialty comprehensive weight management centers found that with the introduction of telemedicine, there was a significant 27.2% reduction in the annual no-show rate and an increase in completed new and follow-up visits (Aras et al., 2021). Telemedicine may improve overall access to care by reducing certain barriers for patients, including access (i.e., childcare limitations, limited transportation or mobility) and time limitations (i.e., inability to take time off from work to travel to an appointment); it can also decrease the challenges related to weight stigma (Kahan et al., 2022; Saunders et al., 2022). There certainly are limitations of telemedicine as well, including technological requirements, less reliable vital signs, the need for privacy-compliant software, and provider training. The use of telemedicine in obesity treatment has promise, although it will take more time and more studies to fully appreciate the long-term impact that it will have on the field of weight management.

Conclusion

The COVID-19 pandemic has exposed the vulnerability of patients with overweight and obesity and the need to treat excess weight. Obesity is a known risk factor for severe disease and death from COVID-19. A weight loss of 5–10% can lead to a clinically relevant reduction in the risk of overall comorbid disease, and a weight loss of 15% or more has been shown to reduce mortality in individuals with obesity (Ryan & Yockey, 2017; Wing et al., 2011). Given the intricate link between obesity and COVID-19 pathophysiology, significant weight loss can lead to improved outcomes in COVID-19. This was clearly illustrated in a matched cohort study of patients with prior bariatric surgery and nonsurgical controls that showed a 60% reduction in overall COVID-19 severity in the surgical cohort (Aminian et al., 2022). Early in the pandemic, there was a call to treat obesity. Unfortunately, the pandemic itself may have accelerated the global trends of weight gain due to major disruptions and changes in all aspects of daily life, although it will take more time and more studies to fully understand its impact. Finally, the treatment of weight during the height of the pandemic proved challenging but potentially successful at weight management centers. Telemedicine has additionally paved a path not only for treatment in the short term but could also be a long-term solution to improving access to care. The COVID-19 pandemic has highlighted the critical need to train more physicians to treat obesity as a chronic disease and the crucial need to increase access to all modalities of treatment for overweight and obesity.

Policies and procedures

Patients with overweight and obesity deserve to receive standard treatment for excess weight, and must feel comfortable in doing so. One hypothesis during the early pandemic of why patients with obesity presented to hospitals with more advanced COVID-19 infections was that patients delayed seeking care due to stigma and bias associated with obesity (Townsend et al., 2020b). The implicit and explicit bias in the healthcare setting is particularly harmful, and all patients must be treated with respect (including using person-first language when discussing obesity as a disease). Policies and procedures must be put into place to help reduce the existing stigma and bias, both in the medical community and society at large. This starts with the global medical community accepting obesity as a disease and not a lack of willpower.

Furthermore, unfortunately there are not enough providers trained to treat obesity, and it is essential that more physicians receive training. Policies must be adopted to incentivize not only the treatment of obesity, but additionally the prevention of overweight and obesity. In addition, these policies must increase access to care for patients. This includes the expansion of insurance coverage for telemedicine, which can significantly improve patient follow-up and can expand the reach of obesity medicine treatment, and the expansion of insurance coverage for antiobesity medications. Finally, patients with obesity must also be educated on their increased risk for severe disease from COVID-19, and for their increased risk for obesity-related comorbidities. Patients must also be directed toward resources to help them lose weight and maintain the loss.

Applications to other areas

In an increasingly globalized world, it is possible that more emerging pathogens will herald additional global pandemics. Obesity will continue to be a risk factor for severe disease with other infectious diseases. Preparation for future pandemics can start now—hospitals and doctors' offices must modernize their equipment to be fully capable of caring for patients with obesity. The existing stigma and bias surrounding obesity must be addressed such that patients feel comfortable seeking care. Clinicians must also be educated on how to care for patients with obesity, and access to care for patients with obesity must be expanded. For outpatient physician practices who have reverted to (or who never stopped) in-person care, there must be a backup option to transition to telemedicine quickly so that there is no delay in ongoing care.

The expansion of telemedicine is additionally something that should be considered in other specialties. Initially adopted to help reduce viral exposures, telemedicine helped encourage regular patient follow-up in patients who were receiving treatment for obesity. This may be for a variety of reasons, including that patients may take their appointments from anywhere. This means that they do not need to request additional time off work to go to an appointment, to find childcare, or to find a means to travel to a provider. This may be particularly helpful, for example, for rural populations that do not have easy access to primary care to help encourage regular follow-up of any chronic condition. Telemedicine could also help improve access to and follow-up with mental health providers, among many other fields and specialties.

Mini-dictionary of terms

- American Board of Obesity Medicine (ABOM): A nonprofit organization that maintains standards for the assessment and credentialing of obesity medicine physicians. ABOM diplomates are physicians who have demonstrated competency in obesity care.
- Antiobesity pharmacotherapy: Pharmacological agents used to treat obesity. Antiobesity medications target a number of metabolic pathways to help patients lose and maintain weight loss. There are currently six medications approved by the US Food and Drug Administration.
- Obesity: A chronic, progressive, relapsing disease. While there are a number of ways to define and quantify obesity, the Centers for Disease Control and Prevention and the World Health Organization use body mass index (BMI), which is a person's weight in kilograms divided by the person's height in meters squared. Obesity is typically defined as a BMI $\geq 30\,kg/m^2$, though certain populations have lower BMI thresholds to qualify as obesity.
- Overweight: Typically defined as a BMI $\geq 25\,kg/m^2$, though certain populations have lower BMI thresholds to qualify as overweight.
- Telemedicine: Remote clinical services.

Summary points

- The COVID-19 pandemic has exposed the vulnerability of patients with overweight and obesity.
- Obesity is a risk factor for severe disease and death from COVID-19.
- Once the increased risk was recognized, many experts in obesity medicine called for expanded access to treatment for those with overweight and obesity.
- Although the full impact may not be known for years, the COVID-19 pandemic may have accelerated global trends of obesity.
- There is a critical shortage of obesity medicine providers.
- Telemedicine was a critical advancement in treating patients during the COVID-19 pandemic, and continued expansion could help improve access issues.

References

ABOM. (2023). https://www.Abom.org.

Abou-Arab, O., Huette, P., Berna, P., & Mahjoub, Y. (2020). Tracheal trauma after difficult airway management in morbidly obese patients with COVID-19. *British Journal of Anaesthesia, 30214-30216*. S0007091220302142.

American Psychological Association. (2021). *Stress in America: One year later, a new wave of pandemic health concerns*. https://www.apa.org/news/press/releases/stress/2021/sia-pandemic-report.pdf.

Aminian, A., Milinovich, A., Wolski, K., et al. (2022). Association of weight loss achieved through metabolic surgery with risk and severity of COVID-19 infection. *JAMA Surgery*, *157*(3), 221–230.

Aras, M., Tchang, B., & Aronne, L. (2021). Impact of telemedicine during the COVID-19 pandemic on patient attendance. *Obesity*, *29*(7), 1093–1094.

Bakaloudi, D. R., Barazzoni, R., Bischoff, S. C., et al. (2021). Impact of the first COVID-19 lockdown on body weight: A combined systematic review and a meta-analysis. *Clinical Nutrition*. S0261-5614:00207-7.

Barenbaum, S., Saunders, H., Chan, K., et al. (2022). Medical weight management protects against weight gain during the COVID-19 pandemic. *Obesity Science and Practice*, *8*(5), 682–687.

Bhutani, S., vanDellen, M., & Cooper, J. (2021). Longitudinal weight gain and related risk behaviors during the COVID-19 pandemic in adults in the US. *Nutrients*, *13*(2), 671.

Centers for Disease Control. (2022). *Underlying medical conditions associated with high risk for severe COVID-19: Information for healthcare providers*. https://www.cdc.gov/coronavirus/2019-ncov/hcp/clinical-care/underlyingconditions.html. Accessed 12 May 2022.

FINRA. (2020). *State "Shelter-in-place" and "stay-at-home" orders*. https://www.finra.org/rules-guidance/key-topics/covid-19/shelter-in-place.

Freedman, D., Kompaniyets, L., Daymont, C., et al. (2022). Weight gain among US adults during the COVID-19 pandemic through May 2021. *Obesity*, *30*, 2064–2070.

Gao, M., Piernas, C., Astbury, N., et al. (2021). Associations between body-mass index and COVID-19 severity in 6.9 million people in England: A prospective, community-based, cohort study. *The Lancet Diabetes and Endocrinology*, *9*(6), 350–359.

Goyal, P., Choi, J., Pinheiro, L., et al. (2020). Clinical characteristics of COVID-19 in New York City. *The New England Journal of Medicine*, *382*.

Hamer, M., Gale, C. R., Kivimäki, M., & Batty, G. D. (2020). Overweight, obesity, and risk of hospitalization for COVID-19: A community-based cohort study of adults in the United Kingdom. *Proceedings of the National Academy of Sciences of the United States of America*, *117*, 21011–21013.

Hotamisligil, G. S., Arner, P., Caro, J. F., et al. (1995). Increased adipose tissue expression of tumor necrosis factor-alpha in human obesity and insulin resistance. *The Journal of Clinical Investigation*, *95*(5), 2409–2415.

Kahan, S., Look, M., & Fitch, A. (2022). The benefit of telemedicine in obesity care. *Obesity*, *30*, 577–586.

Klang, E., Kassim, G., & Soffer, S. (2020). Severe obesity as an independent risk factor for COVID-19 mortality in hospitalized patients younger than 50. *Obesity*, *28*(9), 1595–1599.

Lin, A., Vittinghoff, E., Olgin, J., et al. (2021). Body weight changes during pandemic-related shelter-in-place in a longitudinal cohort study. *JAMA Network Open*, *4*(3), e212536.

McNelis, J. C., & Macrophages, O. J. M. (2014). Immunity, and metabolic disease. *Immunity*, *41*, 36–48.

Minihane, A. M., Vinoy, S., Russell, W. R., et al. (2015). Low-grade inflammation, diet composition and health: Current research evidence and its translation. *The British Journal of Nutrition*, *114*(7), 999–1012.

Must, A., Spandano, J., Coakley, E. H., et al. (1999). The disease burden associated with overweight and obesity. *JAMA*, *282*, 1523–1529.

Ouchi, N., Parker, J. L., Lugus, J. J., & Walsh, K. (2011). Adipokines in inflammation and metabolic disease. *Nature Reviews. Immunology*, *11*(2), 85–97.

Popkin, B. M., Du, S., Green, W. D., et al. (2020). Individuals with obesity and COVID-19: A global perspective on the epidemiology and biological relationships. *Obesity Reviews*, *21*, e131128.

Rizzo, S., Chawla, D., Zalocusky, K., et al. (2020). Descriptive epidemiology of 16,780 hospitalized COVID-19 patients in the United States. *medRxiv*. https://doi.org/10.1101/2020.07.17.20156265. Preprint.

Robinson, E., Gillespie, S., & Jones, A. (2020). Weight-related lifestyle behaviours and the COVID-19 crisis: An online survey study of UK adults during social lockdown. *Obesity Science and Practice*, *6*, 735–740.

Ryan, D., & Yockey, S. (2017). Weight loss and improvement in comorbidity: Differences at 5%, 10%, 15% and over. *Current Obesity Reports*, *6*(2), 187–194.

Rychter, A., Zawada, A., Ratajczak, A., et al. (2020). Should patients with obesity be more afraid of COVID-19? *Obesity Reviews*, *21*, e13083.

Sanchis-Gomar, F., Lavie, C., Mehra, M., et al. (2020). Obesity and outcomes in COVID-19: When an epidemic and pandemic collide. *Mayo Clinic Proceedings*, *95*(7), 1445–1453.

Saunders, K., Igel, L., & Aronne, L. (2022). Telemedicine could be the solution to scaling obesity treatment. *Obesity*, *30*, 573–574.

Schoeller, D. (2014). The effect of holiday weight gain on body weight. *Physiology & Behavior*, *134*, 66–69.

Simonnet, A., Chetboun, M., Poissy, J., et al. (2020). High prevalence of obesity in severe acute respiratory syndrome Coronavirus-2 (SARS-CoV-2) requiring invasive mechanical ventilation. *Obesity*, *28*, 1195–1199.

Stierman, B., Afful, J., Carrol, M., et al. (2021). *National Health and nutrition examination survey 2017–March 2020 prepandemic data files development of files and prevalence estimates for selected health outcomes*. National Health Statistics Reports *Vol. 158*. https://stacks.cdc.gov/view/cdc/106273.

Thomas, C., Mauer, E., Shukla, A., et al. (2016). Low adoption of weight loss medications: A comparison of prescribing patterns of antiobesity pharmacotherapies and SGLT2s. *Obesity*, *24*(9), 1955–1961.

Townsend, M., Kyle, T., & Stanford, F. (2020a). Commentary: COVID-19 and obesity: Exploring biologic vulnerabilities, structural disparities, and weight stigma. *Metabolism*, *110*, 154316.

Townsend, M., Kyle, T., & Stanford, F. (2020b). Commentary: COVID-19 and obesity: Exploring biologic vulnerabilities, structural disparities, and weight stigma. *Metabolism, 110*, 154316.

Wing, R. R., Lang, W., Wadden, T. A., et al. (2011). Benefits of modest weight loss in improving cardiovascular risk factors in overweight and obese individuals with type 2 diabetes. *Diabetes Care, 34*, 1481–1486.

World Health Organization Fact Sheet. (2021). Found at: https://www.who.int/news-room/fact-sheets/detail/obesity-and-overweight.

Yanovski, J., Yanovski, S., Sovik, K., et al. (2000). A prospective study of holiday weight gain. *The New England Journal of Medicine, 342*, 861–867.

Yu, L., Zhang, X., Ye, S., et al. (2022). Obesity and COVID-19: Mechanistic insights from adipose tissue. *The Journal of Clinical Endocrinology & Metabolism, 107*, 1799–1811.

Zeigler, Z. (2021). COVID-19 self-quarantine and weight gain risk factors in adults. *Current Obesity Reports, 12*, 1–11.

Chapter 16

COVID-19 and the management of heart failure using telemedicine

Maria Margarida Andrade[a], Diogo Cruz[a,b], and Marta Afonso Nogueira[c]
[a]*Department of Internal Medicine, Hospital de Cascais, Dr. José de Almeida, Alcabideche, Portugal,* [b]*Faculty of Medicine, Lisbon University, Lisbon, Portugal,* [c]*Department of Cardiology, Hospital de Cascais, Dr. José de Almeida, Alcabideche, Portugal*

Abbreviations

AF	atrial fibrillation
BP	blood pressure
CD	cardiology department
CIED	cardiac implanted electronic device
COVID-19	Coronavirus disease 2019
CRT-D	cardiac resynchronization therapy defibrillator
DMP	disease management program
ECG	electrocardiogram
ED	emergency department
F2F	face to face
GP	general practitioner
HR	hazard ratio
HF	heart failure
HR	heart rate
ICD	implanted cardioverter defibrillator
LAP	left atrial pressure
LVEF	left ventricular ejection fraction
NYHA	New York Heart Association
PA	pulmonary artery
RPM	remote patient monitoring
VV	virtual visits

Introduction

Heart failure (HF) is a chronic disease characterized by high morbidity and mortality. Its prevalence in the European population is estimated to be 0.4%–2%. In the United States, HF affected 5.7 million adults in 2012, and this is predicted to rise to 8.4 million by 2030 (Alvarez et al., 2021; Krzesiński et al., 2021). Hospitalizations account for more than half of the direct and indirect costs associated with HF. The rate of readmission for HF deterioration is approximately 10% 30 days after the initial hospitalization, increasing to 40% within 6 months. Each successive HF readmission worsens the prognosis of the patient (Krzesiński et al., 2021).

The COVID-19 pandemic has affected the care of HF patients (Alvarez et al., 2021; Krzesiński et al., 2021; Palazzuoli et al., 2022). First, due to their older age, overall frailty, reduced immunity, comorbidities, and limited hemodynamic response to cope with more severe infections, they are at high risk of complications (Bertagnin et al., 2021; Tersalvi et al., 2020). It was reported that in HF patients, monocytes seem to produce more tumor necrosis factor-α and less interleukin-10 than healthy subjects, which, in combination with the widespread systemic inflammatory response associated with severe COVID-19 infections, requires enhanced cardiac performance and high cardiac output (that HF patients are generally incapable of) (Bader et al., 2021; D'Amario et al., 2020). Some authors have proposed the name *acute cardiovascular syndrome by COVID-19* to describe the cardiovascular system changes associated with SARS-CoV-2 infection

Management, Body Systems, and Case Studies in COVID-19. https://doi.org/10.1016/B978-0-443-18703-2.00041-4
Copyright © 2024 Elsevier Inc. All rights are reserved, including those for text and data mining, AI training, and similar technologies.

such as arterial and venous thrombotic complications or even acute myocarditis (Afonso Nogueira et al., 2021; Goldraich et al., 2020). New onset or worsening HF is a common complication in patients with COVID-19, with documented poor survival rates of ~60%. Furthermore, nearly 50% of patients who die from COVID-19 with a diagnosis of HF have no previous history of either hypertension or cardiovascular disease (Afonso Nogueira et al., 2021; Charman et al., 2021).

Second, HF patients are among the highest consumers of healthcare services (Oseran et al., 2021), so overcoming psychosocial, economic, and geographical barriers and assuring effective management at distance presented new and unique challenges (Alvarez et al., 2021; Oseran et al., 2021). Healthcare systems have developed disease management programs to overcome these limitations. Telemedicine and telemonitoring are integral parts of those programs and their application increased exponentially in 2020–21 during the COVID-19 pandemic (Alvarez et al., 2021; De Peralta et al., 2021).

Telemedicine and telemonitoring have revolutionized the delivery of healthcare. Despite the common use of telemonitoring technologies in the management of chronic diseases such as HF, these interventions have been used as an adjunct to a consultative face-to-face (F2F) visit, usually in the context of a disease management program (DMP) (Alvarez et al., 2021). Sensor telemonitoring technologies and telehealth programs for HF patients have shown mixed results in the reduction of morbidity and mortality (De Peralta et al., 2021).

The current COVID-19 pandemic has spurred a paradigm shift in the delivery of cardiac consultation for HF patients. To reduce the risk of exposure to SARS-CoV-2, while preventing the patient's baseline health from deteriorating at the same time, telemedical follow-up of stable HF patients was recommended. Direct patient-provider contact was used in emergent/urgent situations only. Although this change was forcibly and rapidly introduced, it is likely to transform the future delivery of consultative care (Afonso Nogueira et al., 2021; De Peralta et al., 2021; Tersalvi et al., 2020).

Telemedicine and telemonitoring in heart failure

Definition and indication

Telemedicine implies the utilization of technology data to monitor patients at distance, enabling communication between a provider and a patient. Telemedicine appointments can be asynchronous or synchronous, continuous or intermittent, and occur in an inpatient or outpatient setting. Examples of asynchronous interventions are a review of an echocardiogram, laboratory evaluation, or answering a patient call. On the other hand, synchronous telemedicine appointments happen when a real-time interaction is established such as a video consultation. Continuous telemedicine medical care is possible nowadays due to the cardiologic devices monitoring a patient's heart (Alvarez et al., 2021).

HF is a chronic disorder for which additional telemedical support was generally shown to be advantageous (Oseran et al., 2021). Multidisciplinary care plans that comprise professionals from different levels of care optimize the diagnosis and treatment of patients with HF (Mazón-Ramos et al., 2022). However, telemedical support is an expensive intervention and HF is one of the most prevalent chronic diseases, so an efficient allocation is desirable (Oseran et al., 2021; Pigorsch et al., 2022). Providers must first identify patients appropriate for a virtual visit and retriage those better served by traditional face-to-face evaluation (Oseran et al., 2021).

Patients with a sufficiently stable disease state will most likely not profit from telemedical intervention. Therefore, it would be ideal if the subgroup of potential telemedical profiters—the actual population of interest—could be identified in advance by an appropriate biomarker (Pigorsch et al., 2022). According to a randomized patient group trial by Pigorsch et al., several biomarkers that were known to be associated with the probability of an unfavorable event were measured, but there were no established cut-off values, allowing a prospective definition of the population of interest (Pigorsch et al., 2022).

During the COVID-19 pandemic, telemedicine programs for HF patients made the continuity of care easier without increasing mortality or the need for in-person consultations (Mazón-Ramos et al., 2022; Qureshi et al., 2020).

The great challenge in patients with HF during COVID-19 was not to provide superior care than the standard but to offer patients with HF a health maintenance strategy that kept them safe from infection, but also continuing with strict monitoring to prevent hospitalizations (Tersalvi et al., 2020).

Many hospital management strategies were implemented in patients with HF during COVID-19. Home monitoring consisted of the remote monitoring of vital parameters and transmission (via devices, telephone, apps) to a care center for interpretation and management to therapy optimization. On the order hand, virtual visits were used to assess symptoms but also for therapy optimization and to meet new HF patients. Appointments to forward triage, in-hospital telemedicine, and telerehabilitation were also developed (Tersalvi et al., 2020).

Regarding clinical practice guidelines for follow-up visits, it is recommended that one happens 7–14 days after hospital discharge for patients who were hospitalized due to acute HF to avoid early hospital readmissions (Mazón-Ramos et al.,

2022). Koehler et al. showed in the TIM-HF2 study that telemedicine appointments are effective for the follow-up of HF patients, as they allow clinical management at each moment, which is fundamental at the start of decompensation to reduce hospitalizations and mortality (Koehler et al., 2018). This is also suitable for the follow-up of HF patients in whom their general practitioner (GP) has identified clinical signs of decompensation or other issues related to patient care, meaning they need to be referred to a cardiology department (CD) (Mazón-Ramos et al., 2022; Tersalvi et al., 2020).

Despite the fact that some patients might prefer in-person visits over virtual encounters (Fraser et al., 2021), the benefits of telemedicine appointments are clear: diminished infection risk in the clinical setting, reduced patient stress and anxiety, continued patient-provider partnership, lower hospital admission rates, and preserving beds for the critically ill. However, some difficulties were also reported such as striving to obtain accurate vital signs, challenging and limited physical examinations, and difficulties related to the use of virtual technology (Tersalvi et al., 2020).

Virtual visits

Synchronous audio/video interactions between the provider and the patient consist of virtual visits (VV). During the COVID-19 pandemic, the increase in this type of visit was exponential and it was found that the benefits overcame the difficulties (Alvarez et al., 2021).

There are several types of virtual visits, each with its own particularities, as shown in Table 1 (Qureshi et al., 2020). Although demanding in terms of human resources, VV for HF patients during a pandemic are proven to be effective in the decline of adverse clinical outcomes (Bertagnin et al., 2021).

The frequency of VV is flexible and determined by clinical judgment. As shown in the randomized trial of telephone intervention in chronic HF (DIAL), the phone call frequency went from weekly in patients with New York Heart Association (NYHA) III–IV symptoms, recent hospitalization, weight gain of more than 2 kg, and severe edema to monthly in patients with NYHA I, not hospitalized within the previous year, <75 years old, and not living alone. Risk stratification tools may guide the intensity of telemedicine interventions (Alvarez et al., 2021; Ferrante et al., 2010).

Lack of familiarity with technology as well as regulatory, legal, and reimbursement concerns were some of the barriers previously impeding this type of consultation (Alvarez et al., 2021).

Interventions

VV should begin with a detailed history. The virtual physical exam should focus on obtaining an accurate assessment of vital signs and fluid balance (Oseran et al., 2021). Patients should be encouraged to monitor their blood pressure, heart rate, heart rhythm, and daily standing weight at home. Home blood pressure guidance detects hypotension that prevents the titration of medical therapy and allows the adjustment of antihypertensive therapy that may impact filling pressures and exercise capacity (Alvarez et al., 2021).

This type of visit allows the assessment of any signs or symptoms suggestive of a decompensated state such as dyspnea on exertion, orthopnea, elevated pressure on the jugular vein, and lower extremity oedema (De Peralta et al., 2021; Oseran et al., 2021). The early detection of clinical features such as congestion associated with increased decompensation risk is one of the main goals of telemedicine (Alvarez et al., 2021). The early diagnosis of mild decompensations allows timely medication dosage adjustment by telephone that avoids the need to resort to healthcare services (Barrios et al., 2020).

TABLE 1 Virtual visits.

Type of virtual visit	Characteristics
Electronic consult	Asynchronous. Ideal for simple clinical questions such as usual prescription renewal
Clinic video telehealth	Synchronous. Ideal if the patient requires evaluation but is unable to or was advised not to come in for a face-to-face visit
Remote monitoring	Asynchronous. Monitor heart rate, blood pressure, pulse oximetry, weight, and pedometer steps Ideal for patients with HF that have had a recent hospitalization (due to worsening heart failure)

Types of virtual visits and their characteristics—type of contact, collected data, and types of patients it is intended for.

VV may be best utilized for medication titration and optimization in stable patients with chronic HF. A review of medications should be performed to ensure a complete and accurate list, as such a review of the patient's adherence to the medication is equally important (Albert & Prasun, 2020). The adequate assessment of volume status or congestion in an asymptomatic patient at distance is quite a challenge (Tersalvi et al., 2020). Daily weight allows adjusting diuretic therapy and the detection of volume overload by the provider during the appointment. However, significant congestion can develop in patients without significant weight gain due to sympathetically mediated volume redistribution in acute HF (Alvarez et al., 2021).

Forward triage was a useful telemedicine strategy for patients with HF during COVID-19. It consisted of the stratification of patients before arriving in the emergency department (ED), as many respiratory symptoms as well as functional decline and fatigue might be present in early COVID-19 and/or decompensated HF. Through a structured telemedicine program, detailed medical and exposure histories might be easily obtained, complemented with the use of screening algorithms. The goal is to guide patients to the right diagnostic—therapeutic pathway while protecting them from unnecessary risk and exposure (Tersalvi et al., 2020).

Telerehabilitation strategies were also described and involved the delivery of rehabilitation services via information and communication technologies. Videos, website information, and emails were the most frequently used ways (Tersalvi et al., 2020). Patients with advanced HF might benefit from telemedicine platforms, as they allow multidisciplinary assessments without delay (Tersalvi et al., 2020).

Regarding possible interventions in VV, overall 51% of contacts led to a clinical decision (adjustment of diuretic doses, change in blood pressure-lowering drugs, rate controls, anticoagulation management, and others) (Salzano et al., 2020). So, as reported by Charman et al., VVs are so much more efficient than standard consultations in primary care practice as they save time. Thus, in-person visits should be reserved for patients approaching or with advanced HF, those who had a recent left ventricular assist device (LVAD) implantation or heart transplant, and those with new-onset HF (Tersalvi et al., 2020). Telemedical interventions provide a wide field of treatment support options that has the potential to better address the patient's individual needs and thereby improve long-term outcomes (Pigorsch et al., 2022).

Wearables and cardiac implantable electronic devices

Sensors that capture continuous functional or physiological data incorporated in patches, clothes, or smartwatches are considered wearables. They collect data such as heart rate, blood pressure, activity, lung water content, and arrhythmia, which, when interpreted, can allow the early diagnosis of chronic disease decompensation (Alvarez et al., 2021).

Some trials were dedicated to evaluating the sensitivity of the data collected as predisposing factors for cardiac decompensation. In the Apple Heart Study, the authors showed that an irregular pulse detected by a commercial smartwatch had a positive predictive value of 84% for detecting atrial fibrillation (AF) on an electrocardiogram (ECG) simultaneously with a subsequent irregular pulse notification (Alvarez et al., 2021; Turakhia et al., 2019). Despite the fact that only a minority of HF patients have AF, we know that a rapid and arrhythmic pulse is a known factor that predisposes to HF decompensation (Turakhia et al., 2019).

Stehlik et al. analyzed the accuracy of a wearable multiparametric sensor to predict HF hospitalization in the LINK-HF study (Stehlik et al., 2020). The model had sensitivity between 76% and 88% and a specificity of 87% to detect HF hospitalization precursors with a median time between alert and admission of 6.5 days (Alvarez et al., 2021; Stehlik et al., 2020).

There is an emergent armamentarium of remote sensor technology for HF patients that might have been beneficial during the time of limited F2F clinical encounters (Bertagnin et al., 2021). Regarding cardiac implanted electronic devices (CIDE), these are recommended for HF patients with severe systolic dysfunction. If the device is intended to prevent sudden death, then it consists of an implantable cardioverter defibrillator (ICD), whereas it is a cardiac resynchronization therapy device (CRT) if the goal is to improve cardiac contractile function by resynchronizing left and right ventricles (Alvarez et al., 2021; Bader et al., 2021; Oseran et al., 2021). In addition to the positive effect on the ejection fraction, these devices allow the collection of data about the patient's condition (Bertagnin et al., 2021; Oseran et al., 2021).

Arya et al. evaluated the difference among daily review CIED data from NYHA functional class II or III, left ventricular ejection fraction (LVEF) <35%, and indication for dual-chamber ICD or CRT patients in addition to standard of care or standard of care alone in HF hospitalizations. The IN-IME study concluded that daily monitoring might improve outcomes because it allowed the early detection of suboptimal device function and clinical worsening (Arya et al., 2008).

There are other examples of technological devices helpful for HF patients. The CardioMEMS (Abbott) device is an implantable pulmonary artery (PA) pressure sensor that reduced hospitalizations and improved quality of life in HF patients, especially those in NYHA class III functional capacity with at least one HF hospitalization within the previous 12 months. It had benefits across the entire ejection fraction spectrum, as it measures and transmits PA pressure variations favoring the early diagnosis of cardiac decompensation (Oseran et al., 2021).

Some CIEDs, such as the Optivol (Medtronic) system, act through the measurement of thoracic impedance that is inversely related to volume overload, so if the pulmonary fluid increases, the intrathoracic impedance decreases. Data showing a reduction in HF hospitalizations have been inconsistent; however, the clinical information collected from this device might predict HF hospitalization at a time when F2F encounters are not possible and patient access is limited (Oseran et al., 2021). Another device, HeartLogic (Boston Scientific), consists of an ICD or CRT with an algorithm that includes intrathoracic impedance, nocturnal heart rate, the presence of a third heart sound, respiration rate, and patient activity. In the MULTI-SENSE study, HeartLogic predicted HF events with 70% sensitivity (Gardner et al., 2018; Oseran et al., 2021). Importantly, nearly 90% of patients enrolled in this study had a HeartLogic alert that preceded a true HF event by at least 2 weeks. This type of technology allows further optimization of HF management and potentially prevents an F2F visit or hospitalization.

Recent telemedicine and telemonitoring trials

Since 2000, most of the telemedicine clinical trials in HF have been focused on home telemonitoring. Due to the heterogeneity of the study groups, implemented devices, and telemedicine systems, their results are inconsistent and incomparable. Nevertheless, a few have demonstrated their superiority to controls for the primary endpoints.

In July 2005, Galinier et al. showed that telemonitoring patients with a previous (12 months) HF hospitalization had failed in reducing all-cause death or unplanned hospitalization for HF. This kind of system provided an alert that nurses evaluated and if appropriate, they advised the patient to contact their general practitioner or cardiologist. They concluded that only some groups benefited from telemonitoring: patients in NYHA class III/IV (HR = 0.71; P = 0.02), socially isolated subjects (hazard ratio (HR) = 0.62; P = 0.043), and those strictly adherent to body mass measurement (HR = 0.63; P = 0.006) (Galinier et al., 2020).

The Tele-HF study (telemonitoring to improve HF outcomes) (Chaudhry et al., 2007) and the WISH trial (weight monitoring in patients with severe HF) (Lyng et al., 2012) were neutral for nurse-supervised teletransmission of body mass and clinical symptoms.

Also, Ong et al., in the BEAT-HF (better effectiveness after transition) study, analyzed 180 days of teleintervention based on remote patient monitoring (blood pressure, heart rate, body mass, and symptoms). It failed to reduce all-cause hospital readmission when compared with standard care. However, the telemonitoring strategy was not integrated with physician care to allow rapid medical treatment changes (Ong et al., 2016). Still, in a secondary analysis, Haynes et al. showed that every day of good patient adherence resulted in a 19% decrease in mortality and an 11% decrease in the rate of hospitalization in the following week (Haynes et al., 2020).

The perspective changed in 2015 when a Cochrane Library systematic review of 41 randomized controlled trials about structured telephone support or noninvasive telemonitoring for patients with HF concluded that those interventions reduced the risk of all-cause mortality and HF-related hospitalizations (Inglis et al., 2017).

A breakthrough in this field was the publication of the results of the TIM-HF2 (telemedical interventional monitoring in HF) trial (Koehler et al., 2018). It was a randomized, controlled, parallel-group unmasked trial that tested a multifactorial telemonitoring intervention compared with usual care. In what concerns TIM-HF results (Koehler et al., 2021), remote patient monitoring compared with standard of care was not associated with reducing all-cause mortality. Unplanned cardiovascular hospital admissions or all-cause death were the primary endpoints. The remote patient monitoring (RPM) was possible due to four Bluetooth-equipped measuring devices: a three-channel ECG, a pulse oximeter, blood pressure monitors, and a digital scale. It revealed that RPM was beneficial in reducing the percentage of days lost to unplanned cardiovascular hospitalizations (4.9% vs 6.6%, HR = 0.80; P = 0.046) and all-cause mortality (HR = 0.70; P = 0.028). The value of these results was so significant that the expert consensus report of the Heart Failure Association of the European Society of Cardiology indicated home monitoring like the one used in TIM-HF2 as worthy of consideration for patients with HF (Koehler et al., 2018).

Further, Salzano et al., in the early months of the COVID-19 pandemic, evaluated the usefulness of a DMP using different modalities of telemedicine such as telephonic consultations, online chats, and video consultations. They carried out a comparative study with a similar population from the same period in 2019 with access to F2F consultations. The team concluded that telemedicine was associated with a lower risk of HF-related hospitalization but a similar risk of death. The authors argued that telemedicine is a valuable tool in HF management and showed feasibility during the COVID-19 outbreak (Salzano et al., 2020).

These results are like those described by Sammour et al. and Zhu et al., which concluded that telematic consultations avoid patient and caregiver journeys without increasing the risk of urgent visits or mortality. The authors of this study also suggested that DMP with the capability of patient risk stratification should be developed to reduce the healthcare burden (Sammour et al., 2021; Zhu et al., 2020).

Finally, more recently, we have seen trials carried out on HF management through remote telemonitoring of left atrial pressure (LAP) using the V-LAP device. The study population consisted of patients in NYHA III with LVEF >15% who had a previous hospitalization for worsening of HF or elevated ambulatory levels of brain natriuretic peptide. Earlier detection of disease progression due to remote analysis of LAP curves prevents clinically relevant decompensation (D'Amario et al., 2020).

COVID-19 and heart failure clinics

Because chronic HF has a tremendous impact on a patient's quality of life, establishing and supporting multidisciplinary HF clinics are key mechanisms to improve the health status and clinical outcomes for patients with HF. There has been growing interest in developing high-performing HF clinics to efficiently provide comprehensive disease management interventions improving the quality of outpatient HF care (Greene et al., 2021). The European Society of Cardiology HF guidelines promote the use of telephone and remote monitoring in the management and follow-up of HF patients. The COVID-19 pandemic revitalized the interest in telemedicine and home monitoring of HF patients (Charman et al., 2021; Greene et al., 2021).

A reduction of HF hospitalizations was observed in early 2020, which was interpreted secondary to the patient's fear of infection. During the same period, a 154% increase in VVs was documented. VVs and telemedicine changed from being sporadic to becoming one of the principal forms of healthcare delivery (Alvarez et al., 2021). Standardizing HF telemedicine is essential, and there is potential to improve patient care and reduce costs to healthcare services through HF-organized clinics (Charman et al., 2021). Gorodeski et al. evaluated the absenteeism rates of HF patients in the HF clinic 7 days post-hospitalization and compared personal with VV rates of 51% versus 34.6%, respectively. No significant differences in hospital readmission, ED visits, or death were observed between the two groups (Bader et al., 2021; Gorodeski et al., 2020).

The Portuguese perspective on the subject is based on previously published data that proved the benefit of HF clinics in improving the quality of life of patients and in diminishing the ED visits/hospitalizations and mortality rates (Nogueira et al., 2020). During the first 3 months of the COVID-19 pandemic, the HF clinic of a district hospital was reorganized. All the prescheduled appointments were converted to teleconsultations (over the phone appointments including drug prescriptions via email or a short message service, with adjustments of drugs in the ambulatory setting) to identify which patients would need in-person care. Personal appointments were limited to urgent situations and to patients in NYHA functional classes III/ambulatory IV. This HF clinic was able to maintain a low rate of admissions due to HF decompensation, without an increase in mortality, despite the increase in the telemedical management of patients (Afonso Nogueira et al., 2021).

Based on evidence that the use of telemedicine technology through a structured HF clinic was safe and effective, one might encourage the incremental use of these tools in HF patients in the context of this or future pandemics and in situations in which physical consultation might not be possible due to logistic issues (Afonso Nogueira et al., 2021).

Conclusion

Telemedical interventions in HF patients have as the main goal avoiding adverse events through early, individualized care to fulfill patient needs (Pigorsch et al., 2022). VVs have exponentially improved during the COVID-19 pandemic. Telemonitoring interventions can range from low to high complexity and should equal the risk profile of the patient (Alvarez et al., 2021).

By reviewing the results of previous trials regarding telemedicine programs during COVID-19, we must be aware that a system that is cost-effective, accessible, and person-centered only has to prove that it is not inferior to traditional ways of delivering care and thus allows the safe maintenance of the patient (Tersalvi et al., 2020). According to De Peralta et al., telemedicine was found to provide an adequate platform for the delivery of HF consultation and to the point studied, was a reliable determinant of patients needing emergency/hospital-level care (De Peralta et al., 2021). Also, Alvarez et al. agreed with that principle as they stated that telehealth was not associated with increased subsequent emergency department visits, hospitalizations, or mortality in comparison with in-person visits (Alvarez et al., 2021).

The evidence regarding the benefit of telemedicine interventions in the management of chronic HF, in both screening of early signs of congestion or in follow-up in advanced HF patients, is clear. All this home monitoring and implementation of strategies requires patient collaboration, not only in being attentive to one's signs and symptoms but also in contacts with the health team. Adherence to medications and a low-salt diet can be the difference maker in this regard (Albert & Prasun, 2020; Bader et al., 2021). As Albert et al. concluded, patients need to understand that they are in control of their health (Albert & Prasun, 2020). Patient-directed self-care is nowadays a mandatory objective regarding health literacy.

Two years have gone by since the beginning of the COVID-19 pandemic and the need for patient surveillance and follow-up favored the use of telemedicine. It is, after all, a safe, reliable method to maintain continuity of care, especially as providers and patients became more experienced and comfortable using telehealth.

Policies and procedures

In this chapter, we reviewed the evidence regarding the benefit of telemedicine interventions in the management of chronic HF. We reviewed in detail its utility in both screening of early signs of congestion and follow-up in advanced HF patients. According to De Peralta and Alvarez, telemedicine was found to provide an adequate platform for the delivery of HF consultation and, to the point studied, was a reliable determinant of patients needing emergency/hospital-level care. Also, telehealth was not associated with increased subsequent emergency department visits, hospitalizations, or mortality in comparison with in-person visits (Alvarez et al., 2021; De Peralta et al., 2021).

Over the last few decades, there has been an emergent armamentarium of remote sensor technology for HF patients that has shown greater benefit during periods when face-to-face consultations were limited (Bertagnin et al., 2021).

The implementation of standard HF telemedicine is essential and there is potential to improve patient care as well as to reduce costs related to healthcare services through HF-organized clinics (Charman et al., 2021). Based on the Portuguese experience with HF clinics, the benefit of organized and accessible healthcare allows an improvement in the quality of life of patients and a decrease of the ED visits/hospitalizations and even mortality rates (Nogueira et al., 2020).

In addition to a lower rate of absenteeism from consultations, we also observed greater adherence to therapy as well as superior patient collaboration, not only in being attentive to one's signs and symptoms but also in contacts with the health team (Albert & Prasun, 2020; Bader et al., 2021). Now more than ever, patient-directed self-care must be a mandatory goal regarding health literacy (Albert & Prasun, 2020).

Applications to other areas

In this review, we highlighted the benefits of using telemedicine in HF patients, a monitoring strategy largely used during the COVID-19 pandemic. Telemedicine implies the utilization of technology data to monitor patients at a distance, enabling communication between a provider and a patient (Alvarez et al., 2021).

Wearables consist of sensors that capture continuous functional or physiological data and are nowadays incorporated in patches, clothes, or smartwatches. They collect data such as heart rate, blood pressure, activity, lung water content, and arrhythmia, which, when interpreted, can allow the early diagnosis of chronic disease decompensation (Alvarez et al., 2021).

Despite the common use of telemonitoring technologies in the management of chronic diseases such as HF, the significant dissemination of this intervention has shown unequivocal advantages in the follow-up of patients (Alvarez et al., 2021). The use of telemedicine may also present advantages in monitoring patients, for example, those with chronic respiratory disease. It may even be possible to use the same technology.

Technologies such as the one used in the LINK-HF study might be adapted to other diseases, including respiratory (Stehlik et al., 2020). Another possibility is the Optivol (Medtronic) system, which tat acts through the measurement of thoracic impedance that is inversely related to volume overload, so if pulmonary fluid increases, intrathoracic impedance decreases. It showed benefits in predicting HF hospitalization and perhaps might give some important information regarding the respiratory function (Oseran et al., 2021).

With the knowledge acquired during this pandemic as well as the ability we now have to interchange F2F consultations and virtual VV, we are better prepared for future pandemics or even periods of extreme climate change that prevent free human circulation.

Mini-dictionary of terms

- **Synchronous visits**: A type of telemedicine appointment in which there is an interaction between the health team and the patient in real-time at distance.
- **Acute cardiovascular syndrome by COVID-19**: Cardiovascular system changes associated with SARS-CoV-2 infection such as arterial and venous thrombotic complications or even acute myocarditis.

- **Home blood pressure guidance**: Titration of medical therapy and/or adjustment of antihypertensive therapy based on blood pressure values recorded by the patient or by telemedicine sensors.
- **Disease management program**: Structured treatment plans that aim to help people better manage their chronic disease and to maintain and improve the quality of life. They may include F2F visits and telemedicine appointments.
- **Wearables**: Sensors that capture continuous functional or physiological data (e.g., patches, smartwatches, and/or devices attached to clothes).

Summary points

- Healthcare providers must identify patients who are appropriate for a VV and retriage those better served by traditional face-to-face evaluation to use the telemedicine resources in the most cost-effective way possible.
- Patients with a sufficiently stable disease state will most likely not profit from telemedical intervention.
- The frequency of VV is flexible and should be determined by clinical judgment.
- The early detection of clinical features such as congestion is one of the main goals of telemedicine.
- 51% of VV contacts led to a clinical decision.
- HF clinic benefits include improvement in the quality of life of patients and a decrease in ED visits/hospitalizations and mortality rates.
- Telemedicine programs are not inferior to traditional ways of delivering care and thus allow the safe maintenance of the patient's health status.

References

Afonso Nogueira, M., Ferreira, F., Raposo, A. F., Mónica, L., Simões Dias, S., Vasconcellos, R., et al. (2021). Impact of telemedicine on the management of heart failure patients during coronavirus disease 2019 pandemic. *ESC Heart Failure*, *8*(2), 1150–1155. https://doi.org/10.1002/ehf2.13157.

Albert, N. M., & Prasun, M. A. (2020). Telemedicine in heart failure during COVID-19: Like it, love it or lose it? *Heart and Lung*, *49*(6), A11–A12. https://doi.org/10.1016/j.hrtlng.2020.10.014.

Alvarez, P., Sianis, A., Brown, J., Ali, A., & Briasoulis, A. (2021). Chronic disease management in heart failure: Focus on telemedicine and remote monitoring. *Reviews in Cardiovascular Medicine*, *22*(2), 403–413. https://doi.org/10.31083/j.rcm2202046.

Arya, A., Block, M., Kautzner, J., Lewalter, T., Mörtel, H., Sack, S., et al. (2008). Influence of home monitoring on the clinical status of heart failure patients: Design and rationale of the IN-TIME study. *European Journal of Heart Failure*, *10*(11), 1143–1148. https://doi.org/10.1016/j.ejheart.2008.08.004.

Bader, F., Manla, Y., Atallah, B., & Starling, R. (2021). Heart failure and COVID-19. *Heart Failure Reviews*, *26*(1), 1–10. https://doi.org/10.1007/s10741-020-10008-2.

Barrios, V., Cosín-Sales, J., Bravo, M., Escobar, C., Gámez, J. M., Huelmos, A., et al. (2020). Telemedicine consultation for the clinical cardiologists in the era of COVID-19: Present and future. Consensus document of the Spanish Society of Cardiology. *Revista Española de Cardiología*, *73*(11), 910–918. https://doi.org/10.1016/j.recesp.2020.06.027.

Bertagnin, E., Greco, A., Bottaro, G., Zappulla, P., Romanazzi, I., Russo, M. D., et al. (2021). Remote monitoring for heart failure management during COVID-19 pandemic. *IJC Heart and Vasculature*, *32*, 100724. https://doi.org/10.1016/j.ijcha.2021.100724.

Charman, S. J., Velicki, L., Okwose, N. C., Harwood, A., McGregor, G., & Ristic, A. (2021). Insights into heart failure hospitalizations, management, and services during and beyond COVID-19. *ESC Heart Failure*, *8*(1), 175–182. https://doi.org/10.1002/ehf2.13061.

Chaudhry, S., Barton, B., & Mattera, J. (2007). Randomized trial of telemonitoring to improve heart failure outcomes (Tele-HF): study design. *Journal of Cardiac Failure*, *13*(9), 709–714. https://doi.org/10.1016/j.cardfail.2007.06.720.

D'Amario, D., Restivo, A., Canonico, F., Rodolico, D., Mattia, G., Francesco, B., et al. (2020). Experience of remote cardiac care during the COVID-19 pandemic: The V-LAP^{TM} device in advanced heart failure. *European Journal of Heart Failure*, *22*(6), 1050–1052. https://doi.org/10.1002/ejhf.1900.

De Peralta, S. S., Ziaeian, B., Chang, D. S., Goldberg, S., Vetrivel, R., & Fang, Y. M. (2021). Leveraging telemedicine for management of veterans with heart failure during COVID-19. *Journal of the American Association of Nurse Practitioners*, *34*(1), 182–187. https://doi.org/10.1097/JXX.0000000000000573.

Ferrante, D., Varini, S., MacChia, A., Soifer, S., Badra, R., Nul, D., et al. (2010). Long-term results after a telephone intervention in chronic heart failure: DIAL (randomized trial of phone intervention in chronic heart failure) follow-up. *Journal of the American College of Cardiology*, *56*(5), 372–378. https://doi.org/10.1016/j.jacc.2010.03.049.

Fraser, M., Mutschler, M., Newman, C., Sackman, K., Mehdi, B., Wick, L., et al. (2021). Heart failure care delivery in the COVID-19 era: The patients' perspective. *Healthcare (Basel, Switzerland)*, *9*(3), 245. https://doi.org/10.3390/healthcare9030245.

Galinier, M., Roubille, F., Berdague, P., Brierre, G., Cantie, P., Dary, P., et al. (2020). Telemonitoring versus standard care in heart failure: A randomised multicentre trial. *European Journal of Heart Failure*, *22*(6), 985–994. https://doi.org/10.1002/ejhf.1906.

Gardner, R. S., Singh, J. P., Stancak, B., Nair, D. G., Cao, M., Schulze, C., et al. (2018). HeartLogic multisensor algorithm identifies patients during periods of significantly increased risk of heart failure events: Results from the MultiSENSE Study. *Circulation. Heart Failure*, *11*(7), e004669. https://doi.org/10.1161/CIRCHEARTFAILURE.117.004669.

Goldraich, L., Silvestre, O., Gomes, E., Biselli, B., & Montera, M. (2020). Emerging topics in heart failure: COVID-19 and heart failure. *Arquivos Brasileiros de Cardiologia*, *115*(5), 942–944. https://doi.org/10.36660/abc.20201081.

Gorodeski, E. Z., Goyal, P., Cox, Z. L., Thibodeau, J. T., Reay, R. E., Rasmusson, K., et al. (2020). Virtual visits for care of patients with heart failure in the era of COVID-19: A statement from the heart failure society of America. *Journal of Cardiac Failure*, *26*(6), 448–456. https://doi.org/10.1016/j.cardfail.2020.04.008.

Greene, S. J., Adusumalli, S., Albert, N. M., Hauptman, P. J., Rich, M. W., Heidenreich, P. A., et al. (2021). Building a heart failure clinic: A practical guide from the Heart Failure Society of America. *Journal of Cardiac Failure*, *27*(1), 2–19. https://doi.org/10.1016/j.cardfail.2020.10.008.

Haynes, S. C., Tancredi, D. J., Tong, K., Hoch, J. S., Ong, M. K., Ganiats, T. G., et al. (2020). Association of adherence to weight telemonitoring with health care use and death: A secondary analysis of a randomized clinical trial. *JAMA Network Open*, *3*(7), e2010174. https://doi.org/10.1001/jamanetworkopen.2020.10174.

Inglis, S. C., Clark, R. A., Dierckx, R., Prieto-Merino, D., & Cleland, J. G. F. (2017). Structured telephone support or noninvasive telemonitoring for patients with heart failure. *Heart*, *103*(4), 255–257. https://doi.org/10.1136/heartjnl-2015-309191.

Koehler, F., Koehler, K., Deckwart, O., Prescher, S., Wegscheider, K., Winkler, S., et al. (2018). Telemedical interventional management in heart failure II (TIM-HF2), a randomised, controlled trial investigating the impact of telemedicine on unplanned cardiovascular hospitalisations and mortality in heart failure patients: Study design and description. *European Journal of Heart Failure*, *20*(10), 1485–1493. https://doi.org/10.1002/ejhf.1300.

Koehler, J., Stengel, A., Hofmann, T., Wegscheider, K., Koehler, K., Sehner, S., et al. (2021). Telemonitoring in patients with chronic heart failure and moderate depressed symptoms: Results of the telemedical interventional monitoring in heart failure (TIM-HF) study. *European Journal of Heart Failure*, *23*(1), 186–194. https://doi.org/10.1002/ejhf.2025.

Krzesiński, P., Siebert, J., Jankowska, E. A., Banasiak, W., Piotrowicz, K., Stańczyk, A., et al. (2021). Rationale and design of the AMULET study: A new model of telemedical care in patients with heart failure. *ESC Heart Failure*, *8*(4), 2569–2579. https://doi.org/10.1002/ehf2.13330.

Lyng, P., Persson, H., Hgg-Martinell, A., Hgglund, E., Hagerman, I., Langius-Eklf, A., et al. (2012). Weight monitoring in patients with severe heart failure (WISH). A randomized controlled trial. *European Journal of Heart Failure*, *14*(4), 438–444. https://doi.org/10.1093/eurjhf/hfs023.

Mazón-Ramos, P., Álvarez-Álvarez, B., Ameixeiras-Cundins, C., Portela-Romero, M., Garcia-Vega, D., Rigueiro-Veloso, P., et al. (2022). An electronic consultation program impacts on heart failure patients' prognosis: Implications for heart failure care. *ESC Heart Failure*, *9*(6), 4150–4159. https://doi.org/10.1002/ehf2.14134.

Nogueira, M., Ferreira, F., Raposo, A., Mónica, L., Cruz, L., Guimarães, M., et al. (2020). Impact of a heart failure clinic on morbidity, mortality and quality of life. *European Journal of Heart Failure*, *22*(S1), 71.

Ong, M. K., Romano, P. S., Edgington, S., Aronow, H. U., Auerbach, A. D., Black, J. T., et al. (2016). Effectiveness of remote patient monitoring after discharge of hospitalized patients with heart failure the better effectiveness after transition-heart failure (BEAT-HF) randomized clinical trial. *JAMA Internal Medicine*, *176*(3), 310–318. https://doi.org/10.1001/jamainternmed.2015.7712.

Oseran, A. S., Afari, M. E., Barrett, C. D., Lewis, G. D., & Thomas, S. S. (2021). Beyond the stethoscope: Managing ambulatory heart failure during the COVID-19 pandemic. *ESC Heart Failure*, *8*(2), 999–1006. https://doi.org/10.1002/ehf2.13201.

Palazzuoli, A., Metra, M., Collins, S. P., Adamo, M., Ambrosy, A. P., Antohi, L. E., et al. (2022). Heart failure during the COVID-19 pandemic: Clinical, diagnostic, management, and organizational dilemmas. *ESC Heart Failure*, *9*(6), 3713–3736. https://doi.org/10.1002/ehf2.14118.

Pigorsch, M., Möckel, M., Gehrig, S., Wiemer, J. C., Koehler, F., & Rauch, G. (2022). Performance evaluation of a new prognostic-efficacy-combination design in the context of telemedical interventions. *ESC Heart Failure*, *9*(6), 4030–4042. https://doi.org/10.1002/ehf2.14122.

Qureshi, R. O., Kokkirala, A., & Wu, W.-C. (2020). Review of telehealth solutions for outpatient heart failure care in a veterans health affairs hospital in the COVID-19 era. *Rhode Island Medical Journal*, *103*(9), 22–25. http://www.ncbi.nlm.nih.gov/pubmed/33126782.

Salzano, A., D'Assante, R., Stagnaro, F. M., Valente, V., Crisci, G., Giardino, F., et al. (2020). Heart failure management during the COVID-19 outbreak in Italy: A telemedicine experience from a heart failure university tertiary referral centre. *European Journal of Heart Failure*, *22*(6), 1048–1050. https://doi.org/10.1002/ejhf.1911.

Sammour, Y., Spertus, J. A., Austin, B. A., Magalski, A., Gupta, S. K., Shatla, I., et al. (2021). Outpatient management of heart failure during the COVID-19 pandemic after adoption of a telehealth model. *JACC. Heart Failure*, *9*(12), 916–924. https://doi.org/10.1016/j.jchf.2021.07.003.

Stehlik, J., Schmalfuss, C., Bozkurt, B., Nativi-Nicolau, J., Wohlfahrt, P., Wegerich, S., et al. (2020). Continuous wearable monitoring analytics predict heart failure hospitalization: The link-HF multicenter study. *Circulation. Heart Failure*, *13*(3), e006513. https://doi.org/10.1161/CIRCHEARTFAILURE.119.006513.

Tersalvi, G., Winterton, D., Cioffi, G. M., Ghidini, S., Roberto, M., Biasco, L., et al. (2020). Telemedicine in heart failure during COVID-19: A step into the future. *Frontiers in Cardiovascular Medicine*, *7*, 612818. https://doi.org/10.3389/fcvm.2020.612818.

Turakhia, M. P., Desai, M., Hedlin, H., Rajmane, A., Talati, N., Ferris, T., et al. (2019). Rationale and design of a large-scale, app-based study to identify cardiac arrhythmias using a smartwatch: The apple heart study. *American Heart Journal*, *207*, 66–75. https://doi.org/10.1016/j.ahj.2018.09.002.

Zhu, Y., Gu, X., & Xu, C. (2020). Effectiveness of telemedicine systems for adults with heart failure: A meta-analysis of randomized controlled trials. *Heart Failure Reviews*, *25*(2), 231–243. https://doi.org/10.1007/s10741-019-09801-5.

Section C

Guidelines for different treatment clinics

Chapter 17

Guidelines for breast imaging in the COVID-19 pandemic

Daniele Ugo Tari
Department of Breast Imaging, District 12, Caserta Local Health Authority, Caserta, Italy

Abbreviations

ABUS	automated breast ultrasound
ACR	American College of Radiology
ACS	American Cancer Society
BC	breast cancer
BI-RADS	breast imaging reporting and data system 5th edition
CAR	Canadian Association of Radiologists
CESM	contrast-enhanced spectral mammography
CNB	core-needle biopsy
CSBI	Canadian Society of Breast Imaging
DBT	digital breast tomosynthesis
EUSOBI	European Society of Breast Imaging
FNAC	fine-needle aspiration cytology
FFDM	full-field digital mammography
FFP	filtering face piece
HCWs	healthcare workers
HHUS	hand-held ultrasound
IARC	International Agency for Research on Cancer
MRI	magnetic resonance imaging
PPE	personal protective equipment
SIRM	Italian Society of Medical Radiology
SBI	society of breast imaging
USPS	United States Preventive Service
VABB	vacuum-assisted breast biopsy

Introduction

Breast cancer (BC) represents the most frequent cancer in women of all ages, with more than 2 million diagnoses every year and 685,000 deaths globally in 2020. It is also the world's most prevalent cancer with 7.8 million women suffering from BC in the past 5 years (WHO, 2021). Early diagnosis is the key to increase the survival rate, and its efficacy has been widely confirmed in both high-income and low-income countries (Dewi et al., 2022; Foti & Mancuso, 2005; Møller et al., 1999; Simpson, 1986; Xie et al., 2022).

Since the COVID-19 emergency outbreak was declared, all nonurgent services, such as breast cancer screening, were temporarily suspended, with potential worse outcomes for BC patients. Although the literature provided conflicting results, a short delay (e.g., 6–12 weeks) should not affect the overall outcome but periodic interruption could have a considerable effect (Bleicher, 2018; Pediconi, Galati, et al., 2020; Pediconi, Mann, et al., 2020). Several studies evaluating breast unit activity in the COVID-19 era compared to previous years reported an increased number of referrals for both diagnostic exams in suspected BC patients (28%) and patients who received their first treatment for a BC diagnosis (16%).

Management, Body Systems, and Case Studies in COVID-19. https://doi.org/10.1016/B978-0-443-18703-2.00015-3
Copyright © 2024 Elsevier Inc. All rights are reserved, including those for text and data mining, AI training, and similar technologies.

Consequently, a reduction in new cancer diagnoses (from −26% to −52.1%) was revealed, especially in the first part of 2020 compared to the same period in 2019 in the BC screening age range (De Vincentiis et al., 2021; Gathani et al., 2021; Jacob et al., 2021; Kaufman et al., 2020; Tari et al., 2022). Moreover, the fear of contracting COVID-19 drove patients away from hospitals, probably due to the belief that they were infection reservoirs (Vanni et al., 2020). Finally, the pandemic not only affected patient decision making but might have also had a psychological impact on healthcare workers (HCWs) (Tari, Santarsiere, et al., 2021). Given these considerations, during the pandemic, the need for specific guidelines was essential to optimize the management of patients and HCWs, perform strict COVID-19 screening, and avoid any impairment of survival of BC patients.

Breast imaging guidelines

Breast cancer screening guidelines

Population-based screening is widely recognized as the best way to reduce deaths by detecting BC when treatment is more effective and less harmful. In a wider framework involving the whole world, there is an extensive debate about the modalities of BC screening. Indeed, there is no universal consensus on age selection and screening interval. Furthermore, the risk of overdiagnosis, which is the detection of a cancer that would not have progressed within a woman's lifetime and that does not require treatment (Independent UK Panel on Breast Cancer Screening, 2012), is considered one of the main potential harms in addition to increased anxiety and discomfort caused by the screening itself.

The incidence of newly diagnosed breast cancer increased with age (NIH, 2022). It is most frequently diagnosed among women aged 50–74, with a median age at diagnosis of 63 in the United States. According to the International Agency for Research on Cancer (IARC), the reduction in mortality for women aged 50–69 attending population-based screening is 40% (Lauby-Secretan et al., 2015). The European Society of Breast Imaging (EUSOBI) recommended biennial screening for average-risk women aged 50–69. As a secondary priority, once sufficient coverage of the aforementioned age group has been attained, it is possible to extend biennial screening to include women up to 75 years old (Sardanelli et al., 2017). Instead, the extension to women aged 40–49 should be evaluated only as a third priority, according to national demographics and local priorities. In this case, the yearly interval should be adopted in consideration of a potential higher speed of BC growth.

In the United States, the US Preventive Services (USPS) Task Force, the American Cancer Society (ACS), and the American College of Radiology (ACR) provide different guidelines. The ACR recommended annual screening from 40 to 74 years; the age where screening stops should be based on a woman's health status rather than an age-based determination (Monticciolo et al., 2017). The USPS Task Force stated that the decision to start screening prior to 50 should be an individual one by considering the potential benefits in comparison with the potential harms. Thus, they suggested performing biennial screening between 40–49 and recommended biennial screening also for women 50–74 (Siu and U.S. Preventive Services Task Force, 2016). The ACS strongly recommended annual screening from 45 to 54. At 55, screening should be biennial or continuing annually. Furthermore, the ACS suggested but didn't recommend annual screening from 40 to 44 and proposed to continue screening as long as a woman's overall health was good and she had a life expectancy of ≥10 years (Oeffinger et al., 2015).

In the United Kingdom, BC screening is currently offered to women aged 50–71 every 3 years as part of a national BC screening program (Public Health England, 2021).

Finally, dedicated pathways should be considered for women at high risk, such as women with high familiarity for BC and with a BRCA1 of BRCA2 mutation; magnetic resonance imaging (MRI) should also be offered in according with national or international guidelines and recommendations (Mann et al., 2015; Sardanelli et al., 2010; Saslow et al., 2007).

Breast cancer screening: Imaging modalities, benefits, and potential harms

Breast imaging requires highly qualified radiologists. Despite the technological evolution in the last decades, two-dimensional full-field digital mammography (FFDM) is still considered the main tool for population-based mass screening with demonstrated effectiveness in reducing mortality and allowing for conservative treatment (Sardanelli et al., 2012). In the near future, digital breast tomosynthesis (DBT), which is a new technology also known as three-dimensional mammography, will probably replace FFDM as a screening tool. Actually, several prospective studies supported the use of DBT as an adjunct or alternative to standard imaging techniques due to increasing the detection rate from 0.5 to 2.7 per 1000 screened women (Ciatto et al., 2013; Houssami, 2015; Lång et al., 2016; Skaane et al., 2013). At the same time, there is still no statistically significant and clinically relevant reduction in the interval cancer rate, so more studies are needed to consolidate the wide use of DBT in a population-based screening program.

Furthermore, BC screening could be affected by some potential harms, such as the aforementioned overdiagnosis, false positive and false negative results, and radiation-induced BCs. With regard to overdiagnosis, the IARC working group accepted the estimate provided by the Euroscreen working group, equal to 6.5% (range 1%–10%) (Paci and EUROSCREEN Working Group, 2012). The average cumulative risk for a false positive is about 20% for women aged 50–69 years (Lauby-Secretan et al., 2015). The risk of radiation-induced BC has been estimated to be 1 per 100,000 screened women, at least 100 times lower than the probability of avoiding a BC death (Hauge et al., 2014). Given these considerations, the potential benefits of an organized BC screening seem to outweigh any potential harms.

Nevertheless, there are other aspects to be considered. In particular, one of the most important limitations to early diagnosis in women, especially women younger than 50, is increased breast density (Fig. 1). Breast density is an independent BC risk factor (Freer, 2015), strongly impacting the sensitivity of a screening mammography. That reduction could go from 86%–89% to 62%–68% due to the masking effect of suspicious lesions. The adoption of DBT could potentially address this issue, along with the incorporation of ultrasonography as a complement to FFDM. However, the integration of all these techniques must contend with the implications for the sustainability of screening programs, especially in terms of costs, given that they might be time-consuming and more expensive than FFDM alone.

None of the aforementioned scientific societies recommended clinical breast examination as a screening tool, and they discouraged the use of hand-held ultrasound (HHUS) as a primary screening tool in asymptomatic women over 40 at average risk.

Automated breast ultrasound (ABUS) and contrast-enhanced spectral mammography (CESM), which represent the most recent advancements in breast imaging, require further studies to evaluate their impact on BC screening and on the assessment of suspicious lesions, especially in comparison with MRI (Bicchierai et al., 2021; Bougias & Stogiannos, 2022; Feng et al., 2022).

Finally, according to the concept of triple assessment (Tari, Morelli, et al., 2021), the needle sampling of a breast lesion should be performed using a core needle biopsy (CNB) or a vacuum-assisted biopsy (VABB) instead of fine needle aspiration cytology (FNAC), unless in strict cooperation with a cytologist.

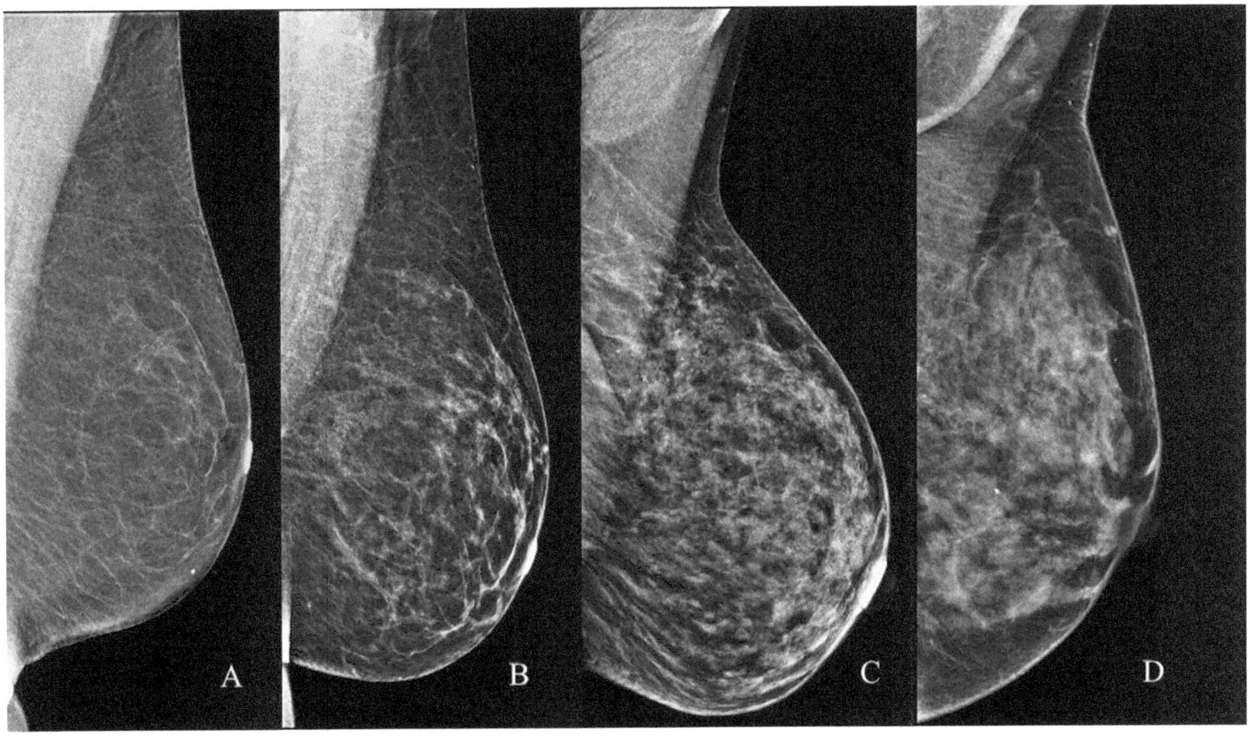

FIG. 1 Density category according to ACR-BIRADS classification 5th edition (D'Orsi et al., 2013). *ACR*, American College of Radiology; *BIRADS*, breast imaging and reporting and data system 5th edition.

Breast imaging during pandemic

COVID-19 and breast imaging: A new challenge

The COVID-19 pandemic imposed new challenges in the management of women needing periodic imaging. In many cases, routine activities were drastically reduced. Furthermore, breast imaging examinations require very close contact with patients. There is no option of physical distancing when performing a mammogram or breast ultrasound exam where the patient's face may be as close as 20–30 cm to the face of the radiologist or the radiographer performing the study. Similarly, during breast intervention procedures, a radiologist may be working within 30 cm of the patient's face (Maio et al., 2021). Accordingly, triage might be necessary to identify which patients should undergo immediate care versus further delay, being aware that medical decisions must be taken balancing individual and community safety as well as the safety of HCWs. Indeed, the risks of disease progression and worse outcomes for patients need to be compared to the risk of patients and staff exposure to SARS-CoV-2. Given these considerations, both patients and HCWs have to be protected from infection during procedures and everyone should avoid inappropriate, unsolicited, or redundant examinations.

Recommendations for breast cancer imaging during a pandemic

Conscientiousness regarding how to improve safety at work for both patients and clinicians is just as important as knowing the biological aspects of pathology. Because patients are forced to choose between seeking treatment and increasing the risk of contracting COVID-19 or postponing therapy and minimizing the risk of contracting the infection, more accessible information for the general population could be useful to clarify that the risk of contamination is relatively low when all effective protective measures are respected (Szmuda et al., 2020).

Many scientific societies have produced guidelines for safely doing daily activities, taking into account that these recommendations are not intended to supersede individual physician judgment or institutional policies and that, as the pandemic rapidly evolves, they might be updated.

In the United States, the COVID-19 Pandemic Breast Cancer Consortium (Dietz et al., 2020) identified three priority categories based on the severity of an individual patient's condition and the potential efficacy of treatments (Table 1). Evolving local conditions and resources (e.g., supply and equipment inventory, availability of intensive care, etc.) will

TABLE 1 Priority categories according to COVID-19 pandemic breast cancer consortium (Dietz et al., 2020).

Category	Cases	Recommendation
A	Severe breast abscess formation Postoperative complications Oncologic emergencies	Imaging not to be delayed Urgent treatment
B	Patients with breast cancer – Suspicious breast symptoms – BI-RADS 4 and 5 – Women at high risk	Imaging delayed to max 6–12 weeks
B1	Inflammatory breast cancer Triple negative BC or HER2+ Neoadjuvant therapy already started Stage 1 or 2 ER+/HER2− BC	Higher priority
B2	Stage 1 HER2+	Mid-priority
B3	DCIS Stage 4 BC	Lower priority
C	BI-RADS 3 short-term follow-up Patients in long-term follow-up for early BC Screening for BC	Imaging suspended and delayed after pandemic

BC, breast cancer; BI-RADS, breast imaging reporting and data system 5th edition; DCIS, ductal carcinoma in situ; ER, estrogen receptors; HER2, human epidermal growth factor receptor 2. Category B is subdivided in categories B1, B2, and B3.

influence which priority category receives treatment. In particular, patients in the Priority A category required urgent treatment to save lives or control the progressing disease while patients in the Priority C category could be deferred without adversely affecting outcomes. Instead, BC patients, who mostly fall in the Priority B category, would not have immediate life-threatening conditions but treatment or services could be delayed for a maximum 6–12 weeks because longer delays could affect outcomes. In these cases, a triage is necessary to justify which patients should undergo treatment. Prioritization has been done also for breast imaging itself (Table 1). Diagnostic imaging for a severe breast abscess formation or to evaluate a serious postoperative complication should be included in Priority A. The Priority B category should include imaging for an abnormal mammogram or for suspicious breast symptoms, biopsies for BI-RADS 4 or 5 lesions, and breast MRIs to evaluate the extent of the disease or for a prechemotherapy assessment. Biopsies for less-suspicious lesions (BI-RADS 4a) may be postponed or biopsied on a case-by-case basis. BI-RADS category 3 patients returning for a short-term follow-up diagnostic mammogram and/or ultrasound and routine breast examinations should have been postponed until the COVID-19 pandemic was over; they would be Priority C. All screening examinations including mammography, ultrasound, and MRI should be placed in Priority C and suspended until after the pandemic. BRCA mutation carriers under 40 may be considered for screening if delays of more than 6 months are expected.

Nevertheless, it was unrealistic that examinations for Priority C patients could be indefinitely suspended. Indeed, BC screening had to be resumed to avoid long-lasting effects on both population and healthcare systems. According to this, Curigliano et al. (2020) suggested correlating the priority category (urgent, high, medium and low) to the pandemic scenario, as defined by the European Center for Disease Prevention and Control (ECDC, 2020). They also proposed risk stratifications considering the supplementary risk factors as being over 60, having a preexisting cardiovascular or respiratory disease, being a smoker, and being male. For each category compared to the pandemic scenario, they defined the priority of screening and diagnostic visits (Table 2).

EUSOBI offered recommendations based on three priorities and established that breast imaging is not justified in a patient with an active COVID-19 infection (Table 3) (Pediconi, Mann, et al., 2020). These recommendations revised and adapted the general schemes proposed by the French health professional societies (Ceugnart et al., 2020), the Dutch working group for breast surgery (Vrancken, 2020), the United States Society of Breast Imaging (SBI) (SBI, 2020), and both the Canadian Society of Breast Imaging (CSBI) and the Canadian Association of Radiologists (CAR) (Seely & Barry, 2020; Seely et al., 2020).

In particular, women who have suspicious BC symptoms (new onset palpable nodule; skin or nipple retraction; orange peel skin; unilateral nipple discharge) and lesions categorized as BI-RADS 4 or 5 should undergo a regular diagnostic workup and needle biopsy as soon as possible. Women with BC requiring staging examinations or the evaluation of neoadjuvant therapy should undergo these studies without further delay. Instead, asymptomatic women at increased risk for BC and asymptomatic women performing annual mammographic follow-up should undergo a screening examination preferably within 15 months from the previous appointment. Asymptomatic women, who did not respond to the invitation for a screening mammography after the onset of the COVID-19 pandemic or who were informed of the screening suspension, should schedule the check preferably within 3–6 months of the due date. In women at average risk, a delay of up to 1 year for biennial screening may be acceptable.

To protect patients and HCWs from infection, the Italian Society of Medical Radiology (SIRM) has recommended the necessary personal protective equipment (PPE) (Table 4) (Pediconi, Galati, et al., 2020), following the guidelines provided by the Italian Experimental Institute Lazzaro Spallanzani (Pianura et al., 2020).

In conclusion, most scientific societies involved in BC management agreed on the necessity to prioritize patients. By summarizing the available guidelines, Tari, Santarsiere, et al. (2021) proposed a model to manage patients and HCWs in both screening and outpatient settings, triaging patients into three different clinical scenarios by performing a telephone questionnaire before each diagnostic exam or a nasopharyngeal swab before every recovery or invasive diagnostic procedure. Due to the proven efficacy, it could be considered a good way to apply guidelines and recommendations to BC diagnosis in both screening and outpatient settings (**see Policies and Procedure paragraph**).

Furthermore, during the pandemic, the majority of encounters should be conducted remotely via telemedicine. Decisions to conduct in-person visits must carefully weigh the risk of viral transmission to patients and HCWs. Increased precautions should be taken surrounding in-person visits/treatments for patients with comorbidities and a high risk of COVID-19 complications.

Axillary lymphadenopathy after COVID-19 vaccination

Alongside preventive measures, the key point to contain the SARS-CoV-2 pandemic was an effective mass vaccination campaign. In this setting, it is important to be aware of a reactive unilateral axillary lymphadenopathy (UAL) as a possible

TABLE 2 Priority category and COVID-19 scenarios (Curigliano et al., 2020).

Modality	Pandemic scenario	Urgent	High priority	Medium priority	Low priority
Outpatient visits	1–2	Infection Hematoma	Newly diagnosed invasive BC Postoperative complications	High-risk patients	Low-risk patients
	3–4	Acute/progressive infection or hemorrhage	None	Postoperative complications	High-risk patients
Diagnostics and screening	1–2	None	BI-RADS 4b BI-RADS 4c BI-RADS 5 BC suspected recurrence	None	Screening for High-risk patients (BRCA1-2) Screening deferred
	3–4	None	None	BI-RADS 4b BI-RADS 4c BI-RADS 5 BC suspected recurrence Newly diagnosed invasive BC	Screening for high-risk patients (BRCA1-2) Screening suspended
Prioritization in patients with BC					
I	Breast cancer patients recently suspected or recently diagnosed				
II	Breast cancer patients on active treatment (e.g., chemotherapy, immunotherapy, anti-HER2 therapy, endocrine therapy with or without targeted therapies)				
III	Breast cancer patients in follow-up (nonactive treatment) or on adjuvant endocrine therapy alone				
Pandemic scenarios according to ECDC					
Scenario 1	Semiurgent setting		Severe disease with no community spread		
Scenario 2	Urgent setting		Severe disease with contained community spread		
Scenario 3	Lockdown		Severe disease with uncontained community spread		
Scenario 4	Uncontrolled pandemic		Severe disease with uncontrolled spread		

Prioritization of outpatient, screening, and diagnostic visits for patients with breast cancer, compared to scenarios to describe progression of COVID-19 outbreaks, according to ECDC (Curigliano et al., 2020). *BC*, breast cancer; *BI-RADS*, breast imaging reporting and data system 5th edition; *ECDC*, European Center for Disease Prevention and Control.

TABLE 3 EUSOBI recommendations for breast imaging and cancer diagnosis during and after the COVID-19 pandemic (Pediconi, Galati, et al., 2020; Pediconi, Mann, et al., 2020).

Priority	Cases	Recommendation
High	Imaging in women presenting with suspicious breast symptoms or suspicious axillary findings Diagnostic imaging in women with an abnormal screening examination MRI screening in women at very high risk for breast cancer Exploration of incidentally detected abnormalities in other imaging modalities (e.g., chest CT) Search for occult primary cancer	Rapid appointment
Medium	Follow-up imaging in women with BI-RADS 3 findings on a previous examination Mammographic screening in women at high risk for breast cancer Systematic follow-up after breast cancer	Appointment within 3 months
Low	Breast MRI for breast implant evaluation Screening mammography in healthy women at average risk for breast cancer	Appointment within 6 months

CT, computed tomography; *MRI*, magnetic resonance imaging; *BI-RADS*, breast imaging and reporting system 5th edition.

TABLE 4 Personal protective equipment for patients and HCWs.

Clinical scenario	Definition	Patients PPE	HCWs PPE[a]
Non-COVID-19 patient	RT-PCR test: negative	FFP2 or surgical mask Physical distancing (at least 1 m) Hand sanitization	FFP2 or surgical mask Visor or goggle protection Disposable gown Disposable cap Disposable gloves Ultrasound probe should be covered with a probe cover or glove
Suspected COVID-19 patient	Under investigation waiting for RT-PCR test result	FFP2 or surgical mask	FFP2 mask Visor or goggle protection Disposable gown Disposable cap Disposable gloves Ultrasound probe should be covered with a probe cover or glove Washable or plastic film on the keyboard and the ultrasound machine
Confirmed COVID-19 patient	RT-PCR test: positive	FFP2 or surgical mask	FFP2 of FFP3 mask Visor or goggle protection Disposable gown Disposable cap Disposable gloves Complete coverage of ultrasound probe and machine Sanitization of all the areas after undressing at the end of procedure

Personal protective equipment (PPE) for both patients and healthcare workers (HCWs), in according to guidelines provided by the Italian Experimental Institute Lazzaro Spallanzani (Pianura et al., 2020; Pediconi, Galati, et al., 2020 SIRM) and Italian Ministry of Health in 2022. *FFP*, filtering face piece; *RT-PCR*, reverse transcriptase-polymerase chain reaction.
[a]*Sanitization of devices is mandatory before every patient.*

mild side effect of vaccination, and both the population and clinicians should know that this could be related to a robust immune response (Shirone et al., 2012; Studdiford et al., 2008). Nevertheless, an isolated lymphadenopathy could be of relevance during a screening for BC, even with negative conventional imaging, due to the presence of the cancer of unknown primary syndrome, which is a metastatic malignancy with an unknown primary tumor that accounts for up to 1% of all breast cancers (Gorkem & Oconnell, 2012).

Before COVID-19, the identification of an isolated UAL without abnormal breast findings would have been classified as BI-RADS category 4 (D'Orsi et al., 2013) and further examinations, included FNAC or CNB, should have been performed. The same evaluation should be done during a pandemic, taking into account that UAL was also reported after vaccination in from 0.3% to 16% of recipients, occurring from 2 to 4 days after inoculation and lasting approximately 10 days (Tu et al., 2021).

Guidelines produced in 2021 by SBI (Grimm et al., 2021), EUSOBI (Schiaffino et al., 2021), CSBI, and CAR (CSBI and CAR, 2021) recommended rescheduling breast imaging from 6 to 12 weeks before or after the vaccination.

Nevertheless, because referrals are no longer considered reasonable, some scientific societies have revised their guidelines to no longer delay screening mammograms around COVID-19 vaccinations (Grimm et al., 2022; NHS, 2022).

Consequently, it is widely recommended to collect information on vaccination status, including the date and side (left vs right), on patient intake forms (Table 5). In the absence of suspicious findings, a BI-RADS category 2 assessment should be given to a UAL after recent vaccination in the ipsilateral arm. For patients who previously had a short term follow-up for presumed vaccine-induced lymphadenopathy, the interval recommended should be 12 or more weeks. In persistent unmodified UAL, an additional follow-up of 6 months could be considered. If UAL was increased, a lymph node sampling should be performed. Finally, to minimize patient anxiety, it is suggested to include a statement that vaccines of all types can induce a temporary swelling of the lymph nodes, which could be considered a sign of response to vaccination.

TABLE 5 Telephonic questionnaire.

Question	Answer	Recommendations
Do you have fever?	Yes/No	If yes access forbidden
Do you have cough or dyspnea?	Yes/No	If yes access forbidden
Do you have anosmia or dysgeusia symptoms?	Yes/No	If yes access forbidden
Have you done COVID-19 serology test	Yes/No	If yes, indicate when and result
Have you done a nasopharyngeal swab?	Yes/No	If yes, indicate when and result
Are you vaccinated against COVID-19?	Yes/No	If yes, indicate when you have done last vaccination and which side (left or right)

Questions to be submitted to patients the day before the appointment with a telephonic questionnaire.

The multidisciplinary approach during a pandemic

It is well known that early detection contributes to a decrease in BC mortality and outcomes are dependent on timely and high-quality multidisciplinary interventions (Duffy et al., 2020; Pediconi & Galati, 2020). During a pandemic, the multidisciplinary team (MDT) meetings should be restricted to one professional for each discipline (Biganzoli et al., 2020) or done online with all the members for cases requiring a multidisciplinary decision.

The basic tenets of cancer care coordination should be followed as much as possible. The American College of Surgeons issued a statement that elective surgeries should be canceled (American College of Surgeons, 2020), but it is not possible to defer these treatments indefinitely. Thus, multidisciplinary discussions documenting priority categories for surgery and/or adjuvant treatments are necessary to ensure the best outcomes for patients. If feasible, tumor board discussions should include both standard and COVID-19 recommendations based on an institution's level of pandemic severity. Documentation of these discussions in the medical record is highly recommended.

Policies and procedures

As shown in Table 2, the literature and guidelines have identified four phases of SARS-CoV-2 infection, depending on the country's COVID-19 status (ECDC, 2020; Ng et al., 2020). The comparison of these phases with a patient's possible COVID-19 infection or BC stage (Tables 2 and 4) is fundamental to optimize human and technological resources. Nevertheless, because laboratory tests are not still validated for screening purposes, and considering the high number of asymptomatic or paucisymptomatic cases, HCWs must treat all patients as if they were infected (Too et al., 2020).

Furthermore, because the most common symptoms of the disease are fever, coughing, and shortness of breath, we suggested performing a SARS-CoV-2 infection telephone questionnaire (Table 5) before access to facilities and a nasopharyngeal swab before a procedure, according to the stage of the diagnostic-therapeutic care pathways (DTCP) (Fig. 2). Both positive and clinically suspected COVID-19 patients have to be postponed and rescheduled 2 weeks after the first asymptomatic day, as established by a complete health evaluation made by the general practitioner (GP) (Tari, Santarsiere, et al., 2021).

If the patient is admitted to screening procedures, hand sanitization and a body temperature check should be performed before the exam, and HCWs have to follow the specific procedures for the clinical scenario identified. The time lapse between the exams should be about 10–20 min, taking into account the patient accommodation, performing diagnostic imaging, accurate machine and room disinfection, and air ventilation.

Fast and awake surgery should be preferred. If possible, patients should be discharged as soon as they are clinically stable, usually 1–3 days after surgery. Postdischarge visits should be reduced to the minimum required. If a patient developed COVID-19-related symptoms, a breast-care nurse or a trained radiographer, both identified as case managers, informed public health officers and family doctors to follow up on the infection's clinical course. They should also monitor all the established care pathways after hospital discharge.

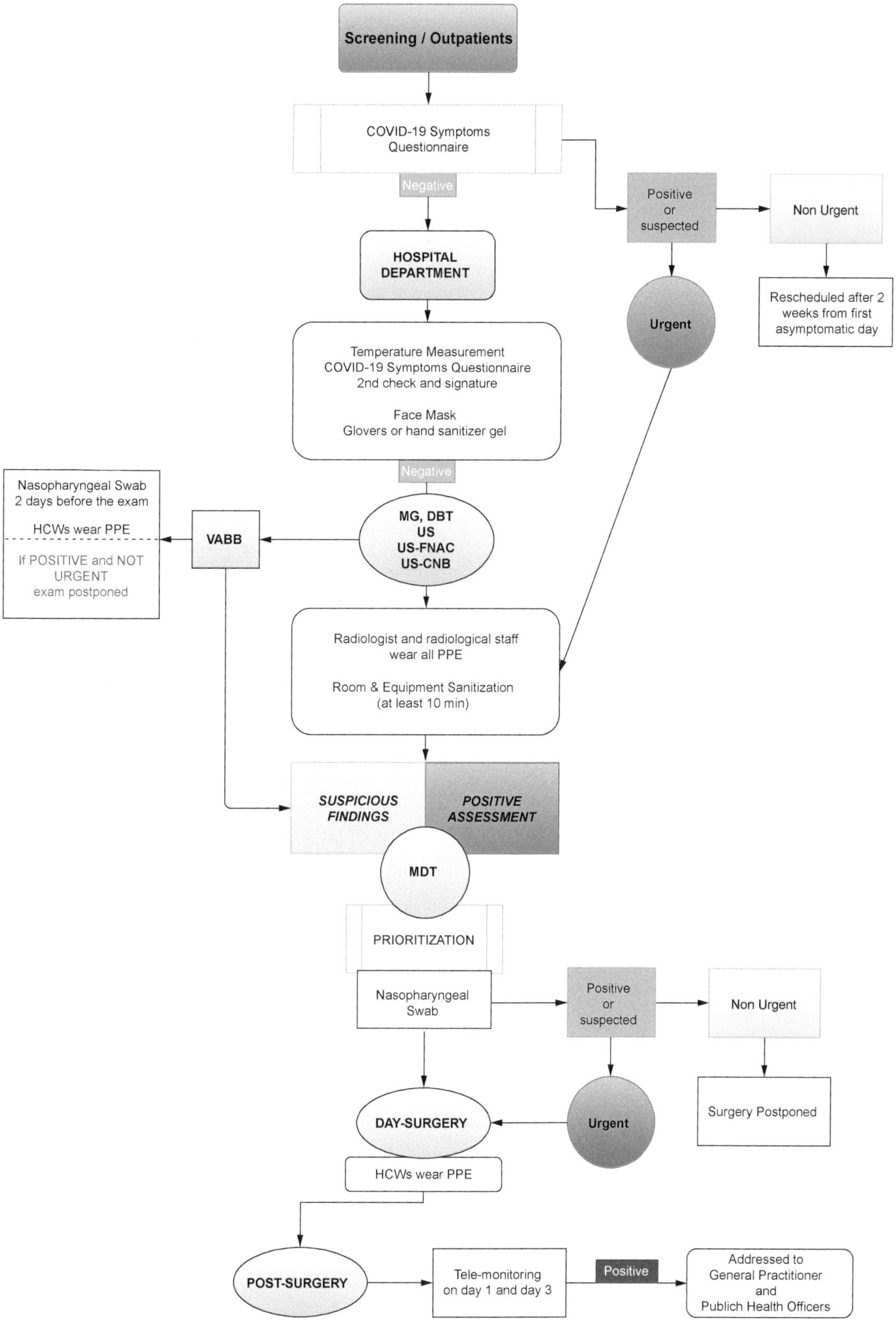

FIG. 2 Breast imaging workflow. Diagnostic-therapeutic care pathway (DTCP). *CNB*, core needle biopsy; *DBT*, digital breast tomosynthesis; *FNAC*, fine-needle aspiration cytology; *HCWs*, healthcare workers; *MG*, mammography; *MDT*, multidisciplinary team; *PPE*, personal protective equipment; *US*, ultrasound; *VABB*, stereotactic vacuum-assisted breast biopsy or tomobiopsy.

Applications to other areas

BC is a disease with a significant social and economic impact on the entire community, due to its long-lasting effects for both population and health services. The COVID-19 pandemic has created long-term and permanent challenges that will change the practice of medicine in the future. One of the major challenges was the maintenance of social distancing, which may not be possible in various settings such as waiting rooms and when performing common procedures.

In this review, we have evaluated how triaging both patients' clinical status and urgent clinical cases should be mandatory to optimize the allocation of the limited resources during higher peaks, increasing the number of treated patients and reducing hospitalizations as well as the risk of cross-infection. This could lead to the implementation of better DTCP.

The telephonic questionnaire (Table 5) and the proposed workflow for resuming BC screening activities (Fig. 2) could be useful also during regular screening activities for other oncologic screenings. For example, during colorectal cancer screening procedures, it could be useful to triage patients who need an urgent colonoscopy due to symptoms or high-risk factors (Mazidimoradi et al., 2021) as well as sanitization procedures that have already been proposed in cardiovascular interventional radiology (Too et al., 2020).

The pandemic rapidly evolved and more challenges will likely come in the future; nevertheless, no effort will be wasted if done with commitment and dedication.

Mini-dictionary of terms

- **Automated breast ultrasound**. A new imaging technique approved by the US Food and Drug Administration in 2012 as a supplemental screening tool for women with heterogeneously and extremely dense breasts. ABUS separates the moment of image acquisition (made by the radiographer) from the moment of image interpretation, thus reducing the operator dependence as well as the time spent by the physician.
- **Contrast-enhanced spectral mammography**. A special type of mammogram that could be performed on patients whose standard breast imaging tests are inconclusive. CESM uses an intravenous injection of iodinated contrast to reveal areas of increased blood supply within the breast. It is proposed as an alternative to magnetic resonance imaging.
- **Digital breast tomosynthesis**. It is also called three-dimensional mammography or breast tomosynthesis. It is an advanced form of breast imaging where multiple images of the breast from different angles are captured and reconstructed into a three-dimensional image set. For some machinery, it is also possible to "synthesize" a 2D reconstructed image that is very similar to standard mammography.
- **Hand-held ultrasound**. It is the standard modality to perform a breast ultrasound examination. It is operator-dependent (physician) and time-consuming in a screening setting. It is fundamental for the evaluation of dense breasts in younger women and it is the second level examination in a BC screening.
- **Mammography**. Specialized medical imaging that uses a low-dose X-ray system to see inside the breasts. A mammography exam, called a mammogram, is the gold standard for the early detection and diagnosis of breast diseases in women over 40. It is the first level examination in a BC screening.
- **Magnetic resonance imaging**. Medical imaging technique that uses a magnetic field and computer-generated radio waves to create detailed images of the organs and tissues in the body. It is useful, particularly in women who have dense breast tissue or who might be at high risk of the disease.
- **Vacuum-assisted breast biopsy**. A percutaneous safe and minimally invasive procedure in which a sample of breast tissue is removed for examination. It could be performed with or without a very small cutaneous incision and only a local anesthetic is used. The biopsy is performed under imaging guidance: mammography, digital breast tomosynthesis, magnetic resonance imaging, or ultrasound. It is useful for very small abnormalities of the breast as an alternative to a surgical biopsy.

Summary points

- After the COVID-19 emergency outbreak was declared, BC screening was temporarily suspended and appointments were deferred. Even though a short delay (e.g., 6–12 weeks) should not affect the overall outcome, longer delays or periodic interruptions could worsen the outcomes for BC patients.
- There is an extensive open debate about the modalities of BC screening with no universal consensus on age selection and screening interval.
- A mammography is the gold standard for breast cancer diagnosis in women over 40.

- DBT is considered an adjunct or alternative to standard imaging techniques, especially in younger women with dense breasts.
- Many scientific societies have produced guidelines to continue to safely perform daily activities during a pandemic.
- Prioritization and triage of patients according to the country's COVID-19 scenario is fundamental to continue performing breast imaging examinations and screening.
- Isolated unilateral axillary lymphadenopathy is a mild effect of COVID-19 vaccination but most of the scientific societies no longer recommend delaying breast imaging around COVID-19 vaccinations.
- The multidisciplinary approach is still the best way to manage breast cancer patients but, during a pandemic, the meetings should be performed online and members should establish the patient's treatment priority.
- A telephonic questionnaire before appointments and a nasopharyngeal swab before invasive examinations or surgery are good ways to triage COVID-19 patients and to safely perform daily activities.
- Some of the proposed recommendations could be useful also in colorectal cancer screening and cardiovascular interventional radiology.

References

American College of Surgeons. (2020). *COVID-19 guidelines for triage of breast cancer patients.* https://www.facs.org/covid-19/clinicalguidance/elective-case/breast-cancer. (Accessed 20 July 2022).

Bicchierai, G., Di Naro, F., De Benedetto, D., Cozzi, D., Pradella, S., Miele, V., & Nori, J. (2021). A review of breast imaging for timely diagnosis of disease. *International Journal of Environmental Research and Public Health, 18*(11), 5509.

Biganzoli, L., Cardoso, F., Beishon, M., Cameron, D., Cataliotti, L., Coles, C. E., et al. (2020). The requirements of a specialist breast centre. *Breast, 51*, 65–84.

Bleicher, R. J. (2018). Timing and delays in breast cancer evaluation and treatment. *Annals of Surgical Oncology, 25*(10), 2829–2838.

Bougias, H., & Stogiannos, N. (2022). Breast MRI: Where are we currently standing? *Journal of Medical Imaging and Radiation Sciences, 53*(2), 203–211.

Ceugnart, L., Delaloge, S., Balleyguier, C., Deghaye, M., Veron, L., Kaufmanis, A., et al. (2020). Dépistage et diagnostic du cancer du sein à la fin de période de confinement COVID-19, aspects pratiques et hiérarchisation des priorités [Breast cancer screening and diagnosis at the end of the COVID-19 confinement period, practical aspects and prioritization rules: Recommendations of 6 French health professionals societies]. *Bulletin du Cancer, 107*(6), 623–628. French.

Ciatto, S., Houssami, N., Bernardi, D., et al. (2013). Integration of 3D digital mammography with tomosynthesis for population breast cancer screening (STORM): A prospective comparison study. *The Lancet. Oncology, 14*(7), 583–589.

CSBI and CAR. (2021). *The Canadian Society of Breast Imaging and Canadian Association of Radiologists' recommendations for the management of axillary adenopathy in patients with recent COVID-19 vaccination— Update.* https://csbi.ca/wp-content/uploads/2021/03/2021_03_24_CSBI_CAR_Axillary-Adenopathy-Update.pdf. (Accessed 20 July 2022).

Curigliano, G., Cardoso, M. J., Poortmans, P., Gentilini, O., Pravettoni, G., Mazzocco, K., … Editorial Board of The Breast. (2020). Recommendations for triage, prioritization and treatment of breast cancer patients during the COVID-19 pandemic. *Breast, 52*, 8–16.

D'Orsi, C. J., Sickles, E. A., Mendelson, E. B., & Morris, E. A. (2013). *ACR BI-RADS atlas, breast imaging reporting and data system* (5th ed.). Reston, VA: American College of Radiology.

De Vincentiis, L., Carr, R. A., Mariani, M. P., & Ferrara, G. (2021). Cancer diagnostic rates during the 2020 "lockdown", due to COVID-19 pandemic, compared with the 2018–2019: An audit study from cellular pathology. *Journal of Clinical Pathology, 74*, 187–189.

Dewi, T. K., Ruiter, R. A. C., Diering, M., Ardi, R., & Massar, K. (2022). Breast self-examination as a route to early detection in a lower-middle-income country: Assessing psychosocial determinants among women in Surabaya, Indonesia. *BMC Womens Health, 22*(1), 179.

Dietz, J. R., Moran, M. S., Isakoff, S. J., Kurtzman, S. H., Willey, S. C., Burstein, H. J., et al. (2020). Recommendations for prioritization, treatment, and triage of breast cancer patients during the COVID-19 pandemic. The COVID-19 pandemic breast cancer consortium. *Breast Cancer Research and Treatment, 181*(3), 487–497.

Duffy, S. W., Tabár, L., Yen, A. M., Dean, P. B., Smith, R. A., Jonsson, H., et al. (2020). Mammography screening reduces rates of advanced and fatal breast cancers: Results in 549,091 women. *Cancer, 126*(13), 2971–2979.

European Centre for Disease Prevention and Control (ECDC). (2020). *Coronavirus disease 2019 (COVID-19) pandemic: Increased transmission in the EU/EEA and the UK—Seventh update.* https://www.ecdc.europa.eu/sites/default/files/documents/RRA-seventh-update-Outbreak-of-coronavirus-disease-COVID-19.pdf. (Accessed 20 July 2022).

Feng, L., Sheng, L., Zhang, L., Li, N., & Xie, Y. (2022). Comparison of contrast-enhanced spectral mammography and contrast-enhanced MRI in screening multifocal and multicentric lesions in breast cancer patients. *Contrast Media and Molecular Imaging, 2022*, 4224701.

Foti, E., & Mancuso, S. (2005). Early breast cancer detection. *Minerva Ginecologica, 57*(3), 269–292. English, Italian.

Freer, P. E. (2015). Mammographic breast density: Impact on breast cancer risk and implications for screening. *Radio Graphics, 35*(2), 302–315.

Gathani, T., Clayton, G., MacInnes, E., & Horgan, K. (2021). The COVID-19 pandemic and impact on breast cancer diagnoses: What happened in England in the first half of 2020. *British Journal of Cancer, 124*, 710–712.

Gorkem, S. B., & Oconnell, A. (2012). Abnormal axillary lymphnodes on negative mammograms: Causes except breast cancer. *Diagnostic and Interventional Radiology, 18*, 473–479.

Grimm, L., Destounis, S., Dogan, B., et al. (2021). *SBI recommendations for the management of axillary adenopathy in patients with recent COVID-19 vaccination.* Available at: https://splf.fr/wp-content/uploads/2021/02/Society-of-breast-imaging-Recommendations-for-the-Management-of-Axillary-Adenopathy-in-Patients-with-Recent-COVID-19-Vaccination-Mis-en-ligne-par-la-Societe-francaise-de-radiologie-le-10-02-21.pdf. (Accessed 20 July 2022).

Grimm, L., Srini, A., Dontchos, B., Daly, C., Tuite, C., Sonnenblick, E., et al. (2022). *Revised SBI recommendations for the management of axillary adenopathy in patients with recent COVID-19 vaccination.* Available from: https://www.sbi-online.org/Portals/0/Position-Statements/2022/SBI-recommendations-for-managing-axillary-adenopathy-post-COVID-vaccination_updatedFeb2022.pdf. (Accessed 20 July 2022).

Hauge, I. H., Pedersen, K., Olerud, H. M., Hole, E. O., & Hofvind, S. (2014). The risk of radiation-induced breast cancers due to biennial mammographic screening in women aged 50-69 years is minimal. *Acta Radiological*, *55*(10), 1174–1179.

Houssami, N. (2015). Digital breast tomosynthesis (3D-mammography) screening: Data and implications for population screening. *Expert Review of Medical Devices*, *12*(4), 377–379.

Independent UK Panel on Breast Cancer Screening. (2012). The benefits and harms of breast cancer screening: An independent review. *Lancet*, *380*(9855), 1778–1786.

Jacob, L., Loosen, S. H., Kalder, M., Luedde, T., Roderburg, C., & Kostev, K. (2021). Impact of the COVID-19 pandemic on cancer diagnoses in general and specialized practices in Germany. *Cancers*, *13*, 408.

Kaufman, H. W., Chen, Z., Niles, J., & Fesko, Y. (2020). Changes in the number of US patients with newly identified cancer before and during the coronavirus disease 2019 (COVID-19) pandemic. *JAMA Network Open*, *3*(8), e2017267.

Lång, K., Andersson, I., Rosso, A., Tingberg, A., Timberg, P., & Zackrisson, S. (2016). Performance of one-view breast tomosynthesis as a standalone breast cancer screening modality: Results from the Malmö breast tomosynthesis screening trial, a population-based study. *European Radiology*, *26*(1), 184–190.

Lauby-Secretan, B., Scoccianti, C., Loomis, D., International Agency for Research on Cancer Handbook Working Group, et al. (2015). Breast cancer screening—Viewpoint of the IARC working group. *The New England Journal of Medicine*, *372*(24), 2353–2358.

Maio, F., Tari, D. U., Granata, V., Fusco, R., Grassi, R., Petrillo, A., & Pinto, F. (2021). Breast cancer screening during COVID-19 emergency: Patients and department management in a local experience. *Journal of Personalized Medicine*, *11*(5), 380.

Mann, R. M., Balleyguier, C., Baltzer, P. A., European Society of Breast Imaging (EUSOBI), with language review by Europa Donna–The European Breast Cancer Coalition, et al. (2015). Breast MRI: EUSOBI recommendations for women's information. *European Radiology*, *25*(12), 3669–3678.

Mazidimoradi, A., Tiznobaik, A., & Salehiniya, H. (2021). Impact of the COVID-19 pandemic on colorectal cancer screening: A systematic review. *Journal of Gastrointestinal Cancer*, 1–15.

Møller, P., Reis, M. M., Evans, G., Vasen, H., Haites, N., Anderson, E., et al. (1999). Efficacy of early diagnosis and treatment in women with a family history of breast cancer. European Familial Breast Cancer Collaborative Group. *Disease Markers*, *15*(1-3), 179–186.

Monticciolo, D. L., Newell, M. S., Hendrick, R. E., Helvie, M. A., Moy, L., Monsees, B., et al. (2017). Breast cancer screening for average-risk women: Recommendations from the ACR commission on breast imaging. *Journal of the American College of Radiology*, *14*(9), 1137–1143.

Ng, C. W. Q., Tseng, M., Lim, J. S. J., & Chan, C. W. (2020). Maintaining breast cancer care in the face of COVID-19. *The British Journal of Surgery*, *107*(10), 1245–1249.

NHS. (2022). *National Health Service, axillary adenopathy after COVID vaccination position statement.* Available from: https://www.nhs.uk/conditions/breast-screening-mammogram/coronavirus-covid-19-updates/. (Accessed 20 July 2022).

NIH. (2022). *National Cancer Institute, cancer stat facts: Female breast cancer.* Available from: https://seer.cancer.gov/statfacts/html/breast.html. (Accessed 20 July 2022).

Oeffinger, K. C., Fontham, E. T., Etzioni, R., Herzig, A., Michaelson, J. S., Shih, Y. C., … American Cancer Society. (2015). Breast cancer screening for women at average risk: 2015 guideline update from the American Cancer Society. *JAMA*, *314*(15), 1599–1614.

Paci, E., & EUROSCREEN Working Group. (2012). Summary of the evidence of breast cancer service screening outcomes in Europe and first estimate of the benefit and harm balance sheet. *Journal of Medical Screening*, *19*(Suppl 1), 5–13.

Pediconi, F., & Galati, F. (2020). Breast cancer screening programs: Does one risk fit all? *Quantitative Imaging in Medicine and Surgery*, *10*(4), 886–890.

Pediconi, F., Galati, F., Bernardi, D., Belli, P., Brancato, B., Calabrese, M., et al. (2020). Breast imaging and cancer diagnosis during the COVID-19 pandemic: Recommendations from the Italian College of Breast Radiologists by SIRM. *La Radiologia Medica*, *125*(10), 926–930 (1).

Pediconi F., Mann R.M., Gilbert F.J., Forrai G., Sardanelli S., Camps Herrero J., on behalf of the EUSOBI Executive Board (2020). EUSOBI recommendations for breast imaging and cancer diagnosis during and after the COVID-19 pandemic. Available online on: https://www.eusobi.org/news/recommendations-breast-covid19/ (2) (Accessed 20 July 2022).

Pianura, E., Stefano, F. D., Cristofaro, M., et al. (2020). COVID-19: A review of the literature and the experience of INMI Lazzaro Spallanzani two months after the epidemic outbreak. *Journal of Radiological Review*, *7*(3), 196–207.

Public Health England. (2021). *NHS breast screening (BSP) programme: Detailed information.* Available from: https://www.gov.uk/topic/population-screening-programmes/breast. (Accessed 20 July 2022).

Sardanelli, F., Aase, H. S., Álvarez, M., Azavedo, E., Baarslag, H. J., Balleyguier, C., et al. (2017). Position paper on screening for breast cancer by the European Society of Breast Imaging (EUSOBI) and 30 national breast radiology bodies from Austria, Belgium, Bosnia and Herzegovina, Bulgaria, Croatia, Czech Republic, Denmark, Estonia, Finland, France, Germany, Greece, Hungary, Iceland, Ireland, Italy, Israel, Lithuania, Moldova, The Netherlands, Norway, Poland, Portugal, Romania, Serbia, Slovakia, Spain, Sweden, Switzerland and Turkey. *European Radiology*, *27*(7), 2737–2743.

Sardanelli, F., Boetes, C., Borisch, B., et al. (2010). Magnetic resonance imaging of the breast: recommendations from the EUSOMA working group. *European Journal of Cancer*, *46*(8), 1296–1316.

Sardanelli, F., Helbich, T. H., & European Society of Breast Imaging (EUSOBI). (2012). Mammography: EUSOBI recommendations for women's information. *Insights Imaging*, *3*(1), 7–10.

Saslow, D., Boetes, C., Burke, W., American Cancer Society Breast Cancer Advisory Group, et al. (2007). American Cancer Society guidelines for breast screening with MRI as an adjunct to mammography. *CA: A Cancer Journal for Clinicians*, *57*(2), 75–89.

SBI. (2020). *Society of breast imaging statement on breast imaging during the COVID-19 pandemic*. Available from: https://www.sbi-online.org/Portals/0/Position%20Statements/2020/society-of-breast-imaging-statement-on-breast-imaging-during-COVID19-pandemic.pdf. (Accessed 20 July 2022).

Schiaffino, S., Pinker, K., Magni, V., Cozzi, A., Athanasiou, A., Baltzer, P. A. T., et al. (2021). Axillary lymphadenopathy at the time of COVID-19 vaccination: Ten recommendations from the European Society of Breast Imaging (EUSOBI). *Insights Imaging*, *12*(1), 119.

Seely, J. M., & Barry, M. (2020). *Canadian Society of Breast Imaging and Canadian Association of radiologists joint position statement on COVID-19*. Available from: https://csbi.ca/wp-content/uploads/2020/03/Covid-19-statement-CSBI_CAR-1.pdf. (Accessed 20 July 2022).

Seely, J. M., Scaranelo, A. M., Yong-Hing, C., Appavoo, S., Flegg, C., Kulkarni, S., et al. (2020). COVID-19: Safe guidelines for breast imaging during the pandemic. *Canadian Association of Radiologists Journal*, *71*(4), 459–469.

Shirone, N., Shinkai, T., Yamane, T., et al. (2012). Axillary lymph node accumulation on FDG-PET/CT after influenza vaccination. *Annals of Nuclear Medicine*, *26*, 248–252.

Simpson, J. S. (1986). The early diagnosis of breast cancer. *The New Zealand Medical Journal*, *99*(798), 188–190.

Siu, A. L., & U.S. Preventive Services Task Force. (2016). Screening for breast cancer: U.S. Preventive Services Task Force recommendation statement. *Annals of Internal Medicine*, *164*(4), 279–296.

Skaane, P., Bandos, A. I., Gullien, R., et al. (2013). Comparison of digital mammography alone and digital mammography plus tomosynthesis in a population-based screening program. *Radiology*, *267*(1), 47–56.

Studdiford, J., Lamb, K., Horvath, K., Altshuler, M., & Stonehouse, A. (2008). Development of unilateral cervical and supraclavicular lymphadenopathy after human papilloma virus vaccination. *Pharmacotherapy*, *28*, 1194–1197.

Szmuda, T., Ozdemir, C., Ali, S., et al. (2020). Readability of online patient education material for the novel coronavirus disease (COVID-19): A cross-sectional health literacy study. *Public Health*, *185*, 21–25.

Tari, D. U., Morelli, L., Guida, A., & Pinto, F. (2021). Male breast cancer review. A rare case of pure DCIS: Imaging protocol, radiomics and management. *Diagnostics (Basel)*, *11*(12), 2199.

Tari, D. U., Santarsiere, A., Palermo, F., Morelli, C. D., & Pinto, F. (2021). The management of a breast unit during the COVID-19 emergency: A local experience. *Future Oncology*, *17*(34), 4757–4767.

Tari, D. U., Santonastaso, R., & Pinto, F. (2022). Consequences of the impact of COVID-19 pandemic on breast cancer at a single Italian Institution. *Exploration of Targeted Anti-Tumor Therapy*, *3*(4), 414–422.

Too, C. W., Wen, D. W., Patel, A., et al. (2020). Interventional radiology procedures for COVID-19 patients: How we do it. *Cardiovascular and Interventional Radiology*, *43*(6), 827–836.

Tu, W., Gierada, D. S., & Joe, B. N. (2021). COVID-19 vaccination-related lymphadenopathy: What to be aware of. *Radiology. Imaging Cancer*, *3*, e210038.

Vanni, G., Materazzo, M., Pellicciaro, M., Ingallinella, S., Rho, M., Santori, F., et al. (2020). Breast cancer and COVID-19: The effect of fear on patients' decision-making process. *In Vivo*, *34*(Suppl. 3), 1651–1659.

Vrancken, M. (2020). *NVCO, mammachirurgie*. Document: Follow-up in corona-time.

WHO. (2021). *World Health Organization, Cancer*. Available from: https://www.who.int/health-topics/cancer#tab=tab_2. (Accessed 20 July 2022).

Xie, E., Colditz, G. A., Lian, M., Greever-Rice, T., Schmaltz, C., Lucht, J., & Liu, Y. (2022). Timing of medicaid enrollment, late-stage breast cancer diagnosis, treatment delays, and mortality. *JNCI Cancer Spectrum*, *6*(3). pkac031.

Chapter 18

Vaccine-induced (immune) thrombotic thrombocytopenia (VITT): Diagnosis, guidelines, and reporting

Emmanuel J. Favaloro[a,b,c], Leonardo Pasalic[a,b,d], and Giuseppe Lippi[e]

[a]*Haematology, Institute of Clinical Pathology and Medical Research (ICPMR), Sydney Centres for Thrombosis and Haemostasis, NSW Health Pathology, Westmead Hospital, Westmead, NSW, Australia,* [b]*School of Dentistry and Medical Sciences, Faculty of Science and Health, Charles Sturt University, Wagga, NSW, Australia,* [c]*School of Medical Sciences, Faculty of Medicine and Health, University of Sydney, Westmead Hospital, Westmead, NSW, Australia,* [d]*Westmead Clinical School, Faculty of Medicine and Health, University of Sydney, Westmead, NSW, Australia,* [e]*Section of Clinical Biochemistry, University of Verona, Verona, Italy*

Abbreviations

APS	antiphospholipid [antibody] syndrome
CDC	Centers for Disease Control and Prevention
COVID-19	Coronavirus disease 2019
DVT	deep vein thrombosis
ELISA	enzyme linked immunosorbent assay
HIT	heparin induced thrombocytopenia
HITT	heparin induced (immune) thrombotic thrombocytopenia
Ig	immunoglobulin
IV	intravenous
PE	pulmonary thrombosis
PF4	platelet factor 4
SARS-CoV-2	severe acute respiratory syndrome coronavirus 2
sHIT	spontaneous-HIT (or HIT-like syndrome)
SRA	serotonin release assay
TGA	therapeutic goods agency
THANZ	Thrombosis and Haemostasis Society of Australia and New Zealand
TTP	thrombotic thrombocytopenic purpura
TTS	thrombotic thrombocytopenia syndrome
VATT	vaccine associated (immune) thrombotic thrombocytopenia
VIPIT	vaccine induced prethrombotic immune thrombocytopenia
VITT	vaccine induced (immune) thrombotic thrombocytopenia
VTE	venous thromboembolism

Introduction

Coronavirus disease 2019 (COVID-19) occurs as a result of infection by the severe acute respiratory syndrome coronavirus (SARS-CoV-2) virus. COVID-19 is believed to have originated in Wuhan, China, in November 2019, and was declared a pandemic by the World Health Organization in early March 2020. These events identify the highly virulent nature of SARS-CoV-2. COVID-19 also shows a high morbidity and mortality rate, in part determined by various factors such as sex (males more adversely affected), age (older people suffer greater morbidity and mortality), and the presence of other comorbidities (Chang et al., 2022; Lippi et al., 2020). The main pathophysiological events that cause morbidity and

Management, Body Systems, and Case Studies in COVID-19. https://doi.org/10.1016/B978-0-443-18703-2.00012-8
Copyright © 2024 Elsevier Inc. All rights are reserved, including those for text and data mining, AI training, and similar technologies.

mortality are difficulty breathing or shortness of breath (dyspnea) and chest pain (all reflective of lung pathology), loss of mobility, confusion (central nervous system effects), immunoinflammation (systemic effects), and increased risk of thrombosis (procoagulant effects).

To reduce infection rates, help prevent COVID-19, and reduce COVID-19 related morbidity and mortality, a number of COVID-19 vaccines were fast tracked to development, clinical trials, and emergency use authorization, with the first few vaccines available in December 2020 (Zidan et al., 2023). As with all vaccination events, there is potential for side effects after immunization. Most of the side effects are mild, and may include such generalized symptoms as pain, redness and swelling at the injection site, tiredness, headache, muscle pain, chills, fever, and nausea (https://www.cdc.gov/coronavirus/2019-ncov/vaccines/expect/after.html). More severe adverse side effects are rare, but include anaphylaxis, Guillain-Barré syndrome (GBS), myocarditis, pericarditis, and a condition called thrombosis with thrombocytopenia syndrome (TTS) by government reporting agencies, but otherwise called vaccine induced (immune) thrombotic thrombocytopenia (VITT) by scientific researchers. VITT has only been confirmed in people vaccinated using adenovirus-based vaccines, thus mostly including the AstraZeneca (Vaxzevria), Johnson and Johnson/Janssen (Jcovden), and Sputnik V (Gam-COVID-Vac) vaccines (Favaloro & Pasalic, 2021; Greinacher, Thiele, et al., 2021a, 2021b; Schultz et al., 2021; Scully et al., 2021; See et al., 2021; Zidan et al., 2023).

VITT vs TTS

VITT and TTS are in fact similar entities that are just being described by different experts, but due to varying case definitions may end up identifying different patient subpopulations. VITT is a term originally defined by several teams of scientific researchers publishing the original case series in the New England Journal of Medicine (NEJM) (Greinacher, Thiele, et al., 2021a; Schultz et al., 2021; Scully et al., 2021). It is meant to reflect a variation of the term "HITT," which describes heparin-induced (immune) thrombotic thrombocytopenia (Favaloro & Pasalic, 2021; Lippi & Favaloro, 2022). The term VITT likely represents some sort of harmonization led by the NEJM editorial team because the term VIPIT (for vaccine induced prothrombotic immune thrombocytopenia) was originally coined by the German group who published the first case series in a preprint in April 2021 (Greinacher, Thiele, et al., 2021b). The term VATT (for vaccine associated (immune) thrombotic thrombocytopenia) has also been used by other researchers. All four terms (VIPIT, VITT, VATT, and TTS) are descriptive in so far as they identify the main events reflective of the arising pathology, being platelet activation leading to platelet aggregation and clearance from circulation (and thus thrombocytopenia) as well as consequent thrombosis. However, the term TTS does not identify a specific association with vaccination, perhaps used intentionally by government agencies (including the Centers for Disease Control and Prevention (CDC) in the United States and the Therapeutic Goods Agency (TGA) in Australia) so as to not cause vaccine fear or hesitancy. Nevertheless, scientific researchers are unlikely to use the term TTS in VITT research studies because the term TTS can encompass any syndrome where thrombosis is associated with thrombocytopenia, including HITT, catastrophic antiphospholipid [antibody] syndrome (APS), and thrombotic thrombocytopenic purpura (TTP) (Favaloro & Pasalic, 2021; Lippi & Favaloro, 2022).

VITT vs HITT—Similar but different

Both VITT and HITT have similar pathologies, representing an autoimmune formation of antibodies against platelet factor 4 (PF4), which then bind to this receptor and ultimately trigger platelet activation (Favaloro & Pasalic, 2021; Lippi & Favaloro, 2022; Warkentin & Greinacher, 2021). These platelets then aggregate, forming blood clots that can cause vascular thrombosis. Despite the involvement of platelets, the arising thrombosis primarily occurs in the venous vascular bed, but occasional arterial thrombosis may also occur. In these respects, HITT and VITT are similar (Table 1). In addition, both VITT and HITT are associated with high levels of plasma D-dimers, resulting from the generation of thrombi and their subsequent breakdown (Favaloro & Pasalic, 2021; Lippi & Favaloro, 2022; Thachil et al., 2022).

However, there are also notable differences between HITT and VITT. First, in HITT, the trigger for the formation of antibodies is a complex comprised of heparin and PF4 (i.e., heparin/PF4). Heparin is an anticoagulant that is given to patients to otherwise prevent or treat thrombosis (Carpenè et al., 2022); it is counterintuitive then that an agent given to prevent thrombosis ends up causing thrombosis in a small proportion (up to 1%–3%) of treated patients. The antibodies are thought to bind to a region of PF4 where heparin also binds (Huynh et al., 2021; Warkentin & Greinacher, 2022a, 2022b). In VITT, the antibodies formed are not against a complex of heparin/PF4, but rather against PF4 or PF4 in complex with some other component (either of the vaccine or a component that may arise following vaccination). In this way, the VITT antibodies bind to a different part of the PF4, and thus are different to HITT antibodies. Although the pathogenesis of VITT is only partially elucidated, it is now commonly believed that vaccine constituents and PF4 may trigger the generation of so-called neoantigens that can then become the targets of an immunological response. The resulting antigenic complexes

TABLE 1 VITT vs HITT—Similar but different.

Similarities	Differences
• Both represent antibodies against platelet factor 4 (PF4) • Both lead to platelet activation, platelet aggregation, platelet clearance (thrombocytopenia), and thrombosis • Both can show high levels of plasma D-dimer • Both can be detected immunologically by anti-PF4 antibody assays • Both can be confirmed using functional platelet activation assays, such as serotonin release assays (SRA)	• VITT and HITT antibodies bind to different PF4 complexes and different epitopes on PF4 • VITT can only be detected immunologically by certain anti-PF4 antibody assays, most notably those using enzyme linked immunosorbent assay (ELISA) methods, with most rapid anti-PF4 antibody assays unable to identify VITT antibodies. Instead, HITT anti-PF4 antibodies can be detected by most anti-PF4 antibody assays, including ELISA and rapid assays • Although both can be confirmed using functional platelet activation assays, such as SRA, the test patterns differ, with therapeutic levels of heparin typically augmenting platelet activation in HITT, whereas in VITT, platelet activation should occur in the absence of therapeutic levels of heparin, with added heparin at the therapeutic level potentially inhibiting platelet activation • D-dimer levels in VITT appear to be higher than those described in HITT

generate a sustained B cell response with production of high-avidity anti-PF4 antibodies, which then bind to the PF4/vaccine complex on platelets, ultimately leading to platelet activation (Greinacher, Selleng, et al., 2021). Concomitantly, anti PF-4 antibodies may also be capable of stimulating neutrophils to release neutrophil extracellular traps (NETs), which would then contribute to boost the abnormal propensity toward clotting seen in VITT (Greinacher, Selleng, et al., 2021).

In addition to the above, and due to the presence of thrombosis, high plasma D-dimers are present in both VITT and HITT, although higher levels appear to occur in VITT.

Immunological detection of anti-PF4 antibodies in VITT (vs HITT)

Anti-PF4 antibodies in both VITT and HITT can be detected immunologically by anti-PF4 antibody assays. However, VITT can only be detected immunologically by certain anti-PF4 antibody assays, most notably enzyme linked immunosorbent assay (ELISA) methods, with most rapid anti-PF4 antibody assays (including the chemiluminescence procedure on the AcuStar instrument) mostly unable to identify VITT antibodies (Favaloro, 2021; Favaloro, Clifford, Leitinger, et al., 2022; Favaloro, Pasalic, Henry, & Lippi, 2022; Platton et al., 2021; Warkentin & Greinacher, 2022a, 2022b) (Fig. 1). Instead, HITT anti-PF4 antibodies can be detected by most anti-PF4 antibody assays, including ELISA and rapid assays (Favaloro, 2018; Favaloro et al., 2018, 2017; Favaloro, Mohammed, et al., 2021; Hvas et al., 2021). Thus, it is important to precharacterize patients clinically as either HITT (i.e., heparin associated) vs VITT (i.e., vaccine associated), and for clinicians to not just request an "anti-PF4 antibody assay" or "HIT" test to ensure that the laboratory performs the correct assay (e.g., laboratory can use rapid chemiluminescence procedure [AcuStar] if HITT, but only ELISA for VITT).

However, there also appears to be some variability in reactivity to different anti-PF4 antibody assays, and in an Australian cohort, up to one-third of VITT cases confirmed by functional assay were reported to be anti-PF4 antibody negative (Favaloro, Clifford, Leitinger, et al., 2022). Thus, although a negative anti-PF4 antibody ELISA result is generally taken to rule out VITT, its exclusion cannot be guaranteed, and functional testing is still warranted in patients with high clinical suspicion of VITT.

Functional testing for platelet activation in VITT (vs HITT)

Similarly, it is important for laboratories to know how to differentially perform platelet activation assays when assessing for VITT vs HITT, or what procedure to use for follow-up of a positive anti-PF4 ELISA assay result. There are various platelet activation assays available, of which the serotonin release assay (SRA) is often considered the gold standard (Favaloro et al., 2017; Warkentin et al., 2021). In this assay, platelets are preloaded with radioactive serotonin, and platelet activating anti-PF4 antibodies will cause release of this radioactive serotonin into the supernatant. This can then be subjected to radioisotypic counting to determine the level of release, which is proportional to the extent of platelet activation. Other methods include platelet aggregation by light transmittance aggregometry (although this is usually insufficiently sensitive (Favaloro et al., 2018)) or by whole blood impedance aggregometry (e.g., multiplate method (Lau et al., 2017; Morel-Kopp et al., 2016)) as well as a variety of flow cytometry-based methods (Lee, Liang, et al., 2022; Lee, Selvadurai, et al., 2022).

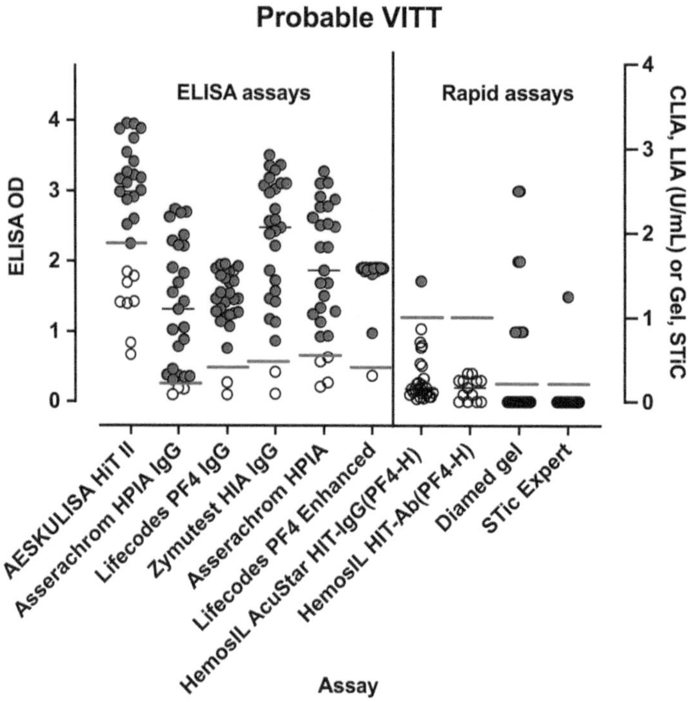

FIG. 1 Summary of reported ELISA optical density readings for PF4 antibodies (left side of figure; left *y*-axis), and results of rapid assays (right side of figure; right *y*-axis), according to the study from Platton et al. (2021). For rapid assays, chemiluminescence and latex immunoassay results given in U/mL; for the Diamed gel method, the numbers refer to the grade of response, and for the Stago STiC method, a value of 0 represents negative and a value of 1.5 represents positive. The *red* horizontal lines indicate the negative/positive cut-off values. Positive values are shown as *red*-filled dots, and negative values as *white*-filled dots.

In classical HITT testing, the tests are performed in the presence of a therapeutic level of heparin (usually between 0.1 and 0.5 U/mL) to assess for heparin induced activation as well as a high (supertherapeutic) level of heparin (usually between 10 and 100 U/mL), which is expected to inhibit platelet activation due to the disruption of heparin/PF4 complex formation. Thus, a high level of platelet activation (or serotonin release in the SRA) in the therapeutic heparin level with inhibition of activation (or release) in the high heparin level is a hallmark of pathological HITT (Favaloro, 2018; Favaloro et al., 2018, 2017; Favaloro, Henry, & Lippi, 2021; Favaloro, Mohammed, et al., 2021; Warkentin et al., 2021). In classical VITT, the addition of a therapeutic level of heparin is not expected to augment platelet activation (or serotonin release) because the antibodies are not generated against the heparin/PF4 complex. Indeed, the addition of a therapeutic level of heparin may instead inhibit platelet activation (or serotine release in the SRA) in VITT (Favaloro, 2021; Favaloro, Clifford, Leitinger, et al., 2022; Vayne et al., 2021). Thus, in VITT, the usual approach entails an assessment of platelet activation (or serotonin release) in the absence of heparin, and the use of a high level of heparin to show inhibition of platelet activation (or serotonin release). Thus, positive platelet activation (or serotonin release) in the absence of heparin, with inhibition in the presence of both therapeutic and high levels of heparin, may be considered a positive confirmatory finding for VITT.

Of course, not all patient test results are so clear cut, and combinations or variations of these findings are possible. For example, although not formally described to our knowledge, it is theoretically possible to have both HITT and VITT antibodies in the same patient because patients may be immunized against COVID-19 by an adenovirus-based vaccine, and then be exposed to heparin should they be hospitalized for surgery or illness.

Finally, it needs to be recognized that not all patients who show evidence of anti-PF4 antibodies and/or thrombocytopenia after vaccination will have VITT. Indeed, there is a background rate of these anti-PF4 antibodies in otherwise healthy individuals after both adenovirus and mRNA-based COVID-19 vaccines, but these "background" antibodies do not trigger platelet activation nor any adverse pathophysiological event (Sørvoll et al., 2021; Thiele et al., 2021).

Guidelines for the diagnosis and management of VITT (or HITT)

There have been a large number of guidelines developed to help clinicians diagnose and/or manage patients with VITT as well as previously with HITT. We recently evaluated those guidelines for the diagnosis of VITT (Favaloro, Pasalic,

& Lippi, 2022a), and some of us have also been in part involved in the development of guidelines for the diagnosis and management of HITT (Joseph et al., 2019). Similar to the situation with VITT vs TTS, the use of different guidelines for VITT diagnosis may yield slightly different subpopulations of cases, depending on the emphasis of various criteria. It should also be recognized that guidelines reflect historical context, with some emerging soon after the recognition of VITT to manage a possible emerging crisis of VITT, and others acting as "living guidelines" that were adapted over time (Favaloro, Pasalic, & Lippi, 2022a).

There are many similarities but also some inconsistencies and variations in the VITT guidelines for the diagnosis of VITT (Favaloro, Pasalic, & Lippi, 2022a). The early published cohorts recognized the onset of symptoms as vaccine association between 2 and 28 days after vaccination as well as thrombocytopenia, the characteristic presentation of thrombosis at unusual sites, high morbidity, and very high levels of plasma D-dimer. While characteristic of early VITT cohorts, limiting the investigation of VITT patients to such criteria may have inadvertently missed some VITT patients who presented after the 28-day limit, or who had more classic or common forms of venous thromboembolism (VTE) such as pulmonary thrombosis (PE) or deep vein thrombosis (DVT), or who did not present with thrombocytopenia. Notably, two of the authors of this review (EJF and LP) were members of the Thrombosis and Haemostasis Society of Australia and New Zealand (THANZ) VITT advisory group, formed to advise on Australian and New Zealand VITT diagnosis and management processes. As the group has recently published (Chen et al., 2021; Favaloro, Clifford, Leitinger, et al., 2022), the Australian cohort had several VITT serologically confirmed cases that presented after the 28-day limit (up to 43 days postvaccination), and/or some patients did not present with thrombocytopenia and/or had PE and/or DVT as the presenting thrombosis.

In brief, most guidelines for diagnosis indicate the performance of common laboratory tests at the site of presentation (especially platelet count and plasma D-dimer level) and investigations into the possible presentation of thrombosis (according to symptoms and the site of symptoms) (Favaloro, Pasalic, & Lippi, 2022a). Subsequently, on the basis of the likelihood or otherwise of VITT, tests for anti-PF4 antibodies by ELISA, and if indicated, platelet activation assays using a VITT-modified protocol would be indicated (Favaloro, Pasalic, & Lippi, 2022a). In terms of management, all guidelines indicated immediate anticoagulation, typically using a nonheparin anticoagulant (although the specific requirement for a nonheparin anticoagulant was aligned to the historical approach used for HITT, and this approach may not be required for all patients with VITT); most (especially the more recent) guidelines also indicate the use of intravenous (IV) immunoglobulin (Ig) to minimize anti-PF4 driven platelet activation (Bourguignon et al., 2021; Chen et al., 2021; Lee, Al Moosawi, et al., 2022; Rizk et al., 2021).

Policies and procedures

In this chapter, we reviewed the pathophysiological event that is VITT as well as the various guidelines available on its diagnosis and management and its differentiation from the historically better-known syndrome called HITT.

For the diagnosis or exclusion of VITT, we recommended following the most recently updated guidelines available locally or else as supported by expert organizations such as the International Society on Thrombosis and Haemostasis, the American Society of Hematology), and THANZ (Favaloro, Pasalic, & Lippi, 2022a; Gresele et al., 2021; Nazy et al., 2021; Oldenburg et al., 2021). However, consideration for the most recent version of the available guideline is also critical. As an example, in Australia we would follow the THANZ guidelines, which for diagnosis/exclusion would recommend assessment upon patient presentation of platelet count and plasma D-dimer, plus investigation of possible thrombosis according to patient symptoms (e.g., major headache not controlled by analgesics would mandate the exclusion of cerebral thrombosis; leg calf pain would mandate the exclusion of DVT; breathing difficulties and chest pain, exclusion of PE; abdominal pain, exclusion of splenic vein thrombosis (Chen et al., 2021) (Fig. 2)). These guidelines were regularly updated online at the THANZ website (https://www.thanz.org.au/).

Subsequently, a plasma/serum sample should be tested for anti-PF4 antibodies by ELISA, and if further indicated, platelet activation should be checked using a VITT modified assay (Chen et al., 2021) (Fig. 2). In Australia, we employed three different platelet activation assays, as performed at three different laboratory sites, primarily to manage the burden of this specialized testing given the complexity and labor intensiveness of these assays (Favaloro, Clifford, Leitinger, et al., 2022). These comprised VITT modified assays by SRA, by multiplate, and by a novel flow cytometry assay. Interestingly, all assays showed a similar relationship to the immunologically performed ELISA assay, although the SRA and flow assays appeared most sensitive to VITT (Favaloro, Clifford, Leitinger, et al., 2022).

In regard to management, if VITT is strongly suspected, then anticoagulation and IV Ig should be given. THANZ favored a nonheparin anticoagulant on the basis of a lack of evidence around the safety of heparin anticoagulation in VITT patients; however, the group also recognized that heparin anticoagulant might be safe in some patients (Chen et al., 2021).

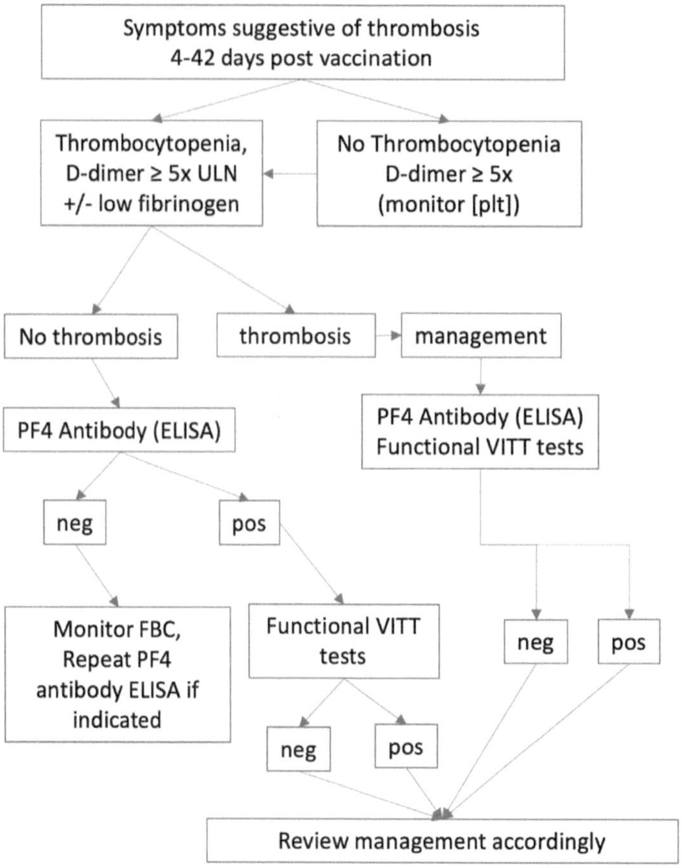

FIG. 2 The VITT diagnostic pathway proposed by THANZ based on the Australian case series (Chen et al., 2021; Favaloro, Clifford, Leitinger, et al., 2022). Abbreviations: *ELISA*, enzyme linked immunosorbent assay; *FBC*, full blood count; *neg*, negative; *PF4*, platelet factor 4; *pos*, positive; *VITT*, vaccine induced thrombotic thrombocytopenia.

Applications to other areas

Although the VITT "crisis" seems to have dissipated in our own geographical localities (Australia and Italy), with reduced use or even elimination of adenovirus-based COVID-19 vaccines, such vaccines are still used in many other counties, particularly in low-income regions. Also, although VITT is primarily identified as a COVID-19 vaccine event, VITT is also theoretically possible in a patient given another adenovirus-based vaccine unrelated to COVID-19. Thus, VITT has implications beyond COVID-19 vaccination because adenovirus-based vaccines can be employed against agents other than SARS-CoV-2. In addition, what we have learned about VITT may also help expand our knowledge about related pathophysiological events, including but not limited to HITT. For example, there is a related condition that some workers call spontaneous-HIT, but which we prefer to call HIT-like syndrome, and which arises either due to infection or after some surgical procedures, especially knee surgery (Favaloro, Pasalic, & Lippi, 2022b; Warkentin & Greinacher, 2021). This condition can thus arise as a result of COVID-19 itself, unrelated to any vaccine use (Favaloro, Henry, & Lippi, 2021). All these conditions share some similarities, but also reflect different origins and potentially different outcomes. There is also an impact of VITT on individuals who do not directly suffer from the conditions discussed. For example, the emergence of VITT had a potentially catastrophic impact on COVID-19 vaccine uptake, essentially causing vaccine hesitancy in a large number of people. Some countries entirely abandoned the use of adenovirus-based vaccines in favor of perceived "safer" alternatives, such as mRNA-based vaccines (Zidan et al., 2023). For example, the AstraZeneca COVID-19 vaccine is no longer available in Italy; although it was still available in Australia at time of writing, most healthcare providers favored the alternate vaccines, and so very few people were being vaccinated with the AstraZeneca COVID-19 vaccine in Australia after the emergence of VITT. It is unclear if the VITT experience will also lead manufacturers to abandon the adenovirus-based vaccine approach in general, or instead lead to improved adenovirus-based formulations that will not lead to VITT. In context, the risk of getting VITT (1 in 50,000–100,000 vaccine doses), or dying from VITT up to 50% of affected patients in the early VITT cohorts (Favaloro, 2021; Greinacher, Thiele, et al., 2021b; Schultz et al., 2021; Scully et al., 2021), but is <5% in the latest Australian cohort; so, in total, the risk is 1 in 1–2 million doses of vaccine (Favaloro, Clifford, Leitinger, et al., 2022), which is much less than the risk of getting COVID-19, or dying from COVID-19 (1%–3% of SARS-CoV-2 infected people) (Elalamy et al., 2021; Whiteley et al., 2022).

Mini-dictionary of terms

COVID-19: A pathology caused by the severe acute respiratory syndrome coronavirus 2, mostly causing respiratory symptoms, but with a significant risk of becoming a systemic disorder accompanied by thrombosis and multiple organ failure in predisposed subjects.

Heparin (unfractionated): A naturally occurring glycosaminoglycan that has been used extensively for more than 70 years as an anticoagulant indicated for both the prevention and treatment of thrombotic events such as deep vein thrombosis. Heparin achieves its anticoagulant effect through its interaction with plasma antithrombin, which enhances the inactivation of the coagulation enzymes thrombin, Factor Xa, and Factor IXa.

HIT-like syndrome: Heparin induced thrombocytopenia-like syndrome, also called spontaneous HIT by some workers. A condition not associated with heparin (or vaccine) use, but still associated with the development of antiplatelet factor 4 (PF4) antibodies, and somewhat pathophysiologically similar to both HITT and VITT.

Heparin induced (immune) thrombotic thrombocytopenia. A condition associated with heparin use, with the development of antiplatelet factor 4 (PF4) antibodies, and somewhat pathophysiologically similar to VITT.

Platelet factor 4; a cytokine of the CXC chemokine family, mostly released from activated platelets, which promotes blood coagulation. PF4 is the main target of anti-PF4 antibodies in HITT and VITT.

Thrombotic thrombocytopenia syndrome. Although used by government reporting agencies as a surrogate for the term VITT (but without mentioning any vaccine association), the term TTS can actually describe any pathology that arises and causes thrombotic thrombocytopenia.

Vaccine induced (immune) thrombotic thrombocytopenia. This is due to the generation of antiplatelet factor 4 (PF4) antibodies that cause platelet activation. VITT can arise following immunization with an adenovirus-based vaccine, most typically against COVID-19, in around 1 in 100,000 doses of vaccine.

Summary points

- VITT arises in a small proportion of patients exposed to vaccination using an adenovirus-based vaccine (incidence around 1 in 100,000 vaccine doses).
- VITT pathophysiology is similar to HITT and HIT-like syndrome, and is caused by the development of antiplatelet factor 4 (PF4) antibodies capable of platelet activation.
- The platelet activation caused by anti-PF4 antibodies leads to platelet aggregation and thrombosis.
- Diagnosis of VITT requires evidence of anti-PF4 antibodies using a sensitive assay (typically ELISA), and potentially confirmed using a VITT modified platelet activation assay.
- Additional markers of VITT include an initial low platelet count (thrombocytopenia) and a high level of plasma D-dimer upon clinical presentation.
- Treatment of VITT requires initiation of anticoagulation, most typically applied using a nonheparin anticoagulant, as well as intravenous immunoglobulin.

Acknowledgments/sources of funding

No funding was received for this work. The views expressed herein are those of the authors and are not necessarily those of NSW Health Pathology or other affiliated institutions.

Conflicts of interest

The authors have no conflicts of interest to disclose.

References

Bourguignon, A., Arnold, D. M., Warkentin, T. E., Smith, J. W., Pannu, T., Shrum, J. M., Al Maqrashi, Z. A. A., Shroff, A., Lessard, M. C., Blais, N., Kelton, J. G., & Nazy, I. (2021). Adjunct immune globulin for vaccine-induced immune thrombotic thrombocytopenia. *The New England Journal of Medicine*, *385*, 720–728. https://doi.org/10.1056/NEJMoa2107051.

Carpenè, G., Negrini, D., Lippi, G., Favaloro, E. J., & Montagnana, M. (2022). Heparin: The journey from parenteral agent to nasal delivery. *Seminars in Thrombosis and Hemostasis*, *48*, 949–954. https://doi.org/10.1055/s-0042-1749395.

Chang, D., Chang, X., He, Y., & Tan, K. J. K. (2022). The determinants of COVID-19 morbidity and mortality across countries. *Scientific Reports*, *12*, 5888. https://doi.org/10.1038/s41598-022-09783-9.

Chen, V. M., Curnow, J. L., Tran, H. A., & Choi, P. Y. (2021). Australian and New Zealand approach to diagnosis and management of vaccine-induced immune thrombosis and thrombocytopenia. *The Medical Journal of Australia*, *215*(6), 245–249. https://doi.org/10.5694/mja2.51229.

Elalamy, I., Gerotziafas, G., Alamowitch, S., Laroche, J. P., Van Dreden, P., Ageno, W., Beyer-Westendorf, J., Cohen, A. T., Jimenez, D., Brenner, B., Middeldorp, S., Cacoub, P., & Scientific Reviewer Committee. (2021). SARS-CoV-2 vaccine and thrombosis: An expert consensus on vaccine-induced immune thrombotic thrombocytopenia. *Thrombosis and Haemostasis*, *121*, 982–991. https://doi.org/10.1055/a-1499-0119.

Favaloro, E. J. (2018). Laboratory tests for identification or exclusion of heparin induced thrombocytopenia—HIT or miss? *American Journal of Hematology*, *93*, 308–314. https://doi.org/10.1002/ajh.24979.

Favaloro, E. J. (2021). Laboratory testing for suspected COVID-19 vaccine-induced (immune) thrombotic thrombocytopenia. *International Journal of Laboratory Hematology*, *43*, 559–570. https://doi.org/10.1111/ijlh.13629.

Favaloro, E. J., Clifford, J., Leitinger, E., Parker, M., Sung, P., Chunilal, S., Tran, H., Kershaw, G., Fu, S., Passam, F., Ahuja, M., Ho, S. J., Duncan, E., Yacoub, O., Tan, C. W., Kaminskis, L., Modica, N., Pepperell, D., Ballard, L., … Chen, V. (2022). Assessment of immunological anti-platelet factor 4 antibodies for vaccine-induced thrombotic thrombocytopenia (VITT) in a large Australian cohort: A multicentre study comprising 1284 patients. *Journal of Thrombosis and Haemostasis*, *20*, 2896–2908. https://doi.org/10.1111/jth.15881.

Favaloro, E. J., Henry, B. M., & Lippi, G. (2021). The complicated relationships of heparin-induced thrombocytopenia and platelet factor 4 antibodies with COVID-19. *International Journal of Laboratory Hematology*, *43*, 547–558. https://doi.org/10.1111/ijlh.13582.

Favaloro, E. J., McCaughan, G., Mohammed, S., Lau, K. K. E., Gemmell, R., Cavanaugh, L., Donikian, D., Kondo, M., Brighton, T., & Pasalic, L. (2018). HIT or miss? A comprehensive contemporary investigation of laboratory tests for heparin induced thrombocytopenia. *Pathology*, *50*, 426–436. https://doi.org/10.1016/j.pathol.2017.11.089.

Favaloro, E. J., McCaughan, G., & Pasalic, L. (2017). Clinical and laboratory diagnosis of heparin induced thrombocytopenia: An update. *Pathology*, *49*, 346–355. https://doi.org/10.1016/j.pathol.2017.02.005.

Favaloro, E. J., Mohammed, S., Donikian, D., Kondo, M., Duncan, E., Yacoub, O., Zebeljan, D., Ng, S., Malan, E., Yuen, A., Beggs, J., Moosavi, S., Coleman, R., Klose, N., Chapman, K., Cavanaugh, L., Pasalic, L., Motum, P., Tan, C. W., & Brighton, T. (2021). A multicentre assessment of contemporary laboratory assays for heparin induced thrombocytopenia. *Pathology*, *53*, 247–256. https://doi.org/10.1016/j.pathol.2020.07.012.

Favaloro, E. J., & Pasalic, L. (2021). COVID-19 vaccine induced (immune) thrombotic thrombocytopenia (VITT)/thrombosis with thrombocytopenia syndrome (TTS): An update. *Australian Journal of Medical Science*, *42*, 86–93.

Favaloro, E. J., Pasalic, L., Henry, B., & Lippi, G. (2022). Laboratory testing for platelet factor 4 antibodies: Differential utility for diagnosis/exclusion of heparin induced thrombocytopenia versus suspected vaccine induced thrombotic thrombocytopenia. *Pathology*, *54*, 254–261. https://doi.org/10.1016/j.pathol.2021.10.008.

Favaloro, E. J., Pasalic, L., & Lippi, G. (2022a). Review and evolution of guidelines for diagnosis of COVID-19 vaccine induced thrombotic thrombocytopenia (VITT). *Clinical Chemistry and Laboratory Medicine*, *60*, 7–17. https://doi.org/10.1515/cclm-2021-1039.

Favaloro, E. J., Pasalic, L., & Lippi, G. (2022b). Antibodies against platelet factor 4 and their associated pathologies: From HIT/HITT to spontaneous HIT-like syndrome, to COVID-19, to VITT/TTS. *Antibodies (Basel)*, *11*, 7. https://doi.org/10.3390/antib11010007.

Greinacher, A., Selleng, K., Palankar, R., Wesche, J., Handtke, S., Wolff, M., Aurich, K., Lalk, M., Methling, K., Völker, U., Hentschker, C., Michalik, S., Steil, L., Reder, A., Schönborn, L., Beer, M., Franzke, K., Büttner, A., Fehse, B., … Renné, T. (2021). Insights in ChAdOx1 nCoV-19 vaccine-induced immune thrombotic thrombocytopenia. *Blood*, *138*(22), 2256–2268. https://doi.org/10.1182/blood.2021013231.

Greinacher, A., Thiele, T., Warkentin, T. E., Weisser, K., Kyrle, P. A., & Eichinger, S. (2021a). Thrombotic thrombocytopenia after ChAdOx1 nCov-19 vaccination. *The New England Journal of Medicine*, *384*, 2092–2101. https://doi.org/10.1056/NEJMoa2104840.

Greinacher, A., Thiele, T., Warkentin, T. E., Weisser, K., Kyrle, P., & Eichinger, S. (2021b). *A prothrombotic thrombocytopenic disorder resembling heparin-induced thrombocytopenia following coronavirus-19 vaccination*. Preprint:. https://doi.org/10.21203/rs.3.rs-362354/v1. https://www.researchsquare.com/article/rs-362354/v1.

Gresele, P., Marietta, M., Ageno, W., Marcucci, R., Contino, L., Donadini, M. P., Russo, L., Tiscia, G. L., Palareti, G., Tripodi, A., Mannucci, P. M., & De Stefano, V. (2021). Management of cerebral and splanchnic vein thrombosis associated with thrombocytopenia in subjects previously vaccinated with Vaxzevria (AstraZeneca): A position statement from the Italian Society for the study of haemostasis and thrombosis (SISET). *Blood Transfusion*, *19*, 281–283.

Huynh, A., Kelton, J. G., Arnold, D. M., Daka, M., & Nazy, I. (2021). Antibody epitopes in vaccine-induced immune thrombotic thrombocytopaenia. *Nature*, *596*, 565–569. https://doi.org/10.1038/s41586-021-03744-4.

Hvas, A. M., Favaloro, E. J., & Hellfritzsch, M. (2021). Heparin-induced thrombocytopenia: Pathophysiology, diagnosis and treatment. *Expert Review of Hematology*, *14*, 335–346.

Joseph, J., Rabbolini, D., Enjeti, A. K., Favaloro, E., Kopp, M. C., McRae, S., Pasalic, L., Chee Wee, T., Ward, C. M., & Chong, B. H. (2019). Diagnosis and management of heparin-induced thrombocytopenia: a consensus statement from the Thrombosis and Haemostasis Society of Australia and New Zealand HIT Writing Group. *The Medical Journal of Australia*, *210*, 509–516. https://doi.org/10.5694/mja2.50213.

Lau, K. K. E., Mohammed, S., Pasalic, L., & Favaloro, E. J. (2017). Laboratory testing protocols for heparin-induced thrombocytopenia (HIT) testing. *Methods in Molecular Biology*, *1646*, 227–243. https://doi.org/10.1007/978-1-4939-7196-1_19.

Lee, A. Y., Al Moosawi, M., Peterson, E. A., McCracken, R. K., Wong, S. K. W., Nicolson, H., Chan, V., Smith, T., Wong, M. P., Lee, L. J., Griffiths, C., Rahal, B., Parkin, S., Afra, K., Ambler, K., Chen, L. Y. C., Field, T. S., Lindsay, H. C., Lavoie, M., … Sweet, D. (2022). Clinical care pathway for suspected vaccine-induced immune thrombotic thrombocytopenia after ChAdOx1 nCoV-19 vaccination. *Blood Advances*, *6*, 3315–3320. https://doi.org/10.1182/bloodadvances.2021006862.

Lee, C. S. M., Liang, H. P. H., Connor, D. E., Dey, A., Tohidi-Esfahani, I., Campbell, H., Whittaker, S., Capraro, D., Favaloro, E. J., Donikian, D., Kondo, M., Hicks, S. M., Choi, P. Y., Gardiner, E. E., Clarke, L. J., Tran, H., Passam, F. H., Brighton, T. A., & Chen, V. M. (2022). A novel flow cytometry procoagulant assay for diagnosis of vaccine-induced immune thrombotic thrombocytopenia. *Blood Advances*, *6*, 3494–3506. https://doi.org/10.1182/bloodadvances.2021006698.

Lee, C. S. M., Selvadurai, M. V., Pasalic, L., Yeung, J., Konda, M., Kershaw, G. W., Favaloro, E. J., & Chen, V. M. (2022). Measurement of procoagulant platelets provides mechanistic insight and diagnostic potential in heparin-induced thrombocytopenia. *Journal of Thrombosis and Haemostasis*, *20*, 975–988. https://doi.org/10.1111/jth.15650.

Lippi, G., & Favaloro, E. J. (2022). Cerebral venous thrombosis developing after COVID-19 vaccination: VITT, VATT, TTS, and more. *Seminars in Thrombosis and Hemostasis*, *48*, 8–14. https://doi.org/10.1055/s-0041-1736168.

Lippi, G., Sanchis-Gomar, F., & Henry, B. M. (2020). COVID-19: Unravelling the clinical progression of nature's virtually perfect biological weapon. *Annals of Translational Medicine*, *8*, 693. https://doi.org/10.21037/atm-20-3989.

Morel-Kopp, M. C., Mullier, F., Gkalea, V., Bakchoul, T., Minet, V., Elalamy, I., Ward, C. M., & Subcommittee on Platelet Immunology. (2016). Heparin-induced multi-electrode aggregometry method for heparin-induced thrombocytopenia testing: Communication from the SSC of the ISTH. *Journal of Thrombosis and Haemostasis*, *14*, 2548–2552. https://doi.org/10.1111/jth.13516.

Nazy, I., Sachs, U. J., Arnold, D. M., McKenzie, S. E., Choi, P., Althaus, K., Ahlen, M. T., Sharma, R., Grace, R. F., & Bakchoul, T. (2021). Recommendations for the clinical and laboratory diagnosis of VITT against COVID-19: Communication from the ISTH SSC subcommittee on platelet immunology. *Journal of Thrombosis and Haemostasis*, *19*, 1585–1588. https://doi.org/10.1111/jth.15341.

Oldenburg, J., Klamroth, R., Langer, F., Albisetti, M., von Auer, C., Ay, C., Korte, W., Scharf, R. E., Pötzsch, B., & Greinacher, A. (2021). Diagnosis and management of vaccine-related thrombosis following AstraZeneca COVID-19 vaccination: Guidance statement from the GTH. *Hämostaseologie*, *41*, 184–189. https://doi.org/10.1055/a-1469-7481.

Platton, S., Bartlett, A., MacCallum, P., Makris, M., McDonald, V., Singh, D., Scully, M., & Pavord, S. (2021). Evaluation of laboratory assays for anti-platelet factor 4 antibodies after ChAdOx1 nCOV-19 vaccination. *Journal of Thrombosis and Haemostasis*, *19*, 2007–2013. https://doi.org/10.1111/jth.15362.

Rizk, J. G., Gupta, A., Sardar, P., Henry, B. M., Lewin, J. C., Lippi, G., & Lavie, C. J. (2021). Clinical characteristics and pharmacological management of COVID-19 vaccine-induced immune thrombotic thrombocytopenia with cerebral venous sinus thrombosis: A review. *JAMA Cardiology*, *6*, 1451–1460. https://doi.org/10.1001/jamacardio.2021.3444.

Schultz, N. H., Sørvoll, I. H., Michelsen, A. E., Munthe, L. A., Lund-Johansen, F., Ahlen, M. T., Wiedmann, M., Aamodt, A. H., Skattør, T. H., Tjønnfjord, G. E., & Holme, P. A. (2021). Thrombosis and thrombocytopenia after ChAdOx1 nCoV-19 vaccination. *The New England Journal of Medicine*, *384*, 2124–2130. https://doi.org/10.1056/NEJMoa2104882.

Scully, M., Singh, D., Lown, R., Poles, A., Solomon, T., Levi, M., Goldblatt, D., Kotoucek, P., Thomas, W., & Lester, W. (2021). Pathologic antibodies to platelet factor 4 after ChAdOx1 nCoV-19 vaccination. *The New England Journal of Medicine*, *384*, 2202–2211. https://doi.org/10.1056/NEJMoa2105385.

See, I., Su, J. R., Lale, A., Woo, E. J., Guh, A. Y., Shimabukuro, T. T., Streiff, M. B., Rao, A. K., Wheeler, A. P., Beavers, S. F., Durbin, A. P., Edwards, K., Miller, E., Harrington, T. A., Mba-Jonas, A., Nair, N., Nguyen, D. T., Talaat, K. R., Urrutia, V. C., … Broder, K. R. (2021). US case reports of cerebral venous sinus thrombosis with thrombocytopenia after Ad26.COV2.S vaccination, March 2 to April 21, 2021. *JAMA*, *325*, 2448–2456. https://doi.org/10.1001/jama.2021.7517.

Sørvoll, I. H., Horvei, K. D., Ernstsen, S. L., Laegreid, I. J., Lund, S., Grønli, R. H., Olsen, M. K., Jacobsen, H. K., Eriksson, A., Halstensen, A. M., Tjønnfjord, E., Ghanima, W., & Ahlen, M. T. (2021). An observational study to identify the prevalence of thrombocytopenia and anti-PF4/polyanion antibodies in Norwegian health care workers after COVID-19 vaccination. *Journal of Thrombosis and Haemostasis*, *19*, 1813–1818. https://doi.org/10.1111/jth.15352.

Thachil, J., Favaloro, E. J., & Lippi, G. (2022). D-dimers—"Normal" levels versus elevated levels due to a range of conditions, including "D-dimeritis," inflammation, thromboembolism, disseminated intravascular coagulation, and COVID-19. *Seminars in Thrombosis and Hemostasis*, *48*, 672–679. https://doi.org/10.1055/s-0042-1748193.

Thiele, T., Ulm, L., Holtfreter, S., Schönborn, L., Kuhn, S. O., Scheer, C., Warkentin, T. E., Bröker, B. M., Becker, K., Aurich, K., Selleng, K., Hübner, N. O., & Greinacher, A. (2021). Frequency of positive anti-PF4/polyanion antibody tests after COVID-19 vaccination with ChAdOx1 nCoV-19 and BNT162b2. *Blood*, *138*, 299–303. https://doi.org/10.1182/blood.2021012217.

Vayne, C., Rollin, J., Gruel, Y., Pouplard, C., Galinat, H., Huet, O., Mémier, V., Geeraerts, T., Marlu, R., Pernod, G., Mourey, G., Fournel, A., Cordonnier, C., & Susen, S. (2021). PF4 immunoassays in vaccine-induced thrombotic thrombocytopenia. *The New England Journal of Medicine*, *385*, 376–378.

Warkentin, T. E., & Greinacher, A. (2021). Spontaneous HIT syndrome: Knee replacement, infection, and parallels with vaccine-induced immune thrombotic thrombocytopenia. *Thrombosis Research*, *204*, 40–51. https://doi.org/10.1016/j.thromres.2021.05.018.

Warkentin, T. E., & Greinacher, A. (2022a). Laboratory testing for VITT antibodies. *Seminars in Hematology*, *59*, 80–88. https://doi.org/10.1053/j.seminhematol.2022.03.003.

Warkentin, T. E., & Greinacher, A. (2022b). Laboratory testing for heparin-induced thrombocytopenia and vaccine-induced immune thrombotic thrombocytopenia antibodies: A narrative review. *Seminars in Thrombosis and Hemostasis*. https://doi.org/10.1055/s-0042-1758818 (in press).

Warkentin, T. E., Smythe, M. A., Ali, M. A., Aslam, N., Sheppard, J. I., Smith, J. W., Moore, J. C., Arnold, D. M., & Nazy, I. (2021). Serotonin-release assay-positive but platelet factor 4-dependent enzyme-immunoassay negative: HIT or not HIT? *American Journal of Hematology*, *96*, 320–329. https://doi.org/10.1002/ajh.26075.

Whiteley, W. N., Ip, S., Cooper, J. A., Bolton, T., Keene, S., Walker, V., Denholm, R., Akbari, A., Omigie, E., Hollings, S., Di Angelantonio, E., Denaxas, S., Wood, A., Sterne, J. A. C., Sudlow, C., & CVD-COVID-UK Consortium. (2022). Association of COVID-19 vaccines ChAdOx1 and BNT162b2 with major venous, arterial, or thrombocytopenic events: A population-based cohort study of 46 million adults in England. *PLoS Medicine*, *19*(2), e1003926. https://doi.org/10.1371/journal.pmed.1003926.

Zidan, A., Noureldin, A., Kumar, S. A., Elsebaie, A., & Othman, M. (2023). COVID-19 vaccine-associated immune thrombosis and thrombocytopenia (VITT): Diagnostic discrepancies and global implications. *Seminars in Thrombosis and Hemostasis*, *49*(1), 9–14. https://doi.org/10.1055/s-0042-1759684. 36603593.

Chapter 19

Lung cancer in the era of COVID-19

Shehab Mohamed[a], Monica Casiraghi[a,b], Lorenzo Spaggiari[a,b], and Luca Bertolaccini[a]
[a]*Department of Thoracic Surgery, IEO, European Institute of Oncology IRCCS, Milan, Italy,* [b]*Department of Oncology and Hemato-Oncology, University of Milan, Milan, Italy*

Abbreviations

COPD	chronic obstructive pulmonary disease
COVID-19	coronavirus disease 2019
ECMO	extracorporeal membrane oxygenation
ERAS	enhanced recovery after surgery
ESMO	European Society for Medical Oncology
ICU	intensive care unit
NSCLC	nonsmall cell lung cancer
SARS-Cov2	severe acute respiratory syndrome coronavirus 2
TERAVOLT	Thoracic Cancers International COVID-19 Collaboration
THORN	Thoracic Surgery Outcomes Research Network
UKLCC	United Kingdom Lung Cancer Coalition
WHO	World Health Organization

Introduction

In March 2020, coronavirus disease 2019 (COVID-19) was declared a pandemic by the World Health Organization (Zhu et al., 2020). More than 756 million cases have been diagnosed with COVID-19, causing more than 6.8 million related deaths worldwide (World Health Organization, 2020). In the last 3 years, the severe acute respiratory syndrome coronavirus (SARS-CoV-2) has changed the preoperative and postoperative management of thoracic surgery patients as well as many other surgical specialties. Older patients or those with relevant chronic comorbidities such as cardiovascular disease, chronic respiratory disease, diabetes, or cancer are considered high-risk patients with an increased risk of mortality after COVID-19 infection (Onder et al., 2020). Indeed, oncological patients treated with systemic therapy or surgical operations tend to be more susceptible to infections due to an immunosuppressed state. The general thoracic surgeon had to work more often with implementing life-saving and life-sustaining therapies, including extracorporeal membrane oxygenation (ECMO) and tracheostomies, used mainly for the more compromised patients. In the last 3 years, especially during outbreaks, the surgical volumes have been substantially reduced, and many elective resections have been postponed to avoid viral transmission during hospitalization.

Moreover, surgical resections that need the intensive care unit (ICU) after operations were more postponed than minor elective surgeries due to the need for ICU beds for COVID-19 patients. Indeed, this new organization of resources was chosen to preserve personnel and material for COVID-19 patient care (Hanna et al., 2020). Of course, patients affected initially by early stage lung cancer may progress to an advanced or inoperable stage due to delayed treatment. For this reason, we reviewed the current data to give practical guidelines on the management of nonsmall cell lung cancer (NSCLC) patients during the COVID-19 pandemic.

Risk factors in NSCLC patients

Oncological patients are considered high-risk patients for greater susceptibility to SARS-Cov-2 infection and severe events or even death associated with the disease (Zhang et al., 2020). Patients with lung cancer may present in an immunocompromised state due to malignancy or systemic treatments (Dingemans et al., 2020). Garassino et al. assessed the outcomes of

Management, Body Systems, and Case Studies in COVID-19. https://doi.org/10.1016/B978-0-443-18703-2.00054-2
Copyright © 2024 Elsevier Inc. All rights are reserved, including those for text and data mining, AI training, and similar technologies.

patients affected by NSCLC and COVID-19 through the Thoracic Cancers International COVID-19 Collaboration (TERAVOLT) registry and reported an increased mortality rate. In this study, most of the patients had a stage IV disease under systemic treatment. However, the ICU admission rate was only 9% despite the high hospitalization rate, perhaps because most of the patients were from Europe, which was highly affected during the first pandemic wave (Garassino et al., 2020). The authors found that older age and other potential comorbidities, such as chronic obstructive pulmonary disease (COPD) and smoking history, are independent risk factors of death (Garassino et al., 2020). Unfortunately, lung cancer symptoms such as cough, dyspnea, or fatigue may mimic SARS-CoV-2 infection, resulting in a delayed diagnosis. Seitlinger et al. reported a multicenter study that analyzed the relationship between COVID-19 and 731 thoracic surgery patients. They described nine cases of COVID-19 and only one COVID-19-related death. The conclusions of this study, which our institution also took into account, were that oncological patients could undergo surgical procedures safely even during the COVID-19 pandemic to avoid delays (Seitlinger et al., 2020).

On the other hand, a few case series reported a significant increase in COVID-19-related deaths following anatomical lung resections (Cai et al., 2020; Huang et al., 2020; Peng et al., 2020). According to these results, the United Kingdom Lung Cancer Coalition (UKLCC) estimated an increased mortality risk in this group of patients of 40%–50% (Gourd, 2020). Another group of patients that needs to be highlighted is those who receive immune checkpoint inhibitors (ICIs) in the neoadjuvant and adjuvant settings. Even though this subgroup of patients may be less immunocompromised than patients receiving systemic chemotherapy, this treatment may have pneumological adverse effects that can be added to the SARS-CoV-2 infection. Because there is no clear evidence of interaction between the SARS-CoV-2 infection and systemic therapy, a brief suspension of oncological therapy may be considered in infected patients with stable cancer disease (Russano et al., 2020). Based on the TERAVOLT results, the type of systemic therapy (chemotherapy or ICIs) did not influence patient outcomes (Garassino et al., 2020).

Risks and benefits

Elective surgeries and procedures during the COVID-19 pandemic had to be suspended temporarily, especially during the first wave (Antonoff et al., 2020). The precise management of all kinds of patients is essential to help the decision-making process. Moreover, a good strategy for oncological patients is to consider the risks and benefits of providing cancer treatment and at what time despite the limited hospital resources (Dingemans et al., 2020). Dingemans et al. and many others reported the importance of maintaining oncological surgical activity during the pandemic to avoid delayed cancer treatment because thoracic surgeons perform procedures for lung cancer and other thoracic malignancies (Dingemans et al., 2020; Passaro et al., 2020). Moreover, the Thoracic Surgery Outcomes Research Network (ThORN) recommended that oncological patients should be treated with priority according to the phase of the COVID-19 pandemic, especially when hospital resources are limited (Table 1). At the same time, the prioritization should also consider the tumor's stage,

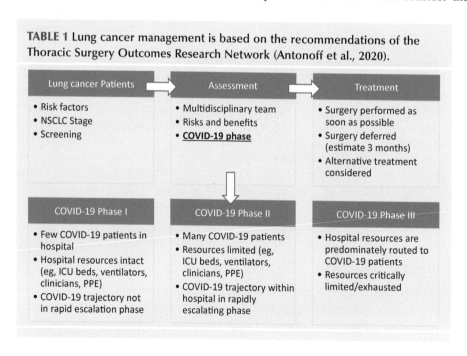

TABLE 1 Lung cancer management is based on the recommendations of the Thoracic Surgery Outcomes Research Network (Antonoff et al., 2020).

Lung cancer Patients →	Assessment →	Treatment
• Risk factors • NSCLC Stage • Screening	• Multidisciplinary team • Risks and benefits • <u>**COVID-19 phase**</u> ↓	• Surgery performed as soon as possible • Surgery deferred (estimate 3 months) • Alternative treatment considered
COVID-19 Phase I	**COVID-19 Phase II**	**COVID-19 Phase III**
• Few COVID-19 patients in hospital • Hospital resources intact (eg, ICU beds, ventilators, clinicians, PPE) • COVID-19 trajectory not in rapid escalation phase	• Many COVID-19 patients • Resources limited (eg, ICU beds, ventilators, clinicians, PPE) • COVID-19 trajectory within hospital in rapidly escalating phase	• Hospital resources are predominately routed to COVID-19 patients • Resources critically limited/exhausted

the symptoms, and any other associated complications (Antonoff et al., 2020). As described by Antonoff et al., several characteristics can make an NSCLC more urgent, meaning that it cannot be postponed. In this paper, the authors reported guidance for the triage of patients with thoracic malignancies, analyzing the different prioritizations based on the pandemic phase:

- Surgery should be performed as soon as possible.
- Surgery is deferred (estimated 3 months).
- Alternative treatment considered.

Few infected patients in the hospital, intact resources, and no rapid escalation characterized the first phase of COVID-19. In this setting, surgeries that should be performed as soon as possible are NSCLC with predominantly solid appearance (>50%), NSCLC greater than 2 cm, postinduction therapy cancer, node-positive NSCLC, and staging interventions (mediastinoscopy, bronchoscopy, and endobronchial ultrasound). Surgery can be deferred up to 3 months in the case of predominantly ground glass (<50% solid) nodules or cancers, a solid nodule or lung cancer <2 cm, indolent histology (e.g., carcinoid, slowly enlarging nodule), pulmonary oligometastases unless clinically necessary for pressing therapeutic or diagnostic indications (i.e., surgery will impact treatment), and patients likely to require prolonged ICU needs (i.e., particularly high-risk patients) (Antonoff et al., 2020). Alternative treatments, the availability of which may vary across health systems, may be considered for NSCLC patients that are eligible for adjuvant therapy. In these cases, neoadjuvant therapy can be used. Also, other possibilities are stereotactic ablative radiotherapy, ablation such as cryotherapy or radiofrequency, a stent for obstructing cancers followed by chemoradiation, and debulking for endobronchial tumors and only in circumstances where alternative therapy is not an option due to the increased risk of aerosolization. Also, the staging interventions may be nonsurgical if not available, such as endobronchial ultrasound (EBUS), imaging, and interventional radiology biopsy (Antonoff et al., 2020). The second phase of COVID-19 is characterized by many infected patients; limited resources such as ICU beds, ventilators, or clinicians; and a rapidly escalating phase. In this phase, surgery is restricted to patients who cannot survive if surgery is not performed within the next few days. The thoracic surgery procedures that must be performed as soon as possible are for the treatment of tumor-associated infections, such as debulking for postobstructive pneumonia or a tumor associated with a hemorrhage not amenable to nonsurgical treatment, and the management of surgical complications (hemothorax, empyema, infected mesh) in a hemodynamically stable patient. All the thoracic procedures initially scheduled as electives should be deferred for up to 3 months. Also, in this phase, there are several alternative treatments recommended, such as transferring the patient to a hospital that is in Phase I; if the patient is eligible for adjuvant therapy, then giving neoadjuvant therapy to save time; stereotactic ablative radiotherapy; ablation (e.g., cryotherapy, radiofrequency ablation); and in specific cases reconsidering a neoadjuvant as the definitive chemoradiation and monitoring patients for salvage surgery (Antonoff et al., 2020). Limited resources characterize the third phase of COVID-19, and they are predominately routed to COVID-19 patients. Surgeries are restricted to patients who cannot survive if surgery is not performed within the next few hours. In this setting, surgeries that should be performed as soon as possible are procedures for threatened airways, tumor-associated sepsis, and the management of surgical complications in an unstable patient, such as active bleeding not curable with nonsurgical management or dehiscence of the airway. Surgery can be deferred up to 3 months for all nonemergent operations. As in the second phase, the alternative treatments recommended are to transfer the patient to a hospital that is in Phase I; if the patient is eligible for adjuvant therapy, then give neoadjuvant therapy to save time; stereotactic ablative radiotherapy; ablation (e.g., cryotherapy, radiofrequency ablation); and in specific cases reconsidering a neoadjuvant as the definitive chemoradiation and monitoring patients for salvage surgery. In all these cases, transfer the patient, whenever feasible, to another hospital (Antonoff et al., 2020). Alternate therapies are suggested, especially for early stage lung cancer. On the other hand, as Banna et al. described in locally advanced lung cancer, neoadjuvant systemic therapy may allow the patient to start surgical treatment, even in a hospital with minimal resources (Banna et al., 2020). Furthermore, when indicated, Dingemans et al. preferred to treat locally advanced lung cancer with upfront surgery followed by chemotherapy to reduce the risk of infection during the neoadjuvant therapy at the hospital (Dingemans et al., 2020). Of course, every step taken to manage lung cancer patients should be discussed in a multidisciplinary setting (Bertolaccini et al., 2022). In a treatment delay and when resources are limited, follow-up imaging with computed tomography should be performed to keep an eye on the disease's evolution; eventually the case should be discussed at the tumor board and then rescheduled for surgery (Antonoff et al., 2020).

Surgical approach

Peng et al. described a significant increase in mortality in 11 patients affected by SARS-CoV-2 after thoracic surgery operations (Peng et al., 2020). After this paper, Scarci et al. stated that a sublobar resection might be considered for lung cancer and tumors larger than 2 cm to decrease the mortality risk due to COVID-19 infection compared to cancer recurrence

(Scarci & Raveglia, 2020). Moreover, a sublobar resection can be considered for patients with decreased preoperative functional respiratory reserve, in which the limit between the indication of lobectomy vs. sublobar resection may lean more often toward the sublobar resection (Muñoz-Largacha & Wei, 2020). During the procedure, it is essential to minimize the risk of infection to the medical staff and to use a minimally invasive approach such as video-assisted thoracic surgery or robotic-assisted thoracic surgery along with the correct respiratory mask use by everyone regardless of the COVID-19 status of the patient, thus minimizing intraoperative exposure to aerosols (Seco et al., 2020). Moreover, to further decrease the spread of COVID-19, especially during the pandemic, enhanced recovery after surgery (ERAS) assumes a more critical role in the postoperative setting. To date, better postoperative management can reduce the complications and hospital stay and increase resources for COVID-19 patients. Not only can the ICU and ward protocols reduce the hospital stay, but the minimally invasive approach can also do this, reducing postoperative complications and pain (Falcoz et al., 2016).

Furthermore, the ERAS society also recommended smoking cessation, prehabilitation, preoperative carbohydrate loading, intra- and postoperative (regional) anesthesia, opioid-sparing analgesia, and early postoperative mobilization to reduce hospital stay and thus the spread of infection (Mina et al., 2020). COVID-19 may infect patients before they undergo thoracic surgeries, even if they tested negative at admission, due to the low viral RNA counts during the incubation period where the infection could be asymptomatic (Mina et al., 2020). Thus, patients admitted to the ward and undergoing surgery can easily transmit the infection to medical staff and other patients. With a positive preoperative test for COVID-19, the operation should be postponed for at least 2–3 weeks, and then the test should be repeated (Scarci & Raveglia, 2020). Also, at our institution, we agree with the same management; during the first wave, we tested patients 48 h before surgery and again in case of any symptoms suspicious for SARS-CoV-2. To date and after the first wave, the COVID-19 swab tests were performed no more than 72 h after hospitalization for elective surgery. COVID-19 symptoms such as shortness of breath, cough, fever, and fatigue are very similar to the postoperative symptoms after lung resections, thus requiring close attention by all the medical staff and, if in doubt, the COVID-19 test should be repeated (Peng et al., 2020). Compared to pneumonia in postoperative patients, COVID-19 pneumonia lymphopenia is more commonly seen and may help in the differential diagnosis. It is essential to consider COVID-19, especially in heavily affected regions with many cases (Antonoff et al., 2020; Chan et al., 2020).

Treatment delay and consequences

Lung cancer has always been a disease that needs treatment immediately. Tsai et al. reported the effects and the better survival rates of minimizing the interval from diagnosis to treatment, especially in early stage NSCLC. The authors noted that initiation of treatment 7 days after diagnosis in early stage NSCLC was associated with a 10% increase in the 5-year survival rate (Tsai et al., 2020). Usually, in non-COVID-19 times, performing a complete imaging and staging in 7 days is very hard to imagine in every hospital, even more when resources are limited due to the pandemic. Based on the US National Database, Merritt et al. stated that a treatment delay of more than 8 weeks is an independent risk factor for disease progression. Thus surgical procedures should not be postponed for more than 2 months (Merritt & Kneuertz, 2020). On the other hand, Passaro et al. stated that surgical procedures for NSCLC should not be postponed for more than 6–8 weeks, according to the European Society for Medical Oncology (ESMO) (Passaro et al., 2020). The negative effect on lung cancer patient management started with the first wave of the pandemic in 2020. In Europe and North America, a decrease in NSCLC diagnoses was reported by several authors such as Dingemans et al., Seitlinger et al., and Werner et al. (Dingemans et al., 2020; Seitlinger et al., 2020; Werner et al., 2022). Moreover, misdiagnosis between lung cancer and COVID-19 for similar symptoms as well as a reduction in screening appointments due to the pandemic can complicate the early detection of lung cancer (Gourd, 2020). In several countries, a decrease in the early detection of lung cancer and an increase in advanced stages during the pandemic have been reported (Bertolaccini et al., 2020; Gourd, 2020; Park et al., 2020). A delayed lung cancer diagnosis decreases survival rates based on histology and volume doubling time (Goldstraw et al., 2016). It is imperative as far as possible, even during the pandemic, to maintain surgical activity and avoid treatment delays to prevent survival rate reduction.

International experiences

Werner et al. reported 50 NSCLC patients who underwent surgical resections in Switzerland and developed COVID-19 symptoms in the postoperative course. A total of 6 patients (12%) experienced COVID-19-like symptoms, including fever, coughing, or shortness of breath. COVID-19 tests were performed on 25 patients, of whom 19 were asymptomatic and none were positive (Werner et al., 2022). Hoyos Mejia et al. reported that out of 101 surgical procedures, including 57 primary oncological resections, 6 lung transplants, and 18 emergency procedures, there were only 5 cases of COVID-19, 3 in the immediate postoperative period and two as outpatients. Although 80% of patients with positive test results for COVID-19 required in-hospital care, none were considered severe or critical, and none died. The authors concluded that a surgical procedure could be performed with a relatively low risk for the patient if, in selected cases, managed by a dedicated multidisciplinary team (Mejía et al., 2020). Dziodzio et al. analyzed and reviewed COVID-19 management recommendations from 14 thoracic societies. The consensus recommended suspending elective procedures or procedures for benign conditions temporarily during the pandemic and prioritizing patients with symptomatic or advanced cancer. Most societies recommend screening all patients for COVID-19 prior to admission. In a few cases, serology tests and CT scans were recommended for the diagnosis of infection. In a confirmed COVID-19 infection, elective surgeries should be suspended and rescheduled after reevaluation. This study also summarized recommendations on operating room precautions and the management of chest drains (Dziodzio et al., 2021). In Greece, Tomos et al. reported their experience with NSCLC patients undergoing surgical resection during the first wave of COVID-19 compared with patients treated before the pandemic. A total of 61 patients were described in the study, 28 (median age 67 years, SD: 7.1) during the pandemic and 33 (median age 67.1 years, SD: 7.5) 1 year earlier (Tomos et al., 2023). A statistically significant more prolonged period of waiting for treatment and an increase in tumor size were observed during the pandemic compared to before, with a median time of 47 days vs. a median time of 18 days (Tomos et al., 2023). No significant increase in lung cancer stage between the subgroups was found (Tomos et al., 2023).

Moreover, the Asian Society for Cardiovascular and Thoracic Surgery published a consensus statement from nine hospitals regarding thoracic surgery during the pandemic. Lung cancer surgeries should not be stopped and should be considered based on disease stage, the availability of nonsurgical treatment options, or patient condition. Any aerosol-forming procedures should be avoided unless extremely necessary. The authors also stated that lung cancer patients should be treated only in hospitals with isolation wards for patients with COVID-19 (Jheon et al., 2020).

Conclusion

Lung cancer patients are at an increased risk of COVID-19 infection, which can quickly lead to severe complications compared to nononcological patients. Unfortunately, especially during the first wave, COVID-19 had a very negative effect, significantly delaying lung cancer treatment. Considering the risks and benefits of treating lung cancer when it is still in a resectable stage, every hospital and department are responsible for taking every measure to protect all the medical staff and other patients to limit the risk of the spread of infection. Moreover, based on the studies reported in this review and our experience, surgical resections can be safely performed while a minimally invasive approach and the ERAS protocol can help with early discharge and a decreased risk of contracting COVID-19. Thoracic oncology activity is safe and feasible.

Policies and procedures

The COVID-19 pandemic has impacted every aspect of healthcare, including surgical management. As a result, new guidelines were established to reduce the risk of transmission and ensure patient and healthcare worker safety. The following policies and procedures outline the guidelines for surgical management in the COVID-19 pandemic.

Policy

This healthcare facility's policy was to follow the guidelines on surgical management in the COVID-19 pandemic to minimize the risk of transmission and ensure the safety of all patients and healthcare workers.

Procedures

Preoperative screening

All patients scheduled for surgery must be screened for COVID-19 before their procedure. This screening will include a medical history, a physical examination, and a COVID-19 test. The test should be performed within 72 h of the scheduled surgery, and patients who test positive should have their surgery delayed until they have recovered.

Personal protective equipment (PPE)

All healthcare workers involved in surgical procedures must wear appropriate PPE. This includes an N95 respirator or a surgical mask, a face shield or goggles, a gown, and gloves. The PPE must be worn throughout the surgical procedure and disposed of appropriately afterward.

Operating room set-up

The operating room must be set up to minimize the risk of COVID-19 transmission. This includes appropriate ventilation to ensure adequate air exchange and negative pressure rooms for patients known or suspected to have COVID-19. All surfaces and equipment must be thoroughly disinfected before and after each procedure.

Surgical procedure

All healthcare workers must follow standard infection prevention and control measures during the surgical procedure. This includes hand hygiene, using sterile techniques, and minimizing the number of personnel in the operating room. Procedures that generate aerosols, such as intubation or extubation, should be cautiously performed.

Postoperative care

Patients undergoing surgery must be monitored for signs and symptoms of COVID-19 during their hospital stay. They should be placed in appropriate isolation if they develop symptoms or test positive for COVID-19. Discharge planning should also include appropriate guidance on self-isolation and monitoring for symptoms.

Conclusion

These policies and procedures provide guidelines on surgical management in the COVID-19 pandemic to help reduce the risk of transmission and ensure the safety of all patients and healthcare workers. Adherence to these guidelines will help mitigate the impact of COVID-19 on surgical procedures and improve patient outcomes.

Applications to other areas

The guidelines on surgical management in the COVID-19 pandemic can be applied to other areas of healthcare to help reduce the risk of transmission and ensure the safety of all patients and healthcare workers. Here are some examples:

1. Outpatient clinics:

   Outpatient clinics can apply these guidelines by implementing prescreening measures for all patients, including temperature checks and screening for COVID-19 symptoms. They can also enforce social distancing measures in waiting rooms and require all patients and staff to wear appropriate PPE. The clinic should also be thoroughly disinfected after each patient.

2. Emergency departments:

   Emergency departments can follow these guidelines by implementing prescreening measures for all patients, including temperature checks and screening for COVID-19 symptoms. They can also use negative pressure rooms for patients known or suspected to have COVID-19 and ensure that all staff wears appropriate PPE. The department should also be thoroughly disinfected after each patient.

3. Long-term care facilities:

   Long-term care facilities can apply these guidelines by implementing prescreening measures for all staff and residents, including temperature checks and screening for COVID-19 symptoms. They can also require all staff to wear appropriate PPE and enforce social distancing measures in common areas. The facility should also be thoroughly disinfected regularly.

4. Dental offices:

   Dental offices can follow these guidelines by implementing prescreening measures for all patients, including temperature checks and screening for COVID-19 symptoms. They can also require all staff and patients to wear appropriate PPE and use high-volume evacuation systems to reduce the spread of aerosols. The office should also be thoroughly disinfected after each patient.

5. Laboratories:

   Laboratories can apply these guidelines by implementing prescreening measures for all staff and visitors, including temperature checks and screening for COVID-19 symptoms. They can also enforce social distancing measures in common areas and require all staff to wear appropriate PPE. The laboratory should also be thoroughly disinfected regularly.

Conclusion

The guidelines on surgical management in the COVID-19 pandemic can be applied to other areas of healthcare to help reduce the risk of transmission and ensure the safety of all patients and healthcare workers. By following these guidelines, healthcare facilities can help mitigate the impact of COVID-19 on their operations and improve patient outcomes.

Mini-dictionary of terms

- Aerosols—Tiny particles that can be suspended in the air and spread the COVID-19 virus; they are generated by specific medical procedures.
- Biohazard—A biological agent threatening living organisms' health, including viruses such as COVID-19.
- Disinfection—The process of killing or removing harmful microorganisms, including the SARS-CoV-2 virus, from surfaces and equipment in medical facilities.
- Endotracheal intubation—A medical procedure that involves the insertion of a tube through the mouth and into the trachea to help a patient breathe, which can generate COVID-19 aerosols.
- Face shield—A protective device worn over the face to prevent exposure to infectious droplets and aerosols during surgical procedures.
- Gown—A protective garment worn over clothing to prevent contamination during surgical procedures.
- Hand hygiene—The practice of cleaning hands with soap and water or an alcohol-based hand sanitizer to prevent the spread of infection, including COVID-19.
- Isolation—The separation of a patient with a known or suspected COVID-19 infection from others to prevent the spread of the virus.
- Joint Commission—An independent organization that accredits and certifies healthcare organizations and programs in the United States, which issued guidelines on surgical management during the COVID-19 pandemic.
- KN95 respirator—A mask that filters out at least 95% of airborne particles, including the SARS-CoV-2 virus.
- Negative pressure room—A hospital room designed to prevent the spread of airborne infections, including COVID-19, by using air pressure to draw contaminated air out of the room and prevent it from spreading.

Summary points

- Lung cancer patients are at an increased risk of COVID-19 infection that can quickly develop severe complications compared to nononcological patients.
- During the first wave, COVID-19 had a very negative effect, with significant delays in lung cancer treatment.
- Surgical resections can be safely performed.
- A minimally invasive approach and ERAS protocol can help for an early discharge and decrease the risk of contracting COVID-19, and thoracic oncology activity is safe and feasible.

References

Antonoff, M., Backhus, L., Boffa, D. J., Broderick, S. R., Brown, L. M., Carrott, P., et al. (2020). Thoracic surgery outcomes research network, Inc. COVID-19 guidance for triage of operations for thoracic malignancies: A consensus statement from thoracic surgery outcomes research network. *The Journal of Thoracic and Cardiovascular Surgery*, *160*(2), 601–605. https://doi.org/10.1016/j.jtcvs.2020.03.061. PubMed: 1097-685X.

Banna, G., Curioni-Fontecedro, A., Friedlaender, A., & Addeo, A. (2020). How we treat patients with lung cancer during the SARS-CoV-2 pandemic: primum non nocere. *ESMO Open*, *5*(2), e000765. PubMed: 2059-7029.

Bertolaccini, L., Mohamed, S., Bardoni, C., Lo Iacono, G., Mazzella, A., Guarize, J., & Spaggiari, L. (2022). The interdisciplinary management of lung cancer in the European Community. *Journal of Clinical Medicine*. https://doi.org/10.3390/jcm11154326.

Bertolaccini, L., Sedda, G., & Spaggiari, L. (2020). Paying another tribute to the COVID-19 pandemic: The decrease of early lung cancers. *The Annals of Thoracic Surgery*, *111*, 745–746. PubMed: 0003-4975.

Cai, Y., Hao, Z., Gao, Y., Ping, W., Wang, Q., Peng, S., et al. (2020). Coronavirus disease 2019 in the perioperative period of lung resection: A brief report from a Single Thoracic Surgery Department in Wuhan, People's Republic of China. *Journal of Thoracic Oncology*, *15*(6), 1065–1072. https://doi.org/10.1016/j.jtho.2020.04.003. PubMed: 1556-1380.

Chan, J. F., Yuan, S., Kok, K. H., To, K. K., Chu, H., Yang, J., et al. (2020). A familial cluster of pneumonia associated with the 2019 novel coronavirus indicating person-to-person transmission: A study of a family cluster. *Lancet*, *395*(10223), 514–523. https://doi.org/10.1016/S0140-6736(20)30154-9. PubMed: 1474-547X.

Dingemans, A. C., Soo, R. A., Jazieh, A. R., Rice, S. J., Kim, Y. T., Teo, L. L., et al. (2020). Treatment guidance for patients with lung cancer during the coronavirus 2019 pandemic. *Journal of Thoracic Oncology*, *15*(7), 1119–1136. https://doi.org/10.1016/j.jtho.2020.05.001. PubMed: 1556-1380.

Dziodzio, T., Knitter, S., Wu, H. H., Ritschl, P. V., Hillebrandt, K.-H., Jara, M., Juraszek, A., Öllinger, R., Pratschke, J., Rückert, J., et al. (2021). Thoracic surgery in the COVID-19 pandemic: A novel approach to reach guideline consensus. *Journal of Clinical Medicine*, *10*, 2769. https://doi.org/10.3390/jcm10132769.

Falcoz, P. E., Puyraveau, M., Thomas, P. A., Decaluwe, H., Hürtgen, M., Petersen, R. H., et al. (2016). Video-assisted thoracoscopic surgery versus open lobectomy for primary non-small-cell lung cancer: A propensity-matched analysis of outcome from the European Society of Thoracic Surgeon database. *European Journal of Cardio-Thoracic Surgery*, *49*(2), 602–609. https://doi.org/10.1093/ejcts/ezv154. PubMed: 1873-734X.

Garassino, M. C., Whisenant, J. G., Huang, L. C., Trama, A., Torri, V., Agustoni, F., et al. (2020). TERAVOLT investigators. COVID-19 in patients with thoracic malignancies (TERAVOLT): First results of an international, registry-based, cohort study. *The Lancet Oncology*, *21*(7), 914–922. https://doi.org/10.1016/S1470-2045(20)30314-4. PubMed: 1474-5488.

Goldstraw, P., Chansky, K., Crowley, J., Rami-Porta, R., Asamura, H., Eberhardt, W. E., et al. (2016). The IASLC lung cancer staging project: Proposals for revision of the TNM stage groupings in the forthcoming (eighth) edition of the TNM classification for lung cancer. *Journal of Thoracic Oncology*, *11*(1), 39–51. https://doi.org/10.1016/j.jtho.2015.09.009. PubMed: 1556-1380.

Gourd, E. (2020). Lung cancer control in the UK hit badly by COVID-19 pandemic. *The Lancet Oncology*, *21*(12), 1559. https://doi.org/10.1016/S1470-2045(20)30691-4. PubMed: 1474-5488.

Hanna, T. P., King, W. D., Thibodeau, S., Jalink, M., Paulin, G. A., Harvey-Jones, E., et al. (2020). Mortality due to cancer treatment delay: Systematic review and meta-analysis. *BMJ*, *371*, m4087. https://doi.org/10.1136/bmj.m4087. PubMed: 1756-1833.

Huang, J., Wang, A., Kang, G., Li, D., & Hu, W. (2020). Clinical course of patients infected with severe acute respiratory syndrome coronavirus 2 soon after thoracoscopic lung surgery. *The Journal of Thoracic and Cardiovascular Surgery*, *160*(2), e91–e93. https://doi.org/10.1016/j.jtcvs.2020.04.026. PubMed: 1097-685X.

Jheon, S., Ahmed, A. D. B., Fang, V. W. T., Jung, W., Khan, A. Z., Lee, J.-M., Sihoe, A. D. L., Thongcharoen, P., Tsuboi, M., Turna, A., & Nakajima, J. (2020). Thoracic cancer surgery during the COVID-19 pandemic: A consensus statement from the Thoracic Domain of the Asian Society for Cardiovascular and Thoracic Surgery. *Asian Cardiovascular & Thoracic Annals*, *28*(6), 322–329.

Mejía, L. H., Román, A. R., Barturen, M. G., del Mar Córdoba Pelaez, M., de la Cruz, J. L. C.-C., Naranjo, J. M., et al. (2020). Thoracic surgery during the coronavirus disease 2019 (COVID-19) pandemic in Madrid, Spain: Single-centre report. *European Journal of Cardio-Thoracic Surgery*, *58*, 991–996.

Merritt, R. E., & Kneuertz, P. J. (2020). Considerations for the surgical management of early stage lung cancer during the COVID-19 pandemic. *Clinical Lung Cancer*, *22*, 156–160. PubMed: 1525-7304.

Mina, M. J., Parker, R., & Larremore, D. B. (2020). Rethinking Covid-19 test sensitivity—A strategy for containment. *The New England Journal of Medicine*, *383*(22), e120. https://doi.org/10.1056/NEJMp2025631. PubMed: 1533-4406.

Muñoz-Largacha, J. A., & Wei, B. (2020). Commentary: Lung surgery in the time of COVID-19. *The Journal of Thoracic and Cardiovascular Surgery*, *160*(2), e97–e98. https://doi.org/10.1016/j.jtcvs.2020.04.088. PubMed: 1097-685X.

Onder, G., Rezza, G., & Brusaferro, S. (2020). Case-fatality rate and characteristics of patients dying in relation to COVID-19 in Italy. *Journal of the American Medical Association*, *323*(18), 1775–1776. https://doi.org/10.1001/jama.2020.4683. PubMed: 1538-3598.

Park, J. Y., Lee, Y. J., Kim, T., Lee, C. Y., Kim, H. I., Kim, J. H., et al. (2020). Collateral effects of the coronavirus disease 2019 pandemic on lung cancer diagnosis in Korea. *BMC Cancer*, *20*(1), 1040. https://doi.org/10.1186/s12885-020-07544-3. PubMed: 1471-2407.

Passaro, A., Addeo, A., Von Garnier, C., Blackhall, F., Planchard, D., Felip, E., et al. (2020). ESMO management and treatment adapted recommendations in the COVID-19 era: Lung cancer. *ESMO Open*, *5*(Suppl 3), 5. https://doi.org/10.1136/esmoopen-2020-000820. PubMed: 2059-7029.

Peng, S., Huang, L., Zhao, B., Zhou, S., Braithwaite, I., Zhang, N., et al. (2020). Clinical course of coronavirus disease 2019 in 11 patients after thoracic surgery and challenges in diagnosis. *The Journal of Thoracic and Cardiovascular Surgery*, *160*(2), 585–592.e2. https://doi.org/10.1016/j.jtcvs.2020.04.005. PubMed: 1097-685X.

Russano, M., Citarella, F., Vincenzi, B., Tonini, G., & Santini, D. (2020). Coronavirus disease 2019 or lung cancer: What should we treat? *Journal of Thoracic Oncology*, *15*(7), e105–e106. https://doi.org/10.1016/j.jtho.2020.04.001. PubMed: 1556-1380.

Scarci, M., & Raveglia, F. (2020). Commentary: The double responsibility of the thoracic surgeon at the time of the pandemic: A perspective from the North of Italy. *The Journal of Thoracic and Cardiovascular Surgery*, *160*(2), 595–596. https://doi.org/10.1016/j.jtcvs.2020.04.003. PubMed: 1097-685X.

Seco, M., Wood, J., & Wilson, M. K. (2020). COVIDSafe thoracic surgery: Minimizing intraoperative exposure to aerosols. *JTCVS Techniques*, *3*, 412–414. https://doi.org/10.1016/j.xjtc.2020.05.017. PubMed: 2666-2507.

Seitlinger, J., Wollbrett, C., Mazzella, A., Schmid, S., Guerrera, F., Nkomo, D. B., et al. (2020). Safety and feasibility of thoracic malignancy surgery during the COVID-19 pandemic. *The Annals of Thoracic Surgery*, *112*, 1870–1876. PubMed: 0003-4975.

Tomos, I., Kapetanakis, E. I., Dimakopoulou, K., Raptakis, T., Kampoli, K., Karakatsani, A., Koumarianou, A., Papiris, S., & Tomos, P. (2023). The impact of COVID-19 pandemic on surgical treatment of resectable non-small cell lung cancer in Greece. *Life*, *13*, 218. https://doi.org/10.3390/life13010218.

Tsai, C. H., Kung, P. T., Kuo, W. Y., & Tsai, W. C. (2020). Effect of time interval from diagnosis to treatment for non-small cell lung cancer on survival: A national cohort study in Taiwan. *BMJ Open*, *10*(4), e034351. https://doi.org/10.1136/bmjopen-2019-034351. PubMed: 2044-6055.

Werner, R. S., Lörtscher, A., Kirschner, M. B., Lauk, O., Furrer, K., Caviezel, C., Schneiter, D., Inci, I., Hillinger, S., Curioni-Fontecedro, A., & Opitz, I. (2022). Surgical management of lung cancer during the COVID-19 pandemic—A narrative review and single-centre report. *Swiss Medical Weekly*, *152*, w30109.

World Health Organization. (2020). *Coronavirus disease*. https://www.Who.int/emergencies/diseases/novel-coronavirus-2019 (21 February 2023, date last accessed).

Zhang, L., Zhu, F., Xie, L., Wang, C., Wang, J., Chen, R., et al. (2020). Clinical characteristics of COVID-19-infected cancer patients: A retrospective case study in three hospitals within Wuhan, China. *Annals of Oncology*, *31*(7), 894–901. https://doi.org/10.1016/j.annonc.2020.03.296. PubMed: 1569-8041.

Zhu, N., Zhang, D., Wang, W., Li, X., Yang, B., Song, J., et al. (2020). China novel coronavirus investigating and research team. A novel coronavirus from patients with pneumonia in China, 2019. *The New England Journal of Medicine*, *382*(8), 727–733. https://doi.org/10.1056/NEJMoa2001017. PubMed: 1533-4406.

Chapter 20

Multisystem inflammatory syndrome in children/pediatric inflammatory multisystem syndrome: Clinical guidelines

Arthur J. Chang[a], Ramesh Kordi[b,c], and Mark D. Hicar[b,c]

[a]*Department of Pediatrics, University of Nebraska Medical Center, Omaha, NE, United States,* [b]*Department of Pediatrics, Jacobs School of Medicine and Biomedical Sciences, University at Buffalo, State University of New York, Buffalo, NY, United States,* [c]*John R. Oishei Children's Hospital, Buffalo, NY, United States*

Abbreviations

ACR	American College of Rheumatology
ASO	anti-streptolysin O
BNP	brain-natriuretic peptide
CDC	Centers for Disease Control and Prevention
COVID-19	coronavirus disease 2019
CRP	C-reactive protein
ECMO	extracorporeal membrane oxygenation
EKG	electrocardiogram
GI	gastrointestinal
HLH	hemophagocytic lymphohistiocytosis
IVIG	intravenous immunoglobulin G
KD	Kawasaki disease
LDH	lactate dehydrogenase
LFTs	liver function tests
MIS-C	multisystem inflammatory syndrome in children
PIMS	pediatric inflammatory multisystem syndrome
PCR	polymerase chain reaction
PT	prothrombin time
PTT	partial thromboplastin time
TE	thromboembolic events
TNF	tumor necrosis factor
TSS	toxic shock syndrome

Introduction

Shortly after the COVID-19 pandemic spread throughout Europe, a pediatric postinfectious inflammatory condition was described worldwide. It was referred to as pediatric inflammatory multisystem syndrome (PIMS) in European publications and multisystem inflammatory syndrome in children (MIS-C) by the US Centers for Disease Control and Prevention (CDC) as well as throughout this document (Fig. 1). All case definitions involved either proven or presumed prior SARS-CoV-2 infection with multisystem inflammation being present without other probable explanation (Whittaker et al., 2020). The case definitions are broad enough that children with Kawasaki disease (KD), toxic shock syndrome, and viral myocarditis likely were initially diagnosed with MIS-C.

Management, Body Systems, and Case Studies in COVID-19. https://doi.org/10.1016/B978-0-443-18703-2.00052-9
Copyright © 2024 Elsevier Inc. All rights are reserved, including those for text and data mining, AI training, and similar technologies.

MIS-C CDC Case Definition

Fever (>24 hours reported or documented ≥38.0°C)
AND laboratory evidence of inflammation*
AND Illness requiring hospitalization
AND Multisystem (≥2) organ involvement
(*e.g.* cardiac, renal, respiratory, gi, heme, derm or neuro)
AND No alternative plausible diagnoses
AND COVID 19 positivity/exposure
(*i.e.* SARS-CoV-2 RT-PCR positive currently
or recently positive on Antibody testing
or COVID-19 exposure within the 4 weeks prior to the onset of symptoms)

*Laboratory evidence supportive but non-specific of MIS-C:
Lymphopenia **or** albuminemia
or elevation of any of the following: Neutrophil count, C-reactive protein (CRP), erythrocyte sedimentation rate (ESR), fibrinogen, procalcitonin, d-dimer, ferritin, lactic acid dehydrogenase (LDH), and interleukin 6 (IL-6)).

FIG. 1 MIS-C CDC case definition as of 2022 (CDC-MIS-C, 2022; Godfred-Cato et al., 2020).

Epidemiology and case definitions

After initial reports from Europe (Belhadjer et al., 2020; Riphagen et al., 2020; Verdoni et al., 2020), other countries reported on the global nature of this condition (Lima-Setta et al., 2020; Mamishi et al., 2020; Okarska-Napierala et al., 2020; Riollano-Cruz et al., 2021; Torres et al., 2020). In contrast with expectations from KD, Asian countries reported a significantly lower incidence of MIS-C (Choe et al., 2021; Li et al., 2020; Mohri et al., 2022), with unpublished reports of low hundreds of cases from Taiwan being an exception. In the United States, the cumulative MIS-C incidence in persons younger than 21 is 2 per 100,000, with approximately 1 case per 4000 following acute infections by an average of 2–5 weeks (Belay et al., 2021). As of August 1, 2022, the CDC reported 8798 MIS-C cases with a near 1% mortality rate (77 deaths) and 98% with positive testing (by nucleic acid, antigen, and/or antibody) for SARS-CoV-2 (CDC-MIS-C). The patients have a median age of 9 years and are predominantly 5–13 years. Of reported cases, 57% occurred in children who are Hispanic/Latino or non-Hispanic Black and 61% of cases are male. These data are similar to reports of age groups and ethnicities from other studies with lower socio-economic status also a risk factor (Goyal et al., 2020; Javalkar et al., 2021). A similar syndrome in young adults is termed MIS-A (Adult) (Morris et al., 2020). A report of a twin boy with MIS-C with no issues occurring in his identical twin after both were acutely infected implies that genetics alone are not predictive (Chang et al., 2022).

Generally, acute COVID-19 has distinct clinical signs and symptoms as well as laboratory findings (Reiff et al., 2021), but some acute respiratory COVID-19 presentations may fulfill MIS-C criteria (Verdoni et al., 2020). Interestingly, younger children with MIS-C are more likely than adolescents to be SARS-CoV-2 IgG positive (86% for cases 0–9 years vs. 70% for 15–20 years), whereas the opposite is seen for polymerase chain reaction (PCR) positivity (48% vs. 64%, respectively) (Belay et al., 2021). Strain variation affects the incidence of MIS-C. In an Israeli study, omicron-associated MIS-C was more than 10-fold less than the preceding alpha (14.3-fold less) and delta (12.9-fold less) waves (Levy et al., 2022). These data are similar to those reported to the CDC (Fig. 2) (CDC-MIS-C).

Clinical presentation

As the incidences of cases are falling with new strains, this diagnosis may become very sporadic and will be a continued challenge. Fever should be reported or confirmed, as a lack of fever would rule out MIS-C. MIS-C should be considered for septic febrile children presenting for care. MIS-C can present with multiorgan system failure with cardiogenic or vasoplegic shock (Godfred-Cato et al., 2020), and cardiac dysfunction is the most concerning and striking feature of this disorder. Patients in whom there is an initial low index of suspicion but who present with some features of MIS-C should be considered for screening for inflammation with a complete blood count (CBC) with differential and C-reactive protein (CRP), as a normal CBC and CRP virtually eliminate this diagnosis. Patients with a high index of suspicion should have broader investigations (see Table 1).

MIS-C frequently presents with prominent gastrointestinal (GI) signs and symptoms (70%–90% of reported cases) (Belhadjer et al., 2020; Lima-Setta et al., 2020; Mamishi et al., 2020; Okarska-Napierala et al., 2020; Riollano-Cruz

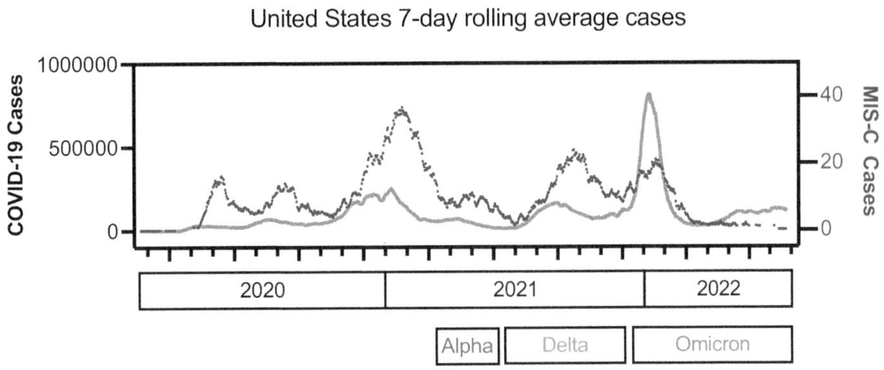

FIG. 2 Variation in MIS-C incidence is influenced by circulating strain. Publicly available data from Centers of Disease Control were used (CDC-MIS-C). *Solid line* represents rolling average COVID-19 cases with *purple dots* representing reported rolling average MIS-C cases. *Ticks* represent each month and major US circulating strains are shown underneath. Higher incidence of MIS-C with lower COVID-19 incidence early in 2020 likely related to lack of widespread testing for acute COVID-19.

TABLE 1 Work-up for cases of MIS-C.

Common labs for initial approach	Other labs to consider depending on lab results and presentation
[a]CBC with differential [a]CRP CMP (albumin and LFTs) ESR COVID-19 PCR *Limited workup* Chest x-ray Echocardiogram EKG COVID-19 PCR COVID-19 IgG (unless known +) ESR CBC with differential CMP (albumin and LFTs) CRP Ferritin D-dimer Troponin BNP or NT-proBNP LDH Procalcitonin Blood cultures	VBG Lactate Creatine kinase PT/PTT Blood type and screen Antithrombin III Von Willebrand panel Fibrinogen Amylase/lipase Abdominal x-ray/CT/ultrasound Complement levels, ANA Triglycerides Soluble interleukin 2 receptor ADAMTS13 and haptoglobin Stool cultures Throat culture/rapid strep Antistreptolysin O (ASO) Specific viral titers Respiratory viral panel Urinalysis with microscopic Urine culture Immunoglobulins Cytokine panel

Based on prior publications (Henderson et al., 2021; Hennon et al., 2021).
[a]*Minimal screening labs.*

et al., 2021; Riphagen et al., 2020; Torres et al., 2020; Verdoni et al., 2020) including abdominal pain, diarrhea, nausea, vomiting, constipation, GI bleeding (Sahn et al., 2021), and pseudoappendicitis (Lishman et al., 2020). Respiratory complaints can be mild or absent (Belay et al., 2021; Dufort et al., 2020); however, many present with diffuse infiltrates on a chest x-ray, which may or may not be correlative of symptomatic oxygen requirement or PCR positivity, particularly in cases presenting early after acute infection. The CDC definition of multisystem involvement includes $\geq$2 organ systems (Fig. 1) (CDC-MIS-C). The level of dysfunction that qualifies for a system to be included for MIS-C has not been specifically defined (Henderson et al., 2021), but reports are consistent with organ dysfunction defined in guidelines for severe COVID-19 (Feldstein et al., 2021). Neurologic manifestations, including encephalitis, meningitis, necrotizing encephalopathy, demyelinating disorders, intracerebral hemorrhage, and stroke, have been described (Verrotti et al., 2021). Although initially described as KD-like, there are distinguishing features of these presentations and fulfilling complete KD criteria is relatively rare for MIS-C patients (Table 2).

TABLE 2 Comparison of demographic and clinical characteristics of MIS-C to common pediatric inflammatory disorders.

Variable	Diagnosis			
	COVID-19	**MIS-C**	**Kawasaki**	**TSS**
Demographic				
SEX Predominance (M/F ratio)	No predominance	Male ~ 1.2:1	Male ~1.6:1	Female ~1:1.4
Age, year; median (IQR)	15 (3–17)	9 (5–13)	3 (2–5)	13 (9–16)
Ethnicity (Highest incidence)	–	African Hispanic	East Asia	–
Clinical manifestations				
Pneumonia	Common	Rare	Rare	Intermediate
Cough, shortness of breath	Very common	Intermediate	Intermediate	Intermediate
Skin rash	Rare	Common	Very common	Very common
Oral mucosal changes	Rare	Uncommon	Very common	Common
Conjunctival injection	Rare	Common	Very common	Intermediate
Neurological symptoms (headache, neck pain)	Uncommon	Intermediate	Rare	Intermediate
Diarrhea	Intermediate	Common	Intermediate	Intermediate
Vomiting	Intermediate	Common	Intermediate	Common
Decreased cardiac function (EF<55%)	Rare	Intermediate	Rare	Rare
Mitral regurgitation	Rare	Intermediate	Rare	Rare
Myocarditis	Rare	Intermediate	Rare	Rare
Coronary artery aneurism	Rare	Uncommon	Intermediate	Rare
Acute kidney injury	Uncommon	Rare	Rare	Common
Cervical adenopathy	Rare	Rare	Intermediate	Rare
Shock (requiring fluids, vasoactives)	Rare	Intermediate	Rare	Very common
Hypotension	Uncommon	Common	Rare	Very common

Very common: > 80%; common: 50%–80%; intermediate: 30%–50%; uncommon: 15%–30%; rare: <15%.
Summarized from Feldstein et al. (2021), Godfred-Cato et al. (2022), Lee et al. (2021), and Sharma et al. (2021).

Cardiac manifestations

Complete transthoracic echocardiograms should be performed urgently in all patients with clinical or laboratory evidence of cardiac injury and/or shock. Initial published reports of COVID-19-associated MIS-C cases described predominantly ventricular dysfunction with some coronary artery changes, atrioventricular valve regurgitation, and pericardial effusions (Belhadjer et al., 2020; Riphagen et al., 2020; Verdoni et al., 2020; Whittaker et al., 2020). Recent studies suggest that persistent aneurysms are rare in MIS-C (Reiff et al., 2021). When coronary arteries are noted to be enlarged during MIS-C, the findings correspond to the peak of fever. In contrast, the aneurysms of KD usually are most prominent 2–3 weeks after the onset of fevers, generally after a fever has resolved (McCrindle et al., 2017).

Coagulopathy

Children with MIS-C are at risk for thromboembolic events (TEs), including deep venous thrombosis, pulmonary embolism, thrombotic microangiopathy, and arterial events such as stroke (Sharathkumar et al., 2021; Whitworth et al., 2021). MIS-C cases exhibited significantly higher levels of D-dimer in comparison to acute COVID-19 cases (median; 0.36 mg/L vs. 2.07 mg/L, respectively, *P*-value <0.001) (Godfred-Cato et al., 2022). In a multicenter retrospective study, TEs were found in 6.5% of patients with MIS-C, as compared to 2.1% and 0.7% in those with symptomatic and asymptomatic COVID-19, respectively. More than two-thirds of TEs occurred in MIS-C despite receiving thromboprophylaxis, especially in those with certain risk factors (Table 3) (Whitworth et al., 2021).

TABLE 3 Risk factors favoring use of prophylactic anticoagulation.

Central line	Cystic fibrosis exacerbation
Mechanical ventilation	Prolonged hospital stay >3 days
Obesity	Thrombophilia
Immobility	Congenital/acquired cardiac disease with impaired venous return
Active malignancy	Personal or first-degree relative with a history of venous thromboembolism
Sickle cell exacerbation	Inflammatory diseases such as lupus or juvenile idiopathic arthritis
Nephrotic syndrome	Inflammatory bowel disease with exacerbation

Summarized from expert consensus guidelines (Goldenberg et al., 2020).

Addressing other etiologies in the differential

Some guidelines recommend many up-front laboratory evaluations (Henderson et al., 2021; Hennon et al., 2021). Currently, there is no specific test or scoring system that confirms the diagnosis of MIS-C. For ill-appearing children being hospitalized with MIS-C high on the differential, a chest x-ray, an electrocardiam (ECG), and an echocardiogram should be performed. Plain abdominal x-ray and abdominal computed tomography (CT) will be helpful for acute abdomen presentations due to symptomatic overlap with inflammatory bowel disease-like disorder, appendicitis, or bowel perforation. The addition of a laboratory evaluation of pancreatitis and gallbladder disease may also be warranted. For general laboratory values, lymphopenia, thrombocytopenia, excessive CRP elevation, hyperferritinemia, elevated troponins, and elevated D-dimers correlated with MIS-C. Particularly, troponin and D-dimer elevations were more specific when compared to children with KD and KD shock syndrome (Davies et al., 2020). Thrombocytosis was more common with KD, low albumin, and relative anemia, although supplemental criteria for diagnosing KD were in fact lower in MIS-C children overall.

As sepsis is a top consideration, investigations such as blood cultures following blood gases and lactates should be considered while starting empiric antibiotics. A work-up for urosepsis is commonly performed on all septic children, but it's unlikely there will be a symptomatic overlap. A work-up for streptococcal infection is often considered, as polymorphic skin rashes and strawberry tongue combined with fever and pharyngitis in patients can be scarlet fever or MIS-C (Barut et al., 2022; Halepas et al., 2021). MIS-C and staphylococcal or streptococcal toxic shock syndrome (TSS) share multiple clinical characteristics including fever, hypotension, shock, rash, and conjunctivitis. However, cardiac dysfunction, prominent with MIS-C, rarely occurs in TSS. Additionally, prominent abdominal and respiratory symptoms are seen less frequently in TSS compared to MIS-C (Godfred-Cato et al., 2022; Vogel et al., 2021). Children with MIS-C may present predominantly with abdominal pain, but fever, nausea, vomiting, and high inflammatory markers also raise the suspicion of an acute abdominal condition such as acute appendicitis (Manz et al., 2021). Even diagnostic imaging procedures, such as ultrasound and CT, cannot clearly differentiate MIS-C and acute appendicitis because appendiceal thickening and ascites have been reported in MIS-C (Morparia et al., 2021). Moreover, encephalopathy has been reported in almost 15% of patients with MIS-C (Fink et al., 2022), and this can be confused with sepsis-induced altered mental status.

Alternative infectious pathogen testing will be warranted in many patients, guided by exposures and specific clinical findings. Skin rashes, conjunctivitis, and cardiac manifestations can occur with such common infections as mycoplasma, Epstein-Barr virus, parvovirus B 19, cytomegalovirus, adenovirus, and enteroviruses (Vogel et al., 2021). Lyme disease and rickettsia infections as well as leptospirosis may also exhibit persistent fever and multiorgan involvement (Son & Friedman, 2021). Gastroenteritis including yersiniosis, mesenteric lymphadenitis, inflammatory bowel syndrome, and typhoid might be considered as a differential diagnosis with MIS-C presenting with GI symptoms. Both typhoid fever and MIS-C may present with polymorphous skin rashes and mucosal changes that make the differential diagnosis more challenging in travelers and persons in endemic areas (Lasheen et al., 2022). Underlying immune deficiencies should be considered with HIV and immune testing performed in select patients.

With thrombocytopenia and anemia, thrombotic microangiopathies such as thrombotic thrombocytopenic purpura and hemolytic uremic syndrome should be considered. ADAMTS13, haptoglobin, and stool cultures should be considered in the appropriate presentation (George & Nester, 2014). The hyperferritinemia in the setting of systemic symptoms is also similar to hemophagocytic lymphohistiocytosis (HLH). HLH is an excessive immune response secondary to malignant or infectious stimulus (Al-Samkari & Berliner, 2018) that is characterized by persistent elevated proinflammatory cytokines, such as interferon (IFN)-gamma, IL-18, and IL-1. Continuous stimulation of immune cells such as cytotoxic T lymphocytes, natural killer cells, and macrophages leads to characteristic features of hemophagocytosis and immune-mediated organ damage. MIS-C shares many characteristics of HLH including high level of cytokines, elevated inflammatory markers such as CRP and ferritin, coagulopathy, thrombocytopenia, and immune-mediated organ damage (Nakra et al., 2020). Many MIS-C children will present with very high ferritin levels, which is a hallmark of HLH, so overlapping of diagnoses should be considered in the hyperinflamed child (Feld et al., 2020). Macrophage activation syndrome is a similar presentation related to autoimmune flares (Henderson et al., 2021). Antinuclear antibody (ANA), complement levels, and specific autoantibodies can be done if severe lupus presentations, juvenile idiopathic arthritis, or other autoimmune presentations are being considered. Notably, patients with MIS-C exhibit autoreactive B cells and a high level of autoantibodies against tissue-specific antigens in the GI tract as well as cardiovascular system, skeletal muscle, and brain tissue (Porritt et al., 2021), especially in more severe cases (Consiglio et al., 2020; Gruber et al., 2020; Porritt et al., 2021). Extensive systemic tissue damage and cell death may contribute to these autoantibody responses. A number of the more specialized tests may not be clinically relevant. Although a number of groups publish on cytokine levels, circulating interleukin levels are poor correlates in other literature, as noted in the American College of Rheumatology (ACR) guideline (Henderson et al., 2021).

With the continued evolution of the pandemic, viral genetic variants may influence the diagnostic accuracy of certain tests. Interpretation of results will depend on several factors including the type of antibody testing locally performed, country of residence, recent travel history, known exposures, and specific SARS-CoV-2 vaccine formulation if obtained.

Criteria for hospitalization

Proposing detailed criteria for hospital admission, the hospital approach, and follow-up are challenging as data are lacking in many areas and local standards vary (Fig. 3). Current epidemiologic definitions include hospitalization as a necessary part of the case definition, but there seems to be a mild form that may not need hospitalization (our experience and reported (Godfred-Cato et al., 2020)), making true incidence estimations problematic. In some relatively well-appearing children, a rapid decline in clinical status can occur, so children should be stabilized and transferred to a tertiary medical center for appropriate consult service involvement. Often, pediatric infectious disease consultation is warranted, as bacterial and viral illness may present similarly. Pediatric cardiology consultation is active if there are abnormalities in brain-natriuretic peptides (BNP), troponin levels, ECG, or the echocardiogram. As the name implies, the multisystem inflammation seen in these cases may require the involvement of a number of other subspecialties. For children sent home after initial evaluation, if MIS-C was at all a consideration, they should be provided strict instructions to seek care if they clinically worsen and they should follow up with their normal physician within days of the first evaluation.

Treatment

A number of guidelines have been developed by the ACR (Henderson et al., 2021), the Italian Rheumatologic Society (Cattalini et al., 2021), and academic centers (Hennon et al., 2021). MIS-C presentation among children admitted to the hospital falls under three distinctive categories for the purposes of treatment recommendations: (i) MIS-C with features of acute COVID-19; (ii) MIS-C with predominant KD features; and (iii) MIS-C with shock, myocarditis, or severe multisystem involvement. As we learn more about this condition, more specific treatments may be developed. However, cardiopulmonary supportive care and a broad antiinflammatory regimen are the backbones of treatment.

Empiric antibiotics

Due to the inflamed state, broad-spectrum antibiotics are often empirically begun. The addition of anaerobic coverage should be considered in cases where ileitis or colitis are on the differential. In cases of severe illness or shock, coverage should target methicillin-resistant *Staphylococcus aureus* and resistant Gram-negatives while including antibiotics that inhibit toxin production because toxic shock will likely be in the differential diagnoses. Once the MIS-C diagnosis is solidified, antibiotics are quickly deescalated.

Antivirals

The antiviral drug remdesivir should be considered for those known to be PCR-positive and/or with a presentation consistent with typical COVID-19, as acute severe COVID-19 can have a clinical overlap with MIS-C. This should be guided by a pediatric infectious disease consultant and by current treatment guidelines. In particular, remdesivir is recommended in

Clinical approach for COVID-19 Associated Multisystem Inflammatory Syndrome in Children (MIS-C)

This condition will remain a challenge as many of the symptoms and laboratory findings are not specific. This should be considered in children with fever **without alternative explanation if:**

- there is a known recent history of COVID-19 or following future local COVID-19 case clusters
- and presenting with severe COVID-19 OR shock OR systemic illness affecting two or more organs systems OR symptoms of Kawasaki Disease (KD) (rash, conjunctivitis, oral/mucosal inflammation), or significant vomiting/diarrhea/abdominal pain.

Initial Evaluation

Admit and treat as MIS-C if concerning labs and lacking a more appropriate diagnosis if:

- Hemodynamically unstable
- Stable with multi-organ system inflammation

There is a lack of clinical scoring systems, but 'concerning labs' may include: neutrophilia, lymphopenia, thrombocytopenia, hypoalbuminemia, hyponatremia, elevated inflammatory markers (see Figure 1).

Decision to not admit children who you considered, but did not meet criteria:

If discharging a case that was 'too mild' or potentially early with relatively benign labs, provide anticipatory guidance and urging to return to care if worsen.

Approach to admission

- Start IV fluids (judicious use with frequent reevaluation).
- Perform any labs not done on initial approach (ferritin, coagulation panel, etc.).
- Echocardiography early in course and repeated as needed.
- Empiric antibiotics as differential includes sepsis and bacterial gastrointestinal disease.
- Notify local health authorities tracking cases
- Consulting services: often include: Rheumatology, Infectious Diseases, Cardiology.
- Appropriate therapies: currently data supports methylprednisolone and many centers use IVIG and adjunct anakinra.
- Anticoagulation: many centers us aspirin if children present similar to Kawasaki Disease and low molecular weight heparin is often used.

Flavors of MIS-C

Kawasaki-disease like

- Methylprednisolone
- Use IVIG and Aspirin

Features of Acute COVID-19 (may be PCR +)

- Methylprednisolone
- Often use IVIG
- Remdesivir

Shock or myocarditis

- Critical care admission
- Methylprednisolone
- Often use IVIG
- Anakinra
- Low molecular weight heparin

All of these can have increasing methylprednisolone or additive/increased anakinra dosing if severe.

Discharge if:

- Resolving inflammation
- stable vital signs
- benign physical exam
- reassuring labs

Follow up care:

- within one week with primary physician.
- other specialist follow-up early may be needed to trend labs.
- cardiology should follow on all patients at four to six weeks.

FIG. 3 Guidelines for evaluation of a child with suspected MIS-C. Guideline was modified from prior publication (Hennon et al., 2021) and based on CDC case definition, published cases, and other guidelines.

COVID-19 patients requiring supplemental oxygen, oxygen through a high-flow device, or noninvasive ventilation, but not in patients needing invasive mechanical ventilation or extracorporeal membrane oxygenation (ECMO) (WHO Solidarity Trial Consortium, 2022). Remdesivir is not recommended in patients with abnormal liver function or severe renal compromise. There does not appear to be a role for convalescent serum or recombinant antibodies for these presentations, even for patients presenting with acute SARS-CoV-2 (Wolf et al., 2021).

IVIG and steroids

For those who qualify as a case of KD, including concern for coronary artery involvement on the echocardiogram, the standard KD approach of using IV immunoglobulin G (IVIG) with initial methylprednisolone should be used. This follows current KD guidelines as methylprednisolone is encouraged if aneurysms are noted on the first echocardiogram or for high-risk individuals (McCrindle et al., 2017). Additionally, data from a retrospective cohort study that showed improvement in fever, left ventricular dysfunction, and intensive care unit stay support the use of corticosteroids in MIS-C cases (Ouldali et al., 2021). A recently published randomized controlled trial and retrospective trials suggest the superiority of methylprednisolone over dexamethasone for acute/respiratory COVID-19 (Ko et al., 2021; Ranjbar et al., 2021), although dexamethasone is commonly used on acute COVID-19 protocols. Generally, methylprednisolone should be used in all cases of MIS-C, unless there is a high index of suspicion of focal bacterial infection or enteritis; then, this is used sparingly or held until other conditions are ruled out.

Antithrombolytics

Expert consensus guidelines (Goldenberg et al., 2020) support prophylactic anticoagulation for pediatric patients with COVID-19 or MIS-C and specific risk factors (Table 3). For those individuals with KD high on the differential, aspirin should be included. This dosing varies by country with a high dose of 80–100 mg/kg/day being used for a variable time on initiation. Daily low-dose aspirin of 3–5 mg/kg/day (maximum 81 mg) is then prescribed after initial improvement and maintained until coronary arteries are assessed during outpatient cardiology follow-up. For other MIS-C phenotypes, it is unclear if aspirin is warranted, although it is recommended by at least one consensus guideline (Henderson et al., 2021). Notably, the American Heart Association KD guidelines allow for the coadministration of heparin and low-dose aspirin, so patients with expanding or large coronary artery aneurysms should be treated following these guidelines (McCrindle et al., 2017).

Other antiinflammatories

The IL-1 receptor antagonist anakinra is often used if additional immune modulation is required (Rajasekaran et al., 2014). Anakinra dosing ranges from 2 to 20 mg/kg/day, depending on the severity of disease (Henderson et al., 2021). Anakinra has the advantages of a quick onset, a large therapeutic window, and a favorable safety profile (Jesus & Goldbach-Mansky, 2014; Sota et al., 2018). It's short half-life also is advantageous in case a contraindication (e.g., systemic reaction, concomitant bacteremia, or alternative diagnosis) is found. If the clinical presentation is most consistent with KD, and there is a failure of first-line treatment, a second dose of IVIG or infliximab (a TNF-α inhibitor) could be considered, per KD guidelines (McCrindle et al., 2017). Other immunomodulators are not recommended (Henderson et al., 2021), but can be used if a patient is refractory to optimized methylprednisolone, anakinra, and IVIG treatment. Notably, anakinra has been successfully used in a small number of patients with IVIG-resistant KD (Kone-Paut et al., 2018).

Inotropes

Early reports showed the potential severity of cases with inotropic required in 80% of cases, while veno-arterial ECMO was initiated in 28% in one study (Belhadjer et al., 2020) and was used in 6.3% in a metaanalysis (Yasuhara et al., 2021). Prior to the COVID-19 pandemic, it was very uncommon for children presenting with KD to have shock or myocardial involvement (Ma et al., 2018). Although their use is common and necessary, currently, there is no specific inotropic regimen that is superior in the treatment of these children.

Hospital course

Children should be continuously monitored for fevers or evidence of ongoing or increasing inflammation. Critical care evaluation and management are warranted in many cases, as up to one-third of cases required intubation and mechanical ventilation (Belhadjer et al., 2020; Davies et al., 2020; Riphagen et al., 2020; Yasuhara et al., 2021).

Early recognition of a shock state (vasoplegic vs. cardiogenic) with judicious fluid resuscitation and appropriate initiation of inotropes and vasopressors are key factors to successful and favorable outcomes (Alhazzani et al., 2020; Belhadjer et al., 2020; Weiss et al., 2020). Extracorporeal therapies such as plasmapheresis and ECMO have a role in severe presentations (Badulak et al., 2021). In a metaanalysis of 917 cases early in the pandemic, 6.3% received ECMO. Left ventricular dysfunction or myocarditis was present in 55.3% (95% CI, 42.4–68.2) and the pooled prevalence of coronary artery aneurysm or dilatation was 21.7% (95% CI, 12.8–30.1) (Yasuhara et al., 2021). With treatment, reports support the rapid resolution of systolic myocardial function (Belhadjer et al., 2020; Yasuhara et al., 2021).

In children with a reassuring initial cardiac evaluation or mild to moderate disease, continued cardiac screening and serial troponins and BNPs throughout hospitalization are warranted due to reports of rapid decompensation. GI symptoms may evolve during admission, initially being mild before progressing to more severe signs including those concerning for an acute abdomen or appendicitis (Meyer et al., 2021; Place et al., 2020). Acute hepatitis has also been reported in MIS-C, generally secondary to the severity of MIS-C rather than specific hepatocyte dysfunction (Cantor et al., 2020). There is an increasing recognition of neurological manifestations including seizure, headache, meningitis, encephalopathy, muscle weakness, and brain stem and cerebellar signs that also support critical care involvement (Abdel-Mannan et al., 2020).

If clinical inflammation continues or worsens, retreatment or alternative therapy should be considered, as guided by infectious diseases and rheumatology. Many centers use aspirin in all children, which follows the ACR MIS-C guidelines (Henderson et al., 2021). As many cases of MIS-C present with initial thrombocytopenia and never show thrombocytosis, it is unclear if that is a necessary therapy, particularly in those who are continued on low molecular weight heparin.

Discharge and follow-up

Improvement in all clinical and concerning laboratory parameters can guide discharge decision making. Primary care follow-up is recommended and other disciplines are warranted. A repeat of pertinent laboratory markers, and if corticosteroids were used, tapering off guided by clinical and laboratory parameters can be managed by primary care, rheumatology, or in infectious disease follow-up. Rheumatology or infectious disease follow-up can review similarities to KD convalescence (such as peeling rash or reactive thrombocytosis) and for the consideration of the measurement of convalescent titers for any alternative diagnoses. Continuation of low-dose aspirin, if used, through follow up with pediatric cardiology is recommended. Generally, aspirin is not added solely due to reactive thrombocytosis in other conditions (Alberio, 2016). Thrombolytic management if prophylaxis is continued or if treatment for an active clot is ongoing should be managed by a specialist in hematology. Posthospital discharge prophylaxis is less clear and can be considered for patients with existing risk factors, continued elevated D-dimers, or those who have not returned to the previous level of functioning. Patients should have a therapeutic antifactor-Xa level prior to discharge home on low molecular weight heparin.

All children with MIS-C should have outpatient follow up with pediatric cardiology, minimally 4 weeks after discharge (Henderson et al., 2021). Recommendations from the American Academy of Pediatrics are to treat all children with MIS-C similarly to children who have had viral myocarditis (AAP, 2022). This includes exercise restriction for 3–6 months and obtaining specific cardiology clearance prior to resuming training or competition. Limited data suggest a reduction in functional exercise capacity, but few long-term organ-specific morbidities after recovery (Penner et al., 2021). Repeat cardiologic testing may include an ECG, cardiac magnetic resonance imaging, and stress testing to evaluate for any evidence of myocardial injury or scarring depending on the individual case and available testing modalities.

Vaccination

Recent studies suggest vaccination will prevent cases of MIS-C (Levy et al., 2022; Zambrano et al., 2022). The paucity of reports of MIS-C or MIS-A postvaccination and our prior data argue against a pure association with immunity to the spike protein (Chang et al., 2022). The recent reports of MIS-C cases directly from vaccination are rare (1 in 3 million) and subject to bias due to the methods of capture (surveillance survey) (Yousaf et al., 2022). Analysis of MIS-A cases diagnosed postvaccination support that natural infection was the impetus rather than vaccination (Belay et al., 2022).

There are few data to guide the optimal timing for vaccination after recovery from a case of MIS-C and only limited published reports on vaccination in those who have recovered from MIS-C (Chang et al., 2022). Currently, the CDC suggests delaying vaccination until 3 months after MIS-C diagnosis if full recovery has been achieved. Rare cases of transient myocarditis in adolescents and young adults following the second dose of mRNA-based vaccines do not appear to have symptomatic overlap with MIS-C. The diagnosis of MIS-C cases will become even more challenging as vaccination and natural immunity lead to lower circulating rates of SARS-CoV-2 infection as cases are likely to be more sporadic.

Summary points

- Fever and multisystem inflammation are universal in children with MIS-C/PIMS.
- There are three general phenotypes of MIS-C/PIMS.
- MIS-C/PIMS may become more difficult to diagnose as herd immunity grows.
- Steroid use is warranted, but current data on other therapies are lacking.
- The preponderance of data supports that MIS-C is distinct from KD.

References

AAP. (2022). *COVID-19 interim guidance: Return to sports and physical activity*. https://www.aap.org/en/pages/2019-novel-coronavirus-covid-19-infections/clinical-guidance/covid-19-interim-guidance-return-to-sports/. Accessed October 24, 2022, Updated 9/9/2022.

Abdel-Mannan, O., Eyre, M., Lobel, U., Bamford, A., Eltze, C., Hameed, B., … Hacohen, Y. (2020). Neurologic and radiographic findings associated with COVID-19 infection in children. *JAMA Neurology*. https://doi.org/10.1001/jamaneurol.2020.2687.

Alberio, L. (2016). Do we need antiplatelet therapy in thrombocytosis? Pro. Diagnostic and pathophysiologic considerations for a treatment choice. *Hämostaseologie*, *36*(4), 227–240. https://doi.org/10.5482/HAMO-14-11-0074.

Alhazzani, W., Moller, M. H., Arabi, Y. M., Loeb, M., Gong, M. N., Fan, E., … Rhodes, A. (2020). Surviving sepsis campaign: Guidelines on the management of critically ill adults with coronavirus disease 2019 (COVID-19). *Intensive Care Medicine*, *46*(5), 854–887. https://doi.org/10.1007/s00134-020-06022-5.

Al-Samkari, H., & Berliner, N. (2018). Hemophagocytic lymphohistiocytosis. *Annual Review of Pathology: Mechanisms of Disease*, *13*, 27–49.

Badulak, J., Antonini, M. V., Stead, C. M., Shekerdemian, L., Raman, L., Paden, M. L., … Brodie, D. (2021). ECMO for COVID-19: Updated 2021 guidelines from the Extracorporeal Life Support Organization (ELSO). *ASAIO Journal*. https://doi.org/10.1097/MAT.0000000000001422.

Barut, K., Özkoca, D., & Kutlubay, Z. (2022). 5 Year old with fever and perioral and periorbital erythema. In *Clinical cases in early-years pediatric dermatology* (pp. 1–5). Springer.

Belay, E. D., Abrams, J., Oster, M. E., Giovanni, J., Pierce, T., Meng, L., … Godfred-Cato, S. (2021). Trends in geographic and temporal distribution of US children with multisystem inflammatory syndrome during the COVID-19 pandemic. *JAMA Pediatrics*. https://doi.org/10.1001/jamapediatrics.2021.0630.

Belay, E. D., Godfred Cato, S., Rao, A. K., Abrams, J., Wyatt Wilson, W., Lim, S., … Bamrah Morris, S. (2022). Multisystem inflammatory syndrome in adults after severe acute respiratory syndrome coronavirus 2 (SARS-CoV-2) infection and coronavirus disease 2019 (COVID-19) vaccination. *Clinical Infectious Diseases*, *75*(1), e741–e748. https://doi.org/10.1093/cid/ciab936.

Belhadjer, Z., Meot, M., Bajolle, F., Khraiche, D., Legendre, A., Abakka, S., … Bonnet, D. (2020). Acute heart failure in multisystem inflammatory syndrome in children (MIS-C) in the context of global SARS-CoV-2 pandemic. *Circulation*. https://doi.org/10.1161/CIRCULATIONAHA.120.048360.

Cantor, A., Miller, J., Zachariah, P., DaSilva, B., Margolis, K., & Martinez, M. (2020). Acute hepatitis is a prominent presentation of the multisystem inflammatory syndrome in children: A single-center report. *Hepatology*, *72*(5), 1522–1527. https://doi.org/10.1002/hep.31526.

Cattalini, M., Taddio, A., Bracaglia, C., Cimaz, R., Paolera, S. D., Filocamo, G., … Rheumatology Study Group of the Italian Society of Pediatrics. (2021). Childhood multisystem inflammatory syndrome associated with COVID-19 (MIS-C): A diagnostic and treatment guidance from the Rheumatology Study Group of the Italian Society of Pediatrics. *Italian Journal of Pediatrics*, *47*(1), 24. https://doi.org/10.1186/s13052-021-00980-2.

CDC-MIS-C. (2022). *Health department-reported cases of multisystem inflammatory syndrome in children (MIS-C) in the United States*. https://www.cdc.gov/mis/mis-c/hcp/index.html?CDC_AA_refVal=https%3A%2F%2Fwww.cdc.gov%2Fmis%2Fhcp%2Findex.htmlaccesssed. Accessed August 28, 2022.

Chang, A. J., Baron, S., & Hicar, M. D. (2022). Robust humoral immune response after boosting in children with multisystem inflammatory syndrome in children. *IDCases*, *29*, e01569. https://doi.org/10.1016/j.idcr.2022.e01569.

Choe, Y. J., Choi, E. H., Choi, J. W., Eun, B. W., Eun, L. Y., Kim, Y. J., … Lee, S. W. (2021). Surveillance of COVID-19-associated multisystem inflammatory syndrome in children, South Korea. *Emerging Infectious Diseases*, *27*(4), 1196–1200. https://doi.org/10.3201/eid2704.210026.

Consiglio, C. R., Cotugno, N., Sardh, F., Pou, C., Amodio, D., Rodriguez, L., … Pascucci, G. R. (2020). The immunology of multisystem inflammatory syndrome in children with COVID-19. *Cell*, *183*(4), 968–981 (e967).

Davies, P., Evans, C., Kanthimathinathan, H. K., Lillie, J., Brierley, J., Waters, G., … Ramnarayan, P. (2020). Intensive care admissions of children with paediatric inflammatory multisystem syndrome temporally associated with SARS-CoV-2 (PIMS-TS) in the UK: A multicentre observational study. *The Lancet. Child & Adolescent Health*, *4*(9), 669–677. https://doi.org/10.1016/S2352-4642(20)30215-7.

Dufort, E. M., Koumans, E. H., Chow, E. J., Rosenthal, E. M., Muse, A., Rowlands, J., … New York State and Centers for Disease Control and Prevention Multisystem Inflammatory Syndrome in Children Investigation Team. (2020). Multisystem inflammatory syndrome in children in New York State. *The New England Journal of Medicine*, *383*(4), 347–358. https://doi.org/10.1056/NEJMoa2021756.

Feld, J., Tremblay, D., Thibaud, S., Kessler, A., & Naymagon, L. (2020). Ferritin levels in patients with COVID-19: A poor predictor of mortality and hemophagocytic lymphohistiocytosis. *International Journal of Laboratory Hematology*, *42*(6), 773–779.

Feldstein, L. R., Tenforde, M. W., Friedman, K. G., Newhams, M., Rose, E. B., Dapul, H., … Overcoming COVID-19 Investigators. (2021). Characteristics and outcomes of US children and adolescents with multisystem inflammatory syndrome in children (MIS-C) compared with severe acute COVID-19. *JAMA*, *325*(11), 1074–1087.

Fink, E. L., Robertson, C. L., Wainwright, M. S., Roa, J. D., Lovett, M. E., Stulce, C., … Holloway, A. (2022). Prevalence and risk factors of neurologic manifestations in hospitalized children diagnosed with acute SARS-CoV-2 or MIS-C. *Pediatric Neurology*, *128*, 33–44.

George, J. N., & Nester, C. M. (2014). Syndromes of thrombotic microangiopathy. *The New England Journal of Medicine*, *371*(19), 1847–1848. https://doi.org/10.1056/NEJMc1410951.

Godfred-Cato, S., Abrams, J. Y., Balachandran, N., Jaggi, P., Jones, K., Rostad, C. A., … Belay, E. D. (2022). Distinguishing multisystem inflammatory syndrome in children from COVID-19, Kawasaki Disease and Toxic Shock Syndrome. *The Pediatric Infectious Disease Journal*, *41*(4), 315.

Godfred-Cato, S., Bryant, B., Leung, J., Oster, M. E., Conklin, L., Abrams, J., … California MIS-C Response Team. (2020). COVID-19-associated multisystem inflammatory syndrome in children—United States, March-July 2020. *MMWR. Morbidity and Mortality Weekly Report*, *69*(32), 1074–1080. https://doi.org/10.15585/mmwr.mm6932e2.

Goldenberg, N. A., Sochet, A., Albisetti, M., Biss, T., Bonduel, M., Jaffray, J., … Thrombosis Subcommittee of the ISTH SSC. (2020). Consensus-based clinical recommendations and research priorities for anticoagulant thromboprophylaxis in children hospitalized for COVID-19-related illness. *Journal of Thrombosis and Haemostasis*, *18*(11), 3099–3105. https://doi.org/10.1111/jth.15073.

Goyal, M. K., Simpson, J. N., Boyle, M. D., Badolato, G. M., Delaney, M., McCarter, R., & Cora-Bramble, D. (2020). Racial and/or ethnic and socioeconomic disparities of SARS-CoV-2 infection among children. *Pediatrics*, *146*(4). https://doi.org/10.1542/peds.2020-009951.

Gruber, C. N., Patel, R. S., Trachtman, R., Lepow, L., Amanat, F., Krammer, F., … Tuballes, K. (2020). Mapping systemic inflammation and antibody responses in multisystem inflammatory syndrome in children (MIS-C). *Cell*, *183*(4), 982–995. e914.

Halepas, S., Lee, K. C., Myers, A., Yoon, R. K., Chung, W., & Peters, S. M. (2021). Oral manifestations of COVID-2019–related multisystem inflammatory syndrome in children: A review of 47 pediatric patients. *The Journal of the American Dental Association*, *152*(3), 202–208.

Henderson, L. A., Canna, S. W., Friedman, K. G., Gorelik, M., Lapidus, S. K., Bassiri, H., … Mehta, J. J. (2021). American College of Rheumatology clinical guidance for multisystem inflammatory syndrome in children associated with SARS-CoV-2 and hyperinflammation in pediatric COVID-19: Version 2. *Arthritis & Rhematology*, *73*(4), e13–e29. https://doi.org/10.1002/art.41616.

Hennon, T. R., Yu, K. O. A., Penque, M. D., Abdul-Aziz, R., Chang, A. C., McGreevy, M. B., … Hicar, M. D. (2021). COVID-19 associated multisystem inflammatory syndrome in children (MIS-C) guidelines; revisiting the Western New York approach as the pandemic evolves. *Progress in Pediatric Cardiology*, *62*, 101407. https://doi.org/10.1016/j.ppedcard.2021.101407.

Javalkar, K., Robson, V. K., Gaffney, L., Bohling, A. M., Arya, P., Servattalab, S., … Dionne, A. (2021). Socioeconomic and racial/ethnic disparities in multisystem inflammatory syndrome. *Pediatrics*. https://doi.org/10.1542/peds.2020-039933.

Jesus, A. A., & Goldbach-Mansky, R. (2014). IL-1 blockade in autoinflammatory syndromes. *Annual Review of Medicine*, *65*, 223–244. https://doi.org/10.1146/annurev-med-061512-150641.

Ko, J. J., Wu, C., Mehta, N., Wald-Dickler, N., Yang, W., & Qiao, R. (2021). A comparison of methylprednisolone and dexamethasone in intensive care patients with COVID-19. *medRxiv*. https://doi.org/10.1101/2021.02.03.21251088.

Kone-Paut, I., Cimaz, R., Herberg, J., Bates, O., Carbasse, A., Saulnier, J. P., … Piram, M. (2018). The use of interleukin 1 receptor antagonist (anakinra) in Kawasaki disease: A retrospective cases series. *Autoimmunity Reviews*, *17*(8), 768–774. https://doi.org/10.1016/j.autrev.2018.01.024.

Lasheen, R. A., ElTohamy, A., & Salaheldin, E. O. (2022). MIS-C frenzy: The importance of considering a broad differential diagnosis. *SAGE Open Medical Case Reports*, *10*. 2050313X221088397.

Lee, M.-S., Liu, Y.-C., Tsai, C.-C., Hsu, J.-H., & Wu, J.-R. (2021). Similarities and differences between COVID-19-related multisystem inflammatory syndrome in children and Kawasaki disease. *Frontiers in Pediatrics*, *9*, 640118.

Levy, N., Koppel, J. H., Kaplan, O., Yechiam, H., Shahar-Nissan, K., Cohen, N. K., & Shavit, I. (2022). Severity and incidence of multisystem inflammatory syndrome in children during 3 SARS-CoV-2 pandemic waves in Israel. *JAMA*, *327*(24), 2452–2454. https://doi.org/10.1001/jama.2022.8025.

Li, W., Tang, Y., Shi, Y., Chen, Y., & Liu, E. (2020). Why multisystem inflammatory syndrome in children has been less commonly described in Asia? *Translational Pediatrics*, *9*(6), 873–875. https://doi.org/10.21037/tp-20-151.

Lima-Setta, F., Magalhaes-Barbosa, M. C., Rodrigues-Santos, G., Figueiredo, E., Jacques, M. L., Zeitel, R. S., … Brazilian Research Network in Pediatric Intensive Care (BRnet-PIC). (2020). Multisystem inflammatory syndrome in children (MIS-C) during SARS-CoV-2 pandemic in Brazil: A multicenter, prospective cohort study. *Jornal de Pediatria*. https://doi.org/10.1016/j.jped.2020.10.008.

Lishman, J., Kohler, C., de Vos, C., van der Zalm, M. M., Itana, J., Redfern, A., … Rabie, H. (2020). Acute appendicitis in multisystem inflammatory syndrome in children with COVID-19. *The Pediatric Infectious Disease Journal*, *39*(12), e472–e473. https://doi.org/10.1097/INF.0000000000002900.

Ma, L., Zhang, Y. Y., & Yu, H. G. (2018). Clinical manifestations of Kawasaki disease shock syndrome. *Clinical Pediatrics (Phila)*, *57*(4), 428–435. https://doi.org/10.1177/0009922817729483.

Mamishi, S., Movahedi, Z., Mohammadi, M., Ziaee, V., Khodabandeh, M., Abdolsalehi, M. R., … Pourakbari, B. (2020). Multisystem inflammatory syndrome associated with SARS-CoV-2 infection in 45 children: A first report from Iran. *Epidemiology and Infection*, *148*, e196. https://doi.org/10.1017/S095026882000196X.

Manz, N., Höfele-Behrendt, C., Bielicki, J., Schmid, H., Matter, M. S., Bielicki, I., … Gros, S. J. (2021). MIS-C-implications for the pediatric surgeon: An algorithm for differential diagnostic considerations. *Children*, *8*(8), 712.

McCrindle, B. W., Rowley, A. H., Newburger, J. W., Burns, J. C., Bolger, A. F., Gewitz, M., … Council on Epidemiology and Prevention. (2017). Diagnosis, treatment, and long-term management of Kawasaki disease: A scientific statement for health professionals from the American Heart Association. *Circulation*, *135*(17), e927–e999. https://doi.org/10.1161/CIR.0000000000000484.

Meyer, J. S., Robinson, G., Moonah, S., Levin, D., McGahren, E., Herring, K., … Shirley, D. A. (2021). Acute appendicitis in four children with SARS-CoV-2 infection. *Journal of Pediatric Surgery Case Reports*, *64*, 101734. https://doi.org/10.1016/j.epsc.2020.101734.

Mohri, Y., Shimizu, M., Fujimoto, T., Nishikawa, Y., Ikeda, A., Matsuda, Y., … Kawaguchi, C. (2022). A young child with pediatric multisystem inflammatory syndrome successfully treated with high-dose immunoglobulin therapy. *IDCases*, *28*, e01493. https://doi.org/10.1016/j.idcr.2022.e01493.

Morparia, K., Park, M. J., Kalyanaraman, M., McQueen, D., Bergel, M., & Phatak, T. (2021). Abdominal imaging findings in critically ill children with multisystem inflammatory syndrome associated with COVID-19. *The Pediatric Infectious Disease Journal*, *40*(2), e82–e83.

Morris, S. B., Schwartz, N. G., Patel, P., Abbo, L., Beauchamps, L., Balan, S., … Godfred-Cato, S. (2020). Case series of multisystem inflammatory syndrome in adults associated with SARS-CoV-2 infection—United Kingdom and United States, March-August 2020. *MMWR. Morbidity and Mortality Weekly Report*, *69*(40), 1450–1456. https://doi.org/10.15585/mmwr.mm6940e1.

Nakra, N. A., Blumberg, D. A., Herrera-Guerra, A., & Lakshminrusimha, S. (2020). Multi-system inflammatory syndrome in children (MIS-C) following SARS-CoV-2 infection: Review of clinical presentation, hypothetical pathogenesis, and proposed management. *Children*, *7*(7), 69.

Okarska-Napierala, M., Ludwikowska, K. M., Szenborn, L., Dudek, N., Mania, A., Buda, P., … Kuchar, E. (2020). Pediatric inflammatory multisystem syndrome (PIMS) did occur in Poland during months with low COVID-19 prevalence, preliminary results of a nationwide register. *Journal of Clinical Medicine*, *9*(11). https://doi.org/10.3390/jcm9113386.

Ouldali, N., Toubiana, J., Antona, D., Javouhey, E., Madhi, F., Lorrot, M., … French Covid-19 Paediatric Inflammation Consortium. (2021). Association of intravenous immunoglobulins plus methylprednisolone vs immunoglobulins alone with course of fever in multisystem inflammatory syndrome in children. *Journal of the American Medical Association*. https://doi.org/10.1001/jama.2021.0694.

Penner, J., Abdel-Mannan, O., Grant, K., Maillard, S., Kucera, F., Hassell, J., … GOSH PIMS-TS MDT Group. (2021). 6-month multidisciplinary follow-up and outcomes of patients with paediatric inflammatory multisystem syndrome (PIMS-TS) at a UK tertiary paediatric hospital: A retrospective cohort study. *The Lancet. Child & Adolescent Health*. https://doi.org/10.1016/S2352-4642(21)00138-3.

Place, R., Lee, J., & Howell, J. (2020). Rate of pediatric appendiceal perforation at a children's hospital during the COVID-19 pandemic compared with the previous year. *JAMA Network Open*, *3*(12), e2027948. https://doi.org/10.1001/jamanetworkopen.2020.27948.

Porritt, R. A., Binek, A., Paschold, L., Rivas, M. N., McArdle, A., Yonker, L. M., … Fasano, A. (2021). The autoimmune signature of hyperinflammatory multisystem inflammatory syndrome in children. *The Journal of Clinical Investigation*, *131*(20).

Rajasekaran, S., Kruse, K., Kovey, K., Davis, A. T., Hassan, N. E., Ndika, A. N., … Birmingham, J. (2014). Therapeutic role of anakinra, an interleukin-1 receptor antagonist, in the management of secondary hemophagocytic lymphohistiocytosis/sepsis/multiple organ dysfunction/macrophage activating syndrome in critically ill children*. *Pediatric Critical Care Medicine*, *15*(5), 401–408. https://doi.org/10.1097/PCC.0000000000000078.

Ranjbar, K., Moghadami, M., Mirahmadizadeh, A., Fallahi, M. J., Khaloo, V., Shahriarirad, R., … Gholampoor Saadi, M. H. (2021). Methylprednisolone or dexamethasone, which one is superior corticosteroid in the treatment of hospitalized COVID-19 patients: A triple-blinded randomized controlled trial. *BMC Infectious Diseases*, *21*(1), 337. https://doi.org/10.1186/s12879-021-06045-3.

Reiff, D. D., Mannion, M. L., Samuy, N., Scalici, P., & Cron, R. Q. (2021). Distinguishing active pediatric COVID-19 pneumonia from MIS-C. *Pediatric Rheumatology Online Journal*, *19*(1), 21. https://doi.org/10.1186/s12969-021-00508-2.

Riollano-Cruz, M., Akkoyun, E., Briceno-Brito, E., Kowalsky, S., Reed, J., Posada, R., … Paniz-Mondolfi, A. (2021). Multisystem inflammatory syndrome in children related to COVID-19: A New York City experience. *Journal of Medical Virology*, *93*(1), 424–433. https://doi.org/10.1002/jmv.26224.

Riphagen, S., Gomez, X., Gonzalez-Martinez, C., Wilkinson, N., & Theocharis, P. (2020). Hyperinflammatory shock in children during COVID-19 pandemic. *Lancet*. https://doi.org/10.1016/S0140-6736(20)31094-1.

Sahn, B., Eze, O. P., Edelman, M. C., Chougar, C. E., Thomas, R. M., Schleien, C. L., & Weinstein, T. (2021). Features of intestinal disease associated with COVID-related multisystem inflammatory syndrome in children. *Journal of Pediatric Gastroenterology and Nutrition*, *72*(3), 384–387. https://doi.org/10.1097/MPG.0000000000002953.

Sharathkumar, A. A., Faustino, E. V. S., & Takemoto, C. M. (2021). How we approach thrombosis risk in children with COVID-19 infection and MIS-C. *Pediatric Blood & Cancer*, *68*(7), e29049. https://doi.org/10.1002/pbc.29049.

Sharma, C., Ganigara, M., Galeotti, C., Burns, J., Berganza, F. M., Hayes, D. A., … Bayry, J. (2021). Multisystem inflammatory syndrome in children and Kawasaki disease: A critical comparison. *Nature Reviews Rheumatology*, *17*(12), 731–748.

Son, M. B. F., & Friedman, K. (2021). COVID-19: Multisystem inflammatory syndrome in children (MIS-C) clinical features, evaluation, and diagnosis. In *upToDate*. Waltham. Accessed August 19, 2021.

Sota, J., Vitale, A., Insalaco, A., Sfriso, P., Lopalco, G., Emmi, G., … "Working Group" of Systemic Autoinflammatory Diseases of SIR (Italian Society of Rheumatology). (2018). Safety profile of the interleukin-1 inhibitors anakinra and canakinumab in real-life clinical practice: A nationwide multicenter retrospective observational study. *Clinical Rheumatology*, *37*(8), 2233–2240. https://doi.org/10.1007/s10067-018-4119-x.

Torres, J. P., Izquierdo, G., Acuna, M., Pavez, D., Reyes, F., Fritis, A., … Tapia, L. I. (2020). Multisystem inflammatory syndrome in children (MIS-C): Report of the clinical and epidemiological characteristics of cases in Santiago de Chile during the SARS-CoV-2 pandemic. *International Journal of Infectious Diseases*, *100*, 75–81. https://doi.org/10.1016/j.ijid.2020.08.062.

Verdoni, L., Mazza, A., Gervasoni, A., Martelli, L., Ruggeri, M., Ciuffreda, M., … D'Antiga, L. (2020). An outbreak of severe Kawasaki-like disease at the Italian epicentre of the SARS-CoV-2 epidemic: An observational cohort study. *Lancet*. https://doi.org/10.1016/S0140-6736(20)31103-X.

Verrotti, A., Mazzocchetti, C., & Iannetti, P. (2021). Definitive pathognomonic signs and symptoms of paediatric neurological COVID-19 are still emerging. *Acta Paediatrica*. https://doi.org/10.1111/apa.15827.

Vogel, T. P., Top, K. A., Karatzios, C., Hilmers, D. C., Tapia, L. I., Moceri, P., … Klein, N. P. (2021). Multisystem inflammatory syndrome in children and adults (MIS-C/A): Case definition & guidelines for data collection, analysis, and presentation of immunization safety data. *Vaccine*, *39*(22), 3037–3049.

Weiss, S. L., Peters, M. J., Alhazzani, W., Agus, M. S. D., Flori, H. R., Inwald, D. P., … Tissieres, P. (2020). Surviving sepsis campaign international guidelines for the management of septic shock and sepsis-associated organ dysfunction in children. *Intensive Care Medicine*, *46*(Suppl 1), 10–67. https://doi.org/10.1007/s00134-019-05878-6.

Whittaker, E., Bamford, A., Kenny, J., Kaforou, M., Jones, C. E., Shah, P., … Consortia, P. (2020). Clinical characteristics of 58 children with a pediatric inflammatory multisystem syndrome temporally associated with SARS-CoV-2. *JAMA*, *324*(3), 259–269. https://doi.org/10.1001/jama.2020.10369.

Whitworth, H., Sartain, S. E., Kumar, R., Armstrong, K., Ballester, L., Betensky, M., … Raffini, L. (2021). Rate of thrombosis in children and adolescents hospitalized with COVID-19 or MIS-C. *Blood*, *138*(2), 190–198. https://doi.org/10.1182/blood.2020010218.

WHO Solidarity Trial Consortium. (2022). Remdesivir and three other drugs for hospitalised patients with COVID-19: Final results of the WHO solidarity randomised trial and updated meta-analyses. *The Lancet*, *399*(10339), 1941–1953. https://doi.org/10.1016/S0140-6736(22)00519-0.

Wolf, J., Abzug, M. J., Wattier, R. L., Sue, P. K., Vora, S. B., Zachariah, P., … Nakamura, M. M. (2021). Initial guidance on use of monoclonal antibody therapy for treatment of COVID-19 in children and adolescents. *Journal of the Pediatric Infectious Diseases Society*. https://doi.org/10.1093/jpids/piaa175.

Yasuhara, J., Watanabe, K., Takagi, H., Sumitomo, N., & Kuno, T. (2021). COVID-19 and multisystem inflammatory syndrome in children: A systematic review and meta-analysis. *Pediatric Pulmonology*, *56*(5), 837–848. https://doi.org/10.1002/ppul.25245.

Yousaf, A. R., Cortese, M. M., Taylor, A. W., Broder, K. R., Oster, M. E., Wong, J. M., & MIS-C Investigation Authorship Group. (2022). Reported cases of multisystem inflammatory syndrome in children aged 12–20 years in the USA who received a COVID-19 vaccine, December, 2020, through August, 2021: A surveillance investigation. *The Lancet. Child & Adolescent Health*, *6*(5), 303–312. https://doi.org/10.1016/S2352-4642(22)00028-1.

Zambrano, L. D., Newhams, M. M., Olson, S. M., Halasa, N. B., Price, A. M., Orzel, A. O., … Overcoming, C.-I. (2022). BNT162b2 mRNA vaccination against COVID-19 is associated with decreased likelihood of multisystem inflammatory syndrome in U.S. Children ages 5-18 years. *Clinical Infectious Diseases*. https://doi.org/10.1093/cid/ciac637.

Section D

Impact on the respiratory system

Chapter 21

Ground-glass nodules in the lungs of COVID-19 patients

Noel Roig-Marín
Independent Researcher, Alicante, Spain

Abbreviations

Chest CT scan	chest computed tomography scan
CXR	chest x-ray
DLCO	diffusing capacity of the lungs for carbon monoxide
ED	emergency department
FEV1	forced expiratory volume in one second
FVC	forced vital capacity
GGF	ground-glass finding
GGN	ground-glass nodule
GGO	ground-glass opacity
TLC	total lung capacity

Introduction

COVID-19 can mimic other lung diseases because its lesions can be found in other pathological processes (Arslan & Ünal, 2021). The most frequent finding in the chest computed tomography (CT) scan is "bilateral ground-glass opacity (GGO) with accompanying consolidation" (Arslan & Ünal, 2021). The presence of ground-glass findings (GGF) is a classic and characteristic lesion of COVID-19, which is mainly formed by fibrin, although some authors believe that there could be amyloid structures in them (Kell et al., 2022). Within the ground-glass findings are the ground-glass nodules (GGN) and the ground-glass opacities. According to Shivananda and Amini (2008), GGN are included within GGO. Therefore, GGN can be considered a type of GGO.

GGOs were one of the first initial manifestations of COVID-19 described, being the most frequent radiographic finding in the pediatric population (Chang et al., 2020). These GGOs usually present a typically bilateral, peripheral distribution with a predominance in the lower fields (Martínez Chamorro et al., 2021). Other findings, in addition to GGO/GGN, are "reticular and linear opacities, residual crazy paving pattern, melted sugar sign, and parenchymal fibrotic bands" (Shaw et al., 2021) that can be found in remission or late phase (Shaw et al., 2021).

Besides, GGOs also occur as sequelae or residual or persistent lesions after the acute infection phase of SARS-CoV-2 (Frija-Masson et al., 2021; Migliore, 2021). Special care should be taken with the observation of GGN/GGO because a differential diagnosis of lung cancer must be made (Yamanaka et al., 2021; Zhang et al., 2020). These elements are highly relevant because they determine the prognosis and evolution of patients.

This chapter will analyze the different implications of GGN/GGO in patients with COVID-19. Data on their mortality and evolution will be attached. It will also discuss what other topics are related, such as the diagnostic, immunological, and possible tumor degeneration section, among others.

GGN/GGO and mortality

In this chapter, the implications of GGN/GGO are reviewed. The main and most relevant outcome measure is mortality, which was studied (Roig-Marín & Roig-Rico, 2022d). The presence of GGFs diagnosed after a chest x-ray (CXR) is infrequent (29.7% of case). GGNs/GGOs observed in the first CXR are associated with a higher mortality rate

Management, Body Systems, and Case Studies in COVID-19. https://doi.org/10.1016/B978-0-443-18703-2.00024-4
Copyright © 2024 Elsevier Inc. All rights are reserved, including those for text and data mining, AI training, and similar technologies.

(36% vs. 22.3%; $p = 0.01$). A history of congestive heart failure was not a confounding factor because its frequency in both groups (with GGN/GGO and without GGN/GGO) was very similar. It is relevant to mention this fact due to heart failure being a pathology that can produce radiographic findings of GGN/GGO, such as those that appear in the context of COVID-19. In addition, during the study, the radiologists in charge of reporting CXRs did not detect that any finding was associated with congestive heart failure, but rather that the GGFs were associated with SARS-CoV-2 infection.

Most of the GGFs studied by Roig-Marín and Roig-Rico (2022d) were peripheral bilateral ground-glass, which coincides with the descriptions made by other authors (Martínez Chamorro et al., 2021; Schmitt & Marchiori, 2020). Bilateral ground-glass was the most frequent (78.7%) and the one associated with a higher mortality rate. Unilateral ground-glass was much less frequent and the associated case fatality rate was not significant. This has only been published in the article by Roig-Marín and Roig-Rico (2022d). The mathematical/statistical results are corroborated by the radiographic images.

Based on a review of all the literature, it is reasonable to consider that the ground-glass variable could be included in COVID-19 prognostic calculators, such as those of different authors (Liang et al., 2020; Wongvibulsin et al., 2021). This variable can be studied in all patients because all of them are initially in the emergency department and a chest x-ray is performed. Prognostic calculators that use variables available in all patients and the vast majority of hospital centers should be made. This would make it possible to know more rigorously if the patient should be admitted to the ward or the intensive care unit (ICU), or even if they could be discharged more quickly than another patient.

Evolution after acute phase of COVID-19 and GGO/GGN

There are different publications about the evolution of patients with COVID-19 (Frija-Masson et al., 2021; Han et al., 2021). On the one hand, Han et al. (2021) performed a 6-month follow-up chest CT scan of 114 patients with severe COVID-19. In this imaging test, fibrotic changes were observed in 35% and residual GGO or "interstitial thickening" in 27% while no findings were found in the rest because they had resolved. Therefore, GGN/GGO may remain residual and is not just a finding found in the early stages. The fibrotic forms seem to be associated more with late phases and not with acute phases, unlike ground-glass, which appears in the early and chronic phases.

On the other hand, Frija-Masson et al. (2021) conducted a similar study in which they followed 137 patients 3 months after an acute episode of pneumonia. The follow-up chest CT scan showed us, in 75% of the GGO patients, a reticular pattern/reticulation in 30% and fibrotic changes in 13%. Thus, in this work, GGN/GGO turned out to be the most frequent finding after the acute phase of COVID-19. According to Migliore (2021), GGO was shown to be the most frequently detected pattern on a chest CT scan after 6 months of the acute phase. Consequently, both studies coincide, but differ from Han et al. (2021) who showed that in their cohort the most frequent changes were fibrotic rather than GGO.

Frija-Masson et al. (2021) stated that patients with ground-glass findings had very clearly lower levels of diffusing capacity of the lungs for carbon monoxide (DLCO) and total lung capacity (TLC) ($p < 0.0001$) and also lower levels of forced vital capacity (FVC) and forced expiratory volume in one second (FEV1). The fibrotic and reticular pattern associated changes showed significant changes, such as lower levels of DLCO and TLC, but with a considerably higher *p*-value. That is to say, GGOs were the findings with the most important statistically significant implications. In this cohort, the majority of patients had alterations in pulmonary function tests, including mild cases. The use of glucocorticoids did not produce notable improvements in terms of an increase in respiratory indices or a reduction/disappearance of radiographic findings. With these two studies mentioned, the need for adequate follow-up of patients with COVID-19 is reflected to observe their evolution and verify the presence of GGN/GGO.

Need for differential diagnosis of GGN/GGO from COVID-19 and lung cancer?

The determination of GGN/GGO can be suggestive of cancer, so it is important to perform a differential diagnosis with tumor pathology (Yamanaka et al., 2021; Zhang et al., 2020). According to Yamanaka et al. (2021), when a radiographic pulmonary nodule is observed in patients with COVID-19, the possibility of a tumor should be taken into account if the GGN shows nonspecific characteristics for viral etiology, such as well-defined borders, nodule characteristics in a solitary lung, relative roundness, etc.

Migliore (2021) considered that the initial identification of GGO is mandatory for proper evaluation and decision making. However, the author believes that the differential diagnosis with cancer pathology is neither useful nor necessary. Migliore (2021) alleges three reasons: GGOs tend to worsen, those patients with COVID-19 and lung cancer can be operated on later, and GGOs associated with lung cancer have a 5-year survival rate that is greater than 90%. The first of the arguments, that "GGOs commonly tend to deteriorate quickly in COVID-19 patients causing the need for rapid hospitalization and oxygen therapy" (Migliore, 2021) can generate some confusion. It can be assumed that Migliore is referring to the fact that in patients with GGOs, these tend to worsen with radiographic changes of the GGOs, so cancer can be more easily ruled out as lung cancer usually has a more progressive course.

The author also admits that although it is not initially profitable to carry out the differential diagnosis of GGOs, there is a long-term problem (Migliore, 2021). The importance and meaning of the residual GGOs observed in the chest CT scan at 6 months of evolution are not known. The GGOs could be scar lesions or related to cancerous pathology (Migliore, 2021). That is to say, they could be lesions on which cancer could settle or degenerate into tumor pathology. It is also not known whether they are lesions that will disappear or if they will remain throughout the patient's life. The Italian researcher also states that when there is diagnostic doubt, such as in an asymptomatic patient with GGO, a bronchoscopy with sample collection for analysis should be performed.

"GGO TNM" and differences between GGOs from COVID-19 and lung cancer

It is noteworthy that the Italian thoracic surgeon also talks about an update of the classification of malignant tumors, which takes into account tumors, nodes, and metastases (TNM), due to GGO should be included as a new tumor (T) categorization. The author proposes to add a prefix or particle of "ggo" or "g" to the TNM staging ("ggo TNM" or "gTNM") (Migliore, 2021). In the TNM section, it is also mentioned that solid GGOs must be differentiated from pure and mixed ones, which should be agreed upon internationally (Migliore, 2021).

On the other hand, Zhang et al. (2020) did a study in which they described the characteristics presented by patients with GGOs in the initial stage and those with incipient lung cancer. The justification for this study was based on the fact that the early stages of COVID-19 can generate confusing patterns similar to those of cancer and it can be difficult to differentiate between one and the other. Thus, the objective of these authors was to clarify which elements are useful to opt for a tumor or COVID-19 diagnosis. The researchers found that COVID-19 patients were more often male, with a higher body mass index (BMI), and were younger. Single lesions correlated more frequently with cancerous pathology (89% vs. 17%) than with COVID-19. In addition, the involvement of COVID-19 involved more lung lobes and segments, with a patchy shape in 54% of cases. Cancer produced rather oval shapes in most cases. With this, it can be verified that COVID-19 with GGO and lung cancer have similarities, but are distinguishable thanks to a series of characteristics such as the type of patient, radiographic image, the evolutionary course, and the epidemiological history of the patient, apart from other tests such as laboratory tests (Zhang et al., 2020).

As a reflection of these last two subsections, initial GGOs in COVID-19 can be confused with lung cancer. Some elements can differentiate, such as the fact that COVID-19 lesions are multiple and are more widely distributed in the lungs while the tumor lesion is usually more circumscribed with defined borders and a more oval/round appearance. The differential diagnosis becomes more complex when the GGOs are residual by dint of the fact that their behavior is far from an acute viral etiology and they become persistent lesions that may suggest the presence of malignant pathology. However, the evolution of the natural history of late GGOs is unknown, so one possibility is that they may be precancerous lesions or that they facilitate the proliferation of tumor tissue. Prospective follow-up studies lasting years are required to clarify these doubts.

Diagnostic difficulties and CXR vs. chest CT scan

The study of CXR lung nodules and opacities is one of the first elements taught in the radiodiagnosis study subject of the medical degree. Today, in universities, the finding of ground-glass is taught rather with a chest CT scan and not with the CXR. This is because detecting GGN/GGO on CXR is a diagnostic challenge that can be solved by expert radiologists. A physician who has not specialized in thoracic radiodiagnosis may have great difficulty discerning the presence or absence of GGFs. Our work on GGOs (Roig-Marín & Roig-Rico, 2022d) was carried out thanks to the findings recorded in reports made by expert radiologists.

A review of the literature was done in which the vast majority of GGOs described in patients with COVID-19 were diagnosed and described from the performance of a chest CT scan (Frija-Masson et al., 2021; Han et al., 2021; Saha et al., 2022). Few studies focus on or defend the diagnostic utility of CXR for the assessment of patients with SARS-CoV-2 infection (Balbi et al., 2021; Hurt et al., 2020; Kaleemi et al., 2021; Litmanovich et al., 2020; Schiaffino et al., 2020; Yasin & Gouda, 2020). However, despite being an imaging test that is performed in all patients, most studies focus on the diagnostic and prognostic possibilities of the findings obtained on chest CT.

Therefore, the predominant works are based on the high sensitivity and specificity of chest CT scans for the diagnosis of radiographic lesions due to COVID-19 (Borges da Silva Teles et al., 2021; Goyal et al., 2020; Hu et al., 2020; Lang et al., 2020; Lu & Pu, 2020; Simpson et al., 2020; Zhu et al., 2020). However, this chapter is postulated in defense of CXR due to its easy access and high availability in different health centers, in addition to being one of the first tests performed on the patient that also offers a pragmatic overview of the initial state.

CXR may be more useful in the context of COVID-19, as it is a simple, fast, and cheap test that reports the initial status of the patient. Indeed, it has less sensitivity than the chest CT scan, but greater access and ease of use. Moreover, in our cohort (Roig-Marín & Roig-Rico, 2022d), there was a total correlation between the radiographic findings.

Confirmation of CXR results by chest CT scan and pragmatism of CXR

In our study, the results of the chest CT scan and the CXR were always correlated. Bilateral peripheral GGOs were observed and the chest CT scan showed the same findings. As a notable novelty, a useful tool that could increase the accuracy of the chest CT scan diagnostic test for the detection and evaluation of GGO is artificial intelligence (Saha et al., 2022).

In short, when the chest CT scan was performed on the ground-glass patients, they confirmed the GGFs 100% of the time. However, only a few patients underwent chest CT scans. In 13 patients with GGOs was a chest CT scan performed and in these 13, the findings were confirmed. Of the few times that a CT scan of the chest was performed, in most cases it was observed as ground-glass. As a consequence, the data in the table refer to the fact that a chest CT scan has high sensitivity and specificity for the detection of ground-glass.

Finally, despite the differences in diagnostic accuracy between the chest CT scan and CXR, the initial CXR is more pragmatic in the context of COVID-19 patients, some of whom are elderly, making it more difficult to mobilize them to perform a CT scan. The findings between both radiographic tests were correlated, so CXR was not as limited as might have been thought before designing our study (Roig-Marín & Roig-Rico, 2022d). For all that has been mentioned, the importance of defending CXR in the initial evaluation of the patient with COVID-19 is reiterated.

Policies and procedures

In this chapter, the impact of the coronavirus pandemic on patients with the presence of ground-glass findings was reviewed. The prognostic, evolutionary, and mortality implications were reviewed. The main variable taken into account is the presence of radiographic findings and the predominant outcome variable is mortality. Additionally, in one of the articles (Frija-Masson et al., 2021), other variables, such as FEV1, FVC, TLC, and DLCO, are used. Of all the literature, the study that best reflects the intrinsic relationship between the presence of ground-glass findings and mortality Roig-Marín & Roig-Rico, 2022d. This study reflects that the presence of ground-glass findings is a prognostic factor that should be taken into account.

Our policy is that the variable GGF is a parameter that must be considered in the emergency department and that it is used, among other criteria, to classify the risk status and severity of the patient to know what degree of attention is required. Proposing risk stratification standards in patients with COVID-19 in the emergency department could be a measure promoted by the Spanish Ministry of Health to be applied in different national public hospitals.

About the ground-glass nodules, reviews of these in the context of COVID-19 have been found in multiple articles (Cardoso et al., 2021; Kong et al., 2020; Mogami et al., 2021; Niu et al., 2021; Ramdani et al., 2021; Shivananda & Amini, 2008; Wang et al., 2021; Xia et al., 2020; Yamanaka et al., 2021; Zhang et al., 2022). According to Shivananda and Amini, ground-glass nodules are included within the focal and diffuse ground-glass opacities, so the well-known and classic GGOs of COVID-19 were added to the review, which is the main topic of different studies (such as Schmitt & Marchiori, 2020; Wu et al., 2021; Zhang et al., 2020). It has also been observed that COVID-19 not only manifests itself in the form of GGNs but also as a single/solitary pulmonary nodule, for which a bibliographic search was also carried out (Arslan & Ünal, 2021; Vaccarello et al., 2021; Varona Porres et al., 2021). Furthermore, SARS-CoV-2 infection can also start as a small solitary ground-glass nodule on CT as an initial manifestation of coronavirus disease 2019 (Xia et al., 2020).

Applications to other areas

Taking into account the infinity of pathologies that have been related to COVID-19 and the study of these (Gold et al., 2020; Imanova Yaghji et al., 2021; López-Candales et al., 2021; Para et al., 2022; Pettus & Skolnik, 2021; Rotondo et al., 2021; White-Dzuro et al., 2021), in addition to the different discoveries related to it (the six articles published by Roig-Marín & Roig-Rico, 2022a, 2022b, 2022c, 2022d, 2022e, 2022f), a little explored topic in relation to GGN/GGO has been chosen for review. In this chapter, the review subject is how immunology is related to the pathophysiology of ground-glass. The best work that has answered this question was written by Wu et al. (2021).

Wu et al. (2021) conducted a retrospective study comparing patients with severe or moderate COVID-19. It was differentiated into two groups: a group in which GGFs were observed, formed mainly by fibrin (Kell et al., 2022), and another group without GGOs. The authors identified that the GGO group had significantly higher levels of IFN-γ, IL-4, and IL-2. Based on multivariate analysis, IL-2 was found to be an independent predictor of GGO (Wu et al., 2021). This information is relevant because Ma et al. (2021) determined that elevated levels of soluble IL-2R (IL-2 receptor) are associated with lengthened disease in patients with severe COVID-19 (Ma et al., 2021). An elevation of IL-2 contributes to the known cytokine storm of COVID-19 (Bagheri-Hosseinabadi et al., 2021) that induces a proinflammatory state; the higher the inflammation, the higher the associated mortality rate (Callejas Rubio et al., 2020). However, Zhu et al. (2021) investigated the possibility of recombinant IL-2 as a therapeutic alternative for COVID-19 because it recovers lymphocyte levels in severely ill patients, making it a highly relevant element in the pathophysiology of COVID-19.

With all this, it is possible to observe the existing correlation between the presence of GGN/GGO and the increase in IL-2 levels associated with the cytokine storm and a more prolonged evolution of the disease. In turn, this proinflammatory state leads to higher mortality. In this way, what the authors Roig-Marín and Roig-Rico (2022d) studied about the existence of GGFs as a mortality risk factor is corroborated. Consequently, the observed results present a consistent immunological and biological basis.

Although recombinant IL-2 has been studied as a therapy in severe patients, it would probably not be the most indicated treatment in patients with GGFs because, according to our study about GGO, patients with GGO had rather higher leukocytosis (which includes the level of lymphocytes) than patients without GGO. That is to say, patients with GGN/GGO had higher leukocytosis and, therefore, less lymphopenia than patients without GGN/GGO. Thus, recombinant IL-2 therapy, whose objective is to reduce lymphopenia in severely ill patients, would not be a treatment especially indicated in these patients, but rather in patients without GGO.

Mini-dictionary of terms

Ground-glass: Radiographic finding that consists of a slight increase in pulmonary attenuation/density. That is to say, it is shown as a density change that is commonly observed in a chest CT scan. Only expert radiologists can easily detect these changes in a chest x-ray. It is a relevant finding in COVID-19.

Ground-glass nodule: It is a round or oval lesion objectified in the lungs in which a slight change in density is observed. This initial finding should be studied to rule out lung cancer.

Ground-glass opacity: It is a finding similar to the previous term. In this case, it is not an oval or rounded structure, but follows a more irregular structure/pattern in the form of patches. This is a more common finding than the ground-glass nodule. It is sometimes the earliest and most predominant finding in the context of COVID-19. If they do appear, they are usually bilateral and peripheral, which worsens the prognosis. The unilateral and reduced forms usually present a better evolution.

Lung nodule: It is an oval or round spot (lesion) that is diagnosed with a chest CT scan or a chest x-ray.

Solitary pulmonary nodule: A solitary pulmonary nodule is a separate lesion that is <3 cm in diameter, which is circumscribed entirely by lung parenchyma. This is a finding that is often detected incidentally when a chest x-ray or chest CT scan is performed for other reasons. It can also be observed during lung cancer screening. There are case reports that described a solitary/single pulmonary nodule related to COVID-19.

Summary points

- The radiographic diagnosis of GGN/GGO is a risk factor for in-hospital mortality.
- The increase in IL-2 levels is an independent predictor of the presence of GGO.
- GGNs/GGOs are findings of COVID-19 that may appear initially, but they can also be observed in the long term.
- GGN can generate diagnostic doubts between COVID-19 and lung cancer.
- Cancerous lesions tend to be single and oval/rounded, but COVID-19 GGOs are patchy, multiple, and with greater lung spread.
- It has been proposed to add GGO to TNM staging.
- The finding of GGN/GGO has been described mainly in studies in which chest CT scans were performed.
- Studies that focused on describing ground-glass findings using CXR were in the minority.
- Diagnosing GGO with a CXR presents high difficulty for physicians who are not expert thoracic radiologists.
- CXR has a lower diagnostic accuracy than the chest CT scan, although CXR is more accessible and offers information on the initial state of the patient.

Acknowledgments

This chapter, titled *Ground-glass nodules in the lungs of COVID-19 patients*, is derived in part from an article published in POSTGRADUATE MEDICINE, Published online: 06 Jan 2022, <copyright Taylor & Francis>, available online: https://www.tandfonline.com/doi/full/10.1080/00325481.2021.2021741

Only some tables that have been modified and the three existing figures and their legends have been taken from this article: Noel Roig-Marín and Pablo Roig-Rico (2022) Ground-glass opacity on emergency department chest X-ray: a risk factor for in-hospital mortality and organ failure in elderly admitted for COVID-19, Postgraduate Medicine, DOI: https://doi.org/10.1080/00325481.2021.2021741. The terms and conditions of the assignment of copyright are fulfilled because what has been done is within the rights retained by the author. Therefore, no permission is required.

References

Arslan, S., & Ünal, E. (2021). One of the many faces of COVID-19 infection: An irregularly shaped pulmonary nodule. *Insights Into Imaging*, *12*(1), 48. https://doi.org/10.1186/s13244-021-00987-7.

Bagheri-Hosseinabadi, Z., Ostad Ebrahimi, H., Bahrehmand, F., Taghipour, G., & Abbasifard, M. (2021). The relationship between serum levels of interleukin-2 and IL-8 with circulating microRNA-10b in patients with COVID-19. *Iranian Journal of Immunology*, *18*(1), 65–73. https://doi.org/10.22034/iji.2021.88780.1904.

Balbi, M., Caroli, A., Corsi, A., Milanese, G., Surace, A., Di Marco, F., Novelli, L., Silva, M., Lorini, F. L., Duca, A., Cosentini, R., Sverzellati, N., Bonaffini, P. A., & Sironi, S. (2021). Chest X-ray for predicting mortality and the need for ventilatory support in COVID-19 patients presenting to the emergency department. *European Radiology*, *31*(4), 1999–2012. https://doi.org/10.1007/s00330-020-07270-1.

Borges da Silva Teles, G., Nunes Fonseca, E. K. U., Yokoo, P., Marques Almeida Silva, M., Yanata, E., Shoji, H., Bastos Duarte Passos, R., Caruso Chate, R., & Szarf, G. (2021). Performance of chest computed tomography in differentiating coronavirus disease 2019 from other viral infections using a standardized classification. *Journal of Thoracic Imaging*, *36*(1), 31–36. https://doi.org/10.1097/RTI.0000000000000563.

Callejas Rubio, J. L., Aomar Millán, I., Moreno Higueras, M., Muñoz Medina, L., López, M., & Ceballos Torres, Á. (2020). Tratamiento y evolución del síndrome de tormenta de citoquinas asociados a infección por SARS-CoV-2 en pacientes octogenarios [Evolution and treatment of storm cytoquine syndrome associated to SARS-CoV-2 infection among octogenarians]. *Revista Española de Geriatría y Gerontología*, *55*(5), 286–288. https://doi.org/10.1016/j.regg.2020.05.004.

Cardoso, A., Gonçalves, L., Inácio, J. R., Cunha, F., Freitas, J. V., Soares, R., Branco, C., Branco, T., Jacinto, N., Santos, L. R., Alvoeiro, L., & Lacerda, A. P. (2021). Bird Fancier's lung diagnosis in times of COVID-19. *Archivos de Bronconeumología*, *57*, 90–91. https://doi.org/10.1016/j.arbres.2020.09.016.

Chang, T. H., Wu, J. L., & Chang, L. Y. (2020). Clinical characteristics and diagnostic challenges of pediatric COVID-19: A systematic review and meta-analysis. *Journal of the Formosan Medical Association = Taiwan yi zhi*, *119*(5), 982–989. https://doi.org/10.1016/j.jfma.2020.04.007.

Frija-Masson, J., Debray, M. P., Boussouar, S., Khalil, A., Bancal, C., Motiejunaite, J., Galarza-Jimenez, M. A., Benzaquen, H., Penaud, D., Laveneziana, P., Malrin, R., Redheuil, A., Donciu, V., Lucidarme, O., Taillé, C., Guerder, A., Arnoult, F., Vidal-Petiot, E., Flamant, M., … Gonzalez-Bermejo, J. (2021). Residual ground glass opacities three months after Covid-19 pneumonia correlate to alteration of respiratory function: The post Covid M3 study. *Respiratory Medicine*, *184*, 106435. https://doi.org/10.1016/j.rmed.2021.106435.

Gold, M. S., Sehayek, D., Gabrielli, S., Zhang, X., McCusker, C., & Ben-Shoshan, M. (2020). COVID-19 and comorbidities: A systematic review and meta-analysis. *Postgraduate Medicine*, *132*(8), 749–755. https://doi.org/10.1080/00325481.2020.1786964. Stals, M., Kaptein, F. H. J., Kroft, L. J. M. et al. (2021). Challenges in the diagnostic approach of suspected pulmonary embolism in COVID-19 patients. *Postgraduate Medicine, 133* (sup1), 36–41.

Goyal, N., Chung, M., Bernheim, A., Keir, G., Mei, X., Huang, M., Li, S., & Kanne, J. P. (2020). Computed tomography features of coronavirus disease 2019 (COVID-19): A review for radiologists. *Journal of Thoracic Imaging*, *35*(4), 211–218. https://doi.org/10.1097/RTI.0000000000000527.

Han, X., Fan, Y., Alwalid, O., Li, N., Jia, X., Yuan, M., Li, Y., Cao, Y., Gu, J., Wu, H., & Shi, H. (2021). Six-month follow-up chest CT findings after severe COVID-19 pneumonia. *Radiology*, *299*(1), E177–E186. https://doi.org/10.1148/radiol.2021203153.

Hu, Y., Zhan, C., Chen, C., Ai, T., & Xia, L. (2020). Chest CT findings related to mortality of patients with COVID-19: A retrospective case-series study. *PLoS One*, *15*(8), e0237302. https://doi.org/10.1371/journal.pone.0237302.

Hurt, B., Kligerman, S., & Hsiao, A. (2020). Deep learning localization of pneumonia: 2019 coronavirus (COVID-19) outbreak. *Journal of Thoracic Imaging*, *35*(3), W87–W89. https://doi.org/10.1097/RTI.0000000000000512.

Imanova Yaghji, N., Kan, E. K., Akcan, S., Colak, R., & Atmaca, A. (2021). Hydroxychloroquine sulfate related hypoglycemia in a non-diabetic COVİD-19 patient: A case report and literature review. *Postgraduate Medicine*, *133*(5), 548–551. https://doi.org/10.1080/00325481.2021.1889820.

Kaleemi, R., Hilal, K., Arshad, A., Martins, R. S., Nankani, A., Tu, H., Basharat, S., & Ansar, Z. (2021). The association of chest radiographic findings and severity scoring with clinical outcomes in patients with COVID-19 presenting to the emergency department of a tertiary care hospital in Pakistan. *PLoS One*, *16*(1), e0244886. https://doi.org/10.1371/journal.pone.0244886.

Kell, D. B., Laubscher, G. J., & Pretorius, E. (2022). A central role for amyloid fibrin microclots in long COVID/PASC: Origins and therapeutic implications. *The Biochemical Journal*, *479*(4), 537–559. https://doi.org/10.1042/BCJ20220016.

Kong, M., Yang, H., Li, X., Shen, J., Xu, X., & Lv, D. (2020). Evolution of chest CT manifestations of COVID-19: A longitudinal study. *Journal of Thoracic Disease*, *12*(9), 4892–4907. https://doi.org/10.21037/jtd-20-1363.

Lang, M., Som, A., Mendoza, D. P., Flores, E. J., Li, M. D., Shepard, J. O., & Little, B. P. (2020). Detection of unsuspected coronavirus disease 2019 cases by computed tomography and retrospective implementation of the Radiological Society of North America/Society of Thoracic Radiology/American College of Radiology Consensus Guidelines. *Journal of Thoracic Imaging*, *35*(6), 346–353. https://doi.org/10.1097/RTI.0000000000000542.

Liang, W., Liang, H., Ou, L., Chen, B., Chen, A., Li, C., Li, Y., Guan, W., Sang, L., Lu, J., Xu, Y., Chen, G., Guo, H., Guo, J., Chen, Z., Zhao, Y., Li, S., Zhang, N., Zhong, N., … China Medical Treatment Expert Group for COVID-19. (2020). Development and validation of a clinical risk score to predict the occurrence of critical illness in hospitalized patients with COVID-19. *JAMA Internal Medicine*, *180*(8), 1081–1089. https://doi.org/10.1001/jamainternmed.2020.2033.

Litmanovich, D. E., Chung, M., Kirkbride, R. R., Kicska, G., & Kanne, J. P. (2020). Review of chest radiograph findings of COVID-19 pneumonia and suggested reporting language. *Journal of Thoracic Imaging*, *35*(6), 354–360. https://doi.org/10.1097/RTI.0000000000000541.

López-Candales, A., Mathur, P., & Xu, J. (2021). Managing multiple myeloma during COVID-19: an ongoing saga. *Postgraduate Medicine*, *133*(6), 589–591. https://doi.org/10.1080/00325481.2021.1885882.

Lu, T., & Pu, H. (2020). Computed tomography manifestations of 5 cases of the novel coronavirus disease 2019 (COVID-19) pneumonia from patients outside Wuhan. *Journal of Thoracic Imaging*, *35*(3), W90–W93. https://doi.org/10.1097/RTI.0000000000000508.

Ma, A., Zhang, L., Ye, X., Chen, J., Yu, J., Zhuang, L., Weng, C., Petersen, F., Wang, Z., & Yu, X. (2021). High levels of circulating IL-8 and soluble IL-2R are associated with prolonged illness in patients with severe COVID-19. *Frontiers in Immunology*, *12*, 626235. https://doi.org/10.3389/fimmu.2021.626235.

Martínez Chamorro, E., Díez Tascón, A., Ibáñez Sanz, L., Ossaba Vélez, S., & Borruel Nacenta, S. (2021). Radiologic diagnosis of patients with COVID-19. Diagnóstico radiológico del paciente con COVID-19. *Radiología*, *63*(1), 56–73. https://doi.org/10.1016/j.rx.2020.11.001.

Migliore, M. (2021). Ground glass opacities of the lung before, during and post COVID-19 pandemic. *Annals of Translational Medicine*, *9*(13), 1042. https://doi.org/10.21037/atm-21-2095.

Mogami, R., Lopes, A. J., Araújo Filho, R. C., Almeida, F. C. S. D., Messeder, A. M. D. C., Koifman, A. C. B., Guimarães, A. B., & Monteiro, A. (2021). Chest computed tomography in COVID-19 pneumonia: A retrospective study of 155 patients at a university hospital in Rio de Janeiro, Brazil. *Radiologia Brasileira*, *54*(1), 1–8. https://doi.org/10.1590/0100-3984.2020.0133.

Niu, Y., Huang, S., Zhang, H., Li, S., Li, X., Lv, Z., Yan, S., Fan, W., Zhai, Y., Wong, E., Wang, K., Zhang, Z., Chen, B., Xie, R., & Xian, J. (2021). Optimization of imaging parameters in chest CT for COVID-19 patients: An experimental phantom study. *Quantitative Imaging in Medicine and Surgery*, *11*(1), 380–391. https://doi.org/10.21037/qims-20-603.

Para, O., Caruso, L., Pestelli, G., Tangianu, F., Carrara, D., Maddaluni, L., Tamburello, A., Castelnovo, L., Fedi, G., Guidi, S., Pestelli, C., Pennella, B., Ciarambino, T., Nozzoli, C., & Dentali, F. (2022). Ferritin as prognostic marker in COVID-19: The FerVid study. *Postgraduate Medicine*, *134*(1), 58–63. https://doi.org/10.1080/00325481.2021.1990091.

Pettus, J., & Skolnik, N. (2021). Importance of diabetes management during the COVID-19 pandemic. *Postgraduate Medicine*, *133*(8), 912–919. https://doi.org/10.1080/00325481.2021.1978704.

Ramdani, H., Allali, N., Chat, L., & El Haddad, S. (2021). Covid-19 imaging: A narrative review. *Annals of Medicine and Surgery*, *69*, 102489. https://doi.org/10.1016/j.amsu.2021.102489.

Roig-Marín, N., & Roig-Rico, P. (2022a). Cardiac auscultation predicts mortality in elderly patients admitted for COVID-19. *Hospital Practice (1995)*, *50*(3), 228–235. https://doi.org/10.1080/21548331.2022.2069772.

Roig-Marín, N., & Roig-Rico, P. (2022b). Elderly hospitalized for COVID-19 and fever: A retrospective cohort study. *Experimental Aging Research*, *48*(4), 328–335. https://doi.org/10.1080/0361073X.2021.1973824.

Roig-Marín, N., & Roig-Rico, P. (2022c). Elderly people with dementia admitted for COVID-19: How different are they? *Experimental Aging Research*, *48*(2), 177–190. https://doi.org/10.1080/0361073X.2021.1943794.

Roig-Marín, N., & Roig-Rico, P. (2022d). Ground-glass opacity on emergency department chest X-ray: A risk factor for in-hospital mortality and organ failure in elderly admitted for COVID-19. *Postgraduate Medicine*, 1–8. Advance online publication https://doi.org/10.1080/00325481.2021.2021741.

Roig-Marín, N., & Roig-Rico, P. (2022e). In elderly patients with COVID-19 early onset of cough indicates a more severe disease development. *Infectious Diseases (London, England)*, *54*(2), 159–161. https://doi.org/10.1080/23744235.2021.1978538.

Roig-Marín, N., & Roig-Rico, P. (2022f). The deadliest lung lobe in COVID-19: A retrospective cohort study of elderly patients hospitalized for COVID-19. *Postgraduate Medicine*, *134*(5), 533–539. https://doi.org/10.1080/00325481.2022.2069356.

Rotondo, C., Corrado, A., Colia, R., Maruotti, N., Sciacca, S., Lops, L., Cici, D., Mele, A., Trotta, A., Lacedonia, D., Foschino Barbaro, M. P., & Cantatore, F. P. (2021). Possible role of higher serum level of myoglobin as predictor of worse prognosis in Sars-Cov 2 hospitalized patients. A monocentric retrospective study. *Postgraduate Medicine*, *133*(6), 688–693. https://doi.org/10.1080/00325481.2021.1949211.

Saha, M., Amin, S. B., Sharma, A., Kumar, T., & Kalia, R. K. (2022). AI-driven quantification of ground glass opacities in lungs of COVID-19 patients using 3D computed tomography imaging. *PLoS One*, *17*(3), e0263916. https://doi.org/10.1371/journal.pone.0263916.

Schiaffino, S., Tritella, S., Cozzi, A., Carriero, S., Blandi, L., Ferraris, L., & Sardanelli, F. (2020). Diagnostic performance of chest X-ray for COVID-19 pneumonia during the SARS-CoV-2 pandemic in Lombardy, Italy. *Journal of Thoracic Imaging*, *35*(4), W105–W106. https://doi.org/10.1097/RTI.0000000000000533.

Schmitt, W., & Marchiori, E. (2020). Covid-19: Round and oval areas of ground-glass opacity. *Pulmonology*, *26*(4), 246–247. https://doi.org/10.1016/j.pulmoe.2020.04.011.

Shaw, B., Daskareh, M., & Gholamrezanezhad, A. (2021). The lingering manifestations of COVID-19 during and after convalescence: Update on long-term pulmonary consequences of coronavirus disease 2019 (COVID-19). *La Radiologia Medica*, *126*(1), 40–46. https://doi.org/10.1007/s11547-020-01295-8.

Shivananda, A., & Amini, B. (2008). *Ground-glass opacification*. Radiopaedia.org. https://doi.org/10.53347/rid-1404.

Simpson, S., Kay, F. U., Abbara, S., Bhalla, S., Chung, J. H., Chung, M., Henry, T. S., Kanne, J. P., Kligerman, S., Ko, J. P., & Litt, H. (2020). Radiological Society of North America expert consensus statement on reporting chest CT findings related to COVID-19. Endorsed by the Society of Thoracic Radiology, the American College of Radiology, and RSNA—Secondary publication. *Journal of Thoracic Imaging*, *35*(4), 219–227. https://doi.org/10.1097/RTI.0000000000000524.

Vaccarello, A., Charley, E., Tharumia Jagadeesan, C., Talon, A., Munoz, J., Irandost, M., Varda, B., & Saeed, A. I. (2021). *An unusual case of a COVID19-associated solitary pulmonary nodule*. TP100. TP100 Unexpected COVID-19 case reports.

Varona Porres, D., Simó, M., Sánchez, A. L., Cabanzo, L., & Andreu, J. (2021). Single pulmonary nodule with reverse halo sign in COVID-19 infection: Incidental finding on FDG PET/CT scan. Nódulo pulmonar único con signo del halo inverso en infección COVID-19: hallazgo incidental en PET/TC. *Medicina Clínica*, *156*(2), 102. https://doi.org/10.1016/j.medcli.2020.10.011.

Wang, L., Jiaerken, Y., Li, Q., Huang, P., Shen, Z., Zhao, T., Zheng, H., Ji, W., Gao, Y., Xia, J., Cheng, J., Ma, J., Liu, J., Liu, Y., Su, M., Ruan, G., Shu, J., Ren, D., Zhao, Z., … Zhang, M. (2021). An illustrated guide to the imaging evolution of COVID in non-epidemic areas of Southeast China. *Frontiers in Molecular Biosciences*, *8*, 648180. https://doi.org/10.3389/fmolb.2021.648180Wa.

White-Dzuro, G., Gibson, L. E., Zazzeron, L., White-Dzuro, C., Sullivan, Z., Diiorio, D. A., Low, S. A., Chang, M. G., & Bittner, E. A. (2021). Multisystem effects of COVID-19: A concise review for practitioners. *Postgraduate Medicine*, *133*(1), 20–27. https://doi.org/10.1080/00325481.2020.1823094.

Wongvibulsin, S., Garibaldi, B. T., Antar, A., Wen, J., Wang, M. C., Gupta, A., Bollinger, R., Xu, Y., Wang, K., Betz, J. F., Muschelli, J., Bandeen-Roche, K., Zeger, S. L., & Robinson, M. L. (2021). Development of severe COVID-19 adaptive risk predictor (SCARP), a calculator to predict severe disease or death in hospitalized patients with COVID-19. *Annals of Internal Medicine*, *174*(6), 777–785. https://doi.org/10.7326/M20-6754.

Wu, Z., Liu, X., Liu, J., Zhu, F., Liu, Y., Liu, Y., & Peng, H. (2021). Correlation between ground-glass opacity on pulmonary CT and the levels of inflammatory cytokines in patients with moderate-to-severe COVID-19 pneumonia. *International Journal of Medical Sciences*, *18*(11), 2394–2400. https://doi.org/10.7150/ijms.56683.

Xia, T., Li, J., Gao, J., & Xu, X. (2020). Small solitary ground-glass nodule on CT as an initial manifestation of coronavirus disease 2019 (COVID-19) pneumonia. *Korean Journal of Radiology*, *21*(5), 545–549. https://doi.org/10.3348/kjr.2020.0240.

Yamanaka, S., Ota, S., Yoshida, Y., & Shinkai, M. (2021). COVID-19 pneumonia and an indelible ground-glass nodule. *Respirology Case Reports*, *9*(5), e00751. https://doi.org/10.1002/rcr2.751.

Yasin, R., & Gouda, W. (2020). Chest X-ray findings monitoring COVID-19 disease course and severity. *Egyptian Journal of Radiology and Nuclear Medicine*, *51*, 193. https://doi.org/10.1186/s43055-020-00296-x.

Zhang, W., Wei, X., Li, Y., & Chen, Y. (2022). Atypical characteristics of COVID-19 infection on CT imaging with preexisting combined pulmonary fibrosis and emphysema: A case report. *Medicine: Case Reports and Study Protocols*, *3*(3), e0225. https://doi.org/10.1097/md9.0000000000000225.

Zhang, Y. J., Yang, W. J., Liu, D., Cao, Y. Q., Zheng, Y. Y., Han, Y. C., Jin, R. S., Han, Y., Wang, X. Y., Pan, A. S., Dai, J. Y., Sun, Q. F., Zhao, F. Q., Yang, Q. Y., Zhang, J. H., Liu, S. J., Da, Q., Guo, W., Li, C. Q., … Qu, J.M. (2020). COVID-19 and early-stage lung cancer both featuring ground-glass opacities: A propensity score-matched study. *Translational Lung Cancer Research*, *9*(4), 1516–1527. https://doi.org/10.21037/tlcr-20-892.

Zhu, M. E., Wang, Q., Zhou, S., Wang, B., Ke, L., & He, P. (2021). Recombinant interleukin-2 stimulates lymphocyte recovery in patients with severe COVID-19. *Experimental and Therapeutic Medicine*, *21*(3), 227. https://doi.org/10.3892/etm.2021.9658.

Zhu, T., Wang, Y., Zhou, S., Zhang, N., & Xia, L. (2020). A comparative study of chest computed tomography features in young and older adults with Corona virus disease (COVID-19). *Journal of Thoracic Imaging*, *35*(4), W97–W101. https://doi.org/10.1097/RTI.0000000000000513.

Chapter 22

Cardiothoracic imaging in patients affected by COVID-19

Tommaso D'Angelo[a,b], Ludovica R.M. Lanzafame[a], M. Ludovica Carerj[a,c], Antonino Micari[a], Silvio Mazziotti[a], and Christian Booz[d]

[a]*Radiology Unit, BIOMORF Department, University Hospital Messina, Messina, Italy,* [b]*Department of Radiology and Nuclear Medicine, Erasmus MC, Rotterdam, The Netherlands,* [c]*Centro Cardiologico Monzino IRCCS, Milan, Italy,* [d]*Division of Experimental Imaging, Department of Diagnostic and Interventional Radiology, University Hospital Frankfurt, Frankfurt am Main, Germany*

Abbreviations

ACR	American College of Radiology
ACS	acute coronary syndrome
AP	anteroposterior
CAD	coronary artery disease
CCTA	coronary computed tomography angiography
CMR	cardiac magnetic resonance
CO-RADS	COVID-19 reporting and data system
COVID-19	coronavirus disease-19
CT	computed tomography
CT-SS	CT-severity score
CXR	chest x-ray
ECV	extracellular volume fraction
GGO	ground-glass opacity
LGE	late gadolinium enhancement
LIE	late iodine enhancement
LUS	lung ultrasound
MinIP	minimum intensity projection
MRI	magnetic resonance imaging
PA	posteroanterior
PE	pulmonary embolism
PPE	personal protective equipment
RSNA	Radiological Society of North America
RT-PCR	real-time reverse-transcriptase PCR
SARS-CoV-2	severe acute respiratory syndrome coronavirus 2

Introduction

Coronavirus Disease 19 (COVID-19) has significantly impacted hospital services. Radiology departments played a key role in the initial management of symptomatic patients, helping with the early detection of pulmonary involvement and risk stratification (Gaia et al., 2020).

However, the pandemic reduced the overall volume of radiological examinations, especially during the first wave (Fleckenstein et al., 2022). This was due to different factors from the overall reduction in general in-hospital assistance to the lower number of outpatient examinations due to a reduced tendency of the population to move. Despite fewer examinations, departments had to reorganize quickly to face the demanding challenge of managing infected patients while continuing to provide essential services.

Management, Body Systems, and Case Studies in COVID-19. https://doi.org/10.1016/B978-0-443-18703-2.00020-7
Copyright © 2024 Elsevier Inc. All rights are reserved, including those for text and data mining, AI training, and similar technologies.

Management of COVID-19 infection in radiology department

The SARS-CoV-2 pandemic represented an important challenge for radiological departments because these environments are often characterized by high flow and mobilization of patients, making these spaces a potential "hot spot" for spreading infectious diseases.

X-rays and computed tomography (CT) examinations played a pivotal role in the diagnostic workup of COVID-19 patients since the first days of the pandemic. However, one of the main priorities has always been ensuring employee health as well as continuity of department services.

Monitoring the spread of respiratory infectious diseases can be challenging. In fact, it does not depend only on the prevention of contact with infected patients but also with people or staff members who may be infected but presenting no symptoms.

Thanks to the adoption of practices such as telephone screening prior to examinations, temperature control at the department entrance, and quick anamnestic questionnaires, it was possible to mitigate this risk by filtering the access of both patients and employees.

In addition to the early detection of potentially infected subjects, it was also necessary to optimize the safety of the department. Personal protective equipment (PPE) can help reduce the risk of spreading the infection among staff. In particular, to protect from droplets, both patients and staff were required to wear protective masks. N95 respirators, with a higher filtration capacity, are preferred to conventional surgical masks. Indeed, according to a US Centers for Disease Control and Prevention study, wearing an N95 respirator is associated with a lower odds ratio (i.e., 0.17) of testing positive for SARS-CoV-2 in comparison with both surgical and cloth masks (Andrejko et al., 2022).

Furthermore, one of the greatest opportunities for radiology services has been the use of telemedicine. Indeed, all noninvasive radiological technique such as ultrasound, x-rays, computed tomography, and magnetic resonance imaging can be interpreted and reported remotely, thus further reducing the risk for of infection for healthcare workers.

Diagnostic work might be rapidly shifted to an out-of-hospital setting thanks to the use of home workstations (Sammer et al., 2020), guaranteeing all the services of the radiology department and facilitating social distancing. The internet performance available today can guarantee sending high-resolution radiological images in a very short time to other regions, countries, or continents. This may also help provide high diagnostic standards in remote regions with a shortage of experienced radiologists.

Thoracic imaging in SARS-COV-2

Chest imaging plays an essential role in the diagnosis and management of patients affected by SARS-CoV-2 infection (Kwee & Kwee, 2020).

Even though the first-line tool for the diagnosis of COVID-19 is a real-time reverse-transcriptase PCR (RT-PCR) test from a nasopharyngeal swab or respiratory secretions, chest imaging, and particularly a chest x-ray (CXR) and computed tomography (CT), are routinely used for early recognition and risk stratification as well as for guiding and evaluating the treatment response (Martínez Chamorro et al., 2021; Rubin et al., 2020; Varadarajan et al., 2021).

Currently, scientific and radiological associations do not recommend imaging tests as screening tools to detect COVID-19. On the other hand, imaging should be performed to monitor disease progression or complications (Martínez Chamorro et al., 2021).

The Fleischner Society published a consensus statement to guide medical practitioners in the use of chest imaging in COVID-19. According to the statement, imaging is recommended to establish patient pulmonary condition and to detect any underlying cardiorespiratory anomaly. This can be useful for stratifying the risk for clinical worsening and progression of disease in mildly symptomatic patients (e.g., age $>$ 65 years, cardiovascular disease, diabetes, chronic respiratory disease, hypertension, etc.). Moreover, imaging is recommended for monitoring patients with moderate to severe symptoms of COVID-19.

The statement does not specify whether radiography or CT should be performed:

- Main recommendations
  - Imaging is not routinely indicated as a screening test for COVID-19 in asymptomatic individuals.
  - Imaging is not indicated for patients with mild symptoms of COVID-19 unless they are at risk for disease progression.
  - Imaging is indicated for patients with moderate to severe symptoms regardless of COVID-19 test results.
  - Imaging is indicated for patients with COVID-19 and evidence of worsening respiratory status.
  - In a resource-constrained environment where access to CT is limited, a CXR may be preferred for patients with COVID-19 unless features of respiratory worsening warrant the use of CT.

- Additional recommendations
  - Daily chest radiographs are not indicated in stable intubated patients with COVID-19.
  - CT is indicated in patients with functional impairment and/or hypoxemia after recovery from COVID-19.
  - COVID-19 testing is indicated in patients incidentally found to have findings suggestive of COVID-19 on a CT scan (Rubin et al., 2020).

Chest x-ray (CXR)

Chest radiography is generally the first-line imaging modality used for patients with suspected COVID-19 because of its large accessibility and low cost. Even though CXR is considered less sensitive in the detection of pulmonary involvement in the early stage of the disease, it is useful for monitoring the rapid progression of lung abnormalities in COVID-19, especially in critical patients admitted to intensive care units; it is also valuable for clinicians to improve risk stratification (Sadiq et al., 2021).

An optimal CXR should include posteroanterior (PA) and lateral projections with the patient in a standing position (Martínez Chamorro et al., 2021).

Nonetheless, anteroposterior (AP) projection performed in a supine position at the bedside using a portable radiography unit is the first-line radiological test recommended by the American College of Radiology (ACR). This helps to reduce the risk of transmission and limits the need for moving potentially infected patients to the radiology department (American College of Radiology, 2020).

Although a CXR is limited, it enables the detection of possible complications (such as pneumothorax, subcutaneous emphysema, and pneumomediastinum), monitoring disease progression, and suggesting alternative diagnoses such as lobar pneumonia suggestive of bacterial superinfection, pneumothorax, and pleural effusion (Manna et al., 2020; Rubin et al., 2020).

Chest radiographs can appear normal, especially in the early stage of disease. Typical features of COVID-19 pneumonia include airspace opacities, described as consolidation or ground-glass opacity (GGO) (Figs. 1 and 2) with bilateral and peripheral distribution; these are more commonly detected in the lower zone (Rubin et al., 2020).

Consolidation denotes the occupation of the air alveolar spaces by pathological products such as water, pus, and blood. This determines an increase of parenchymal attenuation that obscures the margins of the vessels and the airways. GGOs are caused by partial filling of airspaces and appear as areas of mildly increased lung opacity while the pulmonary vessels remain discerned from the affected parenchyma (Hansell et al., 2008; Sun et al., 2020).

Atypical imaging findings consist of pleural effusion and pneumothorax. Pleural effusion may indicate bacterial pneumonia superimposition and presents as blunting of the costophrenic angle with an area of higher attenuation at the lung base (Sadiq et al., 2021).

Pneumothorax can be spontaneous or iatrogenic as a consequence of barotrauma induced by the invasive ventilation or the placement of a subclavian central venous line in COVID-19 patients. Pneumothorax presents as a thin, sharp white line at the visceral pleural edge (Carerj et al., 2022; Steinberger et al., 2022; Sun et al., 2020).

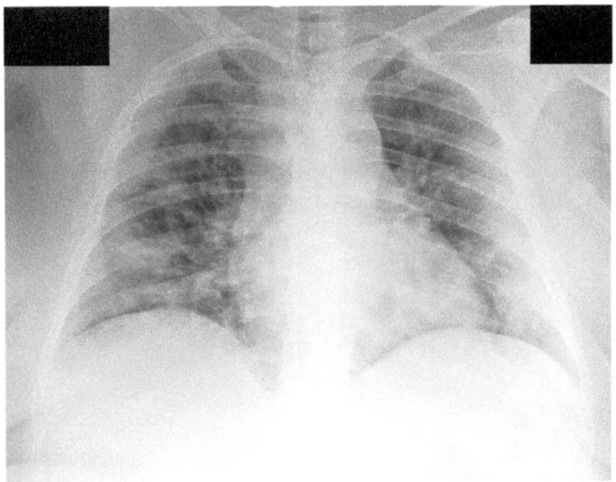

FIG. 1 49-year-old female patient affected by SARS-CoV-2 infection. The anteroposterior CXR reveals bilateral GGOs with peripheral distribution, and tending to consolidation.

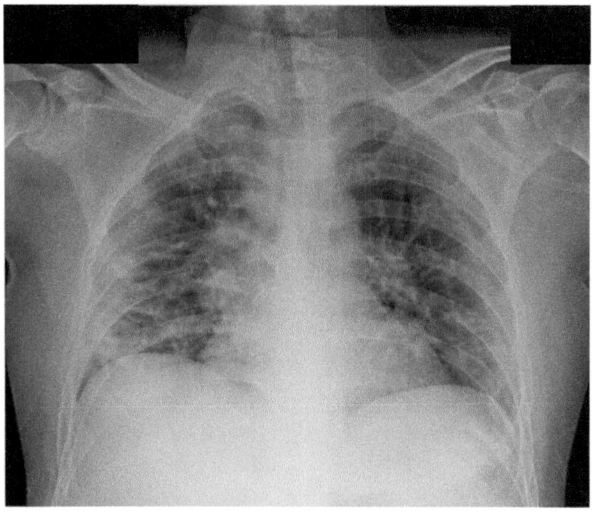

FIG. 2 52-year-old male patient. The anteroposterior CXR shows bilateral GGOs with peripheral distribution.

Chest computed tomography (CT)

A chest CT is considered the most accurate imaging modality to evaluate lung involvement in patients affected by SARS-CoV-2 infection.

Several studies have reported the sensitivity of CT with values ranging between 61% and 99% using RT-PCR as the reference standard (Pontone et al., 2021).

A chest CT should be performed with meticulous safety measures to decrease the risk of contamination. The scan protocol should include an unenhanced acquisition during a single inspiratory breath hold. The acquisition of an expiratory phase increases the radiation dose and is not recommended. The administration of intravenous contrast material should be reserved in case of suspicion of complications such as a pulmonary thromboembolism (Rubin et al., 2020).

The typical early features (0–4 days) of COVID-19 on CT imaging are GGOs (Fig. 3), whose incidence in the metaanalysis of Bao et al. was 83%. Lesion distribution is often bilateral (87%), peripheral (76%), and multilobar (78%) (Salehi et al., 2020).

The hallmark of the subsequent phase (5–8 days) is represented by the increasing number and size of GGOs; the gradual conversion of GGOs into multifocal, consolidative areas (Fig. 4); and the development of a "crazy-paving" pattern (incidence 14%), described as the presence of thickened interlobular septa and intralobular lines overlaid on a background of GGO (Bao et al., 2020; Pan et al., 2020).

The detection of GGOs in the earliest stages of disease is essential to enable a patient severity assessment. However, the smallest opacities are often hardly recognizable due to overlying bronchovascular structures. In these cases, minimum intensity projection (MinIP) reconstructions have been shown to improve the diagnostic accuracy in the detection of these anomalies (Booz et al., 2021).

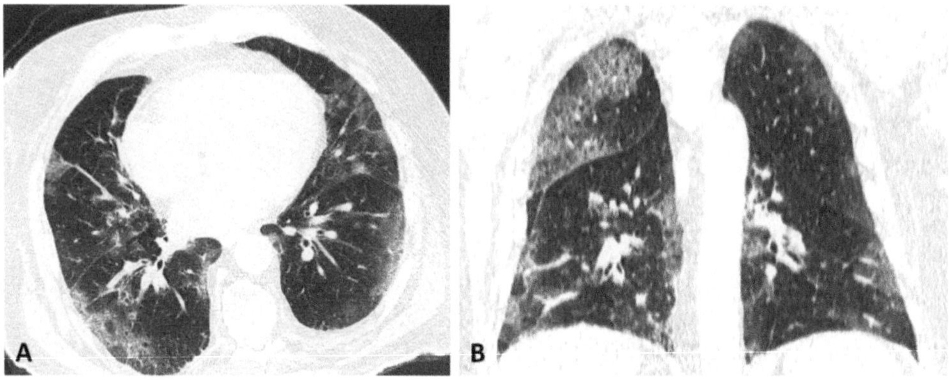

FIG. 3 Axial (A) and coronal (B) chest CT reconstructions showing GGO with peripheral distribution.

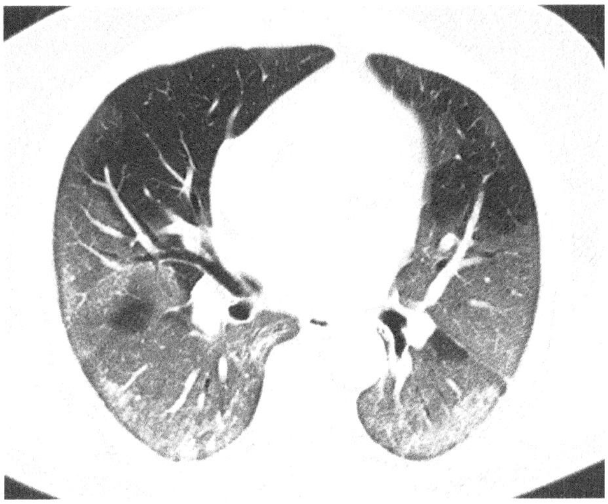

FIG. 4 Chest CT of a 61-year-old male patient demonstrating diffuse GGOs and consolidation areas mainly distributed to lower lobes.

Other less common findings include a "reversed halo sign" (also known as the "atoll sign" and characterized by a central GGO surrounded by a ring of consolidation) and a "spider web sign" (a triangular subpleural GGO with thickened interlobular septa conforming a web) (Godoy et al., 2012; Lomoro et al., 2020; Wu et al., 2020).

The peak COVID-19 phase (9–13 days) is characterized by greater lung involvement and high-density consolidations. Eventually, in the absorption stage, consolidations slowly tend to fade with the appearance of repaired lung signs, such as fibrotic bands (Bao et al., 2020).

However, none of these findings are specific for COVID-19 pneumonia for the noteworthy overlap with other diseases responsible for interstitial damage. In fact, CT specificity in the diagnosis of COVID-19 ranges between 24% and 94% (Pontone et al., 2021).

The Radiological Society of North America (RSNA) endorsed by the Society of Thoracic Radiology and the American College of Radiology (ACR) released a consensus statement to help radiologists distinguish among the findings potentially imputable to COVID-19 than other lung infections or diseases. Four categories of signs have been identified based on the likelihood of being ascribed to COVID-19 infection, including:

- Typical appearance:
  - Peripheral, bilateral, GGO with or without consolidation or visible intralobular lines ("crazy paving" pattern).
  - Multifocal GGOs of rounded morphology with or without consolidation or visible intralobular lines ("crazy paving" pattern).
  - Reverse halo sign or other findings of organizing pneumonia.
- Indeterminate appearance:
  - Absence of typical CT findings and the presence of:
    - Multifocal, diffuse, perihilar, or unilateral GGO with or without consolidation lacking a specific distribution and are nonrounded or nonperipheral.
    - Few small GGO with a nonrounded and nonperipheral distribution.
- Atypical appearance:
  - Absence of typical or indeterminate features and the presence of:
    - Isolated lobar or segmental consolidation without GGOs.
    - Discrete small nodules (e.g., centrilobular, tree-in-bud).
    - Lung cavitation.
    - Smooth interlobular septal thickening with pleural effusion.
- Negative for pneumonia: no CT features to suggest pneumonia (Simpson et al., 2020).

Late in 2020, the Dutch Radiological Society introduced a categorical system to evaluate suspected lung involvement by COVID-19 on chest CT scans, named the COVID-19 Reporting and Data System (CO-RADS).

CO-RADS includes categories from very low (CO-RADS category 1) to very high (CO-RADS category 5) on suspicion of COVID-19 as well as two supplementary categories, one for a technically insufficient examination (CO-RADS category 0) and one for RT-PCR-proven SARS-CoV-2 infection at the time of examination (CO-RADS category 6) (Prokop et al., 2020).

A chest CT was proposed as an important tool to stratify the risk of progression; together with clinical evaluation and laboratory results, it can be useful for monitoring therapy response (D'Angelo & Booz, 2022).

Several semiquantitative scores have been proposed to estimate the extension of pulmonary disease. Pan et al. (2020) used a semiquantitative scoring system to measure the pulmonary involvement scoring of each of the five lung lobes from 0 to 5 (0, no involvement; 1, <5% involvement; 2, 25% involvement; 3, 26%–49% involvement; 4, 50%–75% involvement; 5, >75% involvement). The total CT score is represented by the amount of the individual lobar scores ranging from 0 (no involvement) to 25 (maximum involvement).

The Yang et al. (2020) CT-severity score (CT-SS) divides lungs into 20 regions, assigning scores from 0 to 2 if the parenchymal opacification involved is 0%, less than 50%, or equal to or more than 50% of each region.

Lung ultrasound

Bedside real-time lung ultrasound (LUS) is a fast, noninvasive instrument to evaluate COVID-19 pneumonia that can be used as an added tool or when access to other imaging modalities is limited. LUS allows checking the severity of lung involvement in patients admitted to the emergency and intensive care units (Ji et al., 2020). LUS can also be used to detect lung involvement and monitor disease progression in children and pregnant women, reducing the exposure to ionizing radiation (Booz & D'Angelo, 2022; Nair et al., 2022).

LUS should be performed using convex transducers for the visualization of the parenchyma and the pleural line exploring at least 12 zones, six zones for the right lung and six zones for the left lung (Buonsenso et al., 2020; Ji et al., 2020).

The main reported findings associated with COVID-19 are B lines and small consolidations. B1 lines are associated with interstitial syndrome and reduced lung aeration while B2 lines are confluent lines configuring the "white lung" pattern, which correspond to CT ground-glass opacities. Consolidation appears as a subpleural hypoechoic region compared to a normal lung, mimicking the solid appearance of the liver. The presence of large consolidation should suggest superimposed bacterial pneumonia (Millington et al., 2021; Vetrugno et al., 2020).

Chest magnetic resonance (MR)

The role of magnetic resonance imaging (MRI) for lung visualization is limited due to several disadvantages. In particular, the modest presence of hydrogen protons in the lungs reduces the signal-to-noise ratio while unavoidable artifacts, due to physiologic respiratory and cardiac motion, further deteriorate the image quality. Nonetheless, some authors have suggested conducting a chest MRI for patients with suspected or confirmed COVID-19 to assess the extension of lung parenchymal lesions and monitor the disease progression in case CT is not available (Torkian et al., 2021; Vasilev et al., 2021).

The main MRI finding related to COVID-19 pneumonia is the rise of proton density due to the presence of lung consolidations or edema, showing an increase of signal intensity in T2-weighted images (Barreto et al., 2013; Torkian et al., 2021).

Cardiovascular imaging in SARS-COV-2 disease

Myocardial injury may be a consequence of COVID-19. The mechanism of myocardial injury in COVID-19 depends on numerous phenomena: from direct damage of cardiomyocytes (Hanson et al., 2022) to indirect cellular damage, caused by hypercoagulability or immune-mediated processes (Siripanthong et al., 2022).

A recent metaanalysis (Abate et al., 2021) conducted on 21,204 patients showed that acute myocardial injury occurs in 22.3% of patients with COVID-19. This happens especially in male patients with a history of smoking, diabetes mellitus, hypertension, coronary artery disease, or chronic obstructive pulmonary disease. Among these patients, mortality was approximately nine times higher compared to COVID-19 patients without acute myocardial injury.

Despite acute myocardial damage, patients who develop COVID-19 are at increased risk of long-term cardiovascular disease (Xie et al., 2022), including ischemic, nonischemic, and inflammatory heart disease as well as dysrhythmia and cerebrovascular and thrombotic disorders.

Imaging techniques such as ultrasound, CT, and MRI play essential roles in the noninvasive assessment of acute and long-term cardiovascular involvement in patients with COVID-19.

Echocardiography

An ultrasound is a quick, relatively inexpensive, and widely available technique that can be used to evaluate cardiac structure and function in patients with COVID-19. The examination can be performed at the bedside without any need for patients to be moved, thus reducing the infectious risk for hospital staff. In addition to the ultrasound examination of the heart, performed with a phased array probe, it's also possible to assess the structure and functionality of the whole cardiovascular system with linear and convex probes.

As demonstrated by a prospective international survey carried out on 1216 patients, an ultrasound examination was able to directly change the clinical management of 33% of patients with COVID-19. In particular, it's possible to initiate or modulate heart failure therapy, to optimize the hemodynamic support, or to change the level of patient care, based on cardiac ultrasound findings (Dweck et al., 2020).

The majority of patients with COVID-19, both in critical and noncritical care settings, presented nonspecific patterns of segmental or global ventricular disfunction. However, in a small group of these patients, ultrasound turned out to be essential in the diagnosis of conditions such as myocarditis, Takotsubo cardiomyopathy, or tamponade with a change in the therapeutic management of the patient.

Ultrasound may also be useful to evaluate long-term cardiac abnormalities in adults who recovered from COVID-19 (Garcia-Zamora et al., 2022). In a small percentage of patients, COVID-19 may cause a lasting reduction in left ventricle ejection fraction below 50%, without significant differences between asymptomatic and symptomatic patients. Another rare finding in patients with a previous SARS-CoV-2 infection is pericardial effusion, which can persist even if the patient is asymptomatic. However, it is usually mild and does not compromise the hemodynamic balance. (Fig. 5).

Cardiac computed tomography (CCT)

COVID-19 can be associated with a higher risk of acute coronary syndrome (ACS) during the acute phase of the infection (Xiong et al., 2020), mainly due to systemic inflammation that may promote endothelial damage and platelet activation, increasing the risk of rupture of the atherosclerotic plaques in the coronary arteries.

According to European Society of Cardiology (ESC) guidelines for the management of cardiovascular disease during the COVID-19 pandemic, low and intermediate risk non-ST elevation (NSTE) ACS patients should undergo noninvasive testing to select who can benefit from invasive strategies, favoring coronary CT angiography among all noninvasive diagnostic techniques if the expertise and equipment are available (Baigent et al., 2022). The CCT protocols in case of suspected acute myocardial injury should include an unenhanced CT scan to assess both pneumonia severity and calcium score; a retrospectively acquired angiographic triple rule-out scan to exclude coronary artery disease (CAD), pulmonary embolism (PE), and acute aortic injury; and optionally a late contrast enhancement scan (5–10 min after contrast injection) for myocardial fibrosis and extracellular volume fraction (ECV) evaluation (Palmisano et al., 2022).

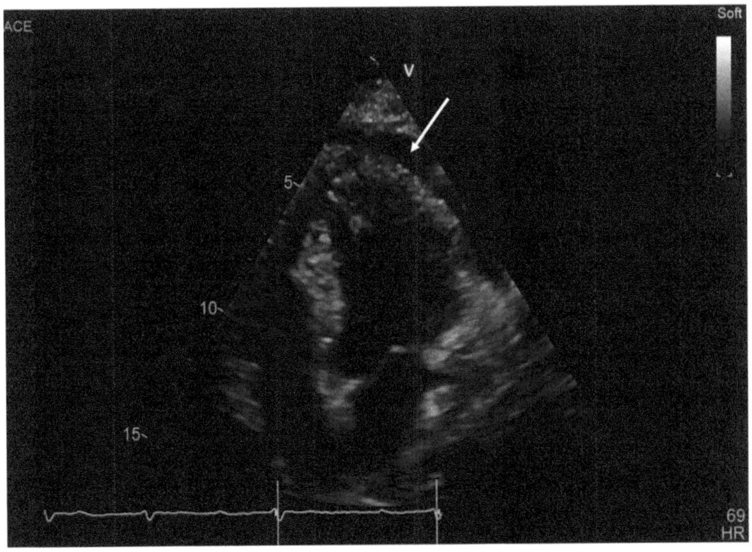

FIG. 5 Apical four-chamber view showing pericardial effusion surrounding left ventricular lateral wall (*arrow*) in a COVID-19 patient.

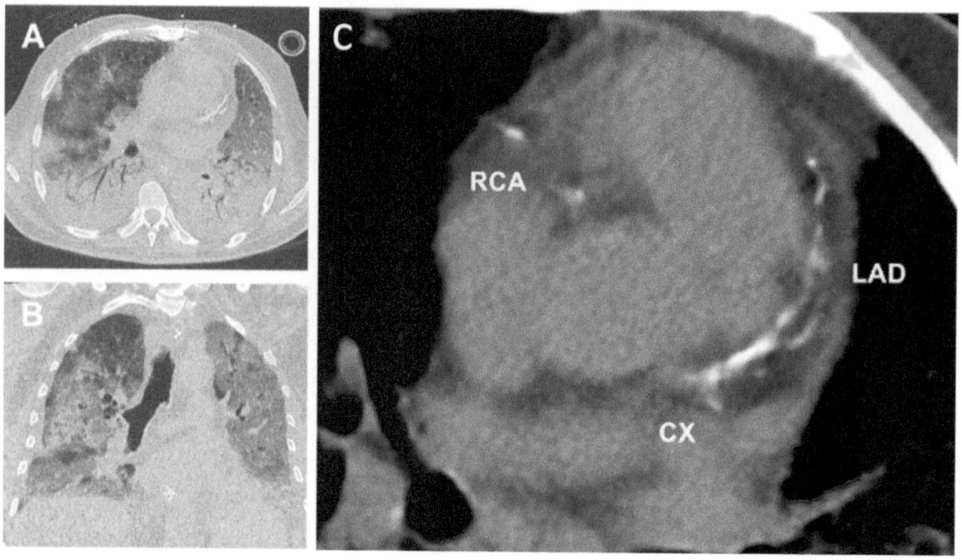

FIG. 6 Chest CT of a 67-year-old patient with interstitial SARS-CoV-2 related pneumonia. Extensive involvement of lung parenchyma as demonstrated on axial (A) and coronal (B) views correlates with high calcium load in the coronary arteries (C). *(Adapted from Koch, V., Gruenewald, L. D., Albrecht, M. H., Eichler, K., Gruber-Rouh, T., Yel, I., Alizadeh, L. S., Mahmoudi, S., Scholtz, J. E., Martin, S. S., Lenga, L., Vogl, T. J., Nour-Eldin, N. E. A., Bienenfeld, F., Hammerstingl, R. M., Graf, C., Sommer, C. M., Hardt, S. E., Mazziotti, S., … Booz, C. (2022). Lung opacity and coronary artery calcium score: A combined tool for risk stratification and outcome prediction in COVID-19 patients.* Academic Radiology, *29(6), 861–870. https://doi.org/10.1016/j.acra.2022.02.019.)*

CCT allows identifying the presence of clinically relevant CAD and recognizing signs of vulnerable plaques (low attenuation, positive remodeling, spotty calcifications, and napkin-ring sign), avoiding futile invasive coronary angiographies, shortening the hospital stay, reducing the spread of infection, and optimizing medical therapies in those patients without critical coronary stenosis (Alasnag et al., 2021).

Ccoronary artery calcification (CAC) scoring is a useful tool for risk stratification in patients with suspected CAD (Hecht et al., 2018). Recently, several studies showed that the coronary plaque burden was strongly associated with SARS-CoV-2 pulmonary involvement and adverse outcomes in patients with COVID-19 (Fig. 6) (Giannini et al., 2021; Koch et al., 2022; Lee et al., 2022; Scoccia et al., 2021).

Late iodine enhancement (LIE) images allow for the assessment of myocardial inflammation. The evaluation of the lung parenchyma, coronary arteries, pulmonary artery, and myocardium late enhancement may be used for a "quadruple rule-out" strategy for interstitial pneumonia, acute coronary syndrome, pulmonary embolism, and myocarditis and/or pericarditis (Pontone et al., 2020). LIE imaging can also be an additional diagnostic opportunity for those patients for whom MRI is contraindicated or not available or to reduce the need to sanitize the MRI room.

Cardiac magnetic resonance (CMR)

CMR plays a pivotal role in the noninvasive assessment of the myocardium, allowing for the evaluation of ventricular function, myocardial edema, and fibrosis (Palmisano et al., 2022).

Since the beginning of the COVID-19 pandemic, CMR has been shown to be a powerful imaging technique to assess both the acute and long-term myocardial involvement related to SARS-CoV-2 infection.

A CMR diagnostic workup should include cine-sequences for functional assessment, T2-based imaging (T2w- STIR or T2 mapping), and T1-based imaging (T1 mapping and late gadolinium enhancement) (Palmisano et al., 2022).

Acute myocarditis is a possible complication in patients hospitalized for COVID-19 (Ammirati et al., 2022; Tanacli et al., 2021). CMR allows for the detection of edema, capillary leakage, permanent myocardium damage, and fibrosis (Muscogiuri et al., 2022).

Ojha et al. (2021), in a metaanalysis on COVID-19 patients who underwent CMR, found myocarditis to be the most frequent diagnosis (40.2%), with patients presenting T1 and T2 mapping anomalies and myocardial edema as the most commonly reported findings.

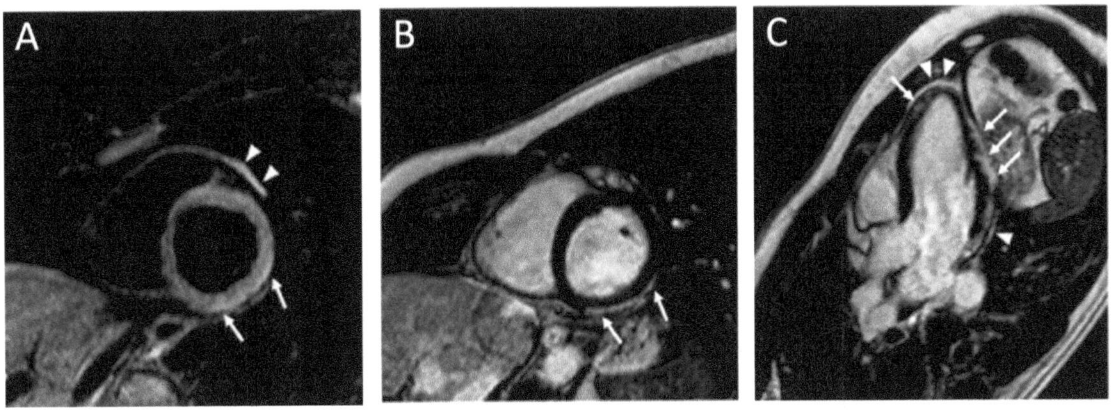

FIG. 7 Cardiac MRI of a 30-year-old man who received his second dose of an mRNA SARS-CoV-2 vaccine 72h earlier. Images show increased subepicardial signal intensity in the inferolateral myocardial segments (*arrows*) and in the pericardium (*arrowheads*) on T2-weighted short-tau inversion recovery images (A). Phase-sensitive inversion recovery sequences acquired along basal short-axis view (B) and three-chamber view (C) show myocardial late gadolinium enhancement on the inferolateral myocardial segments (*arrows*) and of the pericardium (*arrowheads*). *(Adapted from D'Angelo, T., Cattafi, A., Carerj, M. L., Booz, C., Ascenti, G., Cicero, G., Blandino, A., & Mazziotti, S. (2021). Myocarditis after SARS-CoV-2 vaccination: A vaccine-induced reaction?* Canadian Journal of Cardiology, *37(10), 1665–1667. https://doi.org/10.1016/j.cjca.2021.05.010.)*

Vaccination for SARS-CoV-2 has also been associated with the potential development of myocarditis, especially in young adults and adolescent males (Fig. 7) (Bozkurt et al., 2021; D'Angelo et al., 2021; Marshall et al., 2021; Watanabe et al., 2022).

However, vaccine-associated myocarditis usually tends to complete resolution, leaving no or minor residual myocardial scarring on late gadolinium enhancement imaging (Kravchenko et al., 2022).

Less common diagnoses include inducible ischemia and Takotsubo syndrome (Ojha et al., 2021).

Policies and procedures

In this chapter, our goal was to bring together the main scientific evidence regarding the radiological assessment and diagnosis of thoracic and cardiac involvement in SARS-CoV-2. We focused on all available noninvasive imaging methods, namely ultrasound, x-rays, computed tomography, and nuclear magnetic resonance of the chest and heart.

We decided to separate the thoracic and cardiovascular sections because of the different diagnostic techniques used to evaluate each system. Moreover, we included a section regarding the management of the SARS-CoV-2 pandemic inside radiological departments. In fact, given the crucial role of radiology departments in the evaluation of patients with COVID-19, one of the main aims from the beginning of the pandemic was to keep the services active and operational. For this reason, it hasn't been necessary to reorganize spaces and habits to cope with a highly transmissible virus responsible for a pandemic.

To collect all the resources used in our chapter, we performed a review of the scientific literature available until the end of 2022 within the PubMed, Google Scholar, EMBASE, and WHO databases. All the articles included were scientific papers written in English and available online with full text.

To facilitate the reader's consultation, we preferred to refer to the original articles, and not to reviews that directly cited them.

Most of the works cited were published in journals specializing in cardiovascular and thoracic radiology with a high impact factor and the relevance of each article cited was assessed by all authors.

We acknowledge that scientific research on this topic is only at an initial stage, and some of the information may change or evolve in the future. However, we believe that a summary chapter with the evidence available to date can be a benefit for readers interested in expanding their knowledge on this subject.

Applications to other areas

Our chapter is not of interest only to radiologists and imaging experts. In fact, the assessment of regions such as the chest and the cardiovascular system during SARS-CoV-2 infection involves many other figures dedicated to patient care. Indeed, noninvasive imaging has important implications in the stratification and follow-up of the patient and allows the clinician to change the therapeutical management of the patient and the level of care.

In addition, the reorganization of the radiology environments, as we described at the beginning of the chapter, could serve as a model for every department that offers a diagnostic service (i.e., echocardiography or nuclear medicine).

We have included references to the pathophysiology of the infection to explain some of the described findings of noninvasive imaging in COVID-19. Moreover, some of these findings (interstitial pneumonia, myocarditis, pericarditis) also represent common findings in other viral infections not caused by SARS-CoV-2. Therefore, part of the content of this chapter can also be applied transversally in other clinical scenarios.

It's certainly possible to think that all the new evidence in the field of radiological imaging that emerged during this pandemic could serve, in the future, to give us a better understanding of the presentation of other infectious diseases. Furthermore, the experiences with remote workstations that allow diagnosticians to work from home may also help in the future to provide high diagnostic standards in remote regions.

Mini-dictionary of terms

Ultrasound imaging: A noninvasive imaging technique that uses high-frequency sound waves to assess organs and structures. An ultrasound probe emits sound waves that pass through tissues in different ways depending on their composition or that can be reflected. Based on this phenomenon, images of organs and tissues are reconstructed in scales of gray.

Computed tomography: A noninvasive technique that allows acquiring tridimensional images of inorganic and organic materials. CT systems translate the attenuated radiation collected by a detector or system of detectors into gray-scale values. The attenuation takes place because when an x-ray beam is passed through a material, it is either absorbed or scattered differently depending on its energy level and on the nature of the material. Modern CT systems use an x-ray beam and a row of detectors that rotates around the patient and allows acquiring a section of the body in less than a second.

Magnetic resonance imaging: A noninvasive medical imaging test that uses magnetic fields and radio waves to produce images of the body. It doesn't involve the use of x-rays or other types of ionizing radiation. Compared to CT imaging, it offers a better assessment of soft tissues (i.e., muscles, tendons, brain, myocardium, etc.) but with a longer acquisition time. As we have described, in COVID-19 patients it may be particularly useful in the evaluation of myocardial edema in case of suspected myocarditis.

Interstitial pneumonia: It may be the result of infection of the lung parenchyma by SARS-CoV-2. It is characterized by an inflammatory process that occurs mainly within the interstitial walls rather than inside the alveoli. The radiological aspect foresees the presence of ground-glass opacities and consolidations, mainly basal or subpleural. However, it can also be the result of other infectious and noninfectious conditions.

Myocarditis: A condition where an inflammatory process takes place inside the myocardium. It can be caused by various infectious, toxic, or immunological agents. It's a possible complication in patients with COVID-19, and can occur asymptomatically. It has also been very rarely reported after mRNA vaccination for SARS-CoV-2.

Summary points

- At the initial stage of the COVID-19 pandemic, radiological departments played a key role in the early identification of symptomatic patients with pulmonary involvement highly suggestive for SARS-CoV-2 infection.
- One of the main priorities in the pandemic was to ensure the safety of healthcare employees. For this reason, implementing measures that guaranteed a reduction in the virus spread in hospitals was crucial to keep all services active.
- COVID-19 has contributed to the adoption of systems that allow remote reporting in radiology ("teleradiology"). This could be useful in the future, beyond the pandemic, to offer high diagnostic standards to remote locations.
- Noninvasive imaging techniques are a cornerstone in the assessment, diagnosis, and stratification of the involvement of the respiratory and cardiovascular systems in patients with SARS-CoV-2 infection.
- Computed tomography of the chest allows for the evaluation of pulmonary parenchyma and to assess the presence and the extent of interstitial pneumonia. The addition of contrast medium and cardiosynchronization can further expand the information derived from CT imaging, allowing for the evaluation of coronary arteries and the myocardium.
- COVID-19 can lead to myocardial inflammation, which is optimally evaluated with MRI. However, the long-term implications of COVID-19 related myocarditis, especially in asymptomatic subjects, have to be further investigated.

References

Abate, S. M., Mantefardo, B., Nega, S., Chekole, Y. A., Basu, B., Ali, S. A., & Taddesse, M. (2021). Global burden of acute myocardial injury associated with COVID-19: A systematic review, meta-analysis, and meta-regression. In *Vol. 68. Annals of Medicine and Surgery* Elsevier Ltd. https://doi.org/10.1016/j.amsu.2021.102594.

Alasnag, M., Ahmed, W., Al-Nasser, I., & Al-Shaibi, K. (2021). Role of cardiovascular computed tomography in acute coronary syndromes during the COVID-19 pandemic-single center snapshot study. *Frontiers in Cardiovascular Medicine*, *8*. https://doi.org/10.3389/fcvm.2021.665735.

American College of Radiology (2020). ACR recommendations for the use of chest radiography and computed tomography (CT) for suspected COVID-19 infection. ACR website. www.acr.org/Advocacy-and-Economics/ACR-Position-Statements/Recommendations-for-Chest-Radiography-and-CT-for-Suspected-COVID19-Infection. Updated 22 March 2020 (Accessed 10 December 2022).

Ammirati, E., Lupi, L., Palazzini, M., Hendren, N. S., Grodin, J. L., Cannistraci, C.v., Schmidt, M., Hekimian, G., Peretto, G., Bochaton, T., Hayek, A., Piriou, N., Leonardi, S., Guida, S., Turco, A., Sala, S., Uribarri, A., van de Heyning, C. M., Mapelli, M., … Metra, M. (2022). Prevalence, characteristics, and outcomes of COVID-19-associated acute myocarditis. *Circulation*, *145*(15), 1123–1139. https://doi.org/10.1161/CIRCULATIONAHA.121.056817.

Andrejko, K. L., Pry, J. M., Myers, J. F., Fukui, N., Deguzman, J. L., Openshaw, J., Watt, J. P., Lewnard, J. A., & Jain, S. (2022). Effectiveness of face mask or respirator use in indoor public settings for prevention of SARS-CoV-2 infection—California, February-December 2021. *Morbidity and Mortality Weekly Report*. https://doi.org/10.1093/cid/ciab640/6324500#supplementary-data.

Baigent, C., Windecker, S., Andreini, D., Arbelo, E., Barbato, E., Bartorelli, A. L., Baumbach, A., Behr, E. R., Berti, S., Bueno, H., Capodanno, D., Cappato, R., Chieffo, A., Collet, J. P., Cuisset, T., de Simone, G., Delgado, V., Dendale, P., Dudek, D., … Williams, B. (2022). ESC guidance for the diagnosis and management of cardiovascular disease during the COVID-19 pandemic: Part 2-care pathways, treatment, and follow-up. *European Heart Journal*, *43*(11), 1059–1103. https://doi.org/10.1093/eurheartj/ehab697.

Bao, C., Liu, X., Zhang, H., Li, Y., & Liu, J. (2020). Coronavirus disease 2019 (COVID-19) CT findings: A systematic review and meta-analysis. *Journal of the American College of Radiology*, *17*(6), 701–709. https://doi.org/10.1016/j.jacr.2020.03.006.

Barreto, M. M., Rafful, P. P., Rodrigues, R. S., Zanetti, G., Hochhegger, B., Souza, A. S., Guimarães, M. D., & Marchiori, E. (2013). Correlation between computed tomographic and magnetic resonance imaging findings of parenchymal lung diseases. *European Journal of Radiology*, *82*(9). https://doi.org/10.1016/j.ejrad.2013.04.037.

Booz, C., & D'Angelo, T. (2022). Lung ultrasound in pregnant COVID-19 patients: An added tool to expand patient care. *Journal of Clinical Ultrasound*. https://doi.org/10.1002/jcu.23283.

Booz, C., Vogl, T. J., Joseph Schoepf, U., Caruso, D., Inserra, M. C., Yel, I., Martin, S. S., Bucher, A. M., Lenga, L., Caudo, D., Schreckenbach, T., Schoell, N., Huegel, C., Stratmann, J., Vasa-Nicotera, M., Rachovitsky-Duarte, D. E., Laghi, A., de Santis, D., Mazziotti, S., … Albrecht, M. H. (2021). Value of minimum intensity projections for chest CT in COVID-19 patients. *European Journal of Radiology*, *135*. https://doi.org/10.1016/j.ejrad.2020.109478.

Bozkurt, B., Kamat, I., & Hotez, P. J. (2021). Myocarditis with COVID-19 mRNA vaccines. *Circulation*, *144*(6), 471–484. Lippincott Williams and Wilkins https://doi.org/10.1161/CIRCULATIONAHA.121.056135.

Buonsenso, D., Raffaelli, F., Tamburrini, E., Biasucci, D. G., Salvi, S., Smargiassi, A., Inchingolo, R., Scambia, G., Lanzone, A., Testa, A. C., & Moro, F. (2020). Clinical role of lung ultrasound for diagnosis and monitoring of COVID-19 pneumonia in pregnant women. *Ultrasound in Obstetrics and Gynecology*, *56*(1), 106–109. https://doi.org/10.1002/uog.22055.

Carerj, M. L., Bucolo, G. M., Mazziotti, S., Blandino, A., Booz, C., Cicero, G., & D'Angelo, T. (2022). Pulmonary barotrauma in patient suffering from COVID-19. *Heliyon*, *8*(1). https://doi.org/10.1016/j.heliyon.2022.e08745.

D'Angelo, T., & Booz, C. (2022). High correlation between CT and clinical data in COVID-19 progression. *Journal of Clinical Ultrasound*, *50*(3), 383–384. John Wiley and Sons Inc https://doi.org/10.1002/jcu.23163.

D'Angelo, T., Cattafi, A., Carerj, M. L., Booz, C., Ascenti, G., Cicero, G., Blandino, A., & Mazziotti, S. (2021). Myocarditis after SARS-CoV-2 vaccination: A vaccine-induced reaction? *Canadian Journal of Cardiology*, *37*(10), 1665–1667. https://doi.org/10.1016/j.cjca.2021.05.010.

Dweck, M. R., Bularga, A., Hahn, R. T., Bing, R., Lee, K. K., Chapman, A. R., White, A., di Salvo, G., Sade, L. E., Pearce, K., Newby, D. E., Popescu, B. A., Donal, E., Cosyns, B., Edvardsen, T., Mills, N. L., & Haugaa, K. (2020). Global evaluation of echocardiography in patients with COVID-19. *European Heart Journal Cardiovascular Imaging*, *21*(9), 949–958. https://doi.org/10.1093/ehjci/jeaa178.

Fleckenstein, F. N., Maleitzke, T., Böning, G., Kahl, V., Petukhova-Greenstein, A., Kucukkaya, A. S., Gebauer, B., Hamm, B., & Aigner, A. (2022). Changes of radiological examination volumes over the course of the COVID-19 pandemic: A comprehensive analysis of the different waves of infection. *Insights Into Imaging*, *13*(1). https://doi.org/10.1186/s13244-022-01181-z.

Gaia, C., Maria Chiara, C., Silvia, L., Chiara, A., Maria Luisa, D. C., Giulia, B., Silvia, P., Lucia, C., Alessandra, T., Annarita, S., Cristina, V., Maria, A., Maria Rosaria, D. A., Giacinta, A., Riccardo, G., Zaher, K., Andrea, L., Maddalena, B., Catalano, C., & Paolo, R. (2020). Chest CT for early detection and management of coronavirus disease (COVID-19): A report of 314 patients admitted to emergency department with suspected pneumonia. *Radiologia Medica*, *125*(10), 931–942. https://doi.org/10.1007/s11547-020-01256-1.

Garcia-Zamora, S., Picco, J. M., Lepori, A. J., Galello, M. I., Saad, A. K., Ayón, M., Monga-Aguilar, N., Shehadeh, I., Manganiello, C. F., Izaguirre, C., Fallabrino, L. N., Clavero, M., Mansur, F., Ghibaudo, S., Sevilla, D., Cado, C. A., Priotti, M., Liblik, K., Gastaldello, N., & Merlo, P. M. (2022). Abnormal echocardiographic findings after COVID-19 infection: A multicenter registry. *International Journal of Cardiovascular Imaging*. https://doi.org/10.1007/s10554-022-02706-9.

Giannini, F., Toselli, M., Palmisano, A., Cereda, A., Vignale, D., Leone, R., Nicoletti, V., Gnasso, C., Monello, A., Manfrini, M., Khokhar, A., Sticchi, A., Biagi, A., Turchio, P., Tacchetti, C., Landoni, G., Boccia, E., Campo, G., Scoccia, A., … Esposito, A. (2021). Coronary and total thoracic calcium scores predict mortality and provides pathophysiologic insights in COVID-19 patients. *Journal of Cardiovascular Computed Tomography, 15*(5), 421–430. https://doi.org/10.1016/j.jcct.2021.03.003.

Godoy, M., Viswanathan, C., Marchiori, E., Truong, M. T., Benveniste, M. F., Rossi, S., & Marom, E. M. (2012). The reversed halo sign: Update and differential diagnosis. *British Journal of Radiology, 85*(1017), 1226–1235. https://doi.org/10.1259/bjr/54532316.

Hansell, D. M., Bankier, A. A., MacMahon, H., McLoud, T. C., Müller, N. L., & Remy, J. (2008). Fleischner society: Glossary of terms for thoracic imaging. *Radiology, 246*(3), 697–722. https://doi.org/10.1148/radiol.2462070712.

Hanson, P. J., Liu-Fei, F., Ng, C., Minato, T. A., Lai, C., Hossain, A. R., Chan, R., Grewal, B., Singhera, G., Rai, H., Hirota, J., Anderson, D. R., Radio, S. J., & McManus, B. M. (2022). Characterization of COVID-19-associated cardiac injury: Evidence for a multifactorial disease in an autopsy cohort. *Laboratory Investigation, 102*(8), 814–825. https://doi.org/10.1038/s41374-022-00783-x.

Hecht, H. S., Blaha, M. J., Kazerooni, E. A., Cury, R. C., Budoff, M., Leipsic, J., & Shaw, L. (2018). CAC-DRS: Coronary artery calcium data and reporting system. An expert consensus document of the Society of Cardiovascular Computed Tomography (SCCT). *Journal of Cardiovascular Computed Tomography, 12*(3), 185–191. https://doi.org/10.1016/j.jcct.2018.03.008.

Ji, L., Cao, C., Gao, Y., Zhang, W., Xie, Y., Duan, Y., Kong, S., You, M., Ma, R., Jiang, L., Liu, J., Sun, Z., Zhang, Z., Wang, J., Yang, Y., Lv, Q., Zhang, L., Li, Y., Zhang, J., & Xie, M. (2020). Prognostic value of bedside lung ultrasound score in patients with COVID-19. *Critical Care, 24*(1). https://doi.org/10.1186/s13054-020-03416-1.

Koch, V., Gruenewald, L. D., Albrecht, M. H., Eichler, K., Gruber-Rouh, T., Yel, I., Alizadeh, L. S., Mahmoudi, S., Scholtz, J. E., Martin, S. S., Lenga, L., Vogl, T. J., Nour-Eldin, N. E. A., Bienenfeld, F., Hammerstingl, R. M., Graf, C., Sommer, C. M., Hardt, S. E., Mazziotti, S., … Booz, C. (2022). Lung opacity and coronary artery calcium score: A combined tool for risk stratification and outcome prediction in COVID-19 patients. *Academic Radiology, 29*(6), 861–870. https://doi.org/10.1016/j.acra.2022.02.019.

Kravchenko, D., Isaak, A., Mesropyan, N., Bischoff, L. M., Pieper, C. C., Attenberger, U., Kuetting, D., Zimmer, S., Hart, C., & Luetkens, J. A. (2022). Cardiac magnetic resonance follow-up of COVID-19 vaccine associated acute myocarditis. *Frontiers in Cardiovascular Medicine, 9*. https://doi.org/10.3389/fcvm.2022.1049256.

Kwee, T. C., & Kwee, R. M. (2020). Chest CT in COVID-19: What the radiologist needs to know. *Radiographics, 40*(7), 1848–1865. https://doi.org/10.1148/rg.2020200159.

Lee, K. K., Rahimi, O., Lee, C. K., Shafi, A., & Hawwass, D. (2022). A meta-analysis: Coronary artery calcium score and COVID-19 prognosis. *Medical Science, 10*(1), 5. https://doi.org/10.3390/medsci10010005.

Lomoro, P., Verde, F., Zerboni, F., Simonetti, I., Borghi, C., Fachinetti, C., Natalizi, A., & Martegani, A. (2020). COVID-19 pneumonia manifestations at the admission on chest ultrasound, radiographs, and CT: Single-center study and comprehensive radiologic literature review. *European Journal of Radiology Open, 7*. https://doi.org/10.1016/j.ejro.2020.100231.

Manna, S., Wruble, J., Maron, S. Z., Toussie, D., Voutsinas, N., Finkelstein, M., Cedillo, M. A., Diamond, J., Eber, C., Jacobi, A., Chung, M., & Bernheim, A. (2020). COVID-19: A multimodality review of radiologic techniques, clinical utility, and imaging features. *Radiology: Cardiothoracic Imaging, 2*(3). https://doi.org/10.1148/ryct.2020200210.

Marshall, M., Ferguson, I. D., Lewis, P., Jaggi, P., Gagliardo, C., Collins, J. S., Shaughnessy, R., Caron, R., Fuss, C., Corbin, K. J. E., Emuren, L., Faherty, E., Hall, E. K., di Pentima, C., Oster, M. E., Paintsil, E., Siddiqui, S., Timchak, D. M., & Guzman-Cottrill, J. A. (2021). Symptomatic acute myocarditis in 7 adolescents after pfizer-biontech covid-19 vaccination. *Pediatrics, 148*(3). https://doi.org/10.1542/peds.2021-052478.

Martínez Chamorro, E., Díez Tascón, A., Ibáñez Sanz, L., Ossaba Vélez, S., & Borruel Nacenta, S. (2021). Radiologic diagnosis of patients with COVID-19. *Radiología, 63*(1), 56–73. https://doi.org/10.1016/j.rx.2020.11.001.

Millington, S. J., Koenig, S., Mayo, P., & Volpicelli, G. (2021). Lung ultrasound for patients with coronavirus disease 2019 pulmonary disease. *Chest, 159*(1), 205–211. Elsevier Inc https://doi.org/10.1016/j.chest.2020.08.2054.

Muscogiuri, G., Guaricci, A. I., Cau, R., Saba, L., Senatieri, A., Chierchia, G., Pontone, G., Volpato, V., Palmisano, A., Esposito, A., Basile, P., Marra, P., D'angelo, T., Booz, C., Rabbat, M., & Sironi, S. (2022). Multimodality imaging in acute myocarditis. *Journal of Clinical Ultrasound, 50*(8), 1097–1109. John Wiley and Sons Inc https://doi.org/10.1002/jcu.23310.

Nair, A. V., Ramanathan, S., & Venugopalan, P. (2022). Chest imaging in pregnant patients with COVID-19: Recommendations, justification, and optimization. *Acta Radiologica Open, 11*(2). https://doi.org/10.1177/20584601221077394. 205846012210773.

Ojha, V., Verma, M., Pandey, N. N., Mani, A., Malhi, A. S., Kumar, S., Jagia, P., Roy, A., & Sharma, S. (2021). Cardiac magnetic resonance imaging in coronavirus disease 2019 (COVID-19): A systematic review of cardiac magnetic resonance imaging findings in 199 patients. *Journal of Thoracic Imaging, 36*(2), 73–83. Lippincott Williams and Wilkins https://doi.org/10.1097/RTI.0000000000000574.

Palmisano, A., Gambardella, M., D'Angelo, T., Vignale, D., Ascione, R., Gatti, M., Peretto, G., Federico, F., Shah, A., & Esposito, A. (2022). Advanced cardiac imaging in the spectrum of COVID-19 related cardiovascular involvement. *Clinical Imaging, 90*, 78–89. Elsevier Inc https://doi.org/10.1016/j.clinimag.2022.07.009.

Pan, F., Ye, T., Sun, P., Gui, S., Liang, B., Li, L., Zheng, D., Wang, J., Hesketh, R. L., Yang, L., & Zheng, C. (2020). Time course of lung changes at chest CT during recovery from coronavirus disease 2019 (COVID-19). *Radiology, 295*(3), 715–721. https://doi.org/10.1148/radiol.2020200370.

Pontone, G., Baggiano, A., Conte, E., Teruzzi, G., Cosentino, N., Campodonico, J., Rabbat, M. G., Assanelli, E., Palmisano, A., Esposito, A., & Trabattoni, D. (2020). "Quadruple rule-out" with computed tomography in a COVID-19 patient with equivocal acute coronary syndrome presentation. *JACC: Cardiovascular Imaging, 13*(8), 1854–1856. https://doi.org/10.1016/J.JCMG.2020.04.012.

Pontone, G., Scafuri, S., Mancini, M. E., Agalbato, C., Guglielmo, M., Baggiano, A., Muscogiuri, G., Fusini, L., Andreini, D., Mushtaq, S., Conte, E., Annoni, A., Formenti, A., Gennari, A. G., Guaricci, A. I., Rabbat, M. R., Pompilio, G., Pepi, M., & Rossi, A. (2021). Role of computed tomography in COVID-19. *Journal of Cardiovascular Computed Tomography, 15*(1), 27–36. Elsevier Inc https://doi.org/10.1016/j.jcct.2020.08.013.

Prokop, M., van Everdingen, W., van Rees Vellinga, T., van Ufford, H. Q., Stöger, L., Beenen, L., Geurts, B., Gietema, H., Krdzalic, J., Schaefer-Prokop, C., van Ginneken, B., & Brink, M. (2020). CO-RADS: A categorical CT assessment scheme for patients suspected of having COVID-19-definition and evaluation. *Radiology*, *296*(2), E97–E104. https://doi.org/10.1148/radiol.2020201473.

Rubin, G. D., Ryerson, C. J., Haramati, L. B., Sverzellati, N., Kanne, J. P., Raoof, S., et al. (2020). The role of chest imaging in patient management during the COVID-19 pandemic: A multinational consensus statement from the Fleischner Society. *Radiology*, *296*(1), 172–180. https://doi.org/10.1148/radiol.2020201365.

Sadiq, Z., Rana, S., Mahfoud, Z., & Raoof, A. (2021). Systematic review and meta-analysis of chest radiograph (CXR) findings in COVID-19. *Clinical Imaging*, *80*, 229–238. Elsevier Inc https://doi.org/10.1016/j.clinimag.2021.06.039.

Salehi, S., Abedi, A., Balakrishnan, S., & Gholamrezanezhad, A. (2020). Coronavirus disease 2019 (COVID-19): A systematic review of imaging findings in 919 patients. *American Journal of Roentgenology*, *215*(1), 87–93. American Roentgen Ray Society https://doi.org/10.2214/AJR.20.23034.

Sammer, M. B. K., Sher, A. C., Huisman, T. A. G. M., & Seghers, V. J. (2020). Response to the COVID-19 pandemic: Practical guide to rapidly deploying home workstations to guarantee radiology services during quarantine, social distancing, and stay home orders. *American Journal of Roentgenology*, *215*(6), 1417–1420. American Roentgen Ray Society https://doi.org/10.2214/AJR.20.23297.

Scoccia, A., Gallone, G., Cereda, A., Palmisano, A., Vignale, D., Leone, R., Nicoletti, V., Gnasso, C., Monello, A., Khokhar, A., Sticchi, A., Biagi, A., Tacchetti, C., Campo, G., Rapezzi, C., Ponticelli, F., Danzi, G. B., Loffi, M., Pontone, G., … Toselli, M. (2021). Impact of clinical and subclinical coronary artery disease as assessed by coronary artery calcium in COVID-19. *Atherosclerosis*, *328*, 136–143. https://doi.org/10.1016/j.atherosclerosis.2021.03.041.

Simpson, S., Kay, F. U., Abbara, S., Bhalla, S., Chung, J. H., Chung, M., Henry, T. S., Kanne, J. P., Kligerman, S., Ko, J. P., & Litt, H. (2020). Radiological society of North America expert consensus document on reporting chest CT findings related to COVID-19: Endorsed by the society of thoracic radiology, the American college of radiology, and RSNA. *Radiology: Cardiothoracic Imaging*, *2*(2). https://doi.org/10.1148/ryct.2020200152.

Siripanthong, B., Asatryan, B., Hanff, T. C., Chatha, S. R., Khanji, M. Y., Ricci, F., Muser, D., Ferrari, V. A., Nazarian, S., Santangeli, P., Deo, R., Cooper, L. T., Mohiddin, S. A., & Chahal, C. A. A. (2022). The pathogenesis and long-term consequences of COVID-19 cardiac injury. *JACC: Basic to Translational Science*, *7*(3P1), 294–308. Elsevier Inc https://doi.org/10.1016/j.jacbts.2021.10.011.

Steinberger, S., Finkelstein, M., Pagano, A., Manna, S., Toussie, D., Chung, M., Bernheim, A., Concepcion, J., Gupta, S., Eber, C., Dua, S., & Jacobi, A. H. (2022). Barotrauma in COVID 19: Incidence, pathophysiology, and effect on prognosis. *Clinical Imaging*, *90*, 71–77. https://doi.org/10.1016/j.clinimag.2022.06.014.

Sun, Z., Zhang, N., Li, Y., & Xu, X. (2020). A systematic review of chest imaging findings in COVID-19. *Quantitative Imaging in Medicine and Surgery*, *10*(5), 1058–1079. AME Publishing Company 10.21037/QIMS-20-564.

Tanacli, R., Doeblin, P., Götze, C., Zieschang, V., Faragli, A., Stehning, C., Korosoglou, G., Erley, J., Weiss, J., Berger, A., Pröpper, F., Steinbeis, F., Kühne, T., Seidel, F., Geisel, D., Cannon Walter-Rittel, T., Stawowy, P., Witzenrath, M., Klingel, K., … Kelle, S. (2021). COVID-19 vs. classical myocarditis associated myocardial injury evaluated by cardiac magnetic resonance and endomyocardial biopsy. *Frontiers in Cardiovascular Medicine*, *8*. https://doi.org/10.3389/fcvm.2021.737257.

Torkian, P., Rajebi, H., Zamani, T., Ramezani, N., Kiani, P., & Akhlaghpoor, S. (2021). Magnetic resonance imaging features of coronavirus disease 2019 (COVID-19) pneumonia: The first preliminary case series. *Clinical Imaging*, *69*, 261–265. https://doi.org/10.1016/j.clinimag.2020.09.002.

Varadarajan, V., Shabani, M., Ambale Venkatesh, B., & Lima, J. A. C. (2021). Role of imaging in diagnosis and management of COVID-19: A multiorgan multimodality imaging review. *Frontiers in Medicine*, *8*. https://doi.org/10.3389/fmed.2021.765975. Frontiers Media S.A.

Vasilev, Y. A., Sergunova, K. A., Bazhin, A.v., Masri, A. G., Vasileva, Y. N., Semenov, D. S., Kudryavtsev, N. D., Panina, O. Y., Khoruzhaya, A. N., Zinchenko, V.v., Akhmad, E. S., Petraikin, A.v., Vladzymyrskyy, A.v., Midaev, A.v., & Morozov, S. P. (2021). Chest MRI of patients with COVID-19. *Magnetic Resonance Imaging*, *79*, 13–19. https://doi.org/10.1016/j.mri.2021.03.005.

Vetrugno, L., Bove, T., Orso, D., Barbariol, F., Bassi, F., Boero, E., Ferrari, G., & Kong, R. (2020). Our Italian experience using lung ultrasound for identification, grading and serial follow-up of severity of lung involvement for management of patients with COVID-19. *Echocardiography*, *37*(4), 625–627. Pharmacotherapy Publications Inc https://doi.org/10.1111/echo.14664.

Watanabe, K., Ashikaga, T., Maejima, Y., Tao, S., Terui, M., Kishigami, T., Kaneko, M., Nakajima, R., Okata, S., Lee, T., Horie, T., Nagase, M., Nitta, G., Miyazaki, R., Nagamine, S., Nagata, Y., Nozato, T., Goya, M., & Sasano, T. (2022). Case report: Importance of MRI examination in the diagnosis and evaluation of COVID-19 mRNA vaccination induced myocarditis: Our experience and literature review. *Frontiers in Cardiovascular Medicine*, *9*. https://doi.org/10.3389/fcvm.2022.844626.

Wu, J., Wu, X., Zeng, W., Guo, D., Fang, Z., Chen, L., Huang, H., & Li, C. (2020). Chest CT findings in patients with coronavirus disease 2019 and its relationship with clinical features. *Investigative Radiology*, *55*(5), 257–261. Lippincott Williams and Wilkins https://doi.org/10.1097/RLI.0000000000000670.

Xie, Y., Xu, E., Bowe, B., & Al-Aly, Z. (2022). Long-term cardiovascular outcomes of COVID-19. *Nature Medicine*, *28*(3), 583–590. https://doi.org/10.1038/s41591-022-01689-3.

Xiong, T. Y., Redwood, S., Prendergast, B., & Chen, M. (2020). Coronaviruses and the cardiovascular system: Acute and long-term implications. *European Heart Journal*, *41*(19), 1798–1800. https://doi.org/10.1093/eurheartj/ehaa231.

Yang, R., Li, X., Liu, H., Zhen, Y., Zhang, X., Xiong, Q., Luo, Y., Gao, C., & Zeng, W. (2020). Chest ct severity score: An imaging tool for assessing severe covid-19. *Radiology: Cardiothoracic Imaging*, *2*(2). https://doi.org/10.1148/ryct.2020200047.

Chapter 23

Bronchial epithelial cells in cystic fibrosis: What happens in SARS-CoV-2 infection?

Anna Lagni, Erica Diani, Davide Gibellini, and Virginia Lotti
Department of Diagnostics and Public Health, Division of Microbiology and Virology, University of Verona, Verona, Italy

Abbreviations

ABC	ATP-binding cassette
ACE2	angiotensin-converting enzyme 2
ALI	air-liquid interface
ARDS	acute respiratory distress syndrome
ASL	airway surface liquid
BEC	bronchial epithelial cells
CF	cystic fibrosis
CFTR	cystic fibrosis transmembrane conductance regulator
COVID-19	coronavirus disease 2019
CPE	cytopathic effect
ELF	epithelial lining fluid
ENaC	epithelial sodium channel
HAE	human airway epithelial
HBE	human bronchial epithelial
IFN	interferon
IL-10	interleukin 10
IL-1beta	interleukin 1beta
IL-6	interleukin 6
MCC	mucociliary clearance
PAMP	pathogen-associated molecular patterns
PRR	pattern recognition receptors
pwCF	people with cystic fibrosis
SARS-CoV-2	severe acute respiratory syndrome coronavirus 2
SV40	simian virus 40
TEER	trans-epithelial electrical resistance
TGF-β	transforming growth factor beta
TMPRSS2	transmembrane serine protease 2
TNF	tumor necrosis factor
WHO	World Health Organization

Introduction

The human bronchial epithelium represents a primary target for SARS-CoV-2 pathogenesis and bronchi infection may cause increasing hypoxia and pulmonary infiltrates, which can proceed to acute respiratory distress syndrome (ARDS).

Physiologically, bronchial epithelial cells (BECs) serve as an important innate immunological barrier against airborne pollutants and bacteria/viral infections. Mucous and periciliary fluid secreted by apical goblet cells form a layer on the luminal surface of the epithelium, protecting the airway by collecting foreign particles and pathogens and allowing them to pass through the mucociliary elevator via the coordinated beating of ciliated apical airway epithelial cells. In cystic

Management, Body Systems, and Case Studies in COVID-19. https://doi.org/10.1016/B978-0-443-18703-2.00004-9
Copyright © 2024 Elsevier Inc. All rights are reserved, including those for text and data mining, AI training, and similar technologies.

fibrosis (CF) patients, a loss of cystic fibrosis transmembrane conductance regulator (CFTR) channel function and/or expression causes the lack of epithelial sodium channel (ENaC) inhibition, leading to sodium (Na^+) hyperabsorption with concomitant intracellular chloride (Cl^-) accumulation and water inflow. Thus, mucus becomes dry, viscous, and sticky, resulting in reduced airway surface liquid (ASL) height, mucus stasis, and obstruction.

Given these differences in healthy and cystic fibrosis bronchial epithelial cells, it is critical to understand how distinct bronchial epithelial cell properties affect SARS-CoV-2 infection.

Cystic fibrosis

Cystic fibrosis (CF) is the most prevalent autosomal recessive disease affecting Caucasians, and it is caused by mutations in the so-called cystic fibrosis transmembrane conductance regulator (*CFTR*) gene, which was identified for the first time in 1989 (Collins, 1992; Riordan et al., 1989; Rowe et al., 2005). Today, there are about 70,000 reported cases of CF worldwide, and 1000 new cases are diagnosed each year. The mutations in the *CFTR* gene, located on chromosome 7, decrease the synthesis of the functional CFTR ion channel protein, a member of the ATP-binding cassette (ABC) family of proteins with a transmembrane transport function (Rommens et al., 1989). These genetic mutations cause the synthesis of dysfunctional CFTR protein, which fails to perform normal chloride (Cl^-) and bicarbonate (HCO_3^-) ion transport (Anderson et al., 1991; Saint-Criq & Gray, 2017). As a result, water builds up inside the cells of a diverse range of epithelial tissues where the CFTR protein is expressed (Riordan, 2008), dehydrating extracellular mucus and leading to the most common symptoms characterizing CF: mucus accumulation and recurrent bacterial infections (Ratjen et al., 2015). The lungs, pancreas, and even other vital organs are negatively impacted by the mutant CFTR protein (Elborn, 2016; O'Sullivan & Freedman, 2009), and because of this CF is considered a multiorgan disease. The respiratory tract, digestive tract, urogenital tract, and sweat glands are the major areas where CF manifests its main symptoms (Callaghan et al., 2020; Reynaert et al., 2000). The dysfunction of the CFTR ion channel in these organs causes fluid and electrolyte transportation abnormalities as well as clogged ducts, atrophic epithelia, gland enlargement, and finally inflammation and fibrosis (Khan et al., 1995; Puchelle et al., 2002). Until recently, CFTR was assumed to be expressed only by epithelial cells, but recent research has shown that CFTR is even expressed in neurons of the human central nervous system, including the brain and spinal cord (Guo et al., 2009; Mulberg et al., 1994), and is also extended to the peripheral nervous system. These findings imply that CFTR is also important in the physiology of the central and peripheral nervous systems.

There are currently more than 2000 known CFTR variants, although the most prevalent mutation (classified as CFTR F508del) is characterized by a phenylalanine loss at position 508 as a result of a chromosomal deletion of three nucleotides, affecting the 70% of CF patients in homozygosis (Bobadilla et al., 2002; Brennan & Schrijver, 2016). According to the underlying cause of malfunction, CFTR mutations are currently divided into six groups (I–VI) (Marson et al., 2016; Stanke & Tümmler, 2016): Class I has flaws in protein synthesis, Class II has errors in CFTR processing, Class III has errors in CFTR channel regulation, Class IV has errors in channel conductance, Class V has errors in channel number, and Class VI has errors in CFTR stability at the plasma membrane. Because a residual CFTR function was observed, the less-severe CF forms are those from classes IV to VI. Chloride and bicarbonate transport across epithelial tissues is significantly altered by CFTR protein functional derangement and the subsequent reduced expression onto the cell membrane (Lukasiak & Zajac, 2021), which has an adverse effect on biological activity in the intracellular compartments.

Most CF-related morbidity and mortality are due to lung disease caused by chronic respiratory infections. In addition to the most common bacterial infections, several viruses have been implicated in the development of respiratory illnesses. Influenza viruses, parainfluenza viruses, respiratory syncytial virus, and others, in particular, can invade the lower and/or upper airways, causing serious respiratory alterations in some cases. Although viral infection may have a significant role in the disruption of the respiratory system in pwCF, the pathogenic processes in the respiratory apparatus remain mainly unclear (Billard et al., 2017; Frickmann et al., 2012).

Bronchial epithelial cells in normal and cystic fibrosis tissues

Bronchial epithelial cells (BEC) provide a crucial innate immune barrier against airborne toxins such as cigarette smoke, particulate matter, bacteria, and viruses. It is more than just a passive barrier because mucous and periciliary fluid released by apical goblet cells form a layer about 30 μm thick on the luminal surface of the epithelium. Through the coordinated beating of ciliated apical airway epithelial cells, mucous plays a protective role by trapping foreign particles and pathogens and allowing their clearance through the mucociliary elevator (Amatngalim & Hiemstra, 2018).

BECs quickly identify pathogen-associated molecular patterns (PAMP) by expressing pattern recognition receptors (PRR) such as toll-like receptors, and after this recognition, they can interact with, regulate, stimulate, and modulate other immune cells. Upon viral detection, BECs also release chemokines and cytokines that are crucial for both innate and adaptive immunity mechanisms (Amatngalim & Hiemstra, 2018; Hewitt & Lloyd, 2021).

A variety of specialized cell types make up the highly organized human bronchial epithelium. BEC can be divided into three groups: basal, ciliated, and secretory cells. Basal cells are ubiquitous in the large (50%) and small airways (81%), but the absolute cell count decreases with airway size (Bigot et al., 2020; Davis & Wypych, 2021). Basal epithelial cells show an important role in cell adhesion and because of their ability to self-renew and differentiate in response to epithelial injury, they are thought to be progenitor or stem cells. In addition, these cells produce various bioactive molecules, such as cytokines (Ruysseveldt et al., 2021).

Ciliated epithelial cells make up more than 50% of all epithelial cells; the importance of their involvement in clearing mucus from the airways by directional ciliary beating is shown by the fact that each cell possesses a significant number of energy-producing mitochondria near its apical surface and can contain up to 300 cilia (Bigot et al., 2020; Yaghi & Dolovich, 2016).

Mucus (goblet) cells secrete mucus to prevent foreign items from entering the airway lumen. A healthy balance between mucus production and clearance provides a crucial protection barrier and prevents the desiccation of the airway surface. Mucus cells also possess the capacity to self-renew and develop into ciliated epithelial cells (Adler et al., 2013; Rogers, 1994).

Club cells, a type of nonciliated secretory cell, are found in the trachea and small airways. These can be morphologically defined by their unique dome-shaped apical protrusions, and molecularly characterized by the production of club cell secretory protein. By producing bronchiolar surfactants and specific antiproteases, they control the integrity of the bronchiolar epithelium and immune response. They also have an important stem cell function as progenitors for both ciliated and mucus-secreting cells (Barnes, 2015; Rokicki et al., 2016).

In many acute and chronic respiratory disorders as well as in cystic fibrosis, the human bronchial epithelium is harmed or altered pathologically. In patients with CF, mucus becomes dehydrated, viscous, and sticky, which then leads to decreased airway surface liquid (ASL) height as well as mucus stasis and obstruction (Elborn, 2016; O'Sullivan & Freedman, 2009). Loss of functional CFTR in airway epithelial cells also promotes depletion and increased oxidation of the airway surface liquid. Neutrophil infiltrates are unable to clear bacterial infection and, as an adverse effect, contribute to mucosal tissue damage by releasing proteases and reactive oxygen species (Saint-Criq & Gray, 2017). Together, these changes are associated with diminished mucociliary clearance of bacteria, activation of epithelial cell signaling through multiple pathways, and subsequent hyperinflammatory responses in CF airways (Filkins & O'Toole, 2015; Stelzer-Braid et al., 2012) (Fig. 1).

Early pulmonary alterations in CF patients, such as bronchiectasis, squamous metaplasia, or goblet cell hyperplasia, have been documented by autopsy studies. Various investigations examined the epithelial specification in CF bronchial biopsies or lung explants, highlighting the presence of more proliferative zones, basal cell hyperplasia, glandular hyperplasia, and cells expressing mucins; ciliated cell defect and several signatures of impairment of the airway epithelium integrity, including epithelial to mesenchymal transition due to an increase in transforming growth factor beta (TGF-β), were observed in CF patients (Collin et al., 2021; Jacquot et al., 2008).

In people with cystic fibrosis (pwCF), also the immune system is deregulated: respiratory epithelial lining fluid (ELF) from patients with CF contains significantly less soluble interleukin 10 (IL-10) than the ELF of healthy subjects. This can contribute to enhancing local inflammation and tissue damage (Aldallal et al., 2012).

Cellular models of bronchial epithelial cells for in vitro studies (stabilized versus primary cells) of respiratory diseases

Physiological models are essential for understanding the interaction between respiratory diseases and epithelial cells (Feng et al., 2015).

Immortalized bronchial cell lines are a significant benefit to the study of respiratory human diseases. CF and non-CF immortalized cell lines are relevant to physiological, biochemical, and genetical analysis. Among the non-CF immortalized cell lines, Calu-3 and 16HBE14o− are well established.

The immortalized bronchial cell line 16HBE14o− retains different epithelial functions, such as the presence of both tight junctions and adherens junctions. Although these cells were immortalized from primary isolated bronchial epithelial cells, they did not have the capacity to differentiate to the three main bronchial phenotypes and do not form an organized epithelial layer (Callaghan et al., 2020; Hermanns et al., 2017).

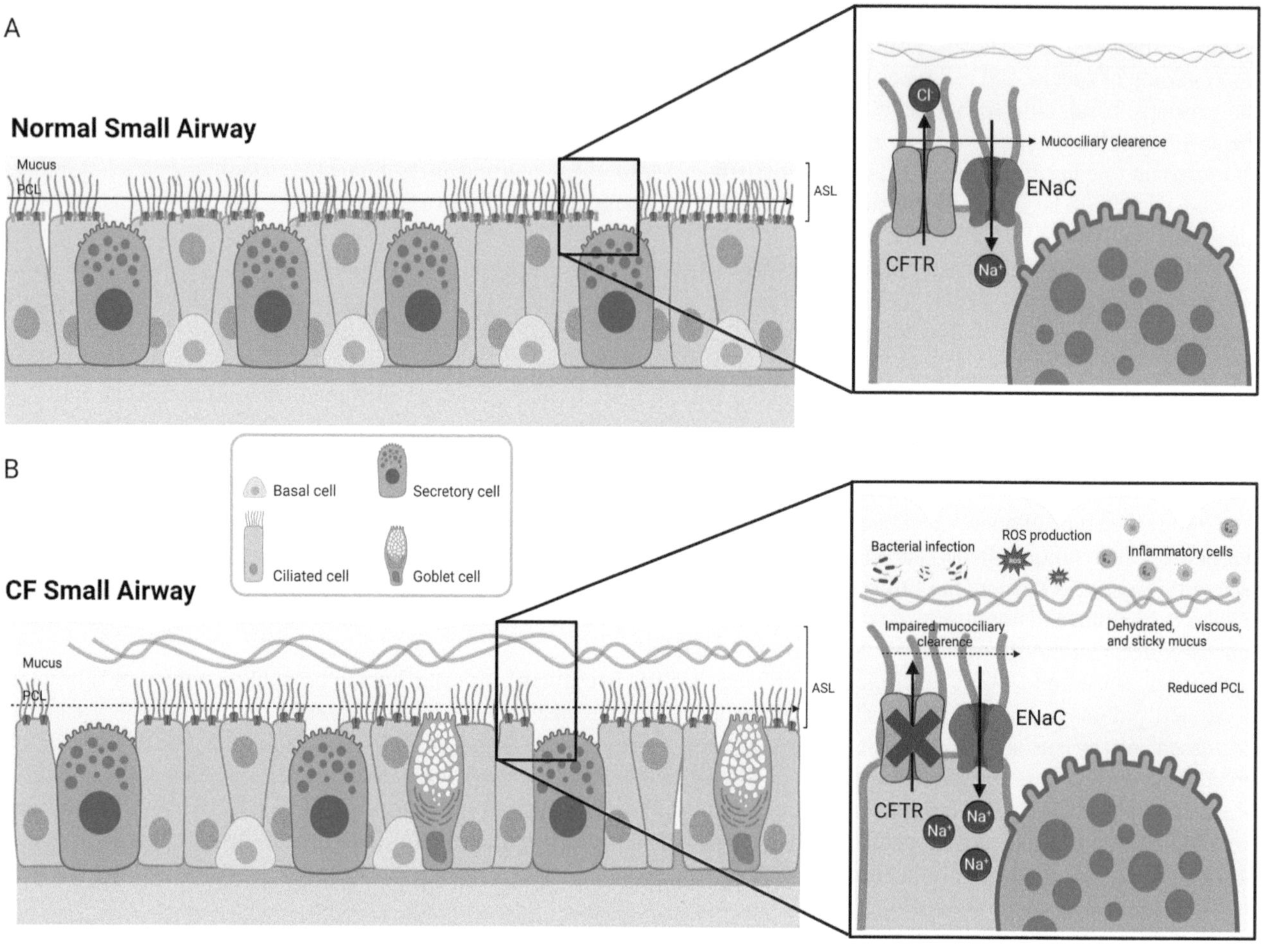

FIG. 1 Normal vs CF small airway epithelium. Normal and CF airway schematic drawings of epithelial apical ion transport. (A) In healthy airways, sodium (Na^+) absorption and chloride (Cl^-) secretion are regulated by the ENaC and CFTR channels, respectively, and this balanced transport largely controls the hydration of the airway surface layer (ASL). This layer is normally transported by effective beating of cilia in the periciliary layer (PCL). (B) In CF airways, the CFTR absence or loss of function leads to impaired Cl^- secretion and dysregulated Na^+ absorption by ENaC, resulting in ASL inadequate hydration and PCL reduction, affecting mucociliary transport. Accumulation of thick and sticky mucus resulted in a low pH environment that favors chronic bacterial infection and inflammation.

Calu-3 cells are a submucosal gland cell line that was generated from a bronchial adenocarcinoma. Under air-liquid interface (ALI) conditions, Calu-3 cells grow in monolayers, producing mucus and cilia-like microvilli. This cell line has gained a lot of interest because of its capacity to form tight monolayers in culture and produce the CFTR protein (Zhu et al., 2010).

Regarding CF immortalized cell lines, the quality of a cell line for CF research is determined by both the stability and level of CFTR expression as well as its ion transport characteristics; one of the most used cell lines is CFBE41o− (Molenda et al., 2014). The CFBE41o− human bronchial epithelial cell line was created from a CF patient who was homozygous for the CFTR mutation F508 and immortalized with the SV40 plasmid (pSVori−), which has a defective origin of replication. All the cystic fibrosis-specific ion transport characteristics, such as impaired calcium-dependent chloride transport and intact cAMP-dependent chloride transport, are displayed by CFBE14o− (Illek et al., 2008). The CFBE41o− cell line has been utilized to produce subclones with a restored wild-type phenotype or overexpressing F508del CFTR mutation (Gottschalk et al., 2016).

Immortalized cell lines have the advantage of being well defined, homogeneous, and genetically similar, which helps produce consistent and repeatable results with a rapid and continuous growth. The main drawback of using these cells is that they cannot be considered "normal," as they divide endlessly and may exhibit distinct gene patterns that are not present in any other cell type in vivo such as the capability of differentiating (Duell et al., 2011; Gazdar et al., 2010; Maqsood et al., 2013). If it is necessary and preferable to have a cell culture model closer to the human in vivo environment, human primary cells are a better choice. Because they are directly extracted from tissues, human primary cells keep the morphological and functional traits of their origin. However, human primary cells experience senescence and do not live indefinitely in passages. Primary cells also begin to exhibit morphological and functional changes as the number of passages increases (Awatade et al., 2018; BéruBé et al., 2010).

Primary human bronchial epithelial (HBE) cells are isolated from the surface epithelium of human bronchi and are conventionally cultured as monolayers (2D cultures). To achieve fully differentiated HBE cells that show remarkable phenotypic similarity to bronchial epithelium, an ALI 3D cell culture is needed. Bronchial epithelial cells cultured using ALI cell culture techniques can form polarized cell layers with the presence of functional beating cilia and mucus secretion, tight junctions, and the development of robust trans-epithelial electrical resistance (TEER) values (Neuberger et al., 2011; Rijsbergen et al., 2021).

Because the CFTR protein is largely expressed at the apical surface of ciliated cells, the polarized, differentiated phenotype is essential for the in vitro evaluation of CFTR function. The use of HBE cell cultures also allows investigating the effects of CFTR correctors and potentiators on CFTR function and epithelial cell biology in the native pathological environment (Awatade et al., 2018; Neuberger et al., 2011).

HBEs are considered the gold standard to study CF pathogenesis and CFTR functional response. However, the extensive damage into the tissue, particularly to the cell layer, presents technical challenges to establishing successful ex vivo cultures. Therefore, the supply of CF patient-derived HBE cells is often limited.

SARS-CoV-2 infection impact in bronchial epithelial cells

After its breakout in December 2019, the coronavirus disease 2019 (COVID-19) outbreak brought on by severe acute respiratory syndrome coronavirus 2 (SARS-CoV-2) expanded around the world. On March 11, 2020, the World Health Organization (WHO) proclaimed it to be a pandemic (Zhu, Wang, Yang, et al., 2020).

SARS-CoV-2 is an enveloped, single-stranded, positive-sense RNA virus with a genomic size of 29.9 kb. Virion particles are assembled from four proteins: spike (S), envelope (E), membrane (M), and nucleocapsid (N). These glycoproteins play roles in host cell receptor recognition, virion assembly and release as well as virion and viral RNA encapsulation. Furthermore, 16 nonstructural proteins and several auxiliary proteins have been identified as being involved in RNA processing, replication, and host cell survival (Guan et al., 2020; Huang et al., 2020).

Viral entrance is mediated by the SARS-CoV-2 spike protein, which is composed of two subunits, S1 and S2, separated by the S1–S2 site, which contains a furin cleavage motif that is cleaved by the host cell. The S1 component mediates attachment to the host cell membrane and engages angiotensin-converting enzyme 2 (ACE2) as the entrance receptor, resulting in membrane fusion or virion endocytosis. Cellular proteases transmembrane serine protease 2 (TMPRSS2), which is present at the cell surface, and endosomal cathepsins then cleave the furin cleavage motif. The S2 subunit is responsible for the fusion of viral and cellular membranes, which results in the release of the viral package into the host cytoplasm (Hoffmann et al., 2020; Pizzato et al., 2022). The expression levels of these molecules differ depending on the type, function, and location of airway epithelial cells such as ciliated, secretory, olfactory epithelial, and alveolar epithelial cells as well as host to host depending on age, sex, or comorbid diseases.

In the lower conducting airways, ciliated cells, mucus-secreting cells, and club cells all express ACE2 and TMPRSS2, with ciliated cells acting as the main infected cell type. Ciliated and secretory epithelial cells in pathological specimens appear infected while progenitor basal cells are unaffected. This is supported by the primary bronchial epithelial cell culture model, in which ciliated cells are the first to become infected, and the infection then spreads to secretory cells (Ryu & Shin, 2021).

Infection in the bronchi produces progressive hypoxia associated with pulmonary infiltrates that can progress to ARDS with diffuse alveolar damage, epithelial and microvascular injury, hyaline membranes, and, in some cases, thrombi of small and large vessels (Mulay et al., 2020; Ravindra et al., 2021). A study was conducted on 16HBE to understand the clinical features of SARS-CoV-2 related to the infection process in bronchial epithelial cells. In particular, it was noticed that SARS-CoV-2 did not induce damage to epithelial cells, maintaining their structure for at least 1 week postinfection, still allowing continuous virus release (Liao et al., 2020). This process is probably facilitated by the low level of IFN stimulation reported in different studies in which type I and type III interferons (IFNs) were not detectable (Chen et al., 2021). A study on primary human airway epithelial cells (HAE) revealed significant areas of disordered cilia, showing that SARS-CoV-2 infection impaired cilia synchronization. Cilium shrinking in the middle of CPE plaque was initially identified as no ciliary beating under the light microscope could be found. The irregular ciliary beating and disruption of cilia synchronization result in poor mucociliary clearance (MCC), which can contribute to secondary infections. Another notable ultrastructural change was the accumulation of organelles toward the apical surface, which included a substantial number of mitochondria with an aberrant shape. Apoptosis was shown to be the cause of the widespread cell death observed in the CPE zone. In addition, unique plaque-like CPEs on the apical surface of HAE were found consistently across multiple propagations. During the incubation period, the size and quantity of these plaques grew, as did the creation of syncytial cells and the breakdown of cell tight junctions (Zhu, Wang, Liu, et al., 2020).

SARS-CoV-2 infection in CFTR-modified bronchial epithelial cells: What is known so far?

pwCFs should be considered at high risk of developing severe symptoms in case of SARS-CoV-2 infection because, as seen above, viral infections can play a significant role in lung disease. Several reports, however, show that SARS-CoV-2 did not result in worse outcomes in pwCF.

Recent in vitro studies indicate cellular and molecular processes to be significant drivers in SARS-CoV-2 having a lower impact on pwCF than expected, ranging from cytokine releases to biochemical alterations leading to morphological rearrangements inside the cells associated with impaired CFTR (Lotti et al., 2023). Some of these in vitro studies focused on the analysis of the SARS-CoV-2 infection in CFTR-modified bronchial epithelial cells. In a study by Lotti et al., SARS-CoV-2 infection experiments were conducted on both immortalized CFBE41o− WT and F508del cell lines, and on the 3D cell culture MucilAir (fully differentiated HBE cells from both healthy patients and patients with cystic fibrosis). This study demonstrated a significantly lower SARS-CoV-2 infection in CFTR-defective cells than in normal cells, importantly correlated with a minor loss of epithelium integrity, thus demonstrating that altering or completely deleting the CFTR gene resulted in a significantly lower SARS-CoV-2 infection impact (Lotti et al., 2022).

A further in vitro study, conducted to investigate the effect of CFTR inhibitors on SARS-CoV-2 replication in human bronchial epithelial cells, showed that CFTR inhibitors had no significant impact on SARS-CoV-2 cell entry, whereas they were significantly more effective against SARS-CoV-2 in the postentry phase (Lagni et al., 2023). Bezzerri et al. showed that the CFTR channel regulates ACE2 production and localization and, depending on the ACE2 pattern of expression, the SARS-CoV-2 S protein induced a very weak IL-6 response in primary CF cells (Bezzerri et al., 2023).

Merigo et al. focused on the morphological effects of SARS-CoV-2 infection on WT and F508del CFBE41o− at the ultrastructural level. In particular, it was seen that WT cells produced double membrane vesicles, characteristic replicative structures with granular and vesicular material, early in infection, and vesicles containing virus particles later in infection. CF cells revealed double-membrane vesicles with an irregular form and degenerative changes as well as vesicles harboring viruses with no regular structure and no organized distribution late in infection (Merigo et al., 2022).

In WT cells, ACE2 was expressed at the plasma membrane and present in the cytoplasm just early in the infection, whereas it remained in the cytoplasm even late in the infection in CF cells. The autophagosome content also differed across the cells: vesicles associated with virus-containing structures were found in WT cells, suggesting that SARS-CoV-2 could subvert autophagy to accelerate its replication, whereas ingested material for lysosomal processing was found in CF cells. An intriguing aspect that distinguishes CF cells from WT cells was the presence of lipid droplets solely in CF cells following SARSCoV-2 infection. It was believed that lipids have a function in energy supply and cell membrane production because viral particles were not found within lipid droplets, hence lipophagy may not act as a scaffold in the replication of SARS-CoV-2 in infected CF cells. Thus, the build-up of lipid droplets in these cells is most likely the result of different cellular activities, such as the requirement for food supplementation and thus energy, or it might be the result of a deficient autophagic breakdown mechanism.

At present, few studies of SARS-CoV-2 infection in CFTR-modified bronchial epithelial cells have been conducted. However, further studies are certainly needed to fully understand the dynamics of replication in these cell types.

Policies and procedures

To write this chapter, we collected papers using the following search terms: "bronchial epithelial cells," "cystic fibrosis," "SARS-CoV-2," "COVID-19," and "CFTR" on the MEDLINE/PubMed database. Papers were also identified via the reference lists of examined papers. Only published papers written in English were reviewed, and they were chosen according to content relevance. This chapter examines the role of SARS-CoV-2 in bronchial epithelial cells, with a focus on cystic fibrosis bronchial cells. First, we discussed the structure and function of bronchial epithelial cells, also comparing the characteristics of bronchial cells in healthy and cystic fibrosis tissues. Then we concentrated on the in vitro bronchial cellular models, particularly focusing on the most common healthy and cystic fibrosis models. Afterward, we turned our attention to the dynamics of SARS-CoV-2 infection of bronchial epithelial cells and the impact on these cells at pathogenic, immunological, and morphological levels. Starting then from retrospective studies that suggested a milder infection in people with cystic fibrosis, we searched for in vitro studies that could explain the impact of SARS-CoV-2 in cystic fibrosis bronchial epithelial cells at both the molecular and ultrastructural levels. Few studies were done, and all were concordant in demonstrating a significant lower SARS-CoV-2 infection in CFTR-defective cells.

Applications to other areas

In this chapter, we reviewed the impact of SARS-CoV-2 infection on human bronchial epithelial cells, focusing our attention specifically on the response of this type of cell in cystic fibrosis patients.

It is well known that in humans, respiratory tract infections are a leading source of morbidity and mortality. In particular, newborns, the elderly, and immunocompromised patients experienced more severe disease in case of viral infection. Thus, this chapter emphasizes the need to study the pathogenesis of viral respiratory viruses in bronchial epithelial cells and how they may affect the at-risk population, as could be the case for pwCF. In this work, we explored the different bronchial epithelial cellular models, representing reproducible and scalable in vitro culture models that more closely reproduce the human respiratory system, which could serve in studying interactions between respiratory viruses and the host as well as testing novel therapeutic approaches. Moreover, because the main CF symptoms are related to the respiratory tract, it could be interesting to explore the response of bronchial epithelial cells of pwCF in case of SARS-CoV-2 infection with and without CFTR modulators and/or corticosteroids, which are the common therapies for pwCF. Concluding, it could be interesting to extend the study of SARS-CoV-2 impact in different cell types as COVID-19 is not only a respiratory disease but also a multisystem disease because gastrointestinal symptoms and renal, cardiac, or central nervous system manifestations were reported, both in normal and CF cells.

Mini-dictionary of terms

- **ACE-2.** The enzyme considered to be the receptor mainly involved in SARS-CoV-2 entrance into host cells.
- **Bronchial epithelial cells.** The innate immune barrier against airborne toxins, with mucous and periciliary fluid released by apical goblet cells to form a layer on the luminal surface of the epithelium. These cells play a fundamental protective role by trapping foreign particles and pathogens in mucus and allowing their clearance through the coordinated beating of ciliated apical airway epithelial cells.
- ***Cystic fibrosis transmembrane conductance regulator (CFTR)* gene,** located on chromosome 7, codes for the homonymous ion channel protein with a transmembrane transport function.
- **Cystic fibrosis.** The most common autosomal recessive disease affecting Caucasians, caused by mutations in the CFTR gene. The main consequences occur at the respiratory level, caused by thick and sticky mucus.
- **Immortalized cell lines:** Immortalized cell lines are either tumorous cells that do not stop dividing or cells that have been artificially modified to proliferate endlessly and can, thus, be grown over numerous generations. These cells are well defined, homogeneous, and genetically similar, enabling consistent and repeatable results.
- **Primary cells:** Primary cultures are cells that have been freshly extracted from organ tissue and are being grown in vitro. They represent a close replicate of the physiological condition of cells in vivo, thus generating the most biologically relevant data.
- **Human bronchial epithelial cells:** Cells isolated from the surface epithelium of human bronchi that are conventionally cultured as monolayers (2D cultures). To achieve fully differentiated HBE cells that show remarkable phenotypic similarity to the bronchial epithelium, the air-liquid interface (ALI) 3D cell culture is needed to form polarized cell layers with the presence of functional beating cilia and mucus secretion, tight junctions, and the development of robust transepithelial electrical resistance values.

Summary points

- BECs serve as an important innate immunological barrier against airborne pollutants such as cigarette smoke, particulate matter, bacteria, and viruses.
- In CF airways, there is a diminished mucociliary clearance, the activation of epithelial cell signaling through multiple pathways, and subsequent hyperinflammatory responses.
- In the bronchial epithelium, ciliated cells act as the main SARS-CoV-2 infected cell type.
- After SARS-CoV-2 infection, primary human airway epithelial cells revealed significant areas of disordered cilia with impaired synchronization, resulting in poor mucociliary clearance. Retrospective studies showed no worse outcome in SARS-CoV-2 infected pwCF.
- In vitro studies demonstrated lower SARS-CoV-2 infection in CFTR-defective cells than in normal cells.

References

Adler, K. B., Tuvim, M. J., & Dickey, B. F. (2013). Regulated mucin secretion from airway epithelial cells. *Frontiers in Endocrinology*, *4*(SEP), 129. https://doi.org/10.3389/FENDO.2013.00129/BIBTEX.

Aldallal, N., McNaughton, E. E., Manzel, L. J., Richards, A. M., Zabner, J., Ferkol, T. W., & Look, D. C. (2012). Inflammatory response in airway epithelial cells isolated from patients with cystic fibrosis. *American Journal of Respiratory and Critical Care Medicine*, *166*(9), 1248–1256. https://doi.org/10.1164/Rccm.200206-627OC.

Amatngalim, G., & Hiemstra, P. (2018). Airway epithelial cell function and respiratory host defense in chronic obstructive pulmonary disease. *Chinese Medical Journal*, *131*(9), 1099. https://doi.org/10.4103/0366-6999.230743.

Anderson, M. P., Gregory, R. J., Thompson, S., Souza, D. W., Paul, S., Mulligan, R. C., Smith, A. E., & Welsh, M. J. (1991). Demonstration that CFTR is a chloride channel by alteration of its anion selectivity. *Science (New York, N.Y.)*, *253*(5016), 202–205. https://doi.org/10.1126/SCIENCE.1712984.

Awatade, N. T., Wong, S. L., Hewson, C. K., Fawcett, L. K., Kicic, A., Jaffe, A., & Waters, S. A. (2018). Human primary epithelial cell models: Promising tools in the era of cystic fibrosis personalized medicine. *Frontiers in Pharmacology*, *9*, 1429. https://doi.org/10.3389/FPHAR.2018.01429/BIBTEX.

Barnes, P. J. (2015). Club cells, their secretory protein, and COPD. *Chest*, *147*(6), 1447–1448. https://doi.org/10.1378/chest.14-3171.

BéruBé, K., Prytherch, Z., Job, C., & Hughes, T. (2010). Human primary bronchial lung cell constructs: The new respiratory models. *Toxicology*, *278*(3), 311–318. https://doi.org/10.1016/J.TOX.2010.04.004.

Bezzerri, V., Gentili, V., Api, M., Finotti, A., Papi, C., Tamanini, A., Boni, C., Baldisseri, E., Olioso, D., Duca, M., Tedesco, E., Leo, S., Borgatti, M., Volpi, S., Pinton, P., Cabrini, G., Gambari, R., Blasi, F., Lippi, G., … Cipolli, M. (2023). SARS-CoV-2 viral entry and replication is impaired in cystic fibrosis airways due to ACE2 downregulation. *Nature Communications*, *14*(1), 1–15. https://doi.org/10.1038/s41467-023-35862-0.

Bigot, J., Guillot, L., Guitard, J., Ruffin, M., Corvol, H., Balloy, V., & Hennequin, C. (2020). Bronchial epithelial cells on the front line to fight lung infection-causing aspergillus fumigatus. *Frontiers in Immunology*, *11*, 1041. https://doi.org/10.3389/FIMMU.2020.01041/BIBTEX.

Billard, L., Le Berre, R., Pilorgé, L., Payan, C., Héry-Arnaud, G., & Vallet, S. (2017). Viruses in cystic fibrosis patients' airways. *Critical Reviews in Microbiology*, *43*(6), 690–708. https://doi.org/10.1080/1040841X.2017.1297763.

Bobadilla, J. L., Macek, M., Fine, J. P., & Farrell, P. M. (2002). Cystic fibrosis: A worldwide analysis of CFTR mutations—Correlation with incidence data and application to screening. *Human Mutation*, *19*(6), 575–606. https://doi.org/10.1002/humu.10041.

Brennan, M. L., & Schrijver, I. (2016). Cystic fibrosis: A review of associated phenotypes, use of molecular diagnostic approaches, genetic characteristics, progress, and dilemmas. *The Journal of Molecular Diagnostics*, *18*(1), 3–14. https://doi.org/10.1016/J.JMOLDX.2015.06.010.

Callaghan, P. J., Ferrick, B., Rybakovsky, E., Thomas, S., & Mullin, J. M. (2020). Epithelial barrier function properties of the 16HBE14o- human bronchial epithelial cell culture model. *Bioscience Reports*, *40*(10), 20201532. https://doi.org/10.1042/BSR20201532.

Chen, H., Liu, W., Wang, Y., Liu, D., Zhao, L., & Yu, J. (2021). SARS-CoV-2 activates lung epithelial cell proinflammatory signaling and leads to immune dysregulation in COVID-19 patients. *eBioMedicine*, *70*, 103500. https://doi.org/10.1016/J.EBIOM.2021.103500.

Collin, A. M., Lecocq, M., Detry, B., Carlier, F. M., Bouzin, C., de Sany, P., Hoton, D., Verleden, S., Froidure, A., Pilette, C., & Gohy, S. (2021). Loss of ciliated cells and altered airway epithelial integrity in cystic fibrosis. *Journal of Cystic Fibrosis*, *20*(6), e129–e139. https://doi.org/10.1016/J.JCF.2021.09.019.

Collins, F. S. (1992). Cystic fibrosis: Molecular biology and therapeutic implications. *Science (New York, N.Y.)*, *256*(5058), 774–779. https://doi.org/10.1126/SCIENCE.1375392.

Davis, J. D., & Wypych, T. P. (2021). Cellular and functional heterogeneity of the airway epithelium. *Mucosal Immunology*, *14*(5), 978–990. https://doi.org/10.1038/s41385-020-00370-7.

Duell, B. L., Cripps, A. W., Schembri, M. A., & Ulett, G. C. (2011). Epithelial cell coculture models for studying infectious diseases: Benefits and limitations. *Journal of Biomedicine and Biotechnology*, *2011*. https://doi.org/10.1155/2011/852419.

Elborn, J. S. (2016). Cystic fibrosis. *The Lancet*, *388*(10059), 2519–2531. https://doi.org/10.1016/S0140-6736(16)00576-6.

Feng, W., Guo, J., Huang, H., Xia, B., Liu, H., Li, J., Lin, S., Li, T., Liu, J., & Li, H. (2015). Human normal bronchial epithelial cells: A novel in vitro cell model for toxicity evaluation. *PLoS ONE*, *10*(4). https://doi.org/10.1371/JOURNAL.PONE.0123520.

Filkins, L. M., & O'Toole, G. A. (2015). Cystic fibrosis lung infections: Polymicrobial, complex, and hard to treat. *PLoS Pathogens*, *11*(12), e1005258. https://doi.org/10.1371/JOURNAL.PPAT.1005258.

Frickmann, H., Jungblut, S., Hirche, T. O., Groß, U., Kuhns, M., & Zautner, A. E. (2012). Spectrum of viral infections in patients with cysticfibrosis. *European Journal of Microbiology & Immunology*, *2*(3), 161. https://doi.org/10.1556/EUJMI.2.2012.3.1.

Gazdar, A. F., Gao, B., & Minna, J. D. (2010). Lung cancer cell lines: Useless artifacts or invaluable tools for medical science? *Lung Cancer*, *68*(3), 309–318. https://doi.org/10.1016/J.LUNGCAN.2009.12.005.

Gottschalk, L. B., Vecchio-Pagan, B., Sharma, N., Han, S. T., Franca, A., Wohler, E. S., Batista, D. A. S., Goff, L. A., & Cutting, G. R. (2016). Creation and characterization of an airway epithelial cell line for stable expression of CFTR variants. *Journal of Cystic Fibrosis: Official Journal of the European Cystic Fibrosis Society*, *15*(3), 285. https://doi.org/10.1016/J.JCF.2015.11.010.

Guan, W., Ni, Z., Hu, Y., Liang, W., Ou, C., He, J., Liu, L., Shan, H., Lei, C., Hui, D. S. C., Du, B., Li, L., Zeng, G., Yuen, K.-Y., Chen, R., Tang, C., Wang, T., Chen, P., Xiang, J., … Zhong, N. (2020). Clinical characteristics of coronavirus disease 2019 in China. *New England Journal of Medicine*, *382*(18), 1708–1720. https://doi.org/10.1056/NEJMOA2002032.

Guo, Y., Su, M., McNutt, M. A., & Gu, J. (2009). Expression and distribution of cystic fibrosis transmembrane conductance regulator in neurons of the human brain. *Journal of Histochemistry and Cytochemistry*, *57*(12), 1113–1120. https://doi.org/10.1369/JHC.2009.953455/ASSET/IMAGES/LARGE/10.1369_JHC.2009.953455-FIG 2.JPEG.

Hermanns, M. I., Freese, C., Anspach, L., Grützner, C., Pohl, C., Unger, R. E., & Kirkpatrick, C. J. (2017). Cell culture systems for studying biomaterial interactions with biological barriers. *Comprehensive Biomaterials*, *II*, 295–334. https://doi.org/10.1016/B978-0-12-803581-8.09821-0.

Hewitt, R. J., & Lloyd, C. M. (2021). Regulation of immune responses by the airway epithelial cell landscape. *Nature Reviews Immunology*, *21*(6), 347–362. https://doi.org/10.1038/s41577-020-00477-9.

Hoffmann, M., Kleine-Weber, H., Schroeder, S., Mü, M. A., Drosten, C., & Pö, S. (2020). SARS-CoV-2 cell entry depends on ACE2 and TMPRSS2 and is blocked by a clinically proven protease inhibitor. *Cell*, *181*, 271–280. https://doi.org/10.1016/j.cell.2020.02.052.

Huang, C., Wang, Y., Li, X., Ren, L., Zhao, J., Hu, Y., Zhang, L., Fan, G., Xu, J., Gu, X., Cheng, Z., Yu, T., Xia, J., Wei, Y., Wu, W., Xie, X., Yin, W., Li, H., Liu, M., … Cao, B. (2020). Clinical features of patients infected with 2019 novel coronavirus in Wuhan, China. *The Lancet*, *395*(10223), 497–506. https://doi.org/10.1016/S0140-6736(20)30183-5.

Illek, B., Maurisse, R., Wahler, L., Kunzelmann, K., Fischer, H., & Gruenert, D. C. (2008). Cl transport in complemented CF bronchial epithelial cells correlates with CFTR mRNA expression levels. *Cellular Physiology and Biochemistry: International Journal of Experimental Cellular Physiology, Biochemistry, and Pharmacology*, *22*(1–4), 57–68. https://doi.org/10.1159/000149783.

Jacquot, J., Tabary, O., Le Rouzic, P., & Clement, A. (2008). Airway epithelial cell inflammatory signalling in cystic fibrosis. *The International Journal of Biochemistry & Cell Biology*, *40*(9), 1703–1715. https://doi.org/10.1016/J.BIOCEL.2008.02.002.

Khan, T. Z., Wagener, J. S., Bost, T., Martinez, J., Accurso, F. J., & Riches, D. W. (1995). Early pulmonary inflammation in infants with cystic fibrosis. *American Journal of Respiratory and Critical Care Medicine*, *151*(4), 1075–1082. https://doi.org/10.1164/AJRCCM.151.4.7697234.

Lagni, A., Lotti, V., Diani, E., Rossini, G., Concia, E., Sorio, C., & Gibellini, D. (2023). CFTR inhibitors display in vitro antiviral activity against SARS-CoV-2. *Cells*, *12*, *776*. https://doi.org/10.3390/CELLS12050776.

Liao, Y., Li, X., Mou, T., Zhou, X., Li, D., Wang, L., Zhang, Y., Dong, X., Zheng, H., Guo, L., Liang, Y., Jiang, G., Fan, S., Xu, X., Xie, Z., Chen, H., Liu, L., & Li, Q. (2020). Distinct infection process of SARS-CoV-2 in human bronchial epithelial cell lines. *Journal of Medical Virology*, *92*(11), 2830. https://doi.org/10.1002/JMV.26200.

Lotti, V., Lagni, A., Diani, E., Sorio, C., & Gibellini, D. (2023). Crosslink between SARS-CoV-2 replication and cystic fibrosis hallmarks. *Frontiers in Microbiology*, *14*, 1460. https://doi.org/10.3389/FMICB.2023.1162470.

Lotti, V., Merigo, F., Lagni, A., Di Clemente, A., Ligozzi, M., Bernardi, P., Rossini, G., Concia, E., Plebani, R., Romano, M., Sbarbati, A., Sorio, C., & Gibellini, D. (2022). CFTR modulation reduces SARS-CoV-2 infection in human bronchial epithelial cells. *Cells*, *11*(8). https://doi.org/10.3390/CELLS11081347.

Lukasiak, A., & Zajac, M. (2021). The distribution and role of the CFTR protein in the intracellular compartments. *Membranes*, *11*(11), 804. https://doi.org/10.3390/MEMBRANES11110804.

Maqsood, M. I., Matin, M. M., Bahrami, A. R., & Ghasroldasht, M. M. (2013). Immortality of cell lines: Challenges and advantages of establishment. *Cell Biology International*, *37*(10), 1038–1045. https://doi.org/10.1002/CBIN.10137.

Marson, F. A. L., Bertuzzo, C. S., & Ribeiro, J. D. (2016). Classification of CFTR mutation classes. *The Lancet Respiratory Medicine*, *4*(8), e37–e38. https://doi.org/10.1016/S2213-2600(16)30188-6.

Merigo, F., Lotti, V., Bernardi, P., Conti, A., Di Clemente, A., Ligozzi, M., Lagni, A., Sorio, C., Sbarbati, A., & Gibellini, D. (2022). Ultrastructural characterization of human bronchial epithelial cells during SARS-CoV-2 infection: Morphological comparison of wild-type and CFTR-modified cells. *International Journal of Molecular Sciences*, *23*(17), 9724. https://doi.org/10.3390/IJMS23179724.

Molenda, N., Urbanova, K., Weiser, N., Kusche-Vihrog, K., Guñzel, D., & Schillers, H. (2014). Paracellular transport through healthy and cystic fibrosis bronchial epithelial cell lines—Do we have a proper model? *PLoS One*, *9*(6), e100621. https://doi.org/10.1371/JOURNAL.PONE.0100621.

Mulay, A., Konda, B., Garcia, G., Jr., Yao, C., Beil, S., Sen, C., Purkayastha, A., Kolls, J. K., Pociask, D. A., Pessina, P., de Aja, J. S., Garcia-de-Alba, C., Kim, C. F., Gomperts, B., Arumugaswami, V., & Stripp, B. R. (2020). SARS-CoV-2 infection of primary human lung epithelium for COVID-19 modeling and drug discovery. *BioRxiv*. https://doi.org/10.1101/2020.06.29.174623.

Mulberg, A. E., Wiedner, E. B., Bao, X., Marshall, J., Jefferson, D. M., & Altschuler, S. M. (1994). Cystic fibrosis transmembrane conductance regulator protein expression in brain. *Neuroreport*, *5*(13), 1684–1688. https://doi.org/10.1097/00001756-199408150-00035.

Neuberger, T., Burton, B., Clark, H., & Van Goor, F. (2011). Use of primary cultures of human bronchial epithelial cells isolated from cystic fibrosis patients for the pre-clinical testing of CFTR modulators. *Methods in Molecular Biology (Clifton, N.J.)*, *741*, 39–54. https://doi.org/10.1007/978-1-61779-117-8_4.

O'Sullivan, B. P., & Freedman, S. D. (2009). Cystic fibrosis. *The Lancet*, *373*(9678), 1891–1904. https://doi.org/10.1016/S0140-6736(09)60327-5.

Pizzato, M., Baraldi, C., Boscato Sopetto, G., Finozzi, D., Gentile, C., Gentile, M. D., Marconi, R., Paladino, D., Raoss, A., Riedmiller, I., Ur Rehman, H., Santini, A., Succetti, V., & Volpini, L. (2022). SARS-CoV-2 and the host cell: A tale of interactions. *Frontiers in Virology*, *1*, 46. https://doi.org/10.3389/FVIRO.2021.815388.

Puchelle, E., Bajolet, O., & Abély, M. (2002). Airway mucus in cystic fibrosis. *Paediatric Respiratory Reviews*, *3*(2), 115–119. https://doi.org/10.1016/S1526-0550(02)00005-7.

Ratjen, F., Bell, S. C., Rowe, S. M., Goss, C. H., Quittner, A. L., & Bush, A. (2015). Cystic fibrosis. *Nature Reviews Disease Primers*, *1*(1), 15010. https://doi.org/10.1038/nrdp.2015.10.

Ravindra, N. G., Alfajaro, M. M., Gasque, V., Huston, N. C., Wan, H., Szigeti-Buck, K., Yasumoto, Y., Greaney, A. M., Habet, V., Chow, R. D., Chen, J. S., Wei, J., Filler, R. B., Wang, B., Wang, G., Niklason, L. E., Montgomery, R. R., Eisenbarth, S. C., Chen, S., … Wilen, C. B. (2021). Single-cell longitudinal analysis of SARS-CoV-2 infection in human airway epithelium identifies target cells, alterations in gene expression, and cell state changes. *PLoS Biology*, *19*(3), e3001143. https://doi.org/10.1371/JOURNAL.PBIO.3001143.

Reynaert, I., Van Der Schueren, B., Gise', G., Degeest, G., Miche', M., Manin, M., Cuppens, H., Scholte, B., & Cassiman, J.-J. (2000). Morphological changes in the vas deferens and expression of the cystic fibrosis transmembrane conductance regulator (CFTR) in control, F508 and Knock-out CFTR mice during postnatal life. *Genetics, Gene Regulation, and Expression*, *55*(2), 125–135. https://doi.org/10.1002/(SICI)1098-2795(200002)55:2.

Rijsbergen, L. C., van Dijk, L. L. A., Engel, M. F. M., de Vries, R. D., & de Swart, R. L. (2021). In vitro modelling of respiratory virus infections in human airway epithelial cells—A systematic review. *Frontiers in Immunology*, *12*, 3301. https://doi.org/10.3389/FIMMU.2021.683002/BIBTEX.

Riordan, J. R. (2008). CFTR function and prospects for therapy. *Annual Review*, *77*, 701–726. https://doi.org/10.1146/ANNUREV.BIOCHEM.75.103004.142532.

Riordan, J. R., Rommens, J. M., Kerem, B. S., Alon, N. O. A., Rozmahel, R., Grzelczak, Z., Zielenski, J., Lok, S. I., Plavsic, N., Chou, J. L., Drumm, M. L., Iannuzzi, M. C., Collins, F. S., & Tsui, L. C. (1989). Identification of the cystic fibrosis gene: Cloning and characterization of complementary DNA. *Science (New York, N.Y.)*, *245*(4922), 1066–1073. https://doi.org/10.1126/SCIENCE.2475911.

Rogers, D. F. (1994). Airway goblet cells: Responsive and adaptable front-line defenders. *European Respiratory Journal*, *7*(9), 1690–1706. https://doi.org/10.1183/09031936.94.07091690.

Rokicki, W., Rokicki, M., Wojtacha, J., & Dzeljijli, A. (2016). The role and importance of club cells (Clara cells) in the pathogenesis of some respiratory diseases. *Kardiochirurgia i Torakochirurgia Polska = Polish Journal of Cardio-Thoracic Surgery*, *13*(1), 26. https://doi.org/10.5114/KITP.2016.58961.

Rommens, J. M., Iannuzzi, M. C., Kerem, B. S., Drumm, M. L., Melmer, G., Dean, M., Rozmahel, R., Cole, J. L., Kennedy, D., Hidaka, N., Zsiga, M., Buchwald, M., Riordan, J. R., Tsui, L. C., & Collins, F. S. (1989). Identification of the cystic fibrosis gene: Chromosome walking and jumping. *Science (New York, N.Y.)*, *245*(4922), 1059–1065. https://doi.org/10.1126/SCIENCE.2772657.

Rowe, S. M., Miller, S., & Sorscher, E. J. (2005). Mechanisms of disease cystic fibrosis. *The New England Journal of Medicine*, *352*.

Ruysseveldt, E., Martens, K., & Steelant, B. (2021). Airway basal cells, protectors of epithelial walls in health and respiratory diseases. *Frontiers in Allergy*, *2*, 88. https://doi.org/10.3389/FALGY.2021.787128.

Ryu, G., & Shin, H. W. (2021). SARS-CoV-2 infection of airway epithelial cells. *Immune Network*, *21*(1), 1–16. https://doi.org/10.4110/IN.2021.21.E3.

Saint-Criq, V., & Gray, M. A. (2017). Role of CFTR in epithelial physiology. *Cellular and Molecular Life Sciences*, *74*(1), 93–115. https://doi.org/10.1007/s00018-016-2391-y.

Stanke, F., & Tümmler, B. (2016). Classification of CFTR mutation classes. *The Lancet Respiratory Medicine*, *4*(8), e36. https://doi.org/10.1016/S2213-2600(16)30147-3.

Stelzer-Braid, S., Johal, H., Skilbeck, K., Steller, A., Alsubie, H., Tovey, E., Van Asperen, P., McKay, K., & Rawlinson, W. D. (2012). Detection of viral and bacterial respiratory pathogens in patients with cystic fibrosis. *Journal of Virological Methods*, *186*(1–2), 109–112. https://doi.org/10.1016/J.JVIROMET.2012.08.008.

Yaghi, A., & Dolovich, M. B. (2016). Airway epithelial cell cilia and obstructive lung disease. *Cell*, *5*(4). https://doi.org/10.3390/CELLS5040040.

Zhu, Y., Chidekel, A., & Shaffer, T. H. (2010). Cultured human airway epithelial cells (calu-3): A model of human respiratory function, structure, and inflammatory responses. *Critical Care Research and Practice*, *2010*, 1–8. https://doi.org/10.1155/2010/394578.

Zhu, N., Wang, W., Liu, Z., Liang, C., Wang, W., Ye, F., Huang, B., Zhao, L., Wang, H., Zhou, W., Deng, Y., Mao, L., Su, C., Qiang, G., Jiang, T., Zhao, J., Wu, G., Song, J., & Tan, W. (2020). Morphogenesis and cytopathic effect of SARS-CoV-2 infection in human airway epithelial cells. *Nature Communications*, *11*(1), 1–8. https://doi.org/10.1038/s41467-020-17796-z.

Zhu, Z., Wang, L., Yang, S., Zhao, X., Huang, B., Shi, W., Lu, R., Niu, P., Zhan, F., Ma, X., Wang, D., Xu, W., Wu, G., Gao, G. F., & Tan, W. (2020). A novel coronavirus from patients with pneumonia in China, 2019. *New England Journal of Medicine*, *382*(8), 727–733. https://doi.org/10.1056/NEJMOA2001017.

Chapter 24

Biomechanics and mechanobiology of the lung parenchyma following SARS-CoV-2 infection

Béla Suki[a], András Lorx[b], and Erzsébet Bartolák-Suki[a]
[a]Department of Biomedical Engineering, Boston University, Boston, MA, United States, [b]Department of Anesthesiology and Intensive Therapy, Semmelweis University, Budapest, Hungary

Abbreviations

ACE2	angiotensin converting enzyme 2
AET2	alveolar epithelial type 2
ARDS	acute respiratory distress syndrome
ARF	acute respiratory failure
C	compliance
C_{dyn}	dynamic compliance
C_L	lung compliance
COPD	chronic obstructive pulmonary disease
C_{rs}	respiratory system compliance
CSE	cigarette smoke extract
CT	computed tomography
ECM	extracellular matrix
FA	focal adhesion
FAK	focal adhesion kinase
FDM	fluctuation-driven mechanotransduction
FiO_2	fractional inspired oxygen
FRC	functional residual capacity
HS	high stretch
IPAP	inspiratory peak airway pressure
LS	low stretch
MAPK	mitogen-activated protein kinase
MV	mechanical ventilation
NIV	noninvasive ventilation
Orf9b	alternative reading frame
PaO_2	alveolar oxygen tension
PCLS	precision cut lung slice
PEEP	positive end-expiratory pressure
P_{tp}	transpulmonary pressure
P-V	pressure-volume
R_{aw}	airway resistance
RV	residual volume
TLC	total lung capacity
TMPRSS2	transmembrane serine protease 2
US	unstretched
V_T	tidal volume
VV	variable ventilation

Management, Body Systems, and Case Studies in COVID-19. https://doi.org/10.1016/B978-0-443-18703-2.00003-7
Copyright © 2024 Elsevier Inc. All rights are reserved, including those for text and data mining, AI training, and similar technologies.

Introduction

In 2020, the world was hit by a major pandemic, coronavirus disease 19 (COVID-19), due to the rapid spread of severe acute respiratory syndrome coronavirus 2 (SARS-CoV-2) (Kumar et al., 2021). The main site of infection is the lung, causing primary symptoms such as fever, cough, fatigue, and dyspnea. COVID-19 often leads to early hypoxia (Bos et al., 2020) followed by acute respiratory distress syndrome (ARDS) (Ziehr et al., 2020). In severe cases, mechanical ventilation (MV) was invoked for ARDS patients (Lentz et al., 2020), many of whom developed pulmonary fibrosis (Wendisch et al., 2021) or multiple organ failure (Sun et al., 2020) and eventually died or were left with long-term consequences (Bazdyrev et al., 2021).

The lung is a mechanically highly active organ due to cyclic respiration. Virtually all cellular processes depend on various mechanical aspects of the lung. For instance, surfactant secretion is exquisitely sensitive to mechanical stretch: a single deep inspiration triggers long-term surfactant release (Wirtz & Dobbs, 1990). Dynamic mechanical forces acting on alveolar epithelial type 2 (AET2) cells stimulate both the secretion and production of surfactant (Bartolak-Suki et al., 2017). This is just one example of a general phenomenon called mechanotransduction, a cellular process that converts mechanical signals to biochemical messages (Bartolak-Suki et al., 2015; Hoffman et al., 2011). Mechanotransduction plays a key role in normal tissue homeostasis (Torday & Rehan, 2003; Tschumperlin et al., 2010), which is crucial for maintaining the healthy biomechanical properties of the lung. However, aberrant mechanotransduction can lead to a breakdown of normal homeostasis and contribute to disease progression (Chen et al., 2016; Waters et al., 2012). Because of altered mechanotransduction, the extracellular matrix (ECM) of the lung can be remodeled by, for example, excessive collagen deposition, which increases tissue stiffness, loosely defined as the resistance of the material to deformation. With respect to COVID-19, the receptor that allows SARS-CoV-2 entry into AET2 cells, the angiotensin converting enzyme-2 (ACE2) mediated by the transmembrane serine protease 2 (TMPRSS2) (Hoffmann et al., 2020), was shown to depend on cell stretch (Bartolak-Suki et al., 2022).

Cells attached to the ECM also respond to the tissue's biomechanical properties, such as the stiffness, via another mechanobiological process called mechanosensing (Guo et al., 2022). When stiffness is outside the homeostatic range, cells respond by moving away from homeostasis. For example, ECM stiffening is a hallmark of pulmonary fibrosis to which quiescent lung fibroblasts respond by progressive proliferation and matrix synthesis (Liu et al., 2010) through the α6-integrin mechanosensitive receptor that stimulates the myofibroblast to develop fibrosis (Chen et al., 2016). Toward the end stage, regional ECM stiffening percolates through the tissue to impair whole lung mechanics (Bates et al., 2007), and lung functions, including gas exchange, deteriorate. Thus, the lung's biomechanical properties are fundamentally linked with the mechanobiology of its cells, and diseases that alter one or the other will break the normal biomechanical homeostasis. This must be valid for COVID-19, especially when normal breathing is substituted with lung stretch provided by MV.

The purpose of this chapter is to review the literature on the biomechanical properties of the COVID-19 lung and the corresponding cellular mechanobiological processes. We start with a brief description of lung mechanical properties and how COVID-19 influences them. Next, we discuss why these mechanical properties play an enormous role in how COVID-19 patients should be mechanically ventilated. Lastly, we review mechanotransduction related to ACE2 and surfactant proteins and provide a coherent picture of how the aberrant mechanosensitivity of these two systems may contribute to the emergence of COVID-19-induced pulmonary fibrosis.

Biomechanics of the non-COVID-19 respiratory system

A detailed account of ECM biomechanics in terms of the ECM's major load-bearing constituents such as collagen, elastin, and proteoglycans can be found elsewhere (Suki, 2021). The lung-specific contributions of these ECM constituents, including surface film lining the lungs, are also reviewed elsewhere (Suki et al., 2011). Here, we provide a brief summary of how these constituents contribute to bulk tissue stiffness or its inverse, the compliance, and the pressure-volume (*P-V*) curve in the normal and diseased lung.

Lung structure and volumes

The complex internal structure of the lung allows efficient gas exchange: the internal surface area is maximized while the distance traveled by O_2 and CO_2 between the alveolar gas and capillary blood is minimized (Weibel, 2009). Capillaries should fit into the alveolar septal walls and have the proper strength and stiffness required for stretching and recoiling with each breath without rupture. To satisfy these constraints, alveoli are connected to the airway opening via the airway tree, a fractal structure with a small dead space and fluid flow resistance (Kitaoka et al., 1999). The terminal airways carry fresh air to the acini, which are comprised of many alveoli where the gas exchange occurs via diffusion (Sapoval et al., 2002). The septal wall thickness is only 6–8 μm and the average diameter of the human alveolus at total lung capacity (TLC) is approximately 200 μm.

In diseases such as asthma, the airways are heterogeneously constricted due to inflammation (Lutchen et al., 2001), leading to uneven ventilation distribution (Downie et al., 2007). In fibrosis, collagen deposition increases the stiffness (Liu & Tschumperlin, 2011) as well as the thickness (Tukiainen et al., 1983) of the alveolar septal walls. The former limits the amount of air drawn into the lung by the respiratory muscles and reduces lung volumes (Selvi et al., 2013) such as functional residual capacity (FRC, lung volume at end-expiration) while the latter slows the diffusion of O_2 and CO_2 (Hempleman & Hughes, 1991). Alternatively, in chronic obstructive pulmonary disease (COPD), the septal walls become weak and rupture (Kononov et al., 2001), leading to elevated residual volume (RV, lowest lung volume achieved by forced expiration), FRC, and hyperinflation (Morris et al., 1996).

Compliance and the *P-V* curve

FRC results from the balance between lung recoil and the outward stress generated by the chest and diaphragm. The recoil pressure is a consequence of the biomechanical properties of the lung, which are reflected in its *P-V* curve. Fig. 1 shows a schematic *P-V* curve of the lung from the degassed state to TLC and back to RV, whereas the small loop demonstrates a typical small tidal volume (V_T) breathing cycle (Suki et al., 2011). In region 1, the increase in transpulmonary pressure (P_{tp}) produces little change in *V* because most airways and alveoli are collapsed in the degassed state and the slight increase reflects the expansion of the larger airways. As P_{tp} further increases (region 2), the critical opening pressures of collapsed airways are gradually exceeded (Gaver 3rd et al., 1990) and airways pop open in avalanches (Suki et al., 1994), recruiting substantial airspaces; hence, *V* increases much faster. The slope of the inflation limb can be much steeper, even taking negative values implying instability depending on the inflation rate (Alencar et al., 2002). The open regions begin to expand and surface film stretching and possibly proteoglycan compression contribute to the curve. In region 3, surface film and tissue elasticity due to elastin contribute to the curve with some straightening of wavy fibers (Maksym & Bates, 1997). Region 4 is mostly due to collagen fiber recruitment with a sharp increase in the stiffness of the lung. During deflation, these processes are reversed, although with different surfactant adsorption-desorption kinetics (Smith & Stamenovic, 1986) giving rise to lower closing pressures than opening pressures (Melville et al., 1978). This phenomenon, together with fiber viscoelasticity, produces visible hysteresis. The closures lead to air trapping that determines RV. The compliance *C* of the lung (C_L) is the local slope of the *P-V* curve. Examining the *P-V* curve in Fig. 1, C_L obviously depends on the lung volume at which it is measured, V_T, as well as volume history (whether on the inspiratory or expiratory limb). In healthy human adults, static inspiratory C_L is approximately 0.25 L/cmH_2O (Suki et al., 1989) and if measured dynamically, it slightly decreases with frequency (Dosman et al., 1975). Total respiratory system compliance (C_{rs}) in healthy adults gradually decreases from its quasistatic value of approximately 0.1 to 0.035 L/cmH_2O as frequency increases to 2 Hz (Hantos et al., 1986).

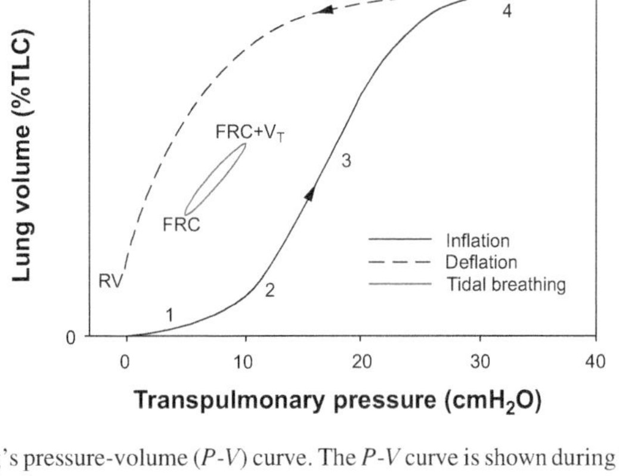

FIG. 1 Schematic representation of the lung's pressure-volume (*P-V*) curve. The *P-V* curve is shown during inflation from the collapsed state *(black solid line)* to total lung capacity (TLC), deflation *(dashed line)* to residual volume (RV), and during breathing *(red)* with tidal volume (V_T) from functional residual capacity (FRC). The regions labeled 1, 2, 3, 4 correspond approximately to regimes of different mechanisms (see text for explanation). *From Suki, B., Stamenovic, D., & Hubmayr, R. D. (2011). Lung parenchymal mechanics. In J. J. Fredberg, G. C. Sieck, & W. T. Gerthoffer (Eds.),* Comprehensive physiology, the respiratory system, respiration mechanics: Organ, cell, molecule *(Vol. 1, pp. 1317–1351). Wiley-Blackwell with permission.*

Compared to the normal lung, in COPD, the recoil force of the parenchyma is reduced due to ruptured septal walls. Consequently, the *P-V* curve shifts upward and C_L increases. In pulmonary fibrosis, the stiffened septal walls increase tissue recoil force, and the *P-V* curve shifts downward, becoming similar in shape to regions 3 and 4, which reduces C_L. In ARDS, the surfactant system is compromised, which elevates surface tension. The end result is edema leading to a *P-V* curve that is similar to the inspiratory *P-V* curve from the degassed state that has a characteristic recruitment phase (regions 1 through 3). The amount of recruitable alveolar space for gas exchange determines the effectiveness of the applied positive end-expiratory pressure (PEEP) during MV.

Biomechanics of the COVID-19 respiratory system

The compliance

While the number of publications on COVID-19 is astounding (>300,000), there is little data on the detailed biomechanics of the lung with the majority of publications limited to simple indexes such as C_{rs}. At the beginning of the pandemic, many patients showed compromised but near normal C_{rs} values (mean ± SD: 0.05 ± 0.014 L/cmH_2O) in the early phase of the disease with minimal parenchymal involvement on CT images that, however, was accompanied by strong hypoxemia (Gattinoni, Coppola, et al., 2020). These patients were called Type L indicating low elastance, or, equivalently, high compliance, of the lung (Gattinoni, Chiumello, et al., 2020). As the disease progressed, Type L could transition to Type H with high elastance or low compliance due to partial collapse of the lung (Gattinoni, Chiumello, et al., 2020). One study reported moderately low C_{rs} values (0.0395 L/cmH_2O) that were accompanied by increased biological markers of vascular damage and thrombosis (Diehl et al., 2020). Another study found similar C_{rs} values (0.041 L/cmH_2O) in patients receiving MV and showed that placing the patient in the prone position improved gas exchange (Zacchetti et al., 2022). In a multicenter study including a cohort of 372 patients with moderate to severe COVID-19, C_{rs} reached a mean of 0.0375 ± 0.0013 L/cmH_2O (Vandenbunder et al., 2021), comparable to that found in regular ARDS. The low C_{rs} values were mostly a consequence of reduced C_L (Roesthuis et al., 2020). To shed light on the controversy as to whether C_{rs} can be used to classify patients with COVID-19, a systematic metaanalysis was carried out that included 11,356 patients from 37 studies (Reddy et al., 2022). The overall C_{rs} assessed around the time of endotracheal intubation was 0.0358 L/cmH_2O. These results did not support the idea of the compliance-based clinical phenotyping of COVID-19 patients. This is important because compliance is often used for selecting MV settings. Based on their observations, the authors concluded that "no change in conventional lung-protective ventilation strategies is warranted" (Reddy et al., 2022). As we shall see below, at least in some patients this conclusion is questionable.

Fig. 2 compares the time course of compliance for COVID-19 survivals and nonsurvivals (Ge et al., 2020). Although the authors stated that the data represent lung compliance, the values are so small that they are most likely respiratory compliance. For simplicity, we refer to their values as "compliance." The compliance of survivals suddenly started to increase

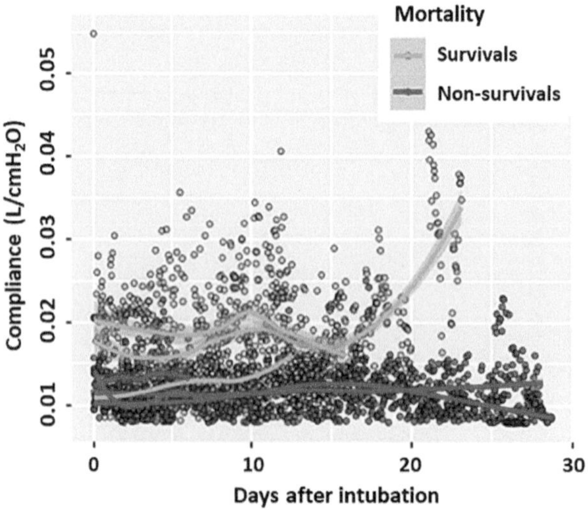

FIG. 2 Time course of compliance in COVID-19 survivals and nonsurvivals. There is a clear separation between the two groups. *Adapted from Ge, H., Pan, Q., Zhou, Y., Xu, P., Zhang, L., Zhang, J., … Zhang, Z. (2020). Lung mechanics of mechanically ventilated patients with COVID-19: Analytics with high-granularity ventilator waveform data.* Frontiers in Medicine(Lausanne), 7, *541. https://doi.org/10.3389/fmed.2020.00541 with permission.*

after about 16 days compared to that of nonsurvivals. A statistical comparison demonstrated that survivors had significantly higher compliance values (0.015 vs 0.011 L/cmH_2O; $P < .001$), suggesting that the mechanical properties of the lung may be a useful predictor of long-term outcome. This issue appears to be somewhat controversial as one study agreed with the above predictive ability of compliance (Longino et al., 2021) whereas another study found a lack of association between compliance and outcome (Vandenbunder et al., 2021). A possible resolution is the effect of time: C_{rs} in survivors started to become significantly higher compared to nonsurvivors after 96–120 h following initiation of MV (Farina-Gonzalez et al., 2022).

Airway resistance

While compliance mostly reflects tissue biomechanical properties, one study also reported airway resistance (R_{aw}) values (Koppurapu et al., 2021), defined as the pressure drop through the airways divided by the inlet flow. Specifically, R_{aw} substantially increased in COVID-19 patients under MV with a mean of 20 ± 4 cmH_2O s/L (Koppurapu et al., 2021). Having excluded artifacts, the authors concluded that "increased airways resistance is an intrinsic feature of severe COVID-19 lung disease." Indeed, this value is extremely high given that in healthy subjects, R_{aw} taken during inspiration is only about 2.5 cmH_2O s/L (Aldrich et al., 1989). Another study found R_{aw} values of 12 cmH_2O s/L (Nezami et al., 2022). A likely explanation for the lower R_{aw} is that the time from intubation to measurement was 7 days in the latter study versus 1.7 days in the former study. In addition, the medications given to patients were different. The most likely candidate mechanisms for the high R_{aw} are heterogeneous closure and massive constriction of small airways, perhaps similar to that in severe asthma due to inflammation, airway wall thickening, and smooth muscle hypertrophy (Lutchen et al., 2001). Indeed, airway wall thickening (Li et al., 2020) as well as extensive cellular and ECM infiltration into bronchioles (Copin et al., 2020) have been reported. Surprisingly, however, the survival rate did not depend on airway obstruction (Nezami et al., 2022).

Lung recruitability

A retrospective metaanalysis suggested no difference between COVID-19-related ARDS and non-COVID-19-related ARDS (Reddy et al., 2022). Several additional studies arrived at similar conclusions with subtle differences. For example, in a matched study (Chiumello et al., 2020), three groups of patients were compared: group 1 included mechanically ventilated COVID-19 patients, group 2 included a historical ARDS population with matched oxygenation, and group 3 was another historical ARDS population with matched C_{rs}. Comparing the oxygenation-matched groups, patients with COVID-19 had higher C_{rs} than those with ARDS (0.050 ± 15 vs 0.039 ± 11 L/cmH_2O, $P = .03$). Comparing the compliance-matched groups, patients with COVID-19 had lower PaO_2/FiO_2 (O_2 tension over fractional inspired O_2) than those with ARDS (106.5 ± 59.6 vs 160 ± 62 mmHg, $P < .001$). It was concluded that COVID-19-associated ARDS is a subset of regular ARDS. There was an additional interesting observation: hypoxemia was not exclusively due to the amount of collapsed tissue because elevating PEEP from 5 to 15 cmH_2O achieved no recruitment and lung mechanics could be even worse in the COVID-19 group (Chiumello et al., 2020). Other studies also reported that increasing PEEP decreased compliance with little to no lung recruitment (Ball et al., 2021; Chiumello et al., 2021; Roesthuis et al., 2020; Sang et al., 2020), albeit with high heterogeneity among patients (Chiumello et al., 2021). Of note is that at higher PEEPs, compliance decreases without a clearcut effect on the ventilation-perfusion mismatch and without reducing shunt (Scaramuzzo et al., 2022). However, the low recruitability appears to be coupled with overdistension even at intermediate PEEPs and V_Ts (Yaroshetskiy et al., 2022).

The lung's *P-V* curve in COVID-19

The need for *P-V* curves

What is the explanation for the main findings that the COVID-19 lung exhibits low recruitability? Assessing how compliance varies with PEEP is usually done by adding V_T at one PEEP to get a first compliance value and then raising the PEEP and adding V_T again to get a second compliance value. This procedure relies on single endpoints that can, however, occur in many ways, linearly, or along a curve. If the curve is convex as in region 2 of the *P-V* curve (Fig. 1), the compliance will increase with PEEP, and it is associated with recruitment. Alternatively, if the curve is concave as in region 4, the compliance will decrease with PEEP, and it represents stiffening. Thus, measuring the entire *P-V* curve can provide more insight into the biomechanical properties than single compliance values. Indeed, as explained in Fig. 1, the *P-V* curve provides detailed information on the mechanisms contributing to lung behavior during breathing or MV.

Unfortunately, very little is known about how COVID-19 alters the lung's *P-V* curve. A recent study presented a series of *P-V* curves of the lung, using the esophageal balloon to measure P_{tp}, in a patient during SARS-CoV-2 infection and the subsequent recovery (Lorx et al., 2022). To our knowledge, this is the only study that has reported lung *P-V* data in COVID-19. Below, we discuss the results and show that the detailed biomechanical properties of the COVID-19 lung offer a deeper understanding of how compliance changes with PEEP with implications for MV.

Biomechanics of the COVID-19 lung during recovery (Lorx et al., 2022)

An unvaccinated 63-year-old woman was admitted to the hospital with respiratory failure due to bilateral pneumonia associated with confirmed SARS-CoV-2 infection. The initial CT image (day 1) showed massive consolidation and some ground-glass opacity and airway dilation (Fig. 3A). She was admitted to the intensive care unit on day 25 with an alveolar-arterial (A-a) gradient of 530 mmHg, and the next CT image demonstrated almost 100% involvement of the lung (Fig. 3B). The patient was then placed on pressure-controlled noninvasive ventilation (NIV) using a PEEP of 8 cmH_2O, and an inspiratory peak airway pressure (IPAP) of 24 cmH_2O. On day 35, still on NIV, the PEEP was lowered to 6 cmH_2O and IPAP was between 24 and 30 cmH_2O. Although the patient received considerable ventilator support, her respiratory efforts were substantial ($P_{tp} > 40$ cmH_2O), resulting in severe dyspnea and fatigue. Elevating IPAP to 35 cmH_2O reduced the patient's inspiratory effort and this more stable condition led to gradual improvements in C_L. On day 95, the patient was discharged.

Fig. 4 shows a series of maximal spontaneous P_{tp}-*V* loops recorded between the 36th and 301st days. During maximum efforts, P_{tp} swings were enormous, reaching values between 70 and 80 cmH_2O (day 36) while the corresponding inhaled volume was only 0.66 L (May 7 curve). Also note the negligible hysteresis of the P_{tp}-*V* curve and the lack of recruitment with the shape of the curve being similar to that in region 4 in Fig. 1. This shape is fully consistent with *P-V* curves reported for pulmonary fibrosing alveolitis (Thompson & Colebatch, 1989). By day 67 (June 7), the maximum lung volume reached 1.03 L and the corresponding CT (Fig. 3C) displayed considerable improvements with little remaining ground-glass

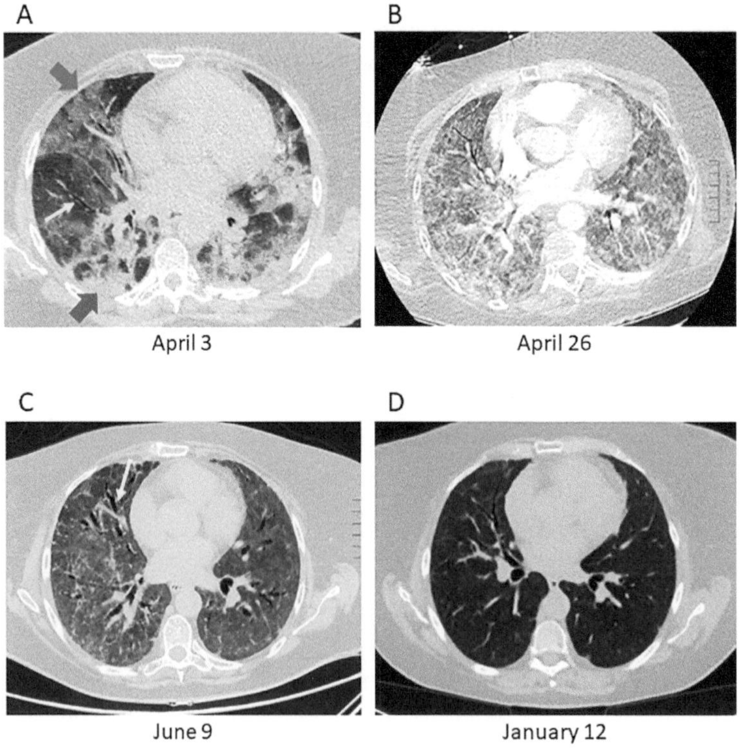

FIG. 3 CT images of a patient with COVID-19. (A) Image taken at the time of hospital admission (Day 1). *Red*, *blue*, and *yellow arrows* mark regions of consolidation, ground-glass opacity, and airway dilation, respectively. (B) The patient was moved to the ICU (Day 25); the corresponding CT image (using contrast) shows massive bilateral ground-glass opacity throughout the lung. (C) The image on Day 69 shows substantial improvement in ground-glass opacity. (D) Nearly normal CT image 7 months later (Day 301). *Reproduced from Lorx, A., Baglyas, S., Podmaniczky, E., Valko, L., Gal, J., & Suki, B. (2022). Lung mechanics during recovery of a non-invasively ventilated patient with severe COVID-19 pneumonia.* Respiratory Physiology & Neurobiology, 306, *103960. https://doi.org/10.1016/j.resp.2022.103960 with permission.*

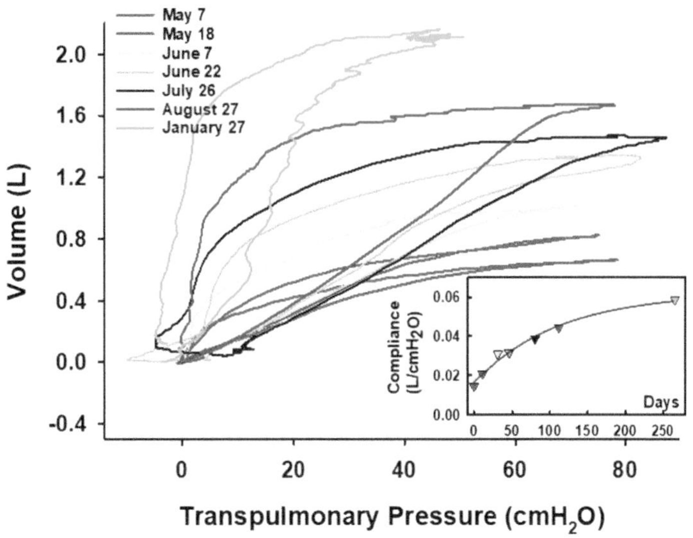

FIG. 4 Transpulmonary pressure-lung volume curves during the recovery of a patient from COVID-19 infection. Different colors correspond to measurements taken on different dates. Inset: Dynamic lung compliance during spontaneous breathing *(triangles)* and a nonlinear regression *(solid line)* as a function of the number of days from hospital admission. *Reproduced from Lorx, A., Baglyas, S., Podmaniczky, E., Valko, L., Gal, J., & Suki, B. (2022). Lung mechanics during recovery of a non-invasively ventilated patient with severe COVID-19 pneumonia.* Respiratory Physiology & Neurobiology, 306, *103960. https://doi.org/10.1016/j.resp.2022.103960 with permission.*

opacities. By day 148 (Aug. 27), the maximum inhaled volume further increased to 1.66 L and the A-a gradient decreased to 60.1 mmHg. At a follow-up visit on day 301 (Jan. 27), the P_{tp}-V curve became normal so that maximum efforts generated $P_{tp} < 50\,cmH_2O$ with a 2.11 L of inhaled air and the CT image (Fig. 3D) presented a healthy lung. The time course of dynamic lung compliance (C_{dyn}) revealed a 3.6-fold increase between day 36 and 301 (inset, Fig. 4). These improvements in lung biomechanics were also accompanied by a gradual restoration of a large hysteresis loop of the P_{tp}-V curve, suggesting normal surfactant function (Smith & Stamenovic, 1986). The relation between pulmonary fibrosis and surfactant is further discussed in the next section. The evolution of the biomechanical and structural features of the lung over a 10-month-long period demonstrates how lung healing returns to normal homeostasis.

Implications for mechanical ventilation

The discussion above favors the notion that acute respiratory failure (ARF) in COVID-19 patients differs from conventional ARDS. To summarize, at the early phase of ARF due to SARS-CoV2 infection, C_L is close to the normal range, but due to lung perfusion issues, severe hypoxemia develops (Herrmann et al., 2020) with a remarkable increase in minute ventilation. Because FRC is still acceptable, the elevated V_Ts do not cause excessive mechanical stretching of the lung tissue. As the disease progresses, both C_L and FRC silently drop, which in turn substantially elevate the intrapleural pressure swings, resulting in a heterogeneous increase of local mechanical stresses. This process is further augmented by alveolar infiltration and collagen deposition, which prevent the formation of uniform negative intrapleural pressure swings along the pleural surface. The end result is a significant amplification of local mechanical stresses that influence cellular mechanotransduction (see below). The clinical course of the disease can last from weeks to months and hence adequate MV support is required to prevent fatigue and ARF. However, improper MV can exacerbate mechanical stresses, and optimization of its setting is crucial for avoiding further injury while maintaining patient condition to facilitate weaning and survival.

How to best set MV conditions is a hotly debated question. Oxygenation is provided mostly by the elevated FiO_2 and increasing the mean alveolar pressure with PEEP helps avoid regional collapse, providing a larger surface area for gas exchange. However, the effects of high PEEP on lung physiology are complex (Cronin et al., 2021) and may not be beneficial in COVID-19 patients (Bhatt et al., 2021). Consider the severe case when C_L is $<0.03\,L/cmH_2O$, and the P-V curve is nonlinear (curves in April and May in Fig. 4): a P_{tp} swing as large as $40\,cmH_2O$ would only yield $\sim$0.4 L of inhaled air, and much of this volume change would occur at lower pressures where C_L is higher. Hence, raising the PEEP to levels consistent with regular ARDS would actually lower C_L. This is consistent with earlier publications that found no recruitment with increasing PEEP and suggested the application of lower PEEPs (Ball et al., 2021; Roesthuis et al., 2020). Furthermore, the traditional high PEEP-low V_T ARDS strategy cannot be delivered to an awake patient with high respiratory demand:

when inspiratory pressure is reduced, V_T should drop but patient effort restores it, eventually leading to fatigue and failure of MV. Nevertheless, a relatively high PEEP may be appropriate in the early stages of the disease with near normal C_L. As the disease worsens, the biomechanical properties of the lung may limit MV to low PEEPs and high inspiratory pressures to compensate for the low C_L. This approach may in fact provide more uniform regional ventilation than the patient's own vigorous spontaneous efforts that can generate self-inflicted lung injury (Cruces et al., 2020).

Lung mechanobiology in COVID-19

Mechanotransduction related to ACE2

CT images revealed that SARS-CoV-2 preferentially affects the base of the lung (Pan et al., 2020), which experiences larger stretches, or volumetric strains, than the apex during inhalation (Napadow et al., 2001). However, the apex is also exposed to a larger static stress, hence strain, due to the weight of the lung (West, 1971). A possible explanation for the spatial distribution of SARS-CoV-2 is that the expression of ACE2 on AET2 cells is stretch-dependent. In vascular smooth muscle cells, ACE2 expression has indeed been found to respond to stretch (Song et al., 2020). Furthermore, we recently reported that the expressions of both ACE2 and TMPRSS2 in the lung are mechanosensitive (Bartolak-Suki et al., 2022).

Briefly, rat precision-cut lung slices (PCLS) were subject to mechanical stimuli utilizing a 12-well stretcher device that reproduced in vivo physiologic conditions (Mondonedo et al., 2020). Samples were sinusoidally stretched for 12 h at 1 Hz (breathing rate for rats). Three stretch groups were set up: (1) unstretched (US); (2) low-stretch (LS) with 5% peak-to-peak area strain mimicking the lung base; and (3) high-stretch (HS), 5% area strain superimposed on 10% static stretch mimicking the lung apex. Additionally, PCLSs were either control or exposed to cigarette smoke extract (CSE) to mimic acute inflammatory effects of cigarette smoking. Western blots were used to evaluate ACE2 and TMPRSS2 expressions from homogenates. The effect of stretch was strong ($P < .001$) independent of CSE (Fig. 5A), but CSE did not affect ACE2 under

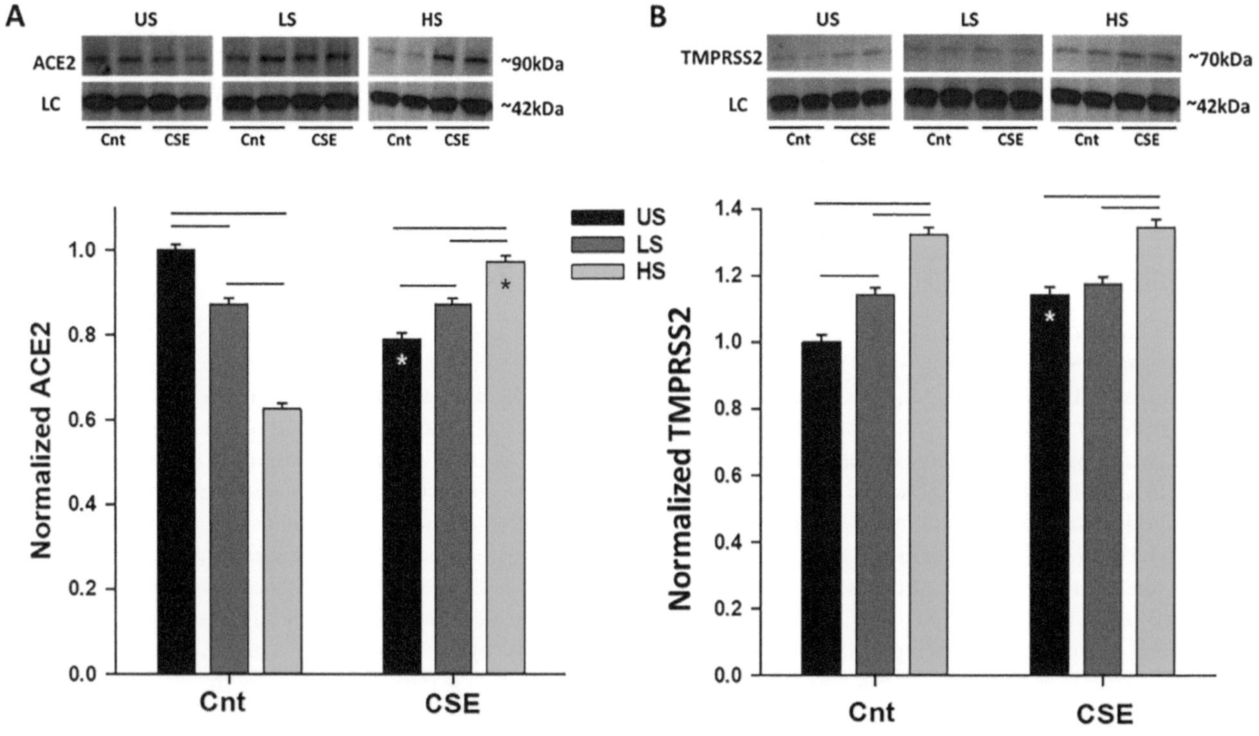

FIG. 5 Expressions of ACE2 and TMPRSS2 as a function of stretch pattern and cigarette smoke exposure. ACE2 (A) and TMPRSS2 (B) protein levels in homogenized precision cut lung slices normalized with the mean of the unstretched (US) control (Cnt) samples measured at three different conditions (*US*, unstretched, *LS*, low stretch with 5% dynamic area strain, *HS*, high stretch with 5% dynamic area strain superimposed on 10% static strain) stretched for 12 h with or without cigarette smoke extract (CSE) exposure. Error bars represent measurement variability in eight PCLS samples from eight rats. Western blots above graphs are representative examples from each condition. *LC*, loading control. Horizontal bars with Cnt or CSE denote statistically significant stretch dependence; * denotes differences due to CSE exposure compared to Cnt at the same stretch level. *Reproduced with permission from Bartolak-Suki, E., Mondonedo, J. R., & Suki, B. (2022). Mechano-inflammatory sensitivity of ACE2: Implications for the regional distribution of SARS-CoV-2 injury in the lung.* Respiratory Physiology & Neurobiology, 296, *103804. https://doi.org/10.1016/j.resp.2021.103804.*

LS (red bars). In contrast, CSE changed the mechanosensitivity of ACE2 by increasing its expression with HS to above the value with LS ($P<.001$). Stretch also influenced TMPRSS2 in both the control and CSE groups ($P<.001$; Fig. 5B). However, no difference was found between TMPRSS2 levels under the US and LS conditions in the CSE group. The effect of CSE was only significant in the US case ($P<.001$).

These results demonstrate that both ACE2 and TMPRSS2 are sensitive to stretch in vitro. Hence, they will likely exhibit regional differences in vivo given the known spatial distribution of stretch magnitudes in the lung (Napadow et al., 2001). With high probability, this mechanosensitivity comes from ACE2 on AET2 cells because they express more ACE2 than endothelial cells (Hikmet et al., 2020). Also, ACE2 is either downregulated in the presence of the larger static stretch at the apex or responds to the larger dynamic stretch in the lung base. Further experiments are required to distinguish between these scenarios. Nevertheless, the mechanosensitivity of ACE2 in lung cells increases the susceptibility of the lung base to SARS-CoV-2, which may, in part, explain the characteristic spatial pattern of lung injury in COVID-19 (Pan et al., 2020). Additionally, prone positioning has shown benefits for patients with severe COVID-19 (Karpov et al., 2020) and the improvement in gas exchange after proning was associated with a reduced duration of MV and death rate (Scaramuzzo et al., 2021). It is possible that the redistribution of gravity-related mechanical stresses following proning protects at-risk lung regions by changing the regional stretch experienced by lung cells. Interestingly, CSE exposure inverted the stretch-dependence of ACE2. This may have implications for the severity of COVID-19 among patients with COPD who have an increased risk of MV and worse outcomes, as demonstrated by the higher rate of ground-glass opacities on CT images (Wu et al., 2020). Fig. 5 also implies that inflammation due to a short-term CSE exposure can alter how ACE2 responds to stretch. Because CSE also increased IL-1β expression in PCLS in a stretch-dependent manner (Mondonedo et al., 2020), the associated inflammation can exacerbate the effects of COVID-19 in COPD patients. Despite the limitations of the rat PCLS, the mechanoinflammatory sensitivity of ACE2 and TMPRSS2 could play a role in the observed spatial distribution of COVID-19-mediated lung injury.

Possible mechanisms in the normal lung

Due to a lack of data, we will speculate that there is similarity between the mechanosensitivities of ACE2 and pulmonary surfactant proteins in AET2 cells. First, recall that ACE2 in vascular smooth muscle cells is mechanosensitive: 10% monotonous sinusoidal strain for 24h upregulated ACE2 (Song et al., 2020). Our data in Fig. 5 demonstrate that not just stretch but the stretch pattern (dynamic strain in LS vs dynamic+static strain in HS) also affects ACE2 expression in lung cells. Second, the surfactant metabolism is highly sensitive to mechanical stimuli, including single large stretches (Wirtz & Dobbs, 1990) and cyclic but not monotonous stretches produced by MV programmed to deliver cycle-to-cycle variations in V_T called variable ventilation (VV) that mimics natural variability in breathing (Bartolak-Suki et al., 2017). The phenomenon by which the variable stretch pattern regulates mechanotransduction is called fluctuation-driven mechanotransduction (FDM) (Suki et al., 2016). Accordingly, we predict that ACE2 expression is also influenced by FDM. The question is, what pathways are involved in ACE2 regulation?

The above study in vascular smooth muscle cells proposed a simple mechanosensitive pathway for ACE2 (Song et al., 2020). Specifically, c-Jun kinase 1/2 and protein kinase C βII, and their downstream transcription factors including the nuclear factor κB, are involved in the stretch-induced upregulation of ACE2 (Song et al., 2020). However, there remain many missing links in the signaling diagram (see question marks in Fig. 6). For example, how are mechanical forces sensed by cells? Mechanotransduction starts with the transmission of forces through cell-ECM connections, usually at focal adhesions (FAs), and the regulation of mechanosensitive integrin activation in lipid rafts, which are specific membrane domains (Lietha & Izard, 2020). Integrin subunits form clusters (Loftus & Liddington, 1997) and undergo conformational changes from low to high affinity states (Hughes et al., 1997). G-protein coupled receptors and Ras-related small GTPases and their effectors modulate the affinity of integrin bindings (Hughes et al., 1997; Smyth et al., 1993). At FAs, talin and α-actinin bind to intracellular domains connecting them to vinculin, paxillin, and tensin as well as FA kinase (FAK), which regulate actin filament formation (Brancaccio et al., 2006). Next, FAK is rapidly phosphorylated to recruit Src, which controls FA assembly/disassembly with additional proteins such as Rho and mitogen-activated protein kinase (MAPK). While such a pathway is likely, this needs verification for ACE2. Furthermore, during breathing-induced FDM in vivo, there are sweeping effects of the variable stretch pattern on signaling that can be different from classical mechanotransduction (Suki et al., 2016). For example, subcortical actin tension due to myosin motors is critically involved in FDM, which in turn regulates mitochondrial ATP production, making the mitochondria exquisitely mechanosensitive (Bartolak-Suki et al., 2015). Although most FDM-related findings are from vascular smooth muscle cells, they should apply to AET2 cells because surfactant and cytokine productions in vivo are significantly influenced by FDM during VV (Bartolak-Suki et al., 2017).

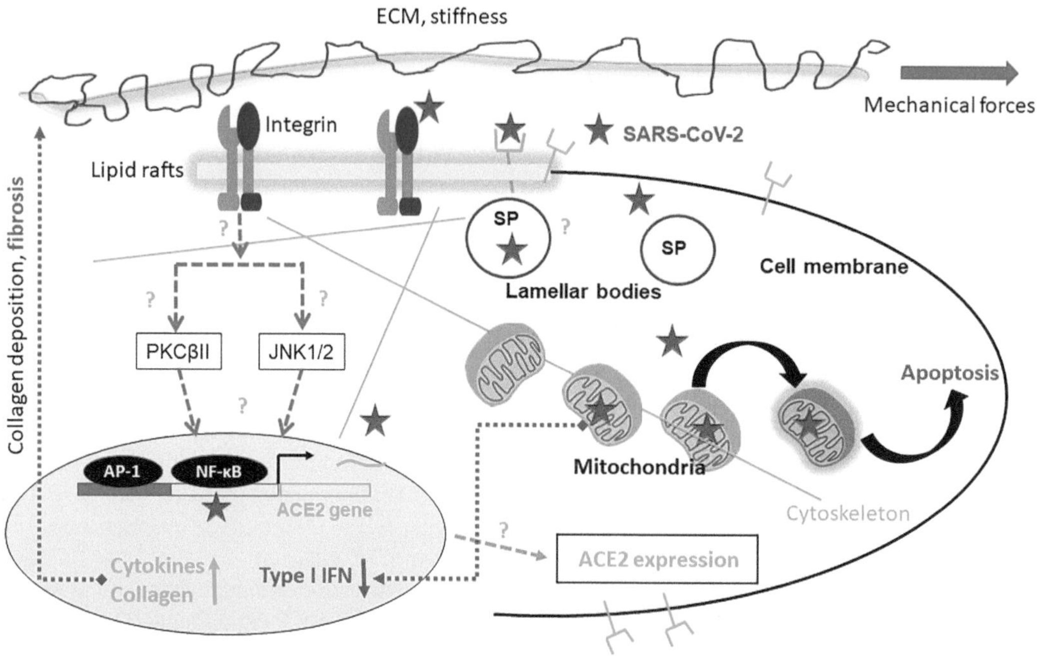

FIG. 6 Schematic mechanotransduction pathways in SARS-CoV-2-infected alveolar epithelial type 2 cells. Mechanical forces of breathing are transmitted through the extracellular matrix, mostly collagen, to AET2 cells that are attached to the ECM via mechanosensitive integrin receptors in lipid rafts. Integrins mediate uptake of the virus *(red stars)* by ACE2 receptors *(green forks)*. *Dashed dark blue arrows* are components of the mechanotransduction pathway proposed by Song et al. (2020). *Green question marks* represent unknown components of this pathway. Lamellar bodies containing both surfactant protein (SP) and viral particles are also shown. The cytoskeleton *(light blue lines)* also participates in viral entry and mechanotransduction to the nucleus *(gray ellipse)* and the mitochondria, which can be normal *(orange)*, infected, or dead *(gray)*. Inflammatory cytokines and collagen are upregulated *(green up arrow)* leading to collagen deposition and fibrosis. Type I interferons (IFNs) are downregulated *(down red arrow)* compromising the cell's host defense.

Possible mechanisms in the COVID-19 lung

How does SARS-CoV-2 affect this mechanotransduction pathway? Although the virus interferes with many cell functions, here we limit the discussion to a few possibilities. The lipid rafts can accumulate ACE2 on the host cell membrane, which may promote the interaction of ACE2 with the viral spike protein (Sorice et al., 2020). These membrane domains also include mechanosensitive integrin receptors. Because the spike protein contains an RGD motif to which integrins can bind, integrins serve as coreceptors to ACE2 and successful cell infection requires integrin signaling through talin (Simons et al., 2021). The entry of the virus through ACE2 is also mediated by clathrin-dependent endocytosis (Bayati et al., 2021), which requires proper structural organization of the cortical actin. Due to the universal nature of how stretch modulates the cytoskeleton (Trepat et al., 2007), we therefore suggest that actin should play a role in viral entry. Additionally, the actin network is involved in many steps of viral replication and transfer (Kloc et al., 2022), most of which awaits experimental support for SARS-CoV-2. As another example, a small viral protein called Orf9b (alternative reading frame) was found to localize on the mitochondria, suppressing type I interferons (Jiang et al., 2020), which are specialized cytokines that coordinate host cell defenses against viral infections (Stetson & Medzhitov, 2006). Such hijacking of the mitochondria has far-reaching consequences because mitochondria control apoptosis and can release DNA into the cytoplasm, activating the inflammasome and suppressing innate and adaptive immunity (Singh et al., 2020). How the mechanosensitivity of mitochondria contributes to these processes is not yet known.

Lastly, we note that mechanosensitive integrins may promote viral entry into cells that do not express ACE2 (Liu et al., 2022) with implications for pulmonary fibrosis because integrins promote fibrotic processes via the activation of transforming growth factor beta (Henderson & Sheppard, 2013). However, SARS-CoV-2 provided an unexpected surprise relevant for pulmonary fibrosis: the virus was found to hide within surfactant lamellar bodies (Huang et al., 2020), suggesting that it interferes with surfactant storage and secretion. Gene regulatory network analysis found that following SARS-CoV-2 infection, the surfactant metabolism was dysregulated (Islam & Khan, 2020), which lowered the amount of surfactant lipids in the bronchoalveolar lavage fluid of COVID-19 patients (Schousboe et al., 2022). Furthermore, these patients had elevated levels of immunoglobulin A, which hinders the normal function of surfactant proteins B and C (Sinnberg et al., 2022).

Interfering with surfactant metabolism and function increases surface tension at the air-liquid interface, leading to alveolar collapse accompanied by poor oxygenation. Interestingly, aberrant surfactant processing through activated Notch1 signaling in AET2 cells from lungs of patients with idiopathic pulmonary fibrosis primes the alveoli for fibrosis (Wasnick et al., 2022), a characteristic manifestation of COVID-19 (Wendisch et al., 2021). It is perhaps not surprising that the Notch1 pathway is upregulated in COVID-19 patients (Islam & Khan, 2020), providing a mechanistic link between the viral impairment of surfactant function and pulmonary fibrosis. Fig. 6 provides a schematic summary of the mechanisms discussed here.

To conclude, mechanobiological processes in the lung are key for the homeostatic regulation of ACE2, surfactant, and many other molecules, organelles, and immune cells that are mechanosensitive but not covered here. By interfering with basic cell functions, SARS-CoV-2 breaks down cellular homeostasis in COVID-19 patients, which causes perfusion-related hypoxemia, compromised lung biomechanics, and potentially pulmonary fibrosis. Once the lung's biomechanical properties deteriorate to a level that MV becomes necessary, a balanced approach is needed that allows gas exchange and facilitates recovery without overdistension. Because VV stimulates the upregulation of surfactant production, we speculate that FDM has the potential to reduce morbidity and perhaps mortality associated with the mechanical ventilation of severe COVID-19 patients. Although mechanobiological processes are an essential part of how the lung responds to SARS-CoV-2 infection, they await further experimental investigation.

Policies and procedures

In this chapter, we reviewed the impact of COVID-19 on the biomechanical properties of the lung and respiratory system. We provided a detailed account of the compliance of the respiratory system and the lung following SARS-CoV-2 infection. We also touched on the fluid flow resistance of the airways. We linked these simple biomechanical indexes to pathobiological processes in the virus-infected lung such as surfactant downregulation and deactivation, cellular and extracellular infiltration, and collagen deposition into the small airways and alveoli. We have associated these processes with mechanosensitive cellular behavior such as mechanotransduction in alveolar epithelial type II cells. We also provided a detailed account of how the lung recovers in a COVID-19 patient. The unique biomechanical behavior observed in this case motivated us to propose specific approaches to noninvasive mechanical ventilation. Specifically, as many studies have reported, the lung is not recruitable in COVID-19 due to the massive alveolar damage, consolidation, lack of surfactant, and collagen deposition. Hence, the application of a high PEEP that is part of the accepted ARDS protocol does not improve oxygenation; rather, it may overstretch the lung and reduce the potential for proper air inhalation. The pressure swings should be sufficiently large to allow gas exchange and avoid the patient fighting ventilation. Because of the high stiffness of the tissue, these large pressures are unlikely to produce overdistension. This strategy may allow lung healing and slow weaning from the ventilator. We also proposed that variable ventilation may improve surfactant function and hence reduce morbidity and healing time.

Applications to other areas

In this chapter, we provided a brief review of the biomechanical properties of the normal and the SARS-CoV-2-infected lung. We identified a simple index, the compliance, that has invariably been used to guide mechanical ventilation procedures. We also argued that the pressure-volume curve provides more information that is not only relevant but can guide physicians setting up mechanical ventilation. We suggest that measuring and utilizing these biomechanical characteristics of the lung should be used in other diseases such as fibrosis, COPD, and viral/bacterial infections when mechanical ventilation becomes necessary. We have reviewed the literature on the mechanobiology related to ACE2 and pulmonary surfactant. We identified mechanosensitive processes that regulate both ACE2 and surfactant in alveolar epithelial type 2 cells. Knowledge of the corresponding pathways can aid studies in other fields such as molecular biology, pathology, and genetic network analysis to reveal further details of how SARS-CoV-2 infects the lung and hijacks pathways such as the surfactant metabolism and organelles such as the mitochondria. Additionally, we discussed the notion of fluctuation-driven mechanotransduction that should apply to all lung cells. The idea is that fluctuations, not stretches per se, influence signal transduction, which leads us to the world of complexity science. We believe that the science of complexity can help unravel gene regulations related to mechanosensitive proteins such as ACE2 and surfactant proteins. Computational modeling of the actin and microtubule cytoskeleton as well as the mitochondria utilizing tools from network science should advance our understanding of the effects of viral infection on overall cell and tissue health.

Mini-dictionary of terms

- **Biomechanics.** A field that characterizes and often quantitatively describes with mathematical models the mechanical properties of biological tissues.
- **Mechanobiology.** A subfield of biology in which mechanical forces play a role in cellular processes.
- **Mechanotransduction.** The conversion of mechanical stimuli to biochemical signals.
- **Mechanosensitivity.** A process by which cells respond to mechanical conditions such as the stiffness of the extracellular matrix.
- **Stiffness.** Change in mechanical stress divided by the corresponding change in deformation such as strain.
- **Pressure-volume curve.** Transpulmonary or transrespiratory pressure as a function of volume during an inhalation/exhalation cycle.
- **Recruitment.** Opening and adding collapsed alveolar regions to the lung's total alveolar airspace available for gas exchange.
- **Variable Ventilation.** A relatively new ventilation mode in which tidal volume changes from cycle to cycle.
- **Fluctuation-driven mechanotransduction.** A specific mode of mechanotransduction in which the magnitude of mechanical stimuli to cells varies from cycle to cycle, but is always present during normal breathing and often delivered by variable ventilation.

Summary points

- COVID-19 alters the biomechanical properties of the lung by reducing its compliance and increasing its airway resistance.
- Surfactant dysfunction and fibrotic alterations are responsible for the changes in biomechanical properties.
- When the biomechanical properties reach a critical level, mechanical ventilation is required to sustain gas exchange.
- The pressure-volume curve of the COVID-19 lung is similar to that of fibrotic patients with little recruitment available.
- Many cellular processes in COVID-19 are mechanosensitive, including the expressions of ACE2 and surfactant proteins.
- Variable ventilation that delivers cycle-by-cycle variations in tidal volume may help restore surfactant function and reduce morbidity.

References

Aldrich, T. K., Shapiro, S. M., Sherman, M. S., & Prezant, D. J. (1989). Alveolar pressure and airway resistance during maximal and submaximal respiratory efforts. *The American Review of Respiratory Disease*, *140*(4), 899–906. https://doi.org/10.1164/ajrccm/140.4.899.

Alencar, A. M., Arold, S. P., Buldyrev, S. V., Majumdar, A., Stamenovic, D., Stanley, H. E., & Suki, B. (2002). Physiology: Dynamic instabilities in the inflating lung. *Nature*, *417*(6891), 809–811. Retrieved from http://www.ncbi.nlm.nih.gov/entrez/query.fcgi?cmd=Retrieve&db=PubMed&dopt=Citation&list_uids=12075340.

Ball, L., Robba, C., Maiello, L., Herrmann, J., Gerard, S. E., Xin, Y., … GECOVID (GEnoa COVID-19) group. (2021). Computed tomography assessment of PEEP-induced alveolar recruitment in patients with severe COVID-19 pneumonia. *Critical Care*, *25*(1), 81. https://doi.org/10.1186/s13054-021-03477-w.

Bartolak-Suki, E., Imsirovic, J., Parameswaran, H., Wellman, T. J., Martinez, N., Allen, P. G., … Suki, B. (2015). Fluctuation-driven mechanotransduction regulates mitochondrial-network structure and function. *Nature Materials*, *14*(10), 1049–1057. https://doi.org/10.1038/nmat4358.

Bartolak-Suki, E., Mondonedo, J. R., & Suki, B. (2022). Mechano-inflammatory sensitivity of ACE2: Implications for the regional distribution of SARS-CoV-2 injury in the lung. *Respiratory Physiology & Neurobiology*, *296*, 103804. https://doi.org/10.1016/j.resp.2021.103804.

Bartolak-Suki, E., Noble, P. B., Bou Jawde, S., Pillow, J. J., & Suki, B. (2017). Optimization of variable ventilation for physiology, immune response and surfactant enhancement in preterm lambs. *Frontiers in Physiology*, *8*, 425. https://doi.org/10.3389/fphys.2017.00425.

Bates, J. H., Davis, G. S., Majumdar, A., Butnor, K. J., & Suki, B. (2007). Linking parenchymal disease progression to changes in lung mechanical function by percolation. *American Journal of Respiratory and Critical Care Medicine*, *176*(6), 617–623. Retrieved from http://www.ncbi.nlm.nih.gov/entrez/query.fcgi?cmd=Retrieve&db=PubMed&dopt=Citation&list_uids=17575096.

Bayati, A., Kumar, R., Francis, V., & McPherson, P. S. (2021). SARS-CoV-2 infects cells after viral entry via clathrin-mediated endocytosis. *The Journal of Biological Chemistry*, *296*, 100306. https://doi.org/10.1016/j.jbc.2021.100306.

Bazdyrev, E., Rusina, P., Panova, M., Novikov, F., Grishagin, I., & Nebolsin, V. (2021). Lung fibrosis after COVID-19: Treatment prospects. *Pharmaceuticals (Basel)*, *14*(8). https://doi.org/10.3390/ph14080807.

Bhatt, A., Deshwal, H., Luoma, K., Fenianos, M., Hena, K., Chitkara, N., … Mukherjee, V. (2021). Respiratory mechanics and association with inflammation in COVID-19-related ARDS. *Respiratory Care*, *66*(11), 1673–1683. https://doi.org/10.4187/respcare.09156.

Bos, L. D. J., Paulus, F., Vlaar, A. P. J., Beenen, L. F. M., & Schultz, M. J. (2020). Subphenotyping acute respiratory distress syndrome in patients with COVID-19: Consequences for ventilator management. *Annals of the American Thoracic Society*, *17*(9), 1161–1163. https://doi.org/10.1513/AnnalsATS.202004-376RL.

Brancaccio, M., Hirsch, E., Notte, A., Selvetella, G., Lembo, G., & Tarone, G. (2006). Integrin signalling: The tug-of-war in heart hypertrophy. *Cardiovascular Research*, *70*(3), 422–433. https://doi.org/10.1016/j.cardiores.2005.12.015.

Chen, H., Qu, J., Huang, X., Kurundkar, A., Zhu, L., Yang, N., … Zhou, Y. (2016). Mechanosensing by the alpha6-integrin confers an invasive fibroblast phenotype and mediates lung fibrosis. *Nature Communications*, *7*, 12564. https://doi.org/10.1038/ncomms12564.

Chiumello, D., Bonifazi, M., Pozzi, T., Formenti, P., Papa, G. F. S., Zuanetti, G., & Coppola, S. (2021). Positive end-expiratory pressure in COVID-19 acute respiratory distress syndrome: The heterogeneous effects. *Critical Care*, *25*(1), 431. https://doi.org/10.1186/s13054-021-03839-4.

Chiumello, D., Busana, M., Coppola, S., Romitti, F., Formenti, P., Bonifazi, M., … Gattinoni, L. (2020). Physiological and quantitative CT-scan characterization of COVID-19 and typical ARDS: A matched cohort study. *Intensive Care Medicine*, *46*(12), 2187–2196. https://doi.org/10.1007/s00134-020-06281-2.

Copin, M. C., Parmentier, E., Duburcq, T., Poissy, J., Mathieu, D., Lille, C.-I., & Anatomopathology, G. (2020). Time to consider histologic pattern of lung injury to treat critically ill patients with COVID-19 infection. *Intensive Care Medicine*, *46*(6), 1124–1126. https://doi.org/10.1007/s00134-020-06057-8.

Cronin, J. N., Camporota, L., & Formenti, F. (2021). Mechanical ventilation in COVID-19: A physiological perspective. *Experimental Physiology*. https://doi.org/10.1113/EP089400.

Cruces, P., Retamal, J., Hurtado, D. E., Erranz, B., Iturrieta, P., Gonzalez, C., & Diaz, F. (2020). A physiological approach to understand the role of respiratory effort in the progression of lung injury in SARS-CoV-2 infection. *Critical Care*, *24*(1), 494. https://doi.org/10.1186/s13054-020-03197-7.

Diehl, J. L., Peron, N., Chocron, R., Debuc, B., Guerot, E., Hauw-Berlemont, C., … Smadja, D. M. (2020). Respiratory mechanics and gas exchanges in the early course of COVID-19 ARDS: A hypothesis-generating study. *Annals of Intensive Care*, *10*(1), 95. https://doi.org/10.1186/s13613-020-00716-1.

Dosman, J., Bode, F., Urbanetti, J., Antic, R., Martin, R., & Macklem, P. T. (1975). Role of inertia in the measurement of dynamic compliance. *Journal of Applied Physiology*, *38*(1), 64–69. https://doi.org/10.1152/jappl.1975.38.1.64.

Downie, S. R., Salome, C. M., Verbanck, S., Thompson, B., Berend, N., & King, G. G. (2007). Ventilation heterogeneity is a major determinant of airway hyperresponsiveness in asthma, independent of airway inflammation. *Thorax*, *62*(8), 684–689. https://doi.org/10.1136/thx.2006.069682.

Farina-Gonzalez, T. F., Nunez-Reiz, A., Yordanov-Zlatkov, V., Latorre, J., Calle-Romero, M., Alonso-Martinez, P., … Sanchez-Garcia, M. (2022). Hourly analysis of mechanical ventilation parameters in critically ill adult Covid-19 patients: Association with mortality. *Journal of Intensive Care Medicine*, *37*(12), 1606–1613. https://doi.org/10.1177/08850666221105423.

Gattinoni, L., Chiumello, D., Caironi, P., Busana, M., Romitti, F., Brazzi, L., & Camporota, L. (2020). COVID-19 pneumonia: Different respiratory treatments for different phenotypes? *Intensive Care Medicine*, *46*(6), 1099–1102. https://doi.org/10.1007/s00134-020-06033-2.

Gattinoni, L., Coppola, S., Cressoni, M., Busana, M., Rossi, S., & Chiumello, D. (2020). COVID-19 does not lead to a "typical" acute respiratory distress syndrome. *American Journal of Respiratory and Critical Care Medicine*, *201*(10), 1299–1300. https://doi.org/10.1164/rccm.202003-0817LE.

Gaver, D. P., 3rd, Samsel, R. W., & Solway, J. (1990). Effects of surface tension and viscosity on airway reopening. *Journal of Applied Physiology*, *69*(1), 74–85. Retrieved from http://www.ncbi.nlm.nih.gov/entrez/query.fcgi?cmd=Retrieve&db=PubMed&dopt=Citation&list_uids=2394665.

Ge, H., Pan, Q., Zhou, Y., Xu, P., Zhang, L., Zhang, J., … Zhang, Z. (2020). Lung mechanics of mechanically ventilated patients with COVID-19: Analytics with high-granularity ventilator waveform data. *Frontiers in Medicine(Lausanne)*, *7*, 541. https://doi.org/10.3389/fmed.2020.00541.

Guo, T., He, C., Venado, A., & Zhou, Y. (2022). Extracellular matrix stiffness in lung health and disease. *Comprehensive Physiology*, *12*(3), 3523–3558. https://doi.org/10.1002/cphy.c210032.

Hantos, Z., Daroczy, B., Suki, B., Galgoczy, G., & Csendes, T. (1986). Forced oscillatory impedance of the respiratory system at low frequencies. *Journal of Applied Physiology (Bethesda, MD: 1985)*, *60*(1), 123–132. https://doi.org/10.1152/jappl.1986.60.1.123.

Hempleman, S. C., & Hughes, J. M. (1991). Estimating exercise DLO2 and diffusion limitation in patients with interstitial fibrosis. *Respiration Physiology*, *83*(2), 167–178. https://doi.org/10.1016/0034-5687(91)90026-f.

Henderson, N. C., & Sheppard, D. (2013). Integrin-mediated regulation of TGFbeta in fibrosis. *Biochimica et Biophysica Acta*, *1832*(7), 891–896. https://doi.org/10.1016/j.bbadis.2012.10.005.

Herrmann, J., Mori, V., Bates, J. H. T., & Suki, B. (2020). Modeling lung perfusion abnormalities to explain early COVID-19 hypoxemia. *Nature Communications*, *11*(1), 4883. https://doi.org/10.1038/s41467-020-18672-6.

Hikmet, F., Mear, L., Edvinsson, A., Micke, P., Uhlen, M., & Lindskog, C. (2020). The protein expression profile of ACE2 in human tissues. *Molecular Systems Biology*, *16*(7), e9610. https://doi.org/10.15252/msb.20209610.

Hoffman, B. D., Grashoff, C., & Schwartz, M. A. (2011). Dynamic molecular processes mediate cellular mechanotransduction. *Nature*, *475*(7356), 316–323. https://doi.org/10.1038/nature10316.

Hoffmann, M., Kleine-Weber, H., Schroeder, S., Kruger, N., Herrler, T., Erichsen, S., … Pohlmann, S. (2020). SARS-CoV-2 cell entry depends on ACE2 and TMPRSS2 and is blocked by a clinically proven protease inhibitor. *Cell*, *181*(2), 271–280. e278 https://doi.org/10.1016/j.cell.2020.02.052.

Huang, J., Hume, A. J., Abo, K. M., Werder, R. B., Villacorta-Martin, C., Alysandratos, K. D., … Kotton, D. N. (2020). SARS-CoV-2 infection of pluripotent stem cell-derived human lung alveolar type 2 cells elicits a rapid epithelial-intrinsic inflammatory response. *Cell Stem Cell*, *27*(6), 962–973. e967 https://doi.org/10.1016/j.stem.2020.09.013.

Hughes, P. E., Renshaw, M. W., Pfaff, M., Forsyth, J., Keivens, V. M., Schwartz, M. A., & Ginsberg, M. H. (1997). Suppression of integrin activation: A novel function of a Ras/Raf-initiated MAP kinase pathway. *Cell*, *88*(4), 521–530. https://doi.org/10.1016/s0092-8674(00)81892-9.

Islam, A., & Khan, M. A. (2020). Lung transcriptome of a COVID-19 patient and systems biology predictions suggest impaired surfactant production which may be druggable by surfactant therapy. *Scientific Reports*, *10*(1), 19395. https://doi.org/10.1038/s41598-020-76404-8.

Jiang, H. W., Zhang, H. N., Meng, Q. F., Xie, J., Li, Y., Chen, H., … Tao, S. C. (2020). SARS-CoV-2 Orf9b suppresses type I interferon responses by targeting TOM70. *Cellular & Molecular Immunology*, *17*(9), 998–1000. https://doi.org/10.1038/s41423-020-0514-8.

Karpov, A., Mitra, A. R., Crowe, S., & Haljan, G. (2020). Prone position after liberation from prolonged mechanical ventilation in COVID-19 respiratory failure. *Critical Care Research and Practice*, *2020*, 6688120. https://doi.org/10.1155/2020/6688120.

Kitaoka, H., Takaki, R., & Suki, B. (1999). A three-dimensional model of the human airway tree. *Journal of Applied Physiology*, *87*(6), 2207–2217. Retrieved from http://www.ncbi.nlm.nih.gov/entrez/query.fcgi?cmd=Retrieve&db=PubMed&dopt=Citation&list_uids=10601169.

Kloc, M., Uosef, A., Wosik, J., Kubiak, J. Z., & Ghobrial, R. M. (2022). Virus interactions with the actin cytoskeleton-what we know and do not know about SARS-CoV-2. *Archives of Virology*, *167*(3), 737–749. https://doi.org/10.1007/s00705-022-05366-1.

Kononov, S., Brewer, K., Sakai, H., Cavalcante, F. S., Sabayanagam, C. R., Ingenito, E. P., & Suki, B. (2001). Roles of mechanical forces and collagen failure in the development of elastase-induced emphysema. *American Journal of Respiratory and Critical Care Medicine*, *164*(10 Pt 1), 1920–1926. Retrieved from http://www.ncbi.nlm.nih.gov/htbin-post/Entrez/query?db=m&form=6&dopt=r&uid=11734447.

Koppurapu, V. S., Puliaiev, M., Doerschug, K. C., & Schmidt, G. A. (2021). Ventilated patients with COVID-19 show airflow obstruction. *Journal of Intensive Care Medicine*, *36*(6), 696–703. https://doi.org/10.1177/0885066621 1000601.

Kumar, A., Singh, R., Kaur, J., Pandey, S., Sharma, V., Thakur, L., … Kumar, N. (2021). Wuhan to world: The COVID-19 pandemic. *Frontiers in Cellular and Infection Microbiology*, *11*, 596201. https://doi.org/10.3389/fcimb.2021.596201.

Lentz, S., Roginski, M. A., Montrief, T., Ramzy, M., Gottlieb, M., & Long, B. (2020). Initial emergency department mechanical ventilation strategies for COVID-19 hypoxemic respiratory failure and ARDS. *The American Journal of Emergency Medicine*, *38*(10), 2194–2202. https://doi.org/10.1016/j.ajem.2020.06.082.

Li, K., Wu, J., Wu, F., Guo, D., Chen, L., Fang, Z., & Li, C. (2020). The clinical and chest CT features associated with severe and critical COVID-19 pneumonia. *Investigative Radiology*, *55*(6), 327–331. https://doi.org/10.1097/RLI.0000000000000672.

Lietha, D., & Izard, T. (2020). Roles of membrane domains in integrin-mediated cell adhesion. *International Journal of Molecular Sciences*, *21* (15). https://doi.org/10.3390/ijms21155531.

Liu, J., Lu, F., Chen, Y., Plow, E., & Qin, J. (2022). Integrin mediates cell entry of the SARS-CoV-2 virus independent of cellular receptor ACE2. *The Journal of Biological Chemistry*, *298*(3), 101710. https://doi.org/10.1016/j.jbc.2022.101710.

Liu, F., Mih, J. D., Shea, B. S., Kho, A. T., Sharif, A. S., Tager, A. M., & Tschumperlin, D. J. (2010). Feedback amplification of fibrosis through matrix stiffening and COX-2 suppression. *The Journal of Cell Biology*, *190*(4), 693–706. https://doi.org/10.1083/jcb.201004082.

Liu, F., & Tschumperlin, D. J. (2011). Micro-mechanical characterization of lung tissue using atomic force microscopy. *Journal of Visualized Experiments*, (54). https://doi.org/10.3791/2911.

Loftus, J. C., & Liddington, R. C. (1997). New insights into integrin-ligand interaction. *The Journal of Clinical Investigation*, *100*(11 Suppl), S77–S81. Retrieved from https://www.ncbi.nlm.nih.gov/pubmed/9413406.

Longino, A., Riveros, T., Risa, E., Hebert, C., Krieger, J., Coppess, S., … Johnson, N. J. (2021). Respiratory mechanics in a cohort of critically ill subjects with COVID-19 infection. *Respiratory Care*, *66*(10), 1601–1609. https://doi.org/10.4187/respcare.09064.

Lorx, A., Baglyas, S., Podmaniczky, E., Valko, L., Gal, J., & Suki, B. (2022). Lung mechanics during recovery of a non-invasively ventilated patient with severe COVID-19 pneumonia. *Respiratory Physiology & Neurobiology*, *306*, 103960. https://doi.org/10.1016/j.resp.2022.103960.

Lutchen, K. R., Jensen, A., Atileh, H., Kaczka, D. W., Israel, E., Suki, B., & Ingenito, E. P. (2001). Airway constriction pattern is a central component of asthma severity: The role of deep inspirations. *American Journal of Respiratory and Critical Care Medicine*, *164*(2), 207–215. Retrieved from http://www.ncbi.nlm.nih.gov/pubmed/11463589.

Maksym, G. N., & Bates, J. H. (1997). A distributed nonlinear model of lung tissue elasticity. *Journal of Applied Physiology*, *82*(1), 32–41. Retrieved from http://www.ncbi.nlm.nih.gov/entrez/query.fcgi?cmd=Retrieve&db=PubMed&dopt=Citation&list_uids=9029195.

Melville, G. N., Iravani, J., & Richter, H. G. (1978). Closing and opening pressures in the intrapulmonary airways of rats. *Respiration*, *35*(1), 22–29. https://doi.org/10.1159/000193855.

Mondonedo, J. R., Bartolak-Suki, E., Bou Jawde, S., Nelson, K., Cao, K., Sonnenberg, A., … Suki, B. (2020). A high-throughput system for cyclic stretching of precision-cut lung slices during acute cigarette smoke extract exposure. *Frontiers in Physiology*, *11*, 566. https://doi.org/10.3389/fphys.2020.00566.

Morris, M. J., Madgwick, R. G., & Lane, D. J. (1996). Difference between functional residual capacity and elastic equilibrium volume in patients with chronic obstructive pulmonary disease. *Thorax*, *51*(4), 415–419. https://doi.org/10.1136/thx.51.4.415.

Napadow, V. J., Mai, V., Bankier, A., Gilbert, R. J., Edelman, R., & Chen, Q. (2001). Determination of regional pulmonary parenchymal strain during normal respiration using spin inversion tagged magnetization MRI. *Journal of Magnetic Resonance Imaging*, *13*(3), 467–474. https://doi.org/10.1002/jmri.1068.

Nezami, B., Tran, H. V., Zamora, K., Lowery, P., Kantrow, S. P., Lammi, M. R., & deBoisblanc, B. P. (2022). Inspiratory airway resistance in respiratory failure due to COVID-19. *Critical Care Explorations*, *4*(4), e0669. https://doi.org/10.1097/CCE.0000000000000669.

Pan, F., Ye, T., Sun, P., Gui, S., Liang, B., Li, L., … Zheng, C. (2020). Time course of lung changes at chest CT during recovery from coronavirus disease 2019 (COVID-19). *Radiology*, *295*(3), 715–721. https://doi.org/10.1148/radiol.2020200370.

Reddy, M. P., Subramaniam, A., Chua, C., Ling, R. R., Anstey, C., Ramanathan, K., … Shekar, K. (2022). Respiratory system mechanics, gas exchange, and outcomes in mechanically ventilated patients with COVID-19-related acute respiratory distress syndrome: A systematic review and meta-analysis. *The Lancet Respiratory Medicine*. https://doi.org/10.1016/S2213-2600(22)00393-9.

Roesthuis, L., van den Berg, M., & van der Hoeven, H. (2020). Advanced respiratory monitoring in COVID-19 patients: Use less PEEP! *Critical Care*, *24*(1), 230. https://doi.org/10.1186/s13054-020-02953-z.

Sang, L., Zheng, X., Zhao, Z., Zhong, M., Jiang, L., Huang, Y., … Zhang, D. (2020). Lung recruitment, individualized PEEP, and prone position ventilation for COVID-19-associated severe ARDS: A single center observational study. *Frontiers in Medicine (Lausanne)*, *7*, 603943. https://doi.org/10.3389/fmed.2020.603943.

Sapoval, B., Filoche, M., & Weibel, E. R. (2002). Smaller is better—But not too small: A physical scale for the design of the mammalian pulmonary acinus. *Proceedings of the National Academy of Sciences of the United States of America*, *99*(16), 10411–10416. https://doi.org/10.1073/pnas.122352499.

Scaramuzzo, G., Gamberini, L., Tonetti, T., Zani, G., Ottaviani, I., Mazzoli, C. A., … ICU-RER COVID-19 Collaboration. (2021). Sustained oxygenation improvement after first prone positioning is associated with liberation from mechanical ventilation and mortality in critically ill COVID-19 patients: A cohort study. *Annals of Intensive Care*, *11*(1), 63. https://doi.org/10.1186/s13613-021-00853-1.

Scaramuzzo, G., Karbing, D. S., Fogagnolo, A., Mauri, T., Spinelli, E., Mari, M., … Spadaro, S. (2022). Heterogeneity of ventilation/perfusion mismatch at different levels of PEEP and in mechanical phenotypes of COVID-19 ARDS. *Respiratory Care*. https://doi.org/10.4187/respcare.10242.

Schousboe, P., Ronit, A., Nielsen, H. B., Benfield, T., Wiese, L., Scoutaris, N., … Plovsing, R. R. (2022). Reduced levels of pulmonary surfactant in COVID-19 ARDS. *Scientific Reports*, *12*(1), 4040. https://doi.org/10.1038/s41598-022-07944-4.

Selvi, E. C., Rao, K. K. V., & Malathi. (2013). Should the functional residual capacity be ignored? *Journal of Clinical and Diagnostic Research*, *7*(1), 43–45. https://doi.org/10.7860/JCDR/2012/4876.2666.

Simons, P., Rinaldi, D. A., Bondu, V., Kell, A. M., Bradfute, S., Lidke, D. S., & Buranda, T. (2021). Integrin activation is an essential component of SARS-CoV-2 infection. *Scientific Reports*, *11*(1), 20398. https://doi.org/10.1038/s41598-021-99893-7.

Singh, K. K., Chaubey, G., Chen, J. Y., & Suravajhala, P. (2020). Decoding SARS-CoV-2 hijacking of host mitochondria in COVID-19 pathogenesis. *American Journal of Physiology. Cell Physiology*, *319*(2), C258–C267. https://doi.org/10.1152/ajpcell.00224.2020.

Sinnberg, T., Lichtensteiger, C., Hasan Ali, O., Pop, O. T., Jochum, A. K., Risch, L., … Flatz, L. (2022). Pulmonary surfactant proteins are inhibited by IgA autoantibodies in severe COVID-19. *American Journal of Respiratory and Critical Care Medicine*. https://doi.org/10.1164/rccm.202201-0011OC.

Smith, J. C., & Stamenovic, D. (1986). Surface forces in lungs. I. Alveolar surface tension-lung volume relationships. *Journal of Applied Physiology*, *60*(4), 1341–1350. Retrieved from http://www.ncbi.nlm.nih.gov/entrez/query.fcgi?cmd=Retrieve&db=PubMed&dopt=Citation&list_uids=3754553.

Smyth, S. S., Joneckis, C. C., & Parise, L. V. (1993). Regulation of vascular integrins. *Blood*, *81*(11), 2827–2843. Retrieved from https://www.ncbi.nlm.nih.gov/pubmed/8499621.

Song, J., Qu, H., Hu, B., Bi, C., Li, M., Wang, L., … Zhang, M. (2020). Physiological cyclic stretch up-regulates angiotensin-converting enzyme 2 expression to reduce proliferation and migration of vascular smooth muscle cells. *Bioscience Reports*, *40*(6). https://doi.org/10.1042/BSR20192012.

Sorice, M., Misasi, R., Riitano, G., Manganelli, V., Martellucci, S., Longo, A., … Mattei, V. (2020). Targeting lipid rafts as a strategy against coronavirus. *Frontiers in Cell and Development Biology*, *8*, 618296. https://doi.org/10.3389/fcell.2020.618296.

Stetson, D. B., & Medzhitov, R. (2006). Type I interferons in host defense. *Immunity*, *25*(3), 373–381. https://doi.org/10.1016/j.immuni.2006.08.007.

Suki, B. (2021). *Structure and function of the extracellular matrix: A multiscale quantitative approach*. Academic Press.

Suki, B., Barabasi, A. L., Hantos, Z., Petak, F., & Stanley, H. E. (1994). Avalanches and power-law behaviour in lung inflation. *Nature*, *368*(6472), 615–618. Retrieved from http://www.ncbi.nlm.nih.gov/entrez/query.fcgi?cmd=Retrieve&db=PubMed&dopt=Citation&list_uids=8145846.

Suki, B., Parameswaran, H., Imsirovic, J., & Bartolak-Suki, E. (2016). Regulatory roles of fluctuation-driven mechanotransduction in cell function. *Physiology (Bethesda)*, *31*(5), 346–358. https://doi.org/10.1152/physiol.00051.2015.

Suki, B., Peslin, R., Duvivier, C., & Farre, R. (1989). Lung impedance in healthy humans measured by forced oscillations from 0.01 to 0.1 Hz. *Journal of Applied Physiology*, *67*(4), 1623–1629. Retrieved from http://www.ncbi.nlm.nih.gov/entrez/query.fcgi?cmd=Retrieve&db=PubMed&dopt=Citation&list_uids=2793762.

Suki, B., Stamenovic, D., & Hubmayr, R. D. (2011). Lung parenchymal mechanics. In J. J. Fredberg, G. C. Sieck, & W. T. Gerthoffer (Eds.), *Vol. 1. comprehensive physiology, the respiratory system, respiration mechanics: Organ, cell, molecule* (pp. 1317–1351). Wiley-Blackwell.

Sun, X., Wang, T., Cai, D., Hu, Z., Chen, J., Liao, H., … Wang, A. (2020). Cytokine storm intervention in the early stages of COVID-19 pneumonia. *Cytokine & Growth Factor Reviews*, *53*, 38–42. https://doi.org/10.1016/j.cytogfr.2020.04.002.

Thompson, M. J., & Colebatch, H. J. (1989). Decreased pulmonary distensibility in fibrosing alveolitis and its relation to decreased lung volume. *Thorax*, *44*(9), 725–731. https://doi.org/10.1136/thx.44.9.725.

Torday, J. S., & Rehan, V. K. (2003). Mechanotransduction determines the structure and function of lung and bone: A theoretical model for the pathophysiology of chronic disease. *Cell Biochemistry and Biophysics*, *37*(3), 235–246. Retrieved from http://www.ncbi.nlm.nih.gov/entrez/query.fcgi?cmd=Retrieve&db=PubMed&dopt=Citation&list_uids=12625629.

Trepat, X., Deng, L., An, S. S., Navajas, D., Tschumperlin, D. J., Gerthoffer, W. T., … Fredberg, J. J. (2007). Universal physical responses to stretch in the living cell. *Nature*, *447*(7144), 592–595. https://doi.org/10.1038/nature05824.

Tschumperlin, D. J., Boudreault, F., & Liu, F. (2010). Recent advances and new opportunities in lung mechanobiology. *Journal of Biomechanics*, *43*(1), 99–107. https://doi.org/10.1016/j.jbiomech.2009.09.015.

Tukiainen, P., Taskinen, E., Holsti, P., Korhola, O., & Valle, M. (1983). Prognosis of cryptogenic fibrosing alveolitis. *Thorax*, *38*(5), 349–355. https://doi.org/10.1136/thx.38.5.349.

Vandenbunder, B., Ehrmann, S., Piagnerelli, M., Sauneuf, B., Serck, N., Soumagne, T., … COVADIS Study Group. (2021). Static compliance of the respiratory system in COVID-19 related ARDS: An international multicenter study. *Critical Care*, *25*(1), 52. https://doi.org/10.1186/s13054-020-03433-0.

Wasnick, R., Korfei, M., Piskulak, K., Henneke, I., Wilhelm, J., Mahavadi, P., … Guenther, A. (2022). Notch1 induces defective epithelial surfactant processing and pulmonary fibrosis. *American Journal of Respiratory and Critical Care Medicine*. https://doi.org/10.1164/rccm.202105-1284OC.

Waters, C. M., Roan, E., & Navajas, D. (2012). Mechanobiology in lung epithelial cells: Measurements, perturbations, and responses. *Comprehensive Physiology*, *2*(1), 1–29. https://doi.org/10.1002/cphy.c100090.

Weibel, E. R. (2009). What makes a good lung? *Swiss Medical Weekly*, *139*(27–28), 375–386. https://doi.org/10.4414/smw.2009.12270.

Wendisch, D., Dietrich, O., Mari, T., von Stillfried, S., Ibarra, I. L., Mittermaier, M., … Sander, L. E. (2021). SARS-CoV-2 infection triggers profibrotic macrophage responses and lung fibrosis. *Cell*, *184*(26), 6243–6261 e6227. https://doi.org/10.1016/j.cell.2021.11.033.

West, J. B. (1971). Distribution of mechanical stress in the lung, a possible factor in localisation of pulmonary disease. *Lancet*, *1*(7704), 839–841. Retrieved from http://www.ncbi.nlm.nih.gov/entrez/query.fcgi?cmd=Retrieve&db=PubMed&dopt=Citation&list_uids=4102531.

Wirtz, H. R., & Dobbs, L. G. (1990). Calcium mobilization and exocytosis after one mechanical stretch of lung epithelial cells. *Science*, *250*(4985), 1266–1269. Retrieved from http://www.ncbi.nlm.nih.gov/entrez/query.fcgi?cmd=Retrieve&db=PubMed&dopt=Citation&list_uids=2173861.

Wu, F., Zhou, Y., Wang, Z., Xie, M., Shi, Z., Tang, Z., … Medical Treatment Expert Group for COPD and COVID-19. (2020). Clinical characteristics of COVID-19 infection in chronic obstructive pulmonary disease: A multicenter, retrospective, observational study. *Journal of Thoracic Disease*, *12*(5), 1811–1823. https://doi.org/10.21037/jtd-20-1914.

Yaroshetskiy, A. I., Avdeev, S. N., Politov, M. E., Nogtev, P. V., Beresneva, V. G., Sorokin, Y. D., … Yavorovskiy, A. G. (2022). Potential for the lung recruitment and the risk of lung overdistension during 21 days of mechanical ventilation in patients with COVID-19 after noninvasive ventilation failure: The COVID-VENT observational trial. *BMC Anesthesiology*, *22*(1), 59. https://doi.org/10.1186/s12871-022-01600-0.

Zacchetti, L., Longhi, L., Bianchi, I., Di Matteo, M., Russo, F., Gandini, L., … Bergamo, C.-G. (2022). Characterization of compliance phenotypes in COVID-19 acute respiratory distress syndrome. *BMC Pulmonary Medicine*, *22*(1), 296. https://doi.org/10.1186/s12890-022-02087-8.

Ziehr, D. R., Alladina, J., Petri, C. R., Maley, J. H., Moskowitz, A., Medoff, B. D., … Hardin, C. C. (2020). Respiratory pathophysiology of mechanically ventilated patients with COVID-19: A cohort study. *American Journal of Respiratory and Critical Care Medicine*, *201*(12), 1560–1564. https://doi.org/10.1164/rccm.202004-1163LE.

Chapter 25

Long noncoding RNA profiling in respiratory specimens from COVID-19 patients

Marta Molinero[a,b], Carlos Rodríguez-Muñoz[a,b], Silvia Gómez[a,b], Ángel Estella[c], Ferran Barbé[a,b], and David de Gonzalo-Calvo[a,b]
[a]*Translational Research in Respiratory Medicine, University Hospital Arnau de Vilanova and Santa Maria, IRBLleida, Lleida, Spain,* [b]*CIBER of Respiratory Diseases (CIBERES), Institute of Health Carlos III, Madrid, Spain,* [c]*Department of Medicine, Intensive Care Unit University Hospital of Jerez, University of Cádiz, INIBiCA, Cádiz, Spain*

Abbreviations

AUC	area under the ROC Curve
BALF	bronchoalveolar lavage fluid
BAS	bronchial aspirate
ICU	intensive care unit
IMV	invasive mechanical ventilation
lncRNAs	long noncoding RNAs
miRNAs	microRNAs
ncRNA	noncoding RNA
NPS	nasopharyngeal swab
OPS	oropharyngeal swab
PBMC	peripheral blood mononuclear cells
TA	tracheal aspirate

Introduction

Although vaccines against SARS-CoV-2 infection have led to a turning point in this pandemic, COVID-19 is still a major concern. Adverse outcomes are challenging for healthcare systems, especially in the elderly, immunosuppressed, and comorbid subjects (Motos et al., 2022). Additional tools providing further information on the disease course could positively impact the care of patients with COVID-19, guiding timely decision making regarding therapeutic strategies and ultimately improving the allocation of resources.

In this scenario, omics approaches, such as transcriptomics, constitute an indispensable tool not only to improve decision making but also to decipher the pathological mechanisms implicated in the disease. Because these techniques can be applied in respiratory samples, transcriptomic profiling is an interesting approach to identify novel factors associated with COVID-19 pathology and to define the local pathological processes within the respiratory system (García-Hidalgo et al., 2022). Here, recent advances in noncoding RNA (ncRNA) profiling in respiratory specimens, particularly long noncoding RNAs (lncRNAs), are discussed in the context of COVID-19 patients.

Respiratory samples

Airway compartmentalization has been described in specific biological processes, such as inflammation (Bendib et al., 2021). Therefore, an analysis of molecular information directly reflecting pathophysiological mechanisms within the respiratory system is mandatory.

Management, Body Systems, and Case Studies in COVID-19. https://doi.org/10.1016/B978-0-443-18703-2.00039-6
Copyright © 2024 Elsevier Inc. All rights are reserved, including those for text and data mining, AI training, and similar technologies.

A number of respiratory specimens are routinely collected from patients and submitted to the laboratory for clinical decision-making. Depending on the sampling location, the specimens are divided into upper and lower respiratory tract samples. Upper respiratory tract samples come from the oral cavity, nasal cavity, oropharynx, or nasopharynx. The most common are saliva, oropharyngeal swab (OPS), and nasopharyngeal swab (NPS). For the lower respiratory tract, specimens are obtained from the trachea, bronchi, bronchioles, or alveoli. Lower respiratory specimens can be divided according to the invasiveness required for their collection. Samples obtained via noninvasive procedures include sputum and tracheal aspirate (TA). Samples collected using invasive procedures, such as fibrobronchoscopy, include bronchial aspirate (BAS) samples and bronchoalveolar lavage fluid (BALF). Bronchoscopic brushing and transbronchial biopsy samples are also obtained through invasive procedures, but their usefulness is more focused on the diagnosis of bacterial pneumonias as well as the analysis of localized lesions or suspected noninfectious etiologies.

Upper respiratory tract

Upper airway specimens are collected using noninvasive procedures. Saliva is obtained from the oral cavity, OPS samples from the oropharynx, and NPS samples from the nasopharynx through the nasal cavity. Saliva is the fluid produced by the salivary glands while an OPS consists mainly of buccal mucosa obtained by swabbing with a swab or brush.

These types of samples have a high yield for diagnostic testing for respiratory viruses, including SARS-CoV-2 (Lee et al., 2021). The collection of upper respiratory specimens via an NPS or OPS has been proposed as the preferred method for SARS-CoV-2 testing (Wahidi et al., 2020). Although an OPS requires less training, is less challenging to collect, and shows a high concordance with NPS samples, the latter have been described as more sensitive specimens in more advanced phases of the disease (Patel et al., 2021). Salivary diagnostic platforms, a less-invasive procedure that does not require specialized training for healthcare professionals, have also been reported as a feasible alternative to detect SARS-CoV-2 (Caixeta et al., 2021; Tee et al., 2022). Previous data suggest that, when compared to NPS samples, saliva-based sampling for SARS-CoV-2 detection shows a similar sensitivity and a lower cost (Bastos et al., 2021).

The use of upper tract samples for molecular phenotyping has been applied in the context of SARS-CoV-2 infection. For instance, Biji et al. (2021) proposed elevated mRNA levels of S100 family genes, a group of genes with a variety of functions, as markers of severe COVID-19 disease in NPS samples. Nevertheless, some limitations should be taken into account. The potential impact of environmental confounding factors in saliva, such as oral intake of food or medication, should not be disregarded when working with this type of sample (Spick et al., 2022). Furthermore, the presence of virus in the upper respiratory tract does not necessarily lead to a diagnosis of pneumonia.

Lower respiratory tract

During an intensive care unit (ICU) stay, airway hypersecretion and coinfections often occur in mechanically ventilated COVID-19 patients. In this context, flexible bronchoscopy can be used to improve clinical management in patients under invasive mechanical ventilation (IMV) when there is a suspicion of secondary infections, atelectasis, or airway stenosis, among other issues (Wahidi et al., 2020). BAS samples are obtained from the distal bronchus while BALF is obtained from the alveolar space. Although a safe procedure in intubated patients, BALF extraction is more invasive and can lead to more complications than BAS collection, such as hypoxemia or fever. In contrast, BALF provides information directly from the epithelial lining fluid from the alveoli. Furthermore, BAS samples are more susceptible to contamination from the upper respiratory tract. Regarding samples obtained noninvasively, both sputum and TA come from the trachea, but while sputum is obtained by the expectoration of the patient, TAs are collected by aspiration through an endotracheal tube.

At the beginning of the pandemic, bronchoscopy was not recommended due to the suspicion that aerosols generated during collection would increase the risk of viral transmission to healthcare workers (Wahidi et al., 2020). Nonetheless, lower respiratory tract samples are useful for COVID-19 diagnosis and associated infectious complications in critically ill patients, such as COVID-19-associated aspergillosis or ventilator-associated pneumonia (Estella et al., 2021). Yu et al. (2020) suggested that sputum is a better indicator of viral load than upper respiratory tract samples such as OPS and NPS samples. Similar findings were reported in independent investigations (Tian et al., 2021).

This type of specimen can also provide valuable molecular information on pathological mechanisms linked to the disease. For example, Sarma et al. (2021) identified distinct immunological features of COVID-19 ARDS using RNA sequencing in TA samples. Recent studies have also proposed BAS-based molecular testing as a novel tool for risk assessment in the ICU and a useful matrix to define the biological factors associated with the pathology of SARS-CoV-2-induced ARDS and its adverse outcomes, including postacute pulmonary sequelae after hospital discharge (Molinero, Benítez, et al., 2022; Molinero, Gómez, et al., 2022).

The noncoding transcriptome

Omic technologies constitute a useful tool to develop innovative biomarkers, to understand the underlying mechanisms of COVID-19 pathophysiology, and consequently to identify potential therapeutic targets (Shen et al., 2020). Among omic technologies, transcriptomics has gained great attention in the last decade.

The transcriptome consists of all RNA transcripts, including coding and noncoding transcripts, present in a cell in a given biological state. Although ~80% of the human genome is transcribed (Dunham et al., 2012), only 1%–3% codes for proteins (Pennisi, 2012). Therefore, RNA can be divided into RNAs with coding potential and RNAs without coding potential, that is, ncRNAs. In the last decade, huge advances in techniques such as RNA sequencing have led to the discovery of thousands of ncRNAs. Initially, the biological functions of these noncoding transcripts were unclear. Nevertheless, increasing evidence suggests a key role in biology and pathology (Beermann et al., 2016). Of note, ncRNAs are currently considered key elements in the development of complex life (Liu et al., 2013).

ncRNAs can be classified into two subclasses according to an arbitrary size threshold. Transcripts smaller than 200 nucleotides are designated short ncRNAs, including microRNAs (miRNAs), which are probably the most studied group of ncRNAs; transcripts longer than 200 nucleotides are called lncRNAs. Here, the chapter focuses on the second class of noncoding transcripts.

Long noncoding RNAs

lncRNAs are a heterogeneous group of transcripts that encompasses one of the largest classes of ncRNAs. The number of human lncRNAs is estimated to be more than 19,000 (www.gencodegenes.org, Release 42, GRCh38.p13). According to their genomic organization or their relation to protein-coding genes, lncRNAs can be classified as sense, antisense, bidirectional, intergenic, intronic, or enhancer lncRNAs (Nojima & Proudfoot, 2022). Compared to mRNAs, lncRNAs are more tissue specific, poorly expressed, and preferentially nuclear (Ransohoff et al., 2018). In addition, they are relatively poorly conserved at the nucleotide sequence level in comparison to other ncRNAs, such as miRNAs.

lncRNAs play diverse functional and structural roles in almost all biological processes (de Gonzalo-Calvo & Bär, 2021; Garratt et al., 2021). These transcripts modulate gene expression at multiple levels through complex epigenetic mechanisms, including the regulation of chromatin architecture and remodeling, the integrity and function of nuclear bodies, transcription, and translation (Mattick et al., 2023).

In addition to their intracellular localization, lncRNAs have been detected in the extracellular space and in different body fluids (de Gonzalo-Calvo et al., 2016a). Prior evidence supports the notion that lncRNAs, similar to other ncRNAs, are actively secreted into the extracellular space and encapsulated in lipid vesicles (Kenneweg et al., 2019), which protect them against degradation and provide high extracellular stability. A possible role in intercellular crosstalk, at least at the paracrine level, has been described for extracellular lncRNAs (Kenneweg et al., 2019).

Long noncoding RNAs in viral infections and the immunological response

lncRNAs are implicated in human health and the onset and development of various diseases (Issler et al., 2022; Viereck et al., 2020). Experimental evidence has suggested that lncRNAs are critical components in viral infections and host defense. Transcriptome analysis of cells infected with different viruses has shown that the levels of cellular lncRNAs are altered in response to viral replication and viral protein expression (Fortes & Morris, 2016). For instance, the host lncRNA DREH is downregulated by hepatitis B virus X protein in hepatocellular carcinoma (Huang et al., 2013). Furthermore, host lncRNAs can function as regulators of the innate immune response. Ouyang et al. (2014) identified a lncRNA named NRAV that suppresses the transcription of interferon-stimulated genes and whose downregulation has been proposed as part of the host antiviral response. Overall, these transcripts seem to be crucial in the regulation of inflammatory and antiviral gene expression programs.

Long noncoding RNAs as novel biomarkers

lncRNAs have the optimal biochemical properties to become excellent biomarkers in clinical practice. They can be quantified in samples currently used in clinical laboratories. Indeed, lncRNAs have been isolated from sputum and bronchial brushings (Vencken et al., 2015). Furthermore, previous investigations have reported that lncRNAs can be stably detected in exosomes isolated from respiratory specimens, such as BALF (Song et al., 2022). As nucleic acids, lncRNAs are generally much easier to detect and quantitate than peptide-based biomarkers, even at low amounts. lncRNAs can be quantified at a relatively low cost and with high accuracy through techniques already available in clinical laboratories, such as qPCR. This technique and other relatively accessible techniques, such as microarray or sequencing, allow the analysis of global lncRNA profiles in a single experiment.

The profile of extracellular lncRNAs reflects the physiological and pathological status of a patient, leading to the possibility of using these molecules as attractive transcriptional biomarkers in liquid biopsies (Meessen et al., 2021). As a consequence, a substantial number of studies have supported the potential role of lncRNAs as biomarkers with clinical application (de Gonzalo-Calvo et al., 2016b). Extracellular lncRNAs are currently available for clinical practice. In 2012, the FDA approved a lncRNA prostate cancer antigen 3 assay (the Progensa PCA3 assay) as a sensitive and specific test for prostate cancer (de Kok et al., 2002).

Long noncoding RNAs and COVID-19

Since the beginning of the pandemic, a number of well-performed investigations have used transcriptome sequencing of respiratory specimens from COVID-19 patients, mostly focused on immune cells isolated from BALF samples, to understand disease pathogenesis, develop antiviral strategies, and identify novel biomarkers (Liao et al., 2020; Xiong et al., 2020; Zhou et al., 2020). The public availability of the data from these pioneering investigations has allowed the development of multiple studies. This is also the case for investigations focused on lncRNAs.

Shaath et al. (2020) published one of the first studies on the topic discussed here (Table 1). The authors used publicly available datasets from a total of 68,873 single cells derived from healthy subjects, patients with mild COVID-19, and

TABLE 1 Studies focused on long noncoding RNAs and respiratory samples in COVID-19.

Article	Findings	Type of sample	Specimen	Reference
Shaath H, *Cells*, 2020	MALAT1, NEAT1, and SNHG25 were downregulated in patients with mild and severe COVID-19.	Publicly available RNA sequencing data from BALF and PBMCs.	BALF dataset: severe COVID-19 patients, mild COVID-19 patients, and healthy patients.	Shaath et al. (2020)
Mukherjee S, *Viruses*, 2021	WAKMAR2, EGOT, EPB41L4A-AS1, and ENSG00000271646 were significantly upregulated in SARS-CoV-2-infected samples.	Publicly available RNA sequencing data from cell lines and BALF.	BALF dataset: healthy and COVID-19 patients.	Mukherjee et al. (2021)
Moazzam-Jazi M, *J Cell Mol Med*, 2021	MALAT1 and NEAT1 were significantly upregulated in BALF samples from COVID-19 patients. HOTAIRM1, PVT1, and AL392172.1 established more than half of lncRNA-protein coding gene interactions and exhibited high affinity for binding the SARS-CoV-2 genome.	Publicly available RNA sequencing data from BALF and PBMCs.	BALF dataset: healthy and COVID-19 patients.	Moazzam-Jazi et al. (2021)
Rodrigues AC, *Mol Oral Microbiol*, 2021	NEAT1 and MALAT1 were upregulated in COVID-19 patients compared to non-COVID-19 patients.	NPS and saliva	COVID-19 patients and non-COVID-19 patients.	Rodrigues et al. (2021)
Huang K, *PLoS One*, 2022	MALAT1 was overexpressed in BALF monocytes/macrophages from severe versus mild COVID-19 patients and in CD4+ T cells from mild patients. NEAT1 was overexpressed in nine different cell types in BALF samples from severe COVID-19 patients.	Publicly available RNA sequencing data from BALF and PBMCs.	BALF dataset: severe COVID-19 patients, mild COVID-19 patients, and healthy patients.	Huang et al. (2022)
Devadoss D, *iScience*, 2022	LASI, WAKMAR2, and NEAT1 were significantly upregulated in COVID-19 patients with a high viral load.	NPS	COVID-19 patients and non-COVID-19 patients.	Devadoss et al. (2022)

BALF, bronchoalveolar lavage fluid; *lncRNA*, long noncoding RNA; *NPS*, nasopharyngeal swab.

patients with severe COVID-19. The analysis of data from BALF cells suggested downregulation of the lncRNAs MALAT1, NEAT1, and SNHG25 in mild and severe cases of COVID-19 compared to healthy controls. According to previous findings (Wei et al., 2019), the deregulation of MALAT1 and NEAT1 levels might be associated with the regulation of mechanisms such as neutrophil chemotaxis and viral infection, respectively.

This regulatory role of lncRNAs in the host response to viral infection has also been proposed in independent investigations. Interestingly, contradictory findings have been reported concerning MALAT1 and NEAT1. Using publicly available RNA sequencing data of BALF and peripheral blood mononuclear cell (PBMC) samples from COVID-19 patients and healthy individuals, Moazzam-Jazi et al. (2021) reported that the transcript levels of MALAT1 and NEAT1 were upregulated in BALF samples from COVID-19 patients compared to controls. The perturbation in lncRNA levels was linked to the regulation of inflammation in response to the infection. The mechanisms were not completely defined. NEAT1, which is increased in red blood cell-depleted whole blood from patients with severe COVID-19 compared to moderate COVID-19 and healthy subjects, has been described as a key player in the regulation of miRNA networks and subsequent inflammatory processes (Tang et al., 2020). MALAT1 may function as an miRNA sponge in COVID-19 patients, particularly for miRNAs with antiinflammatory functions, such as miR-142-3p and miR-146a-5p (Tang et al., 2020). Due to their role as upstream "sponges" that inhibit the function of miRNAs implicated in COVID-19 pathogenesis, both lncRNAs may constitute a treatment strategy for COVID-19.

More recently, Huang et al. (2022) integrated BALF and PBMC datasets and analyzed the combined cell population. The authors reported MALAT1 and NEAT1 as highly differentially expressed between mild and severe COVID-19 patients. MALAT1 was overexpressed in BALF monocytes/macrophages from severe versus mild COVID-19 patients and in CD4+ T cells from mild patients. NEAT1 was overexpressed in nine different cell types in BALF samples from severe COVID-19 patients, including M1, M2, and intermediate monocytes/macrophages; NK cells; CD4+ T cells; CD8+ memory T cells; naïve B cells; myeloid dendritic cells; and epithelium/basal cells. According to their findings, NEAT1 and MALAT1 constitute key components of immune dysregulation in COVID-19 and therefore provide targets for severity biomarkers or therapeutic targets. In support of this potential role of lncRNAs as biomarkers, Rodrigues et al. (2021) observed that both lncRNAs were highly expressed in saliva and NPS samples from individuals infected with SARS-CoV-2 compared to noninfected individuals. When evaluating their value as potential clinical indicators, the authors proposed NEAT1 levels in the respiratory tract as a possible biomarker with a high value for discriminating positive and negative saliva samples [the area under the ROC curve (AUC) for NEAT1 was 0.81].

Devadoss et al. (2022) assessed the acute mucoinflammatory response in a three-dimensional (3D) airway tissue model of SARS-CoV-2 infection and in COVID-19 patient NPS samples. The authors found upregulation of NEAT1 and other lncRNAs, such as LASI and WAKMAR2, in samples with a high viral load compared with low viral load samples. No differences were observed in MALAT1 levels. Furthermore, the expression levels of NEAT1, LASI, and WAKMAR2 were correlated with SARS-CoV-2 viral RNA levels. Then, the authors focused on the lncRNA LASI because RNA microscopy and molecular modeling indicated a possible interaction between viral RNA and this lncRNA. Using interference RNA technology in 3D cultured cells, blocking LASI was found to reduce SARS-CoV-2 replication and attenuate the CoV-2-altered antiviral interferon response. Therefore, LASI has emerged as a modulator of SARS-CoV-2 infection and the respiratory mucoinflammatory response.

Additional data support the critical regulatory role of lncRNAs in the host response to viral infection. A biological network of trans-acting lncRNAs interacting with protein-coding genes within BALF samples revealed that AL392172.1, HOTAIRM1, and PVT1 are key elements during SARS-CoV-2 infection (Moazzam-Jazi et al., 2021). Functional enrichment of these lncRNAs identified a number of molecular mechanisms implicated in chemokine activity, chemokine receptor binding, viral transcription, cytokine-cytokine receptor interaction, and the IL-17 signaling pathway, among others. A similar approach was used by Mukherjee et al. (2021). The authors performed an mRNA-lncRNA coexpression network analysis with publicly available SARS-CoV-2-infected transcriptome data of human lung epithelial cell lines and BALF from COVID-19 patients. The analysis identified four lncRNAs—WAKMAR2, EGOT, EPB41L4A-AS1, and ENSG00000271646—whose aberrant expression is associated with cytokine storms and antiviral responses during severe SARS-CoV-2 infection of the lungs. A correlation with genes involved in immune-related pathways crucial for cytokine signaling was also reported. These results suggest a possible role of lncRNAs in hyperinflammatory responses in critical COVID-19 patients.

The possible roles of lncRNAs in COVID-19 pathophysiology are not limited to regulation of the host response to SARS-CoV-2 infection. Moazzam-Jazi et al. (2021) identified 56 lncRNAs detected in BALF samples, in particular PVT1 and HOTAIRM1, with a high affinity for binding to the SARS-CoV-2 genome. These results are supported by independent investigations. Natarelli et al. (2021) identified endogenous lncRNAs able to interact with the SARS-CoV-2 genome.

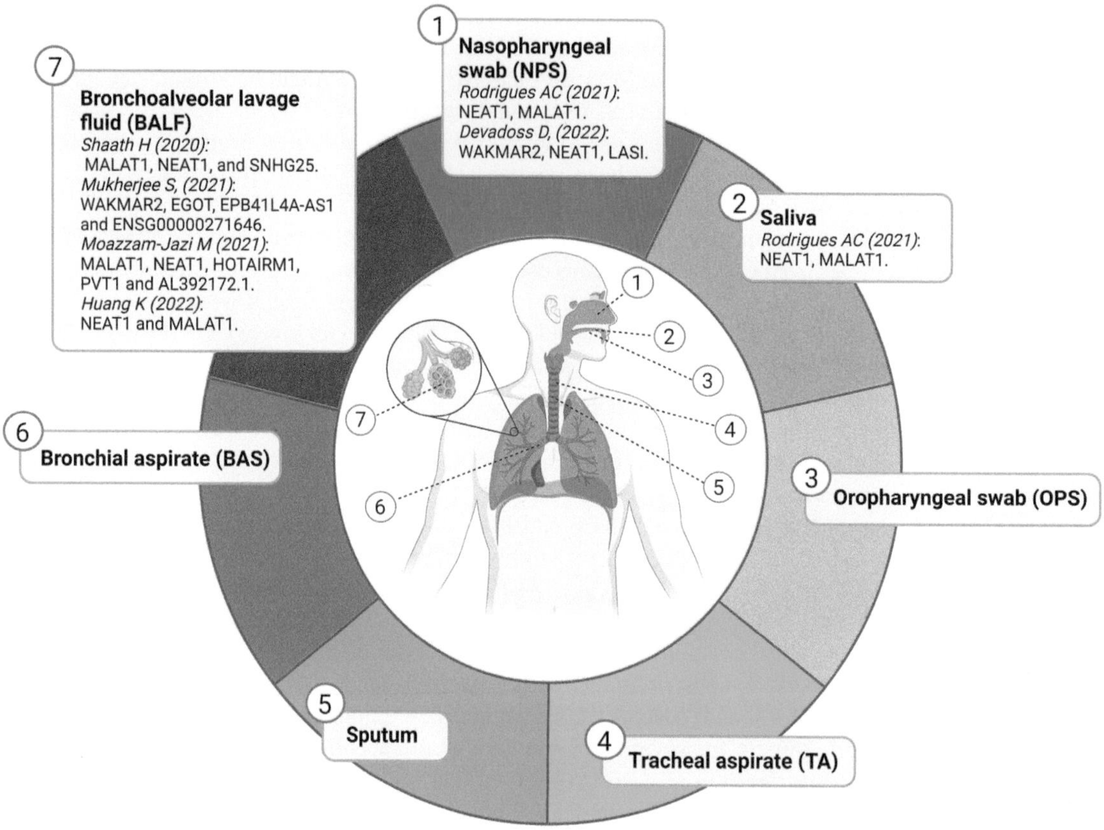

FIG. 1 Summary of long noncoding RNAs in respiratory samples associated with COVID-19 pathology. Overview of the long noncoding RNAs identified in respiratory samples. *BAS*, bronchial aspirate; *BALF*, bronchoalveolar lavage fluid; *NPS*, nasopharyngeal swab; *OPS*, oropharyngeal swab; *TA*, tracheal aspirate.

The lncRNAs ENDRR, HOTAIR, and LINC01505 were found to potentially interact with spike mRNA. In summary, lncRNAs constitute an interesting strategy for designing RNA-based agents with antiviral effects.

Conclusions

SARS-CoV-2 infection induces a unique lncRNA profile in respiratory samples from COVID-19 patients (Fig. 1). Accumulating studies have provided compelling evidence of the crucial roles of lncRNAs in the pathophysiology of COVID-19. Investigation of the host lncRNA fingerprint could allow a better understanding of SARS-CoV-2 pathogenesis and the host biological response. lncRNA profiling in respiratory samples holds interesting potential to improve medical decision-making and identify new therapeutic agents. Further experimental validations are required to corroborate the promising findings discussed here.

Policies and procedures

As previously discussed, some contradictory results have been reported even after a comparison of similar groups of patients. The causes of the disparity in the results among diverse studies could be associated with a number of factors. Low sample size, lncRNA isoforms, use of preamplification protocols, lack of normalization, and poor standardization of the methodology could impact the findings reported by independent laboratories [discussed in more detail in de Gonzalo-Calvo et al. (2022)].

Additional efforts directed at improving reproducibility are fundamental to fully exploring the potential application of lncRNAs as biomarkers and therapeutics. As a consequence, the analysis of lncRNAs in respiratory samples requires that the procedures and methodology used, from RNA extraction to ncRNA quantification, are precise and well defined. To this end, it is crucial to: (i) generate best practice guidelines and standard operating procedures; (ii) provide a detailed description of the methodology used for lncRNA quantification in publications on the topic; and (iii) provide specialized training to healthcare personnel. In this scenario, large, international, and collaborative approaches that include basic researchers, clinicians, and industry partners are of paramount importance.

The use of quantified lncRNAs in respiratory specimens as a source of information to improve medical decision making has been poorly explored. This is surprising because studies using other types of samples have provided interesting results. Using RNA sequencing of PBMCs, Cheng et al. (2021) identified specific lncRNA profiles in patients with severe COVID-19, nonsevere COVID-19, and healthy controls. They constructed a 7-lncRNA panel using LASSO regression with an optimal discrimination value between severe and nonsevere COVID-19 patients. Furthermore, they provided evidence on the use of lncRNAs in defining specific subphenotypes of the disease. Indeed, the lncRNA panel in combination with an immune score described two subtypes of COVID-19 patients, which supports the heterogenicity of the disease. Additional investigations should evaluate whether these findings could also be translated to respiratory specimens.

Applications to other areas

Since the beginning of the century, growing interest in ncRNAs has led to the discovery and functional analysis of many short ncRNAs, which appear to be largely involved in the regulation of physiological pathways in virtually every organ and in disease development. The long counterparts of miRNAs called lncRNAs caught the attention of the biomedical research community at a later stage (Robinson et al., 2022).

A potential clinical application in medical decision making has been demonstrated for lncRNAs under a wide array of conditions. An advantage of the use of lncRNAs as biomarkers is that they can be quantified at relatively low cost using standardized techniques in clinical laboratories, such as qPCR. Therefore, the analysis of lncRNAs in respiratory specimens may have applications at different scales.

In the context of COVID-19, the establishment of a prognostic test based on a lncRNA analyzed in respiratory samples that can predict the presence of postacute respiratory sequelae could improve the clinical management of survivors of the disease. Furthermore, it could provide valuable information on a topic poorly explored: the mechanistic pathways that mediate pulmonary postacute sequelae.

The results discussed here may also have an impact on the management of other respiratory infections, particularly other viral respiratory infections. A prognostic test based on quantified lncRNAs in respiratory specimens could be useful to improve medical decision making in patients infected with syncytial virus or *Influenzavirus A*. These viruses have a strong impact on society due to their high incidence in cold seasons. Therefore, early detection through a lncRNA-based test in a relatively easily accessible respiratory sample constitutes an interesting and innovative approach. The molecular information provided by lncRNA profiling in the context of other viral respiratory infections should also facilitate the development of novel therapeutic approaches.

Mini-dictionary of terms

- **Bronchoalveolar lavage fluid**: Lower respiratory tract specimen obtained from the alveolus via flexible bronchoscopy after instillation of variable volumes of physiological saline solution into a subsegment of the lung. It has high diagnostic power for pulmonary diseases.
- **Sputum**: Lower respiratory tract sample obtained by either induced or spontaneous expectoration. It usually comes from the bronchi and lungs and is produced in response to an inflammatory process.
- **Bronchial aspirate**: Lower respiratory tract specimen obtained from the bronchial tree via flexible bronchoscopy. It is commonly collected for diagnostic purposes in mechanically ventilated patients.
- **Long noncoding**: Noncoding RNAs with a length of 200 nucleotides or more that are involved in posttranscriptional gene regulation.
- **Nasopharyngeal swab**: Upper respiratory tract specimen obtained from the nasopharyngeal cavity by swabbing. It is commonly used for the diagnosis of respiratory infections.
- **Noncoding RNA**: Group of RNA molecules derived from DNA transcription but not translated into proteins. They play an important role in posttranscriptional gene regulation.
- **Oropharyngeal swab**: Upper respiratory tract specimen obtained from the oropharyngeal cavity by smearing with a swab. It may be contaminated by the microbiota of the oral cavity.
- **Tracheal aspirate**: Lower respiratory tract sample obtained from the trachea via aspiration through an endotracheal tube or tracheostomy cannula.

Summary points

- Respiratory samples represent a useful matrix for lncRNA profiling.
- lncRNAs are crucial elements in the host response to SARS-CoV-2 infection.
- lncRNAs have emerged as an interesting tool for antiviral therapies.
- lncRNAs hold great potential to improve medical decision making.
- Additional functional studies are needed to corroborate the current findings.
- Additional efforts should be made to improve the reproducibility of data.

Funding

DdGC received financial support from Instituto de Salud Carlos III (Miguel Servet 2020: CP20/00041), cofunded by the European Union. MM is the recipient of a predoctoral fellowship (PFIS: FI21/00187) funded by Instituto de Salud Carlos III, cofunded by the European Union. CR is supported by Departament de Salut (Pla Estratègic de Recerca i Innovació en Salut (PERIS): SLT028/23/000191). CIBERES is an initiative of the Instituto de Salud Carlos III.

References

Bastos, M. L., Perlman-Arrow, S., Menzies, D., & Campbell, J. R. (2021). The sensitivity and costs of testing for SARS-CoV-2 infection with saliva versus nasopharyngeal swabs: A systematic review and meta-analysis. *Annals of Internal Medicine*, *174*, 501–510.

Beermann, J., Piccoli, M. T., Viereck, J., & Thum, T. (2016). Non-coding RNAs in development and disease: Background, mechanisms, and therapeutic approaches. *Physiological Reviews*, *96*, 1297–1325.

Bendib, I., Beldi-Ferchiou, A., Schlemmer, F., Surenaud, M., Maitre, B., Plonquet, A., ... de Prost, N. (2021). Alveolar compartmentalization of inflammatory and immune cell biomarkers in pneumonia-related ARDS. *Critical Care*, *25*, 23.

Biji, A., Khatun, O., Swaraj, S., Narayan, R., Rajmani, R. S., Sardar, R., ... Tripathi, S. (2021). Identification of COVID-19 prognostic markers and therapeutic targets through meta-analysis and validation of Omics data from nasopharyngeal samples. *eBioMedicine*, *70*, 103525.

Caixeta, D. C., Oliveira, S. W., Cardoso-Sousa, L., Cunha, T. M., Goulart, L. R., Martins, M. M., ... Sabino-Silva, R. (2021). One-year update on salivary diagnostic of COVID-19. *Frontiers in Public Health*, *9*, 589564.

Cheng, J., Zhou, X., Feng, W., Jia, M., Zhang, X., An, T., ... Zhou, C. (2021). Risk stratification by long non-coding RNAs profiling in COVID-19 patients. *Journal of Cellular and Molecular Medicine*, *25*, 4753–4764.

de Gonzalo-Calvo, D., & Bär, C. (2021). Going the long noncoding RNA way toward cardiac regeneration: Mapping candidate long noncoding RNA controllers of regeneration. *The Canadian Journal of Cardiology*, *37*, 374–376.

de Gonzalo-Calvo, D., Kenneweg, F., Bang, C., Toro, R., van der Meer, R. W., Rijzewijk, L. J., ... Thum, T. (2016a). Circulating long-non coding RNAs as biomarkers of left ventricular diastolic function and remodelling in patients with well-controlled type 2 diabetes. *Scientific Reports*, *6*, 37354.

de Gonzalo-Calvo, D., Kenneweg, F., Bang, C., Toro, R., van der Meer, R. W., Rijzewijk, L. J., ... Thum, T. (2016b). Circulating long noncoding RNAs in personalized medicine: Response to pioglitazone therapy in type 2 diabetes. *Journal of the American College of Cardiology*, *68*, 2914–2916.

de Gonzalo-Calvo, D., Sopić, M., & Devaux, Y. (2022). Methodological considerations for circulating long noncoding RNA quantification. *Trends in Molecular Medicine*, *28*, 616–618.

de Kok, J. B., Verhaegh, G. W., Roelofs, R. W., Hessels, D., Kiemeney, L. A., Aalders, T. W., ... Schalken, J. A. (2002). DD3 PCA3, a very sensitive and specific marker to detect prostate tumors. *Cancer Research*, *62*, 2695–2698.

Devadoss, D., Acharya, A., Manevski, M., Houserova, D., Cioffi, M. D., Pandey, K., ... Chand, H. S. (2022). Immunomodulatory LncRNA on antisense strand of ICAM-1 augments SARS-CoV-2 infection-associated airway mucoinflammatory phenotype. *iScience*, *25*, 104685.

Dunham, I., Kundaje, A., Aldred, S. F., Collins, P. J., Davis, C. A., Doyle, F., ... Lochovsky, L. (2012). An integrated encyclopedia of DNA elements in the human genome. *Nature*, *489*, 57–74.

Estella, A., Vidal-Cortés, P., Rodríguez, A., Andaluz Ojeda, D., Martín-Loeches, I., Díaz, E., ... Zaragoza, R. (2021). Management of infectious complications associated with coronavirus infection in severe patients admitted to ICU. *Medicina Intensiva*, *45*, 485–500.

Fortes, P., & Morris, K. V. (2016). Long noncoding RNAs in viral infections. *Virus Research*, *212*, 1–11.

García-Hidalgo, M. C., Peláez, R., González, J., Santisteve, S., Benítez, I. D., Molinero, M., ... Larráyoz, I. M. (2022). Genome-wide transcriptional profiling of pulmonary functional sequelae in ARDS- secondary to SARS-CoV-2 infection. *Biomedicine and Pharmacotherapy*, *154*, 113617.

Garratt, H., Ashburn, R., Sopić, M., Nogara, A., Caporali, A., & Mitić, T. (2021). Long non-coding RNA regulation of epigenetics in vascular cells. *Non-Coding RNA*, *7*, 62.

Huang, J. F., Guo, Y. J., Zhao, C. X., Yuan, S. X., Wang, Y., Tang, G. N., ... Sun, S. H. (2013). Hepatitis B virus X protein (HBx)-related long noncoding RNA (lncRNA) down-regulated expression by HBx (Dreh) inhibits hepatocellular carcinoma metastasis by targeting the intermediate filament protein vimentin. *Hepatology (Baltimore, Md.)*, *57*, 1882–1892.

Huang, K., Wang, C., Vagts, C., Raguveer, V., Finn, P. W., & Perkins, D. L. (2022). Long non-coding RNAs (lncRNAs) NEAT1 and MALAT1 are differentially expressed in severe COVID-19 patients: An integrated single-cell analysis. *PLoS One*, *17*, e0261242.

Issler, O., van der Zee, Y. Y., Ramakrishnan, A., Xia, S., Zinsmaier, A. K., Tan, C., ... Nestler, E. J. (2022). The long noncoding RNA FEDORA is a cell type- and sex-specific regulator of depression. *Science Advances*, *8*, eabn9494.

Kenneweg, F., Bang, C., Xiao, K., Boulanger, C. M., Loyer, X., Mazlan, S., … Thum, T. (2019). Long noncoding RNA-enriched vesicles secreted by hypoxic cardiomyocytes drive cardiac fibrosis. *Molecular Therapy- -Nucleic Acids*, *18*, 363–374.

Lee, R. A., Herigon, J. C., Benedetti, A., Pollock, N. R., & Denkinger, C. M. (2021). Performance of saliva, oropharyngeal swabs, and nasal swabs for SARS-CoV-2 molecular detection: A systematic review and meta-analysis. *Journal of Clinical Microbiology*, *59*, e02881-20.

Liao, M., Liu, Y., Yuan, J., Wen, Y., Xu, G., Zhao, J., … Zhang, Z. (2020). Single-cell landscape of bronchoalveolar immune cells in patients with COVID-19. *Nature Medicine*, *26*, 842–844.

Liu, G., Mattick, J. S., & Taft, R. J. (2013). A meta-analysis of the genomic and transcriptomic composition of complex life. *Cell Cycle (Georgetown, Tex.)*, *12*, 2061–2072.

Mattick, J. S., Amaral, P. P., Carninci, P., Carpenter, S., Chang, H. Y., Chen, L.-L., … Wu, M. (2023). Long non-coding RNAs: Definitions, functions, challenges and recommendations. *Nature Reviews Molecular Cell Biology*, *24*, 430–447.

Meessen, J. M. T. A., Bär, C., di Dona, F. M., Staszewsky, L. I., di Giulio, P., di Tano, G., … Latini, R. (2021). LIPCAR Is increased in chronic symptomatic HF patients. A sub-study of the GISSI-HF trial. *Clinical Chemistry*, *67*, 1721–1731.

Moazzam-Jazi, M., Lanjanian, H., Maleknia, S., Hedayati, M., & Daneshpour, M. S. (2021). Interplay between SARS-CoV-2 and human long non-coding RNAs. *Journal of Cellular and Molecular Medicine*, *25*, 5823–5827.

Molinero, M., Benítez, I. D., González, J., Gort-Paniello, C., Moncusí-Moix, A., Rodríguez-Jara, F., … de Gonzalo-Calvo, D. (2022). Bronchial aspirate-based profiling identifies microRNA signatures associated with COVID-19 and fatal disease in critically Ill patients. *Frontiers in Medicine*, *8*, 756517.

Molinero, M., Gómez, S., Benítez, I. D., Vengoechea, J. J., González, J., Polanco, D., … de Gonzalo-Calvo, D. (2022). Multiplex protein profiling of bronchial aspirates reveals disease-, mortality- and respiratory sequelae-associated signatures in critically ill patients with ARDS secondary to SARS-CoV-2 infection. *Frontiers in Immunology*, *13*, 942443.

Motos, A., López-Gavín, A., Riera, J., Ceccato, A., Fernández-Barat, L., Bermejo-Martin, J. F., … Zapatero, A. (2022). Higher frequency of comorbidities in fully vaccinated patients admitted to the ICU due to severe COVID-19: A prospective, multicentre, observational study. *European Respiratory Journal*, *59*, 2102275.

Mukherjee, S., Banerjee, B., Karasik, D., & Frenkel-Morgenstern, M. (2021). mRNA-lncRNA co-expression network analysis reveals the role of lncRNAs in immune dysfunction during severe SARS-CoV-2 Infection. *Viruses*, *13*, 402.

Natarelli, L., Parca, L., Mazza, T., Weber, C., Virgili, F., & Fratantonio, D. (2021). MicroRNAs and long non-coding RNAs as potential candidates to target specific motifs of SARS-CoV-2. *Non-Coding RNA*, *7*, 1–16.

Nojima, T., & Proudfoot, N. J. (2022). Mechanisms of lncRNA biogenesis as revealed by nascent transcriptomics. *Nature Reviews. Molecular Cell Biology*, *23*, 389–406.

Ouyang, J., Zhu, X., Chen, Y., Wei, H., Chen, Q., Chi, X., … Chen, J. L. (2014). NRAV, a long noncoding RNA, modulates antiviral responses through suppression of interferon-stimulated gene transcription. *Cell Host & Microbe*, *16*, 616–626.

Patel, M. R., Carroll, D., Ussery, E., Whitham, H., Elkins, C. A., Noble-Wang, J., … Brooks, J. T. (2021). Performance of oropharyngeal swab testing compared with nasopharyngeal swab testing for diagnosis of coronavirus disease 2019-United States, January 2020-February 2020. *Clinical Infectious Diseases*, *72*, 482–485.

Pennisi, E. (2012). Genomics. ENCODE project writes eulogy for junk DNA. *Science (New York, N.Y.)*, *337*, 1159–1161.

Ransohoff, J. D., Wei, Y., & Khavari, P. A. (2018). The functions and unique features of long intergenic non-coding RNA. *Nature Reviews. Molecular Cell Biology*, *19*, 143–157.

Robinson, E. L., Baker, A. H., Brittan, M., Mccracken, I., Condorelli, G., Emanueli, C., … Martelli, F. (2022). Dissecting the transcriptome in cardiovascular disease. *Cardiovascular Research*, *118*, 1004–1019.

Rodrigues, A. C., Adamoski, D., Genelhould, G., Zhen, F., Yamaguto, G. E., Araujo-Souza, P. S., … Carvalho de Oliveira, J. (2021). NEAT1 and MALAT1 are highly expressed in saliva and nasopharyngeal swab samples of COVID-19 patients. *Molecular Oral Microbiology*, *36*, 291–294.

Sarma, A., Christenson, S. A., Byrne, A., Mick, E., Pisco, A. O., DeVoe, C., … Langelier, C. R. (2021). Tracheal aspirate RNA sequencing identifies distinct immunological features of COVID-19 ARDS. *Nature Communications*, *12*, 5152.

Shaath, H., Vishnubalaji, R., Elkord, E., & Alajez, N. M. (2020). Single-cell transcriptome analysis highlights a role for neutrophils and inflammatory macrophages in the pathogenesis of severe COVID-19. *Cell*, *9*, 2374.

Shen, B., Yi, X., Sun, Y., Bi, X., Du, J., Zhang, C., … Guo, T. (2020). Proteomic and metabolomic characterization of COVID-19 patient sera. *Cell*, *182*, 59–72.e15.

Song, M., Zhang, X., Gao, Y., Wan, B., Wang, J., Li, J., … Wang, X. (2022). RNA sequencing reveals the emerging role of bronchoalveolar lavage fluid exosome lncRNAs in acute lung injury. *PeerJ*, *10*, e13159.

Spick, M., Lewis, H. M., Frampas, C. F., Longman, K., Costa, C., Stewart, A., … Bailey, M. J. (2022). An integrated analysis and comparison of serum, saliva and sebum for COVID-19 metabolomics. *Scientific Reports*, *12*, 11867.

Tang, H., Gao, Y., Li, Z., Miao, Y., Huang, Z., Liu, X., … Su, W. (2020). The noncoding and coding transcriptional landscape of the peripheral immune response in patients with COVID-19. *Clinical and Translational Medicine*, *10*, e200.

Tee, M. L., Abrilla, A. A., Tee, C. A., Dalmacio, L. M. M., Villaflor, V. J. P., Abubakar, A. Z. A., … Matias, R. R. (2022). Saliva as alternative to naso-oropharyngeal swab for SARS-CoV-2 detection by RT-qPCR: A multicenter cross-sectional diagnostic validation study. *Scientific Reports*, *12*, 12612.

Tian, R. R., Yang, C. X., Zhang, M., Feng, X. L., Luo, R. H., Duan, Z. L., … Zheng, Y. T. (2021). Lower respiratory tract samples are reliable for severe acute respiratory syndrome coronavirus 2 nucleic acid diagnosis and animal model study. *Zoological Research*, *42*, 161–169.

Vencken, S. F., Greene, C. M., & McKiernan, P. J. (2015). Non-coding RNA as lung disease biomarkers. *Thorax*, *70*, 501–503.

Viereck, J., Bührke, A., Foinquinos, A., Chatterjee, S., Kleeberger, J. A., Xiao, K., ... Thum, T. (2020). Targeting muscle-enriched long non-coding RNA H19 reverses pathological cardiac hypertrophy. *European Heart Journal*, *41*, 3462–3474.

Wahidi, M. M., Lamb, C., Murgu, S., Musani, A., Shojaee, S., Sachdeva, A., ... Eapen, G. (2020). American Association for Bronchology and Interventional Pulmonology (AABIP) statement on the use of bronchoscopy and respiratory specimen collection in patients with suspected or confirmed COVID-19 infection. *Journal of Bronchology & Interventional Pulmonology*, *27*, e52–e54.

Wei, L., Li, J., Han, Z., Chen, Z., & Zhang, Q. (2019). Silencing of lncRNA MALAT1 prevents inflammatory injury after lung transplant ischemia-reperfusion by downregulation of IL-8 via p300. *Molecular Therapy- -Nucleic Acids*, *18*, 285–297.

Xiong, Y., Liu, Y., Cao, L., Wang, D., Guo, M., Jiang, A., ... Chen, Y. (2020). Transcriptomic characteristics of bronchoalveolar lavage fluid and peripheral blood mononuclear cells in COVID-19 patients. *Emerging Microbes & Infections*, *9*, 761–770.

Yu, F., Yan, L., Wang, N., Yang, S., Wang, L., Tang, Y., ... Zhang, F. (2020). Quantitative detection and viral load analysis of SARS-CoV-2 in infected patients. *Clinical Infectious Diseases*, *71*, 793–798.

Zhou, Z., Ren, L., Zhang, L., Zhong, J., Xiao, Y., Jia, Z., ... Wang, J. (2020). Heightened innate immune responses in the respiratory tract of COVID-19 patients. *Cell Host & Microbe*, *27*, 883–890.e2.

Chapter 26

COVID-19 and acute pulmonary embolism

Marco Zuin[a] and Gianluca Rigatelli[b]
[a]*Department of Translational Medicine, University of Ferrara, Ferrara, Italy,* [b]*Department of Cardiology, Schiavonia General Hospital, Padova, Italy*

Abbreviations

aPTT	active partial thromboplastin time
CTPA	computed tomography pulmonary angiography
HIT	heparin-induced thrombocytopenia
ICU	intensive care unit
NETs	neutrophil extracellular traps
NOACs	novel oral anticoagulants
PE	pulmonary embolism
PT	prothrombin time
RVD	right ventricular dysfunction
TTE	transthoracic echocardiography

Introduction

Acute pulmonary embolism (PE) has been recognized as a frequent and relevant complication of SARS-CoV-2 infection, influencing the clinical course and outcomes of these subjects (Zuin, Rigatelli, et al., 2022). Frequently, the PE diagnosis in these patients is incidental or is made when patients exhibit a deterioration of their hemodynamic or respiratory function. In this chapter, we comprehensively review the epidemiology, pathophysiology, diagnosis, and use of prognostic biomarkers as well as the current available treatment of acute PE in COVID-19 patients during the acute phase of the disease.

Epidemiology

From an epidemiological perspective, the rates of patients diagnosed with acute PE range between 0.5% and 61% of cases across all risk groups (Bilaloglu et al., 2020; Bompard et al., 2020; Ceriani et al., 2010; Zuin, Engelen, et al., 2022). However, the prevalence of PE Is generally higher among critically ill patients, such as those hospitalized in intensive care units (ICU). However, the real incidence and prevalence of such complications is difficult to assess. The gold-standard technique for diagnosis is computational tomography angiography (CTPA) (Moore et al., 2018). Unfortunately, some patients with coronavirus disease 2019 (COVID-19) may require hospitalization due to the infection and/or occurrence of concomitant venous thromboembolism as those with hemodynamic and/or respiratory instability. Subjects who die early after admission or who cannot be transported to the radiology for CTPA may have been underdiagnosed for acute PE. Furthermore, specific recommendations or guidelines on the appropriate use of CTPA in the diagnostic pathway of PE in COVID-19 are still lacking (Porfidia et al., 2020; Suh et al., 2021). Age has been reported to be the main determinant of COVID-19 related in-hospital mortality (Henkens et al., 2022). All these aspects as well as the rate of patients evaluated using CTPA represent critical factors to determine the real incidence of acute PE in these patients, Finally, some cases may have been underdiagnosed during the first phase of the pandemic.

Pathophysiology of acute pulmonary embolism in COVID-19 patients

From a pathophysiological point of view, the real pathogenetic mechanisms underlying the onset of acute PE in COVID-19 patients have not yet been determined. However, available data indicate a multicausal pathogenesis (Fig. 1). COVID-19

Management, Body Systems, and Case Studies in COVID-19. https://doi.org/10.1016/B978-0-443-18703-2.00036-0
Copyright © 2024 Elsevier Inc. All rights are reserved, including those for text and data mining, AI training, and similar technologies.

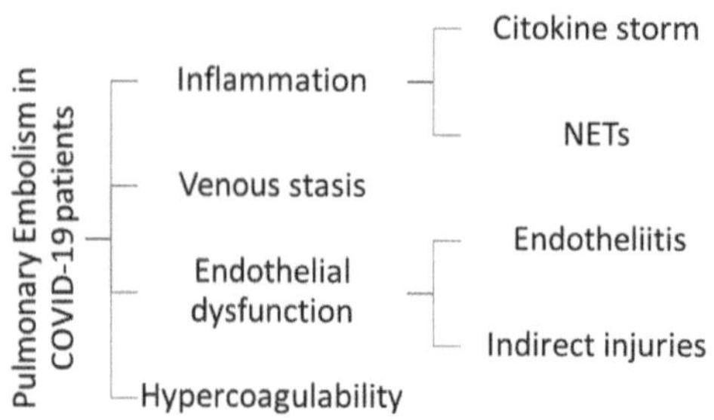

FIG. 1 Pathogenetic mechanisms involved in the genesis of acute pulmonary embolism in COVID-19 patients: *NETs*, neutrophil extracellular traps. Schematic representation of pathophysiological mechanisms involved in the genesis of acute pulmonary embolism in COVID-19 patients.

infection is characterized by hypercoagulability (Zhang et al., 2021). Indeed, since the first cases of COVID-19 infection were reported, researchers identified the Virchow triad, represented by endothelial dysfunction, venous stasis, and hypercoagulability, as a mainstay of the SARS-CoV-2 infection. The endothelial injury is due to either direct or indirect damage produced by the virus, which results in the release of thrombotic factors and the inhibition of fibrinolysis (Varga et al., 2020). Conversely, the inflammatory states beyond the cytokine storm are related to the neutrophil extracellular traps (NETS) used by neutrophils to target pathogens. Moreover, a direct relationship between NETs and hypercoagulability has been demonstrated (Pugh & Ratcliffe, 2017; Zuo et al., 2020). Infected patients generally show a normal or prolonged prothrombin time (PT), an active partial thromboplastin time (aPTT), normal or elevated platelets, elevated D-dimer levels, and elevated von Willebrand factor antigen (Libby & Lüscher, 2020; Lowenstein & Solomon, 2020; Panigada et al., 2020). Concomitantly, there may be a slight decrease in the values of antithrombin and protein S (Ranucci et al., 2020). All these aspects generate a significant hypercoagulable state that may trigger the onset of acute PE or more general venous thromboembolism events (Fig. 2).

Diagnosis of acute PE in COVID-19 patients

The diagnosis of acute PE often remains a challenge. The overlap between COVID-19 and acute PE makes the differential diagnosis demanding, and leads to a significant diagnostic delay in patients with concomitant PE. Symptoms of COVID-19 and PE are nonspecific, including dyspnea, hypoxemia, tachycardia, pleuritic chest pain, and fever (Alimohamadi et al., 2020; Morrone & Morrone, 2018; Righini & Robert-Ebadi, 2018). Therefore, it is important to suspect acute PE in case of sudden respiratory and/or hemodynamic decompensation, refractory hypoxemia, or in a specific subgroup of patients with an intrinsic higher risk of thromboembolic events such as elderly, obese and neoplastic subjects. Unfortunately, a chest

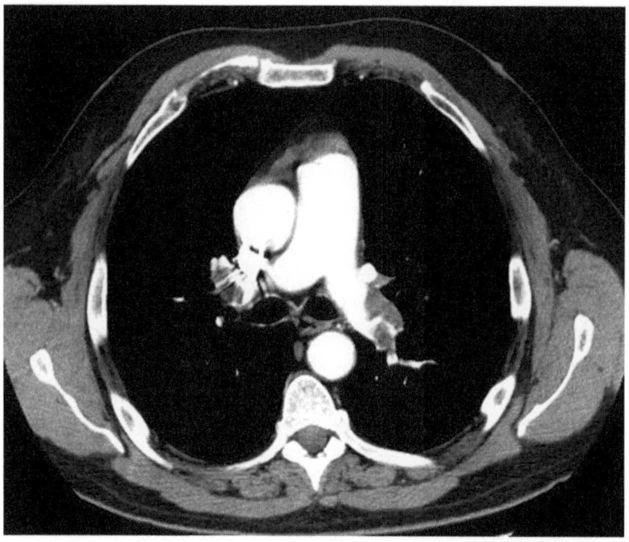

FIG. 2 Pulmonary embolism in a patient with a COVID-19 infection. Computed tomography angiography of 53-year-old women hospitalized for respiratory distress during COVID-19 infection. Multiple filling defects are visible, especially involving the left main pulmonary artery branch.

TABLE 1 Computed tomography signs observed in COVID-19 patients with acute pulmonary embolism.

Signs
Ground-glass opacities
Consolidation interlobular septal thickening
Crazy paving pattern
Air bronchogram
Pleural alterations
Subpleural curvilinear changes
Fibrous stripes
Air bubble sign
Nodules
Halo sign
Vascular enlargement
Lymphadenopathy
Pleural and pericardial effusion

This table describes the concomitant CT findings observed in COVID-19 patients with acute pulmonary embolism.

X-ray has poor accuracy in diagnosing acute PE and is typically used to exclude another concomitant acute conditions such as pneumonia, pulmonary edema, or pneumothorax that could simulate the clinical presentation of thromboembolic disease. CTPA remains the gold standard to diagnose acute PE, also in patients with SARS-CoV-2 infection. The possible radiological signs associated with acute PE in CT are presented in Table 1. CTPA should be considered in patients with worsening respiratory or hemodynamic status, especially in the presence of a limited lung disease or the detection of abnormal coagulation parameters and right ventricular dysfunction (RVD) through transthoracic echocardiography (TTE). Conversely, the use of ventilation-perfusion (VQ) scintigraphy should be reserved for those patients having contraindications for intravenous contrast medium administration and pregnancy. However, the respiratory function of COVID-19 patients may worsen after the inhalation of radiopharmaceuticals.

Prognostic biomarkers

Since the beginning of the pandemic, several biomarkers have been studied in the effort to detect acute PE promptly and accurately in COVID-19 patients. For example, some investigations have identified an association between inflammatory biomarkers or markers of thrombin generation such as D-dimer or the product of fibrin degradation and acute PE (Thachil et al., 2020; Woller et al., 2022; Yu et al., 2020). In daily clinical practice, the D-dimer is widely used in combination with a low pretest probability of thromboembolic disease to exclude the diagnosis and reduce the number of unnecessary CTPAs (Kraaijenhagen et al., 2003; Ten Cate-Hoek & Prins, 2005). Unfortunately, D-dimer is generally increased in patients with concomitant infection, including those with COVID-19 disease. Moreover, the conventional D-dimer thresholds adopted in clinical practice produce a high false positive rate (Righini et al., 2008). Some investigators reported that a D-dimer cut-off of $<2\,\mu$/mL was associated with a high negative predictive value (NPV) of 98.5% (95% confidence interval –CI–: 98.0%−98.9%) with a sensitivity of 70.3% (95% CI: 62.6%−77.2%), and a specificity of 82.4% (95% CI: 81.1%−83.6%) for PE among patients with SARS-CoV-2 infection (Bledsoe et al., 2022). Conversely, Choi et al. discovered that progressively increasing D-dimer thresholds increased the likelihood of PE, but a posttest probability of PE of 3% was present among subjects with a D-dimer <1000 ng/mL (Choi et al., 2020). However, the use of a fixed D-dimer cut-off level in COVID-19 patients may be problematic and incorrect. Indeed, the mortality rate as well as the prevalence of severe forms of infection increase with aging (Onder et al., 2020). Therefore, it would be more appropriate to adopt an age-adjusted cut-off in these subjects, as also recommended by the current guidelines from the European Society of Cardiology (Konstantinides et al., 2020). Notably, the adoption of an age-adjusted threshold can significantly reduce the need for imaging tests compared to a fixed cut-off (Righini et al., 2014). Furthermore, the higher sensitivity and specificity obtained using an age-adjusted D-dimer cut-off also limits the need to transfer an infectious patient to the radiology ward and/or to perform an unnecessary CTPA to exclude acute PE as well as to guide anticoagulant treatments in these subjects (Machowski et al., 2022; Zuin, Rigatelli, & Roncon, 2021; Zuin, Rigatelli, Zuliani, & Roncon, 2021).

Anticoagulation

Due to the elevated prothrombotic risk, COVID-19 patients may require the administration of anticoagulants. According to the recent American guidelines and other international recommendations, critically ill patients should receive a prophylactic dose of anticoagulants, whereas moderately ill subjects if not contraindicated should start therapeutic doses (Moores et al., 2022; Spyropoulos et al., 2020). These recent recommendations were derived from several randomized controlled trials, which demonstrated that the initial treatment with heparin at a therapeutic dose increased the survival probability of moderately ill patients in medical wards (Lawler et al., 2021; Spyropoulos et al., 2021), but not those in ICU because this therapeutic approach was potentially harmful compared to the standard prophylaxis (Goligher et al., 2021). However, in a recent metaanalysis based on both randomized-controlled trials (RCTs) and cohort studies, no differences regarding the all-cause mortality of the high/therapeutic dose compared to the low/prophylactic dose of anticoagulants in all hospitalized COVID-19 patients were observed at the cost of an increased bleeding risk when the higher doses were administered (Flumignan et al., 2022). As evidenced, the optimal prophylactic treatment in COVID-19 patients remains currently unknown. D-dimer levels have been suggested for categorized patients who may benefit from higher doses of anticoagulants, sometimes achieving conflicting results (Kow et al., 2022; Ortega-Paz et al., 2021; Sholzberg et al., 2021). In general, COVID-19 subjects should be monitored for thromboembolic events. Patients who aren't critically ill should receive heparin at a prophylactic dose during the acute phase of the infection or until discharge (Ceccato et al., 2022). Moreover, as evidenced by the Medically Ill hospitalized Patients for Covid - THrombosis Extended ProphyLaxis with rivaroxaban ThErapy (MICHELLE) trial, rivaroxaban 10 mg/daily could be administered after discharge in patients with an increased risk of venous thromboembolic events within 35 days after discharge (Ramacciotti et al., 2022). Conversely, severe COVID-19 patients should be treated with a prophylactic dose of heparin until discharge or during the acute phase according to the local standard of care (Ceccato et al., 2022). Intriguingly, between 20% and 40% of COVID-19 patients developed thromboembolic events, despite therapeutic or prophylactic anticoagulation (de la Morena-Barrio et al., 2021). This phenomenon has raised the hypothesis of heparin resistance. The exact underlying pathophysiological mechanisms remain unknown. This condition of resistance may be diagnosed due to a so-called pseudoresistance, which is characterized by therapeutic values of anti-Xa activity and a concomitant decreased aPTT due to low heparin concentrations or to antithrombin deficiency (de la Morena-Barrio et al., 2021). If the patient reported a previous history of heparin-induced thrombocytopaenia (HIT), prophylaxis with fondaparinux is recommended. Moreover, in the case of new HIT diagnosis, the thrombocytopenia, timing of platelet count fall, thrombosis or other sequelae, and other causes for thrombocytopenia (4Ts score) must be used to confirm the diagnosis and then the use of novel oral anticoagulants (NOACs), argatroban, bivalirudin, danaparoid, or fondaparinux should be considered (Streng et al., 2020).

Systemic thrombolysis

Systemic thrombolysis in PE patients with concomitant SARS-CoV-2 infection has been mainly reported by isolated case reports (Alharthy et al., 2021; Salam et al., 2020; Philippe et al., 2021). Indeed, no guidelines or general recommendations exist for the management of hemodynamically unstable PE patients with COVID-19 infection. Moreover, the use of fibrinolytics in COVID-19 subjects may be problematic considering that a nonnegligible proportion of these patients have thrombocytopenia (Lippi et al., 2020) and the administration of systemic thrombolysis in these subjects could significantly deteriorate an already precarious hematological balance (Roncon et al., 2020). Therefore, in this scenario, systemic reperfusion must be tailored according to the severity of thrombocytopenia.

Percutaneous treatments

Some isolated case reports and studies have analyzed the potential benefit of percutaneous treatment in COVID-19 patients with acute PE (Pendower et al., 2020). In particular, the use of fixed-dose ultrasound-assisted catheter-directed thrombolysis (USAT) in COVID-19 patients with acute PE led to a rapid and clinically relevant reduction in pulmonary artery pressures as well as an improvement of right ventricular function and hemodynamic parameters. This interventional approach appeared to be particularly effective in those infected subjects having a centrally located acute PE associated with RVD (Carlsson et al., 2020; Galastri et al., 2020; Voci et al., 2022).

Mini-dictionary of terms

- **Computed tomography angiography**: A technique that combines a CT scan with an intravenous injection of a special dye to visualize blood vessels and tissues in a part of the body.
- **Neutrophil extracellular traps**: Complexes of chromosomal DNA, histones, and granule proteins that are released by neutrophils and ensnare extracellular microorganisms.
- **Novel oral anticoagulants**: Direct thrombin inhibitor or direct factor Xa inhibitors that have largely replaced warfarin.
- **Percutaneous treatments**: Invasive treatments able to resolve acute thrombotic obstructions and reestablish flow in the main pulmonary arteries.
- **Right ventricular dysfunction**: A sudden increase in right ventricle afterload, both through direct increases in peripheral vascular resistance due to the clot burden as well as through neurohormonal and hypoxia-mediated feedback mechanisms.
- **Ultrasound-assisted thrombolysis**: A pharmacomechanical thrombolysis that consists of a combination of catheter-direct treatment with a catheter system that employs ultrasound.

Summary points

- Pulmonary embolism is a frequent and relevant complication of SARS-CoV-2 infection.
- Pulmonary embolism influences the clinical course and outcomes of COVID-19 patients.
- PE is frequently detected by maintaining a high suspicion of the disease.
- Prompt diagnosis and risk stratification of COVID-19 patients with acute PE are fundamental.
- Appropriate thromboprophylaxis regimens must be administered early in these patients.

References

Alharthy, A., Faqihi, F., Papanikolaou, J., Balhamar, A., Blaivas, M., Memish, Z. A., & Karakitsos, D. (2021). Thrombolysis in severe COVID-19 pneumonia with massive pulmonary embolism. *The American Journal of Emergency Medicine*, *41*, 261.e1–261.e3. https://doi.org/10.1016/j.ajem.2020.07.068. PMID: 32763101. PMCID: PMC7392155.

Alimohamadi, Y., Sepandi, M., Taghdir, M., & Hosamirudsari, H. (2020). Determine the most common clinical symptoms in COVID-19 patients: A systematic review and meta-analysis. *Journal of Preventive Medicine and Hygiene*, *61*, E304–E312.

Bilaloglu, S., Aphinyanaphongs, Y., Jones, S., et al. (2020). Thrombosis in hospitalized patients with COVID-19 in a New York City health system. *JAMA*, *324*, 799–801.

Bledsoe, J. R., Knox, D., Peltan, I. D., Woller, S. C., Lloyd, J. F., Snow, G. L., Horne, B. D., Connors, J. M., & Kline, J. A. (2022). D-dimer thresholds to exclude pulmonary embolism among COVID-19 patients in the emergency department: Derivation with independent validation. *Clinical and Applied Thrombosis/Hemostasis*, *28*. 10760296221117997.

Bompard, F., Monnier, H., Saab, I., et al. (2020). Pulmonary embolism in patients with COVID-19 pneumonia. *The European Respiratory Journal*, *56*, 2001365.

Carlsson, T. L., Walton, B., & Collin, G. (2020). Pulmonary artery thrombectomy—A life-saving treatment in a patient with presumed COVID-19 complicated by a massive pulmonary embolus. *British Journal of Haematology*, *190*, e143–e146.

Ceccato, A., Camprubí-Rimblas, M., Campaña-Duel, E., Areny-Balagueró, A., Morales-Quinteros, L., & Artigas, A. (2022). Anticoagulant treatment in severe ARDS COVID-19 patients. *Journal of Clinical Medicine*, *11*, 2695.

Ceriani, E., Combescure, C., Le Gal, G., et al. (2010). Clinical prediction rules for pulmonary embolism: A systematic review and meta-analysis. *Journal of Thrombosis and Haemostasis*, *8*, 957–970.

Choi, J. J., Wehmeyer, G. T., Li, H. A., et al. (2020). D-dimer cut-off points and risk of venous thromboembolism in adult hospitalized patients with COVID-19. *Thrombosis Research*, *196*, 318–321.

de la Morena-Barrio, M. E., Bravo-Pérez, C., de la Morena-Barrio, B., Orlando, C., Cifuentes, R., Padilla, J., Miñano, A., Herrero, S., Marcellini, S., Revilla, N., Bernal, E., Gómez-Verdú, J. M., Jochmans, K., Herranz, M. T., Vicente, V., Corral, J., & Lozano, M. L. (2021). A pilot study on the impact of congenital thrombophilia in COVID-19. *European Journal of Clinical Investigation*, *51*, e13546.

Flumignan, R. L., Civile, V. T., de Sa Tinôco, J. D., Pascoal, P. I., Areias, L. L., Matar, C. F., et al. (2022). Anticoagulants for people hospitalised with COVID-19. *Cochrane Database of Systematic Reviews*, *3*. CD013739.

Galastri, F. L., Valle, L. G. M., Affonso, B. B., Silva, M. J., Garcia, R. G., Junior, M. R., Ferraz, L. J. R., de Matos, G. F. J., de la Cruz Scarin, F. C., & Nasser, F. (2020). COVID-19 complicated by pulmonary embolism treated with catheter directed thrombectomy. *VASA*, *49*, 333–337.

Goligher, E. C., Bradbury, C. A., McVerry, B. J., REMAP-CAP Investigators, ACTIV-4a Investigators, ATTACC Investigators, et al. (2021). Therapeutic anticoagulation with heparin in critically Ill patients with Covid-19. *The New England Journal of Medicine*, *385*, 777–789.

Henkens, M. T. H. M., Raafs, A. G., Verdonschot, J. A. J., Linschoten, M., van Smeden, M., Wang, P., van der Hooft, B. H. M., Tieleman, R., Janssen, M. L. F., Ter Bekke, R. M. A., Hazebroek, M. R., van der Horst, I. C. C., Asselbergs, F. W., Magdelijns, F. J. H., Heymans, S. R. B., & CAPACITY-COVID collaborative consortium. (2022). Age is the main determinant of COVID-19 related in-hospital mortality with minimal impact of pre-existing comorbidities, a retrospective cohort study. *BMC Geriatrics*, *22*, 184.

Konstantinides, S. V., Meyer, G., Becattini, C., Bueno, H., Geersing, G. J., Harjola, V. P., et al. (2020). 2019 ESC guidelines for the diagnosis and management of acute pulmonary embolism developed in collaboration with the European Respiratory Society (ERS). *European Heart Journal*, *41*, 543–603.

Kow, C. S., Ramachandram, D. S., & Hasan, S. S. (2022). The effect of higher-intensity dosing of anticoagulation on the clinical outcomes in hospitalized patients with COVID-19: A meta-analysis of randomized controlled trials. *Journal of Infection and Chemotherapy*, *28*, 257–265.

Kraaijenhagen, R. A., Wallis, J., Koopman, M. M., de Groot, M. R., Piovella, F., Prandoni, P., et al. (2003). Can causes of false-normal D-dimer test [SimpliRED] results be identified? *Thrombosis Research*, *111*, 155–158.

Lawler, P. R., Goligher, E. C., Berger, J. S., ATTACC Investigators, ACTIV-4a Investigators, REMAP-CAP Investigators, et al. (2021). Therapeutic anticoagulation with heparin in noncritically Ill patients with COVID-19. *The New England Journal of Medicine*, *385*, 790–802.

Libby, P., & Lüscher, T. (2020). COVID-19 is, in the end, an endothelial disease. *European Heart Journal*, *41*, 3038–3044.

Lippi, G., Plebani, M., & Henry, B. M. (2020). Thrombocytopenia is associated with severe coronavirus disease 2019 (COVID-19) infections: A meta-analysis. *Clinica Chimica Acta*, *506*, 145–148.

Lowenstein, C. J., & Solomon, S. D. (2020). Severe COVID-19 Is a microvascular disease. *Circulation*, *142*, 1609–1611.

Machowski, M., Polańska, A., Gałecka-Nowak, M., Mamzer, A., Skowrońska, M., Perzanowska-Brzeszkiewicz, K., Zając, B., Ou-Pokrzewińska, A., Pruszczyk, P., & Kasprzak, J. D. (2022). Age-adjusted D-dimer levels may improve diagnostic assessment for pulmonary embolism in COVID-19 patients. *Journal of Clinical Medicine*, *11*, 3298.

Moore, A. J. E., Wachsmann, J., Chamarthy, M. R., Panjikaran, L., Tanabe, Y., & Rajiah, P. (2018). Imaging of acute pulmonary embolism: An update. *Cardiovascular Diagnosis and Therapy*, *8*, 225–243.

Moores, L. K., Tritschler, T., Brosnahan, S., Carrier, M., Collen, J. F., Doerschug, K., et al. (2022). Thromboprophylaxis in patients with COVID-19: A brief update to the CHEST guideline and expert panel report. *Chest*. S0012–3692: 00250–251.

Morrone, D., & Morrone, V. (2018). Acute pulmonary embolism: Focus on the clinical picture. *Korean Circulation Journal*, *48*, 365–381.

Onder, G., Rezza, G., & Brusaferro, S. (2020). Case-fatality rate and characteristics of patients dying in relation to COVID-19 in Italy. *JAMA*, *323*, 1775–1776.

Ortega-Paz, L., Galli, M., Capodanno, D., Franchi, F., Rollini, F., Bikdeli, B., Mehran, R., Montalescot, G., Gibson, C. M., Lopes, R. D., et al. (2021). Safety and efficacy of different prophylactic anticoagulation dosing regimens in critically and non-critically ill patients with COVID-19: A systematic review and meta-analysis of randomized controlled trials. *European Heart Journal - Cardiovascular Pharmacotherapy*. pvab070.

Panigada, M., Bottino, N., Tagliabue, P., Grasselli, G., Novembrino, C., Chantarangkul, V., Pesenti, A., Peyvandi, F., & Tripodi, A. (2020). Hypercoagulability of COVID-19 patients in intensive care unit: A report of thromboelastography findings and other parameters of hemostasis. *Journal of Thrombosis and Haemostasis*, *18*, 1738–1742.

Pendower, L., Benedetti, G., Breen, K., & Karunanithy, N. (2020). Catheter-directed thrombolysis to treat acute pulmonary thrombosis in a patient with COVID-19 pneumonia. *BMJ Case Reports*, *13*, e237046.

Philippe, J., Cordeanu, E. M., Leimbach, M. B., Greciano, S., & Younes, W. (2021). Acute pulmonary embolism and systemic thrombolysis in the era of COVID-19 global pandemic 2020: A case series of seven patients admitted to a regional hospital in the French epidemic cluster. *European Heart Journal - Case Reports*, *5*. ytaa522.

Porfidia, A., Valeriani, E., Pola, R., et al. (2020). Venous thromboembolism in patients with COVID-19: Systematic review and meta-analysis. *Thrombosis Research*, *196*, 67–74.

Pugh, C. W., & Ratcliffe, P. J. (2017). New horizons in hypoxia signaling pathways. *Experimental Cell Research*, *356*, 116–121.

Ramacciotti, E., Barile, A. L., Calderaro, D., Aguiar, V. C. R., Spyropoulos, A. C., de Oliveira, C. C. C., dos Santos PharmD, J. L., Volpiani, G. G., Sobreira, M. L., Joviliano, E. E., et al. (2022). Rivaroxaban versus no anticoagulation for post-discharge thromboprophylaxis after hospitalisation for COVID-19 (MICHELLE): An open-label, multicentre, randomised, controlled trial. *Lancet*, *399*, 50–59.

Ranucci, M., Ballotta, A., Di Dedda, U., Bayshnikova, E., Dei Poli, M., Resta, M., Falco, M., Albano, G., & Menicanti, L. (2020). The procoagulant pattern of patients with COVID-19 acute respiratory distress syndrome. *Journal of Thrombosis and Haemostasis*, *18*, 1747–1751.

Righini, M., Perrier, A., De Moerloose, P., & Bounameaux, H. (2008). D-Dimer for venous thromboembolism diagnosis: 20 years later. *Journal of Thrombosis and Haemostasis*, *6*, 1059–1071.

Righini, M., & Robert-Ebadi, H. (2018). Diagnosis of acute pulmonary embolism. *Hämostaseologie*, *38*, 11–21.

Righini, M., Van Es, J., Den Exter, P. L., Roy, P. M., Verschuren, F., Ghuysen, A., et al. (2014). Age-adjusted D-dimer cutoff levels to rule out pulmonary embolism: The ADJUST-PE study. *JAMA*, *311*, 1117–1124.

Roncon, L., Zuin, M., & Zonzin, P. (2020). Fibrinolysis in COVID-19 patients with hemodynamic unstable acute pulmonary embolism: Yes or no? *Journal of Thrombosis and Thrombolysis*, *50*, 221–222.

Salam, S., Mallat, J., & Elkambergy, H. (2020). Acute high-risk pulmonary embolism requiring thrombolytic therapy in a COVID-19 pneumonia patient despite intermediate dosing deep vein thromboprophylaxis. *Respiratory Medicine Case Reports*, *31*, 101263.

Sholzberg, M., da Costa, B. R., Tang, G. H., Rahhal, H., AlHamzah, M., Baumann, K. L., Ní, Á. F., Almarshoodi, M. O., James, P. D., Lillicrap, D., et al. (2021). Randomized trials of therapeutic heparin for COVID-19: A meta-analysis. *Research and Practice in Thrombosis and Haemostasis*, *5*, e12638.

Spyropoulos, A. C., Goldin, M., Giannis, D., Diab, W., Wang, J., Khanijo, S., et al. (2021). Efficacy and safety of therapeutic-dose heparin vs standard prophylactic or intermediate-dose heparins for thromboprophylaxis in high-risk hospitalized patients with COVID-19: The HEP-COVID randomized clinical trial. *JAMA Internal Medicine*, *181*, 1612–1620.

Spyropoulos, A. C., Levy, J. H., Ageno, W., Connors, J. M., Hunt, B. J., Iba, T., et al. (2020). Scientific and standardization committee communication: Clinical guidance on the diagnosis, prevention, and treatment of venous thromboembolism in hospitalized patients with COVID-19. *Journal of Thrombosis and Haemostasis*, *18*, 1859–1865.

Streng, A. S., Delnoij, T. S. R., Mulder, M. M. G., Sels, J. W. E. M., Wetzels, R. J. H., Verhezen, P. W. M., Olie, R. H., Kooman, J. P., van Kuijk, S. M. J., Brandts, L., Ten Cate, H., Lorusso, R., van der Horst, I. C. C., van Bussel, B. C. T., & Henskens, Y. M. C. (2020). Monitoring of unfractionated heparin in severe COVID-19: An observational study of patients on CRRT and ECMO. *TH Open*, *4*, e365–e375.

Suh, Y. J., Hong, H., Ohana, M., et al. (2021). Pulmonary embolism and deep vein thrombosis in COVID-19: A systematic review and meta-analysis. *Radiology*, *298*, E70–E80.

Ten Cate-Hoek, A. J., & Prins, M. H. (2005). Management studies using a combination of D-dimer test result and clinical probability to rule out venous thromboembolism: A systematic review. *Journal of Thrombosis and Haemostasis*, *3*, 2465–2470.

Thachil, J., Longstaff, C., Favaloro, E. J., et al. (2020). The need for accurate D-dimer reporting in COVID-19: Communication from the ISTH SSC on fibrinolysis. *Journal of Thrombosis and Haemostasis*, *18*, 2408–2411.

Varga, Z., Flammer, A. J., & Steiger, P. (2020). Endothelial cell infection and endotheliitis in COVID-19. *Lancet*, *395*, 1417–1418.

Voci, D., Zbinden, S., Micieli, E., Kucher, N., & Barco, S. (2022). Fixed-dose ultrasound-assisted catheter-directed thrombolysis for acute pulmonary embolism associated with COVID-19. *Viruses*, *14*, 1606.

Woller, S. C., De Wit, K., Hansen, J. B., et al. (2022). Predictive and diagnostic variables scientific standardization committee podium presentation: Biomarkers predictive for VTE in hospitalized COVID-19 patients: A systematic review paper presented at: International Society of Thrombosis and Haemostasis (ISTH2021). *Research and Practice in Thrombosis and Haemostasis*, *6*, e12786.

Yu, B., Li, X., Chen, J., et al. (2020). Evaluation of variation in D-dimer levels among COVID-19 and bacterial pneumonia: A retrospective analysis. *Journal of Thrombosis and Thrombolysis*, *50*, 548–557.

Zhang, S., Zhang, J., & Wang, C. (2021). COVID-19 and ischemic stroke: Mechanisms of hypercoagulability (review). *International Journal of Molecular Medicine*, *47*, 21.

Zuin, M., Engelen, M. M., Bilato, C., Vanassche, T., Rigatelli, G., Verhamme, P., Vandenbriele, C., Zuliani, G., & Roncon, L. (2022). Prevalence of acute pulmonary embolism at autopsy in patients with COVID-19. *The American Journal of Cardiology*, *171*, 159–164.

Zuin, M., Rigatelli, G., Bilato, C., Quadretti, L., Roncon, L., & Zuliani, G. (2022). COVID-19 patients with acute pulmonary embolism have a higher mortality risk: Systematic review and meta-analysis based on Italian cohorts. *Journal of Cardiovascular Medicine (Hagerstown, Md.)*. https://doi.org/10.2459/JCM.0000000000001354. Epub ahead of print.

Zuin, M., Rigatelli, G., & Roncon, L. (2021). Age-adjusted d-dimer cut-off levels to exclude venous thromboembolism in COVID-19 patients. *Anaesthesia, Critical Care and Pain Medicine*, *40*, 100941.

Zuin, M., Rigatelli, G., Zuliani, G., & Roncon, L. (2021). Age-adjusted D-dimer cutoffs to guide anticoagulation in COVID-19. *Lancet*, *398*, 1303–1304.

Zuo, Y., Yalavarthi, S., & Shi, H. (2020). Neutrophil extracellular traps in COVID-19. *JCI Insight*, *5*, 138999.

Chapter 27

The usefulness of the alveolar-arteriolar gradient during the COVID-19 pandemic

Giuseppe Pipitone[a], Miriam De Michele[b], Massimo Sartelli[c], Francesco Onorato[a], Claudia Imburgia[a], Antonio Cascio[d], and Chiara Iaria[a]

[a]*Infectious Disease Unit, ARNAS Civico-Di Cristina Hospital, Palermo, Italy,* [b]*Provincial Health Authority of Palermo, Palermo, Italy,* [c]*Department of Surgery, Macerata Hospital, Macerata, Italy,* [d]*Infectious Disease Unit, P. Giaccone University Hospital, Palermo, Italy*

Abbreviations

ARDS	acute respiratory distress syndrome
AUC	area under the curve
BGA	blood gas analysis
CAP	community acquired pneumonia
COPD	chronic obstructive pulmonary disease
COVID-19	coronavirus disease2019
CPAP	continuous positive airway pressure
$D(A\text{-}a)O_2$	alveolar-arterial gradient
ED	emergency department
FiO_2	fractional inspired oxygen
HR	hazard ratio
ICU	intensive care unit
IDSA	Infectious Disease Society of America
NIMV	noninvasive mechanical ventilation
NIV	noninvasive ventilation
$PACO_2$	alveolar carbon dioxide partial pressure
Palv	alveolar pressure
PAO_2	alveolar oxygen partial pressure
PaO_2	arterial oxygen partial pressure
Patm	barometric pressure or atmospheric pressure
PE	pulmonary embolism
PH_2O	airway water vapor pressure
R	respiratory quotient
ROC	receiver operating characteristics
SARS-CoV-2	severe acute respiratory syndrome coronavirus 2
V/Q	ventilation/perfusion ratio

Introduction

Coronavirus disease 2019 (COVID-19), caused by the severe acute respiratory syndrome coronavirus 2 (SARS-CoV-2), spread rapidly worldwide. COVID-19 had high mortality, mainly in the first pandemic phase. Approximately 14% of infected patients had a severe disease requiring hospitalization and oxygen therapy. Moreover, 5% required admission to an intensive care unit (ICU) (Yang et al., 2020). The severity of illness is often due to impaired respiratory function. Clinicians had to deal with a new pathology with little in the way of pathophysiological and therapeutic efficacy data. During the first months of the pandemic, COVID-19 represented an unknown factor: apparently healthy patients could suddenly deteriorate and be moved to the ICU a few hours later. At hospital admission, a patient's proper stratification, according to their severity, is crucial for providing early treatment and improving clinical management (Prediletto et al., 2021). Patients with severe COVID-19 often have a type I respiratory failure (hypoxemia and hypocapnia).

Management, Body Systems, and Case Studies in COVID-19. https://doi.org/10.1016/B978-0-443-18703-2.00009-8
Copyright © 2024 Elsevier Inc. All rights are reserved, including those for text and data mining, AI training, and similar technologies.

The pathophysiological basis could be an intrapulmonary shunt, due to alveolar occlusion or atelectasis, and alteration of the ventilation/perfusion ratio (V/Q), due to severe interstitial pneumonia and sometimes pulmonary embolism (PE). Hypoxia and hypocapnia during PE could be due to multiple mechanisms. Occlusion of one or more pulmonary vessels causes a V/Q mismatch (some alveolus could be ventilated but not perfused). There will be a blood hyperflow to the remaining parenchyma, creating a *de facto* "shunt." This could justify the low responsivity to oxygen-therapy during a massive PE. The result will be hypoxia. The compensative hyperventilation of the remaining lung will determine hypocapnia.

PaO_2/FiO_2

The arterial oxygen partial pressure (PaO_2) to fractional inspired oxygen (FiO_2) ratio (PaO_2/FiO_2) is used to evaluate the severity of hypoxia in patients requiring oxygen supplementation. A $PaO_2/FiO_2 > 300$ mmHg identifies a normal lung function (ARDS Definition Task Force et al., 2012). The PaO_2/FiO_2 ratio was included in the definition of acute respiratory distress syndrome (ARDS) in COVID-19 according to the Berlin criteria and was globally used to stratify the severity of respiratory insufficiency during the pandemic (Zhou et al., 2020).

Worldwide, when a COVID-19 patient is admitted to the emergency department (ED), a blood gas analysis (BGA) is performed and the PaO_2/FiO_2 is calculated. The patient can be stratified basing on Berlin criteria for ARDS, but it unknown how patients will clinically evolve. PaO_2/FiO_2 is easy to calculate and it relates to mortality, so it was used since early 2020 to identify patients with the worst outcomes. So, when patients arrived at the hospital, clinicians calculated the PaO_2/FiO_2; the lower the PaO_2/FiO_2, the worse the prognosis.

However, the PaO_2/FiO_2 ratio has some limitations. It depends on the FiO_2 percentage, so when oxygen flow is increased, the PaO_2/FiO_2 ratio can dramatically decrease, even if the pulmonary ventilation is little impaired. PaO_2/FiO_2 doesn't reflect respiratory insufficiency severity at all because it does not include alveolar $PACO_2$, an indirect measure of the patient's respiratory effort (e.g., during tachypnea). Finally, PaO_2/FiO_2 cannot provide information on pulmonary V/Q.

$D(A\text{-}a)O_2$ or alveolar-arterial gradient

Another index of pulmonary function is the alveolar-arteriolar oxygen gradient ($D(A\text{-}a)O_2$).

$D(A\text{-}a)O_2$ is the difference between the alveolar and arteriolar concentrations of oxygen. Before the pandemic, $D(A\text{-}a)O_2$ was frequently used to discriminate between types of respiratory insufficiency.

$D(A\text{-}a)O_2$ remains low in normal condition or in case of "pump failure" and it increases in case of "lung failure" such as alveolar-capillary membrane alteration (e.g., interstitial pneumonia) or V/Q impairment (e.g., PE, ARDS, severe pneumonia). As noted, all the above could be aspects of severe COVID-19 pneumonia (Chu et al., 2018). So, $D(A\text{-}a)O_2$ has been proposed as an early marker of respiratory insufficiency in COVID-19. $D(A\text{-}a)O_2$ is automatically calculated by a blood gas analyzer, and normally its value is between 5 and 10 mmHg. The equation is: $D(A\text{-}a)O_2 = PAO_2 - PaO_2$.

It is a simple difference between PAO_2 (the alveolar oxygen pressure) and PaO_2 (the arterial oxygen pressure, calculated in the arterial blood gas analysis).

The PAO_2 value should be similar to PaO_2 in a standard situation: the more oxygen supplied to the alveoli (PAO_2), the more oxygen that reaches the artery (PaO_2), and the alveolar-arteriolar gradient remains low (normally 5–10). This is what happens in a normal condition: the lower PAO_2 is, the lower $D(A\text{-}a)O_2$ is; the higher PaO_2 is, the lower $D(A\text{-}a)O_2$ is.

PAO_2 is calculated as: $[(FiO_2 \times (Patm - PH_2O)) - (PACO_2/R)]$.

The formula appears a bit difficult to learn, but it isn't.

FiO_2 is the fractional inspired oxygen, a percentage of inspired oxygen. It is 0.21 (or 21%) in room air. This value is set by the clinician when using systems capable of supplying oxygen (venturi masks, ventilators etc.) and can be increased up to 100%.

FiO_2 can be thought of as the percentage of oxygen reaching the airways, depending on the alveolar pressure (Palv).

"$Patm - PH_2O$," also called Palv, is the difference between Patm (atmospheric pressure, 760 mmHg at sea level) and PH_2O (airways water vapor pressure, 37 mmHg at 37°C). It could be considered as the ambient pressure, 723 mmHg in standard condition.

FiO_2 could be imagined as strictly related to Palv because the oxygen reaching the airway (FiO_2) with room air or Venturi mask/CPAP/NIMV exerts a pressure into the alveolus proportional to Palv: the higher the atmospheric pressure (Patm), the higher the pressure of O_2 into the alveolus; the higher the airway's water vapor pressure (PH_2O), the lower the pressure of O_2. For informative purposes only, one can imagine gaseous water competing with O_2 along the airways. My chemist friends will forgive me for this inaccuracy.

$PACO_2$ is the alveolar CO_2 pressure. It is approximated to the arterial pressure of CO_2 ($PaCO_2$) because CO_2 crosses the alveolar-arteriolar membrane in almost every situation.

R is respiratory quotient, a ratio between the CO_2 produced by the organism and the O_2 consumed during redox processes. The median value is 0.8 (to note that a human produces less CO_2 than O_2 consumed).

Once in the alveolus, oxygen with its pressure ($FiO_2 \times (Patm - PH_2O)$) could be imagined as competing with $PACO_2$ into the alveolus to cross the alveolar-capillary membrane.

The complete formula for the alveolar-arterial gradient is $D(A\text{-}a)O_2 = [(FiO_2 \times (Patm - PH_2O)) - (PACO_2/R)] - PaO_2$.

$D(A\text{-}a)O_2$ describes the amount of oxygen administered to the patient (FiO_2), the atmospheric pressure (Patm), the partial oxygen pressure (PaO_2) in arterial blood, the airway's pressure of gaseous water (PH_2O), and the alveolar pressure of CO_2 ($PACO_2$).

Also in COVID-19, a high $D(A\text{-}a)O_2$ could identify patients who need mechanical ventilation such as patients with PE, ARDS, or chronic obstructive pulmonary disease (COPD) instead of a gradual incrementation of oxygen flow (Alhazzani et al., 2021).

Some studies have evaluated $D(A\text{-}a)O_2$ on identifying severe COVID-19 pneumonia (Carlino et al., 2020; de Roos et al., 2021; Kamran et al., 2020; Secco et al., 2021), and only one study compared the performance of $D(A\text{-}a)O_2$ and PaO_2/FiO_2 in this context (Pipitone et al., 2022).

PaO_2/FiO_2 was early used to identify patients with severe COVID-19

Multiple studies show an association between low PaO_2/FiO_2 and severe COVID-19.

One of the first studies published (Guan et al., 2020) showed a similar P/F value between severe and nonsevere COVID-19 patients. The study evaluated 1099 COVID-19 patients admitted to more than 500 hospitals in different Chinese provinces. Patients were divided into two groups, severe and nonsevere, following the 2019 Infectious Disease Society of America (IDSA) criteria for community-acquired pneumonia (CAP). The population was well distributed among the two groups. Demographic, clinical, and laboratory data were obtained. Sadly, data on PaO_2/FiO_2 at arrival in ED were missed for 81% of patients (maybe due to the emergency). No difference among the two groups was found.

The study from Yang et al. (2020) was conducted among critical ill patients. The study enrolled 52 such patients out of 710 with SARS-CoV-2 pneumonia. Data on PaO_2/FiO_2 were collected at admission to the ICU, and patients were divided into survivors and nonsurvivors. This study showed a difference in PaO_2/FiO_2 ratio among the two groups: survivors had a higher PaO_2/FiO_2 than nonsurvivors (100 vs 62.5 mmHg). No statistical analysis was performed, but the difference could be considered significant. However, this study has some limitations, for example, the inclusion and exclusion criteria are lacking and the baseline population characteristics weren't statistically analyzed, meaning there was no information available on population distribution among the two groups. Patients were enrolled if they were admitted to the ICU or if they had an $FiO_2 \geq 60\%$. As can be noted, ICU admission depends on a patient's clinical condition, and the fraction of inspired oxygen depends on a medical decision, so this could be a bias. However, this study could confirm that the worse the PaO_2/FiO_2, the worse the condition. Those data are confirmed by another study among COVID-19 patients admitted to the ICU by Grasselli et al. (2020), which showed that a 100-point increase in PaO_2/FiO_2 ratio decreased by 44% the hazard for mortality.

PaO_2/FiO_2 has been hypothesized to be useful in urinary tract infection (UTI) patients requiring invasive mechanical ventilation (IMV). Hueda-Zavaleta et al. (2022) analyzed the data from 200 patients admitted to the ICU and requiring IMV. The PaO_2/FiO_2 value 24 h after IMV and its variation after IMV were higher among survivors (260 vs 190, $P = .001$ and 152 vs 106, $P = 0.001$, respectively). The area under the curve (AUC) receiver operating characteristic (ROC) of PaO_2/FiO_2 24 h after IMV was higher than PaO_2/FiO_2 24 h, but no statistically significant differences were observed between both curves ($P = .054$)

The PaO_2/FiO_2 ratio is simple to use and gives immediate information about respiratory failure, but this information is incomplete because it doesn't reflect hyperventilation, the respiratory muscle effort, and the V/Q mismatch. This information should be given by $PaCO_2$. Mays (1973) suggested that an estimation of V/Q mismatch may be optimized by "standardizing" PaO_2 for $PaCO_2$ by using the formula ${}_{ST}PaO_2 = PaO_2 + 1.66^{*}PaCO_2 - 66.4$. in their study of 349 COVID-19 patients admitted to respiratory unit, Prediletto et al. demonstrated that ${}_{ST}PaO_2$ better relates than PaO_2 to predict mortality, AUC ROC 0.710 versus 0.688, $P = 0.012$. This study opens the door to the usefulness of $D(A\text{-}a)O_2$ in COVID-19.

$D(A\text{-}a)O_2$ and its use among COVID-19 patients

The alveolar-arterial gradient may give some additional information about the V/Q mismatch and respiratory efforts. Since 2020, some authors focused on its use in COVID-19. Kamran et al. (2020) analyzed data on 252 COVID-19 patients. They observed that $D(A\text{-}a)O_2$ has a hazard ratio (HR) of 2.14 (95% CI: 1.04–4.39, $P = .038$) in predicting disease progression and death. Furthermore, the alveolar-arterial gradient, oxygen saturation, and respiratory rate were considered statistically significant high-risk predictors of death or disease progression in the formed scoring model.

Pipitone et al. (2022) showed the usefulness of $D(A\text{-}a)O_2$ in identifying patients with severe pneumonia. This multicentric study was conducted among patients admitted to the ED for COVID-19. Patients were divided into two groups, severe and nonsevere, based on IDSA CAP criteria. Data were collected at admission and showed that the alveolar-arterial gradient is more appropriate than PaO_2/FiO_2 to identify patients at risk of developing severe pneumonia. $D(A\text{-}a)O_2$, compared to PaO_2/FiO_2, had a higher sensitivity (77.8% vs 66.7%), positive predictive value (75% vs 71.4%), and negative predictive value (94% vs 91%), but similar specificity (94.4% vs 95.5%). A $D(A\text{-}a)O_2$ cut-off >60 mmHg at hospital admission has an AUC-ROC of 0.877 (95% CI: 0.675-1) in identifying patients at risk for developing severe pneumonia.

Two studies analyzed the usefulness of the alveolar-arterial gradient in association with radiological examination. The study by de Roos et al. (2021) conducted among 72 COVID-19 patients showed that the alveolar-arterial gradient should be a rapid and accurate tool to diagnose COVID-19 pneumonia and to select patients in need of hospitalization.

The study by Secco et al. (2021) showed that the lung ultrasound and alveolar-arterial gradient are useful tools in predicting the need for oxygen support and survival in patients with a normal PaO_2/FiO_2 ratio.

Also, among patients admitted to the ICU, the alveolar-arterial gradient is considered an accurate marker of ICU admission. The study conducted by Carlino et al. (2020) demonstrated that $D(A\text{-}a)O_2$ could be useful in identifying patients in need of ICU admission.

Another study by Singh et al. (2022) stated that the alveolar-arterial gradient at ICU admission in COVID-19 was a poor predictor of survival. The study was conducted among 300 patients admitted to the ICU, and baseline characteristics were collected. $D(A\text{-}a)O_2$ was statistically different among survivors and nonsurvivors ($P = 0.024$), with an AUC ROC of 0.602 and a cut-off of 0.28. This is in line with the study from Gupta et al. (2021) that showed the usefulness of $D(A\text{-}a)O_2$ in predicting mortality among patients in need of NIMV.

Conclusion

The epidemiology of COVID-19 is still changing, so the early identification of patients that will develop severe pneumonia is mandatory. PaO_2/FiO_2 is still used, and it will be used worldwide because it gives rapid information on hypoxia and its fast calculation. However, PaO_2/FiO_2 isn't so precise because of it doesn't give information on the V/Q mismatch and respiratory muscle efforts.

The alveolar-arterial gradient may be a more accurate and earlier marker to identify patients at risk, given the importance of $PaCO_2$ in its formula. So, $D(A\text{-}a)O_2$ should be an appropriate and early marker of pneumonia severity in COVID-19 patients at admission.

Policies and procedures

In this chapter, we reviewed the usefulness of the alveolar-arterial gradient to identify patients at high risk, then compared the alveolar-arterial gradient to PaO_2/FiO_2 as a marker to identify patients that will develop severe pneumonia (sec. IDSA community acquired pneumonia severity criteria, Metlay et al., 2019) or patients at high risk of death.

The alveolar-arterial gradient identifies patients who will develop severe pneumonia better and earlier (Pipitone et al., 2022). The alveolar-arterial gradient should be used with radiological examination to identify patients in need of hospitalization (in association with a chest CT scan, de Roos et al., 2021) or in need of oxygen support also with a normal PaO_2/FiO_2 (in association with a lung ultrasound, Secco et al., 2021).

Our policy is to use the alveolar-arterial gradient in interstitial pneumonia as well as in normal PaO_2/FiO_2 to earlier identify patients at risk of developing severe disease and in need of oxygen support or ventilation. An alveolar-arterial gradient cut-off >60 mmHg at hospital admission has an AUC-ROC of 0.877 (95% CI: 0.675-1) to identify patients at risk for developing severe pneumonia.

$D(A\text{-}a)O_2$, compared to PaO_2/FiO_2, had a higher sensibility (77.8% vs 66.7%), positive predictive value (75% vs 71.4%), and negative predictive value (94% vs 91%), but similar specificity (94.4% vs 95.5%) to identify patients with severe pneumonia.

Applications to other areas

In this review, we highlighted the increased alveolar-arterial gradient in COVID-19 pneumonia. The alveolar-arterial gradient is an indirect measurement of respiratory muscle effort and alveolar-arterial membrane alteration. It is often used to differentiate hypoxemic patients with "pump failure" and "lung failure." During the pandemic, many authors highlighted the usefulness of the alveolar-arterial gradient in COVID-19, often in normal PaO_2/FiO_2 ratio. This is one of the most important values of the alveolar-arterial gradient. The alveolar-arterial gradient should be used also in new variants or new pandemics with new respiratory viruses. In case of any infectious pneumonia, especially if the pathophysiological aspect involves interstitial space, massive pneumonia, vascular damage, or pulmonary thromboembolism, the alveolar-arterial gradient should be useful in identifying patients at risk for developing severe pneumonia, in need of oxygen support or ventilation, or necessitating ICU admission.

Mini-dictionary of terms

- **ARDS**. Acute respiratory distress syndrome, defined with the Berlin criteria as PaO_2/FiO_2 <300 with a CPAP $\geq$5 cmH_2O. It's a hypoxemic respiratory insufficiency not responding to simple oxygen support due to pulmonary alveolar and arteriolar damage.
- **Berlin criteria**. A validated definition and stratification for ARDS.
- **D(A-a)O_2**. Alveolar-arterial gradient, an indirect measurement of ventilation/perfusion ratio, respiratory effort, and shunt. It helps in differentiating "lung failure" and "pump failure." Automatically calculated by the blood gas analyzer.
- **PaO_2/FiO_2**. Arterial partial pressure of oxygen to fraction of inspired oxygen, used to evaluate the hypoxemic state and to stratify patients according to the Berlin criteria.
- **Ventilation/perfusion ratio (V/Q)**. An indirect measurement of blood oxygen concentration reaching systemic concentration from pulmonary vessels. It depends on the efficacy of ventilated and perfused areas in the lung. An alteration in V/Q should be observed in case of ARDS, severe pneumonia, pulmonary edema, etc.

Summary points

- COVID-19 often leads to severe pneumonia with ARDS, EP, V/Q mismatch.
- PaO_2/FiO_2, used to stratify patients with ARDS, is used worldwide to identify COVID-19 hypoxemic patients.
- D(A-a)O_2, used to differentiate "lung failure" and "pump failure," showed a high correlation with a risk of developing severe pneumonia in COVID-19 patients.
- D(A-a)O_2 may be used with chest CT or lung ultrasound to identify patients in need of hospitalization for oxygen support.
- D(A-a)O_2 is better than PaO_2/FiO_2 at identifying patients with the worst outcomes.
- A D(A-a)O_2 >60 mmHg is the best cut-off to identify patients who will develop severe pneumonia, also in the case of normal or slightly low PaO_2/FiO_2.

Acknowledgment

Dedicated to little Vincenzo who came to overturn and fill our lives, and thanks to my wife Miriam who gave me the love of science.

References

Alhazzani, W., Evans, L., Alshamsi, F., Møller, M. H., Ostermann, M., Prescott, H. C., Arabi, Y. M., Loeb, M., Ng Gong, M., Fan, E., Oczkowski, S., Levy, M. M., Derde, L., Dzierba, A., Du, B., Machado, F., Wunsch, H., Crowther, M., Cecconi, M., … Rhodes, A. (2021). Surviving sepsis campaign guidelines on the management of adults with coronavirus disease 2019 (COVID-19) in the ICU: First update. *Critical Care Medicine*, *49*(3), e219–e234. https://doi.org/10.1097/CCM.0000000000004899.

ARDS Definition Task Force, Ranieri, V. M., Rubenfeld, G. D., Thompson, B. T., Ferguson, N. D., Caldwell, E., Fan, E., Camporota, L., & Slutsky, A. S. (2012). Acute respiratory distress syndrome: The Berlin definition. *JAMA*, *307*(23), 2526–2533. https://doi.org/10.1001/jama.2012.5669.

Carlino, M. V., Valenti, N., Cesaro, F., Costanzo, A., Cristiano, G., Guarino, M., & Sforza, A. (2020). Predictors of intensive care unit admission in patients with coronavirus disease 2019 (COVID-19). *Monaldi Archives for Chest Disease = Archivio Monaldi per le malattie del torace*, *90*(3). https://doi.org/10.4081/monaldi.2020.1410.

Chu, D. K., Kim, L. H., Young, P. J., Zamiri, N., Almenawer, S. A., Jaeschke, R., Szczeklik, W., Schünemann, H. J., Neary, J. D., & Alhazzani, W. (2018). Mortality and morbidity in acutely ill adults treated with liberal versus conservative oxygen therapy (IOTA): A systematic review and meta-analysis. *Lancet (London, England)*, *391*(10131), 1693–1705. https://doi.org/10.1016/S0140-6736(18)30479-3.

de Roos, M. P., Kilsdonk, I. D., Hekking, P. W., Peringa, J., Dijkstra, N. G., Kunst, P. W. A., Bresser, P., & Reesink, H. J. (2021). Chest computed tomography and alveolar-arterial oxygen gradient as rapid tools to diagnose and triage mildly symptomatic COVID-19 pneumonia patients. *ERJ Open Research*, *7*(1), 00737–02020. https://doi.org/10.1183/23120541.00737-2020.

Grasselli, G., Greco, M., Zanella, A., Albano, G., Antonelli, M., Bellani, G., Bonanomi, E., Cabrini, L., Carlesso, E., Castelli, G., Cattaneo, S., Cereda, D., Colombo, S., Coluccello, A., Crescini, G., Forastieri Molinari, A., Foti, G., Fumagalli, R., Iotti, G. A., ... COVID-19 Lombardy ICU Network. (2020). Risk factors associated with mortality among patients with COVID-19 in intensive care units in lombardy, Italy. *JAMA Internal Medicine*, *180*(10), 1345–1355. https://doi.org/10.1001/jamainternmed.2020.3539.

Guan, W. J., Ni, Z. Y., Hu, Y., Liang, W. H., Ou, C. Q., He, J. X., Liu, L., Shan, H., Lei, C. L., Hui, D. S. C., Du, B., Li, L. J., Zeng, G., Yuen, K. Y., Chen, R. C., Tang, C. L., Wang, T., Chen, P. Y., Xiang, J., ... China Medical Treatment Expert Group for Covid-19. (2020). Clinical characteristics of coronavirus disease 2019 in China. *The New England Journal of Medicine*, *382*(18), 1708–1720. https://doi.org/10.1056/NEJMoa2002032.

Gupta, B., Jain, G., Chandrakar, S., Gupta, N., & Agarwal, A. (2021). Arterial blood gas as a predictor of mortality in COVID pneumonia patients initiated on noninvasive mechanical ventilation: A retrospective analysis. *Indian Journal of Critical Care Medicine*, *25*(8), 866–871. https://doi.org/10.5005/jp-journals-10071-23917.

Hueda-Zavaleta, M., Copaja-Corzo, C., Miranda-Chávez, B., Flores-Palacios, R., Huanacuni-Ramos, J., Mendoza-Laredo, J., Minchón-Vizconde, D., Gómez de la Torre, J. C., & Benites-Zapata, V. A. (2022). Determination of PaO_2/FiO_2 after 24 h of invasive mechanical ventilation and $\Delta PaO_2/FiO_2$ at 24 h as predictors of survival in patients diagnosed with ARDS due to COVID-19. *PeerJ*, *10*, e14290. https://doi.org/10.7717/peerj.14290.

Kamran, S. M., Mirza, Z. E., Moeed, H. A., Naseem, A., Hussain, M., Fazal, I., Sr., Saeed, F., Alamgir, W., Saleem, S., & Riaz, S. (2020). CALL score and RAS score as predictive models for coronavirus disease 2019. *Cureus*, *12*(11), e11368. https://doi.org/10.7759/cureus.11368.

Mays, E. E. (1973). An arterial blood gas diagram for clinical use. *Chest*, *63*(5), 793–800.

Metlay, J. P., Waterer, G. W., Long, A. C., Anzueto, A., Brozek, J., Crothers, K., ... Whitney, C. G. (2019). Diagnosis and treatment of adults with community-acquired pneumonia. An official clinical practice guideline of the American Thoracic Society and Infectious Diseases Society of America. *American Journal of Respiratory and Critical Care Medicine*, *200*(7), e45–e67. https://doi.org/10.1164/rccm.201908-1581ST.

Pipitone, G., Camici, M., Granata, G., Sanfilippo, A., Di Lorenzo, F., Buscemi, C., Ficalora, A., Spicola, D., Imburgia, C., Alongi, I., Onorato, F., Sagnelli, C., & Iaria, C. (2022). Alveolar-arterial gradient is an early marker to predict severe pneumonia in COVID-19 patients. *Infectious Disease Reports*, *14* (3), 470–478. https://doi.org/10.3390/idr14030050.

Prediletto, I., D'Antoni, L., Carbonara, P., Daniele, F., Dongilli, R., Flore, R., Pacilli, A. M. G., Pisani, L., Tomsa, C., Vega, M. L., Ranieri, V. M., Nava, S., & Palange, P. (2021). Standardizing PaO_2 for $PaCO_2$ in P/F ratio predicts in-hospital mortality in acute respiratory failure due to Covid-19: A pilot prospective study. *European Journal of Internal Medicine*, *92*, 48–54. https://doi.org/10.1016/j.ejim.2021.06.002.

Secco, G., Salinaro, F., Bellazzi, C., La Salvia, M., Delorenzo, M., Zattera, C., Barcella, B., Resta, F., Vezzoni, G., Bonzano, M., Cappa, G., Bruno, R., Casagranda, I., & Perlini, S. (2021). Can alveolar-arterial difference and lung ultrasound help the clinical decision making in patients with COVID-19? *Diagnostics (Basel, Switzerland)*, *11*(5), 761. https://doi.org/10.3390/diagnostics11050761.

Singh, A., Soni, K. D., Singh, Y., Aggarwal, R., Venkateswaran, V., Ashar, M. S., & Trikha, A. (2022). Alveolar arterial gradient and respiratory index in predicting the outcome of COVID-19 patients; a retrospective cross-sectional study. *Archives of Academic Emergency Medicine*, *10*(1), e28. https://doi.org/10.22037/aaem.v10i1.1543.

Yang, X., Yu, Y., Xu, J., Shu, H., Xia, J., Liu, H., Wu, Y., Zhang, L., Yu, Z., Fang, M., Yu, T., Wang, Y., Pan, S., Zou, X., Yuan, S., & Shang, Y. (2020). Clinical course and outcomes of critically ill patients with SARS-CoV-2 pneumonia in Wuhan, China: A single-centered, retrospective, observational study. *The Lancet Respiratory Medicine*, *8*(5), 475–481. https://doi.org/10.1016/S2213-2600(20)30079-5.

Zhou, F., Yu, T., Du, R., Fan, G., Liu, Y., Liu, Z., Xiang, J., Wang, Y., Song, B., Gu, X., Guan, L., Wei, Y., Li, H., Wu, X., Xu, J., Tu, S., Zhang, Y., Chen, H., & Cao, B. (2020). Clinical course and risk factors for mortality of adult inpatients with COVID-19 in Wuhan, China: A retrospective cohort study. *Lancet (London, England)*, *395*(10229), 1054–1062. https://doi.org/10.1016/S0140-6736(20)30566-3.

Section E

Effects on cardiovascular and hematological systems

Chapter 28

Cardiac manifestations of COVID-19: An overview

Naveed Rahman[a], Mirza H. Ali[a], Aanchal Sawhney[b], Apurva Vyas[c], and Rahul Gupta[c]
[a]*Department of Internal Medicine, Lehigh Valley Health Network, Allentown, PA, United States,* [b]*Department of Internal Medicine, Crozer Chester Medical Center, Upland, PA, United States,* [c]*Lehigh Valley Heart Institute, Lehigh Valley Health Network, Allentown, PA, United States*

Abbreviations

ACE	angiotensin converting enzyme
ACE2R	angiotensin converting enzyme 2 receptor
ACS	acute coronary syndrome
Akt	protein serine-threonine kinase
ARB	angiotensin receptor blocker
BNP	brain natriuretic peptide
CK	creatinine kinase
CMR	cardiac magnetic resonance
CSVT	cerebral sinus venous thrombosis
CXCL10	CXC motif chemokine ligand 10
ECG	electrocardiogram
EUA	emergency use authorization
HFpEF	heart failure with preserved ejection fraction
ICU	intensive care unit
IFN	interferon
IL	interleukin
KD	Kawasaki disease
KS	Kounis syndrome
LGE	late gadolinium enhancement
LV	left ventricle
MI	myocardial infarction
MINOCA	myocardial infarction with nonobstructive coronary arteries
MIS-A	multisystem inflammatory syndrome in adults
MIS-C	multisystem inflammatory syndrome in children
MRI	magnetic resonance imaging
ORF	open reading frames
PI3K	phosphatidylinositol 3-kinase
POTS	postural orthostatic tachycardia syndrome
RAAS	renin-angiotensin-aldosterone system
ROS	reactive oxygen species
RV	right ventricle
TAPSE	tricuspid annular plane systolic excursion
Th	T-helper
TNF	tumor necrosis factor
VA-ECMO	venoarterial extracorporeal membrane oxygenation

Management, Body Systems, and Case Studies in COVID-19. https://doi.org/10.1016/B978-0-443-18703-2.00034-7
Copyright © 2024 Elsevier Inc. All rights are reserved, including those for text and data mining, AI training, and similar technologies.

Introduction

Since the identification of the novel coronavirus now known as SARS-CoV-2 in late 2019, there have been more than 600 million confirmed cases and nearly 6.5 million deaths due to COVID-19 (Isath et al., 2022). The treatment and management of this disease were extensively studied throughout the pandemic; research and development into treatment as well as long-term manifestations continue to this day, now years later. We continue to learn more about the long-term complications of this devastating virus regularly, and new insights will undoubtedly continue to be made in the future at all levels from basic science to the bedside. The state of knowledge has grown tremendously, although the primary focus was and often still revolves around the virus's pulmonary manifestations, and numerous extrapulmonary manifestations are known to exist. Here, we will discuss some of the effects that are relevant to the cardiovascular system. We will review the cardiac complications of the virus itself as well as some of the complications associated with the drugs and vaccines used for treatment and prevention.

Cardiac manifestations of COVID-19

Before discussing the specific manifestations of COVID-19, it is important to discuss the possible mechanisms proposed for how the virus damages the body. Etiology remains multifactorial. These mechanisms include direct viral myocardial damage, hypoxia, hypotension, systemic inflammation, mismatch of myocardial supply and demand, plaque and rupture, coronary thrombosis, disseminated intravascular coagulation secondary to sepsis, electrolyte imbalances, endothelitis, hypercoagulability, angiotensin converting enzyme 2 receptor (ACE2R) downregulation, drug toxicity, and endogenous adrenergic state (Kochi et al., 2020).

Possible mechanisms of cardiac effects

Direct myocardial injury

Direct myocardial injury is one of the ways in which this virus can affect the cardiovascular system. Autopsies of patients infected with SARS-CoV-2 showed a viral presence within cardiac myocytes, cardiac vascular endothelial cells, and cardiac fibroblasts. Active replication was also observed as shown by the presence of negative sense mRNA in postmortem studies (Almamlouk et al., 2022). As with all coronaviruses, ACE2R plays a key role in viral entry into cells, and cardiac myocytes are no exception to this. Cardiac myocytes have been observed to express ACE2 receptors (Bargehr et al., 2021). Additionally, ACE2 receptor levels have been observed to be elevated in hearts with preexisting disease due to the pharmacologic inhibition of the renin-angiotensin-aldosterone system (RAAS) (Talasaz et al., 2021). This can lead to myocarditis and direct myocardial injury as a result of infection. Exactly what role ACE inhibitors or angiotensin receptor blockers (ARBs) play in this process is still uncertain.

In addition to direct viral entry by the SARS-CoV-2 virus, myocardial injury can occur as a result of the cytokine storm. Specifically, it is an imbalance in T helper cells, types 1 and 2 (Th1/Th2), that has been associated with excess cytokine expression and poor outcomes. Excess Th2-type response has been associated with increased circulating cytokines such as IL-1, IL-6, IL-12, TNF-alpha, and IFN-gamma. These proinflammatory cytokines have been implicated in myocardial injury (Bugert et al., 2021).

Hypoxia

It is well known that the major manifestation of COVID-19 is a pulmonary disease that results in impaired gas exchange and ultimately oxygen delivery. Tissue oxygen delivery is directly proportional to cardiac output, hemoglobin concentration, and arterial oxygen saturation. With the pulmonary involvement of COVID-19, oxygen saturation can be dramatically reduced. This can result in reduced oxygen delivery and ultimately tissue hypoxia, even in the absence of ischemia. Because the heart is one of the most metabolically active organs, tissue hypoxia can consequently lead to myocardial injury.

Thrombogenesis

COVID-19 is highly associated with extensive microvascular and macrovascular thrombosis. Both arterial thrombosis and venous thrombosis have been described. A cohort study in England demonstrated a hazard ratio of 21.7 for arterial thrombosis and 33.2 for venous thrombosis (Knight et al., 2022). Within the intensive care unit (ICU) cohort, the incidence of thrombotic complications in COVID-19 patients has been observed to be greater than 30% (Klok et al., 2020). As in the case of non-COVID-19 related thrombosis, thrombi can occur within the coronary vasculature, resulting in acute coronary syndrome; venous thromboembolisms can occur within the pulmonary vasculature, which can lead to right ventricular strain and ultimately right ventricle failure. An increased incidence of left ventricular thrombus has also been observed, even within patients who had no underlying comorbidities (Russell et al., 2022).

The cause of the increased thrombosis is likely multifactorial. Virchow's triad describes three major contributing factors to thrombosis: endothelial injury, hypercoagulability, and stasis of flow. Endotheliitis has been observed to be one of the many adverse effects of COVID-19 (Calabretta et al., 2021). It is also well known that inflammation results in hypercoagulability. In addition, the nature of being hospitalized or critically ill results in a greater degree of venous stasis, fulfilling the triad.

Viral infection

Viral infection innately can also predispose a patient to cardiac complications. Chronic cardiovascular disease may become exacerbated in the setting of acute viral infection secondary to imbalances between the infection-induced increase in metabolic demand and reduced overall cardiac reserve (Kochi et al., 2020). This imbalance can lead to possible acute coronary syndrome, heart failure, or arrhythmias. A study showed that cardiac injury was present in 19.7% of patients admitted with COVID-19 (Shi et al., 2020). This study also showed the mortality rate in patients with cardiac injury was higher than those without cardiac involvement (Shi et al., 2020).

Viral RNA and protein localize to mitochondria by regulating the ACE2 receptors and open reading frames (ORF) and act by altering the Bax/BCL-2 ratio, which leads to the release of apoptogenic factors. This cellular level damage in the presence of cardiovascular risk factors such as hypertension, diabetes, age, obesity, and dyslipidemia can potentiate cardiac mitochondrial damage, associated with decreased ATP levels, increased reactive oxygen species (ROS), and an increase in circulating mitochondrial DNA. These cellular level changes cause cardiac mitochondrial damage, which can lead to heart failure and cardiac arrest (Ryback & Eirin, 2021).

Arrhythmias

An arrhythmia is when the heart beats in an abnormal pattern or timing. The causes are usually enhanced automaticity, triggered activity, or reentry. Many things can lead to the development of arrhythmias, including COVID-19. Although the exact mechanism is still unknown, various studies have shown associations between COVID-19 and its treatments with the development of arrhythmias (Maitz et al., 2022). Drugs known to treat COVID-19 have been shown to have QTc prolongation effects, one mechanism by which they increase the propensity for arrhythmias (Kochi et al., 2020).

Sinus tachycardia

Sinus tachycardia is not an arrhythmia per se, but rather an abnormality in the heart's rate. This occurs when the heart rate becomes elevated but the heart's rhythm remains normal. This can occur as a consequence of numerous conditions in which the heart naturally increases its automaticity in response to an insult. Most of the time, sinus tachycardia is a physiologic response to a demand for greater cardiac output, increasing sympathetic drive, or decreased parasympathetic drive. Many infections can lead to sinus tachycardia, COVID-19 being one of them. In a study, 71.9% of patients with COVID-19 were found to have sinus tachycardia during hospitalization, and 38.8% remained tachycardic on outpatient follow-up even after discharge (Yu et al., 2006). Tachycardia was noticed to be the most common rhythm or rate abnormality of the heart seen in COVID-19 patients (Yu et al., 2006). In another prospective observational study of hospitalized and monitored patients with COVID-19, sinus tachycardia was found to be associated with higher in-hospital mortality (Cho et al., 2020).

Sinus bradycardia

Sinus bradycardia is also an abnormality in the heart's rate in which the rate is decreased compared to the normal range, but the rhythm remains normal. Just like sinus tachycardia, sinus bradycardia can be physiologic and appropriate or benign. However, some diseases and drugs can cause this rhythm as well, which can lead to lightheadedness, syncope, worsening angina, heart failure, cognitive slowing, exercise intolerance, and more. In the same study that tracked the number of patients with COVID-19 who had sinus tachycardia, 14.9% of patients were found to have sinus bradycardia; however, this was noted to be transient and self-limiting (Cho et al., 2020).

Atrial fibrillation

Atrial fibrillation is the most common cardiac arrhythmia seen in clinical practice, and according to a systematic review on the global burden of atrial fibrillation, it appears the overall burden, incidence, and prevalence continue to rise. It is important to be aware of this arrhythmia as it can increase the risk of thromboembolic events, hospitalization, heart failure, and death. Atrial fibrillation is an arrhythmia in which the rhythm is irregularly irregular due to either rapid local ectopic firing, single-circuit reentry, or multiple-circuit reentry (Iwasaki et al., 2011). In an analysis of more than 30,000 patients from 120 institutions across the United States hospitalized with COVID-19, 5.4% were found to develop new-onset atrial fibrillation during their hospitalization (Rosenblatt et al., 2022). The cause of this new-onset atrial fibrillation is likely multifactorial with direct myocardial damage from the virus, increased adrenergic stimulation in the setting of acute illness, and inflammatory response to the virus all playing roles. Another systematic review and metaanalysis that tried to further elucidate the clinical impact of atrial fibrillation and new-onset atrial fibrillation in patients with COVID-19 included 21,653 hospitalized patients, and found the prevalence of atrial fibrillation to be 11% overall, and approximately sixfold higher in patients who had severe COVID-19 vs. nonsevere COVID-19 (19% vs. 3%) (Li et al., 2021). Both preexisting atrial fibrillation and new-onset atrial fibrillation were significantly associated with an increased risk of all-cause mortality among patients with COVID-19 (Li et al., 2021). Another analysis of 31 studies with a total of 187,716 COVID-19 patients found the prevalence of atrial fibrillation to be as high as 8%, and also found COVID-19 patients with atrial fibrillation were older, hypertensive, and critically ill compared to those without atrial fibrillation (Romiti et al., 2021). This study also showed COVID-19 patients with atrial fibrillation had a fourfold higher risk of death than their counterparts (Romiti et al., 2021).

Ventricular arrhythmias

Ventricular arrhythmias, both tachycardia and fibrillation, are life-threatening arrhythmias that can occur for a variety of reasons such as an electrolyte imbalance, autonomic nervous system dysfunctions, ischemia, damaged myocardium, and more. These arrhythmias result from abnormal automaticity, triggered activity, or reentry. In a prospective observational study, both ventricular tachycardia and ventricular fibrillation were found to be relatively rare in COVID-19 patients, but more common in patients found to have elevated troponins compared to patients who had normal troponin levels (Cho et al., 2020).

Cardiac complications of COVID-19

Cardiac arrest

Suggested etiologies for cardiac arrest in COVID-19 patients include lung injury from COVID-19 leading to hypoxemia, which can alter myocardial electric signaling; direct myocardial cell injury from the virus itself; viral injury causing worsening of preexisting conduction defects; and extreme anxiety related to illness augmenting catecholamine release (Pan et al., 2003). A study in Italy showed an increase in cardiac arrests by around 58% from 2019 to 2020 during the same time period and region; of those, about 77.4% of the increase in cases of out-of-hospital cardiac arrests observed were suspected to have been related to COVID-19 (Baldi et al., 2020a). An increase in out-of-hospital cardiac arrests in 2020 was found to be significantly correlated to the COVID-19 pandemic and coupled with a reduction in short-term outcomes (Baldi et al., 2020b). Another study from Sweden showed that COVID-19 was involved in at least 10% of all outside-of-hospital cardiac arrests and 16% of in-hospital cardiac arrests, and among those patients with COVID-19, the 30-day mortality was increased 3.4-fold in out-of-hospital cardiac arrests and 2.3-fold in in-hospital cardiac arrests (Sultanian et al., 2021).

Myocardial damage

Myocardial damage can be identified with the use of cardiac biomarkers such as cardiac troponin. Although acute coronary syndrome is classically thought of as the cause of elevated cardiac troponin, a multitude of other conditions can lead to elevations. In patients with COVID-19, the mechanisms proposed for troponin elevation are direct damage by the virus, systemic inflammatory responses, destabilized coronary plaque, and aggravated hypoxia (Guo et al., 2020). A case series study of 187 patients with COVID-19 found that 27.8% of patients had a myocardial injury resulting in cardiac dysfunction or arrhythmias, and myocardial injury was significantly associated with fatal outcomes in these patients (Guo et al., 2020). This study also showed that those patients who had underlying cardiovascular disease were more prone to experiencing myocardial injury while infected with COVID-19. Another notable finding was that patients with underlying cardiovascular disease that did not experience any myocardial injury while infected with COVID-19 were found to have a relatively favorable outcome. This study showed a significantly positive linear correlation between plasma troponin levels and plasma C-reactive protein levels, indicating that myocardial injury may be closely associated with inflammatory pathogenesis during disease progression. This inflammatory cytokine release precipitated by acute COVID-19 infection may also be the trigger for coronary blood flow reduction, decreases in oxygen supply, destabilization of coronary plaque, and microthrombosis, which are all potential mechanisms of myocardial injury (Guo et al., 2020).

Myocardial infarction

Myocardial infarction (MI) is defined as a clinical event with associated myocardial damage. Patients with myocardial infarction often have symptoms typical of a cardiac event, electrocardiogram (ECG) changes consistent with ischemia, or imaging evidence of myocardial damage such as a loss of viable myocardium or new regional wall motion abnormalities. Myocardial infarctions can be further categorized into different types, with type 1 and 2 being the more common forms. Type 1 myocardial infarction is due to acute atherothrombotic coronary artery disease, often secondary to atherosclerotic plaque rupture. Type 2 myocardial infarction is more often related to a mismatch between oxygen supply and demand; multiple mechanisms can precipitate this disturbance. A study of 86,742 COVID-19 patients in Sweden showed that COVID-19 was a risk factor for acute myocardial infarction and ischemic stroke (Katsoularis et al., 2021). Additionally, a retrospective study in Lombardy looked at all patients with confirmed COVID-19 who underwent urgent coronary angiography because of an ST-elevation myocardial infarction and showed that out of 28 selected patients, only 60.7% had evidence of a culprit lesion requiring revascularization, whereas 39.3% did not have obstructive coronary artery disease (Stefanini et al., 2020). This is unusual as ST-elevation myocardial infarction patients often have an identifiable culprit lesion. Elevated troponin levels in patients with COVID-19 infection on day 2 of admission with normal values on day 1 are reported to be the strongest independent predictor of increased mortality in a cohort by Nuzzi et al. (Nuzzi et al., 2021).

Kounis syndrome

Mast cell and inflammatory cell-mediated cardiac symptoms (including chest pain, coronary vasospasm, MI, and sudden cardiac death) following an allergic insult, drugs, and postinflammation are referred to as Kounis syndrome (KS). The first type of KS involves vasospasm of nondiseased coronary arteries with no prior coronary artery disease secondary to an acute rise in inflammatory markers, now referred to as myocardial infarction with nonobstructive coronary arteries (MINOCA), and represents endothelial dysfunction or microvascular angina (Kounis et al., 2019). The second type of KS occurs in patients with diseased vessels or prior coronary artery disease in which inflammation induces plaque erosion presenting as myocardial infarction. The third type of KS involves coronary stent thrombosis or restenosis due to allergic inflammation (Kounis et al., 2019). As reported in some case reports, KS has been observed not only in and post-COVID-19 infections, but also 15 min after CoronaVac vaccine administration and 30 min after vaccination with Pfizer (Tajstra et al., 2021).

Myocarditis

Myocarditis is by definition inflammation of the myocardium, and there are many potential causes of myocarditis, with one of the more common ones being viral infections. Although a definitive diagnosis of myocarditis requires an endomyocardial biopsy, the diagnosis of clinically suspected myocarditis can be made if other criteria are met. Several mechanisms are thought to be possibilities in the pathogenesis of COVID-19-induced myocarditis, including the direct destruction of cardiomyocytes through the occurrence of virus-mediated lysis with damage to cardiac structures, resulting in myocyte injury and cardiac dysfunction (Kawakami et al., 2021; Maitz et al., 2022). However, this has been deemed a less likely possibility as an analysis of 39 autopsy cases in Germany showed that the presence of COVID-19 was not associated with increased infiltration of mononuclear cells in the myocardium. Another potential mechanism of cardiac injury that has been proposed is the direct entry of the virus into endothelial cells in the heart without entering the myocytes themselves (Dhaduk et al., 2022). A third possible mechanism is a cardiac injury induced by hyperactivation of the immune system characterized by the release of multiple inflammatory mediators. Overall, COVID-19-induced myocarditis appears to be a relatively uncommon phenomenon (Kawakami et al., 2021). A case report described a COVID-19-positive patient who presented with mild flu-like symptoms with rapid clinical deterioration into respiratory distress and cardiogenic shock requiring mechanical ventilation and venoarterial extracorporeal membrane oxygenation (VA-ECMO) (Tavazzi et al., 2020). This is the most emergent cardiovascular presentation, fulminant myocarditis defined as ventricular dysfunction and heart failure within 2–3 weeks of COVID-19 infection (Esfandiarei & McManus, 2008). In this case, clinical manifestations were similar to those of fulminant myocarditis, but an endomyocardial biopsy showed evidence of low-grade myocardial inflammation with an absence of myocyte necrosis (Tavazzi et al., 2020). This case shows that COVID-19 can lead to presentations similar to that of myocarditis, even if the pathology does not fulfill the official diagnostic criteria of myocarditis. Another study established evidence of SARS-CoV-2 genome detection in endomyocardial biopsies of patients with suspected myocarditis, although the exact mechanism of myocarditis or implications of this finding still needs to be evaluated (Escher et al., 2020). Additionally, other cases have shown biopsy-proven myocarditis in patients with no other identifiable etiology other than a recent COVID-19 infection (Nicol et al., 2020).

Cardiomyopathy

Cardiomyopathy is essentially any disease of the myocardium that results in a structural or functional abnormality. The mechanisms known to cause cardiomyopathy with COVID-19 infection include sepsis with associated cytokine storm and stress-induced cardiomyopathy (Takotsubo) (Siripanthong et al., 2020; Venkata et al., 2020). One case reported reverse Takotsubo cardiomyopathy in a COVID-19-infected patient with the recovery of systolic function on day 7 as noted on cardiac magnetic resonance imaging with virus-negative immune-mediated myocarditis noted on endomyocardial biopsy (Sala et al., 2020).

Another case report discussed a COVID-19 patient who came in with acute systolic heart failure who recovered within days of receiving COVID-19 treatment with dexamethasone and tocilizumab (Schreiber et al., 2022). This case supports the hypothesis that a cytokine storm is the mechanism by which COVID-19 induces cardiomyopathy and myocardial stunning. In a systematic review of 29 articles with 1460 patients, cardiomyopathies secondary to COVID-19 were analyzed. This review showed that both the exacerbation of underlying cardiomyopathy and the emergence of new cardiomyopathy were common in COVID-19 patients (Omidi et al., 2021). Based on the included studies, the mortality rate of patients with cardiomyopathy who developed COVID-19 was 25%. In this review, cardiomyopathy was defined as any evidence of new left ventricle (LV) systolic dysfunction on echocardiography in addition to signs of cardiogenic shock, an increase in creatinine kinase (CK) or troponin, or a reduction in central venous oxygen saturation below 70% (Omidi et al., 2021).

Another study showed that in individuals with COVID-19, while LV systolic function was often preserved in the majority of patients, LV diastolic function and overall right ventricle (RV) function were often impaired (Szekely et al., 2020). In addition, elevated troponin levels were associated with worse RV function. The most common cardiac abnormality found on echocardiography in this study of these patients with COVID-19 was RV dilation (Szekely et al., 2020).

Pericarditis/pericardial effusion

A study that evaluated pericardial effusion and pericarditis in hospitalized patients with COVID-19 showed that the prevalence of pericardial effusion is around 14% in these patients and is associated with COVID-19 severity, worse RV systolic function as measured by tricuspid annular plane systolic excursion (TAPSE), lower cardiac index, elevated brain natriuretic peptide (BNP), and excess mortality scoring (Ghantous et al., 2022). Interestingly, while pericardial effusion was common and associated with myocardial dysfunction and excess mortality, it was found to be rarely attributable to pericarditis or myocarditis (Ghantous et al., 2022).

Myocardial fibrosis

Myocardial fibrosis is characterized by an increase in collagen type 1 deposition and fibroblast activation. COVID-19 has been implicated in the development of cardiac fibrosis. A German study noted that nearly one-half of recovered patients had persistent evidence of myocardial inflammation, and one-third of patients who had recovered had late gadolinium enhancement on follow-up cardiac magnetic resonance imaging (MRI). Interestingly, similar findings were seen even in the case of mild symptoms (Puntmann et al., 2020). Similar results were seen in a single-center study at Beijing Anzhen Hospital (Wang et al., 2021). This suggests that some degree of inflammation is present even after clinical recovery. It is currently unclear what, if any, long-term sequelae these findings may entail.

Multisystem inflammatory syndrome

Multisystem inflammatory syndrome in children (MIS-C) is a hyperinflammatory condition associated with COVID-19 that was first described early in the COVID-19 pandemic. It exhibits similarities to Kawasaki disease (KD) as well as features similar to toxic shock syndrome, and typically occurs 2–6 weeks after acute SARS-CoV-2 infection (Kounis et al., 2021). It is characterized by fever, gastrointestinal symptoms, respiratory failure, polymorphic rashes, multiorgan failure, hypotension, conjunctivitis, and cardiac abnormalities. Cardiac abnormalities include coronary artery aneurysms and cardiomyopathy. The mechanism is not clearly understood; hypotheses include inflammatory cascades resulting from inappropriate t-cell activation and cytokine release, similar to the changes seen during the cytokine storm associated with

COVID-19 (Roarty & Waterfield, 2022). MIS-C differs from Kawasaki disease in the population characteristics in that it commonly affects older children (median age 9–10 years) of African descent in comparison to KD, which is often observed in Asian children younger than 5 years. Also, KD is an inflammatory disorder whereas MIS-C is described as an autoinflammatory and autoimmune disorder with cytokine storm syndrome (Kounis et al., 2021). Soon after its description, a similar syndrome was noted to occur in adult patients as well called multisystem inflammatory syndrome in adults (MIS-A). Compared to its pediatric counterpart, MIS-A was less likely to result in mucocutaneous manifestations and patients were more likely to require mechanical ventilation (Patel et al., 2021).

The most common abnormal echocardiogram pattern noted among deteriorating patients was RV dilation and dysfunction due to its proximity to the pulmonary vasculature and secondary to conditions increasing pulmonary vascular resistance or pulmonary pressure in hospitalized patients (Szekely et al., 2020). Right ventricular pressure overload in the setting of pulmonary embolism leads to compensatory RV dilation, which in turn decreases elasticity, resulting in a fall in cardiac output and further decreased perfusion of the RV. The decrease in the trans-septal pressure gradient between RV and LV results in septal bowing toward the LV, resulting in cardiac myofibrils remodeling and a reduction in function (Szekely et al., 2020).

In a case report published in 2020, COVID-19 was associated with a transient Brugada-like pattern on ECG that reverted after the infection subsided; as the patient was asymptomatic, further investigations including cardiac MRI were not warranted (Vidovich, 2020). In comparison, a patient who presented with syncope was diagnosed with COVID-19 infection and noted to have a Brugada pattern on ECG. In this case, cardiac MRI was performed, and an ICD was placed; however, the Brugada pattern resolved afterward (Chang et al., 2020).

Cardiac effects of COVID-19 treatments

Remdesivir

Remdesivir is an adenosine nucleotide prodrug that inhibits viral RNA synthesis by binding to RNA polymerase and terminating RNA transcription prematurely (Eastman et al., 2020). Remdesivir is approved by the US Food and Drug Administration for the treatment of COVID-19. Unfortunately, remdesivir has cardiotoxic and proarrhythmic effects that are especially pronounced in patients with preexisting cardiovascular disease (Nabati & Parsaee, 2022). There have been cases of patients with COVID-19 who received remdesivir and developed sinus bradycardia, hypotension, T-wave abnormalities, atrial fibrillation, prolonged QT interval, cardiac arrest, and complete heart block (Table 1). Case reports have suggested that bradycardia and heart block resolve after discontinuation of the drug (Selvaraj et al., 2021).

TABLE 1 Treatments for COVID-19 with known adverse cardiac effects.

Treatment	Mechanism	Indication	Adverse effects
Remdesivir	Inhibits RNA synthesis	Hospitalized patients as well as nonhospitalized patients with high risk of progression to severe disease	The most frequently described cardiac adverse effects include arrhythmias, particularly bradyarrhythmia and heart block. Resolves after discontinuation
Dexamethasone	Glucocorticoid	Patients with hypoxemia and/or critical illness due to COVID-19	Glucocorticoids are known to predispose to volume retention and overload. In addition, those who experience MI on glucocorticoids have an increased incidence of free wall rupture
Nirmatrelvir/ritonavir Lopinavir/ritonavir	Inhibition of viral protease	Patients with high risk of progression to severe disease	Several drug interactions have been described, including those used in patients with cardiovascular disease such as statins and antiplatelet drugs. Arrhythmias have also been described, including prolonging of QT and PR intervals and torsades de pointes
Hydroxychloroquine	Lysosomal disruption Toll-like receptor suppression	No longer recommended	Commonly associated with fascicular blocks and QT prolongation. In combination with other QT prolong agents, can result in torsades de pointes and cardiac arrest

Dexamethasone

Dexamethasone was found to be effective in treating COVID-19 patients relatively early in the pandemic, and today it remains recommended to administer dexamethasone in COVID-19 patients who develop hypoxemia and/or are critically ill (Bouadma et al., 2022). Common cardiovascular effects of corticosteroids such as dexamethasone include hypertension and dyslipidemia; however, these are more typically associated with long-term use. Corticosteroids also result in some degree of salt and volume retention, which may result in heart failure exacerbations in those with preexisting congestive heart failure. Corticosteroid use is also associated with myocardial free wall rupture in patients who experience MI, likely due to a delay in scar formation due to the suppression of fibroblasts.

In patients receiving antiviral treatment for COVID-19 within 30 days of an acute coronary syndrome (ACS) and stent placement, it is important to note whether the patient is concomitantly receiving clopidogrel and lopinavir/ritonavir, as this is known to diminish antiplatelet activity, which can lead to stent restenosis or thrombosis. Due to its inhibitory activity of CYP3A4, lopinavir/ritonavir can increase the risk of bleeding with ticagrelor as well (Talasaz et al., 2021). Adverse effects associated with its use include prolonged PR and QT intervals, torsades de pointes, conduction disorders, and cardiac arrhythmias. Lopinavir/ritonavir is known to cause drug-mediated cardiotoxicity by its actions on the phosphatidylinositol 3-kinase (PI3K) and protein serine-threonine kinase (Akt) signaling pathways, regulation of pyroptosis, connexin 43 redistribution, and calcium signaling pathway (Reyskens et al., 2013).

Hydroxychloroquine

Although hydroxychloroquine is not recommended to treat COVID-19, there was much discussion on its possible effectiveness in treating COVID-19 early on in the pandemic. Chloroquine use may increase depolarization length duration and the Purkinje fiber refractory period, leading to conduction abnormalities, most commonly fascicular blocks (Kochi et al., 2020). The combination of azithromycin with hydroxychloroquine increased the risk of ventricular arrhythmias, torsades de pointes, and sudden cardiac death. Chloroquine use can also cause QT prolongation, leading to possible ventricular tachycardia or sudden cardiac death (Kochi et al., 2020).

Cardiac effects of COVID-19 vaccines

As noted in a systematic review of the adverse cardiovascular effects of COVID-19 vaccinations, the reported cardiac injury mechanisms postvaccination include myocarditis, myopericarditis, myocardial infarction, stress cardiomyopathy, and Kounis syndrome (Al-Ali et al., 2022). Smadja et al. reported cases of MI associated with arterial thrombosis after patients received the Pfizer, Moderna, and AstraZeneca vaccines (Smadja et al., 2021).

Both Pfizer and Moderna vaccines are mRNA vaccines, which have been associated with a short-term increased risk for myocarditis and pericarditis, as seen in many observational studies and further established with case-control studies, cohort studies, and self-controlled case series (Table 2). The highest risk is among men 18–25 years of age. The absolute risk is 1.7–2.1 per 100,000. This primarily occurs within 1 week of vaccination. The exact mechanism has not yet been elucidated. It is worth noting that myocarditis is not a novel association with vaccines, and thus the association may not be related to the

TABLE 2 Sars-CoV2 Vaccines, mechanisms of action, and associated adverse cardiac effects.

Vaccine (*Trade name*)	Mechanism	Notable adverse effects
Moderna (*Spikevax*)	mRNA Vaccine	Associated with an increased risk of myocarditis most pronounced in men 18–25 years. Typically occurs within 1 week of vaccination
Pfizer-BioNTech (*Comirnaty*)	mRNA Vaccine	Associated with myocarditis similar to the Moderna vaccine. Has additionally been demonstrated to increase expression of chemokines including CXCL10, which has been associated with angiostasis and atherosclerosis
Jansen/Johnson and Johnson (*JCovden*)	Adenovirus vector	HIT-like syndrome resulting in thrombocytopenia and thrombosis. Greatest risk is within reproductive age women
Novavax (*Nuvaxovid*)	Subunit vaccine	Six cases of myocarditis reported during clinical trial; causal link not established as of this writing

mRNA delivery or encoded spike protein. Long-term effects, if any, from vaccine-associated myocarditis/pericarditis have not been established.

The Pfizer (BNT162b2 mRNA) vaccine is associated with an increase in the number of chemokines in the blood such as CXC motif chemokine ligand 10 (CXCL10), IL-15, and IFN gamma (Karimabad et al., 2021). In the pre-COVID-19 era, CXCL 10 was studied as a marker of atherosclerotic plaque formation and coronary artery disease. It is also associated with angiostasis, inhibiting collateral formation in chronic ischemic heart disease. CXCL 10 increases in concentration during severe COVID-19 infection and postvaccination, which is currently being studied to be used as a possible marker of mRNA vaccine efficacy (Bergamaschi et al., 2021).

Adenovirus vaccines such as the Jansen/Johnson and Johnson vaccine are associated with thrombosis. This is associated with thrombocytopenia, and the pathophysiology is thought to resemble heparin-induced thrombocytopenia. This effect is thought to be related to the adenovirus vector used for these vaccines. Cerebral sinus venous thrombosis (CSVT) has been the most notable form of thrombosis with this vaccine; however, other forms of thrombosis can manifest, potentially resulting in cardiac injury as described earlier. The population at greatest risk for thrombosis in this setting is reproductive-age women. The Jansen vaccine is not currently recommended as the first line and should only be administered if the patient would otherwise be unable to obtain an alternative COVID-19 vaccine (FSFHPAVV, 2022).

The Novavax vaccine was given emergency use authorization (EUA). It contains the SARS-CoV-2 spike protein and the Matrix-M adjuvant. In the Novavax clinical trial of about 40,000 people, there were six reported cases of myocarditis (Document FB, 2022). Clinical data did not support a causal relationship, as the myocarditis rates between the vaccine and placebo groups were found to be very similar.

There are several other novel therapies for COVID-19, including antivirals and monoclonal antibodies. Some of these antiviral medications being used are nirmatrelvir/ritonavir (*Paxlovid*), tixagevimab/cilgavimab (*Evusheld*), molnupiravir, and bebtelovimab. These agents have been given EUA by the FDA for the treatment or prophylaxis of COVID-19. Cardiovascular effects are not strongly associated with these drugs; however, it is worth noting that data at this time remain overly sparse due to the relatively recent authorization and limited study period.

Long-term effects of COVID-19

COVID-19 long haulers (long-COVID-19) and post-COVID-19 conditions are described as the persistence of symptoms ≥ 4 to 12 weeks after COVID-19 diagnosis. The persistence of cardiovascular symptoms including chest pain, palpitations, dyspnea, and tachycardia irrespective of the severity of COVID-19 falls under this umbrella definition (Satterfield et al., 2022). The presence of excessive tachycardia (heart rate increase of >30 beats per minute in adults) within 10 min of standing in the absence of orthostatic hypotension is diagnosed as postural orthostatic tachycardia syndrome (POTS). Patients with these symptoms >12 weeks post-COVID-19 infection are termed to have long-COVID-19 POTS (Raj et al., 2021).

In a study including cardiac magnetic resonance (CMR) imaging in 26 competitive adults who were asymptomatic or experienced mild COVID-19 symptoms, 15% of individuals displayed features of myocarditis, including myocardial edema and myocardial injury, as evidenced by the presence of nonischemic late gadolinium enhancement (LGE); two athletes from this study had evidence of pericardial effusion with myocarditis (Rajpal et al., 2021). In a large cohort study of 1597 competitive athletes at major universities in America, CMR screening revealed the prevalence of subclinical and clinical evidence of myocarditis in 2.3% of individuals, which previously has been recognized as a risk factor of sudden cardiac death in young athletes (Daniels et al., 2021).

Structural changes noted on cardiac MRI performed 37–71 days post-COVID-19 infection irrespective of the severity of the illness revealed elevated T1 values (suggestive of fibrosis and inflammation) in up to 73% of the patients, T2 enhancement (suggestive of edema) in 60% of the patients, and LGE in 32%–45% of the patients (Puntmann et al., 2020). A recent study by Joy et al. studied 74 seropositive healthcare workers 6 months post-COVID-19 infection and observed only 4% LGE compared to matched healthy controls (Joy et al., 2021). LGE is used as a gold standard for the quantification of myocardial fibrosis and is elevated in patients with diastolic dysfunction along with elevated filling pressures (E/e′ ratio). The presence of LGE on cardiac MRI is prudent to predict the diagnosis of heart failure with preserved ejection fraction (HFpEF) in patients with no cardiovascular risk factors other than a history of COVID-19 infection.

Policies and procedures

In this chapter, we reviewed the cardiovascular effects of the SARS-CoV-2 virus. We reviewed in detail the pathophysiology, manifestations, and complications associated with the treatment and vaccination of COVID-19, in addition to the cardiovascular effects of the disease itself. No policies or guidelines have mentioned routine cardiac testing in patients with COVID-19 infection or those who have recovered from infection with persistent symptoms.

We have compiled different hypotheses that explain the cardiovascular manifestations of COVID-19 infection. A metaanalysis using various databases with a symmetric distribution of different continents revealed a sixfold higher prevalence of atrial fibrillation in severe COVID-19 infections (Baldi et al., 2020a). Another study mentioned in this chapter is a case series of 187 patients with COVID-19 infection, where 27.8% of patients had a myocardial injury resulting in cardiac dysfunction or arrhythmias, and the myocardial injury was significantly associated with fatal outcomes (Nuzzi et al., 2021).

The article has been segregated under headings including the introduction, the cardiac manifestations of COVID-19 infection, possible mechanisms of cardiac effects, cardiovascular complications of COVID-19 infection, the cardiovascular effects of various treatment modalities and vaccination, and the long-term effects of COVID-19. Within the heading of cardiac manifestations, this article emphasized the different hypotheses mentioned in the literature related to COVID-19 infection including hypoxia, direct myocardial injury, thrombogenesis, and viral infection.

The literature review from published case reports, case series, metaanalyses, and reviews revealed atrial and ventricular arrhythmias, angina, and heart failure as the presenting clinical manifestations. The cardiac complications associated with increased mortality include cardiac arrest, direct myocardial injury, myocardial infarction, Kounis syndrome, myocarditis, cardiomyopathy, pericardial or pleural effusion, myocardial fibrosis, and multisystem inflammatory syndrome studied in children as well as adults. We have observed a continual evolution in the management of COVID-19 infection and this chapter mentions the cardiac adverse effects with the current guidelines of medical management, including steroids and remdesivir. We also discussed the cardiovascular adverse effects of the vaccines for COVID-19, a controversial topic that has been a talking point since the vaccines were first developed. Our aim was to summarize all the COVID-19 infection-related cardiovascular effects currently known in this chapter and open discussion on future routine cardiac assessments in symptomatic patients with COVID-19 infection.

Applications to other areas

In this chapter, we detailed the various cardiac effects of COVID-19 infection, the pathophysiology that contributes to these effects, and the effects from treatments and vaccines for COVID-19. COVID-19 is a virus and many of its effects could be applied to other viruses. It is very likely that many other viruses also can cause adverse effects to the cardiovascular system, including myocarditis, arrhythmias, myocardial injury, and infarction, among others. In addition, the pathophysiology behind how the damage occurs, such as direct myocardial injury and inflammation, is likely to be the same in other viruses. Similarly, the effects of the vaccine for COVID-19 are likely also seen from other vaccines for different viruses as well. We can speculate that a lot of the pathophysiology behind how COVID-19 affects the cardiovascular system is likely similar to how it affects other organ systems as well. Much research has been done on the numerous effects of COVID-19 on the pulmonary system, but not enough on most other organ systems. The information we have gathered in this article about COVID-19 can reasonably be generalized to many other viruses as well. The effects of COVID-19 and its related treatments and vaccines continue to expand and evolve with time, and future reviews will likely include additional effects that have not been identified yet. In addition, our findings generate the question of how we should deal with this knowledge. Should we make certain management guidelines for cardiac assessments in certain patient populations who have been infected with COVID-19 and would be at risk for certain adverse effects? Currently, there is no specific such plan, but as we continue to learn more about the long-term effects of this deadly virus, things may change in the future.

Mini-dictionary of terms

- **Endothelitis**: Immune response of endothelial cells leading to generalized endothelial cell damage and inflammation.
- **Hypoxia**: Condition in which the oxygen levels within tissue are inadequate.
- **Thrombogenesis**: Process of forming a thrombus or clot.
- **Troponin**: A protein involved in muscle contraction found in myocardial cells.
- **QTc**: ECG measurement corrected for heart rate that represents the total time from ventricular depolarization to complete repolarization.
- **VA-ECMO**: A temporary mechanical circulatory support system that enables complete and immediate cardiopulmonary support useful in setting of cardiogenic shock or cardiac arrest.

- **MINOCA**: Condition in which there is clinical evidence of a myocardial infarction but coronary arteries are angiographically normal or only minimally obstructive (stenosis less than 50%).
- **BNP**: A hormone secreted by cardiac myocytes in the ventricles in response to stretching due to increased blood volume.

Summary points

- COVID-19 infection can involve the cardiovascular system via direct myocardial injury, hypoxia, thrombogenesis, and the viral infection itself.
- Common cardiovascular manifestations of COVID-19 infection studied include arrhythmias, angina, and pulmonary vascular congestion secondary to congestive heart failure, among others.
- The more severe and often fatal cardiac manifestations of COVID-19 infection include cardiac arrest, acute coronary syndrome, Kounis syndrome, myocarditis, and myocardial fibrosis.
- Remdesivir can be a proarrhythmic, cardiotoxic drug with QTc-prolonging effects.
- Corticosteroids worsen the prognosis following acute coronary syndrome in patients with COVID-19 infection by increasing the risk of free wall rupture, diminishing the antiplatelet effect, and increasing the risk of bleeding with ticagrelor.
- Vaccines against COVID-19 infection such as Pfizer and Moderna are known to increase the risk of myocarditis and pericarditis within 1 week of administration.
- Men 18–25 are known to be at the highest risk of COVID-19 vaccine-associated adverse effects.
- Adenovirus vaccines such as the Jansen/Johnson and Johnson vaccine are associated with thrombosis more commonly in women of reproductive age.

Disclosures

The authors have no conflicts of interest to disclose.

(1) This paper is not under consideration elsewhere; (2) All authors have read and approved the manuscript; (3) All authors take responsibility for all aspects of the reliability and freedom from bias of the data presented and their discussed interpretation.

References

Al-Ali, D., Elshafeey, A., Mushannen, M., Kawas, H., Shafiq, A., Mhaimeed, N., et al. (2022). Cardiovascular and haematological events post COVID-19 vaccination: A systematic review. *Journal of Cellular and Molecular Medicine*, *26*(3), 636–653.

Almamlouk, R., Kashour, T., Obeidat, S., Bois, M. C., Maleszewski, J. J., Omrani, O. A., et al. (2022). COVID-19-associated cardiac pathology at the postmortem evaluation: A collaborative systematic review. *Clinical Microbiology and Infection*, *28*(8), 1066–1075.

Baldi, E., Sechi, G. M., Mare, C., Canevari, F., Brancaglione, A., Primi, R., et al. (2020a). Out-of-hospital cardiac arrest during the Covid-19 outbreak in Italy. *The New England Journal of Medicine*, *383*(5), 496–498.

Baldi, E., Sechi, G. M., Mare, C., Canevari, F., Brancaglione, A., Primi, R., et al. (2020b). COVID-19 kills at home: The close relationship between the epidemic and the increase of out-of-hospital cardiac arrests. *European Heart Journal*, *41*(32), 3045–3054.

Bargehr, J., Rericha, P., Petchey, A., Colzani, M., Moule, G., Malgapo, M. C., et al. (2021). Cardiovascular ACE2 receptor expression in patients undergoing heart transplantation. *ESC Heart Failure*, *8*(5), 4119–4129.

Bergamaschi, C., Terpos, E., Rosati, M., Angel, M., Bear, J., Stellas, D., et al. (2021). Systemic IL-15, IFN-gamma, and IP-10/CXCL10 signature associated with effective immune response to SARS-CoV-2 in BNT162b2 mRNA vaccine recipients. *Cell Reports*, *36*(6), 109504.

Bouadma, L., Mekontso-Dessap, A., Burdet, C., Merdji, H., Poissy, J., Dupuis, C., et al. (2022). High-dose dexamethasone and oxygen support strategies in intensive care unit patients with severe COVID-19 acute hypoxemic respiratory failure: The COVIDICUS randomized clinical trial. *JAMA Internal Medicine*, *182*(9), 906–916.

Bugert, C. L., Kwiat, V., Valera, I. C., Bugert, J. J., & Parvatiyar, M. S. (2021). Cardiovascular injury due to SARS-CoV-2. *Current Clinical Microbiology Reports*, *8*(3), 167–177.

Calabretta, E., Moraleda, J. M., Iacobelli, M., Jara, R., Vlodavsky, I., O'Gorman, P., et al. (2021). COVID-19-induced endotheliitis: Emerging evidence and possible therapeutic strategies. *British Journal of Haematology*, *193*(1), 43–51.

Chang, D., Saleh, M., Garcia-Bengo, Y., Choi, E., Epstein, L., & Willner, J. (2020). COVID-19 infection unmasking Brugada syndrome. *HeartRhythm Case Reports*, *6*(5), 237–240.

Cho, J. H., Namazi, A., Shelton, R., Ramireddy, A., Ehdaie, A., Shehata, M., et al. (2020). Cardiac arrhythmias in hospitalized patients with COVID-19: A prospective observational study in the western United States. *PLoS ONE*, *15*(12), e0244533.

Daniels, C. J., Rajpal, S., Greenshields, J. T., Rosenthal, G. L., Chung, E. H., Terrin, M., et al. (2021). Prevalence of clinical and subclinical myocarditis in competitive athletes with recent SARS-CoV-2 infection: Results from the big ten COVID-19 cardiac registry. *JAMA Cardiology*, *6*(9), 1078–1087.

Dhaduk, K., Khosla, J., Hussain, M., Mangaroliya, V., Chauhan, S., Ashish, K., et al. (2022). COVID-19 vaccination and myocarditis: A review of current literature. *World Journal of Virology*, *11*(4), 170–175.

Document FB. (June 7, 2022). *Novavax COVID-19 vaccine (NVX-CoV2373) VRBPAC briefing document*. Available from: https://www.fda.gov/media/158912/download.

Eastman, R. T., Roth, J. S., Brimacombe, K. R., Simeonov, A., Shen, M., Patnaik, S., et al. (2020). Remdesivir: A review of its discovery and development leading to emergency use authorization for treatment of COVID-19. *ACS Central Science*, *6*(5), 672–683.

Escher, F., Pietsch, H., Aleshcheva, G., Bock, T., Baumeier, C., Elsaesser, A., et al. (2020). Detection of viral SARS-CoV-2 genomes and histopathological changes in endomyocardial biopsies. *ESC Heart Failure*, *7*(5), 2440–2447.

Esfandiarei, M., & McManus, B. M. (2008). Molecular biology and pathogenesis of viral myocarditis. *Annual Review of Pathology*, *3*, 127–155.

FSFHPAVV. (May 05, 2022). *Emergency use authorization (EUA) of the janssen COVID-19 vaccine to prevent coronavirus disease 2019 (COVID-19)*. PROVIDERS.

Ghantous, E., Szekely, Y., Lichter, Y., Levi, E., Taieb, P., Banai, A., et al. (2022). Pericardial involvement in patients hospitalized with COVID-19: Prevalence, associates, and clinical implications. *Journal of the American Heart Association*, *11*(7), e024363.

Guo, T., Fan, Y., Chen, M., Wu, X., Zhang, L., He, T., et al. (2020). Cardiovascular implications of fatal outcomes of patients with coronavirus disease 2019 (COVID-19). *JAMA Cardiology*, *5*(7), 811–818.

Isath, A., Malik, A. H., Goel, A., Gupta, R., Shrivastav, R., & Bandyopadhyay, D. (2022). Nationwide analysis of the outcomes and mortality of hospitalized COVID-19 patients. *Current Problems in Cardiology*, *48*(2), 101440.

Iwasaki, Y. K., Nishida, K., Kato, T., & Nattel, S. (2011). Atrial fibrillation pathophysiology: Implications for management. *Circulation*, *124*(20), 2264–2274.

Joy, G., Artico, J., Kurdi, H., Seraphim, A., Lau, C., Thornton, G. D., et al. (2021). Prospective case-control study of cardiovascular abnormalities 6 months following mild COVID-19 in healthcare workers. *JACC: Cardiovascular Imaging*, *14*(11), 2155–2166.

Karimabad, M. N., Kounis, N. G., Hassanshahi, G., Hassanshahi, F., Mplani, V., Koniari, I., et al. (2021). The involvement of CXC motif chemokine ligand 10 (CXCL10) and its related chemokines in the pathogenesis of coronary artery disease and in the COVID-19 vaccination: A narrative review. *Vaccines (Basel)*, *9*(11).

Katsoularis, I., Fonseca-Rodriguez, O., Farrington, P., Lindmark, K., & Fors Connolly, A. M. (2021). Risk of acute myocardial infarction and ischaemic stroke following COVID-19 in Sweden: A self-controlled case series and matched cohort study. *Lancet*, *398*(10300), 599–607.

Kawakami, R., Sakamoto, A., Kawai, K., Gianatti, A., Pellegrini, D., Nasr, A., et al. (2021). Pathological evidence for SARS-CoV-2 as a cause of myocarditis: JACC review topic of the week. *Journal of the American College of Cardiology*, *77*(3), 314–325.

Klok, F. A., Kruip, M., van der Meer, N. J. M., Arbous, M. S., Gommers, D., Kant, K. M., et al. (2020). Incidence of thrombotic complications in critically ill ICU patients with COVID-19. *Thrombosis Research*, *191*, 145–147.

Knight, R., Walker, V., Ip, S., Cooper, J. A., Bolton, T., Keene, S., et al. (2022). Association of COVID-19 with major arterial and venous thrombotic diseases: A population-wide cohort study of 48 million adults in England and Wales. *Circulation*, *146*(12), 892–906.

Kochi, A. N., Tagliari, A. P., Forleo, G. B., Fassini, G. M., & Tondo, C. (2020). Cardiac and arrhythmic complications in patients with COVID-19. *Journal of Cardiovascular Electrophysiology*, *31*(5), 1003–1008.

Kounis, N. G., Koniari, I., de Gregorio, C., Assimakopoulos, S. F., Velissaris, D., Hung, M. Y., et al. (2021). COVID-19 disease, Women's predominant non-heparin vaccine-induced thrombotic thrombocytopenia and Kounis syndrome: A passepartout cytokine storm interplay. *Biomedicine*, *9*(8).

Kounis, N. G., Koniari, I., Velissaris, D., Tzanis, G., & Hahalis, G. (2019). Kounis syndrome-not a single-organ arterial disorder but a multisystem and multidisciplinary disease. *Balkan Medical Journal*, *36*(4), 212–221.

Li, Z., Shao, W., Zhang, J., Ma, J., Huang, S., Yu, P., et al. (2021). Prevalence of atrial fibrillation and associated mortality among hospitalized patients with COVID-19: A systematic review and meta-analysis. *Frontiers in Cardiovascular Medicine*, *8*, 720129.

Maitz, T., Parfianowicz, D., Vojtek, A., Rajeswaran, Y., Vyas, A. V., & Gupta, R. (2022). COVID-19 cardiovascular connection: A review of cardiac manifestations in COVID-19 infection and treatment modalities. *Current Problems in Cardiology*, 101186.

Nabati, M., & Parsaee, H. (2022). Potential cardiotoxic effects of Remdesivir on cardiovascular system: A literature review. *Cardiovascular Toxicology*, *22*(3), 268–272.

Nicol, M., Cacoub, L., Baudet, M., Nahmani, Y., Cacoub, P., Cohen-Solal, A., et al. (2020). Delayed acute myocarditis and COVID-19-related multisystem inflammatory syndrome. *ESC Heart Failure*, *7*(6), 4371–4376.

Nuzzi, V., Merlo, M., Specchia, C., Lombardi, C. M., Carubelli, V., Iorio, A., et al. (2021). The prognostic value of serial troponin measurements in patients admitted for COVID-19. *ESC Heart Failure*, *8*(5), 3504–3511.

Omidi, F., Hajikhani, B., Kazemi, S. N., Tajbakhsh, A., Riazi, S., Mirsaeidi, M., et al. (2021). COVID-19 and cardiomyopathy: A systematic review. *Frontiers in Cardiovascular Medicine*, *8*, 695206.

Pan, S. F., Zhang, H. Y., Li, C. S., & Wang, C. (2003). Cardiac arrest in severe acute respiratory syndrome: Analysis of 15 cases. *Zhonghua Jie He He Hu Xi Za Zhi*, *26*(10), 602–605.

Patel, P., DeCuir, J., Abrams, J., Campbell, A. P., Godfred-Cato, S., & Belay, E. D. (2021). Clinical characteristics of multisystem inflammatory syndrome in adults: A systematic review. *JAMA Network Open*, *4*(9), e2126456.

Puntmann, V. O., Carerj, M. L., Wieters, I., Fahim, M., Arendt, C., Hoffmann, J., et al. (2020). Outcomes of cardiovascular magnetic resonance imaging in patients recently recovered from coronavirus disease 2019 (COVID-19). *JAMA Cardiology*, *5*(11), 1265–1273.

Raj, S. R., Arnold, A. C., Barboi, A., Claydon, V. E., Limberg, J. K., Lucci, V. M., et al. (2021). Long-COVID postural tachycardia syndrome: An American autonomic society statement. *Clinical Autonomic Research*, *31*(3), 365–368.

Rajpal, S., Tong, M. S., Borchers, J., Zareba, K. M., Obarski, T. P., Simonetti, O. P., et al. (2021). Cardiovascular magnetic resonance findings in competitive athletes recovering from COVID-19 infection. *JAMA Cardiology*, *6*(1), 116–118.

Reyskens, K. M., Fisher, T. L., Schisler, J. C., O'Connor, W. G., Rogers, A. B., Willis, M. S., et al. (2013). Cardio-metabolic effects of HIV protease inhibitors (lopinavir/ritonavir). *PLoS ONE*, *8*(9), e73347.

Roarty, C., & Waterfield, T. (2022). Review and future directions for PIMS-TS (MIS-C). *Archives of Disease in Childhood*.

Romiti, G. F., Corica, B., Lip, G. Y. H., & Proietti, M. (2021). Prevalence and impact of atrial fibrillation in hospitalized patients with COVID-19:- A systematic review and meta-analysis. *Journal of Clinical Medicine*, *10*(11).

Rosenblatt, A. G., Ayers, C. R., Rao, A., Howell, S. J., Hendren, N. S., Zadikany, R. H., et al. (2022). New-onset atrial fibrillation in patients hospitalized with COVID-19: Results from the American Heart Association COVID-19 cardiovascular registry. *Circulation. Arrhythmia and Electrophysiology*, *15*(5), e010666.

Russell, J., Wagoner, M., DuPont, J., Myers, D., Muthu, K., & Thotakura, S. (2022). Left ventricular thrombus of unknown etiology in a patient with COVID-19 disease with no significant medical history. *Cardiovascular Revascularization Medicine*, *40S*, 329–331.

Ryback, R., & Eirin, A. (2021). Mitochondria, a missing link in COVID-19 heart failure and arrest? *Frontiers in Cardiovascular Medicine*, *8*, 830024.

Sala, S., Peretto, G., Gramegna, M., Palmisano, A., Villatore, A., Vignale, D., et al. (2020). Acute myocarditis presenting as a reverse Tako-Tsubo syndrome in a patient with SARS-CoV-2 respiratory infection. *European Heart Journal*, *41*(19), 1861–1862.

Satterfield, B. A., Bhatt, D. L., & Gersh, B. J. (2022). Cardiac involvement in the long-term implications of COVID-19. *Nature Reviews. Cardiology*, *19*(5), 332–341.

Schreiber, A., Elango, K., Soussu, C., Fakhra, S., Asad, S., & Ahsan, C. (2022). COVID-19 induced cardiomyopathy successfully treated with Tocilizumab. *Case Reports in Cardiology*, *2022*, 9943937.

Selvaraj, V., Bavishi, C., Patel, S., & Dapaah-Afriyie, K. (2021). Complete heart block associated with Remdesivir in COVID-19: A case report. *European Heart Journal - Case Reports*, *5*(7), ytab200.

Shi, S., Qin, M., Shen, B., Cai, Y., Liu, T., Yang, F., et al. (2020). Association of cardiac injury with mortality in hospitalized patients with COVID-19 in Wuhan, China. *JAMA Cardiology*, *5*(7), 802–810.

Siripanthong, B., Nazarian, S., Muser, D., Deo, R., Santangeli, P., Khanji, M. Y., et al. (2020). Recognizing COVID-19-related myocarditis: The possible pathophysiology and proposed guideline for diagnosis and management. *Heart Rhythm*, *17*(9), 1463–1471.

Smadja, D. M., Yue, Q. Y., Chocron, R., Sanchez, O., & Lillo-Le, L. A. (2021). Vaccination against COVID-19: Insight from arterial and venous thrombosis occurrence using data from VigiBase. *The European Respiratory Journal*, *58*(1).

Stefanini, G. G., Montorfano, M., Trabattoni, D., Andreini, D., Ferrante, G., Ancona, M., et al. (2020). ST-elevation myocardial infarction in patients with COVID-19: Clinical and angiographic outcomes. *Circulation*, *141*(25), 2113–2116.

Sultanian, P., Lundgren, P., Stromsoe, A., Aune, S., Bergstrom, G., Hagberg, E., et al. (2021). Cardiac arrest in COVID-19: Characteristics and outcomes of in- and out-of-hospital cardiac arrest. A report from the Swedish registry for cardiopulmonary resuscitation. *European Heart Journal*, *42*(11), 1094–1106.

Szekely, Y., Lichter, Y., Taieb, P., Banai, A., Hochstadt, A., Merdler, I., et al. (2020). Spectrum of cardiac manifestations in COVID-19: A systematic echocardiographic study. *Circulation*, *142*(4), 342–353.

Tajstra, M., Jaroszewicz, J., & Gasior, M. (2021). Acute coronary tree thrombosis after vaccination for COVID-19. *JACC. Cardiovascular Interventions*, *14*(9), e103–e104.

Talasaz, A. H., Kakavand, H., Van Tassell, B., Aghakouchakzadeh, M., Sadeghipour, P., Dunn, S., et al. (2021). Cardiovascular complications of COVID-19: Pharmacotherapy perspective. *Cardiovascular Drugs and Therapy*, *35*(2), 249–259.

Tavazzi, G., Pellegrini, C., Maurelli, M., Belliato, M., Sciutti, F., Bottazzi, A., et al. (2020). Myocardial localization of coronavirus in COVID-19 cardiogenic shock. *European Journal of Heart Failure*, *22*(5), 911–915.

Venkata, V. R. S., Gupta, R., Aedma, S. K., & Andrus, B. W. (2020). Covid-19 and cardiovascular complications: Pooled analysis of observational studies. *Circulation*, *142*(Suppl_3), A15097.

Vidovich, M. I. (2020). Transient Brugada-like electrocardiographic pattern in a patient with COVID-19. *JACC. Case Reports*, *2*(9), 1245–1249.

Wang, H., Li, R., Zhou, Z., Jiang, H., Yan, Z., Tao, X., et al. (2021). Cardiac involvement in COVID-19 patients: Mid-term follow up by cardiovascular magnetic resonance. *Journal of Cardiovascular Magnetic Resonance*, *23*(1), 14.

Yu, C. M., Wong, R. S., Wu, E. B., Kong, S. L., Wong, J., Yip, G. W., et al. (2006). Cardiovascular complications of severe acute respiratory syndrome. *Postgraduate Medical Journal*, *82*(964), 140–144.

Chapter 29

Limb ischemia and COVID-19

Raffaello Bellosta, Sara Allievi, Luca Attisani, Luca Luzzani, and Matteo Alberto Pegorer
Division of Vascular Surgery, Department of Cardiovascular Surgery, Fondazione Poliambulanza, Brescia, Italy

Abbreviations

ABI	ankle brachial index
ALI	acute limb ischemia
AP	ankle pressure
aPTT	activated partial thromboplastin time
AT	antithrombin
ATIII	antithrombin III
BMI	body mass index
CDT	catheter-directed thrombolysis
CE-MRA	contrast enhanced magnetic resonance angiography
CK	creatine kinase
COVID-19	coronavirus disease 2019
CTA	computed tomography angiography
DSA	digital subtraction angiography
DUS	duplex ultrasound
HIT	heparin-induced thrombocytopenia
non-CoV-2-RV	non-CoV-2 respiratory virus
PAD	peripheral artery disease
PAT	percutaneous thrombus aspiration
PCD	phlegmasia cerulea dolens
PT	prothrombin time
rtPA	recombinant tissue plasminogen activator
TP	toe pressure
UFH	unfractionated heparin
VTE	venous thromboembolism
vWF	von Willebrand factor

Epidemiology and etiopathogenesis

Acute limb ischemia (ALI) is defined as a sudden decrease in arterial perfusion of an extremity associated with a threat to its survival, requiring urgent evaluation and management. ALI is considered when the symptom duration is less than 2 weeks. The most common causes of ALI include embolism, in situ thrombosis of atherosclerotic or previously treated arteries, aneurysms, dissection and traumatic arterial injuries (Björck et al., 2020).

During the coronavirus disease 2019 (COVID-19) pandemic in Brescia, Italy, ALI occurred approximately five times more frequently in COVID-19-positive patients; the increase in patients with ALI was from 1.8% in 2019 to 16.3% in 2020 (Bellosta et al., 2020). At the regional level, ALI was observed in 64% of COVID-19-positive admissions to the vascular service and only in 23% of COVID-19-negative admissions from March 9 to April 28, 2020 (Kahlberg et al., 2021). This was similar to the experience in New York City: Ilonzo et al. reported a relative increase in lower extremity revascularization cases, likely secondary to ALI, from 10.7% of the total volume of cases to 29.8% (Ilonzo, Koleilat, et al., 2021). Indes et al. reported a higher incidence of aortoiliac thrombosis in COVID-19-positive patients compared with COVID-19-negative patients admitted during the same period (Indes et al., 2021).

Management, Body Systems, and Case Studies in COVID-19. https://doi.org/10.1016/B978-0-443-18703-2.00027-X
Copyright © 2024 Elsevier Inc. All rights are reserved, including those for text and data mining, AI training, and similar technologies.

Based on these case series (Bellosta et al., 2020; Bilaloglu et al., 2020; Etkin et al., 2021; Ilonzo, Rao, et al., 2021; Indes et al., 2021), patients with COVID-19-associated ALI were, on average, older than 60, with a body mass index (BMI) >25, and had typical cardiovascular risk factors including hypertension, peripheral artery disease (PAD), and diabetes. The COVID-19 outbreak limited accessibility to patients and the chance for elective surgery in cases of chronic artery disease, leading to more severe vascular complications at the initial presentation. However, it is noteworthy that thrombosis developed even in patients without previous evidence of atherosclerosis. Similar to the general population, the lower extremity was more commonly affected than the upper extremity (Bellosta et al., 2020; Etkin et al., 2021; Fournier et al., 2021; Ilonzo, Rao, et al., 2021).

Thrombotic complications are not unique to COVID-19. Other non-CoV-2 respiratory viral (non-CoV-2-RV) infections have also been associated with both arterial and venous thrombosis. However, there appears to be an increased hypercoagulability in severely ill COVID-19 patients compared with non-CoV-2-RV patients (Tan et al., 2021). Two factors seem to be associated with arterial thrombosis: endothelitis, with diffuse endothelial damage and infiltration by inflammatory cells, and hypercoagulability, characterized by the elevation of D-dimer, prothrombin, and fibrinogen. These two factors, together with prolonged immobilization in critically ill COVID-19 patients, could complete the Virchow's triad, providing a plausible explanation for the mechanism of arterial thrombosis. Furthermore, hyperviscosity resulting from systemic inflammation and hypercytokinemia, as well as the age-associated increase in plasma concentration of coagulation factors, could activate the coagulation cascade (Cheruiyot et al., 2021).

Clinical presentation

The majority of COVID-19 patients with severe and moderate respiratory symptoms develop ALI during hospitalization (De Hous et al., 2021; Etkin et al., 2021), and ALI could be the first clinical manifestation in up to 45% of asymptomatic COVID-19 patients (Attisani et al., 2021; Etkin et al., 2021). The involvement of other vascular districts (visceral, cerebral, and coronary), secondary to the underlying hypercoagulable state, has also been described, ranging from 12% to 18% (Jongkind et al., 2022).

The clinical presentation of ALI depends on the time course of vessel occlusion, the location of the affected vessels, the ability to recruit collateral circulation to provide flow around the occlusion, and the presence of underlying vascular disease. The occlusion of large proximal vessels may result in ischemia of the entire limb, while small vessel occlusion could result in toe/digital ischemia. In patients with known PAD, symptoms can develop insidiously over a period of hours to days; in young and nonatherosclerotic patients, ALI occurs acutely due to the absence of collateral blood vessels.

The typical manifestations of ALI were first described by Pratt in 1954 as the "six Ps" (Pratt & Krahl, 1954):

- **Pain**: Usually located in the distal extremities, it increases in severity and progresses proximally over time. Later, it may decrease due to ischemic sensory loss. The patient is often prostrated and describes the pain as something never experienced before.
- **Pulses** deficit: The absence of pulses in the affected extremity is typical but not sufficient for the diagnosis because patients with PAD may have weak or absent pulses. Normal pulses in the controlateral extremity suggest the absence of chronic arterial occlusive disease and advocate for an embolic etiology.
- **Pallor** and **poikilothermia**: First, the skin becomes pale and cool (poikilothermic) compared to the controlateral side as a consequence of reduced perfusion; afterward, it becomes mottled with delayed capillary filling as a consequence of capillary venodilatation.
- **Paresthesia** and **paralysis**: At an early stage, patients experience numbness or paresthesia, that gradually evolves into deep anesthesia. Muscle weakness and paralysis are signs of advanced limb-threatening ischemia.

Rarely, these signs and symptoms are present at the same time. According to the Society for Vascular Surgery/International Society for Clinical Vascular Surgery (SVS/ISCVS) (Rutherford et al., 1997), the severity of ALI should be determined by the presence and degree of sensory-motor deficits and Doppler findings (Table 1):

- **Grade I** (limb not immediately threatened)—no deficit, arterial and venous Doppler signals audible.
- **Grade IIa** (salvageable limb if promptly treated)—minimal sensory loss (toes), no impaired motor function, absence of arterial Doppler signals, presence of venous Doppler signals.
- **Grade IIb** (salvageable limb if immediate treatment)—sensory loss not just confined to toes, mild-to-moderate muscle weakness, absence of arterial Doppler signals, presence of venous Doppler signals.
- **Grade III** (inevitable major tissue loss or permanent nerve damage)—profound sensory and motor deficits (anesthesia and paralysis), absence of arterial and venous Doppler signals.

TABLE 1 Classification of ALI according to the Society for Vascular Surgery and the International Society for Cardiovascular Surgery.

		Findings		Doppler signals	
Category	Description/prognosis	Sensory loss	Muscle weakness	Arterial	Venous
I. Viable	Not immediately threatened	None	None	Audible	Audible
II. Threatened					
a. Marginally	Salvageable if promptly treated	Minimal (toes) or none	None	Inaudible	Audible
b. Immediately	Salvageable with immediate revascularization	More than toes, associated with rest pain	Mild, moderate	Inaudible	Audible
III. Irreversible	Major tissue loss or permanent nerve damage inevitable	Profound, anesthetic	Profound, paralysis (rigor)	Inaudible	Inaudible

Phlegmasia cerulea dolens (PCD) has been described in COVID-19 patients (Akkari & Schwartz, 2020; Alghamdi et al., 2022; Moraes et al., 2021; Orso et al., 2021). It is a rare and peculiar cause of ALI resulting from massive thrombosis of the deep venous axis of the limb (Fig. 1). It usually affects the lower limb and is associated with an amputation rate ranging from 20% to 50% (Brockman & Vasko, 1965; Haimovici, 1965). The limb's edema, caused by venous outflow obstruction, leads to compartment syndrome, resulting in impaired arterial flow. Clinical manifestations include initial limb swelling as well as pain and cyanosis of extremities; if not promptly treated, it could evolve into venous gangrene at a later stage.

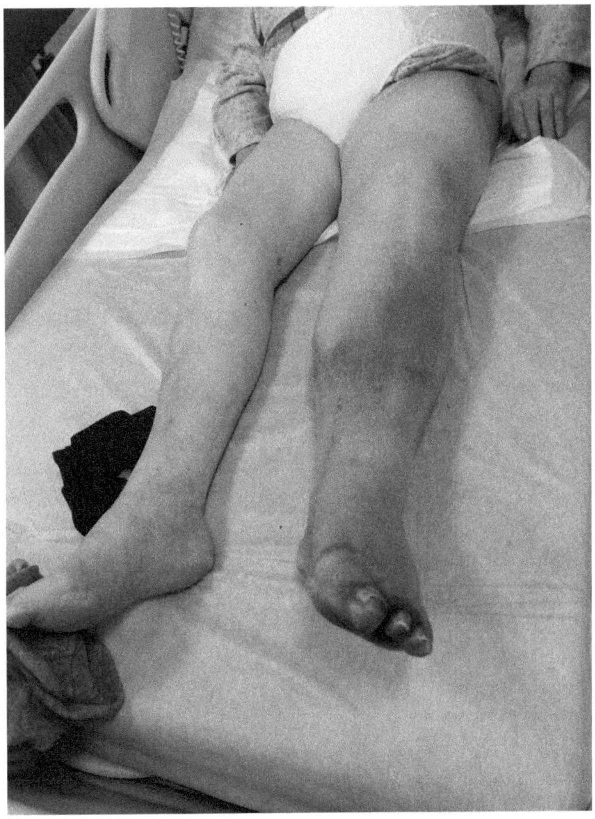

FIG. 1 Digital gangrene in a COVID-19 patient affected by phlegmasia cerulea dolens.

Diagnosis

ALI diagnosis is mostly clinical, based on signs and symptoms summarized as the "six Ps." Rarely, these signs and symptoms are present at the same time; in mild and uncertain forms, ankle pressure (AP), toe pressure (TP), and the ankle-brachial index (ABI) could be helpful in determining the correct diagnosis and assessing the severity of the ischemia. Since these investigations are time-consuming, they should be performed without interfering with or delaying the treatment of ALI.

Noninvasive imaging techniques are crucial to confirm the diagnosis and plan an appropriate treatment. In the emergency setting during a pandemic, duplex ultrasound (DUS) is a readily available tool that allows bedside evaluation and an accurate assessment of the femoro-popliteal district (Bellosta et al., 2020). Given that its accuracy is lower for the evaluation of the aortoiliac segment and tibial arteries, DUS should not be used as the sole modality to rule out arterial occlusion.

The latest guidelines (Björck et al., 2020) suggest computed tomography angiography (CTA) as the preferred imaging modality. It helps in seeking potential proximal embolic sources, looking for other emboli, and determining the anatomical location and extension of the occlusion (Fig. 2). COVID-19 patients with ALI should undergo a comprehensive CTA, with imaging extending from the aortic arch through the limbs, to rule out intracardiac or aortic thrombogenic nidus and concomitant venous thrombosis (Jongkind et al., 2022). This has become of great importance, because a more proximal anatomical location of the clot nidus has been pointed out in COVID-19 patients compared with non-COVID-19 patients (Goldman et al., 2020).

Contrast enhanced magnetic resonance angiography (CE-MRA) only plays a minor role in ALI, because lengthy examination times, scarce availability, and potential artifacts limit its use in any emergency setting. Digital subtraction angiography (DSA) represents the gold standard for ALI, and the first and only diagnostic method when clear and advanced signs of ischemia are present and any delay in further investigations could compromise limb salvage. DSA is commonly used in combination with endovascular revascularization techniques and vasodilators, which can enhance the visualization of the distal arterial bed.

Laboratory tests are useful for the early detection of patients with hypercoagulable states and, consequently, at higher risk for thromboembolic events (Cheruiyot et al., 2021). Routine blood tests should include platelet count, PT, aPTT, and fibrinogen. Additionally, D-dimer, which was found to be a predictor of severe outcomes with moderate sensitivity and specificity, serves as a screening tool for venous thromboembolism (VTE) and helps assess the efficacy of anticoagulant

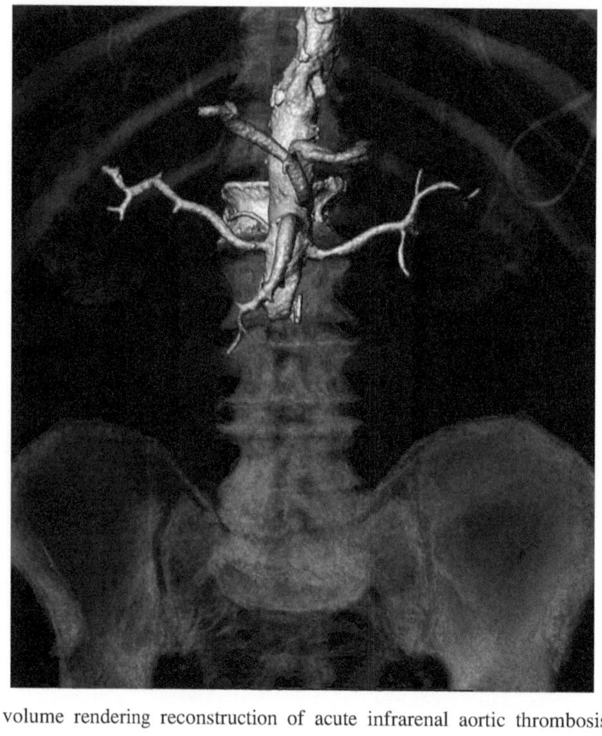

FIG. 2 Preoperative CTA with 3D volume rendering reconstruction of acute infrarenal aortic thrombosis (3mensio Medical Imaging BV, the Netherlands).

therapy in hospitalized COVID-19 patients (Baram et al., 2020; Li et al., 2020; Mouhat et al., 2020; Zhan et al., 2021). Myoglobin, creatine kinase (CK), and serum lactate are well-known markers of ischemic damage, especially for skeletal muscle, and may play a role in the perioperative management of ALI. However, studies supporting their routine use as prognostic factors are lacking.

Investigations of genetic factors, such as protein C, protein S, and antithrombin III (ATIII) deficiencies, and factor V Leiden mutations, could identify inherited forms of thrombophilia to guide anticoagulation therapy and secondary thromboprophylaxis (Elbadry et al., 2022).

Operative techniques

Although advanced stages of ALI require prompt treatment, several factors should be taken into account when deciding whether an intervention is appropriate. Severely ill COVID-19 patients often present systemic complication, which could represent a contraindication for any treatment. Vasoactive drugs, administered in case of shock, could induce vasospasm, exacerbating ischemia. For all patients, therapeutic anticoagulation should be provided as soon as ALI is diagnosed. Bleeding, recent surgery, and a history of heparin-induced thrombocytopenia (HIT) may require dosage adjustment. Local anesthesia and endovascular techniques could offer less-invasive options. However, due to the association of the infection with more severe and extended thrombosis, a hybrid approach may be required, increasing the total procedure time and surgical risks.

If the affected limb is nonviable at presentation or the patient's general conditions are too compromised, primary amputation may be necessary. Anesthesiological support is fundamental to prevent the worsening of multiorgan damage due to the ischemic/reperfusion injury.

Surgical revascularization

The balloon catheter thrombectomy or embolectomy, first introduced by Fogarty in 1963 (Fogarty et al., 1963), has been the cornerstone of therapy for acute limb ischemia, even in COVID-19 patients.

During the COVID-19 pandemic, open surgical interventions were performed in the usual operating room with routine antibiotic prophylaxis and preferably with locoregional or local anesthesia combined with intravenous sedation.

The procedure began with the dissection of the target vessel: a longitudinal groin incision in case of aortoiliac or femoropopliteal thrombosis, or a below-the-knee incision in case of popliteal and/or tibial thrombosis. Tibial arteries and/or forearm vessels were selectively approached in case of distal occlusion (Fig. 3). Vessel loops were used to facilitate proximal and distal control and, depending on the vessel size and associated atherosclerotic disease, either a transverse or longitudinal arteriotomy was performed. Fogarty catheters, sized according to the location of the occlusion, were gently passed proximally and distally until no more thrombus was collected or a good pulse or back bleeding was achieved (Khryshchanovich et al., 2021; Topcu et al., 2021). Intravenous UFH (80 UI/kg) was administered at arterial clamping. Interestingly, thrombus specimens macroscopically appeared more adhesive and viscous in comparison to those obtained from non-COVID-19 patients (Wichmann et al., 2020).

At the end of the procedure, a completion angiography was performed (Fig. 4) in the operating room through the surgical access to exclude residual thrombus and distal embolization, to assess the patency of distal arteries, and to eventually allow immediate reintervention (Karnabatidis et al., 2011). In fact, COVID-19 infection might promote a procoagulant state, leading to the production of both microthrombi and macrothrombi (Mietto et al., 2020; Wichmann et al., 2020).

In the most severe cases, angiography showed a typical pattern known as "desert foot" (Fig. 5), characterized by the absence of forefoot microcirculation, probably due to virus-related acute microvascular thrombosis. In these cases, selective regional intraarterial thrombolysis was performed with the administration of a bolus of recombinant tissue plasminogen activator (rtPA) (20mg in 20min) through a catheter in the lower popliteal artery or selectively in a tibial vessel (Bellosta et al., 2020). Anticoagulation therapy with UFH infusion was initiated immediately in the operating room, and continued for 72/96h, with a target aPTT range of 2.5–3.0s. Effective anticoagulation was difficult to achieve despite increasing the UFH dose. Heparin resistance in COVID-19 patients has been associated with high levels of fibrinogen, factor VIII, and von Willebrand factor (vWF) (Connors & Levy, 2020). In this case, blood tests were taken to measure antithrombin (AT) levels; if AT levels were less than 80% of the baseline level of ATIII, 500 UI of ATIII were infused to reach the correct therapeutic range (Bellosta et al., 2020).

The use of prostanoids after revascularization in COVID-19 patients has been described (Manhanelli et al., 2020) with good results in terms of pain relief, without side effects.

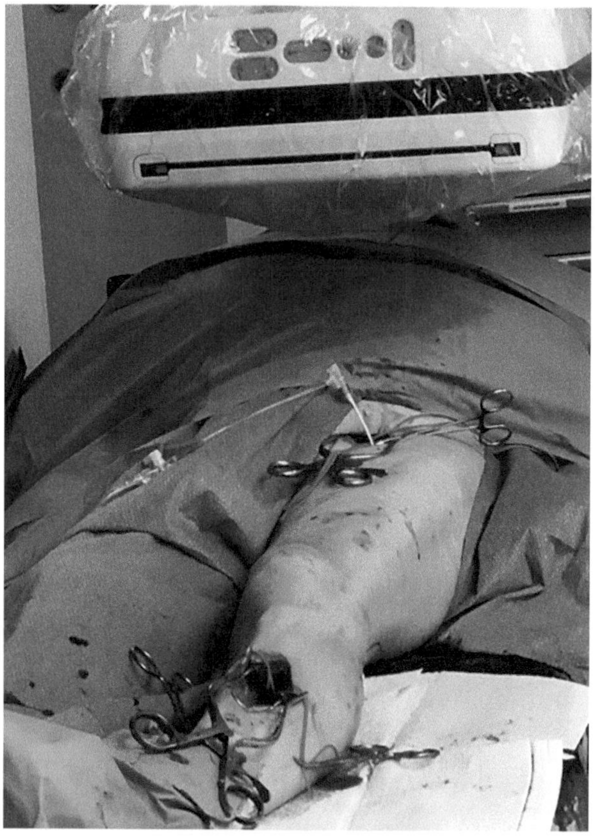

FIG. 3 Simultaneous femoral and tibial surgical accesses for extensive femoro-popliteal-tibial thrombosis.

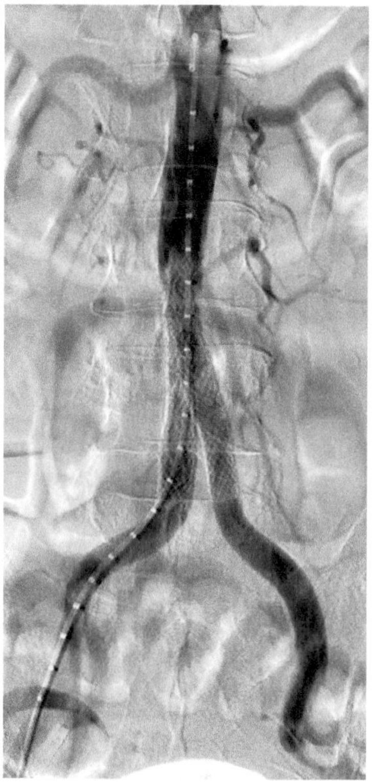

FIG. 4 Final completion angiography after aortoiliac endovascular recanalization.

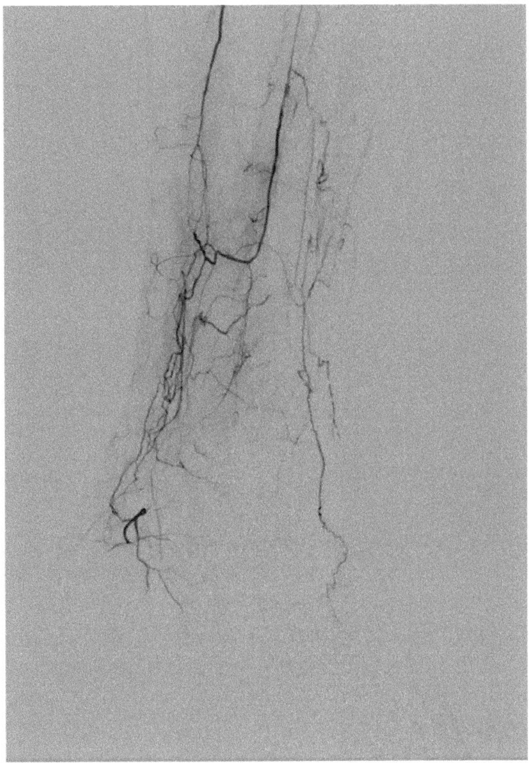

FIG. 5 DSA showing the typical "desert foot" appearance.

Endovascular treatment

Endovascular revascularization of the affected limb colud be achieved using a thrombolytic agent alone or in conjunction with a mechanical device for aspiration, fragmentation of the thrombus, angioplasty, or stenting (Gonzalez-Urquijo et al., 2021; Valle & Waldo, 2017). In COVID-19 patients, the literature on ALI reports a more frequent use of multiple techniques, probably due to the extent and severity of the disease.

Percutaneous thrombus aspiration (PAT) served as a cheap and fast therapeutic option, using large lumen end-hole catheters (6–8 French), especially in below-the-knee arteries. The catheter was delivered over a guidewire to the thrombosed arterial segment and connected to a negative pressure syringe to aspirate clots and thrombus fragments. The introducer sheath was equipped with a removable hemostatic valve to fully remove the thrombus and decrease the risk of thromboembolism.

Neuroendovascular catheters, more atraumatic and flexible, allow thrombosuction even from smaller arteries. However, their widespread use is limited due to decreased effectiveness in cases of organized thrombus, large vessels, and disseminated atherosclerotic lesions and to the risk of inadvertent vessel injury if multiple passages are required. In this cases, a long sheath with a removable hemostatic valve should be placed as close as possible to the occlusion (Gonzalez-Urquijo et al., 2021; Lukasiewicz, 2016).

Follow-up

There is limited evidence regarding postoperative management after ALI treatment in COVID-19 patients, because only small cohort studies are available. Several case series suggest that perioperative and postoperative systemic heparin might improve both primary patency and overall survival (Bellosta et al., 2020; Mascia et al., 2020). However, there is no consensus regarding treatment after the acute phase. It seems reasonable, according to the latest guidelines for ALI, to maintain long-term anticoagulation therapy after thrombectomy or endovascular treatment (Björck et al., 2020), either with oral anticoagulation or low molecular weight heparin. Patient outcomes are largely impacted by concomitant COVID-19-related systemic disorders. Several case series report high perioperative mortality, ranging from 14% to 50% (Behrendt et al., 2021; Etkin et al., 2021), mainly due to cardiac arrest, pneumonia, and multiple-organ failure. Data on long-term outcomes are still sparse, as only short-term and mid-term outcomes have been reported so far. Although limb salvage is initially successful, the patency rate is poor after either surgical or endovascular revascularization. Faries et al.

(Faries et al., 2022) reported a 6-month limb salvage rate of 89.2% [±7.2], similar to non-COVID-19 patients, and 6-month patency rates for open surgical and endovascular interventions of 66.7% [±13.6] and 55.6% [±24.9], respectively. The discrepancy between mid-term patency and amputation rates might be driven by persistent endothelial dysfunction and the subsequent prothrombotic state that increases the risk of recurrent thrombosis (Mosleh et al., 2020). The high readmission rate reported mostly reflects postoperative complications, mainly wound infections, hemorrhagic complications, and cardiovascular events.

Given the high rate of recurrent thrombosis, close follow-up with physical examination and DUS evaluation is crucial. Despite the absence of solid evidence, a follow-up at 1, 3, 6, and 12 months should be performed for the early identification of complications and prompt reintervention if needed.

Mini dictionary of terms

- Acute limb ischemia—Sudden decrease in arterial perfusion of an extremity associated with a threat to its survival with symptoms fewer than 14 days in duration.
- Hypercoagulable state—Also known as thrombophilia, an inherited or acquired disorder, characterized by an increased tendency to develop blood clots (thrombosis).
- Virchow's triad—Triad that consists of three factors (stasis, endothelial injury, and hypercoagulable state) used to describe the etiology and assess the risk of thrombosis, especially of deep veins.
- "Desert Foot"—Advanced pedal artery occlusive disease with occlusion of all the main foot arteries.
- Thromboembolectomy—Surgical removal of a thrombus or an embolus that caused arterial occlusion.
- Percutaneous thrombus aspiration—Endovascular procedure that requires large lumen end-hole catheters connected to a vacuum system to extract the thrombus.
- Thrombolysis—Either mechanical or chemical, a procedure used to break up clots that restrict blood flow.
- Heparin-induced thrombocytopenia—Immune complication that could occur in patients exposed to any form and amount of heparin, where antibodies against heparin and platelet factor 4 complex lead to a decrease in platelet count (thrombocytopenia).
- Phlegmasia cerulea dolens—Acute and extensive DVT characterized by marked swelling that may lead to arterial ischemia and gangrene.

Summary points

- During the COVID-19 pandemic in Brescia, Italy, the incidence of ALI was five times higher compared to the same period of the previous year.
- The systemic inflammation associated with COVID-19 led to a hypercoagulable and hyperviscosity state, responsible for arterial and venous thrombosis.
- Arterial thrombosis mainly involved the femoro-popliteal-tibial district, but a more severe and extended thrombosis was also typical.
- Clinical manifestation depended on the time course of vessel occlusion and the location of affected vessels. If not promptly treated, the patient developed advanced limb ischemia with a high risk of major amputation.
- Diagnosis was mostly clinical. DUS was usually the first and most readily available diagnostic tool to confirm the arterial occlusion and to plan the intervention.
- CTA was the preferred imaging modality in case of aortoiliac involvement to determine the anatomical location and extension of the occlusion, and to seek a potential embolic source.
- Surgical treatment included the traditional balloon catheter thrombectomy along with endovascular techniques, such as angioplasty and percutaneous thrombus aspiration. Procedures were performed in the hybrid suite.
- Intraarterial catheter-directed thrombolysis was used in case of microvascular thrombosis, with good results.
- Postoperative management was mainly represented by intravenous infusion of unfractioned heparin for 72/96 h, with an aPTT therapeutic range of 2.5–3.0 s. Effective anticoagulation was difficult to achieve due to the high rate of heparin resistance.
- Given the hypercoagulable state responsible for the high rate of recurrent thrombosis, the patency rate was poor.
- Long-term anticoagulation therapy improved both primary patency and overall survival.

Policies and procedures

During the COVID-19 pandemic, in-hospital transmission was directly related to the degree of contact between patients and healthcare providers. To prevent the spread of the virus and protect healthcare providers, several precautions were implemented:

1. In case of suspected ALI, a bedside ultrasound evaluation was preferred to limit unnecessary patients mobilization.
2. After diagnosis, surgeons assessed the urgency of revascularization. If the procedure could be delayed, patients were hospitalized and treated only after two negative molecular swabs.
3. In cases requiring treatment, prompt communication allowed the anesthesia care team, the scrub nurse, and the radiology technologist to prepare the operating room:
   a. All unnecessary equipment was moved out and the remaining pieces were wrapped to protect them from droplets.
   b. Entrance and exit areas were provided with alcohol-based hand sanitizers, containers for goggles, sodium hypochlorite solution, surgical masks, and gloves. Disposable drapes soaked with sodium hypochlorite were placed in front of each door.
4. Personal protective equipment recommended for procedures on known or suspected COVID-19 patients included a double surgical cap, N95/FFP2 mask, eye protection (face shield or goggles), full length long-sleeved gown, shoe coverage, and double gloves. Team members dressed in designated areas where posters displayed detailed instructions on appropriate dressing.
5. Patients were directly taken into the operating room, avoiding stays in common areas.
6. At the end of the procedure, a quick transfer through a designated dirty path was arranged to move the patients without any risk of contact.
7. Meanwhile, all operators moved into the filter area. First, they removed gloves and gowns within the room. On the drape soaked with sodium hypochlorite, they removed all the other PPE, while doing hand hygiene with alcohol gel. After complete doffing, operators proceeded to change into new scrub suits.
8. Two nurses, before doffing, dumped everything in biohazard plastic bags and removed plastic sheets by rolling them down to prevent the spread of droplets. All equipment and furniture used were left in the room.
9. A dedicated staff member wiped down everything with disinfectant and sodium hypochlorite before treating the room with hydrogen peroxide vapor. Due to its toxicity, the room had to remain sealed and unused for the next 3 h. If there was another procedure on a COVID-19-positive patient, the same room was used before undergoing treatment with hydrogen peroxide vapor.

Applications to other areas

In this chapter, we highlighted the importance of a prompt diagnosis of ALI, especially in COVID-19 patients, who were affected by more severe and extended arterial thrombosis. If ALI was suspected by any physician, heparin was immediately administered, and the patient, if transferable, was promtly referred to a "hub" hospital.

Lombardy faced the highest and most unexpected increase of COVID-19 patients. To address this massive and unanticipated surge of critically ill patients and ensure assistance continuity for urgencies and emergencies, national and regional healthcare systems had to rapidly reorganize into a "hub" and "spoke" hospital network. Hub hospitals were considered referral centers to cope with vascular emergencies and urgencies. Spoke hospitals served as satellites, dedicated to supporting COVID-19 patient in terms of intensive care and/or internal medicine, with vascular activity reduced mainly to transferable patients. A weekly briefing was held to share ongoing strategies, and a new surgical waiting list, graded according to the National Health Care System, was shared between the hub and its spokes. Only class A (mandatory hospitalization within 30 days) deferrable cases were included on the waiting list, and all patients referred to hub hospitals were tested for SARS-CoV-2 with a nasopharyngeal swab and a chest x-ray. A mixed team of spoke and hub surgeons was established, alongside a backup shift in case of a sick surgeon. Records of all admitted patients were directly collected by the involved physicians.

During the pandemic, the hub and spoke hospital network proved successful, ensuring prompt evaluation and treatment for all patients affected by vascular urgencies or emergencies.

References

Akkari, R., & Schwartz, B. (2020). Phlegmasia cerulea dolens: An atypical COVID-19 presentation. *Chest, 158*(4), A2090. https://doi.org/10.1016/j.chest.2020.08.1805.

Alghamdi, L., Alattab, N., Alwohaibi, A., Alotaibi, Y. H., & AlSheef, M. (2022). Phlegmasia Cerulea Dolens secondary to COVID-19 and May-Thurner syndrome: A case report. *Cureus, 14*(1), e21301. https://doi.org/10.7759/cureus.21301.

Attisani, L., Pucci, A., Luoni, G., et al. (2021). COVID-19 and acute limb ischemia: A systematic review. *The Journal of Cardiovascular Surgery, 62*(6), 542–547. https://doi.org/10.23736/S0021-9509.21.12017-8.

Baram, A., Kakamad, F. H., Abdullah, H. M., et al. (2020). Large vessel thrombosis in patient with COVID-19, a case series. *Annals of Medicine and Surgery, 60*, 526–530. https://doi.org/10.1016/j.amsu.2020.11.030.

Behrendt, C.-A., Seiffert, M., Gerloff, C., L'Hoest, H., Acar, L., & Thomalla, G. (2021). How does SARS-CoV-2 infection affect survival of emergency cardiovascular patients? A cohort study from a German insurance claims database. *European Journal of Vascular and Endovascular Surgery: The Official Journal of the European Society for Vascular Surgery, 62*(1), 119–125. https://doi.org/10.1016/j.ejvs.2021.03.006.

Bellosta, R., Luzzani, L., Natalini, G., et al. (2020). Acute limb ischemia in patients with COVID-19 pneumonia. *Journal of Vascular Surgery, 72*(6), 1864–1872. https://doi.org/10.1016/j.jvs.2020.04.483.

Bilaloglu, S., Aphinyanaphongs, Y., Jones, S., Iturrate, E., Hochman, J., & Berger, J. S. (2020). Thrombosis in hospitalized patients with COVID-19 in a New York city health system. *Journal of the American Medical Association, 324*(8), 799–801. https://doi.org/10.1001/jama.2020.13372.

Björck, M., Earnshaw, J. J., Acosta, S., et al. (2020). Editor's choice—European Society for Vascular Surgery (ESVS) 2020 clinical practice guidelines on the management of acute Limb Ischaemia. *European Journal of Vascular and Endovascular Surgery: The Official Journal of the European Society for Vascular Surgery, 59*(2), 173–218. https://doi.org/10.1016/j.ejvs.2019.09.006.

Brockman, S. K., & Vasko, J. S. (1965). Phlegmasia cerulea dolens. *Surgery, Gynecology & Obstetrics, 121*(6), 1347–1356.

Cheruiyot, I., Kipkorir, V., Ngure, B., Misiani, M., Munguti, J., & Ogeng'o, J. (2021). Arterial thrombosis in coronavirus disease 2019 patients: A rapid systematic review. *Annals of Vascular Surgery, 70*, 273–281. https://doi.org/10.1016/j.avsg.2020.08.087.

Connors, J. M., & Levy, J. H. (2020). COVID-19 and its implications for thrombosis and anticoagulation. *Blood, 135*(23), 2033–2040. https://doi.org/10.1182/blood.2020006000.

De Hous, N., Hollering, P., Van Looveren, R., Tran, T., De Roover, D., & Vercauteren, S. (2021). Symptomatic arterial thrombosis associated with novel coronavirus disease 2019 (COVID-19): Report of two cases. *Acta Chirurgica Belgica*, 1–4. https://doi.org/10.1080/00015458.2021.1911751. Published online June.

Elbadry, M. I., Tawfeek, A., Abdellatif, M. G., et al. (2022). Unusual pattern of thrombotic events in young adult non-critically ill patients with COVID-19 may result from an undiagnosed inherited and acquired form of thrombophilia. *British Journal of Haematology, 196*(4), 902–922. https://doi.org/10.1111/bjh.17986.

Etkin, Y., Conway, A. M., Silpe, J., et al. (2021). Acute arterial thromboembolism in patients with COVID-19 in the New York city area. *Annals of Vascular Surgery, 70*, 290–294. https://doi.org/10.1016/j.avsg.2020.08.085.

Faries, C. M., Rao, A., Ilonzo, N., et al. (2022). Follow-up after acute thrombotic events following COVID-19 infection. *Journal of Vascular Surgery, 75*(2), 408–415.e1. https://doi.org/10.1016/j.jvs.2021.08.092.

Fogarty, T. J., Cranley, J. J., Krause, R. J., Strasser, E. S., & Hafner, C. D. (1963). A method for extraction of arterial emboli and thrombi. *Surgery, Gynecology & Obstetrics, 116*, 241–244.

Fournier, M., Faille, D., Dossier, A., et al. (2021). Arterial thrombotic events in adult inpatients with COVID-19. *Mayo Clinic Proceedings, 96*(2), 295–303. https://doi.org/10.1016/j.mayocp.2020.11.018.

Goldman, I. A., Ye, K., & Scheinfeld, M. H. (2020). Lower-extremity arterial thrombosis associated with COVID-19 is characterized by greater thrombus burden and increased rate of amputation and death. *Radiology, 297*(2), E263–E269. https://doi.org/10.1148/radiol.2020202348.

Gonzalez-Urquijo, M., Gonzalez-Rayas, J. M., Castro-Varela, A., et al. (2021). Unexpected arterial thrombosis and acute limb ischemia in COVID-19 patients. Results from the Ibero-Latin American acute arterial thrombosis registry in COVID-19: (ARTICO-19). *Vascular*. https://doi.org/10.1177/17085381211052033. Published online December. 17085381211052032.

Haimovici, H. (1965). The ischemic forms of venous thrombosis. 1. Phlegmasia cerulea dolens. 2. Venous gangrene. *The Journal of Cardiovascular Surgery, 5*(6 (Suppl.)), 164–173.

Ilonzo, N., Koleilat, I., Prakash, V., et al. (2021). The effect of COVID-19 on training and case volume of vascular surgery trainees. *Vascular and Endovascular Surgery, 55*(5), 429–433. https://doi.org/10.1177/1538574420985775.

Ilonzo, N., Rao, A., Safir, S., et al. (2021). Acute thrombotic manifestations of coronavirus disease 2019 infection: Experience at a large New York city health care system. *Journal of Vascular Surgery, 73*(3), 789–796. https://doi.org/10.1016/j.jvs.2020.08.038.

Indes, J. E., Koleilat, I., Hatch, A. N., et al. (2021). Early experience with arterial thromboembolic complications in patients with COVID-19. *Journal of Vascular Surgery, 73*(2), 381–389.e1. https://doi.org/10.1016/j.jvs.2020.07.089.

Jongkind, V., Earnshaw, J. J., Bastos Gonçalves, F., et al. (2022). Editor's choice—Update of the European Society for Vascular Surgery (ESVS) 2020 clinical practice guidelines on the management of Acute Limb Ischaemia in light of the COVID-19 pandemic, based on a scoping review of the literature. *European Journal of Vascular and Endovascular Surgery: The Official Journal of the European Society for Vascular Surgery, 63*(1), 80–89. https://doi.org/10.1016/j.ejvs.2021.08.028.

Kahlberg, A., Mascia, D., Bellosta, R., et al. (2021). Vascular surgery during COVID-19 emergency in hub hospitals of Lombardy: Experience on 305 patients. *European Journal of Vascular and Endovascular Surgery: The Official Journal of the European Society for Vascular Surgery, 61*(2), 306–315. https://doi.org/10.1016/j.ejvs.2020.10.025.

Karnabatidis, D., Spiliopoulos, S., Tsetis, D., & Siablis, D. (2011). Quality improvement guidelines for percutaneous catheter-directed intra-arterial thrombolysis and mechanical thrombectomy for acute lower-limb ischemia. *Cardiovascular and Interventional Radiology*, *34*(6), 1123–1136. https://doi.org/10.1007/s00270-011-0258-z.

Khryshchanovich, V. Y., Rogovoy, N. A., & Nelipovich, E. V. (2021). Arterial thrombosis and acute limb ischemia as a complication of COVID-19 infection. *The American Surgeon*. https://doi.org/10.1177/00031348211023416. Published online May. 31348211023416.

Li, C., Hu, B., Zhang, Z., et al. (2020). D-dimer triage for COVID-19. *Academic Emergency Medicine: Official Journal of the Society for Academic Emergency Medicine*, *27*(7), 612–613. https://doi.org/10.1111/acem.14037.

Lukasiewicz, A. (2016). Treatment of acute lower limb ischaemia. *VASA*, *45*(3), 213–221. https://doi.org/10.1024/0301-1526/a000527.

Manhanelli, M. A. B., Duarte, E. G., Mariuba, J. V.d. O., et al. (2020). Alprostadil associated with low molecular weight heparin to treat limb ischemia caused by SARS-CoV2. *Jornal Vascular Brasileiro*, *19*, e20200072. https://doi.org/10.1590/1677-5449.200072.

Mascia, D., Kahlberg, A., Melloni, A., Rinaldi, E., Melissano, G., & Chiesa, R. (2020). Single-center vascular hub experience after 7 weeks of COVID-19 pandemic in Lombardy (Italy). *Annals of Vascular Surgery*, *69*, 90–99. https://doi.org/10.1016/j.avsg.2020.07.022.

Mietto, C., Salice, V., Ferraris, M., et al. (2020). Acute lower limb ischemia as clinical presentation of COVID-19 infection. *Annals of Vascular Surgery*, *69*, 80–84. https://doi.org/10.1016/j.avsg.2020.08.004.

Moraes, B., Hashemi, A., Mancheno, K., ObanDo, M., & Marra, E. (2021). Hypercoagulability due to COVID-19 leading to impending phlegmasia cerulea dolens and sub-massive bilateral pulmonary embolism. *Cureus*, *13*(8), e17351. https://doi.org/10.7759/cureus.17351.

Mosleh, W., Chen, K., Pfau, S. E., & Vashist, A. (2020). Endotheliitis and endothelial dysfunction in patients with COVID-19: Its role in thrombosis and adverse outcomes. *Journal of Clinical Medicine*, *9*(6). https://doi.org/10.3390/jcm9061862.

Mouhat, B., Besutti, M., Bouiller, K., et al. (2020). Elevated D-dimers and lack of anticoagulation predict PE in severe COVID-19 patients. *The European Respiratory Journal*, *56*(4). https://doi.org/10.1183/13993003.01811-2020.

Orso, D., Mattuzzi, L., Scapol, S., Delrio, S., Vetrugno, L., & Bove, T. (2021). Phlegmasia cerulea dolens superimposed on disseminated intravascular coagulation in COVID-19. *Acta Bio-Medica*, *92*(4), e2021101. https://doi.org/10.23750/abm.v92i4.11478.

Pratt, G. H., & Krahl, E. (1954). Surgical therapy for the occluded artery. *American Journal of Surgery*, *87*(5), 722–729. https://doi.org/10.1016/0002-9610(54)90171-3.

Rutherford, R. B., Baker, J. D., Ernst, C., et al. (1997). Recommended standards for reports dealing with lower extremity ischemia: Revised version. *Journal of Vascular Surgery*, *26*(3), 517–538. https://doi.org/10.1016/s0741-5214(97)70045-4.

Tan, C. W., Tan, J. Y., Wong, W. H., et al. (2021). Clinical and laboratory features of hypercoagulability in COVID-19 and other respiratory viral infections amongst predominantly younger adults with few comorbidities. *Scientific Reports*, *11*(1), 1793. https://doi.org/10.1038/s41598-021-81166-y.

Topcu, A. C., Ozturk-Altunyurt, G., Akman, D., Batirel, A., & Demirhan, R. (2021). Acute limb ischemia in hospitalized COVID-19 patients. *Annals of Vascular Surgery*, *74*, 88–94. https://doi.org/10.1016/j.avsg.2021.03.003.

Valle, J. A., & Waldo, S. W. (2017). Current endovascular management of acute Limb Ischemia. *Interventional Cardiology Clinics*, *6*(2), 189–196. https://doi.org/10.1016/j.iccl.2016.12.003.

Wichmann, D., Sperhake, J.-P., Lütgehetmann, M., et al. (2020). Autopsy findings and venous thromboembolism in patients with COVID-19: A prospective cohort study. *Annals of Internal Medicine*, *173*(4), 268–277. https://doi.org/10.7326/M20-2003.

Zhan, H., Chen, H., Liu, C., et al. (2021). Diagnostic value of D-dimer in COVID-19: A meta-analysis and meta-regression. *Clinical and Applied Thrombosis/Hemostasis: Official Journal of the International Academy of Clinical and Applied Thrombosis/Hemostasis*, *27*, 10760296211010976. https://doi.org/10.1177/10760296211010976.

Chapter 30

Thrombosis and coagulopathy in COVID-19: A current narrative

Alejandro Lazo-Langner[a] and Mateo Porres-Aguilar[b]

[a]*Division of Hematology, Department of Medicine and Department of Epidemiology and Biostatistics, Western University, London, ON, Canada,*
[b]*Department of Internal Medicine, Divisions of Hospital and Adult Clinical Thrombosis Medicine, Texas Tech University Health Sciences Center and Paul L. Foster School of Medicine, El Paso, TX, United States*

Abbreviations

ACE2	angiotensin-converting enzyme 2
BAL	Bronchoalveolar lavage
CAC	COVID-19-associated coagulopathy
CI	confidence interval
COVID-19	coronavirus disease 2019
DIC	disseminated intravascular coagulation
DVT	deep vein thrombosis
ECMO	extracorporeal membrane oxygenation
ICU	intensive care unit
IL	interleukin
IMPROVE	International Medical Prevention Registry on Venous Thromboembolism
ISTH	International Society on Thrombosis and Hemostasis
JAK	Janus kinase
LMWH	low molecular weight heparin
MAPK	mitogen-activated protein kinase
NETs	neutrophil extracellular traps
OR	odds ratio
PBMC	peripheral blood mononuclear cells
PE	pulmonary embolism
RR	relative risk
SARS-CoV-2	severe acute respiratory syndrome coronavirus 2
STAT	signal transducer and activator of transcription protein
TFPA	tissue factor pathway inhibitor
UFH	unfractionated heparin
VTE	venous thromboembolism

Introduction

Since the early days of the coronavirus 2019 (COVID-19) pandemic, it was rapidly recognized that many patients with the disease frequently developed a derangement of the coagulation system and had an increased occurrence of thrombotic events. It was also noticed that the risk of these complications was higher in patients with a severe presentation necessitating hospital admission, in particular those admitted to intensive care units (ICUs) (Al-Samkari et al., 2020; Klok et al., 2020a, 2020b; Mansory et al., 2022). Particularly noteworthy was the fact that, although these patients developed both deep vein thrombosis (DVT) and pulmonary emboli (PE), the latter was much more frequent than what is usually observed in hospitalized patients (Mansory et al., 2021).

The previous observations led to a rapid explosion of studies assessing the epidemiology and pathophysiology of these abnormalities. A quick search of the PubMed database revealed more than 10,000 publications in this area since the onset of

Management, Body Systems, and Case Studies in COVID-19. https://doi.org/10.1016/B978-0-443-18703-2.00006-2
Copyright © 2024 Elsevier Inc. All rights are reserved, including those for text and data mining, AI training, and similar technologies.

the pandemic, with that number rapidly increasing. In this chapter, we review and summarize the current knowledge of the pathophysiology of hemostatic alterations and the epidemiology and management of thromboembolic disease.

Pathophysiology of coagulopathy in COVID-19

Initial autopsy reports noted the presence of diffuse alveolar damage and microvascular thrombi in the lungs (Fox et al., 2020; Wichmann et al., 2020) and these findings were confirmed in subsequent studies (Bryce et al., 2021; Menter et al., 2020). Concurrently, several studies reported that elevated D-dimer levels were associated with disease severity, pointing to a marked activation of the coagulation system (Guan et al., 2020; Liang et al., 2020). Although some subsequent studies reported contradictory results (Liang et al., 2020), two recently published systematic reviews concluded that D-dimer levels are higher in patients with severe disease (Len et al., 2022; Zhang et al., 2022). Other studies reported the presence of a mildly prolonged prothrombin time and modestly decreased platelet counts in COVID-19 patients, again particularly in those with severe disease (Huang et al., 2020; Wang et al., 2020; Zhou et al., 2020). Taken together, these observations strongly pointed to the occurrence of a marked derangement in the coagulation system in patients with acute COVID-19 infections and especially in severe cases.

Despite these alterations in the hemostatic system, it was also noticed that these patients do not tend to show a marked bleeding tendency, at least initially. Initial studies reported the development of disseminated intravascular coagulation (DIC) in COVID-19 patients using the diagnostic criteria of the International Society on Thrombosis and Hemostasis (ISTH) (Tang et al., 2020; Taylor Jr. et al., 2001), but subsequent reports were contradictory. Other studies have suggested that the patterns of prothrombotic coagulopathy in COVID-19 and DIC are different from that in sepsis, where the platelet count is usually decreased, and the prothrombin time is prolonged with an associated hemorrhagic tendency (Ranucci et al., 2020). These key differences, which include a modest drop in the platelet count, prolongation of the prothrombin time, and higher fibrinogen levels, point to the fact that this condition represents a separate entity that has been denominated as COVID-19-associated coagulopathy (CAC). This is supported by the observation that, in contrast to patients with consumptive coagulopathies such as DIC, in critically ill COVID-19 patients the fibrinogen levels are markedly elevated and correlate with higher mortality, probably from an acquired state of resistance to fibrinolysis contributing to significant thrombosis in situ (Asakura & Ogawa, 2021; Weiss et al., 2020).

Other observations have shown that in COVID-19 patients, there is evidence of extensive tissue damage and endothelial injury that is mediated by an exaggerated T-cell response with marked release of cytokines, including interleukins IL-6, IL-2R, and IL-10 as well as tumor necrosis factor alpha (Cao et al., 2020; Chen et al., 2020; Xu & Gao, 2004). There is also strong evidence of the presence of a marked inflammatory state in these patients with strikingly elevated ferritin and c-reactive protein levels, which are related to clinical prognosis (Burke et al., 2022).

These observations have led to considering CAC as the result of an exaggerated immunothrombosis response. Immunothrombosis is a phenomenon arising as a result of the immune response to a pathogen that activates the coagulation system, in particular in small vessels, to contain and destroy such pathogens (Engelmann & Massberg, 2013). It consists of a complex interplay among many components, including neutrophils, monocytes, platelets, and endothelial cells. This process involves the release and exposure of tissue factor by monocytes and monocyte-derived microparticles as a response to certain patterns of molecular or pathogen-induced damage, resulting in the activation of the initiation phase of the coagulation system and the subsequent formation of thrombin and fibrin. Additionally, the exposure of the tissue factor leads to the activation of platelets, subsequently leading to the recruitment and activation of neutrophils and the release of highly thrombogenic neutrophil extracellular traps (NETs). The latter have several prothrombotic effects: a) they increase platelet recruitment by binding to the von Willebrand factor, b) they lead to increased thrombin generation by binding to the tissue factor, c) the presence of histones within NETs triggers the activation of platelets, d) several neutrophil-derived proteases lead to the inactivation of natural anticoagulant pathways, and e) they can activate the contact pathway by directly activating factor XII (Jayarangaiah et al., 2020).

Studies have shown that SARS-CoV-2 has multiple effects on these mechanisms, leading to endothelial damage and subsequent thrombosis. It has been demonstrated that NET levels are significantly elevated in COVID-19 patients, and that SARS-CoV-2 can induce the release of NETs in healthy neutrophils. It also has been shown that in COVID-19 patients, the infection induces an increased number of a subset of granulocytes known as low-density granulocytes, which are more prone to NET formation (Veras et al., 2020). These studies have also found that in COVID-19, NETs are associated with damage of the lung epithelium and endothelium by inducing the apoptosis of such cells.

A particularly interesting fact is that SARS-CoV-2 uses the receptor for angiotensin-converting enzyme 2 (ACE2) as its host cell surface receptor, which might induce endothelial cell dysfunction by additional mechanisms (Conway et al., 2022). A recent study analyzed publicly available transcriptomics datasets of SARS-CoV-2-infected normal human bronchial epithelial cells, COVID-19 patient bronchoalveolar lavage (BAL) fluid, and COVID-19 peripheral blood mononuclear cells (PBMC) (FitzGerald & Jamieson, 2020). The study found that infection with SARS-CoV-2 results in the activation of tissue factor signaling, but not on its inhibitor tissue factor pathway inhibitor (TFPA). It also results in the downregulation of protein S in the lung and an upregulation of the mechanisms of plasminogen inactivation. These changes were found in bronchial epithelial cells and BAL fluid but not on PBMC, suggesting that, in addition to the inflammatory changes, SARS-CoV-2 infection results in a prothrombotic state through an enhanced procoagulant response in the pulmonary milieu.

In addition to the effect of SARS-CoV-2 on NETs and coagulation proteins, studies have suggested that the virus can trigger complement activation by the classical, alternative, and lectin pathways (Conway et al., 2022). It can also induce the synthesis of complement factors from infected epithelial and endothelial cells by activation of the JAK-STAT signaling pathway (Yan et al., 2020).

Finally, it has also been found that COVID-19 patients have an augmented platelet reactivity, probably because of changes to the platelet transcriptome induced by the virus, mediated at least in part by the activation of the MAPK pathway as well as thromboxane generation (Manne et al., 2020).

In summary, SARS-CoV-2 infection induces a complex inflammatory and thrombotic response mediated by an intricate interplay of different pathways (Fig. 1), resulting in a vicious circle that amplifies the initial insult and propagates and perpetuates the imbalance between the procoagulant and anticoagulant mechanisms. Considering the advancing knowledge on the pathophysiology of immunothrombosis in these patients, it is now believed that many of them develop a pulmonary thrombosis in situ rather than a true embolism. This could potentially explain, at least in part, the clinical observations about a higher than expected frequency of PE in these patients compared to other critical patients without COVID-19. Recent advances in the understanding of these mechanisms have opened potential avenues for targeted interventions in this disease.

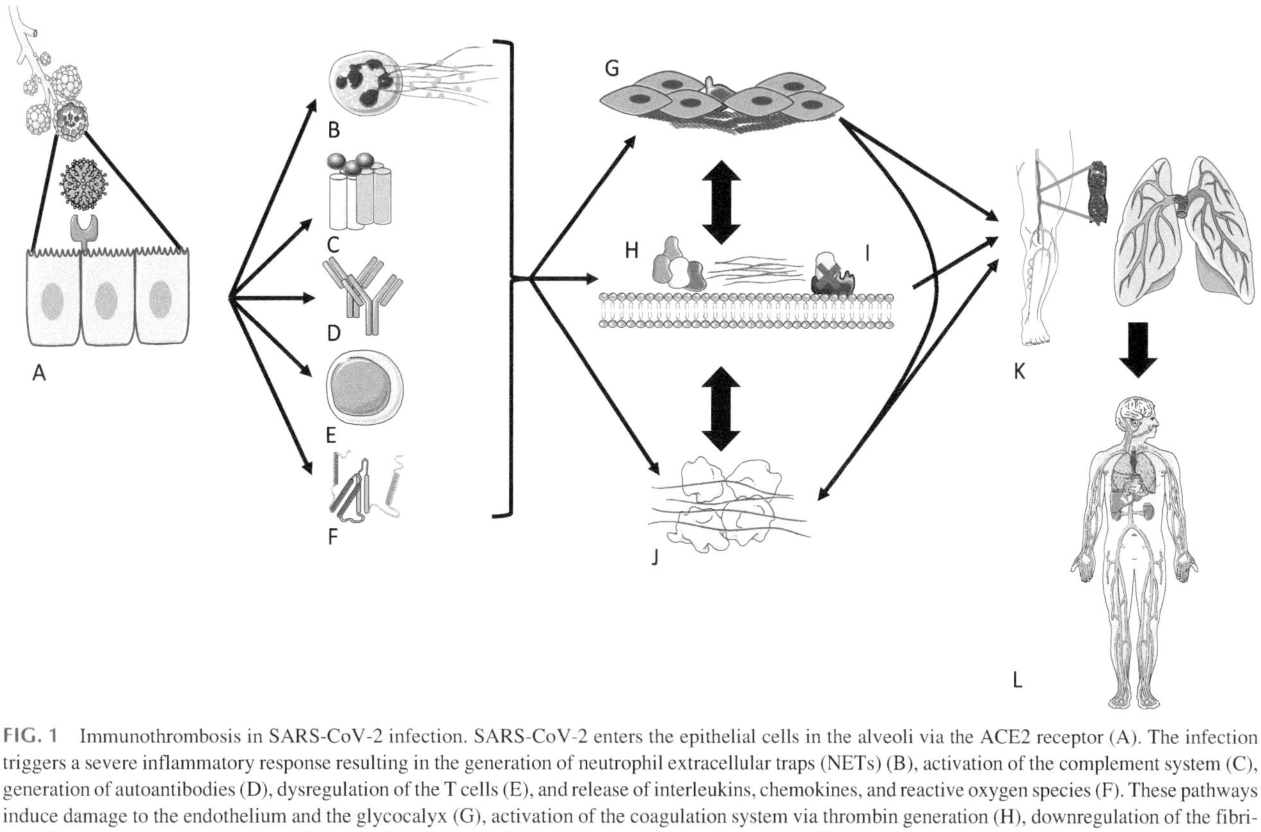

FIG. 1 Immunothrombosis in SARS-CoV-2 infection. SARS-CoV-2 enters the epithelial cells in the alveoli via the ACE2 receptor (A). The infection triggers a severe inflammatory response resulting in the generation of neutrophil extracellular traps (NETs) (B), activation of the complement system (C), generation of autoantibodies (D), dysregulation of the T cells (E), and release of interleukins, chemokines, and reactive oxygen species (F). These pathways induce damage to the endothelium and the glycocalyx (G), activation of the coagulation system via thrombin generation (H), downregulation of the fibrinolytic system (I), and activation of platelets (J). These mechanisms lead to the development of COVID-19-associated coagulopathy with a significant increase in fibrin, microthrombi, and further endothelial damage. The interaction of these pathways leads eventually to the development of thrombosis (K) and multiorgan dysfunction (L). *Adapted from Conway, E. M., Mackman, N., Warren, R. Q., Wolberg, A. S., Mosnier, L. O., Campbell, R. A., . . . Morrissey, J. H. (2022). Understanding COVID-19-associated coagulopathy. Nature Reviews. Immunology, 22(10), 639–649. https://doi.org/10.1038/s41577-022-00762-9. Parts of the figure were drawn by using pictures with modifications from Servier Medical Art. Servier Medical Art by Servier is licensed under a Creative Commons Attribution 3.0 Unported License (https://creativecommons.org/licenses/by/3.0/).*

Venous thromboembolism in COVID-19

Early studies suggested a high frequency of venous thromboembolism (VTE) in COVID-19 patients. Case series of fatal cases reported VTE in 58% of the patients and PE as the cause of death in 33% (Wichmann et al., 2020). A small Chinese retrospective study found that DVT was diagnosed in around 20% of critically ill patients despite the use of thromboprophylaxis (Xu et al., 2020). Subsequent studies from France, Italy, the United Kingdom, China, and the Netherlands reported similar findings (Cui et al., 2020; Helms et al., 2020; Klok et al., 2020a, 2020b; Llitjos et al., 2020; Lodigiani et al., 2020; Middeldorp, Coppens, van Haaps, Foppen, Vlaar, Müller, et al., 2020a; Middeldorp, Coppens, van Haaps, Foppen, Vlaar, Muller, & van Es, 2020b; Poissy et al., 2020; Thomas et al., 2020). Most importantly, all initial studies suggested that, although the incidence of VTE in hospitalized COVID-19 patients was high, the frequency was particularly elevated in critically ill patients, compared to that observed in critically ill patients without COVID-19 (Helms et al., 2020). Additionally, all these studies reported an unusually high frequency of PE rather than DVT.

However, the interpretation of the results has been limited by different study designs, the use or not of screening strategies, and different outcome definitions. We recently reported a systematic review and meta-analysis of 91 studies including more than 35,000 hospitalized patients (Mansory et al., 2021). The results of the meta-analysis found that the percentage of patients with a VTE was 13% in all populations, but much higher in those patients admitted to an ICU compared to those who were not (24% vs. 8%). However, sensitivity analyses suggested that the frequency of VTE may be somewhat overestimated, as there was evidence of significant statistical heterogeneity associated with the study sample size. The larger studies (those in the upper quintile of sample size) included more than 27,000 patients and reported a lower percentage of patients with VTE in all hospitalized (5%), ICU (16%), and non-ICU patients (6%). In any case, even accounting for this, the frequency of VTE, in particular PE, in these patients is quite significant and related to both the pathophysiology of the disease and its prognosis.

Given the high incidence of VTE in hospitalized patients, other studies have sought to evaluate the risk in nonhospitalized patients. Our group recently reported a systematic review and meta-analysis including studies looking at venous events in adult patients with COVID-19 either posthospital discharge or in ambulatory patients with no history of hospitalization (Mansory et al., 2022). The review included 16 observational studies comprising 102,779 patients, of which 85,051 were outpatients and the rest had been recently discharged from a hospital. The pooled estimates for VTE prevalence were 0.28% and 1.16%, respectively. For studies with available data, the pooled estimates for the VTE incidence rates per 1000 patient-days of observation were 0.06 for outpatients and 0.12 for patients discharged from a. hospital. Finally, we found that, compared to outpatients who were never hospitalized, the pooled relative risk of VTE for discharged patients was 3.87, but this was not statistically significant (95%CI 0.38 to 39.18; $p = 0.252$).

Arterial thrombosis in COVID-19

Arterial thrombotic events have also been reported in COVID-19 patients with varying frequencies. Early studies reported the occurrence of stroke in approximately 3.5% of patients (Helms et al., 2020; Klok et al., 2020a, 2020b; Lodigiani et al., 2020; Mao et al., 2020; Thomas et al., 2020). A meta-analysis of 61 studies enrolling more than 108,000 patients found that the pooled incidence of stroke in patients with was 1.4% (Nannoni et al., 2021). This study also found that COVID-19 patients developing a stroke were older and had a higher frequency of cardiovascular risk factors such as hypertension, diabetes, and coronary artery disease, compared to those patients who did not develop a stroke. Importantly, when comparing patients with COVID-19 and stroke and patients with a stroke without the infection, the former were younger, less likely to have hypertension or a previous stroke, and there was no difference in other cardiovascular risk factors. Also, they had large vessel occlusions more frequently, the strokes were more severe, and in-hospital mortality was five times higher (Nannoni et al., 2021).

In addition, the frequency of other arterial events in hospitalized COVID-19 patients has been reported to be nonnegligible. The largest review published to date reported information on more than 100,000 patients included in 36 studies. Overall, the frequency of arterial events was 2% with frequencies for acute myocardial infarction, acute limb ischemia, and other arterial events being 0.8%, 0.2%, and 0.5%, respectively. These numbers are similar to what is observed in patients hospitalized for other infectious conditions(Candeloro & Schulman, 2022).

In the case of nonhospitalized patients, our systematic review found an overall proportion of arterial events of 0.75%. In outpatients who were never hospitalized, arterial events were reported in 0.23% and in those discharged from a hospital, the frequency was 1.46%.The estimated incidence rates per 1000 patient-days of observation were 0.1 and 0.26 for outpatients and postdischarge patients, respectively (Mansory et al., 2022).

Anticoagulant use in COVID-19 patients

The observation of the increased frequency of thrombotic events prompted a fast adoption of anticoagulant therapy in hospitalized COVID-19 patients, despite a lack of sound evidence of benefit. More than 50 studies have been published evaluating different doses of anticoagulants in these patients, including at least 13 randomized trials; these are summarized in Table 1. Initial studies recommended the use of anticoagulant therapy, mainly with unfractionated heparin (UFH) or low molecular weight heparins (LMWHs), with many of them suggesting that the use of these agents at intermediate or therapeutic doses was superior to prophylactic doses. As a result, many groups implemented the latter strategies in hospitalized patients, despite the highly conflicting low-quality evidence. It was not until 2021 when the results of the first large randomized trials were published. In general, the results support the generalized use of prophylactic anticoagulation in all hospitalized patients and are less clear regarding the use of higher doses of anticoagulants.

In general, studies have indicated that the benefit from therapeutic anticoagulation may differ according to the severity of the disease. In patients admitted to the ICU, all trials failed to show a benefit in mortality or a reduction of VTE while increasing the risk of bleeding. Two main trials have been published exploring either intermediate or therapeutic doses of low molecular weight heparin compared to prophylactic doses in COVID-19 patients admitted to the ICU. The multiplatform trial was a pragmatic randomized controlled trial comprised of three different platforms in five continents using harmonized protocols. The study randomized more than 1000 patients to either therapeutic dose anticoagulation or prophylactic anticoagulation according to local institutional practices. The main outcome was the number of days free of organ support. The study was terminated early due to futility as therapeutic doses did not confer any benefit regarding the primary outcome and were associated with a significant increase in the risk of major bleeding (3.8% vs. 2.3% in therapeutic and prophylactic groups, respectively) (Goligher et al., 2021).

The second study compared conventional thromboprophylaxis with intermediate doses of low molecular weight heparin in patients admitted to the ICU in 10 centers in Iran. The study included 562 patients and showed no difference in the primary outcome, which was a composite of VTE, arterial events, the need for extracorporeal membrane oxygenation (ECMO), or death with major bleeding occurring in 2.5% of patients in the intermediate-dose group compared to 1.4% in the prophylactic-dose group (Bikdeli et al., 2022; Sadeghipour et al., 2021).

Anticoagulant use in noncritical patients with COVID-19 has been assessed in several studies. In the multiplatform trial, as with the trial in critical patients, the primary composite outcome was the probability of the intervention resulting in less organ support-free days, combining in-hospital death and number of days free of cardiopulmonary support up to day 21 among patients who survived until hospital discharge. The study showed that the probability of a benefit of therapeutic anticoagulation by increasing the number of organ support-free days was 98.5%. However, the absolute difference in survival until hospital discharge was only 4% and the rates of major bleeding were 1.9% in the therapeutic arm, compared to 0.9% in the nontherapeutic group (Lawler et al., 2021).

The ACTION trial randomized patients to receive therapeutic doses of rivaroxaban (20 mg daily or 15 mg daily if renally adjusted) or conventional thromboprophylaxis with either LMWH or UFH. While there was no difference in the primary efficacy outcome (death, duration of hospitalization, and duration of supplemental oxygen), the therapeutic group had a higher bleeding risk compared to the conventional thromboprophylaxis group (8% vs. 2%).(Lopes et al., 2021).

The RAPID trial randomly assigned patients to heparin at therapeutic or prophylactic doses. The primary composite outcome included death, use of mechanical ventilation, and VTE. There were no differences in the primary outcome but there seemed to be a reduction in mortality at 28 days (Sholzberg et al., 2021).

Finally, the HEP-COVID trial randomized patients with elevated D-dimer or objective evidence of CAC to either conventional thromboprophylaxis/intermediate doses or therapeutic doses of heparins. The efficacy outcomes were VTE, arterial thrombosis, or death. (Spyropoulos et al., 2021) This study showed that therapeutic anticoagulation was associated with a 32% risk reduction in the primary outcome. However, the benefit was observed only in patients who were not admitted to the ICU.

In addition to the previous studies, more than 40 observational studies and at least seven additional randomized trials comparing low versus nonlow doses of anticoagulants have been published to date. A recent meta-analysis concluded that the use of intermediate compared to prophylactic doses of anticoagulants does not result in significant differences in VTE or major bleeding (Reis et al., 2022). However, when comparing therapeutic versus nontherapeutic anticoagulation, there might be a benefit from the former but only in moderately ill patients. The authors reported that in such patients, therapeutic anticoagulation resulted in a reduction in 28-day mortality (RR 0.39, 95% CI 0.16–0.96) but with a higher risk of bleeding (RR 1.78, 95% CI 1.15–2.74) and no difference in hospital mortality, thrombotic events, or a composite of VTE and mortality (Reis et al., 2022). Nevertheless, the reported benefit in mortality was based on only two studies that enrolled 635 patients, and further studies are needed to confirm this finding (Sholzberg et al., 2021; Spyropoulos et al., 2021).

TABLE 1 Randomized trials evaluating anticoagulation in hospitalized and ambulatory COVID-19.

Study/reference	Study design	Number of patients	Patient population	Intervention	Comparator	Outcomes
2020						
HESACOVID, Lemos-2020 (Lemos et al., 2020)	RCT open-label,	20	Hospitalized+ intensive care unit	Enoxaparin 1 mg/kg BID for at least 96 h and up to 14 days	Enoxaparin 40 mg OD, weight and CrCl adjusted	28-day mortality, hospital mortality, thrombotic events
2021						
ACTION, Lopes-2021 (Lopes et al., 2021)	RCT open-label	614	Hospitalized+ intensive care unit	Rivaroxaban 20 mg OD (280 patients, 90%) for 30 days	Enoxaparin 40 mg OD, weight and CrCl adjusted	30-day mortality, survival until discharge from hospital (30 days), thrombotic events, thrombotic events or deaths, major bleeding
ACTIV 4B, Connors-2021 (Connors et al., 2021)	RCT, double-blind	657	Outpatients	Apixaban 2.5 mg BID for 45 days	Placebo	45-day mortality, hospitalization due to cardiovascular events or death within 45 days, thrombotic events within 45 days, severe bleeding within 45 days
ATTAC, ACTIV-4a, REMAP-CAP, Goligher-2021 (Goligher et al., 2021)	RCT open-label	1207	Intensive care unit	Enoxaparin 1 mg/kg minus 10% BID, weight and CrCl adjusted	Low/intermediate dose low molecular weight heparin (52.1%: enoxaparin, 32.8%: dalteparin; Low dose: 40.4%, Intermediate dose: 51.7%)	Mortality in hospital, thrombotic events, thrombotic events or deaths, major bleeding
ATTACC, ACTIV-4a, REMAP-CAP, Lawler-2021 (Lawler et al., 2021)	RCT open-label	2244	Hospitalized	Enoxaparin 1 mg/kg sc minus 10% BID, weight and CrCl adjusted	Low/intermediate dose low molecular weight heparin (78.7%: enoxaparin, 9.6%: dalteparin. Low dose: 71.7%, Intermediate dose: 26.5%)	Mortality in hospital, clinical improvement: discharge without organ support, thrombotic event, thrombotic event or death, major bleeding
HEP-COVID, Spyropoulos-2021 (Spyropoulos et al., 2021)	RCT open-label	257	Hospitalized	Enoxaparin 1 mg/kg BID or 40 mg OD/BID weight and CrCl adjusted, until discharge	Enoxaparin 40 mg OD/BID weight and CrCl adjusted until discharge	30-day mortality, thrombotic events, thrombotic events or deaths, major bleeding
INSPIRATION, Sadeghipour-2021 (Sadeghipour et al., 2021)	RCT, open-label	600	Hospitalized+ intensive care unit	Enoxaparin 1 mg/kg OD weight and CrCl adjusted, for 30 days	Enoxaparin 40 mg OD; Weight and CrCl adjusted	30-day mortality, 90-day mortality, venous thrombotic events, venous thrombotic events or death, major bleeding

Oliynyk-2021 (Oliynyk et al., 2021)	RCT double-blind	126	Intensive care unit	Enoxaparin: 100 Anti-Xa IU/kg BID or UFH adjusted to APTT	Enoxaparin 50 Anti-Xa IU/kg QD × 28 days	28-day mortality
Perepu-2021 (Perepu et al., 2021)	RCT open-label	173	Hospitalized+ intensive care unit	Enoxaparin 1 mg/kg OD weight and CrCl adjusted until discharge from hospital	Enoxaparin 40 mg OD weight and CrCl adjusted	30-day mortality, venous thrombotic events, major bleeding
RAPID, Sholzberg-2021 (Sholzberg et al., 2021)	RCT open-label	465	Hospitalized	Enoxaparin 1 mg/kg BID; weight and CrCl adjusted	Enoxaparin 40 mg OD, weight and CrCl adjusted	28-day mortality, thrombotic events, major bleeding
2022						
Ananworanich-2022 (Ananworanich et al., 2022)	RCT, double-blind	497	Outpatients	Rivaroxaban 10 mg OD for 21 days	Placebo	35-day mortality, hospitalization within 28 days, severe adverse events within 35 days, adverse events within 35 days, severe bleeding within 35 days
MICHELLE, Ramacciotti-2022 (Ramacciotti et al., 2022)	RCT open-label	318	Post discharge	Rivaroxaban 10 mg OD for 35 days	No anticoagulation	Any thrombotic event and cardiovascular death, any symptomatic venous thrombotic event or all causes, major bleeding, clinically relevant nonsevere bleeding, other bleeding
X-COVID, Morici-2021 (Morici et al., 2022)	RCT open-label	186	Hospitalized+ Intensive Care Unit	Enoxaparin 40 mg BID until discharge from the hospital	Enoxaparin 40 mg OD until discharge	30-day mortality, venous thrombotic events, venous thrombotic events or deaths, major bleeding

APTT, activated partial thromboplastin time; *BID*, twice daily; *CrCl*, creatinine clearance; *OD*, once daily; *RCT*, randomized controlled trial; *UFH*, unfractionated heparin.
Adapted in part from Reis, S., Popp, M., Schießer, S., Metzendorf, M. I., Kranke, P., Meybohm, P., & Weibel, S. (2022). Anticoagulation in COVID-19 patients—An updated systematic review and meta-analysis. Thrombosis Research, 219, 40–48. https://doi.org/10.1016/j.thromres.2022.09.001.

Of note, three of the published randomized trials included patients with either prophylactic or intermediate doses of LMWH in the same arm (Goligher et al., 2021; Lawler et al., 2021; Spyropoulos et al., 2021). Considering this, another systematic review excluded these studies from the overall estimates because the authors were unable to obtain granular data, and concluded that the use of therapeutic doses of anticoagulants does not reduce mortality (OR 0.67, 95% CI 0.18–2.54) in randomized studies. The same review included 40 observational studies and in these, the pooled estimated mortality was higher among patients receiving therapeutic doses (OR 1.74, 95% CI 1.27–2.40) (Elsebaie et al., 2022). Regarding the occurrence of VTE, neither observational studies nor randomized trials showed a difference between therapeutic vs. prophylactic doses (OR 0.8, 95% CI 0.51–1.46; OR 0.65. 95% CI 0.39–1.09).

Studies in ambulatory patients are scarce. The MICHELLE trial assessed the use of prophylactic anticoagulation after hospital discharge in patients hospitalized for COVID-19 (Ramacciotti et al., 2022). Patients with an elevated risk of VTE assessed by the International Medical Prevention Registry on Venous Thromboembolism (IMPROVE VTE) were randomized to rivaroxaban 10 mg daily or no anticoagulation for 35 days. The primary efficacy was a composite of symptomatic or fatal venous thromboembolism, asymptomatic venous thromboembolism on mandatory screening with bilateral lower-limb venous ultrasound and CT pulmonary angiogram performed in all patients at day 35, symptomatic arterial thromboembolism, and cardiovascular death. Half the patients had been admitted to the ICU. The study reported a 67% risk reduction in the intervention arm (RR 0.33. 95% CI 0.12–0.90) with no increased bleeding.

The ACTIV-4B trial included symptomatic outpatients and randomized patients to aspirin 81 mg daily, apixaban 2.5 mg twice daily, apixaban 5 mg twice daily, or a placebo. The primary end point was a composite of all-cause mortality, symptomatic venous and arterial thrombosis, myocardial infarction, or stroke. The study did not find differences in the number of events but was terminated early due to a low event rate (Connors et al., 2021). Finally, a phase 2b study assessed the use of rivaroxaban on disease progression in patients with mild COVID-19 symptoms and at high risk for COVID-19 progression based on age, body mass index, or comorbidity (Ananworanich et al., 2022). The study was stopped for futility.

Based on the previous studies, the most recent guidelines from the American College of Chest Physicians, the American Society of Hematology, the International Society on Thrombosis and Hemostasis, and the National Institutes of Health recommend that all hospitalized patients should receive anticoagulant prophylaxis. They also suggest that the use of therapeutic anticoagulation with heparins (UFH or LMWH) over conventional thromboprophylaxis can be considered in hospitalized COVID-19 patients who are moderately ill and have a low bleeding risk (COVID-19 Treatment Guidelines Panel and National Institutes of Health, 2022; Cuker et al., 2022; Moores et al., 2022; Schulman et al., 2022). The use of intermediate doses of heparins was not recommended, nor was the routine use of anticoagulation in ambulatory patients or the use of extended anticoagulant prophylaxis after hospital discharge.

Conclusions and future directions

Thrombotic events, in particular VTE, are a frequent complication of infection by SARS-CoV-2, especially in hospitalized patients. This is in part related to the severity of the disease and is a consequence of the intense immune response, which is closely related to the activation of several mechanisms resulting in a state of coagulopathy unique to this infection. Considering the unusually high frequency of PE in these patients, it is believed that many of these patients develop a pulmonary thrombosis in situ rather than a true embolism.

Given this, the use of anticoagulants has been extensively studied and, while there is a general agreement that all patients admitted to the hospital with COVID-19 require anticoagulant prophylaxis with low dose heparins, there is less agreement on whether patients who are hospitalized but not critically ill may benefit from therapeutic anticoagulation. In fact, most studies have shown that the use of therapeutic anticoagulation results in a higher bleeding risk while providing little to no benefit. To this point, there are many limitations precluding a solid conclusion, including the lack of a standard definition of moderately ill and critical cases, the lack of validated tools to assess the risk of bleeding, different study designs, and concerns for selection bias in published studies.

Additionally, given the changes in the dynamics of the pandemic and the surge of new SARS-CoV-2 variants, the prothrombotic profile of the infection could potentially change, thus raising questions about the future benefit of these interventions. Other unresolved issues include the impact that other therapies such as antivirals, janus kinase inhibitors, monoclonal antibodies, and others, may have on the risk of thrombosis. Also, there is a need to conduct more studies in hospitalized patients and after hospital discharge to identify patients at higher risk of developing VTE and who have a low bleeding risk who might potentially benefit from extended prophylaxis because it is not possible to recommend this conclusively based on the information available. Finally, it is crucial to develop standard definitions, protocols, and outcomes to be able to analyze and facilitate future research.

Policies and procedures

In this chapter, we reviewed the pathophysiology of SARS-CoV-2 infection in relation to its thrombogenic potential. We provided an overview of the immunothrombosis process triggered by acute infection and how it leads to a self-perpetuating cycle that augments tissue damage. We also reviewed the frequency of thromboembolic complications, both venous and arterial, in patients with COVID-19. Finally, we provided an overview of the studies assessing the use of different anticoagulation strategies and their indications. These findings may be of interest to clinicians involved in the management of these patients and for scientists developing new therapeutic interventions.

Mini-dictionary of terms

- **D-dimer**: Soluble degradation product of cross-linked fibrin that, when elevated, may indicate activation of the clotting system.
- **Deep vein thrombosis**: Development of a blood clot (thrombus) in the deep veins, usually in the legs.
- **Disseminated intravascular coagulation**: Systemic activation of blood coagulation, resulting in the consumption of clotting proteins and platelets.
- **Low molecular weight heparin**: A common injectable anticoagulant.
- **Pulmonary embolism**: Development of a thrombus in the pulmonary arteries, usually as a result of embolization of a detached deep vein thrombosis.
- **Tissue factor**: The major trigger of the coagulation system.
- **Venous thromboembolism**: Occurrence of either a deep vein thrombosis or a pulmonary embolism.

Summary points

- Venous thromboembolism is very frequent in hospitalized patients with COVID-19.
- The risk of thrombosis is higher in critically ill patients.
- The thrombotic risk in COVID-19 is a result of a severe immune inflammatory response leading to the activation of coagulation through multiple mechanisms.
- COVID-19 patients admitted to the hospital should be considered for anticoagulant prophylaxis with heparins (low molecular weight or unfractionated) using standard doses.
- Indiscriminate use of intermediate or therapeutic doses in all hospitalized patients is not recommended.
- Hospitalized patients who are not critically ill may benefit from therapeutic anticoagulation if there is no contraindication.
- There is insufficient evidence to recommend the routine use of anticoagulants in ambulatory patients or patients discharged from a hospital in the absence of other indications.

Acknowledgments

AL-L is an investigator of the Canadian Venous Thromboembolism Research Network (CanVECTOR). The CanVECTOR Network receives grant funding from the Canadian Institutes of Health Research (Funding Reference: CDT-142654).

Conflict of interest

The authors declare no conflicts.

Funding

This work did not receive any specific grant from funding agencies in the public, commercial, or not-for-profit sectors.

References

Al-Samkari, H., Karp Leaf, R. S., Dzik, W. H., Carlson, J. C. T., Fogerty, A. E., Waheed, A., & Rosovsky, R. P. (2020). COVID-19 and coagulation: Bleeding and thrombotic manifestations of SARS-CoV-2 infection. *Blood*, *136*(4), 489–500. https://doi.org/10.1182/blood.2020006520.

Ananworanich, J., Mogg, R., Dunne, M. W., Bassyouni, M., David, C. V., Gonzalez, E., & Heaton, P. (2022). Randomized study of rivaroxaban vs placebo on disease progression and symptoms resolution in high-risk adults with mild coronavirus disease 2019. *Clinical Infectious Diseases*, *75*(1), e473–e481. https://doi.org/10.1093/cid/ciab813.

Asakura, H., & Ogawa, H. (2021). COVID-19-associated coagulopathy and disseminated intravascular coagulation. *International Journal of Hematology*, *113*(1), 45–57. https://doi.org/10.1007/s12185-020-03029-y.

Bikdeli, B., Talasaz, A. H., Rashidi, F., Bakhshandeh, H., Rafiee, F., Rezaeifar, P., & Sadeghipour, P. (2022). Intermediate-dose versus standard-dose prophylactic anticoagulation in patients with COVID-19 admitted to the intensive care unit: 90-day results from the INSPIRATION randomized trial. *Thrombosis and Haemostasis*, *122*(1), 131–141. https://doi.org/10.1055/a-1485-2372.

Bryce, C., Grimes, Z., Pujadas, E., Ahuja, S., Beasley, M. B., Albrecht, R., & Fowkes, M. E. (2021). Pathophysiology of SARS-CoV-2: The Mount Sinai COVID-19 autopsy experience. *Modern Pathology*, *34*(8), 1456–1467. https://doi.org/10.1038/s41379-021-00793-y.

Burke, H., Freeman, A., O'Regan, P., Wysocki, O., Freitas, A., Dushianthan, A., & Wilkinson, T. (2022). Biomarker identification using dynamic time warping analysis: A longitudinal cohort study of patients with COVID-19 in a UK tertiary hospital. *BMJ Open*, *12*(2), e050331. https://doi.org/10.1136/bmjopen-2021-050331.

Candeloro, M., & Schulman, S. (2022). Arterial thrombotic events in hospitalized COVID-19 patients: A short review and meta-analysis. *Seminars in Thrombosis and Hemostasis*. https://doi.org/10.1055/s-0042-1749661.

Cao, Y. C., Deng, Q. X., & Dai, S. X. (2020). Remdesivir for severe acute respiratory syndrome coronavirus 2 causing COVID-19: An evaluation of the evidence. *Travel Medicine and Infectious Disease*, *35*, 101647. https://doi.org/10.1016/j.tmaid.2020.101647.

Chen, G., Wu, D., Guo, W., Cao, Y., Huang, D., Wang, H., & Ning, Q. (2020). Clinical and immunological features of severe and moderate coronavirus disease 2019. *The Journal of Clinical Investigation*, *130*(5), 2620–2629. https://doi.org/10.1172/jci137244.

Connors, J. M., Brooks, M. M., Sciurba, F. C., Krishnan, J. A., Bledsoe, J. R., Kindzelski, A., & Investigators, A.-B. (2021). Effect of antithrombotic therapy on clinical outcomes in outpatients with clinically stable symptomatic COVID-19: The ACTIV-4B randomized clinical trial. *JAMA*, *326*(17), 1703–1712. https://doi.org/10.1001/jama.2021.17272.

Conway, E. M., Mackman, N., Warren, R. Q., Wolberg, A. S., Mosnier, L. O., Campbell, R. A., & Morrissey, J. H. (2022). Understanding COVID-19-associated coagulopathy. *Nature Reviews. Immunology*, *22*(10), 639–649. https://doi.org/10.1038/s41577-022-00762-9.

COVID-19 Treatment Guidelines Panel, & National Institutes of Health. (2022). *Coronavirus disease 2019 (COVID-19) treatment guidelines*. Retrieved from: https://www.covid19treatmentguidelines.nih.gov/.

Cui, S., Chen, S., Li, X., Liu, S., & Wang, F. (2020). Prevalence of venous thromboembolism in patients with severe novel coronavirus pneumonia. *Journal of Thrombosis and Haemostasis*. https://doi.org/10.1111/jth.14830.

Cuker, A., Tseng, E. K., Nieuwlaat, R., Angchaisuksiri, P., Blair, C., Dane, K., & Schünemann, H. J. (2022). American Society of Hematology living guidelines on the use of anticoagulation for thromboprophylaxis in patients with COVID-19: July 2021 update on postdischarge thromboprophylaxis. *Blood Advances*, *6*(2), 664–671. https://doi.org/10.1182/bloodadvances.2021005945.

Elsebaie, M. A. T., Baral, B., Elsebaie, M., Shrivastava, T., Weir, C., Kumi, D., & Birch, N. W. (2022). Does high-dose thromboprophylaxis improve outcomes in COVID-19 patients? A meta-analysis of comparative studies. *TH Open*, *6*(4), e323–e334. https://doi.org/10.1055/a-1930-6492.

Engelmann, B., & Massberg, S. (2013). Thrombosis as an intravascular effector of innate immunity. *Nature Reviews Immunology*, *13*(1), 34–45. https://doi.org/10.1038/nri3345.

FitzGerald, E. S., & Jamieson, A. M. (2020). Unique transcriptional changes in coagulation cascade genes in SARS-CoV-2-infected lung epithelial cells: A potential factor in COVID-19 coagulopathies. *bioRxiv*. https://doi.org/10.1101/2020.07.06.182972.

Fox, S. E., Akmatbekov, A., Harbert, J. L., Li, G., Brown, J. Q., & Vander Heide, R. S. (2020). Pulmonary and cardiac pathology in COVID-19: The first autopsy series from New Orleans. *medRxiv*. https://doi.org/10.1101/2020.04.06.20050575.

Goligher, E. C., Bradbury, C. A., McVerry, B. J., Lawler, P. R., Berger, J. S., Gong, M. N., & Zarychanski, R. (2021). Therapeutic anticoagulation with heparin in critically ill patients with COVID-19. *The New England Journal of Medicine*, *385*(9), 777–789. https://doi.org/10.1056/NEJMoa2103417.

Guan, W. J., Ni, Z. Y., Hu, Y., Liang, W. H., Ou, C. Q., He, J. X., & Zhong, N. S. (2020). Clinical characteristics of coronavirus disease 2019 in China. *The New England Journal of Medicine*. https://doi.org/10.1056/NEJMoa2002032.

Helms, J., Tacquard, C., Severac, F., Leonard-Lorant, I., Ohana, M., & Delabranche, X. (2020). High risk of thrombosis in patients in severe SARS-CoV-2 infection: A multicenter prospective cohort study. *Intensive Care Medicine*. https://doi.org/10.1007/s00134-020-06062-x.

Huang, C., Wang, Y., Li, X., Ren, L., Zhao, J., Hu, Y., & Cao, B. (2020). Clinical features of patients infected with 2019 novel coronavirus in Wuhan, China. *Lancet*, *395*(10223), 497–506. https://doi.org/10.1016/s0140-6736(20)30183-5.

Jayarangaiah, A., Kariyanna, P. T., Chen, X., Jayarangaiah, A., & Kumar, A. (2020). COVID-19-associated coagulopathy: An exacerbated immunothrombosis response. *Clinical and Applied Thrombosis/Hemostasis*, *26*. https://doi.org/10.1177/1076029620943293.

Klok, F. A., Kruip, M., van der Meer, N. J. M., Arbous, M. S., Gommers, D., Kant, K. M., & Endeman, H. (2020a). Confirmation of the high cumulative incidence of thrombotic complications in critically ill ICU patients with COVID-19: An updated analysis. *Thrombosis Research*. https://doi.org/10.1016/j.thromres.2020.04.041.

Klok, F. A., Kruip, M., van der Meer, N. J. M., Arbous, M. S., Gommers, D., Kant, K. M., & Endeman, H. (2020b). Incidence of thrombotic complications in critically ill ICU patients with COVID-19. *Thrombosis Research*. https://doi.org/10.1016/j.thromres.2020.04.013.

Lawler, P. R., Goligher, E. C., Berger, J. S., Neal, M. D., McVerry, B. J., Nicolau, J. C., & Zarychanski, R. (2021). Therapeutic anticoagulation with heparin in noncritically ill patients with COVID-19. *The New England Journal of Medicine*, *385*(9), 790–802. https://doi.org/10.1056/NEJMoa2105911.

Lemos, A. C. B., do Espírito Santo, D. A., Salvetti, M. C., Gilio, R. N., Agra, L. B., Pazin-Filho, A., & Miranda, C. H. (2020). Therapeutic versus prophylactic anticoagulation for severe COVID-19: A randomized phase II clinical trial (HESACOVID). *Thrombosis Research*, *196*, 359–366. https://doi.org/10.1016/j.thromres.2020.09.026.

Len, P., Iskakova, G., Sautbayeva, Z., Kussanova, A., Tauekelova, A. T., Sugralimova, M. M., & Barteneva, N. S. (2022). Meta-analysis and systematic review of coagulation disbalances in COVID-19: 41 studies and 17,601 patients. *Frontiers in Cardiovascular Medicine*, *9*, 794092. https://doi.org/10.3389/fcvm.2022.794092.

Liang, W., Liang, H., Ou, L., Chen, B., Chen, A., Li, C., & He, J. (2020). Development and validation of a clinical risk score to predict the occurrence of critical illness in hospitalized patients with COVID-19. *JAMA Internal Medicine*. https://doi.org/10.1001/jamainternmed.2020.2033.

Llitjos, J. F., Leclerc, M., Chochois, C., Monsallier, J. M., Ramakers, M., Auvray, M., & Merouani, K. (2020). High incidence of venous thromboembolic events in anticoagulated severe COVID-19 patients. *Journal of Thrombosis and Haemostasis*. https://doi.org/10.1111/jth.14869.

Lodigiani, C., Iapichino, G., Carenzo, L., Cecconi, M., Ferrazzi, P., Sebastian, T., & Humanitas COVID-19 Task Force. (2020). Venous and arterial thromboembolic complications in COVID-19 patients admitted to an academic hospital in Milan, Italy. *Thrombosis Research*. https://doi.org/10.1016/j.thromres.2020.04.024.

Lopes, R. D., de Barros, E. S. P. G. M., Furtado, R. H. M., Macedo, A. V. S., Bronhara, B., Damiani, L. P., & Berwanger, O. (2021). Therapeutic versus prophylactic anticoagulation for patients admitted to hospital with COVID-19 and elevated D-dimer concentration (ACTION): An open-label, multicentre, randomised, controlled trial. *Lancet*, *397*(10291), 2253–2263. https://doi.org/10.1016/s0140-6736(21)01203-4.

Manne, B. K., Denorme, F., Middleton, E. A., Portier, I., Rowley, J. W., Stubben, C., & Campbell, R. A. (2020). Platelet gene expression and function in patients with COVID-19. *Blood*, *136*(11), 1317–1329. https://doi.org/10.1182/blood.2020007214.

Mansory, E. M., Abu-Farhaneh, M., Iansavitchene, A., & Lazo-Langner, A. (2022). Venous and arterial thrombosis in ambulatory and discharged COVID-19 patients: A systematic review and meta-analysis. *TH Open*, *6*(3), e276–e282. https://doi.org/10.1055/a-1913-4377.

Mansory, E. M., Srigunapalan, S., & Lazo-Langner, A. (2021). Venous thromboembolism in hospitalized critical and noncritical COVID-19 patients: A systematic review and meta-analysis. *TH Open*, *5*(3), e286–e294. https://doi.org/10.1055/s-0041-1730967.

Mao, L., Jin, H., Wang, M., Hu, Y., Chen, S., He, Q., & Hu, B. (2020). Neurologic manifestations of hospitalized patients with coronavirus disease 2019 in Wuhan, China. *JAMA Neurology*. https://doi.org/10.1001/jamaneurol.2020.1127.

Menter, T., Haslbauer, J. D., Nienhold, R., Savic, S., Hopfer, H., Deigendesch, N., & Tzankov, A. (2020). Postmortem examination of COVID-19 patients reveals diffuse alveolar damage with severe capillary congestion and variegated findings in lungs and other organs suggesting vascular dysfunction. *Histopathology*, *77*(2), 198–209. https://doi.org/10.1111/his.14134.

Middeldorp, S., Coppens, M., van Haaps, T. F., Foppen, M., Vlaar, A. P., Muller, M. C., & van Es, N. (2020a). Incidence of venous thromboembolism in hospitalized patients with COVID-19. *Journal of Thrombosis and Haemostasis*. https://doi.org/10.20944/preprints202004.0345.v1.

Middeldorp, S., Coppens, M., van Haaps, T. F., Foppen, M., Vlaar, A. P., Müller, M. C. A., … van Es, N. (2020b). Incidence of venous thromboembolism in hospitalized patients with COVID-19. *Journal of Thrombosis and Haemostasis*. https://doi.org/10.1111/jth.14888.

Moores, L. K., Tritschler, T., Brosnahan, S., Carrier, M., Collen, J. F., Doerschug, K., & Wells, P. (2022). Thromboprophylaxis in patients with COVID-19: A brief update to the CHEST guideline and expert panel report. *Chest*, *162*(1), 213–225. https://doi.org/10.1016/j.chest.2022.02.006.

Morici, N., Podda, G., Birocchi, S., Bonacchini, L., Merli, M., Trezzi, M., & Cattaneo, M. (2022). Enoxaparin for thromboprophylaxis in hospitalized COVID-19 patients: The X-COVID-19 randomized trial. *European Journal of Clinical Investigation*, *52*(5), e13735. https://doi.org/10.1111/eci.13735.

Nannoni, S., de Groot, R., Bell, S., & Markus, H. S. (2021). Stroke in COVID-19: A systematic review and meta-analysis. *International Journal of Stroke*, *16*(2), 137–149. https://doi.org/10.1177/1747493020972922.

Oliynyk, O., Barg, W., Slifirczyk, A., Oliynyk, Y., Dubrov, S., Gurianov, V., & Rorat, M. (2021). Comparison of the effect of unfractionated heparin and enoxaparin sodium at different doses on the course of COVID-19-associated coagulopathy. *Life (Basel)*, *11*(10). https://doi.org/10.3390/life11101032.

Perepu, U. S., Chambers, I., Wahab, A., Ten Eyck, P., Wu, C., Dayal, S., & Lentz, S. R. (2021). Standard prophylactic versus intermediate dose enoxaparin in adults with severe COVID-19: A multi-center, open-label, randomized controlled trial. *Journal of Thrombosis and Haemostasis*, *19*(9), 2225–2234. https://doi.org/10.1111/jth.15450.

Poissy, J., Goutay, J., Caplan, M., Parmentier, E., Duburcq, T., Lassalle, F., & Susen, S. (2020). Pulmonary embolism in COVID-19 patients: Awareness of an increased prevalence. *Circulation*. https://doi.org/10.1161/circulationaha.120.047430.

Ramacciotti, E., Barile Agati, L., Calderaro, D., Aguiar, V. C. R., Spyropoulos, A. C., de Oliveira, C. C. C., & Barbosa Santos, M. V. (2022). Rivaroxaban versus no anticoagulation for post-discharge thromboprophylaxis after hospitalisation for COVID-19 (MICHELLE): An open-label, multicentre, randomised, controlled trial. *The Lancet*, *399*(10319), 50–59. https://doi.org/10.1016/s0140-6736(21)02392-8.

Ranucci, M., Ballotta, A., Di Dedda, U., Bayshnikova, E., Dei Poli, M., Resta, M., & Menicanti, L. (2020). The procoagulant pattern of patients with COVID-19 acute respiratory distress syndrome. *Journal of Thrombosis and Haemostasis*. https://doi.org/10.1111/jth.14854.

Reis, S., Popp, M., Schießer, S., Metzendorf, M. I., Kranke, P., Meybohm, P., & Weibel, S. (2022). Anticoagulation in COVID-19 patients—An updated systematic review and meta-analysis. *Thrombosis Research*, *219*, 40–48. https://doi.org/10.1016/j.thromres.2022.09.001.

Sadeghipour, P., Talasaz, A. H., Rashidi, F., Sharif-Kashani, B., Beigmohammadi, M. T., Farrokhpour, M., & Bikdeli, B. (2021). Effect of intermediate-dose vs standard-dose prophylactic anticoagulation on thrombotic events, extracorporeal membrane oxygenation treatment, or mortality among patients with COVID-19 admitted to the intensive care unit: The INSPIRATION randomized clinical trial. *JAMA*, *325*(16), 1620–1630. https://doi.org/10.1001/jama.2021.4152.

Schulman, S., Sholzberg, M., Spyropoulos, A. C., Zarychanski, R., Resnick, H. E., Bradbury, C. A., & Haemostasis. (2022). ISTH guidelines for antithrombotic treatment in COVID-19. *Journal of Thrombosis and Haemostasis*, *20*(10), 2214–2225. https://doi.org/10.1111/jth.15808.

Sholzberg, M., Tang, G. H., Rahhal, H., AlHamzah, M., Kreuziger, L. B., Áinle, F. N., & Jüni, P. (2021). Effectiveness of therapeutic heparin versus prophylactic heparin on death, mechanical ventilation, or intensive care unit admission in moderately ill patients with covid-19 admitted to hospital: RAPID randomised clinical trial. *BMJ*, *375*, n2400. https://doi.org/10.1136/bmj.n2400.

Spyropoulos, A. C., Goldin, M., Giannis, D., Diab, W., Wang, J., Khanijo, S., & Weitz, J. I. (2021). Efficacy and safety of therapeutic-dose heparin vs standard prophylactic or intermediate-dose heparins for thromboprophylaxis in high-risk hospitalized patients with COVID-19: The HEP-COVID randomized clinical trial. *JAMA Internal Medicine*, *181*(12), 1612–1620. https://doi.org/10.1001/jamainternmed.2021.6203.

Tang, N., Li, D., Wang, X., & Sun, Z. (2020). Abnormal coagulation parameters are associated with poor prognosis in patients with novel coronavirus pneumonia. *Journal of Thrombosis and Haemostasis*, *18*(4), 844–847. https://doi.org/10.1111/jth.14768.

Taylor, F. B., Jr., Toh, C. H., Hoots, W. K., Wada, H., & Levi, M. (2001). Towards definition, clinical and laboratory criteria, and a scoring system for disseminated intravascular coagulation. *Thrombosis and Haemostasis*, *86*(5), 1327–1330.

Thomas, W., Varley, J., Johnston, A., Symingotn, E., Robinson, M., Sheares, K., & Besser, M. (2020). Thrombotic complications of patients admitted to intensive care with COVID-19 at a teaching hospital in the United Kingdom. *Thrombosis Research*. https://doi.org/10.1016/j.thromres.2020.04.028.

Veras, F. P., Pontelli, M. C., Silva, C. M., Toller-Kawahisa, J. E., de Lima, M., Nascimento, D. C., & Cunha, F. Q. (2020). SARS-CoV-2-triggered neutrophil extracellular traps mediate COVID-19 pathology. *The Journal of Experimental Medicine*, *217*(12). https://doi.org/10.1084/jem.20201129.

Wang, D., Hu, B., Hu, C., Zhu, F., Liu, X., Zhang, J., & Peng, Z. (2020). Clinical characteristics of 138 hospitalized patients with 2019 novel coronavirus-infected pneumonia in Wuhan, China. *JAMA*. https://doi.org/10.1001/jama.2020.1585.

Weiss, E., Roux, O., Moyer, J. D., Paugam-Burtz, C., Boudaoud, L., Ajzenberg, N., & de Raucourt, E. (2020). Fibrinolysis resistance: A potential mechanism underlying COVID-19 coagulopathy. *Thrombosis and Haemostasis*, *120*(9), 1343–1345. https://doi.org/10.1055/s-0040-1713637.

Wichmann, D., Sperhake, J. P., Lutgehetmann, M., Steurer, S., Edler, C., Heinemann, A., & Kluge, S. (2020). Autopsy findings and venous thromboembolism in patients with COVID-19: A prospective cohort study. *Annals of Internal Medicine*. https://doi.org/10.7326/m20-2003.

Xu, X., & Gao, X. (2004). Immunological responses against SARS-coronavirus infection in humans. *Cellular & Molecular Immunology*, *1*(2), 119–122.

Xu, J.-F., Wang, L., Zhao, L., Li, F., Liu, J., Zhang, L., & Liu, J. (2020). Risk assessment of venous thromboembolism and bleeding in COVID-19 patients. *ResearchSquare*. https://doi.org/10.21203/rs.3.rs-18340/v1.

Yan, B., Freiwald, T., Chauss, D., Wang, L., West, E., Bibby, J., & Kazemian, M. (2020). SARS-CoV2 drives JAK1/2-dependent local and systemic complement hyper-activation. *Research Square*. https://doi.org/10.21203/rs.3.rs-33390/v1.

Zhang, H., Wu, H., Pan, D., & Shen, W. (2022). D-dimer levels and characteristics of lymphocyte subsets, cytokine profiles in peripheral blood of patients with severe COVID-19: A systematic review and meta-analysis. *Frontiers in Medicine (Lausanne)*, *9*, 988666. https://doi.org/10.3389/fmed.2022.988666.

Zhou, F., Yu, T., Du, R., Fan, G., Liu, Y., Liu, Z., & Cao, B. (2020). Clinical course and risk factors for mortality of adult inpatients with COVID-19 in Wuhan, China: A retrospective cohort study. *Lancet*, *395*(10229), 1054–1062. https://doi.org/10.1016/s0140-6736(20)30566-3.

Chapter 31

COVID-19 myocarditis: Features of echocardiography

Antonello D'Andrea[a], Dario Fabiani[b], Francesco Sabatella[b], Carmen Del Giudice[b], Luigi Cante[b], Adriano Caputo[b], Stefano Palermi[c], Francesco Giallauria[d], and Vincenzo Russo[b]

[a]*Cardiology Unit, Umberto I Hospital, University of Campania "Luigi Vanvitelli", Nocera Inferiore, Italy,* [b]*Cardiology Unit, Department of Medical Translational Sciences, University of Campania "Luigi Vanvitelli", Naples, Italy,* [c]*Public Health Department, University of Naples "Federico II", Naples, Italy,* [d]*Department of Translational Medical Sciences, University of Naples "Federico II", Naples, Italy*

Abbreviations

2D	two-dimensional
3D-RVEF	right ventricle ejection fraction measured with three-dimensional echocardiography
ASE	American Society of Echocardiography
CE	contrast echocardiography
CMR	cardiac magnetic resonance
COVID-19	coronavirus disease 2019
CRP	C-reactive protein
ESC	European Society of Cardiology
FAC	fractional area change
GLS	global longitudinal strain
GWI	global work index
Hs-TnI	high sensitive troponin I
LGE	late gadolinium enhancement
LV	left ventricle
LVEF	left ventricular ejection fraction
LVGLS	left ventricular global longitudinal strain
MB	myocardial biopsy
MW	myocardial work
NT-proBNP	amino-terminal fragment of B-type natriuretic propeptide
PAPs	estimate of systolic pulmonary pressure
PE	pulmonary embolism
PH	pulmonary hypertension
PLAX	parasternal long axis
POCUS	point of care ultrasound
PVR	pulmonary vascular resistance
RIMP	right ventricle index of myocardial performance
RV	right ventricle
RVLS	right ventricular longitudinal strain
RV-PA	right ventricle–pulmonary artery
S′	tricuspid lateral annular systolic velocity wave
SARS-CoV2	severe acute respiratory syndrome coronavirus 2
TAPSE	tricuspid annular plane systolic excursion
TDI	tissue doppler imaging
TTE	transthoracic echocardiography

Management, Body Systems, and Case Studies in COVID-19. https://doi.org/10.1016/B978-0-443-18703-2.00030-X
Copyright © 2024 Elsevier Inc. All rights are reserved, including those for text and data mining, AI training, and similar technologies.

Introduction

Coronavirus disease 2019 (COVID-19) myocarditis constitutes an important cardiac complication in the intensive care setting. In this context, bedside transthoracic echocardiography (TTE) represents the first-line imaging exam both to diagnose myocarditis and to stratify prognosis through targeted therapeutic interventions in COVID-19 patients (Picard & Weiner, 2020). During the COVID-19 pandemic, TTE was characterized by a high contagion risk due to closer contact to patients; preventive measures should have been adopted in critical care units (Cameli et al., 2020). This chapter will discuss the role of TTE in COVID-19 myocarditis from two-dimensional (2D) echocardiographic features to the most recent tools, such as speckle tracking echocardiography (STE) and myocardial work (MW) analysis. The last part of this chapter will focus on the right ventricle (RV) involvement in COVID-19 patients and the use of conventional 2D parameters complemented by new echocardiographic applications, such as STE or 3D ejection fraction, to detect RV dysfunction.

Role of transthoracic echocardiography in COVID-19 myocarditis

TTE is a rapid, noninvasive, low-cost, and widely available technique that represents the first-line imaging exam in COVID-19 myocarditis evaluation (Picard & Weiner, 2020). The European Society of Cardiology (ESC) Working Group on Myocardial and Pericardial Diseases recommended TTE as part of the initial diagnostic workup for all COVID-19 patients with suspected myocarditis (Citro et al., 2021). The advantages and limitations of TTE are summarized in Table 1. Nevertheless, during the COVID-19 pandemic, a TTE examination was associated with a higher risk of infection. Due to contagion risk, all operators should adhere to specific measures, such as wearing masks, plastic glasses, gloves, and scrubs (Klompas et al., 2020). Also, after each examination, the ultrasound machines, probes, and rooms should have been accurately disinfected (Agricola et al., 2020; Cameli et al., 2020). Additionally, many units have ultrasound machines dedicated to COVID-19 patients (Drake et al., 2020). According to the ESC Task Force document (The Task Force for the management of COVID-19 of the European Society of Cardiology, 2022a) and the American Society of Echocardiography (ASE) (Kirkpatrick et al., 2020), in COVID-19 patients with cardiac involvement, TTE should be based on point of care ultrasound (POCUS) both to answer specific clinical questions and to minimize patient contact with machines and operators. Despite the advantages of POCUS examination, a comprehensive TTE exam could be useful to characterize systematic COVID-19 cardiac involvement (Carrizales-Sepúlveda et al., 2021). In a critical care setting, TTE allows determining the causes of hemodynamic instability and monitoring treatment response, providing an immediate impact on the clinical management and prognosis (Dweck et al., 2020). Conversely, other diagnostic exams, such as cardiac magnetic resonance (CMR) and myocardial biopsy (MB), are less useful and available than TTE in the COVID-19-related intensive care scenario (Cameli et al., 2020).

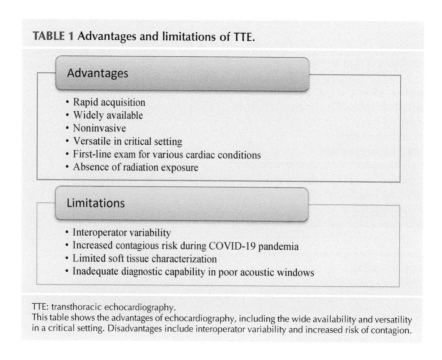

TABLE 1 Advantages and limitations of TTE.

Advantages

- Rapid acquisition
- Widely available
- Noninvasive
- Versatile in critical setting
- First-line exam for various cardiac conditions
- Absence of radiation exposure

Limitations

- Interoperator variability
- Increased contagious risk during COVID-19 pandemia
- Limited soft tissue characterization
- Inadequate diagnostic capability in poor acoustic windows

TTE: transthoracic echocardiography.
This table shows the advantages of echocardiography, including the wide availability and versatility in a critical setting. Disadvantages include interoperator variability and increased risk of contagion.

Two-dimensional transthoracic echocardiography and other techniques

Viral infection of the myocardium leads to acute inflammatory cell infiltration, interstitial edema, thickening of the left ventricle (LV) myocardial wall, and reduced LV contractility (Felker et al., 2000). The pathophysiological mechanisms of COVID-19-related myocardial damage are not fully elucidated, contributing several factors such as hypoxia, direct cytopathic damage (Atri et al., 2020), enhanced immune response with cytokine storm, and atherosclerotic plaque ruptures with coronary thrombosis (De Lorenzo et al., 2020). A comprehensive TTE examination can be performed through various echographic views. Among these, in a critical care setting, the subcostal four-chamber view (Fig. 1) provides a global vision of the heart and is the easiest to obtain in the forced supine position of the patient, as in mechanical ventilation. If the subcostal four-chamber view is difficult or not possible to obtain, other echocardiographic views should be used, such as the parasternal long axis (PLAX) view (Fig. 2) or the apical four-chamber view (Fig. 3) (Altersberger et al., 2021). 2D echocardiographic findings in COVID-19 myocarditis include LV or RV increased endocavitary dimensions, normal or increased wall thickness, global or regional RV or LV systolic dysfunction, LV diastolic dysfunction, LV thrombus, and pericardial involvement with effusion (Purdy et al., 2021; Rathore et al., 2021). These echocardiographic features are not specific only for COVID-19 but are common with myocarditis due to other viral agents. These characteristics are shown in Table 2. In fulminant myocarditis, a dilated and hypocontractile LV is seen, resulting clinically in cardiogenic shock status (Felker et al., 2000). Tissue Doppler imaging (TDI) and contrast echocardiography (CE) are other TTE techniques that could be used to further stratify COVID-19 myocarditis prognosis (Goerlich, Gilotra, et al., 2021; Goerlich, Minhas, et al., 2021). TDI is a useful application to evaluate LV diastolic function through a noninvasive estimate of left ventricular filling pressures (D'Andrea et al., 2018). TDI measures are usually impaired in patients with acute myocarditis (Escher et al., 2011). In addition, CE represents an echocardiographic application allowing the detection of a more accurate regional LV systolic function and also finding LV thrombi (Olszewski et al., 2007). However, further research is needed to evaluate the role of TDI and CE in COVID-19 myocarditis.

Speckle tracking echocardiography and myocardial work

STE is an echocardiographic tool that provides a quantitative, objective and nondoppler-related assessment of systolic function, evaluating myocardial deformation on 2D images (Mondillo et al., 2011). This technique owns a higher diagnostic accuracy in the examination of early LV systolic dysfunction compared to conventional TTE, also in case of COVID-19

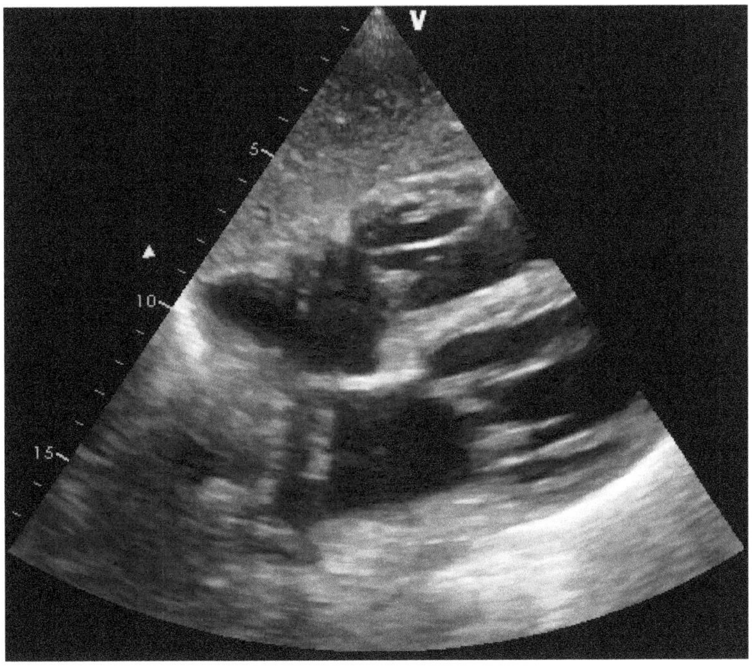

FIG. 1 Subcostal four-chamber view. Subcostal four-chamber view provides a global vision of the heart and is the easiest echocardiographic view to obtain in the forced supine position of the patient. This view also allows detecting associated pericardial effusion. This figure shows right and left cardiac chambers surrounded by pericardium.

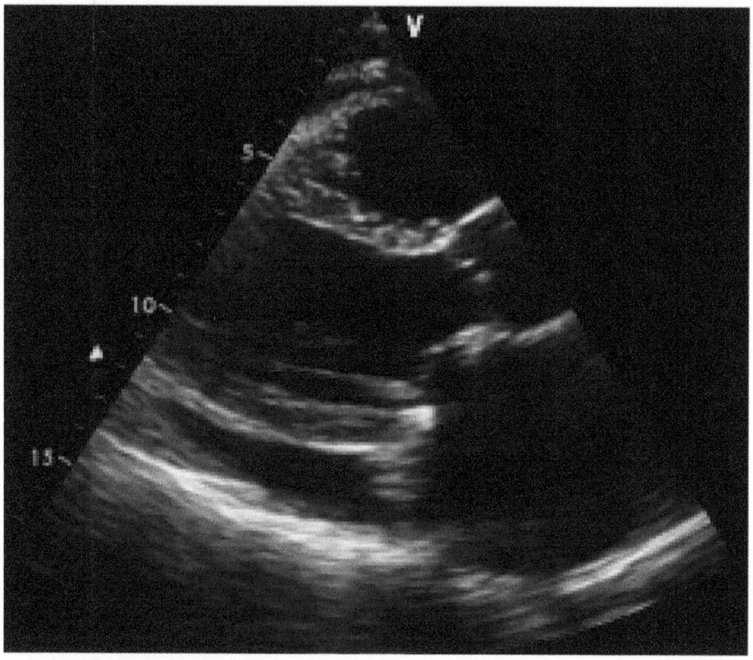

FIG. 2 Parasternal long axis view. It shows dilated left atrium and ventricle with associated pericardial effusion.

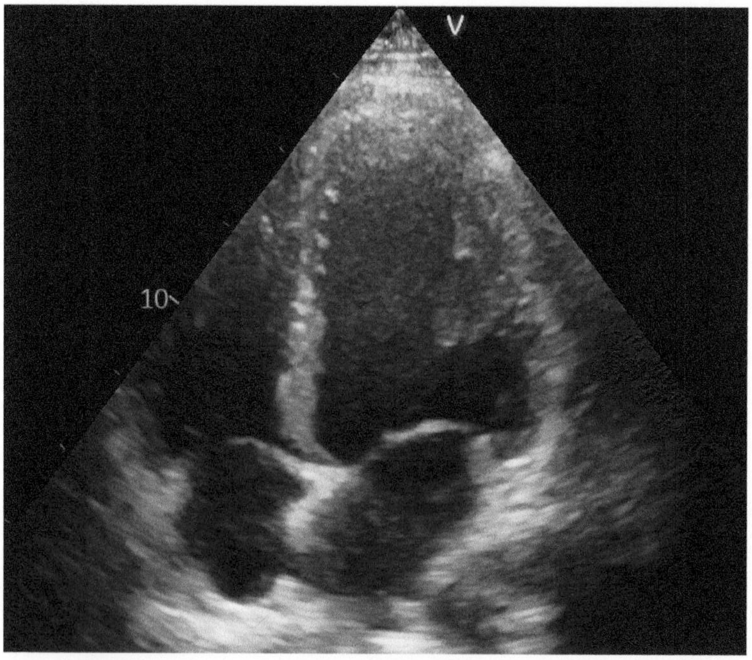

FIG. 3 Apical four-chamber view. The apical four-chamber reported in the figure shows dilated left atrium and ventricle.

myocarditis (Adeboye et al., 2022; Basar et al., 2022; Croft et al., 2021). Indeed, global longitudinal strain (GLS) analysis, performed in COVID-19 hospitalized patients with an average LV ejection fraction (LVEF) value of 52.1%, has showed a discordant mean GLS value (.12.9%). Weckbach et al. (2021) revealed that a preserved or mildly impaired LVEF was associated to impaired GLS values (−7.6% to −15.1%) in COVID-19 patients with cardiac involvement, emphasizing STE role in the early diagnosis of myocardial damage. In COVID-19 myocarditis, GLS value alterations are strictly related to both serum markers of inflammation, such as C-reactive protein (CRP) and cytokines, and interstitial edema distribution detected with CMR on T2-weighted sequences (Li et al., 2021). These data have demonstrated a strong correlation between STE abnormalities and myocardial inflammation. Furthermore, in COVID-19 myocarditis, STE has shown a typical

TABLE 2 Spectrum of two-dimensional TTE characteristics in COVID-19 myocarditis.

	Echocardiographic parameters	Most common findings	
Morphological characteristics	LV	Dimensions	Normal or increased endocavitary dimensions
		Septal thickness	Normal or increased due to pseudohypertrophy
	RV	Could be dilated	
	Pericardium	Pericardial effusion could be present	
Functional characteristics	Systolic function	LV global hypokinesis	
		LV segmentary hypokinesia not corresponding to specific coronary perfusion territories	
		Decreased LVGLS on STE	
		Global RV systolic function could be reduced	
	Diastolic function	LV diastolic dysfunction with increased ventricular filling pressures (↑E/E′)	

LV: left ventricle; RV: right ventricle, LVGLS: left ventricle global longitudinal strain; STE: speckle tracking echocardiography.
TTE in COVID-19 myocarditis could detect both morphological (alterations of cardiac chambers dimensions, increased septal thickness, pericardial effusion) and functional characteristics (systolic and/or diastolic dysfunction). The two-dimensional echocardiographic features are in common with myocarditis due to other viral agents.

deformation pattern caused by the involvement of the subepicardial layer of the myocardial wall (D'Andrea et al., 2022). The subendocardial region sparing allows excluding ischemic myocardial damage and making a differential diagnosis with acute coronary syndrome, particularly in the subset of myocarditis simulating this clinical presentation (D'Andrea et al., 2022). Furthermore, Stöbe et al. (2020) reported the presence of early alterations in the basal regional radial strain (RS) and the basal regional circumferential strain (CS). These data are associated with a subepicardial injury starting from basal LV segments, called the "reversal tako-tsubo" pattern (Fig. 4). This STE pattern appears to be specific to COVID-19 myocarditis (Goerlich, Gilotra, et al., 2021). The pathognomonic localization of STE abnormalities indicated that LV basal segments are more vulnerable than other myocardial regions, probably for a heterogeneous distribution of the viral receptor ACE-2 (Cecchetto et al., 2022). Moreover, STE provides both an early identification of subclinical LV systolic impairment and the evaluation of residual damage after the acute phase of myocarditis. Indeed, several studies have noted that LV STE alterations after the acute phase are suggestive of scar distribution identified with late gadolinium enhancement (LGE) on T1-weighted CMR, representing a potential substrate for ventricular arrhythmias (Stöbe et al., 2020). STE impairment

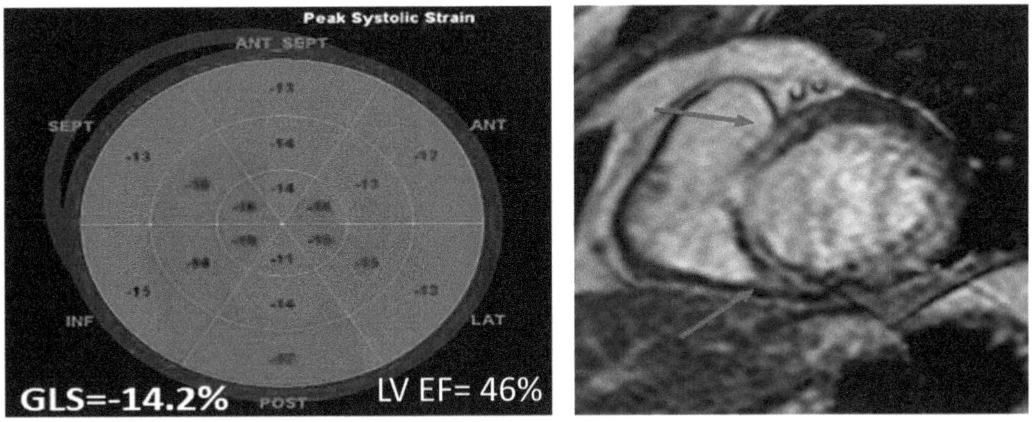

FIG. 4 To the left, a reduced global longitudinal strain mainly localized into basal segments ("reversal tako-tsubo" pattern). To the right, areas of late gadolinium enhancement, suggestive of nonischemic myocardial fibrosis, on cardiac magnetic resonance images.

correlates with the myocardial fibrosis burden detected on CMR with higher sensitivity compared to LVEF measured with Simpson's biplane technique (D'Andrea et al., 2022). MW is a recent TTE application, utilizing LVGLS values indexed to systolic blood pressure, that allows achieving a load-independent LV systolic function evaluation (Roemer et al., 2021). In fact, the global work index (GWI) seems to be an estimate of COVID-19-related systolic dysfunction that is more sensible than LVEF and LVGLS (Goerlich, Minhas, et al., 2021). However, further data are needed to clarify the MW and GWI roles in COVID-19 myocarditis. Although, according to the ESC Task Force (The Task Force for the management of COVID-19 of the European Society of Cardiology, 2022b), CMR represents the main noninvasive exam to detect COVID-19-related myocardial damage. Recently STE took on an emerging role for its ability to identify subclinical COVID-19-related myocardial injuries in an intensive care setting, where a bedside assessment is more feasible (D'Andrea et al., 2022).

Role of transthoracic echocardiography in right ventricle dysfunction

RV damage is prevalent in COVID-19 patients and is caused by several mechanisms, including increased RV afterload in pulmonary diseases (acute respiratory distress syndrome (ARDS) or pulmonary embolism (PE), hypoxic vasoconstriction, direct myocardial damage, and the negative inotropic effect of inflammatory cytokines (Almarawi et al., 2020). RV involvement leads to poor clinical outcomes in COVID-19 patients and TTE represents an important diagnostic exam to detect RV failure. Moreover, several echocardiographic parameters are strongly associated with prognosis and risk stratification (Lan et al., 2021). RV dysfunction could be defined using the following conventional echocardiographic indices: fractional area change (FAC) $<35\%$; tricuspid lateral annular systolic velocity wave (S′) by TDI <9.5 cm/s; tricuspid annular plane systolic excursion (TAPSE) <17 mm; and RV index of myocardial performance (RIMP) by TDI >0.43 (Lang et al., 2015). However, new echocardiographic parameters, such as the ratio between TAPSE and the estimate of systolic pulmonary pressure (PAPs), RV longitudinal strain (RVLS), and RV ejection fraction measured with three-dimensional echocardiography (3D-RVEF) are emerging prognostic predictors (Figs. 5 and 6).

Involvement of right ventricle in COVID-19 patients

Mortality in COVID-19 patients who develop RV dysfunction, RV dilatation, or pulmonary hypertension (pH) is high (Paternoster et al., 2021), and RV function evaluation should always be considered an important predictor of prognosis (Szekely et al., 2020). In COVID-19 patients with acute onset of hypotension and hemodynamical deterioration associated with elevated D-dimer serum levels, TTE can help rule out PE according to ESC guidelines (Ezhilkugan et al., 2021;

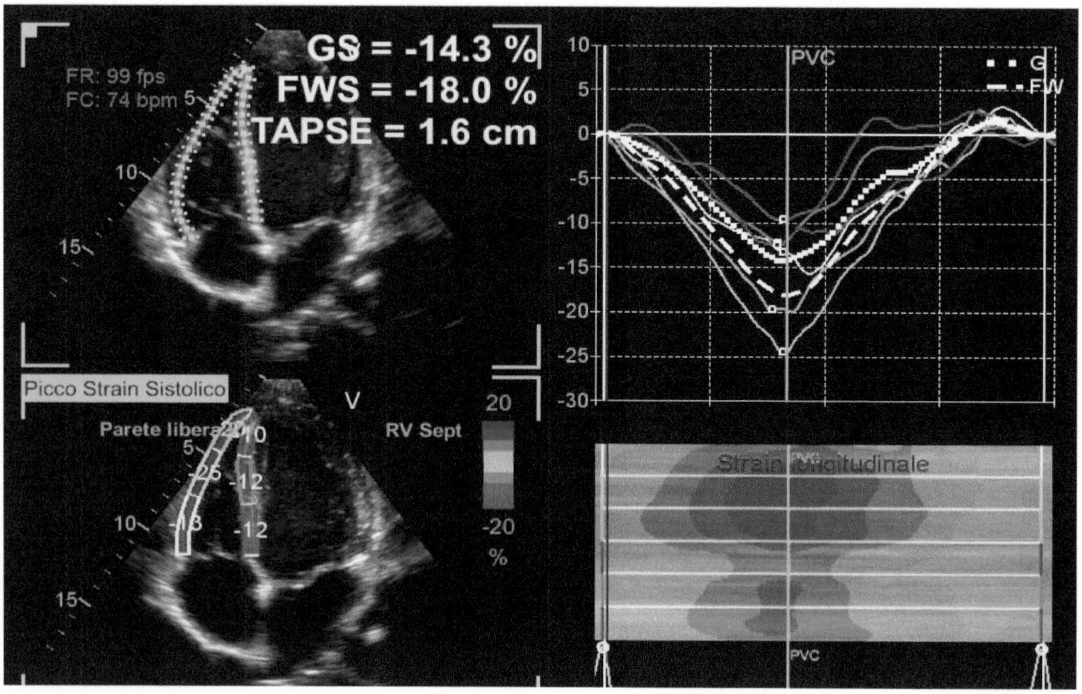

FIG. 5 Right ventricular longitudinal strain. The image shows a reduced right ventricular longitudinal strain. This latter constitutes a new echocardiographic marker of right ventricle dysfunction.

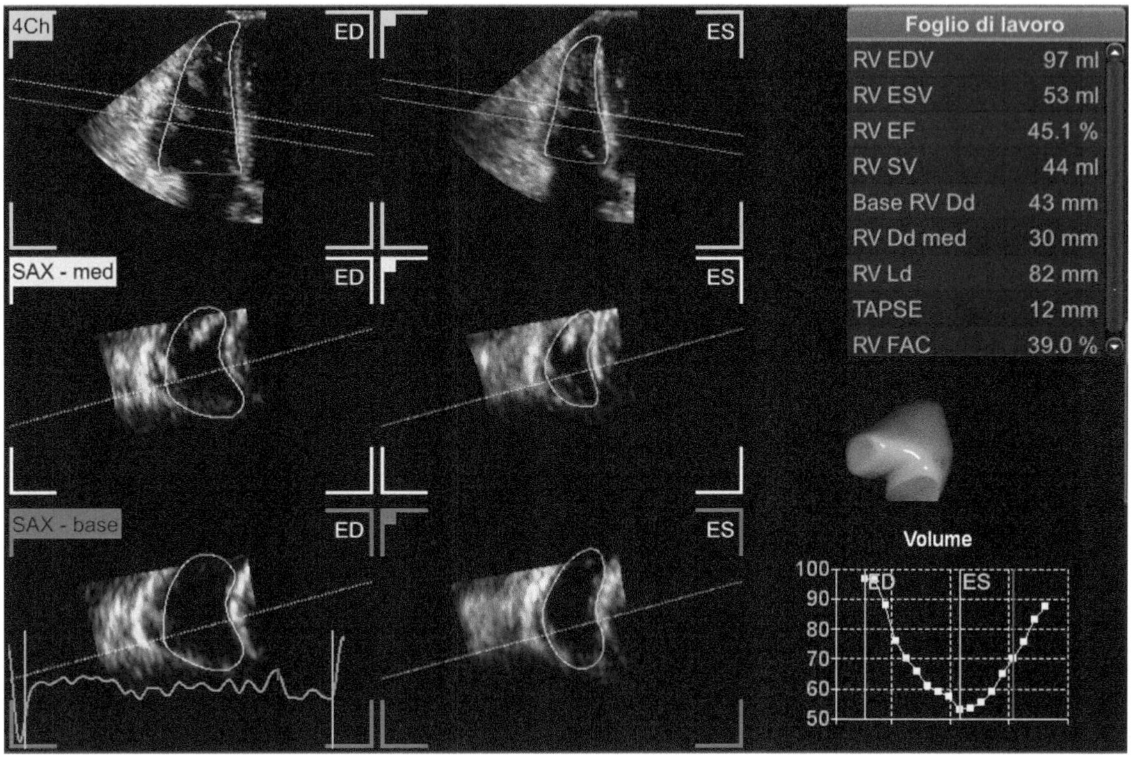

FIG. 6 Right ventricular ejection fraction measured with three-dimensional echocardiography. 3D-RVEF is a new echocardiographic parameter. It is reported to be decreased in COVID-19 patients with cardiac involvement and strongly associated with higher mortality.

Konstantinides et al., 2020). Mahmoud-Elsayed et al. (2020) showed that RV involvement in COVID-19 patients was associated with a marked reduction in radial RV systolic function indices, such as FAC, without longitudinal shortening involvement. However, studies revealed that TAPSE was an independent predictor of in-hospital death both in COVID-19 (D'Andrea et al., 2020; Li, Zheng, et al., 2020) and in ARDS patients (Shah et al., 2016). Bleakley et al. (2021) showed that RV dysfunction in COVID-19 patients affected by ARDS was characterized by a significantly reduced RV FAC compared to longitudinal systolic parameters. Hence, a significantly higher proportion of patients were identified as having RV dysfunction by RV FAC than by TAPSE and S′. Moreover, RV radial impairment was associated with higher serum levels of the amino-terminal fragment of the B-type natriuretic propeptide (NT-proBNP) and high sensitive Troponin I (hs-TnI) (Bleakley et al., 2021). Right ventricle-pulmonary artery (RV-PA) coupling indicated RV function according to its afterload and is derived noninvasively from the RV FAC/right ventricular systolic pressure (RVSP) ratio or by the TAPSE/PAPs ratio (Guazzi et al., 2017). RV-PA coupling was also associated with the P/F ratio, PEEP, and NT-proBNP (Bleakley et al., 2021). Furthermore, D'Alto et al. (2020) demonstrated that the TAPSE/PAPs ratio is strongly associated with clinical outcomes and mortality in COVID-19 patients complicated by ARDS. RVLS analysis has been recently demonstrated to be accurate and sensitive to detect RV dysfunction, even if subclinical (Longobardo et al., 2017). Li, Li, et al. (2020) showed a strong correlation between RVLS and clinical outcomes in COVID-19 patients. The 3D-RVEF has been used to evaluate RV function, providing additive prognostic information in several cardiovascular diseases (Focardi et al., 2015). Moreover, 3D-RVEF has been shown to be significantly decreased in COVID-19 patients compared to a control group, and is strongly associated with higher mortality (Zhang et al., 2021).

Policies and procedures

In this chapter, we reviewed the echocardiographic features of COVID-19 myocarditis. Bedside TTE has acquired a consolidated role in intensive care units for its capacity to diagnose COVID-19-associated myocardial injury and to stratify patient outcome. We investigated in detail the indication of TTE in COVID-19 myocarditis, based on ASE and ESC guidelines, describing advantages and limitations associated with this imaging technique. During the COVID-19 pandemic, the higher contagion risk represented a major health public problem for which several preventive measures were adopted. The discussion in this chapter has focused on the main echocardiographic techniques to evaluate LV: conventional 2D echocardiography, STE, and MW. RV involvement in COVID-19 patients has an emerging role. We also discussed conventional and new echocardiographic tools to diagnose RV dysfunction. In this chapter, our policy was based on a review of guidelines, clinical studies, and narrative review articles regarding COVID-19 myocarditis. The research was carried out on open-access medical portals such as PubMed.

Applications to other areas

In this chapter, the major indications and features of TTE in COVID-19 myocarditis were discussed. TTE is a noninvasive imaging technique representing the first-line exam for several conditions. Bedside TTE has a key role in a critical care setting, and not just for COVID-19 myocarditis. In fact, the 2D echocardiographic features are not specific only for COVID-19 but also for myocarditis due to other viral agents or, in general, for other major cardiovascular diseases. Moreover, TTE is applied to these clinical contexts as specified in international guidelines and allows noninvasively monitoring the hemodynamic state and treatment response. STE analysis has revealed a peculiar aspect of COVID-19 myocarditis, the "reverse tako-tsubo" pattern, and is similarly used to characterize specific patterns of other cardiac conditions, allowing differential diagnosis. RV involvement requires TTE evaluation similar to other clinical contexts, such as ARDS due to COVID-19 or other causes, pH, and PE. Emerging new echocardiographic applications seem to be predictive of a poor prognosis in COVID-19 patients and could also be applied in other cardiac conditions.

Mini-dictionary of terms

- **Myocarditis:** An inflammatory disease of the myocardium associated with the necrosis of cardiomyocytes. It can be caused by various etiological agents, such as SARS-CoV2.
- **POCUS:** A targeted ultrasound examination, performed and interpreted by an experiencing physician in the intensive care unit, to answer a specific clinical question.
- **Transthoracic echocardiography:** An imaging technique that uses ultrasound to evaluate the morphologic and functional features of the heart.
- **Speckle tracking echocardiography:** A noninvasive echocardiographic application allowing the assessment of global and regional ventricular function, independently of in-plane translation movements and angle of ultrasonographic insonation.
- **Acute distress respiratory syndrome:** A type of respiratory insufficiency caused by various etiological agents, responsible for the accumulation of fluid in the lungs with subsequent reduced gas exchanges.

Summary points

- COVID-19 myocarditis represents a cardiac complication, characterized by a heterogeneous presentation from subclinical forms to cardiogenic shock.
- TTE represents the first-line noninvasive imaging exam in COVID-19 myocarditis evaluation.
- During the COVID-19 pandemic, all operators should have adhered to specific preventive measures.
- The subcostal four-chamber TTE view plays a key role in critical care unit.
- The 2D echocardiographic features are common with myocarditis due to other viral agents.
- STE is characterized by the specific "reversal tako-tsubo" pattern of LV and allowed the identification of subclinical COVID-19-related myocardial injury.
- RV dysfunction is associated with a poor outcome in COVID-19 patients.
- The 2D conventional parameters with associated new echocardiographic applications, such as STE or 3D RV ejection fraction, allowed diagnosing RV dysfunction.

References

Adeboye, A., Alkhatib, D., Butt, A., Yedlapati, N., & Garg, N. (2022). A review of the role of imaging modalities in the evaluation of viral myocarditis with a special focus on COVID-19-related myocarditis. *Diagnostics*, *12*(2), 549.

Agricola, E., Beneduce, A., Esposito, A., Ingallina, G., Palumbo, D., Palmisano, A., Ancona, F., Baldetti, L., Pagnesi, M., Melisurgo, G., Zangrillo, A., & De Cobelli, F. (2020). Heart and lung multimodality imaging in COVID-19. *JACC: Cardiovascular Imaging*, *13*, 1792–1808.

Almarawi, H. A., Alwani, M. M., & Arabi, A. (2020). Right ventricular involvement in COVID-19 patients: A rapid review. *Heart Views*, *21*(3), 193–194.

Altersberger, M., Schneider, M., Schiller, M., Binder-Rodriguez, C., Genger, M., Khafaga, M., Binder, T., & Prosch, H. (2021). Point of care echocardiography and lung ultrasound in critically ill patients with COVID-19. *Wiener Klinische Wochenschrift*, *133*(23–24), 1298–1309.

Atri, D., Siddiqi, H. K., Lang, J. P., Nauffal, V., Morrow, D. A., & Bohula, E. A. (2020). COVID-19 for the cardiologist: Basic virology, epidemiology, cardiac manifestations, and potential therapeutic strategies. JACC: Basic to translational. *Science*, *5*, 518–536.

Başar, E. Z., Usta, E., Akgün, G., Güngör, H. S., Sönmez, H. E., & Babaoğlu, K. (2022). Is strain echocardiography a more sensitive indicator of myocardial involvement in patients with multisystem inflammatory syndrome in children (MIS-C) associated with SARS-CoV-2? *Cardiology in the Young*, *32*(10), 1657–1667.

Bleakley, C., Singh, S., Garfield, B., Morosin, M., Surkova, E., Mandalia, M. S., Dias, B., Androulakis, E., Price, L. C., McCabe, C., Wort, S. J., West, C., Li, W., Khattar, R., Senior, R., Patel, B. V., & Price, S. (2021). Right ventricular dysfunction in critically ill COVID-19 ARDS. *International Journal of Cardiology*, *327*, 251–258.

Cameli, M., Pastore, M. C., Henein, M., Aboumarie, H. S., Mandoli, G. E., D'Ascenzi, F., Cameli, P., Franchi, F., Mondillo, S., & Valente, S. (2020). Safe performance of echocardiography during the COVID-19 pandemic: A practical guide. *Reviews in Cardiovascular Medicine*, *21*, 217–223.

Carrizales-Sepúlveda, E. F., Vera-Pineda, R., Flores-Ramírez, R., Hernández-Guajardo, D. A., Pérez-Contreras, E., Lozano-Ibarra, M. M., & Ordaz-Farías, A. (2021). Echocardiographic manifestations in COVID-19: A review. *Heart, Lung & Circulation*, *30*(8), 1117–1129.

Cecchetto, A., Nistri, S., Baroni, G., Torreggiani, G., Aruta, P., Pergola, V., Baritussio, A., Previtero, M., Palermo, C., Iliceto, S., & Mele, D. (2022). Role of cardiac imaging modalities in the evaluation of COVID-19-related cardiomyopathy. *Diagnostics*, *12*(4), 896.

Citro, R., Pontone, G., Bellino, M., Silverio, A., Iuliano, G., Baggiano, A., Manka, R., Iesu, S., Vecchione, C., Asch, F. M., Ghadri, J. R., & Templin, C. (2021). Role of multimodality imaging in evaluation of cardiovascular involvement in COVID-19. *Trends in Cardiovascular Medicine*, *31*, 8–16.

Croft, L. B., Krishnamoorthy, P., Ro, R., Anastasius, M., Zhao, W., Buckley, S., Goldman, M., Argulian, E., Sharma, S. K., Kini, A., & Lerakis, S. (2021). Abnormal left ventricular global longitudinal strain by speckle tracking echocardiography in COVID-19 patients. *Future Cardiology*, *17*(4), 655–661.

D'Alto, M., Marra, A. M., Severino, S., Salzano, A., Romeo, E., De Rosa, R., Stagnaro, F. M., Pagnano, G., Verde, R., Murino, P., Farro, A., Ciccarelli, G., Vargas, M., Fiorentino, G., Servillo, G., Gentile, I., Corcione, A., Cittadini, A., Naeije, R., & Golino, P. (2020). Right ventricular-arterial uncoupling independently predicts survival in COVID-19 ARDS. *Critical Care*, *24*(1), 670.

D'Andrea, A., Cante, L., Palermi, S., Carbone, A., Ilardi, F., Sabatella, F., Crescibene, F., Di Maio, M., Giallauria, F., Messalli, G., Russo, V., & Bossone, E. (2022). COVID-19 myocarditis: Prognostic role of bedside speckle-tracking echocardiography and association with Total scar burden. *International Journal of Environmental Research and Public Health*, *19*(10), 5898.

D'Andrea, A., Scarafile, R., Riegler, L., Liccardo, B., Crescibene, F., Cocchia, R., & Bossone, E. (2020). Right ventricular function and pulmonary pressures as independent predictors of survival in patients with COVID-19pneumonia. *JACC: Cardiovascular Imaging*, *13*(11), 2467–2468.

D'Andrea, A., Vriz, O., Ferrara, F., Cocchia, R., Conte, M., Di Maio, M., Driussi, C., Scarafile, R., Martone, F., Sperlongano, S., Tocci, G., Citro, R., Caso, P., Bossone, E., & Golino, P. (2018). Reference ranges and physiologic variations of left *E/e'* ratio in healthy adults: Clinical and echocardiographic correlates. *Journal of Cardiovascular Echography*, *28*(2), 101–108.

De Lorenzo, A., Kasal, D. A., Tura, B. R., Lamas, C. C., & Rey, H. C. (2020). Acute cardiac injury in patients with COVID-19. *American Journal of Cardiovascular Disease*, *10*, 28–33.

Drake, D. H., De Bonis, M., Covella, M., Agricola, E., Zangrillo, A., Zimmerman, K. G., & Cobey, F. C. (2020). Echocardiography in pandemic: Front-line perspective, expanding role of ultrasound, and ethics of resource allocation. *Journal of the American Society of Echocardiography*, *33*, 683–689.

Dweck, M. R., Bularga, A., Hahn, R. T., Bing, R., Lee, K. K., Chapman, A. R., White, A., Salvo, G. D., Sade, L. E., Pearce, K., Newby, D. E., Popescu, B. A., Donal, E., Cosyns, B., Edvardsen, T., Mills, N. L., & Haugaa, K. (2020). Global evaluation of echocardiography in patients with COVID-19. *European Heart Journal Cardiovascular Imaging*, 949–958.

Escher, F., Westermann, D., Gaub, R., Pronk, J., Bock, T., Al-Saadi, N., Kühl, U., Schultheiss, H. P., & Tschöpe, C. (2011). Development of diastolic heart failure in a 6-year follow-up study in patients after acute myocarditis. *Heart*, *97*, 709–714.

Ezhilkugan, G., Balamurugan, N., Vivekanandan, M., Ajai, R., & Dorje, N. (2021). Unexplained acute right ventricular dilatation and dysfunction in COVID-19. *Journal of Global Infectious Diseases*, *13*(4), 200–201.

Felker, G. M., Boehmer, J. P., Hruban, R. H., Hutchins, G. M., Kasper, E. K., Baughman, K. L., & Hare, J. L. (2000). Echocardiographic findings in fulminant and acute myocarditis. *Journal of the American College of Cardiology*, *36*(1), 227–232.

Focardi, M., Cameli, M., Carbone, S. F., Massoni, A., De Vito, R., & Lisi, M. (2015). Traditional and innovative echocardiographic parameters for the analysis of right ventricular performance in comparison with cardiac magnetic resonance. *European Heart Journal Cardiovascular Imaging*, *16*(1), 47–52.

Goerlich, E., Gilotra, N. A., Minhas, A. S., Bavaro, N., Hays, A. G., & Cingolani, O. H. (2021). Prominent longitudinal strain reduction of basal left ventricular segments in patients with coronavirus Disease-19. *Journal of Cardiac Failure*, *27*(1), 100–104.

Goerlich, E., Minhas, A. S., Mukherjee, M., Sheikh, F. H., Gilotra, N. A., Sharma, G., Michos, E. D., & Hays, A. G. (2021). Multimodality imaging for cardiac evaluation in patients with COVID-19. *Current Cardiology Reports*, *23*(5), 44.

Guazzi, M., Dixon, D., Labate, V., Beussink-Nelson, L., Bandera, F., & Cuttica, M. J. (2017). RV contractile function and its coupling to pulmonary circulation in heart failure with preserved ejection fraction: Stratification of clinical phenotypes and outcomes. *JACC: Cardiovascular Imaging*, *10*, 1211–1221.

Kirkpatrick, J. N., Mitchell, C., Taub, C., Kort, S., Hung, J., & Swaminathan, M. (2020). ASE statement on protection of patients and echocardiography service providers during the 2019 novel coronavirus outbreak: Endorsed by the American College of Cardiology. *Journal of the American Society of Echocardiography*, *33*(6), 648–653.

Klompas, M., Morris, C. A., Sinclair, J., Pearson, M., & Shenoy, E. S. (2020). Universal masking in hospitals in the Covid-19 era. *New England Journal of Medicine*, *382*(21), e63.

Konstantinides, S. V., Meyer, G., Becattini, C., Bueno, H., Geersing, G. J., Harjola, V. P., Huisman, M. V., Humbert, M., Jennings, C. S., Jiménez, D., Kucher, N., Lang, I. M., Lankeit, M., Lorusso, R., Mazzolai, L., Meneveau, N., Ní Áinle, F., Prandoni, P., Pruszczyk, P., … Zamorano, J. L. (2020).

2019 ESC guidelines for the diagnosis and management of acute pulmonary embolism developed in collaboration with the European Respiratory Society (ERS). *European Heart Journal*, *41*(4), 543–603.

Lan, Y., Liu, W., & Zhou, Y. (2021). Right ventricular damage in COVID-19: Association between myocardial injury and COVID-19. *Frontiers in Cardiovascular Medicine*, *8*, 606318.

Lang, R. M., Badano, L. P., Mor-Avi, V., Afilalo, J., Armstrong, A., Ernande, L., Flachskampf, F. A., Foster, E., Goldstein, S. A., Kuznetsova, T., Lancellotti, P., Muraru, D., Picard, M. H., Rietzschel, E. R., Rudski, L., Spencer, K. T., Tsang, W., & Voigt, J. U. (2015). Recommendations for cardiac chamber quantification by echocardiography in adults: An update from the American Society of Echocardiography and the European Association of Cardiovascular Imaging. *Journal of the American Society of Echocardiography*, *28*, 1–39.

Li, Y., Li, H., Zhu, S., Xie, Y., Wang, B., He, L., Zhang, D., Zhang, Y., Yuan, H., Wu, C., Sun, W., Zhang, Y., Li, M., Cui, L., Cai, Y., Wang, J., Yang, Y., Lv, Q., Zhang, L., & Xie, M. (2020). Prognostic value of right ventricular longitudinal strain in patients with COVID-19. *JACC: Cardiovascular Imaging*, *13*(11), 2287–2299.

Li, R., Wang, H., Ma, F., Cui, G. L., Peng, L. Y., Li, C. Z., Zeng, H. S., Marian, A. J., & Wang, D. W. (2021). Widespread myocardial dysfunction in COVID-19 patients detected by myocardial strain imaging using 2-D speckle-tracking echocardiography. *Acta Pharmacologica Sinica*, *42*(10), 1567–1574.

Li, Y. L., Zheng, J. B., Jin, Y., Tang, R., Li, M., Xiu, C. H., Dai, Q. Q., Zuo, S., Wang, H. Q., Wang, H. L., Zhao, M. Y., Ye, M., & Yu, K. J. (2020). Acute right ventricular dysfunction in severe COVID-19 pneumonia. *Reviews in Cardiovascular Medicine*, *21*(4), 635–641.

Longobardo, L., Suma, V., Jain, R., Carerj, S., Zito, C., Zwicke, D. L., & Khandheria, B. K. (2017). Role of two-dimensional speckle-tracking echocardiography strain in the assessment of right ventricular systolic function and comparison with conventional parameters. *Journal of the American Society of Echocardiography*, *30*(10), 937–946.

Mahmoud-Elsayed, H. M., Moody, W. E., Bradlow, W. M., Khan-Kheil, A. M., Senior, J., Hudsmith, L. E., & Steeds, R. P. (2020). *Echocardiographic* findings in patients with COVID-19 pneumonia. *Canadian Journal of Cardiology*, *36*, 1203–1207.

Mondillo, S., Galderisi, M., Mele, D., Cameli, M., Lomoriello, V. S., Zacà, V., Ballo, P., D'Andrea, A., Muraru, D., Losi, M., Agricola, E., D'Errico, A., Buralli, S., Sciomer, S., Nistri, S., & Badano, L. (2011). Speckle-tracking echocardiography: A new technique for assessing myocardial function. *Journal of Ultrasound in Medicine*, *30*(1), 71–83.

Olszewski, R., Timperley, J., Szmigielski, C., Monaghan, M., Nihoyannopoulos, P., Senior, R., & Becher, H. (2007). The clinical applications of contrast echocardiography. *European Journal of Echocardiography*, *8*, S13–S23.

Paternoster, G., Bertini, P., Innelli, P., Trambaiolo, P., Landoni, G., Franchi, F., Scolletta, S., & Guarracino, F. (2021). Right ventricular dysfunction in patients with COVID-19: A systematic review and Meta-analysis. *Journal of Cardiothoracic and Vascular Anesthesia*, *35*(11), 3319–3324.

Picard, M. H., & Weiner, R. B. (2020). Echocardiography in the time of COVID-19. *Journal of the American Society of Echocardiography*, *33*, 674–675.

Purdy, A., Ido, F., Sterner, S., Tesoriero, E., Matthews, T., & Singh, A. (2021). Myocarditis in COVID-19 presenting with cardiogenic shock: A case series. *European Heart Journal - Case Reports*, *5*(2), ytab028.

Rathore, S. S., Rojas, G. A., Sondhi, M., Pothuru, S., Pydi, R., Kancherla, N., Singh, R., Ahmed, N. K., Shah, J., Tousif, S., Baloch, U. T., & Wen, Q. (2021). Myocarditis associated with Covid-19 disease: A systematic review of published case reports and case series. *International Journal of Clinical Practice*, *75*(11), e14470.

Roemer, S., Jaglan, A., Santos, D., Umland, M., Jain, R., Tajik, A. J., & Khandheria, B. K. (2021). The utility of myocardial work in clinical practice. *Journal of the American Society of Echocardiography*, *34*(8), 807–818.

Shah, T. G., Wadia, S. K., Kovach, J., Fogg, L., & Tandon, R. (2016). Echocardiographic parameters of right ventricular function predict mortality in acute respiratory distress syndrome: A pilot study. *Pulmonary Circulation*, *6*(2), 155–160.

Stöbe, S., Richter, S., Seige, M., Stehr, S., Laufs, U., & Hagendorff, A. (2020). Echocardiographic characteristics of patients with SARS-CoV-2 infection. *Clinical Research in Cardiology: Official Journal of the German Cardiac Society*, *109*(12), 1549–1566.

Szekely, Y., Lichter, Y., Taieb, P., Banai, A., Hochstadt, A., Merdler, I., Gal Oz, A., Rothschild, E., Baruch, G., Peri, Y., Arbel, Y., & Topilsky, Y. (2020). Spectrum of cardiac manifestations in COVID-19: A systematic echocardiographic study. *Circulation*, *142*, 342–353.

The Task Force for the management of COVID-19 of the European Society of Cardiology. (2022a). European Society of Cardiology guidance for the diagnosis and management of cardiovascular disease during the COVID-19 pandemic: Part 1—Epidemiology, pathophysiology, and diagnosis. *European Heart Journal*, *43*(11), 1033–1058.

The Task Force for the management of COVID-19 of the European Society of Cardiology. (2022b). ESC guidance for the diagnosis and management of cardiovascular disease during the COVID-19 pandemic: Part 2—Care pathways, treatment, and follow-up. *European Heart Journal*, *43*, 1059–1103.

Weckbach, L. T., Curta, A., Bieber, S., Kraechan, A., Brado, J., Hellmuth, J. C., Muenchhoff, M., Scherer, C., Schroeder, I., Irlbeck, M., Maurus, S., Ricke, J., Klingel, K., Kääb, S., Orban, M., Massberg, S., Hausleiter, J., & Grabmaier, U. (2021). Myocardial inflammation and dysfunction in COVID-19-associated myocardial injury. *Circulation: Cardiovascular Imaging*, *14*(1), e012220.

Zhang, Y., Sun, W., Wu, C., Zhang, Y., Cui, L., Xie, Y., Wang, B., He, L., Yuan, H., Zhang, Y., Cai, Y., Li, M., Zhang, Y., Yang, Y., Li, Y., Wang, J., Yang, Y., Lv, Q., Zhang, L., & Xie, M. (2021). Prognostic value of right ventricular ejection fraction assessed by 3D echocardiography in COVID-19 patients. *Frontiers in Cardiovascular Medicine*, *8*, 641088.

Chapter 32

COVID-19 lockdown and impact on arrhythmias

Valentino Ducceschi and Giovanni Domenico Ciriello
Electrophysiology and Cardiac Pacing Unit, Pellegrini Hospital, Naples, Italy

Introduction

The COVID-19 pandemic caused a worldwide emergency that spread rapidly throughout Asia, Europe, and the Americas. Most countries dealt with a dramatic rise in infection rates, leading to the establishment of specific rules to restrict social contact by following the example of China, where the outbreak first began. These lockdown measures were not synchronously taken throughout the world, given that the pandemic wave hit the Asian, European, and American nations, peaking in each one from the beginning of winter to spring, according to the spread of the SARS-CoV-2 virus.

In Italy, for example, the lockdown period lasted from March 9 to May 4, 2021, with very tight restrictions aimed at reducing social contact as much as possible (Italian Civil Protection Department et al., 2020). As a consequence, several public health and hospital services such as outpatient clinics and routine hospital admissions for chronic illnesses were severely limited (Boriani et al., 2020; Holt et al., 2020; Russo et al., 2021; Schmitt et al., 2022). The different healthcare systems had to face an entirely new situation. The usual population screening to detect heart rhythm disorders was not possible based on common strategies and resources, given that only patients with "acute" arrhythmias manifesting severe symptoms were able to access emergency departments (de Almeida Fernandes et al., 2021; Ducceschi et al., 2022; Harding et al., 2021). The impact of lockdowns on cardiac arrhythmias has shown sometimes conflicting results, so our intention is to clarify the reasons for apparently different reports from the medical literature (Dell'Era et al., 2022; Diemberger et al., 2021; Galand et al., 2021; Mascioli et al., 2021; Matteucci et al., 2021; Sassone et al., 2021).

Our aim is to review the different effects of lockdown restrictions established during the COVID-19 pandemic on the incidence of cardiac rhythm disorders. First, we will analyze how the lockdown period affected the implantation of cardiac implantable electronic devices (CIEDs) in terms of urgent or scheduled indications. Second, we will focus on cardiac tachyarrhythmias, reporting their incidence in emergency departments and their detection in patients with CIEDs through home monitoring technology.

COVID-19 lockdown and cardiac arrhythmias

Generally speaking, in comparing the two months of lockdown with the same time frame in 2019 or with an equal period of 2020, we see that the total number of CIED implantations declined, mainly due to a significant decrease in scheduled procedures, but also hospital admissions for urgent cardiac pacing operations were reduced (de Almeida Fernandes et al., 2021; Dell'Era et al., 2022; Russo et al., 2021). We might have expected a remarkable fall in the number of scheduled procedures owing to a very strict policy aimed at limiting social contacts. Indeed, all activities involving general population screening such as outpatient clinics or chronic patient follow-up visits were prohibited. However, also urgent cardiac pacing interventions diminished during the lockdown, even when comparing the data with the two immediately subsequent months in which the pandemic wave was still peaking (de Almeida Fernandes et al., 2021). A possible explanation could be the fear of infection with the deadly virus with all its consequences on hospitalization time, social distancing from family members, etc.

The real impact of the COVID-19 lockdown on the incidence of tachyarrhythmias in the general population is difficult to assess. Only one study conducted in Denmark analyzed atrial fibrillation occurrence during this time frame (Holt et al., 2020). A 47% reduction in total new-onset cases was registered, with a greater prevalence of unfavorable events such as death or ischemic stroke (2.7% vs 1.3% and 5.3% vs 4.3%, respectively). With these data, the authors hypothesized that the

Management, Body Systems, and Case Studies in COVID-19. https://doi.org/10.1016/B978-0-443-18703-2.00023-2
Copyright © 2024 Elsevier Inc. All rights are reserved, including those for text and data mining, AI training, and similar technologies.

inflexible measures taken to reduce social contact, thus limiting access to the healthcare system, resulted in a certain amount of undiagnosed atrial fibrillation cases. According to the same authors, this might have led to poorer outcomes, especially in a prolonged or repeated lockdown (Holt et al., 2020).

In another study comprising 1098 patients with an insertable loop recorder followed by remote monitoring, it emerged that during the lockdown, with respect to the pre-COVID-19 era, there was a fivefold increase in arrhythmic episodes, mainly atrial fibrillation, but also narrow complex tachycardias and ventricular arrhythmias (Harding et al., 2021) (Figs. 1 and 2). In the same paper, however, a very significant increase in total patient-initiated events was recorded, the majority of which were not arrhythmias. The authors attributed these data to the severe psychological strain imposed by COVID-19 restrictions that encouraged bad habits such as sedentariness and alcohol consumption (Harding et al., 2021).

If data regarding the general population are very poor and derive from just the two cited studies, arrhythmia occurrence in patients with CIEDs during the lockdown has been investigated quite deeply by exploiting the remote monitoring platforms.

While some authors confirmed what was highlighted in the aforementioned papers, demonstrating an increase in the incidence of atrial (Diemberger et al., 2021; Ducceschi et al., 2022; Schmitt et al., 2022) and ventricular arrhythmias (Ducceschi et al., 2022; Schmitt et al., 2022), other studies (9 Mascioli et al., 2021; Matteucci et al., 2021; Sassone et al., 2021) did not endorse similar results. The reason for this apparent discrepancy might be due to differences in the patient populations examined. Matteucci et al. (2021) considered fewer sick patients with CIEDs, including those with pacemakers. Mascioli et al. (2021) selected a small cohort of patients (just 180) and what they defined as the postlockdown period was, in fact, still part of the lockdown established in Italy. Sassone et al. (2021) likewise conducted a study that was

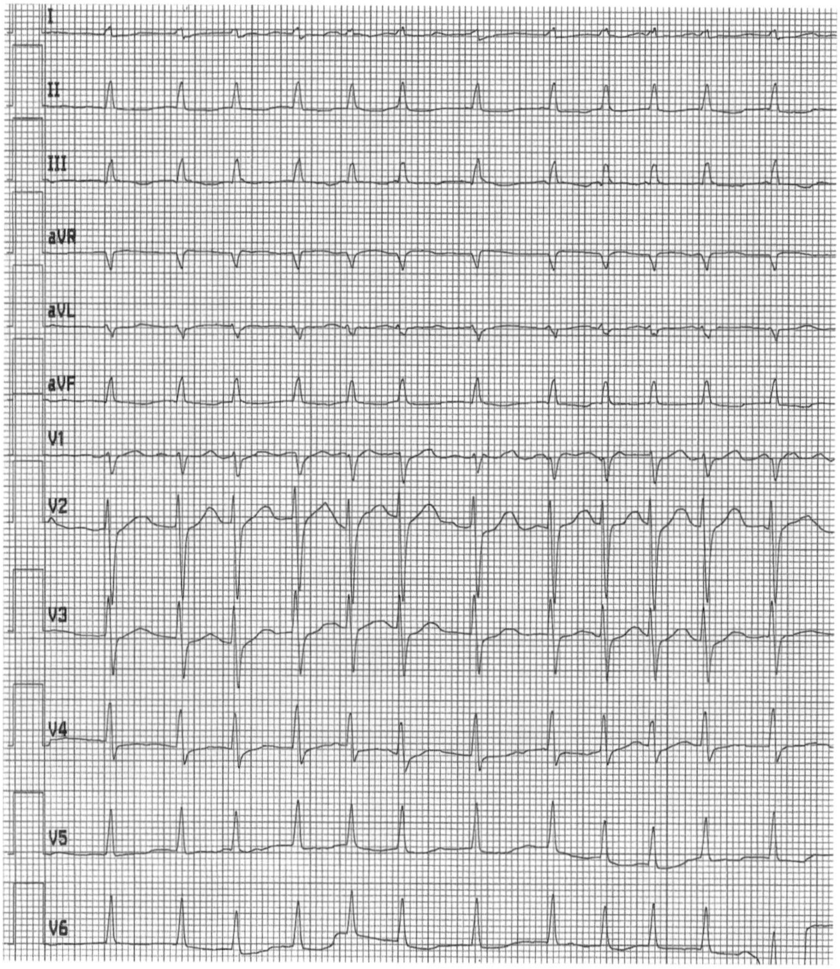

FIG. 1 Atrial fibrillation with rapid ventricular response.

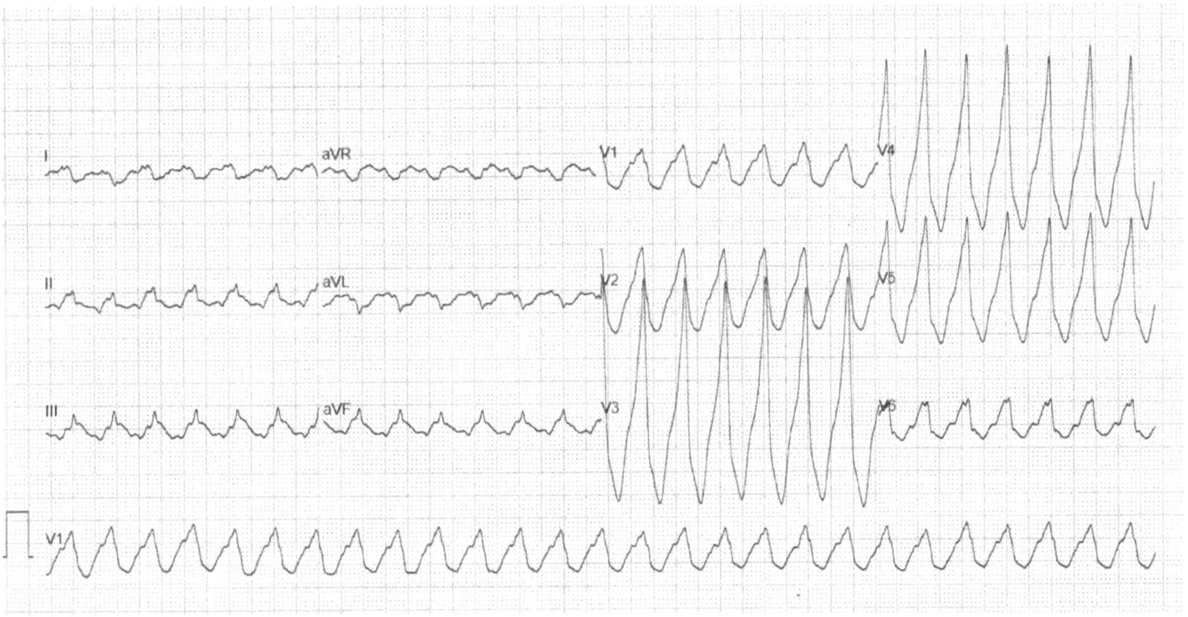

FIG. 2 Ventricular tachycardia.

geographically quite limited, involving only one city. Besides, the observation of a neglectable impact of the lockdown on cardiac arrhythmias reported in this latter paper might be attributable to a too-wide programmed arrhythmia detection window that also included a monitor zone.

On the contrary, other studies embracing a larger geographical area and therefore a greater number of hospitals reported a significant increase in atrial fibrillation (Diemberger et al., 2021; Ducceschi et al., 2022; Schmitt et al., 2022) and ventricular tachyarrhythmias during the lockdown (Ducceschi et al., 2022; Schmitt et al., 2022). These authors precisely respected the time frame and enrolled a more homogeneous population of patients, that is, those with defibrillators with or without resynchronization capability (Diemberger et al., 2021; Ducceschi et al., 2022; Schmitt et al., 2022). In addition, the detection zones for ventricular tachyarrhythmia recognition were the same that delivered therapies, so that only significant episodes were considered (Diemberger et al., 2021; Ducceschi et al., 2022). Other data regarding the mean heart rate, daily activity, and heart rate variability reflected a more homogeneous response in patients with CIEDs, showing a similar decrease (Diemberger et al., 2021; Ducceschi et al., 2022; Schmitt et al., 2022).

The reason can be attributed to the strict restrictions imposed during the lockdown, which forced people to stay at home as much as possible. This led to unprecedented psychological stress on the population, reflected by an increased sympathetic tone, as evidenced by the finding of reduced heart rate variability (Diemberger et al., 2021).

Such a strain was the unavoidable consequence of limiting leisure activities and unnecessary social contacts. Furthermore, mass media pressure played a key role by reporting precise bulletins regarding the number of victims and the severe crisis of the healthcare system. Hospitals had to cope with overcrowded intensive care units inadequate to accommodate the everyday increasing number of critically ill infected patients.

Several papers from the literature underline the importance of heart-brain interactions in the occurrence of cardiac arrhythmias (Buckley & Shivkumar, 2016; Di Renzo et al., 2020; Peacock & Whang, 2013; Taggart et al., 2011) Mental stress and emotions may alter the sympathetic-parasympathetic balance, affecting the spatial distribution of autonomic input to the heart and even causing coronary artery vasoconstriction and ischemia, favoring the onset of tachyarrhythmias (Buckley & Shivkumar, 2016; Rossi et al., 2020; Ziegelstein, 2007).

It is conceivable that this negative heart-brain interaction might cause stressful conditions in populations that have experienced lockdown, especially those with advanced cardiac diseases such as patients with CIEDs (Ducceschi et al., 2022). Recent evidence suggests that asymmetric brain activity can contribute to heart susceptibility to arrhythmias (Buckley & Shivkumar, 2016; Ziegelstein, 2007). Emotional stress may lateralize cerebral activity, thus stimulating the cardiac muscle asymmetrically and resulting in areas of inhomogeneous repolarization. This condition provokes electrical instability and therefore promotes the onset of arrhythmias (Buckley & Shivkumar, 2016; Rossi et al., 2020; Ziegelstein, 2007).

Conclusions

The measures adopted during the lockdown aimed at containing the pandemic spread negatively affected the health of patients with heart disease, making difficult the screening of chronically ill patients in outpatient clinics. This not only led to a decline in urgent pacemaker implantations and likewise urgent arrhythmia ablations, with obvious deleterious effects on public health, but also to a decreased number of scheduled procedures such as CIED operations and other cardiac ablations. The important consequences this caused on public health are still being evaluated.

Finally, the huge psychosocial stress that affected the world population contributed to a rise in the occurrence of cardiac tachyarrhythmias, and this was manifest especially in patients with CIEDs with remote monitoring function. This is because the widespread fear of being infected by COVID-19 together with a lack of hospital beds for pathologies other than COVID-19 strongly discouraged people with rhythm disorders to ask for medical assistance.

Mini-dictionary of terms

- **Catheter ablation**: Minimally invasive interventional procedure performed to treat different types of cardiac arrhythmias.
- **CIEDs**: Devices that are implanted to treat cardiac rhythm disturbances; they are represented by pacemakers and cardiac defibrillators.
- **Remote monitoring**: Last-generation pacemakers or defibrillators are capable of daily transmissions of data to a central server, allowing nurses and physicians to review automatically received alert notifications about arrhythmias and technical issues, avoiding unnecessary hospital admissions for patients.
- **Atrial fibrillation**: The most common form of sustained arrhythmia in the general population. It is characterized by irregular atrial electrical activity with a variable heart rate. More common symptoms of atrial fibrillation are palpitations, feeling of an irregular heartbeat, or a pounding heart.
- **Ventricular arrhythmias**: The scariest cardiac arrhythmias are those that arise from the right and left ventricles of the heart, such as ventricular tachycardia or ventricular fibrillation. These types of arrhythmias are sometimes life-threatening, leading to sudden cardiac death if not treated.

Summary points

- A significant increase in arrhythmia episodes occurred during the COVID-19 lockdown.
- Data collected from cardiac devices showed an increase in both atrial and ventricular arrhythmias, compared with the same time frame of previous years.
- The number of cardiac device implantations and ablation procedures declined during the COVID-19 lockdown.
- Remote monitoring technology was useful to follow patients during the COVID-19 outbreak, reducing hospital admissions.
- Psychological stress due to COVID-19 restrictions played a key role in promoting susceptibility to arrhythmias.

References

Boriani, G., Palmisano, P., Guerra, F., Bertini, M., Zanotto, G., Lavalle, C., Notarstefano, P., Accogli, M., Bisignani, G., Forleo, G. B., Landolina, M., D'Onofrio, A., Ricci, R., De Ponti, R., & AIAC Ricerca Network Investigators. (2020). Impact of COVID-19 pandemic on the clinical activities related to arrhythmias and electrophysiology in Italy: Results of a survey promoted by AIAC (Italian Association of Arrhythmology and Cardiac Pacing). *Internal and Emergency Medicine*, *15*(8), 1445–1456. https://doi.org/10.1007/s11739-020-02487-w.

Buckley, U., & Shivkumar, K. (2016). Stress-induced cardiac arrhythmias: The heart-brain interaction. *Trends in Cardiovascular Medicine*, *26*(1), 78–80. https://doi.org/10.1016/j.tcm.2015.05.001.

de Almeida Fernandes, D., Cadete, R., António, N., Ventura, M., Cristóvão, J., Elvas, L., & Gonçalves, L. (2021). Impact of the COVID-19 lockdown in urgent pacemaker implantations: A cross-sectional study. *Journal of Arrhythmia*, *38*(1), 137–144. https://doi.org/10.1002/joa3.12658.

Dell'Era, G., Colombo, C., Forleo, G. B., Curnis, A., Marcantoni, L., Racheli, M., Sartori, P., Notarstefano, P., De Salvia, A., Guerra, F., Ziacchi, M., Tondo, C., Gandolfi, E., De Vecchi, F., Mascioli, G., Coppolino, A., Catuzzo, B., Amellone, C., Mantica, M., … Patti, G. (2022). Reduction of admissions for urgent and elective pacemaker implant during the COVID-19 outbreak in Northern Italy. *Journal of Cardiovascular Medicine (Hagerstown, Md.)*, *23*(1), 22–27. https://doi.org/10.2459/JCM.0000000000001189.

Di Renzo, L., Gualtieri, P., Pivari, F., Soldati, L., Attinà, A., Cinelli, G., Leggeri, C., Caparello, G., Barrea, L., Scerbo, F., Esposito, E., & De Lorenzo, A. (2020). Eating habits and lifestyle changes during COVID-19 lockdown: An Italian survey. *Journal of Translational Medicine*, *18*(1), 229. https://doi.org/10.1186/s12967-020-02399-5.

Diemberger, I., Vicentini, A., Cattafi, G., Ziacchi, M., Iacopino, S., Morani, G., Pisanò, E., Molon, G., Giovannini, T., Dello Russo, A., Boriani, G., Bertaglia, E., Biffi, M., Bongiorni, M. G., Rordorf, R., & Zucchelli, G. (2021). The impact of COVID-19 pandemic and lockdown restrictions on cardiac implantable device recipients with remote monitoring. *Journal of Clinical Medicine*, *10*(23), 5626. https://doi.org/10.3390/jcm10235626.

Ducceschi, V., de Divitiis, M., Bianchi, V., Calvanese, R., Covino, G., Rapacciuolo, A., Russo, V., Canciello, M., Volpicelli, M., Ammirati, G., Sangiuolo, R., Papaccioli, G., Ciardiello, C., Innocenti, S., & D'Onofrio, A. (2022). Effects of COVID-19 lockdown on arrhythmias in patients with implantable cardioverter-defibrillators in southern Italy. *Journal of Arrhythmia*, *38*(3), 439–445. https://doi.org/10.1002/joa3.12713.

Galand, V., Hwang, E., Gandjbakhch, E., Sebag, F., Marijon, E., Boveda, S., Leclercq, C., Defaye, P., Rosier, A., & Martins, R. P. (2021). Impact of COVID-19 on the incidence of cardiac arrhythmias in implantable cardioverter defibrillator recipients followed by remote monitoring. *Archives of Cardiovascular Diseases*, *114*(5), 407–414. https://doi.org/10.1016/j.acvd.2021.02.005.

Harding, I., Khan, P., Alves, K., Weerasinghe, N., Daily, T., Arumugam, P., Kamdar, R., Sohal, M., Murgatroyd, F., & Scott, P. A. (2021). Remote monitoring of arrhythmias in the COVID lockdown era: A multicentre experience. *Circulation. Arrhythmia and Electrophysiology*, *14*(1), e008932. https://doi.org/10.1161/CIRCEP.120.008932.

Holt, A., Gislason, G. H., Schou, M., Zareini, B., Biering-Sørensen, T., Phelps, M., Kragholm, K., Andersson, C., Fosbøl, E. L., Hansen, M. L., Gerds, T. A., Køber, L., Torp-Pedersen, C., & Lamberts, M. (2020). New-onset atrial fibrillation: Incidence, characteristics, and related events following a national COVID-19 lockdown of 5.6 million people. *European Heart Journal*, *41*(32), 3072–3079. https://doi.org/10.1093/eurheartj/ehaa494.

Italian Civil Protection Department, Morettini, M., Sbrollini, A., Marcantoni, I., & Burattini, L. (2020). COVID-19 in Italy: Dataset of the Italian civil protection department. *Data in Brief*, *30*, 105526. https://doi.org/10.1016/j.dib.2020.105526.

Mascioli, G., Lucca, E., Napoli, P., & Giacopelli, D. (2021). Impact of COVID-19 lockdown in patients with implantable cardioverter and cardiac resynchronization therapy defibrillators: Insights from daily remote monitoring transmissions. *Heart and Vessels*, *36*(11), 1694–1700. https://doi.org/10.1007/s00380-021-01843-w.

Matteucci, A., Bonanni, M., Centioni, M., Zanin, F., Geuna, F., Massaro, G., & Sangiorgi, G. (2021). Home management of heart failure and arrhythmias in patients with cardiac devices during pandemic. *Journal of Clinical Medicine*, *10*(8), 1618. https://doi.org/10.3390/jcm10081618.

Peacock, J., & Whang, W. (2013). Psychological distress and arrhythmia: Risk prediction and potential modifiers. *Progress in Cardiovascular Diseases*, *55*(6), 582–589. https://doi.org/10.1016/j.pcad.2013.03.001.

Rossi, R., Socci, V., Talevi, D., Mensi, S., Niolu, C., Pacitti, F., Di Marco, A., Rossi, A., Siracusano, A., & Di Lorenzo, G. (2020). COVID-19 pandemic and lockdown measures impact on mental health among the general population in Italy. *Frontiers in Psychiatry*, *11*, 790. https://doi.org/10.3389/fpsyt.2020.00790.

Russo, V., Pafundi, P. C., Rapacciuolo, A., de Divitiis, M., Volpicelli, M., Ruocco, A., Rago, A., Uran, C., Nappi, F., Attena, E., Chianese, R., Esposito, F., Del Giorno, G., D'Andrea, A., Ducceschi, V., Russo, G., Ammendola, E., Carbone, A., Covino, G., … D'Onofrio, A. (2021). Cardiac pacing procedures during coronavirus disease 2019 lockdown in Southern Italy: Insights from Campania Region. *Journal of Cardiovascular Medicine (Hagerstown, Md.)*, *22*(11), 857–859. https://doi.org/10.2459/JCM.0000000000001156.

Sassone, B., Virzì, S., Bertini, M., Pasanisi, G., Manzoli, L., Myers, J., Grazzi, G., & Muser, D. (2021). Impact of the COVID-19 lockdown on the arrhythmic burden of patients with implantable cardioverter-defibrillators. *Pacing and Clinical Electrophysiology: PACE*, *44*(6), 1033–1038. https://doi.org/10.1111/pace.14280.

Schmitt, J., Wenzel, B., Brüsehaber, B., Anguera, I., de Sousa, J., Nölker, G., Bulava, A., Marques, P., Hatala, R., Golovchiner, G., Meyhöfer, J., Ilan, M., & On behalf of the BIO|STREAM.HF investigators. (2022). Impact of lockdown during COVID-19 pandemic on physical activity and arrhythmia burden in heart failure patients. *Pacing and Clinical Electrophysiology: PACE*, *45*(4), 471–480. https://doi.org/10.1111/pace.14443.

Taggart, P., Boyett, M. R., Logantha, S., & Lambiase, P. D. (2011). Anger, emotion, and arrhythmias: From brain to heart. *Frontiers in Physiology*, *2*, 67. https://doi.org/10.3389/fphys.2011.00067.

Ziegelstein, R. C. (2007). Acute emotional stress and cardiac arrhythmias. *JAMA*, *298*(3), 324–329. https://doi.org/10.1001/jama.298.3.324.

Chapter 33

Cardiometabolic disease and COVID-19: A new narrative

Mohamad B. Taha[a], Bharat Narasimhan[a], Eleonora Avenatti[a], Aayush Shah[a], and Wilbert S. Aronow[b]
[a]*Department of Cardiology, Houston Methodist DeBakey Heart & Vascular Center, Houston, TX, United States,* [b]*Cardiology Division, Department of Medicine, Westchester Medical Center and New York Medical College, Valhalla, NY, United States*

Abbreviations

ACE2	angiotensin-converting enzyme 2
ACEIs	angiotensin-converting enzyme inhibitors
ARBs	angiotensin 2 receptor blockers
BMI	body mass index
COVID-19	coronavirus disease 2019
CVD	cardiovascular disease
HTN	hypertension
SARS-CoV-2	severe acute respiratory syndrome coronavirus 2
T2D	type 2 diabetes mellitus

Introduction

Cardiometabolic disease refers to a cluster of common but often preventable related conditions that increases the risk of cardiovascular diseases (CVDs) and type 2 diabetes (T2D), including insulin resistance, hyperglycemia, abdominal obesity, nonalcoholic fatty liver disease, dyslipidemia, inflammation, and hypertension (HTN), in addition to environmental risk factors such as sedentary lifestyle, poor dietary habits, and smoking (Mechanick et al., 2020). On the spectrum of disease progression to CVD and T2D is the metabolic syndrome, which describes a common cooccurrence of metabolic risk factors for CVD and T2D. Based on the National Cholesterol Education Program's Adult Treatment Panel III, metabolic syndrome is diagnosed if patients have any three of the following five clinical traits: abdominal obesity, high serum triglycerides, low serum high-density lipoprotein, elevated blood pressure, and evidence of impaired glucose tolerance (Grundy et al., 2005). While it is unclear whether cardiometabolic disease has a unique pathophysiology or is simply a collection of the risk factors of its individual components, cardiometabolic disease and its components are important risk factors of atherosclerotic CVD and T2D and, therefore, must be identified and managed throughout a patient's life. Thus, to reduce the chances of complications in at-risk patients, physicians must identify the signs of cardiometabolic disease and prescribe aggressive lifestyle changes centered around healthy nutrition, physical activity, and weight loss (Fig. 1).

The current coronavirus disease 2019 (COVID-19) pandemic caused by severe acute respiratory syndrome coronavirus 2 (SARS-CoV-2) infection has affected 628 million people resulting in more than 6.5 million deaths. The interlinkage between CMD and COVID-19 has been increasingly recognized over the past years. This began with the observation that patients with cardiometabolic disease are at higher risk of poor outcomes with COVID-19. On the other hand, the pandemic in itself has impacted overall cardiometabolic disease via a multitude of biological, behavioral, and healthcare policy-based mechanisms. In this chapter, we delve into both the aforementioned issues while also exploring potential strategies to mitigate the long-term impact of COVID-19.

Management, Body Systems, and Case Studies in COVID-19. https://doi.org/10.1016/B978-0-443-18703-2.00013-X
Copyright © 2024 Elsevier Inc. All rights are reserved, including those for text and data mining, AI training, and similar technologies.

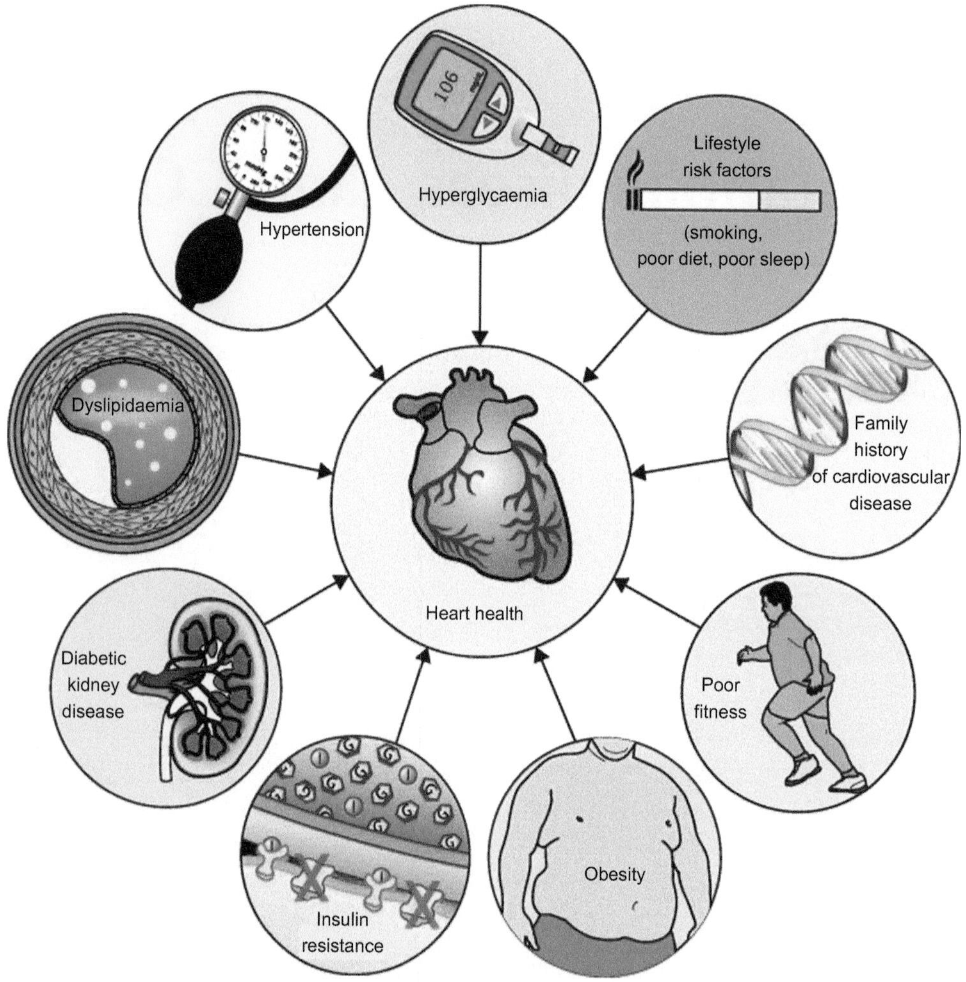

FIG. 1 Various components of cardiometabolic disease increase the risk of cardiovascular diseases. *Reprinted with permission Bjornstad, P., Donaghue, K. C., & Maahs, D. M. (2018). Macrovascular disease and risk factors in youth with type 1 diabetes: Time to be more attentive to treatment? The Lancet Diabetes and Endocrinology, 6(10), 809–820. https://doi.org/10.1016/S2213-8587(18)30035-4.*

Impact of cardiometabolic disease on COVID-19

Susceptibility to COVID-19 infection

True estimates of COVID-19 infection risks in patients with underlying cardiometabolic disease are limited by the observational design of the majority of studies on the subject as well as the inability to capture asymptomatic cases, thus complicating an estimation of global prevalence of the infection. Hence, most of our understanding arises from patients hospitalized with COVID-19 (Fig. 2).

Since the early reports from China in 2020, the high prevalence of cardiovascular disease and risk factors in patients hospitalized with COVID-19 has become clear. The earliest data from Wuhan on hospitalized patients in January 2020 show the prevalence of T2D, HTN, and other CVDs at 20%, 15%, and 15%, respectively (Huang et al., 2020). This finding has been extensively confirmed in larger cohorts (Guan, Ni, et al., 2020; Wu & McGoogan, 2020; Zhou et al., 2020). In the United Kingdom (UK), a prospective observational study of 16,749 patients hospitalized with COVID-19 between February and April 2020–accounting for 14.7% of all people who tested positive for COVID-19 in the UK-reported a prevalence of CVD, T2D, and obesity at 29%, 19%, and 11%, respectively (Docherty et al., 2020). Similar findings were reported from the United States; in a study of 393 patients hospitalized for COVID-19, 50% of patients had HTN, 36% had obesity, and 25% had T2D (Goyal et al., 2020).

Obesity was introduced at that point as a major risk factor that increases susceptibility to COVID-19. This was not as highlighted in the Chinese population, likely due to baseline differences in the rates of obesity between the two areas.

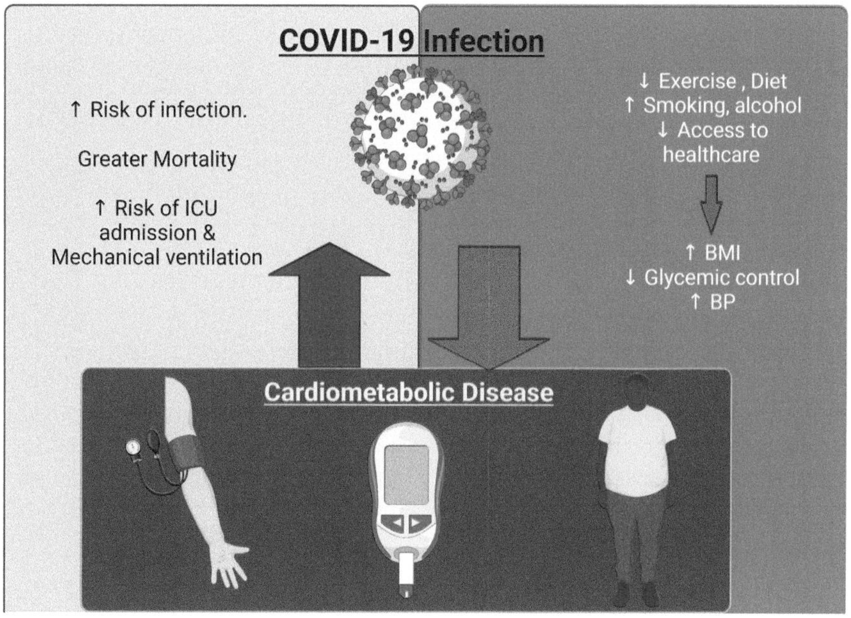

FIG. 2 Relationship between COVID-19 and cardiometabolic disease.

Impact on COVID-19 disease severity

The global case fatality rate of COVID-19 peaked at approximately 10% and is currently estimated at between 2% and 3% (Abou Ghayda et al., 2022). The aforementioned comorbidities have been associated with increased risks of severe disease such as the need for intensive care unit (ICU) level of care and mechanical ventilation in addition to mortality (Hendren et al., 2021; Mahamat-Saleh et al., 2021). It is often difficult to tease out the relative impact of singular cardiovascular risk factors on COVID-19 outcome, which often coexist in the same patient. An analysis of 44,672 COVID-19 cases in China revealed a mortality rate for patients with CVD, T2D, and HTN at 10.5%, 7.3%, and 6%, respectively (Wu & McGoogan, 2020). Moreover, in a meta-analysis of 186 studies representing more than 200,000 deaths among more than 1.3 million patients with COVID-19, the absolute risk of COVID-19 mortality was increased by 14%, 11%, 12%, and 7% for T2D, HTN, obesity, and smoking, respectively (Mahamat-Saleh et al., 2021).

In addition to mortality, numerous studies of patients with COVID-19 support the increased severity of illness in patients with cardiometabolic disease. Data from 1590 patients with confirmed COVID-19 across China showed that COVID-19 patients with HTN were three times more likely to require the ICU and invasive mechanical ventilation (Guan, Liang, et al., 2020). Similarly, patients with T2D have a significantly higher risk of more critical illness. In a multicentered study of 7337 COVID-19 patients, the presence of preexisting T2D was associated with significantly higher rates of ICU care (Zhu et al., 2020). Additional studies report obesity as a major and independent prognostic factor for severe COVID-19. Studies from China showed that patients who are overweight and have obesity are 1.8-fold and 3.4-fold more likely to develop severe COVID-19 disease, respectively, compared with normal-weight patients (Cai et al., 2020). Data from the United States report that COVID-19 patients with a body mass index (BMI) $\geq 35\,kg/m^2$ were 3.6-fold more likely to require ICU admission as well as invasive mechanical ventilation (Cai et al., 2020). Moreover, progressively higher obesity classes were associated with a progressively higher risk of in-hospital death or need for mechanical ventilation. A significant interaction with age was described, such that the association of BMI with severe outcomes from COVID-19 was strongest in adults ≤ 50 years, and progressively weaker for the strata of 51–70 years, and >70 years old (Hendren et al., 2021).

These findings are particularly concerning when considering the alarmingly high prevalence of cardiometabolic disease in the US general population. In the United States from 1999 to 2018, the prevalence of obesity increased from 30.5% to 42.4% while the prevalence of severe obesity increased from 4.7% to 9.2% (Hales et al., 2020). In 2018, the age-adjusted prevalence of all types of heart diseases was 11.2% (Tsao et al., 2022). Additionally, national reports estimated that 37.3 million people have T2D (11.3% of the US population), of which only 77% are diagnosed, and 96 million adults have prediabetes (38% of the adult US population) (Centers for Disease Control and Prevention, 2022).

The impact of aggressive cardiometabolic disease management on the mitigation of infection rates and severity is yet to be determined.

Mechanism

The exact mechanisms underlying the increased susceptibility and higher risk of morbidity and mortality with COVID-19 in patients with cardiometabolic disease remain poorly understood; however, there are several key factors to consider. Cardiovascular tissues are a direct target for the SARS-CoV2 virus due to the expression of angiotensin-converting enzyme 2 (ACE2), which is the main viral receptor binding site mediating viral entry and cell invasion (Ziegler et al., 2020). ACE2 is highly upregulated in patients with T2D, HTN, and obesity. These patients also have increased levels of key proinflammatory markers, such as C-reactive protein, tumor necrosis factor-alpha, interleukin-6, and many other adipokines, indicative of immune dysregulation (Zhu et al., 2021). Hyperactivation of the immune system, interleukin-6 in particular, has been linked to the severe inflammation that is the hallmark of the "chemokine storm," the severe third phase of COVID-19 disease associated with systemic damage (Siddiqi & Mehra, 2020).

Adipose tissue also has a high expression of ACE2; in patients with obesity, this likely contributes to the increased susceptibility to and severity of the disease (Al-Benna, 2020). The excess adipose tissue in these patients could represent a reservoir for viral replication (Al-Benna, 2020; Kassir, 2020). Additionally, obesity and its associated cardiovascular, metabolic, respiratory, and thrombotic complications can impair the body's ability to overcome the infection. Examples of these complications include compromised cardiorespiratory reserve and fitness and increased susceptibility to immune-driven vascular and thrombotic effects.

Management consideration of COVID-19

Due to the heightened susceptibility of patients with cardiometabolic disease to SARS-CoV-2 infection, it is necessary to consider the multiple variables that must be effectively managed in such patients. Individuals with T2D or other metabolic disorders who have not been infected with the SARS-CoV-2 virus should intensify their metabolic control as a means to prevent infection and poor COVD-19-related outcomes; this includes adequate glycemic and blood pressure control.

Management of hypertension and hyperglycemia in patients at risk of or with COVID-19 is of critical importance. Patients with T2D and HTN, who are at increased risk of severe COVID-19, are often treated with angiotensin-converting enzyme inhibitors (ACEIs) and angiotensin 2 receptor blockers (ARBs) due to their efficacy in the reduction of cardiovascular outcomes and renal preservation. Treatment with these medications was believed to potentially increase the expression of ACE2, potentially increasing risk of infection. As such, calcium-channel blockers (CCBs) were proposed as the preferred antihypertensive agents during COVID-19 illness. However, no studies to date have demonstrated a significant association between ACEI and ARB use and the risk of developing severe COVID-19 (Gnanenthiran et al., 2022). Similarly, studies have shown no substantial increase in the risk of infection or developing severe COVID-19 symptoms in association with five common classes of antihypertensive medications: ACEi, ARBs, CCBs, beta-blockers, or thiazide diuretics (Reynolds et al., 2020). Given the large benefits of these drugs on cardiometabolic disease, ACEIs and ARBs should not be discontinued.

The management of COVID-19 patients with T2D is another important topic of consideration. In recently published multisociety recommendations for the care of COVID-19 patients with T2D, emphasis was placed on the continuous and reliable control of the blood glucose level (Bornstein et al., 2020). Studies have shown that in individuals with T2D, poorly controlled blood glucose (glycemic variability >10mmol/L) yielded a markedly higher mortality rate compared with well-controlled blood glucose. Currently, insulin is the mainstay of glycemic control in hospitalized COVID-19 patients, with basal or intermediate-acting insulin given daily, along with preprandial doses of short-acting insulin in patients who tolerate oral intake. Studies have shown that insulin-mediated blood glucose control with continuous glucose monitoring significantly improves the prognosis for hospitalized patients with COVID-19 and hyperglycemia (Wong et al., 2022; Zhu et al., 2020). For critically ill patients in the ICU, continuous intravenous insulin infusion is recommended with a target blood glucose of 140–180mg/dL. The concurrent use of steroids in the management of severe COVID-19 adds a layer of complexity and significant challenges to the achievement of stable target blood glucose levels. Noninsulin treatments require reevaluation in the hospitalized patient. Metformin should be discontinued in patients with acute illness and dehydration due to the risk of lactic acidosis, and pioglitazone should be stopped in critically severe COVID-19 cases. Sodium-glucose transporter 2 inhibitors (SGLT2i), which are known to exert both cardio- and reno-protective effects in patients, have been hypothesized to prevent cardiovascular and renal complications in COVID-19 patients with T2D. However, the current recommendation is to withhold their use during acute illness due to the increased risk of euglycemic ketoacidosis and dehydration. Dipeptidyl peptidase-4 inhibitors and glucagon-like peptide-1 receptor agonists (GLP-1RA) may be continued in noncritically ill patients. Ultimately, there is still much to be learned about the short- and long-term management of T2D in patients with COVID-19. Treatment guidelines shared by experts emphasize the

TABLE 1 Management consideration of potential interfering effects of medication in COVID-19 patients.

Hypoglycemic medications	
Insulin	• Insulin therapy should not be stopped as it's the mainstay of glycemic treatment during severe illness • Regular self-monitoring of blood-glucose every 2–4h should be encouraged, or continuous glucose monitoring • Carefully adjust regular therapy if appropriate to reach therapeutic goals
Metformin	• Potentially leads to dehydration and lactic acidosis, more likely if patients are dehydrated, therefore should be stopped during acute illness • There is a risk of acute kidney injury or chronic kidney disease, and renal function should be carefully monitored
Dipeptidyl peptidase-4 inhibitors	• Generally well tolerated and can be continued
Glucagon-like peptide-1 receptor agonists	• Generally well tolerated and can be continued • Adequate fluid intake and regular meals should be encouraged
Sodium-glucose-co-transporter 2 inhibitors	• Risk of dehydration and euglycemic ketoacidosis during illness, so patients should stop taking the medication during acute illness • Renal function should be carefully monitored for acute kidney injury
Blood pressure medication	
Angiotensin-converting enzyme inhibitors	• Can be continued and has no significant association with a risk of developing severe COVID-19 disease • There is a risk of acute kidney injury or chronic kidney disease, and renal function should be carefully monitored • Avoid during critical illness
Angiotensin 2 receptor blockers	• Can be continued and has no significant association with a risk of developing severe COVID-19 disease • There is a risk of acute kidney injury or chronic kidney disease, and renal function should be carefully monitored • Avoid during critical illness
Calcium-channel blockers	• Can be continued and has no significant association with a risk of developing severe COVID-19 disease
Thiazide diuretics	• Can be continued and has no significant association with a risk of developing severe COVID-19 disease • There is a risk of dehydration, acute kidney injury, or chronic kidney disease, and renal function should be carefully monitored • Avoid during critical illness
Beta-blockers	• Can be continued and has no significant association with a risk of developing severe COVID-19 disease

key importance of effective glycemic control. Table 1 highlights management consideration of potential interfering effects of medication in COVID-19 patients.

Lastly, statins are lipid-lowering therapies with antiinflammatory and immunomodulating properties that constitute the cornerstone of cardiometabolic treatment. Several observational studies have suggested that antecedent and in-hospital statin use is associated with a lower risk of mortality in COVID-19 (Lohia et al., 2021; Permana et al., 2021). Therefore, it is recommended that statins be continued in patients with COVID-19 and optimal control of lipid concentrations be achieved.

Vaccination consideration of COVID-19

Vaccines have proven a true game-changer in the management of the COVID-19 pandemic. Given the close links among cardiometabolic disease and COVID-19 outcomes as well as the issues of vaccine hesitancy, it is imperative to discuss any potential relationship between cardiometabolic disease and vaccination. With regard to CVD, vaccination against COVID-19 is not a contraindication, as there has been no evidence of increased major adverse cardiovascular events with COVID-19 vaccination (Ye et al., 2022). Recent data suggest a reduced risk of acute myocardial injury and ischemic stroke in patients with COVID-19 infection who were vaccinated, as compared to the unvaccinated population (Kim et al., 2022). Moreover, vaccination also prevents the progression to severe COVID-19 infection (Tenforde et al., 2021). Thus, evidence suggests that the benefits of vaccination against COVID-19 infection heavily outweigh any potential adverse effects.

High blood pressure and a history of hypertension are associated with lower antibody titer response, higher breakthrough infection, and a temporary increase in blood pressure after receiving two doses of the inactivated viral vaccine (Angeli et al., 2022; Soegiarto et al., 2022). Currently, there are no contraindications to COVID-19 vaccination for uncontrolled hypertension, and data encourages tighter blood pressure control in patients with a history of uncontrolled hypertension prior to receiving COVID-19 vaccination. Similarly, patients with a smoking history may experience lower antibody titer response (Ferrara et al., 2022). Another important consideration is the relationship between obesity and reduced COVID-19 vaccine efficacy. Although recent studies suggest a significantly lower antibody response in patients with obesity, currently approved COVID-19 vaccines are all highly efficacious in reducing COVID-19-associated hospitalization and death and were found to be equally efficacious in individuals with obesity compared with normal-weight individuals. Current recommendations do not support the hypothesis of impaired humoral immune responses to the SARS-CoV-2 vaccine in individuals with obesity (Butsch et al., 2021).

Given the clear and well-established benefits of COVID-19 vaccination, current data support proceeding with any of the currently approved vaccines, regardless of, and especially in, patients with cardiometabolic disease.

Impact of COVID-19 on cardiometabolic health

The profound impact of the pandemic on health-related behaviors as well as disruptions of routine non-COVID-19 healthcare cannot be understated. During the acute stages of COVID-19 disease, a 1.8-fold higher incidence of T2D was reported that can persist for up to 23 weeks postinfection (Rezel-Potts et al., 2022). A meta-analysis of 3700 patients reported a more modest 14.4% rate of newly diagnosed diabetes mellitus in patients hospitalized with COVID-19 (Tang et al., 2021). This has been attributed to relative insulin resistance during the hyperinflammatory state, direct pancreatic toxicity, and the impact of steroid exposure. This risk remains elevated in the postacute phase (i.e., long COVID-19), where a composite of incident T2D or new use of medication for glucose control was reported to be 46% higher in patients who had a positive COVID-19 test compared to the control group (Xie & Al-Aly, 2022). A study from Italy reported a mean increase in glycated hemoglobin (HbA1c) levels from 7% to 7.3% (Karatas et al., 2021). Of note, a similar degree of worsened glycemic control was reported in countries without significant national lockdowns such as Japan (Tanji et al., 2021). Similarly, modest increases in blood pressure were reported worldwide, with a mean rise in systolic blood pressure ranging from 1 to 3 mmHg (Laffin et al., 2022; Shah et al., 2022).

In the United States, significant increases in mean BMI (+0.6%) as well as the prevalence of obesity (+3%) were reported through the pandemic (Restrepo, 2022). Among Chinese adults, a weight gain of more than 2.5 kg was reported by 19% of men and 16% of women (Xu et al., 2021). A survey of adults in Brazil noted a dramatic decline in self-reported rates of physical activity from 69% prepandemic to 39% during periods of lockdown (Puccinelli et al., 2021). A similar 30% reduction in physical activity was reported by a large survey of 74,430 respondents in England (Strain et al., 2022). This was associated with greater rates of anxiety and depression in these groups. In the longer term, modeling studies and group-based trajectory models indicate a consistently negative impact of the pandemic on multiple facets of cardiometabolic health, including worsening BMI and physical activity (Beydoun et al., 2022).

While a direct causal relationship is yet to be established, a multitude of biological mechanisms, including an overexuberant inflammatory response, ACE-2 overexpression, changes in immune status, and altered estrogen dynamics, have been proposed to contribute to those negative changes. Changes in diet, sleep, and exercise habits have been highly varied in different individuals, with an overall heterogeneous impact on health. Access to healthcare for routine preventive care and non-COVID-19-related issues was overall negatively impacted by the pandemic. This potentially curtailed routine screening practices as well as follow-up visits for medication optimization. Furthermore, economic constraints during the pandemic likely had a negative impact on medication compliance among patients. Finally, the adverse impacts of the aforementioned changes during the pandemic were disproportionately borne by certain subpopulations, thus further widening preexisting disparities in healthcare.

Policies and procedures

Worldwide, innumerable policies and strategies were implemented against COVID-19 outbreaks. Such preventive measures had a largely negative impact on lifestyle patterns, resulting in overall worse cardiometabolic health, as described above. The focus now and for the future is on finding a balance between healthcare policies to limit the spread of an infection while also minimizing the damage to overall cardiometabolic health. Fortunately, numerous innovative policies and tools are available to address this issue (Table 2).

Physicians should routinely discuss with their patients their level of activity and encourage a level of activity that is in concordance with the recommendations of at least 150–300 min of moderate-intensity activity or 75–150 min of vigorous-intensity aerobic activity weekly. High-intensity exercise combined with high-load resistance can help with weight loss for those who have gained weight during the pandemic (Arnett et al., 2019). Exercises can be in the form of traditional in-person training or online home exercise classes that can be potentially more accessible and less expensive; these have become more widely accepted. In regard to the challenges with obesity during the COVID-19 pandemic, they mostly stemmed from stress and anxiety, the home food environment and reduced access to fresh foods, and challenges associated with physical activity (Caldwell et al., 2022). Therefore, any obesity treatment program (either in-person or online) should consider those aspects. Additional tools such as telehealth-based nutrition counseling programs can provide further support.

Furthermore, public health policy changes are needed to provide healthcare opportunities for individuals with or at risk of cardiometabolic disease. Improving health insurance coverage and financial assistance programs can enhance prevention efforts, improve risk factor control, and narrow disparities in cardiometabolic health. Importantly, governmental support of various innovative in-person and virtual programs that promote a healthier lifestyle, home self-monitoring, and increased equitable access to telehealth might result in improved cardiometabolic health outcomes through increased utilization and adherence.

Application to other areas

Given the high prevalence of cardiometabolic diseases worldwide, the monitoring and management of patients with T2D, HTN, dyslipidemia, obesity, and metabolic syndrome during the COVID-19 pandemic was very important. Several innovative models for delivering cardiometabolic healthcare can be implemented and utilized (Taha et al., 2022). Population healthcare delivery models focus on improving health outcomes for large patient groups within a system and frequently use electronic health records to identify potential candidates for therapies. Digital solutions can identify high-risk groups and be leveraged to guide management in select populations. Such innovative digital models led to improvements in multiple cardiometabolic diseases and improved medication prescription and adherence (Scirica et al., 2021). Additionally, multidisciplinary cardiometabolic centers are an opportunity for patients and healthcare providers to provide a holistic approach to cardiometabolic disease. Results from such an approach showed that patients receiving care at a cardiometabolic center had a greater degree of weight loss, reductions in glycated hemoglobin and low-density lipoprotein, and were more likely to be

TABLE 2 Suggested procedures and policies to help improve cardiometabolic health during endemic COVID-19.

Physical activity	• Physicians should routinely discuss with their patients their level of activity • Follow the recommendations of at least 150–300 min of moderate-intensity activity or 75–150 min of vigorous-intensity aerobic activity weekly, and high-intensity exercise combined with high-load resistance can help with weight loss for those who have gained weight during the pandemic • Exercises can be in the form of traditional in-person training or online home-based exercise classes
Weight management	• Obesity treatment program (either in-person or online) should consider level of physical activity, stress and anxiety, home food environment, and access to fresh foods • Telehealth-based nutrition counseling programs • Utilizing obesity medication with favorable cardiometabolic health profile such as GLP-1RA
Digital health	• Implementing and utilizing innovative models for delivering cardiometabolic health care such as population healthcare delivery models and promoting cardiometabolic centers
Technology	• Teleconsultation, virtual follow-up and communication, remote monitoring, and wearables
Public health policy	• Improving health insurance coverage and financial assistance programs • Governmental support of various innovative in-person and virtual programs that promote a healthier lifestyle, home self-monitoring, and increased equitable access to telehealth

treated with medications that have a favorable cardiometabolic impact such as SGLT2i and GLP-1RA (Sammour et al., 2021). Moreover, studies have suggested that self-assessment at home using a self-monitoring kit has the potential to facilitate the identification of individuals at risk for cardiometabolic disease in low-income settings (Calvert et al., 2022).

Digital health has been accelerated by the COVID-19 pandemic. Such digital solutions provide evidence as to how implementing and utilizing newer technologies can be used to support care and improve outcomes. These digital solutions include teleconsultation, more frequent virtual follow-ups and communication, remote monitoring, and wearables. These technologies can be applied to other healthcare areas outside cardiometabolic health, such as heart failure and patients who are immunosuppressed.

Mini-dictionary of terms

Cardiometabolic disease: A spectrum of interconnected pathophysiological changes that increases the risk of type 2 diabetes and cardiovascular diseases, including insulin resistance, hyperglycemia, abdominal obesity, dyslipidemia, inflammation, and hypertension, in addition to environmental risk factors such as sedentary lifestyle, poor dietary habits, and smoking.

Metabolic syndrome: Having any three of the following five traits: (1) abdominal obesity (waist circumference $\geq$40 in. in men and $\geq$35 in. in women), (2) serum triglycerides $\geq$150 mg/dL or drug treatment for elevated triglyceride, (3) serum high-density lipoprotein <40 mg/dL in men and <50 mg/dL in women, (4) blood pressure $\geq$130/85 mmHg or drug treatment for hypertension, and (5) fasting plasma glucose $\geq$100 mg/dL or drug treatment for hyperglycemia.

Cardiometabolic outcomes: Obesity, hypertension, type 2 diabetes mellitus, dyslipidemia, and cardiovascular disease (including atherosclerotic cardiovascular disease, heart failure, and atrial fibrillation).

Population healthcare delivery model: Healthcare delivery system aimed at improving healthcare outcomes for large patient groups within a system across the continuum of care that frequently utilizes electronic health records to identify potential candidates for therapies.

Summary points

- Individuals with cardiometabolic disease are particularly vulnerable to both infection by SARS-CoV-2 and severe outcomes from COVID-19 disease, including the need for intensive care units, mechanical ventilation, and death.
- Mechanisms underlying the susceptibility to and more severe outcomes of COVID-19 disease are still unclear. High levels of expression of ACE2, elevated levels of key proinflammatory markers, and obesity-related pathophysiological complications are key factors.
- The alarmingly high prevalence of cardiometabolic disease in the general population implies the need for more aggressive cardiometabolic disease prevention and management to mitigate COVID-19 infection rates and severity.
- No data support the need to discontinue ACE or ARBs in patients with COVID-19 disease. Glucose control in patients with COVID-19 disease improves outcomes and should be achieved with the preferential use of insulin during hospitalization. Certain oral hypoglycemic medications that increase the risk of complications during more severe illnesses should be avoided, including metformin and sodium-glucose transporter 2 inhibitors.
- Vaccination against SARS-CoV-2 is safe in patients with cardiometabolic disease and should be prioritized given the increased susceptibility to and worse outcome in this population.
- Lessons learned from the care of patients with cardiometabolic disease during the COVID-19 pandemic could guide the implementation of effective strategies for the treatment of other subgroups of patients.
- Adopting and implementing effective healthcare public policies and innovation tools and technologies to improve cardiometabolic health is of critical importance; these include telehealth, virtual health-promoting programs, digital healthcare solutions, and cardiometabolic clinics.

References

Abou Ghayda, R., Lee, K. H., Han, Y. J., Ryu, S., Hong, S. H., Yoon, S., Jeong, G. H., Yang, J. W., Lee, H. J., Lee, J., Lee, J. Y., Effenberger, M., Eisenhut, M., Kronbichler, A., Solmi, M., Li, H., Jacob, L., Koyanagi, A., Radua, J., & Shin, J. I. (2022). The global case fatality rate of coronavirus disease 2019 by continents and national income: A meta-analysis. *Journal of Medical Virology*, *94*(6), 2402–2413. https://doi.org/10.1002/JMV.27610.

Al-Benna, S. (2020). Association of high level gene expression of ACE2 in adipose tissue with mortality of COVID-19 infection in obese patients. *Obesity Medicine*, *19*. https://doi.org/10.1016/J.OBMED.2020.100283.

Angeli, F., Reboldi, G., Trapasso, M., Santilli, G., Zappa, M., & Verdecchia, P. (2022). Blood pressure increase following COVID-19 vaccination: A systematic overview and meta-analysis. *Journal of Cardiovascular Development and Disease*, *9*(5). https://doi.org/10.3390/JCDD9050150.

Arnett, D. K., Blumenthal, R. S., Albert, M. A., Buroker, A. B., Goldberger, Z. D., Hahn, E. J., Himmelfarb, C. D., Khera, A., Lloyd-Jones, D., McEvoy, J. W., Michos, E. D., Miedema, M. D., Muñoz, D., Smith, S. C., Virani, S. S., Williams, K. A., Yeboah, J., & Ziaeian, B. (2019). 2019 ACC/AHA guideline on the primary prevention of cardiovascular disease: A report of the American College of Cardiology/American Heart Association task force on clinical practice guidelines. *Circulation*, *140*(11), e596–e646. https://doi.org/10.1161/CIR.0000000000000678.

Beydoun, H. A., Beydoun, M. A., Gautam, R. S., Alemu, B. T., Weiss, J., Hossain, S., & Zonderman, A. B. (2022). COVID-19 pandemic impact on trajectories in cardiometabolic health, physical activity, and functioning among adults from the 2006-2020 health and retirement study. *The Journals of Gerontology. Series A, Biological Sciences and Medical Sciences*, *77*(7), 1371–1379. https://doi.org/10.1093/GERONA/GLAC028.

Bornstein, S. R., Rubino, F., Khunti, K., Mingrone, G., Hopkins, D., Birkenfeld, A. L., Boehm, B., Amiel, S., Holt, R. I., Skyler, J. S., DeVries, J. H., Renard, E., Eckel, R. H., Zimmet, P., Alberti, K. G., Vidal, J., Geloneze, B., Chan, J. C., Ji, L., & Ludwig, B. (2020). Practical recommendations for the management of diabetes in patients with COVID-19. *The Lancet Diabetes and Endocrinology*, *8*(6). https://doi.org/10.1016/S2213-8587(20)30152-2.

Butsch, W. S., Hajduk, A., Cardel, M. I., Donahoo, W. T., Kyle, T. K., Stanford, F. C., Zeltser, L. M., Kotz, C. M., & Jastreboff, A. M. (2021). COVID-19 vaccines are effective in people with obesity: A position statement from The Obesity Society. *Obesity*, *29*(10), 1575–1579. https://doi.org/10.1002/OBY.23251.

Cai, Q., Chen, F., Wang, T., Luo, F., Liu, X., Wu, Q., He, Q., Wang, Z., Liu, Y., Liu, L., Chen, J., & Xu, L. (2020). Obesity and COVID-19 severity in a designated Hospital in Shenzhen, China. *Diabetes Care*, *43*(7), 1392–1398. https://doi.org/10.2337/DC20-0576.

Caldwell, A. E., Thomas, E. A., Rynders, C., Holliman, B. D., Perreira, C., Ostendorf, D. M., & Catenacci, V. A. (2022). Improving lifestyle obesity treatment during the COVID-19 pandemic and beyond: New challenges for weight management. *Obesity Science & Practice*, *8*(1), 32–44. https://doi.org/10.1002/OSP4.540.

Calvert, C., Kolkenbeck-Ruh, A., Crouch, S. H., Soepnel, L. M., & Ware, L. J. (2022). Reliability, usability and identified need for home-based cardiometabolic health self-assessment during the COVID-19 pandemic in Soweto, South Africa. *Scientific Reports*, *12*(1). https://doi.org/10.1038/S41598-022-11072-4.

Centers for Disease Control and Prevention. (2022). *National diabetes statistics report website*. Retrieved November 13, 2022, from: https://www.cdc.gov/diabetes/data/statistics-report/index.html.

Docherty, A. B., Harrison, E. M., Green, C. A., Hardwick, H. E., Pius, R., Norman, L., Holden, K. A., Read, J. M., Dondelinger, F., Carson, G., Merson, L., Lee, J., Plotkin, D., Sigfrid, L., Halpin, S., Jackson, C., Gamble, C., Horby, P. W., Nguyen-Van-Tam, J. S., & Semple, M. G. (2020). Features of 20 133 UK patients in hospital with covid-19 using the ISARIC WHO clinical characterisation protocol: Prospective observational cohort study. *BMJ (Clinical Research Ed.)*, *369*. https://doi.org/10.1136/BMJ.M1985.

Ferrara, P., Ponticelli, D., Agüero, F., Caci, G., Vitale, A., Borrelli, M., Schiavone, B., Antonazzo, I. C., Mantovani, L. G., Tomaselli, V., & Polosa, R. (2022). Does smoking have an impact on the immunological response to COVID-19 vaccines? Evidence from the VASCO study and need for further studies. *Public Health*, *203*, 97–99. https://doi.org/10.1016/J.PUHE.2021.12.013.

Gnanenthiran, S. R., Borghi, C., Burger, D., Caramelli, B., Charchar, F., Chirinos, J. A., Cohen, J. B., Cremer, A., Di Tanna, G. L., Duvignaud, A., Freilich, D., Gommans, D. H. F., Gracia-Ramos, A. E., Murray, T. A., Pelorosso, F., Poulter, N. R., Puskarich, M. A., Rizas, K. D., Rothlin, R., & Schutte, A. E. (2022). Renin-angiotensin system inhibitors in patients with COVID-19: A meta-analysis of randomized controlled trials led by the International Society of Hypertension. *Journal of the American Heart Association*, *11*(17), 26143. https://doi.org/10.1161/JAHA.122.026143.

Goyal, P., Choi, J. J., Pinheiro, L. C., Schenck, E. J., Chen, R., Jabri, A., Satlin, M. J., Campion, T. R., Nahid, M., Ringel, J. B., Hoffman, K. L., Alshak, M. N., Li, H. A., Wehmeyer, G. T., Rajan, M., Reshetnyak, E., Hupert, N., Horn, E. M., Martinez, F. J., & Safford, M. M. (2020). Clinical characteristics of COVID-19 in New York City. *The New England Journal of Medicine*, *382*(24), 2372–2374. https://doi.org/10.1056/NEJMC2010419.

Grundy, S. M., Cleeman, J. I., Daniels, S. R., Donato, K. A., Eckel, R. H., Franklin, B. A., Gordon, D. J., Krauss, R. M., Savage, P. J., Smith, S. C., Spertus, J. A., & Costa, F. (2005). Diagnosis and management of the metabolic syndrome. *Circulation*, *112*(17), 2735–2752. https://doi.org/10.1161/CIRCULATIONAHA.105.169404.

Guan, W.-J., Liang, W.-H., Zhao, Y., Liang, H.-R., Chen, Z.-S., Li, Y.-M., Liu, X.-Q., Chen, R.-C., Tang, C.-L., et al. (2020). Comorbidity and its impact on 1590 patients with COVID-19 in China: A nationwide analysis. *The European Respiratory Journal*, *55*(5), 640. https://doi.org/10.1183/13993003.00547-2020.

Guan, W., Ni, Z., Hu, Y., Liang, W., Ou, C., He, J., Liu, L., Shan, H., Lei, C., Hui, D. S. C., Du, B., Li, L., Zeng, G., Yuen, K.-Y., Chen, R., Tang, C., Wang, T., Chen, P., Xiang, J., & Zhong, N. (2020). Clinical characteristics of coronavirus disease 2019 in China. *The New England Journal of Medicine*, *382*(18), 1708–1720. https://doi.org/10.1056/NEJMOA2002032.

Hales, C. M., Carroll, M. D., Fryar, C. D., & Ogden, C. L. (2020). Prevalence of obesity and severe obesity among adults: United States, 2017-2018 key findings data from the National Health and Nutrition Examination survey. *NCHS Data Brief*, 1–8.

Hendren, N. S., de Lemos, J. A., Ayers, C., Das, S. R., Rao, A., Carter, S., Rosenblatt, A., Walchok, J., Omar, W., Khera, R., Hegde, A. A., Drazner, M. H., Neeland, I. J., & Grodin, J. L. (2021). Association of body mass index and age with morbidity and mortality in patients hospitalized with COVID-19: Results from the American Heart Association COVID-19 cardiovascular disease registry. *Circulation*, *143*(2), 135–144. https://doi.org/10.1161/CIRCULATIONAHA.120.051936.

Huang, C., Wang, Y., Li, X., Ren, L., Zhao, J., Hu, Y., Zhang, L., Fan, G., Xu, J., Gu, X., Cheng, Z., Yu, T., Xia, J., Wei, Y., Wu, W., Xie, X., Yin, W., Li, H., Liu, M., & Cao, B. (2020). Clinical features of patients infected with 2019 novel coronavirus in Wuhan, China. *The Lancet*, *395*(10223), 497–506. https://doi.org/10.1016/S0140-6736(20)30183-5.

Karatas, S., Yesim, T., & Beysel, S. (2021). Impact of lockdown COVID-19 on metabolic control in type 2 diabetes mellitus and healthy people. *Primary Care Diabetes*, *15*(3), 424–427. https://doi.org/10.1016/J.PCD.2021.01.003.

Kassir, R. (2020). Risk of COVID-19 for patients with obesity. *Obesity Reviews : An Official Journal of the International Association for the Study of Obesity*, *21*(6). https://doi.org/10.1111/OBR.13034.

Kim, Y. E., Huh, K., Park, Y. J., Peck, K. R., & Jung, J. (2022). Association between vaccination and acute myocardial infarction and ischemic stroke after COVID-19 infection. *JAMA*, *328*(9), 887–889. https://doi.org/10.1001/JAMA.2022.12992.

Laffin, L. J., Kaufman, H. W., Chen, Z., Niles, J. K., Arellano, A. R., Bare, L. A., & Hazen, S. L. (2022). Rise in blood pressure observed among US adults during the COVID-19 pandemic. *Circulation*, *145*(3), 235–237. https://doi.org/10.1161/CIRCULATIONAHA.121.057075.

Lohia, P., Kapur, S., Benjaram, S., & Mir, T. (2021). Association between antecedent statin use and severe disease outcomes in COVID-19: A retrospective study with propensity score matching. *Journal of Clinical Lipidology*, *15*(3). https://doi.org/10.1016/j.jacl.2021.03.002.

Mahamat-Saleh, Y., Fiolet, T., Rebeaud, M. E., Mulot, M., Guihur, A., El Fatouhi, D., Laouali, N., Peiffer-Smadja, N., Aune, D., & Severi, G. (2021). Diabetes, hypertension, body mass index, smoking and COVID-19-related mortality: A systematic review and meta-analysis of observational studies. *BMJ Open*, *11*(10), e052777. https://doi.org/10.1136/BMJOPEN-2021-052777.

Mechanick, J. I., Farkouh, M. E., Newman, J. D., & Garvey, W. T. (2020). Cardiometabolic-based chronic disease, adiposity and dysglycemia drivers. *Journal of the American College of Cardiology*, *75*(5), 525. https://doi.org/10.1016/J.JACC.2019.11.044.

Permana, H., Huang, I., Purwiga, A., Kusumawardhani, N. Y., Sihite, T. A., Martanto, E., Wisaksana, R., & Soetedjo, N. N. M. (2021). In-hospital use of statins is associated with a reduced risk of mortality in coronavirus-2019 (COVID-19): Systematic review and meta-analysis. *Pharmacological Reports*, *73*(3). https://doi.org/10.1007/s43440-021-00233-3.

Puccinelli, P. J., da Costa, T. S., Seffrin, A., de Lira, C. A. B., Vancini, R. L., Nikolaidis, P. T., Knechtle, B., Rosemann, T., Hill, L., & Andrade, M. S. (2021). Reduced level of physical activity during COVID-19 pandemic is associated with depression and anxiety levels: An internet-based survey. *BMC Public Health*, *21*(1). https://doi.org/10.1186/S12889-021-10470-Z.

Restrepo, B. J. (2022). Obesity prevalence among U.S. adults during the COVID-19 pandemic. *American Journal of Preventive Medicine*, *63*(1), 102–106. https://doi.org/10.1016/J.AMEPRE.2022.01.012.

Reynolds, H. R., Adhikari, S., Pulgarin, C., Troxel, A. B., Iturrate, E., Johnson, S. B., Hausvater, A., Newman, J. D., Berger, J. S., Bangalore, S., Katz, S. D., Fishman, G. I., Kunichoff, D., Chen, Y., Ogedegbe, G., & Hochman, J. S. (2020). Renin–angiotensin–aldosterone system inhibitors and risk of COVID-19. *New England Journal of Medicine*, *382*(25). https://doi.org/10.1056/nejmoa2008975.

Rezel-Potts, E., Douiri, A., Sun, X., Chowienczyk, P. J., Shah, A. M., & Gulliford, M. C. (2022). Cardiometabolic outcomes up to 12 months after COVID-19 infection. A matched cohort study in the UK. *PLoS Medicine*, *19*(7), e1004052. https://doi.org/10.1371/JOURNAL.PMED.1004052.

Sammour, Y., Nassif, M., Gunta, P., Tang, F., Magwire, M., O'Keefe, J. H., & Kosiborod, M. (2021). Cardiometabolic Center of Excellence: Analysis of two-year outcomes. *American Heart Journal*, *242*, 168–169. https://doi.org/10.1016/J.AHJ.2021.10.058.

Scirica, B. M., Cannon, C. P., Fisher, N. D. L., Gaziano, T. A., Zelle, D., Chaney, K., Miller, A., Nichols, H., Matta, L., Gordon, W. J., Murphy, S., Wagholikar, K. B., Plutzky, J., & MacRae, C. A. (2021). Digital care transformation: Interim report from the first 5000 patients enrolled in a remote algorithm-based cardiovascular risk management program to improve lipid and hypertension control. *Circulation*, *143*(5), 507–509. https://doi.org/10.1161/CIRCULATIONAHA.120.051913.

Shah, N. P., Clare, R. M., Chiswell, K., Navar, A. M., Shah, B. R., & Peterson, E. D. (2022). Trends of blood pressure control in the U.S. during the COVID-19 pandemic. *American Heart Journal*, *247*, 15–23. https://doi.org/10.1016/J.AHJ.2021.11.017.

Siddiqi, H. K., & Mehra, M. R. (2020). COVID-19 illness in native and immunosuppressed states: A clinical-therapeutic staging proposal. *The Journal of Heart and Lung Transplantation: the Official Publication of the International Society for Heart Transplantation*, *39*(5), 405–407. https://doi.org/10.1016/J.HEALUN.2020.03.012.

Soegiarto, G., Wulandari, L., Purnomosari, D., Dhia Fahmita, K., Ikhwan Gautama, H., Tri Hadmoko, S., Edwin Prasetyo, M., Aulia Mahdi, B., Arafah, N., Prasetyaningtyas, D., Prawiro Negoro, P., Rosita Sigit Prakoeswa, C., Endaryanto, A., Gede Agung Suprabawati, D., Tinduh, D., Basuki Rachmad, E., Astha Triyono, E., Wahyuhadi, J., Budi Keswardiono, C., & Oceandy, D. (2022). Hypertension is associated with antibody response and breakthrough infection in health care workers following vaccination with inactivated SARS-CoV-2. *Vaccine*, *40*(30), 4046–4056. https://doi.org/10.1016/J.VACCINE.2022.05.059.

Strain, T., Sharp, S. J., Spiers, A., Price, H., Williams, C., Fraser, C., Brage, S., Wijndaele, K., & Kelly, P. (2022). Population level physical activity before and during the first national COVID-19 lockdown: A nationally representative repeat cross-sectional study of 5 years of Active Lives data in England. *The Lancet Regional Health*, *12*. https://doi.org/10.1016/J.LANEPE.2021.100265.

Taha, M. B., Rao, N., Vaduganathan, M., Cainzos-Achirica, M., Nasir, K., & Patel, K. V. (2022). Implementation of cardiometabolic centers and training programs. *Current Diabetes Reports*, *22*(5), 203–212. https://doi.org/10.1007/S11892-022-01459-Y/FIGURES/1.

Tang, X., Uhl, S., Zhang, T., Xue, D., Li, B., Vandana, J. J., Acklin, J. A., Bonnycastle, L. L., Narisu, N., Erdos, M. R., Bram, Y., Chandar, V., Chong, A. C. N., Lacko, L. A., Min, Z., Lim, J. K., Borczuk, A. C., Xiang, J., Naji, A., & Chen, S. (2021). SARS-CoV-2 infection induces beta cell transdifferentiation. *Cell Metabolism*, *33*(8), 1577–1591.e7. https://doi.org/10.1016/J.CMET.2021.05.015.

Tanji, Y., Sawada, S., Watanabe, T., Mita, T., Kobayashi, Y., Murakami, T., Metoki, H., & Akai, H. (2021). Impact of COVID-19 pandemic on glycemic control among outpatients with type 2 diabetes in Japan: A hospital-based survey from a country without lockdown. *Diabetes Research and Clinical Practice*, *176*. https://doi.org/10.1016/J.DIABRES.2021.108840.

Tenforde, M. W., Self, W. H., Adams, K., Gaglani, M., Ginde, A. A., McNeal, T., Ghamande, S., Douin, D. J., Talbot, H. K., Casey, J. D., Mohr, N. M., Zepeski, A., Shapiro, N. I., Gibbs, K. W., Files, D. C., Hager, D. N., Shehu, A., Prekker, M. E., Erickson, H. L., & Patel, M. M. (2021). Association between mRNA vaccination and COVID-19 hospitalization and disease severity. *JAMA*, *326*(20), 2043–2054. https://doi.org/10.1001/JAMA.2021.19499.

Tsao, C. W., Aday, A. W., Almarzooq, Z. I., Alonso, A., Beaton, A. Z., Bittencourt, M. S., Boehme, A. K., Buxton, A. E., Carson, A. P., Commodore-Mensah, Y., Elkind, M. S. V., Evenson, K. R., Eze-Nliam, C., Ferguson, J. F., Generoso, G., Ho, J. E., Kalani, R., Khan, S. S., Kissela, B. M., & Martin,

S. S. (2022). Heart disease and stroke statistics—2022 update: A report from the American Heart Association. *Circulation*, *145*(8), E153–E639. https://doi.org/10.1161/CIR.0000000000001052.

Wong, R., Hall, M., Vaddavalli, R., Anand, A., Arora, N., Bramante, C. T., Garcia, V., Johnson, S., Saltz, M., Tronieri, J. S., Yoo, Y. J., Buse, J. B., Saltz, J., Miller, J., Moffitt, R., Bennett, T., Casiraghi, E., Chute, C., DeWitt, P., … Wooldridge, J. (2022). Glycemic control and clinical outcomes in U.S. patients with COVID-19: Data from the national COVID cohort collaborative (N3C) database. *Diabetes Care*, *45*(5), 1099. https://doi.org/10.2337/DC21-2186.

Wu, Z., & McGoogan, J. M. (2020). Characteristics of and important lessons from the coronavirus disease 2019 (COVID-19) outbreak in China: Summary of a report of 72 314 cases from the Chinese Center for Disease Control and Prevention. *JAMA*, *323*(13), 1239–1242. https://doi.org/10.1001/JAMA.2020.2648.

Xie, Y., & Al-Aly, Z. (2022). Risks and burdens of incident diabetes in long COVID: A cohort study. *The Lancet. Diabetes and Endocrinology*, *10*(5), 311–321. https://doi.org/10.1016/S2213-8587(22)00044-4.

Xu, X., Yan, A. F., Wang, Y., & Shi, Z. (2021). Dietary patterns and changes in weight status among Chinese men and women during the COVID-19 pandemic. *Frontiers in Public Health*, *9*. https://doi.org/10.3389/FPUBH.2021.709535.

Ye, X., Ma, T., Blais, J. E., Yan, V. K. C., Kang, W., Chui, C. S. L., Lai, F. T. T., Li, X., Wan, E. Y. F., Wong, C. K. H., Tse, H. F., Siu, C. W., Wong, I. C. K., & Chan, E. W. (2022). Association between BNT162b2 or CoronaVac COVID-19 vaccines and major adverse cardiovascular events among individuals with cardiovascular disease. *Cardiovascular Research*, *118*(10), 2329–2338. https://doi.org/10.1093/CVR/CVAC068.

Zhou, F., Yu, T., Du, R., Fan, G., Liu, Y., Liu, Z., Xiang, J., Wang, Y., Song, B., Gu, X., Guan, L., Wei, Y., Li, H., Wu, X., Xu, J., Tu, S., Zhang, Y., Chen, H., & Cao, B. (2020). Clinical course and risk factors for mortality of adult inpatients with COVID-19 in Wuhan, China: A retrospective cohort study. *Lancet*, *395*(10229), 1054–1062. https://doi.org/10.1016/S0140-6736(20)30566-3.

Zhu, J., Pang, J., Ji, P., Zhong, Z., Li, H., Li, B., & Zhang, J. (2021). Elevated interleukin-6 is associated with severity of COVID-19: A meta-analysis. *Journal of Medical Virology*, *93*(1), 35–37. https://doi.org/10.1002/JMV.26085.

Zhu, L., She, Z. G., Cheng, X., Qin, J. J., Zhang, X. J., Cai, J., Lei, F., Wang, H., Xie, J., Wang, W., Li, H., Zhang, P., Song, X., Chen, X., Xiang, M., Zhang, C., Bai, L., Xiang, D., Chen, M. M., & Li, H. (2020). Association of Blood Glucose Control and Outcomes in patients with COVID-19 and pre-existing type 2 diabetes. *Cell Metabolism*, *31*(6), 1068–1077.e3. https://doi.org/10.1016/J.CMET.2020.04.021.

Ziegler, C. G. K., Allon, S. J., Nyquist, S. K., Mbano, I. M., Miao, V. N., Tzouanas, C. N., Cao, Y., Yousif, A. S., Bals, J., Hauser, B. M., Feldman, J., Muus, C., Wadsworth, M. H., Kazer, S. W., Hughes, T. K., Doran, B., Gatter, G. J., Vukovic, M., Taliaferro, F., & Zhang, K. (2020). SARS-CoV-2 receptor ACE2 is an interferon-stimulated gene in human airway epithelial cells and is detected in specific cell subsets across tissues. *Cell*, *181*(5), 1016–1035.e19. https://doi.org/10.1016/J.CELL.2020.04.035.

Chapter 34

Postrecovery COVID-19 and interlinking diabetes and cardiovascular events

Giuseppe Seghieri[a,b]

[a]*Epidemiology Unit, Regional Health Agency of Tuscany, Florence, Italy,* [b]*Department of Experimental and Clinical Biomedical Sciences, University of Florence, Florence, Italy*

Introduction

The postacute sequelae of COVID-19 still remain not fully clarified due to several aspects that make analysis of this issue complex. First, the burden of late consequences of exposure to SARS-CoV-2 infection is strongly influenced by the prior clinical history of those who have recovered from COVID-19, whether they were previously hospitalized, and, in case of hospitalization, whether they were discharged from ordinary inpatient units or intensive care units (ICU) (Huang et al., 2021; Xie et al., 2022). In addition, the period of the epidemic to which the observational study refers is important, considering that the occurrence of viral variants and the different vaccination rates are strictly dependent on the period of the COVID-19 epidemic and, at the same time, portend different prognostic outcomes at follow-up. Finally, the expression of the late consequences of prior exposure to SARS-CoV-2 infection is obviously dependent on the length of the observation period.

In this context, it has been widely shown that COVID-19, during the acute phase of the disease, increases the risk of acute cardiovascular diseases such as stroke and myocardial infarction. It has also been hypothesized that the relationship among COVID-19 cardiovascular events is due to the direct impact of acute viral infection, in association with other accompanying risk factors for atherosclerosis such as aging, diabetes, hypertension, and obesity. Several studies agree that the concurrence of these factors leads to a worse prognosis regarding adverse outcomes, often associated with the reduced efficiency of the overall healthcare system during the epidemic (Belani et al., 2020; Boulos et al., 2023; Klok et al., 2020; Kumar et al., 2021; Liu et al., 2020; Magadum & Kishore, 2020; Primessnig et al., 2021; Schmid et al., 2021; Stefanou et al., 2023).

Post-COVID-19 epidemiological studies

What has been shown over time is that the excess risk of cardiovascular events extends into post-COVID-19 recovery for an indefinite period, as demonstrated by a very large retrospective cohort study using national healthcare databases from the US Department of Veterans Affairs (Xie et al., 2022). This study has additionally shown that the risk of incident acute cardiovascular disease is increased also in individuals who were not hospitalized during the acute phase of the disease, quantifying the excess risk in about the 60% hazard ratio (HR): 1.63 (95% CI 1.51–1.75) for acute myocardial infarction and in about 50% for stroke: HR 1.52 (95% CI 1.43–1.62). Interestingly, the excess burden of incident cardiovascular events was significantly higher in the group of nonpreviously hospitalized patients (Xie et al., 2022). A second study addressing this issue concerned a large cohort of the US population aged 18–65. Also in this case, the excess burdens of coronary acute diseases and ischemic stroke were significantly higher in post-COVID-19 follow-up, independently of any prior hospitalization (Daugherty et al., 2021). In conclusion, there is enough evidence that the rate of incidence and the risk of cardiovascular events are increased after exposure to SARS-CoV-2 viral infection, independent of the acute phase hospitalization. The reasons for this, however, remains to be clarified. A first hypothesis should be that some risk factors for COVID-19 and cardiovascular events, such as obesity or diabetes, cluster together while the viral infection initiating a process that needs time and the copresence of risk factors to find full occurrence (Belani et al., 2020; Klok et al., 2020; Kumar et al., 2021; Magadum & Kishore, 2020; Primessnig et al., 2021; Schmid et al., 2021).

Management, Body Systems, and Case Studies in COVID-19. https://doi.org/10.1016/B978-0-443-18703-2.00056-6
Copyright © 2024 Elsevier Inc. All rights are reserved, including those for text and data mining, AI training, and similar technologies.

COVID-19 and diabetes

In this respect, the association between COVID-19 and diabetes and of both with cardiovascular events is complex and seems to be bidirectional (Harding et al., 2023). There are many studies suggesting that COVID-19 constitutes a risk factor for new cases of diabetes (Ssentongo et al., 2022) while, on the other hand, preexisting diabetes augments, as expected, the risk of acute cardiovascular events such as ischemic stroke or acute myocardial infarction. However, while there are many studies identifying diabetes as an important risk factor for mortality and severity during the acute phase of SARS-CoV-2 infection (Barron et al., 2020; Khunti et al., 2023; Ran et al., 2021), no studies have adequately addressed the question of whether diabetes specifically increases the risk of subsequent cardiovascular events later in the short, medium, or long term after SARS-CoV-2 infection.

Recovery from COVID-19 and incident new cases of cardiovascular diseases in diabetes

The eventual link between diabetes and the risk of incident cardiovascular diseases appears to be widely modified by several factors that are intertwined, including aging, gender, prevalent comorbidities, and accompanying therapies.

With these premises, at the end of the first epidemic wave, the Regional Health Agency of Tuscany designed a retrospective observational study based in Tuscany, an Italian region of approximatively 3.7 million residents. The study used the administrative databases regarding the first phase of the COVID-19 epidemic (Profili et al., 2022).

This study retrospectively evaluated people with or without diabetes, exempt from previous hospitalizations for cardiovascular diseases, previously tested positive for SARS-CoV-2 by March 1, 2020, and who subsequently presented a negative swab for SARS-CoV-2 infection. Groups recovering from prior SARS-CoV-2 infection were compared with age- and gender-matched people, apparently not previously exposed to viral infection because there was no trace of them in the regional registry of individuals undergoing swab tests for SARS-CoV-2 over this same period. The main purpose of the study was to test whether diabetes modified the relative risk of mortality and hospitalizations for the first acute cardiovascular events (myocardial infarction or stroke) by 6 months after a negative SARS-CoV-2 swab. This study had some strengths: 1. It involved the onset of the first COVID-19 wave, thereby reducing at minimum the possible confounding effect of vaccination on postrecovery outcomes. 2. The study considered, besides information about demographic characteristics and socioeconomic status, the current glucose-lowering therapy as well as two drug classes, antiplatelets and statins, that are widely used in the primary or secondary prevention of acute cardiovascular events.

Looking at the characteristics of the population involved in the study, we had a relatively young population well balanced for gender and socioeconomic status, with about 70% of individuals younger than 65. This latter characteristic was also probably also due to the progressive loss, over time, of more elderly people that didn't survive the first pandemic wave. Further characteristics were a low prevalence of insulin treatment in people with diabetes (about 17%) and the low burden of comorbidities as testified by no prior hospitalizations (Charlson index = 0) in about the totality of cases. It was also considered that the basal recruitment requirement was being exempt of prior hospitalizations for acute cardiovascular events. Therefore, we were dealing with a population that could be categorized as at a low risk of mortality and with a reduced risk of developing first incident cardiovascular events over a short follow-up time. Moreover, a further analysis was undertaken to compare the post-COVID-19 risk for cardiovascular events with the background diabetes-associated risk in our population, pairing the post-COVID-19 cohort with a cohort followed up in 2019 consisting of people strictly matched for age, gender, comorbidities, and socioeconomic status. In this historic pre-COVID-19 cohort, diabetes was associated with a higher adjusted incidence rate ratio of death (IRR) at 6 months of about the 17%: 1.17 (95% CI 1.13–1.22) while the IRR of being hospitalized for acute myocardial infarction or stroke rose to about 70%: 1.67 (95% CI 1.56–1.78). Comparing these incidence rates with the rates in the actual post-COVID-19 cohort, we observed that while independently of diabetes the IRR of death increased by about twofold: 1.92 (95% CI 1.63–2.25), in those who recovered from SARS-CoV-2 infection the IRR of incident cardiovascular events (AMI or stroke) remained almost neutral: 0.80 (95% CI 0.58–1.10). Interestingly, however, in presence of diabetes, the trend of IRR values appeared completely reversed. The IRR of death in people recovering from COVID-19 was no more significant: IRR: 1.23 (95% CI 0.74–1.78), also due to the scarce number of events while, at the same time, the IRR of cardiovascular events (myocardial infarction or stroke) remained significantly higher: 2.24 (95% CI 2.18–4.25). In conclusion, according to this study, in people without diabetes the SARS-CoV-2 infection increased the risk of death at 6 months but not of incident cardiovascular events while diabetes maintained the excess risk of cardiovascular events as in the historic cohort, conversely reducing or nullifying the excess risk of mortality. These results were further statistically reinforced after introducing the interaction term diabetes × COVID-19 into regression models. In conclusion from these data, after recovery from SARS-CoV-2 infection the presence of diabetes maintained or even reinforced the risk of incident cardiovascular events such as acute myocardial infarction or stroke, while statistically reduced or nullified its background excess risk of mortality. The effect on mortality could be explained with the low number of cases, also related by the background low basal risk of death in this population, or even by other reasons such as current medication with statins as mentioned below.

The main limitation of this real-world study, however, was the low number of cardiovascular events (stroke and/or myocardial infarctions, $n = 54$) and a not too long observation period (6 months after being tested negative for SARS-CoV-2). Further characteristics of this survey were that the study was performed in a young stratum of the population, recruited for the absence of any prior cardiovascular event at basal; given these conditions, the adjunctive effect of prior COVID-19 resulted as nonsignificant. In these conditions, however, the real effect of diabetes, especially aimed at its excess risk of incident cardiovascular events independently of the exposure to SARS-CoV-2 infection, stands out even more evidently. In addition, this study provides further confirmation that it is not easy, in observational studies of the real world, to disentangle the eventual role of COVID-19 on the global cardiovascular risk associated with diabetes. The risk of cardiovascular events associated with diabetes in relation with COVID-19 is heterogeneous and may likewise be heterogeneously distributed in the real world. Our study involved a very particular stratum of people with diabetes and the expected effect of diabetes on the cardiovascular risk may be even higher when considering older populations with a greater basal burden of comorbidities. However, even in this low-risk population, diabetes maintains or even increases its basal effect on cardiovascular risk. Whether SARS-CoV-2 infection will trigger further cardiovascular events after recovery has been addressed by a recent study. According to this study, the risk of cardiovascular sequelae and death after hospital discharge for COVID-19 greatly increases when myocardial injury is evidenced during hospitalization by a rise in the plasma concentration of high sensitivity cardiac troponin I (hs-cTnI). This suggests that the excess risk of cardiovascular events post-COVID-19 may be triggered by SARS-CoV-2 infection during the acute phase of the disease remaining clinically evident over time (Rinaldi et al., 2022). In this case, however, there is no data to evaluate the role of diabetes in this initial process of myocardial injury.

Relation between gender and risk of cardiovascular events in diabetes

As mentioned before, the relation interlinking postrecovery of COVID-19 with diabetes and cardiovascular events must consider further factors: gender, current therapy with glucose-lowering drugs, and, among other currently used drugs, treatment with statins.

As to the first point, while females are at higher risk of all the post-COVID-19 sequelae (Tsampasian et al., 2023), men seem to be at a higher risk of post-COVID-19 cardiovascular events (Xie et al., 2022). According to our study in Tuscany among people who recovered from COVID-19, women outnumbered men (54% vs 46%) even if men were at a higher risk of mortality (by 14%) and of cardiovascular events (by 44%) after recovering from COVID-19 (Profili et al., 2022). According to the prior literature, women seem to be more sensitive to SARS-Cov-2 infection, even if men are at a higher risk of post-COVID-19 complications (Bienvenu et al., 2020; Newson et al., 2021; Seeland et al., 2020).

No previous study, however, has to date addressed the specific question of whether gender impacts the relationship between diabetes and the risk of cardiovascular events at a longer follow-up. This could be an interesting issue for investigation in light of the greater relative risk of cardiovascular diseases associated with diabetes in women compared to men (Kautzky-Willer et al., 2023), especially in the postmenopausal period (Policardo et al., 2017). Interestingly, a study carried out during the first wave of the pandemic in the United Kingdom has shown that the risk of mortality post-COVID-19 was significantly higher among men while both type 1 and type 2 diabetes were associated with a greater relative risk of mortality among women compared to men (Barron et al., 2020).

Post-COVID-19 complications and glucose-lowering therapy in diabetes

Insulin therapy seems to be related with a worse prognostic outcome after COVID-19 in people with diabetes. The more plausible hypotheses to explain the relation between insulin therapy with worse prognostic outcomes could be found in likely prior worse metabolic control or in a longer duration of diabetes among people treated with insulin. A further hypothesis could be related with the need of associating insulin treatment with the use of glucocorticoids in people with more severe COVID-19 during the acute phase of the disease, even if some have suggested a direct effect of insulin in worsening the prognosis of COVID-19 (Kan et al., 2021; Wiyarta & Wisnu, 2022). No specific effect was found in the cohort of people with diabetes in our study, however, even if the low percentage of insulin treatment and the clinical characteristics of our population don't allow evaluating any possible link between insulin treatment and the incidence of cardiovascular diseases after recovery from COVID-19 in patients with diabetes.

Statins and post-COVID-19 events

Another point that remains to be further elucidated regarding post-COVID-19 cardiovascular events in people with or without diabetes is the role of statins. This class of drugs could in fact have a double impact, reducing the risk of cardiovascular events both in primary and in secondary prevention, and, at the same time having a protective effect against infectious diseases, including SARS-CoV-2 (Barkas et al., 2020; Ganjali et al., 2020; Policardo et al., 2018). It has been calculated that the number needed to treat (NNT) for preventing one hospitalization for infectious diseases is within the 95% CIs of the NNT previously found for the primary prevention of one coronary ischemic event (Policardo et al., 2018). The protective effect of statins in SARS-CoV-2 infection, nonetheless, is a matter of debate. In this regard, our epidemiological study found an independent reduction in the risk of mortality by about 30% associated with statin use after recovery from COVID-19; a recent metaanalysis showed a reduction approximately at the same extent (Zein et al., 2022). Interestingly, however, according to our study the protective effect of statins didn't extend to the risk of hospitalization for first incident cardiovascular events. Statins in the post-COVID-19 cohort of our study were prescribed in about 45% of people with diabetes, and only in 11% of people without diabetes, remembering that our study cohort excluded people with prior hospitalizations for cardiovascular diseases. It is therefore plausible that the greater protective effect against mortality from statins, eventually used in primary prevention, may be more evident in people with diabetes, and interestingly presence of diabetes was not associated with a rise in risk of mortality in our population. Finally, a further intriguing factor in the interaction linking COVID-19, diabetes, and statins is that it has been demonstrated that both COVID-19 and statins increase the risk of new cases of diabetes, even if the mechanisms, duration, and implied risk cofactors are still unclear (Carfì et al., 2020; Khunti et al., 2021; Singh et al., 2022).

In summary, COVID-19 has a biunivocal link with diabetes: it seems to elicit an increase of new cases of diabetes as previously reported and, after recovery, preexisting diabetes seems to maintain its background excess risk of incident cardiovascular events. However, the full manifestation of adverse events is associated with many factors, including advanced age, duration of diabetes, comorbidities, male gender, and prevailing therapy with insulin or statins. The latter, mainly used in primary prevention in people with diabetes, could even reduce the risk of death in people recovering from prior COVID-19.

Summary points

- Epidemiological studies suggest that postrecovery COVID-19 is associated with a rise in the risk of mortality and of cardiovascular events after long-term follow-ups, independently of prior hospitalizations.
- According to all population observational studies, the excess risk for mortality or incident cardiovascular events after recovery from COVID-19 is modified by several confounders such as comorbidities, aging, male gender, or the copresence of risk factors for cardiovascular disease such as obesity, smoking, hypertension, or diabetes.
- COVID-19 has a biunivocal link with diabetes. It elicits an increase of new cases of diabetes and, after recovery from COVID-19, preexisting diabetes is associated with a significant rise in the risk of incident cardiovascular events.
- No conclusive data exist about the effect of gender on the post-COVID-19 incidence of cardiovascular events in diabetes, even if women seem to be at higher risk of SARS-CoV-2 infection and men at higher risk of post-COVID-19 adverse outcomes.
- Prior insulin therapy seems to portend a worse prognosis during the acute phase of COVID-19, seemingly depending on comorbidities, advanced age, metabolic control and duration of diabetes.
- Prior statin therapy seems to reduce the risk of post-COVID-19 mortality. Whether statins, after recovery from COVID-19, further protect from new incident cardiovascular events, in association or not with diabetes, remains, however, to be ascertained.

References

Barkas, F., Milionis, H., Anastasiou, G., & Liberopoulos, E. (2020). Statins and PCSK9 inhibitors: What is their role in coronavirus disease 2019? *Medical Hypotheses*, *146*, 110452F.

Barron, E., Bakhai, C., Kar, P., Weaver, A., Bradley, D., Ismail, H., Knighton, P., Holman, N., Khunti, K., Sattar, N., Wareham, N. J., Young, B., & Valabhji, J. (2020). Associations of type 1 and type 2 diabetes with COVID-19-related mortality in England: A whole-population study. *The Lancet Diabetes and Endocrinology*, *8*, 813–822.

Belani, P., Schefflein, J., Kihira, S., Rigney, B., Delman, B. N., Mahmoudi, K., et al. (2020). COVID-19 is an independent risk factor for acute ischemic stroke. *AJNR. American Journal of Neuroradiology*, *4*, 1361–1364.

Bienvenu, L. A., Noonan, J., Wang, X., & Peter, K. (2020). Higher mortality of COVID-19 in males: Sex differences in immune response and cardiovascular comorbidities. *Cardiovascular Research*, *116*, 2197–2206.

Boulos, P. K., Freeman, S. V., Henry, T. D., Mahmud, E., & Messenger, J. C. (2023). Interaction of COVID-19 with common cardiovascular disorders. *Circulation Research*, *132*, 1259–1271.

Carfì, A., Bernabei, R., Landi, F., & Gemelli Against COVID-19 Post-Acute Care Study Group. (2020). Persistent symptoms in patients after acute COVID-19. *JAMA*, *324*, 603–605.

Daugherty, S. E., Guo, Y., Heath, K., Dasmariñas, M. C., Jubilo, K. G., Samranvedhya, J., Lipsitch, M., & Cohen, K. (2021). Risk of clinical sequelae after the acute phase of SARS-CoV-2 infection: Retrospective cohort study. *BMJ*, *373*, n1098.

Ganjali, S., Bianconi, V., Penson, P. E., Pirro, M., Banach, M., Watts, G. F., et al. (2020). Commentary: Statins, COVID-19, and coronary artery disease: Killing two birds with one stone. *Metabolism*, *113*, 154375.

Harding, J. L., Oviedo, S. A., Ali, M. K., Ofotokun, I., Gander, J. C., Patel, S. A., Magliano, D. J., & Patzer, R. E. (2023). The bidirectional association between diabetes and long-COVID-19—A systematic review. *Diabetes Research and Clinical Practice*, *195*, 10202.

Huang, C., Huang, L., Wang, Y., Li, X., Ren, L., Gu, X., Kang, L., Guo, L., Liu, M., Zhou, X., Luo, J., Huang, Z., Tu, S., Zhao, Y., Chen, L., Xu, D., Li, Y., Li, C., Peng, L., … Cao, B. (2021). 6-month consequences of COVID-19 in patients discharged from hospital: A cohort study. *Lancet*, *397*, 220–232.

Kan, C., Zhang, Y., Han, F., Xu, Q., Ye, T., Hou, N., & Sun, X. (2021). Mortality risk of antidiabetic agents for type 2 diabetes with COVID-19: A systematic review and meta-analysis. *Frontiers in Endocrinology (Lausanne)*, *12*, 708494.

Kautzky-Willer, A., Leutner, M., & Harreiter, J. (2023). Sex differences in type 2 diabetes. *Diabetologia*, *66*, 986–1002.

Khunti, K., Del Prato, S., Mathieu, C., Kahn, S. E., Gabbay, R. A., & Buse, J. B. (2021). COVID-19, hyperglycemia, and new-onset diabetes. *Diabetes Care*, *44*, 2645–2655.

Khunti, K., Valabhji, J., & Misra, S. (2023). Diabetes and the COVID-19 pandemic. *Diabetologia*, *66*, 255–266.

Klok, F. A., Kruip, M. J. H. A., van der Meer, N. J. M., Arbous, M. S., Gommers, D., Kant, K. M., et al. (2020). Confirmation of the high cumulative incidence of thrombotic complications in critically ill ICU patients with COVID-19: An updated analysis. *Thrombosis Research*, *191*, 148–150.

Kumar, N., Verma, R., Lohana, P., Lohana, A., & Ramphul, K. (2021). Acute myocardial infarction in COVID-19 patients. A review of cases in the literature. *Archives of Medical Science – Atherosclerotic Diseases*, *6*, 169–175.

Liu, P. P., Blet, A., Smyth, D., & Li, H. (2020). The science underlying COVID-19: Implications for the cardiovascular system. *Circulation*, *42*, 68–78.

Magadum, A., & Kishore, R. (2020). Cardiovascular manifestations of COVID-19 infection. *Cells*, *9*, 2508.

Newson, L., Manyonda, I., Lewis, R., Preissner, R., Preissner, S., & Seeland, U. (2021). Sensitive to infection but strong in defense-female sex and the power of oestradiol in the COVID-19 pandemic. *Frontiers in Global Women's Health*, *2*, 651752.

Policardo, L., Seghieri, G., Francesconi, P., Anichini, R., Franconi, F., & Del Prato, S. (2017). Gender difference in diabetes related excess risk of cardiovascular events: When does the 'risk window' open? *Journal of Diabetes and its Complications*, *31*, 74–79.

Policardo, L., Seghieri, G., Gualdani, E., & Franconi, F. (2018). Effect of statins in preventing hospitalizations for infections: A population study. *Pharmacoepidemiology and Drug Safety*, *27*, 878–884.

Primessnig, U., Pieske, B. M., & Sherif, M. (2021). Increased mortality and worse cardiac outcome of acute myocardial infarction during the early COVID-19 pandemic. *ESC Heart Failure*, *8*, 333–343.

Profili, F., Seghieri, G., & Francesconi, P. (2022). Effect of diabetes on short-term mortality and incidence of first hospitalizations for cardiovascular events after recovery from SARS-CoV-2 infection. *Diabetes Research and Clinical Practice*, *87*, 109872.

Ran, J., Zhao, S., Han, L., Ge, Y., Chong, M. K. C., Cao, W., & Sun, S. (2021). Increase in diabetes mortality associated with COVID-19 pandemic in the U.S. *Diabetes Care*, *44*, e146–e147.

Rinaldi, R., Basile, M., Salzillo, C., Grieco, D. L., Caffè, A., Masciocchi, C., Lilli, L., Damiani, A., La Vecchia, G., Iannaccone, G., Bonanni, A., De Pascale, G., Murri, R., Fantoni, M., Liuzzo, G., Sanna, T., Massetti, M., Gasbarrini, A., Valentini, V., … On Behalf Of The Gemelli Against Covid Group. (2022). Myocardial injury portends a higher risk of mortality and long-term cardiovascular sequelae after hospital discharge in COVID-19 survivors. *Journal of Clinical Medicine*, *11*, 5964.

Schmid, A., Petrovic, M., Akella, K., Pareddy, A., Velavan, S. S., & Bozzani, A. (2021). Getting to the heart of the matter: Myocardial injury, coagulopathy, and other potential cardiovascular implications of COVID-19. *International Journal of Vascular Medicine*, *2021*, 1–16.

Seeland, U., Coluzzi, F., Simmaco, M., Mura, C., Bourne, P. E., Heiland, M., Preissner, R., & Preissner, S. (2020). Evidence for treatment with estradiol for women with SARS-CoV-2 infection. *BMC Medicine*, *18*, 369.

Singh, H., Sikarwar, P., Khurana, S., & Sharma, J. (2022). Assessing the incidence of new-onset diabetes mellitus with statin use: A systematic review of the systematic reviews and meta-analyses. *touchREVIEWS in Endocrinology*, *18*, 96–101.

Ssentongo, P., Zhang, Y., Witmer, L., Chinchilli, V. M., & Ba, D. M. (2022). Association of COVID-19 with diabetes: A systematic review and meta-analysis. *Scientific Reports*, *12*, 20191.

Stefanou, E., Karvelas, N., Bennett, S., & Kole, C. (2023). Cerebrovascular manifestations of SARS-COV-2-CoV-2: A comprehensive review. *Current Treatment Options in Neurology*, *25*, 71–92.

Tsampasian, V., Elghazaly, H., Chattopadhyay, R., Debski, M., Naing, T. K. P., Garg, P., Clark, A., Ntatsaki, E., & Vassiliou, V. S. (2023). Risk factors associated with post-COVID-19 condition: A systematic review and meta-analysis. *JAMA Internal Medicine*, *23*, e230750.

Wiyarta, E., & Wisnu, W. (2022). Does insulin use worsen the prognosis of COVID-19 patients with type 2 diabetes mellitus? A current update. *Current Diabetes Reviews*, *18*, e171121197988.

Xie, Y., Xu, E., Bowe, B., & Al-Aly, Z. (2022). Long-term cardiovascular outcomes of COVID-19. *Nature Medicine*, *28*, 583–590.

Zein, A. F. M. Z., Sulistiyana, C. S., Khasanah, U., Wibowo, A., Lim, M. A., & Pranata, R. (2022). Statin and mortality in COVID-19: A systematic review and meta-analysis of pooled adjusted effect estimates from propensity-matched cohorts. *Postgraduate Medical Journal*, *98*, 503–508.

Chapter 35

COVID-19 patients and extracorporeal membrane oxygenation

Mario Castano[a], Pasquale Maiorano[a], Laura Castillo[a], Gregorio Laguna[a], Guillermo Muniz-Albaiceta[b], Victor Sagredo[c], Elio Martín-Gutiérrez[a], and Javier Gualis[a]

[a]*Cardiac Surgery Department, University Hospital of León, León, Spain,* [b]*Intensive Care Medicine Department, University Central Hospital of Asturias, Oviedo, Spain,* [c]*Intensive Care Medicine Department, University Hospital of Salamanca, Salamanca, Spain*

Abbreviations

AVWS	acquired von-Willebrand syndrome
APTT	activated partial thromboplastin time
AKI	acute kidney injury
ARDS	acute respiratory distress syndrome
COVID-19	Coronavirus Disease 2019
BMI	body mass index
ECMO	extracorporeal membrane oxygenation
ICU	intensive care unit
PEEP	positive end-expiratory pressure
RV	right ventricular
VV	venovenous
VA	venoarterial
vWF	Von Willebrand factor
VILI	ventilator-induced lung injury

Introduction

Extracorporeal membrane oxygenation (ECMO) has been widely used for the treatment of patients with viral-related acute respiratory distress syndrome (ARDS) since 2009, when patients with refractory hypoxemia and/or hypercapnia despite optimal medical treatment and invasive ventilatory support during the H1N1 pandemic were successfully treated with these devices.

In this chapter, we review the evidence about rationale, indications, contraindications, management, results, and resource optimization of ECMO treatment in patients with refractory ARDS due to Coronavirus Disease 2019 (COVID-19).

Rationale, risk factors for mortality, and indications for ECMO in COVID-19 patients

As previously described in other chapters of this book, a significant percentage of COVID-19 patients develop refractory hypoxemia and/or hypercapnia despite optimal medical treatment and invasive ventilatory support. Less frequently, concomitant viral myocarditis or pulmonary thromboembolic events induce severe heart failure. In both situations, subsequent multiorgan failure and concomitant bacterial infections occur, with extremely low survival rates.

In these patients, ECMO support may improve outcomes because it allows adequate perfusion and oxygenation/ventilation with optimized protective lung ventilation while viral-induced lung damage heals.

There is a lack of controlled studies comparing the results of ECMO and conventional treatment in these groups of patients. Therefore, the identification of risk factors for mortality is paramount to select those patients who may benefit

Management, Body Systems, and Case Studies in COVID-19. https://doi.org/10.1016/B978-0-443-18703-2.00048-7
Copyright © 2024 Elsevier Inc. All rights are reserved, including those for text and data mining, AI training, and similar technologies.

most from ECMO to improve mortality, functional recovery, and resource optimization. Risk factors of mortality for veno-venous (VV)-ECMO are as follows:

- Increasing age (Barbaro et al., 2020; Castaño et al., 2022; Lebreton et al., 2021; Riera et al., 2022). Relative risk ranges from 1.37 to 6.81 in VV ECMO patients >60 years and >70 years, respectively (Friedrichson et al., 2022; Herrmann et al., 2022) and cut-off values as low as 48 years for increased mortality have been described (Castaño et al., 2022). Although age should not be considered an absolute contraindication, and comorbidities and functional preprocedural status must be weighted, an increase in mortality over the age of 60 years and a poor survival over 70 years should be expected.
- A shorter duration of pre-ECMO mechanical ventilation enhances survival (Lebreton et al., 2021; Supady, DellaVolpe, et al., 2021). However, there is no clearly defined cut-off point and other conflicting results have been reported (Barbaro et al., 2020; Riera et al., 2022), even when patients were ventilated for more than 10 days before ECMO (Hermann et al., 2022). Therefore, longer mechanical ventilation times are not an absolute contraindication but when longer than 7–10 days, lung mechanics and elasticity, the aggressiveness of ventilation, and tomographic morphological findings must be carefully assessed.
- Pre-ECMO noninvasive ventilatory support time and long periods of high-flow oxygen therapy may be related with increased mortality (Barbaro et al., 2020) because they may similarly induce greater lung injury than invasive protective ventilation (Riera et al., 2022).
- Obesity (a body mass index (BMI) of at least 30 kg/m^2) shows a positive association with improved survival in patients undergoing ECMO for COVID-19 in meta-analyses (Tran et al., 2022) and registries (Castaño et al., 2022), especially when compared with significantly low BMI (Herrmann et al., 2022). Therefore, obesity is no longer a limiting factor in most centers, except for morbid obesity, due to technical issues during cannulation.
- Driving pressure prior to ECMO has the highest level of evidence among the multiple ventilatory parameters studied as a predictor of mortality, similar to ARDS from other causes (Riera et al., 2022). Values greater than 16 cmH_2O induce a greater probability of death on ECMO.
- Both severe hypoxemia and pre-ECMO hypercapnia correlate with mortality. Patients with greater pre-ECMO PaO_2/FiO_2 ratios (Warren et al., 2022) or lower PCO_2 (Tran et al., 2022) had significantly lower mortality, even after adjusting for potential confounding factors.
- Concomitant comorbidities such as immunosuppression, acute kidney injury (AKI), chronic respiratory disease, and ischemic cardiomyopathy have been consistently described as risk factors of mortality (Barbaro et al., 2020; Hermann et al., 2022; Lebreton et al., 2021).
- Prognostic scores. A high SOFA Score and a renal component ≥3 have been identified as risk factors for mortality (Lebreton et al., 2021). Contrarily, the respiratory ECMO survival prediction (RESP) score, developed from data from the Extracorporeal Life Support Organization registry, is a poor predictor of survival in COVID-19 patients undergoing ECMO. The prediction of survival on the ECMO therapy score (PRESET-Score), developed before the COVID-19 pandemic, includes extrapulmonary variables (mean arterial pressure, lactate, arterial pH, platelets, and days of hospital stay pre-ECMO). It has been validated in COVID-19 and identifies patients with the greatest ECMO support benefits (Powell et al., 2022). Some multicenter experiences suggest that these scores cannot be recommended for definite therapeutic decision making, but can be considered as an additional valuable tool to predict prognosis (Supady, DellaVolpe, et al., 2021).
- Treatment in centers with a high volume of cases (especially if >30 cases/year) is consistently associated with increased probability of survival (Herrmann et al., 2022; Lebreton et al., 2021; Riera et al., 2022), and this benefit persists after 90 days of follow-up. However, some experiences suggest that patients treated in the admitting center has less mortality than those transferred to other centers (Castaño et al., 2022).

As predicted, indications differ widely among centers and publications, but considering these prognostic factors and the possible deleterious effects of pre-ECMO hypoxia and hypercapnia, the modified EOLIA trial criteria (Combes et al., 2018) are considered appropriate:

- $PO_2/FiO_2 < 50$ for more than 3 h.
- $PO_2/FiO_2 < 80$ for more than 6 h.
- pH <7.25 and $PCO_2 \geq 60$ mmHg for more than 6 h.

Before indicating ECMO, all patients must receive protective mechanical ventilation and adjuvant therapies, as discussed later in this chapter.

A practical guideline for VV-ECMO recommendations in COVID-19 patients is described in the "Policies and Procedures" section of this chapter.

Some patients with COVID-19 develop myocarditis, massive pulmonary embolism, stress cardiomyopathy, arrhythmias, or acute coronary syndrome that may require venoarterial (VA) ECMO. VA-ECMO benefits in COVID-19 patients are much less clear, with significantly higher mortality and less established indications and contraindications.

Management of COVID-19 patients treated with ECMO

Cannulation

In brief, for isolated respiratory support, venous drainage for ECMO from the inferior vena cava is oxygenated and usually returned into the superior vena cava through an internal jugular vein cannula or the contralateral femoral vein cannula (VV-ECMO). When circulatory support is required, oxygenated venous blood is returned into a femoral artery or, alternatively, a subclavian artery (VA-ECMO). In some cases with initial VV-ECMO that develop heart failure, an additional arterial cannula may be implanted for arterial return (VV-A and V-VA-ECMO configurations).

Isolated VA-ECMO support is required in only 3%–5% of COVID-19 cases treated with ECMO (Barbaro et al., 2021; Castaño et al., 2022). However, pulmonary vascular resistance can increase during ARDS in COVID-19 by vasoconstriction due to hypoxia/hypercapnia, inflammatory mediators, interstitial edema, and thromboembolic events. In fact, right ventricular (RV) failure is present in 13%–51% of COVID-19 cases (Rojas-Velasco et al., 2022). Therefore, bedside echocardiography must be performed in every COVID-19 patient before ECMO mode selection to rule out heart failure. These patients need a VA-ECMO configuration or, alternatively, venopulmonary ECMO for RV support.

In every COVID-19 patient, cannulation is performed in the intensive care unit (ICU) by the strictly necessary staff. A percutaneous Seldinger technique is selected for arterial and venous access under echography or fluoroscopic guidance for adequate cannulae position. Initial cannulation and configuration must ensure adequate return flow and minimize potential configuration changes. For this purpose, multiperforated and two large cannulae (separate inflow and outflow) are recommended (Shekar et al., 2020).

In general, femoro-femoral cannulation is preferred in COVID-19 patients. These accesses minimize cannulation time, prevent neck manipulation, and keep operators away from the airway (Shekar et al., 2020). Additionally, this technique doesn't need patient transfer for dedicated fluoroscopy. Although some groups have reported good results with single dual-lumen cannulation with the Avalon cannula (Getinge, Göteborg, Sweden) for cavo-caval VV configuration (Calcaterra et al., 2020), ELSO guidelines discourage this approach in COVID-19 patients due to the potential increase in thrombotic complications and longer procedural times (Shekar et al., 2020). In VA-ECMO, an axillary arterial return may be beneficial for Harlequin syndrome prevention and, therefore, minimizing the incidence of configuration changes (Akar et al., 2019).

In associated RV failure, some centers suggest good results with V-P-ECMO in which a specific dual-stage cannula is inserted through the internal jugular vein to the main pulmonary artery with the ProtekDuo cannula (Fig. 1) (LivaNova, London, UK) (Rojas-Velasco et al., 2022). Venous return from the right atrium is sent directly to the pulmonary artery for RV support. RV dysfunction may be extremely difficult to diagnose. Data from retrospective noncontrolled studies suggest a survival benefit with these modes, even as the default ECMO configuration (Smith, Park, et al., 2022). Although this is promising, controlled studies are needed to confirm the benefits of this cannulation strategy.

Respiratory management

Ventilatory strategies during ECMO

The main goal of mechanical ventilation in ARDS patients (irrespective of the level of severity or therapeutic support) is to ensure an adequate gas exchange without inflicting ventilator-induced lung injury (VILI) (Albaiceta & Amado-Rodríguez, 2022). Therefore, the delicate balance between gas exchange goals and the cost of ventilation must be weighed. During standard protective ventilation, a tidal volume of 6mL/kg and moderate positive end-expiratory pressure (PEEP) values (in the 8–15 cmH_2O range) with a driving pressure below 15 cmH_2O and a plateau pressure below 28 cmH_2O are considered standard markers of protective ventilation; however, even with these settings, some lung areas may be exposed to overdistension (Terragni et al., 2007). Increasing tidal volumes or PEEP may improve gas exchange, but usually increase VILI.

Extracorporeal gas exchange directly improves oxygenation and CO_2 washout, and indirectly allows the use of less aggressive ventilatory strategies. As VILI is one of the major determinants of outcome in ARDS patients, the onset of ECMO must lead to an immediate adjustment of the ventilatory strategy to achieve the goals of lung protection. Tidal volumes can be gradually decreased to values around 3mL/kg of predicted body weight, avoiding large shifts in arterial CO_2 that have been linked to worse outcomes (Cavayas et al., 2020). Given the range of lung compliances in COVID-19,

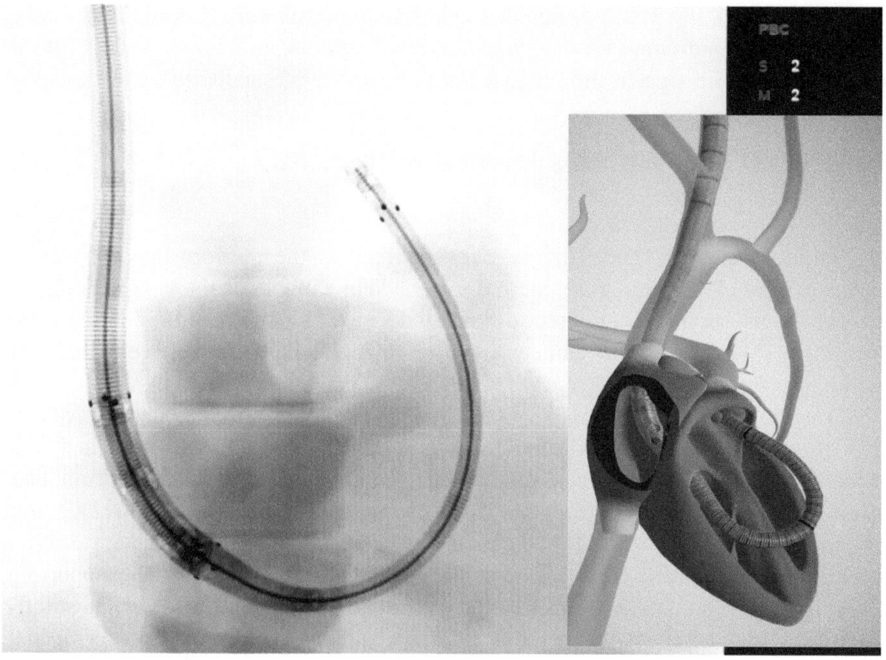

FIG. 1 ProtekDuo cannula fluoroscopy. *With permission from LivaNova, London, UK.*

this reduction in tidal volume usually results in airway pressures within the safe range. However, these ultralow tidal volumes may cause also alveolar collapse due to the lack of end-inspiratory recruitment (Amado-Rodríguez et al., 2021). Ventilation at a low end-expiratory lung volume may also cause damage due to cyclic changes in aeration (atelectrauma) and increased pulmonary vascular resistance (Duggan et al., 2003). Detection of this collapse may be difficult without advanced monitoring techniques (such as impedance tomography or measurement of static lung volumes), but conventional radiology or standard respiratory mechanics (i.e., a large decrease in lung compliance or an increase in airway resistance) may be useful in this setting. Increasing PEEP levels and decreasing inspired oxygen may help preserve aeration. The safety of ventilatory strategies that allow large areas of nonventilated lung is an ongoing debate in current critical care.

Extracorporeal support also allows fine tuning the other ventilatory settings that contribute to VILI. Decreasing respiratory rate, inspiratory time, and/or flow results in a less aggressive ventilatory pattern that ultimately decreases mechanical power (the energy transfer between the ventilator and the lung), thus avoiding VILI (Cressoni et al., 2016). Although alternative modes of ventilatory support, such as airway pressure release ventilation, have been proposed (Rozé et al., 2017), there are no data on their superiority to improve patient-centered outcomes.

In spite of the potential benefits of ECMO to facilitate a protective ventilation strategy, the indiscriminate use of ultraprotective ventilation has failed to improve both long-term outcomes or short-term markers of alveolar damage (Amado-Rodríguez et al., 2021). This lack of positive results highlights the need to identify subgroups of patients in which this intervention may be beneficial, such as those with clear markers of VILI (keeping in mind that currently there is no clear gold standard method to identify alveolar overdistension at the bedside) or with high-risk subphenotypes (such as those with hyperinflammatory systemic responses, as with some COVID-19 patients) (López-Martínez et al., 2022). Personalization of ECMO therapy, in terms of patient selection and fine tuning, is mandatory to provide efficient care of critically ill patients.

Coadjuvant respiratory measures

Several strategies complementary to protective ventilation may help improve gas exchange or lung protection. Among these, only the prone position has shown a mortality benefit in randomized control trials (Guérin et al., 2013). However, their role once the patient is on ECMO is still debatable.

Prone position in ECMO patients may be a logistic challenge, especially in COVID-19 patients, but is feasible and safe with experienced teams (Massart et al., 2023). As in patients without ECMO, there is consistent evidence that prone position improves survival in COVID-19 and other ARDS patients (Giani et al., 2021; Massart et al., 2023).

Although many other strategies (mainly recruitment maneuvers and inhaled nitric oxide) can be applied in ECMO patients, no strong evidence is available as to their benefits. Recruitment maneuvers may potentially improve gas exchange in ECMO patients, but increasing concerns about their safety in terms of VILI and right ventricle overload exist (Mercado et al., 2018).

When ventilation goals cannot be achieved despite ECMO support and coadjuvant maneuvers, optimization of flows, control of recirculation, or changes in the ECMO configuration to more complex settings (including VV-A-ECMO) must be attempted.

Weaning from ECMO. Post-ECMO ventilatory settings

Removal of ECMO must be considered if protective ventilation goals can be achieved without extracorporeal gas exchange. This implies not only adequate PaO2 and $PaCO_2$ levels, but also airway and transpulmonary pressures below the accepted safety thresholds. As most of the patients at this stage receive only partial support from the ventilator, hidden inspiratory efforts and dyssynchronies that can contribute to the so-called patient self-inflicted lung injury must also be considered. A recent study considered ECMO weaning in patients under pressure support ventilation with tidal esophageal pressure swings lower than 15 cmH_2O and a respiratory rate below 30/min with $PaO_2 > 70\,mmHg$, $PaCO_2 < 60$, and $pH > 7.25$.

Following experiences of weaning off other life support systems, such as mechanical ventilation itself, it has been suggested that a systematic approach to ECMO weaning, including daily trials, may speed up recovery and discharge in COVID-19 patients (Teijeiro-Paradis et al., 2021). In this setting, physiological markers such as dead space may be useful to predict a successful weaning (Lazzari et al., 2022).

Lung transplant in COVID-19

Lung transplantation has been performed in selected cases of severe COVID-19, either during the acute phase or in late fibrotic stages (Avella & Bharat, 2022). Most of the published series include severe patients with refractory hypoxemia and massive lung destruction, usually supported with ECMO. In this setting, and keeping in mind the high risk of selection bias, published data show mortality rates similar to non-COVID-19 ARDS cases or other lung chronic diseases (Florissi et al., 2022).

Anticoagulation and bleeding management

Anticoagulation management is critical during ECMO treatment because thrombotic, thromboembolic, and bleeding events are extremely common, as described later. Many factors are associated with these complications such as anticoagulation treatments, coagulation factor consumption, platelet activation and dysfunction, increased fibrinolysis, and exaggerated inflammation. Blood contact with nonendothialized ECMO components induces intense activation of inflammatory and coagulation cascades. Meanwhile, SARS-COV-2 interaction with endothelial cells induces a systemic endotheliopathy that enhances thombosis and bleeding while synergistic effects occur that significantly increase the rate of bleeding and thromboembolic events (Arachchillage et al., 2022). Additionally, during the acute phase of COVID-19, systemic endothelial cell activation induces a depletion of Weibel-Palade bodies with secretion of large amounts of von Willebrand factor (vWF), with increased serum concentrations and subsequent prothombotic state (Kalbhenn & Zieger, 2022).

However, the optimal anticoagulant strategy during ECMO has not yet been established in COVID-19 patients. The most commonly used treatment is intravenous unfractionated heparin, which can be monitored with different methods such as activated clotting time (180–200 s), activated partial thromboplastin time (APTT, 60–75 s), and anti-Xa determination (0.3–0.5 IU/mL), that is more aggressive than in the EOLIA trial criteria (M. Schmidt et al., 2020), or even thromboelastography. In COVID-19 patients, a significant discordance between APTT and anti-Xa values up to 50% has been described, and some authors say that routine anti-Xa determination is recommended (Rhoades et al., 2021). Similarly, antithrombin III deficiency is common and serial determinations have been suggested to maintain an activity of at least 70%.

Contrarily, some authors recommend minimal anticoagulation with low molecular weight heparin (for an anti-Xa activity of 0.15–0.35 UI/mL) for bleeding prevention because the shear forces of ECMO flows induce vWF levels to decrease in all patients, and an acquired von Willebrand syndrome (AVWS) with associated thrombocytopenia and platelet dysfunction develops with subsequent increased bleeding risk. When bleeding occurs and INR, APTT, and calcium levels are corrected with conventional treatments, platelet concentrate transfusion is administered for a platelet count of >100.000 platelets/mm^3. If bleeding persists, the vWF:A/vWF:Ag ratio is determined. If reduced, AVWS is suspected and vWF concentrates are infused. In these cases, desmopressin is useless because the vWF content of Weibel-Palade bodies is significantly reduced in COVID-19 patients (Kalbhenn & Zieger, 2022).

Heparin-induced thrombocytopenia favors thrombosis and bleeding and appears to be more frequent in COVID-19 patients under ECMO support than in other indications for ECMO. Bivalirudin appears to be safe in COVID-19 during ECMO support with bleeding and thromboembolic event rates similar compared with non-COVID-19 patients (Trigonis et al., 2022). Additionally, a higher bivalirudin dose requirement has been observed in these patients, although anticoagulation levels were more stable than in non-COVID-19 patients.

A few hours after patient decannulation, vWF and factor VIII levels significantly rebound and a prothrombotic state is induced that increases the risk of thrombosis. Therefore, early subcutaneous enoxaparin at therapeutic ranges is recommended (or, alternatively, direct oral anticoagulants when possible) for an anti-Xa of 0.5–0.8 UI/ml, which is maintained for 6 weeks after ECMO weaning (Kalbhenn & Zieger, 2022).

Adjuvant therapies during ECMO

Several adjuvant therapies have been implemented in ECMO systems in severe COVID-19 patients to blunt the deleterious clinical effects of high viral loads and a hyperinflammatory cytokine storm.

As mentioned in other chapters of this book, high viral SARS-CoV-2 RNAemia has been related to disease severity.

The Seraph-100 Microbind Affinity Blood Filter (ExThera, California, United States) is a sorbent hemoperfusion filter of ultrahigh molecular weight polyethylene beads coated with endpoint-attached heparin with the capacity for the adsorption of microorganisms and toxins that have been clinically tested for viral load reduction. Viruses irreversibly bind to the immobilized heparin of the system, similar to the heparan sulphate of the endothelium (Schmidt et al., 2022), and are removed from the bloodstream.

Evidence regarding Seraph-100 treatment in severe COVID-19 patients is very limited and, especially, experience during ECMO is virtually absent. So, controlled randomized studies are warranted.

Hyperinflammation and increased cytokine (IL-6) levels have been associated with poor prognosis in COVID-19 patients (Stockmann et al., 2022).

Some studies have tested the clinical effect of cytokine removal during ECMO treatment in this subgroup of patients with CytoSorb (CytoSorbents Corporation, Monmouth Junction, New Jersey, United States), a recent hemoadsorption biocompatible sorbent that contains polystyrene divinylbenzene beads coated with polyvinylpyrrolidone. This device removes molecules of medium weight (5–55 kDa) from the treated blood, including cytokines and drugs. Some noncontrolled studies have observed a significant decrease in IL-6 and other proinflammatory molecules during CytoSorb treatment (Friesecke et al., 2019). However, two recent randomized studies in ECMO-treated COVID-19 patients failed to find beneficial effects. Stockmann et al. described that the time until resolution of vasoplegic shock, the incidence of shock resolution, fluid balance, and catecholamine requirements were similar between groups (Stockmann et al., 2022), although intervention might have been initiated too late regarding both the time from ICU admission and the poor clinical state. Supady et al. observed an unexpected reduced survival after 30 days in the CytoSorb group (Fig. 2) and IL-6 levels were comparable (Supady, Weber, et al., 2021). Protective (antiinflammatory) molecules or drugs may be cleared from the blood (specially antibiotics and antivirals) and impact the results.

Results and complications of ECMO in COVID-19 patients

Hospital and mid-term mortality

The in-hospital mortality of COVID-19 patients treated with ECMO was initially very high (80%–90%) and some authors suggested that it could be even deleterious by increasing the hyperinflammatory response and prothrombosis (Henry, 2020).

Subsequently, numerous studies have shown good survival rates that differ greatly among centers due to heterogeneity in patient selection and clinical characteristics, management protocols, and experience in ECMO treatment. In-hospital mortality at 90 days varies between 35% and 65% of cases (Castaño et al., 2022; Schmidt et al., 2020), similar to those reported in the treatment of ARDS from other pathologies. However, no controlled studies have demonstrated the superiority of ECMO treatment over optimal medical treatment in patients with severe COVID-19 and, given the declining incidence of the disease, this level of evidence will unlikely be reached in the near future. Remarkably, mortality during the second wave of the pandemic was consistently higher in multiple experiences than that observed in earlier and later phases (Fig. 3)(Barbaro et al., 2021; Castaño et al., 2022; Riera et al., 2021).

After discharge, these patients show excellent survival at 6–12 months, with very low mortality during the first year of follow-up (Biancari et al., 2021; Castaño et al., 2022; Smith, Chang, et al., 2022) (Fig. 4).

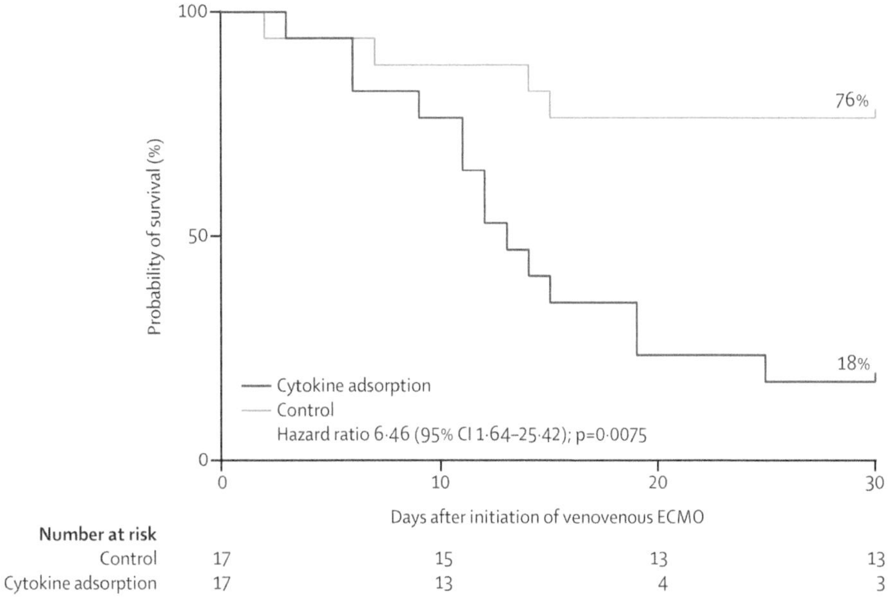

FIG. 2 Increased mortality of patients treated with Cytosorb. *With permission from Supady, A., DellaVolpe, J., Taccone, F. S., Scharpf, D., Ulmer, M., Lepper, P. M., … Staudacher, D. L. (2021). Outcome prediction in patients with severe COVID-19 requiring extracorporeal membrane oxygenation-A retrospective international multicenter study. Membranes (Basel), 11(3). https://doi.org/10.3390/membranes11030170.*

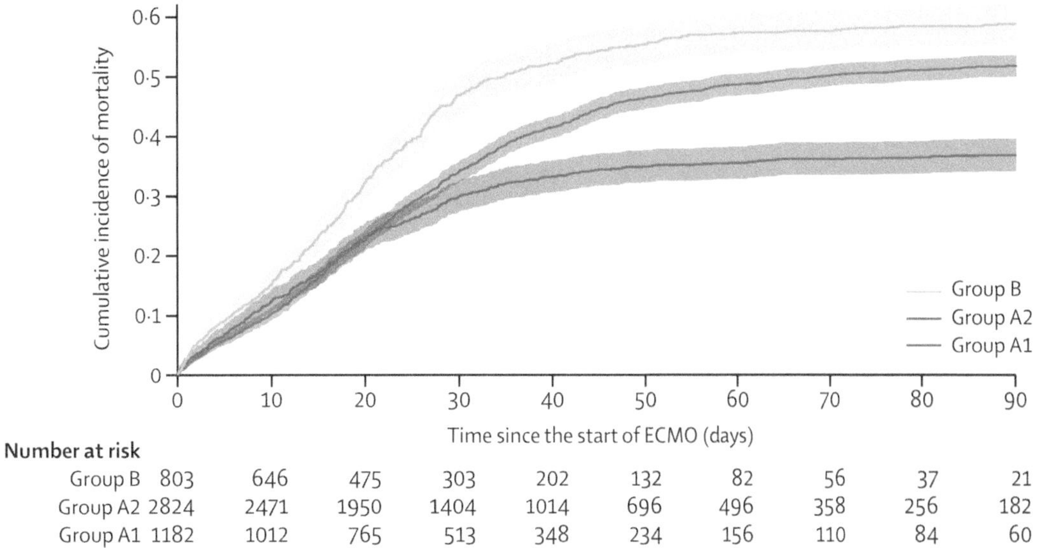

FIG. 3 Augmented mortality during the second wave of the pandemic. *With permission from Barbaro, R. P., MacLaren, G., Boonstra, P. S., Combes, A., Agerstrand, C., Annich, G., . . . Brodie, D. (2021). Extracorporeal membrane oxygenation for COVID-19: Evolving outcomes from the international extracorporeal life support organization registry. Lancet. https://doi.org/10.1016/s0140-6736(21)01960-7.*

Morbidity

Bleeding and thromboembolic complications

Bleeding and thrombotic events are the most common complications in these patients. Significant bleeding occurs in 30%–40% of ECMO-treated COVID-19 patients (Arachchillage et al., 2022), with the cannulation site the most common form (Nunez et al., 2022) and intracraneal bleeding the most lethal. Central nervous system hemorrhage occurs in 5%–10% of patients and increases mortality by 5- to 14-fold (Nunez et al., 2022; Saeed et al., 2022). An autopsy registry study reported a 21% rate of intracraneal bleeding, and it was identified as the main cause of death in 78% of the cases (von Stillfried et al., 2022). Digestive

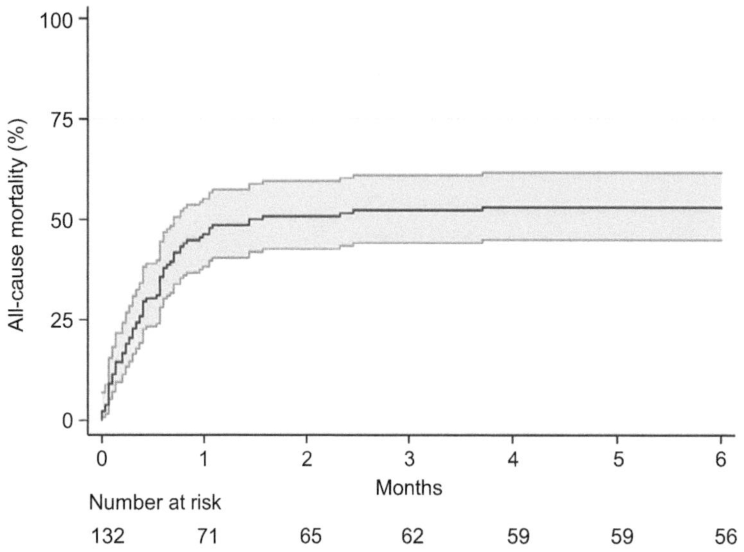

FIG. 4 Six-month excellent survival rate of ECMO-treated COVID-19 patients. *With permission from Biancari, F., Mariscalco, G., Dalén, M., Settembre, N., Welp, H., Perrotti, A., . . . Fiore, A. (2021). Six-month survival after extracorporeal membrane oxygenation for severe COVID-19. Journal of Cardiothoracic and Vascular Anesthesia, 35(7), 1999–2006. https://doi.org/10.1053/j.jvca.2021.01.027.*

tract and pulmonary bleeding account for 10% and 25%, respectively, of the total bleeding events in these patients (Arachchillage et al., 2022).

Thrombotic events are extremely common, with incidences between 20% and 65% (Arachchillage et al., 2022; Helms et al., 2020). The most frequent event is the pulmonary thromboembolism (66%), followed by arterial thrombosis (18%–20%) and thrombosis of the ECMO circuit (8%–10%). Compared with non-COVID-19 patients, ECMO-treated COVID-19 patients present more commonly with pulmonary thromboembolism and total thrombotic events (Helms et al., 2020), including ECMO circuit thrombosis/exchange (Fig. 5). When routine venous echography is performed, deep venous thrombosis is present in more than 80% of VV-ECMO patients. Importantly in these patients, early circuit exchange may significantly improve hemostasis and prevent bleeding complications by reducing the coagulation factor consumption and fibrinolysis induced by subclinical circuit microthrombosis (Bemtgen et al., 2021).

Other complications

AKI occurs in 20%–50% of COVID-19 patients treated with ECMO, with most of them requiring renal replacement therapy (Castaño et al., 2022; Pathangey et al., 2021), that is usually inserted in the ECMO circuit. The etiology of AKI is multifactorial. Vasoplegia originated by the septic/inflammatory state and enhanced by the ECMO system, vasoactive therapy,

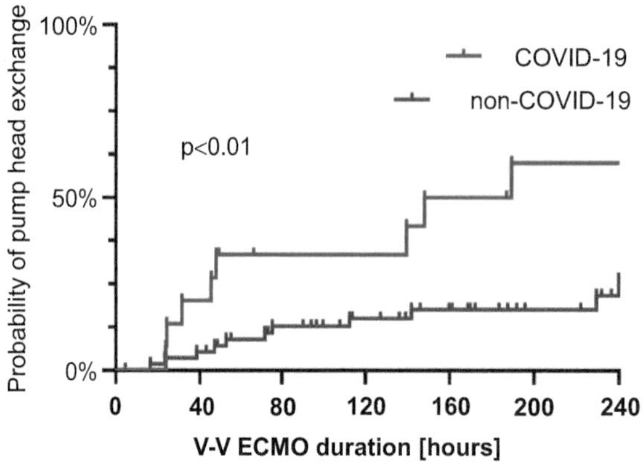

FIG. 5 Increased incidence of ECMO-circuit thrombosis in COVID-19 patients. *With permission from Bemtgen, X., Zotzmann, V., Benk, C., Rilinger, J., Steiner, K., Asmussen, A., . . . Staudacher, D. L. (2021). Thrombotic circuit complications during venovenous extracorporeal membrane oxygenation in COVID-19. Journal of Thrombosis and Thrombolysis, 51(2), 301–307. https://doi.org/10.1007/s11239-020-02217-1.*

parenchymal microembolism, and the tropism of SARS Cov-2 for renal cells contribute to AKI development (Pathangey et al., 2021). Patients with AKI present significantly increased mortality over patients with normal renal function.

Other neurological complications in these patients include meningoencephalitis, encephalopathy, and encephalomyelitis. Their etiology is frequently multifactorial, a combination of inflammatory response, neuronal trophism of SARS COV-2, and coinfection/sepsis of the critical patient (Gabelloni et al., 2022).

After pulmonary thromboembolism, bacterial coinfection is the most frequent pulmonary complication in these patients. Meticillin-resistant *Staphylococcus aureus* and pseudomona aeruginosa are the most frequent etiological agents in early (<48 h from ICU admission) and late (>7 days) coinfections, respectively. The timing of support initiation and, especially, a concurrent reduced pulmonary compliance, may influence coinfection incidence because they imply more delayed support weaning and longer invasive ventilator support (Yang et al., 2020).

Hypertransaminasemia, paralytic ileus, and mesenteric ischemia are the most frequent gastrointestinal complications in severely ill COVID-19 patients, especially when supported with ECMO. Terminal ileitis and colitis, predominantly related to vasopressor treatment in critical patients, have also been reported.

Post-ECMO functional recovery

Chronic respiratory disorders are relatively common after ARDS, predominantly asthma, which was observed in 10% during the first year when functional tests were performed (Oh et al., 2021). This complication significantly increases mid-term mortality rates. However, Kanji et al. (Kanji et al., 2021), in a cohort of 67 COVID-19 patients treated with ECMO, observed median values of FEV1, FVC, and FEV/FVC at 6 months of discharge of 86.0%, 89%, and 95%, respectively, with similar results in other series (Smith, Chang, et al., 2022). Moreover, oxygen therapy dependence is uncommon (Smith, Chang, et al., 2022).

The contagious nature of the disease and the need for isolation limit family and social support during recovery as well as rehabilitation treatment during and after ECMO therapy, which increases the functional limitation and disability of these patients. In fact, functional recovery of ECMO-treated COVID-19 patients is slow. At 6–12 months, more than 70% have variable degrees of disability and only 25%–50% have returned to work, in some cases part-time. On the other hand, 1 year after discharge, around 40%–60% have anxiety/depressive symptoms and/or signs of posttraumatic stress (Lorusso et al., 2023; Rajajee et al., 2021).

When cognitive impairment was specifically evaluated early after hospital discharge, abnormalities were found in up to 87% of patients, significantly more frequently than in patients not treated with ECMO (Taylor et al., 2022). Fortunately, prolonged cognitive impairment is rare when quantified with dedicated tests, although patients subjectively more commonly perceive such impairment, as occurs usually after ECMO treatment. This discrepancy is most likely due to the frequent association with mood disorders in these patients (Rajajee et al., 2021).

Several factors have been associated with improved functional recovery, such as patient selection and early ECMO implantation. When the device is implanted in young patients with good prior clinical condition and the "only-rescue therapy" concept during ECMO indication is discarded, functional results are excellent (Smith, Chang, et al., 2022). Similarly, specific multidisciplinary early rehabilitation programs, including patients during ECMO therapy, have been shown to improve functional outcomes in small series (Mayer et al., 2021).

Policies and procedures

Recommendations and contraindication for ECMO treatment in COVID-19 patients

(1) Justified, strongly recommended:
- Less than 3 days of mechanical ventilation and driving pressure < 16cmH_2O.
- Absence of extrapulmonary organ failure.
- Absence of previous cardiac arrest.
- Age < 40 years.
- PRESET score $\leq$ 6.
- BMI $\leq$ 40 kg/m^2.
- Absence of chronic respiratory disease.
- Absence of comorbidity or single low-risk comorbidity, absence of immunosuppression, malignant neoplastic disease, or chronic disease with short life expectancy.

- Favorable patient will.
- Highly experienced center (>30 cases/year).

(2) Indication discussed. Individual assessment.
- 3–10 days of mechanical ventilation pre-ECMO or driving pressure > 16cmH_2O.
- Single, treatable extrapulmonary organ failure.
- Short duration cardiac arrest (flow time restoration in less than 15 min).
- Age between 40 and 69.
- PRESET Score ≥ 7.
- BMI >40 kg/m^2.
- Moderate chronic respiratory disease.
- Single high-risk or multiple low-risk comorbidity, moderate immunosuppression, stable or in remission malignant neoplastic disease, or chronic disease with unknown life expectancy.
- Unknown patient will.
- Medium experience center (20–30 cases/year).

(3) High probability of nonindication, contraindication:
- More than 10 days of mechanical ventilation and driving pressure > 16cmH_2O.
- Multiple organ failure with little chance of recovery.
- Prolonged previous cardiac arrest.
- Age ≥ 70 years.
- PRESET Score 8–9.
- BMI >50 kg/m^2.
- Severe chronic respiratory disease.
  - Multiple high-risk comorbidities, severe immunosuppression, advanced malignant neoplastic disease, or chronic disease with short life expectancy.
  - Severe brain injury or concomitant brain hemorrhage.
  - Contraindication or impossibility to perform adequate anticoagulation.
  - Clinical frailty scale ≥3.
  - Blood products refusal.
  - Clear will of the patient against.
  - Center with little experience (<20 cases/year).

Applications to other areas

ECMO treatment is a very specialized, uncommon, and complex treatment. Its application is restricted to ICU environments in advanced ARDS patients. On one side, the extreme increase of patients treated during the COVID-19 pandemic has been a breakthrough in this field that has boosted center experience and, more importantly, the resource optimization process. However, ECMO in COVID-19 patient management requires specific approaches that are difficult to extrapolate to other ARDS or ECMO patients.

Mini-dictionary

- **Extracorporeal membrane oxygenation**. Systems that allow adequate oxygenation, systemic perfusion, and CO_2 removal by draining venous blood through appropriate cannulae that is oxygenated by an extracorporeal membrane oxygenator and reinfused into the body by a pump.
- **Ventilator-induced lung injury**. Lung damage originated by excessive volumes or pressures during invasive ventilator support.
- **Prone position**. It may improve survival in ECMO patients because it enhances oxygenation, homogenizes ventilation along lung parenchyma, and improve aerations due to mitigation of atelectasic areas caused by very low tidal volumes.
- **Second wave mortality increase**. It probably was due to fewer restricted indications and patient selection after the good results obtained during the first wave and the dispersion of cases among a greater number of centers with less experience.

- **"Only-rescue therapy" concept**. When ECMO indications are restricted to critically sick patients with poor clinical condition and advanced disease, and indications are stablished later in the process, predicted survival and functional recovery are seriously worsened.

Summary points

- ECMO probably improves the outcomes of COVID-19 patients with refractory hypoxemia, hypercapnia, and/or low-output syndrome despite optimal medical treatment and invasive ventilatory support, although no controlled studies are available.
- Indications for ECMO in these patients are not fully established, but some typical scenarios may be correctly outlined for correct patient selection and resource optimization.
- Younger age, shorter duration of pre-ECMO mechanical and noninvasive ventilatory support, obesity, low driving pressure, absence of comorbidities, and high-volume experience of the treating center are associated with improved survival.
- Bleeding and thromboembolic events are extremely frequent complications of ECMO support in these patients and meticulous anticoagulation protocols must be implemented.
- Cytokine adsorption is not beneficial when used in the early phases of the management of severe COVID-19 patients, and maybe even deleterious in ECMO-treated patients.
- COVID-19 patients that require prolonged ICU stays and, especially those treated with ECMO, need early multimodal (physical and neuropsychological) rehabilitation to improve functional recovery and reduce mid-term depression, anxiety, and disability.

References

Akar, A. R., Ertugay, S., Kervan, Ü., İnan, M. B., Sargın, M., Engin, Ç., & Özatik, M. A. (2019). Turkish Society of Cardiovascular Surgery (TSCVS) Proposal for use of ECMO in respiratory and circulatory failure in COVID-19 pandemic era. *Türk Göğüs Kalp Damar Cerrahisi Dergisi*, *28*(2), 229–235. https://doi.org/10.5606/tgkdc.dergisi.2020.09293.

Albaiceta, G. M., & Amado-Rodríguez, L. (2022). Ventilator-induced lung injury and lung protective ventilation. In G. Bellani (Ed.), *Mechanical ventilation from pathophysiology to clinical evidence* (pp. 165–176). Springer International Publishing.

Amado-Rodríguez, L., Del Busto, C., López-Alonso, I., Parra, D., Mayordomo-Colunga, J., Arias-Guillén, M., & Albaiceta, G. M. (2021). Biotrauma during ultra-low tidal volume ventilation and venoarterial extracorporeal membrane oxygenation in cardiogenic shock: A randomized crossover clinical trial. *Annals of Intensive Care*, *11*(1), 132. https://doi.org/10.1186/s13613-021-00919-0.

Arachchillage, D. J., Rajakaruna, I., Scott, I., Gaspar, M., Odho, Z., Banya, W., & Yusuff, H. (2022). Impact of major bleeding and thrombosis on 180-day survival in patients with severe COVID?19 supported with veno-venous extracorporeal membrane oxygenation in the United Kingdom: A multicentre observational study. *British Journal of Haematology*, *196*(3), 566–576. https://doi.org/10.1111/bjh.17870.

Avella, D., & Bharat, A. (2022). Lung transplantation in patients with lung disease secondary to coronavirus disease 2019 infection. *Current Opinion in Critical Care*, *28*(6), 681–685. https://doi.org/10.1097/mcc.0000000000000996.

Barbaro, R. P., MacLaren, G., Boonstra, P. S., Combes, A., Agerstrand, C., Annich, G., & Brodie, D. (2021). Extracorporeal membrane oxygenation for COVID-19: Evolving outcomes from the international extracorporeal life support organization registry. *Lancet*. https://doi.org/10.1016/s0140-6736(21)01960-7.

Barbaro, R. P., MacLaren, G., Boonstra, P. S., Iwashyna, T. J., Slutsky, A. S., Fan, E., & Brodie, D. (2020). Extracorporeal membrane oxygenation support in COVID-19: An international cohort study of the extracorporeal life support organization registry. *Lancet*, *396*(10257), 1071–1078. https://doi.org/10.1016/s0140-6736(20)32008-0.

Bemtgen, X., Zotzmann, V., Benk, C., Rilinger, J., Steiner, K., Asmussen, A., & Staudacher, D. L. (2021). Thrombotic circuit complications during veno-venous extracorporeal membrane oxygenation in COVID-19. *Journal of Thrombosis and Thrombolysis*, *51*(2), 301–307. https://doi.org/10.1007/s11239-020-02217-1.

Biancari, F., Mariscalco, G., Dalén, M., Settembre, N., Welp, H., Perrotti, A., & Fiore, A. (2021). Six-month survival after extracorporeal membrane oxygenation for severe COVID-19. *Journal of Cardiothoracic and Vascular Anesthesia*, *35*(7), 1999–2006. https://doi.org/10.1053/j.jvca.2021.01.027.

Calcaterra, D., Heather, B., Kohl, L. P., Erickson, H. L., & Prekker, M. E. (2020). Bedside veno-venous ECMO cannulation: A pertinent strategy during the COVID-19 pandemic. *Journal of Cardiac Surgery*, *35*(6), 1180–1185. https://doi.org/10.1111/jocs.14641.

Castaño, M., Sbraga, F., Pérez Sota, E., Arribas, J. M., Cámara, M. L., Voces, R., & Donado, A. (2022). Extracorporeal membrane oxigenation in COVID-19 patients: Results of the ECMO-COVID Registry of the Spanish Society of Cardiovascular and Endovascular Surgery. *Cirugía Cardiovascular*, *29*, 89–102.

Cavayas, Y. A., Munshi, L., Del Sorbo, L., & Fan, E. (2020). The early change in Pa(CO(2)) after extracorporeal membrane oxygenation initiation is associated with neurological complications. *American Journal of Respiratory and Critical Care Medicine*, *201*(12), 1525–1535. https://doi.org/10.1164/rccm.202001-0023OC.

Combes, A., Hajage, D., Capellier, G., Demoule, A., Lavoué, S., Guervilly, C., & Mercat, A. (2018). Extracorporeal membrane oxygenation for severe acute respiratory distress syndrome. *New England Journal of Medicine*, *378*(21), 1965–1975. https://doi.org/10.1056/NEJMoa1800385.

Cressoni, M., Gotti, M., Chiurazzi, C., Massari, D., Algieri, I., Amini, M., & Gattinoni, L. (2016). Mechanical power and development of ventilator-induced lung injury. *Anesthesiology*, *124*(5), 1100–1108. https://doi.org/10.1097/aln.0000000000001056.

Duggan, M., McCaul, C. L., McNamara, P. J., Engelberts, D., Ackerley, C., & Kavanagh, B. P. (2003). Atelectasis causes vascular leak and lethal right ventricular failure in uninjured rat lungs. *American Journal of Respiratory and Critical Care Medicine*, *167*(12), 1633–1640. https://doi.org/10.1164/rccm.200210-1215OC.

Florissi, I. S., Etchill, E. W., Barbur, I., Verdi, K. G., Merlo, C., & Bush, E. L. (2022). Lung transplantation in patients with COVID-19-The early national experience. *Seminars in Thoracic and Cardiovascular Surgery*. https://doi.org/10.1053/j.semtcvs.2022.08.008.

Friedrichson, B., Kloka, J. A., Neef, V., Mutlak, H., Old, O., Zacharowski, K., & Piekarski, F. (2022). Extracorporeal membrane oxygenation in coronavirus disease 2019: A nationwide cohort analysis of 4279 runs from Germany. *European Journal of Anaesthesiology*, *39*(5), 445–451. https://doi.org/10.1097/eja.0000000000001670.

Friesecke, S., Träger, K., Schittek, G. A., Molnar, Z., Bach, F., Kogelmann, K., & Brunkhorst, F. M. (2019). International registry on the use of the CytoSorb adsorber in ICU patients: Study protocol and preliminary results. *Medizinische Klinik, Intensivmedizin und Notfallmedizin*, *114*(8), 699–707. https://doi.org/10.1007/s00063-017-0342-5.

Gabelloni, M., Faggioni, L., Cioni, D., Mendola, V., Falaschi, Z., Coppola, S., & Neri, E. (2022). Extracorporeal membrane oxygenation (ECMO) in COVID-19 patients: A pocket guide for radiologists. *La Radiologia Medica*, *127*(4), 369–382. https://doi.org/10.1007/s11547-022-01473-w.

Giani, M., Martucci, G., Madotto, F., Belliato, M., Fanelli, V., Garofalo, E., & Grasselli, G. (2021). Prone positioning during venovenous extracorporeal membrane oxygenation in acute respiratory distress syndrome. A multicenter cohort study and propensity-matched analysis. *Annals of the American Thoracic Society*, *18*(3), 495–501. https://doi.org/10.1513/AnnalsATS.202006-625OC.

Guérin, C., Reignier, J., Richard, J. C., Beuret, P., Gacouin, A., Boulain, T., & Ayzac, L. (2013). Prone positioning in severe acute respiratory distress syndrome. *New England Journal of Medicine*, *368*(23), 2159–2168. https://doi.org/10.1056/NEJMoa1214103.

Helms, J., Tacquard, C., Severac, F., Leonard-Lorant, I., Ohana, M., Delabranche, X., & Meziani, F. (2020). High risk of thrombosis in patients with severe SARS-CoV-2 infection: A multicenter prospective cohort study. *Intensive Care Medicine*, *46*(6), 1089–1098. https://doi.org/10.1007/s00134-020-06062-x.

Henry, B. M. (2020). COVID-19, ECMO, and lymphopenia: A word of caution. *The Lancet Respiratory Medicine*, *8*(4), e24. https://doi.org/10.1016/s2213-2600(20)30119-3.

Hermann, M., Laxar, D., Krall, C., Hafner, C., Herzog, O., Kimberger, O., & Hermann, A. (2022). Duration of invasive mechanical ventilation prior to extracorporeal membrane oxygenation is not associated with survival in acute respiratory distress syndrome caused by coronavirus disease 2019. *Annals of Intensive Care*, *12*(1), 6. https://doi.org/10.1186/s13613-022-00980-3.

Herrmann, J., Lotz, C., Karagiannidis, C., Weber-Carstens, S., Kluge, S., Putensen, C., & Meybohm, P. (2022). Key characteristics impacting survival of COVID-19 extracorporeal membrane oxygenation. *Critical Care (London, England)*, *26*(1), 190. https://doi.org/10.1186/s13054-022-04053-6.

Kalbhenn, J., & Zieger, B. (2022). Bleeding during veno-venous ECMO: Prevention and treatment. *Frontiers in Medicine (Lausanne)*, *9*(879), 579. https://doi.org/10.3389/fmed.2022.879579.

Kanji, H. D., Chouldechova, A., Harris-Fox, S., Ronco, J. J., O'Dea, E., Harvey, C., & Peek, G. J. (2021). Quality of life and functional status of patients treated with venovenous extracorporeal membrane oxygenation at 6 months. *Journal of Critical Care*, *66*, 26–30. https://doi.org/10.1016/j.jcrc.2021.07.010.

Lazzari, S., Romitti, F., Busana, M., Vassalli, F., Bonifazi, M., Macrí, M. M., & Gattinoni, L. (2022). End-tidal to arterial Pco(2) ratio as guide to weaning from venovenous extracorporeal membrane oxygenation. *American Journal of Respiratory and Critical Care Medicine*, *206*(8), 973–980. https://doi.org/10.1164/rccm.202201-0135OC.

Lebreton, G., Schmidt, M., Ponnaiah, M., Folliguet, T., Para, M., Guihaire, J., & Leprince, P. (2021). Extracorporeal membrane oxygenation network organization and clinical outcomes during the COVID-19 pandemic in Greater Paris, France: A multicentre cohort study. *The Lancet Respiratory Medicine*, *9*(8), 851–862. https://doi.org/10.1016/s2213-2600(21)00096-5.

López-Martínez, C., Martín-Vicente, P., Gómez de Oña, J., López-Alonso, I., Gil-Peña, H., Cuesta-Llavona, E., & Amado-Rodríguez, L. (2022). Transcriptomic clustering of critically ill COVID-19 patients. *The European Respiratory Journal*. https://doi.org/10.1183/13993003.00592-2022.

Lorusso, R., De Piero, M. E., Mariani, S., Di Mauro, M., Folliguet, T., Taccone, F. S., & Belohlavek, J. (2023). In-hospital and 6-month outcomes in patients with COVID-19 supported with extracorporeal membrane oxygenation (EuroECMO-COVID): A multicentre, prospective observational study. *The Lancet Respiratory Medicine*, *11*(2), 151–162. https://doi.org/10.1016/s2213-2600(22)00403-9.

Massart, N., Guervilly, C., Mansour, A., Porto, A., Flécher, E., Esvan, M., & Nesseler, N. (2023). Impact of prone position in COVID-19 patients on extracorporeal membrane oxygenation. *Critical Care Medicine*, *51*(1), 36–46. https://doi.org/10.1097/ccm.0000000000005714.

Mayer, K. P., Jolley, S. E., Etchill, E. W., Fakhri, S., Hoffman, J., Sevin, C. M., & Rove, J. Y. (2021). Long-term recovery of survivors of coronavirus disease (COVID-19) treated with extracorporeal membrane oxygenation: The next imperative. *JTCVS Open*, *5*, 163–168. https://doi.org/10.1016/j.xjon.2020.11.006.

Mercado, P., Maizel, J., Kontar, L., Nalos, M., Huang, S., Orde, S., & Slama, M. (2018). Moderate and severe acute respiratory distress syndrome: Hemodynamic and cardiac effects of an open lung strategy with recruitment maneuver analyzed using echocardiography. *Critical Care Medicine*, *46*(10), 1608–1616. https://doi.org/10.1097/ccm.0000000000003287.

Nunez, J. I., Gosling, A. F., O'Gara, B., Kennedy, K. F., Rycus, P., Abrams, D., & Grandin, E. W. (2022). Bleeding and thrombotic events in adults supported with venovenous extracorporeal membrane oxygenation: An ELSO registry analysis. *Intensive Care Medicine*, *48*(2), 213–224. https://doi.org/10.1007/s00134-021-06593-x.

Oh, T. K., Cho, H. W., Lee, H. T., & Song, I. A. (2021). Chronic respiratory disease and survival outcomes after extracorporeal membrane oxygenation. *Respiratory Research*, *22*(1), 195. https://doi.org/10.1186/s12931-021-01796-8.

Pathangey, G., Fadadu, P. P., Hospodar, A. R., & Abbas, A. E. (2021). Angiotensin-converting enzyme 2 and COVID-19: Patients, comorbidities, and therapies. *American Journal of Physiology. Lung Cellular and Molecular Physiology*, *320*(3), L301–l330. https://doi.org/10.1152/ajplung.00259.2020.

Powell, E. K., Lankford, A. S., Ghneim, M., Rabin, J., Haase, D. J., Dahi, S., & Tabatabai, A. (2022). Decreased PRESET-Score corresponds with improved survival in COVID-19 veno-venous extracorporeal membrane oxygenation. *Perfusion*. https://doi.org/10.1177/02676591221128237.

Rajajee, V., Fung, C. M., Seagly, K. S., Park, P. K., Raghavendran, K., Machado-Aranda, D. A., & Napolitano, L. M. (2021). One-year functional, cognitive, and psychological outcomes following the use of extracorporeal membrane oxygenation in coronavirus disease 2019: A prospective study. *Critical Care Explorations*, *3*(9), e0537. https://doi.org/10.1097/cce.0000000000000537.

Rhoades, R., Leong, R., Kopenitz, J., Thoma, B., McDermott, L., Dovidio, J., & Al-Rawas, N. (2021). Coagulopathy monitoring and anticoagulation management in COVID-19 patients on ECMO: Advantages of a heparin anti-Xa-based titration strategy. *Thrombosis Research*, *203*, 1–4. https://doi.org/10.1016/j.thromres.2021.04.008.

Riera, J., Alcántara, S., Bonilla, C., Fortuna, P., Blandino Ortiz, A., Vaz, A., & Roncon-Albuquerque, R., Jr. (2022). Risk factors for mortality in patients with COVID-19 needing extracorporeal respiratory support. *European Respiratory Journal*, *59*(2). https://doi.org/10.1183/13993003.02463-2021.

Riera, J., Roncon-Albuquerque, R., Jr., Fuset, M. P., Alcántara, S., & Blanco-Schweizer, P. (2021). Increased mortality in patients with COVID-19 receiving extracorporeal respiratory support during the second wave of the pandemic. *Intensive Care Medicine*, *1–4*. https://doi.org/10.1007/s00134-021-06517-9.

Rojas-Velasco, G., Carmona-Levario, P., Manzur-Sandoval, D., Lazcano-DÌaz, E., & Damas de Los Santos, F. (2022). Pulmonary artery cannulation during venovenous extracorporeal membrane oxygenation: An alternative to manage refractory hypoxemia and right ventricular dysfunction. *Respiratory Medicine Case Reports*, *38*(101), 704. https://doi.org/10.1016/j.rmcr.2022.101704.

Rozé, H., Richard, J. M., Thumerel, M., & Ouattara, A. (2017). Spontaneous breathing (SB) using airway pressure-release ventilation (APRV) in patients under extracorporeal-membrane oxygenation (ECMO) for acute respiratory distress syndrome (ARDS). *Intensive Care Medicine*, *43*(12), 1919–1920. https://doi.org/10.1007/s00134-017-4892-z.

Saeed, O., Stein, L. H., Cavarocchi, N., Tatooles, A. J., Mustafa, A., Jorde, U. P., & Silvestry, S. (2022). Outcomes by cannulation methods for venovenous extracorporeal membrane oxygenation during COVID-19: A multicenter retrospective study. *Artificial Organs*, *46*(8), 1659–1668. https://doi.org/10.1111/aor.14213.

Schmidt, J. J., Borchina, D. N., Van Klooster, M., Bulhan-Soki, K., Okioma, R., Herbst, L., & Kielstein, J. T. (2022). Interim analysis of the COSA (COVID-19 patients treated with the Seraph® 100 Microbind Affinity filter) registry. *Nephrology, Dialysis, Transplantation*, *37*(4), 673–680. https://doi.org/10.1093/ndt/gfab347.

Schmidt, M., Hajage, D., Lebreton, G., Monsel, A., Voiriot, G., Levy, D., & Combes, A. (2020). Extracorporeal membrane oxygenation for severe acute respiratory distress syndrome associated with COVID-19: A retrospective cohort study. *The Lancet Respiratory Medicine*, *8*(11), 1121–1131. https://doi.org/10.1016/s2213-2600(20)30328-3.

Shekar, K., Badulak, J., Peek, G., Boeken, U., Dalton, H. J., Arora, L., & Pellegrino, V. (2020). Extracorporeal life support organization coronavirus disease 2019 interim guidelines: A consensus document from an International Group of Interdisciplinary Extracorporeal Membrane Oxygenation Providers. *ASAIO Journal*, *66*(7), 707–721. https://doi.org/10.1097/mat.0000000000001193.

Smith, D. E., Chang, S. H., Geraci, T. C., James, L., Kon, Z. N., Carillo, J. A., & Galloway, A. C. (2022). One-year outcomes with venovenous extracorporeal membrane oxygenation support for severe COVID-19. *The Annals of Thoracic Surgery*. https://doi.org/10.1016/j.athoracsur.2022.01.003.

Smith, N. J., Park, S., Zundel, M. T., Dong, H., Szabo, A., Cain, M. T., & Durham, L. A., 3rd. (2022). Extracorporeal membrane oxygenation for COVID-19: An evolving experience through multiple waves. *Artificial Organs*, *46*(11), 2257–2265. https://doi.org/10.1111/aor.14381.

Stockmann, H., Thelen, P., Stroben, F., Pigorsch, M., Keller, T., Krannich, A., & Lehner, L. J. (2022). CytoSorb rescue for COVID-19 patients with vasoplegic shock and multiple organ failure: A prospective, open-label, randomized controlled pilot study. *Critical Care Medicine*, *50*(6), 964–976. https://doi.org/10.1097/ccm.0000000000005493.

Supady, A., DellaVolpe, J., Taccone, F. S., Scharpf, D., Ulmer, M., Lepper, P. M., & Staudacher, D. L. (2021). Outcome prediction in patients with severe COVID-19 requiring extracorporeal membrane oxygenation-A retrospective international multicenter study. *Membranes (Basel)*, *11*(3). https://doi.org/10.3390/membranes11030170.

Supady, A., Weber, E., Rieder, M., Lother, A., Niklaus, T., Zahn, T., & Duerschmied, D. (2021). Cytokine adsorption in patients with severe COVID-19 pneumonia requiring extracorporeal membrane oxygenation (CYCOV): A single centre, open-label, randomized, controlled trial. *The Lancet Respiratory Medicine*, *9*(7), 755–762. https://doi.org/10.1016/s2213-2600(21)00177-6.

Taylor, L. J., Jolley, S. E., Ramani, C., Mayer, K. P., Etchill, E. W., Mart, M. F., & Rove, J. Y. (2022). Early posthospitalization recovery after extracorporeal membrane oxygenation in survivors of COVID-19. *The Journal of Thoracic and Cardiovascular Surgery*. https://doi.org/10.1016/j.jtcvs.2021.11.099.

Teijeiro-Paradis, R., Tiwari, P., Spriel, A., Del Sorbo, L., & Fan, E. (2021). Standardized liberation trials in patients with COVID-19 ARDS treated with venovenous extracorporeal membrane oxygenation: When ready, let them breathe! *Intensive Care Medicine*, *47*(12), 1494–1496. https://doi.org/10.1007/s00134-021-06523-x.

Terragni, P. P., Rosboch, G., Tealdi, A., Corno, E., Menaldo, E., Davini, O., & Ranieri, V. M. (2007). Tidal hyperinflation during low tidal volume ventilation in acute respiratory distress syndrome. *American Journal of Respiratory and Critical Care Medicine*, *175*(2), 160–166. https://doi.org/10.1164/rccm.200607-915OC.

Tran, A., Fernando, S. M., Rochwerg, B., Barbaro, R. P., Hodgson, C. L., Munshi, L., & Brodie, D. (2022). Prognostic factors associated with mortality among patients receiving venovenous extracorporeal membrane oxygenation for COVID-19: A systematic review and meta-analysis. *The Lancet Respiratory Medicine*. https://doi.org/10.1016/s2213-2600(22)00296-x.

Trigonis, R., Smith, N., Porter, S., Anderson, E., Jennings, M., Kapoor, R., & Rahman, O. (2022). Efficacy of bivalirudin for therapeutic anticoagulation in COVID-19 patients requiring ECMO support. *Journal of Cardiothoracic and Vascular Anesthesia*, *36*(2), 414–418. https://doi.org/10.1053/j.jvca.2021.10.026.

von Stillfried, S., Bülow, R. D., Röhrig, R., Meybohm, P., & Boor, P. (2022). Intracranial hemorrhage in COVID-19 patients during extracorporeal membrane oxygenation for acute respiratory failure: A nationwide register study report. *Critical Care (London, England)*, *26*(1), 83. https://doi.org/10.1186/s13054-022-03945-x.

Warren, A., McKie, M. A., Villar, S. S., Camporota, L., & Vuylsteke, A. (2022). Effect of hypoxemia on outcome in respiratory failure supported with extracorporeal membrane oxygenation: A cardinality matched cohort study. *ASAIO Journal*, *68*(12), e235–e242. https://doi.org/10.1097/mat.0000000000001835.

Yang, X., Cai, S., Luo, Y., Zhu, F., Hu, M., Zhao, Y., & Peng, Z. (2020). Extracorporeal membrane oxygenation for coronavirus disease 2019-induced acute respiratory distress syndrome: A multicenter descriptive study. *Critical Care Medicine*, *48*(9), 1289–1295. https://doi.org/10.1097/ccm.0000000000004447.

Chapter 36

Sars-CoV-2 infection in different hematological patients

Saša Anžej Doma[a,b]
[a]University Medical Centre Ljubljana, Hematology Department, Ljubljana, Slovenia, [b]Faculty of Medicine, University of Ljubljana, Ljubljana, Slovenia

Abbreviations

ALL acute lymphoblastic leukemia
AML acute myeloid leukemia
APTT activated partial thromboplastin time
FLT3 FMS-like tyrosine kinase 3
G-CSF granulocyte colony stimulating factor
HMAs hypomethylating agents
HSCT hematopoietic stem cell transplantation
ICU intensive care unit
ISTH International Society of Thrombosis and Haemostasis
LMWH low molecular weight heparin
MDS myelodysplastic syndrome
MM multiple myeloma
MPN myeloproliferative neoplasm
PT prothrombin time
RT-PCR reverse transcription polymerase chain reaction

Introduction

COVID-19 predominantly affects the respiratory system but several hematopoietic system and hemostasis abnormalities have also been identified. These are variable and nonspecific but more prominent in patients with severe COVID-19 (Al-Saadi & Abdulnabi, 2022; Gajendra, 2022; Goyal et al., 2020; Youssry et al., 2022; Zhou et al., 2020). Common hematological abnormalities associated with SARS-CoV-2 infection are lymphopenia, thrombocytopenia, and elevated D-dimer (Goyal et al., 2020; Terpos, Ntanasis-Stathopoulos, et al., 2020).

Blood counts and other laboratory parameters

Lymphopenia is the most common laboratory finding, present in 25%–80% of patients at admission (Gajendra, 2022; Goyal et al., 2020; Zhou et al., 2020). It is common even in mild disease (Al-Saadi & Abdulnabi, 2022). The underlying pathophysiological process of lymphopenia is the invasion of lymphocytes by the virus, resulting in cell lysis, cytokines causing lymphocyte apoptosis, and atrophy of lymphoid organs, further impairing lymphocyte turnover and reducing lymphocyte proliferation due to raised lactic acid (Gajendra, 2022). Lymphopenia was reported to correlate with the severity of the disease (Terpos, Ntanasis-Stathopoulos, et al., 2020). Neutrophilia was observed in severe COVID-19 and was associated with a poor outcome (Mehrpouri, 2021; Rahman et al., 2021; Zhou et al., 2020).

Thrombocytopenia has been reported in 5%–54% of patients (Cheung et al., 2019; Gajendra, 2022). According to some studies (Mehrpouri, 2021; Rahman et al., 2021), it correlates with the severity of COVID-19; however, it is usually mild (Fig. 1). It is supposed to be a direct effect of the SARS-CoV-2 virus on platelet production, autoimmune destruction of platelets, or increased platelet consumption (Rahman et al., 2021).

Management, Body Systems, and Case Studies in COVID-19. https://doi.org/10.1016/B978-0-443-18703-2.00007-4
Copyright © 2024 Elsevier Inc. All rights are reserved, including those for text and data mining, AI training, and similar technologies.

Test name	*result*	*units*	*Reference range*
WBC	7.8	10^9/L	4.0 – 10.0
RBC	4.12	10^{12}/L	3.80 – 4.80
hemoglobin	123	g/L	120-150
hematocrit	0.379	1	0.360 – 0.460
MCV	92.0	fL	83.0 – 101.0
MCH	29.9	pg	27.0 – 32.0
MCHC	325	g/L	315 - 345
RDW	14.2	%	11.6 – 14.0
Platelet count	**118**	10^9/L	150 - 410
MPV	9.2	fL	7.8 – 11.0
Differential:			
Total neutrophils, %	83	%	40-80
Total lymphocytes,%	**8.1**	%	20-40
Monocytes, %	7.6	%	2-10
Eosinophils, %	0.8	%	1.0-6.0
Basophils, %	0.5	%	0.0-2.0
Total neutrophils, absolute	6.47	10^9/L	1.50 – 7.40
Total lymphocytes, absolute	**0.63**	10^9/L	1.10 – 3.50
monocytes, absolute	0.59	10^9/L	0.21 – 0.92
Eosinophils, absolute	0.06	10^9/L	0.02 – 0.67
Basophils, absolute	0.04	10^9/L	0.00 – 0.13
Coagulation tests:			
Prothrombin time	1.06	1	0.70-1.30
Activated partial thromboplastin time	35.5	s	25.9 – 36.6
fibrinogen	**7.1**	g/L	1.8 - 3.5
Thrombin time	17.5	s	< 21.0
D-dimer	**1760**	μg/L	< 500
Frequent hematological findings in COVID-19 are lymphopenia, mild thrombocytopenia, elevated D-dimer and fibrinogen (bolded). Among biochemical parameters common findings are: elevated liver enzymes, elevated lactate dehydrogenase, elevated inflammatory markers (C-reactive protein, ferritin), elevated troponin and creatine phosphokinase, acute kidney injury.			

FIG. 1 **Typical hematological findings in a patient with COVID-19.** *MCH*, mean corpuscular hemoglobin; *MCHC*, mean corpuscular hemoglobin concentration; *MCV*, mean corpuscular volume; *MPV*, mean platelet volume; *RBC*, red blood cell count; *RDW*, red cell distribution width; *WBC*, total white blood cell count

Coagulation abnormalities are frequently encountered, especially in patients with severe COVID-19, and are most probably a consequence of antiviral inflammatory response shifting the balance in the anticoagulant and procoagulant pathways (Rahman et al., 2021; Terpos, Ntanasis-Stathopoulos, et al., 2020). Among the most frequently reported are elevated D-dimer levels (Rahman et al., 2021; Terpos, Ntanasis-Stathopoulos, et al., 2020; Youssry et al., 2022; Zhou et al., 2020). Elevated D-dimer levels or their gradual increase during the course of the disease was associated with disease worsening (Goyal et al., 2020; Terpos, Ntanasis-Stathopoulos, et al., 2020; Zhou et al., 2020). An admission D-dimer > 2.0 μg/mL was predictive of in-hospital mortality (Zhang et al., 2020). Prolongation of prothrombin time (PT) and activated partial thromboplastin time (APTT) was reported in some patients with severe COVID-19, including those presenting with cardiac injury in the context of the disease (Terpos, Ntanasis-Stathopoulos, et al., 2020). Increased endothelial activation markers such as von Willebrand factor, factor VIII, P-selectin, and a fibrinolytic inhibitor plasminogen activator inhibitor 1 (PAI-1) were also reported in COVID-19 (Gajendra, 2022; Rahman et al., 2021). Altered coagulability was shown to be a poor prognostic sign (Al-Saadi & Abdulnabi, 2022). However, coagulopathy associated with COVID-19 (CAC) differs from proper disseminated intravascular coagulation in that fibrinogen is often increased (Gajendra, 2022; Sahu & Cerny, 2021). The rate of symptomatic venous thromboembolism complications in hospitalized COVID-19 patients was around 10% (Terpos, Ntanasis-Stathopoulos, et al., 2020) and even higher (25%–31%) among COVID-19 patients admitted to the intensive care unit (ICU) (Rahman et al., 2021). According to the International Society of Thrombosis and Haemostasis, the routine use of low molecular weight heparin is recommended for all in-patients with COVID-19; for selected patients it is even recommended in therapeutic doses (Kreuziger et al., 2022; Schulman et al., 2022). Anticoagulants should be given with caution in patients with hemophilia and allied disorders (Hermans et al., 2020).

Hematological diseases

As hematological malignancies directly affect the immune system, these patients are at a significantly higher risk for severe infections, including COVID-19 (Langerbeins & Hallek, 2022; Wang et al., 2021). The other reason for their impaired immunity is the cytotoxic therapy that compromises not only malignant blood cells but also healthy ones. Classical risk factors for severe disease and death, such as older age and comorbidities, apply to this population as well (Langerbeins & Hallek, 2022). Based on international registries and metaanalyses, the mortality of COVID-19 among hematological patients is high, with 14%–51% of patients dying with a documented SARS-CoV-2 infection (Buske et al., 2022; Langerbeins & Hallek, 2022). This was the rationale for hematological patients to be a high-priority subgroup for SARS-CoV-2 vaccination. Importantly, in patients with lymphoid malignancies and in patients receiving B-cell depleting agents, anti-CD38 monoclonal antibodies, Bruton's tyrosine kinase inhibitors, JAK inhibitors, and venetoclax therapy, the vaccine-induced antibody response is significantly impaired. Additionally, many patients will lose their immunity after hematopoietic stem cell transplantation (Buske et al., 2022). In addition to vaccination, which should preferably be administered before the initiation of hematological treatment, especially before anti-CD20 antibodies (Buske et al., 2022), antiviral drugs and monoclonal antibodies for preexposure and postexposure prophylaxis and for early treatment of COVID-19 have recently become available (Langerbeins & Hallek, 2022). The oral protease inhibitor nirmatrelvir in combination with ritonavir can be prescribed to patients with hematological malignancies and a confirmed SARS-CoV-2 infection to prevent severe COVID-19, keeping in mind relevant CYP3A4 drug interactions (kinase inhibitors, venetoclax). Molnupiravir has a similar indication. As the first approved drug to treat COVID-19, remdesivir received emergency use authorization to treat COVID-19 patients with pneumonia requiring supplemental oxygen. However, it can also be prescribed to patients with hematological malignancies with asymptomatic or mild COVID-19 to alleviate the course of the infection. Dexamethasone is indicated in the treatment of COVID-19 adults who require supplemental oxygen therapy. Several monoclonal antibodies against the virus have been developed and used (etesevimab-bamlanivimab, imdevimab-casirivimab, sotrovimab, tixagevimab-cilgavimab) but are no longer effective against the new variants of the virus. Tixagevimab-cilgavimab was authorized also as preexposure prophylaxis to prevent COVID-19 in moderately to severely immunocompromised patients or for whom an active COVID-19 vaccination was not possible (patients with active hematological malignancy or receiving immunosuppressive treatment and transplant recipients). (Langerbeins & Hallek, 2022; Montgomery et al., 2022) (Table 1)

Important elements to managing hematological patients include common preventive strategies and frequent screening, preferably with reverse transcription polymerase chain reaction (RT-PCR) before in-hospital stay and at presentation of any symptoms of COVID-19. Hematology units should be SARS-CoV-2 free zones, dedicated solely to hematological treatment (Buske et al., 2022; Rahman et al., 2021). After a COVID-19 diagnosis, two negative results are recommended before admission to care units. Prolonged infections in immunosuppressed patients, a positive PCR without active infection versus prolonged viral shedding, reactivation, or reinfection (Langerbeins & Hallek, 2022; Zhou et al., 2020) impose significant challenges for the management of this population. It is generally recommended to avoid a hospital stay and reduce or sometimes delay the initiation of immunocompromised treatments, if possible (Buske et al., 2022). While this applies

TABLE 1 Medical conditions where immune response against SARS-CoV-2 vaccine is impaired.

Lymphoid malignancies: CLL, indolent lymphoma; NOT Hodgkin lymphoma, and NOT hairy cell leukemia
Therapy with B-cell depleting agents (anti-CD20 antibodies) within 6–12 months
Anti-CD38 monoclonal antibody therapy
Recent hematopoietic stem cell transplantation
JAK inhibitor therapy (ruxolitinib)
Bruton's tyrosine kinase inhibitor treatment
Venetoclax therapy

CLL, chronic lymphocytic leukemia; *JAK*, Janus kinase.

TABLE 2 Management of hematological malignancies in patients with active SARS-CoV-2 infection.

MM[a]	In symptomatic COVID-19, delay treatment unless acute renal failure or other MM-related complications; closely monitor for COVID-19 progression Asymptomatic patients should stay quarantined at home for at least 14 days, may continue therapy in case of active MM (MM-related symptoms, new diagnosis, recent relapse, suboptimal response to therapy) Steroids and drugs inducing lymphopenia should be deintensified Consider prophylactic G-CSF
AML, ALL, high-risk MDS, blast phase of MPN/CML[a]	Delay any intensive chemotherapy until at least 2 weeks after resolution of symptoms and SARS-CoV-2 PCR negativity (2 negative swabs), whenever possible Best available therapy for COVID-19 should be given Best supportive therapy for a hematological malignancy should be given In newly diagnosed patients, low-intensity therapies to avoid progression (prednisone plus central nervous system prophylaxis in ALL, hydroxyurea, and/or HMAs or *FLT3* inhibitors in AML) can be used AML, ALL, APL consolidation and maintenance therapies could be delayed Thrombosis prophylaxis is recommended if asparaginase is to be used and asparaginase omitted if thrombotic events are present A low-intensity therapy as a bridge to HSCT in patients with a high risk of progression could be considered Patients already under active treatment, especially if prolonged myelosuppression is expected, must be admitted to a COVID-19 unit and closely monitored
Lymphoma[a]	All positive cases should be investigated with lung CT In newly diagnosed patients, defer commencement of treatment until nasopharyngeal swab negativity and resolution of the infection If already on treatment, BTK inhibitor therapy and other novel inhibitors might be continued in mild COVID-19 In hospitalized patients or those requiring oxygen therapy, targeted therapy as well as monoclonal antibodies and/or chemotherapy should be withheld until recovery Therapy could be resumed if patients are asymptomatic for at least 48 h and at least 14 days after symptom onset, and if possible, after two consecutive negative RT-PCR tests In aggressive lymphoma, delay the start of treatment without compromising treatment in a curative setting
MPN (excluding blast phase of MPN/CML)	No adjustment required (continuation of TKI, ruxolitinib etc.)
MDS	Lower-risk MDS patients responding to erythropoiesis-stimulating agents, luspatercept, or lenalidomide and higher-risk MDS patients on HMAs beyond the third cycle without hematological toxicity can continue treatment (preferably if it can be delivered at home)

ALL, acute lymphoblastic leukemia; *AML*, Acute myeloid leukemia; *APL*, acute promyelocytic leukemia; *CML*, chronic myeloid leukemia; *FLT3*, FMS-like tyrosine kinase 3; *G-CSF*, granulocyte colony stimulating factor; *HMAs*, hypomethylating agents; *HSCT*, hematopoietic stem cell transplantation; *MDS*, myelodysplastic syndrome; *MM*, multiple myeloma; *MPN*, myeloproliferative neoplasm; *TKI*, tyrosine kinase inhibitors.
[a]*The decision to continue/delay/start anticancer treatment in patients with aggressive hematological malignancies should be made on a case-by-case basis.*

mainly to indolent lymphomas, there usually isn't much place for modification of the treatment of aggressive lymphomas, acute leukemias, high-risk myelodysplastic syndromes, and multiple myeloma; both the SARS-CoV-2 infection and hematological malignancy risk a patient's life. On the other hand, the therapy of myeloproliferative diseases, including chronic myeloid leukemia, needs no adjustment (Buske et al., 2022). This similarly applies to nonmalignant hematological diseases, although recommendations for management have been issued for hemoglobin disorders (Chowdhury & Anwar, 2020) and hemophilia (Hermans et al., 2020; Pipe et al., 2021). Table 2 sums up the current recommendations for the management of hematological malignancies in patients with active SARS-CoV-2 infection.

Case reports

The following cases illustrate real-life challenges where management must often be individualized. Most of the decision making is based on little evidence (Buske et al., 2022).

Case report 1: Acute myeloid leukemia and COVID-19

A 49-year-old female patient with schizophrenia, hypothyreosis, arterial hypertension, asthma, and obesity was diagnosed with COVID-19 early in December 2020. After a week of respiratory symptoms as well as loss of taste and smell, she was admitted to the hospital due to fatigue. The oxygen saturation was 100%, blood pressure 110/70 mm Hg, pulse 104/min, and she was afebrile. Her laboratory results at admission were CRP 41 mg/L (ref. <5); procalcitonin 0.15 μg/L (ref. ≤0.24); LDH 7.38 μkat/L (ref. ≤4.12); normal bilirubin and transaminases; gamma-GT 2.28 μkat/L (ref. ≤2.28); creatinine 187 μmol/L (ref. 44–97); urea 9.0 mmol/L (ref. 2.8–7.5); ferritin 782 μg/L (ref. 10–120); L 40.2 x10^9/L (ref. 4.0–10.0) with 23% of blast cells; segmented neutrophils were 3.22×10^9/L (ref. 1.5–7.4); lymphocytes 8.04×10^9/L (ref. 1.1–3.5); hb 67 g/L (ref. 120–150); platelets 53×10^9/L (ref. 150–410); prothrombin time 0.74 (ref. 0.70–1.30); INR 1.20 (ref. <1.30); and D-dimer 4074 μg/L (ref. <500). Acute myeloid leukemia (AML) with inv (16) was diagnosed by bone marrow aspirate and hydroxyurea introduced due to high white blood cells. Transfusions of red blood cells and platelets were given to maintain hb above 80 g/L and platelets above 20×10^9/L (Döhner et al., 2022). After a few days, the patient deteriorated due to pneumonia, progressive acute renal failure, and septic shock. She was admitted to the ICU and put on mechanical ventilation. Her laboratory results were as follows: CRP 309 mg/L (ref. <5); procalcitonin 33.33 μg/L (ref. ≤0.24); IL-6 1920 ng/L (ref. ≤7.0); LDH 17.48 μkat/L (ref. ≤4.12); bilirubin 78/56 μmol/L (ref. <22/7); troponin 162 ng/L (ref. <40); gamma-GT 1.62 μkat/L (ref. ≤2.28); creatinine 291 μmol/L (ref. 44-97); urea 21.0 mmol/L (ref. 2.8–7.5); ferritin 1807 μg/L (ref. 10–120); L 30.5×10^9/L (ref. 4.0–10,0) with 2% of blast cells; segmented neutrophils were 16.47×10^9/L (ref. 1.5–7.4); lymphocytes 4.27×10^9/L (ref. 1.1–3.5); hb 72 g/L (ref. 120–150); platelets 38×10^9/L (ref. 150–410); prothrombin time 0.53 (ref. 0.70–1.30); INR 1.46 (ref. <1.30); APTT 54.6 s (ref. 25.9–36.6); fibrinogen 5.8 g/L (ref. 1.8–3.5); antithrombin 0.81 (ref. 0.83–1.18); and D-dimer 2287 μg/L (ref. <500). She was treated with broad-spectrum antibiotics (piperacillin/tazobactam, cefepime, clindamycin, meropenem), methylprednisolone according to the protocol for critically ill COVID-19 patients, an antifungal, and pneumocystis prophylaxis. Due to hemodynamic instability, she received therapy with noradrenaline. Remdesivir could not be given due to renal failure. She received intravenous immunoglobulins. Drainage of her right pleural space due to empyema was required. Prolongation of the coagulation tests was even more pronounced (PT 0.45; INR 1.73; APTT 63.6 s). D-dimer and fibrinogen levels were increased too (persistently above 4000 μg/L and 5.8 g/L, respectively). Despite platelet transfusions, she suffered a subarachnoid hemorrhage. Nimodipin was introduced and platelets intensively substituted. The patient received no thromboprophylaxis. When she was admitted to the hematology department (after having positive nasopharyngeal swabs for SARS-CoV-2 for a whole month), her ECOG performance status was 4. She had the severe myopathy of a critically ill patient and needed intensive physiotherapy. Her blood counts were L 1.43×10^9/L; neutrophil granulocytes 0.40×10^9/L; lymphocytes 0.40×10^9/L; hb 86 g; and platelets 101×10^9/L. We decided to treat her nonintensively with azacitidine. Already after the first cycle, her blood counts improved and after two cycles of azacytidine, bone marrow aspirate demonstrated complete hematological remission. She continued with azacitidine for seven more cycles and slowly rehabilitated. She fully recovered from COVID-19 but had a relapse of AML in August 2021. Again, an AML with inv(16) was diagnosed and this time the patient received induction therapy with daunorubicin and cytarabine as well as three consolidation courses with high doses of cytarabine. She achieved complete hematological and molecular remission.

In May 2021, she was vaccinated with one dose of the Pfizer vaccine. In February 2022, just after finishing chemotherapy, she got COVID-19 again but received sotrovimab and the infection was uneventful.

Case report 1: Discussion

The patient was diagnosed with acute leukemia at the same time of COVID-19 infection. Besides a hematological malignancy, she also had other risk factors for severe COVID-19, namely obesity, asthma, and arterial hypertension (Goyal et al., 2020; Zhou et al., 2020). Studies have shown that patients with hematological malignancies have a higher rate of severe/critical disease compared to the general population (Passamonti et al., 2020), or patients with solid tumors. Particularly, patients with AML have the worst clinical outcome and the highest mortality rate (García-Suárez et al., 2020; Lee et al., 2020; Pagano et al., 2021). Concerning hematological parameters heralding severe COVID-19, our patient did not develop lymphopenia, but instead neutrophilia (and became neutropenic only after 3 weeks). She also had increased CRP, procalcitonin, ferritin, and LDH. All these parameters have been reported to be associated with severe disease (Al-Saadi & Abdulnabi, 2022; Gajendra, 2022; Rahman et al., 2021; Terpos, Ntanasis-Stathopoulos, et al., 2020; Youssry et al., 2022; Zhang et al., 2020). The patient also had renal insufficiency, the incidence of which is reported to be as high as 40% in COVID-19 patients. This is due to direct kidney damage from SARS-CoV-2 or through the production of proinflammatory cytokines, but is also caused by indirect pathogenesis from critical care interventions (Gajendra, 2022). She also developed COVID-19 coagulopathy, characterized by prolonged PT and APTT, elevated fibrinogen, extremely elevated IL-6, and high D-dimer levels (Pipe et al., 2021). Thrombocytopenia was probably a consequence of both AML and COVID-19. It was moderate at admission but became severe over the course of hospitalization. Thrombocytopenia and later the subarachnoid hemorrhage were the reasons the patient did not receive thromboprophylaxis. According to ISTH recommendations, thromboprophylaxis with heparins is applied to all hospitalized patients unless contraindicated (active bleeding and platelet count $<25 \times 10^9$/L) (Thachil et al., 2020). It is known from the literature that not only thrombotic complications but bleedings also occur in patients with severe COVID-19, although at a much lower frequency (4.8% bleeding rate according to Al-Samkari et al., 2020).

It is recommended to delay treatment for acute leukemia until at least 2 weeks after the resolution of symptoms and SARS-CoV-2 PCR negativity (Buske et al., 2022; Cheung et al., 2019). Low-intensity therapies to avoid progression can be used (prednisolone plus central nervous system prophylaxis in ALL and hydroxyurea and/or hypomethylating agents or *FLT3* inhibitors for AML) (Buske et al., 2022). Our patient demonstrated prolonged PCR positivity, which is known to complicate the management of patients with hematological malignancies (Buske et al., 2022; Langerbeins & Hallek, 2022). She couldn't start intensive chemotherapy even after the resolution of COVID-19 as she developed severe myopathy and wasn't fit enough for intensive chemotherapy. Despite the fact that a core binding factor AML is known to be highly susceptible to cytarabine (Borthakur & Kantarjian, 2021), our patient responded surprisingly well to hydroxyurea and azacitidine and stayed in remission for several months.

It has been demonstrated that protection after SARS-CoV-2 infection is short-lasting in patients with hematological malignancies (Langerbeins & Hallek, 2022). Luckily, the second time the patient caught COVID-19, she was in hematological remission, was vaccinated, and received antiviral drugs; the course of the infection was uneventful.

Case report 2: Multiple myeloma and COVID-19 infection

A 71-year-old patient with multiple myeloma (MM) IgA kappa (diagnosed in March 2019), actively treated with lenalidomide, ixazomib, and dexamethasone with very good partial response, was admitted into the ICU in mid-October 2020 for respiratory failure after 2 days of upper respiratory tract symptoms. He tested positive for SARS-CoV-2 infection. His saturation was 85% on Ohio mask, blood pressure 95/65 mm Hg. His laboratory parameters were as follows: L 2.2×10^9/L (ref. 4.0–10.0); lymphocytes 0.13×10^9/L (ref. 1.1–3.5); neutrophils 1.76×10^9/L (ref. 1.5–7.4); hb 121 g/L (ref. 130–170); platelets 57×10^9/L (ref. 150–410); CRP 251 mg/L (ref. <5); procalcitonin 0.40 μg/L (ref. ≤0.24); IL-6 85.7 ng/L (ref. ≤7.0); ferritin 2399 μg/L (ref. 20–300); creatinine 191 μmol/L (ref. 44–97); troponin 305 ng/L (ref. <58); LDH 5.82 (ref. ≤4.13); PT 0.98 (ref. 0.70–1.30); APTT 32.2 s (ref. 25.9–36.6); fibrinogen 6.2 g/L (ref. 1.8–3.5); and D-dimer 909 μg/L (ref. ≤500). He required urgent intubation and vasoactive therapy with noradrenaline. A chest X-ray demonstrated pneumonic infiltrates, and he was empirically treated with piperacillin/tazobactam. Ixazomib and lenalidomide were stopped and methylprednisolone introduced according to the protocol (1 mg/kg body weight). Besides MM, he had diabetes, arterial hypertension, gout, and was overweight (BMI 40 kg/m^2). Prophylactic LMWH was introduced; a dose was adjusted according to the thrombocytopenia and renal insufficiency (dalteparin 2500 IU/24 h). Due to the worsening of his renal function, a dialysis procedure was required, but subsequently his renal function improved (the dose of dalteparin was increased to 5000 IU/12 h). Also, the platelet counts improved and were between 60 and 120×10^9/L. D-dimer levels increased significantly (up to 16,166 μg/L) and this was, together with improvement of the platelet count and renal function, a reason to increase the dose of dalteparin (5000 IE/12 h). Pulmonary embolisms and deep venous thrombosis were excluded by CT angiography and ultrasound, respectively. In the next few days, the patient deteriorated with purulent aspirates, from which *Klebsiella pneumoniae* and *Proteus mirabilis* were isolated, and cefepime was introduced. Due to isolates from the sacral decubitus, metronidazole was added and later cefepime was substituted by meropenem. Respiratory function further worsened and a chest CT showed vast ground-glass opacities in both lungs. After 10 days, his nasopharyngeal swab for SARS-CoV-2 was negative but then positive again after 10 days. Over the course of the disease, neutropenia and anemia developed, and he required a transfusion of red blood cells. After 32 days of mechanical ventilation, the patient died.

Case report 2: Discussion

Patients with MM are at increased risk for severe COVID-19 due to suppressed levels of normal immunoglobulins as well as dysfunctional cellular and innate immunity (Terpos, Engelhardt, et al., 2020). Therapy is immunocompromising as well. Dexamethasone as a key component of the treatment regimen is well known for its inhibitory effects on the immune cells and cytokines (Zen et al., 2011). Proteasome inhibitors and immunomodulatory drugs exert various effects on the immune system and put patients under an increased risk of severe infection. Besides hematological malignancy, our patient had other predictors of adverse outcome, namely increased age, male gender, obesity, arterial hypertension, and diabetes (Gajendra, 2022; Goyal et al., 2020; Zhou et al., 2020). Additionally, renal failure further increases the risk of infection (Terpos, Engelhardt, et al., 2020). Our patient also had laboratory features predicting severe COVID-19: lymphopenia, elevated D-dimer, CRP, and LDH (Rahman et al., 2021; Youssry et al., 2022). While moderate thrombocytopenia was detected at previous hospital visits and was probably caused by the therapy (lenalidomide and ixazomib) (Kumar et al., 2017), as it improved after discontinuation of the drugs, profound lymphopenia was a consequence of the virus.

The patient was in the maintenance phase of treatment, in which an all-oral therapy regimen is preferred in the era of COVID-19 for transplant-ineligible patients, providing it is not inferior to the alternative intravenous schemes. A dexamethasone dose in such a clinical setting is recommended not to exceed 20 mg/week (Buske et al., 2022). It is recommended to interrupt antimyeloma treatment in symptomatic COVID-19 until total clinical recovery, especially in patients in whom MM is under control. Prophylactic G-CSF for the prevention of profound neutropenia is encouraged (Buske et al., 2022) and was also used in our patient who experienced reactive neutropenia during the long ICU stay. The patient also experienced anemia of inflammation, which has also been reported in the literature (Mehrpouri, 2021).

Regarding the well-known thrombotic potential of SARS-CoV-2, antithrombotic prophylaxis is recommended for hospitalized patients with COVID-19 in the ICU (Schulman et al., 2022) and a higher prophylactic dose was speculated to be beneficial in obese patients or in those with a higher thrombotic risk (Susen et al., 2020). The dose of dalteparin was adjusted according to the patient's renal function and thrombocytopenia, according to the local policy.

The patient remained positive for SARS-CoV-2 for 3 weeks. Prolonged infections and prolonged viral shedding are often noticed in hematological patients (Langerbeins & Hallek, 2022; Zhou et al., 2020). Nowadays, vaccination and the early introduction of antivirals should be integrated into the management of patients with MM (Terpos, Engelhardt, et al., 2020). Unfortunately, none of these were available in autumn 2020. As it is known that patients with MM have suboptimal responses to COVID-19 vaccination, due to therapy as well as the disease itself (Ludwig et al., 2021), preexposure prophylaxis with tixagevimab-cilgavimab is recommended for these patients (Langerbeins & Hallek, 2022).

Case report 3: Hemophilia and COVID-19 infection

A 56-year-old male patient with severe hemophilia A on emicizumab prophylaxis was admitted to the traumatological department after a polytrauma and hypovolemic shock in October 2021. He had his left leg amputated under the knee and an osteosynthesis of a periprosthetic fracture of his left femur and his right calf was performed. All surgical procedures were performed under a recombinant extended half-life FVIII product (turoctocog alfa pegol) coverage, which was continued after the operation to keep the activity of FVIII above 80% for the first week (Srivastava et al., 2020). The first day he also needed a lot of transfusions of red blood cells as well as fresh frozen plasma and platelets due to dilutional coagulopathy. Because of the infection of the surgery wounds, he was also treated with broad-spectrum antibiotics. After 2 weeks of hospitalization, he got infected by SARS-CoV-2 from another patient and was referred to the infectious department. He developed only mild symptoms. No lymphopenia and thrombocytopenia were detected. While most biochemical markers were only slightly elevated (LDH 4.70 μkat/L (ref. <4.13); ferritin 358 μg/L (ref. 22–275); CRP 20 mg/L (ref. <5)), D-dimer was highly increased (10,255 μg/L; ref. <500) while PT and APTT were in the normal range. The patient did not have other risk factors for severe COVID-19. Because the wounds were not completely healed and emicizumab was temporarily discontinued, we decided to substitute FVIII for the whole hospital stay, aiming at 30% FVIII trough activity. According to our practice for the surgery setting, we also did not use LMWH prophylaxis. After discharge, the patient continued with emicizumab prophylaxis.

Case report 3: Discussion

Patients with hemophilia A and other inherited coagulation disorders are not at greater risk of contracting COVID-19 or developing a severe form of the disease. Our patient also didn't have any classical risk factors for severe COVID-19. He did not develop any hematological changes, typical of COVID-19. Studies have shown, though, that lymphocytopenia has been reported even in mild COVID-19 in 80% of cases; however, thrombocytopenia and coagulation abnormalities are characteristics of a more severe disease (Al-Saadi & Abdulnabi, 2022). However, the proper management of these patients can be complicated by a need to correct hemostasis before invasive procedures. It is important to know that in patients on emicizumab, APTT can be falsely normal (Jenkins et al., 2020). Patients on emicizumab are effectively protected from spontaneous bleedings but they still require FVIII substitution before major procedures and also before the majority of minor procedures (in case inhibitor bypassing agents are required). In such instances, chromogenic assays containing bovine FIX and FX for the accurate determination of FVIII levels in the presence of emicizumab are used. Thromboprophylaxis is not routinely used in patients with hemophilia undergoing surgery (Badulescu et al., 2020; Srivastava et al., 2020). Concerning COVID-19, thromboprophylaxis is recommended in patients with hemophilia and severe COVID-19 while for patients with a mild disease, the risk of thrombosis must be assessed for the individual patient and so the decision on prophylaxis with LMWH. Prophylaxis with LMWH requires concomitant correction of FVIII/FIX above 30% or prophylaxis with emicizumab (Pipe et al., 2021). In patients admitted to the ICU, FVIII substitution should be even more intensive, aiming to keep trough and peak concentrations of FVIII at 50%–100% (Fig. 2). Arterial blood gases, for example, should not be performed unless FVIII is at least 50%. In all other aspects, the management of COVID-19 should not differ compared to the general population (Hermans et al., 2020). It is also important that patients with inherited coagulation disorders get vaccinated against COVID-19; as the vaccine is administered intramuscularly, the smallest gauge needle available should be used, and pressure should be applied to the site for at least 10min postinjection to reduce bleeding and swelling. Patients on emicizumab can be vaccinated by intramuscular injection at any time without extra hemostatic protection. For patients with severe/moderate hemophilia or type 3 von Willebrand disease, the vaccine injection should be given after an FVIII/FIX/vonWillebrand factor-containing concentrate (routine prophylaxis or on demand treatment), respectively. For patients with basal FVIII/FIX level above 10%, no hemostatic precautions are required (Kaczmarek et al., 2021).

Conclusion

Since the beginning of the COVID-19 pandemic, SARS-CoV-2 has evolved into several variants. A number of vaccines and antivirals have become available, resulting in a decline of hospital admission and mortality. Compared with immunocompetent persons, patients with hematological malignancies still have considerable mortality (Blennow et al., 2022; Langerbeins & Hallek, 2022). The management of those patients requires the incorporation of all available agents and acts to prevent SARS-CoV-2 infection. However, once these patients are infected, they should be quickly admitted for antiviral treatments and carefully monitored for complications. Over the time of the pandemic, different recommendations for nonmalignant hematological diseases, including thromboprophylaxis, were also issued. Further data are needed to improve the hematological patients' outcome.

Policies and procedures

In this chapter, we reviewed the hematological parameters typically observed in patients with COVID-19, namely lymphopenia, thrombocytopenia, D-dimer, and coagulation tests. Additionally, we focused on the impact of COVID-19 on patients with hematological diseases and recommendations on their management, in general and in the setting of acute infection. We presented three case reports and analyzed particular issues arising in a hematological patient facing COVID-19 using the data from the literature. The case studies are presented according to the Helsinki declaration.

Clinical data and laboratory measurements of the patients presented were collected from the medical records of selected patients treated in the university medical center Ljubljana, both in the infectious diseases department and the hematology department. A literature search was performed using PubMed to identify studies on hematological findings of COVID-19 up to Nov. 30, 2022. The keywords used in the search were COVID-19, SARS-CoV-2 infection, hematology, coagulation, and blood counts. The initial selection focused on the article titles and abstracts, following which the full-text articles were read.

Applications to other areas

The topic focused on hematological, coagulation, and biochemical parameters encountered in COVID-19 patients in general, paying attention also to factors predicting a poor outcome. Second, the influence of COVID-19 on different hematological patients, particularly patients with hematological malignancies, was discussed, together with their management. Hematological therapy and its influence on immunity were explained. Modes of prevention and treatment of COVID-19 (vaccines, antiviral therapies) were presented and extrapolated to hematological patients. A special consideration was given to thromboprophylaxis, also in the context of acquired coagulation disorders. The topic thus applies to hematology, infectious diseases, and intensive care medicine.

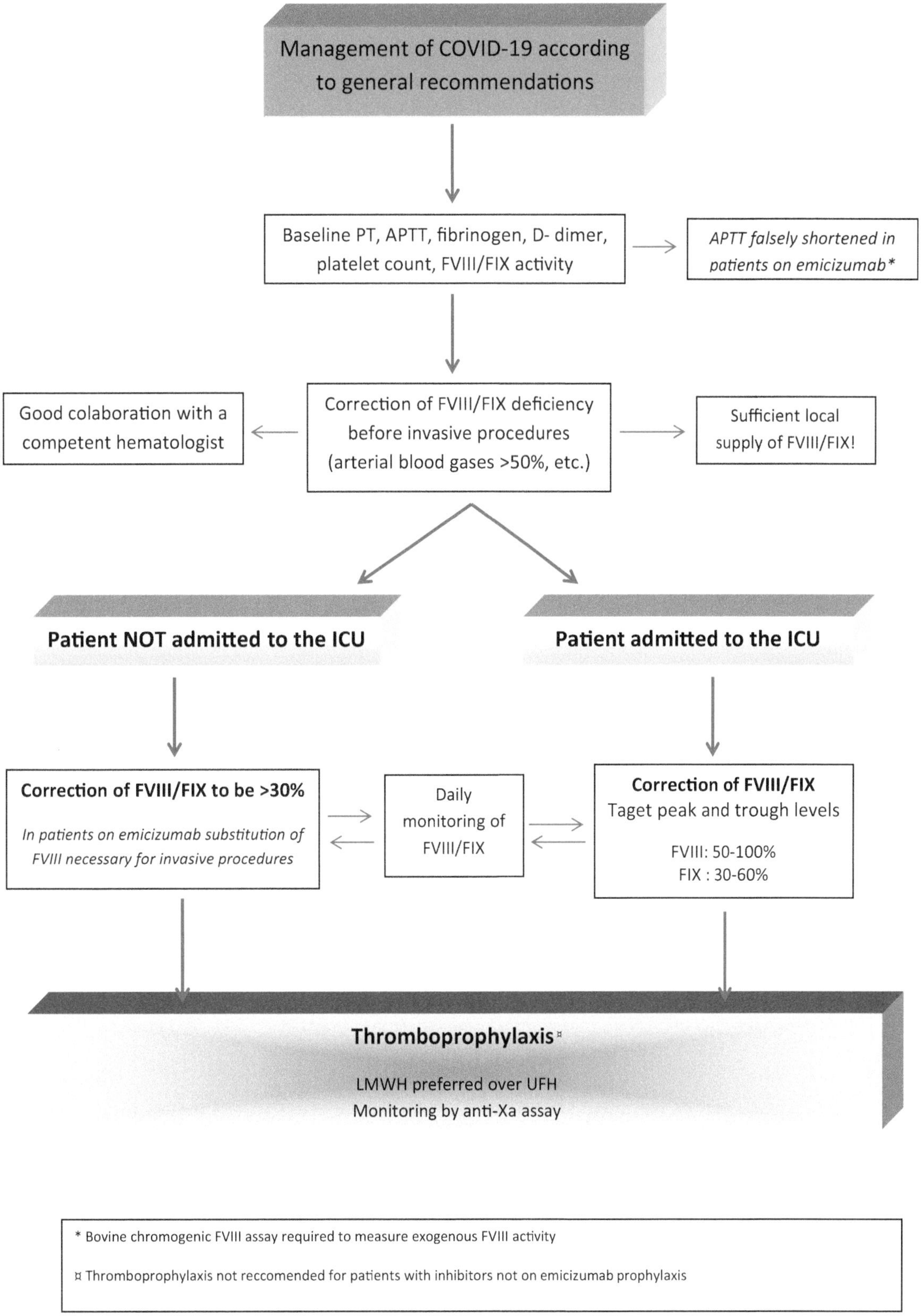

FIG. 2 **Management of hospitalized patients with hemophilia and COVID-19 (Hermans et al., 2020; Pipe et al., 2021).** *APTT*, activated partial thromboplastin time; *F*, factor; *ICU*, intensive care unit; *LMWH*, low molecular weight heparin; *PT*, prothrombin time; *UFH*, unfractioned heparin.

Mini-dictionary of terms

Coagulation tests: Prothrombin time, activated partial thromboplastin time, and thrombin time.

B-cell depleting agents: Used against malignant B lineage cells or autoimmune disease producing B cells in patients with hematological malignancies and autoimmune diseases, for example, rituximab.

Anti-CD38 monoclonal antibodies: They bind to a protein called CD38, which is found on some types of blood cells and in high levels on some cancer cells, including multiple myeloma cells; anti-CD38 monoclonal antibodies are used to treat multiple myeloma.

Bruton's tyrosine kinase inhibitors: Drugs that inhibit the enzyme BTK, which is a crucial part of the B-cell receptor signaling pathway. Examples are ibrutinib, acalabrutinib, and zanubrutinib; they are used to treat chronic lymphocytic leukemia and certain lymphomas, which use B-cell receptor signaling for growth and survival.

JAK inhibitors: Protein tyrosine kinases that bind to transmembrane type 1 and type 2 cytokine receptors and mediate cellular responses to numerous cytokines and growth factors; these mediators are important in immune defense and in immune-mediated disease. JAK inhibitors are used to treat myelofibrosis, polycythemia vera, and graft-versus-host disease as well as for some rheumatological diseases.

Venetoclax: A selective small molecule inhibitor of the antiapoptotic protein Bcl-2, with potential antineoplastic activity, used for acute myeloid leukemia and chronic lymphocytic leukemia.

Summary points

- The most common hematological abnormalities in COVID-19 are lymphopenia, elevated D-dimer, and thrombocytopenia; these are also associated with worse outcomes of the disease.
- Patients with hematological malignancies, particularly acute myeloid leukemia, are prone to severe COVID-19, resulting in a high mortality rate.
- The vaccine-induced antibody response is significantly impaired in patients with lymphoid malignancies and in patients receiving B-cell depleting agents, anti-CD38 monoclonal antibodies, Bruton's tyrosine kinase inhibitors, JAK inhibitors, and venetoclax therapy.
- Apart from vaccination, antiviral drugs and monoclonal antibodies for preexposure and postexposure prophylaxis and for the early treatment of COVID-19 are important in the management of hematological patients.
- Important parts of management of hematological patients are common preventive strategies and frequent screening, preferably with RT-PCR.
- Thromboprophylaxis should be applied also to patients with thrombocytopenia and coagulation disorders and COVID-19.

References

Al-Saadi, E. A. K. D., & Abdulnabi, M. A. (2022). Hematological changes associated with COVID-19 infection. *Journal of Clinical Laboratory Analysis, 36*, e24064.

Al-Samkari, H., Karp Leaf, R. S., Dzik, W. H., Carlson, J. C. T., Fogerty, A. E., Waheed, A., et al. (2020). COVID-19 and coagulation: Bleeding and thrombotic manifestations of SARS-CoV-2 infection. *Blood, 136*, 489–500.

Badulescu, O. V., Filip, N., Sirbu, P. D., Bararu-Bojan, I., Vladeanu, M., Bojan, A., et al. (2020). Current practices in haemophilic patients undergoing orthopedic surgery—A systematic review. *Experimental and Therapeutic Medicine, 20*, 207.

Blennow, O., Salmanton-García, J., Nowak, P., Itri, F., Van Doesum, J., López-García, A., et al. (2022). Outcome of infection with omicron SARS-CoV-2 variant in patients with hematological malignancies: An EPICOVIDEHA survey report. *American Journal of Hematology, 97*, E312–E317.

Borthakur, G., & Kantarjian, H. (2021). Core binding factor acute myelogenous leukemia-2021 treatment algorithm. *Blood Cancer Journal, 11*, 114.

Buske, C., Dreyling, M., Alvarez-Larrán, A., Apperley, J., Arcaini, L., Besson, C., et al. (2022). Managing hematological cancer patients during the COVID-19 pandemic: An ESMO-EHA interdisciplinary expert consensus. *ESMO Open, 7*, 100403.

Cheung, C. K. M., Law, M. F., Lui, G. C. Y., Wong, S. H., & Wong, R. S. M. (2019). Coronavirus disease 2019 (COVID-19): A haematologist's perspective. *Acta Haematologica, 144*, 10–23.

Chowdhury, S. F., & Anwar, S. (2020). Management of hemoglobin disorders during the COVID-19 pandemic. *Frontiers in Medicine, 7*, 306.

Döhner, H., Wei, A. H., Appelbaum, F. R., Craddock, C., DiNardo, C. D., Dombret, H., et al. (2022). Diagnosis and management of AML in adults: 2022 recommendations from an international expert panel on behalf of the ELN. *Blood, 140*, 1345–1377.

Gajendra, S. (2022). Spectrum of hematological changes in COVID-19. *American Journal of Blood Research, 12*, 43–53.

García-Suárez, J., de la Cruz, J., Cedillo, Á., Llamas, P., Duarte, R., Jiménez-Yuste, V., et al. (2020). Impact of hematologic malignancy and type of cancer therapy on COVID-19 severity and mortality: Lessons from a large population-based registry study. *Journal of Hematology and Oncology, 13*, 133.

Goyal, P., Choi, J. J., Pinheiro, L. C., Schenck, E. J., Chen, R., Jabri, A., et al. (2020). Clinical characteristics of COVID-19 in New York City. *New England Journal of Medicine*, *382*, 2372–2374.

Hermans, C., Lambert, C., Sogorb, A., Wittebole, X., Belkhir, L., & Yombi, J. C. (2020). In-hospital management of persons with haemophilia and COVID-19: Practical guidance. *Haemophilia*, *26*, 768–772.

Jenkins, P. V., Bowyer, A., Burgess, C., Gray, E., Kitchen, S., Murphy, P., et al. (2020). Laboratory coagulation tests and emicizumab treatment A United Kingdom haemophilia centre doctors' organisation guideline. *Haemophilia*, *26*, 151–155.

Kaczmarek, R., El Ekiaby, M., Hart, D. P., Hermans, C., Makris, M., Noone, D., et al. (2021). Vaccination against COVID-19: Rationale, modalities and precautions for patients with haemophilia and other inherited bleeding disorders. *Haemophilia*, *27*, 515–518.

Kreuziger, L. B., Sholzberg, M., & Cushman, M. (2022). Anticoagulation in hospitalized patients with COVID-19. *Blood*, *140*, 809–814.

Kumar, S., Moreau, P., Hari, P., Mateos, M. V., Ludwig, H., Shustik, C., et al. (2017). Management of adverse events associated with ixazomib plus lenalidomide/dexamethasone in relapsed/refractory multiple myeloma. *British Journal of Haematology*, *178*, 571–582.

Langerbeins, P., & Hallek, M. (2022). COVID-19 in patients with hematologic malignancy. *Blood*, *140*, 236–252.

Lee, L. Y. W., Cazier, J. B., Starkey, T., Briggs, S. E. W., Arnold, R., Bisht, V., et al. (2020). COVID-19 prevalence and mortality in patients with cancer and the effect of primary tumour subtype and patient demographics: a prospective cohort study. *The Lancet Oncology*, *21*, 1309–1316.

Ludwig, H., Sonneveld, P., Facon, T., San-Miguel, J., Avet-Loiseau, H., Mohty, M., et al. (2021). COVID-19 vaccination in patients with multiple myeloma: A consensus of the European Myeloma network. *The Lancet Haematology*, *8*, e934–e946.

Mehrpouri, M. (2021). Hematological abnormalities in patients with COVID-19: An emerging approach to differentiate between severe COVID-19; Compared with non-severe forms of the disease. *Acta Medica Iranica*, *59*, 126–132.

Montgomery, H., Hobbs, F. D. R., Padilla, F., Arbetter, D., Templeton, A., Seegobin, S., et al. (2022). Efficacy and safety of intramuscular administration of tixagevimab-cilgavimab for early outpatient treatment of COVID-19 (TACKLE): A phase 3, randomised, double-blind, placebo-controlled trial. *The Lancet Respiratory Medicine*, *10*, 985–996.

Pagano, L., Salmanton-García, J., Marchesi, F., Busca, A., Corradini, P., Hoenigl, M., et al. (2021). COVID-19 infection in adult patients with hematological malignancies: A European Hematology Association Survey (EPICOVIDEHA). *Journal of Hematology and Oncology*, *14*, 168.

Passamonti, F., Cattaneo, C., Arcaini, L., Bruna, R., Cavo, M., Merli, F., et al. (2020). Clinical characteristics and risk factors associated with COVID-19 severity in patients with haematological malignancies in Italy: A retrospective, multicentre, cohort study. *The Lancet Haematology*, *7*, e737–e745.

Pipe, S. W., Kaczmarek, R., Srivastava, A., Pierce, G. F., Makris, M., Hermans, C., et al. (2021). Management of COVID-19-associated coagulopathy in persons with haemophilia. *Haemophilia*, *27*, 41–48.

Rahman, A., Niloofa, R., Jayarajah, U., De Mel, S., Abeysuriya, V., & Seneviratne, S. L. (2021). Hematological abnormalities in COVID-19: A narrative review. *American Journal of Tropical Medicine and Hygiene*, *104*, 1188–1201.

Sahu, K. K., & Cerny, J. (2021). A review on how to do hematology consults during COVID-19 pandemic. *Blood Reviews*, *47*, 100777.

Schulman, S., Sholzberg, M., Spyropoulos, A. C., Zarychanski, R., Resnick, H. E., Bradbury, C. A., et al. (2022). International society on thrombosis and haemostasis. ISTH guidelines for antithrombotic treatment in COVID-19. *Journal of Thrombosis and Haemostasis*, *20*, 2214–2225.

Srivastava, A., Santagostino, E., Dougall, A., Kitchen, S., Sutherland, M., Pipe, S. W., et al. (2020). WFH guidelines for the management of hemophilia, 3rd edition. *Haemophilia*, *26*, 1–158.

Susen, S., Tacquard, C. A., Godon, A., Mansour, A., Garrigue, D., Nguyen, P., et al. (2020). Prevention of thrombotic risk in hospitalized patients with COVID-19 and hemostasis monitoring. *Critical Care*, *24*, 364.

Terpos, E., Engelhardt, M., Cook, G., Gay, F., Mateos, M. V., Ntanasis-Stathopoulos, I., et al. (2020). Management of patients with multiple myeloma in the era of COVID-19 pandemic: A consensus paper from the European Myeloma Network (EMN). *Leukemia*, *34*, 2000–2011.

Terpos, E., Ntanasis-Stathopoulos, I., Elalamy, I., Kastritis, E., Sergentanis, T. N., Politou, M., et al. (2020). Hematological findings and complications of COVID-19. *American Journal of Hematology*, *95*, 834–847.

Thachil, J., Tang, N., Gando, S., Falanga, A., Cattaneo, M., Levi, M., et al. (2020). ISTH interim guidance on recognition and management of coagulopathy in COVID-19. *Journal of Thrombosis and Haemostasis*, *18*, 1023–1026.

Wang, Q., Berger, N. A., & Xu, R. (2021). When hematologic malignancies meet COVID-19 in the United States: Infections, death and disparities. *Blood Reviews*, *47*, 100775.

Youssry, I., Abd Elaziz, D., Ayad, N., & Eyada, I. (2022). The cause-effect dilemma of hematologic changes in COVID-19: One year after the start of the pandemic. *Hematology Reports*, *14*, 95–102.

Zen, M., Canova, M., Campana, C., Bettio, S., Nalotto, L., Rampudda, M., et al. (2011). The kaleidoscope of glucorticoid effects on immune system. *Autoimmunity Reviews*, *10*, 305–310.

Zhang, L., Yan, X., Fan, Q., Liu, H., Liu, X., Liu, Z., et al. (2020). D-dimer levels on admission to predict in-hospital mortality in patients with Covid-19. *Journal of Thrombosis and Haemostasis*, *18*, 1324–1329.

Zhou, F., Yu, T., Du, R., Fan, G., Liu, Y., Liu, Z., et al. (2020). Clinical course and risk factors for mortality of adult inpatients with COVID-19 in Wuhan, China: A retrospective cohort study. *Lancet*, *395*, 1054–1062.

Section F

Effects on body systems

Chapter 37

Dermatological reactions associated with personal protective equipment use during the COVID-19 pandemic

Nicholas Herzer[a], Fletcher G. Young[b], Chrystie Nguyen[b], Aniruddha Singh[c], and Doug McElroy[a]

[a]*Department of Biology, Western Kentucky University and Western Kentucky Heart and Lung/Med Center Health Research Foundation, Bowling Green, KY, United States,* [b]*Lake Cumberland Regional Hospital, Somerset, KY, United States,* [c]*Western Kentucky Heart and Lung/Med Center Health Research Foundation and Tower Health Medical Group Cardiology-West Reading, West Reading, PA, United States*

Abbreviations

COVID-19	Coronavirus disease
N95	nonoil, 95% efficiency
PPE	personal protective equipment
SARS	severe acute respiratory syndrome
SARS-CoV-2	severe acute respiratory syndrome coronavirus 2
UVGI	ultraviolet germicidal irradiation

Introduction

Personal protective equipment (PPE) has been used to protect healthcare workers and limit the spread of infectious agents since the early 17th century (Huremović, 2019). As the Black Death ravaged Eurasia, plague doctors were highly sought after—as well as increasingly rare due to profound death among the ranks (Matuschek et al., 2020b). As they traversed some of the most highly infected areas of the continent, these physicians sought protection in the form of barriers and herbs to snuff out the "miasma," or foul/tainted air that was thought to be claiming so many lives (Huremović, 2019; Fig. 1A). Quarantining and isolation also became regular practices to reduce the spread of infection during this time (Huremović, 2019).

During the 1918 influenza pandemic, society began to adopt PPE as an everyday item; however, its use within the medical system was both rudimentary by today's standards (Fig. 1B) and viewed with apprehension (Matuschek et al., 2020b). There was a notable lack of support from surgeons and other medical professionals due to the perception that PPE impeded a surgeon's ability to effectively conduct their surgeries as well as other associated tasks (Adams et al., 2016; Matuschek et al., 2020b).

PPE continued to evolve over time as our understanding of infectious disease grew, and advancements in technology allowed us to progress beyond simple cloth masks that still allowed smaller pathogens to infect those who wore them (Mikulicz, 1897; Matuschek et al., 2020b). Research and development were dedicated to protecting those in harm's way and led to an evolution of PPE, culminating in the surgical masks, gloves, air filtration systems, and other tools that are the current state of the art (Fig. 1C) (Matuschek et al., 2020a). Ultimately, increases in the regular use of PPE coincided with the increased ability to protect healthcare workers, patients, and society at large, though data supporting this claim were insufficient until the 1990s (Matuschek et al., 2020a). In 2003, the outbreak of severe acute respiratory syndrome (SARS) was rapidly contained by proper practices of isolation, quarantine, and PPE use (Lau et al., 2005).

In 2019, a novel version of the SARS coronavirus-SARS-CoV-2–emerged with increased infectious ability, and COVID-19 swept across the globe in a matter of weeks. During the initial phase of the pandemic, PPE use scaled exponentially, and healthcare workers experienced shortages that led to the need for reuse and increased wear time; this in turn

Management, Body Systems, and Case Studies in COVID-19. https://doi.org/10.1016/B978-0-443-18703-2.00045-1
Copyright © 2024 Elsevier Inc. All rights are reserved, including those for text and data mining, AI training, and similar technologies.

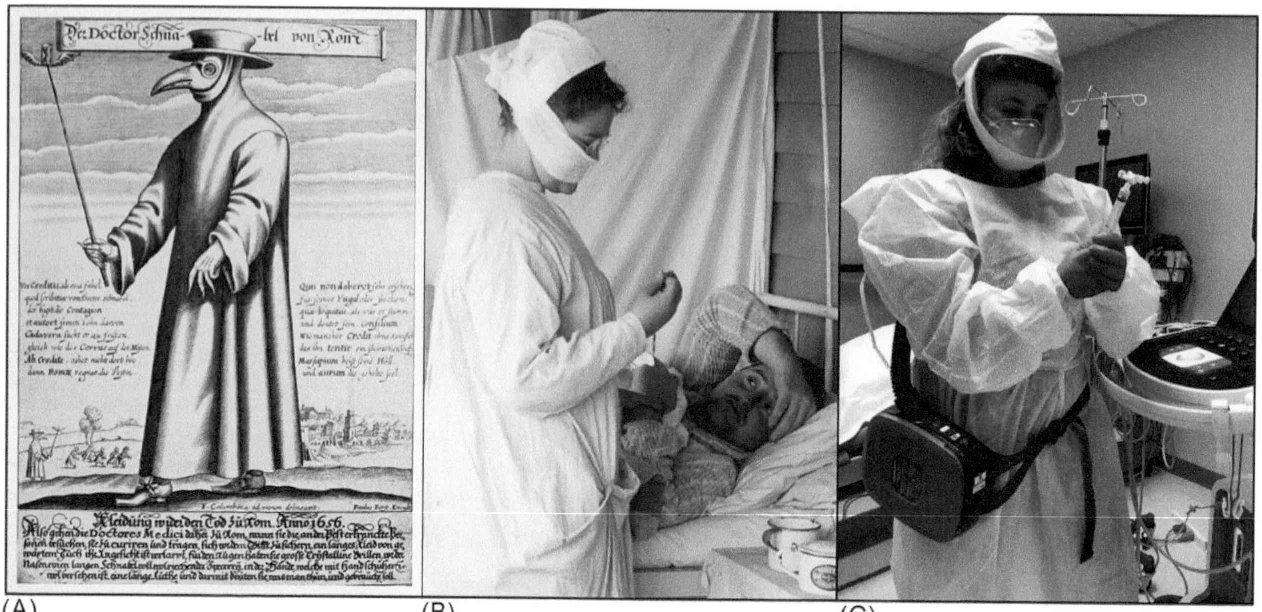

FIG. 1 The evolution of PPE from the Black Death through COVID-19. (A) Plague doctor, Rome, circa 1656; (B) Walter Reed Hospital Flu Ward, Washington, D.C., circa 1918; (C) Med Center Health, Bowling Green, Kentucky, 2022. *Image credits: (A) Eugen Holländer,* Die Karikatur und Satire in der Medizin: Medico-Kunsthistorische Studie von Professor Dr. Eugen Holländer, *2nd edn (Stuttgart:Ferdinand Enke, 1921), public domain; (B) Harris & Ewing via Library of Congress website, public domain; (C) Kendra Jones & Chandler Brown, used by permission.*

led to second- and third-order effects such as dermatological reactions, breathing complications, and pathogenic accumulation (Matuschek et al., 2020a; Nguyen et al., 2022; Pendlebury et al., 2022).

This chapter will review the dermatological effects of PPE use during the COVID-19 pandemic, with the primary focus on the healthcare system and its workers. Rather than highlighting the collective benefits of PPE use, our focus will be on elucidating the dermatological side effects of improper and extended PPE use and identifying best practices for the mitigation of these adverse reactions.

Incidence of adverse reactions

A number of surveys and other studies have been conducted among healthcare workers in Southern Asia, Australia, and the United States that assess the incidence rates of adverse dermatological reactions within and among physicians, nurses, medical assistants, and other hospital workers. In general, these studies found the incidence of such reactions to be between 40% and 90% in this group (Aloweni et al., 2022; Chowdhury et al., 2022; Darnall et al., 2022; Lin et al., 2020; Nguyen et al., 2022). This is a large range, which may reflect the fact that these studies generally allowed respondents to self-identify their adverse reaction(s).

In comparing studies that reported lower-end numbers (40%–50% incidence) to those that reported higher-end ones (80%–90% incidence), two trends were evident, reflecting the duration of PPE use and the gender of the respondent (Aloweni et al., 2022; Chowdhury et al., 2022; Darnall et al., 2022; Lin et al., 2020). On the lower end, the duration of PPE use was less than 6h in most respondents (Aloweni et al., 2022; Chowdhury et al., 2022). On the higher end, the duration of PPE use was generally reported to be more than 6h, with some studies reporting a typical duration of PPE of more than 10h (Darnall et al., 2022; Lin et al., 2020). A number of studies quantified incidence rate across a range of use durations; despite a rather high variation in overall incidence rate among studies, there was a consistent pattern of increased incidence of adverse reactions as a function of hours of PPE per day (Fig. 2).

Studies that reported a lower reaction incidence also had a low percentage of female respondents—typically less than 30%—while those with a higher overall reaction incidence had percentages of female respondents of 80%–90% (Aloweni et al., 2022; Chowdhury et al., 2022; Darnall et al., 2022; Krajewski et al., 2020; Lin et al., 2020; Santoro et al., 2021). This is a potentially important distinction, as multiple studies reported female gender as a risk factor for dermatological reaction (Lin et al., 2020; Santoro et al., 2021; Aloweni et al., 2022; Chowdhury et al., 2022; Darnall et al., 2022; see below).

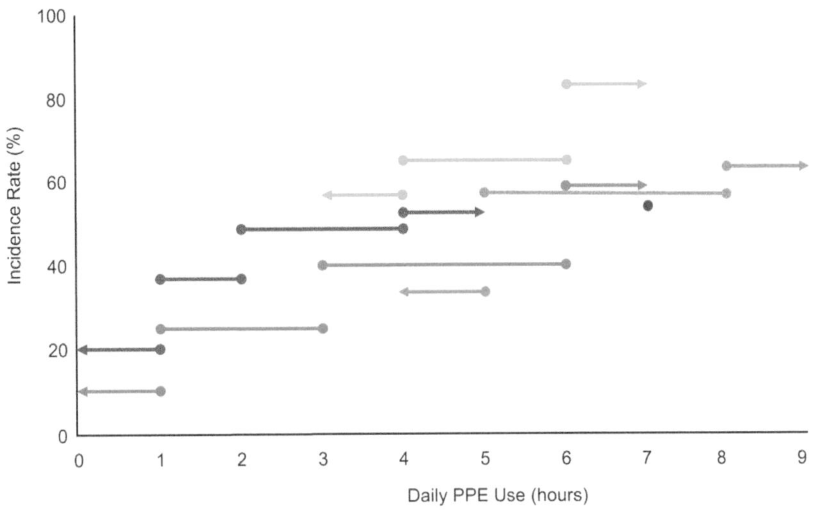

FIG. 2 Incidence rates of adverse dermatological reactions vs. extent of average daily PPE use. Data from individual studies are indicated by color as follows: *Red*, Krajewski et al. (2020); *Green*, Lin et al. (2020); *Blue*, Santoro et al. (2021); *Orange*, Chowdhury et al. (2022); *Purple*, Aloweni et al. (2022). While substantial variation exists in overall incidence rate among studies, all studies support the interpretation that incidence rate increases with the extent of daily PPE use.

Dermatological reactions to PPE use are of definite concern, particularly within the healthcare community, and it is important to explore more fully the nature and severity of such reactions as well as associated risk factors.

Reaction areas and associated incidence

It is clear from the studies conducted to date that dermatological reactions occur on the parts of the body that are most constrained when donning PPE. These areas are most directly subjected to additional pressure, a microenvironment that results from lack of breathability, and/or direct contact with the material itself (Ansari et al., 2022; Barnawi et al., 2021; Darnall et al., 2022; Foo et al., 2006; Hu et al., 2020; Jose et al., 2021; Sarfraz et al., 2022). Reactions were most commonly reported on the face; this is not unexpected given the need to wear highly restrictive masks and goggles for extended periods of time. Specifically, respondents across multiple studies reported physical damage to the nose, the zygomatic arch, and the forehead (Ansari et al., 2022; Darnall et al., 2022; Hu et al., 2020; Jose et al., 2021; Sarfraz et al., 2022). In addition, there were also manifestations of several types of dermatitis in areas directly covered by a mask such as the chin, cheeks, and below the nose (Ansari et al., 2022; Darnall et al., 2022; Justin & Yew, 2022; Sarfraz et al., 2022).

Another area commonly reported to be impacted was the hands, suggested to be due to extended periods of glove wear (Ansari et al., 2022; Sarfraz et al., 2022). The hands are largely affected by maceration, which can lead to subsequent side effects and a weakened skin constitution (Ansari et al., 2022, Sarfraz et al., 2022).

Gowns, goggles, and masks with tight-fitting elastic bands or other hard materials such as metal molds were found to constrain areas behind the ears, around the eyes, and the wrists; these areas were seen to be affected by physical damage rather than an inflammatory reaction (Ansari et al., 2022; Barnawi et al., 2021; Darnall et al., 2022; Foo et al., 2006; Hu et al., 2020; Jose et al., 2021; Sarfraz et al., 2022). It is clear from these studies that the dermatological effects of PPE tend to be localized rather than systemic reactions, specific to areas where the PPE is in contact with the body or too restrictive.

Types of reactions

Studies to date have reported a wide spectrum of reactions in those areas of the body covered by PPE. The most prevalent types of reactions included acne (50%–60% of reports), seborrheic dermatitis (50%–60% of reports), inflammation such as redness/itching (40%–50% of reports), and contact dermatitis (20%–30% of reports) (Cosansu et al., 2022; da Silva et al., 2022; Darnall et al., 2022; Foo et al., 2006; Hu et al., 2020; Justin & Yew, 2022; Sarfraz et al., 2022). Surgical masks and N95 masks were found in several studies to be associated with a high incidence of acne, itching, and contact dermatitis; this was particularly the case around the ears and nose, areas noted to be subjected to high frictional strain from elastic-type materials and metal supports within the masks (da Silva et al., 2022; Foo et al., 2006; Hu et al., 2020). While no causal evidence linked such materials to incidence rates of frictional dermatitis, skin erosion/ulcers, and scarring, it was a common

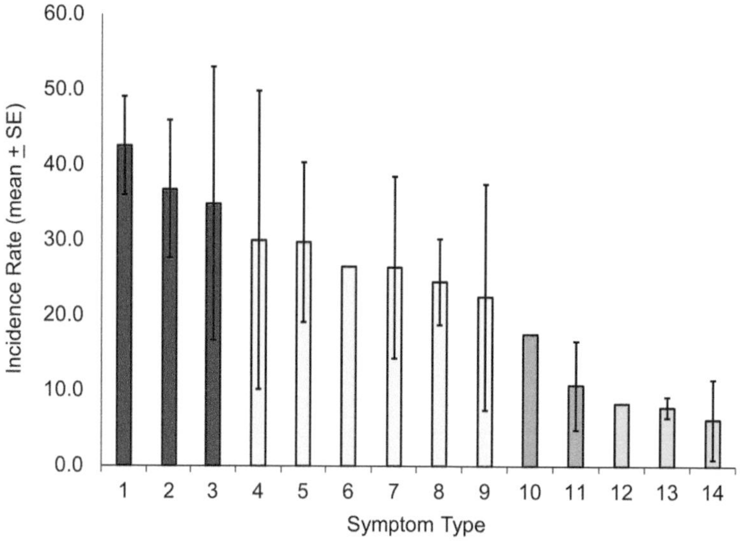

FIG. 3 Mean incidence rate of commonly reported facial reactions resulting from PPE use. Bar heights represent the mean incidence rate across all studies that reported a given symptom. The large standard errors reflect the inherent variation in incidence rate among studies; bars without errors reflect symptoms reported in a single study. Symptoms are ranked from 1 to 14 as follows (number of contributing studies in parentheses): 1, itching (5); 2, acne (6); 3, rosacea/redness/erythema (5); 4, nasal bridge damage/scarring (3); 5, dryness/scaling (4); 6, increased pore size (1); 7, pressure injuries/ulceration (3); 8, contact dermatitis/rash (3); 9, burning sensation (2); 10, moisture injuries/maceration (1); 11, frictional dermatitis/peeling skin/desquamation (3); 12, seborrheic dermatitis (1); 13, pigmentation/melasma (3); 14, wheals/indentations (2). Colors indicate the magnitude of mean incidence rate, as follows: *red*, greater than 30%; *yellow*, 20%–30%; *blue*, 10%–20%; *green*, less than 10%.

inference that they were a contributing factor (da Silva et al., 2022; Foo et al., 2006; Hu et al., 2020). Gloves were reported to be associated with a high incidence of contact dermatitis, xerosis, and itching (da Silva et al., 2022; Foo et al., 2006; Hu et al., 2020). Gowns, face shields, and goggles were found to affect areas of the wrists and around the eyes, where the most common reaction types were pressure-related injuries such as heels, friction injury, and several stages of pressure ulcers related to mask wear as well as contact dermatitis, particularly on the wrist areas (da Silva et al., 2022). Fig. 3 indicates the most commonly reported dermatological symptoms.

Effects of materials and improper wear

The materials that make PPE effective are those that also contribute to a lack of breathability, generate microclimates, and afford direct contact with abrasive conditions, all of which contribute to the incidence of dermatological reactions (Lin et al., 2022; Yu et al., 2021). Nonwoven materials found in surgical and N95 masks are highly effective at preventing the spread of infectious disease by decreasing droplet spread; however, these materials were found to induce site-specific temperature changes as well as increase localized humidity (Lin et al., 2022; Rashid et al., 2022; Yin et al., 2020). Such microenvironments are especially troublesome, as the skin naturally supports a microbiome that includes both symbiotic and pathogenic bacterial populations, and the balance of these populations often shifts—in as little as 2 h—when subjected to extended changes in moisture and temperature (Belova et al., 2022; Lin et al., 2022). In addition, these environments can lead to a buildup of oils, debris, and moisture, which contribute to dermatological reactions (Yu et al., 2021).

Beyond environmental changes, the materials of the masks themselves—including elastic banding, metal molds, and nonwoven materials—were shown in multiple studies to contribute to frictional damage (Yin et al., 2020; Yu et al., 2021). Nonwoven materials, though highly effective at protecting against disease transmission, have been shown to cause allergic reactions and frictional damage (Yin et al., 2020; Yu et al., 2021). Elastic bands are meant to ensure an airtight seal between the PPE and the body and be both unisex and size independent in nature; however, multiple studies have shown that over-tightening or other improper use can increase the risk for frictional dermatitis and erosion (Dowdle et al., 2021; Jose et al., 2021; Zhou et al., 2020). Several types of latex and some nitrile gloves are manufactured using accelerators that can cause allergic reactions, and a preexisting allergy to these accelerators could be exacerbated with extended use (Crepy et al., 2018; Dowdle et al., 2021).

The need to reuse PPE was prevalent during the COVID-19 pandemic due to shortages across the globe (Ayton et al., 2022; Cohen & van der Rodgers, 2020). Cohen and van der Rodgers (2020) specifically cited the strain COVID-19 placed on hospital systems and necessitated the regular practice of reusing PPE. Lack of sufficient rotation through fresh PPE led to improper wear, increased duration of wear, and increased PPE contamination, thus contributing to increased rates of adverse reactions (Doos et al., 2022; Jose et al., 2021; Yu et al., 2021). In summary, the materials that make PPE effective also contribute to incidence rates of dermatological reactions, and improper wear can exacerbate these effects.

Effects of gender and healthcare position

Multiple studies attributed the incidence rate of dermatological manifestations to gender and proximity to COVID-19 patients by healthcare profession (Chowdhury et al., 2022; Daye et al., 2020; Etgu & Onder, 2021; Nguyen et al., 2022; Proietti et al., 2022; Santoro et al., 2021). Females reported a significantly higher reaction incidence rate, with 50%–60% of female participants reporting some form of dermatological reaction compared to 30%–40% of males (Chowdhury et al., 2022; Daye et al., 2020; Etgu & Onder, 2021; Nguyen et al., 2022; Proietti et al., 2022; Santoro et al., 2021). A number of these studies supported a higher incidence of reactions among females relative to their male counterparts, independent of other factors (Chowdhury et al., 2022; Daye et al., 2020; Etgu & Onder, 2021; Nguyen et al., 2022; Proietti et al., 2022; Santoro et al., 2021). Notably, one study (Daye et al., 2020) found that even though their dataset included a lower number of female participants compared to males, analyses strongly supported gender as a risk factor for dermatological reactions. Additionally, this same study found strong correlations between gender and specific reactions such as dryness, itching, flaking, and cracking (Daye et al., 2020).

Healthcare workers assigned to areas of a hospital more prone to regular contact with COVID-19 patients—such as the emergency department, intensive care unit, or respiratory wards—were found to have a higher incidence of adverse dermatological reactions (Daye et al., 2020; Nguyen et al., 2022; Proietti et al., 2022; Santoro et al., 2021). This was presumably associated with a need for increased duration of PPE wear in elevated risk areas as well as stricter sanitation protocols (Santoro et al., 2021; Nguyen et al., 2022). In the study by Nguyen et al. (2022), data were collected to examine relationships among healthcare professions, hospital units, wear time, and reaction severity. Hospital workers interacting directly with COVID-19 patients reported significantly higher rates and severity of adverse reactions (70.8%) compared to those who did not (57.4%). Healthcare providers working in a dedicated COVID-19 unit or long-term care facility reported the highest incidence of adverse reactions (>70%), followed by those in the emergency department and intensive care unit (>60%); those in surgical and outpatient units reported the lowest incidence rates in this study (>50%) (Nguyen et al., 2022). Medical assistants and nurses reported the highest mean hours of PPE use per day as well as the highest incidence of adverse reactions (66%–70%) (Nguyen et al., 2022). The relationship among reaction incidence rates vs. both gender and healthcare position may be related; for example, Nguyen et al. (2022) inferred that those underlying differences in the gender ratio of workers in different positions (e.g., physician vs. nurse) could lead to the observed relationships with incidence rate in different groups.

Effects of hygiene

During the COVID-19 pandemic, common hygienic practices were emphasized to a higher degree, including handwashing, use of gel-based sanitizers, surface disinfectants, and ultraviolet (UV) sterilization and autoclaving of PPE (Graça et al., 2022; Grinshpun et al., 2020; Nguyen et al., 2022; Santoro et al., 2021). Maintaining proper hygiene can prevent the build-up of oils, dirt, bacteria, and moisture, thus reducing the risk for various types of infection; however, increased emphasis on hygiene was found to be associated with increased incidence of dermatological reactions (Graça et al., 2022; Nguyen et al., 2022; Santoro et al., 2021). These reactions often included dry skin, itching, and other irritant dermatitis and were found to be related to the overcleansing of areas such as the hands and face (Graça et al., 2022; Nguyen et al., 2022; Santoro et al., 2021). In a study by Malik and English (2015), dermatitis on the hands was found to be highly associated with frequent hand washing. According to Darlenski and Tsankov (2020), overcleansing can disrupt the integrity of the skin, thus contributing further to dermatitis. Hand washing during COVID-19 was reported to be as high as 10 times per day (Santoro et al., 2021); workers were found to cleanse every time they donned and removed PPE even if it was for reuse (Graça et al., 2022). This frequent cleansing repetitively stripped the skin of natural oils and subjected them to frictional erosion as well as harsh cleansing products necessary for disinfection (Graça et al., 2022).

Associated with these reports was a lack of moisturizer use, via application of a humectant or emollients, which can help to maintain skin integrity (Darlenski & Tsankov, 2020; Dowdle et al., 2021). When dry irritated skin is then exposed to the hot humid conditions created when donning PPE, this results in a risk factor for dermatological injury (Graça et al., 2022). By contrast, the use of UV germicidal irradiation (UVGI) was associated with fewer days of adverse reaction per month (Nguyen et al., 2022).

Time of use and other predictors of adverse reactions

The most profound and obvious effect that COVID-19 had on PPE use was in the duration and frequency of daily use. Many workers were required to wear PPE for as many as 10–12h per day (Chowdhury et al., 2022; Nguyen et al., 2022; Santoro et al., 2021), and for as many as 30 days in a row (Chowdhury et al., 2022). Chowdhury et al. (2022) found that the average weekly mask use increased from less than 32h to more than 56h per week, and incidence rates of adverse reactions increased 30%. This was also reflected in the data of Nguyen et al. (2022), where healthcare workers in positions requiring higher frequency or duration of mask wear reported increased incidence of reactions. In general, studies agreed that donning PPE for more than 6h per day was associated with increased incidence of some form of dermatitis (Chowdhury et al., 2022; Etgu & Onder, 2021; Justin & Yew, 2022; Proietti et al., 2022; Santoro et al., 2021). Nguyen et al. (2022) found a significant linear relationship between hours of PPE use per day and both reaction incidence and severity (Table 1). Furthermore, they found that the total days of PPE use per month were associated with increased incidence of adverse reactions while the number of consecutive days of wear was also associated with a higher severity of reactions (Nguyen et al., 2022; Table 1). Several additional studies (Chowdhury et al., 2022; Proietti et al., 2022; Santoro et al., 2021) identified time of use and other significant factors affecting the likelihood and/or severity of adverse dermatological reactions resulting

TABLE 1 Predictive models describing the relationship between patterns of PPE use and aspects of adverse dermatological reactions. In most cases, significant effect variables were associated with increased incidence or severity of dermatological reactions; however, several mitigating factors were identified, which are identified below as (protective).

Study	Statistical method	Dependent variable	Independent variables (significant effects in bold)
Santoro et al. (2021)	Multiple Logistic Regression	Adverse Dermatological Reactions	**Age 31–40 Years (protective)** **Age Over 50 (protective)** **Female Gender** **Occupation as Nurse/Midwife** **Day Hospital Setting (protective)**
Santoro et al. (2021)	Multiple Logistic Regression	Adverse Dermatological Reactions on Hands	Age Gender **Working With COVID-19 Patients** **History of Acne/Dermatitis (protective)** **3–6h Glove Use Per Day** Hand Washing Use of Hand Cream
Santoro et al. (2021)	Multiple Logistic Regression	Adverse Dermatological Reactions on Face	Age Gender **Working With COVID-19 Patients** **History of Acne** **3–6h Glove Use Per Day** Time of Mask Use Recycling of Masks **Use of Face Cream** **Use of Moisturizer (protective)**
Chowdhury et al. (2022)	Multiple Logistic Regression	Dermatological Problems	Age **Female Gender** **Occupation as Nurse (protective)** **N95 Mask Use** Mask Use/Cleaning Pattern **Mask Reuse** **>32h Mask Use Per Week** **History of COVID-19**
Nguyen et al. (2022)	Stepwise Linear Regression	Severity of Adverse Reactions	**Female Gender** **Hours Per Day of PPE Use** **Consecutive Days of PPE Use** Hygiene Practices

TABLE 1 Predictive models describing the relationship between patterns of PPE use and aspects of adverse dermatological reactions. In most cases, significant effect variables were associated with increased incidence or severity of dermatological reactions; however, several mitigating factors were identified, which are identified below as (protective)—cont'd

Study	Statistical method	Dependent variable	Independent variables (significant effects in bold)
Nguyen et al. (2022)	Stepwise Linear Regression	Total days of Adverse Reaction per Month	Gender **Total Days of PPE Use Per Month** **Frequency of Handwashing** **Hours Per Day of PPE Use** Hygiene Practices **Use of UVGI (protective)**
Proietti et al. (2022)	Multiple Logistic Regression	Adverse Effects of Mask Use	**Younger Age** **Female Gender** **>6 h Per Day of PPE Use** **Job** **Sector** Type of PPE
Proietti et al. (2022)	Multiple Logistic Regression	Adverse Effects Related to Bonnet Use	Age **Female Gender** **>6 h Per Day of PPE Use** Job Sector Type of PPE

from PPE use (Table 1). These findings offer predictive models for understanding how PPE wear can affect dermatological reactions in the longer term (Nguyen et al., 2022; Santoro et al., 2021), suggesting that dermatological reactions that would occur during use—if not provided the proper chance to resolve—might increase in severity with time.

Mitigation practices

Several potential mitigation practices were identified in our review. First, replacing PPE as needed has been suggested to be in the best interest of the individual and those they are around (Grinshpun et al., 2020). Reuse of PPE has been found to compromise the material composition, thus increasing the risk of exposure to viral agents (Grinshpun et al., 2020). Reuse may also contribute to increased risk of dermatological issues through bacterial contamination as well as improper wear (Grinshpun et al., 2020). Second, wearing properly fitting PPE can decrease the incidence of dermatological reactions by preserving skin integrity as well as preventing physical wear through friction (Nguyen et al., 2022).

The finding that increasing hygiene protocols between periods of extended PPE use can increase rates of reactions suggests that cleansing strategies might best be limited to the use of only soap and water when visual filth is noted (Darlenski & Tsankov, 2020; Desai et al., 2020; Dowdle et al., 2021; Nguyen et al., 2022; Zhou et al., 2020); if additional measures are needed, it may be best to use an alcohol-based cleanser and/or UVGI (Dowdle et al., 2021; Nguyen et al., 2022). Other recommendations have included supplementing cleansing protocols with an emollient or humectant that can rehydrate the skin, thus reducing risks such as contact dermatitis associated with dry skin (Desai et al., 2020; Navarro-Triviño & Ruiz-Villaverde, 2020; Nguyen et al., 2022; Zhou et al., 2020). Finally, barrier protection on areas most subject to friction or harsh materials has been shown to be effective at mitigating dermatological reactions (Coyer et al., 2020; Desai et al., 2020; Nguyen et al., 2022).

With regard to patterns of use, studies found that taking a 15-min break at least every 2 h and removing PPE in low-risk areas could reduce rates of dermatological reactions (Desai et al., 2020; Dowdle et al., 2021; Nguyen et al., 2022). This practice comes with a need to ensure adequate staffing of high-risk areas (Dowdle et al., 2021; Nguyen et al., 2022) as well as the challenge of ensuring adequate resourcing of PPE (Cohen & van der Rodgers, 2020; Dowdle et al., 2021).

TABLE 2 Recommended practices for mitigation of adverse dermatological effects resulting from PPE use.

Category	Recommendation	Source
General	Limit daily PPE use to less than 6h per day	Santoro et al. (2021); Chowdhury et al. (2022); Etgu and Onder (2021); Justin and Yew (2022); Nguyen et al. (2022); Proietti et al. (2022)
	Allow skin to rest between PPE donning for 15min every 2h	Dowdle et al. (2021)
	Use gentle cleansers with warm water	Dowdle et al. (2021)
	Apply an emollient or humectant regularly	Desai et al. (2020); Navarro-Triviño and Ruiz-Villaverde (2020); Zhou et al. (2020)
	Avoid direct contact with surface disinfectants	Dowdle et al. (2021)
	Use UVGI as a means of disinfection	Nguyen et al. (2022)
	Use barrier protection where skin may contact abrasive materials as well as those with little breathability	Navarro-Triviño and Ruiz-Villaverde (2020); Zhou et al. (2020); Dowdle et al. (2021)
	Limit consecutive and total days per month of PPE use	Nguyen et al. (2022)
Hands	Wash hands only if visible filth is noted, otherwise sanitize with a 0.5% glycerin sanitizer	Menegueti et al. (2019); Navarro-Triviño and Ruiz-Villaverde (2020); Zhou et al. (2020); Dowdle et al. (2021)
	Use emollients and humectant as part of your regular hygienic routine and allow proper absorption	Desai et al. (2020); Navarro-Triviño and Ruiz-Villaverde (2020); Zhou et al. (2020)
	Wear cotton liner under gloves for reduced maceration	Navarro-Triviño and Ruiz-Villaverde (2020); Zhou et al. (2020); Dowdle et al. (2021)
Face	Wear PPE using proper donning instructions and avoid overtightening	Dowdle et al. (2021); Navarro-Triviño and Ruiz-Villaverde (2020); Zhou et al. (2020)
	Dry skin thoroughly before donning after washing	Dowdle et al. (2021)
	Apply a barrier film for reduced friction and moisture	Dowdle et al. (2021); Navarro-Triviño and Ruiz-Villaverde (2020); Zhou et al. (2020)

Policies and procedures

In this chapter, we reviewed dermatological reactions associated with COVID-19 PPE use, and considered specific causes and methods of mitigation. Given these data, and a number of mitigation practices highlighted above, we can put forward a series of recommendations intended to preserve a balance between safety and overall health. Table 2 provides a list of best practices for individuals to prevent long-term dermatological damage. In addition to individual practices, those agencies that create PPE and/or the material for the creation of PPE should invest time and funding in identifying alternative materials that do not compromise efficacy in withstanding viral spread while providing additional comfort to users who may need to wear PPE for extended periods of time and under strenuous conditions.

Applications to other areas

It is easy to see how the dermatological effects identified within the context of COVID-19 and this chapter may be relevant to multiple scenarios where the donning of PPE is required for extended periods of time. Within the medical field, COVID-19 highlighted PPE in both a positive and a negative sense. On the one hand, its efficacy in preventing illness is well-established; however, a lack of supporting data for long-term usage led medical professionals to be ill-equipped to anticipate and thus mitigate associated negative effects such as adverse dermatological reactions.

PPE is regularly used in the medical field on an hourly/daily basis outside pandemic conditions. As such, adoption of the practices provided within this chapter may be advisable and applicable to a broad range of medical professionals and settings. Beyond the medical field, there are a variety of specialties that are required to wear PPE on a regular basis and for extended periods of time. The food industry, for example, necessitates PPE and regular hygiene to protect both workers and consumers; the above practices are highly applicable in this context as well. Similarly, chemical plant workers, miners, construction workers, laboratory technicians, and those who work in the cosmetics industry are also likely to benefit from practices that mitigate the side effects associated with long-term PPE wear. Finally, should countries continue to adopt long-term PPE use protocols, these measures may offer a reprieve and/or confidence to the general public.

Mini-dictionary of terms

- Dermatitis: Inflammation of the skin.
- Contact dermatitis: Dermatitis caused by direct contact with an irritant or allergen.
- Seborrheic dermatitis: Dermatitis typically affecting the face or areas with many sebaceous glands.
- Frictional dermatitis: Dermatitis caused by excessive friction from materials.
- Xerosis: Excessive dryness of a body part.
- Maceration: When the skin becomes weak due to excessive moisture.

Summary points

- PPE has been used for decades to protect healthcare workers and patients in the medical industry.
- COVID-19 dramatically increased daily PPE use and required long-term duration of use.
- Excessive use of PPE leads to dermatological reactions based on several risk factors, including duration of use, types of materials, improper use, and associated hygiene practices.
- Types of reactions include acne, xerosis, maceration, and dermatitis.
- Mitigation practices that can reduce the incidence and severity of dermatological reactions include wearing PPE appropriately, allowing the skin to rest between periods of PPE use, using gentle emollients and limiting hand washing, and limiting total and/or consecutive hours/days of PPE use.

References

Adams, L. W., Aschenbrenner, C. A., Houle, T. T., & Roy, R. C. (2016). Uncovering the history of operating room attire through photographs. *Anesthesiology*, *124*(1), 19–24. https://doi.org/10.1097/ALN.0000000000000932.

Aloweni, F., Bouchoucha, S. L., Hutchinson, A., Ang, S. Y., Toh, H. X., Bte Suhari, N. A., Bte Sunari, R. N., & Lim, S. H. (2022). Health care workers' experience of personal protective equipment use and associated adverse effects during the COVID-19 pandemic response in Singapore. *Journal of Advanced Nursing*, *78*(8), 2383–2396. https://doi.org/10.1111/jan.15164.

Ansari, R. T., Farooq, N., Nasreen, S., & Faisal, D. (2022). Adverse effects of personal protective equipment used on healthcare workers' skin during COVID-19 outbreak. *Journal of the Dow University of Health Sciences*, *16*(1). https://doi.org/10.36570/jduhs.2022.1.1215.

Ayton, D., Soh, S.-E., Berkovic, D., Parker, C., Yu, K., Honeyman, D., Manocha, R., MacIntyre, R., & Ananda-Rajah, M. (2022). Experiences of personal protective equipment by Australian healthcare workers during the COVID-19 pandemic, 2020: A cross-sectional study. *PLoS One*, *17*(6), e0269484. https://doi.org/10.1371/journal.pone.0269484.

Barnawi, G. M., Barnawi, A. M., & Samarkandy, S. (2021). The association of the prolonged use of personal protective equipment and face mask during COVID-19 pandemic with various dermatologic disease manifestations: A systematic review. *Cureus*, *13*(7), e16544. https://doi.org/10.7759/cureus.16544.

Belova, E., Shashina, E., Zhernov, Y., Zabroda, N., Sukhov, V., Gruzdeva, O., Khodykina, T., Laponova, E., Makarova, V., Simanovsky, A., Zhukova, A., Isiutina-Fedotkova, T., Shcherbakov, D., & Mitrokhin, O. (2022). Assessment of hygiene indicators when using gloves by transport workers in Russia during the COVID-19 pandemic. *International Journal of Environmental Research and Public Health*, *19*(3), 1198. https://doi.org/10.3390/ijerph19031198.

Chowdhury, S., Roy, S., Iktidar, M. A., Rahman, S., Liza, M. M., Islam, A. M. K., Akhter, S., Medha, M. B., Tasnim, A., Gupta, A. D., Deb, A., Chowdhury, S., & Hawlader, M. D. H. (2022). Prevalence of dermatological, oral and neurological problems due to face mask use during COVID-19 and its associated factors among the health care workers of Bangladesh. *PLoS One*, *17*(4), e0266790. https://doi.org/10.1371/journal.pone.0266790.

Cohen, J., & van der Rodgers, Y. (2020). Contributing factors to personal protective equipment shortages during the COVID-19 pandemic. *Preventive Medicine*, *141*(106263), 106263. https://doi.org/10.1016/j.ypmed.2020.106263.

Cosansu, N. C., Yuksekal, G., Kutlu, O., Umaroglu, M., Yaldız, M., & Dikicier, B. S. (2022). The change in the frequency and severity of facial dermatoses and complaints in healthcare workers during the COVID-19. *Journal of Cosmetic Dermatology*, *21*(8), 3200–3205. https://doi.org/10.1111/jocd.15044.

Coyer, F., Coleman, K., Hocking, K., Leong, T., Levido, A., & Barakat-Johnson, M. (2020). Maintaining skin health and integrity for staff wearing personal protective equipment for prolonged periods: A practical tip sheet. *Wound Practice & Research*, *28*(2), 75–83. https://doi.org/10.33235/wpr.28.2.75-83.

Crepy, M.-N., Lecuen, J., Ratour-Bigot, C., Stocks, J., & Bensefa-Colas, L. (2018). Accelerator-free gloves as alternatives in cases of glove allergy in healthcare workers. *Contact Dermatitis*, *78*(1), 28–32. https://doi.org/10.1111/cod.12860.

da Silva, L. F. M., de Almeida, A. G., Pascoal, L. M., Santos Neto, M., Lima, F. E. T., & Santos, F. S. (2022). Skin injuries due to personal protective equipment and preventive measures in the COVID-19 context: An integrative review. *Revista Latino-Americana de Enfermagem*, *30*. https://doi.org/10.1590/1518-8345.5636.3522.

Darlenski, R., & Tsankov, N. (2020). COVID-19 pandemic and the skin: What should dermatologists know? *Clinics in Dermatology*, *38*(6), 785–787. https://doi.org/10.1016/j.clindermatol.2020.03.012.

Darnall, A. R., Sall, D., & Bay, C. (2022). Types and prevalence of adverse skin reactions associated with prolonged N95 and simple mask usage during the COVID-19 pandemic. *Journal of the European Academy of Dermatology and Venereology: JEADV*, *36*(10), 1805–1810. https://doi.org/10.1111/jdv.18365.

Daye, M., Cihan, F. G., & Durduran, Y. (2020). Evaluation of skin problems and dermatology life quality index in health care workers who use personal protection measures during COVID-19 pandemic. *Dermatologic Therapy*, *33*(6), e14346. https://doi.org/10.1111/dth.14346.

Desai, S. R., Kovarik, C., Brod, B., James, W., Fitzgerald, M. E., Preston, A., & Hruza, G. J. (2020). COVID-19 and personal protective equipment: Treatment and prevention of skin conditions related to the occupational use of personal protective equipment. *Journal of the American Academy of Dermatology*, *83*(2), 675–677. https://doi.org/10.1016/j.jaad.2020.05.032.

Doos, D., Barach, P., Sarmiento, E., & Ahmed, R. (2022). Reuse of personal protective equipment: Results of a human factors study using fluorescence to identify self-contamination during donning and doffing. *The Journal of Emergency Medicine*, *62*(3), 337–341. https://doi.org/10.1016/j.jemermed.2021.12.010.

Dowdle, T. S., Thompson, M., Alkul, M., Nguyen, J. M., & Sturgeon, A. L. E. (2021). COVID-19 and dermatological personal protective equipment considerations. *Proceedings (Baylor University Medical Center)*, *34*(4), 469–472. https://doi.org/10.1080/08998280.2021.1899730.

Etgu, F., & Onder, S. (2021). Skin problems related to personal protective equipment among healthcare workers during the COVID-19 pandemic (online research). *Cutaneous and Ocular Toxicology*, *40*(3), 207–213. https://doi.org/10.1080/15569527.2021.1902340.

Foo, C. C. I., Goon, A. T. J., Leow, Y.-H., & Goh, C.-L. (2006). Adverse skin reactions to personal protective equipment against severe acute respiratory syndrome—A descriptive study in Singapore. *Contact Dermatitis*, *55*(5), 291–294. https://doi.org/10.1111/j.1600-0536.2006.00953.x.

Graça, A., Martins, A. M., Ribeiro, H. M., & Marques Marto, J. (2022). Indirect consequences of coronavirus disease 2019: Skin lesions caused by the frequent hand sanitation and use of personal protective equipment and strategies for their prevention. *The Journal of Dermatology*, *49*(9), 805–817. https://doi.org/10.1111/1346-8138.16431.

Grinshpun, S. A., Yermakov, M., & Khodoun, M. (2020). Autoclave sterilization and ethanol treatment of re-used surgical masks and N95 respirators during COVID-19: Impact on their performance and integrity. *The Journal of Hospital Infection*, *105*(4), 608–614. https://doi.org/10.1016/j.jhin.2020.06.030.

Hu, K., Fan, J., Li, X., Gou, X., Li, X., & Zhou, X. (2020). The adverse skin reactions of health care workers using personal protective equipment for COVID-19. *Medicine*, *99*(24), e20603. https://doi.org/10.1097/MD.0000000000020603.

Huremović, D. (2019). Brief history of pandemics (pandemics throughout history). In *Psychiatry of pandemics* (pp. 7–35). Springer International Publishing.

Jose, S., Cyriac, M. C., & Dhandapani, M. (2021). Health problems and skin damages caused by personal protective equipment: Experience of frontline nurses caring for critical COVID-19 patients in intensive care units. *Indian Journal of Critical Care Medicine: Peer-Reviewed, Official Publication of Indian Society of Critical Care Medicine*, *25*(2), 134–139. https://doi.org/10.5005/jp-journals-10071-23713.

Justin, L. Y. S., & Yew, Y. W. (2022). Facial dermatoses induced by face masks: A systematic review and meta-analysis of observational studies. *Contact Dermatitis*, *87*(6), 473–484. https://doi.org/10.1111/cod.14203.

Krajewski, P. K., Matusiak, Ł., Szepietowska, M., Białynicki-Birula, R., & Szepietowski, J. C. (2020). Increased prevalence of face mask-induced itch in health care workers. *Biology*, *9*(12), 451. https://doi.org/10.3390/biology9120451.

Lau, J. T. F., Yang, X., Pang, E., Tsui, H. Y., Wong, E., & Wing, Y. K. (2005). SARS-related perceptions in Hong Kong. *Emerging Infectious Diseases*, *11*(3), 417–424. https://doi.org/10.3201/eid1103.040675.

Lin, X., Li, Y. Z., Chen, T., Min, S. H., Wang, D. F., Ding, M. M., & Jiang, G. (2022). Effects of wearing personal protective equipment during COVID-19 pandemic on composition and diversity of skin bacteria and fungi of medical workers. *Journal of the European Academy of Dermatology and Venereology: JEADV*, *36*(9), 1612–1622. https://doi.org/10.1111/jdv.18216.

Lin, P., Zhu, S., Huang, Y., Li, L., Tao, J., Lei, T., Song, J., Liu, D., Chen, L., Shi, Y., Jiang, S., Liu, Q., Xie, J., Chen, H., Duan, Y., Xia, Y., Zhou, Y., Mei, Y., Zhou, X., & Li, H. (2020). Adverse skin reactions among healthcare workers during the coronavirus disease 2019 outbreak: A survey in Wuhan and its surrounding regions. *The British Journal of Dermatology*, *183*(1), 190–192. https://doi.org/10.1111/bjd.19089.

Malik, M., & English, J. (2015). Irritant hand dermatitis in health care workers. *Occupational Medicine (Oxford, England)*, *65*(6), 474–476. https://doi.org/10.1093/occmed/kqv067.

Matuschek, C., Moll, F., Fangerau, H., Fischer, J. C., Zänker, K., van Griensven, M., Schneider, M., Kindgen-Milles, D., Knoefel, W. T., Lichtenberg, A., Tamaskovics, B., Djiepmo-Njanang, F. J., Budach, W., Corradini, S., Häussinger, D., Feldt, T., Jensen, B., Pelka, R., Orth, K., & Haussmann, J. (2020a). Face masks: Benefits and risks during the COVID-19 crisis. *European Journal of Medical Research*, *25*(1), 32. https://doi.org/10.1186/s40001-020-00430-5.

Matuschek, C., Moll, F., Fangerau, H., Fischer, J. C., Zänker, K., van Griensven, M., Schneider, M., Kindgen-Milles, D., Knoefel, W. T., Lichtenberg, A., Tamaskovics, B., Djiepmo-Njanang, F. J., Budach, W., Corradini, S., Häussinger, D., Feldt, T., Jensen, B., Pelka, R., Orth, K., & Haussmann, J. (2020b). The history and value of face masks. *European Journal of Medical Research*, *25*(1), 23. https://doi.org/10.1186/s40001-020-00423-4.

Menegueti, M. G., Laus, A. M., Ciol, M. A., Auxiliadora-Martins, M., Basile-Filho, A., Gir, E., Pires, D., Pittet, D., & Bellissimo-Rodrigues, F. (2019). Glycerol content within the WHO ethanol-based handrub formulation: Balancing tolerability with antimicrobial efficacy. *Antimicrobial Resistance and Infection Control*, *8*(1), 109. https://doi.org/10.1186/s13756-019-0553-z.

Mikulicz, J. (1897). Das operieren in sterilisierten Zwirnhandschuhen und mit Mundbinde. *Centralblatt für Chirurgie*, *24*, 713–717.

Navarro-Triviño, F. J., & Ruiz-Villaverde, R. (2020). Therapeutic approach to skin reactions caused by personal protective equipment (PPE) during COVID-19 pandemic: An experience from a tertiary hospital in Granada, Spain. *Dermatologic Therapy*, *33*(6), e13838. https://doi.org/10.1111/dth.13838.

Nguyen, C., Young, F. G., McElroy, D., & Singh, A. (2022). Personal protective equipment and adverse dermatological reactions among healthcare workers: Survey observations from the COVID-19 pandemic: Survey observations from the COVID-19 pandemic. *Medicine*, *101*(9), e29003. https://doi.org/10.1097/MD.0000000000029003.

Pendlebury, G. A., Oro, P., Haynes, W., Merideth, D., Bartling, S., & Bongiorno, M. A. (2022). The impact of COVID-19 pandemic on dermatological conditions: A novel, comprehensive review. *Dermatopathology (Basel, Switzerland)*, *9*(3), 212–243. https://doi.org/10.3390/dermatopathology9030027.

Proietti, I., Borrelli, I., Skroza, N., Santoro, P. E., Gualano, M. R., Bernardini, N., Mambrin, A., Tolino, E., Marchesiello, A., Marraffa, F., Michelini, S., Rossi, G., Volpe, S., Ricciardi, W., Moscato, U., & Potenza, C. (2022). Adverse skin reactions to personal protective equipment during COVID-19 pandemic in Italian health care workers. *Dermatologic Therapy*, *35*(6), e15460. https://doi.org/10.1111/dth.15460.

Rashid, T. U., Sharmeen, S., & Biswas, S. (2022). Effectiveness of N95 masks against SARS-CoV-2: Performance efficiency, concerns, and future directions. *Journal of Chemical Health and Safety*, *29*(2), 135–164. https://doi.org/10.1021/acs.chas.1c00016.

Santoro, P. E., Borrelli, I., Gualano, M. R., Proietti, I., Skroza, N., Rossi, M. F., Amantea, C., Daniele, A., Ricciardi, W., Potenza, C., & Moscato, U. (2021). The dermatological effects and occupational impacts of personal protective equipment on a large sample of healthcare workers during the COVID-19 pandemic. *Frontiers in Public Health*, *9*, 815415. https://doi.org/10.3389/fpubh.2021.815415.

Sarfraz, Z., Sarfraz, A., Sarfraz, M., Felix, M., Bernstein, J. A., Fonacier, L., & Cherrez-Ojeda, I. (2022). Contact dermatitis due to personal protective equipment use and hygiene practices during the COVID-19 pandemic: A systematic review of case reports. *Annals of Medicine and Surgery*, *74*, 103254. https://doi.org/10.1016/j.amsu.2022.103254.

Yin, L., Shim, E., & DenHartog, E. (2020). A study of skin physiology, sensation and friction of nonwoven fabrics used in absorbent hygiene products in neutral and warm environments. *Biotribology*, *24*, 100149. https://doi.org/10.1016/j.biotri.2020.100149.

Yu, J., Chen, J. K., Mowad, C. M., Reeder, M., Hylwa, S., Chisolm, S., Dunnick, C. A., Goldminz, A. M., Jacob, S. E., Wu, P. A., Zippin, J., & Atwater, A. R. (2021). Occupational dermatitis to facial personal protective equipment in health care workers: A systematic review. *Journal of the American Academy of Dermatology*, *84*(2), 486–494. https://doi.org/10.1016/j.jaad.2020.09.074.

Zhou, N.-Y., Yang, L., Dong, L.-Y., Li, Y., An, X.-J., Yang, J., Yang, L., Huang, C.-Z., & Tao, J. (2020). Prevention and treatment of skin damage caused by personal protective equipment: Experience of the first-line clinicians treating SARS-CoV-2 infection: Experience of the first-line clinicians treating 2019-nCoV infection. *International Journal of Dermatology and Venereology*, *3*(2), 70–75. https://doi.org/10.1097/jd9.0000000000000085.

Chapter 38

Hemodialysis patients, effects of infections by SARS-CoV-2 and vaccine response

Diana Rodríguez-Espinosa, Elena Cuadrado-Payán, and José Jesús Broseta
Department of Nephrology and Renal Transplantation, Hospital Clínic of Barcelona, Barcelona, Spain

Abbreviations

CKD	chronic kidney disease
COVID-19	Coronavirus disease 2019
CYP	Cytochrome P450
FDA	Food and Drug Administration
HD	hemodialysis
mAbs	monoclonal antibodies
RdRp	RNA-dependent RNA polymerase
rRT-PCR	real-time reverse-transcription polymerase chain reaction
S	Spike
SARS-CoV-2	severe acute respiratory syndrome coronavirus 2

Clinical presentation and outcomes

Infections are the second most frequent cause of death among dialysis patients (Betjes, 2013). Due to the inherent respiratory transmission risk of SARS-CoV-2, and their dysfunctional immune system, chronic kidney disease (CKD) patients and those on in-center hemodialysis (HD) were of particular interest during the coronavirus disease-2019 (COVID-19) pandemic (Alberici et al., 2020; Ke et al., 2020; Wölfel et al., 2020). HD patients are exposed more to the virus and thus have a reported higher prevalence of infection (Broseta et al., 2020). Moreover, once this population is infected, their clinical outcomes are dismal as they require hospital admissions more frequently; once admitted, they have longer hospital stays and have a greater mortality risk than infected individuals from the general population (Goicoechea et al., 2020; Trujillo et al., 2020; Xiong et al., 2020). During the first COVID-19 wave, the most common clinical manifestation among symptomatic patients was fever in almost 50%–70% of them. Other common symptoms were shortness of breath and cough (around 50%). Gastrointestinal symptoms—mainly diarrhea or vomiting—occurred in 22% of patients. Notably, the mortality rate was strikingly high, ranging from 30% to 41% according to different series, much higher than in the general population, which approached 10% (Grasselli et al., 2020; Garcia-Vidal et al., 2021; Ronda et al., 2020; Sánchez-Álvarez et al., 2020). In fact, data obtained during the disease's first wave in 2020 in a single center show that 85% of infected HD patients were hospitalized and that 38% of them died (Broseta et al., 2020); these figures differ significantly from the overall 9.7% mortality seen in all COVID-19 admitted patients from the same hospital (Garcia-Vidal et al., 2021). Increased age, history of hypertension, cardiovascular disease, and CKD are some of the factors associated with worse clinical outcomes in COVID-19-infected patients. However, HD patients are at a higher risk of poorer outcomes because they mostly already share these comorbid conditions. Indeed, those who did not survive were older, had more comorbidities, and had a longer time on dialysis. It is worth mentioning that risk scales routinely calculated for community-acquired pneumonia (i.e., CURB-65 and PSI) failed to discern whether admitted patients would develop a poor outcome (Broseta et al., 2020).

Prevention

HD patients are of particular epidemiological interest given their increased risk of contagion, which led to numerous COVID-19 outbreaks in HD units across the globe (Clarke et al., 2020; Ke et al., 2020). According to the World Health

Management, Body Systems, and Case Studies in COVID-19. https://doi.org/10.1016/B978-0-443-18703-2.00051-7
Copyright © 2024 Elsevier Inc. All rights are reserved, including those for text and data mining, AI training, and similar technologies.

Organization, SARS-CoV-2 is transmitted through direct, indirect, or close contact with saliva, respiratory secretions, or respiratory droplets of infected patients (Chan et al., 2020; Liu et al., 2020). This can lead to outbreaks in indoor spaces (Leclerc et al., 2020), which is why social distancing, using face masks, and avoiding crowded indoor places were encouraged (European Centre for Disease Prevention and Control, 2020).

Unfortunately, in-center HD therapy's inherent characteristics make it impossible for patients to avoid high-risk situations such as sharing transportation, dressing rooms, and other center facilities with more patients, at least for triweekly 4 h HD sessions. Therefore, rigorous preventive measures became essential in this population. While the overall prevalence in the general population ranged from 7% to 14% at the pandemic's peak (Ministerio de Ciencia e Innovación de España & Ministerio de Sanidad de España, 2020), the prevalence among in-center HD patients was significantly higher, more than doubling the general population's prevalence in some cases. For instance, even though there was a much higher overall prevalence in the city of London than in Barcelona (14.5% and 7%, respectively), the prevalence in the studied population of in-center HD patients was significantly higher (36.2%) than in the former (Rodríguez-Espinosa et al., 2020). In this particular case, the infection precaution measures adopted by the London cohort were based on the NICE guidelines and did not make the use of face masks mandatory during the totality of the HD sessions (COVID-19 Rapid Guideline: Dialysis Service Delivery (NG160), 2020).

Therefore, educating patients on how and when to wash their hands is recommended. Body temperature should be measured upon arrival and before entering the main dialysis room. Medical masks are mandatory for every person in the center during outbreaks; this includes patients as well as for medical, nursing, and cleaning staff. Alcoholic disinfectants should be placed next to every door and next to every dialysis machine. Every trash can in the center should have a foot pedal. And, if any patient had a body temperature higher than 37.3° Celsius (99° Fahrenheit) during an outbreak, that patient is to be dialyzed in isolation while awaiting either a COVID-19 antigen test or a real-time reverse-transcription polymerase chain reaction (rRT-PCR) result.

Vaccination

Considering the combination of high-risk factors in CKD patients, this population was promptly and effectively immunized against SARS-CoV-2 as recommended by different nephrology societies. Since the emergence of SARS-CoV-2 vaccines, authors recommended the use of mRNA vaccines in HD patients, given that these were the most potent vaccines on the market (Windpessl et al., 2021). The two mRNA vaccines tested in CKD are the mRNA-1273 (Moderna) and the BNT162b2 (Pfizer-BioNTech) (Broseta, Rodríguez-Espinosa, Rodríguez, et al., 2021; Grupper et al., 2021). Both consist of a lipid-encapsulated nanoparticle that encodes the prefusion-stabilized full-length SARS-CoV-2 spike that is recognized and induces T-helper, cytotoxic T-cell, and antibody immune responses (Teijaro & Farber, 2021).

Immunological response

Many groups generated data on the effectiveness of mRNA vaccines in HD patients (Agur et al., 2021; Attias et al., 2021; Frantzen et al., 2021; Grupper et al., 2021). Seroconversion occurs in 82%–98% of patients that develop either humoral or cellular responses after two doses of either mRNA-1273 or BNT162b2, resembling results reported in the vaccine clinical trials. The strongest predictors of humoral vaccine response and antibody levels are similar to those reported across other publications, such as age, immunosuppressive treatment, and previous SARS-CoV-2 infection. However, when comparing age-matched dialysis-dependent CKD patients with the general population, what stands out is that those on dialysis have a slower pace of seroconversion with progressive increments as the weeks pass, finally reaching 94.9% by the seventh week when, in otherwise healthy subjects, this percentage is achieved by the second week (Broseta, Rodríguez-Espinosa, Soruco, & Maduell, 2021). This slowdown may be related to the widely described impaired immune response of dialysis-dependent CKD patients reported in other immunization programs (Eleftheriadis et al., 2007, 2014). As expected, in a cohort of HD patients followed postvaccination, seroconversion went from 95.4% after two doses to 81.25% 6 months afterward, accompanied by a drop in antibody levels already observed as early as 3 months after vaccination (Broseta et al., 2014; Doria-Rose et al., 2021; Favresse et al., 2021). In view of this drop, HD patients clearly benefit from booster doses to maintain antibody levels in those who still had them and to restimulate an immune response in those who had lost them, at least while COVID-19 continues to be a threat.

Clinical response to vaccination

The vaccine's success relied on the fact that SARS-CoV-2 breakthrough infections became less frequent and had much milder presentations (i.e., fewer hospital admissions and mortality) among those vaccinated over those who are not. Among the affected population during COVID-19 waves, HD patients had a higher susceptibility to infection and poorer outcomes. However, vaccination, together with proper confinement and personal sanitary measures, decreased the incidence of SARS-CoV-2 infection in comparison to the incidence reported in the same population during COVID-19's first wave in 2020 (Broseta et al., 2020; Goicoechea et al., 2020). Even though a lower infection rate remains an important outcome, the real success of vaccination is that it led to a reduction in significant clinical outcomes, such as fewer and shorter hospital admissions, as well as less severe cases with fewer ICU admissions and lower mortality associated with COVID-19. Two-dose vaccinated HD patients with COVID-19 required significantly fewer hospital admissions (22.2%), fewer cases of severe disease (5.5%), and most importantly, mortality dropped to zero in some cohorts (Rodríguez-Espinosa et al., 2022).

Pharmacological management

After the COVID-19 pandemic that began in 2020, treatment guidelines evolved rapidly, and various treatment regimens were approved (Brogan & Ross, 2022). The problem physicians face today is that data on the dialysis population are scarce, given their exclusion from every clinical trial, with insufficient and contradictory information regarding the dosage, frequency, and safety of these drugs in CKD patients (Dr et al., 2020; El Karoui & De Vriese, 2022; Kale et al., 2023).

We will discuss the different pharmacological groups (Kale et al., 2023; Marra et al., 2021) (Table 1) used with evidence from the general population and the available data on patients on dialysis.

Anticoagulants

Patients affected by COVID-19 experience a higher frequency of thrombotic events. Therefore, dialysis patients are at risk of presenting arteriovenous fistula thrombosis, mechanical catheter dysfunction, or circuit clotting. For this reason, as in patients without CKD, prophylactic anticoagulation therapy is recommended (El Karoui & De Vriese, 2022).

Corticosteroids

Systemic corticosteroids improve clinical outcomes and reduce mortality in hospitalized patients with COVID-19 with supplemental oxygen needs. In contrast, their use has not proven beneficial in hospitalized patients with no oxygen

TABLE 1 COVID-19 treatments available.

Corticosteroids		Dexamethasone Hydrocortisone Methylprednisolone
Antiviral	RNA-dependent RNA polymerase (RdRp) inhibitor:	Remdesivir Molnupiravir Famciclovir Favipiravir
	Protease inhibitors	Atazanavir Darunavir Nirmatrelvir/ritonavir combination Lopinavir/ritonavir combination
	Nucleoside reverse transcriptase inhibitors	Azvudine
	HIV integrase inhibitors/ nonnucleoside reverse transcriptase inhibitors	Dolutegravir/rilpivirine
Monoclonal antibodies		Sotrovimab Bamlanivimab/etesevimab Casirivimab/imdevimab
Immunomodulators	IL-6 receptor antagonist	Tocilizumab, Sarilumab
	Janus kinase inhibitor	Baricitinib
	IL-1 receptor antagonist	Anakinra

requirement and may cause harm (The RECOVERY Collaborative Group, 2021; Tomazini et al., 2020). There is no need to adjust the corticosteroid dose in dialysis patients.

Remdesivir

Remdesivir was the first antiviral drug approved by the US Food and Drug Administration (FDA) for emergency use in moderate-to-severe COVID-19 in May 2020 (Cuadrado-Payán et al., 2022). This drug acts by inhibiting the viral RNA-dependent RNA polymerase (RdRp) interfering in RNA replication and should be administered within 7 days of the onset of infection (Allen et al., 2022;Brogan & Ross, 2022; Kale et al., 2023).

This antiviral is not recommended for patients with an eGFR less than 30 mL/min/1.73 m^2 due to the presumed toxicity of the drug itself and the accumulation of its solubilizing excipient sulfobutyl ether β-cyclodextrin (Cuadrado-Payán et al., 2022; Kale et al., 2023). Despite the lack of data on remdesivir safety in patients with severe CKD, different case series and prospective studies with small sample sizes have documented a good experience in this population, reporting liver function stability with no abnormalities before or after initiating treatment nor drug discontinuation due to side effects (Brogan & Ross, 2022; Cuadrado-Payán et al., 2020; Grein et al., 2020).

Remdesivir has been proven to reduce recovery times and shorten the hospital stay (Beigel et al., 2020; Cuadrado-Payán et al., 2022; WHO Solidarity Trial Consortium, 2021); however, more recent observational data have not shown any benefit in dialysis patients, though it may have benefit in dialysis patients who are on immunosuppressive treatment (Cacho et al., 2022; Cuadrado-Payán et al., 2022).

Molnupiravir

Molnupiravir is another RdRp inhibitor with clinical benefit in adults with moderate COVID-19 or those unvaccinated at risk of progressing to a more severe illness. Their initiation has to be within 5 days of symptom onset, when it has been demonstrated to reduce all-cause mortality and the risk of hospitalization (Brogan & Ross, 2022; Fischer et al., 2021; Lingscheid et al., 2022). Because there is no kidney clearance, molnupiravir can be safely used in kidney disease and dialysis patients without dose adjustment (Brogan & Ross, 2022; Kale et al., 2023). Moreover, it has also been shown to improve clinical outcomes in vaccinated dialysis patients (Chen et al., 2023).

Nirmatrelvir/ritonavir combination

Nirmatrelvir/ritonavir is a protease inhibitor combination with good results against mild to severe SARS-CoV-2 infection, including the omicron variant, although several drug interactions limit its use (El Karoui & De Vriese, 2022; Kale et al., 2023). It inhibits the 3-chymotrypsin-like cysteine protease enzyme responsible for viral replication (Kale et al., 2023). When given alone, nirmatrelvir is rapidly metabolized via cytochrome P450 (CYP) 3A4 and mainly eliminated by renal clearance (Lingscheid et al., 2022), and when administered with the CYP3A4 inhibitor, ritonavir's half-life is prolonged (Brogan & Ross, 2022).

In mild to moderate kidney disease patients (eGFR 30–60 mL/min/1.73 m^2), these drugs are safe though they require a reduction of 50% of the dosage (Brogan & Ross, 2022; Kale et al., 2023; Lingscheid et al., 2022). However, this drug combination has not been approved for patients with advanced kidney disease (eGFR $<$ 30 mL/min/1.73 m^2) or on dialysis (El Karoui & De Vriese, 2022). However, a clinical trial is currently under way in this population (NCT05487040) on this matter (Kale et al., 2023).

Monoclonal antibodies (mAbs)

The FDA has approved several mAbs in patients with mild to moderate COVID-19. They include bamlanivimab plus etesevimab, casirivimab plus imdevimab, sotrovimab, and bebtelovimab (Kale et al., 2023). Also, in December 2021, tixagevimab plus cilgavimab (Evusheld) received authorization to be used as preexposure prophylaxis in patients at risk (El Karoui & De Vriese, 2022).

They target the SARS-CoV-2 spike (S) protein. Hence, the efficacy is affected by mutations in S, causing the recommendations for its use to be modified as new SARS-CoV-2 variants appear (Kale et al., 2023). The data available show that these treatments are implausible to be active against the omicron variant. Sotrovimab is the only one that retains efficacy against this widely circulating variant (Destras et al., 2022).

All of them can be used in dialysis patients without dose adjustment (Kale et al., 2023; Mambelli et al., 2022).

Tocilizumab and sarilumab

Both are interleukin-6 receptor antagonists approved for COVID-19 treatment in hospitalized adults who require systemic corticosteroids, supplemental oxygen, noninvasive ventilation, mechanical ventilation, or extracorporeal membrane oxygenation (Rosas et al., 2021; Salama et al., 2021).

As they are not eliminated through the kidneys, they can be safely administered to patients with mild to moderate kidney disease without dose modification (Chamlagain et al., 2021). However, their effectiveness in patients with severe kidney impairment has not been evaluated (Marra et al., 2021).

Baricitinib and tofacitinib

These Janus kinase inhibitors can prevent the phosphorylation of essential proteins involved in signal transduction, leading to immune activation and inflammation in COVID-19 (Guimarães et al., 2021;Kalil et al., 2021; Marconi et al., 2021). Furthermore, baricitinib has direct antiviral activity through interference with SARS-CoV-2 endocytosis in susceptible cells (Kalil et al., 2021; Marconi et al., 2021).

Both are recommended with systemic corticosteroids in hospitalized patients with evidence of inflammation and increased oxygen needs (Kalil et al., 2021; Marconi et al., 2021).

CKD patients with an eGFR $< 60 mL/min/1.73 m^2$ require a dose adjustment as the kidneys mainly clear these drugs. Therefore, they are not recommended for dialysis patients, given the lack of safety data in this population (Brogan & Ross, 2022).

Policies and procedures

In this chapter, we reviewed the impact of the coronavirus pandemic on patients with in-center hemodialysis, the results of the immunization programmes in them and the treatment options available considering their impaired kidney function.

Applications to other areas

In this chapter, we reviewed the epidemiological, management, immunization, and therapeutical particularities of COVID-19 in hemodialysis patients. The lessons learned from this pandemic are essential and must be taken into account in case of a new breakthrough infection of a new variant of SARS-CoV-2 or even a new virus.

Mini-dictionary of terms

- **RNA-dependent RNA polymerase inhibitors:** Drugs that inhibit this enzyme that catalyzes the replication of RNA.
- **Protease inhibitors:** Drugs that bind proteolytic enzymes and, thus, viral replication.
- **Nucleoside reverse transcriptase inhibitor:** Drugs that competitively block the reverse transcriptase and, thus, the viral reverse transcription.
- **HIV integrase inhibitors/nonnucleoside reverse transcriptase inhibitors:** Drugs that noncompetitively block the reverse transcriptase and, thus, the viral reverse transcription.
- **Janus kinase inhibitor:** Drugs that inhibit these nonreceptor tyrosine kinases that are part of signal transduction pathways for the regulation of gene expression.

Summary points

- COVID-19 in HD patients, although its effect has been attenuated by vaccines, has been a major problem. Infections are the second cause of death in this population.
- HD patients are at a higher risk of poorer outcomes in case of infection with SARS-CoV-2 because they usually have comorbid conditions.
- The fact that HD patients have to continue with their treatments makes them more vulnerable in pandemic situations such as COVID-19.
- Although HD patients have poorer responses to vaccines, the immunization programs are effective in laboratory and clinical terms and are recommended in them.
- Although HD patients are often excluded from clinical trials and some drugs are not recommended for them due to kidney impairment, some data suggest the safety and efficacy of drugs in COVID-19 infections.

References

Agur, T., Ben-Dor, N., Goldman, S., Lichtenberg, S., Herman-Edelstein, M., Yahav, D., Rozen-Zvi, B., & Zingerman, B. (2021). Antibody response to mRNA SARS-CoV-2 vaccine among dialysis patients—a prospective cohort study. *Nephrology, Dialysis, Transplantation: Official Publication of the European Dialysis and Transplant Association - European Renal Association*, *36*(7), 1347–1349. https://doi.org/10.1093/NDT/GFAB155.

Alberici, F., Delbarba, E., Manenti, C., Econimo, L., Valerio, F., Pola, A., Maffei, C., Possenti, S., Gaggia, P., Movilli, E., Bove, S., Malberti, F., Bracchi, M., Costantino, E. M., Bossini, N., Gaggiotti, M., & Scolari, F. (2020). Management of patients on dialysis and with kidney transplant during Covid-19 Coronavirus infection. *Brescia Renal Covid Task Force*, 1–15. https://www.era-edta.org/en/wp-content/uploads/2020/03/COVID_guidelines_finale_eng-GB.pdf.

Allen, R., Turner, M., & deSouza, I. S. (2022). Remdesivir for the treatment of COVID-19. *American Family Physician*, *105*(2), 131–132. https://doi.org/10.1002/14651858.cd014962.pub2.

Attias, P., Sakhi, H., Rieu, P., Soorkia, A., Assayag, D., Bouhroum, S., Nizard, P., & el Karoui, K. (2021). Antibody response to the BNT162b2 vaccine in maintenance hemodialysis patients. *Kidney International*, *99*(6), 1490–1492. https://doi.org/10.1016/J.KINT.2021.04.009.

Beigel, J. H., Tomashek, K. M., Dodd, L. E., Mehta, A. K., Zingman, B. S., Kalil, A. C., Hohmann, E., Chu, H. Y., Luetkemeyer, A., Kline, S., Lopez de Castilla, D., Finberg, R. W., Dierberg, K., Tapson, V., Hsieh, L., Patterson, T. F., Paredes, R., Sweeney, D. A., Short, W. R., … Lane, H. C. (2020). Remdesivir for the Treatment of Covid-19—Final Report. *New England Journal of Medicine*, *383*(19), 1813–1826. https://doi.org/10.1056/nejmoa2007764.

Betjes, M. G. H. (2013). Immune cell dysfunction and inflammation in end-stage renal disease. *Nature Reviews Nephrology*, *9*(5), 255–265. https://doi.org/10.1038/nrneph.2013.44.

Brogan, M., & Ross, M. J. (2022). Annual review of medicine COVID-19 and kidney disease. *Annual Review of Medicine*. https://doi.org/10.1146/annurev-med-042420.

Broseta, J. J., Rodríguez-Espinosa, D., Bedini, J. L., Rodríguez, N., & Maduell, F. (2014). Antibody maintenance 3 months after complete mRNA COVID-19 vaccination in hemodialysis. *Nephrology, Dialysis, Transplantation*, *16*(3), 518–524. https://doi.org/10.1093/NDT/GFAB272.

Broseta, J. J., Rodríguez-Espinosa, D., Cuadrado, E., et al. (2020). SARS-CoV-2 infection in a Spanish cohort of CKD-5D patients: prevalence, clinical presentation, outcomes, and de-isolation results. *Blood Purification*, *50*(4–5), 531–538. https://doi.org/10.1159/000510557.

Broseta, J. J., Rodríguez-Espinosa, D., Rodríguez, N., del Mosquera, M. M., Marcos, M.Á., Egri, N., Pascal, M., Soruco, E., Bedini, J. L., Bayés, B., & Maduell, F. (2021). Humoral and cellular responses to mRNA-1273 and BNT162b2 SARS-CoV-2 vaccines administered to hemodialysis patients. *American Journal of Kidney Diseases*. https://doi.org/10.1053/j.ajkd.2021.06.002.

Broseta, J. J., Rodríguez-Espinosa, D., Soruco, E., & Maduell, F. (2021). Weekly seroconversion rate of the mRNA-1273 SARS-CoV-2 vaccine in haemodialysis patients. *Nephrology, Dialysis, Transplantation*, *36*(9), 1754–1755. https://doi.org/10.1093/ndt/gfab195.

Cacho, J., Nicolás, D., Bodro, M., Cuadrado-Payán, E., Torres-Jaramillo, V., Gonzalez-Rojas, Á., Ventura-Aguiar, P., Montagud-Marrahi, E., Herrera, S., Rico, V., Cofàn, F., Oppenheimer, F., Revuelta, I., Diekmann, F., & Cucchiari, D. (2022). Use of remdesivir in kidney transplant recipients with SARS-CoV-2 Omicron infection. *Kidney International*, *102*(4), 917–921. https://doi.org/10.1016/J.KINT.2022.08.001.

Chamlagain, R., Shah, S., Sharma Paudel, B., Dhital, R., & Kandel, B. (2021). Efficacy and safety of sarilumab in COVID-19: A systematic review. In *2021. Interdisciplinary Perspectives on Infectious Diseases* Hindawi Limited. https://doi.org/10.1155/2021/8903435.

Chan, J. F. W., Yuan, S., Kok, K. H., Tso, K. K. W., Chu, H., Yang, J., Xing, F., Liu, J., Yip, C. C. Y., Poon, R. W. S., Tsoi, H. W., Lo, S. K. F., Chan, K. H., Poon, V. K. M., Chan, W. M., Ip, J. D., Cai, J. P., Cheng, V. C. C., Chen, H., … Yuen, K. Y. (2020). A familial cluster of pneumonia associated with the 2019 novel coronavirus indicating person-to-person transmission: a study of a family cluster. *The Lancet*, *395*(10223), 514–523. https://doi.org/10.1016/S0140-6736(20)30154-9.

Chen, P.-C., Huang, C.-C., Fu, C.-M., Chang, Y.-C., Wu, P.-J., Lee, W.-C., Lee, C.-T., Tsai, K.-F., Chen, P.-C., Huang, C.-C., Fu, C.-M., Chang, Y.-C., Wu, P.-J., Lee, W.-C., Lee, C.-T., & Tsai, K.-F. (2023). Real-world effectiveness of SARS-CoV-2 vaccine booster in hemodialysis patients with COVID-19 receiving Molnupiravir. *Viruses*, *15*(2), 543. https://doi.org/10.3390/V15020543.

Clarke, C., Prendecki, M., Dhutia, A., Ali, M. A., Sajjad, H., Shivakumar, O., … Willicombe, M. (2020). High prevalence of asymptomatic COVID-19 infection in hemodialysis patients detected using serologic screening. *Journal of the American Society of Nephrology*, *31*(9), 1969–1975. https://doi.org/10.1681/ASN.2020060827.

Cuadrado-Payán, E., Montagud-Marrahi, E., Torres-Elorza, M., Bodro, M., Blasco, M., Poch, E., Soriano, A., & Piñeiro, G. J. (2020). SARS-CoV-2 and influenza virus co-infection. *The Lancet*, *395*(10236). https://doi.org/10.1016/S0140-6736(20)31052-7. p. e84. Lancet Publishing Group.

Cuadrado-Payán, E., Rodríguez-Espinosa, D., Broseta, J. J., Guillén-Olmos, E., & Maduell, F. (2022). Safety profile and clinical results of Remdesivir in Hemodialysis patients infected with SARS-CoV-2. A single-center Spanish cohort study. *Journal of Nephrology*. https://doi.org/10.1007/s40620-022-01364-3.

Destras, G., Bal, A., Simon, B., Lina, B., & Josset, L. (2022). *Sotrovimab drives SARS-CoV-2 omicron variant evolution in immunocompromised patients*. https://doi.org/10.1101/2021.03.09.434607.

Doria-Rose, N., Suthar, M. S., Makowski, M., O'Connell, S., McDermott, A. B., Flach, B., Ledgerwood, J. E., Mascola, J. R., Graham, B. S., Lin, B. C., O'Dell, S., Schmidt, S. D., Widge, A. T., Edara, V.-V., Anderson, E. J., Lai, L., Floyd, K., Rouphael, N. G., Zarnitsyna, V., … Kunwar, P. (2021). Antibody persistence through 6 months after the second dose of mRNA-1273 vaccine for Covid-19. *New England Journal of Medicine*, *384*(23), 2259–2261. https://doi.org/10.1056/nejmc2103916.

Dr, R. M., Selvaskandan, H., Makkeyah, Y. M., Hull, K., Kuverji, A., & Graham-Brown, M. (2020). The exclusion of patients with CKD in prospectively registered interventional trials for COVID-19-a rapid review of international registry data. *Journal of the American Society of Nephrology*, *31*(10), 2250–2252. https://doi.org/10.1681/ASN.2020060877.

El Karoui, K., & De Vriese, A. S. (2022). COVID-19 in dialysis: clinical impact, immune response, prevention, and treatment. *Kidney International*, *101* (5), 883–894. https://doi.org/10.1016/j.kint.2022.01.022. Elsevier B.V.

Eleftheriadis, T., Antoniadi, G., Liakopoulos, V., Kartsios, C., & Stefanidis, I. (2007). Disturbances of acquired immunity in hemodialysis patients. *Seminars in Dialysis*, *20*(5), 440–451. https://doi.org/10.1111/j.1525-139X.2007.00283.x.

Eleftheriadis, T., Pissas, G., Antoniadi, G., Liakopoulos, V., & Stefanidis, I. (2014). Factors affecting effectiveness of vaccination against hepatitis B virus in hemodialysis patients. *World Journal of Gastroenterology*, *20*(34), 12018–12025. https://doi.org/10.3748/wjg.v20.i34.12018. WJG Press.

European Centre for Disease Prevention and Control. (2020). *Guidelines for the implementation of non-pharmaceutical interventions against COVID-19*. September.

Favresse, J., Bayart, J.-L., Mullier, F., Elsen, M., Eucher, C., van Eeckhoudt, S., Roy, T., Wieers, G., Laurent, C., Dogné, J.-M., Closset, M., & Douxfils, J. (2021). Antibody titres decline 3-month post-vaccination with BNT162b2. *Emerging Microbes & Infections*, *10*(1), 1495–1498. https://doi.org/10.1080/22221751.2021.1953403.

Fischer, W., Eron, J. J., Holman, W., Cohen, M. S., Fang, L., Szewczyk, L. J., Sheahan, T. P., Baric, R., Mollan, K. R., Wolfe, C. R., Duke, E. R., Azizad, M. M., Borroto-Esoda, K., Wohl, D. A., Loftis, A. J., Alabanza, P., Lipansky, F., & Painter, W. P. (2021). Molnupiravir, an Oral Antiviral Treatment for COVID-19. *Science Translational Medicine*. https://doi.org/10.1101/2021.06.17.21258639.

Frantzen, L., Cavaillé, G., Thibeaut, S., & El-Haik, Y. (2021). Efficacy of the BNT162b2 mRNA Covid-19 Vaccine in a hemodialysis cohort. *Nephrology, Dialysis, Transplantation : Official Publication of the European Dialysis and Transplant Association - European Renal Association*, *36*(9), 1756–1757. https://doi.org/10.1093/NDT/GFAB165.

Garcia-Vidal, C., Cózar-Llistó, A., Meira, F., Dueñas, G., Puerta-Alcalde, P., Cilloniz, C., Garcia-Pouton, N., Chumbita, M., Cardozo, C., Hernández, M., Rico, V., Bodro, M., Morata, L., Castro, P., Almuedo-Riera, A., García, F., Mensa, J., Antonio Martínez, J., Sanjuan, G., ... Soriano, A. (2021). Trends in mortality of hospitalised COVID-19 patients: A single centre observational cohort study from Spain. *The Lancet Regional Health - Europe*, *3*, 100041. https://doi.org/10.1016/j.lanepe.2021.100041.

Goicoechea, M., Sánchez Cámara, L. A., Macías, N., Muñoz de Morales, A., González Rojas, Á., Bascuñana, A., Arroyo, D., Vega, A., Abad, S., Verde, E., García Prieto, A. M., Verdalles, U., Barbieri, D., Felipe Delgado, A., Carbayo, J., Mijaylova, A., Pérez de José, A., Melero, R., Tejedor, A., ... Aragoncillo, I. (2020). COVID-19: Clinical course and outcomes of 36 maintenance hemodialysis patients from a single center in Spain. *Kidney International*. https://doi.org/10.1016/j.kint.2020.04.031.

Grasselli, G., Zangrillo, A., Zanella, A., Antonelli, M., Cabrini, L., Castelli, A., Cereda, D., Coluccello, A., Foti, G., Fumagalli, R., Iotti, G., Latronico, N., Lorini, L., Merler, S., Natalini, G., Piatti, A., Ranieri, M. V., Scandroglio, A. M., Storti, E., ... Pesenti, A. (2020). Baseline Characteristics and Outcomes of 1591 Patients Infected with SARS-CoV-2 Admitted to ICUs of the Lombardy Region, Italy. *JAMA: The Journal of the American Medical Association*, *323*(16), 1574–1581. https://doi.org/10.1001/jama.2020.5394.

Grein, J., Ohmagari, N., Shin, D., Diaz, G., Asperges, E., Castagna, A., Feldt, T., Green, G., Green, M. L., Lescure, F.-X., Nicastri, E., Oda, R., Yo, K., Quiros-Roldan, E., Studemeister, A., Redinski, J., Ahmed, S., Bernett, J., Chelliah, D., ... Flanigan, T. (2020). Compassionate use of remdesivir for patients with severe Covid-19. *New England Journal of Medicine*, *382*(24), 2327–2336. https://doi.org/10.1056/nejmoa2007016.

Grupper, A., Sharon, N., Finn, T., Cohen, R., Israel, M., Agbaria, A., Rechavi, Y., Schwartz, I. F., Schwartz, D., Lellouch, Y., & Shashar, M. (2021). Humoral response to the pfizer bnt162b2 vaccine in patients undergoing maintenance hemodialysis. *Clinical Journal of the American Society of Nephrology*, *16*(7), 1037–1042. https://doi.org/10.2215/CJN.03500321.

Guimarães, P. O., Quirk, D., Furtado, R. H., Maia, L. N., Saraiva, J. F., Antunes, M. O., Kalil Filho, R., Junior, V. M., Soeiro, A. M., Tognon, A. P., Veiga, V. C., Martins, P. A., Moia, D. D. F., Sampaio, B. S., Assis, S. R. L., Soares, R. V. P., Piano, L. P. A., Castilho, K., Momesso, R. G. R. A. P., ... Berwanger, O. (2021). Tofacitinib in patients hospitalized with Covid-19 pneumonia. *New England Journal of Medicine*, *385*(5), 406–415. https://doi.org/10.1056/nejmoa2101643.

Kale, A., Shelke, V., Dagar, N., Anders, H. J., & Gaikwad, A. B. (2023). How to use COVID-19 antiviral drugs in patients with chronic kidney disease. In *14. Frontiers in Pharmacology*. https://doi.org/10.3389/fphar.2023.1053814. Frontiers Media S.A.

Kalil, A. C., Patterson, T. F., Mehta, A. K., Tomashek, K. M., Wolfe, C. R., Ghazaryan, V., Marconi, V. C., Ruiz-Palacios, G. M., Hsieh, L., Kline, S., Tapson, V., Iovine, N. M., Jain, M. K., Sweeney, D. A., El Sahly, H. M., Branche, A. R., Regalado Pineda, J., Lye, D. C., Sandkovsky, U., ... Beigel, J. H. (2021). Baricitinib plus Remdesivir for Hospitalized Adults with Covid-19. *New England Journal of Medicine*, *384*(9), 795–807. https://doi.org/10.1056/nejmoa2031994.

Ke, C., Wang, Y., Zeng, X., Yang, C., & Hu, Z. (2020). 2019 Novel coronavirus disease (COVID-19) in hemodialysis patients: A report of two cases. *Clinical Biochemistry*. https://doi.org/10.1016/j.clinbiochem.2020.04.008.

Leclerc, Q. J., Fuller, N. M., Knight, L. E., Funk, S., & Knight, G. M. (2020). What settings have been linked to SARS-CoV-2 transmission clusters? *Wellcome Open Research*, *5*. https://doi.org/10.12688/wellcomeopenres.15889.2.

Lingscheid, T., Kinzig, M., Krüger, A., Müller, N., Bölke, G., Tober-Lau, P., Münn, F., Kriedemann, H., Witzenrath, M., Sander, L. E., Sörgel, F., & Kurth, F. (2022). Pharmacokinetics of nirmatrelvir and ritonavir in COVID-19 patients with end-stage renal disease on intermittent hemodialysis. *Antimicrobial Agents and Chemotherapy*, *66*(11). https://doi.org/10.1128/aac.01229-22.

Liu, J., Liao, X., Qian, S., Yuan, J., Wang, F., Liu, Y., ... Zhang, Z. (2020). Community transmission of severe acute respiratory. *Emerging Infectious Diseases*, *26*(6), 1320–1323. https://doi.org/10.3201/eid2606.200239.

Mambelli, E., Gasperoni, L., Maldini, L., Biagetti, C., & Rigotti, A. (2022). Sotrovimab in SARS-COV-2 chronic hemodialysis patients in the Omicron era. Is intradialytic administration feasible? Report of 4 cases. *Journal of Nephrology*. https://doi.org/10.1007/s40620-022-01449-z.

Marconi, V. C., Ramanan, A. V., de Bono, S., Kartman, C. E., Krishnan, V., Liao, R., Piruzeli, M. L. B., Goldman, J. D., Alatorre-Alexander, J., de Cassia Pellegrini, R., Estrada, V., Som, M., Cardoso, A., Chakladar, S., Crowe, B., Reis, P., Zhang, X., Adams, D. H., Ely, E. W., ... Zirpe, K. (2021). Efficacy and safety of baricitinib for the treatment of hospitalised adults with COVID-19 (COV-BARRIER): a randomised, double-blind, parallel-group, placebo-controlled phase 3 trial. *The Lancet Respiratory Medicine*, *9*(12), 1407–1418. https://doi.org/10.1016/S2213-2600(21)00331-3.

Marra, F., Smolders, E. J., El-Sherif, O., Boyle, A., Davidson, K., Sommerville, A. J., ... Back, D. (2021). Recommendations for dosing of repurposed COVID-19 medications in patients with renal and hepatic impairment. *Drugs in R and D*, *21*(1), 9–27. https://doi.org/10.1007/s40268-020-00333-0. Adiws.

Ministerio de Ciencia e Innovación de España, & Ministerio de Sanidad de España. (2020). *ESTUDIO ENE-COVID : INFORME FINAL confianza y la generosidad de más de 68.000 participantes que han entendido el interés de proporcionar tiempo, información y muestras.*

COVID-19 Rapid Guideline: Dialysis Service Delivery (NG160). (2020). *NICE Guidelines 2020*. https://www.nice.org.uk/guidance/ng160.

Rodríguez-Espinosa, D., Broseta, J. J., Cuadrado, E., & Maduell, F. (2020). Prevalence of COVID-19 infection in hemodialysis patients detected using serologic screening. *Journal of the American Society of Nephrology*, *31*(12), 2967. https://doi.org/10.1681/ASN.2020081193.

Rodríguez-Espinosa, D., Montagud-Marrahi, E., Cacho, J., Arana, C., Taurizano, N., Hermida, E., del Risco-Zevallos, J., Casals, J., Rosario, A., Cuadrado-Payán, E., Molina-Andújar, A., Rodríguez, N., Vilella, A., Bodro, M., Ventura-Aguiar, P., Revuelta, I., Cofàn, F., Poch, E., Oppenheimer, F., ... Cucchiari, D. (2022). Incidence of severe breakthrough SARS-CoV-2 infections in vaccinated kidney transplant and haemodialysis patients. *Journal of Nephrology*, *1*, 3. https://doi.org/10.1007/s40620-022-01257-5.

Ronda, E. E. C., Infección, L. A., & En, P. O. R. S. O. V. (2020). *Estudio Ene-COVID: cuarta ronda* (pp. 2–7).

Rosas, I. O., Bräu, N., Waters, M., Go, R. C., Hunter, B. D., Bhagani, S., Skiest, D., Aziz, M. S., Cooper, N., Douglas, I. S., Savic, S., Youngstein, T., Del Sorbo, L., Cubillo Gracian, A., De La Zerda, D. J., Ustianowski, A., Bao, M., Dimonaco, S., Graham, E., ... Malhotra, A. (2021). Tocilizumab in hospitalized patients with severe Covid-19 pneumonia. *New England Journal of Medicine*, *384*(16), 1503–1516. https://doi.org/10.1056/nejmoa2028700.

Salama, C., Han, J., Yau, L., Reiss, W. G., Kramer, B., Neidhart, J. D., Criner, G. J., Kaplan-Lewis, E., Baden, R., Pandit, L., Cameron, M. L., Garcia-Diaz, J., Chávez, V., Mekebeb-Reuter, M., Lima de Menezes, F., Shah, R., González-Lara, M. F., Assman, B., Freedman, J., & Mohan, S. V. (2021). Tocilizumab in patients hospitalized with Covid-19 pneumonia. *New England Journal of Medicine*, *384*(1), 20–30. https://doi.org/10.1056/nejmoa2030340.

Sánchez-Álvarez, J. E., Fontán, M. P., Martín, C. J., Pelícano, M. B., Reina, C. J. C., Prieto, Á. M. S., Melilli, E., Barrios, M. C., Heras, M. M., & del Pino, M. D. P.y. (2020). Status of SARS-CoV-2 infection in patients on renal replacement therapy. Report of the COVID-19 Registry of the Spanish Society of Nephrology (SEN). *Nefrología*, *40*(3), 272–278. https://doi.org/10.1016/j.nefroe.2020.04.002.

Teijaro, J. R., & Farber, D. L. (2021). COVID-19 vaccines: modes of immune activation and future challenges. *Nature Reviews Immunology*, *21*(4), 195–197. https://doi.org/10.1038/s41577-021-00526-x.

The RECOVERY Collaborative Group. (2021). Dexamethasone in hospitalized patients with Covid-19. *New England Journal of Medicine*, *384*(8), 693–704. https://doi.org/10.1056/NEJMoa2021436.

Tomazini, B. M., Maia, I. S., Cavalcanti, A. B., Berwanger, O., Rosa, R. G., Veiga, V. C., Avezum, A., Lopes, R. D., Bueno, F. R., Silva, M. V. A. O., Baldassare, F. P., Costa, E. L. V., Moura, R. A. B., Honorato, M. O., Costa, A. N., Damiani, L. P., Lisboa, T., Kawano-Dourado, L., Zampieri, F. G., ... Azevedo, L. C. P. (2020). Effect of dexamethasone on days alive and ventilator-free in patients with moderate or severe acute respiratory distress syndrome and COVID-19: The CoDEX randomized clinical trial. *JAMA: The Journal of the American Medical Association*, *324*(13), 1307–1316. https://doi.org/10.1001/jama.2020.17021.

Trujillo, H., Caravaca-Fontán, F., Sevillano, Á., Gutiérrez, E., Caro, J., Gutiérrez, E., Yuste, C., Andrés, A., & Praga, M. (2020). SARS-CoV-2 infection in hospitalized patients with kidney disease. *Kidney International Reports*, 1–5. https://doi.org/10.1016/j.ekir.2020.04.024.

WHO Solidarity Trial Consortium. (2021). Repurposed antiviral drugs for Covid-19—Interim WHO solidarity trial results. *New England Journal of Medicine*, *384*(6), 497–511. https://doi.org/10.1056/NEJMoa2023184.

Windpessl, M., Bruchfeld, A., Anders, H.-J., Kramer, H., Waldman, M., Renia, L., Ng, L. F. P., Xing, Z., & Kronbichler, A. (2021). COVID-19 vaccines and kidney disease. *Nature Reviews Nephrology*, *17*(5), 291–293. https://doi.org/10.1038/s41581-021-00406-6.

Wölfel, R., Corman, V. M., Guggemos, W., Seilmaier, M., Zange, S., Müller, M. A., Niemeyer, D., Jones, T. C., Vollmar, P., Rothe, C., Hoelscher, M., Bleicker, T., Brünink, S., Schneider, J., Ehmann, R., Zwirglmaier, K., Drosten, C., & Wendtner, C. (2020). Virological assessment of hospitalized patients with COVID-2019. *Nature*, *581*(7809), 465–469. https://doi.org/10.1038/s41586-020-2196-x.

Xiong, F., Tang, H., Liu, L., Tu, C., Tian, J. B., Lei, C. T., Liu, J., Dong, J. W., Chen, W. L., Wang, X. H., Luo, D., Shi, M., Miao, X. P., & Zhang, C. (2020). Clinical characteristics of and medical interventions for COVID-19 in hemodialysis patients in Wuhan, China. *Journal of the American Society of Nephrology: JASN*, 1–11. https://doi.org/10.1681/ASN.2020030354.

Chapter 39

Coinfections with COVID-19: A focus on tuberculosis (TB)

Chijioke Obiwe Onyeani[a,b], Precious Chisom Dimo[a], Emmanuel Ebuka Elebesunu[a], Malachy Ekene Ezema[c], Samuel Ogunsola[d], and Ademola Aiyenuro[e,f]

[a]*Department of Medical Laboratory Sciences, University of Nigeria, Nsukka, Nigeria,* [b]*Department of Microbiology and Physiological Systems, University of Massachusetts Chan Medical School, Worcester, MA, United States,* [c]*Faculty of Pharmaceutical Sciences, University of Nigeria, Nsukka, Nigeria,* [d]*Department of Physiology and Pathophysiology, Max Rady College of Medicine, Rady Faculty of Health Science, University of Manitoba, Winnipeg, MB, Canada,* [e]*Research4Knowledge, Lagos, Nigeria,* [f]*University of Cambridge, Cambridge, United Kingdom*

Abbreviations

ALT	alanine aminotransferase
CK-MB	creatinine kinase MB
CPK	creatinine phosphokinase
CRP	C-reactive protein
ESR	erythrocyte sedimentation rate
HRCT	high-resolution computerized tomography
IGRA	interferon-gamma release assay
IL	interleukin
INF	interferon
LDH	serum lactate dehydrogenase
LTBI	latent tuberculosis infection
MERS	Middle East respiratory syndrome
NT pro-BNP	NT-proB-type natriuretic peptide
SAR	severe acute respiratory syndrome
TB	tuberculosis

Introduction

COVID-19, also known as coronavirus disease 2019, was referred to as "public enemy number one" by the World Health Organization (WHO). The virus caused widespread disruption to healthcare systems around the world, as it is highly contagious and can spread rapidly through both individuals and communities. In certain regions, such as Southeast Asia, South America, and Africa, the COVID-19 pandemic may have been exacerbated by the convergence of the virus with tuberculosis (TB), another lethal disease. Tuberculosis, another airborne disease that has affected humans for at least 70,000 years, is the leading cause of death from infectious diseases. However, since April 2020, COVID-19 has caused similar numbers of daily deaths worldwide. The convergence of these two diseases is particularly concerning in countries where TB is already prevalent. The COVID-19 pandemic has caused significant challenges for TB control programs, particularly in terms of diagnosis and treatment. In countries with inadequate diagnostic systems, it is difficult to accurately identify TB and COVID-19 infections, which can impact therapeutic decision-making and worsen the prognosis of both diseases. The pandemic has also led to a shortage of resources, drugs, and medical supplies as well as reduced mobility for patients and healthcare professionals, all of which can contribute to treatment failure and an increase in the incidence of multidrug-resistant TB. The impacts of COVID-19 on TB go beyond logistical and administrative issues and are already being felt in some countries.

Much is still unknown about the interaction between TB and the SARS-CoV-2 virus that causes COVID-19. A comprehensive survey that examines various aspects of TB and COVID-19 coinfections, including their incidence, mortality,

Management, Body Systems, and Case Studies in COVID-19. https://doi.org/10.1016/B978-0-443-18703-2.00026-8
Copyright © 2024 Elsevier Inc. All rights are reserved, including those for text and data mining, AI training, and similar technologies.

clinical symptoms, diagnosis, treatment, and laboratory evidence, could be useful for managing both diseases. This article discusses the pathology, diagnosis, treatment, and management of TB and COVID-19 coinfections.

Pathology of COVID-19 and tuberculosis coinfection

Clinical manifestations

The clinical presentation of coinfection with COVID-19 and TB can vary depending on the individual's overall health, the severity of the infections, and the presence of other comorbidities (Sahu, Lal, & Mishra, 2020; Sahu, Mishra, & Lal, 2020; Sahu, Mishra, Lal, & Sahu, 2020; Sahu et al., 2020a, 2020b). In general, individuals with coinfection may experience more severe symptoms and complications than those who have only one of the infections (Goyal et al., 2020).

The symptoms of TB and COVID-19 can be similar, which can make diagnosis challenging. However, the onset of TB symptoms is usually more gradual, and the duration of the illness is typically longer than that of COVID-19, lasting from weeks to months (Mishra et al., 2020). COVID-19 can present with a range of symptoms, including respiratory and non-respiratory symptoms, and can range from asymptomatic to severe or fatal. Common clinical symptoms of COVID-19 include fever, dry cough, and difficulty breathing while other nonspecific symptoms may include fatigue, muscle pain, headache, and gastrointestinal symptoms such as loss of appetite, nausea, vomiting, diarrhea, abdominal pain, and bleeding. Respiratory symptoms of COVID-19, including a productive cough and hemoptysis, can also be similar to those of TB (Stochino et al., 2020).

There is evidence that the features of lung imaging in COVID-19 patients can include bilateral involvement, peripheral distribution, a mixture of ground-glass opacity and consolidation, and vascular thickening while the most common computed tomography (CT) findings of coinfection with COVID-19 and TB include bilateral lesions, cavities, infiltrates, ground-glass opacity, nodules, pleural effusion, and fibrosis (He et al., 2020). Previous research suggests that TB can increase the body's susceptibility to other airborne infections and may exacerbate the severity of COVID-19, potentially due to immune impairment and cytokine overexpression (Prompetchara et al., 2020). COVID-19 may also increase the risk of latent TB becoming active and worsen the progression of TB. In COVID-19 and TB coinfection, dyspnea and CT findings such as bilateral lesion infiltrates and the "tree in bud" appearance may be good predictors of disease severity (Kumar et al., 2020).

Morbidity and mortality rates

The severity of disease and symptoms in individuals with COVID-19 and TB coinfection can vary widely, with some patients requiring transfer to the intensive care unit after hospitalization (Sun et al., 2020). Complications that have been reported in these patients include hypoxemia, respiratory failure, acute respiratory distress syndrome (ARDS), the need for noninvasive ventilation, abnormal glucose levels, multiorgan failure, extended hospital stays (up to 130 days), superimposed bacterial infections, and an increased risk of death (Chopra et al., 2020; Mishra et al., 2020).

As of Jan. 3, 2023, the WHO had reported 661 million confirmed cases of COVID-19 and 6.69 million deaths (approximately 1%) globally (World Health Organization, 2023). Studies have found that the case fatality rates for patients with coinfection with TB and COVID-19 are higher than those for COVID-19 alone, with rates of 27.3%, 12.3%, 11.6%, and 23.6% reported by Sy et al. (2020), Gupta et al. (2020), Tadolini et al. (2020), and Motta et al. (2020), respectively. Older age, particularly over 65 years, may be a risk factor for death from COVID-19 and TB coinfection, which is consistent with previous findings in that the mortality rate from COVID-19 increases significantly with age (Promislow, 2020). A model-based analysis found that the estimated overall death rate for COVID-19 was 0.66%, but this rate increased to 7.8% for patients over 80 and decreased to 0.0016% for children under 9 years (Verity et al., 2020). These differences may be due to the higher prevalence of preexisting comorbidities, immune system dysregulation, and chronic subclinical systemic inflammation in older adults compared to younger people (Zhou et al., 2020). Therefore, the elderly should be a primary focus of COVID-19 and COVID-19-TB prevention efforts due to their higher mortality risk (Sahu, Lal, & Mishra, 2020; Sahu, Mishra, & Lal, 2020; Sahu, Mishra, Lal, & Sahu, 2020; Sahu et al., 2020a, 2020b).

COVID-19 and TB coinfection patients also have a higher rate of comorbidities than COVID-19 patients alone (56.41% vs. 25.1%) (Kang & Jung, 2020). The most common comorbidities among COVID-19 patients are hypertension (21.1%), diabetes (9.7%), cardiovascular disease (8.4%), and respiratory system disease (1.5%) while the most common comorbidities among COVID-19-TB patients are diabetes (24.36%), hypertension (17.95%), HIV infection (6.41%), hepatitis (3.85%), epilepsy (3.85%), and cancer (2.56%) (Guan et al., 2020). COVID-19-TB patients who died were more likely to have hypertension (47.06% vs. 9.84%) and cancer (11.76% vs. 0%) compared to those who survived (Zheng et al.,

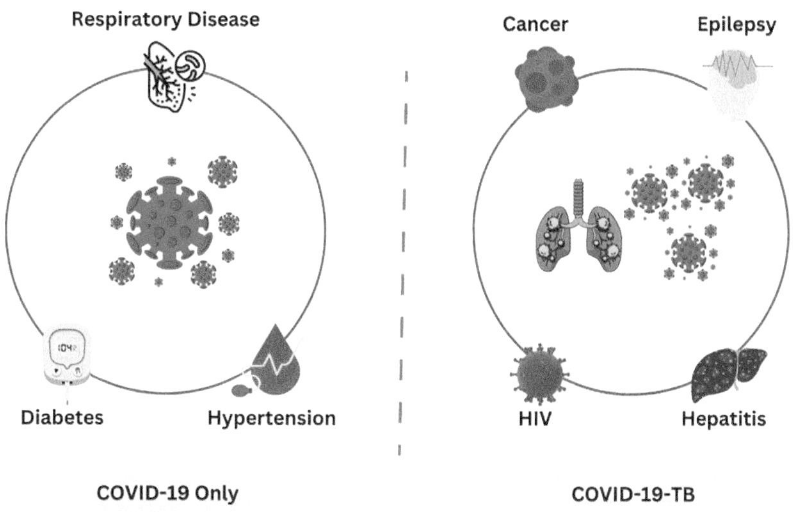

FIG. 1 Common comorbidities among COVID-19 and COVID-19-TB patients.

2020). Predisposing factors for COVID-19 in TB patients have included medical comorbidities such as chronic obstructive pulmonary disease (COPD), diabetes, HIV, renal failure, liver disease, alcohol abuse, and smoking as well as demographic factors such as male gender and being part of a migrant population (Tham et al., 2020). Both COVID-19 and TB tend to spread in overcrowded areas and among poor and malnourished populations, and these comorbidities and circumstances can have a synergistic impact on the severity of the diseases (Mishra et al., 2020) (Fig. 1).

Diagnosis of COVID-19 and tuberculosis

The symptoms of active TB and COVID-19 often overlap (Dheda et al., 2020). The study by Tadolini et al. (2020) showed that 74% of patients with COVID-19 and TB coinfection had TB diagnosed before COVID-19 (including 234 patients with previous TB, corresponding to 31.3% of the whole cohort) while 16.5% were diagnosed with both diseases within the same week (after COVID-19 symptoms prompted further testing that revealed preexisting TB), and 9.5% were diagnosed with COVID-19 first. However, it can be difficult to differentiate between the two diseases based on symptoms alone. The radiography of active tuberculosis and COVID-19 is also often similar, despite the fact that isolated upper-lobe pulmonary infiltrates may indicate active tuberculosis while lower infiltrates may be indicative of another bacterial infection (Sahu, Lal, & Mishra, 2020; Sahu, Mishra, & Lal, 2020; Sahu, Mishra, Lal, & Sahu, 2020; Sahu et al., 2020a, 2020b). It is therefore misleading to base a clinic-radiological diagnosis on the characteristic patterns of active tuberculosis and COVID-19 presentation, even in tuberculosis-endemic environments (Fig. 2). It is important to test for both diseases separately, or at least consider testing for TB when testing for COVID-19, to accurately diagnose patients (Song et al., 2021).

Microbiology

In most reported cases of COVID-19 and pulmonary TB coinfection, the TB was caused by *Mycobacterium tuberculosis*. Various diagnostic tools were used to identify TB, including sputum culture (45%), the nucleic acid amplification test (NAAT, 26.5%), and sputum smear positivity (18.4%) (Motta et al., 2020). There have also been reports of TB caused by *Mycobacterium bovis*. Both drug-susceptible and drug-resistant strains of mycobacteria have been identified, with drug-susceptible strains being more common (82%). Approximately 18% of patients with coinfections had TB that was resistant to a single anti-TB drug, and 9% had multidrug-resistant TB (MDR TB) (Tadolini et al., 2020).

Laboratory panel

Initial reports of COVID-19 and TB coinfection have indicated the presence of leukopenia, lymphocytopenia, and elevated inflammatory markers such as ESR, CRP, and LDH (Liu et al., 2020). Other laboratory abnormalities that have been reported in this subset of patients include low serum albumin, elevated ALT, abnormal glucose levels, and elevated CPK. Studies on COVID-19 patients have consistently identified multiple laboratory abnormalities in patients with severe

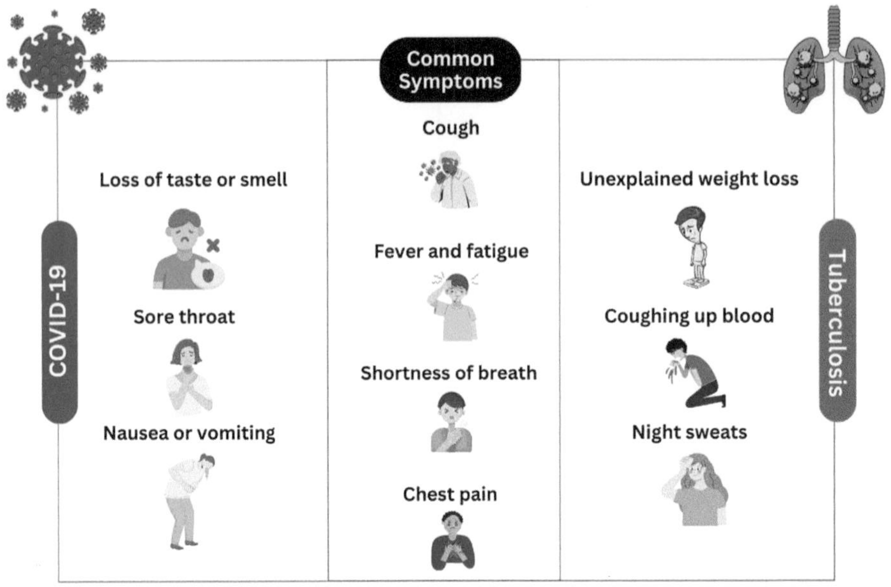

FIG. 2 Symptoms of COVID-19 and tuberculosis.

disease, including elevated levels of procalcitonin, ferritin, IL-2, IL-7, and tumor necrosis factor in patients experiencing a cytokine storm (Mishra et al., 2020). Severe COVID-19 patients with cardiac injury have also been reported to have elevated levels of troponin, CK-MB, myoglobin, D-dimer, highly sensitive troponin, and NT pro-BNP (Fu et al., 2020).

Imaging

High-resolution computerized tomography (HRCT) and chest radiography have been used to identify patterns of pulmonary involvement in COVID-19 and TB coinfection. HRCT has been the initial modality of imaging in more than 40% of patients. In most cases of coinfection (98%), imaging details were available (Lal, Mishra, & Sahu, 2020; Lal, Mishra, Sahu, & Abraham, 2020). Imaging findings that have been reported in TB patients with COVID-19 include the development of multiple, bilateral ground-glass opacities and consolidations with air bronchograms. Among patients with both TB and COVID-19, 33% had unilateral pulmonary infiltrates and 19% had bilateral infiltrates. Chest CT findings that may suggest the diagnosis of pulmonary TB include cavitating lung lesions. In patients with TB, bilateral cavitary lesions were more common (27%) compared to unilateral cavitary lesions (21%). Other patterns seen on imaging in these patients include a miliary pattern, a crazy paving pattern, and a tree in bud pattern (Sahu, Lal, & Mishra, 2020; Sahu, Mishra, & Lal, 2020; Sahu, Mishra, Lal, & Sahu, 2020; Sahu et al., 2020a, 2020b). The pattern of pulmonary involvement in TB patients with COVID-19 is similar to that seen in non-TB patients with COVID-19, with ground-glass opacities being the most common finding (Lal, Mishra, & Sahu, 2020; Lal, Mishra, Sahu, & Abraham, 2020). Interestingly, ground-glass opacities have also been reported to resolve in TB patients following improvement in their COVID-19 illness, similar to what has been seen in non-TB patients (Khurana & Aggarwal, 2020).

The impact of COVID-19 on TB

Challenges and opportunities

The COVID-19 pandemic had a significant effect on TB patient care in 2020. According to the WHO, a report from 84 countries showed an estimated deficit of 1.4 million in the number of persons receiving TB care between 2019 and 2020. The outcome of this deficit resulted in a high morbidity rate of TB, as a staggering 1.5 million TB-related deaths were recorded in 2020 (Boulle et al., 2021).

Policies associated with the COVID-19 pandemic such as restrictions on movement stalled public health efforts as access to medical care was impaired. The effect goes beyond just lack of early detection, as the disruption of essential services for TB patients also posed a challenge to their quality of life and hence, their prognosis. The combination of TB and COVID-19 can lead to more severe illness and an increased risk of death compared to either condition alone. TB is already a leading cause of death worldwide, and the added burden of COVID-19 can further strain healthcare systems.

Biological, clinical, and public health effects

Coinfection with TB and COVID-19 can complicate the diagnosis and treatment of both diseases (Fig. 3). The symptoms of TB and COVID-19 can be similar, making it difficult to distinguish between the two infections. The implications of this would be misdiagnoses of TB, which presents with a great transmission risk to the immediate environment. Also, we know that early diagnosis of TB and immediate initiation of pharmaceutical care are vital for effective TB control. Patients who remain undiagnosed with pulmonary TB mostly provide a reservoir for infecting others, and of course, late diagnosis will likely worsen the disease severity as well as elevate the mortality rate and the likelihood of TB transmission in the community, as each infectious case will result in 10–15 secondary infections. (Stochino et al., 2020).

The respiratory system is commonly affected in both TB and COVID-19, and coinfection can lead to more severe respiratory symptoms, such as difficulty in breathing and a worsening of TB-related lung damage. This ultimately increases the severity of the illness, especially in individuals with underlying health conditions or a weakened immune system.

Studies have shown that patients with COVID-19-TB coinfection possess a low capacity to build an immune response to SARS-COV-2. The patients have also demonstrated low blood lymphocyte counts. The decline in response to SARS-COV-2 antigens from patients with TB-COVID-19 coinfection might be because of the huge partitioning of the specific-T-cells in the infectious foci or, as seen in other infectious diseases, by the removal of effector T-cells when in contact with high amounts of antigens (Prompetchara et al., 2020). It is unclear if the lack of a SARS-COV-2-specific response correlates to a worse clinical outcome.

While the story of COVID-19-TB coinfection is not a happy one, it still provides an opportunity to reevaluate TB care and possibly find areas of improvement. This provides an opportunity to scale up simultaneous testing for TB and COVID-19, taking into consideration the similarity of symptoms (cough, fever, and difficulty breathing), and based on the exposure or presence of risk factors (Falasca et al., 2020). Maintaining current molecular diagnostic services for TB patients will be essential as countries prepare to share existing molecular platforms for COVID-19 testing.

At a point when movement might still be restricted in some countries, there is a need to promote access to people-centered prevention and care services. Home-based and community-based prevention and care should be the preferred choice over hospital treatment for TB patients (unless serious conditions require hospitalization). This is to reduce opportunities for transmission (Goudiaby et al., 2022). Treatments for extensively drug-resistant TB and multidrug-resistant TB are all oral, as recommended by the WHO. Digital adherence technologies can also help bridge the gap in communication. Preventive treatment for TB should be provided for household contacts due to the increased risk of exposure. These needed shifts in the paradigm of TB treatment that will impact TB care can be regarded as opportunities and lessons that can be taken from the challenge of COVID-19-TB coinfection.

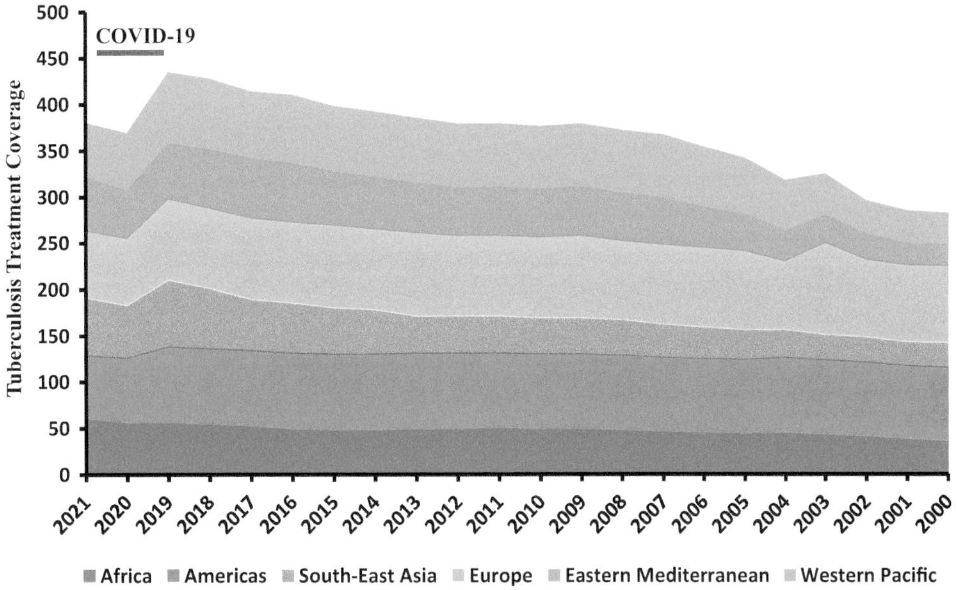

FIG. 3 Tuberculosis treatment coverage based on World Health Organization region (Data adapted from WHO website).

Effects of COVID-19 and tuberculosis coinfection on the pulmonary system

Both COVID-19 and tuberculosis are known to have a high affinity for the respiratory system and lung tissue. Hence, radiological findings play an important role in the diagnostic process for distinguishing both diseases and for detecting a case of coinfection. The most prevalent feature in primary TB is the presence of hilar and mediastinal lymphadenopathy, often with pleural effusion and parenchymal consolidations that resolve slowly (Nachiappan et al., 2017). Irregular consolidations and thick-walled cavitations occur in TB reactivation, and in miliary tuberculosis, diffuse nodules of about 1–3 mm are usually seen (Osejo-Betancourt et al., 2021). On the other hand, the radiological abnormalities common to COVID-19 are bilateral and peripheral ground-glass opacities that are multifocal, with progressive consolidations, a crazy paving pattern, and pneumonia (Kwee & Kwee, 2020). It is worth noting that the above radiological findings are progressive on HRCT, and depict the severe pulmonary inflammation caused by SARS-CoV-2 (Kwee & Kwee, 2020). An important factor that delays the investigation and diagnosis of coinfection with COVID-19 and TB is that the associated symptoms of both diseases are often similar, and their radiological manifestations tend to overlap in the affected patient, which often results in a misdiagnosis or a diagnosis of just one of the two diseases in a case of coinfection (Musso et al., 2021).

In several case studies involving COVID-19 and TB coinfection, the pulmonary abnormalities commonly reported included symptoms such as persistent cough, dyspnea, and widespread crackling upon lung auscultation (Musso et al., 2021; Niazkar et al., 2022). The radiological manifestations from chest x-ray and CT results that were regarded as tuberculosis sequelae included thick-walled cavitary lesions, numerous nodular opacities, and ill-defined airspace opacities, which were more frequent in the apical portions of the right lung (Mishra et al., 2020; Musso et al., 2021; Niazkar et al., 2022), but also occurred bilaterally in some cases (Adzic-Vukicevic et al., 2022). The manifestations that were suggestive of COVID-19-associated pneumonia included diffuse areas of consolidation and bilateral ground-glass opacities, which were present in the lungs of almost all diagnosed patients, along with signs of left basal lung disease (Nyanti et al., 2022; Song et al., 2021). A number of case studies reported some form of mix-up or delay in detecting the coinfection of both diseases, and the affected patients often had worse lung parenchymal damage than was observed in singular cases of either of the diseases, with some cases resulting in mortality (Biswas et al., 2021; Musso et al., 2021) while other patients eventually recovered after continued therapy (Al Lawati et al., 2021).

Effect of COVID-19 and tuberculosis coinfection on the immune system

Prior to COVID-19, there had been reports of coinfection with pulmonary TB and the Middle East respiratory syndrome (MERS) or severe acute respiratory syndrome (SARS). Several such cases had been reported in countries such as Taiwan, Singapore, and China (Chen, 2008; Liu et al., 2006). It was noted that coinfections of TB and MERS-CoV or SARS-CoV had the tendency to augment each other and result in significant immunosuppression (Alfaraj et al., 2017), which is quite similar to what happens in SARS-CoV-2 and TB coinfection. Reports show that the presence of TB in a patient makes them more vulnerable to SARS-CoV-2 infection, and both diseases may significantly affect the immune system, resulting in a severe clinical picture (Song et al., 2021). In COVID-19 infection, a cytokine storm is described among the major pathophysiological mechanisms, and the cytokines involved also have a critical role in host resistance to TB (Crisan-Dabija et al., 2020). In addition, it has been reported that TB-associated lung damage is a factor that makes one more susceptible to airborne diseases like COVID-19 (Mousquer et al., 2021). Research also showed that *Mycobacterium tuberculosis* and SARS-CoV-2 share several similar host protein interactomes (same interaction partner proteins), which is a major critical factor in coinfections because both diseases have a strong affinity for pulmonary tissue (Gordon et al., 2020).

The immune response to TB is typically cell-mediated, mainly involving alveolar macrophages, CD4+ T-cells, and sometimes CD8+ T-cells, which function to limit the bacterial infection and give rise to a latent tuberculosis infection (LTBI). On the other hand, in the COVID-19 immune response, a humoral response involving type 1 interferon (INF-1) is triggered, leading to a rise in the levels of cytokines such as INF-γ, IL-6, and IL-8, which function to recruit neutrophils, macrophages, natural killer (NK) cells, and CD4+ T-cells; this is accompanied by elevated levels of inflammatory markers such as C-reactive protein (CRP) and ferritin (Osuchowski et al., 2021). Eventually, the combined impact of SARS-COV-2 and TB infection on the immune system results in a dysregulated immune response that yields pronounced lymphocytopenia and triggers the activation of LTBI into an active tuberculosis infection. Marked lymphopenia is often an indicator of the severity of COVID-19 infection, and an indeterminate or negative score for the interferon-gamma release assay (IGRA) as a TB test indicates an increased risk of disease dissemination and mortality after therapy initiation in TB patients (Auld et al., 2016) (Fig. 4).

Test results depicting the immune response in case studies of COVID-19 and TB coinfection showed various similarities, such as low lymphocyte and IFN-γ levels from lymphocyte counts and QuantiFERON-tests, which indicated

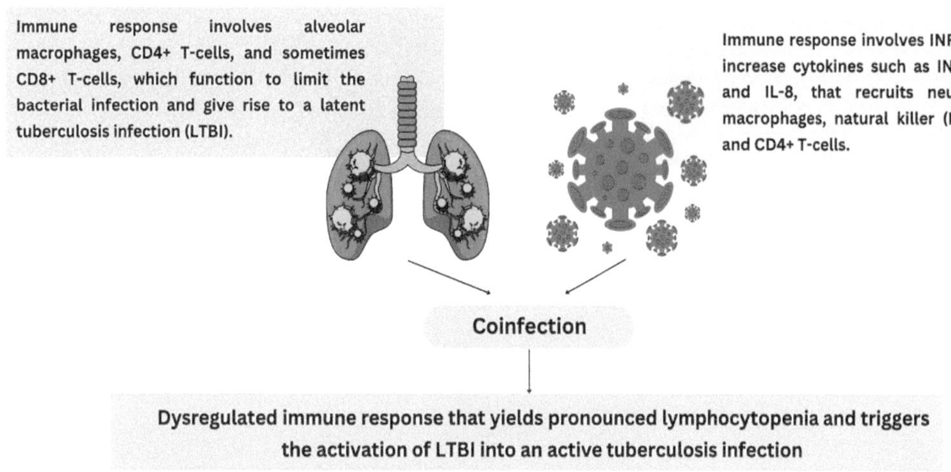

FIG. 4 Immune response to tuberculosis and COVID-19 coinfection.

immunosuppression. The results from general inflammation indicator tests such as erythrocyte sedimentation rate (ESR) and C-reactive protein (CRP) were also significantly elevated, including lactate dehydrogenase levels. Coagulation studies also showed elevated levels of D-dimer (Al Lawati et al., 2021; Musso et al., 2021; Niazkar et al., 2022). Reports from case studies of coinfection showed that TB infection came first in most patients, which may have had immunosuppressive effects, prompting the subsequent infection with COVID-19. In turn, the COVID-19 infection further dampened the body's immune response, resulting in increased TB disease progression and worse prognosis in the patients. Hence, both diseases had a synergistic effect in worsening the health outcomes and disease severity in patients. However, a significant proportion of the patients recovered fully after consistent therapy (Al Lawati et al., 2021; Niazkar et al., 2022) while the outcome in some patients further worsened and resulted in mortality (Biswas et al., 2021; Musso et al., 2021).

Treatment of COVID-19 and TB coinfection

All the patients with active tuberculosis in the study by Sahu et al. received a combination of drugs known as multidrug regimen antitubercular therapy (Sahu, Lal, & Mishra, 2020; Sahu, Mishra, & Lal, 2020; Sahu, Mishra, Lal, & Sahu, 2020; Sahu et al., 2020a, 2020b). For patients with pan-susceptible organisms, this treatment included isoniazid, rifampin, pyrazinamide, and ethambutol (Sahu, Lal, & Mishra, 2020; Sahu, Mishra, & Lal, 2020; Sahu, Mishra, Lal, & Sahu, 2020; Sahu et al., 2020a, 2020b). In patients with multidrug-resistant TB who were found to be resistant to rifampicin through Gene Xpert PCR testing, a combination of bedaquiline, levofloxacin, linezolid, and clofazimine was used in conjunction with pyrazinamide (Rakotosamimanana et al., 2020). Other drugs used to treat COVID-19 in patients with tuberculosis coinfection included hydroxychloroquine, azithromycin, lopinavir/ritonavir, and a combination of darunavir/cobicistat. Glucocorticoids such as methylprednisolone and dexamethasone have also been given to patients with TB and COVID-19 coinfection (Lai et al., 2015). There are also reports of the use of anticoagulants such as enoxaparin and parnaparin (Tadolini et al., 2020). While the tuberculosis treatment regimens in most studies were similar, the therapies used for COVID-19 varied among all patients (Sadanshiv et al., 2018). Most details about the treatment–including the reasons for choosing certain therapies, the doses, the duration of medications, and information about drug interactions, dosage adjustments, and side effects–were not consistently reported for either form of therapy (Mishra et al., 2020).

The management of COVID-19 is similar in patients with both active tuberculosis and COVID-19. The need for steroids and respiratory support is based on measures of oxygenation (such as oxygen saturation and partial pressure of oxygen) and the patient's clinical symptoms (Yang, 2020). If steroids are necessary to treat COVID-19 in patients with active tuberculosis, the dose is usually doubled due to the effect of rifampicin on liver enzymes (van Paassen et al., 2020). Similar considerations may also apply to some antiviral therapies; for example, rifampicin is expected to reduce the effectiveness of remdesivir, though the clinical significance of this reduced exposure is not yet clear. Patients who require mechanical ventilation may need to receive tuberculosis treatment intravenously or through a nasogastric tube. Administration via nasogastric tube may require therapeutic drug monitoring (Esmail et al., 2018). The management of people with rifampicin-resistant active tuberculosis and COVID-19 coinfection is similar to that of people without a coinfection. Given the increasing incidence of fungal infections such as mucormycosis among people who survive COVID-19, which may be related to diabetes and prolonged steroid use, it is important to maintain good glycemic control and avoid the overuse of steroids, especially in people with concurrent active tuberculosis (Ceriello et al., 2020).

In addition to increasing the mortality risk for patients with tuberculosis, COVID-19 has also been found to negatively impact other tuberculosis outcomes, such as treatment failure and loss of follow-up, through several mechanisms (Sahu, Lal, & Mishra, 2020; Sahu, Mishra, & Lal, 2020; Sahu, Mishra, Lal, & Sahu, 2020; Sahu et al., 2020a, 2020b). The risk factors of COVID-19 and tuberculosis coinfection have implications for prioritizing patients for more advanced respiratory support in areas where tuberculosis is prevalent, and given the higher mortality in patients with both active tuberculosis and COVID-19, it supports giving priority to patients with active tuberculosis and tuberculosis survivors for early SARS-CoV-2 vaccination (Mishra et al., 2020). Additionally, testing for tuberculosis should be conducted for all people with COVID-19, and testing for COVID-19 should be conducted for all people with newly diagnosed active tuberculosis when there is still significant community transmission of SARS-CoV-2. Furthermore, people with active tuberculosis and COVID-19 coinfection should be closely monitored with a lower threshold for referral and intervention (Lai et al., 2015).

Conclusion

The COVID-19 pandemic and TB epidemic have had detrimental impacts on and disrupted the lives of people in many countries. Because viral respiratory infections and tuberculosis both weaken the body's immune response, it is likely that the combination of these two conditions will have more severe consequences than they would have individually. As a result, a more robust management strategy is needed due to the risk of delaying TB treatment and worsening the patient's condition. Routine screening for *Mycobacterium tuberculosis* is recommended among suspected or confirmed cases of COVID-19 in countries with a high burden of TB due to the worse prognosis of COVID-19-TB and the confounding clinical symptoms of these two diseases. Patients with TB should be given priority for COVID-19 prevention measures, such as immunization, and TB should be recognized as a risk factor for severe COVID-19 illness.

Mini-dictionary of terms

- **Cavitary lesions:** The presence of hollow spaces or cavities in the lung tissue due to pulmonary damage caused by factors such as infections, inflammation, tumors, etc.
- **Crazy paving pattern:** A radiographic appearance seen on chest x-rays and computed tomography (CT) scans of the lungs, characterized by a network of thickened septa and a hazy appearance of the lung tissue, giving it a "crazy paving" or mosaic-like pattern.
- **Cytokine storm:** A potentially life-threatening immune reaction where the immune system overreacts to an infection by releasing large amounts of proinflammatory cytokines that cause widespread inflammation and tissue damage, leading to organ failure and other complications.
- **Ground-glass opacity:** An abnormal appearance on radiographic images where some areas of the lungs appear hazy and less dense than normal lung tissue while the underlying blood vessels and air spaces can still be visualized.
- **Hilar and mediastinal lymphadenopathy:** The enlargement of lymph nodes located in the hilar and mediastinal spaces. The hilum is the area of the lung where the bronchi, blood vessels, and nerves enter and exit while the mediastinum is the space between the lungs.
- **Latent tuberculosis infection (LTBI):** A condition where an individual is infected with the tubercule bacilli but does not have active tuberculosis disease. However, the active tuberculosis disease may develop at some point in the future if the infection is not treated.
- **Lymphocytopenia:** An abnormally low level of lymphocytes in the blood; these are cells that play a critical role in the immune response.
- **Miliary tuberculosis:** This is when the tubercule bacilli spread through the bloodstream and infect multiple organs and tissues in the body. The name "miliary" refers to the small size of the lesions that develop in the affected organs, as they resemble millet seeds.
- **Mucormycosis:** A serious fungal infection caused by a group of molds called *Mucormycetes*, which can infect the sinuses, lungs, skin, and brain, and can be life-threatening in people with weakened immune systems.
- **Multidrug-resistant tuberculosis (MDR-TB):** A condition where the tubercule bacilli become resistant to at least two of the most powerful first-line TB drugs, isoniazid and rifampicin. This makes treatment more difficult and prolonged, and increases the risk of the disease being spread to others.
- **Parenchymal consolidations:** An abnormal condition where the air spaces in the lungs become filled with fluid or other substances such as pus or blood, resulting in a denser appearance on radiographic images.

- **Pleural effusion:** Accumulation of an excessive amount of fluid in the pleural space, which is the space between the two layers of tissue (pleura) that line the outside of the lungs and the inside of the chest cavity.
- **Pulmonary infiltrates:** The abnormal accumulation of fluid, cells, and other substances in the lung tissue, often visualized as areas of increased density or opacification in the lungs.
- **Sequelae:** The long-term or permanent complications that can arise as a result of an illness, injury, or medical treatment. These complications can occur even after the primary condition has resolved or been treated.
- **"Tree in bud" appearance:** A radiological finding that appears as small branching tubular structures surrounded by small nodules, giving the impression of a tree with buds on its branches. This is caused by the filling of the small airways in the lungs with fluid or mucus and the resulting inflammation.

Summary points

- Individuals with TB and COVID-19 coinfection present with more severe symptoms and complications than in singular infections with either disease.
- The symptoms and radiological manifestations of TB and COVID-19 often overlap, resulting in misdiagnosis or delayed diagnosis of a case of coinfection.
- Delays in diagnosis result in further lung parenchymal damage, worse prognosis, and higher mortality rates.
- The case fatality rates for patients with TB and COVID-19 coinfection (11.6–27.3%) are higher than those for COVID-19 alone (0.0016–7.8%).
- COVID-19 and TB coinfection patients also have a higher rate of comorbidities (56.41%) than COVID-19 patients alone (25.1%).
- In patients with coinfection, 18% had TB that was resistant to a single anti-TB drug, and 9% had multidrug-resistant TB (MDR-TB).
- Radiological manifestations in COVID-19 and TB coinfection include thick-walled cavitary lesions, numerous nodular opacities, diffuse areas of consolidation, bilateral ground-glass opacities, pulmonary infiltrates, and a "tree in bud" appearance.
- Laboratory findings from TB and COVID-19 coinfection include lymphocytopenia, low IFN-γ levels from QuantiFERON-tests, and elevated levels of lactate dehydrogenase (LDH), C-reactive protein (CRP), erythrocyte sedimentation rate (ESR), and D-dimer.
- COVID-19 and TB coinfection patients are managed with steroids and respiratory support, with the administration of isoniazid and rifampin for TB treatment, and hydroxychloroquine, lopinavir/ritonavir, and darunavir/cobicistat for COVID-19 treatment.
- Coinfection patients with multidrug-resistant TB are administered a combination of bedaquiline, levofloxacin, linezolid, and clofazimine together with pyrazinamide.

References

Adzic-Vukicevic, T., Stosic, M., Antonijevic, G., Jevtic, M., Radovanovic-Spurnic, A., & Velickovic, J. (2022). Tuberculosis and COVID-19 co-infection in Serbia: Pandemic challenge in a low-burden country. *Frontiers in Medicine*, *9*, 971008.

Al Lawati, R., Al Busaidi, N., Al Umairi, R., Al Busaidy, M., Al Naabi, H. H., & Khamis, F. (2021). COVID-19 and pulmonary mycobacterium tuberculosis co-infection. *Oman Medical Journal*, *36*(5), e298.

Alfaraj, S. H., Al-Tawfiq, J. A., Altuwaijri, T. A., & Memish, Z. A. (2017). Middle East respiratory syndrome coronavirus and pulmonary tuberculosis co-infection: Implications for infection control. *Intervirology*, *60*(1–2), 53–55.

Auld, S. C., Lee, S. H., Click, E. S., Miramontes, R., Day, C. L., Gandhi, N. R., & Heilig, C. M. (2016). IFN-γ release assay result is associated with disease site and death in active tuberculosis. *Annals of the American Thoracic Society*, *13*(12), 2151–2158.

Biswas, S. S., Awal, S. S., & Awal, S. K. (2021). COVID-19 and pulmonary tuberculosis—A diagnostic dilemma. *Radiology Case Reports*, *16*(11), 3255–3259.

Boulle, A., Davies, M. A., Hussey, H., Ismail, M., Morden, E., Vundle, Z., Zweigenthal, V., Mahomed, H., Paleker, M., Pienaar, D., Tembo, Y., Lawrence, C., Isaacs, W., Mathema, H., Allen, D., Allie, T., Bam, J. L., Buddiga, K., Dane, P., … Tamuhla, T. (2021). Risk factors for Coronavirus Disease 2019 (COVID-19) death in a population cohort study from the Western Cape Province, South Africa. *Clinical Infectious Diseases*, *73*(7), E2005–E2015.

Ceriello, A., De Nigris, V., & Prattichizzo, F. (2020). Why is hyperglycemia worsening COVID-19 and its prognosis? *Diabetes, Obesity & Metabolism*, *22*(10), 1951–1952.

Chen, Y. M. A. (2008). The 2003 SARS outbreaks in Taiwan. *Emerging Infections in Asia*, 117–129.

Chopra, K. K., Arora, V. K., & Singh, S. (2020). COVID 19 and tuberculosis. *The Indian Journal of Tuberculosis*, *67*(2), 149–151.

Crisan-Dabija, R., Grigorescu, C., Pavel, C. A., Artene, B., Popa, I. V., Cernomaz, A., & Burlacu, A. (2020). Tuberculosis and COVID-19: Lessons from the past viral outbreaks and possible future outcomes. *Canadian Respiratory Journal*, *2020*, 1401053.

Dheda, K., Jaumdally, S., Davids, M., Chang, J. W., Gina, P., Pooran, A., Makambwa, E., Esmail, A., Vardas, E., & Preiser, W. (2020). Diagnosis of COVID-19: Considerations, controversies and challenges. *African Journal of Thoracic Critical Care Medicine*, *26*(2), 10.

Esmail, A., Sabur, N. F., Okpechi, I., & Dheda, K. (2018). Management of drug-resistant tuberculosis in special sub-populations including those with HIV co-infection, pregnancy, diabetes, organ-specific dysfunction, and in the critically ill. *Journal of Thoracic Disease*, *10*(5), 3102–3118.

Falasca, L., Nardacci, R., Colombo, D., Lalle, E., Di Caro, A., Nicastri, E., et al. (2020). Postmortem findings in Italian patients with COVID-19: A descriptive full autopsy study of cases with and without comorbidities. *Journal of Infectious Diseases*, *222*, 1807–1815.

Fu, J., Kong, J., Wang, W., Wu, M., Yao, L., Wang, Z., Jin, J., Wu, D., & Yu, X. (2020). The clinical implication of dynamic neutrophil to lymphocyte ratio and D-dimer in COVID-19: A retrospective study in Suzhou China. *Thrombosis Research*, *192*, 3–8.

Gordon, D. E., Jang, G. M., Bouhaddou, M., Xu, J., Obernier, K., White, K. M., O'Meara, M. J., Rezelj, V. V., Guo, J. Z., Swaney, D. L., Tummino, T. A., Hüttenhain, R., Kaake, R. M., Richards, A. L., Tutuncuoglu, B., Foussard, H., Batra, J., Haas, K., Modak, M., … Krogan, N.J. (2020). A SARS-CoV-2 protein interaction map reveals targets for drug repurposing. *Nature*, *583*(7816), 459–468.

Goudiaby, M. S., Gning, L. D., Diagne, M. L., Dia, B. M., Rwezaura, H., & Tchuenche, J. M. (2022). Optimal control analysis of a COVID-19 and tuberculosis co-dynamics model. *Informatics in Medicine Unlocked*, *28*, 100849.

Goyal, P., Choi, J. J., Pinheiro, L. C., Schenck, E. J., Chen, R., Jabri, A., Satlin, M. J., Campion, T. R., Jr., Nahid, M., Ringel, J. B., & Hoffman, K. L. (2020). Clinical characteristics of Covid-19 in New York City. *New England Journal of Medicine*, *382*(24), 2372–2374.

Guan, W. J., Liang, W. H., Zhao, Y., Liang, H. R., Chen, Z. S., Li, Y. M., Liu, X. Q., Chen, R. C., Tang, C. L., & Wang, T. (2020). China Medical Treatment Expert Group for COVID-19. Comorbidity and its impact on 1590 patients with COVID-19 in China: A nationwide analysis. *The European Respiratory Journal*, *55*(5), 2000547.

Gupta, N., Ish, P., Gupta, A., Malhotra, N., Caminero, J. A., Singla, R., Kumar, R., Yadav, S. R., Dev, N., Agrawal, S., Kohli, S., Sen, M. K., Chakrabarti, S., & Gupta, N. K. (2020). A profile of a retrospective cohort of 22 patients with COVID-19 and active/treated tuberculosis. *The European Respiratory Journal*, *56*(5), 2003408.

He, G., Wu, J., Shi, J., Dai, J., Gamber, M., Jiang, X., Sun, W., & Cai, J. (2020). COVID-19 in tuberculosis patients: A report of three cases. *Journal of Medical Virology*, *92*(10), 1802–1806.

Kang, S. J., & Jung, S. I. (2020). Age-related morbidity and mortality among patients with COVID-19. *Infection and Chemotherapy*, *52*(2), 154–164.

Khurana, A., & Aggarwal, D. (2020). The (in) significance of TB and COVID-19 co-infection. *The European Respiratory Journal*, *56*(2), 1–4.

Kumar, D.R., Bhattacharya, D.B., Meena, D.V., Soneja, D.M., Wig, D.N. (2020), COVID-19 and TB co-infection—Finishing touch in perfect recipe to 'severity' or 'death'. Journal of Infectious Diseases, 81(3), e39-e40.

Kwee, T. C., & Kwee, R. M. (2020). Chest CT in COVID-19: What the radiologist needs to know. *Radiographics*, *40*(7), 1848–1865.

Lai, C. C., Lee, M. T., Lee, S. H., Lee, S. H., Chang, S. S., & Lee, C. C. (2015). Risk of incident active tuberculosis and use of corticosteroids. *The International Journal of Tuberculosis and Lung Disease*, *19*(8), 936–942.

Lal, A., Mishra, A. K., & Sahu, K. K. (2020). CT chest findings in coronavirus disease-19 (COVID-19). *Journal of the Formosan Medical Association*, *119*(5), 1000–1001.

Lal, A., Mishra, A. K., Sahu, K. K., & Abraham, G. M. (2020). The return of Koch's: Ineffective treatment or re-infection. *Enfermedades Infecciosas y Microbiología Clínica*, *38*(3), 144–145.

Liu, W., Fontanet, A., Zhang, P. H., Zhan, L., Xin, Z. T., Tang, F., Baril, L., & Cao, W. C. (2006). Pulmonary tuberculosis and SARS, China. *Emerging Infectious Diseases*, *12*(4), 707–709.

Liu, S. L., Wang, S. Y., Sun, Y. F., Jia, Q. Y., Yang, C. L., Cai, P. J., Li, J. Y., Wang, L., & Chen, Y. (2020). Expressions of SAA, CRP, and FERR in different severities of COVID-19. *European Review in Medical Pharmacological Science*, *24*(21), 11386–11394.

Mishra, A., Sahu, K., George, A., & Lal, A. (2020). A review of cardiac manifestations and predictors of outcome in patients with COVID–19. *Heart & Lung*, *49*(6), 848–852.

Motta, I., Centis, R., D'Ambrosio, L., García-García, J. M., Goletti, D., Gualano, G., Lipani, F., & Palmieri, F. (2020). Tuberculosis, COVID-19 and migrants: Preliminary analysis of deaths occurring in 69 patients from two cohorts. *Pulmonology*, *26*(4), 233–240.

Mousquer, G. T., Peres, A., & Fiegenbaum, M. (2021). Pathology of TB/COVID-19 co-infection: The phantom menace. *Tuberculosis*, *126*, 102020.

Musso, M., Di Gennaro, F., Gualano, G., Mosti, S., Cerva, C., Fard, S. N., Libertone, R., Di Bari, V., Cristofaro, M., Tonnarini, R., Castilletti, C., Goletti, D., & Palmieri, F. (2021). Concurrent cavitary pulmonary tuberculosis and COVID-19 pneumonia with in vitro immune cell anergy. *Infection*, *49*(5), 1061–1064.

Nachiappan, A. C., Rahbar, K., Shi, X., Guy, E. S., Mortani Barbosa, E. J., Jr., Shroff, G. S., Ocazionez, D., Schlesinger, A. E., Katz, S. I., & Hammer, M. M. (2017). Pulmonary tuberculosis: Role of radiology in diagnosis and management. *Radiographics*, *37*(1), 52–72.

Niazkar, H. R., Zibaee, B., Razavi, S. B., Ghanaeian, K., Talebzadeh, V., & Vosugh, N. H. (2022). Evaluation of tuberculosis infection in COVID-19 patients: A case of tuberculosis and COVID-19 co-infection. *The Egyptian Journal of Internal Medicine*, *34*(1), 46.

Nyanti, L. E., Wong, Z. H., Sachdev Manjit Singh, B., Chang, A. K. W., Jobli, A. T., & Chua, H. H. (2022). Pulmonary Tuberculosis and COVID-19 co-infection: Hickam's Dictum revisited. *Respiratory Medicine Case Reports*, *37*, 101653.

Osejo-Betancourt, M., Molina-Paez, S., & Rubio-Romero, M. (2021). Pulmonary tuberculosis and COVID-19 co-infection: A new medical challenge. *Monaldi Archives for Chest Disease*, *92*(3).

Osuchowski, M. F., Winkler, M. S., Skirecki, T., Cajander, S., Shankar-Hari, M., Lachmann, G., Monneret, G., Venet, F., Bauer, M., Brunkhorst, F. M., Weis, S., Garcia-Salido, A., Kox, M., Cavaillon, J. M., Uhle, F., Weigand, M. A., Flohé, S. B., Wiersinga, W. J., Almansa, R., … Rubio, I. (2021). The COVID-19 puzzle: Deciphering pathophysiology and phenotypes of a new disease entity. *The Lancet Respiratory Medicine*, *9*(6), 622–642.

Promislow, D. E. L. (2020). A geroscience perspective on COVID-19 mortality. *Journal of Gerontology and Biological Sciences and Medical Sciences*, *75*(9), e30–e33.

Prompetchara, E., Ketloy, C., & Palaga, T. (2020). Immune responses in COVID-19 and potential vaccines: Lessons learned from SARS and MERS epidemic. *Asian Pacific Journal of Allergy and Immunology*, *38*(1), 1–9.

Rakotosamimanana, N., Randrianirina, F., Randremanana, R., Raherison, M. S., Rasolofo, V., Solofomalala, G. D., Spiegel, A., & Heraud, J. M. (2020). GeneXpert for the diagnosis of COVID-19 in LMICs. *The Lancet Global Health*, *8*(12), e1457–e1458.

Sadanshiv, M., George, A. A., Mishra, A. K., & Kuriakose, C. K. (2018). Rifampicin-induced immune allergic reaction. *Tropical Doctor*, *48*(2), 156–159.

Sahu, K. K., Lal, A., & Mishra, A. K. (2020). An update on CT chest findings in coronavirus disease-19 (COVID-19). *Heart & Lung*, *49*(5), 442–443.

Sahu, K. K., Mishra, A. K., & Lal, A. (2020). Trajectory of the COVID-19 pandemic: Chasing a moving target. *Annals of Translational Medicine*, *8*(11), 694.

Sahu, K. K., Mishra, A. K., Lal, A., & Sahu, S. A. (2020). India fights back: COVID-19 pandemic. *Heart & Lung*, *49*(5), 446–448.

Sahu, K. K., Mishra, A. K., Martin, K., & Chastain, I. (2020a). COVID-19 and restrictive lung disease: A deadly combo to trip off the fine balance. *Monaldi Archives of Chest Diseases*, *90*(2). https://doi.org/10.4081/monaldi.2020.1346.

Sahu, K. K., Mishra, A. K., Martin, K., & Chastain, I. (2020b). COVID-19 and clinical mimics. Correct diagnosis is the key to appropriate therapy. *Monaldi Archives for Chest Disease*, *90*(2). https://doi.org/10.4081/monaldi.2020.1327.

Song, W. M., Zhao, J. Y., Zhang, Q. Y., Liu, S. Q., Zhu, X. H., An, Q. Q., Xu, T. T., Li, S. J., Liu, J. Y., Tao, N. N., Liu, Y., Li, Y. F., & Li, H. C. (2021). COVID-19 and tuberculosis co-infection: An overview of case reports/case series and meta-analysis. *Frontiers in Medicine*, *8*, 657006.

Stochino, C., Villa, S., Zucchi, P., Parravicini, P., Gori, A., & Raviglione, M. C. (2020). Clinical characteristics of COVID-19 and active tuberculosis co-infection in an Italian reference hospital. *The European Respiratory Journal*, *56*(1), 2001708.

Sun, Y., Dong, Y., Wang, L., Xie, H., Li, B., Chang, C., & Wang, F. S. (2020). Characteristics and prognostic factors of disease severity in patients with COVID-19: The Beijing experience. *Journal of Autoimmunity*, *112*, 102473.

Sy, K. T. L., Haw, N. J. L., & Uy, J. (2020). Previous and active tuberculosis increases risk of death and prolongs recovery in patients with COVID-19. *Infectious Diseases*, *52*(12), 902–907.

Tadolini, M., Codecasa, L. R., & García-García, J. M. (2020). Active tuberculosis, sequelae and COVID-19 co-infection: First cohort of 49 cases. *The European Respiratory Journal*, *56*(1), 2001398.

Tham, S. M., Lim, W. Y., Lee, C. K., Loh, J., Premkumar, A., Yan, B., Kee, A., Chai, L., Tambyah, P. A., & Yan, G. (2020). Four patients with COVID-19 and tuberculosis, Singapore, April–May 2020. *Emerging Infectious Diseases*, *26*(11), 2764–2766.

van Paassen, J., Vos, J. S., Hoekstra, E. M., Neumann, K. M. I., Boot, P. C., & Arbous, S. M. (2020). Corticosteroid use in COVID-19 patients: A systematic review and meta-analysis on clinical outcomes. *Critical Care*, *24*(1), 696.

Verity, R., Okell, L. C., Dorigatti, I., Winskill, P., Whittaker, C., Imai, N., Cuomo-Dannenburg, G., Thompson, H., Walker, P. G. T., Fu, H., Dighe, A., Griffin, J. T., Baguelin, M., Bhatia, S., & Boonyasiri, A. (2020). Estimates of the severity of coronavirus disease 2019: A model-based analysis. *Lancet Infectious Diseases*, *20*(6), 669–677.

World Health Organization. (2023). *WHO Coronavirus (COVID-19) Dashboard.* Available at: https://covid19.who.int/. (Accessed 3 January 2023).

Yang, K. (2020). What do we know about remdesivir drug interactions? *Clinical and Translational Science*, *13*(5), 842–844.

Zheng, Z., Peng, F., Xu, B., Zhao, J., Liu, H., Peng, J., Li, Q., Jiang, C., Zhou, Y., Liu, S., Ye, C., Zhang, P., Xing, Y., Guo, H., & Tang, W. (2020). Risk factors of critical & mortal COVID-19 cases: A systematic literature review and meta-analysis. *Journal of Infectious Diseases*, *81*(2), e16–e25.

Zhou, F., Yu, T., Du, R., Fan, G., Liu, Y., Liu, Z., Xiang, J., Wang, Y., Song, B., Gu, X., Guan, L., Wei, Y., Li, H., Wu, X., Xu, J., Tu, S., Zhang, Y., Chen, H., & Cao, B. (2020). Clinical course and risk factors for mortality of adult inpatients with COVID-19 in Wuhan, China: A retrospective cohort study. *Lancet*, *395*(10229), 1054–1062.

Chapter 40

Patients with autoimmune liver disease and the impact of Sars-COV-2 infection

Annarosa Floreani[a,b], Sara De Martin[c], and Nora Cazzagon[b]
[a]*Scientific Institute for Research, Hospitalization and Healthcare, Verona, Italy,* [b]*Department of Surgery, Oncology and Gastroenterology, University of Padova, Padova, Italy,* [c]*Department of Pharmaceutical and Pharmacological Sciences, University of Padova, Padova, Italy*

Abbreviations

AASLD	American Association for the Study of Liver Diseases
AIH	autoimmune hepatitis
AILD	autoimmune liver diseases
AISF	Italian Association for the Study of the Liver
ALI	acute liver injury
APASL	Asian-Pacific Association for the Study of the Liver
CLD	chronic liver disease
EASL	European Association for the Study of the Liver
ERN-RARE LIVER	European Reference Network on Rare Hepatological Diseases
ESMID	European Society of Clinical Microbiology and Infectious Diseases
MAFLD	metabolic dysfunction-associated fatty liver
MMF	mycophenolate mofetil
PBC	primary biliary cholangitis
PSC	primary sclerosing cholangitis
SARS-CoV-2	severe acute respiratory syndrome coronavirus-2

Introduction

In late 2019, severe acute respiratory syndrome coronavirus 2 (SARS-CoV-2) emerged in China as a serious threat to public health. Since then, SARS-CoV-2 has become a devastating pandemic that overwhelmed healthcare systems around the world, resulting in 636,440,663 confirmed cases confirmed cases of COVID-19, including 6,606,624 deaths, reported to the World Health Organization (WHO) as of Nov. 23, 2022 (WHO COVID-19 dashboard, https://covid19.who.int/). Since COVID-19's first descriptions, it has emerged that it is not only a respiratory disease, but it can also involve the gastrointestinal tract and the liver. Enteric symptoms are present in up to 20% of patients; these can also precede the development of respiratory symptoms (Guan et al., 2020) and are mostly represented by anorexia, nausea, vomiting, and diarrhea (Perisetti et al., 2020). The most common effect of COVID-19 on the liver is hepatocellular injury, with mild elevations of serum transaminases with occasionally more prominent elevations (Perisetti et al., 2020). In their first systematic description of liver damage in patients hospitalized for COVID-19, Cai et al. observed that around 77% of patients had abnormal liver test results and 21.5% of patients had liver injury while hospitalized; liver injury was predictive of progression to severe pneumonia (Cai et al., 2020). Since then, several reports have explored the presence and type of liver damage in patients with COVID-19 and the possible pathogenic mechanisms (Payus et al., 2022). Notably, since the beginning of the pandemic, several physicians and scientists around the world evaluated the impact of SARS-CoV-2 on patients with preexisting liver diseases (Singh & Khan, 2020), including autoimmune liver diseases (AILD). These diseases, namely autoimmune hepatitis (AIH), primary biliary cholangitis (PBC), and primary sclerosing cholangitis (PSC), are chronic immune mediated liver diseases that primarily affect hepatocytes (AIH) or cholangiocytes (PBC and PSC) and that can progress to cirrhosis and its complications. The aim of this review is to review the current evidence regarding the complex relationship between SARS-COV-2 infections and AILD. More specifically, this review hopes to address the following clinical questions:

Management, Body Systems, and Case Studies in COVID-19. https://doi.org/10.1016/B978-0-443-18703-2.00055-4
Copyright © 2024 Elsevier Inc. All rights are reserved, including those for text and data mining, AI training, and similar technologies.

I. What is the epidemiology of SARS-CoV-2 infection in patients with AILD?
II. What is the course of COVID-19 in patients with AILDs and what is the impact of COVID-19 on liver function in patients with AILDs?
III. What is the efficacy and safety of COVID-19 vaccines in patients with previous AILD?
IV. Is there an increased risk of AILD development after SARS-COV-2 infection?
V. Is there an increased risk of AILD development after COVID-19 vaccination?

What is the epidemiology of SARS-CoV-2 infection in patients with AILD?

Different methodologies have been applied to evaluate the epidemiology of SARS-CoV-2 infection in patients with AILD.

Registry studies

A retrospective observational study was conducted in Barcelona from March 1– May 29, 2020, to describe a multidisciplinary cohort of patients with autoimmune and immune-mediated inflammatory diseases (AI/IMID) and symptomatic SARS-CoV-2 infection, focusing on sociodemographic, clinical, and therapeutic factors associated with a poor prognosis (Sarmiento-Monroy et al., 2021). The study was based on a local SARS-CoV-2 infection database and included a total of 175 patients (58 with AI/IMID and 117 controls matched by age and gender). The prevalence of symptomatic SARS-CoV-2 infection was recorded in 1.3% of patients with AI/IMID. In logistic regression analysis, patients with AI/IMID had a lower risk of developing severe COVID-19 pneumonia, including the need to stay in the intensive care unit (ICU) and mechanical ventilation. Moreover, no differences in mortality were found between the groups. Several factors have been hypothesized to explain the better outcome of patients with autoimmunity compared to controls, including: (i) the potential beneficial effect of chronic immunosuppression; (ii) the possible immunomodulatory effect of vitamin D as part of the medical therapy for patients with autoimmunity; and (iii) a better baseline vaccination status (Sarmiento-Monroy et al., 2021).

Online survey

An online survey was available on the EUSurvey platform (supported by the European Commission, ec.eurpa.eu/eusurvey) in nine languages from June 24–Oct. 14, 2020. The survey was supported by patient representatives of the European Reference Network on Hepatological Diseases (Zecher et al., 2021). The survey included 1779 participants with AILD, of whom 34.9% had PBC, 34.7% had AIH, 20.2% had PSC, 6.3% had PBC-AIH variant syndrome, and 3.8% had PSC-AIH variant syndrome. One-third of patients (29%) had liver cirrhosis and 7% had received previous liver transplantation. Overall, 2.2% of participants had COVID-19, a percentage comparable to the general population (Zecher et al., 2021).

These results suggest that patients with autoimmune liver diseases (AILD) are not at increased risk of SARS-CoV-2 infection compared to the general population.

Does SARS-COV-2 infection affect clinical outcomes of AILDs?

Patients with AILD hospitalized for severe COVID-19 infection did not have an increased risk of adverse outcomes (intensive care unit admission, intubation, or death) compared to age- and sex-matched controls from New York City (Richardson et al., 2020). Interestingly, a nationwide cohort study of all Swedish adults with chronic liver disease (CLD) ($n = 42{,}320$ adults including $n = 2{,}549$ with cirrhosis) was matched with up to five randomly selected general population comparators derived from the total population register (Simon et al., 2021). Between Feb. 1–July 31, 2020, 161 (0.38%) chronic liver disease patients and 435 (0.21%) controls were hospitalized for COVID-19. Although a higher risk of hospitalization in CLD patients compared to the general population was observed, patients with CLD did not have an increased risk of developing severe COVID-19 nor did they have increased mortality. The results were similar in patients with alcoholic liver disease, metabolic dysfunction-associated steatotic liver disease, viral hepatitis, autoimmune hepatitis, and other aetiologies (Simon et al., 2021).

The first case series of 10 Italian patients with AIH under immunosuppressive therapy was published during the first outbreak of SARS-CoV-2 (Gerussi et al., 2020). Eight patients were on biochemical remission at the time of SARS-CoV-2 infection. All subjects had a respiratory syndrome; in seven patients, the dosage of immunosuppressive medication was reduced. In general, the clinical outcome was comparable to the reported cases occurring in nonimmunosuppressed patients (Gerussi et al., 2020).

Interestingly, data from patients with AIH and SARS-CoV-2 were combined from three international reporting registries: the R-Liver COVID-19 registry (coordinated by the European Reference Network on Rare Hepatological Disease [ERN-RARE-LIVER]), the SECURE-cirrhosis registry (coordinated by the University of North Carolina, United States) and the COVID-Hep.net registry (coordinated by the University of Oxford) (Marjot et al., 2021). Data from 70 patients with AIH were compared to data from patients with CLD of different aetiologies (non-AIH CLD) and to patients without CLD (non-CLD). Of patients with AIH, 83% were taking immunosuppressive therapy. There were no differences in rates of major outcomes (i.e., hospitalization, ICU admission, and death) between patients with AIH and non-AIH CLD. Factors associated with death within the AIH cohort included age and advanced liver disease (child B and C), but not immunosuppression. Moreover, patients with AIH compared to non-CLD patients demonstrated an increased risk of hospitalization but an equivalent risk of all other outcomes (Marjot et al., 2021). These observations clearly underline that patients with AIH do not have a higher risk of adverse outcomes, despite being on immunosuppressive treatment, compared to other causes of CLD and to matched cases without liver disease.

An extensive review of the impact of COVID-19 on preexisting liver disease, including AILD, has been recently published (Sharma, Patnaik, et al., 2021b). Although patients with metabolic dysfunction-associated fatty liver (MAFLD) had shown a 4–6-fold increase in the severity of COVID-19, this was not the case for patients with AILD. However, it has been shown that cirrhosis is an independent predictor of severity for COVID-19, causing increased hospitalization and mortality (Sharma, Kumar, et al., 2021a).

The effect of immunosuppressive therapy on viral infection has been evaluated in detail. In fact, it has been hypothesized that the iatrogenic effect of immunosuppressive therapy can complicate the outcome of SARS-CoV-2 infection (Favalli et al., 2020). A large international retrospective study carried out between March 11, 2020, and May 15, 2021, examined the impact of AIH medications, including glucocorticoids, thiopurines, mycophenolate mofetil (MMF), and tacrolimus on the risk of worse COVID-19 severity (Efe et al., 2022a). A total of 254 AIH patients with a median age of 50 years (range 17–85) were included, 92.1% of whom were on treatment with glucocorticoids ($n = 156$), thiopurines ($n = 151$), MMF ($n = 22$), or tacrolimus ($n = 156$), alone or in combination. Overall, 94 patients (37%) were hospitalized and 7.1% died. Baseline treatment with systemic glucocorticoids or thiopurines prior to the onset of COVID-19 was significantly associated with COVID-19 severity (Efe et al., 2022a, 2022b). However, more than half the patients in both groups had cirrhosis, which is a very well-established risk factor for adverse events (APASL Covid-19 Task Force et al., 2020).

COVID-19 outcomes, with a mortality rate as high as 34%. In an interesting editorial, Terziroli Beretta-Piccoli and Lleo underlined that risk factors for severe COVID-19 in AIH largely overlap with those reported in non-AIH hepatic and extrahepatic conditions, so whether chronic corticosteroid and thiopurine use is an additional risk factor remains to be proven (Terziroli Beretta-Piccoli & Lleo, 2022). On the other hand, dose reduction or discontinuation of immunosuppressive drugs may expose patients with AIH to relapse and does not appear to be of any benefit.

How has the management of AILD changed during the pandemic?

In the early period of the pandemic, when data on outcomes of CLD patients were still lacking, a consensus statement from an expert panel of the American Association for the Study of Liver Diseases (AASLD) was published to provide clinical recommendations and policies to mitigate the impact of the pandemic on liver patients and healthcare providers (Fix et al., 2020). In immunosuppressed liver disease patients with COVID-19, the following recommendations were originally stated:

- Reduce high-dose prednisone, but maintain a sufficient dosage to avoid adrenal insufficiency.
- Consider reducing azathioprine or MMF dosage, especially in with lymphopenia, fever, or worsening pneumonia attributed to COVID-19.
- Consider reducing, but not stopping, daily calcineurin inhibitor dosage, especially in lymphopenia, fever, or worsening pulmonary status.
- In AILD patients with an indication for immunosuppressive therapy (AIH, graft rejection), caution should be taken in initiating steroids.
- In general, for liver disease patients, primary prophylaxis with beta-blockers instead of a screening endoscopy should be considered.

Similarly, the European Association for the Study of the Liver (EASL) and the European Society of Clinical Microbiology and Infectious Diseases (ESCMID) published a position paper after 6 months of the pandemic on the care of patients with liver disease (Boettler et al., 2020). In this paper, the authors advised against reducing immunosuppressive therapy in patients with AILD to prevent

SARS-CoV-2 infection. Reductions should only be considered under special circumstances (e.g., medication-induced lymphopenia, or bacterial/fungi superinfection in case of severe COVID-19) after consultation with a specialist. They also recommend considering budesonide as a first-line agent to induce remission in patients without cirrhosis and a flare of AIH to minimize systemic glucocorticoid exposure. Moreover, in patients treated with corticosteroids who develop COVID-19, corticosteroid dosing should be maintained at a sufficient dose to prevent adrenal insufficiency. The addition of, or conversion to, dexamethasone should only be considered in patients with COVID-19 who require hospitalization and respiratory support. Finally, all patients should receive vaccination for *Streptococcus pneumoniae* and *influenza* (Boettler et al., 2020).

The Asian-Pacific Association for the Study of the Liver (APASL) formulated clinical practice guidance during COVID-19. Recommendations for patients with AILD were similar to the AASLD and EASL-ESMID societies. This panel of experts also recommended disease control and follow-up maintained through altered pathways, including telehealth/telephone consultation and local blood testing.

The most recent recommendations from EASL concluded that patients with AIH on immunosuppression do not appear to be at a higher risk of SARS-CoV-2 infection or COVID-19-related mortality. However, the baseline use of glucocorticoids or azathioprine may be associated with more severe COVID-19; however, discontinuing or reducing the dose of these agents should only occur following a careful assessment of all risks and benefits (Marjot et al., 2022).

Finally, we suggest that evidence regarding the management of immunosuppressive treatment in AIH patients with COVID-19 following COVID-19 vaccinations is still missing, although we hope to have it in the near future.

What is the efficacy and safety of COVID-19 vaccines in patients with previous AILD?

The efficacy of COVID-19 vaccinations in patients with autoimmune liver diseases was assessed in a prospective observational study that compared the humoral and T-cellular immune responses to SARS-CoV-2 vaccination in such patients (Duengelhoef et al., 2022). In this study, 103 patients with AIH and 125 patients with PBC or PSC were included. In almost all patients, a humoral vaccination response was detected. Notably, the levels of anti-COVID-19 antibodies were lower in AIH patients under immunosuppression than in those without immunosuppression. Antibody titers significantly declined within 7 months after the second dose of vaccination. In the autoimmune assay of 20 AIH patients, despite a positive serology, a spike-specific T-cell response was undetectable while 85% of the PBC/PSC patients demonstrated a spike-specific T-cell response (Duengelhoef et al., 2022). Consequently, patients with AIH may have an increased risk of being infected by the SARS-CoV-2 virus while patients with PBC or PSC might be more protected from the infection.

Subsequently, the same group conducted a further study to assess the effect of the third COVID-19 vaccination in 83 patients with AIH (Hartl et al., 2022). The authors showed that median antibody titers were significantly lower in AIH compared to healthy controls, and this observation was particularly evident in AIH patients treated with MMF or steroids. Actually, AIH patients had antibody titers below the 10% percentile of the healthy controls. Moreover, in accordance with their previous observations (Duengelhoef et al., 2022), patients with AIH after the third vaccination remained at high risk of failing to develop a spike-specific T-cell response (44% vs. 12% in healthy controls) and showed overall lower frequencies of spike-specific CD4+T-cells compared to healthy individuals (Hartl et al., 2022). However, an analysis conducted in a subgroup of patients with available antibody titers before and after booster vaccination showed a strong, 148-fold increase in antibody titers, especially in those without detectable/low antibody titers (< 100 AU/mL) after the second vaccination (Hartl et al., 2022). This means that a third COVID-19 vaccination efficiently boosts antibody levels and T-cell responses in AIH patients and even seroconversion in patients with absent immune response after two vaccinations, but to a lower level compared to controls. This suggests the need to routinely assess antibody levels in AIH patients and offer additional booster vaccinations to those with a suboptimal response. Similarly, in 1073 patients followed in rheumatology and clinical immunology as well as liver disease outpatient clinics, of whom 16% of patients affected by AILD were under immunosuppressants, the use of MMF significantly reduced both the rate of seroconversion in COVID-19-naïve patients and anti-SARS-CoV-2 antibody titers in seroconverted patients (De Santis et al., 2022).

Safety data on anti-COVID-19 vaccinations in patients with autoimmune diseases have been collected both during the clinical trial and by real-life observation after their approval. The vaccines receiving approval include the mRNA vaccine BNT 162b2 produced by Pfizer-BioNTech and the mRNA-1273 vaccine by Moderna as well as the adenovirus vaccine vector produced by Oxford University/Astra Zeneca, that is, the ChAdOx1 nCOV-19 vaccine (AZD 1222). During 2021, sporadic cases of thrombotic events in young women following vaccination with AZD 1222 were reported, particularly of the rare condition of cavernous sinus thrombosis. This is of particular interest for patients diagnosed with autoimmune diseases because some studies have documented a pathogenic mechanism related to the antibodies against the SARS-CoV-2 spike protein, possibly able to cross-react with platelet factor 4 (PF4/CXLC4). This mechanism resembles that of autoimmune heparin-induced thrombocytopenia. Furthermore, anecdotal cases of different types of autoimmunity triggered by vaccines have been reported, including a subacute cutaneous lupus erythematosus induction, observed in a patient with PBC 10 days after receiving the second dose of the SARS-CoV-2 mRNA Pfizer-BioNTech vaccine (Zengarini et al., 2022).

In summary, despite the scarce data regarding the safety and efficacy of vaccines against COVID-19 in chronic autoimmune liver diseases, no significant concerns regarding the safety of these vaccines in patients with AILD have emerged. For this reason, they are recommended. Presently, clinical trials are ongoing to measure the immunogenicity and safety of the available COVID-19 vaccines in patients with chronic liver diseases, and registries, such as SECURE-cirrhosis and COVID-Hep, are also constantly yielding such data on a large scale (Sharma, Patnaik, et al., 2021b).

The Italian Association for the Study of the Liver (AISF) recently published a clinical update on the risks and efficacy of anti-SARS-CoV-2 vaccines in patients with AIH (Lleo et al., 2022). According to these recommendations, patients with AIH should receive anti-SARS-CoV-2 vaccination without preference for one vaccine over another. Patients with AIH are suggested to undergo vaccination when the disease activity is controlled by immunosuppressive therapy, and routine testing of transaminase levels after vaccination could be suggested in selected patients, although the timing for performing liver enzymes needs to be defined (Lleo et al., 2022).

Is there evidence for an increased incidence of AILDs as a consequence of SARS-CoV-2 infection?

The relationship between autoimmunity and COVID-19 is complex (Knight et al., 2021). In fact, hyperinflammation and macrophage activation observed in patients with COVID-19 can resemble the immunopathology of various autoimmune diseases. Moreover, de novo autoimmunity following SARS-CoV-2 infection is clearly recognized and can express in a range of clinical phenomena, including SLE, immune thrombocytopenic purpura, and AIH (Liu et al., 2021). This could be related to viral-induced molecular mimicry (Knight et al., 2021) determining the development of new-onset autoantibodies targeting traditional autoantigens or cytokines (Chang et al., 2021). Despite the epidemiology of SARS-CoV-2 infection, there are only a few reported cases of de novo AIH following COVID-19. All cases occurred within 1 month of mild COVID-19 and the diagnosis of AIH was based on typical biochemical, serological, and histological criteria. Most cases responded well to immunosuppressive therapy (Marjot et al., 2022). Prospective series have demonstrated a high prevalence of tissue-specific autoantibodies during or soon after recovery from COVID-19, including smooth muscle antibody and antinuclear antibody in up to 30% and 44%, respectively. The long-term clinical significance of these autoantibodies remains unclear and to date, there is no evidence to recommend routine monitoring for AIH development in all patients following SARS-CoV-2 infection (Marjot et al., 2022).

Another recent and interesting observation is the fact that the COVID-19 can also impact in the biliary system, causing a pattern of severe biliary tract injury called COVID-19 cholangiopathy. This entity is characterized by marked elevations of serum alkaline phosphatase, often accompanied by hyperbilirubinemia and radiological evidence of biliary duct changes comparable to that observed in PSC or secondary sclerosing cholangitis (i.e., strictures alternated to dilatation of the intra- and or extrahepatic bile ducts). Different from hepatocellular injury that normally occurs concomitantly to COVID-19, this cholangiopathy occurs weeks or months after the initial diagnosis of SARS-CoV-2 infection and can be associated with a severe prognosis and the need for a liver transplant. The pathophysiology is unknown but several possible mechanisms have been reported, including: (1) biliary ischemia; (2) biliary injury caused by cytokine release syndrome; (3) direct injury caused by the virus that can enter the cholangiocyte via the ACE2 receptor; and (4) ketamine-associated cholangiopathy (Faruqui et al., 2022).

Is there evidence for an increased incidence of AILDs as a consequence of COVID-19 vaccination?

Since the widespread implementation of COVID-19 programs worldwide, cases of adverse events including rare complications affecting the liver in the form of acute liver injury (ALI) mimicking AIH have been described. Moreover, there were also several reports of severe venous thrombosis of unusual sites, such as the portal and splenic vein and immune thrombocytopenic purpura associated with ALI.

A survey evaluating the risk of acute liver injury following the mRNA (BNT162b2) and inactivated (Corona Vac) COVID-19 vaccines was conducted using vaccination records in Hong Kong (Wong et al., 2022). Among 2,343,288 COVID-19 vaccine recipients, 4,677 patients developed acute liver injury (ALI). The number of ALI cases within 56 days after the first and second doses of vaccination were 307 and 521 (335 and 334 per 100,000 person-years). The majority of ALI in mRNA vaccine recipients showed hepatocellular injury along with high titers of autoantibodies, elevated IgG, and response to corticosteroid therapy. The findings suggested that the absolute risk of ALI is very low following SARS-CoV-2 infection. None of the ALI cases were severe or fatal. Although most studies did not suggest direct vaccine-induced liver injury and AIH, vaccine-induced immune-mediated hepatitis was postulated as one of the possible mechanisms. Previous population-based studies have shown that drug-induced hepatotoxicity occurred in 19 cases per 100,000 recipients while hepatotoxicity with autoimmune features was seen in fewer than 1 case per million (LiverTox Database, 2012; Bjornsoon et al., 2013).

Regarding COVID-19 and AILD, several case reports and a recent multicenter series (Efe et al., 2022a, 2022b) (Table 1) described a temporal association between COVID-19 vaccination and the first manifestation of (AIH); for this reason, a possible causal role has been suggested. By Nov. 23, 2022, 12,959,275,260 doses of vaccines had been administered and almost 100 cases of AIH following COVID-19 vaccination were reported; this is shown in Table 1. Patients were more often women ranging from 35 to 89 years old. In 20 patients, a preexisting autoimmune disease was present, including two patients with PBC and one with PSC. Notably, the type of onset was quite unequivocally acute icteric, and fibrosis was described in almost one-third of patients. The latter observation suggests the possibility of a preexisting undiagnosed AIH before COVID-19 vaccination. Finally, almost all patients recovered well after the introduction of immunosuppressive treatment that in most cases was represented by prednisone or prednisolone. Three cases of death were reported: two for sepsis and one for acute liver failure.

Notably, to assess whether AIH post-COVID-19 vaccination differs from classical AIH, Boettler and colleagues analyzed the hepatic tissue of a male who developed acute immune mediated hepatitis after COVID-19 vaccination and responded well to systemic corticosteroids. They found an immune infiltrate quantitatively dominated by activated cytotoxic CD8 T cells with pan-lobular distribution and an enrichment of CD4 T cells, B cells, plasma cells, and myeloid cells compared to the control. Moreover, they observed an enrichment of the intrahepatic infiltrate of CD8 T cells with SARS-CoV-2 specificity compared to the peripheral blood, thus suggesting that these vaccine-induced cells can contribute to liver inflammation (Boettler et al., 2022).

Finally, considering that at that time more than 6 billion people had received at least one dose of a COVID-19 vaccine, an increased incidence of AIH should have been expected (WHO COVID-19 dashboard).

In a very wise analysis, Ruther and colleagues clearly showed that the proportion of new AIH diagnoses per total AIH followed up in one of the European and national reference centers for AIH fell substantially in 2021, that is, the first year after vaccination implementation; this observation does not support the assumption that COVID-19 vaccination induces AIH (Rüther et al., 2022).

It could be said that COVID-19 is not a pandemic, but instead a syndemic (Horton, 2020). Syndemics are characterized by biological and social interactions that increase a person's susceptibility to harm or worsen their health outcomes. In fact, because of the vulnerability of older subjects toward COVID-19 infection, a syndemic approach provides an orientation to clinical medicine and to a public health approach. Thus, the new expectations are devoted not only to treating individual patients but also preventing the infection and its consequences in all ages. Priorities for vaccination and for the availability of new drugs are a special issue of public health and government.

Conclusions

In conclusion, the current evidence suggests that patients with autoimmune liver diseases are not at increased risk of SARS-CoV-2-infection compared to the general population. In particular, patients with AIH do not have a higher risk of adverse outcomes after SARS-CoV-2 infection, despite being on immunosuppressive treatment, compared to other cause of CLD and to matched cases without liver disease. Thus, a reduction of immunosuppressive drugs in patients with AIH and COVID-19 should not be considered in all patients but only in selected cases by balancing the risk of having an AIH flare-up over the benefit of reducing IS treatment to favor COVID-19 resolution. Moreover, we reported that COVID-19 vaccines are safe in patients with AILD, but a lower efficacy has been observed in patients with AIH. This suggests the need for repeated COVID-19 vaccination in AIH patients. Finally, we reported that SARS-CoV-2 infection can be associated with the development of COVID-19 cholangiopathy but not with an increased risk of de novo AIH. Moreover, the current evidence does not support an increased risk of de novo AIH after COVID-19 vaccination.

TABLE 1 Summary of the reported cases of de novo AIH after COVID-19 vaccination.

Reference	Vaccine	Age, gender	Autoimmune comorbidities	Previous SARS-CoV-2 infection	AIH onset	Fibrosis	Therapy	Outcome
Avci and Abasiyanik (2021)	mRNA Pfizer/BioNTech, 1 month before	61, F	Yes	Yes, mild, 8 months before	Acute icteric	F2	Prednisolone + azathioprine add-on	Recovered, 35 days follow-up
Bril (2021)	mRNA Pfizer/BioNTech, 7 days before	35, F	Not reported	No	Acute icteric	Absent	Prednisone	Recovered, 50 days of follow-up
Cao et al. (2022)	Inactivated whole-virion SARS-CoV-2 (CoronaVac)	57, F	Not reported	No	Acute icteric	F2	Methylprednisolone, UDCA + azathioprine add-on	Recovered, 5 months follow-up
Camacho-Domínguez et al. (2022)	ChAdOx1 nCOV-19 vaccine (Oxford-AstraZeneca), 15 days after the I dose	79, M	No	Not reported	Acute icteric	Not reported	Prednisone	Recovered, 40 days of follow-up
Clayton-Chubb et al. (2021)	ChAdOx1 nCoV-19 vaccine (Oxford-AstraZeneca), 26 days before	36, M	No	No	Acute subicteric	Absent	Prednisolone	Recovered, 24 days of follow-up
Efe et al. (2022b)	mRNA Pfizer/BioNTech, Oxford Astra Zeneca, Moderna	45 Cases	15 cases with autoimmune disease	Not reported	40/45 hepatocellular injury, 3/45 mixed injury, 2/45 cholestatic injury	28/45 fibrosis 0–2	32/45 Steroid	54 (15–185) days to recover, no liver transplant
Erard et al. (2022)	mRNA Pfizer/BioNTech, 10 days after the II dose	80, F	Not reported	Not reported	Acute icteric	Absent	Steroid	Recovered
	mRNA Moderna 21 days after the 1st dose	73, F	Not reported	Not reported	Acute icteric	Absent	Steroid	Recovered
	ChAdOx1 nCoV-19 vaccine (Oxford-AstraZeneca) 20 days after the I dose	68, F	Not reported	Not reported	Acute liver failure	Absent	Not reported	Death for liver failure and sepsis

Continued

TABLE 1 Summary of the reported cases of de novo AIH after COVID-19 vaccination—cont'd

Reference	Vaccine	Age, gender	Autoimmune comorbidities	Previous SARS-CoV-2 infection	AIH onset	Fibrosis	Therapy	Outcome
Fimiano et al. (2022)	mRNA Pfizer/ BioNTech	63, F	No	No	Acute icteric	Absent	Methylprednisolone and add-on azathioprine	Recovered
Garrido et al. (2021)	mRNA Moderna, 2 weeks before	65, F	No	No	Acute icteric severe	Absent	Prednisolone	Recovered, 1 month, of follow-up
Ghielmetti et al. (2021)	mRNA-1273, 7 days before	63, M	No	No, unknown but anticardiolipin +	Acute icteric	Absent	Prednisone	Recovered, 14 days of follow-up
Goulas et al. (2022)	mRNA Moderna, 2 weeks before	52, F	No	No	Acute icteric	Absent	Prednisolone, azathioprine add-on	Unknown
Hasegawa et al. (2022)	mRNA Pfizer/BioNTech 7 days after the I dose	82, F	No	No	Acute icteric	Absent	Prednisolone	Recovered
Kang et al. (2022)	Inactivated vaccine CoronaVac(Sinovac) 1 week after the II dose	52, F	No	No	Acute icteric	F3	Prednisolone and azathioprine	Recovered
Kulkarni et al. (2023)	Inactived or nonreplicating viral vector: Coronavac, bBIBP-CorV, covaxinm Ad5-nCOV Median of 18 days after the vaccination (70% after the I dose, 30% after the II dose)	13 cases Median age 42(22–67)	1 Patient with PBC	Not reported	Acute	Not reported	Steroid	Recovered in 39 days (median) in treated and untreated patients and no relapsed observed after a median 265 (20–450) months of follow-up
Lee et al. (2022)	mRNA Pfizer/BioNTech 2 weeks after the II dose	57, F	No	Not reported	PBC-AIH variant	Absent	UDCA	Recovered

Londono et al. (2021)	mRNA Moderna, 7 days after the II dose	41, F	Not reported	No	Acute icteric	Absent	Prednisone	Recovered
Mekritthikrai et al. (2022)	Inactivated CoronaVac vaccine (Sinovac Biotech) 1 week after the II dose	52, F	No	No	Acute icteric	F4	Prednisone	Recovered
McShane et al. (2021)	mRNA Moderna, 4 days after	71, F	No	Not reported	Acute icteric severe	Absent	Prednisone	Recovered
Palla et al. (2022)	mRNA Pfizer/BioNTech 1 month after II dose	40, F			Chronic asymptomatic	F3	Prednisolone	Recovered
Rela et al. (2021)	ChAdOx1 nCoV-19 vaccine (Oxford-AstraZeneca), 20 days before	38, F	No	No	Acute icteric	Absent	Prednisolone and tapering after 4 weeks	Recovered, 1 month of follow-up
	ChAdOx1 nCoV-19 vaccine (Oxford-AstraZeneca), 16 days before	62, M			Acute severe AIH	F1–F2	Prednisolone + plasma exchange 5 cycles	Persistent cholestasis ➔ death in 21 days for economic constraints regarding liver transplantation
Rocco et al. (2021)	Pfizer/BioNTech 1 week before (II dose)	89, F	Yes	No	Acute icteric	Absent	Prednisone	Recovered, 3 months of follow-up
Shahrani et al. (2022)	ChAdOx1 nCoV-19 vaccine (Oxford-AstraZeneca), after the 12 days since the II dose	59, F	Not reported	Not reported	Acute icteric	Absent	Prednisone	Recovered
	ChAdOx1 nCoV-19 vaccine (Oxford-AstraZeneca),14 days after the 1st dose	63, F	PSC and IBD	Not reported	Acute icteric	F3	Prednisone	Death for sepsis
	Inactivated vaccine (CoronaVac), 10 days after the II dose	72, F	No	Not reported	Acute	Absent	Prednisone	Recovered

Continued

TABLE 1 Summary of the reported cases of de novo AIH after COVID-19 vaccination—cont'd

Reference	Vaccine	Age, gender	Autoimmune comorbidities	Previous SARS-CoV-2 infection	AIH onset	Fibrosis	Therapy	Outcome
Suzuki et al. (2022)	Pfizer/BioNTech 10 days before (II dose)	80, F	Not reported	Not reported	Acute icteric	Not reported	Prednisone	Recovered, 50 days of follow-up
	Pfizer/BioNTech 4 days before (II dose)	75, F	Not reported	Not reported	Acute icteric	Not reported	Prednisone	Recovered, 105 days of follow-up
	Pfizer/BioNTech 7 days before (I dose)	78, F	Yes (PBC)	Not reported	Acute	Not reported	Prednisone	Recovered, 103 days of follow-up
Tan et al. (2021)	mRNA Moderna, 6 weeks before	56, F	Not reported	No	Acute icteric	F1	Budesonide	1 week of follow-up
Vuille-Lessard et al. (2021)	mRNA Moderna, 3 days before	76, F	Yes	Yes, 3 months before (mild disease)	Acute icteric	Not evaluable	Prednisolone+ azathioprine add-on	Recovered, 4 months follow-up
Zin Tun et al. (2022)	mRNA Moderna, 3 days before (I dose) and 2 days before (II dose)	47, M	Not reported	No	Acute icteric	F1	Prednisolone	Recovered, 2 weeks of follow-up

Summary points

- Patients with AILD are not at increased risk of SARS-CoV-2 infection compared to the general population.
- Patients with AIH do not have a higher risk of adverse outcomes after SARS-CoV-2 infection.
- Reduction of immunosuppressive drugs in patients with AIH and COVID-19 should be done only in selected cases.
- COVID-19 vaccines are safe and effective in AILD patients.
- In patients with AIH, the type of immunosuppressive treatment can influence the response to COVID-19 vaccination, and repeated vaccinations are needed.
- SARS-CoV-2 infection does not seem to be associated with an increased risk of AIH development.
- No strong evidence exists about an increased risk of de novo AIH after COVID-19 vaccination.

References

APASL Covid-19 Task Force, Lau, G., & Sharma, M. (2020). Clinical practice guidance for hepatology and liver transplant providers during the COVID-19 pandemic: APASL expert panel consensus recommendations. *Hepatology International*, *14*, 415–428.

Avci, E., & Abasiyanik, F. (2021). Autoimmune hepatitis after SARS-CoV-2 vaccine: New-onset or flare-up? *Journal of Autoimmunity*, *125*, 102745.

Bjornsoon, E. S., Bergmann, O. M., Bjornsson, H. K., Kvaran, R. B., & Olafsson, S. (2013). Incidence, presentation, and outcomes in patients with drug-induced liver injury in the general population of Iceland. *Gastroenterology*, *144*, 1419–1425.

Boettler, T., Marjot, T., Newsome, P. N., Mondelli, M. U., Maticic, M., Cordero, E., et al. (2020). Impact of COVID-19 on the care of patients with liver disease: EASL-ESCMID position paper after 6 months of the pandemic. *JHEP Reports*, *2*, 100169.

Boettler, T., Csernalabics, B., Salié, H., Luxenburger, H., Wischer, L., Alizei, E. S., et al. (2022). SARS-CoV-2 vaccination can elicit a CD8 T-cell dominant hepatitis. *Journal of Hepatology*, *77*, 653–659.

Bril, F. (2021). Autoimmune hepatitis developing after coronavirus disease 2019 (COVID-19) vaccine: One or even several swallows do not make a summer. *Journal of Hepatology*, *75*, 1256–1257.

Cai, Q., Huang, D., Yu, H., Zhu, Z., Xia, Z., Sun, Y., et al. (2020). COVID-19: Abnormal liver function tests. *Journal of Hepatology*, *73*, 566–574.

Camacho-Domínguez, L., Rodríguez, Y., Polo, F., Restrepo Gutierrez, J. C., Zapata, E., Rojas, M., & Anaya, J. M. (2022). COVID-19 vaccine and autoimmunity. A new case of autoimmune hepatitis and review of the literature. *Journal of Translational Autoimmunity*, *5*, 100140.

Cao, Z., Gui, H., Sheng, Z., Xin, H., & Xie, Q. (2022). Letter to the editor: Exacerbation of autoimmune hepatitis after COVID-19 vaccination. *Hepatology*, *75*, 757–759.

Chang, S. E., Feng, A., Meng, W., Apostolidis, S. A., Mack, E., Artandi, M., et al. (2021). New-onset IgG autoantibodies in hospitalized patients with COVID-19. *Nature Communications*, *12*, 5417.

Clayton-Chubb, D., Schneider, D., Freeman, E., Kemp, W., & Roberts, S. K. (2021). Autoimmune hepatitis developing after the ChAdOx1 nCoV-19 (Oxford-AstraZeneca) vaccine. *Journal of Hepatology*, *75*, 1249–1250.

De Santis, M., Motta, F., Isailovic, N., Clementi, M., Criscuolo, E., Clementi, N., et al. (2022). Dose-dependent impairment of the immune response to the Moderna-1273 mRNA vaccine by mycophenolate mofetil in patients with rheumatic and autoimmune liver diseases. *Vaccines*, *10*, 801. https://doi.org/10.3390/vaccines10050801.

Duengelhoef, P., Hartl, J., Rüther, D., Steinmann, S., Brehm, T. T., Weltzsch, J. P., Glaser, F., Schaub, G. M., Sterneck, M., Sebode, M., et al. (2022). SARS-CoV-2 vaccination response in patients with autoimmune hepatitis and autoimmune cholestatic liver disease. *United European Gastroenterol Journal*, *10*, 319–329.

Efe, C., Lammert, C., Tascilar, K., Dhanasekaran, R., Ebik, B., Higuera-de la Tijera, F., et al. (2022a). Effects of immunosuppressive drugs on COVID-19 severity in patients with autoimmune hepatitis. *Liver International*, *42*, 607–614.

Efe, C., Kulkarni, A. V., Terziroli Beretta-Piccoli, B., Magro, B., Friedrich Staettermayer, A., Cengiz, M., et al. (2022b). Liver injury after SARS-CoV-2 vaccination: Features of immune-mediated hepatitis, role of corticosteroid therapy and outcome. *Hepatology*. https://doi.org/10.1002/hep.32572.

Erard, D., Villeret, F., Lavrut, P. M., & Dumortier, J. (2022). Autoimmune hepatitis developing after COVID 19 vaccine: Presumed guilty? *Clinics and Research in Hepatology and Gastroenterology*, *46*, 101841.

Faruqui, S., Shanbhogue, L., & Jacobson, I. M. (2022). Biliary tract injury in patients with COVID-19: A review of the current literature. *Gastroenterology & Hepatology*, *18*, 380–387.

Favalli, F. G., Ingegnoli, F., De Lucia, O., Cincinelli, G., Cimaz, R., & Caporali, R. (2020). COVID-19 infection and rheumatoid arthritis: Faraway, so close! *Autoimmunity Reviews*, *19*, 102523.

Fimiano, F., D'Amato, D., Gambella, A., Marzano, A., Saracco, G. M., & Morgando, A. (2022). Autoimmune hepatitis or drug-induced autoimmune hepatitis following Covid-19 vaccination? *Liver International*, *42*, 1204–1205.

Fix, O. K., Hameed, B., Fontana, R. J., Kwok, R. M., McGuire, B. M., Mulligan, D. C., et al. (2020). Clinical best practice advice for hepatology and liver transplant providers during the COVID-19 pandemic: AASLD Expert Panel Consensus Statement. *Hepatology*, *72*, 287–304.

Garrido, I., Lopes, S., Simoes, M. S., Liberal, R., Lopes, J., Carneiro, F., et al. (2021). Autoimmune hepatitis after COVID-19 vaccine—More than a coincidence. *Journal of Autoimmunity*, *125*, 102741.

Gerussi, A., Rigamonti, C., Elia, C., Cazzagon, N., Floreani, A., Pozzi, R., et al. (2020). Coronavirus Disease 2019 in autoimmune hepatitis: A lesson from immunosuppressed patients. *Hepatology Communications*, *4*, 1257–1262.

Ghielmetti, M., Schaufelberger, H. D., Mieli-Vergani, G., Cerny, A., Dayer, E., Vergani, D., et al. (2021). Acute autoimmune-like hepatitis with atypical anti-mitochondrial antibody after mRNA COVID-19 vaccination: A novel clinical entity? *Journal of Autoimmunity*, *123*, 102706.

Goulas, A., Kafiri, G., Kranidioti, H., & Manolakopoulos, S. (2022). A typical autoimmune hepatitis (AIH) case following Covid-19 mRNA vaccination. More than a coincidence? *Liver International*, *42*, 254–255.

Guan, W. J., Ni, Z. Y., Hu, Y., Liang, W. H., Ou, C. Q., He, J. X., et al. (2020). China Medical Treatment Expert Group of Covid-19. Clinical characteristics of coronavirus disease 2019 in China. *The New England Journal of Medicine*, *382*, 1708–1720.

Hartl, J., Ruther, D. F., Duengelhoef, P. M., Brehm, T. T., Steinmann, S., Weltzsch, J. P., et al. (2022). Analysis of the humoral and cellular response after the third COVID-19 vaccination in patients with autoimmune hepatitis. *Liver International*. https://doi.org/10.1111/liv.15368.

Hasegawa, N., Matsuoka, R., Ishikawa, N., Endo, M., Terasaki, M., Seo, E., & Tsuchiya, K. (2022). Autoimmune hepatitis with history of HCV treatment triggered by COVID-19 vaccination: Case report and literature review. *Clinical Journal of Gastroenterology*, *15*, 791–795.

Horton, R. (2020). Offline: COVID-19 is not a pandemic. *Lancet*, *396*, 874.

Kang, S. H., Kim, M. Y., Cho, M. Y., & Baik, S. K. (2022). Autoimmune Hepatitis Following Vaccination for SARS-CoV-2 in Korea: Coincidence or autoimmunity? *Journal of Korean Medical Science*, *18*(37), e116.

Knight, J.S.; Caricchio, R.; Casanova, J-L.; Combes, A.J.; Diamond, B.; Fox, S.E., et al. (2021). The intersection of COVID-19 and autoimmunity. The Journal of Clinical Investigation, 131, e154886.

Kulkarni, A. V., Anders, M., Nazal, L., Ridrujeo, E., & Efe, C. (2023). Cases of severe acute liver injury following inactivated SARS-CoV-2 vaccination. *Journal of Hepatology*, *78*, e56–e74.

Lee, S. K., Kwon, J. K., Yoon, N., Lee, S. H., & Sung, P. S. (2022). Immune-mediated liver injury represented as overlap syndrome after SARS-CoV-2 vaccination. *Journal of Hepatology*, *77*, 1209–1211.

Liu, Y., Sawalha, A. H., & Lu, Q. (2021). COVID-19 and autoimmune diseases. *Current Opinion in Rheumatology*, *33*, 155–162.

LiverTox Database. (2012). *LiverTox: Clinical and research information on drug-induced liver injury*. Bethesda, MD: National Institute of Diabetes and Digestive and Kidney Diseases. Last access: December 2023.

Lleo, A., Cazzagon, N., Rigamonti, C., Muratori, L., Carbone, M., & On behalf of the Italian Association for the Study of the Liver. (2022). Clinical update on risks and efficacy of anti-SARS-CoV-2 vaccines in patients with autoimmune hepatitis and summary of reports on post-vaccination liver injury. *Digestive and Liver Disease*, *54*, 722–726.

Londono, M. C., Gratacos-Gines, J., & Saez-Penataro, J. (2021). Another case of autoimmune hepatitis after SARS-CoV-2 vaccination—still casualty? *Journal of Hepatology*, *75*, 1248–1249.

Marjot, T., Buescher, G., Sebode, M., Barnes, E., Barritt, A. S., Armstrong, M. J., et al. (2021). SARS-CoV-2 infection in patients with autoimmune hepatitis. *Journal of Hepatology*, *74*, 1335–1343.

Marjot, T., Eberhardt, C. S., Boettler, T., Belli, L. S., Berenguer, M., Buti, M., et al. (2022). Impact of COVID-19 on the liver and on the care of patients with chronic liver disease, hepatobiliary cancer, and liver transplantation: An updated EASL position paper. *Journal of Hepatology*, *77*, 161–1197.

Mekritthikrai, K., Jaru-Ampornpan, P., Komolmit, P., & Thanapirom, K. (2022). Autoimmune Hepatitis Triggered by COVID-19 Vaccine: The first case from inactivated vaccine. *ACG Case Reports Journal*, *9*(7), e00811.

McShane, C., Kiat, C., Rigby, J., & Crosbie, O. (2021). The mRNA COVID-19 vaccine - A rare trigger of autoimmune hepatitis? *Journal of Hepatology*, *75*, 1252–1254.

Palla, P., Vergadis, C., Sakellariou, S., & Androutsakos, T. (2022). Letter to the editor: Autoimmune hepatitis after COVID-19 vaccination: A rare adverse effect? *Hepatology*, *75*, 489–490.

Payus, O., Noh, M. M., Azizan, N., & Chettiar, R. M. (2022). SARS-CoV-2-induced liver injury: A review article on the high-risk populations, manifestations, mechanisms, pathological changes, management, and outcomes. *World Journal of Gastroenterology*, *28*, 5723–5730.

Perisetti, A., Goyal, H., Gajendran, M., Boregowda, U., Mann, R., & Sharma, N. (2020). Prevalence, mechanisms, and implications of gastrointestinal symptoms in COVID-19. *Frontiers in Medicine*, *7*, 588711. https://doi.org/10.3389/fmed.2020.588711.

Rela, M., Jothimani, D., Vij, M., Rajakumar, A., & Rammohan, A. (2021). Auto-immune hepatitis following COVID vaccination. *Journal of Autoimmunity*, *123*, 102688.

Richardson, S., Hirsch, J. S., Narasimhan, M., et al. (2020). Presenting characteristics, comorbidities, and outcomes among 5700 patients hospitalized with COVID-19 in the New York City area. *JAMA*, *323*, 2052.

Rocco, A., Sgamato, C., Compare, D., & Nardone, G. (2021). Autoimmune hepatitis following SARS-CoV-2 vaccine: May not be a casuality. *Journal of Hepatology*, *75*, 728–729.

Rüther, D. F., Weltzch, J. P., Schramm, C., Sebode, M., & Lohse, A. W. (2022). Autoimmune hepatitis and COVID-19: No increased risk for AIH after vaccination but reduced care. *Journal of Hepatology*, *77*, 249–268.

Sarmiento-Monroy, J. C., Espinosa, G., Londono, M. C., Meira, F., Caballol, B., Llufrin, S., et al. (2021). A multidisciplinary registry of patients with autoimmune and immune-mediated diseases with symptomatic COVID-19 from a single centre. *Journal of Autoimmunity*, *117*. https://doi.org/10.1016/j.jaut.2020.102580.

Shahrani, S., Sooi, C. Y., Hilmi, I. N., & Mahadeva, S. (2022). Autoimmune hepatitis (AIH) following coronavirus (COVID-19) vaccine-No longer exclusive to mRNA vaccine? *Liver International*, *2022*(42), 2344–2345.

Sharma, P., Kumar, A., Anikhindi, S., Bansal, N., Singla, V., Shivam, K., & Arora, A. (2021a). Effect of COVID-19 on pre-existing liver disease: What hepatologist should know. *Journal of Clinical and Experimental Hepatology*, *11*, 484–493.

Sharma, A., Patnaik, I., Kumar, A., & Gupta, R. (2021b). COVID-19 Vaccines in Patients with Chronic Liver Disease. *Journal of Clinical and Experimental Hepatology*, *11*, 720–726.

Simon, T. G., Hagstrom, H., Sharma, R., Soderling, J., Roelstraete, B., Larsson, E., et al. (2021). Risk of severe COVID-19 and mortality in patients with established chronic liver disease: A nationwide matched cohort study. *BMC Gastroenterology*, *21*, 439.

Singh, S., & Khan, A. (2020). Clinical characteristics and outcomes of coronavirus disease 2019 among patients with pre-existing liver disease in the United States: A multicenter research network study. *Gastroenterology*, *159*, 768–771.e3.

Suzuki, Y., Kakisaka, K., & Takikawa, Y. (2022). Letter to the editor: Autoimmune hepatitis after COVID-19 vaccination: Need for population-based epidemiological study. *Hepatology*, *75*, 759–760.

Tan, C. K., Wong, Y. J., Wang, L. M., Ang, T. L., & Kumar, R. (2021). Autoimmune hepatitis following COVID-19 vaccination: True causality or mere association?. *Journal of Hepatology*, *75*, 1250–1252.

Terziroli Beretta-Piccoli, B., & Lleo, A. (2022). Is immunosuppression truly associated with worse outcomes in autoimmune hepatitis patients with COVID-19? *Liver International*, *42*(2), 274–276. https://doi.org/10.1111/liv.15138.

Vuille-Lessard, E., Montani, M., Bosch, J., & Semmo, N. (2021). Autoimmune hepatitis triggered by SARS-CoV-2 vaccination. *Journal of Autoimmunity*, *123*, 102710. https://doi.org/10.1016/j.jaut.2021.102710.

Wong, C. K. H., Mak, L. Y., Au, I. C. H., Lai, F. T. T., Li, X., Wan, E. Y. F., et al. (2022). Risk of acute liver injury following the mRNA (BNT162b2) and inactivated (Corona Vac) COVID-19 vaccines. *Journal of Hepatology*. https://doi.org/10.1016/J.Jhep.2022.06.032.

Zecher, B. F., Buescher, G., Willemse, J., Walmsley, M., Taylor, A., Leburgue, A., et al. (2021). Prevalence of COVID-19 in patients with autoimmune liver disease in Europe: A patient-oriented online survey. *United European Gastroenterology Journal*, *9*, 797–808.

Zengarini, C., Pileri, A., Salamone, F. P., Piraccini, B. M., Vitale, G., & La Placa, M. (2022). Subacute cutaneous lupus erythematosus induction after SARS-CoV-2 vaccine in a patient with primary biliary cholangitis. *Academy of Dermatology and Venereology*, *36*, e179–e180.

Zin Tun, G. S., Gleeson, D., Al-Joudeh, A., & Dube, A. (2022). Immune-mediated hepatitis with the Moderna vaccine, no longer a coincidence but confirmed. *Journal of Hepatology*, *76*, 747–749.

Chapter 41

COVID-19 severity and nonalcoholic fatty liver disease

Nina Vrsaljko[a], Branimir Gjurašin[b], and Neven Papić[c]
[a]*Department of Infectious Diseases, University Hospital for Infectious Diseases Zagreb, Zagreb, Croatia,* [b]*Department for Intensive Care and Neuroinfections, University Hospital for Infectious Diseases Zagreb, Zagreb, Croatia,* [c]*Department of Infectious Diseases, School of Medicine, University of Zagreb, Zagreb, Croatia*

Abbreviations

ACE2	angiotensin-converting enzyme 2
CRP	C-reactive protein
CXCL10	C-X-C motif chemokine ligand 10
HCC	hepatocellular carcinoma
IFN-γ	interferon gamma
IL-6	interleukin 6
IL-8	interleukin 8
IL-10	interleukin 10
MS	metabolic syndrome
NAFLD	nonalcoholic fatty liver disease
NASH	nonalcoholic steatohepatitis
PACS	postacute COVID-19 syndrome
PT	pulmonary thrombosis
T2DM	type 2 diabetes mellitus
Th17	T helper 17 cells
TMPRSS2	transmembrane protease serine 2
Treg	regulatory T cells

Introduction

Nonalcoholic fatty liver disease (NAFLD) is the most common chronic liver disease in western countries, and is present in up to 30% of the general population (Riazi et al., 2022). NAFLD is closely related with components of metabolic syndrome (MS), with a prevalence of 55% in patients with type 2 diabetes mellitus (T2DM), 60% in patients with obesity, and 75% in patients with both T2DM and obesity. Importantly, the prevalence of both MS and NAFLD significantly increased in the last decade (Riazi et al., 2022). The primary feature of NAFLD is steatosis, followed by nonalcoholic steatohepatitis (NASH) with inflammation, fibrosis, and liver damage, potentially leading to cirrhosis and hepatocellular carcinoma (HCC) (Juanola et al., 2021). Currently, NAFLD is considered a multisystemic disease associated with systemic changes in immune response, chronic low-level inflammation, and endothelial dysfunction, all possibly related to infectious disease outcomes (Juanola et al., 2021; Tilg et al., 2021).

In the pre-COVID era, the association between NAFLD and serious infections was described for community-acquired pneumonia, recurrent urinary tract infections, *Clostridioides difficile* disease, invasive group B streptococcus infections, primary bacteremia of presumed gastrointestinal origin, and general recurrence of bacterial infections (Gjurasin et al., 2022; Krznaric & Vince, 2022; Nseir et al., 2016, 2019; Papic et al., 2020). This association was shown to be independent of the components of MS (i.e., obesity and T2DM).

There is growing evidence that patients with NAFLD have an increased risk of developing SARS-CoV-2 infection, severe disease requiring hospitalization, and more frequent complications that are related to distinct immune responses in this patient group. In this chapter, we will review the current knowledge on the role of NAFLD in COVID-19.

Management, Body Systems, and Case Studies in COVID-19. https://doi.org/10.1016/B978-0-443-18703-2.00001-3
Copyright © 2024 Elsevier Inc. All rights are reserved, including those for text and data mining, AI training, and similar technologies.

The risk of SARS-CoV-2 infection in patients with NAFLD

While it is still uncertain whether patients with chronic liver diseases are more susceptible to SARS-CoV-2 infection, the cumulative incidence of COVID-19 was reported to be higher among patients with components of MS, including NAFLD (Ghoneim et al., 2020; Roca-Fernandez et al., 2021; Yoo et al., 2021). In a large cohort study in the United States, among all components of MS, NASH showed the strongest link with developing COVID-19 (Ghoneim et al., 2020).

A prospective study from the United Kingdom evaluated the impact of preexisting liver disease on the risk of acquiring SARS-CoV-2 infection and developing more severe forms of the disease (Roca-Fernandez et al., 2021). Patients with NAFLD had a 35% increased risk of testing positive over those without NAFLD. Furthermore, obese patients with NAFLD had a fivefold higher risk of hospitalization.

While the prevalence of NAFLD in the general population is 25%–30% (Younossi et al., 2019), overrepresentation of NAFLD in COVID-19 patients has been frequently reported (ranging from 30% in an Israel cohort to 42% in Mexico and 52% in the United States) (Chen et al., 2021; Lopez-Mendez et al., 2021; Mahamid et al., 2021; Zhou et al., 2020).

These data suggest that patients with NAFLD are at increased risk of acquiring SARS-CoV-2 infection. It is a matter of debate as to whether the liver in NAFLD upregulates the expression of SARS-CoV-2 critical entry proteins, such as ACE2 and transmembrane protease serine 2 (TMPRSS2), possibly rendering NAFLD patients more susceptible to SARS-CoV-2 infection (Biquard et al., 2020; Fondevila et al., 2021; Meijnikman et al., 2021; Singh et al., 2021). However, a dysregulated immune response in patients with NAFLD also heightens the risk for severe disease, as described in the following chapter.

Clinical course and outcomes of COVID-19 in patients with NAFLD

Severe COVID-19 predominantly occurs in adults with risk factors, such as advanced age and certain underlying comorbidities, including T2DM, obesity, dyslipidemia, and arterial hypertension, all closely related to NAFLD (Guan et al., 2020; Harrison et al., 2020; Petrilli et al., 2020; Zheng et al., 2020). Moreover, chronic liver disease (such as cirrhosis, alcoholic liver disease, autoimmune hepatitis, and NAFLD) heightens the risk for severe COVID-19 (Mohammed et al., 2021).

In T2DM, this association could be due to higher affinity cellular binding and entry, decreased viral clearance, diminished immune responses, increased susceptibility to hyperinflammation, and the presence of cardiovascular disease (Kazakou et al., 2022). Obese patients could have a higher risk for severe disease due to altered respiratory physiology, impaired immune response, and ACE2 enzyme expression in adipose tissue (Yu et al., 2022). However, most studies examining the impact of MS did not include NAFLD as a variable. As highlighted in a study by Roca-Fernandez et al. (2021), an increased risk of severe disease was not observed in obese patients without NAFLD. Similarly, Ghoneim et al. (2020) reported that NAFLD and NASH have a higher impact on COVID-19 outcomes than T2DM or obesity. This suggests that the impact of components of MS on COVID-19 outcomes might be due to the underlying NAFLD, not MS per se.

With COVID-19, numerous studies reported an association of NAFLD with increased COVID-19 severity and mortality (Ji et al., 2020; Mahamid et al., 2021; Vrsaljko et al., 2022; Wang et al., 2022). It has been reported that patients with NAFLD have a higher risk of disease progression, which is even higher in those with an increased fibrosis-4 index and NAFLD fibrosis score (Chen et al., 2021; Ji et al., 2020; Vrsaljko et al., 2022; Wang et al., 2021; Younossi et al., 2022). Patients with NAFLD more frequently had severe respiratory symptoms upon admission and required advanced respiratory support, intensive care unit treatment, and longer hospital stays than non-NAFLD controls (Vrsaljko et al., 2022; Younossi et al., 2022). Regarding laboratory findings, patients with NAFLD had higher inflammatory biomarkers, such as C-reactive protein (CRP) or interleukin 6 (IL-6); higher lactate-dehydrogenase, aminotransferases, fibrinogen, ferritin, and D-dimer; and lower lymphocyte counts, all shown to be negative predictors of COVID-19 outcomes (Papic et al., 2022; Vrsaljko et al., 2022; Younossi et al., 2022). The risk of mortality in NAFLD patients was reported to be associated with inflammatory markers (Forlano et al., 2020; Vrsaljko et al., 2022; Younossi et al., 2022).

While liver injury is frequently observed in patients with COVID-19, in patients with NAFLD this is even more pronounced. Acute liver injury was observed in 4% of patients with NAFLD (vs. 2.4%), however, the clinical significance is not clear (Younossi et al., 2022).

The risk of pulmonary thrombosis in patients with COVID-19 and NAFLD

Pulmonary thrombosis (PT) is a frequent complication of severe COVID-19 with an estimated incidence ranging from 5% to 35% (Katsoularis et al., 2022; Mocibob et al., 2022). NAFLD was recently recognized as an independent risk factor for

PT in non-COVID-19 patients (Zeina et al., 2022). The pathophysiological mechanism can be explained by the chronic low-grade inflammation associated with endothelial vascular dysfunction, platelet dysfunction, and changes in the coagulation cascade, which characterize NAFLD as a prothrombotic state (Northup et al., 2008; Verrijken et al., 2014).

Several small postmortem analyses showed high steatosis prevalence in COVID-19 patients with pulmonary thrombosis (Falasca et al., 2020). A prospective observational study included 216 hospitalized adult patients with severe COVID-19 and 55.5% of them were diagnosed with NAFLD (Vrsaljko et al., 2022). In total, 20.8% of patients were diagnosed with PT during hospitalization: 26.7% NAFLD patients and 13.5% non-NAFLD patients. In this study, NAFLD was associated with PT in hospitalized COVID-19 patients independently of other components of MS.

This highlights the importance of a more detailed screening process for PT in NAFLD patients that could be relevant in choosing and guiding anticoagulation therapy.

Immune response to SARS-CoV-2 infection in patients with NAFLD

One of the hallmarks of COVID-19 pathogenesis is hyperactivation and dysregulation of the immune system that results in a so called "cytokine storm," a state of increased synthesis of various proinflammatory and antiinflammatory cytokines and chemokines (Laing et al., 2020; Zanza et al., 2022).

The immune changes in NAFLD are complex and include impaired cellular immunity with an imbalance of the Treg/Th17 axis and an increased release of proinflammatory cytokines (He et al., 2017; Paquissi, 2016). Meanwhile, the negative regulators of Th17 response (such as IL-10, IL-4, IL-22, and IFN-γ) are suppressed, which further contributes to the progression of liver inflammation (He et al., 2017; Paquissi, 2016). NAFLD should be viewed in the context of MS, where the immune activity of adipose tissue leads to a cytokine imbalance with an increased release of proinflammatory cytokines (TNF-α, IL-6, IL-1β) and subsequent chronic low-grade inflammation (Han et al., 2020; Lukas & Herbert, 2018). This might exacerbate insulin resistance, another key component of NAFLD (Fontes-Cal et al., 2021). Recently, a "lung-liver axis" model was described that implicated the importance of liver acute phase proteins in host defense modulation and the inflammation of lung parenchyma (Herrero et al., 2020; Young et al., 2016). Therefore, it seems reasonable to suggest that patients with NAFLD have different immune responses to infection.

So far, several studies have shown that patients with NAFLD and COVID-19 have distinct serum cytokine profiles in comparison to patients without NAFLD, including higher levels of IL-6, IL-8, IL-10, and CXCL10, and lower levels of IFN-γ (Papic et al., 2022). Furthermore, patients who progressed to more severe forms of the disease had higher serum concentrations of IL-6, -8, -10, and IFN-γ upon admission (Papic et al., 2022).

Therefore, patients with NAFLD show an impaired liver-derived immune response to SARS-CoV-2 infection. This contributes to the imbalance of the pro- and antiinflammatory mediators with repercussions on the severity and outcome of the disease (Fig. 1).

NAFLD and post-COVID-19 condition

The relationship of NAFLD and post-COVID-19 condition (or long COVID-19) is not fully understood.

A cross-sectional observational Italian study aimed to determine the prevalence of NAFLD in patients with post-COVID-19 (Milic et al., 2022). The prevalence of NAFLD was 37.3% upon admission and 55.3% at follow-up. However, the correlation between NAFLD and specific post-COVID-19 symptoms was weak. Interestingly, due to the significant occurrence of new liver steatosis in the convalescent phase, the authors argued that NAFLD might be a separate symptom of post-COVID-19 as a consequence of metabolic changes and weight gain.

This is further supported by the finding of glycemic abnormalities that can be detected for at least 2 months in patients who recovered from COVID-19 (Montefusco et al., 2021). A systematic review and metaanalysis found that the risk of new onset T2DM increased 1.2-fold after COVID-19 infection compared to patients with general upper respiratory tract infections, and this risk was even higher in patients with severe COVID-19 (Zhang et al., 2022). This risk was highest in the first 3 months after COVID-19.

Similarly, several studies reported the high prevalence of dyslipidemia in patients after COVID-19 recovery (Dennis et al., 2021; Gameil et al., 2021).

In conclusion, NAFLD might be considered as a separate post-COVID-19 symptom with possible implications on metabolic and cardiovascular health.

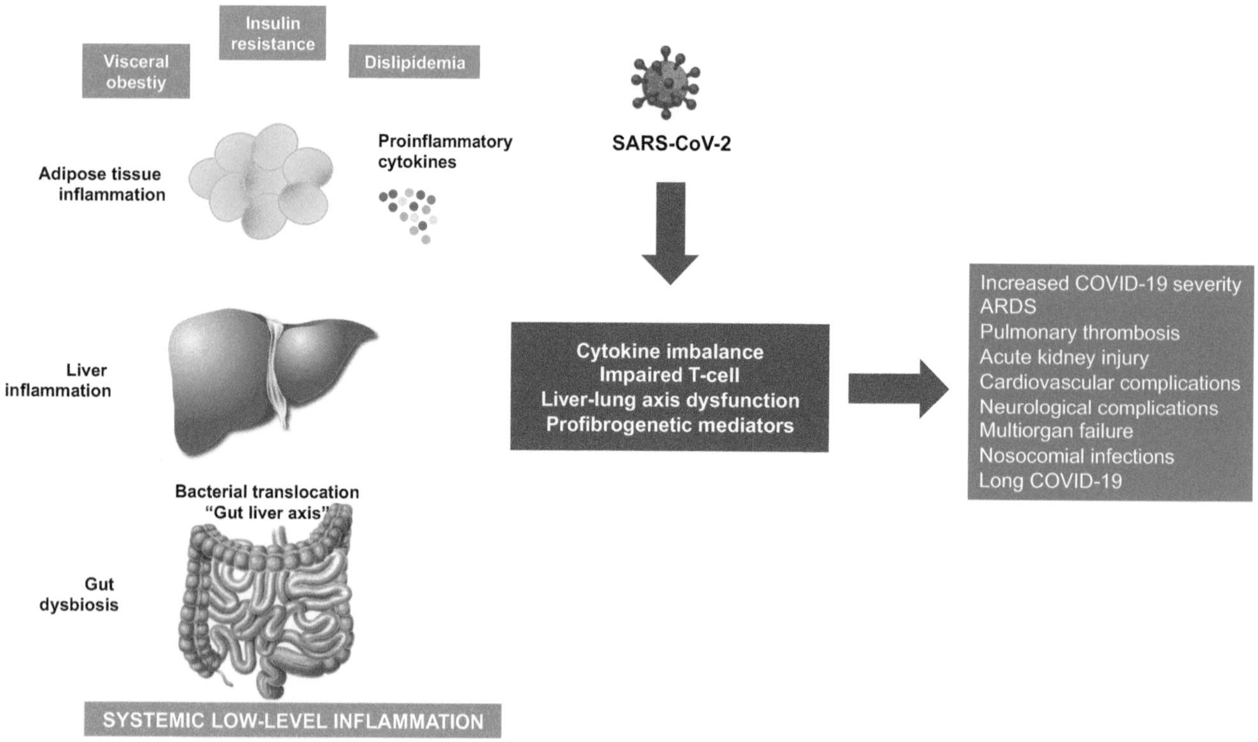

FIG. 1 The proposed interplay between NAFLD and SARS-CoV-2 infection. The simplified overview of immune changes in NAFLD that might contribute to COVID-19 severity.

Policies and procedures

Policy

Nonalcoholic fatty liver disease is highly prevalent and associated with worse COVID-19 outcomes. The rising prevalence and underdiagnosing of NAFLD emphasizes the need for large-scale healthcare campaigns and screening strategies. Patients with NAFLD should be prioritized for vaccination and other preventive strategies. There might be a benefit for screening of hospitalized COVID-19 patients for components of metabolic syndrome and NAFLD, as well as in patients with post-COVID-19 condition.

Applications to other areas

In the pre-COVID-19 era, data on the association of NAFLD and infectious diseases were scarce. The COVID-19 pandemic gave us a broader insight into the prevalence of the metabolic syndrome and NAFLD, and their significant impact on disease outcomes. This raises questions as to the role of NAFLD in other infectious diseases, such as bloodstream infections, flu, or acute respiratory distress syndrome, warranting further research on the effects of NAFLD and other severe infections.

Mini-dictionary of terms

- *Liver steatosis*. The accumulation of abnormal amounts of lipids in 5% or more of liver cells.
- *Lung-liver axis*. A bidirectional interplay model explaining the role of liver acute phase proteins in modulating lung inflammation and vice versa.
- *NAFLD*. Nonalcoholic fatty liver disease is excessive fat accumulation in the liver without another clear cause (such as alcohol). It is frequently associated with other components of metabolic syndrome.
- *Metabolic syndrome*. Metabolic syndrome is a cooccurrence of at least three of the following conditions: abdominal obesity, high blood pressure, high blood glucose levels, high serum triglycerides, and low serum high-density lipoprotein.

Summary points

- Nonalcoholic fatty liver disease is highly prevalent in patients with COVID-19.
- NAFLD increases susceptibility to SARS-CoV-2 infection.
- NAFLD is associated with COVID-19 severity and mortality.
- NAFLD is associated with pulmonary thrombosis in hospitalized patients with COVID-19.
- NAFLD might be considered a manifestation of post-COVID-19 condition.

References

Biquard, L., Valla, D., & Rautou, P. E. (2020). No evidence for an increased liver uptake of SARS-CoV-2 in metabolic-associated fatty liver disease. *Journal of Hepatology*, *73*(3), 717–718. https://doi.org/10.1016/j.jhep.2020.04.035.

Chen, V. L., Hawa, F., Berinstein, J. A., Reddy, C. A., Kassab, I., Platt, K. D., … Sharma, P. (2021). Hepatic steatosis is associated with increased disease severity and liver injury in coronavirus disease-19. *Digestive Diseases and Sciences*, *66*(9), 3192–3198. https://doi.org/10.1007/s10620-020-06618-3.

Dennis, A., Wamil, M., Alberts, J., Oben, J., Cuthbertson, D. J., Wootton, D., … COVERSCAN Study Investigators. (2021). Multiorgan impairment in low-risk individuals with post-COVID-19 syndrome: A prospective, community-based study. *BMJ Open*, *11*(3), e048391. https://doi.org/10.1136/bmjopen-2020-048391.

Falasca, L., Nardacci, R., Colombo, D., Lalle, E., Di Caro, A., Nicastri, E., … Del Nonno, F. (2020). Postmortem findings in Italian patients with COVID-19: A descriptive full autopsy study of cases with and without comorbidities. *The Journal of Infectious Diseases*, *222*(11), 1807–1815. https://doi.org/10.1093/infdis/jiaa578.

Fondevila, M. F., Mercado-Gomez, M., Rodriguez, A., Gonzalez-Rellan, M. J., Iruzubieta, P., Valenti, V., … Nogueiras, R. (2021). Obese patients with NASH have increased hepatic expression of SARS-CoV-2 critical entry points. *Journal of Hepatology*, *74*(2), 469–471. https://doi.org/10.1016/j.jhep.2020.09.027.

Fontes-Cal, T. C. M., Mattos, R. T., Medeiros, N. I., Pinto, B. F., Belchior-Bezerra, M., Roque-Souza, B., … Gomes, J. A. S. (2021). Crosstalk between plasma cytokines, inflammation, and liver damage as a new strategy to monitoring NAFLD progression. *Frontiers in Immunology*, *12*, 708959. https://doi.org/10.3389/fimmu.2021.708959.

Forlano, R., Mullish, B. H., Mukherjee, S. K., Nathwani, R., Harlow, C., Crook, P., … Manousou, P. (2020). In-hospital mortality is associated with inflammatory response in NAFLD patients admitted for COVID-19. *PLoS One*, *15*(10), e0240400. https://doi.org/10.1371/journal.pone.0240400.

Gameil, M. A., Marzouk, R. E., Elsebaie, A. H., & Rozaik, S. E. (2021). Long-term clinical and biochemical residue after COVID-19 recovery. *Egyptian Liver Journal*, *11*(1), 74. https://doi.org/10.1186/s43066-021-00144-1.

Ghoneim, S., Butt, M. U., Hamid, O., Shah, A., & Asaad, I. (2020). The incidence of COVID-19 in patients with metabolic syndrome and non-alcoholic steatohepatitis: A population-based study. *Metabolism Open*, *8*, 100057. https://doi.org/10.1016/j.metop.2020.100057.

Gjurasin, B., Jelicic, M., Kutlesa, M., & Papic, N. (2022). The impact of nonalcoholic fatty liver disease on severe community-acquired pneumonia outcomes. *Life (Basel)*, *13*(1). https://doi.org/10.3390/life13010036.

Guan, W. J., Liang, W. H., Zhao, Y., Liang, H. R., Chen, Z. S., Li, Y. M., … China Medical Treatment Expert Group for COVID-19. (2020). Comorbidity and its impact on 1590 patients with COVID-19 in China: A nationwide analysis. *The European Respiratory Journal*, *55*(5). https://doi.org/10.1183/13993003.00547-2020.

Han, M. S., White, A., Perry, R. J., Camporez, J. P., Hidalgo, J., Shulman, G. I., & Davis, R. J. (2020). Regulation of adipose tissue inflammation by interleukin 6. *Proceedings of the National Academy of Sciences of the United States of America*, *117*(6), 2751–2760. https://doi.org/10.1073/pnas.1920004117.

Harrison, S. L., Fazio-Eynullayeva, E., Lane, D. A., Underhill, P., & Lip, G. Y. H. (2020). Comorbidities associated with mortality in 31,461 adults with COVID-19 in the United States: A federated electronic medical record analysis. *PLoS Medicine*, *17*(9), e1003321. https://doi.org/10.1371/journal.pmed.1003321.

He, B., Wu, L., Xie, W., Shao, Y., Jiang, J., Zhao, Z., … Cui, D. (2017). The imbalance of Th17/Treg cells is involved in the progression of nonalcoholic fatty liver disease in mice. *BMC Immunology*, *18*(1), 33. https://doi.org/10.1186/s12865-017-0215-y.

Herrero, R., Sanchez, G., Asensio, I., Lopez, E., Ferruelo, A., Vaquero, J., … Lorente, J. A. (2020). Liver-lung interactions in acute respiratory distress syndrome. *Intensive Care Medicine Experimental*, *8*(Suppl 1), 48. https://doi.org/10.1186/s40635-020-00337-9.

Ji, D., Qin, E., Xu, J., Zhang, D., Cheng, G., Wang, Y., & Lau, G. (2020). Non-alcoholic fatty liver diseases in patients with COVID-19: A retrospective study. *Journal of Hepatology*, *73*(2), 451–453. https://doi.org/10.1016/j.jhep.2020.03.044.

Juanola, O., Martinez-Lopez, S., Frances, R., & Gomez-Hurtado, I. (2021). Non-alcoholic fatty liver disease: Metabolic, genetic, epigenetic and environmental risk factors. *International Journal of Environmental Research and Public Health*, *18*(10). https://doi.org/10.3390/ijerph18105227.

Katsoularis, I., Fonseca-Rodriguez, O., Farrington, P., Jerndal, H., Lundevaller, E. H., Sund, M., … Fors Connolly, A. M. (2022). Risks of deep vein thrombosis, pulmonary embolism, and bleeding after covid-19: Nationwide self-controlled cases series and matched cohort study. *BMJ*, *377*, e069590. https://doi.org/10.1136/bmj-2021-069590.

Kazakou, P., Lambadiari, V., Ikonomidis, I., Kountouri, A., Panagopoulos, G., Athanasopoulos, S., … Mitrakou, A. (2022). Diabetes and COVID-19; A bidirectional interplay. *Frontiers in Endocrinology (Lausanne)*, *13*, 780663. https://doi.org/10.3389/fendo.2022.780663.

Krznaric, J., & Vince, A. (2022). The role of non-alcoholic fatty liver disease in infections. *Life (Basel)*, *12*(12). https://doi.org/10.3390/life12122052.

Laing, A. G., Lorenc, A., Del Molino Del Barrio, I., Das, A., Fish, M., Monin, L., … Hayday, A. C. (2020). A dynamic COVID-19 immune signature includes associations with poor prognosis. *Nature Medicine*, *26*(10), 1623–1635. https://doi.org/10.1038/s41591-020-1038-6.

Lopez-Mendez, I., Aquino-Matus, J., Gall, S. M., Prieto-Nava, J. D., Juarez-Hernandez, E., Uribe, M., & Castro-Narro, G. (2021). Association of liver steatosis and fibrosis with clinical outcomes in patients with SARS-CoV-2 infection (COVID-19). *Annals of Hepatology*, *20*, 100271. https://doi.org/10.1016/j.aohep.2020.09.015.

Lukas, N., & Herbert, T. (2018). Cytokines and fatty liver diseases. *Liver Research*, *2*(1), 14–20. https://doi.org/10.1016/j.livres.2018.03.003.

Mahamid, M., Nseir, W., Khoury, T., Mahamid, B., Nubania, A., Sub-Laban, K., … Goldin, E. (2021). Nonalcoholic fatty liver disease is associated with COVID-19 severity independently of metabolic syndrome: A retrospective case-control study. *European Journal of Gastroenterology & Hepatology*, *33*(12), 1578–1581. https://doi.org/10.1097/MEG.0000000000001902.

Meijnikman, A. S., Bruin, S., Groen, A. K., Nieuwdorp, M., & Herrema, H. (2021). Increased expression of key SARS-CoV-2 entry points in multiple tissues in individuals with NAFLD. *Journal of Hepatology*, *74*(3), 748–749. https://doi.org/10.1016/j.jhep.2020.12.007.

Milic, J., Barbieri, S., Gozzi, L., Brigo, A., Beghe, B., Verduri, A., … Raggi, P. (2022). Metabolic-associated fatty liver disease is highly prevalent in the postacute COVID syndrome. *Open Forum Infectious Diseases*, *9*(3), ofac003. https://doi.org/10.1093/ofid/ofac003.

Mocibob, L., Susak, F., Situm, M., Viskovic, K., Papic, N., & Vince, A. (2022). COVID-19 and pulmonary thrombosis—An unresolved clinical puzzle: A single-center cohort study. *Journal of Clinical Medicine*, *11*(23). https://doi.org/10.3390/jcm11237049.

Mohammed, A., Paranji, N., Chen, P. H., & Niu, B. (2021). COVID-19 in chronic liver disease and liver transplantation: A clinical review. *Journal of Clinical Gastroenterology*, *55*(3), 187–194. https://doi.org/10.1097/MCG.0000000000001481.

Montefusco, L., Ben Nasr, M., D'Addio, F., Loretelli, C., Rossi, A., Pastore, I., … Fiorina, P. (2021). Acute and long-term disruption of glycometabolic control after SARS-CoV-2 infection. *Nature Metabolism*, *3*(6), 774–785. https://doi.org/10.1038/s42255-021-00407-6.

Northup, P. G., Sundaram, V., Fallon, M. B., Reddy, K. R., Balogun, R. A., Sanyal, A. J., … Coagulation in Liver Disease Group. (2008). Hypercoagulation and thrombophilia in liver disease. *Journal of Thrombosis and Haemostasis*, *6*(1), 2–9. https://doi.org/10.1111/j.1538-7836.2007.02772.x.

Nseir, W., Amara, A., Farah, R., Ahmad, H. S., Mograbi, J., & Mahamid, M. (2019). Non-alcoholic fatty liver disease is associated with recurrent urinary tract infection in premenopausal women independent of metabolic syndrome. *The Israel Medical Association Journal*, *21*(6), 386–389. Retrieved from https://www.ncbi.nlm.nih.gov/pubmed/31280506.

Nseir, W., Artul, S., Nasrallah, N., & Mahamid, M. (2016). The association between primary bacteremia of presumed gastrointestinal origin and non-alcoholic fatty liver disease. *Digestive and Liver Disease*, *48*(3), 343–344. https://doi.org/10.1016/j.dld.2015.10.004.

Papic, N., Jelovcic, F., Karlovic, M., Maric, L. S., & Vince, A. (2020). Nonalcoholic fatty liver disease as a risk factor for Clostridioides difficile infection. *European Journal of Clinical Microbiology & Infectious Diseases*, *39*(3), 569–574. https://doi.org/10.1007/s10096-019-03759-w.

Papic, N., Samadan, L., Vrsaljko, N., Radmanic, L., Jelicic, K., Simicic, P., … Vince, A. (2022). Distinct cytokine profiles in severe COVID-19 and non-alcoholic fatty liver disease. *Life (Basel)*, *12*(6). https://doi.org/10.3390/life12060795.

Paquissi, F. C. (2016). Immune imbalances in non-alcoholic fatty liver disease: From general biomarkers and neutrophils to interleukin-17 axis activation and new therapeutic targets. *Frontiers in Immunology*, *7*, 490. https://doi.org/10.3389/fimmu.2016.00490.

Petrilli, C. M., Jones, S. A., Yang, J., Rajagopalan, H., O'Donnell, L., Chernyak, Y., … Horwitz, L. I. (2020). Factors associated with hospital admission and critical illness among 5279 people with coronavirus disease 2019 in New York City: Prospective cohort study. *BMJ*, *369*, m1966. https://doi.org/10.1136/bmj.m1966.

Riazi, K., Azhari, H., Charette, J. H., Underwood, F. E., King, J. A., Afshar, E. E., … Shaheen, A. A. (2022). The prevalence and incidence of NAFLD worldwide: A systematic review and meta-analysis. *The Lancet Gastroenterology & Hepatology*, *7*(9), 851–861. https://doi.org/10.1016/S2468-1253(22)00165-0.

Roca-Fernandez, A., Dennis, A., Nicholls, R., McGonigle, J., Kelly, M., Banerjee, R., … Sanyal, A. J. (2021). Hepatic steatosis, rather than underlying obesity, increases the risk of infection and hospitalization for COVID-19. *Frontiers in Medicine (Lausanne)*, *8*, 636637. https://doi.org/10.3389/fmed.2021.636637.

Singh, M. K., Mobeen, A., Chandra, A., Joshi, S., & Ramachandran, S. (2021). A meta-analysis of comorbidities in COVID-19: Which diseases increase the susceptibility of SARS-CoV-2 infection? *Computers in Biology and Medicine*, *130*, 104219. https://doi.org/10.1016/j.compbiomed.2021.104219.

Tilg, H., Adolph, T. E., Dudek, M., & Knolle, P. (2021). Non-alcoholic fatty liver disease: The interplay between metabolism, microbes and immunity. *Nature Metabolism*, *3*(12), 1596–1607. https://doi.org/10.1038/s42255-021-00501-9.

Verrijken, A., Francque, S., Mertens, I., Prawitt, J., Caron, S., Hubens, G., … Van Gaal, L. (2014). Prothrombotic factors in histologically proven non-alcoholic fatty liver disease and nonalcoholic steatohepatitis. *Hepatology*, *59*(1), 121–129. https://doi.org/10.1002/hep.26510.

Vrsaljko, N., Samadan, L., Viskovic, K., Mehmedovic, A., Budimir, J., Vince, A., & Papic, N. (2022). Association of nonalcoholic fatty liver disease with COVID-19 severity and pulmonary thrombosis: CovidFAT, a prospective, observational cohort study. *Open Forum Infectious Diseases*, *9*(4), ofac073. https://doi.org/10.1093/ofid/ofac073.

Wang, Y., Wang, Y., Duan, G., & Yang, H. (2022). NAFLD was independently associated with severe COVID-19 among younger patients rather than older patients: A meta-analysis. *Journal of Hepatology*. https://doi.org/10.1016/j.jhep.2022.10.009.

Wang, G., Wu, S., Wu, C., Zhang, Q., Wu, F., Yu, B., … Zhong, Y. (2021). Association between non-alcoholic fatty liver disease with the susceptibility and outcome of COVID-19: A retrospective study. *Journal of Cellular and Molecular Medicine*, *25*(24), 11212–11220. https://doi.org/10.1111/jcmm.17042.

Yoo, H. W., Jin, H. Y., Yon, D. K., Effenberger, M., Shin, Y. H., Kim, S. Y., … Lee, S. W. (2021). Non-alcoholic fatty liver disease and COVID-19 susceptibility and outcomes: A Korean nationwide cohort. *Journal of Korean Medical Science*, *36*(41), e291. https://doi.org/10.3346/jkms.2021.36.e291.

Young, R. P., Hopkins, R. J., & Marsland, B. (2016). The gut-liver-lung axis. Modulation of the innate immune response and its possible role in chronic obstructive pulmonary disease. *American Journal of Respiratory Cell and Molecular Biology*, *54*(2), 161–169. https://doi.org/10.1165/rcmb.2015-0250PS.

Younossi, Z. M., Stepanova, M., Lam, B., Cable, R., Felix, S., Jeffers, T., … Gerber, L. (2022). Independent predictors of mortality among patients with NAFLD hospitalized with COVID-19 infection. *Hepatology Communications*, *6*(11), 3062–3072. https://doi.org/10.1002/hep4.1802.

Younossi, Z., Tacke, F., Arrese, M., Chander Sharma, B., Mostafa, I., Bugianesi, E., … Vos, M. B. (2019). Global perspectives on nonalcoholic fatty liver disease and nonalcoholic steatohepatitis. *Hepatology*, *69*(6), 2672–2682. https://doi.org/10.1002/hep.30251.

Yu, L., Zhang, X., Ye, S., Lian, H., Wang, H., & Ye, J. (2022). Obesity and COVID-19: Mechanistic insights from adipose tissue. *The Journal of Clinical Endocrinology and Metabolism*, *107*(7), 1799–1811. https://doi.org/10.1210/clinem/dgac137.

Zanza, C., Romenskaya, T., Manetti, A. C., Franceschi, F., La Russa, R., Bertozzi, G., … Longhitano, Y. (2022). Cytokine storm in COVID-19: Immunopathogenesis and therapy. *Medicina (Kaunas, Lithuania)*, *58*(2). https://doi.org/10.3390/medicina58020144.

Zeina, A. R., Kopelman, Y., Mari, A., Ahmad, H. S., Artul, S., Khalaila, A. S., … Abu Baker, F. (2022). Pulmonary embolism risk in hospitalized patients with nonalcoholic fatty liver disease: A case-control study. *Medicine (Baltimore)*, *101*(45), e31710. https://doi.org/10.1097/MD.0000000000031710.

Zhang, T., Mei, Q., Zhang, Z., Walline, J. H., Liu, Y., Zhu, H., & Zhang, S. (2022). Risk for newly diagnosed diabetes after COVID-19: A systematic review and meta-analysis. *BMC Medicine*, *20*(1), 444. https://doi.org/10.1186/s12916-022-02656-y.

Zheng, Z., Peng, F., Xu, B., Zhao, J., Liu, H., Peng, J., … Tang, W. (2020). Risk factors of critical & mortal COVID-19 cases: A systematic literature review and meta-analysis. *The Journal of Infection*, *81*(2), e16–e25. https://doi.org/10.1016/j.jinf.2020.04.021.

Zhou, Y. J., Zheng, K. I., Wang, X. B., Sun, Q. F., Pan, K. H., Wang, T. Y., … Zheng, M. H. (2020). Metabolic-associated fatty liver disease is associated with severity of COVID-19. *Liver International*, *40*(9), 2160–2163. https://doi.org/10.1111/liv.14575.

Chapter 42

Changes in obesity and diabetes severity during the COVID-19 pandemic at Virginia Commonwealth University health system

Asmaa M. Namoos[a], Vanessa Sheppard[b], NourEldin Abosamak[c], Martin Lavallee[d], Rana Ramadan[c], Estelle Eyob[e], Chen Wang[d], and Tamas S. Gal[f]

[a]*Department of Social and Behavioral Sciences, School of Medicine, Virginia Commonwealth University, Richmond, VA, United States,* [b]*Department of Health Behavior and Policy, Massey Cancer Center, School of Medicine, Virginia Commonwealth University, Richmond, VA, United States,* [c]*Virginia Commonwealth University, Richmond, VA, United States,* [d]*Department of Biostatistics, Virginia Commonwealth University, Richmond, VA, United States,* [e]*Department of Biology & Public Health, William & Mary, Sadler Center, Williamsburg, VA, United States,* [f]*Department of Research Informatics, Wright Center for Clinical and Translational Research and Massey Cancer Center, Virginia Commonwealth University, Richmond, VA, United States*

Abbreviations

BMI	body mass index
DM	diabetes mellitus
ED	emergency department
HbA1c	hemoglobin A1c

Introduction

Obesity is a global pandemic that negatively affects multiple aspects of individual well-being and is a known risk factor for diabetes and other diseases. According to the National Center for Health Statistics, 42.5% of the US population aged 20 and older were classified as obese in 2018 (American Heart Association, 2021; Fryar et al., 2020). A few studies in different countries documented an increase in body mass index (BMI) after the lockdown caused by the COVID-19 pandemic including in the United States, where 42% of adults surveyed reported an undesired increase in weight (American Psychological Association, 2020; Bakaloudi et al., 2021; Pellegrini et al., 2016). Multiple studies found that the pandemic has led to behavioral changes that increased the tendency of individuals to gain weight, including changes in nutritional choices, financial burdens, psychological alterations, and physical activity patterns (Almandoz et al., 2020; Ashby, 2020). Additionally, food companies have initiated innovative marketing techniques since the start of the pandemic, which added to the increased consumption of unhealthy foods and alcohol (Khan & Moverley Smith, 2020).

Obesity and diabetes mellitus

Obesity has a strong negative effect on diabetes mellitus (DM) prognosis. Obesity is the most prominent risk factor for insulin resistance in the United States, which is the main reason for Type 2 DM (Lee et al., 2006). More than 90% of diabetic patients are classified as having Type 2 DM. Though obesity is less of a risk factor for Type 1 diabetes, it plays an important role in the management of Type 1 DM as well. About 11% of the US population suffers from Type 1 or Type 2 diabetes (CDC, 2020), and more than 75% of DM patients are obese (Bays et al., 2007).

Management, Body Systems, and Case Studies in COVID-19. https://doi.org/10.1016/B978-0-443-18703-2.00047-5
Copyright © 2024 Elsevier Inc. All rights are reserved, including those for text and data mining, AI training, and similar technologies.

Methods for evaluating diabetes control

There is evidence that obesity reduces the glycemic control of diabetic patients in addition to increasing the severity of comorbidities (Scheen, 2000). A common way of evaluating diabetes is measuring the concentration of hemoglobin A1c (HbA1c), which depends on both the concentration of glucose in the blood and the life span of red blood cells. Because the average lifespan of red blood cells is about 120 days, HbA1c reflects the integrated glucose concentration over 8–12 weeks prior to the test. This offers the advantage of eliminating day-to-day fluctuations that occur in blood glucose concentrations (Goldstein et al., 2004). In addition, HbA1c blood samples can be taken at any time of the day without patient preparations (Weykamp, 2013). The HbA1c concentration is frequently used to monitor the glycemic status in both Type 1 and Type 2 diabetes patients over time. This measurement helps to indicate the degree of glycemic control, response to treatment, and the risk of developing of diabetes-related complications. A higher level of HbA1c is strongly suggestive of poorly controlled diabetes (American Diabetes Association, 2011; The Diabetes Control and Complications Trial Research Group, 1993). An HbA1c level of at least 6.5% would be diagnostic of diabetes if confirmed by an elevated blood glucose level (Saudek et al., 2008).

COVID-19 and diabetes mellitus

Uncontrolled glucose levels were found to affect the morbidity and mortality of diabetic patients infected with COVID-19. Diabetic patients infected with COVID-19 are twice as likely to develop severe COVID-19 disease with twice the risk of mortality; therefore, a large portion of diabetics present with COVID-19 infection to the emergency department (ED) with severe respiratory symptoms and are admitted to the hospital (Kumar et al., 2020). A study by Ghosal et al. demonstrated that poor glycemic control increased the risk of mortality in diabetic patients with COVID-19. The authors also found that the duration of lockdown was directly proportional to the deterioration of glycemic control and diabetes-related complications (Ghosal et al., 2020).

African American susceptibility to diabetes complications

About 30% of the patient population of the VCU Health System is African American. African Americans are documented to experience more detrimental effects related to diabetes. The National Center for Health Statistics reports that 24.9% of African American diabetic patients have poor glycemic control, compared to only 8.8% of non-Hispanic White diabetic patients. African Americans are 2.1 times more likely to die from diabetes and 1.7 times more likely to be hospitalized as a result of diabetes compared to non-Hispanic Whites (CDC, 2021; Kochanek et al., 2019). This difference in complication rates can be attributed to many factors, including the disproportionate socioeconomic status of African Americans compared to non-Hispanic Whites. Additionally, blood glucose monitoring rates are significantly lower for African Americans than those for non-Hispanic Whites (Levine et al., 2009), further increasing the risk of uncontrolled diabetes. Minority populations such as African Americans constitute 20% of the rural population, where there is limited access to primary healthcare and lower screening rates of chronic diseases (Johnson, 2012). Death rates from diabetes in African Americans are higher in rural areas of the United States than in urban areas; they are also higher than the rates in the rural White population (Probst et al., 2004; Slifkin et al., 2000; Starfield et al., 2005).

Policies and procedures

Our study analyzed BMI data of the general adult population between January 2018 and December 2020 at the VCU Health System, an academic safety-net healthcare institution serving a large underserved African American population. We also analyzed HbA1c lab results and the number of ED visits from diabetic patients in the same population and timeframe. We hypothesized that the disruption in everyday life caused by the COVID-19 pandemic might have caused an increased obesity rate as well as less effective management of diabetes, manifesting in higher HbA1c values and more frequent ED visits. Our study had mixed results (Fig. 1). While some of the models showed significant differences between the prepandemic and pandemic timeframes, the measured difference was negligible. It is difficult to assess what parts of the changes shown by our analysis were caused by a long-term trend that would have occurred without the COVID-19 pandemic, and what parts of the changes were caused by the pandemic.

Our analysis reaffirms that the pandemic has affected the African American population more severely than the White population (Tables 1–4). Our most significant finding is that the number of ED visits significantly decreased from 2018 and 2019 to 2020 (Tables 5 and 6), which can be easily explained by restrictions that were put in place at medical facilities and by the fear of infection at these facilities. These factors likely prevented necessary care seeking and caused harm and death.

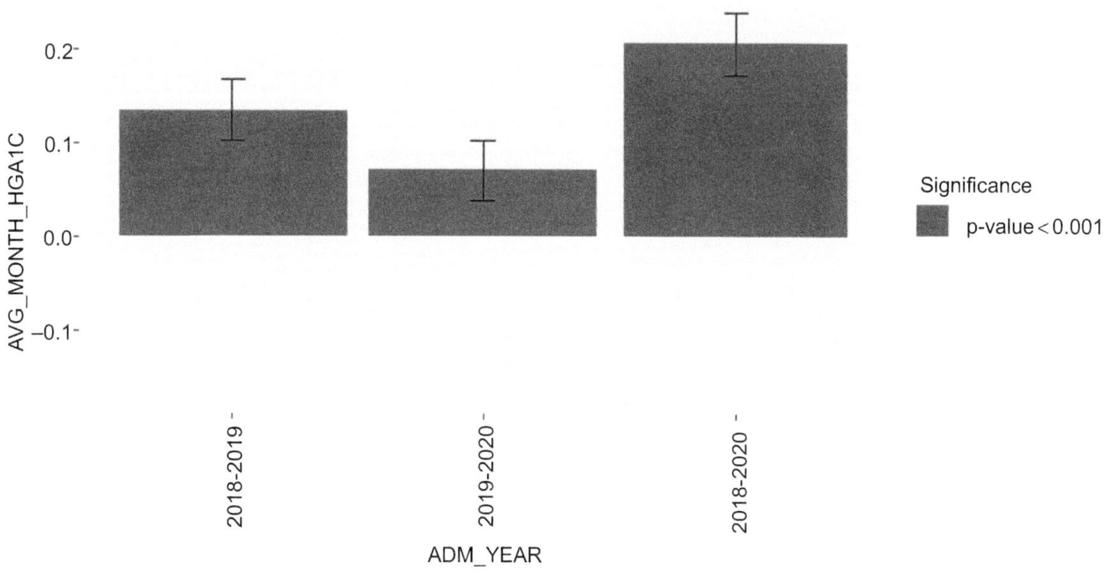

FIG. 1 HbA1c/year contrast.

TABLE 1 Body mass index (BMI) cohort summary table.

Variable	Group	2018	2019	2020	2018	2019	2020
		Total			Percentage		
Total		69,604	89,718	98,475	46.5	53.4	63.4
Age	18–19	1724	2272	1575	2.5	2.5	1.6
Age	20–39	16,451	21,288	24,177	23.6	23.7	24.6
Age	40–59	25,030	31,032	33,181	36.0	34.6	33.7
Age	>60	26,264	34,977	39,542	37.7	39.0	40.2
Age	Missing	135	149	0	0.2	0.2	0
Gender	Female	43,057	54,848	60,034	61.9	61.1	61.0
Gender	Male	26,546	34,869	38,440	38.1	38.9	39.0
Race	American Indian	70	91	103	0.1	0.1	0.1
Race	Asian	744	1022	1090	1.1	1.1	1.1
Race	African American	30,099	37,024	41,024	43.2	41.3	41.7
Race	Multiple	114	148	167	0.2	0.2	0.2
Race	Other/unknown	2811	4584	5041	4.0	5.1	5.1
Race	White	35,766	46,849	51,050	51.4	52.2	51.8
Overweight	Yes (BMI ≥ 25)	51,250	66,019	72,402	73.6	73.6	73.5
Overweight	No (BMI < 25)	18,354	23,699	26,073	26.4	26.4	26.5
		Mean			Standard deviation		
BMI	American Indian	30.52	30.99	30.66	7.63	7.40	6.94

Continued

TABLE 1 Body mass index (BMI) cohort summary table—cont'd

Variable	Group	2018	2019	2020	2018	2019	2020
BMI	Asian	25.64	25.71	25.69	4.93	5.03	4.98
BMI	African American	31.97	32.03	31.98	8.21	8.28	8.36
BMI	Multiple	29.63	29.62	29.54	8.76	8.90	8.40
BMI	Other/unknown	29.11	29.32	29.31	6.58	6.80	6.75
BMI	White	29.49	29.53	29.51	7.19	7.20	7.24
BMI	Overall	30.56	30.57	30.54	7.75	7.78	7.82

TABLE 2 Diabetes cohort.

Variable	Group	2018	2019	2020	2018	2019	2020
		Total			Percentage		
Total		19,482	22,376	27,652	49.4	56.8	70.2
Age	18–19	83	99	119	0.4	0.4	0.4
Age	20–39	1551	1667	1965	8.0	7.4	7.1
Age	40–59	7757	8420	9720	39.8	37.6	35.2
Age	>60	10,084	12,179	15,836	51.8	54.4	57.3
Age	Missing	7	11	12	0.0	0.0	0.0
Gender	Female	11,096	12,544	14,966	57.0	56.1	54.1
Gender	Male	8386	9832	12,686	43.0	43.9	45.9
Race	American Indian	29	31	37	0.1	0.1	0.1
Race	Asian	194	226	274	1.0	1.0	1.0
Race	African American	10,898	12,168	14,580	55.9	54.4	52.7
Race	Multiple	40	44	53	0.2	0.2	0.2
Race	Other/unknown	865	1090	1647	4.4	4.9	6.0
Race	White	7456	8817	11,061	38.3	39.4	40.0
High HbA1c	Yes (HbA1c ≥ 6.4)	4271	5224	6752	54.2	59.3	63.0
High HbA1c	No (HbA1c < 6.4)	3606	3581	3969	45.8	40.7	37.0
Obese	Yes (BMI ≥ 25)	15,473	177,714	20,129	87.3	86.5	84.9
Obese	No (BMI < 25)	2248	2755	3589	12.7	13.5	15.1
		Mean			Standard deviation		
HbA1c	American Indian	7.153	7.785	7.520	1.135	1.468	1.588
HbA1c	Asian	6.773	7.380	7.308	1.361	1.668	1.673
HbA1c	African American	7.288	7.363	7.475	2.017	1.917	1.995
HbA1c	Multiple	7.178	7.152	7.279	1.681	1.444	2.498
HbA1c	Other/unknown	7.095	7.248	7.331	1.749	1.771	1.816

TABLE 2 Diabetes cohort—cont'd

Variable	Group	2018	2019	2020	2018	2019	2020
HbA1c	White	6.967	7.105	7.234	1.651	1.630	1.735
HbA1c	Overall	7.150	7.254	7.370	1.870	1.799	1.886
ED visits	American Indian	0.413	0.129	0.324	0.682	0.428	0.784
ED visits	Asian	0.211	0.248	0.266	0.490	0.626	0.689
ED visits	African American	0.855	0.845	0.727	1.409	1.409	1.270
ED visits	Multiple	2.000	1.977	1.321	1.812	1.759	1.300
ED visits	Other/unknown	0.559	0.530	0.420	1.108	1.198	0.985
ED visits	White	0.487	0.489	0.453	1.088	1.102	0.962
ED visits	Overall	0.696	0.685	0.595	1.291	1.294	1.144

ED, emergency department; *HbA1c*, hemoglobin A1c.

TABLE 3 BMI models using the BMI cohort.

Model	Fixed effects	Estimates	Random effects	Estimate of SD	ICC
Model 1(Linear Model)	Intercept	30.57 (30.540, 30.596)	NA	NA	NA
	Time	−0.013 (−0.034, 0.009)			
Model 2	Intercept	30.22 (30.173, 30.267)	Patient	7.380 (7.347, 7.413)	0.932
	Time	0.0083 (0.002, 0.0143)	Residual	1.999 (1.996, 2.002)	
Model 3	Intercept	30.22 (30.17, 30.26)	Patient	7.380 (7.347, 7.413)	0.932
	yr2018	−0.0174 (−0.029, −0.006)			
	yr2019	0.0003 (−0.010, 0.011)	Residual	1.999 (1.996, 2.002)	
Model 4	Intercept	28.53 (28.35, 28.72)	Patient	7.273 (7.241, 7.306)	0.930
	yr2018	−0.182 (−0.235, −0.130)			
	yr2019	−0.094 (−0.139, −0.049)			
	raceAA	3.174 (2.979, 3.369)			
	raceWhite	0.725 (0.532, 0.917)	Residual	1.995 (1.996, 2.002)	
	2018*AA	0.133 (0.078, 0.189)			
	2019*AA	0.067 (0.019, 0.115)			
	2018*White	0.210 (0.154, 0.265)			
	2019*White	0.129 (0.082, 0.177)			

AA, African Americans; *NA*, not applicable.

TABLE 4 GEE models for hemoglobin A1c (diabetes cohort) and body mass index (BMI cohort).

Model	Effect	Estimates	Sandwich estimator SE	Wald statistic	*p*-Value
High HbA1c (≥6.4)	Inter:No	0.214	0.089	5.826	0.016
	timeYr	0.360	0.090	16.112	0.000
	AA	0.109	0.094	1.334	0.248
	White	0.058	0.097	0.365	0.546
	yr2020	−0.270	0.148	3.341	0.068
	AA:yr2020	0.219	0.158	1.922	0.166
	White:yr2020	0.175	0.161	1.173	0.279
	timeYr:AA	−0.194	0.096	4.122	0.042
	timeYr:White	−0.155	0.098	2.472	0.116
Overweight (BMI ≥ 25)	Inter:No	0.705	0.040	304.134	0.000
	timeYr	0.047	0.033	2.052	0.152
	AA	0.633	0.044	210.624	0.000
	White	0.203	0.043	22.681	0.000
	yr2020	−0.031	0.046	0.456	0.499
	AA:yr2020	−0.011	0.050	0.046	0.830
	White:yr2020	0.042	0.049	0.728	0.393
	timeYr:AA	−0.040	0.035	1.279	0.258
	timeYr:White	−0.054	0.034	2.516	0.113

AA, African Americans; *Yr*, Year.

TABLE 5 Hemoglobin A1c models using the diabetes cohort.

Model	Fixed effects	Estimates	Random effects	Estimate of SD	ICC
Model 1(Linear Model)	Intercept	7.148 (30.540, 30.596)	NA	NA	NA
	Time	0.110 (−0.034, 0.009)			
Model 2	Intercept	7.132 (7.099, 7.165)	Patient	1.559 (1.537, 1.582)	0.668
	Time	0.103 (0.086, 0.119)	Residual	1.100 (1.089, 1.110)	
Model 3	Intercept	7.328 (7.30, 7.36)	Patient	1.559 (1.537, 1.582)	0.668
	yr2018	−0.206 (−0.240, −0.173)			
	yr2019	−0.071 (−0.102, −0.039)	Residual	1.099 (1.089, 1.110)	
Model 4	Intercept	7.28 (7.169, 7.392)	Patient	1.554 (1.532, 1.577)	0.668
	yr2018	−0.285 (−0.409, −0.161)			
	yr2019	−0.022 (−0.139, 0.095)			
	raceAA	0.163 (0.044, 0.283)			
	raceWhite	−0.106 (−0.229, 0.017)	Residual	1.099 (1.089, 1.110)	
	2018*AA	0.099 (−0.033, 0.231)			
	2019*AA	−0.069 (−0.195, 0.055)			
	2018*White	−0.058 (−0.077, 0.193)			
	2019*White	−0.031 (−0.161, 0.094)			

AA, African Americans; *NA*, not applicable.

TABLE 6 Emergency department visit count models using the diabetes cohort.

Model	Fixed effects	Estimates (p-value)	Random effects	Estimate of SD	ICC ($\sigma_e^2 = 2.36$)
Model 1(Linear Model)	Intercept	−0.339 (<0.05)	NA	NA	
	Time	−0.081 (<0.05)			
Model 2	Intercept	−1.075 (<0.05)	Patient (intercept)	1.216	0.386
	Time	−0.067 (<0.05)			
Model 3	Intercept	−1.099 (<0.05)	Patient (intercept)	1.217	0.386
	yr2018	0.129 (<0.05)			
	yr2019	0.133 (<0.05)			
Model 4	Intercept	−1.530 (<0.05)	Patient (intercept)	1.188	0.374
	yr2018	0.086 (<0.05)			
	yr2019	0.099 (<0.05)			
	AA	0.570 (<0.05)			
	2018*AA	0.054 (0.04)			
	2019*AA	0.05 (0.06)			

AA, African Americans; *NA*, not applicable.

Applications to other areas

The respiratory disease COVID-19 has had an enormous effect on global health. An overlooked outcome, however, is its indirect effects on obesity, and how that has impacted diabetes. The prevalence of obesity continues to rise worldwide, and it is often described as being an epidemic. In the United States, the African American population is especially affected by the obesity epidemic. Looking back at a longer period of time before the pandemic, and also having access to more data since the beginning of the pandemic would make it easier to separate the effects of the pandemic from other general trends. Having only one health system's data included in the analysis is also a limitation. Results might be more generalizable if a similar analysis would be performed on larger consolidated datasets, such as the National Covid Cohort Collaborative (N3C) (Haendel et al., 2021). Future research in this field may include analyzing data from multiple health systems, using a longer time period, and adding more comorbidities in the analysis. Exploring patient behavior related to stress, diet, exercise, and other lifestyle factors is also necessary.

Summary points

- Obesity has a major effect on diabetes outcome.
- During the pandemic, HbA1c measurements showed an increasing trend in diabetics.
- Emergency department visits of diabetic patients decreased in 2020.
- African Americans were more negatively affected compared to other races.
- Research into exploring patient behavior related to stress, diet, exercise, and other lifestyle factors is necessary.

References

Almandoz, J. P., Xie, L., Schellinger, J. N., Mathew, M. S., Gazda, C., Ofori, A., … Messiah, S. E. (2020). Impact of COVID-19 stay-at-home orders on weight-related behaviors among patients with obesity. *Clinical Obesity*. https://doi.org/10.1111/cob.12386.

American Diabetes Association. (2011). Standards of medical care in diabetes-2011. *Diabetes Care*, *34*(S1), S11–S61.

American Heart Association. (2021). Obesity and cardiovascular disease: A scientific statement from the American Heart Association. *Circulation*. https://www.ahajournals.org/doi/abs/10.1161/CIR.0000000000000973.

American Psychological Association. (2020). *Stress in America: One year later, a new wave of pandemic health concerns*. Available from: https://www.apa.org/news/press/releases/stress/2021/sia-pandemic-report.pdf.

Ashby, N. J. S. (2020). The impact of the COVID-19 pandemic on unhealthy eating in populations with obesity. *Obesity*. https://doi.org/10.1002/oby.22940.

Bakaloudi, D. R., Barazzoni, R., Bischoff, S. C., Breda, J., Wickramasinghe, K., & Chourdakis, M. (2021). Impact of the first COVID-19 lockdown on body weight: A combined systematic review and a meta-analysis. *Clinical Nutrition*. https://doi.org/10.1016/j.clnu.2021.04.015.

Bays, H. E., Chapman, R. H., Grandy, S., & SHIELD Investigators' Group. (2007). The relationship of body mass index to diabetes mellitus, hypertension and dyslipidaemia: Comparison of data from two national surveys [published correction appears in Int J Clin Pract. 61 (10), 1777-1778]. *International Journal of Clinical Practice*, *61*(5), 737–747. https://doi.org/10.1111/j.1742-1241.2007.01336.x.

CDC. (2021). *US diabetes surveillance system national diabetic medication use*. Retrieved January 8, 2021, from https://gis.cdc.gov/grasp/diabetes/DiabetesAtlas.html.

Centers for Disease Control and Prevention. (2020). *National diabetes statistics report, 2020*. Atlanta, GA: Centers for Disease Control and Prevention, U.S. Dept of Health and Human Services.

Fryar, C. D., Carroll, M. D., & Afful, J. (2020). *Prevalence of overweight, obesity, and severe obesity among adults aged 20 and over: United States, 1960–1962 through 2017–2018*. NCHS Health E-Stats.

Ghosal, S., Sinha, B., Majumder, M., & Misra, A. (2020). Estimation of effects of nationwide lockdown for containing coronavirus infection on worsening of glycosylated haemoglobin and increase in diabetes-related complications: A simulation model using multivariate regression analysis. *Diabetes and Metabolic Syndrome: Clinical Research and Reviews*. https://doi.org/10.1016/j.dsx.2020.03.014.

Goldstein, D. E., Little, R. R., Lorenz, R. A., et al. (2004). Tests of glycemia in diabetes. *Diabetes Care*, *27*, 1761–1773.

Haendel, M. A., Chute, C. G., Bennett, T. D., Eichmann, D. A., Guinney, J., Kibbe, W. A., Gersing, K. R., … N3C Consortium. (2021). The National COVID Cohort Collaborative (N3C): Rationale, design, infrastructure, and deployment. *Journal of the American Medical Informatics Association*, *28* (3), 427–443. https://doi.org/10.1093/jamia/ocaa196. PMID: 32805036.

Johnson, K. M. (2012). *Carsey Institute University of New Hampshire. Rural demographic change in the new century: Slower growth, increased diversity*. Available at: http://scholars.unh.edu/cgi/viewcontent.cgi?article=1158&context=carsey.

Khan, M. A., & Moverley Smith, J. E. (2020). "Covibesity," a new pandemic. *Obesity Medicine*, *100282*. https://doi.org/10.1016/j.obmed.2020.100282.

Kochanek, K. D., Murphy, S. L., Xu, J. Q., & Arias, E. (2019). *Deaths: Final data for 2017. National vital statistics reports. Vol. 68 no 9*. Hyattsville, MD: National Center for Health Statistics.

Kumar, A., Arora, A., Sharma, P., Anikhindi, S. A., Bansal, N., Singla, V., Khare, S., & Srivastava, A. (2020). Is diabetes mellitus associated with mortality and severity of COVID-19? A meta-analysis. *Diabetes & Metabolic Syndrome*, *14*(4), 535–545. https://doi.org/10.1016/j.dsx.2020.04.044.

Lee, J. M., Okumura, M. J., Davis, M. M., Herman, W. H., & Gurney, J. G. (2006). Prevalence and determinants of insulin resistance among U.S. adolescents: A population-based study. *Diabetes Care*, *29*(11), 2427–2432. https://doi.org/10.2337/dc06-0709.

Levine, D. A., Allison, J. J., Cherrington, A., Richman, J., Scarinci, I. C., & Houston, T. K. (2009). Disparities in self-monitoring of blood glucose among low-income ethnic minority populations with diabetes, United States. *Ethnicity and Disease*, *19*(2), 97–103 PMID: 19537217.

Pellegrini, M., Ponzo, V., Rosato, R., Scumaci, E., Goitre, I., Benso, A., Belcastro, S., Crespi, C., De Michieli, F., Ghigo, E., Broglio, F., & Bo, S. (2016). Changes in weight and nutritional habits in adults with obesity during the "lockdown" period caused by the COVID-19 virus emergency. *Nutrients*, *2020*, 12. https://doi.org/10.3390/nu12072016.

Probst, J. C., Moore, C. G., Glover, S. H., & Samuels, M. E. (2004). Person and place: The compounding effects of race/ethnicity and rurality on health. *American Journal of Public Health*, *94*(10), 1695–1703.

Saudek, C. D., Herman, W. H., Sacks, D. B., Bergenstal, R. M., Edelman, D., & Davidson, M. B. (2008). A new look at screening and diagnosing diabetes mellitus. *The Journal of Clinical Endocrinology & Metabolism*, *93*(7), 2447–2453. https://doi.org/10.1210/jc.2007-2174. PMID: 18460560.

Scheen, A. J. (2000). Treatment of diabetes in patients with severe obesity. *Biomedicine & Pharmacotherapy*, *54*(2), 74–79. https://doi.org/10.1016/s0753-3322(00)88855-1.

Slifkin, R., Goldsmith, L., & Ricketts, T. (2000). *Race and place: Urban-rural differences in health for racial and ethnic minorities. NC RHRP working paper series, No. 66*.

Starfield, B., Shi, L., & Macinko, J. (2005). Contribution of primary care to health systems and health. *The Milbank Quarterly*, *83*(3), 457–502. https://doi.org/10.1111/j.1468-0009.2005.00409.x.

The Diabetes Control and Complications Trial Research Group. (1993). The effect of intensive treatment of diabetes on the development and progression of long-term complications in insulin-dependent diabetes mellitus. *The New England Journal of Medicine*, *329*, 977–986.

Weykamp, C. (2013). HbA1c: A review of analytical and clinical aspects. *Annals of Laboratory Medicine*, *33*(6), 393. https://doi.org/10.3343/alm.2013.33.6.393.

Chapter 43

COVID-19 associated rhino-orbital-cerebral mucormycosis in patients with diabetes and comorbid conditions

Caglar Eker
Department of Otolaryngology, Faculty of Medicine, Balcali Hospital, University of Cukurova, Saricam, Adana, Turkey

Abbreviations

CAM	COVID-19-associated mucormycosis
CNS	central nervous system
CotH	fungal ligand spore coating homolog
CT	computed tomography
DM	diabetes mellitus
GRP-78	glucose regulatory protein 78
IL-1	interleukin-1
IL-6	interleukin-6
KOH	potassium hydroxide
L-AmB	liposomal amphotericin B
MRI	magnetic resonance imaging
PCR	polymerase chain reaction
ROCM	rhino-orbital-cerebral mucormycosis
SARS-CoV-2	severe acute respiratory syndrome coronavirus 2

Introduction

A pandemic of viral pneumonia caused by severe acute respiratory syndrome coronavirus 2 (SARS-CoV-2) began in Wuhan, China, in December 2019 and spread rapidly around the world. The disease caused by this virus was named COVID-19 and was named a global health emergency (Wu & McGoogan, 2020). COVID-19 has manifested itself in a wide range from mild rhinitis to life-threatening pneumonia during the pandemic. Since the onset of the pandemic, COVID-19 has shown many transitions in disease presentation, course, and outcomes. Many opportunistic infections have been associated with COVID-19 (Kubin et al., 2021). Mucormycosis is one of these opportunistic infections that has caused serious mortality since the beginning of the pandemic.

Mucormycosis is an opportunistic fungal infection caused by fungi of the order Mucorales. Immunocompromised patients such as those with diabetes mellitus (DM), malignancy, prolonged neutropenia, and hematopoietic system disorders are at increased risk for mucormycosis (Spellberg & Ibrahim, 2018, p. 1538). Mucormycosis is classified according to the parts of the body involved: rhino-orbital-cerebral, pulmonary, gastrointestinal, cutaneous, and disseminated. Rhino-orbital-cerebral mucormycosis (ROCM) is the most common form (39%) (Reid et al., 2020). The main route of transmission is by inhalation, and vascular invasion occurs by ingesting fungal spores that penetrate the sinonasal epithelium. This situation results in tissue ischemia and necrosis. Treatment consists of control of the underlying disease, surgical debridement, and systemic antifungal therapy. Early diagnosis and treatment are of paramount importance because of the high mortality and morbidity.

Management, Body Systems, and Case Studies in COVID-19. https://doi.org/10.1016/B978-0-443-18703-2.00043-8
Copyright © 2024 Elsevier Inc. All rights are reserved, including those for text and data mining, AI training, and similar technologies.

There was a dramatic increase in cases of mucormycosis, particularly ROCM, during the COVID-19 pandemic. These cases were named COVID-19-associated mucormycosis (CAM). Although a definite reason for this increase has not been revealed, immune changes caused by the virus (Peman et al., 2020), high-dose corticosteroid administrations given to severe COVID-19 patients (Ahmadikia et al., 2021), and the presence of uncontrolled DM in the majority of patients (Bhandari et al., 2022) were shown as possible reasons for this increase. It has been considered that conditions such as low oxygen concentration in the mucosa, a favorable acidic environment, a high glucose level, and a decrease in phagocytic activity resulting from immune suppression (COVID-19-mediated and corticosteroid use) played a role in the development of the disease in CAM (Sengupta & Nayak, 2022). When considering the possible mechanisms in the occurrence of CAM, DM, which is a common comorbid condition, and high-dose steroids, which were frequently implemented especially at the beginning of the pandemic, are the most important risk factors. Many reports, especially from India, have been made regarding the coexistence of these three conditions during the pandemic. Taking all these reports and cases into consideration, it is necessary to consider COVID-19, mucormycosis, and DM as an emerging disease association.

Etiology and risk factors

There are multiple possible factors contributing to the development of mucormycosis in patients with COVID-19, including diabetes mellitus, corticosteroid use, hyperglycemia, ketoacidosis, and the development of cytokine storms. Diabetes mellitus can be considered the most important factor that causes CAM development and complicates its management. In addition to the effects of diabetes on the immune system, hyperglycemia that occurs in patients with poor diabetes control also contributes to the development of CAM (Moorthy et al., 2021). In fact, in a study, patients with hyperglycemia of blood glucose >400 mg/dL were qualified as the highest risk group for the development of CAM (Arora et al., 2022). Apart from preexisting DM, hyperglycemia due to COVID-19-mediated pancreatic islet cell destruction contributes to the formation of CAM. The acidic environment that will occur in diabetic ketoacidosis facilitates the germination of fungal spores and their invasion into the tissue.

Although high-dose corticosteroids have proven beneficial in some hospitalized groups of COVID-19 patients, it increases the risk of developing secondary infections such as CAM (Horby et al., 2021). Moreover, this treatment predisposes a patient to hyperglycemia. In the study by John et al., they found that 94% of CAM patients had diabetes. They also found that 88% of them received corticosteroid treatment (John et al., 2021). Singh et al. (2021) found that 80% of 101 cases had diabetes and 76.3% of them took steroids for the treatment of COVID-19. In a series of 235 cases originating from India, 84.3% of the patients received corticosteroid therapy, 44.7% had known DM, and 42.1% had newly diagnosed DM (Bhandari et al., 2022). In a comprehensive review on this subject, a high number of CAM cases were examined, and it was stated that 81.6% of their patients, 82% of Indian cases, 81.6% of Egyptians, and 82.5% of Europeans received corticosteroid treatment (Almyroudi et al., 2022). Even though the number of cases in our clinic was not as high as in the studies mentioned above, all the CAM patients we followed had known or new-onset DM. Furthermore, most of the patients had received corticosteroid treatment due to COVID-19. Both the reports in the literature and our clinical experience demonstrate that these two risk factors play the most important roles in the development of the disease.

Immunosuppression, hematological malignancies, and solid organ or bone marrow transplantation are also well-known but less-common risk factors for CAM. Zinc consumption is another risk factor discussed in the literature. Despite one study claiming zinc to be associated with CAM, another study provided evidence for the protective effect of zinc use (Kumar et al., 2022). Besides all these risk factors, the fact that the cases are especially observed in equatorial regions such as India and Egypt suggests that climatological conditions such as high humidity and temperature are also risk factors.

Possible reasons

Even though it is quite well known that diabetes is the most important factor in the emergence of ROCM, it has been thought that COVID-19 have a synergistic effect on ROCM development directly and indirectly with its certain characteristics and effects. Various scenarios have been suggested for this association. While COVID-19 has direct effects that cause susceptibility to the disease, it also has indirect effects such as predisposing a person to hyperglycemia and immune suppression by the corticosteroids used in its treatment. Some proposed scenarios provide information about the role of COVID-19 in triggering the disease.

Diabetes and corticosteroids cause immunosuppression by causing impaired neutrophilic and macrophage phagocytic functions. COVID-19 triggers thromboembolic phenomena, causing endothelial damage and thrombosis. Moreover, it causes a decrease in the level of CD4+ and CD8+ T lymphocytes, leading to mucormycosis (Anand et al., 1992). In the cytokine storm observed in COVID-19 patients, the increase of interleukin-6 (IL-6) and interleukin-1 (IL-1) especially leads to an increase in serum free iron levels, which is an essential element for the growth of mucor (Iwasaki et al., 2021). COVID-19 itself causes an increase in blood sugar levels by its effect on the pancreas (Kusmartseva et al., 2020) and provides an optimum environment with a low oxygen concentration, which will facilitate the growth of mucor (Singh et al., 2021).

Diabetes, COVID-19, and corticosteroids cause hyperglycemia. Increased blood sugar creates an appropriate environment for mucor spores to germinate. The acidic environment that will occur in the event of ketoacidosis supports the germination of mucor spores. Low pH does not allow free iron to bind to transferrin, and serum free iron levels rise, leading to mucormycosis formation (Artis et al., 1982). Furthermore, elevated blood glucose leads to the glycosylation of transferrin and ferritin, resulting in decreased iron binding and increased free iron levels. Another proposed mechanism is the increased expression of the fungal ligand spore coating homolog (CotH) protein and glucose regulatory protein 78 (GRP-78) in endothelial cells, enabling angio-invasion and tissue necrosis (Gebremariam et al., 2014) (Fig. 1).

Clinical presentation

Infection begins with the inhalation of spores into the oral and nasal cavities. It rarely develops in individuals with an intact immune system, as fungal spores are phagocytosed by macrophages. However, this infection may appear due to severe immune system suppression in patients with uncontrolled diabetes mellitus and COVID-19 and/or given high doses of corticosteroids used in its treatment. This fungal infection begins in the nasal and sinus mucosa. It then spreads to the orbit, usually via vascular or anatomical proximity. Its next stop will be the central nervous system (CNS). Although it has various forms such as pulmonary, gastrointestinal, and disseminated, the most common clinical presentation of mucormycosis is the rhino-orbito-cerebral form (Sugar, 1992).

In most cases, there is a time period between the development of CAM and the diagnosis of COVID-19. According to the review from Hoenigl et al. (2022), this period is a median 10 days. According to Sen et al. (2021), it is 14.5 days, and according to Almyroudi et al. (2022), it is 17.8 days. However, CAM cases cooccurring with acute COVID-19 have also been reported, albeit less frequently. In a review, 24% of included patients had the coexistence of COVID-19 and mucormycosis at the time of hospital admission. In the remaining patients, the time taken for the diagnosis of CAM was correlated with the control status of the patient's diabetes. This period was observed to be shorter in patients with uncontrolled DM compared to those with controlled DM (Hoenigl et al., 2022). It should be kept in mind that the time period for the diagnosis

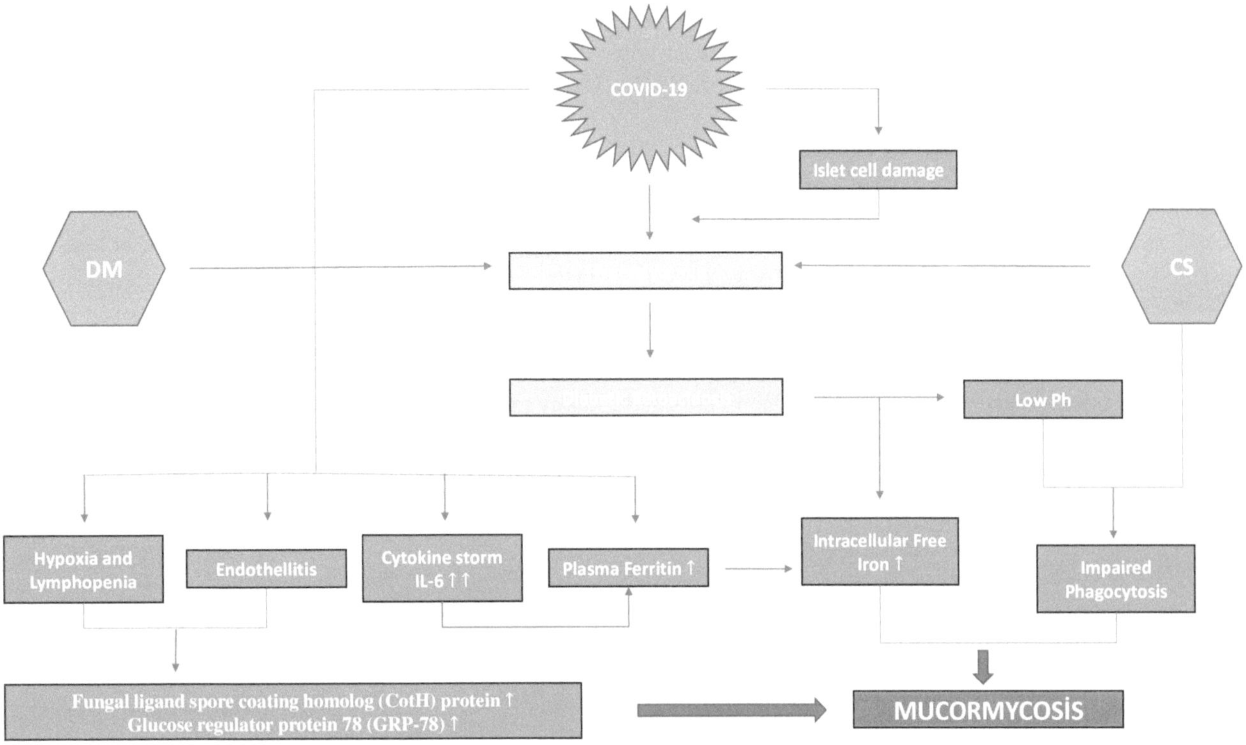

FIG. 1 Putative associations of diabetes, corticosteroids, and COVID-19 with mucormycosis.

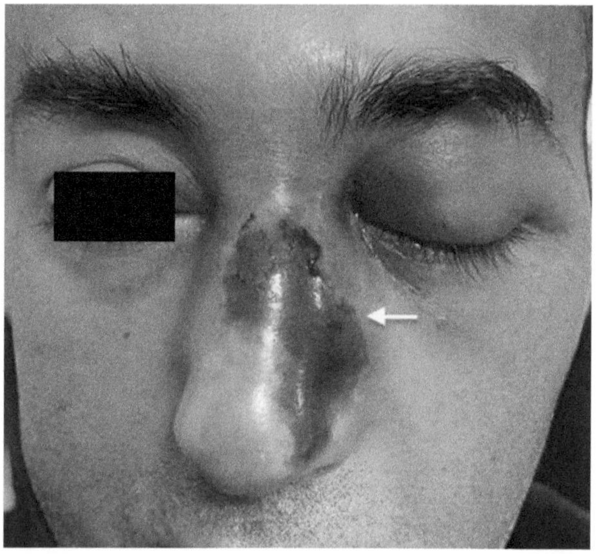

FIG. 2 The clinical picture showing advanced rhino-orbital mucormycosis invading the nasal dorsum skin and causing necrosis (*white arrow*, developed black scar due to mucormycosis).

of CAM may be related to the general clinical condition of the patient (delayed emergence of CAM findings in patients with relatively good clinical status), and may also reflect the delay in diagnosis. The importance of early diagnosis in the management of ROCM is well known. Thus, a delay in diagnosis may result in increased mortality (Fouad et al., 2021).

According to the European Confederation of Medical Mycology, symptoms of ROCM vary depending on where the infection developed. Unilateral facial swelling and/or proptosis, fever, nasal congestion, nasal discharge, facial pain and headache, loss of vision, and dark lesions on the nasal dorsum or palate are the main signs and symptoms (Cornely et al., 2019) (Fig. 2). Symptoms attributed to nasal and oral cavity invasion include epistaxis, nasal congestion, and palate destruction. Orbital extension may cause destruction of the ophthalmic artery and optic nerves, resulting in ptosis, proptosis, visual disturbances, and total vision loss. According to a systematic review conducted before the COVID-19 pandemic, the most common symptom of ROCM is facial swelling (64.5%). This is followed by fever (62.9%) and nasal congestion (52.2%). Ocular symptoms were present in approximately 50% of patients, headache in fewer than 50%, and palatal necrosis symptoms in 20.8% (Turner et al., 2013). Together with COVID-19, some differences were observed in the clinical presentation of ROCM patients. Orbital and ocular signs and symptoms were more prominent in CAM. In a multicenter comprehensive review of 2826 patients from India, the most common primary symptom frequencies of ROCM were orbital/facial pain (23%), orbital/facial edema (21%), visual loss (19%), ptosis (11%), and nasal congestion (9%). The most common primary signs of ROCM were periocular/facial edema (33%), vision loss (21%), ptosis (12%), proptosis (11%), and nasal discharge (10%). Looking at the cumulative incidence of clinical signs of ROCM, vision loss (63%) was the most common symptom, followed by periocular or facial edema (61%), ptosis (54%), and proptosis (38%) (Sen et al., 2021).

Fungal hyphae invading the orbit can involve the cavernous sinus through anatomical passages in the orbital apex. This condition manifests itself with diplopia and ophthalmoplegia (Pai et al., 2021). The disease spreads from here to other cerebral structures. Apart from cavernous sinus involvement, it may present as cerebral abscess, meningitis, or cerebrovascular disease (Mehta et al., 2022). Cerebral involvement was noted in 18.9% of CAM patients, according to a review of 7388 patients (Almyroudi et al., 2022). Cerebral involvement is notable as it is associated with high mortality.

Diagnosis

Early diagnosis is of paramount importance, as delayed CAM treatment is associated with high mortality (Meshram et al., 2021). Diagnosing mucormycosis depends on the availability of experienced personnel and appropriate techniques. Because mucormycosis is a quite challenging disease to manage, suspected cases should be referred to the referral center immediately. The most critical step in the diagnosis of mucormycosis is the clinician's suspicion. If symptoms and signs occur in patients with predisposing factors, the clinician should definitely consider mucormycosis and conduct a detailed examination in this direction. This examination begins with a detailed endoscopy (Fig. 3) and radiological imaging.

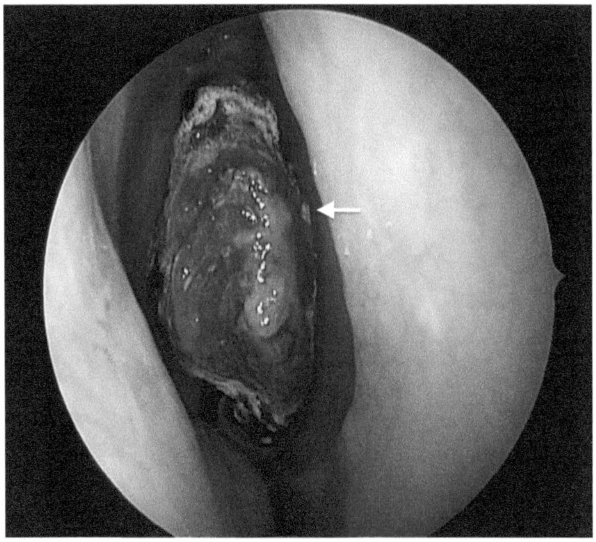

FIG. 3 Nasal endoscopy picture showing early stage mucormycosis involving isolated left middle turbinate (*white arrow*, black eschar).

Maxillofacial and cranial computer tomography (CT) or magnetic resonance imaging (MRI) is highly recommended to determine whether ROCM is present in diabetic patients in the presence of symptoms of facial pain, sinusitis, proptosis, and ophthalmoplegia (Cornely et al., 2019). In addition, endoscopy and radiological imaging are required for mucormycosis in patients with hematological malignancies in case of fever that does not respond to antibacterial therapy.

According to the criteria recommended by the European Confederation of Medical Mycology, the diagnosis of mucormycosis is based on clinical and imaging features and confirmed with direct microscopy, histopathological analysis, and culture of specimens taken by biopsy (Cornely et al., 2019). The diagnosis is challenging as suitable specimens are obtained by invasive procedures and specific dyes are needed for diagnosis. Direct microscopy with potassium hydroxide (KOH) is the most widely used method and is important for the rapid diagnosis of mucormycosis due to delayed results with culture and histopathology. Direct microscopy reveals broad, nonsegmented, ribbon-like hyaline hyphae with irregular right-angle branching characteristic of Mucorales (Samson & Dharne, 2022). Besides KOH, staining with fluorescent dye blankophor or chalcofluor white in direct microscopy is another method (Jung et al., 2015). Although direct microscopy is a rapid and beneficial method for diagnosis, it is difficult for it to distinguish mucorales from other molds. In addition, the fact that the patient received antifungal treatment before the diagnosis alters the structure of the hyphae and complicates the diagnosis by direct microscopy (Fathima et al., 2021). Therefore, it is recommended that the diagnosis of mucormycosis be confirmed and supported by histopathology, culture, or molecular techniques. In the histopathological analysis of the tissue, vascular infarcts, angioinvasion, and perineural invasion are detected. Mucorales has been difficult to cultivate in culture due to the fragility of the hyphae. Even when mucormycosis is recognized on direct microscopic and histopathological analyses, the culture may be negative in 50% of cases (Samson & Dharne, 2022). In a review of 2175 cases with CAM with microbiological evidence, evidence was obtained through direct microscopy in 89% of cases while only 19% of cases had cultivation in culture (Sen et al., 2021). Molecular tests that are not commonly used include monoclonal antibodies or polymerase chain reaction (PCR) and immunohistochemistry techniques applied on paraffin embedded tissue. In addition to studies claiming to be highly specific, there are also studies reporting variations in sensitivity (Drogari-Apiranthitou et al., 2016; Sunagawa et al., 2013).

Radiological imaging

CT and MRI are imaging modalities utilized in the diagnosis of CAM, the determination of the extent of the disease, and presurgical planning. CT evaluates bone structures better while MRI evaluates soft tissues better. In ROCM patients, these examinations should be performed to cover the maxillofacial, orbit, and CNS. If the patient has no contraindications, CT and MRI should be performed with contrast.

Radiological findings can be listed as signs of sinusitis (thickness of the mucosa, opacification of the sinuses, air fluid levels), irregulation of periorbital fatty tissue (Fig. 4), cavernous sinus thrombosis and infiltration (Fig. 5), internal carotid artery infiltration, cerebritis, cerebral infarction, mycotic aneurysms, subarachnoid hemorrhage, and subdural, epidural, or

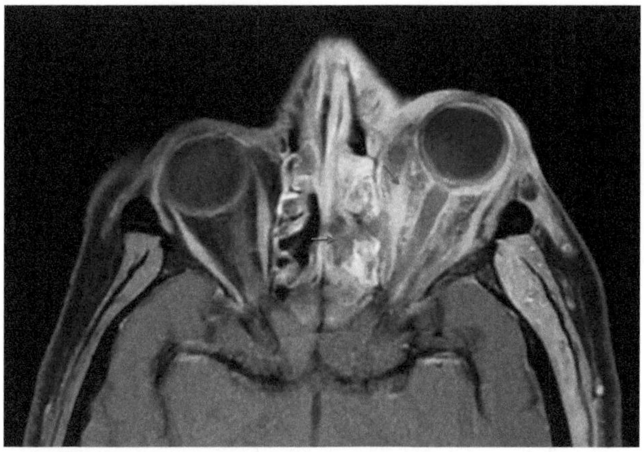

FIG. 4 Axial contrast enhanced MRI (T1) of the orbit and paranasal sinuses showing mucosal thickenings in the ethmoid sinuses, contrast enhancement, and soft tissue images consistent with necrotic material that does not retain contrast in places (*red arrow*), and edema and contrast enhancement in the periorbital tissues consistent with diffuse inflammation in the periorbital tissues (*blue arrow*).

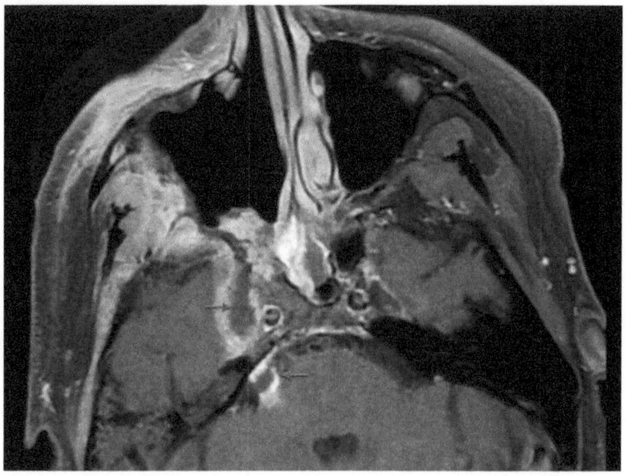

FIG. 5 Axial contrast enhanced MRI (T1) of the paranasal sinuses showing cavernous sinus thrombosis and involvement, adjacent to the right temporal lobe (*blue arrow*), brainstem involvement (*red arrow*).

cerebral abscess formation (Pai et al., 2021). The "black turbinate mark" appears on MRI as a result of concha necrosis due to mucormycosis. However, it is not a specific finding for ROCM. Bone structure erosion such as paranasal sinuses, cribriform plate, and lamina papricea can be seen frequently in CT evaluation (Patel, Adke, et al., 2021).

The early detection of soft tissue involvement and a better evaluation of orbital and cerebral involvement compared to CT are the advantages of MRI (Sen et al., 2021). In addition to these, MRI is the preferred method for early diagnosis because erosions of bone structure are not an expected finding in the early period. However, in cases where MRI is not feasible, the patient should be evaluated with contrast-enhanced CT. Moreover, a paranasal sinus CT yields necessary information for the surgeon in planning before surgery. Positron emission tomography is a beneficial tool for detecting and evaluating response to treatment, but its cost is its major limitation (Liu et al., 2013).

Management and treatment

The treatment approach in CAM patients is similar to that of invasive ROCM patients in other risk groups. However, in patients with hematological malignancies, treatment focuses more on the management of neutropenia, whereas in patients with CAM it usually focuses more on diabetes and corticosteroid exposure. Treatment of mucormycosis in CAM patients mainly includes control of hyperglycemia and other risk factors, optimal surgical debridement, and antifungal agents (Cornely et al., 2019). The importance of the combined treatment approach was shown in a large review of 929 cases. This review indicated that survival was 3% without any intervention, 57% with surgery alone, 61% with amphotericin deoxycholate, and 70% when treated with both amphotericin and surgical debridement (Roden et al., 2005).

First, it is necessary to carefully control the impaired glucose balance. If the patient has ketoacidosis at presentation, it is important to treat quickly with intravenous hydration and insulin. In the long-term use of high-dose corticosteroids, their reduction and elimination should be considered.

Second, systemic antifungal therapy should be administered. High-dose liposomal amphotericin B (L-AmB) is the first-line therapy for mucormycosis (Cornely et al., 2019). It has two forms, amphotericin B deoxycholate and amphotericin B lipid complex. The liposomal form is preferred as it is less nephrotoxic and therefore higher doses can be given over an extended period of time (Lanternier & Lortholary, 2008). According to the Global Mucormycosis Guidelines published in 2019, 5–10 mg/kg L-AmB treatment per day is recommended in various organ involvements (Cornely et al., 2019). In the case of intracranial involvement, the dose can be increased up to 10 mg/kg per day. Increases in serum creatinine and decreases in resistant potassium are the most important dose-limiting side effects.

Posaconazole and isavuconazole have emerged as appropriate alternatives to L-AmB in the treatment of mucormycosis, especially in patients with renal dysfunction (Perfect et al., 2018). Besides, these azoles have indications for salvage treatment (Cornely et al., 2019). While posaconazole can be given intravenously, it can also be given orally with DR tablets with increased bioavailability. However, the serum levels of the drug should be monitored and kept above 1.0 g/mL, and liver functions should be monitored due to possible hepatotoxicity (Khanna & Kayarat, 2021). Isavuconazole is a better-tolerated molecule than posaconazole. Besides exhibiting less hepatotoxicity and fewer drug interactions, therapeutic drug monitoring is not required. It is used both orally and intravenously. Because its safety profile is good, it is well tolerated for use longer than 6 months. Thus, it has been suggested that it is not inferior to L-AmB as a primary therapy in mucormycosis, in refractory cases, and in toxicity to other antifungals (Marty et al., 2016). Combination therapy is controversial. In a study, it was suggested that the combination of L-AmB with echinocandins or posaconazole or isavuconazole would improve survival in resistant cases (Krishna et al., 2022). However, there is insufficient data to support the use of combination therapy. Randomized placebo-controlled studies are required. Another critical issue is that L-AmB, Posaconazole, and isavuconazole are the most effective drugs against mucormycosis while their efficacy varies by species (Divakar, 2022). Therefore, accurate species identification is crucial not only for the correct diagnosis of the disease but also for its treatment, and accurate and rapid species identification is urgently needed.

As the third pillar of treatment, the necrotic tissue from mucormycosis should be carefully debrided to control infection. Surgical debridement is indispensable in the treatment of mucormycosis and it should be kept in mind that repeated interventions may be required. Critical areas affecting surgical decision making include involvement of the premaxillary, periorbital soft tissues, hard and soft palate, pterygopalatine fossa, infratemporal fossa, vidian canal, and skull base. Surgical debridement of these areas is very challenging with conservative methods and aggressive surgery may be required. Perineural invasion is another hallmark of ROCM (Sravani et al., 2014). This feature is responsible for early optic nerve involvement and orbital apex syndrome. In this regard, the decision for orbital exenteration is often controversial and there are no standard guidelines for making this decision (Hargrove et al., 2006). As an alternative treatment method, patients with orbital involvement and good vision can be treated with transcutaneous retrobulbar injections of amphotericin B. Satisfactory responses have been reported, particularly in patients without necrosis (Walia et al., 2021).

Orbital exenteration is performed in the absence of vision and diffuse orbital involvement. Because there is no definite consensus on the indications and timing of orbital exenteration, the surgical decision belongs to the treating physician group. The most critical decision in the treatment of ROCM is whether to exenter the orbit. Orbital mucormycosis not only corresponds to advanced disease, but also poses a high risk of disease progression to the intracranial region due to the anatomical connections directly related to this region. Therefore, in the case of rapidly progressive disease with orbital invasion, emergency exenteration may be necessary to keep the patient alive. However, recent advances in medical therapy, such as low-toxicity L-AmB, may allow the orbit to be preserved in the case of slowly progressing disease. In the published case series, no significant difference was found in survival with or without orbital exenteration (Eker et al., 2023). In contrast, in a large series of patients from India, it was noted that orbital exenteration improved survival in patients with orbital involvement with total vision loss or more extensive involvement. They also suggested that orbital exenteration is beneficial in patients with CNS involvement, contrary to the standard understanding (Sen et al., 2021). There are similar controversial points and opposing views for the approach to be applied in CNS involvement. Although there are publications claiming to the contrary, it is accepted that surgical debridement in CNS involvement has no effect on survival (Hoenigl et al., 2022).

Prognosis and outcome

Mortality rates vary in the literature. In a comprehensive review of a high number patients on CAM, the median all-cause mortality rate was 21.4%. The median mortality was 50% in patients with CNS involvement (Almyroudi et al., 2022). In another comprehensive review, the overall mortality rate for ROCM associated with COVID-19 was 31% (Singh et al., 2021). The mortality rate reported in the patient series from Patel et al. was higher than in previous studies. The 6-week mortality rate was 38.3% and the 12-week mortality rate was 45.7% (Patel, Agarwal, et al., 2021). Possible reasons for this variability in mortality rates could be that the case series were from different geographical regions as well as the heterogeneity in patient groups in terms of age, comorbidity, etc.; the difficulty of diagnosis in concomitant pulmonary mucormycosis; and the variability of the severity of concomitant COVID-19.

The patient's ketoacidosis, CNS involvement, and development of ROCM while under moderate/severe COVID-19 treatment are poor prognostic factors (Dravid et al., 2022). Delay in CAM diagnosis can also be considered a poor prognostic factor. While the median mortality was 33.7% when CAM was diagnosed after 15 days, the rate decreased to 23.4% when it was diagnosed before 15 days (Cag et al., 2022). Furthermore, a history of mechanical ventilation due to COVID-19, older age, HbA1c level > 9.1%, advanced disease, development of renal dysfunction during hospital stay, orbital involvement, and history of tocilizumab use are also poor prognostic factors (Almyroudi et al., 2022).

Policies

The prevalence of diabetes in the community; the uncontrolled use of corticosteroids for the treatment of COVID-19, especially at the beginning of the pandemic; and the inhalation of fungal spores from oxygenated humidifiers by COVID-19 patients treated in the hospital have been effective in the formation of the COVID-19-related mucormycosis epidemic. The inefficiency of hospital sanitation and disinfection processes in developing countries, especially during periods such as pandemics, may have caused hospital waters to be a real reservoir for mucormycosis (Rammaert et al., 2012). Early detection of mucormycosis cases, prompt treatment with antifungal drugs and surgical operations, controlling glycemic levels and corticosteroid use in COVID-19 patients, and taking appropriate hygiene and sanitation measures will limit the increasing incidence of this fungal infection.

Even after patient recovery from COVID-19, active and regular follow-up is necessary, especially for those with an uncontrolled glycemic index and who have received high-dose corticosteroid therapy for a significant period of time. In particular, extreme caution should be exercised during the 15 days following the diagnosis of COVID-19 (Pal et al., 2021).

In addition to the high mortality of the disease, it can cause serious morbidity for surviving patients. These patients should be provided with posttreatment rehabilitation and psychological support. Rhino-orbito-cerebral mucormycosis has demonstrated the potential to be a serious public health concern in the COVID-19 pandemic. Mucormycosis must be dealt with aggressively through a joint effort from multidisciplinary medical teams and governments (Sen et al., 2021).

Another issue that needs to be mentioned is orbital exenteration. It is a substantially radical surgery. Although this intervention is sometimes life-saving, it should be considered repeatedly when making a decision because it is a surgery with high morbidity. High-dose L-AmB and/or transcutaneous retrobulbar L-AmB injection treatment may be given more weight, especially if it has arisen due to conditions that cause temporary immune changes such as COVID-19.

Application to other areas

In December 2019, the Chinese city of Wuhan became the starting point of the viral pneumonia epidemic called COVID-19 (Wu & McGoogan, 2020). Then, the virus spread rapidly around the world. It infected many people and caused their death. Since COVID-19 progressed with immune alterations, it caused an increase in deadly opportunistic infections such as mucormycosis (Rawson et al., 2020). Although the pandemic and the mucormycosis epidemic seem to be under control now, another viral disease or variant that will cause a pandemic in the future may cause a similar scenario, perhaps worse. Currently available therapeutic interventions are not sufficient to reduce the high mortality associated with mucormycosis. However, genetic studies have identified new genes encoding virulence factors such as multidrug resistance in mucor hyphae. These new developments are promising for future antifungal treatments (Trieu et al., 2017). Physicians will be more prepared for a possible future epidemic thanks to these new treatment methods.

Posaconazole and isavuconazole have emerged as appropriate alternatives to L-AmB in the treatment of mucormycosis, especially in patients with renal dysfunction (Perfect et al., 2018). While these two azole derivatives are utilized in cases where L-AmB cannot be used, studies have been conducted in which they were used as an alternative to L-AmB as the primary treatment (Manesh et al., 2016; Marty et al., 2016). In addition to the primary treatment, there are also studies in which it has been used in combination with L-AmB (Reed et al., 2008). Promising results have been obtained, but randomized, placebo-controlled studies are needed to determine a definitive benefit. These new treatment approaches can be used in unrelated cases as well as in mucormycosis associated with COVID-19.

Mini-dictionary of terms

- *Mucormycosis*. Serious but rare fungal infection caused by a group of molds called mucormycetes.
- *Interleukin*. A group of cytokines primarily synthesized by leukocytes that regulates immune responses.
- *Transferrin*. A blood plasma glycoprotein that plays a role in iron metabolism and is responsible for iron ion transport.
- *Ferritin*. A plasma protein containing iron, responsible for its storage.
- *Proptosis*. The swelling of one or both eyes out of their natural position.
- *Exenteration*. A surgical procedure that involves the removal of the entire orbit with its surrounding structures including muscles, fat, nerves, and eyelids.

Summary points

- Together with the COVID-19 pandemic, a dramatic increase was observed in the cases of rhino-orbito-cerebral mucormycosis.
- Immunosuppression due to diabetes, corticosteroid use, and COVID-19 predispose one to mucormycosis.
- Early diagnosis is paramount in the management of mucormycosis.
- Control of comorbid risk factors, surgical debridement, and antifungal agents constitute the basis of treatment.
- Rapid and optimal surgical debridement is one of the factors that improves survival.
- L-AmB is the first choice as an antifungal in the treatment of primary CAM.
- In patients who are not appropriate for L-AmB therapy due to renal dysfunction, posaconazole and isavuconazole can be utilized safely in treatment.
- Because there is no definite consensus on the indications and timing of orbital exenteration, the surgical decision belongs to the treating physician group.
- Although mortality rates of CAM vary, a delay in diagnosis as well as orbital and intracranial involvement are the main indicators of poor prognosis.

References

Ahmadikia, K., Hashemi, S. J., Khodavaisy, S., Getso, M. I., Alijani, N., Badali, H., et al. (2021). The double-edged sword of systemic corticosteroid therapy in viral pneumonia: A case report and comparative review of influenza-associated mucormycosis versus COVID-19 associated mucormycosis. *Mycoses*, *64*(8), 798–808.

Almyroudi, M. P., Akinosoglou, K., Rello, J., Blot, S., & Dimopoulos, G. (2022). Clinical phenotypes of COVID-19 associated mucormycosis (CAM): A comprehensive review. *Diagnostics (Basel)*, *12*(12), 3092.

Anand, V. K., Alemar, G., & Griswold, J. A. (1992). Intracranial complications of mucormycosis: An experimental model and clinical review. *The Laryngoscope*, *102*, 656–662.

Arora, U., Priyadarshi, M., Katiyar, V., Soneja, M., Garg, P., Gupta, I., et al. (2022). Risk factors for coronavirus disease-associated mucormycosis. *The Journal of Infection*, *84*(3), 383–390.

Artis, W. M., Fountain, J. A., Delcher, H. K., & Jones, H. E. (1982). A mechanism of susceptibility to mucormycosis in diabetic ketoacidosis: Transferrin and iron availability. *Diabetes*, *31*(12), 1109–1114.

Bhandari, S., Bhargava, S., Samdhani, S., Singh, S. N., Sharma, B. B., Agarwal, S., et al. (2022). COVID-19, diabetes and steroids: The demonic trident for mucormycosis. *Indian Journal of Otolaryngology and Head and Neck Surgery*, *74*(Suppl 2), 3469–3472.

Cag, Y., Erdem, H., Gunduz, M., Komur, S., Ankarali, H., Ural, S., et al. (2022). Survival in rhino-orbito-cerebral mucormycosis: An international, multicenter ID-IRI study. *European Journal of Internal Medicine*, *100*, 56–61.

Cornely, O. A., Alastruey-Izquierdo, A., Arenz, D., Chen, S. C. A., Dannaoui, E., Hochhegger, B., et al. (2019). Global guideline for the diagnosis and management of mucormycosis: An initiative of the European Confederation of Medical Mycology in cooperation with the Mycoses Study Group Education and Research Consortium. *The Lancet Infectious Diseases*, *19*, e405–e421.

Divakar, P. K. (2022). Fungal taxa responsible for mucormycosis/"black fungus" among COVID-19 patients in India. *Journal of Fungi (Basel)*, *7*(8), 641.

Dravid, A., Kashiva, R., Khan, Z., Bande, B., Memon, D., Kodre, A., et al. (2022). Epidemiology, clinical presentation and management of COVID-19 associated mucormycosis: A single Centre experience from Pune, Western India. *Mycoses*, *65*(5), 526–540.

Drogari-Apiranthitou, M., Panayiotides, I., Galani, I., Konstantoudakis, S., Arvanitidis, G., Spathis, A., et al. (2016). Diagnostic value of a semi-nested PCR for the diagnosis of mucormycosis and aspergillosis from paraffin-embedded tissue: A single center experience. *Pathology, Research and Practice*, *212*, 393–397.

Eker, C., Tarkan, O., Surmelioglu, O., Dagkiran, M., Tanrisever, I., Yucel Karakaya, S. P., et al. (2023). Alternating pattern of rhino-orbital-cerebral mucormycosis with COVID-19 in diabetic patients. *European Archives of Oto-Rhino-Laryngology*, *280*(1), 219–226.

Fathima, A. S., Mounika, V. L., Kumar, V. U., Gupta, A. K., Garapati, P., Ravichandiran, V., et al. (2021). Mucormycosis: A triple burden in patients with diabetes during COVID-19 pandemic. *Health Sciences Review (Oxford)*, *1*, 100005.

Fouad, Y. A., Bakre, H. M., Nassar, M. A., Gad, M. O. A., & Shaat, A. A. K. (2021). Characteristics and outcomes of a series of COVID-associated mucormycosis patients in two different settings in Egypt through the third pandemic wave. *Clinical Ophthalmology*, *15*, 4795–4800.

Gebremariam, T., Liu, M., Luo, G., Bruno, V., Phan, Q. T., Waring, A. J., et al. (2014). CotH3 mediates fungal invasion of host cells during mucormycosis. *The Journal of Clinical Investigation*, *124*(1), 237–250.

Hargrove, R. N., Wesley, R. E., Klippenstein, K. A., Fleming, J. C., & Haik, B. G. (2006). Indications for orbital exenteration in mucormycosis. *Ophthalmic Plastic & Reconstructive Surgery*, *22*(4), 286–291.

Hoenigl, M., Seidel, D., Carvalho, A., Rudramurthy, S. M., Arastehfar, A., Gangneux, J. P., et al. (2022). The emergence of COVID-19 associated mucormycosis: A review of cases from 18 countries. *The Lancet. Microbe*, *3*(7), e543–e552.

Horby, P., Lim, W. S., Emberson, J. R., Mafham, M., Bell, J. L., Linsell, L., et al. (2021). Dexamethasone in hospitalized patients with Covid-19. *The New England Journal of Medicine*, *384*(8), 693–704.

Iwasaki, M., Saito, J., Zhao, H., Sakamoto, A., Hirota, K., & Ma, D. (2021). Inflammation triggered by SARS-CoV-2 and ACE2 augment drives multiple organ failure of severe COVID-19: Molecular mechanisms and implications. *Inflammation*, *44*(1), 13–34.

John, T. M., Jacob, C. N., & Kontoyiannis, D. P. (2021). When uncontrolled diabetes mellitus and severe COVID-19 converge: The perfect storm for mucormycosis. *Journal of Fungi (Basel)*, *7*(4), 298.

Jung, J., Park, Y. S., Sung, H., Song, J. S., Lee, S. O., Choi, S. H., et al. (2015). Using immunohistochemistry to assess the accuracy of histomorphologic diagnosis of aspergillosis and mucormycosis. *Clinical Infectious Diseases*, *61*(11), 1664–1670.

Khanna, P., & Kayarat, B. (2021). Posaconazole in the prevention of COVID-19-associated mucormycosis: A concerning contributor to the rise in antifungal resistance. *Indian Journal of Critical Care Medicine*, *25*, 1209–1210.

Krishna, V., Bansal, N., Morjaria, J., & Kaul, S. (2022). COVID-19-associated pulmonary mucormycosis. *Journal of Fungi (Basel)*, *8*(7), 711.

Kubin, C. J., McConville, T. H., Dietz, D., Zucker, J., May, M., Nelson, B., et al. (2021). Characterization of bacterial and fungal infections in hospitalized patients with coronavirus disease 2019 and factors associated with health care-associated infections. *Open Forum Infectious Diseases*, *8*(6), ofab201.

Kumar, S., Acharya, S., Jain, S., Shukla, S., Talwar, D., Shah, D., et al. (2022). Role of zinc and clinicopathological factors for COVID-19-associated mucormycosis (CAM) in a rural hospital of Central India: A case-control study. *Cureus*, *14*(2), e22528.

Kusmartseva, I., Wu, W., Syed, F., Van Der Heide, V., Jorgensen, M., Joseph, P., et al. (2020). Expression of SARS-CoV-2 entry factors in the pancreas of normal organ donors and individuals with COVID-19. *Cell Metabolism*, *32*(6), 1041–1051.e6.

Lanternier, F., & Lortholary, O. (2008). AMBIZYGO: Phase II study of high dose liposomal amphotericin B (AmBisome) [10 mg/kg/j] efficacy against zygomycosis. *Médecine et Maladies Infectieuses*, *38*(suppl 2), S90–S91.

Liu, Y., Wu, H., Huang, F., Fan, Z., & Xu, B. (2013). Utility of 18F-FDG PET/CT in diagnosis and management of mucormycosis. *Clinical Nuclear Medicine*, *38*, e370–e371.

Manesh, A., John, A. O., Mathew, B., Varghese, L., Rupa, V., Zachariah, A., et al. (2016). Posaconazole: An emerging therapeutic option for invasive rhino-orbito-cerebral mucormycosis. *Mycoses*, *59*, 765–772.

Marty, F. M., Ostrosky-Zeichner, L., Cornely, O. A., Mullane, K. M., Perfect, J. R., Thompson, G. R., 3rd, et al. (2016). Isavuconazole treatment for mucormycosis: A single-arm open-label trial and case-control analysis. *The Lancet Infectious Diseases*, *16*, 828–837.

Mehta, R., Nagarkar, N. M., Jindal, A., Rao, K. N., Nidhin, S. B., Arora, R. D., et al. (2022). Multidisciplinary management of COVID-associated mucormycosis syndemic in India. *The Indian Journal of Surgery*, *84*(5), 934–942.

Meshram, H. S., Kute, V. B., Yadav, D. K., Godara, S., Dalal, S., Guleria, S., et al. (2021). Impact of COVID-19-associated mucormycosis in kidney transplant recipients: A multicenter cohort study. *Transplantation direct*, *8*(1), e1255.

Moorthy, A., Gaikwad, R., Krishna, S., Hegde, R., Tripathi, K. K., Kale, P. G., et al. (2021). SARS-CoV-2, uncontrolled diabetes and corticosteroids-an unholy trinity in invasive fungal infections of the maxillofacial region? A retrospective, multi-centric analysis. *Journal of Maxillofacial and Oral Surgery*, *20*(3), 418–425.

Pai, V., Sansi, R., Kharche, R., Bandili, S. C., & Pai, B. (2021). Rhino-orbito-cerebral mucormycosis: Pictorial review. *Insights Into Imaging*, *12*(1), 167.

Pal, R., Singh, B., Bhadada, S. K., Banerjee, M., Bhogal, R. S., Hage, N., et al. (2021). COVID-19-associated mucormycosis: An updated systematic review of literature. *Mycoses*, *64*(12), 1452–1459.

Patel, D. D., Adke, S., Badhe, P. V., Lamture, S., Marfatia, H., & Mhatre, P. (2021). COVID-19 associated rhino-orbito-cerebral mucormycosis: Imaging spectrum and clinico-radiological correlation—A single centre experience. *Clinical Imaging*, *82*, 172–178.

Patel, A., Agarwal, R., Rudramurthy, S. M., Shevkani, M., Xess, I., Sharma, R., et al. (2021). Multicenter epidemiologic study of coronavirus disease-associated mucormycosis, India. *Emerging Infectious Diseases*, *27*(9), 2349–2359.

Peman, J., Gaitan, A. R., Vidal, C. G., Salavert, M., Ramirez, P., Puchades, F., et al. (2020). Fungal co-infection in COVID-19 patients: Should we be concerned? *Revista Iberoamericana de Micología*, *37*(2), 41–46.

Perfect, J. R., Cornely, O. A., Heep, M., Ostrosky-Zeichner, L., Mullane, K. M., Maher, R., et al. (2018). Isavuconazole treatment for rare fungal diseases and for invasive aspergillosis in patients with renal impairment: Challenges and lessons of the VITAL trial. *Mycoses*, *61*, 420–429.

Rammaert, B., Lanternier, F., Zahar, J. R., Dannaoui, E., Bougnoux, M. E., Lecuit, M., et al. (2012). Healthcare-associated mucormycosis. *Clinical Infectious Diseases*, *54*(Suppl 1), S44–S54.

Rawson, T. M., Moore, L. S. P., Zhu, N., Ranganathan, N., Skolimowska, K., Gilchrist, M., et al. (2020). Bacterial and fungal coinfection in individuals with coronavirus: A rapid review to support COVID-19 antimicrobial prescribing. *Clinical Infectious Diseases*, *71*(9), 2459–2468.

Reed, C., Bryant, R., Ibrahim, A. S., Edwards, J., Jr., Filler, S. G., Goldberg, R., et al. (2008). Combination polyene-caspofungin treatment of rhino-orbital-cerebral mucormycosis. *Clinical Infectious Diseases*, *47*, 364–371.

Reid, G., Lynch, J. P., 3rd, Fishbein, M. C., & Clark, N. M. (2020). Mucormycosis. *Seminars in Respiratory and Critical Care Medicine*, *41*(1), 99–114.

Roden, M. M., Zaoutis, T. E., Buchanan, W. L., Knudsen, T. A., Sarkisova, T. A., Schaufele, R. L., et al. (2005). Epidemiology and outcome of zygomycosis: A review of 929 reported cases. *Clinical Infectious Diseases*, *41*, 634–653.

Samson, R., & Dharne, M. (2022). COVID-19 associated mucormycosis: Evolving technologies for early and rapid diagnosis. *3 Biotech*, *12*(1), 6.

Sen, M., Honavar, S. G., Bansal, R., Sengupta, S., Rao, R., Kim, U., et al. (2021). Epidemiology, clinical profile, management, and outcome of COVID-19-associated rhino-orbital-cerebral mucormycosis in 2826 patients in India—Collaborative OPAI-IJO study on mucormycosis in COVID-19 (COSMIC), report 1. *Indian Journal of Ophthalmology*, *69*(7), 1670–1692.

Sengupta, I., & Nayak, T. (2022). Coincidence or reality behind mucormycosis, diabetes mellitus and Covid-19 association: A systematic review. *Journal of Medical Mycology*, *32*(3), 101257.

Singh, A. K., Singh, R., Joshi, S. R., & Misra, A. (2021). Mucormycosis in COVID-19: A systematic review of cases reported worldwide and in India. *Diabetes and Metabolic Syndrome: Clinical Research and Reviews*, *15*(4), 102146.

Spellberg, B., & Ibrahim, A. S. (2018). Mucormycosis. In J. L. Jameson, A. S. Fauci, D. L. Kasper, S. L. Hauser, D. Longo, & J. Loscalzo (Eds.), *Harrison's principles of internal medicine* (20th ed., p. 1538). New York: McGraw-Hill Education.

Sravani, T., Uppin, S. G., Uppin, M. S., & Sundaram, C. (2014). Rhinocerebral mucormycosis: Pathology revisited with emphasis on perineural spread. *Neurology India*, *62*, 383–386.

Sugar, A. M. (1992). Mucormycosis. *Clinical Infectious Diseases*, *14*(Suppl 1), S126–S129.

Sunagawa, K., Ishige, T., Kusumi, Y., Asano, M., Nisihikawa, E., Kato, M., et al. (2013). Renal abscess involving mucormycosis by immunohistochemical detection in a patient with acute lymphocytic leukemia: A case report and literature review. *Japanese Journal of Infectious Diseases*, *66*, 345–347.

Trieu, T. A., Navarro-Mendoza, M. I., Pérez-Arques, C., Sanchis, M., Capilla, J., Navarro-Rodriguez, P., et al. (2017). RNAi-based functional genomics identifies new virulence determinants in mucormycosis. *PLoS Pathogens*, *13*(1), e1006150.

Turner, J. H., Soudry, E., Nayak, J. V., & Hwang, P. H. (2013). Survival outcomes in acute invasive fungal sinusitis: A systematic review and quantitative synthesis of published evidence. *The Laryngoscope*, *123*(5), 1112–1118.

Walia, S., Bhaisare, V., Rawat, P., Kori, N., Sharma, M., Gupta, N., et al. (2021). COVID-19-associated mucormycosis: Preliminary report from a tertiary eye care centre. *Indian Journal of Ophthalmology*, *69*(12), 3685–3689.

Wu, Z., & McGoogan, J. M. (2020). Characteristics of and important lessons from the coronavirus disease 2019 (COVID-19) outbreak in China: Summary of a report of 72314 cases from the Chinese Center for Disease Control and Prevention. *JAMA*, *323*(13), 1239–1242.

Chapter 44

Tissue location of SARS-CoV-2 RNA: A focus on bone and implications for skeletal health

Edoardo Guazzoni[a,b,*], Luigi di Filippo[c,*], Alberto Castelli[d], Andrea Giustina[c], and Federico Grassi[d]

[a]*IRCCS Humanitas Research Hospital, Rozzano, Milan, Italy,* [b]*Fondazione Livio Sciutto Onlus, Campus Savona, Università Degli Studi di Genova, Savona, Italy,* [c]*Institute of Endocrine and Metabolic Sciences, Università Vita-Salute San Raffaele, IRCCS Ospedale San Raffaele, Milan, Italy,* [d]*Department of Orthopaedics and Traumatology, IRCCS Fondazione Policlinico San Matteo, University of Pavia, Pavia, Italy*

Abbreviations

ACE2	angiotensin-converting enzyme 2
BMD	bone mineral density
BV/TV	trabecular bone volume fraction
CCL-2	chemokine ligands 2
CSS	cytokine storm syndrome
CT	computed tomography
CXCL10	C-X-C motif chemokine ligand 10
CXCR3	C-X-C chemokine receptor 3
Dpi	days post infection
DXA	dual energy X-ray absorptiometry
ED	emergency department
HPCs	hematopoietic progenitor cells
HSCs	hematopoietic stem cells
ICU	intensive care unit
IL-8	Interleukin-8
K18-hACE2	epithelial cell cytokeratin 18-human angiotensin I-converting enzyme 2
MCP-1	monocyte chemoattractant protein 1
MIP-1α	macrophage inhibitory protein 1 α
NIMV	noninvasive mechanical ventilation
OBs	osteoblasts
OCs	osteoclasts
OPG	osteoprotegerin
PTH	parathyroid hormone
RANKL	receptor activator of nuclear factor kappa-B ligand
RCTs	randomized clinical trials
ROS	reactive oxygen species
TLR4	Toll-like receptor 4
TNF-α	tumor necrosis factor-α
VEGF-A	vascular endothelial growth factor A
VFs	vertebral fractures
μCT	microcomputed tomography

* Contributed equally.

Management, Body Systems, and Case Studies in COVID-19. https://doi.org/10.1016/B978-0-443-18703-2.00016-5
Copyright © 2024 Elsevier Inc. All rights are reserved, including those for text and data mining, AI training, and similar technologies.

COVID-19 and bone health

The initial scientific focus on COVID-19 was on the acute disease and the life-threatening complications associated with it. Nowadays, more attention is being paid to the possible negative long-term sequelae of this disease on multiple systems, including the skeletal system. This chapter will describe what is known of the consequences of SARS-CoV-2 infection on the skeletal system and the detrimental implications of skeletal fragility on patients affected by COVID-19, both during the acute disease and during disease recovery.

COVID-19 and bone health: Basic and preclinical implications

Patients infected with severe acute respiratory syndrome-associated coronavirus (SARS-CoV), which belongs to the same family and genus as SARS-CoV-2, were characterized by reduced bone mineral density (BMD) after disease recovery. It was first suggested that the reduction in BMD was mainly due to the corticosteroid treatments used in these patients (Griffith, 2011; Lau et al., 2005). However, other evidence showed a BMD reduction in patients with acute disease (van Niekerk & Engelbrecht, 2018), suggesting new reflections about other possible underlying pathophysiological mechanisms.

Osteoblasts (OBs) and osteoclasts (OCs) are crucial key actors in regulating the homeostatic balance of bone formation and resorption. This system can be dysregulated in various ways during SARS-CoV-2 infection, potentially leading to skeletal complications (Fig. 1). SARS-CoV-2 enters the cells with the help of its spike protein that bonds to angiotensin-converting enzyme 2 (ACE2). ACE2 receptors are known to have an antiinflammatory role in hydrolyzing angiotensin II, a peptide hormone with proinflammatory effects, to angiotensin 1–7, a vasoactive peptide with antiinflammatory effects. After entering cells, SARS-CoV-2 promotes the downregulation of the expression of ACE2 receptors (Kuba et al., 2006), increasing angiotensin II and thus inflammation through the upregulation of interleukin-8 (IL-8), chemokine ligands 2 (CCL-2), CCL-5, and selectin. It has been demonstrated that increasing levels of angiotensin II promote the differentiation of OCs through the upregulation of the receptor activator of the nuclear factor kappa-B ligand (RANKL), accelerating bone reabsorption (Queiroz-Junior et al., 2019). Patients affected by acute COVID-19 are subject to an increase of different cytokines, chemokines, and growth factors such as IL-1β, IL-2, IL-6, IL-7, IL-17, tumor necrosis factor-α (TNF-α), C-X-C motif chemokine ligand 10 (CXCL10), vascular endothelial growth factor A (VEGF-A), macrophage inhibitory protein 1 α (MIP-1α), and monocyte chemoattractant protein 1 (MCP-1). Many of them are involved in the

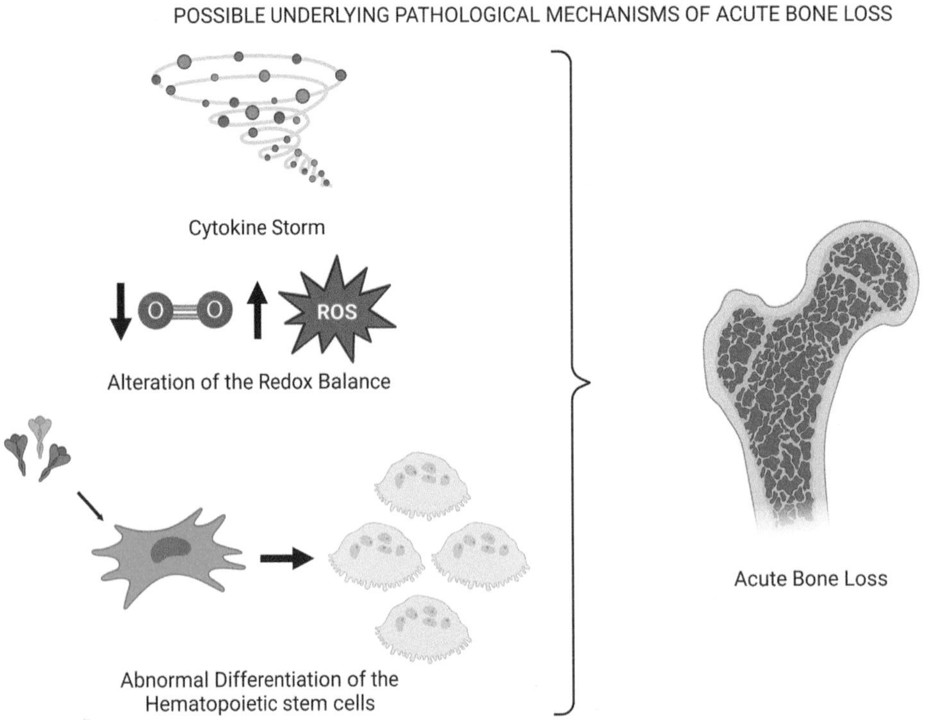

FIG. 1 Possible underlying physiopathological mechanisms of bone loss during SARS-CoV-2 infection.

upregulation of osteoclastogenesis through different pathways and can be involved in skeletal impairment during SARS-CoV-2 infection. IL-1β is involved in OC development at various steps and plays an important role in bone metabolism through the activation of RANK signaling other than inducing RANKL to promote osteoclastogenesis (Lee et al., 2010). IL-6 is known to stimulate OC activity, upregulating the production of RANKL (Kwan Tat et al., 2004). CXCL10 though TLR4 and C-X-C chemokine receptor 3 (CXCR3) upregulated the production of osteoclastogenic cytokines, inducing bone resorption (Lee et al., 2017). TNF-α stimulates OC differentiation by a mechanism independent of the ODF/RANKL-RANK interaction (Kobayashi et al., 2000). VEGF-A has been shown in vitro to promote bone reabsorption though OC recruitment (Niida et al., 1999).

The clinical importance and occurrence of the cytokine storm syndrome (CSS) in critically ill patients have been questioned by multiple studies (Kox et al., 2020; Sinha et al., 2020; Syed et al., 2021). In patients affected by acute infection, there is the activation of innate immunity through the activation of Toll-like receptor 4 (TLR4) by angiotensin II (Biancardi et al., 2017). In contrast, it has been demonstrated, both in vitro and in vivo, that the downregulation of TLR4 by microRNA-4485, upregulated in patients with acute COVID-19 disease, inhibits osteogenic differentiation and bone remodeling, impairing fracture healing (Mi et al., 2021).

The redox balance and reactive oxygen species (ROS) production are commonly altered during COVID-19. In fact, hypoxemia, a common condition in patients affected by acute disease, can increase the production of ROS with respect to the antioxidant defense (Chernyak et al., 2020) and it is associated with the severity of the disease. ROS has been found to favor osteoclastogenesis through the upregulation of RANKL and osteoprotegerin (OPG), inducing the apoptosis of OBs and osteocytes (Domazetovic et al., 2017) and possibly impairing skeletal health.

Another possible underlying mechanism involved in bone impairment in COVID-19 patients is represented by the abnormal differentiation of the hematopoietic lineage exposed to the spike protein of SARS-CoV-2, with a subsequent osteoclastogenesis. After ex vivo exposition to the viral S-protein, the hematopoietic stem cells (HSCs) and progenitor cells (HPCs) expand less effectively and have less functional colony-forming capacity while peripheral blood monocytes upregulate CD14 expression with changes in their size and granularity (Ropa et al., 2021).

A specific impairment of bone health, characterized by a decrease in trabecular structure, has been demonstrated in animal models (Fig. 2).

One of the first studies investigating the role of SARS-CoV-2 infection in bone impairment on a murine model was conducted by Awosanya et al. (2022). They used the epithelial cell cytokeratin 18-human angiotensin I-converting enzyme

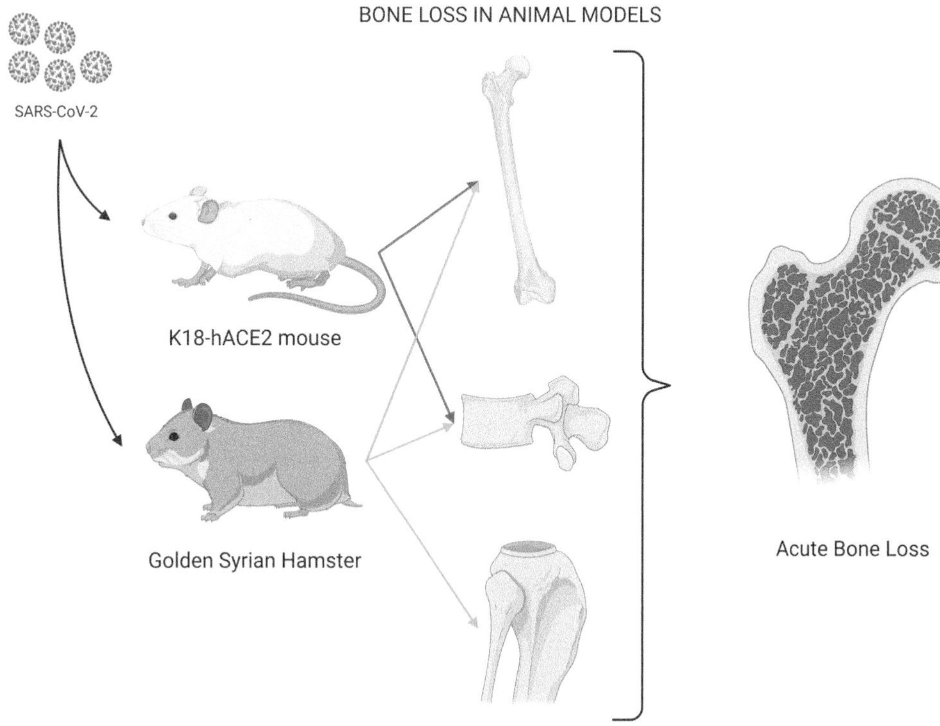

FIG. 2 Bone loss in animal models of COVID-19.

2 (K18-hACE2) transgenic mouse model, which expresses the hACE2 receptor in the epithelial cell, to use an unaltered SARS-CoV-2 virus. It has been shown that this murine model can get infected and develop a severe disease (McCray et al., 2007). Awosanya et al. infected 19-week-old mice with different viral doses and analyzed their femoral tissues 12–14 days post infection (dpi). The microcomputed tomography (μCT) analysis demonstrated a reduction of 24.4% of the trabecular bone volume fraction (BV/TV) and the histomorphometric analysis indicated an increase of OCs of more than 60% (Awosanya et al., 2022). Christiansen et al. infected the same K18-hACE2 mice at the age of 6 weeks and evaluated the L5 vertebra after 1 week, reporting a 10% decrease in BV/TV (Christiansen et al., 2021).

Another well-established animal model that closely emulates COVID-19 in humans is the golden Syrian hamster (Chan et al., 2020), which is a nonlethal model that has the advantage of mimicking a mild to moderate disease. Qiao W et al. infected the golden Syrian hamster and euthanized them at 4 (acute phase), 30 (postrecovery phase), and 60 (chronic phase) dpi to study the loss of trabecular bone in the distal metaphyseal femur, proximal metaphyseal tibia, and lumbar vertebrae (Qiao et al., 2022). A notable reduction in BMD as well as the number and thickness of the trabecula was demonstrated even at 4 dpi. The BV/TV reduction continued through the postrecovery phase, reaching a peak between 30 and 60 dpi with a reduction up to 50%. They reported an upregulation of the OC activity and number while the OB activity and number were unchanged.

Osteo-metabolic findings in COVID-19 patients

In April 2020, during the first pandemic spread in Europe, a case of severe acute hypocalcemia in a previously thyroidectomized Italian patient with COVID-19 was described and, for the first time, SARS-CoV-2 infection was suggested as the possible precipitating cause of a subclinical postsurgical hypoparathyroidism presenting with a severe acute hypocalcemia (Bossoni et al., 2020). Interestingly, previous studies conducted in 2003 and 2016 focused on the SARS and Ebola epidemic emergencies reported low calcium levels as a highly prevalent biochemical abnormality in affected patients (di Filippo, Doga, et al., 2022). Based on these past observations and on the first reported case of acute hypocalcemia in a COVID-19 patient, a first retrospective study was conducted in Italy to identify the prevalence of low calcium levels in a cohort of COVID-19 subjects. A total of 531 patients were included in this study and hypocalcemia was found in about the 80% of them (Di Filippo et al., 2020). Several following studies conducted worldwide confirmed these data, reporting a prevalence of low calcium levels ranging from 62.6% to 87.2% (di Filippo, Doga, et al., 2022). The presence of this biochemical alteration was suggested early on as a distinctive biochemical feature characterizing these patients. Low calcium levels were reported more frequently in patients affected by SARS-CoV-2 infection as compared to those affected by other viral or bacterial pneumonia presenting in the hospital with comparable baseline clinical parameters (di Filippo, Doga, et al., 2022; di Filippo, Formenti, Doga, Frara, et al., 2021; di Filippo, Frara, et al., 2022). Moreover, in the first studies evaluating calcium levels in COVID-19 patients, hypocalcemia was negatively associated with inflammatory, coagulopathy, and organ-injury parameters, and recognized as an independent risk factor for worse COVID-19 outcomes, representing a strongly reliable biomarker of disease severity (di Filippo, Doga, et al., 2022; di Filippo, Frara, et al., 2022). These findings were subsequently confirmed by following systematic reviews and metaanalyses reporting that hypocalcemia was significantly associated with worse disease outcomes (Alemzadeh et al., 2021; Martha et al., 2021; Song et al., 2023). Several underlying pathophysiological mechanisms have been hypothesized to cause and promote the hypocalcemia occurrence in COVID-19 patients (di Filippo, Doga, et al., 2022; di Filippo, Frara, & Giustina, 2021), including the calcium-dependent viral mechanisms of action determining intra- and extracellular cation dysmetabolism (di Filippo, Frara, & Giustina, 2021; Guazzoni et al., 2022), the diffuse hematologic and thrombotic complications altering the calcium key role in coagulation and platelet adhesion, the severe catabolic status typically present in those affected by acute illness, and the increased synthesis of unbound and unsaturated fatty acids during acute inflammatory status that can bind calcium ions and albumin causing and triggering calcium saponification and hypocalcemia (di Filippo, Doga, et al., 2022; di Filippo, Frara, & Giustina, 2021). Finally, central roles in promoting the highly prevalent hypocalcemia occurrence in these patients were also hypothesized early on for the widespread hypovitaminosis D present in the general population of western countries, and, during acute infections, for an impaired compensatory and secondary parathyroid hormone (PTH) response unable to maintain the calcium levels in the normal range in conditions of vitamin D deficiency (di Filippo, Allora, et al., 2021).

Vitamin D plays a key role in intestinal calcium absorption and the prevention of bone diseases. Also, increasing evidence shows the role of vitamin D in other pathological conditions, highlighting its possible extraskeletal effects (Bouillon et al., 2019). Vitamin D is well known to have several regulating properties in the innate response to infectious diseases, influencing the antimicrobial and antiviral mechanisms of action and regulating the adaptive immune response and the immune-system activity (Bouillon et al., 2019). Based on this immunomodulatory role and on the widespread hypovitaminosis D present in western countries heavily affected by the first COVID-19 pandemic spread, vitamin D was implicated early on as a potential factor influencing SARS-CoV-2 infectivity and COVID-19 severity (Giustina & Formenti, 2020). During the first months of the pandemic, mostly cross-sectional observational and retrospective cohort studies suggested that vitamin D deficiency may be associated with an increased susceptibility to SARS-CoV-2 infection and an increased risk of severe COVID-19 (Bilezikian et al., 2023; di Filippo, Allora, et al., 2022). Systematic reviews and metaanalyses of these first studies revealed inverse associations between serum 25(OH) vitamin D levels and the risk of developing severe COVID-19, mortality, intensive care unit (ICU) admission, length of ICU stay, and need for noninvasive mechanical ventilation (NIMV) (Bilezikian et al., 2023). However, it must be considered that these studies published during the first pandemic wave presented several methodological limitations, and it was a matter of debate as to whether the low vitamin D levels observed in COVID-19 patients during the acute illness represented only an epiphenomenon of the disease, included in an "acute phase reaction," or a potential preexisting modifiable risk factor for severe outcomes. This aspect was only subsequently clarified by prospective and interventional studies conducted with better and rigorous study-designs in which the negative role of low vitamin D was consistently confirmed (Bilezikian et al., 2023; di Filippo et al., 2023). In line with these findings, promising but inconsistent results regarding vitamin D supplementation in these patients were reported early on in many retrospective and quasiexperimental studies conducted with heterogeneous methodologies. Subsequently, several randomized clinical trials (RCTs) addressing the effect of vitamin D supplementation in treating COVID-19 as well as metaanalyses evaluating these trials were published. These showed significant benefits of vitamin D supplementation in terms of COVID-19 severity, suggesting that vitamin D administration could positively affect the outcomes of COVID-19 patients, although further robust data in this specific setting are still needed (Bilezikian et al., 2023).

In line with these biochemical findings, diagnostic techniques have shown a possible impact of COVID-19 on the osteometabolic phenotype of these patients. In fact, a retrospective study including 58 patients evaluated slightly less than 3 months after acute COVID-19 observed a BMD decrease by a mean of 8.6% from diagnosis to follow-up, as measured by quantitative computed tomography (CT) scans to evaluate COVID-19 severity (Berktaş et al., 2022). Another study included 56 COVID-19 patients and matched 44 patients without infection. It evaluated BMD before the infection (within 3 months before COVID-19 occurrence) and after 9 months using a dual energy X-ray absorptiometry (DXA) scan; it reported a significant decrease in 9-month BMD at both the lumbar spine and hip regions, only in the infected group (Elmedany et al., 2022).

Moreover, using the morphometric diagnostic technique applied on lateral chest x-rays performed at hospital admission, which allowed a qualitative and quantitative evaluation of vertebral shape, we reported that vertebral fractures (VFs) were highly prevalent in patients with acute COVID-19 (di Filippo, Compagnone, et al., 2022; di Filippo, Formenti, Doga, Pedone, et al., 2021). A following morphometric study on CT in patients seen in emergency departments (ED) confirmed the high prevalence of VFs in SARS-CoV-2 infection (Battisti et al., 2021).

Conclusions

In conclusion, among the several different tissues and organs potentially affected by SARS-CoV-2 infection and COVID-19, the skeletal metabolism has been reported to be negatively influenced by this multisystemic disease (Fig. 3). COVID-19 has been potentially associated with an impairment of bone health in both humans and animal models. In addition, the highly prevalent bone comorbidities and alterations in the calcium-vitamin D metabolism in these patients were reported early on as strong risk factors for severe disease outcomes. These bidirectional relationships must be considered in the acute and long-term management of all COVID-19 and osteoporotic patients.

Policies and procedures

In this chapter, we reviewed the impact of the coronavirus pandemic, SARS-CoV-2 infection, and COVID-19 on skeletal health. First, we outlined the preclinical implications of the SARS-CoV-2 infection on bone tissue, reporting the evidence derived from in vitro and in vivo studies in animal models. These data were evaluated taking into account the preclinical observations reported during previous emergencies caused by other coronavirus infections.

The main focus was on the detrimental role of a hyperreactive immune response, proinflammatory molecules, and cytokine storm syndrome on the finely regulated balance of bone resorption and formation, which is eventually compromised by the altered activity of the main key factors involved in the preservation of skeletal health. In this section, we also discussed the negative effects of the disease on the skeletal tissues of murine models infected with SARS-CoV-2.

Second, we summarized the main clinical findings reported by different studies done during the last 3 years on patients affected by SARS-CoV-2 to investigate the possible impact of COVID-19 on bone health and skeletal complications, and vice versa. We highlighted the main osteo-metabolic findings that were disrupted in COVID-19 patients, including acute

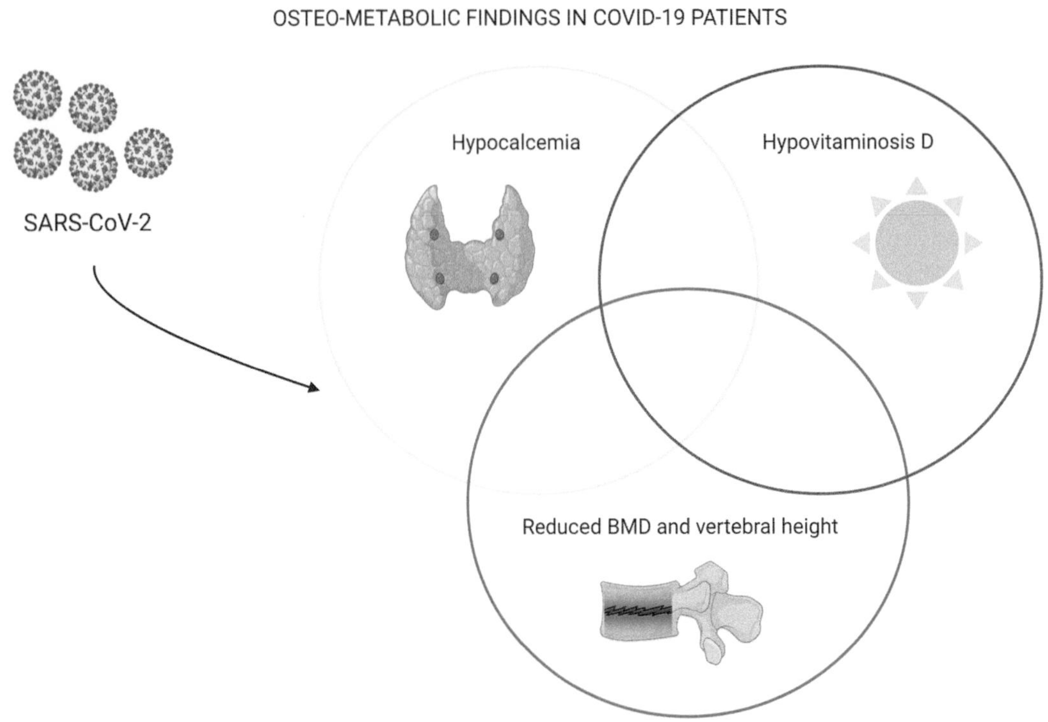

FIG. 3 Osteo-metabolic findings in COVID-19 patients.

alterations in calcium ion homeostasis, particularly widespread in these patients and possibly representing a specific biochemical characteristic of the disease and a risk predictor for worse outcomes, as well as hypovitaminosis D, which has proven to negatively influence the course of the acute disease and recovery after hospitalization.

These data show a strict bidirectional influence between COVID-19 and bone health in patients affected.

Applications to other areas

In the context of the multisystemic involvement of SARS-CoV-2 infection and COVID-19, including also endocrine and metabolic features, this chapter summarizes the recent literature highlighting the occurrence of a distinct osteo-metabolic phenotype in these patients.

The involvement of skeletal tissue in SARS-CoV-2 infection is supported by both preclinical and clinical findings reported during these years in several studies conducted worldwide. The clinical relevance of this phenotype was confirmed in both acute and postacute COVID-19, revealing a strict bidirectional relationship between the disease and the skeletal features influencing bone complications, alterations in osteo-metabolic biochemical findings, and COVID-19 outcomes and recovery. In this osteo-metabolic scenario, the reported preclinical data can expand our knowledge of the pathophysiological mechanisms underlying bone involvement and damage during acute illness conditions. These findings could be transposed to other acute infections as well as inflammatory and immunological disorders. In particular, the data focusing on murine models may represent a seminal and reliable in vivo basis to further investigate the mechanistic patterns involved in skeletal impairment during acute conditions. Also, the data reported in clinical studies are able to show how the acute disease can negatively influence the skeletal health of the patients affected. The osteo-metabolic biochemical, clinical, and radiological findings observed in COVID-19 should also be investigated in patients affected by other acute conditions for the early detection and prevention of these skeletal complications, that can be emergingly under-diagnosed and -treated impairing quality of life and increasing risk of mortality. Moreover, the relevance of this osteo-metabolic phenotype on short- and long-term outcomes of the disease suggests that all its components may represent suitable therapeutic targets to investigate in further studies to prevent SARS-CoV-2 infection, poor COVID-19 outcomes, and long-COVID-19 occurrence as well as to possibly evaluate their efficacy in other acute infections and inflammatory conditions.

Mini-dictionary of terms

- Cytokine storm syndrome: Syndrome caused by the unrestrained release of proinflammatory cytokines.
- Transgenic mouse: Mouse that has its genome modified for scientific research.

Summary points

- SARS-CoV-2 may negatively affect bone metabolism promoting a proinflammatory state, favoring hypoxia and the production of ROS and directly infecting the hematopoietic cell lineage.
- An impairment of bone trabecular structures has been demonstrated in animal models of COVID-19.
- A distinctive osteo-metabolic phenotype of COVID-19 characterized by widespread acute hypocalcemia and hypovitaminosis D, and a high prevalence of skeletal complications such as VFs, was reported early on in affected patients.
- These findings highlight the central role of a multidisciplinary team in evaluating COVID-19 patients for a proactive search of negative skeletal consequences because they may represent suitable therapeutic targets to prevent SARS-CoV-2 infection, poor COVID-19 outcomes, and long COVID-19 occurrence.

References

Alemzadeh, E., et al. (2021). The effect of low serum calcium level on the severity and mortality of Covid patients: A systematic review and meta-analysis. *Immunity, Inflammation and Disease*, *9*(4), 1219–1228. https://doi.org/10.1002/iid3.528.

Awosanya, O. D., et al. (2022). Osteoclast-mediated bone loss observed in a COVID-19 mouse model. *Bone*, *154*, 116227. https://doi.org/10.1016/j.bone.2021.116227.

Battisti, S., et al. (2021). Vertebral fractures and mortality risk in hospitalised patients during the COVID-19 pandemic emergency. *Endocrine*, *74*(3), 461–469. https://doi.org/10.1007/s12020-021-02872-1.

Berktaş, B. M., et al. (2022). COVID-19 illness and treatment decrease bone mineral density of surviving hospitalized patients. *European Review for Medical and Pharmacological Sciences*, *26*(8), 3046–3056. https://doi.org/10.26355/eurrev_202204_28636.

Biancardi, V. C., et al. (2017). The interplay between angiotensin II, TLR4 and hypertension. *Pharmacological Research*, *120*, 88–96. https://doi.org/10.1016/j.phrs.2017.03.017.

Bilezikian, J. P., et al. (2023). Consensus and controversial aspects of vitamin D and COVID-19. *The Journal of Clinical Endocrinology & Metabolism*, *108*(5), 1034–1042. https://doi.org/10.1210/clinem/dgac719.

Bossoni, S., Chiesa, L., & Giustina, A. (2020). Severe hypocalcemia in a thyroidectomized woman with Covid-19 infection. *Endocrine*, *68*(2), 253–254. https://doi.org/10.1007/s12020-020-02326-0.

Bouillon, R., et al. (2019). Skeletal and extraskeletal actions of vitamin D: Current evidence and outstanding questions. *Endocrine Reviews*, *40*(4), 1109–1151. https://doi.org/10.1210/er.2018-00126.

Chan, J. F.-W., et al. (2020). Simulation of the clinical and pathological manifestations of coronavirus disease 2019 (COVID-19) in a golden syrian hamster model: implications for disease pathogenesis and transmissibility. *Clinical Infectious Diseases: An Official Publication of the Infectious Diseases Society of America*, *71*(9), 2428–2446. https://doi.org/10.1093/cid/ciaa325.

Chernyak, B. V., et al. (2020). COVID-19 and oxidative stress. *Biochemistry. Biokhimiia*, *85*(12), 1543–1553. https://doi.org/10.1134/S0006297920120068.

Christiansen, B., Ball, E., Haudenschild, A., Yik, J., Coffey, L., & Haudenschild, D. (2021). Bone loss following SARS-CoV-2 infection in mice. *Journal of Bone and Mineral Research*, *36*(Suppl. 1).

di Filippo, L., Allora, A., et al. (2021). Hypocalcemia in COVID-19 is associated with low vitamin D levels and impaired compensatory PTH response. *Endocrine*, *74*(2), 219–225. https://doi.org/10.1007/s12020-021-02882-z.

di Filippo, L., Allora, A., et al. (2022). Vitamin D levels are associated with blood glucose and BMI in COVID-19 patients, predicting disease severity. *The Journal of Clinical Endocrinology & Metabolism*, *107*(1), e348–e360. https://doi.org/10.1210/clinem/dgab599.

di Filippo, L., Compagnone, N., et al. (2022). Vertebral fractures at hospitalization predict impaired respiratory function during follow-up of COVID-19 survivors. *Endocrine*, *77*(2), 392–400. https://doi.org/10.1007/s12020-022-03096-7.

di Filippo, L., Doga, M., et al. (2022). Hypocalcemia in COVID-19: Prevalence, clinical significance and therapeutic implications. *Reviews in Endocrine and Metabolic Disorders*, *23*(2), 299–308. https://doi.org/10.1007/s11154-021-09655-z.

di Filippo, L., Formenti, A. M., Doga, M., Frara, S., et al. (2021). Hypocalcemia is a distinctive biochemical feature of hospitalized COVID-19 patients. *Endocrine*, *71*(1), 9–13. https://doi.org/10.1007/s12020-020-02541-9.

di Filippo, L., Formenti, A. M., Doga, M., Pedone, E., et al. (2021). Radiological thoracic vertebral fractures are highly prevalent in COVID-19 and predict disease outcomes. *The Journal of Clinical Endocrinology & Metabolism*, *106*(2), e602–e614. https://doi.org/10.1210/clinem/dgaa738.

di Filippo, L., Frara, S., & Giustina, A. (2021). The emerging osteo-metabolic phenotype of COVID-19: Clinical and pathophysiological aspects. *Nature Reviews Endocrinology*, *17*(8), 445–446. https://doi.org/10.1038/s41574-021-00516-y.

di Filippo, L., Frara, S., et al. (2022). The osteo-metabolic phenotype of COVID-19: An update. *Endocrine*, *78*(2), 247–254. https://doi.org/10.1007/s12020-022-03135-3.

Di Filippo, L., et al. (2020). Hypocalcemia is highly prevalent and predicts hospitalization in patients with COVID-19. *Endocrine*, *68*(3), 475–478. https://doi.org/10.1007/s12020-020-02383-5.

di Filippo, L., et al. (2023). Low vitamin D levels predict outcomes of COVID-19 in patients with both severe and non-severe disease at hospitalization. *Endocrine*, *80*(3), 669–683. https://doi.org/10.1007/s12020-023-03331-9.

Domazetovic, V., et al. (2017). Oxidative stress in bone remodeling: Role of antioxidants. *Clinical Cases in Mineral and Bone Metabolism: The Official Journal of the Italian Society of Osteoporosis, Mineral Metabolism, and Skeletal Diseases*, *14*(2), 209–216. https://doi.org/10.11138/ccmbm/2017.14.1.209.

Elmedany, S. H., et al. (2022). Bone mineral density changes in osteoporotic and osteopenic patients after COVID-19 infection. *Egyptian Rheumatology and Rehabilitation*, *49*(1), 64. https://doi.org/10.1186/s43166-022-00165-7.

Giustina, A., & Formenti, A. M. (2020). Does hypovitaminosis D play a role in the high impact of COVID infection in Italy? *British Medical Journal*. Available at: https://www.bmj.com/content/368/bmj.m810/rr-36.

Griffith, J. F. (2011). Musculoskeletal complications of severe acute respiratory syndrome. *Seminars in Musculoskeletal Radiology*, *15*(5), 554–560. https://doi.org/10.1055/s-0031-1293500.

Guazzoni, E., et al. (2022). Detection of SARS-CoV-2 in cancellous bone of patients with COVID-19 disease undergoing orthopedic surgery: Laboratory findings and clinical applications. *International Journal of Environmental Research and Public Health*, *19*(17). https://doi.org/10.3390/ijerph191710621.

Kobayashi, K., et al. (2000). Tumor necrosis factor alpha stimulates osteoclast differentiation by a mechanism independent of the ODF/RANKL-RANK interaction. *The Journal of Experimental Medicine*, *191*(2), 275–286. https://doi.org/10.1084/jem.191.2.275.

Kox, M., et al. (2020). Cytokine levels in critically ill patients with COVID-19 and other conditions. *JAMA*, *324*(15), 1565–1567. https://doi.org/10.1001/jama.2020.17052.

Kuba, K., Imai, Y., & Penninger, J. M. (2006). Angiotensin-converting enzyme 2 in lung diseases. *Current Opinion in Pharmacology*, *6*(3), 271–276. https://doi.org/10.1016/j.coph.2006.03.001.

Kwan Tat, S., et al. (2004). IL-6, RANKL, TNF-alpha/IL-1: Interrelations in bone resorption pathophysiology. *Cytokine & Growth Factor Reviews*, *15*(1), 49–60. https://doi.org/10.1016/j.cytogfr.2003.10.005.

Lau, E. M. C., et al. (2005). Reduced bone mineral density in male severe acute respiratory syndrome (SARS) patients in Hong Kong. *Bone*, *37*(3), 420–424. https://doi.org/10.1016/j.bone.2005.04.018.

Lee, Y.-M., et al. (2010). IL-1 plays an important role in the bone metabolism under physiological conditions. *International Immunology*, *22*(10), 805–816. https://doi.org/10.1093/intimm/dxq431.

Lee, J.-H., et al. (2017). Pathogenic roles of CXCL10 signaling through CXCR3 and TLR4 in macrophages and T cells: Relevance for arthritis. *Arthritis Research & Therapy*, *19*(1), 163. https://doi.org/10.1186/s13075-017-1353-6.

Martha, J. W., Wibowo, A., & Pranata, R. (2021). Hypocalcemia is associated with severe COVID-19: A systematic review and meta-analysis. *Diabetes & Metabolic Syndrome: Clinical Research & Reviews*, *15*(1), 337–342. https://doi.org/10.1016/j.dsx.2021.01.003.

McCray, P. B. J., et al. (2007). Lethal infection of K18-hACE2 mice infected with severe acute respiratory syndrome coronavirus. *Journal of Virology*, *81*(2), 813–821. https://doi.org/10.1128/JVI.02012-06.

Mi, B., et al. (2021). SARS-CoV-2-induced overexpression of miR-4485 suppresses osteogenic differentiation and impairs fracture healing. *International Journal of Biological Sciences*, *17*(5), 1277–1288. https://doi.org/10.7150/ijbs.56657.

Niida, S., et al. (1999). Vascular endothelial growth factor can substitute for macrophage colony-stimulating factor in the support of osteoclastic bone resorption. *The Journal of Experimental Medicine*, *190*(2), 293–298. https://doi.org/10.1084/jem.190.2.293.

Qiao, W., et al. (2022). Author correction: SARS-CoV-2 infection induces inflammatory bone loss in golden Syrian hamsters. *Nature Communications*, *13*(1), 3139. https://doi.org/10.1038/s41467-022-30952-x.

Queiroz-Junior, C. M., et al. (2019). The angiotensin converting enzyme 2/angiotensin-(1–7)/Mas receptor axis as a key player in alveolar bone remodeling. *Bone*, *128*, 115041. https://doi.org/10.1016/j.bone.2019.115041.

Ropa, J., et al. (2021). Human hematopoietic stem, progenitor, and immune cells respond ex vivo to SARS-CoV-2 spike protein. *Stem Cell Reviews and Reports*, *17*(1), 253–265. https://doi.org/10.1007/s12015-020-10056-z.

Sinha, P., Matthay, M. A., & Calfee, C. S. (2020). Is a "cytokine storm" relevant to COVID-19? *JAMA Internal Medicine. United States*, 1152–1154. https://doi.org/10.1001/jamainternmed.2020.3313.

Song, H. J. J. M. D., et al. (2023). Electrolyte imbalances as poor prognostic markers in COVID-19: A systemic review and meta-analysis. *Journal of Endocrinological Investigation*, *46*(2), 235–259. https://doi.org/10.1007/s40618-022-01877-5.

Syed, F., et al. (2021). Excessive matrix Metalloproteinase-1 and hyperactivation of endothelial cells occurred in COVID-19 patients and were associated with the severity of COVID-19. *The Journal of Infectious Diseases*, *224*(1), 60–69. https://doi.org/10.1093/infdis/jiab167.

van Niekerk, G., & Engelbrecht, A.-M. (2018). Inflammation-induced metabolic derangements or adaptation: An immunometabolic perspective. *Cytokine & Growth Factor Reviews*, *43*, 47–53. https://doi.org/10.1016/j.cytogfr.2018.06.003.

Chapter 45

Linking between gastrointestinal tract effects of COVID-19 and tryptophan metabolism

Yoshihiro Yokoyama and Hiroshi Nakase
Department of Gastroenterology and Hepatology, School of Medicine, Sapporo Medical University, Chuo-ku Sapporo, Hokkaido, Japan

Introduction

Coronavirus disease (COVID-19) caused by severe acute respiratory syndrome coronavirus 2 (SARS-CoV-2) is an infectious disease that caused a global pandemic in 2020 and disrupted the world. The main symptoms of infection are pulmonary symptoms, such as fever, sore throat and cough; however, some patients show gastrointestinal (GI) symptoms, such as diarrhea, abdominal pain, and vomiting (Tariq et al., 2020; Wiersinga et al., 2020). The main route of infection is through direct contact, including droplet and close contact transmission (Meyerowitz et al., 2021; Sharma et al., 2021). The detection of SARS-CoV-2 in the feces of some patients suggests that SARS-CoV-2 could be transmitted via a fecal-oral route (Guo et al., 2021; Jiao et al., 2021). It has been speculated that the direct causes of intestinal symptoms due to SARS-CoV-2 are viral infection and cytokine storms (Jin et al., 2021); however, involvement of COVID-19 in the mucosal immune system remains unclear. Angiotensin-converting enzyme 2 (ACE2) regulates tryptophan (Trp) metabolism in the small intestine and prevents intestinal inflammation by regulating the intestinal microbiota (Hashimoto et al., 2012). We elucidated that a decrease in tryptophan metabolism in the intestinal tract of patients with severe COVID-19 and the resulting inflammation in the intestinal tract may contribute to the severity of the disease (Yokoyama et al., 2022). Furthermore, in this study, we found that ACE2 expression markedly decreased in the intestinal mucosa of patients with severe COVID-19, suggesting an inability to absorb tryptophan in the intestinal tract. Blackett et al. reported the association between long-term effects of COVID-19 ("long COVID-19") and decreased Trp pathway mediated with gut dysbiosis (Blackett et al., 2022). Omics analysis of serum from patients with COVID-19 reported that in critical patients, decreased Trp metabolism is associated with impaired T cell dysfunction (Wu et al., 2021). Taken together, understanding the relationship between the gastrointestinal tract and Trp in the pathogenesis of COVID-19 is essential. In this article, we discuss the GI manifestations of COVID-19 based on their mechanisms with a focus on the gut microbiota and Trp metabolism in the intestinal tract.

Gastrointestinal symptoms and their mechanisms in COVID-19

SARS-CoV-2 is named after the Latin word "corona," meaning crown, after the shape of the ring of spikes visible through an electron microscope (Mortaz et al., 2020). When SARS-CoV-2 infects humans and causes COVID-19, symptoms usually appear 4 or 5 days after SARS-CoV-2 infection or as long as 2 weeks later (Wiersinga et al., 2020), and more than 15% of cases have GI symptoms such as nausea, diarrhea, and abdominal pain (Jin et al., 2021). The incidence of gastrointestinal symptoms is increased in severe cases of COVID-19 (Zhang et al., 2021). A meta-analysis has shown that abdominal pain is a risk factor for severe COVID-19 (Hayashi et al., 2021).

The human ACE2 receptor is an essential route of transmission for SARS-CoV-2 and S protein on the virus surface binds to ACE2 receptor, allowing SARS-CoV-2 to enter the human cell (Jackson et al., 2022; Villablanca et al., 2022) (Fig. 1). ACE2 receptors are known to be expressed in the digestive, urinary, respiratory, and vascular endothelium, especially highly expressed in the small or large intestinal mucosa (An et al., 2021). Therefore, it is assumed that the GI tract is easily penetrated by SARS-CoV-2. In a study by Xiao et al., infectious SARS-CoV-2 was isolated from stools in COVID-19 patients, and positive fluorescent immunostaining of gastrointestinal tissue for ACE2 and SARS-CoV-2 was observed in

Management, Body Systems, and Case Studies in COVID-19. https://doi.org/10.1016/B978-0-443-18703-2.00053-0
Copyright © 2024 Elsevier Inc. All rights are reserved, including those for text and data mining, AI training, and similar technologies.

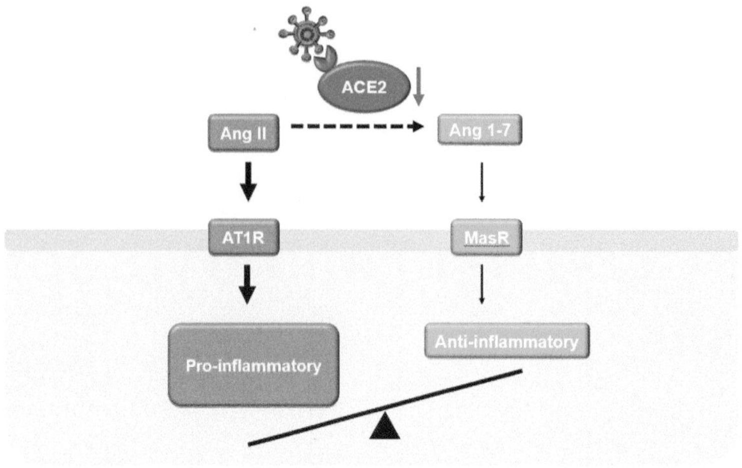

FIG. 1 Inflammatory response due to renin-angiotensin system imbalance with decreased ACE2. This figure shows that SARS-CoV-2 infection causes decreases in ACE2, which disrupts renin-angiotensin (RAS) system homeostasis and induces inflammation. Increased angiotensin II (Ang II) increases angiotensin II receptor type1 (AT1R)-mediated inflammation, while decreased Ang 1–7 attenuates Mas receptor (*R*)-mediated anti-inflammatory effects.

patients with positive stool test (Wong et al., 2020; Xiao et al., 2020). Viscous mucus in the GI tract protects the invasion of viral RNA and viral particles, allowing the virus to remain infectious. The virus is subsequently excreted through the stool. The inflammatory state of the GI tract disrupts the multilayer barrier system and increases ACE2 expression in intestinal epithelium, allowing SARS-CoV-2 to invade intestinal epithelial cells (Lamers et al., 2020; Potdar et al., 2021; Zhou et al., 2020). The infection of intestinal cells with SARS-CoV-2 has two main characteristics: First, the immune response induced by SARS-CoV-2 is distinct; under in vitro conditions, SARS-CoV-2 has been observed to infect human intestinal epithelial and mesenteric cells and cause their apoptosis (Lamers et al., 2020). Second, lung infections may also be accompanied by GI tract infections. Some patients had positive stool tests, even when the pharyngeal swab test was masked. This suggests that SARS-CoV-2 has been present and replicating at low levels in the intestinal tract for a long time. This is because in the early stages of COVID-19 infection, it is difficult for a small number of patients to rapidly acquire immunity to SARS-CoV-2 and produce targeted IgM and IgG, eventually resulting in long-term persistence of the virus in the GI tract (Zhang et al., 2021). The renin angiotensin system (RAS) regulates systemic and local body function. ACE2 is the major bioactive molecule of the RAS, and converts angiotensin (Ang) II to Ang 1–7. Ang II binds to angiotensin II receptors (ATR1 and produces proinflammatory cytokines via nuclear factor-kB. Ang 1–7 is known to bind to the Mas receptor (MasR) and act in an antiinflammatory effect, producing molecules involved in the production of IL-10 and inhibition of neutrophil migration and adhesion (Simões e Silva et al., 2013). After SARS-CoV-2 infection, ACE2 expression in the host is down-regulated, and the ACE2 protein level decreases; this results in an increase in Ang II level and a decrease in Ang 1–7 level, which is associated with systemic inflammation (Bian & Li, 2021) (Fig. 1).

Several reports have reported that the GI manifestations of COVID-19 resemble ischemic enterocolitis (Norsa et al., 2020; Ponzetto et al., 2021). Gastrointestinal hemorrhage, and acute mesenteric thrombosis have also been repeatedly reported, the pathogenesis of which is primarily histological ischemia due to microvascular thrombosis (Azouz et al., 2020). The mechanism of intestinal ischemia in COVID-19 involves ACE2 and transmembrane serine proteases (TMPRSS) expressed in vascular endothelium (Bonaventura et al., 2021). The SARS-CoV-2 S protein downregulates ACE2 and TMPRSS2 expression in the vascular endothelium, inducing the downregulation of ACE2. ACE2 downregulation is associated with a local increase in Ang II levels, which increases blood levels of tumor necrosis factor-α (TNF-α) and interleukin (IL)-6. Both cytokines promote the rupture of the endothelial barrier and decrease Ang-(1–7) levels, leading to sepsis development. A postulated mechanism suggests that vascular endothelial cell damage and increased blood cytokines combine to increase blood coagulation capacity, leading to intestinal injury.

The role of tryptophan and its metabolites in the gastrointestinal tract

Trp was first isolated and discovered by Frederick Gowland Hopkins in 1901 from trypsin hydrolyzed casein, a milk protein (Curzon, 2019). Trp can be broadly classified into two components: proteins and precursors of bioactive substances. The known precursors of bioactive substances include (1) serotonin, a neurotransmitter and smooth muscle contractile substance; (2) melatonin, a hormone related to sleep; and (3) kynurenine, an agonist of the aryl hydrocarbon receptor (AhR) (Agus et al., 2018; Anderson et al., 2021; Stockinger et al., 2021). Trp is also known to be a neurotransmitter, such as for serotonin and melatonin, but most Trp is metabolized in vivo via the kynurenine metabolic pathway. Some intestinal bacteria such as *Escherichia coli* produce indole from Trp. On the other hand, indole-3-acetic acid (IAA), a major Trp metabolite, is known to activate the nuclear receptor AHR, whose activation induces IL-22 production in immune cells such as regulatory T cells (Yang et al., 2020). In addition, histidine exerts a specific inhibitory effect on intestinal inflammation. Previous studies have shown that histidine suppresses TNF1α production by macrophages via suppression of NF-κB signaling (Andou et al., 2009).

Thus, Trp and its metabolites have been shown to be closely related to intestinal immunity.

ACE2 and tryptophan metabolism

Transepithelial absorption of amino acids occurs in intestinal epithelial cells, and the absorption of amino acids is a continuous transport process through the lumen, brush border, and basolateral membrane. The process by which amino acids pass through the lumen requires intervention by various amino acid transport systems. The stage at which amino acids pass through the lumen requires intervention by various amino acid transporters. It has been reported that the amino acid transporter B(0)AT1, a sodium-dependent neutral amino acid transporter, is expressed in cells of the small intestine and plays an important role in amino acid transport and absorption.

The expression and function of B(0)AT1 have been reported to be dependent on the presence of ACE2, and Hashimoto et al. found no changes in the morphology and ultrastructure of the small and large intestines of ACE2 knockout mice. However, after stimulation with sodium dextran sulfate (SDS), the intestines of ACE2 knockout mice showed more obvious inflammation, weight loss, and severe diarrhea. Although B(0)AT1 cannot be expressed in the intestines of ACE2 knockout mice, dietary Trp is absorbed mainly through the B(0)AT1/ACE2 transport pathway. Plasma Trp concentrations were significantly reduced, and deficiency of Trp and its metabolite nicotinamide reduced the activity of the mTOR pathway. The mTOR pathway regulates the expression of antimicrobial peptides that affect the gut microbiota. Alterations in antimicrobial peptides affect the ecology of the microflora in the large and small intestines, leading to localized enteritis and diarrhea. These data are important because they demonstrate that ACE2 regulates intestinal inflammation by modulating tryptophan metabolism through regulation of intestinal bacteria.

COVID-19 and intestinal microbiota

The human body is inhabited by bacteria that have approximately 10 times as many cells as the number of cells in the body (approximately 60 trillion cells), and the "bacterial flora", an ecosystem created by these bacteria, produces beneficial substances that the body cannot produce and suppresses excessive immune responses.

Clinical studies have shown changes in the intestinal microbiota and a decrease in anaerobic bacteria (*Bacteroides*, *Clostridium*, *Prevotella*, etc.) in COVID-19 infected individuals compared to healthy controls (Zuo et al., 2020). In another prospective cohort study analyzing the microbiome patterns of healthy participants, patients infected with COVID-19 and influenza virus (HIN1), it was possible to discriminate between COVID-19 and Influenza virus (HIN1) infections with high accuracy and precision using the abundance ratio of seven species of bacteria (*Streptococcus*, *Fusicatenibacter*, *Collinsella*, *Dorea*, *Agathobacter*, *Eubacterium hallii*, and *Ruminococcus torques* group) (Gu et al., 2020). This result led us to suspect that these seven bacterial species are involved in the pathogenesis of severe COVID-19 infections. In a clinical study investigating the correlation between gut microbiota and cytokines/chemokines, and inflammatory markers in patients with COVID-19 according to the disease severity, the beta diversity of the gut microbiota decreased with worsening severity of COVID-19 (Yeoh et al., 2021). These data suggest that disruption of the gut microbiota can lead to severe COVID-19 disease by producing increased gastrointestinal symptoms and enhanced cytokine-mediated inflammatory responses.

COVID-19 and tryptophan metabolism in the intestinal tract

We discuss the findings of our laboratory (Yokoyama et al., 2022) and those reported to date on Trp metabolism and COVID-19 in the intestinal tract. Our institution has previously reported a case of COVID-19 pneumonia with severe gastrointestinal mucosal damage (Yamakawa et al., 2022) and that abdominal pain is a poor prognostic factor in COVID-19 (Hayashi et al., 2021). Following these reports, we performed a study to clarify the relationship between the severity of COVID-19 and GI damage. We first found that patients with critical COVID-19 had inflammation of the GI tract, as shown by high fecal calprotectin levels. Stool metabolomic analysis showed that the level of indole-3-propionic acid, a ligand for

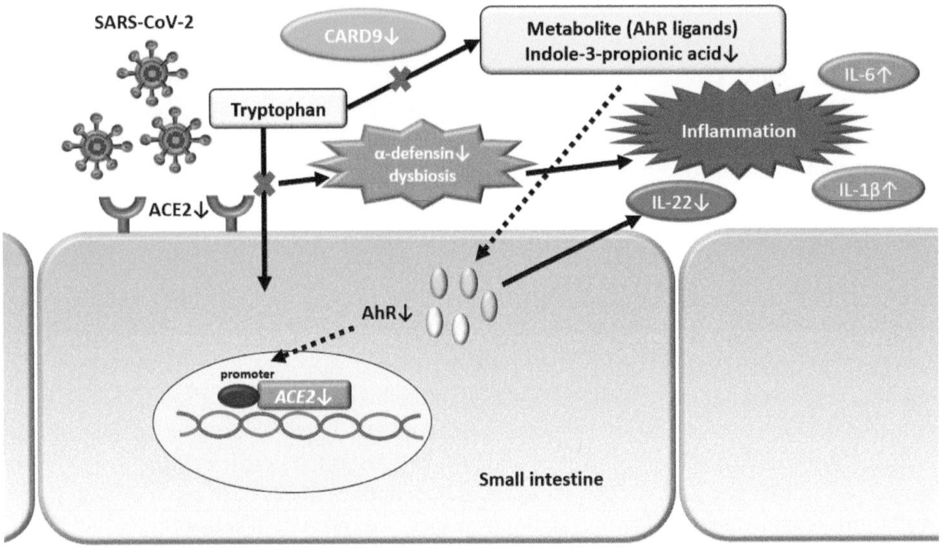

FIG. 2 Possible mechanisms of gastrointestinal inflammation in patients with critical COVID-19. This figure shows the possible mechanism of inflammation within the small intestine of patients with critical COVID-19. When ACE2 in the small intestine is reduced by SARS-CoV-2 infection, tryptophan metabolism via the AhR pathway is impaired and induces inflammation of the gastrointestinal tract. *(From Yokoyama, Y., Ichiki, T., Yamakawa, T., Tsuji, Y., Kuronuma, K., Takahashi, S., Narimatsu, E., & Nakase, H. (2022). Impaired tryptophan metabolism in the gastrointestinal tract of patients with critical coronavirus disease 2019.* Frontiers in Medicine, 9, *941422. https://doi.org/10.3389/fmed.2022.941422 with permission.)*

AhR, was markedly decreased in the critical COVID-19 patients. Moreover, the expression of genes involved in tryptophan metabolism, including ACE2, AhR, caspase recruitment domain-containing protein 9 (CARD9), and IL22, was downregulated in the ileum of critical COVID-19 patients. These results suggested that the decreased ACE2 expression and impaired AhR pathway in the small intestine were responsible for inflammation in the gastrointestinal tracts of the critical COVID-19 patients (Fig. 2).

Blackett et al. investigated Trp metabolism and 5-hydroxytryptamine (5-HT)-based signaling in the gut microbiome of post-COVID-19 patients (Blackett et al., 2022). Using fecal and blood samples from COVID-19 patients during the acute phase, they found that L-Trp biosynthesis in the gut microbiome during the acute phase of COVID-19 was decreased in those who developed more severe GI symptoms. All Trp pathways were less active in the patients with more severe GI symptoms. The same pathways also decreased in patients with the most severe psychiatric symptoms after COVID-19. Furthermore, plasma 5-HT concentrations during COVID-19 were 5.1-fold higher in patients with GI symptoms than in those with mental health symptoms only. Taken together, these findings suggest that an acute decrease in gut bacteria-mediated 5-HT signaling may contribute to long-term GI and mental health symptoms after COVID-19.

OMICS analysis in COVID-19 patients showed that neutrophil overactivation, arginine depletion, and Trp metabolite accumulation in asymptomatic and severe patients correlated with T cell dysfunction in severe patients, supporting the hypothesis that Trp metabolic impairment occurs in severe COVID-19 (Wu et al., 2021). In this study, correlation analysis suggested the possibility of regulatory crosstalk between inflammatory cytokines/chemokines (IL-6, M-CSF, IL-1α, IL-1β, etc.) and arginine, Trp, and purine metabolism. In addition, the release of proinflammatory cytokines by peripheral blood mononuclear cells isolated from SARS-CoV-2-infected rhesus monkeys was markedly modulated by targeting metabolism, suggesting that the exploitation of metabolic changes may cause lethal cytokine storms in COVID-19. In recent years, various findings have suggested that Trp metabolism disorder is associated with COVID-19 severity or post-COVID-19. Therefore, the improvement of Trp metabolism in the intestinal tract needs to prevent COVID-19 severity and sequelae caused by COVID-19.

Conclusion

There are many unresolved issues related to the factors involved in the GI damage and severity of COVID-19. Gut microbiome dysbiosis and impaired tryptophan metabolism observed in COVID-19 patients is an essential key to resolving these issues. Further research on the linking of COVID-19 and tryptophan in the GI tract is expected to overcome COVID-19.

Summary points

- COVID-19 is an infection that causes gastrointestinal symptoms.
- SARS-CoV-2 invades intestinal cells via ACE2 and induces inflammation of the digestive tract.
- Impaired tryptophan metabolism in the gastrointestinal tract of patients with severe COVID-19 has been demonstrated.
- The association of COVID-19 with intestinal dysbiosis and intestinal metabolomic abnormalities has been clarified.
- Improving intestinal gut dysbiosis and tryptophan metabolism in the gastrointestinal tract could prevent the severity of COVID-19.

References

Agus, A., Planchais, J., & Sokol, H. (2018). Gut microbiota regulation of tryptophan metabolism in health and disease. *Cell Host & Microbe*, *23*(6), 716–724. https://doi.org/10.1016/j.chom.2018.05.003.

An, X., Lin, W., Liu, H., Zhong, W., Zhang, X., Zhu, Y., … Sheng, Q. (2021). SARS-CoV-2 host receptor ACE2 protein expression atlas in human gastrointestinal tract. *Frontiers in Cell and Developmental Biology*, *9*, 659809. https://www.frontiersin.org/articles/10.3389/fcell.2021.659809.

Anderson, G., Carbone, A., & Mazzoccoli, G. (2021). Tryptophan metabolites and aryl hydrocarbon receptor in severe acute respiratory syndrome, CORONAVIRUS-2 (SARS-CoV-2) pathophysiology. *International Journal of Molecular Sciences*, *22*(4), 1597. https://doi.org/10.3390/ijms22041597.

Andou, A., Hisamatsu, T., Okamoto, S., Chinen, H., Kamada, N., Kobayashi, T., Hashimoto, M., Okutsu, T., Shimbo, K., Takeda, T., Matsumoto, H., Sato, A., Ohtsu, H., Suzuki, M., & Hibi, T. (2009). Dietary histidine ameliorates murine colitis by inhibition of proinflammatory cytokine production from macrophages. *Gastroenterology*, *136*(2). https://doi.org/10.1053/j.gastro.2008.09.062. 564–574.e2.

Azouz, E., Yang, S., Monnier-Cholley, L., & Arrivé, L. (2020). Systemic arterial thrombosis and acute mesenteric ischemia in a patient with COVID-19. *Intensive Care Medicine*, *46*(7), 1464–1465. https://doi.org/10.1007/s00134-020-06079-2.

Bian, J., & Li, Z. (2021). Angiotensin-converting enzyme 2 (ACE2): SARS-CoV-2 receptor and RAS modulator. *Acta Pharmaceutica Sinica B*, *11*(1), 1–12. https://doi.org/10.1016/j.apsb.2020.10.006.

Blackett, J. W., Sun, Y., Purpura, L., Margolis, K. G., Elkind, M. S. V., O'Byrne, S., Wainberg, M., Abrams, J. A., Wang, H. H., Chang, L., & Freedberg, D. E. (2022). Decreased gut microbiome tryptophan metabolism and serotonergic signaling in patients with persistent mental health and gastrointestinal symptoms after COVID-19. *Clinical and Translational Gastroenterology*. https://doi.org/10.14309/ctg.0000000000000524.

Bonaventura, A., Vecchié, A., Dagna, L., Martinod, K., Dixon, D. L., Van Tassell, B. W., Dentali, F., Montecucco, F., Massberg, S., Levi, M., & Abbate, A. (2021). Endothelial dysfunction and immunothrombosis as key pathogenic mechanisms in COVID-19. *Nature Reviews. Immunology*, *21*(5), 319–329. https://doi.org/10.1038/s41577-021-00536-9.

Curzon, G. (2019). *Hopkins and the discovery of tryptophan* (pp. XXIX–XL). De Gruyter. https://doi.org/10.1515/9783110854657-004.

Gu, S., Chen, Y., Wu, Z., Chen, Y., Gao, H., Lv, L., Guo, F., Zhang, X., Luo, R., Huang, C., Lu, H., Zheng, B., Zhang, J., Yan, R., Zhang, H., Jiang, H., Xu, Q., Guo, J., Gong, Y., … Li, L. (2020). Alterations of the gut microbiota in patients with coronavirus disease 2019 or H1N1 influenza. *Clinical Infectious Diseases: An Official Publication of the Infectious Diseases Society of America*, *71*(10), 2669–2678. https://doi.org/10.1093/cid/ciaa709.

Guo, M., Tao, W., Flavell, R. A., & Zhu, S. (2021). Potential intestinal infection and faecal-oral transmission of SARS-CoV-2. *Nature Reviews. Gastroenterology & Hepatology*, *18*(4), 269–283. https://doi.org/10.1038/s41575-021-00416-6.

Hashimoto, T., Perlot, T., Rehman, A., Trichereau, J., Ishiguro, H., Paolino, M., … Penninger, J. M. (2012). ACE2 links amino acid malnutrition to microbial ecology and intestinal inflammation. *Nature*, *487*(7408), 477–481. https://doi.org/10.1038/nature11228.

Hayashi, Y., Wagatsuma, K., Nojima, M., Yamakawa, T., Ichimiya, T., Yokoyama, Y., Kazama, T., Hirayama, D., & Nakase, H. (2021). The characteristics of gastrointestinal symptoms in patients with severe COVID-19: A systematic review and meta-analysis. *Journal of Gastroenterology*, *56*(5), 409–420. https://doi.org/10.1007/s00535-021-01778-z.

Jackson, C. B., Farzan, M., Chen, B., & Choe, H. (2022). Mechanisms of SARS-CoV-2 entry into cells. *Nature Reviews. Molecular Cell Biology*, *23*(1), 3–20. https://doi.org/10.1038/s41580-021-00418-x.

Jiao, L., Li, H., Xu, J., Yang, M., Ma, C., Li, J., Zhao, S., Wang, H., Yang, Y., Yu, W., Wang, J., Yang, J., Long, H., Gao, J., Ding, K., Wu, D., Kuang, D., Zhao, Y., Liu, J., … Peng, X. (2021). The gastrointestinal tract is an alternative route for SARS-CoV-2 infection in a nonhuman primate model. *Gastroenterology*, *160*(5), 1647–1661. https://doi.org/10.1053/j.gastro.2020.12.001.

Jin, B., Singh, R., Ha, S. E., Zogg, H., Park, P. J., & Ro, S. (2021). Pathophysiological mechanisms underlying gastrointestinal symptoms in patients with COVID-19. *World Journal of Gastroenterology*, *27*(19), 2341–2352. https://doi.org/10.3748/wjg.v27.i19.2341.

Lamers, M. M., Beumer, J., van der Vaart, J., Knoops, K., Puschhof, J., Breugem, T. I., Ravelli, R. B. G., Paul van Schayck, J., Mykytyn, A. Z., Duimel, H. Q., van Donselaar, E., Riesebosch, S., Kuijpers, H. J. H., Schipper, D., van de Wetering, W. J., de Graaf, M., Koopmans, M., Cuppen, E., Peters, P. J., … Clevers, H. (2020). SARS-CoV-2 productively infects human gut enterocytes. *Science (New York, N.Y.)*, *369*(6499), 50–54. https://doi.org/10.1126/science.abc1669.

Meyerowitz, E. A., Richterman, A., Gandhi, R. T., & Sax, P. E. (2021). Transmission of SARS-CoV-2: A review of viral, host, and environmental factors. *Annals of Internal Medicine*, *174*(1), 69–79. https://doi.org/10.7326/M20-5008.

Mortaz, E., Tabarsi, P., Varahram, M., Folkerts, G., & Adcock, I. M. (2020). The immune response and immunopathology of COVID-19. *Frontiers in Immunology*, *11*, 2037. https://doi.org/10.3389/fimmu.2020.02037.

Norsa, L., Bonaffini, P. A., Indriolo, A., Valle, C., Sonzogni, A., & Sironi, S. (2020). Poor outcome of intestinal ischemic manifestations of COVID-19. *Gastroenterology*, *159*(4), 1595–1597.e1. https://doi.org/10.1053/j.gastro.2020.06.041.

Ponzetto, A., Holton, J., & Porta, M. (2021). Intestinal ischemic manifestations of COVID-19. *Gastroenterology*, *160*(6), 2191. https://doi.org/10.1053/j.gastro.2020.10.060.

Potdar, A. A., Dube, S., Naito, T., Li, K., Botwin, G., Haritunians, T., Li, D., Casero, D., Yang, S., Bilsborough, J., Perrigoue, J. G., Denson, L. A., Daly, M., Targan, S. R., Fleshner, P., Braun, J., Kugathasan, S., Stappenbeck, T. S., & McGovern, D. P. B. (2021). Altered intestinal ACE2 levels are associated with inflammation, severe disease, and response to anti-cytokine therapy in inflammatory bowel disease. *Gastroenterology*, *160*(3), 809–822.e7. https://doi.org/10.1053/j.gastro.2020.10.041.

Sharma, A., Ahmad Farouk, I., & Lal, S. K. (2021). COVID-19: A review on the novel coronavirus disease evolution, transmission, detection. *Control and Prevention. Viruses*, *13*(2), 202. https://doi.org/10.3390/v13020202.

Simões e Silva, A. C., Silveira, K. D., Ferreira, A. J., & Teixeira, M. M. (2013). ACE2, angiotensin-(1–7) and Mas receptor axis in inflammation and fibrosis. *British Journal of Pharmacology*, *169*(3), 477–492. https://doi.org/10.1111/bph.12159.

Stockinger, B., Shah, K., & Wincent, E. (2021). AHR in the intestinal microenvironment: Safeguarding barrier function. *Nature Reviews. Gastroenterology & Hepatology*, *18*(8), 559–570. https://doi.org/10.1038/s41575-021-00430-8.

Tariq, R., Saha, S., Furqan, F., Hassett, L., Pardi, D., & Khanna, S. (2020). Prevalence and mortality of COVID-19 patients with gastrointestinal symptoms: A systematic review and meta-analysis. *Mayo Clinic Proceedings*, *95*(8), 1632–1648. https://doi.org/10.1016/j.mayocp.2020.06.003.

Villablanca, E. J., Selin, K., & Hedin, C. R. H. (2022). Mechanisms of mucosal healing: Treating inflammatory bowel disease without immunosuppression? *Nature Reviews Gastroenterology & Hepatology*, *19*(8). https://doi.org/10.1038/s41575-022-00604-y. Article 8.

Wiersinga, W. J., Rhodes, A., Cheng, A. C., Peacock, S. J., & Prescott, H. C. (2020). Pathophysiology, transmission, diagnosis, and treatment of coronavirus disease 2019 (COVID-19): A review. *JAMA*, *324*(8), 782–793. https://doi.org/10.1001/jama.2020.12839.

Wong, M. C., Huang, J., Lai, C., Ng, R., Chan, F. K. L., & Chan, P. K. S. (2020). Detection of SARS-CoV-2 RNA in fecal specimens of patients with confirmed COVID-19: A meta-analysis. *The Journal of Infection*, *81*(2), e31–e38. https://doi.org/10.1016/j.jinf.2020.06.012.

Wu, P., Chen, D., Ding, W., Wu, P., Hou, H., Bai, Y., Zhou, Y., Li, K., Xiang, S., Liu, P., Ju, J., Guo, E., Liu, J., Yang, B., Fan, J., He, L., Sun, Z., Feng, L., Wang, J., … Chen, G. (2021). The trans-omics landscape of COVID-19. *Nature Communications*, *12*(1), 4543. https://doi.org/10.1038/s41467-021-24482-1.

Xiao, F., Tang, M., Zheng, X., Liu, Y., Li, X., & Shan, H. (2020). Evidence for gastrointestinal infection of SARS-CoV-2. *Gastroenterology*, *158*(6). https://doi.org/10.1053/j.gastro.2020.02.055. 1831–1833.e3.

Yamakawa, T., Ishigami, K., Takizawa, A., Takada, Y., Ohwada, S., Yokoyama, Y., … Nakase, H. (2022). Extensive mucosal sloughing of the small intestine and colon in a patient with severe COVID-19. *DEN Open*, *2*(1), e42. https://doi.org/10.1002/deo2.42.

Yang, W., Yu, T., Huang, X., Bilotta, A. J., Xu, L., Lu, Y., Sun, J., Pan, F., Zhou, J., Zhang, W., Yao, S., Maynard, C. L., Singh, N., Dann, S. M., Liu, Z., & Cong, Y. (2020). Intestinal microbiota-derived short-chain fatty acids regulation of immune cell IL-22 production and gut immunity. *Nature Communications*, *11*(1), 4457. https://doi.org/10.1038/s41467-020-18262-6.

Yeoh, Y. K., Zuo, T., Lui, G. C.-Y., Zhang, F., Liu, Q., Li, A. Y., Chung, A. C., Cheung, C. P., Tso, E. Y., Fung, K. S., Chan, V., Ling, L., Joynt, G., Hui, D. S.-C., Chow, K. M., Ng, S. S. S., Li, T. C.-M., Ng, R. W., Yip, T. C., … Ng, S. C. (2021). Gut microbiota composition reflects disease severity and dysfunctional immune responses in patients with COVID-19. *Gut*, *70*(4), 698–706. https://doi.org/10.1136/gutjnl-2020-323020.

Yokoyama, Y., Ichiki, T., Yamakawa, T., Tsuji, Y., Kuronuma, K., Takahashi, S., Narimatsu, E., & Nakase, H. (2022). Impaired tryptophan metabolism in the gastrointestinal tract of patients with critical coronavirus disease 2019. *Frontiers in Medicine*, *9*, 941422. https://doi.org/10.3389/fmed.2022.941422.

Zhang, H., Shao, B., Dang, Q., Chen, Z., Zhou, Q., Luo, H., Yuan, W., & Sun, Z. (2021). Pathogenesis and mechanism of gastrointestinal infection with COVID-19. *Frontiers in Immunology*, *12*, 674074. https://doi.org/10.3389/fimmu.2021.674074.

Zhou, J., Li, C., Liu, X., Chiu, M. C., Zhao, X., Wang, D., Wei, Y., Lee, A., Zhang, A. J., Chu, H., Cai, J.-P., Yip, C. C.-Y., Chan, I. H.-Y., Wong, K. K.-Y., Tsang, O. T.-Y., Chan, K.-H., Chan, J. F.-W., To, K. K.-W., Chen, H., & Yuen, K. Y. (2020). Infection of bat and human intestinal organoids by SARS-CoV-2. *Nature Medicine*, *26*(7), 1077–1083. https://doi.org/10.1038/s41591-020-0912-6.

Zuo, T., Zhang, F., Lui, G. C. Y., Yeoh, Y. K., Li, A. Y. L., Zhan, H., Wan, Y., Chung, A. C. K., Cheung, C. P., Chen, N., Lai, C. K. C., Chen, Z., Tso, E. Y. K., Fung, K. S. C., Chan, V., Ling, L., Joynt, G., Hui, D. S. C., Chan, F. K. L., … Ng, S. C. (2020). Alterations in gut microbiota of patients with COVID-19 during time of hospitalization. *Gastroenterology*, *159*(3), 944–955.e8. https://doi.org/10.1053/j.gastro.2020.05.048.

Chapter 46

Organ damage in SARS-CoV-2 infection in children: A focus on acute kidney injury

Girish Chandra Bhatt[a], Yogendra Singh Yadav[b], and Tanya Sharma[c]
[a]*Division of Pediatrics Nephrology, Department of Pediatrics, AIIMS Bhopal, Bhopal, India,* [b]*Department of Pediatrics, AIIMS Bhopal, Bhopal, India,* [c]*Department of Pathology & Laboratory Medicine, AIIMS Bhopal, Bhopal, India*

Abbreviations

AKI	acute kidney injury
AKIN	acute kidney injury network
APOL1	apolipoprotien L 1
ARDS	acute respiratory distress syndrome
COVID-19	coronavirus disease 2019
CRRT	continuous renal replacement therapy
eGFR	estimated glomerular filtration rate
eCL	estimated creatinine clearance
GFR	glomerular filtration rate
IHD	intermittent hemodialysis
ICU	intensive care unit
KDIGO	kidney disease improving global outcomes
KST	kidney support therapy
MIS-C	multisystem inflammatory syndrome in children
PIHD	prolonged intermittent hemodialysis
PD	peritoneal dialysis
SARS	severe acute respiratory syndrome
SARS-CoV-2	severe acute respiratory syndrome coronavirus 2
sC5b9	soluble complement 5b-9
SLEDD	sustained low-efficiency daily dialysis
SCr	serum creatinine
TNF	tumor necrosis factor
WHO	World Health Organization

Introduction

Coronavirus disease 2019 (COVID-19) is an illness caused by severe acute respiratory syndrome coronavirus 2 (SARS-CoV-2). As with adults, the disease manifestations among children vary, including asymptomatic infection, mild disease with fever, respiratory or gastrointestinal symptoms, more severe disease, and systemic involvement causing multi-organ failure. Although COVID-19 primarily affects the lungs, acute kidney injury (AKI) is also a prevalent complication. Additionally, a rare yet severe manifestation among children, known as multisystem inflammatory syndrome in children (MIS-C), has been identified. This condition is associated with COVID-19 and is characterized by symptoms such as continuous fever, hypotension, gastrointestinal issues, rashes, myocarditis, shock, and elevated inflammation markers (World Health Organization, 2023).

AKI is defined as an abrupt decline in kidney function characterized by an increase in serum creatinine or a decrease in urine output (Kidney Disease Improving Global Outcomes (KDIGO) Acute Kidney Injury Work Group, 2012). Table 1 shows the definition of AKI as proposed by different working groups (Bhatt et al., 2022). AKI occurs in approximately one-fifth of hospitalized children with COVID-19, and one-third of those required PICU admission (Kari et al., 2021).

Management, Body Systems, and Case Studies in COVID-19. https://doi.org/10.1016/B978-0-443-18703-2.00011-6
Copyright © 2024 Elsevier Inc. All rights are reserved, including those for text and data mining, AI training, and similar technologies.

TABLE 1 Pediatric AKI definition by various working groups.

pRIFLE classification			AKIN staging			KDIGO staging		
Class	eGFR/eCL	Urine output	Stage	S creatinine	Urine output	Stage	S creatine	Urine output
Risk	Decrease in eCl≥25%	<0.5 mL/kg/h for 6 h	1	Increase in SCr to 26.5 μmol/L OR Increase in SCr≥150% to 200% from baseline	<0.5 mL/kg/h for >6 h	1	Increase in SCr to ≥1.5 to 1.9 times baseline OR ≥26.5 mmol/L increase	<0.5 mL/kg/h for 6–12 h
Injury	Decrease in eCl≥50%	<0.5 mL/kg/h for 12 h	2	Increase in SCr≥200% to 300% from baseline	<0.5 mL/kg/h for >12 h	2	Increase in SCr to ≥2 to 2.9	<0.5 mL/kg/h for ≥12 h
Failure	Decrease in eCl≥75% OR eCl<35 mL/min/ 1.73 m^2	<0.3 mL/kg/h for 24 h OR anuria for 12 h	3	Increase in SCr≥300% (>threefold) from baseline or if baseline SCr≥353.6 μmol/L or increase in SCr≥44.2 μmol/L; also includes patients requiring KST independent of stage	<0.3 mL/kg/h for >24 h OR anuria for 12 h	3	Increase in SCr to 3.0 times baseline OR increase in SCr to ≥353.6 mmol/L OR Initiation of RRT OR In patients ≤18 years, decrease in eGFR to ≤35 mL/min/1.73 m^2	<0.3 mL/kg/h for ≥24 h OR anuria for ≥12 h
Loss	Complete loss of kidney function >4 weeks	–	–	–	–	–	–	–
ESKD	Complete loss of kidney function >3 months							

eGFR, estimated glomerular filtration rate; *AKIN*, acute kidney injury network (AKIN); *eCL*, estimated creatinine clearance; *SCr*, serum creatinine.

AKI is more commonly seen in younger children and in those with comorbid conditions; it is also associated with increased mortality and morbidity. A small proportion of children with AKI can develop residual renal impairment. Nonetheless, it tends to be milder than in adults, with a lower incidence of oliguria and less need for kidney support therapy (KST) (Kari et al., 2021).

AKI occurs in approximately 25–33% of patients with MIS-C and is associated with a poor prognosis in critically ill children (Basalely et al., 2021). The mechanism of AKI in SARS-CoV-2 patients is multifactorial, including dehydration, poor cardiac output, cytokine storm, the direct cytopathic effect of the virus on renal tubular cells, and the use of nephrotoxic drugs. However, the mechanism implicated in AKI development in MIS-C patients is chiefly due to renal hypoperfusion (Deep et al., 2020).

AKI frequently complicates the course of COVID-19 hospitalizations and is associated with increased severity of illness, prolonged hospitalization stay, and poor prognosis. Thus, the early identification of comorbidities and renal complications including AKI is essential to improve the outcomes of COVID-19 patients.

Pathophysiology

The exact mechanism of kidney damage caused by COVID-19 is not yet clear. The etiology and pathogenesis of AKI are multifactorial. Fig. 1 shows various pathways of pathophysiology that may contribute to COVID-19-associated AKI and renal complications (Legrand et al., 2021). The pathophysiology of COVID-19 AKI is thought to involve local and systemic inflammatory and immune responses, endothelial injury, and the activation of coagulation pathways and the renin-angiotensin system (Ferlicot et al., 2021). However, the role of direct viral infection in the development of AKI remains controversial. The enhanced release of inflammatory mediators by immune and resident kidney cells is likely to be a key mechanism of tissue damage in patients with COVID-19. Inflammatory mediators, such as tumor necrosis factor (TNF) and tumor necrosis factor ligand superfamily (FAS), can bind to their specific receptors expressed by renal endothelial and tubule epithelial cells, causing a direct injury (Cantaluppi et al., 2014).

It has been reported that severely ill patients with COVID-19 exhibit an impaired type I interferon response, which may contribute to the inefficacy in clearing the virus from kidney cells in a specific group of patients. However, collapsing nephropathy in patients with COVID-19 seems to be associated with the high-risk apolipoprotien L 1 (APOL1) genotype and may involve pathogenic pathways linked to interferon-mediated podocyte injury (Wu et al., 2020). One mechanism of renal injury showed that a combination of viral infection of the cell along with complement activation and vascular injury

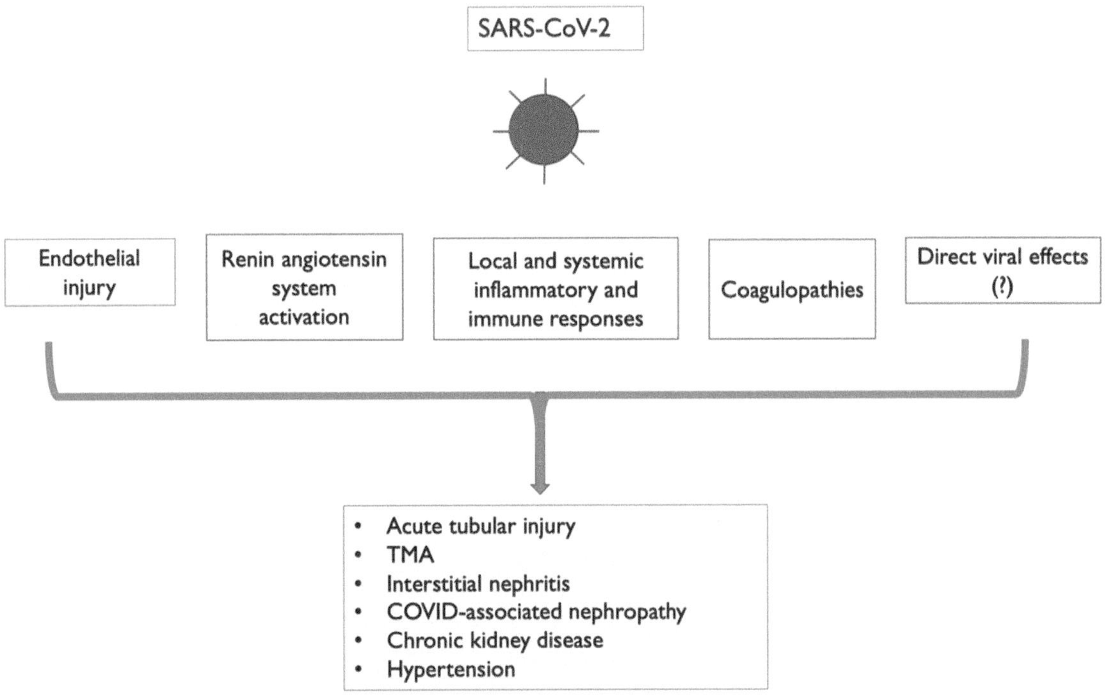

FIG. 1 Pathophysiology of kidney injury in COVID-19. *SARS-CoV-2*, severe acute respiratory syndrome coronavirus 2; *TMA*, thrombotic microangiopathy.

was the cause, and elevations in sC5b9 are independent of other MIS-C markers and associated with renal injury (Diorio et al., 2020).

Viral sequences of the SARS virus are detected in the distal tubular epithelium of the kidney while clusters of viral fragments are observed within the cytoplasm of the same epithelium. Patients infected with SARS are found to have hemorrhagic conditions in their kidneys. COVID-19 AKI patients commonly experience acute tubular injury, but it is typically mild despite significant impairments in kidney function. Evidence has shown that COVID-19 AKI patients frequently experience endothelial injury, microvascular thrombi, inflammation, and immune cell infiltration. Although there is increased knowledge regarding kidney injury processes in COVID-19, there is currently a lack of specific therapeutic strategies targeting the kidneys (Gu et al., 2005).

Clinical manifestations

The most common clinical features of COVID-19 among children include fever, dry cough, and pneumonia, along with an increasing prevalence of multisystem dysfunction. Kidney manifestations from COVID-19 are becoming increasingly prevalent, with AKI contributing to high mortality rates (Gao et al., 2021).

COVID-19 patients with AKI rarely showed clinical symptoms. Hematuria and proteinuria were the predominant manifestation among those who developed AKI and this was notable more prevalent in individuals with severe or critical illness (Chan et al., 2021). More than 50% of patients without AKI as defined by kidney disease improving global outcomes (KDIGO) serum creatinine criteria have hematuria and more than 70% presented with proteinuria. The presence of urinalysis abnormalities in those not meeting the definition of AKI suggests kidney injury without notable acute changes in kidney function. Proteinuria and/or hematuria are indicative of kidney injury, even in the absence of a rise in the serum creatinine (SCr) level or a drop in the glomerular filtration rate (GFR). Further injury is associated with a drop in GFR and a rise in SCr. Underlying chronic kidney disease or a factor such as aging limits the baseline functional reserve and can precipitate the development of AKI (Pei et al., 2020).

More than 10% of children who were hospitalized with COVID-19 or MIS-C experienced AKI, a condition characterized by inflammation and occasionally following SARS-CoV-2 infection (Raina et al., 2022; Rowley, 2020). Children who experience AKI may face challenges in regulating their body fluids and normal urination. While this injury typically clears up on its own, it can potentially develop into chronic kidney disease. It is important to monitor children who have had COVID-19 or MIS-C for persistent symptoms, such as high blood pressure or protein loss in urine, which could indicate an increased risk of chronic kidney disease. Acute renal impairment disrupts prerenal factors, which can lead to multiorgan failure. Age and acute respiratory distress syndrome are the two risk factors determining the progression of acute renal impairment in SARS-affected individuals (Chu et al., 2005).

Laboratory work up

Kidney damage caused by COVID-19 typically manifests as tubular damage with obvious urinalysis abnormalities. Impaired glomerular filtration also occurs, usually manifested by increased blood creatinine and urea nitrogen levels. Proteinuria is common in patients with kidney damage caused by COVID-19. A study was carried out in 13 centers across Argentina involving urinalysis in 106 cases. The results indicated that 18 individuals (16.9%) exhibited leukocyturia, 17 (16.0%) had proteinuria (with one case of massive proteinuria, but without nephrotic syndrome), and 14 (13.2%) showed microscopic hematuria (Martin et al., 2022). Also, various studies have reported a link between renal involvement and hypoalbuminemia, lymphopenia, and leukocytosis. This association could be attributed to systemic inflammation causing an increase in capillary permeability, which could result in hypoalbuminemia as well as alterations in white blood cells (Basalely et al., 2021; Liguoro et al., 2020; Qiu et al., 2020).

Management

The severity of COVID-19 is directly linked to kidney damage in patients. It is imperative to initiate a thorough treatment plan at the earliest convenience. The consistent application of antiviral and oxygen treatments, alongside respiratory support, circulation monitoring, and nutrition support therapy, must be maintained for those with COVID-19.

The estimated incidence of AKI and the need for KST among hospitalized COVID-19 patients vary between studies, ranging from 0.5% to as high as 40% (Fu et al., 2020). The management of AKI caused by COVID-19 as well as the indications for KST and the type of KST used are comparable to those employed in other scenarios. Furthermore, extracorporeal therapies have been investigated as a possible solution for clearing cytokines in patients with sepsis, and may be considered for preventing organ damage in critically ill individuals with COVID-19 who have cytokine release syndrome (Ronco et al., 2020). KST should be used to treat AKI based on established criteria such as restoring immune homeostasis, eliminating inflammatory mediators that cause ARDS, and preventing fluid overload, which is a well-known factor in intensive care unit (ICU) mortality (Bhatt et al., 2021; Garzotto et al., 2016).

AKI requiring dialysis can be managed with a variety of modalities, including continuous renal replacement therapy (CRRT), prolonged intermittent hemodialysis (PIHD), intermittent hemodialysis (IHD), and peritoneal dialysis (PD). The choice of dialysis modality to be used in managing a specific patient is influenced by several factors, including the goals of dialysis, the unique advantages and disadvantages of each modality, and institutional resources (Sabaghian et al., 2022).

CRRT emulates the human glomerular filtration process by introducing arteriovenous blood into a filter made of a semipermeable membrane. This method eliminates both proinflammatory and antiinflammatory transmitters from the blood stream nondiscriminately via convection and diffusion. CRRT blocks the production of inflammatory cascade reactions, regulates the water electrolyte disturbance and acid-base balance, improves the body's immune function, reduces the peak concentration of inflammatory factors in the body, downregulates body inflammation, and prevents the excessive activation of inflammatory transmitters in the body, resulting in a stable and balanced internal environment. The primary reasons for choosing CRRT instead of IHD are hemodynamic instability and high catabolic states (Adapa et al., 2020). Because CRRT can block the inflammatory factor storm in time, it greatly improves the prognosis of COVID-19 patients (Chu et al., 2005).

PIHD, otherwise known as sustained low-efficiency daily dialysis (SLEDD), refers to a type of IHD that is performed for an extended duration, typically around 6–18 h and at least three times a week. The approach behind PIHD involves using a combination of convective (hemofiltration) and diffusion (HD) principles to remove solutes from the blood (Edrees et al., 2016). PIHD is typically more favorable compared to IHD, as it enables a greater volume of ultrafiltration while posing a reduced risk of hemodynamic instability (Kielstein et al., 2004).

The application of IHD for critically ill COVID-19 patients must be restricted to cases where CRRT or PIHD is not available, and should only be considered for ICU patients experiencing severe AKI and are in a hemodynamically stable state. Close monitoring is required during the entire session due to the elevated flow rates necessary to enhance dialysis efficiency to prevent any hemodynamic instabilities (Shemies et al., 2022).

PD may provide a safe and viable alternative to HD for patients with unplanned dialysis needs. Studies have shown that PD is a viable treatment option for critically ill individuals with AKI, and can be utilized effectively in such cases. PD offers several advantages including cost-effectiveness, easy availability in resource-constrained settings, minimal expertise required, and can be performed in patients with coagulation abnormalities (Atlani et al., 2022).

Prognosis

In both adult and pediatric patients, AKI has been identified as a detrimental prognostic factor, which is evidenced by the high incidence rate (30.51%) and mortality rate (2.55%) observed among COVID-19-positive pediatric patients (Raina et al., 2021).

Patients with AKI requiring dialysis are much more likely to die than patients without AKI. Patients with COVID-19-related kidney damage should follow up to ensure kidney function is returning to normal. Therefore, the advanced identification of kidney diseases in patients affected by COVID-19 helps clinicians reduce the mortality rate due to comorbidities such as chronic kidney disease, AKI, proteinuria, and hematuria (Cheng et al., 2020). Early recognition and specific therapy for AKI, including adequate hemodynamic support and avoidance of nephrotoxic drugs, may limit associated complications such as long-term chronic kidney disease or end-stage kidney disease and may help to improve critically ill patients with COVID-19. Furthermore, physicians must be aware that patients who recover from AKI induced by SARS-CoV-2 require monitoring of their kidneys on follow up, as there is rising evidence showing eGFR decreases among patients with a history of COVID-19-associated AKI (Nugent et al., 2021). According to a research study (Kari et al., 2021), approximately 9% of individuals in the study suffered from residual renal impairment upon being discharged. The factors implicated for this were linked to reduced tissue perfusion (such as hypotension, heart failure, acute respiratory syndrome, and hypoxia), sepsis (including high C-reactive protein and positive blood culture), a worsening clinical state (elevated liver enzymes, hypernatremia), or preexisting health conditions.

To minimize mortality and morbidity, it is crucial to closely monitor kidney complications during the diagnosis and treatment of COVID-19. Research has indicated that around 20% of children suffering from MIS-C tend to experience AKI, and this condition is linked with a higher probability of death (4.68 times) (Tripathi et al., 2023).

Summary points

- Approximately 20% of children with MIS-C develop AKI, which is associated with higher mortality.
- The mechanism of AKI in SARS-CoV-2 patients is multifactorial; various factors contribute to COVID-19-associated acute kidney injury.
- COVID-19-associated AKI presents with hematuria and/or proteinuria.
- Kidney damage caused by COVID-19 is usually manifested by increased blood creatinine and urea nitrogen levels.
- The mortality rates are higher in children suffering from COVID-19-induced AKI.

References

Adapa, S., Aeddula, N. R., Konala, V. M., Chenna, A., Naramala, S., Madhira, B. R., et al. (2020). COVID-19 and renal failure: Challenges in the delivery of renal replacement therapy. *Journal of Clinical Medical Research, 12*, 276–285.

Atlani, M. K., Pilania, R. K., & Bhatt, G. C. (2022). Outcomes following peritoneal dialysis for COVID-19-induced AKI: A literature review. *Peritoneal Dialysis International, 42*(6), 554–561.

Basalely, A., Gurusinghe, S., Schneider, J., et al. (2021). Acute kidney injury in pediatric patients hospitalized with acute COVID-19 and multisystem inflammatory syndrome in children associated with COVID-19. *Kidney International, 100*, 138–145.

Bhatt, G. C., Das, R. R., & Satapathy, A. (2021). Early versus late initiation of renal replacement therapy: Have we reached the consensus? An updated meta-analysis. *Nephron, 145*(4), 371–385.

Bhatt, G. C., Esezobor, C. I., Raina, R., Hodson, E. M., & Das, R. R. (2022). Interventions for preventing and treating acute kidney injury in children. *Cochrane Database of Systematic Reviews*, (11), CD015296.

Cantaluppi, V., et al. (2014). Interaction between systemic inflammation and renal tubular epithelial cells. *Nephrology, Dialysis, Transplantation, 29*, 2004–2011.

Chan, L., Chaudhary, K., Saha, A., et al. (2021). AKI in hospitalized patients with COVID-19. *Journal of the American Society of Nephrology, 32*, 151–160.

Cheng, Y., Luo, R., Wang, K., et al. (2020). Kidney disease is associated with in-hospital death of patients with COVID-19. *Kidney International, 97*, 829–838.

Chu, K. H., Tsang, W. K., Tang, C. S., et al. (2005). Acute renal impairment in coronavirus-associated severe acute respiratory syndrome. *Kidney International, 67*, 698–705.

Deep, A., Upadhyay, G., du Pré, P., et al. (2020). Acute kidney injury in pediatric inflammatory multisystem syndrome temporally associated with severe acute respiratory syndrome coronavirus-2 pandemic: Experience from PICUs across United Kingdom. *Critical Care Medicine, 48*, 1809–1818.

Diorio, C., McNerney, K. O., Lambert, M., et al. (2020). Evidence of thrombotic microangiopathy in children with SARS-CoV-2 across the spectrum of clinical presentations. *Blood Advances, 4*, 6051–6063.

Edrees, F., Li, T., & Vijayan, A. (2016). Prolonged intermittent renal replacement therapy. *Advances in Chronic Kidney Disease, 23*, 195–202.

Ferlicot, S., Jamme, M., Gaillard, F., et al. (2021). The spectrum of kidney biopsies in hospitalized patients with COVID-19, acute kidney injury, and/or proteinuria. *Nephrology, Dialysis, Transplantation, 12*, gfab042.

Fu, E. L., Janse, R. J., et al. (2020). Acute kidney injury and kidney replacement therapy in COVID-19: A systematic review and meta-analysis. *Clinical Kidney Journal, 13*(4), 550–563.

Gao, Y. D., Ding, M., Dong, X., et al. (2021). Risk factors for severe and critically ill COVID-19 patients: A review. *Allergy, 76*(2), 428–455.

Garzotto, F., Ostermann, M., Martin-Langerwerf, D., et al. (2016). The dose response multicentre investigation on fluid assessment (DoReMIFA) in critically ill patients. *Critical Care, 20*, 196.

Gu, J., Gong, E., Zhang, B., Zheng, J., & Gao, Z. (2005). Multiple organ infection and the pathogenesis of SARS. *The Journal of Experimental Medicine, 202*, 415–424.

Kari, J. A., Shalaby, M. A., & Albanna, A. S. (2021). Acute kidney injury in children with COVID-19: A retrospective study. *BMC Nephrology, 22*(1), 202.

Kidney Disease Improving Global Outcomes (KDIGO) Acute Kidney Injury Work Group. (2012). KDIGO clinical practice guideline for acute kidney injury. *Kidney International, 2*, 1–138.

Kielstein, J. T., Kretschmer, U., Ernst, T., et al. (2004). Efficacy and cardiovascular tolerability of extended dialysis in critically ill patients: A randomized controlled study. *American Journal of Kidney Diseases, 43*, 342–349.

Legrand, M., Bell, S., & Forni, L. (2021). Pathophysiology of COVID-19-associated acute kidney injury. *Nature Reviews. Nephrology, 17*(11), 751–764.

Liguoro, I., Pilotto, C., Bonanni, M., et al. (2020). SARS-COV-2 infection in children and newborns: A systematic review. *European Journal of Pediatrics, 179*(7), 1029–1046.

Martin, S. M., Meni Battaglia, L., Ferraris, J., et al. (2022). Prevalence of renal involvement among pediatric patients hospitalized due to coronavirus disease 2019: A multicenter study. *Archivos Argentinos de Pediatría, 20*(5), 310–316.

Nugent, J., Aklilu, A., Yamamoto, Y., et al. (2021). Assessment of acute kidney injury and longitudinal kidney function after hospital discharge among patients with and without COVID-19. *JAMA Network Open, 4*, e211095.

Pei, G., et al. (2020). Renal involvement and early prognosis in patients with COVID-19 pneumonia. *Journal of the American Society of Nephrology, 31*, 1157–1165.

Qiu, H., Wu, J., Hong, L., Luo, Y., et al. (2020). Clinical and epidemiological features of 36 children with coronavirus disease 2019 (COVID-19) in Zhejiang, China: An observational cohort study. *The Lancet Infectious Diseases*, *20*(6), 689–696.

Raina, R., Chakraborty, R., Mawby, I., et al. (2021). Critical analysis of acute kidney injury in pediatric COVID-19 patients in the intensive care unit. *Pediatric Nephrology*, *36*(9), 2627–2638.

Raina, R., Mawby, I., Chakraborty, R., et al. (2022). Acute kidney injury in COVID-19 pediatric patients in North America: Analysis of the virtual pediatric systems data. *PLoS One*, *17*(4), e0266737.

Ronco, C., Reis, T., & De Rosa, S. (2020). Coronavirus epidemic and extracorporeal therapies in intensive care: Si vis pacem Para bellum. *Blood Purification*, *49*, 255–258.

Rowley, A. H. (2020). Understanding SARS-CoV-2-related multisystem inflammatory syndrome in children. *Nature Reviews. Immunology*, *20*, 453–454.

Sabaghian, T., Kharazmi, A. B., Ansari, A., et al. (2022). COVID-19 and acute kidney injury: A systematic review. *Frontiers in Medicine (Lausanne)*, *4*(9), 705908.

Shemies, R. S., Nagy, E., Younis, D., et al. (2022). Renal replacement therapy for critically ill patients with COVID-19-associated acute kidney injury: A review of current knowledge. *Therapeutic Apheresis and Dialysis*, *26*(1), 15–23.

Tripathi, A. K., Pilania, R. K., Bhatt, G. C., Atlani, M., Kumar, A., & Malik, S. (2023). Acute kidney injury following multisystem inflammatory syndrome associated with SARS-CoV-2 infection in children: A systematic review and meta-analysis. *Pediatric Nephrology*, *38*(2), 357–370.

World Health Organization. (2023). *WHO coronavirus disease (COVID-19) dashboard*. World Health Organization. https://covid19.who.int/.

Wu, H., et al. (2020). AKI and collapsing glomerulopathy associated with COVID-19 and APOL1 high-risk genotype. *Journal of the American Society of Nephrology*, *31*, 1688–1695.

Chapter 47

Obesity, COVID-19 severity, and mortality

Riecha Joshi[a], Aarushi Sudan[b], Akshat Banga[c], Rahul Kashyap[d], and Vikas Bansal[e]

[a]*Department of Pediatrics, Government Medical College, Kota, India,* [b]*Department of Internal Medicine, Jacobi Medical Center, The Bronx, NY, United States,* [c]*Department of Internal Medicine, Sawai Man Singh Medical College, Jaipur, India,* [d]*Department of Research, WellSpan Health, York, PA, United States,* [e]*Division of Pulmonary and Critical Care Medicine, Department of Internal Medicine, Mayo Clinic, Rochester, MN, United States*

Abbreviations

ACE2	angiotensin-converting enzyme 2ACEIs
ACEIs	angiotensin-converting enzyme inhibitor
ACS	acute coronary syndromes
AF	atrial fibrillation
ALI	acute limb ischemia
APC	antigen-presenting cell
ARB	angiotensin receptor blockers
ARDS	acute respiratory distress syndrome
AT1	angiotensin type 1 receptors
ATII	angiotensin II
CAC	COVID-19-associated coagulopathy
CAD	coronary artery disease
CCI	Charlson Comorbidity Index
CDC	Centers for Disease Control and Prevention
CHD	coronary heart disease
CI	confidence interval
COPD	chronic obstructive pulmonary disease
COVID-19	coronavirus disease 2019
CVD	cardiovascular disease
DIC	disseminated intravascular coagulation
DVT	deep venous thrombosis
ED	emergency department
GFR	glomerular filtration rate
HF	heart failure
HRV	heart rate variability
HT	hypertension
ICU	intensive care unit
IFN	interferons
Ig	immunoglobulin
IL-6	interleukin-6
LDH	lactate dehydrogenase
LMWH	low molecular weight heparin
LPS	lipopolysaccharides
NA	not applicable
NETS	neutrophil extracellular traps
PE	pulmonary embolism
PRPs	pattern recognition receptors
RAAS	renin-angiotensin-aldosterone system
RNA	ribonucleic acid
RT-PCR	reverse-transcriptase-polymerase chain reaction
SARS-COV-2	severe acute respiratory syndrome coronavirus 2

Management, Body Systems, and Case Studies in COVID-19. https://doi.org/10.1016/B978-0-443-18703-2.00044-X
Copyright © 2024 Elsevier Inc. All rights are reserved, including those for text and data mining, AI training, and similar technologies.

TIA	transient ischemic attack
TMPRSS2	transmembrane serine protease-2
TNF	tumor necrosis factor
VTE	venous thromboembolism
WHO	World Health Organization

Introduction

Epidemiology of BMI and COVID-19

Obesity is defined by the World Health Organization (WHO) as having a body mass index (BMI, weight in kilograms divided by height in meters squared) >30 kg/m^2, and overweight is defined as having a BMI >25 kg/m^2 (Hamer et al., 2020). Two-thirds of the United States adult population and at least half of the majority of the populations of developed countries are currently overweight or obese (Berrington de Gonzalez et al., 2010). Obesity is characterized as an abnormal or excessive fat build-up that puts one's health at risk. It has been established that having a BMI >30 kg/m^2 has an increase in mortality rate from certain chronic medical conditions such as metabolic syndromes, heart disease, and strokes as well as certain cancers (Whitlock et al., 2009). An association between obesity and infectious diseases was first noted during the 2009 influenza A (H1N1) pandemic, when it was found that obese patients had higher influenza-associated hospitalizations and deaths.

The respiratory disease known as coronavirus disease-19 (COVID-19) is the source of the global COVID-19 pandemic that the world is currently recovering from. It started in December 2019 when pneumonia of unknown origin was diagnosed in Hubei province, Wuhan, China. Severe acute respiratory syndrome coronavirus-2 (SARS-CoV2), the causative agent, is a stranded RNA virus. It is passed through respiratory droplets in humans. Fever, coughing, exhaustion, shortness of breath, and a loss of taste and smell are some of the typical symptoms. Obesity has been associated with increased severity and mortality in COVID-19 disease, resulting in greater rates of hospitalization and deaths (Jain et al., 2009).

Overview of the impact of obesity on COVID-19

The severity of COVID-19 is significantly impacted by obesity, which also raises the possibility of complications. According to studies, people who are obese are more likely than people who are a healthy weight to die from COVID-19 and experience severe illness that requires hospitalization (Singh et al., 2022). Obesity weakens the immune system, which results in chronic inflammation, a compromised immune response, and decreased lung function, all of which can exacerbate COVID-19's respiratory symptoms. Additionally, angiotensin converting enzyme-2 (ACE2) receptors, the entry points for the SARS-CoV-2 virus, are more highly expressed in adipose tissue, especially in the abdominal region. The higher susceptibility and worse outcomes observed may be related to the higher viral burden in obese people (Cheng et al., 2020). Additionally, known risk factors for cardiovascular disease, diabetes, and hypertension are frequently present alongside obesity as underlying conditions; these are associated with worsening outcomes in COVID-19. For effective strategies to reduce risks and enhance outcomes for obese people, it is essential to recognize and address the impact of obesity on COVID-19. To reduce the risks associated with obesity, improve the general well-being of obese people, fight the growing obesity epidemic, and lessen the effects of contagious diseases such as COVID-19, it is important to promote healthy lifestyles, manage weight, and ensure equal access to healthcare services.

Process of viral infection

SARS-CoV-2 is a positive-sense single-stranded RNA virus. It contains four types of proteins: nucleocapsid, membrane, envelope, and spike. There are two subunits in the spike protein, S1 and S2, that are required for virus entry into cells. The S1 subunit binds to the ACE2 receptor; transmembrane serine protease-2 (TMPRSS2), in contrast, cleaves S2, allowing the virus to fuse with the cell membrane. The spike protein is heavily glycosylated, thus shielding it from the immune system and helping with dynamically conforming the receptor binding on S1, ensuring the ACE2 interaction of the virus. Epithelia of the tracheobronchial tree, type 2 pneumocytes (AT2 cells), endothelial cells, cardiomyocytes, and epithelial cells of the small bowel and colon are among the tissues that express both ACE2 and TMPRSS2 and have the highest viral burden. Heparan sulphate helps viruses attach to cells, and the spike protein's furin-like cleavage sites encourage viral replication in the lungs (Chu et al., 2021). These furin-derived substrates are bound by neuropilin1 (NRP1), which facilitates viral entry into nasal cells. Proinflammatory proteases (such as furin, PCSK5, and PCSK7) that are typically expressed at higher levels in males and increase with aging can coexpress with ACE2 (Cantuti-Castelvetri et al., 2020).

The clinical manifestations of SARS-CoV2 infection are related to the viral effect on the immune system and the resultant injury of tissue. The mechanism by which viral infection causes tissue damage is through aberrant interferon-related response, which leads to perturbation in cytokine production, thus depleting the action of immune cells while at the same time recruiting hyperactive macrophages and altered neutrophils. B-cells, T-cells, and natural killer (NK) cell activity are further suppressed by the excessive activation of these macrophages and neutrophils (Hoffmann et al., 2020). Additionally, this proinflammatory state promotes the coagulation cascade and the development of neutrophil extracellular traps (NETS), both of which cause a state of "thrombo-inflammation" and uncontrolled complement activation (Sungnak et al., 2020).

The global prevalence of individuals with overweight and obesity

According to the WHO, obesity is defined as having a BMI $>30\,kg/m^2$, and overweight is defined as having a BMI $>25\,kg/m^2$. At this time, the majority of developed countries' populations (at least half) and the adult population of the USA are both overweight or obese (Berrington de Gonzalez et al., 2010). Obesity is defined as an abnormal or excessive buildup of fat that endangers one's health. Obesity used to be seen as a problem in high-income countries (Berrington de Gonzalez et al., 2010). According to Molarius et al. (2000), obesity and socioeconomic status were found to have an inverse relationship in those high-income countries, especially for women. In contrast, the obesity prevalence tended to be low and restricted to people with relatively high socioeconomic status in low- and middle-income countries. In 2003, Monteiro et al. (2004) were among the first to demonstrate that this was no longer the case and that obesity had now spread to lower socioeconomic groups, particularly women in middle-income nations.

The rise in obesity has a significant effect on health impairment and the lowered quality of life globally. Obesity, in particular, plays a significant role worldwide in the rise in cardiovascular disease, type 2 diabetes, cancer, osteoarthritis, work disability, and sleep apnea. Morbidity is more severely affected by obesity than is mortality. Increases in the prevalence of obesity have the potential to lengthen the time that people experience morbidity and disability due to obesity. Data from the Global Burden of Disease Study and the World Health Organization demonstrate a strong correlation between obesity and rising rates of ill health, disability, and mortality across diverse geographies (Berrington de Gonzalez et al., 2010; Monteiro et al., 2004).

Obesity and COVID-19: Shared immunological perturbations

Immunological alterations in obesity

Obesity is associated with metabolic disturbances that trigger immune activation in various tissues, including adipose tissue, liver, pancreas, and vasculature, resulting in elevated markers of chronic low-grade inflammation (Winer et al., 2009). Obesity is also linked to an increased risk of complications from infections. The increasing prevalence of obesity throughout the world predisposes individuals to an increased risk of both metabolic and infectious diseases. Immunity in obesity is compromised by more than one mechanism. We briefly discuss the key alterations in the immune system that stem from ectopic lipid accumulation, a proinflammatory state caused by insulin resistance and hypertrophied adipocytes, and dyslipidemia associated with obesity.

The physiologic dysfunction in obesity leads to ectopic lipid accumulation in nonadipose tissue, including tissues of the immune system, bone marrow, and thymus (Kanneganti & Dixit, 2012). The increasing fat content of lymphoid tissues alters the cellular milieu by disrupting tissue integrity, suppressing hematopoiesis, and disproportionately affecting the ratios of progenitor lineages generated in the bone marrow (Kanneganti & Dixit, 2012). Adipose accumulation in the lymphoid tissue is also naturally known to occur with age, adversely affecting immunity in older individuals (Castelo-Branco & Soveral, 2014). Obesity can, thus, cause premature aging of the immune system.

Hypertrophied adipocytes observed in obesity are more prone to activating the endoplasmic reticulum and mitochondrial stress responses. This can lead to adipocyte apoptosis and the release of chemotactic mediators, resulting in inflammatory leukocyte inflammation (Naaz et al., 2004). As a result, a high proportion of adipocytes are observed in hypertrophic obese adipose tissue (Sartipy & Loskutoff, 2003). Both hypertrophic adipocytes and macrophages increase the secretion of TNF-a, IL-6, and monocyte chemoattractant protein-1, ATII (Lagathu et al., 2006). These proinflammatory mediators block the production of antiinflammatory and insulin-sensitizing adipokines such as adiponectin (Guilherme et al., 2008). Insulin resistance observed in obesity can inhibit T cell-mediated resolution of inflammation, as insulin has been shown to promote antiinflammatory TH2 cell differentiation (Viardot et al., 2007).

The low levels and dysfunctional nature of high-density lipoprotein (HDL) in obese populations also contribute to the disruption of leukocyte activation and the coordination of the immune response. HDL acts as an acceptor of leukocyte cholesterol via ATP-binding cassette transporter A1 and ATP-binding cassette transporter G1 (Kellner-Weibel & de la Llera-Moya, 2011), which directly affects leukocyte activity and the proliferative capacity. HDL also alters inflammatory responses by neutralizing lipopolysaccharides (LPS), an immunostimulatory glycolipid located on the outer membrane of Gram-negative bacteria that activates the innate immune response. As a consequence, LPS levels are often high in obesity, and high levels of Gram-negative bacteremia are observed (Soehnlein & Lindbom, 2010).

Studies have shown a high rate of vaccine failure in obese individuals (Hemalatha Rajkumar, 2013). Obesity impairs memory CD8+ T cell response to antigens. This has been seen in several studies where obesity was associated with a failure to maintain antigen-specific CD8+ memory T cells, which are essential to ensure vaccine efficacy. Accordingly, obesity has been shown to increase the risk of vaccine failure, including the vaccines for hepatitis B (Bandaru et al., 2011), tetanus (Eliakim et al., 2006), and influenza (Sheridan et al., 2012).

Immunological alterations in COVID-19

Reports from many countries indicate that COVID-19 shares the mechanism with SARS-Cov-2 by which it enters or invades host cells. It is documented that bats are the source of related viruses, and that human-to-human transmission plays a critical role in the pathogenesis of COVID-19 (Huang et al., 2020).

The spike protein present in the virus interacts with the receptor present in the target cells and gains entry into the cells (Tortorici & Veesler, 2019). The RNA present in the virus is encapsulated and polyadenylated, which encodes various polypeptide genes (Lambeir et al., 2003). In general, pattern recognition receptors (PRRs) recognize invading pathogens, including viruses (Ben Addi et al., 2008). Once viruses enter the body, they stimulate the host to initiate immune responses by augmenting the release of inflammatory factors, the synthesis of type I interferons (IFNs), and the maturation of dendritic cells (DC), which are required to limit the viral spread (Ben Addi et al., 2008). SARS-CoV-2 activates both the innate and acquired immune responses. CD4+ T cells stimulate B cells to produce virus-specific antibodies, including immunoglobulin (Ig) G and IgM, whereas CD8+ T cells directly kill virus-infected cells. T helper cells produce proinflammatory cytokines and mediators to help the other immune cells. T cell function is suppressed by SARS-CoV-2 by a process known as apoptosis, also known as programmed cell death (Lu et al., 2011). Viral-Track, a novel computational approach for screening scRNA-seq data for viral RNAs, identified a significant change in the bronchoalveolar lavage immune cell landscape during severe SARS-CoV-2 infection (Bost et al., 2020).

The pathogenesis of COVID-19 is, therefore, a result of an abnormal host response or overreaction of the immune system in some patients with unknown etiology. The production of extremely high levels of a large number of inflammatory cytokines, chemokines, and free radicals locally causes severe damage to the lungs and other organs. In severe cases, multi-organ failure and even death can result (Collange et al., 2020). Acute respiratory distress syndrome (ARDS) is the main cause of death due to COVID-19, believed to occur because of a cytokine storm. COVID-19 infection also induces pneumonia, which is characterized by bilateral infiltrates on chest imaging (Huang et al., 2020). Vascular congestion, edema with prominent proteinaceous exudates, fibrinoid material, and multinucleated giant cells have also been reported in the lungs of COVID-19-infected patients (Tian et al., 2020).

It is worth mentioning that even though the body's response to SARS-CoV-2 in terms of IFN-I and -III is diminished, recent studies indicate the presence of a consistent chemokine signature (Blanco-Melo et al., 2020). Lymphocyte count is associated with increased disease severity in COVID-19, and it can vary among leukopenia, leukocytosis, and lymphopenia (Huang et al., 2020). Lymphopenia indicates a poor prognosis in COVID-19 patients. The etiology remains unknown, but SARS-CoV RNA has been found in T cells, which suggests that there is a direct effect of SARS virus on T cells, potentially through apoptosis (Liu et al., 2020).

During infection with SARS-CoV, antigen-presenting cell (APC) function is altered, and there is impaired migration of dendritic cells, which results in reduced priming of T cells. This results in a fewer number of virus-specific T cells within the lungs (Yoshikawa et al., 2009). After initial infection with the virus, lung respiratory DCs (rDCs), when they come in contact with the invading pathogen or antigens from infected epithelial cells, get activated and migrate to the draining (mediastinal and cervical) lymph nodes. Once in the lymph nodes, these rDCs present the antigen in the form of an MHC/peptide complex to the circulating T cells. Engagement of the T cell receptor (TCR) with the peptide-MHC complex induces T cell activation, proliferation, and migration to the site of infection (Larsson et al., 2000).

Cytotoxic lymphocytes (CTLs) and natural killer (NK) cells play an important role in limiting viral infection, and a reduction in these cells may increase the severity of the disease, such as in COVID-19. The total number of NK and CTLs is decreased in COVID-19, accompanied by a concurrent decline in their functional capacity and an increase in the upregulation of the inhibitory receptor CD94/NK group 2 member A on NK cells. After the successful recovery of COVID-19 patients, the number of NK and CD8+ T cells may get restored but with reduced expression of NKG2A. Furthermore, there is a lower percentage of CD107a+NK, IFN-γ+NK, IL-2+ NK, and TNF-α+NK cells in COVID-19 patients (Norbury et al., 2002).

Due to increased T cell apoptosis in COVID-19 patients, there is a decreased number of CD4+ and CD8+ T cells in the peripheral blood of SARS-CoV-2-infected patients. In addition, CD4 and CD8 cells have impaired activation, as evidenced by the appearance of CD25, CD28, and CD69 expression on these T cell subsets (Cai et al., 2004). These factors may account for the delayed development of the adaptive immune response and prolonged virus clearance in severe human SARS-CoV infection (Cameron et al., 2008).

The severity of SARS in humans during the acute phase is strongly linked to lower levels of T cells (Li et al., 2004). The S and N proteins present in SARS-CoV are recognized by CD4 and CD8 T cells. The viral S protein results in the formation of neutralizing antibodies, which are able to induce an eosinophilic response in animals (Bolles et al., 2011). To produce these antibodies, the viral antigen needs to be recognized by APC, as these subsequently stimulate the body's humoral immunity with the help of B and plasma cells. In SARS, the antibodies found were IgM and IgG, which were detected in the patient's blood after 3–6 days and 8 days, respectively (Lee et al., 2010). The IgM antibodies specific to SARS disappeared at the end of week 12, whereas the IgG antibody lasted for a longer time, suggesting IgG antibodies may be essential to provide a protective role for a longer time (Li et al., 2003).

In addition to cell and humoral-mediated defense, proinflammatory cytokine release also plays a vital role against COVID-19 infection. Cytokines such as IFN-γ directly inhibit viral replication and enhance antigen presentation (Saha et al., 2010). However, the secretion of a novel short protein by SARS-CoV-2, which is encoded by orf3b, results in the inhibition of expression of IFNβ, which further enhances the pathogenicity of the virus (Chan et al., 2020). Chemokines produced by activated T cells recruit more innate and adaptive cells to control the pathogen burden.

The presence of metabolic balance syndrome, obesity, diabetes, hypertension, or a history of a previous cardiovascular disease is associated with an increased risk of developing a more serious illness, requiring hospital admission and probably invasive ventilation, highlighting the fact that the presence of comorbidities affects the prognosis of the COVID-19 (Li et al., 2020).

Role of ACE-2 in pathogenesis

The angiotensin converting enzyme is a crucial component of the renin-angiotensin system. ACE-2 plays a major role in the negative regulation of RAS by metabolizing AT II into the beneficial peptide Ang 1–7 and is widely distributed in the heart, kidney, lungs, and testes (Cheng et al., 2020). ACE2 also acts as a negative regulator of pulmonary fibrosis. The SARS-COV-2 virus uses ACE-2 as its receptor to invade the human alveolar epithelial cells (Cheng et al., 2020). An overexpression of ACE-2 receptors has been observed in obesity, aiding in infection and serving as a viral reservoir. The binding of the virus to ACE2 leads to RAS activation, a direct loss of ACE2, and an increase in ATII. Higher levels of ATII lead to the progression of lung injury among COVID-19 patients by triggering the NADH/NADPH oxidase (Singh et al., 2022).

ACE-2 overexpression on adipocytes and adipose-like cells can cause trans-differentiation of pulmonary lipofibroblasts into myofibroblasts, contributing to the development of PF and, thus, influencing the clinical severity of COVID-19. The application of TXD and metformin is being studied as adjuvant therapy in COVID-19 to reduce the development of PF and thus attenuate the severity of the course of the disease (Kruglikov & Scherer, 2020).

Endothelial dysfunction and arterial stiffness

COVID-19 may impair the cardiovascular system directly via viral toxicity or indirectly through hypoxemia, thrombogenesis, systemic inflammation, and dysregulated immune and renin-angiotensin systems (Guilherme et al., 2008). SARS-CoV-2 acts on the main receptor for infection, angiotensin-converting enzyme II receptors, inhibiting the conversion of angiotensin II to angiotensin (1–7) and triggering harmful events to the vascular wall such as endothelial dysfunction, increased thrombogenic activity, and platelet aggregation (South et al., 2020). The virus decreases nitric oxide bioavailability by reducing production or increasing degradation by reactive oxygen species and triggering immunological reactions, altering the vascular wall and inhibiting antiproliferative, antithrombotic, and antiatherogenic phenotypes (Siddiqi & Mehra, 2020). Endothelial dysfunction is observed during and after COVID-19 (Kar, 2022). After weeks of infection, an impaired systemic vascular function was also observed in healthy adults (Ratchford et al., 2021). As a result, COVID-19-induced alterations may lead to stiffness of the arteries in the acute phase. AT2 enhances the sympathetic autonomic tones in the central nervous system by interacting with angiotensin 1 receptors (AT1). These dysfunctions were observed after

COVID-19 infection, including palpitations, fatigue, orthostatic intolerance, dizziness, and syncope. Therefore, sympathetic nerve activity, dysregulation of the RAS, and altered ACE-2 expression are believed to cause an adverse structural and functional remodeling of the arteries. Consequently, there exists a mutual relationship between preexisting arterial stiffness and the severity of COVID-19, along with increased arterial stiffness after recovering from COVID-19. Thus, monitoring the heart rate variability (HRV) to assess the cardiac autonomic function of patients with COVID-19 may help diagnose adverse cardiovascular outcomes (Araujo et al., 2023).

Cardiovascular events and thrombosis

While viral pneumonia is a typical presentation of a COVID-19 patient, cardiovascular manifestations such as myocardial infarction, acute coronary syndromes (ACS), venous and arterial thrombosis, and arrhythmia have all been observed.

Myocardial injury

Patients admitted with COVID-19 are frequently found to have myocardial infarction. The prevalence of myocardial injury in patients hospitalized with COVID-19 ranges from 5% to 38%, with an overall crude prevalence of approximately 20%. It is known that myocardial injury can occur in the setting of acute stressors such as hypoxemia, hypotension/shock, acute kidney injury (AKI), and congestive HF, but it has also been found that patients admitted with COVID-19 are also at higher risk of myocardial injury (Zheng et al., 2020). COVID-19 patients with an underlying cardiovascular disease have a higher risk of developing myocardial injury. In a cohort of 416 patients of COVID-19, cardiac injury was associated with chronic hypertension, CAD, and cerebrovascular disease. In the study, patients with myocardial injury were more likely to present with abnormal laboratory results, including higher elevations in acute phase reactants, pro-B-type natriuretic peptide, and creatinine, and with lower levels of platelets and albumin, all of which suggest that those with myocardial injury are more critically ill (Zheng et al., 2020).

The presence of myocardial injury in those with COVID-19 has also been associated with a significant increase in mortality, markedly higher in hospital adverse events, including death, ARDS, malignant arrhythmias, acute coagulopathy, and AKI in patients with elevated troponin levels compared with patients with normal troponin levels was observed (Zhang et al., 2020).

Thrombotic complications

Vascular complications in COVID-19 patients occur due to the interaction between SARS-CoV-2 and platelets, viral-mediated microvascular trauma, proinflammatory cytokine release, and endothelial dysfunction, resulting in micro- and macrovascular complications (Akhmerov & Marban, 2020). Additionally, inflammatory and hypercoagulability markers are also elevated in patients with severe COVID-19, especially patients admitted to the intensive care unit (ICU), and an elevated D-dimer has been shown to predict the likelihood of a possible arterial thrombotic event (Del Prete et al., 2022). Furthermore, baseline elevations and up-trending D-dimer levels throughout a patient's hospital course have been associated with increased mortality risk (Katwa et al., 2022).

Studies have also shown an elevated incidence of venous thromboembolism (VTE) among COVID-19 patients (Katwa et al., 2022). The risk of VTE seems to be higher in hospitalized patients with COVID-19 (Duarte et al., 2021). Arterial thromboembolism, including acute ischemic stroke, ACS, acute limb ischemia (ALI), and mesenteric, renal, and splenic infarcts, have all been documented in patients with COVID-19 (Katwa et al., 2022). In a study of patients with COVID-19 presenting with STEMI, those with COVID-19 were more likely to have multivessel coronary thrombosis and stent thrombosis. Which were further associated with an increased rate of in-hospital death and acute thrombotic event (Chen et al., 2020).

Cardiac arrhythmias

Both types of arrhythmias, atrial and ventricular, have been commonly observed in COVID-19 patients, with palpitations being the most common presenting symptom (Mohammad et al., 2022) and sinus tachycardia being the most common form of rhythm disturbance (Atri et al., 2020).

Atrial fibrillation and flutter (AF) are the most common cardiac dysrhythmias in the United States (Jaffe et al., 2020), and are associated with an increased risk of multiple cardiovascular complications. COVID-19 patients with preexisting AF have been associated with an increased risk of mortality and thromboembolic complications (Henning, 2022). The use of oral anticoagulation in patients with AF admitted with COVID-19 has been associated with decreased rates of thrombotic events and less severe COVID-19.

Paradoxical effect of obesity on different pulmonary conditions and obesity antiparadox effect on COVID-19

Obesity has a heavy, negative impact on the health of the population, especially on cardiovascular health, leading to worsening hypertension, atherogenic dyslipidemia, cardiovascular disease (CVD), insulin resistance or type 2 diabetes, and metabolic syndromes, thus leading to chronic low-grade system inflammation. Despite the effects of obesity, it has been observed that obese patients with CVD and chronic obstructive pulmonary disease (COPD) have better short-term prognoses compared to their leaner counterparts. This is termed the "obesity paradox." There are several theories that explain this phenomenon. The first theory is the muscle mass hypothesis, which explains the improved prognosis of COPD patients who are on long-term hospital treatment (Giri Ravindran et al., 2022). The theory states that obese patients are better tolerant to muscle mass wasting, thus having a higher chance of survival. As advanced congestive heart failure (CHF) is associated with cachexia, having a higher weight and more muscle mass could be a protective or an associative marker of maintained vigor. However, it does not explain the obesity paradox in less severe cases and stable coronary heart diseases (CHD). The second theory that helps us understand the paradox better is based on the effects of systemic inflammation. It suggests that an increase in the body's fat composition is associated with elevated proinflammatory cytokine levels (Lavie et al., 2021). This causes the prognosis of COPD and CHD to be greatly influenced by the body's fat composition. White adipose tissue and brown adipose tissue make up the majority of the adipose tissue in the body. Brown adipose tissue has properties that behave like the lean mass of a body in that it lowers the body's level of lipopolysaccharides, which in turn inhibits the proinflammatory cytokine, effectively lowering the level of systemic inflammation. In COVID-19 patients, an antiparadox effect has been observed. The various biological mechanisms that increase the risk of COVID-19 in obese patients are as follows (Singh et al., 2022):

1. Ectopic fat causes an increased production of proinflammatory cytokines such as interleukin-6 (IL-6) and tumor necrosis factor-alpha (TNF-) as well as angiotensin II, which worsens the inflammation brought on by COVID-19.
2. Obese patients have lower levels of the antiinflammatory adipokine adiponectin, which is associated with a higher level of ATII. Overexpression of ACE2 receptors, which may facilitate infection and act as a viral reservoir, is linked to obesity. Coronavirus decreases the activity of ACE2 inhibitors, which again raises the level of ATII.

In COVID-19 patients, higher levels of ATII accelerate the progression of lung injury by activating the NADH/NADPH oxidase system and encouraging fibrosis, contraction, and vasoconstriction. This is also linked to endothelial dysfunction, which is a major pathogenic factor in COVID-19 and a cause of mortality and morbidity. The production of the cytokines TNF- and IL-6, which are linked to alveolar damage and higher severity and mortality, is also increased by an increased expression of inflammatory adipokine molecules.

COVID-19 and risks related to obesity among adults

A BMI $>30\,kg/m^2$ indicates obesity, which is a global health concern (2021). In developed nations, it affects a sizable portion of adults, resulting in a number of chronic medical conditions and higher mortality rates. Infectious diseases such as the ongoing COVID-19 pandemic and the 2009 H1N1 influenza pandemic have also been linked to obesity.

The COVID-19-causing SARS-CoV-2 virus primarily affects the respiratory system and enters cells by tying up with the ACE2 receptor. Given that adipose tissue expresses ACE2 receptors, obese people frequently have higher viral burdens. Additionally, immunological changes linked to obesity may worsen COVID-19 severity. Immune dysfunction in obese people is a result of chronic inflammation, insulin resistance, and dysregulated immune responses (Singh et al., 2022).

Obesity has been linked to vaccine failure and increases the risk of complications from infectious diseases. Understanding the link between obesity and COVID-19 is essential for putting into practice efficient public health initiatives and giving obese people the care they need (Hamer et al., 2020). Promoting healthy lifestyles, managing weight, and ensuring equal access to healthcare services should be prioritized to reduce the risks related to obesity as well as lessening the effects of contagious illnesses such as COVID-19.

Clinical evidence linking obesity to worse outcomes in patients with COVID-19

- **Being an individual with obesity and the risk of COVID-19**

Obesity is associated with an alteration of immune cells in adipose tissue. It leads to a decrease in Th2 cells, Treg cells, and macrophages while simultaneously increasing proinflammatory cells, including CD8+ T cells. This ultimately results in a state of chronic inflammation both at the local and systemic levels (Lavie et al., 2021). Various studies have indicated that the disease severity and outcome of COVID-19 patients are associated with systemic inflammation and increased proinflammatory cytokines. This suggests that the acute inflammation that arises from COVID-19 amplifies the chronic inflammation due to obesity, thus leading to an increased risk of infection and severity. Paul MacDaragh Ryan and Noel M. Caplice recently put forth a similar theory in a paper. According to these authors, obese people are more likely to overreact to coronavirus infection because they have higher levels of various inflammatory signals (Vaughan et al., 2020).

- **Being an individual with obesity and COVID-19 illness severity**

A BMI >25.0 has been linked to an increased risk of COVID-19 hospitalization but not death; however, obesity (BMI >30.0) and extreme obesity (BMI >40.0) increase the risk of both hospitalization and death. Additionally, there was a linear dose-response relationship between COVID-19 results and obesity categories. Future causation studies should be guided by these dose-response relationships. Obesity and extreme obesity have been linked in a significant way to hospital admission, ICU stays, and fatalities. Age, sex, and comorbidities are among the factors that complicate the relationship between obesity and COVID-19, and accounting for these factors could lower the effect estimate. Furthermore, studies using more recent data revealed a weaker correlation between obesity and COVID-19 outcomes than those conducted during the early stages of the pandemic. The treatment has improved as the disease has advanced, with healthcare facilities better prepared, knowledgeable, and experienced in clinical management in addition to the discovery of new drugs and increased public awareness about the spread and prevention of the disease (Sawadogo et al., 2022).

- **Being an individual with obesity and a COVID-19 prognosis**

According to data from the Centers for Disease Control and Prevention (CDC), 28.3% of 148,494 adults who were diagnosed with COVID-19 during an emergency department (ED) or inpatient visit at 238 US hospitals between March-December 2020 had overweight, and 50.8% had obesity. Obesity was a risk factor for hospitalization and death, especially in adults under the age of 65. Obesity was also a risk factor for invasive mechanical ventilation. Patients with BMIs $<25\,kg/m^2$ had the lowest risks for hospitalization, ICU admission, and death. These risks sharply increased with increasing BMI. Over the entire BMI range, from 15 to $60\,kg/m^2$, the risk of invasive mechanical ventilation increased. Clinicians should make sure to consider the risk of severe outcomes in patients with higher BMIs, particularly in those with severe obesity, as they develop care plans for COVID-19 patients. These findings draw attention to the effects of higher BMIs on clinical and public health, such as the need for intensive COVID-19 illness management as obesity severity rises, the promotion of COVID-19 prevention strategies such as ongoing vaccine prioritization and masking, and policies to ensure community access to nutrition and physical activity that support and promote a healthy BMI.

The results of this analysis by the CDC support the earlier findings that overweight or obese people are at an increased risk of developing severe COVID-19-related illnesses. They also add to the body of knowledge regarding the dose-response relationship between higher BMI and the likelihood of hospitalization, ICU admission, invasive mechanical ventilation, and death. The finding that higher BMI is associated with an increased risk for severe COVID-19-associated illness suggests that patients with more severe obesity may require more intensive COVID-19 management (Kompaniyets et al., 2021).

- **Obesity and Sars-CoV-2 hypercoagulability/thrombosis: the lethal triad of COVID-19**

COVID-19 is associated with having a hypercoagulable status and thrombosis as "COVID-19-associated coagulopathy" (Sarfraz et al., 2022). In severe COVID-19 infection, severe thrombocytopenia, prolongation of prothrombin time and activated partial thromboplastin time, an elevated D-dimer level, and decreased fibrinogen values have been observed, caused by the rapid and severe activation of coagulation and clotting factors, leading to disseminated intravascular coagulation (DIC) and VTE (Sanchis-Gomar et al., 2020). It is important to note that overweight and obesity are associated with hypercoagulability and, thus, an increased risk of developing VTE. The various mechanisms responsible for obesity-related hypercoagulability include the action of adipocytokines such as leptins and adiponectin on coagulation; the hyperactivity of coagulation factors such as fibrinogen, factor VII, factor VIII, and the von Willebrand factor; a chronic proinflammatory state leading to elevated cytokines such as tumor necrosis factor and IL-6; and elevated Ang II levels leading to endothelial dysfunction, venous stasis, and impaired venous return. Therefore, the additive effects of obesity and SARS-CoV-2 infection may be the cause of increased hypercoagulability and thrombosis in patients with COVID-19 infection, thus they are referred to as the lethal triad of COVID-19 (Sarfraz et al., 2022).

Implications for treatment and vaccination strategies for being an individual with obesity

The effectiveness of vaccines relies on the ability of the vaccine to initiate and amplify the immune response. In response to an immune system challenge, innate immune pathways cause the mobilization of effector cells that release cytokines to launch an antigen-specific adaptive immune response. This provides a general response to infection or vaccination. Obese people have a tendency to have a weakened immune response to vaccinations. These groups are more susceptible to infection and complications from natural exposure that can be prevented by vaccination if vaccination fails to elicit a protective immune response. Due to increased body fat and leptin production, obesity may impair a person's ability to mount a successful immune response to vaccination or an infection. However, a study showed that even in people with severe obesity, COVID-19 vaccines successfully target neutralizing spike epitopes. Instead, the relative decrease in the neutralizing power of vaccine-induced antibodies is linked to the absence of high-affinity antibodies. There is evidence that even a 5% weight loss could lower the risk of many metabolic complications brought on by diseases such as diabetes and CHD, which frequently affect obese people. Thus, interventions such as bariatric surgery and lifestyle changes that promote weight loss could similarly reduce the effects of COVID-19 (Painter et al., 2015; van der Klaauw et al., 2023).

Policies and procedures for obesity and COVID-19

I. **Awareness and education**
   - Develop and implement public health campaigns to raise awareness of the risks of obesity and its association with COVID-19.
   - Provide educational materials and resources about healthy lifestyles, weight management, and the importance of vaccination.
   - Collaborate with healthcare professionals and community organizations to disseminate accurate information about obesity and COVID-19.

II. **Prevention and early intervention**
   - Promote healthy eating habits and physical activity through community programs, school initiatives, and workplace wellness programs.
   - Offer accessible and affordable healthy food options in schools, workplaces, and public places.
   - Provide screening and early intervention programs for individuals at risk of obesity and related health conditions.

III. **Accessible healthcare services**
   - Ensure equitable access to healthcare services, including preventive care, screenings, and treatment for obesity and related conditions.
   - Implement telehealth services and digital tools to improve access to healthcare for individuals with obesity, especially during a pandemic such as COVID-19.
   - Train healthcare professionals to provide culturally sensitive and evidence-based care for individuals with obesity.

IV. **Integrated care**
   - Encourage interdisciplinary collaboration among healthcare providers, including primary care physicians, dieticians, psychologists, and exercise specialists, to provide comprehensive care for individuals with obesity and COVID-19.
   - Develop care pathways and referral systems to ensure coordinated and continuous care for individuals with obesity and related health conditions.

V. **Research and data collection**
   - Support research efforts to better understand the relationship between obesity and COVID-19, including the immunological mechanisms and treatment outcomes.
   - Collect and analyze data on obesity prevalence, COVID-19 outcomes, and the effectiveness of interventions to inform evidence-based policies and strategies.
   - Collaborate with international organizations and research institutions to share data and findings on obesity and COVID-19.

VI. **Policy and advocacy**
   - Advocate for policies that promote healthy environments such as restrictions on marketing unhealthy foods and beverages to children.
   - Support policies that ensure access to affordable, nutritious food, including subsidies for fruits and vegetables and regulation of food labeling.
   - Advocate for policies that address social determinants of obesity, such as poverty, education, and food insecurity.

VII. **Collaboration and partnerships**
   - Foster collaboration among government agencies, nonprofit organizations, healthcare systems, and community stakeholders to develop and implement comprehensive strategies to address obesity and COVID-19.

- Engage with international organizations and participate in global initiatives to combat obesity and its associated health risks, including COVID-19.

VIII. **Evaluation and monitoring**
- Establish monitoring systems to track the prevalence of obesity, COVID-19 outcomes, and the impact of interventions over time.
- Regularly evaluate the effectiveness of policies and programs, using data-driven approaches to inform future strategies.
- Use feedback from individuals with obesity, healthcare providers, and community stakeholders to continuously improve policies and procedures.

Directions for future research and management of COVID-19 in the obese population

I. **Epidemiological studies**
- Conduct large-scale epidemiological studies to further investigate the relationship between obesity and COVID-19, including the impact on disease severity, hospitalization rates, and mortality.
- Explore the underlying mechanisms that contribute to increased susceptibility and worse outcomes in obese individuals with COVID-19.
- Investigate the long-term effects of COVID-19 on the metabolic health of individuals with obesity.

II. **Immunological and molecular studies**
- Investigate the immunological alterations in obese individuals with COVID-19, focusing on the dysregulation of immune cells, cytokine profiles, and inflammatory responses.
- Study the molecular mechanisms by which obesity affects viral entry, replication, and immune response to SARS-CoV-2.
- Explore the role of adipose tissue in viral reservoirs, viral persistence, and chronic inflammation in obese individuals with COVID-19.

III. **Clinical management**
- Conduct clinical trials to evaluate the efficacy and safety of antiviral treatments, immunomodulatory agents, and vaccines, specifically in the obese population.
- Develop evidence-based guidelines for the management of COVID-19 in obese individuals, considering the unique challenges and comorbidities associated with obesity.
- Investigate the impact of weight loss interventions, such as lifestyle modifications, bariatric surgery, and pharmacological treatments, on COVID-19 outcomes in individuals with obesity.

IV. **Healthcare system**
- Assess the healthcare system's capacity to effectively manage the healthcare needs of obese individuals with COVID-19, including resources, equipment, and specialized care.
- Explore strategies to improve access to healthcare services for obese individuals during the pandemic, such as telemedicine and home-based care options.
- Evaluate the effectiveness of multidisciplinary care models that involve collaboration among healthcare providers from various specialties to optimize COVID-19 management in obese individuals.

V. **Public health interventions**
- Evaluate the effectiveness of public health interventions targeting obesity prevention and management in reducing the burden of COVID-19 in the population.
- Assess the impact of community-based programs, policies, and environmental changes on promoting healthy behaviors and reducing obesity-related risk factors for COVID-19.
- Investigate the acceptability, feasibility, and effectiveness of interventions aimed at improving vaccine effectiveness in obese individuals.

VI. **Health disparities**
- Investigate the disproportionate impact of COVID-19 on obese individuals from disadvantaged socioeconomic backgrounds, ethnic minorities, and marginalized populations.
- Identify and address barriers to healthcare access and utilization among obese individuals with COVID-19, with a focus on reducing health disparities.
- Develop culturally appropriate and tailored interventions for obesity management and COVID-19 prevention in diverse populations.

VII. **Health communication and education**
- Assess the effectiveness of health communication strategies in raising awareness about the risks of COVID-19 in obese individuals and promoting preventive measures.
- Develop educational materials and resources specifically targeting obese individuals to enhance the understanding of COVID-19 risks, management, and available support services.
- Explore innovative approaches, such as digital health technologies and social media platforms, to disseminate accurate and accessible information to the obese population.
- By addressing these research directions and implementing evidence-based management strategies, we can improve the understanding of COVID-19 in the obese population and develop effective interventions to reduce the burden of the disease in this vulnerable group.

Applications to other areas

I. **Infectious disease research**
- The study of the relationship between obesity and infectious diseases, including COVID-19, can provide valuable insights into the impact of obesity on immune responses and susceptibility to other viral, bacterial, and fungal infections.
- Understanding the immunological alterations associated with obesity can help in the development of targeted interventions and therapies for infectious diseases in obese individuals.

II. **Public health strategies**
- The findings from research on obesity and COVID-19 can inform public health strategies and interventions aimed at reducing the burden of other chronic diseases, such as cardiovascular disease, diabetes, and certain types of cancer, which are also influenced by obesity.
- Lessons learned from managing COVID-19 in obese individuals can be applied to other infectious diseases to develop effective prevention and management strategies for vulnerable populations.

III. **Obesity prevention and management**
- The COVID-19 pandemic has highlighted the need for comprehensive obesity prevention and management strategies. Lessons learned from the pandemic can inform and strengthen initiatives aimed at promoting healthy lifestyles, improving access to healthy food options, and creating supportive environments for physical activity.
- The implementation of telemedicine and digital health solutions during the pandemic can be leveraged to enhance obesity prevention and management programs, providing remote access to healthcare services, personalized support, and behavioral interventions.

IV. **Healthcare delivery**
- The challenges faced in managing COVID-19 in the obese population, such as the need for specialized equipment and healthcare providers with expertise in obesity-related care, can shed light on the importance of integrating obesity management into healthcare delivery systems.
- The development of multidisciplinary care models and collaborations among healthcare providers from different specialties can be extended to other chronic diseases, improving patient outcomes and optimizing healthcare resource utilization.

V. **Health equity and disparities**
- The recognition of health disparities and the disproportionate impact of COVID-19 on obese individuals from marginalized populations can drive efforts to address health equity in various health domains, including obesity prevention, access to healthcare, and health outcomes.
- Lessons learned from COVID-19 can inform initiatives that aim to reduce health disparities and promote health equity, ensuring that vulnerable populations receive equitable access to healthcare services, preventive measures, and treatment options.

VI. **Vaccine development and effectiveness**
- Vaccine failure in obese individuals during the COVID-19 pandemic underscores the importance of considering obesity as a factor in vaccine development and assessing vaccine effectiveness in this population.
- Future research can focus on optimizing vaccine strategies, dosages, and delivery methods to ensure adequate immune responses and protection in obese individuals, not only for COVID-19 but also for other infectious diseases.

Overall, the knowledge gained from studying the impact of obesity on COVID-19 can have broader implications for infectious disease research, public health strategies, obesity prevention and management, healthcare delivery, health equity, and vaccine development. Applying these insights to other areas can contribute to improving population health outcomes and addressing the challenges posed by obesity and infectious diseases in diverse communities.

Mini-dictionary of terms

- Body mass index: This is calculated by dividing a person's weight in kilograms by the square of their height in meters.
- Obesity: The medical condition known as obesity is characterized by an excessive build-up of body fat to the point where it may be harmful to a person's health. Obesity is typically defined as having a BMI of 30 or higher.
- Sleep apnea: A sleep disorder characterized by recurrent episodes of partial or total cessation of breathing while sleeping. A person's overall health and well-being may suffer as a result of this serious condition.
- Dyslipidemia: Characterized by abnormally high blood lipid (fat) levels. The two main lipids found in the bloodstream, triglycerides and cholesterol, are out of balance in this condition.
- Arrhythmias: Conditions in which the heart experiences abnormal electrical activity that interferes with its regular pattern of beating.
- Thrombosis: A blood clot (thrombus) forming inside a blood vessel and blocking the regular flow of blood. The most common cause of it is damage or injury to blood vessels, but it can also happen as a result of certain illnesses or abnormalities in the blood clotting process.

Summary points

- Obesity is associated with worse outcomes and increased mortality in individuals infected with COVID-19.
- Acute kidney injury and increased short- and long-term mortality are more prevalent in obese patients with COVID-19.
- Obese patients have higher hospitalization rates, increased need for mechanical ventilation, and poorer prognosis.
- In obesity, the inflammatory cascade is downregulated, leading to the hyperactivation of inflammatory pathways and imbalances in cytokines, adiponectin, and leptin as well as impaired lung mechanics and gaseous exchange.
- Overexpression of ACE2 receptors in obesity contributes to the severity of COVID-19 infection and acts as a viral reservoir.
- Weight reduction through increased physical activity and calorie restriction is recommended for obese individuals, although long-term outcomes in COVID-19 are still uncertain.
- Metformin use has shown a decrease in COVID-19-related mortality by inhibiting viral attachment, suppressing infectivity, reducing inflammation, and lowering BMI and body weight.
- The COVID-19 pandemic has led to an increase in obesity rates due to factors such as limited physical activity, economic burden, and unhealthy eating habits.
- Clinicians should exercise vigilance and provide preemptive treatment to obese individuals with COVID-19 as well as encourage vaccination despite potential challenges.
- Including obesity or BMI in predicting a prognosis can aid in developing appropriate management strategies and treatment guidelines for COVID-19.

References

Akhmerov, A., & Marban, E. (2020). COVID-19 and the heart. *Circulation Research*, *126*(10), 1443–1455. https://doi.org/10.1161/CIRCRESAHA.120.317055.

Araujo, C., Fernandes, J., Caetano, D. S., Barros, A., de Souza, J. A. F., Machado, M., … Brandao, D. C. (2023). Endothelial function, arterial stiffness and heart rate variability of patients with cardiovascular diseases hospitalized due to COVID-19. *Heart & Lung*, *58*, 210–216. https://doi.org/10.1016/j.hrtlng.2022.12.016.

Atri, D., Siddiqi, H. K., Lang, J. P., Nauffal, V., Morrow, D. A., & Bohula, E. A. (2020). COVID-19 for the cardiologist: Basic virology, epidemiology, cardiac manifestations, and potential therapeutic strategies. *JACC. Basic to Translational Science*, *5*(5), 518–536. https://doi.org/10.1016/j.jacbts.2020.04.002.

Bandaru, P., Rajkumar, H., & Nappanveettil, G. (2011). Altered or impaired immune response to hepatitis B vaccine in WNIN/GR-Ob rat: An obese rat model with impaired glucose tolerance. *ISRN Endocrinology*, *2011*, 980105. https://doi.org/10.5402/2011/980105.

Ben Addi, A., Lefort, A., Hua, X., Libert, F., Communi, D., Ledent, C., … Robaye, B. (2008). Modulation of murine dendritic cell function by adenine nucleotides and adenosine: Involvement of the A(2B) receptor. *European Journal of Immunology*, *38*(6), 1610–1620. https://doi.org/10.1002/eji.200737781.

Berrington de Gonzalez, A., Hartge, P., Cerhan, J. R., Flint, A. J., Hannan, L., MacInnis, R. J., … Thun, M. J. (2010). Body-mass index and mortality among 1.46 million white adults. *The New England Journal of Medicine*, *363*(23), 2211–2219. https://doi.org/10.1056/NEJMoa1000367.

Blanco-Melo, D., Nilsson-Payant, B. E., Liu, W. C., Uhl, S., Hoagland, D., Moller, R., … tenOever, B. R. (2020). Imbalanced host response to SARS-CoV-2 drives development of COVID-19. *Cell*, *181*(5), 1036–1045. e1039 https://doi.org/10.1016/j.cell.2020.04.026.

Bolles, M., Deming, D., Long, K., Agnihothram, S., Whitmore, A., Ferris, M., … Baric, R. S. (2011). A double-inactivated severe acute respiratory syndrome coronavirus vaccine provides incomplete protection in mice and induces increased eosinophilic proinflammatory pulmonary response upon challenge. *Journal of Virology*, *85*(23), 12201–12215. https://doi.org/10.1128/JVI.06048-11.

Bost, P., Giladi, A., Liu, Y., Bendjelal, Y., Xu, G., David, E., … Amit, I. (2020). Host-viral infection maps reveal signatures of severe COVID-19 patients. *Cell*, *181*(7), 1475–1488 e1412. https://doi.org/10.1016/j.cell.2020.05.006.

Cai, C., Zeng, X., Ou, A. H., Huang, Y., & Zhang, X. (2004). Study on T cell subsets and their activated molecules from the convalescent SARS patients during two follow-up surveys. *Xi Bao Yu Fen Zi Mian Yi Xue Za Zhi*, *20*(3), 322–324. Retrieved from https://www.ncbi.nlm.nih.gov/pubmed/15193228.

Cameron, M. J., Bermejo-Martin, J. F., Danesh, A., Muller, M. P., & Kelvin, D. J. (2008). Human immunopathogenesis of severe acute respiratory syndrome (SARS). *Virus Research*, *133*(1), 13–19. https://doi.org/10.1016/j.virusres.2007.02.014.

Cantuti-Castelvetri, L., Ojha, R., Pedro, L. D., Djannatian, M., Franz, J., Kuivanen, S., … Simons, M. (2020). Neuropilin-1 facilitates SARS-CoV-2 cell entry and infectivity. *Science*, *370*(6518), 856–860. https://doi.org/10.1126/science.abd2985.

Castelo-Branco, C., & Soveral, I. (2014). The immune system and aging: A review. *Gynecological Endocrinology*, *30*(1), 16–22. https://doi.org/10.3109/09513590.2013.852531.

Chan, J. F., Kok, K. H., Zhu, Z., Chu, H., To, K. K., Yuan, S., & Yuen, K. Y. (2020). Genomic characterization of the 2019 novel human-pathogenic coronavirus isolated from a patient with atypical pneumonia after visiting Wuhan. *Emerging Microbes & Infections*, *9*(1), 221–236. https://doi.org/10.1080/22221751.2020.1719902.

Chen, L., Li, X., Chen, M., Feng, Y., & Xiong, C. (2020). The ACE2 expression in human heart indicates new potential mechanism of heart injury among patients infected with SARS-CoV-2. *Cardiovascular Research*, *116*(6), 1097–1100. https://doi.org/10.1093/cvr/cvaa078.

Cheng, H., Wang, Y., & Wang, G. Q. (2020). Organ-protective effect of angiotensin-converting enzyme 2 and its effect on the prognosis of COVID-19. *Journal of Medical Virology*, *92*(7), 726–730. https://doi.org/10.1002/jmv.25785.

Chu, H., Hu, B., Huang, X., Chai, Y., Zhou, D., Wang, Y., … Yuen, K. Y. (2021). Host and viral determinants for efficient SARS-CoV-2 infection of the human lung. *Nature Communications*, *12*(1), 134. https://doi.org/10.1038/s41467-020-20457-w.

Collange, O., Tacquard, C., Delabranche, X., Leonard-Lorant, I., Ohana, M., Onea, M., … Mertes, P. M. (2020). Coronavirus disease 2019: Associated multiple organ damage. *Open Forum Infectious Diseases*, *7*(7), ofaa249. https://doi.org/10.1093/ofid/ofaa249.

Del Prete, A., Conway, F., Della Rocca, D. G., Biondi-Zoccai, G., De Felice, F., Musto, C., … Versaci, F. (2022). COVID-19, acute myocardial injury, and infarction. *Cardiac Electrophysiology Clinics*, *14*(1), 29–39. https://doi.org/10.1016/j.ccep.2021.10.004.

Duarte, M., Pelorosso, F., Nicolosi, L. N., Salgado, M. V., Vetulli, H., Aquieri, A., … Rothlin, R. P. (2021). Telmisartan for treatment of Covid-19 patients: An open multicenter randomized clinical trial. *EClinicalMedicine*, *37*, 100962. https://doi.org/10.1016/j.eclinm.2021.100962.

Eliakim, A., Schwindt, C., Zaldivar, F., Casali, P., & Cooper, D. M. (2006). Reduced tetanus antibody titers in overweight children. *Autoimmunity*, *39*(2), 137–141. https://doi.org/10.1080/08916930600597326.

Giri Ravindran, S., Saha, D., Iqbal, I., Jhaveri, S., Avanthika, C., Naagendran, M. S., … Santhosh, T. (2022). The obesity paradox in chronic heart disease and chronic obstructive pulmonary disease. *Cureus*, *14*(6), e25674. https://doi.org/10.7759/cureus.25674.

Guilherme, A., Virbasius, J. V., Puri, V., & Czech, M. P. (2008). Adipocyte dysfunctions linking obesity to insulin resistance and type 2 diabetes. *Nature Reviews. Molecular Cell Biology*, *9*(5), 367–377. https://doi.org/10.1038/nrm2391.

Hamer, M., Gale, C. R., Kivimaki, M., & Batty, G. D. (2020). Overweight, obesity, and risk of hospitalization for COVID-19: A community-based cohort study of adults in the United Kingdom. *Proceedings of the National Academy of Sciences of the United States of America*, *117*(35), 21011–21013. https://doi.org/10.1073/pnas.2011086117.

Hemalatha Rajkumar, P. B. (2013). The impact of obesity on immune response to infection and vaccine: An insight into plausible mechanisms. *Endocrinology & Metabolic Syndrome*, *02*(02). https://doi.org/10.4172/2161-1017.1000113.

Henning, R. J. (2022). Cardiovascular complications of COVID-19 severe acute respiratory syndrome. *American Journal of Cardiovascular Disease*, *12*(4), 170–191. Retrieved from https://www.ncbi.nlm.nih.gov/pubmed/36147783.

Hoffmann, M., Kleine-Weber, H., Schroeder, S., Kruger, N., Herrler, T., Erichsen, S., … Pohlmann, S. (2020). SARS-CoV-2 cell entry depends on ACE2 and TMPRSS2 and is blocked by a clinically proven protease inhibitor. *Cell*, *181*(2), 271–280 e278. https://doi.org/10.1016/j.cell.2020.02.052.

Huang, C., Wang, Y., Li, X., Ren, L., Zhao, J., Hu, Y., … Cao, B. (2020). Clinical features of patients infected with 2019 novel coronavirus in Wuhan, China. *Lancet*, *395*(10223), 497–506. https://doi.org/10.1016/S0140-6736(20)30183-5.

Jaffe, A. S., Cleland, J. G. F., & Katus, H. A. (2020). Myocardial injury in severe COVID-19 infection. *European Heart Journal*, *41*(22), 2080–2082. https://doi.org/10.1093/eurheartj/ehaa447.

Jain, S., Kamimoto, L., Bramley, A. M., Schmitz, A. M., Benoit, S. R., Louie, J., … 2009 Pandemic Influenza A (H1N1) Virus Hospitalizations Investigation Team. (2009). Hospitalized patients with 2009 H1N1 influenza in the United States, April-June 2009. *The New England Journal of Medicine*, *361*(20), 1935–1944. https://doi.org/10.1056/NEJMoa0906695.

Kanneganti, T. D., & Dixit, V. D. (2012). Immunological complications of obesity. *Nature Immunology*, *13*(8), 707–712. https://doi.org/10.1038/ni.2343.

Kar, M. (2022). Vascular dysfunction and its cardiovascular consequences during and after COVID-19 infection: A narrative review. *Vascular Health and Risk Management*, *18*, 105–112. https://doi.org/10.2147/VHRM.S355410.

Katwa, L. C., Mendoza, C., & Clements, M. (2022). CVD and COVID-19: Emerging roles of cardiac fibroblasts and myofibroblasts. *Cells*, *11*(8), 1316. https://doi.org/10.3390/cells11081316.

Kellner-Weibel, G., & de la Llera-Moya, M. (2011). Update on HDL receptors and cellular cholesterol transport. *Current Atherosclerosis Reports*, *13*(3), 233–241. https://doi.org/10.1007/s11883-011-0169-0.

Kompaniyets, L., Goodman, A. B., Belay, B., Freedman, D. S., Sucosky, M. S., Lange, S. J., … Blanck, H. M. (2021). Body mass index and risk for COVID-19-related hospitalization, intensive care unit admission, invasive mechanical ventilation, and death-United States, March-December 2020. *Mmwr-Morbidity and Mortality Weekly Report*, *70*(10), 355–361. https://doi.org/10.15585/mmwr.mm7010e4.

Kruglikov, I. L., & Scherer, P. E. (2020). The role of adipocytes and adipocyte-like cells in the severity of COVID-19 infections. *Obesity (Silver Spring)*, *28* (7), 1187–1190. https://doi.org/10.1002/oby.22856.

Lagathu, C., Yvan-Charvet, L., Bastard, J. P., Maachi, M., Quignard-Boulange, A., Capeau, J., & Caron, M. (2006). Long-term treatment with interleukin-1beta induces insulin resistance in murine and human adipocytes. *Diabetologia*, *49*(9), 2162–2173. https://doi.org/10.1007/s00125-006-0335-z.

Lambeir, A. M., Durinx, C., Scharpe, S., & De Meester, I. (2003). Dipeptidyl-peptidase IV from bench to bedside: An update on structural properties, functions, and clinical aspects of the enzyme DPP IV. *Critical Reviews in Clinical Laboratory Sciences*, *40*(3), 209–294. https://doi.org/10.1080/713609354.

Larsson, M., Messmer, D., Somersan, S., Fonteneau, J. F., Donahoe, S. M., Lee, M., … Bhardwaj, N. (2000). Requirement of mature dendritic cells for efficient activation of influenza A-specific memory CD8+ T cells. *Journal of Immunology*, *165*(3), 1182–1190. https://doi.org/10.4049/jimmunol.165.3.1182.

Lavie, C. J., Coursin, D. B., & Long, M. T. (2021). The obesity paradox in infections and implications for COVID-19. *Mayo Clinic Proceedings*, *96*(3), 518–520. https://doi.org/10.1016/j.mayocp.2021.01.014.

Lee, H. K., Lee, B. H., Seok, S. H., Baek, M. W., Lee, H. Y., Kim, D. J., … Park, J. H. (2010). Production of specific antibodies against SARS-coronavirus nucleocapsid protein without cross reactivity with human coronaviruses 229E and OC43. *Journal of Veterinary Science*, *11*(2), 165–167. https://doi.org/10.4142/jvs.2010.11.2.165.

Li, G., Chen, X., & Xu, A. (2003). Profile of specific antibodies to the SARS-associated coronavirus. *The New England Journal of Medicine*, *349*(5), 508–509. https://doi.org/10.1056/NEJM200307313490520.

Li, T., Qiu, Z., Zhang, L., Han, Y., He, W., Liu, Z., … Wang, A. (2004). Significant changes of peripheral T lymphocyte subsets in patients with severe acute respiratory syndrome. *The Journal of Infectious Diseases*, *189*(4), 648–651. https://doi.org/10.1086/381535.

Li, B., Yang, J., Zhao, F., Zhi, L., Wang, X., Liu, L., … Zhao, Y. (2020). Prevalence and impact of cardiovascular metabolic diseases on COVID-19 in China. *Clinical Research in Cardiology*, *109*(5), 531–538. https://doi.org/10.1007/s00392-020-01626-9.

Liu, J., Li, S., Liu, J., Liang, B., Wang, X., Wang, H., … Zheng, X. (2020). Longitudinal characteristics of lymphocyte responses and cytokine profiles in the peripheral blood of SARS-CoV-2 infected patients. *eBioMedicine*, *55*, 102763. https://doi.org/10.1016/j.ebiom.2020.102763.

Lu, X., Pan, J., Tao, J., & Guo, D. (2011). SARS-CoV nucleocapsid protein antagonizes IFN-beta response by targeting initial step of IFN-beta induction pathway, and its C-terminal region is critical for the antagonism. *Virus Genes*, *42*(1), 37–45. https://doi.org/10.1007/s11262-010-0544-x.

Mohammad, K. O., Rodriguez, J. B. C., & Urey, M. A. (2022). Coronavirus disease 2019 and the cardiologist. *Current Opinion in Cardiology*, *37*(4), 335–342. https://doi.org/10.1097/HCO.0000000000000958.

Molarius, A., Seidell, J. C., Sans, S., Tuomilehto, J., & Kuulasmaa, K. (2000). Educational level, relative body weight, and changes in their association over 10 years: An international perspective from the WHO MONICA project. *American Journal of Public Health*, *90*(8), 1260–1268. https://doi.org/10.2105/ajph.90.8.1260.

Monteiro, C. A., Moura, E. C., Conde, W. L., & Popkin, B. M. (2004). Socioeconomic status and obesity in adult populations of developing countries: A review. *Bulletin of the World Health Organization*, *82*(12), 940–946. Retrieved from https://www.ncbi.nlm.nih.gov/pubmed/15654409.

Naaz, A., Holsberger, D. R., Iwamoto, G. A., Nelson, A., Kiyokawa, H., & Cooke, P. S. (2004). Loss of cyclin-dependent kinase inhibitors produces adipocyte hyperplasia and obesity. *The FASEB Journal*, *18*(15), 1925–1927. https://doi.org/10.1096/fj.04-2631fje.

Norbury, C. C., Malide, D., Gibbs, J. S., Bennink, J. R., & Yewdell, J. W. (2002). Visualizing priming of virus-specific CD8+ T cells by infected dendritic cells in vivo. *Nature Immunology*, *3*(3), 265–271. https://doi.org/10.1038/ni762.

Painter, S. D., Ovsyannikova, I. G., & Poland, G. A. (2015). The weight of obesity on the human immune response to vaccination. *Vaccine*, *33*(36), 4422–4429. https://doi.org/10.1016/j.vaccine.2015.06.101.

Ratchford, S. M., Stickford, J. L., Province, V. M., Stute, N., Augenreich, M. A., Koontz, L. K., … Stickford, A. S. L. (2021). Vascular alterations among young adults with SARS-CoV-2. *American Journal of Physiology. Heart and Circulatory Physiology*, *320*(1), H404–H410. https://doi.org/10.1152/ajpheart.00897.2020.

Saha, B., Jyothi Prasanna, S., Chandrasekar, B., & Nandi, D. (2010). Gene modulation and immunoregulatory roles of interferon gamma. *Cytokine*, *50*(1), 1–14. https://doi.org/10.1016/j.cyto.2009.11.021.

Sanchis-Gomar, F., Lavie, C. J., Mehra, M. R., Henry, B. M., & Lippi, G. (2020). Obesity and outcomes in COVID-19: when an epidemic and pandemic collide. *Mayo Clinic Proceedings*, *95*(7), 1445–1453. https://doi.org/10.1016/j.mayocp.2020.05.006.

Sarfraz, A., Sarfraz, Z., Siddiqui, A., Totonchian, A., Bokhari, S., Hussain, H., … Michel, J. (2022). Hypercoagulopathy in overweight and obese COVID-19 patients: A single-center case series. *Journal of Critical Care Medicine (Targu Mures)*, *8*(1), 41–48. https://doi.org/10.2478/jccm-2021-0032.

Sartipy, P., & Loskutoff, D. J. (2003). Monocyte chemoattractant protein 1 in obesity and insulin resistance. *Proceedings of the National Academy of Sciences of the United States of America*, *100*(12), 7265–7270. https://doi.org/10.1073/pnas.1133870100.

Sawadogo, W., Tsegaye, M., Gizaw, A., & Adera, T. (2022). Overweight and obesity as risk factors for COVID-19-associated hospitalisations and death: Systematic review and meta-analysis. *BMJ Nutrition, Prevention & Health*, *5*(1), 10–18. https://doi.org/10.1136/bmjnph-2021-000375.

Sheridan, P. A., Paich, H. A., Handy, J., Karlsson, E. A., Hudgens, M. G., Sammon, A. B., … Beck, M. A. (2012). Obesity is associated with impaired immune response to influenza vaccination in humans. *International Journal of Obesity*, *36*(8), 1072–1077. https://doi.org/10.1038/ijo.2011.208.

Siddiqi, H. K., & Mehra, M. R. (2020). COVID-19 illness in native and immunosuppressed states: A clinical-therapeutic staging proposal. *The Journal of Heart and Lung Transplantation*, *39*(5), 405–407. https://doi.org/10.1016/j.healun.2020.03.012.

Singh, R., Rathore, S. S., Khan, H., Karale, S., Chawla, Y., Iqbal, K., … Bansal, V. (2022). Association of Obesity with COVID-19 severity and mortality: An updated systemic review, meta-analysis, and meta-regression. *Frontiers in Endocrinology (Lausanne)*, *13*, 780872. https://doi.org/10.3389/fendo.2022.780872.

Soehnlein, O., & Lindbom, L. (2010). Phagocyte partnership during the onset and resolution of inflammation. *Nature Reviews. Immunology*, *10*(6), 427–439. https://doi.org/10.1038/nri2779.

South, A. M., Diz, D. I., & Chappell, M. C. (2020). COVID-19, ACE2, and the cardiovascular consequences. *American Journal of Physiology. Heart and Circulatory Physiology*, *318*(5), H1084–H1090. https://doi.org/10.1152/ajpheart.00217.2020.

Sungnak, W., Huang, N., Becavin, C., Berg, M., Queen, R., Litvinukova, M., … Network, H. C. A. L. B. (2020). SARS-CoV-2 entry factors are highly expressed in nasal epithelial cells together with innate immune genes. *Nature Medicine*, *26*(5), 681–687. https://doi.org/10.1038/s41591-020-0868-6.

Tian, S., Hu, W., Niu, L., Liu, H., Xu, H., & Xiao, S. Y. (2020). Pulmonary pathology of early-phase 2019 novel coronavirus (COVID-19) pneumonia in two patients with lung cancer. *Journal of Thoracic Oncology*, *15*(5), 700–704. https://doi.org/10.1016/j.jtho.2020.02.010.

Tortorici, M. A., & Veesler, D. (2019). Structural insights into coronavirus entry. *Advances in Virus Research*, *105*, 93–116. https://doi.org/10.1016/bs.aivir.2019.08.002.

van der Klaauw, A. A., Horner, E. C., Pereyra-Gerber, P., Agrawal, U., Foster, W. S., Spencer, S., … Thaventhiran, J. E. D. (2023). Accelerated waning of the humoral response to COVID-19 vaccines in obesity. *Nature Medicine*, *29*(5), 1146–1154. https://doi.org/10.1038/s41591-023-02343-2.

Vaughan, C. J., Cronin, H., Ryan, P. M., & Caplice, N. M. (2020). Obesity and COVID-19: A Virchow's triad for the 21st century. *Thrombosis and Haemostasis*, *120*(11), 1590–1593. https://doi.org/10.1055/s-0040-1714216.

Viardot, A., Grey, S. T., Mackay, F., & Chisholm, D. (2007). Potential antiinflammatory role of insulin via the preferential polarization of effector T cells toward a T helper 2 phenotype. *Endocrinology*, *148*(1), 346–353. https://doi.org/10.1210/en.2006-0686.

Whitlock, G., Lewington, S., Sherliker, P., Clarke, R., Emberson, J., Halsey, J., … Prospective Studies Collaboration. (2009). Body-mass index and cause-specific mortality in 900 000 adults: Collaborative analyses of 57 prospective studies. *Lancet*, *373*(9669), 1083–1096. https://doi.org/10.1016/S0140-6736(09)60318-4.

Winer, S., Chan, Y., Paltser, G., Truong, D., Tsui, H., Bahrami, J., … Dosch, H. M. (2009). Normalization of obesity-associated insulin resistance through immunotherapy. *Nature Medicine*, *15*(8), 921–929. https://doi.org/10.1038/nm.2001.

Yoshikawa, T., Hill, T., Li, K., Peters, C. J., & Tseng, C. T. (2009). Severe acute respiratory syndrome (SARS) coronavirus-induced lung epithelial cytokines exacerbate SARS pathogenesis by modulating intrinsic functions of monocyte-derived macrophages and dendritic cells. *Journal of Virology*, *83*(7), 3039–3048. https://doi.org/10.1128/JVI.01792-08.

Zhang, F., Xiong, Y., Wei, Y., Hu, Y., Wang, F., Li, G., … Zhu, W. (2020). Obesity predisposes to the risk of higher mortality in young COVID-19 patients. *Journal of Medical Virology*, *92*(11), 2536–2542. https://doi.org/10.1002/jmv.26039.

Zheng, Y. Y., Ma, Y. T., Zhang, J. Y., & Xie, X. (2020). COVID-19 and the cardiovascular system. *Nature Reviews. Cardiology*, *17*(5), 259–260. https://doi.org/10.1038/s41569-020-0360-5.

Chapter 48

Upper and lower gastrointestinal symptoms and manifestations of COVID-19

Brittany Woods[a], Priyal Mehta[b], Gowthami Sai Kogilathota Jagirdhar[c], Rahul Kashyap[d], and Vikas Bansal[d]

[a]*Department of Hospital Medicine, Advent Health Redmond, Rome, GA, United States,* [b]*Department of Medicine, M.W. Desai Hospital, Mumbai, Maharashtra, India,* [c]*Department of Gastroenterology, Saint Michael's Medical Center, Newark, NJ, United States,* [d]*Department of Critical Care Medicine, Mayo Clinic, Rochester, MN, United States*

Abbreviations

ACE2	angiotensin converting enzyme 2
AP	acute pancreatitis
BUN/creatinine	blood urea nitrogen/creatinine
CT	computed tomography
GI	gastrointestinal
ICU	intensive care unit
IL	interleukin
RAS	renin angiotensin system
ROS	reactive oxygen species
SMA	superior mesenteric artery
SMV	superior mesenteric vein
TMPRSS2	transmembrane serine protease 2
TNF-alpha	tumor necrosing factor-alpha
TRP	transient receptor potential

Introduction

Overview of the impact of COVID-19 on the gastrointestinal tract

The COVID-19 pandemic caused by the novel coronavirus SARS-CoV-2 had a profound impact on global health, leading to almost seven million deaths worldwide since its emergence in late 2019. Though it was most notable for its severe respiratory side effects, emerging studies over the last 3 years have shown that it also causes significant effects to the gastrointestinal (GI) system. Some of the most common gastrointestinal symptoms that patients with COVID-19 experience are diarrhea, nausea, and/or vomiting, which will be discussed later in this chapter.

The virus enters the GI tract by the angiotensin converting enzyme 2 (ACE2) receptors in the colon, esophagus, and parts of the small bowel. For patients with chronic conditions such as IBD, cirrhosis, GERD, or GI malignancies, COVID-19 infection can lead to worse outcomes and serious illness. Many patients with IBD are maintained on immunosuppressive therapy and some regimens such as systemic corticosteroids, sulfasalazine, or 5-aminosalicylate are recommended to be discontinued after a COVID-19 diagnosis. It was found in a small study that these patients are at higher risk for severe COVID-19 infection (Ghazanfar et al., 2022). Patients with cirrhosis or other chronic liver diseases were also found to be at higher risk for poor outcomes in COVID-19 infection (Ghazanfar et al., 2022). In patients with GERD or peptic ulcer disease, the use of PPI, especially within 30 days of COVID-19 infection, led to a higher risk of worse outcomes (Ghazanfar et al., 2022). Lastly, patients with gastrointestinal malignancy were more likely to become infected by the virus due to their immunocompromised state (Ghazanfar et al., 2022).

Management, Body Systems, and Case Studies in COVID-19. https://doi.org/10.1016/B978-0-443-18703-2.00008-6
Copyright © 2024 Elsevier Inc. All rights are reserved, including those for text and data mining, AI training, and similar technologies.

Epidemiology and incidence of COVID-19 on gastrointestinal tract

It is important to understand the epidemiology and incidence of gastrointestinal symptoms in COVID-19, as this is crucial for the accurate and comprehensive management and monitoring of the disease. The incidence of gastrointestinal symptoms in COVID-19 is quite broad. In an April 2023 study by Al-Momani et al., around 40% of patients presented gastrointestinal symptoms. The study also found that the presence of gastrointestinal symptoms was not linked to poor outcomes in these patients. Most patients with gastrointestinal symptoms presented with anorexia, nausea, vomiting, and diarrhea.

When it comes to other gastrointestinal disorders such as acute pancreatitis, bleeding and liver injury are not as prevalent as nausea, vomiting, anorexia, and diarrhea. Acute pancreatitis has about a 0.27% prevalence in patients with COVID-19 (Babajide et al., 2022; de Madaria & Capurso, 2021). Gastrointestinal bleeding is fairly uncommon in COVID-19 patients, with a prevalence around 1.5%–3.0% (Cappell & Friedel, 2023). As for liver injury in COVID-19, a study in February 2023 showed that the incidence of liver injury manifesting as elevated transaminase levels and elevated bilirubin was around 14.8%–53.0% (Vujčić, 2023). Typically, these elevations in laboratories are transient and most return to the baseline following recovery (Vujčić, 2023).

Pathophysiology of COVID-19 and its effects on the gastrointestinal tract

The SARS-CoV-2 virus, more commonly known as COVID-19, is a positive-sense single-stranded RNA virus that has a spike glycoprotein that allows the virus to bind to ACE2. Cells of the respiratory, cardiac, and renal systems and cell membranes of the absorptive enterocytes of the ileum, colon, and duodenum, including the esophageal epithelium, have this receptor (Beyerstedt et al., 2021), (Groff et al., 2021), (Zuo et al., 2020). The expression of ACE2 is four-fold higher in the intestine than that of other tissues (Zhang, Shao, et al., 2021). Tissues of the upper GI tract susceptible to the virus include the glandular cells of the esophagus, stomach, and duodenum. In these tissues, the spike (S) protein of COVID-19 binds to ACE2, facilitating viral entry into the host cell (Zhang, Shao, et al., 2021). The S protein binds with ACE2 in concert with host cell transmembrane serine protease 2 (TMPRSS2) (Zhang, Shao, et al., 2021). TMPRSS2 is a protease that plays a critical part in cleaving the S protein of COVID-19, thus facilitating invasion into host cells (Zhang, Liu, et al., 2021). It is important to understand that only cleavage by TMPRSS2 results in host cell entry of COVID-19 (Xiao et al., 2020).

Though the function of ACE2 in the lungs is well known, it is important to understand that ACE2 functions independently of the renin-angiotensin system (RAS). The difference is that ACE2 stabilizes the amino acid transporters, which ultimately compromises the intestinal uptake of important dietary amino acids (Penninger et al., 2021). Some of these dietary amino acids, including tryptophan and glutamine, are important in signaling the downregulation of lymphoid proinflammatory cytokines, aid in the defense mechanisms of the mucosal cells and tight junction formation, and function in the activation of antimicrobial peptide release (Chua, 2006). Disruption of these important mechanisms leads to gut dysbiosis (Zhang, Garrett, & Sun, 2021). Dysbiosis and a lack of tight junctions leading to the leaky gut can contribute to a cytokine storm in COVID-19 patients who are critically ill (Penninger et al., 2021).

As the virus continues to replicate, the level of ACE2 decreases, leading to the destruction of host cells (Zhang, Shao, et al., 2021). Unfortunately, if the host's immune system cannot defeat the infection, the virus continues to replicate in considerable amounts. The effects of this process can lead to vasoconstriction, inflammation, capillary permeability, fibrosis, and oxidation. The release of the cytokine and chemokine instigates acute intestinal inflammation, thus leading to symptoms of gastroenteritis (Zhang, Garrett, & Sun, 2021). Fig. 1 shows the various gastrointestinal manifestations of COVID-19.

Upper gastrointestinal symptoms in COVID-19

Upper gastrointestinal symptoms present in COVID-19 include nausea, vomiting, anorexia, dysgeusia, or ageusia. Like its binding capacity via the spike protein in the lungs, heart, and kidneys, COVID-19 also binds to the ACE2 receptors of the gastrointestinal tract, including the esophageal epithelium, duodenum, and even the stomach (Groff et al., 2021; Liu et al., 2021). There have been several analyses that correlate the presence of gastrointestinal symptoms with severe symptoms of COVID-19 pneumonia. A study performed by Effenberger et al. (2020) revealed elevated levels of fecal calprotectin and a systemic IL-6 response in patients with COVID-19 during the acute phase of infection. A study in 2022 by Guo et al. (2021) showed that 40% of COVID-19-positive patients with gastrointestinal symptoms had a severe course of infection. In fact, patients who complained of abdominal pain had a 2.8-fold increased risk of progressing to severe COVID-19

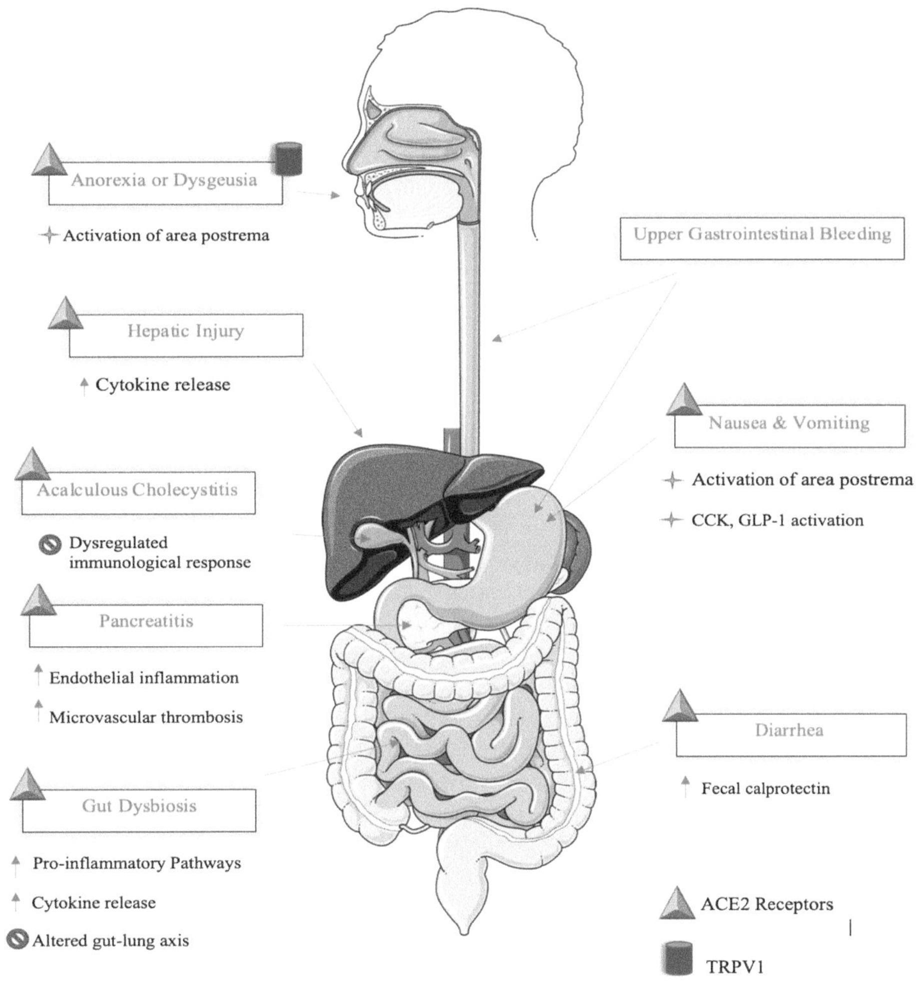

FIG. 1 Various gastrointestinal manifestations and their mechanisms of COVID-19. *Adapted from Servier and created by the authors.*

(Guo et al., 2021). A study in 2022 revealed that there is increased activity of the von Willebrand factor (vWF), WF antigen, factor VIII, and a marker of endothelial activation, thrombomodulin, in patients diagnosed with COVID-19, thus creating endotheliopathy, platelet dysfunction, and thrombosis (McConnell et al., 2022). The following section focuses on the pathophysiology, treatment, and complications of upper gastrointestinal symptoms seen in COVID-19.

Nausea, vomiting, and anorexia in COVID-19 patients

Some studies have reported that nausea or vomiting can be an early presentation of acute COVID-19 infection and are common (Zhang, Shao, et al., 2021). Histopathological studies revealed that the tissue of the gastrointestinal tract is a direct target for COVID-19, and this possibly can explain the damage the virus creates, leading to symptoms of nausea and vomiting (Yao et al., 2022). This study by Yao et al. (2022) also reported that the incidence of nausea ranged from 1.0% to 27.5% and vomiting ranged from 1.0% to 12.5%. In a retrospective analysis by Song et al. (2022), of 750 patients meeting inclusion criteria, 49.7% presented with at least one gastrointestinal symptom.

Few studies have been performed to explain the exact mechanism of nausea and emesis in COVID-19, but some studies have proposed theories. In a study by Andrews et al. (2021), the authors proposed that mediators released from the intestinal epithelium by COVID-19 may activate emetic mechanisms and modulate the vagal afferents of the brainstem. Once these mediators enter the bloodstream, they may also activate the area postrema (Andrews et al., 2021). The study also revealed that data on COVID-19 showed a plasma increase of angiotensin II (Andrews et al., 2021). Angiotensin II receptors are rich in the area postrema, producing emetic effects leading to another potential mechanism for the gastrointestinal symptoms in COVID-19 (Gupta et al., 1989). Andrews et al. (2021) also found that nausea and vomiting are triggered by the binding of

COVID-19 to ACE2 receptors in the digestive tract, leading to a possible increase in neuroactive agents from enteroendocrine cells and inflammatory mediators. These agents and mediators acted upon the area postrema and vagal afferents, leading to nausea and vomiting.

The area postrema is also implicated in the effects of anorexia. A study in the United States by Redd et al. (2020) revealed that patients presenting with gastrointestinal symptoms may experience anorexia 34.8% of the time. A study by Yao et al. (2022) also suggested that another mechanism of anorexia is the transient receptor potential (TRP) channels expressed in the gastrointestinal tract and a person's intake of food. A specific type of TRP channel, TRPV1, is expressed in the "esophageal sensory neurons and stomach-labeled vagal nodose neurons" (Yao et al., 2022). It is a receptor thought to impact appetite via modulation of gastrointestinal vagal afferents (Yao et al., 2022). Because COVID-19 is thought to affect gastrointestinal vagal afferents, this receptor may play a role in the symptoms of anorexia seen in COVID-19-positive patients.

Lastly, a final mechanism proposed to induce nausea and vomiting is a theory regarding neuroactive agents originating from the enteroendocrine cells-cholecystokinin (CCK) and glucagon-like-1 peptide (GLP-1) (Andrews et al., 2021). CCK is a brain-gut peptide that acts on the proximal stomach and pylorus and can inhibit gastric motility and emptying. The mechanism by which CCK functions is through a "capsaicin-sensitive vagal pathway" (Chua, 2006). Glucagon-like-1 peptide is a gastrointestinal hormone involved in the effects of delayed gastric emptying and early satiety. These agents are gastrointestinal hormones that have the capability of inducing symptoms of nausea, vomiting, anorexia, or reduction of appetite (Andrews et al., 2021). This theory is thought to be a possible mechanism of nausea and vomiting, as ACE2 is expressed in enteroendocrine cells, but research is required to further evaluate this concept.

Gut dysbiosis

Now that studies have shown that COVID-19 not only disrupts the respiratory tract but also the gastrointestinal tract, it is important to know that the virus also disrupts the gut microbiome. The gut microbiome is defined as a "collection of bacteria, archaea, and eukarya" that colonizes and coinhabits the gastrointestinal tract, creating eubiosis (Thursby & Juge, 2017). It is said that the number of microorganisms is vast, estimated to exceed 104 (Thursby & Juge, 2017). Alterations in the gut microbiome are referred to as dysbiosis. Natarajan et al. (2022) revealed that up to 49.2% of patients diagnosed with COVID-19 possess viral RNA in their feces in the first week and even up to 7 months postinfection. These patients were also noted to present with gastrointestinal symptoms. Because the virus appears to stay within the GI tract for long periods, studies have found that it can disrupt the gut microbiome. A recently published study in January 2023 by Mańkowska-Wierzbicka et al. (2023) sought to discover whether COVID-19 patients had alterations in their gut microbiome compared to healthy patients. Their study revealed that patients diagnosed with COVID-19 had "statistically greater taxonomic diversity compared to healthy subjects" (Mańkowska-Wierzbicka et al., 2023). In other words, patients with COVID-19 exhibited gut dysbiosis.

There has been evidence to suggest that the microbiota of the lungs and the gastrointestinal tract may play a key role in the maintenance of homeostasis (Mańkowska-Wierzbicka et al., 2023). Dysbiosis of the gut can be related to disease severity and clinical progression (De & Dutta, 2022). A study in 2020 by Zuo et al. (2020) revealed in 15 patients that alterations in the fecal microbiota were associated with "fecal levels of SARS-CoV-2 and COVID-19 severity." This study also revealed that dysbiosis persisted even after the resolution of respiratory symptoms and clearance of COVID-19 (in throat swabs). The study evaluated specific gut bacteria including Clostridium, Coprobacillus, Faecalibacterium, and Bacteroides. In their conclusions, they found that the antiinflammatory bacterium *Faecalibacterium prausnitzii* correlated inversely with COVID-19 disease severity. It was also noted that specific Bacteroides species including *Bacteroides dorei*, *thetaiotaomicron*, *massiliensis*, and *ovatus*-bacteria responsible for the downregulation of ACE2 expression in murine subjects—also correlated inversely with the COVID-19 viral load of fecal samples in hospitalized patients. Overall, multiple studies have concluded that COVID-19 disrupts the microbiome, and this disruption is associated with worse outcomes compared to healthy subjects.

Upper gastrointestinal bleeding

There are studies discussing upper gastrointestinal bleeding in COVID-19. A rapid review study in July 2022 by Negro et al. (2022) sought to find the incidence and risk factors associated with COVID-19 and gastrointestinal bleeding. The study found that after an analysis of 20 relevant studies, the rate of GI bleeding ranged from 1.1% to 13%. Notably, critically ill patients requiring admission to the intensive care unit (ICU) had an increased risk of "stress-related mucosal disease (SRMD) from gastrointestinal bleeding" (Negro et al., 2022).

A prior study in 2020 revealed that the most common etiologies of upper gastrointestinal bleeding were gastric or duodenal ulcers (Martin et al., 2020). In both studies, the conclusion was that the rate of bleeding is related to the severity of critical illness. Because critically ill patients are also prone to venous thromboembolism (VTE), prophylactic anticoagulation is considered on a case-to-case basis depending on the risk-benefit ratio.

Treatment and complications of upper gastrointestinal symptoms

Treatment of upper gastrointestinal symptoms is mostly supportive. There are no treatment-specific medications for nausea and vomiting secondary to COVID-19. The use of antiemetics such as ondansetron or aprepitant is common, as these medications act centrally in the area postrema and peripherally at the vagal afferents. Lastly, as with most patients with vomiting, hydration—whether oral or parenteral—is the mainstay of treatment. As we have learned thus far, gastrointestinal symptoms of COVID-19 are quite common. Many studies have sought to determine whether gastrointestinal symptoms affect the clinical outcomes of patients diagnosed with COVID-19. Over the past 2 years, multiple studies have linked the severity of disease progression and respiratory symptoms to gastrointestinal symptoms while other studies have suggested that GI symptoms are linked to a mild disease course or not at all. Most studies have concluded that the presence of gastrointestinal symptoms is not linked to patient mortality in COVID-19 (Chua, 2006; Raman et al., 2022; Wang et al., 2022). A study published in 2022 found that the presence of isolated gastrointestinal symptoms such as nausea, vomiting, or abdominal pain is not linked to mortality (Wang et al., 2021).

COVID-19-induced pancreatitis

Etiopathogenesis

Acute pancreatitis (AP) is a rapidly progressing inflammatory disorder that primarily impacts the acinar cells until its subsequent spread to the surrounding tissue. The duodenum epithelium expresses abundant ACE2 receptors. Infection-mediated gut dysbiosis can disrupt the mucosal barrier and lead to infection of the ACE2-expressing acinar and islet cells via the pancreatic duct (Brisinda et al., 2022; Khreefa et al., 2023; Memon & Abdelalim, 2021; Neoptolemos & Greenhalf, 2022). SARS-CoV-2 is also able to ignite endothelial inflammation and microvascular thrombosis, leading to diffuse microischemia. The study also underscores the impact of SARS-CoV2 on the pancreatic parenchymal stellate cells, leading to the production of IL-18 and the activation of resident pancreatic macrophages, setting off the cytokine storm and culminating in a severe systemic response. Hyperlipidemia is common in COVID-19 patients, with a pooled incidence of 32.98%. The accumulation of free fatty acids and, thereafter, the activation of the inflammatory response in the pancreas could potentially cause an increase in the levels of inflammatory mediators such as TNF-alpha, interleukin-6, and interleukin-10, which might strengthen the systemic inflammatory response and local pancreatic injury (Tang et al., 2022). The possibility of drug-induced pancreatitis from nonsteroidal antiinflammatory drugs and glucocorticoids should be considered in patients with COVID-19 (Liu et al., 2020).

The gut-liver axis is strongly affected by damage to the intestinal mucosa by SARS CoV-2 replication in the intestinal epithelium as well as the cytokine storm causing dysbiosis. The liver releases bile acids into biliary and systemic circulations. This could lead to pancreatic function damage in COVID-19 patients (Brisinda et al., 2022). Fig. 2 demonstrates in detail the proposed mechanism for the development of acute pancreatitis.

Clinical features

Aziz et al., in their study on COVID-19, underlined that nearly 63% of patients eventually diagnosed with AP presented with acute abdominal pain (Aziz et al., 2022). In patients that displayed no symptoms, elevated serum levels of amylase and lipase (indicative of substantial pancreatic inflammation) have been observed in various surveys (Sinagra et al., 2022). Among COVID-19 patients, the prevalence of hyperamylasemia ranged from 15.7% to 33% while the pooled prevalence of hyperlipidemia in patients with COVID-19 was 11.7%.

True AP, defined by the revised Atlanta criteria, is shown to be less prevalent. A few studies also concluded that AP had a point prevalence of 0.27%, with 69% of the cases being idiopathic (Babajide et al., 2022). Case reports describe uncomplicated self-resolving AP after patients received a dose of the Pfizer/BioNTech mRNA vaccine (Sinagra et al., 2022). Neoptolemos and Greenhalf describe in the Gerasimenko model that the spike protein, or even just part of the spike protein, could cause pancreatitis in a susceptible individual; this could explain reports of acute pancreatitis following vaccination (Neoptolemos & Greenhalf, 2022). Case reports of necrotizing pancreatitis occurring across various age groups in

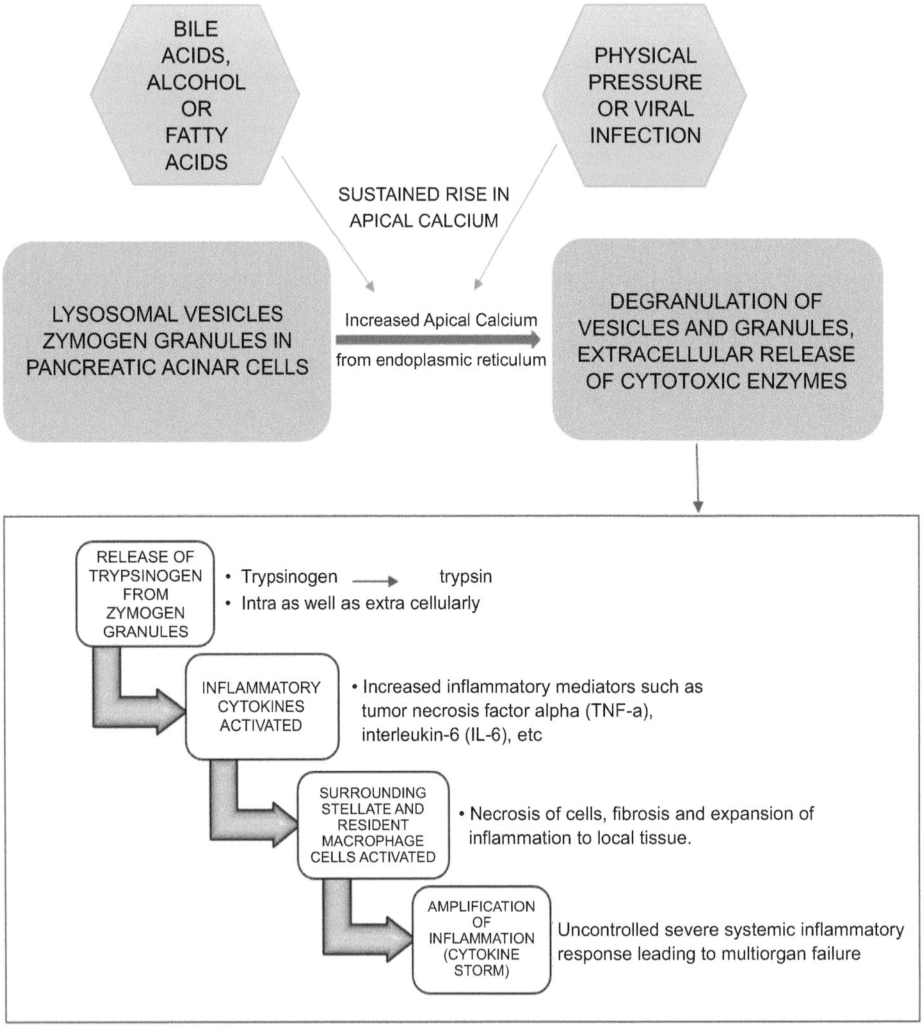

FIG. 2 Etiopathogenesis of acute pancreatitis. *Adapted from Neoptolemos, J. P., & Greenhalf, W. (2022). Linking COVID-19 and acute pancreatitis through the pathogenic effects of the SARS-CoV-2 S protein subunit 1 on pancreatic stellate cells and macrophages. Function, 3(2). doi: https://doi.org/10.1093/function/zqac009 and created by the authors.*

COVID-19 are presented in the literature (Maalouf et al., 2021; Samanta et al., 2020). In the management of necrotizing AP, open operative debridement maintains a role in cases not amenable to less invasive endoscopic and/or laparoscopic procedures. Prompt action is of paramount significance in such cases to prevent mortality (Brisinda et al., 2022).

Management

As shown in the COVIDPAN study, AP in COVID-19-positive patients is more aggressive and requires longer hospital stays, necessitating the use of mechanical ventilation and rapidly precipitating multiorgan failure (Babajide et al., 2022; Pandanaboyana et al., 2021; Sinagra et al., 2022). Therefore, apart from the routine laboratory tests done in a suspected case of AP, it is essential to discern end-organ failure with hemodynamic status monitoring, serial BUN/creatinine ratios, and liver enzymes. There is evidence showing that the lipase/lymphocyte ratio can be useful in estimating mortality and prognosis (Gullu et al., 2022). In all patients with or without SARS-CoV-2 illness, a contrast-enhanced computed tomography (CT) scan is the gold standard for the diagnosis of acute pancreatitis to evaluate both pancreatic and extrapancreatic alterations (Brisinda et al., 2022). Grusova et al. showed that the prevalence of peripancreatic stranding or fluid collection is higher in patients diagnosed with COVID-19 infection (Grusova et al., 2021). Bozdang et al. calculated the average density of four anatomical structures of the pancreas and inferred that a decrease in density on CT is an early finding suggestive of pancreatic involvement in COVID-19 (Bozdag et al., 2021).

Treatment

General principles, including nil by mouth until feeding is tolerated by the patient, are suggested to prevent the release of inflammatory mediators in AP (Su & Chen, 2022). In addition to pain control and nutritional support, early fluid resuscitation is recommended to improve tissue perfusion (Brisinda et al., 2022). Early resumption of enteral nutrition, especially with supplementation of probiotics, may decrease septic complications. In patients with COVID-19 requiring mechanical ventilation, early enteral nutrition is associated with earlier liberation from ventilator support, a shorter ICU and hospital stay, and decreased cost.

Most patients with sterile necrosis can be managed without invasive treatments. If necrotic gas-forming changes or hemorrhage occur, then curettage of the necrotic tissues is necessary (Su & Chen, 2022). The surgical treatment strategy includes the early administration of antibiotics and percutaneous drainage of the collected fluid, followed by surgical resection of the infected necrotic tissue. Patients with COVID-19 who underwent pancreatic surgery had a 3.6-fold higher risk of perioperative mortality, a 2.2-fold higher risk of major complications, a 2.6-fold higher risk of late postoperative bleeding, and a 2.1-fold higher risk of postoperative pancreatic fistula (Brisinda et al., 2022; Su & Chen, 2022).

Liver manifestations in COVID-19

Studies indicate that one in four to five COVID-19 patients develops liver function abnormalities, especially when the infection is severe (Mao et al., 2020; Wijarnpreecha et al., 2021). Elevations of total bilirubin may be accompanied by icterus, and patients with acute liver failure may also complain of acute right upper quadrant pain. Nevertheless, these abnormalities are generally found to be transient, and the majority of patients remain asymptomatic. Intensive monitoring of liver function tests may thus be justified in patients with severe COVID-19 infection because it is associated with disease severity in the prior literature (Wijarnpreecha et al., 2021). Coexisting chronic liver disease was present in up to 37.6% of patients with COVID-19. Chronic liver disease may decompensate precipitously and present as acute-on-chronic liver failure (ACLF) in patients with COVID-19. Aquino-Matus et al. discussed the effect of COVID-19 on patients presenting with ACLF. These patients had higher chances of ICU admission, greater disease severity, and increased mortality in COVID-19 (Aquino-Matus et al., 2022). Patients with metabolic-associated fatty liver disease (MAFLD) were also at higher risk of developing complications related to COVID-19 (Aquino-Matus et al., 2022).

Etiopathogenesis

SARS-CoV-2 infection reaches the liver either via the hematogenous route or by the direct extension of infection from the GI tract. SARS-CoV-2-mediated liver injury is characterized as multifactorial. ACE2 receptors are expressed 20 times higher in cholangiocytes than in hepatocytes; histopathological analysis revealed abundant virus particles within the cytoplasm of hepatocytes, which is indicative of the upregulation of ACE2 following viral infection of those cells (Abdulla et al., 2020; Khreefa et al., 2023). Liver biopsies also revealed prominent Kupffer cells, acidophilic bodies, portal tracts with chronic inflammatory infiltrates, and ballooning degeneration of hepatocytes with frequent mitotic bodies, suggesting damaged/ virally infected COVID-19 cells. Liver endothelial cells were also found to abundantly express ACE2. Liver vascular injury may result in increased vascular permeability and damage, allowing the leakage of inflammatory cytokines and free reactive oxygen species (ROS) into liver tissues (Khreefa et al., 2023). Endothelial damage exposes underlying collagen, which subsequently attracts platelets and neutrophils to promote endothelin and microthrombi formation, causing hypoxic damage to the liver.

Hypoxic hepatic damage is also associated with pneumonia, septic shock, and the correlated hypoxemia encountered in severe disease (Radivojevic et al., 2022). Administration of thrombolytics may lead to a paradoxical rise in liver enzymes (Radivojevic et al., 2022). This is hypothesized to be due to hypoxia reperfusion injury to hepatocytes. Drug-induced liver injury (DILI) and herb-induced liver injury (HILI) are defined as liver dysfunction and/or abnormalities in liver function tests secondary to the use of medications, herbs, or xenobiotics, within the reasonable exclusion of other etiologies (Aquino-Matus et al., 2022; Khreefa et al., 2023; Radivojevic et al., 2022). Worldwide, polypharmacy was used to manage COVID-19 through the use of antiviral agents (e.g., lopinavir, ritonavir, remdesivir, and favipiravir), antibiotics (e.g., azithromycin), antimalarials (e.g., chloroquine and hydroxychloroquine), monoclonal antibodies (e.g., tocilizumab), JAK inhibitors (e.g., baricitinib), complementary alternative medicine (e.g., chlorine dioxide and Ayurvedic Kadha), and home remedies (*Allium sativum*), many of which have been directly linked with hepatotoxicity, especially due to their first-pass metabolism. The cytokine storm and multiorgan dysfunction from severe systemic inflammation also contribute to the liver damage seen in these patients (Aquino-Matus et al., 2022; Radivojevic et al., 2022). The immune dysfunction mediated

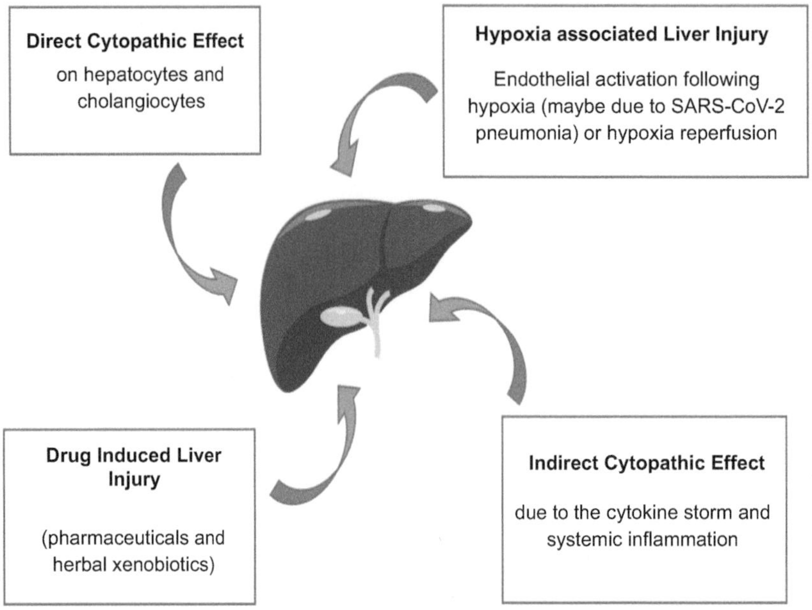

FIG. 3 Pathogenesis of COVID-19 on the liver. *Adapted from Radivojevic, A., Abu Jad, A. A., Ravanavena, A., Ravindra, C., Igweonu-Nwakile, E. O., Ali, S., Paul, S., Yakkali, S., Teresa Selvin, S., Thomas, S., Bikeyeva, V., Abdullah, A., & Balani, P. (2022). A systematic review of SARS-CoV-2-associated hepatic dysfunction and the impact on the clinical outcome of COVID-19. Cureus doi: https://doi.org/10.7759/cureus.26852 and created by the authors.*

by the high circulating levels of interleukin-1β, interferon γ, and monocyte chemoattractant protein-1 upregulates the ACE2 and TMPRSS2 receptors in all tissues, including the liver. Fig. 3 shows the various mechanisms by which COVID-19 acts on the liver.

Lower gastrointestinal symptoms in COVID-19

Diarrhea in COVID-19

Diarrhea is one of the most common symptoms experienced by patients with GI symptoms (Chen et al., 2021; Puli et al., 2020). In a pooled analysis, the percentage of patients who had diarrhea was 6% (Zeng et al., 2022). The appearance of diarrhea symptoms is mainly due to the viral invasion of the absorptive epithelial cells in the intestine, where such cells are the most vulnerable, resulting in malabsorption and intestinal secretion disorders. In addition, the heavy use of antibiotics and antiviral drugs can also cause diarrhea in some patients.

Ischemic colitis

Acute ischemic colitis is a rare intestinal emergency that requires urgent and adequate management (Alratrout & Debroux, 2022). Patients develop ischemic colitis, most likely due to SARS-CoV-2-induced endotheliitis or direct bowel damage due to the expression of ACE-2 (Uhlenhopp et al., 2022). Patients with severe COVID-19 infection have nonocclusive colonic ischemia (CI) due to intense vasoconstriction and decreased mesenteric blood flow secondary to hemodynamic compromise and the use of inotropic agents. From the limited cases of GI ischemia in COVID-19 patients in the literature, almost half the patients died, even after surgical intervention. Colitis manifests itself within 3 days of the onset of a compromised blood supply when mucosal edema is absorbed, resulting in mucosal ulceration. These ulcerations take several months to heal, although the patient may be free of any clinical symptoms (Krejčová et al., 2022). More severe ischemic damage may lead to full-width necrosis of the bowel wall. A rare case of pancolitis secondary to active COVID-19 with simultaneous gastrointestinal involvement was reported (Al-Roubaie & Udayasiri, 2022). In COVID-19-associated intestinal ischemia, bowel gangrene has been reported to occur with or without the involvement of the major arteries, such as the superior mesenteric artery, or veins, such as the superior mesenteric vein and portal vein (Varshney et al., 2021).

Typical clinical presentations are sudden-onset abdominal pain, the urge to defecate, and rectal bleeding within 24 h (Gill & Siau, 2022; Patel et al., 2021; Plut et al., 2023; Uhlenhopp et al., 2022). Colonic ischemia is the cause of 9%–24% of all hospitalizations for acute lower GI bleeding. These symptoms resolve in 2–3 days, and the colon usually heals in 1–2 weeks. The left-sided colon is most affected and includes the watershed areas-splenic flexure and sigmoid colon.

Early colonoscopy is recommended in these patients. Treatment of ischemic colitis depends on the severity of its manifestation. Milder cases can be managed mainly with supportive care—bowel rest, intravenous fluids, antibiotics, and close observation—if there is no evidence of necrosis, gangrene, or perforation of the bowel. Surgical intervention and colonic resection are indicated in cases where imaging proves colon necrosis or for patients with right-sided colon involvement.

Lower gastrointestinal bleeding

The incidence of patients with lower gastrointestinal bleeding (GIB) in the overall COVID-19 population amounts to 0.06%. Among 633 patients with GIB, bleeding was localized in the lower GI tract in 24.8% (Ashktorab et al., 2022). The virus might lead to GIB through direct mucosal damage followed by a strong immune response and the indirect consequences of viral-induced hypoxia and coagulopathy. The most common GIB manifestations and reasons for endoscopy were melena (47.5%), hematochezia (37.8%), and anemia (21.0%) (Ashktorab et al., 2022). Diarrhea was positively associated with GIB. Strong positive correlations with death and hematochezia were found (Chen et al., 2021). GI bleeding is an independent predictor of mortality, ICU admission, and rebleeding in COVID-19 patients. In the lower GIT, cases of rectal ulceration (associated with rectal catheter insertion), colitis, and diverticular bleeding are recorded (Krejčová et al., 2022). Due to COVID-19 restrictions, most patients were managed medically. However, endoscopy is diagnostic as well as therapeutic, making it the preferred mode of intervention in these patients.

Rare gastrointestinal complications of COVID-19

Abdominal compartment syndrome

Al Armashi et al. demonstrated the catastrophic consequences of necrotizing pancreatitis complicated by sepsis and ACS in a COVID-19-positive patient. AP and COVID-19 both increase the likelihood of developing a cytokine storm, which amplifies the local necrosis, thereby causing extensive abdominal inflammation. This increases the risk of abdominal compartment syndrome (ACS), which was confirmed by increased pressure on a trans-bladder measurement of intraabdominal pressure (Al Armashi et al., 2021).

Mesenteric thrombosis

COVID-19 has been plied with coagulopathy, generally in patients with serious ailments. The laboratory results of elevated D-dimer/fibrinogen levels and prolonged prothrombin time found in patients with severe illness as well as the postmortem findings of venous thromboembolism in COVID-19 cases reaffirm the probable correlation between COVID-19 and coagulopathy. Arterial thrombosis, albeit less frequent than venous thrombosis, has also been documented in patients with COVID-19. Hanif et al. (2021) reported a case of acute intestinal ischemia secondary to superior mesenteric thrombosis in a young female patient with mild COVID-19. It reinforces the importance of monitoring for thrombotic complications via serial measurement of D-dimer, C-reactive protein (CRP) levels, platelet counts, and coagulation panels, even in patients without severe COVID-19.

Acute mesenteric ischemia

Bowel wall ischemia may be attributed to hypercoagulability induced by COVID-19, acute embolism occluding arteries and veins supplying the intestine, and nonocclusive hypoxic injury. CT angiography represents the gold standard for the definitive diagnosis of acute mesenteric ischemia (Brandi et al., 2022; Patel et al., 2021). Small bowel ischemia (46.67%) was the most prevalent abdominal CT finding, followed by ischemic colitis (37.3%) (Ojha et al., 2022). Nonocclusive mesenteric ischemia (NOMI; 67.9%), indicating microvascular involvement, was the most common pattern of bowel involvement. Bowel wall thickening/edema (50.9%) was more common than bowel hypoperfusion (20.7%). While the ileum and colon were both equally involved in bowel segments (32.07% each), SMA (24.9%), SMV (14.3%), and the spleen (12.5%) were the most involved arteries, veins, and solid organs, respectively (Ojha et al., 2022). Of the patients receiving conservative/medical management, 50% died, highlighting the high mortality that occurred without surgery (Ojha et al., 2022). Fig. 4 depicts the various causes of acute mesenteric ischemia in the event of a COVID-19 infection.

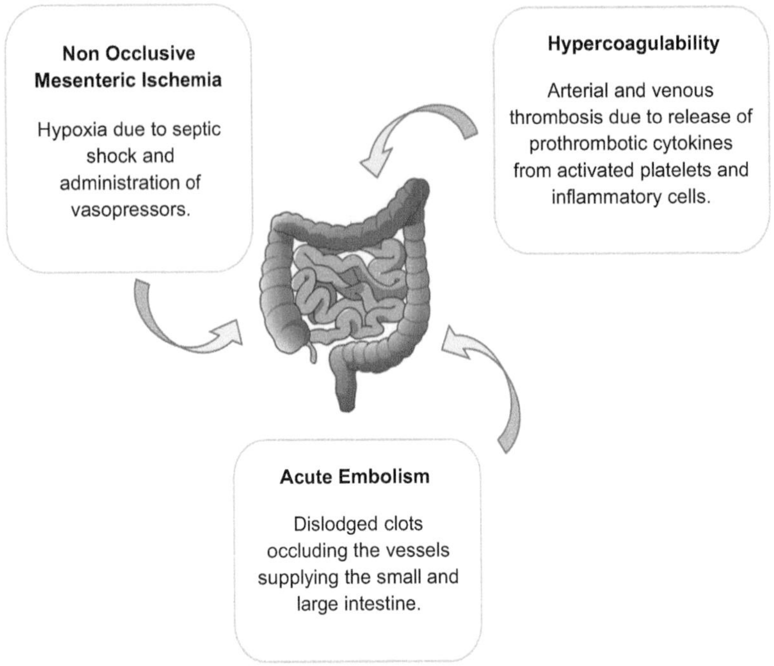

FIG. 4 Etiologies for acute mesenteric ischemia. *Adapted from Patel, S., Parikh, C., Verma, D., Sundararajan, R., Agrawal, U., Bheemisetty, N., Akku, R., Sánchez-Velazco, D., & Waleed, M. S. (2021). Bowel ischemia in COVID-19: A systematic review. In International Journal of Clinical Practice (Vol. 75, Issue 12). doi: https://doi.org/10.1111/ijcp.14930 and created by the authors.*

Acute acalculous cholecystitis

Acute acalculous cholecystitis is secondary to gallbladder stasis in critically ill patients, often requiring mechanical ventilation with severe hemodynamic instability (Abaleka et al., 2021; Alhassan et al., 2020; Kyungu et al., 2022). These risk factors are frequent in severe COVID-19 patients; hence, COVID-19 causing acalculous cholecystitis is a rare but plausible association. It can be attributed to the body's dysregulated immunological response against the virus, resulting in systemic inflammation. Abaleka et al. demonstrated the presence of SARS-CoV-2 in the bile of these patients (Abaleka et al., 2021). Management is decided based on the clinical status of individual patients. However, prompt treatment is validated as the acalculous cholecystitis in COVID-19 infection can have fatal complications such as gall bladder necrosis and perforation, leading to sepsis and even death. Hajebi et al. reported two cases in which patients did not show any common symptoms of COVID-19 infection and only presented as gangrenous acute cholecystitis. Timely diagnosis and treatment with open cholecystectomy prevented mortality in these patients (Hajebi et al., 2022).

Kyungu et al. (2022) reported a case of acute acalculous cholecystitis in a young subject with no COVID-19 infection following the Janssen COVID-19 vaccine.

Postvaccination colitis

Diarrhea symptoms have been reported in 4.61% of cases following COVID-19 vaccination (Doman et al., 2023). This is the first case report of colitis with persistent diarrhea and hypereosinophilia following the BNT162b2 mRNA COVID-19 vaccine. The mRNA-1273 vaccine is shown to have a strong CD4 T-cell response, leading to Th1 cytokine expression (Lee et al., 2022). The mRNA vaccination activates the innate immune system and may inadvertently trigger autoreactive lymphocytes, leading to autoimmune conditions such as flare-ups of IBD or lymphocytic colitis. Vaccine-induced acute colitis with venous thrombosis, thought to be due to a mechanism like that of autoimmune ulcerative colitis, was reported by Unno et al. (2022).

Post-COVID-19 syndrome of the gastrointestinal tract

Gastrointestinal symptoms are not only a part of the acute phase of infection in COVID-19, but also are seen postinfection. Long-COVID-19, now called "post-COVID-19 condition" by the World Health Organization (WHO), is defined as the persistence of COVID-19 symptoms greater than a month from the onset of COVID-19 that last at least 2 months and is unexplained by alternative diagnosis (Castanares-Zapatero et al., 2022; di Gennaro et al., 2022). To date, there is not an exact or well-understood theory as to the pathophysiology of this condition. Castanares-Zapatero et al. (2022) proposed a few theories regarding pathophysiology, including immune dysregulation, occult viral persistence, coagulation activation, and possibly autonomic nervous system damage. A study in November 2022 by di Gennaro et al. (2022) examined more than 120,000 COVID-19-positive patients and found that the incidence of post-COVID-19 conditions was around 56.9%. The study revealed that some of the most common symptoms were anxiety, sleep disturbances, and dyspnea (di Gennaro et al., 2022).

The presence of gastrointestinal symptoms, including abdominal pain, nausea, vomiting, diarrhea, constipation, dyspepsia, and irritable bowel syndrome, have been reported in patients diagnosed with long-COVID-19. A study by Choudhury et al. (2022) included a systematic review and meta-analysis of gastrointestinal symptoms in long-COVID-19. The study found that gastrointestinal symptoms were observed in 22% of patients diagnosed with long-COVID-19. The data revealed that the symptoms reported were abdominal pain, loss of taste, nausea, diarrhea, and vomiting. The most common gastrointestinal symptom in these patients was constipation (Choudhury et al., 2022). A study by Freedberg and Chang (2022) revealed that around 6 months after the initial diagnosis of COVID-19, up to 10%–25% of patients will experience gastrointestinal symptoms. The study also found that around 11% of long-COVID-19 patients rated their gastrointestinal symptoms as bothersome.

The exact pathophysiology of this phenomenon is currently still under investigation. As mentioned in a prior discussion, the gastrointestinal symptoms of acute COVID-19 infection are linked to ACE2 leading to intestinal viral infection. A theory as to why gastrointestinal symptoms appear in long-COVID-19 is the virus's prolonged shedding in the gastrointestinal tract. Because the virus remains in the gastrointestinal tract for prolonged periods, this allows the opportunity to disrupt the gut microbiome, and like acute COVID-19, there are also increased levels of *Bacteroides* and decreased levels of *Faecalibaterium prausnitzii* (Freedberg & Chang, 2022). Despite these findings, there is still more to learn and investigate on the exact mechanism of gastrointestinal symptoms in long-COVID-19 patients.

Mini dictionary of terms

Enterocytes. Cells of the intestinal lining involved in ion, water, and nutrient uptake and absorption.
Protease. A specific enzyme involved in breaking peptide bonds in proteins.
Cytokine. A broad term describing a small, secreted protein released by some immune cells that causes a specific effect on cell signaling.
Dysbiosis. Disruption of the microbiome resulting in alterations in the microbiota of the gut.
Dysgeusia. Impaired or altered sense of taste.
Area Postrema. A paired structure found in the medulla oblongata that detects chemical messengers in the circulating blood and transduces them to neural signals.
Afferents. Carries signals from sensory organs to the central nervous system.
Cholecystokinin. A hormone secreted by the duodenum that stimulates the release of bile.
Venous thromboembolism. A blood clot(s) in the veins.
Acinar cells. Cells in the pancreas that are involved in the synthesis, secretion, and storage of digestive enzymes.
Systemic inflammatory response. A defense response produced by the body when exposed to a noxious stressor.
Enteral. Nutrition delivered via the intestine.
Icterus. Jaundice.
Cholangiocytes. Cells of the bile ducts that aid in bile secretion.
Hepatocytes. Multifunctional cells of the liver.
Melena. Black or tarry stool typically seen in upper GI bleeding.
Hematochezia. Red stool typically seen in lower GI bleeding but can occur in brisk upper GI bleeding.

Summary of key points

Introduction

- Gastrointestinal symptoms are common in patients diagnosed with COVID-19.
- Symptoms that are commonly seen in COVID-19 include nausea, vomiting, anorexia, and diarrhea.

Upper GI Manifestations

- Nausea, vomiting, and anorexia are common upper gastrointestinal symptoms in COVID-19 patients.
- Treatment for upper gastrointestinal symptoms is typically supportive.

Pancreatitis

- Acute pancreatitis is an uncommon disease caused by COVID-19.

Lower GI Manifestations

- Diarrhea is the most common lower gastrointestinal symptom in patients diagnosed with COVID-19.
- The incidence of lower gastrointestinal bleeding in COVID-19 is extremely low.

Complications

- Rare complications from COVID-19 include abdominal compartment syndrome, mesenteric ischemia, and acalculous cholecystitis.
- Post-COVID-19 syndrome is a set of symptoms that is persistent a month following COVID-19 infection and lasting for at least 2 months.

Conclusions

Gastrointestinal manifestations are common in COVID-19 patients. This manuscript provides details on common and uncommon GI features of COVID-19 and its pathogenesis. The chapter also mentions the management of these complications. Further research will help better understand the pathophysiology of various manifestations of COVID-19 and its health impact, including the severe complications that can occur in some patients.

Acknowledgments

Figures were drawn by using pictures from Servier Medical Art. Servier Medical Art by Servier is licensed under a Creative Commons Attribution 3.0 Unported License (https://creativecommons.org/licenses/by/3.0/).

References

Abaleka, F. I., Nigussie, B., Bedanie, G., Mohammed, A., & Galiboglu, S. (2021). Acute acalculous cholecystitis due to COVID-19, an unusual presentation. *Cureus*. https://doi.org/10.7759/cureus.15431.

Abdulla, S., Hussain, A., Azim, D., Abduallah, E. H., Elawamy, H., Nasim, S., Kumar, S., & Naveed, H. (2020). COVID-19-induced hepatic injury: A systematic review and meta-analysis. *Cureus*. https://doi.org/10.7759/cureus.10923.

Al Armashi, A. R., Somoza-Cano, F. J., Patell, K., Al Zubaidi, A., & Ravakhah, K. (2021). COVID-19, necrotizing pancreatitis, and abdominal compartment syndrome: A perfect cytokine storm? *Cureus*. https://doi.org/10.7759/cureus.17230.

Alhassan, S. M., Iqbal, P., Fikrey, L., Mohamed Ibrahim, M. I., Qamar, M. S., Chaponda, M., & Munir, W. (2020). Post COVID 19 acute acalculous cholecystitis raising the possibility of underlying dysregulated immune response, a case report. *Annals of Medicine and Surgery*, *60*, 434–437. https://doi.org/10.1016/j.amsu.2020.11.031.

Alratrout, H., & Debroux, E. (2022). Acute right-sided ischemic colitis in a COVID-19 patient: A case report and review of the literature. *Journal of Medical Case Reports*, *16*(1). https://doi.org/10.1186/s13256-022-03276-z.

Al-Roubaie, A., & Udayasiri, R. (2022). Pancolitis post COVID-19 infection: A case report. *Cureus*. https://doi.org/10.7759/cureus.31384.

Andrews, P. L. R., Cai, W., Rudd, J. A., & Sanger, G. J. (2021). COVID-19, nausea, and vomiting. *Journal of Gastroenterology and Hepatology*, *36*(3), 646–656. https://doi.org/10.1111/jgh.15261.

Aquino-Matus, J., Uribe, M., & Chavez-Tapia, N. (2022). COVID-19: Current status in gastrointestinal, hepatic, and pancreatic diseases—A concise review. In *Vol. 7. Tropical medicine and infectious disease*. https://doi.org/10.3390/tropicalmed7080187. Issue 8.

Ashktorab, H., Russo, T., Oskrochi, G., Latella, G., Massironi, S., Luca, M., Chirumamilla, L. G., Laiyemo, A. O., & Brim, H. (2022). Clinical and endoscopic outcomes in COVID-19 patients with gastrointestinal bleeding. *Gastro Hep Advances*, *1*(4), 487–499. https://doi.org/10.1016/j.gastha.2022.02.021.

Aziz, A. A., Aziz, M. A., Saleem, M., & Haseeb Ul Rasool, M. (2022). Acute pancreatitis related to COVID-19 infection: A systematic review and analysis of data. *Cureus*. https://doi.org/10.7759/cureus.28380.

Babajide, O. I., Ogbon, E. O., Adelodun, A., Agbalajobi, O., & Ogunsesan, Y. (2022). COVID-19 and acute pancreatitis: A systematic review. In *Vol. 6. JGH open* (pp. 231–235). John Wiley and Sons Inc. https://doi.org/10.1002/jgh3.12729. Issue 4.

Beyerstedt, S., Casaro, E. B., & Rangel, É. B. (2021). COVID-19: Angiotensin-converting enzyme 2 (ACE2) expression and tissue susceptibility to SARS-CoV-2 infection. *European Journal of Clinical Microbiology & Infectious Diseases*, *40*(5), 905–919. https://doi.org/10.1007/s10096-020-04138-6.

Bozdag, A., Eroglu, Y., Sagmak Tartar, A., Gundogan Bozdag, P., & Aglamis, S. (2021). Pancreatic damage and radiological changes in patients with COVID-19. *Cureus*. https://doi.org/10.7759/cureus.14992.

Brandi, N., Ciccarese, F., Rimondi, M. R., Balacchi, C., Modolon, C., Sportoletti, C., Renzulli, M., Coppola, F., & Golfieri, R. (2022). An imaging overview of COVID-19 ARDS in ICU patients and its complications: A pictorial review. In *Vol. 12*. *Diagnostics*. https://doi.org/10.3390/diagnostics12040846. Issue 4.

Brisinda, G., Chiarello, M. M., Tropeano, G., Altieri, G., Puccioni, C., Fransvea, P., & Bianchi, V. (2022). SARS-CoV-2 and the pancreas: What do we know about acute pancreatitis in COVID-19 positive patients? *World Journal of Gastroenterology*, *28*(36), 5240–5249. https://doi.org/10.3748/wjg.v28.i36.5240.

Cappell, M. S., & Friedel, D. M. (2023). Gastrointestinal bleeding in COVID-19-infected patients. *Gastroenterology Clinics of North America*, *52*(1), 77–102. https://doi.org/10.1016/j.gtc.2022.10.004. Epub 2022 Nov 1. PMID: 36813432. PMCID: PMC9622379.

Castanares-Zapatero, D., Chalon, P., Kohn, L., Dauvrin, M., Detollenaere, J., Maertens de Noordhout, C., Primus-de Jong, C., Cleemput, I., & van den Heede, K. (2022). Pathophysiology and mechanism of long COVID: A comprehensive review. *Annals of Medicine*, *54*(1), 1473–1487. https://doi.org/10.1080/07853890.2022.2076901.

Chen, H., Tong, Z., Ma, Z., Luo, L., Tang, Y., Teng, Y., Yu, H., Meng, H., Peng, C., Zhang, Q., Zhu, T., Zhao, H., Chu, G., Li, H., Lu, H., & Qi, X. (2021). Gastrointestinal bleeding, but not other gastrointestinal symptoms, is associated with worse outcomes in COVID-19 patients. *Frontiers in Medicine*, *8*. https://doi.org/10.3389/fmed.2021.759152.

Choudhury, A., Tariq, R., Jena, A., Vesely, E. K., Singh, S., Khanna, S., & Sharma, V. (2022). Gastrointestinal manifestations of long COVID: A systematic review and meta-analysis. *Therapeutic Advances in Gastroenterology*, *15*. https://doi.org/10.1177/17562848221118403.

Chua, A. (2006). Cholecystokinin hyperresponsiveness in functional dyspepsia. *World Journal of Gastroenterology*, *12*(17), 2688. https://doi.org/10.3748/wjg.v12.i17.2688.

De, R., & Dutta, S. (2022). Role of the microbiome in the pathogenesis of COVID-19. *Frontiers in Cellular and Infection Microbiology*, *12*. https://doi.org/10.3389/fcimb.2022.736397.

de Madaria, E., & Capurso, G. (2021). COVID-19 and acute pancreatitis: Examining the causality. *Nature Reviews. Gastroenterology & Hepatology*, *18* (1), 3–4. https://doi.org/10.1038/s41575-020-00389-y.

di Gennaro, F., Belati, A., Tulone, O., Diella, L., Fiore Bavaro, D., Bonica, R., Genna, V., Smith, L., Trott, M., Bruyere, O., Mirarchi, L., Cusumano, C., Dominguez, L. J., Saracino, A., Veronese, N., & Barbagallo, M. (2022). Incidence of long COVID-19 in people with previous SARS-Cov2 infection: A systematic review and meta-analysis of 120,970 patients. *Internal and Emergency Medicine*. https://doi.org/10.1007/s11739-022-03164-w.

Doman, T., Saito, H., Tanaka, Y., Hirasawa, D., Endo, M., Togo, D., & Matsuda, T. (2023). Colitis with Hypereosinophilia following the second dose of the BNT162b2 mRNA COVID-19 vaccine: A case report with a literature review. *Internal Medicine*. https://doi.org/10.2169/internalmedicine.0518-22.

Effenberger, M., Grabherr, F., Mayr, L., Schwaerzler, J., Nairz, M., Seifert, M., Hilbe, R., Seiwald, S., Scholl-Buergi, S., Fritsche, G., Bellmann-Weiler, R., Weiss, G., Müller, T., Adolph, T. E., & Tilg, H. (2020). Faecal calprotectin indicates intestinal inflammation in COVID-19. *Gut*, *69*(8), 1543–1544. https://doi.org/10.1136/gutjnl-2020-321388.

Freedberg, D. E., & Chang, L. (2022). Gastrointestinal symptoms in COVID-19: The long and the short of it. *Current Opinion in Gastroenterology*, *38*(6), 555–561. https://doi.org/10.1097/MOG.0000000000000876.

Ghazanfar, H., Kandhi, S., Shin, D., Muthumanickam, A., Gurjar, H., Qureshi, Z. A., Shaban, M. A., Farag, M. A., Haider, A., Budhathoki, P., Bhatt, T., Ghazanfar, A., Jyala, A., & Patel, H. (2022). Impact of COVID-19 on the gastrointestinal tract: A clinical review. *Cureus*. https://doi.org/10.7759/cureus.23333.

Gill, R., & Siau, E. (2022). Colitis after SARS-CoV-2 infection. *Cureus*. https://doi.org/10.7759/cureus.26532.

Groff, A., Kavanaugh, M., Ramgobin, D., McClafferty, B., Aggarwal, C. S., Golamari, R., & Jain, R. (2021). Gastrointestinal manifestations of COVID-19: A review of what we know. *The Ochsner Journal*, *21*(2), 177–180. https://doi.org/10.31486/toj.20.0086.

Grusova, G., Bruha, R., Bircakova, B., Novak, M., Lambert, L., Michalek, P., Tomas, G., & Burgetova, A. (2021). Pancreatic injury in patients with SARS-Cov-2 (COVID-19) infection: A retrospective analysis of CT findings. *Gastroenterology Research and Practice*, *2021*. https://doi.org/10.1155/2021/5390337.

Gullu, H. F., Yavuz, O., & Gamze, A. (2022). Evaluation of patients with acute pancreatitis associated with SARS-CoV-2 (COVID-19); the importance of lipase/lymphocyte ratio in predicting mortality. *Bratislava Medical Journal*, *123*(6), 428–434. https://doi.org/10.4149/BLL_2022_066.

Guo, M., Tao, W., Flavell, R. A., & Zhu, S. (2021). Potential intestinal infection and faecal–oral transmission of SARS-CoV-2. *Nature Reviews Gastroenterology & Hepatology*, *18*(4), 269–283. https://doi.org/10.1038/s41575-021-00416-6.

Gupta, Y. K., Chugh, A., Bhandari, P., & Seth, S. D. (1989). Effect of intracerebroventricular administration of angiotensin II on emetic reflex in dogs. *Indian Journal of Experimental Biology*, *27*(6), 576–577.

Hajebi, R., Habibi, P., Maroufi, S. F., Bahreini, M., & Miratashi Yazdi, S. A. (2022). COVID-19 patients presenting with gangrenous acalculous cholecystitis: Report of two cases. *Annals of Medicine and Surgery*, *76*. https://doi.org/10.1016/j.amsu.2022.103534.

Hanif, M., Ahmad, Z., Khan, A. W., Naz, S., & Sundas, F. (2021). COVID-19-induced mesenteric thrombosis. *Cureus*. https://doi.org/10.7759/cureus.12953.

Khreefa, Z., Barbier, M. T., Koksal, A. R., Love, G., & del Valle, L. (2023). Pathogenesis and mechanisms of SARS-CoV-2 infection in the intestine, liver, and pancreas. *Cell*, *12*(2), 262. https://doi.org/10.3390/cells12020262.

Krejčová, I., Berková, A., Kvasnicová, L., Vlček, P., Veverková, L., Penka, I., Zoufalý, D., & Červeňák, V. (2022). Ischemic colitis in a patient with severe COVID-19 pneumonia. *Case Reports in Gastroenterology*, *16*(2), 526–534. https://doi.org/10.1159/000525840.

Kyungu, F. M., Katumba, A. M., Kamwira, H. L., Mayikuli, A. V., Mala, A., Banza, M. I., Manirou, H. M., Tosali, F. B., & Mulenga, P. C. (2022). Acute acalculous cholecystitis following COVID-19 vaccination: A case report. *The Pan African Medical Journal*, *41*. https://doi.org/10.11604/pamj.2022.41.291.33047.

Lee, P., Wei, M. T., Gubatan, J., Forgó, E., Berry, G. J., Verma, R., & Friedland, S. (2022). De novo diagnosis of lymphocytic colitis after SARS-CoV-2 vaccination. *ACG Case Reports Journal*, *9*(9), e00849. https://doi.org/10.14309/crj.0000000000000849.

Liu, F., Long, X., Zhang, B., Zhang, W., Chen, X., & Zhang, Z. (2020). ACE2 expression in pancreas may cause pancreatic damage after SARS-CoV-2 infection. In *Vol. 18. Clinical gastroenterology and hepatology* (pp. 2128–2130.e2). W.B. Saunders. https://doi.org/10.1016/j.cgh.2020.04.040. Issue 9.

Liu, Y., Wu, Q., Wan, D., He, H., Lin, H., Wang, K., Que, G., Wang, Y., Chen, Y., Tang, X., Wu, L., & Yang, X. (2021). Expression and possible significance of ACE2 in the human liver, esophagus, stomach, and colon. *Evidence-based Complementary and Alternative Medicine*, *2021*, 1–9. https://doi.org/10.1155/2021/6949902.

Maalouf, R. G., Kozhaya, K., & el Zakhem, A. (2021). SARS-CoV-2 induced necrotizing pancreatitis. In *Vol. 156. Medicina clinica* (pp. 629–630). https://doi.org/10.1016/j.medcli.2021.01.005. Issue 12.

Mańkowska-Wierzbicka, D., Zuraszek, J., Wierzbicka, A., Gabryel, M., Mahadea, D., Baturo, A., Zakerska-Banaszak, O., Slomski, R., Skrzypczak-Zielinska, M., & Dobrowolska, A. (2023). Alterations in gut microbiota composition in patients with COVID-19: A pilot study of whole hypervariable 16S rRNA gene sequencing. *Biomedicine*, *11*(2), 367. https://doi.org/10.3390/biomedicines11020367.

Mao, R., Qiu, Y., He, J. S., Tan, J. Y., Li, X. H., Liang, J., Shen, J., Zhu, L. R., Chen, Y., Iacucci, M., Ng, S. C., Ghosh, S., & Chen, M. H. (2020). Manifestations and prognosis of gastrointestinal and liver involvement in patients with COVID-19: A systematic review and meta-analysis. *The Lancet Gastroenterology and Hepatology*, *5*(7), 667–678. https://doi.org/10.1016/S2468-1253(20)30126-6.

Martin, T. A., Wan, D. W., Hajifathalian, K., Tewani, S., Shah, S. L., Mehta, A., Kaplan, A., Ghosh, G., Choi, A. J., Krisko, T. I., Fortune, B. E., Crawford, C., & Sharaiha, R. Z. (2020). Gastrointestinal bleeding in patients with coronavirus disease 2019: A matched case-control study. *American Journal of Gastroenterology*, *115*(10), 1609–1616. https://doi.org/10.14309/ajg.0000000000000805.

McConnell, M. J., Kondo, R., Kawaguchi, N., & Iwakiri, Y. (2022). COVID-19 and liver injury: Role of inflammatory endotheliopathy, platelet dysfunction, and thrombosis. *Hepatology Communications*, *6*(2), 255–269. https://doi.org/10.1002/hep4.1843.

Memon, B., & Abdelalim, E. M. (2021). ACE2 function in the pancreatic islet: Implications for relationship between SARS-CoV-2 and diabetes. In *Vol. 233. Acta physiologica* John Wiley and Sons Inc. https://doi.org/10.1111/apha.13733. Issue 4.

Natarajan, A., Zlitni, S., Brooks, E. F., Vance, S. E., Dahlen, A., Hedlin, H., Park, R. M., Han, A., Schmidtke, D. T., Verma, R., Jacobson, K. B., Parsonnet, J., Bonilla, H. F., Singh, U., Pinsky, B. A., Andrews, J. R., Jagannathan, P., & Bhatt, A. S. (2022). Gastrointestinal symptoms and fecal shedding of SARS-CoV-2 RNA suggest prolonged gastrointestinal infection. *Med*, *3*(6), 371–387.e9. https://doi.org/10.1016/j.medj.2022.04.001.

Negro, A., Villa, G., Rolandi, S., Lucchini, A., & Bambi, S. (2022). Gastrointestinal bleeding in COVID-19 patients: A rapid review. *Gastroenterology Nursing: The Official Journal of the Society of Gastroenterology Nurses and Associates*, *45*(4), 267–275. https://doi.org/10.1097/SGA.0000000000000676.

Neoptolemos, J. P., & Greenhalf, W. (2022). Linking COVID-19 and acute pancreatitis through the pathogenic effects of the SARS-CoV-2 S protein subunit 1 on pancreatic stellate cells and macrophages. *Function*, *3*(2). https://doi.org/10.1093/function/zqac009.

Ojha, V., Mani, A., Mukherjee, A., Kumar, S., & Jagia, P. (2022). Mesenteric ischemia in patients with COVID-19: An updated systematic review of abdominal CT findings in 75 patients. *Abdominal Radiology*, *47*(5), 1565–1602. https://doi.org/10.1007/s00261-021-03337-9.

Pandanaboyana, S., Moir, J., Leeds, J. S., Oppong, K., Kanwar, A., Marzouk, A., Belgaumkar, A., Gupta, A., Siriwardena, A. K., Haque, A. R., Awan, A., Balakrishnan, A., Rawashdeh, A., Ivanov, B., Parmar, C., Halloran, M., Caruana, C., Borg, C. M., Gomez, D., & Nayar, M. (2021). SARS-CoV-2 infection in acute pancreatitis increases disease severity and 30-day mortality: COVID PAN collaborative study. *Gut*, *70*(6), 1061–1069. https://doi.org/10.1136/gutjnl-2020-323364.

Patel, S., Parikh, C., Verma, D., Sundararajan, R., Agrawal, U., Bheemisetty, N., Akku, R., Sánchez-Velazco, D., & Waleed, M. S. (2021). Bowel ischemia in COVID-19: A systematic review. *International Journal of Clinical Practice*, *75*(12). https://doi.org/10.1111/ijcp.14930.

Penninger, J. M., Grant, M. B., & Sung, J. J. Y. (2021). The role of angiotensin converting enzyme 2 in modulating gut microbiota, intestinal inflammation, and coronavirus infection. *Gastroenterology*, *160*(1), 39–46. https://doi.org/10.1053/j.gastro.2020.07.067.

Plut, S., Hanzel, J., & Gavric, A. (2023). COVID-19 associated colitis. *Gastrointestinal Endoscopy*. https://doi.org/10.1016/j.gie.2023.01.046.

Puli, S., Baig, M., & Walayat, S. (2020). Gastrointestinal symptoms and elevation in liver enzymes in COVID-19 infection: A systematic review and meta-analysis. *Cureus*. https://doi.org/10.7759/cureus.9999.

Radivojevic, A., Abu Jad, A. A., Ravanavena, A., Ravindra, C., Igweonu-Nwakile, E. O., Ali, S., Paul, S., Yakkali, S., Teresa Selvin, S., Thomas, S., Bikeyeva, V., Abdullah, A., & Balani, P. (2022). A systematic review of SARS-CoV-2-associated hepatic dysfunction and the impact on the clinical outcome of COVID-19. *Cureus*. https://doi.org/10.7759/cureus.26852.

Raman, B., Bluemke, D. A., Lüscher, T. F., & Neubauer, S. (2022). Long COVID: Post-acute sequelae of COVID-19 with a cardiovascular focus. *European Heart Journal*, *43*(11), 1157–1172. https://doi.org/10.1093/eurheartj/ehac031.

Redd, W. D., Zhou, J. C., Hathorn, K. E., McCarty, T. R., Bazarbashi, A. N., Thompson, C. C., Shen, L., & Chan, W. W. (2020). Prevalence and characteristics of gastrointestinal symptoms in patients with severe acute respiratory syndrome coronavirus 2 infection in the United States: A multicenter cohort study. *Gastroenterology*, *159*(2), 765–767.e2. https://doi.org/10.1053/j.gastro.2020.04.045.

Samanta, J., Gupta, R., Singh, M. P., Patnaik, I., Kumar, A., & Kochhar, R. (2020). Coronavirus disease 2019 and the pancreas. In *Vol. 20. Pancreatology* (pp. 1567–1575). Elsevier B.V. https://doi.org/10.1016/j.pan.2020.10.035. Issue 8.

Sinagra, E., Shahini, E., Crispino, F., Macaione, I., Guarnotta, V., Marasà, M., Testai, S., Pallio, S., Albano, D., Facciorusso, A., & Maida, M. (2022). COVID-19 and the pancreas: A narrative review. *Life*, *12*(9). https://doi.org/10.3390/life12091292.

Song, J., Patel, J., Khatri, R., Nadpara, N., Malik, Z., & Parkman, H. P. (2022). Gastrointestinal symptoms in patients hospitalized with COVID-19. *Medicine*, *101*(25), e29374. https://doi.org/10.1097/MD.0000000000029374.

Su, Y. J., & Chen, T. H. (2022). Surgical intervention for acute pancreatitis in the COVID-19 era. *World Journal of Clinical Cases*, *10*(31), 11292–11298. https://doi.org/10.12998/wjcc.v10.i31.11292.

Tang, Q., Gao, L., Tong, Z., & Li, W. (2022). Hyperlipidemia, COVID-19 and acute pancreatitis: A tale of three entities. *The American Journal of the Medical Sciences*, *364*(3), 257–263. https://doi.org/10.1016/j.amjms.2022.03.007.

Thursby, E., & Juge, N. (2017). Introduction to the human gut microbiota. *Biochemical Journal*, *474*(11), 1823–1836. https://doi.org/10.1042/BCJ20160510.

Uhlenhopp, D. J., Ramachandran, R., Then, E., Parvataneni, S., Grantham, T., & Gaduputi, V. (2022). COVID-19-associated ischemic colitis: A rare manifestation of COVID-19 infection—Case report and review. *Journal of Investigative Medicine High Impact Case Reports*, *10*. https://doi.org/10.1177/23247096211065625.

Unno, S., Yoshizawa, Y., & Hosoda, Y. (2022). Acute entire colitis and vein thrombosis after COVID-19 mRNA-1273 vaccination. In *Vol. 34. Digestive endoscopy* (p. 1069). John Wiley and Sons Inc. https://doi.org/10.1111/den.14307. Issue 5.

Varshney, R., Bansal, N., Khanduri, A., Gupta, J., & Gupta, R. (2021). Colonic gangrene: A sequela of coronavirus disease 2019. *Cureus*. https://doi.org/10.7759/cureus.14687.

Vujčić, I. (2023). Outcomes of COVID-19 among patients with liver disease. *World Journal of Gastroenterology*. https://doi.org/10.3748/wjg.v29.i5.815.

Wang, Y., Li, Y., Zhang, Y., Liu, Y., & Liu, Y. (2022). Are gastrointestinal symptoms associated with higher risk of mortality in COVID-19 patients? A systematic review and meta-analysis. *BMC Gastroenterology*, *22*(1), 106. https://doi.org/10.1186/s12876-022-02132-0.

Wang, M.-K., Yue, H.-Y., Cai, J., Zhai, Y.-J., Peng, J.-H., Hui, J.-F., Hou, D.-Y., Li, W.-P., & Yang, J.-S. (2021). COVID-19 and the digestive system: A comprehensive review. *World Journal of Clinical Cases*, *9*(16), 3796–3813. https://doi.org/10.12998/wjcc.v9.i16.3796.

Wijarnpreecha, K., Ungprasert, P., Panjawatanan, P., Harnois, D. M., Zaver, H. B., Ahmed, A., & Kim, D. (2021). COVID-19 and liver injury: A meta-analysis. *European Journal of Gastroenterology & Hepatology*, *33*(7), 990–995. https://doi.org/10.1097/MEG.0000000000001817.

Xiao, L., Sakagami, H., & Miwa, N. (2020). ACE2: The key molecule for understanding the pathophysiology of severe and critical conditions of COVID-19: Demon or angel? *Viruses*, *12*(5), 491. https://doi.org/10.3390/v12050491.

Yao, Y., Liu, Z.-J., Zhang, Y.-K., & Sun, H.-J. (2022). Mechanism and potential treatments for gastrointestinal dysfunction in patients with COVID-19. *World Journal of Gastroenterology*, *28*(48), 6811–6826. https://doi.org/10.3748/wjg.v28.i48.6811.

Zeng, W., Qi, K., Ye, M., Zheng, L., Liu, X., Hu, S., Zhang, W., Tang, W., Xu, J., Yu, D., & Wei, Y. (2022). Gastrointestinal symptoms are associated with severity of coronavirus disease 2019: A systematic review and meta-analysis. *European Journal of Gastroenterology and Hepatology*, *34*(2), 168–176. https://doi.org/10.1097/MEG.0000000000002072.

Zhang, J., Garrett, S., & Sun, J. (2021). Gastrointestinal symptoms, pathophysiology, and treatment in COVID-19. *Genes & Diseases*, *8*(4), 385–400. https://doi.org/10.1016/j.gendis.2020.08.013.

Zhang, T., Liu, D., Tian, D., & Xia, L. (2021). The roles of nausea and vomiting in COVID-19: Did we miss something? *Journal of Microbiology, Immunology and Infection*, *54*(4), 541–546. https://doi.org/10.1016/j.jmii.2020.10.005.

Zhang, H., Shao, B., Dang, Q., Chen, Z., Zhou, Q., Luo, H., Yuan, W., & Sun, Z. (2021). Pathogenesis and mechanism of gastrointestinal infection with COVID-19. *Frontiers in Immunology*, *12*. https://doi.org/10.3389/fimmu.2021.674074.

Zuo, T., Zhang, F., Lui, G. C. Y., Yeoh, Y. K., Li, A. Y. L., Zhan, H., Wan, Y., Chung, A. C. K., Cheung, C. P., Chen, N., Lai, C. K. C., Chen, Z., Tso, E. Y. K., Fung, K. S. C., Chan, V., Ling, L., Joynt, G., Hui, D. S. C., Chan, F. K. L., & Ng, S. C. (2020). Alterations in gut microbiota of patients with COVID-19 during time of hospitalization. *Gastroenterology*, *159*(3), 944–955.e8. https://doi.org/10.1053/j.gastro.2020.05.048.

Chapter 49

The COVID-19 survivors: Impact on skeletal muscle strength

Renata Gonçalves Mendes[a], Alessandro Domingues Heubel[a], Naiara Tais Leonardi[a], Stephanie Nogueira Linares[a], Vanessa Teixeira do Amaral[b], and Emmanuel Gomes Ciolac[b]

[a]*Cardiopulmonary Physiotherapy Laboratory, Physiotherapy Department, Federal University of São Carlos, São Paulo, Brazil,* [b]*Exercise and Chronic Disease Research Laboratory, Department of Physical Education, School of Sciences, São Paulo State University, São Paulo, Brazil*

Abbreviations

1mSTS	1-min sit to stand test
1RM	1-repetition maximum
30sSTS	30-s sit to stand test
5CRT	5-chair raise test
5STS	5-sit to stand test
6MWD	6-min walk distance
6MWT	6-min walk test
ACE2	angiotensin-converting enzyme 2
AT	aerobic training
ATP	adenosine triphosphate
BBS	Berg balance scale
CG	control group
HGSc	handgrip strength
HRmax	heart rate maximal
HRR	heart rate reserve
ICU	intensive care unit
IG	intervention group
ILD	interstitial lung disease
IME	inspiratory muscle endurance
IMS	inspiratory muscle strength
IMT	inspiratory muscle training
IMV	invasive mechanical ventilation
LICT	light intensity continuous training
LLS	lower limb strength
MEP	maximal expiratory pressure
MIP	maximal inspiratory pressure
MIVT	moderate intensity variable training
n	number
PCS	postacute COVID-19 syndrome
PMS	peripheral muscle strength
PR	pulmonary rehabilitation
RMS	respiratory muscle strength
RMT	respiratory muscle training
RPE	rating of perceived exertion
RT	resistance training
SM	skeletal muscle
SMIP	sustained maximal inspiratory pressure
STS	sit-to stand
TMPRSS2	Transmembrane serine protease 2

Management, Body Systems, and Case Studies in COVID-19. https://doi.org/10.1016/B978-0-443-18703-2.00025-6
Copyright © 2024 Elsevier Inc. All rights are reserved, including those for text and data mining, AI training, and similar technologies.

COVID-19, skeletal muscle impairment and consequences

Muscle damage resulting from SARS-CoV-2 infection has been suggested to occur in two ways: direct and/or indirect (dos Santos et al., 2022). The direct damage is explained by the infiltration of skeletal muscle tissue by SARS-CoV-2, which would be facilitated through viral binding to angiotensin-converting enzyme 2 (ACE2) and/or transmembrane serine protease 2 (TMPRSS2), both expressed on the surface of muscle cells (dos Santos et al., 2022). Postmortem evidence has shown the presence of viral particles in the diaphragm muscle of critically ill COVID-19 patients (Shi et al., 2021). Although direct invasion consequences are unclear, SARS-CoV-2 appears to use the cellular machinery for viral replication, which can downregulate cellular activities, inducing muscle cell injury and death (Silva et al., 2022).

However, the SARS-CoV-2 direct muscle cell invasion hypothesis is not supported by other studies. In a recent case–control study, autopsies of 35 patients who died after COVID-19 demonstrated a negative immunohistochemical analysis for SARS-CoV-2 in all psoas muscle samples (Suh et al., 2021). In another study with 43 nonsurviving COVID-19 patients, although viral RNA was detected in 16% and 5% of the quadriceps and deltoid muscles, respectively, antibodies for the spike protein and analysis with electron microscopy failed to detect SARS-CoV-2 in muscle samples (Aschman et al., 2021).

Although direct infection to muscle cells is not clear, consistent histopathological evidence demonstrates muscle damage resulting from SARS-CoV-2 infection (Aschman et al., 2021; Suh et al., 2021). These findings are mainly attributed to an exacerbated systemic inflammation called a cytokine storm, which occurs due to a dysregulated immune response resulting in the excessive release of cytokines and proinflammatory mediators (Silva et al., 2022; Soares et al., 2022). In muscle tissue, a hyperinflammatory state leads to increased muscle proteolysis, decreased protein synthesis, and, consequently, muscle dysfunction (Silva et al., 2022). Other indirect mechanisms of muscle damage associated with COVID-19 include vascular dysfunction, motoneuron degeneration, hypoxemia, effects of medications, immobility (muscle disuse), and malnutrition (Fig. 1) (Silva et al., 2022; Soares et al., 2022).

As a consequence of muscle damage, COVID-19 can lead to several debilitating musculoskeletal symptoms including myalgia (muscle pain), muscle weakness, and fatigue (dos Santos et al., 2022; Mills et al., 2022). Generally, symptoms occur during the first few days of infection and, in some cases, may begin even before the respiratory manifestations (e.g., dry cough, nasal congestion, sore throat, and dyspnea). Although most patients have complete recovery within a few days or weeks, other patients may have muscle symptoms for a longer period of time. These persistent muscular symptoms are among those that characterize the postacute COVID-19 syndrome (or long-COVID-19), which is known for a wide range of symptoms that last for months after acute infection (Montani et al., 2022).

Muscle pain prevalence after COVID-19 has been reported in several studies. For example, among 1077 COVID-19 patients who were discharged from the hospital, 57.2% presented persistent musculoskeletal pain after 6 months (Evans et al., 2021). In another cohort including 331 patients who were discharged from the hospital, 166 (~50%) reported at least one site of musculoskeletal pain after 45 days (Mills et al., 2022). Ranking from highest to lowest prevalence, the distribution of musculoskeletal pain sites was lower limb (27%), lumbar spine (19%), no location specified (17%), multisite (13%), upper limb (13%), thoracic spine (7%), and neck (4%) (Mills et al., 2022). Although the mechanisms causing COVID-19-associated muscle pain are uncertain, some potential risk factors have already been identified and include female sex, a history of musculoskeletal pain, the presence of myalgia and headache as COVID-19 symptoms at the acute phase, and length of hospital stay (Fernández-de-Las-Peñas et al., 2022).

Fatigue is also a symptom commonly persisting after COVID-19. In a cohort study including 143 hospitalized patients with COVID-19, more than half (~53%) reported fatigue after an average of 60 days from symptom onset (Carfì et al., 2020). In a larger follow-up study including 1733 COVID-19 survivors, fatigue and muscle weakness were the most prevalent symptoms 6 months after hospital discharge, being reported by 63% of the patients (Huang et al., 2021). This marked occurrence of post-COVID-19 fatigue was also recently verified in a systematic review including 127,117 patients, where the prevalence of chronic fatigue syndrome was 45.2% 4 weeks after the onset of COVID-19 symptoms (Salari et al., 2022).

Definitely, post-COVID-19 fatigue is an extremely disabling symptom with a profound negative impact on patients' quality of life (Noujaim et al., 2022). As a possible trigger mechanism, skeletal muscle weakness has been suggested as a determinant for post-COVID-19 fatigue (Azzolino & Cesari, 2022). For example, findings from 16 COVID-19 survivors with persistent fatigue showed muscle weakness in 50% and myopathic electromyography in 75%, with all patients showing histological changes including muscle fiber atrophy (38%), mitochondrial changes (62%), inflammation (62%), and capillary damage (75%) (Hejbøl et al., 2022). Electrophysiological analysis in 20 patients with long-COVID-19 revealed fatigue and abnormalities on quantitative electromyography in the whole sample, suggesting a role for myopathic changes in symptom persistence (Agergaard et al., 2021).

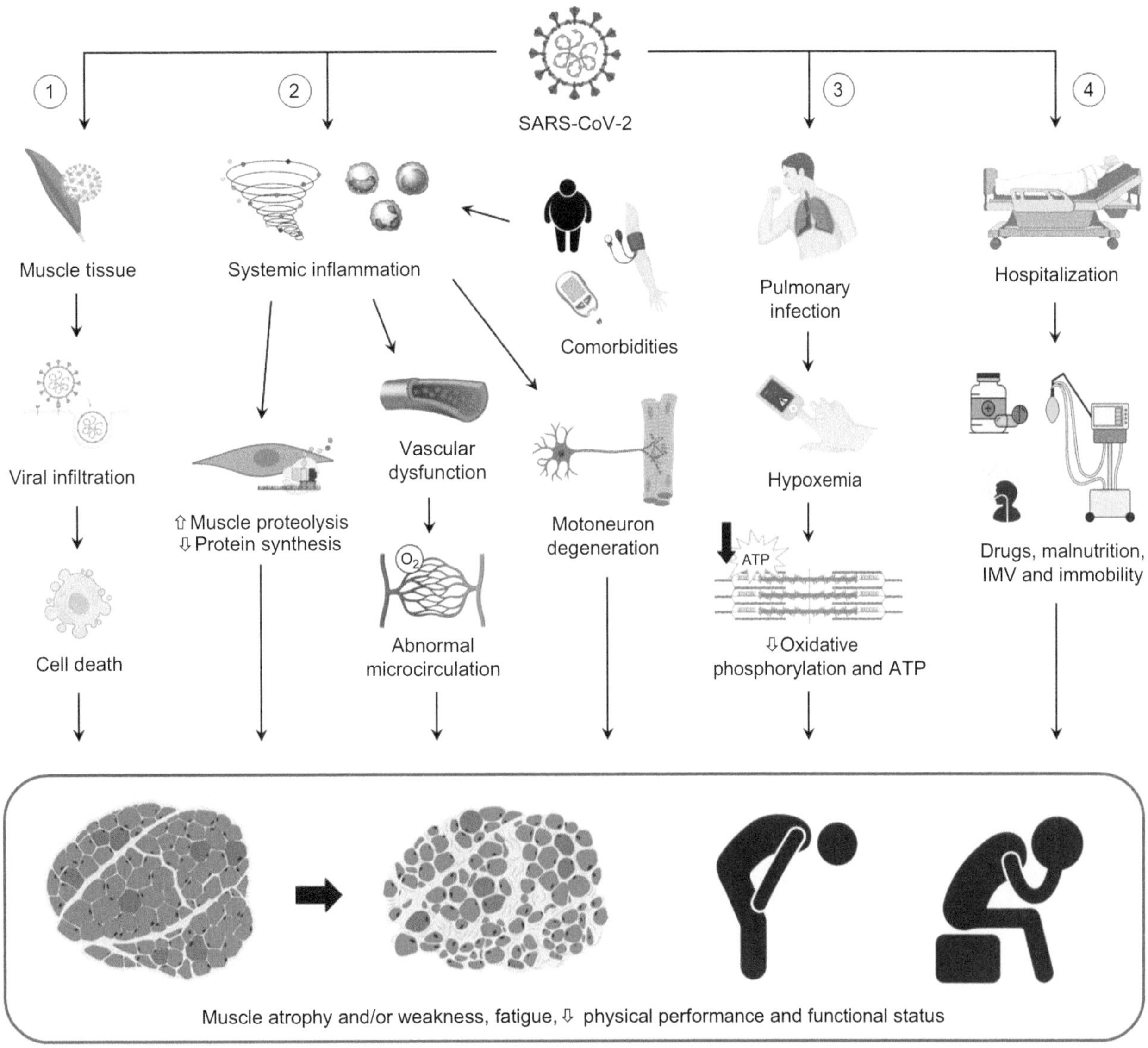

FIG. 1 Direct (1) and indirect (2–4) mechanisms that contribute to muscle weakness, poor physical performance, and impaired functional status in patients with COVID-19. This figure shows the mechanisms of muscle damage in COVID-19 patients: (1) SARS-CoV-2 directly infects the muscle cell and uses the cell machinery for viral replication, resulting in cell death and tissue damage. (2) Systemic inflammation, which can be potentiated by comorbidities (e.g., obesity, diabetes, and hypertension), increases circulating proinflammatory mediators, which are transported by blood to other organs and systems. In the muscle tissue, inflammatory cytokines increase muscle proteolysis and decrease protein synthesis. In the vascular system, the proinflammatory state leads to endothelial damage and dysfunction, which results in a persistent reduction of blood flow through microcirculation, contributing to limiting the supply of nutrients and oxygen to the musculoskeletal tissue. In the peripheral nervous system, antibodies attack nerves, causing damage to the axon and/or myelin, leading to motoneuron degeneration and muscle hypotrophy. (3) Viral pneumonia and diffuse alveolar damage lead to hypoxemia, which negatively interferes with myogenesis, muscle metabolism, and energy production. (4) Hospitalization for COVID-19 can cause muscle damage as a result of noxious stimuli. Corticosteroids and other drugs used in COVID-19 treatment have side effects on the musculoskeletal system, such as myalgia, muscle weakness, and atrophy. Mechanical ventilation, the use of sedatives and neuromuscular blockers, and immobility can lead to neuromyopathy and muscle atrophy. Nutritional deficits during illness can accentuate muscle catabolism, resulting in greater muscle loss and weakness. ATP, adenosine triphosphate; IMV, invasive mechanical ventilation.

It is very important to point out that skeletal muscle dysfunction is a determining factor for exercise intolerance and reduced physical performance in other cardiopulmonary diseases (Gosker et al., 2000). This condition appears to be similar in COVID-19, as there is increasing evidence of muscle damage and musculoskeletal sequelae resulting from the disease (Silva et al., 2022; Soares et al., 2022). In addition to peripheral muscle impairment, respiratory muscle dysfunction is also a muscle damage consequence of COVID-19 (Severin et al., 2020, 2022). However, the importance of respiratory muscle dysfunction is often overlooked, given that its screening is seldom performed in clinical practice. The weakness of inspiratory muscles, especially the diaphragm, directly impairs pulmonary ventilation. In addition, post-COVID-19 patients may have impaired lung function, with a predominance of restrictive disorder as a result of interstitial disease and pulmonary fibrosis (Amin et al., 2022; Torres-Castro et al., 2021). Consequently, both these conditions impair respiratory mechanics, decreasing lung compliance and increasing the work of breathing (Severin et al., 2022). As a result, post-COVID-19 patients may experience an even more significant decrease in exercise capacity and physical performance.

It is important to note that although muscle impairments are frequently reported as a post-COVID-19 symptom, conditions were initially reported from patients' self-report (Huang et al., 2021). However, recent studies have been concerned with assessing muscle strength by using objective measurements. From this perspective, in the next topic we present a synthesis of the most recent evidence focused on assessing peripheral and respiratory muscle strength in patients affected by COVID-19.

Current evidence of muscle strength impairment

Here, we summarize the current findings regarding the impact of COVID-19 on peripheral and respiratory muscle strength. For this purpose, we considered maximal respiratory pressures and handgrip strength, which are measurements widely used in clinical practice to assess respiratory and peripheral muscle strength, respectively.

Respiratory muscle strength (RMS)

Maximal inspiratory and expiratory pressures (MIP and MEP, respectively) were used to assess RMS in six observational studies that investigated patients at a post-COVID-19 condition (Fig. 2A). In a prospective study, including 57 patients, 49% and 23% presented MIP and MEP values, respectively, below 80% of predicted after 30 days of hospital discharge (Huang et al., 2020). MIP and MEP were not different between severe and nonsevere subgroups (Huang et al., 2020). MIP and MEP values below the lower limit of normality were also found in 41% and 96% of individuals, respectively, in another prospective cohort that evaluated 242 COVID-19 survivors at 45 days after hospital discharge (Mancuzo et al., 2021). Additionally, inspiratory muscle strength (IMS) was significantly lower in the subgroup with greater disease severity, characterized by patients who were admitted to the intensive care unit (ICU) and received invasive mechanical ventilation (IMV) (Mancuzo et al., 2021).

In a larger 3-month cohort study with 530 COVID-19 survivors, MIP and MEP were lower than 80% of predicted values in 22% and 76%, respectively (González-Islas et al., 2022). However, RMS was not different between patients who did or did not require mechanical ventilation support (González-Islas et al., 2022). In a small cohort with 50 COVID-19 survivors discharged from the ICU, 48% showed impaired IMS (MIP < 70% predicted) at 3 months, which was reduced to 24% at 6 months, suggesting some natural and gradual recovery over time (Núñez-Seisdedos et al., 2022). In a cross-sectional study assessing 67 patients convalescing from COVID-19, 5 months after symptom onset, reduced MIP (values below sex- and age-specific cutoffs) was found in 88% of symptomatic patients (Hennigs et al., 2022). Surprisingly, MIP impairment was present in 65% of patients who were not hospitalized, suggesting that inspiratory muscle weakness is also a frequent finding after mild to moderate COVID-19 (Hennigs et al., 2022). Finally, in a cohort study assessing the influence of previous levels of physical activity on cardiorespiratory and functional outcomes in 61 survivors from COVID-19 requiring hospitalization, most patients showed impaired RMS (MIP and MEP below 80% of predicted levels) 30–45 days after hospital discharge (Viana et al., 2022). Interestingly, relative MEP (% of predicted) was 26% higher in those physically active prior to hospitalization, suggesting that previous levels of physical activity may be a predictor of RMS (Viana et al., 2022).

Peripheral muscle strength (PMS)

Handgrip strength (HGS) was assessed in six studies with patients in different COVID-19 conditions, including inpatients and outpatients (Fig. 2B). Lower levels of HGS at hospital admission were found in 87% of 118 patients with acute COVID-19 (Pucci et al., 2022). In addition, weight-normalized HGS was an independent and inverse predictor of poor outcomes (death and/or endotracheal intubation) (Pucci et al., 2022). In another prospective cohort assessing HGS at hospital admission in 163 patients, 40% of patients had lower levels of muscle strength (cut-offs: 16 and 27 kg for women and men, respectively) (Piotrowicz et al., 2022). In the same study, analysis of outcomes showed that reduced HGS was associated with a longer hospital stay, but not with mortality rates (Piotrowicz et al., 2022).

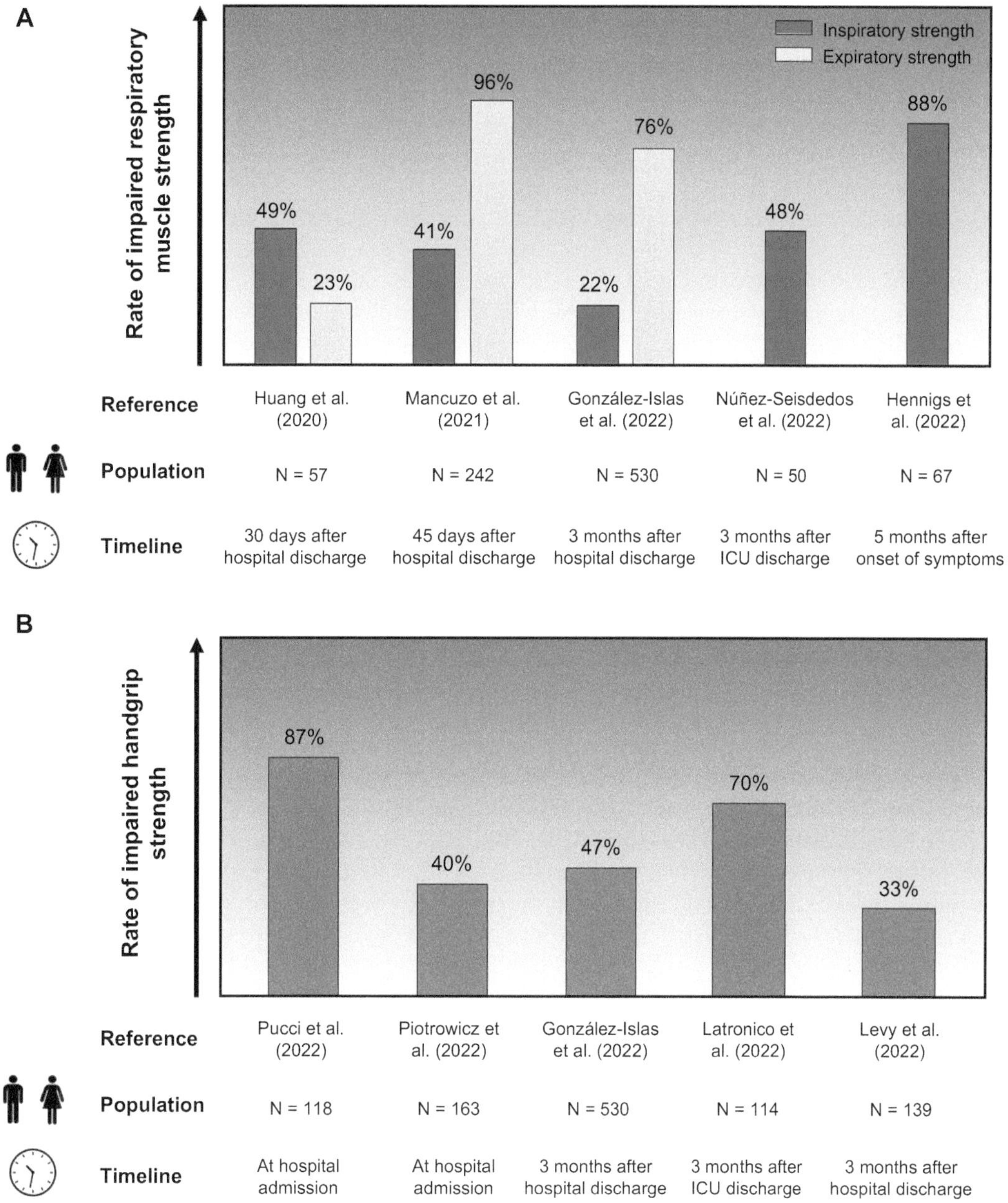

FIG. 2 Rates of impaired muscle strength in different studies. This figure shows results of studies showing rates of COVID-19 patients with impaired respiratory muscle strength (A) and peripheral muscle strength (B).

Another four studies also reported a reduced HGS in COVID-19 patients after months of follow-up. A larger cohort study including 530 COVID-19 patients showed low muscle strength (HGS < 27 kg for men and < 16 kg for women) in 47% of patients 3 months after hospital discharge (González-Islas et al., 2022). For those, the low HGS prevalence was higher in patients who received IMV compared to those who did not (58% vs. 29%, respectively) (González-Islas et al., 2022). In 114 patients discharged from the ICU, reduced HGS (<80% of predicted value) was found in 70% at 3 months of follow-up, with significant improvement over time at 6 and 12 months (Latronico et al., 2022). In a prospective study with 139 COVID-19 survivors, the prevalence of low HGS (<27 kg for men and <16 kg for women) and sarcopenia after 3 months of hospital discharge was 33% and 16%, respectively (Levy et al., 2022). However, these impairments were largely reversed at 6 months, suggesting a natural slow recovery of peripheral muscle strength over time (Levy et al., 2022). Finally, different from what occurred with RMS, patients who were higher, intermediate, or lower physically active before COVID-19 hospital admission presented similar HGS 30–45 days after discharge, suggesting no influence of previous levels of physical activity on HGS after COVID-19 hospitalization (Viana et al., 2022).

Evidence of other consequences of muscle weakness

Muscle weakness in COVID-19 patients is associated with other consequences, such as pulmonary function impairments, hospitalization, use of mechanical ventilation, social isolation, and perhaps activity limitations. For example, walking capacity limitations, assessed by a 6-min walking test (6MWT), below 80% of predicted was found in 48% of patients with severe COVID-19 at 3 months of discharge (van Gassel et al., 2021). A poor 6MW distance (6MWD) was also found in 50 mechanically ventilated COVID-19 survivors after 3 months (71% (66–82) of predicted), which improved 6 months (82% (71.5–90) of predicted) after hospital discharge (Núñez-Seisdedos et al., 2022). Walking limitations were present up to 12 months after discharge, in which 17% of the patients were not able to reach predicted 6MWT distances (Berentschot et al., 2022), suggesting a lower muscle strength/RMS association. In this context, a prospective cohort assessed 46 patients (69% male) at 3 months after hospital discharge. For those with impaired physical performance (6MWD < 80% predicted), there was a lower HGS (% of predicted, 73.7 ± 17.3 vs. 87.4 ± 15.7, $P = 0.008$) and a higher intermuscular adipose tissue area (cm^2, 3.3 (1.4, 4.2) vs. 2.1 (1.4, 2.6), $P = 0.037$) compared to a normal group (6MWD > 80%) (van Gassel et al., 2021). In addition, a strong positive correlation between 6MWD and MIP was found in 50 invasively ventilated COVID-19 survivors (61 (52–68) years, 52% male) at 3 months (rho = 0.657, $P < 0.01$) and 6 months (rho = 0.622, $P < 0.01$) of follow-up, suggesting that respiratory muscle weakness may be related to a reduced walking capacity (Núñez-Seisdedos et al., 2022).

Reduced muscle strength after COVID-19 also appears to be associated with distinct functional limitations. Lower levels of the Berg balance scale (BBS) were found in patients with post-acute COVID-19 syndrome (PCS) when compared with a control group (53 (46–56) vs. 56 (56–56) points, $P = 0.001$), and there was a strong correlation between BBS and HGS (rho = 0.602, $P < 0.0001$) (de Sousa et al., 2022). Also, a significant quadriceps strength decrease (7.4 ± 8.1% lower than baseline levels) and 24% of exercise-induced desaturation during the sit-to-stand test (STS) were shown in 41 patients at hospital discharge (Paneroni et al., 2021).

As a debilitating condition, muscle strength and functional capacities limitations, associated with personal and environmental factors, may lead to restricted performance in daily living activities. Accordingly, a study assessing 100 patients (66 ± 22 years, 66 men, length of stay in acute care = 14 days) from a postacute COVID-19 rehabilitation center reported a reduced daily living score at admission (Barthel score = 77.3 ± 26.7) as well as an association between the Barthel score (measure of activities of daily living performance) and STS and HGS, both at admission (STS: rho = 0.66, $P < 0.001$; HGS: rho = 0.43, $P = 003$) and at discharge (ST: rho = 0.53, $P < 0.001$; HGS: rho = 0.39, $P = 0.007$) (Piquet et al., 2021). Finally, considering the progressive loss of skeletal muscle mass with decreased muscle strength and physical capacity in COVID-19 survivors, mainly in older adults, the sarcopenia screening should be considered (Ali & Kunugi, 2021). The prevalence of sarcopenia may vary from 0.8% to 90.2% with a pooled of 48%, depending on the setting, time of assessment, and criteria used in patients with COVID-19 (Xu et al., 2022). Therefore, it is essential to properly assess the functionality of patients post-COVID-19 to intervene early through a multiprofessional approach and reduce the damage caused by the disease.

Important considerations

In the context of COVID-19, it is important to highlight that the majority of the studies present as a limitation the absence of the previous muscular condition status of the patients (Fig. 2). This lack of data makes it difficult to accurately determine the exact involvement of COVID-19 in muscle dysfunction because the patients may have some previous muscle impairment. In addition, there is a wide heterogeneity of results among studies, which can be attributed to several factors such as the characteristics of the population, disease severity, time point for assessment, methods and devices used, and different prediction equations and cut-off points to define and identify reduced muscle strength (e.g., 70% or 80% of predicted). However, the current body of evidence shows that impaired muscle strength is a common clinical finding in patients with COVID-19. Among the compiled evidence, all reported frequencies are above 20% (Fig. 2), which in practice means that at least 1 in 5 patients experience respiratory and/or peripheral muscle impairment. Given that these conditions are determinants for comprehensive functional status and quality of life, assessing muscle strength and focusing on the appropriate muscle treatment strategies should be a priority of care in post-COVID-19 patients.

Assessing respiratory and peripheral muscle strength in COVID-19 patients

In clinical practice, it is important to assess the potential short- and long-term effects of the disease on muscle health, which is defined as muscle mass and function, such as strength and performance (Montes-Ibarra et al., 2022). HGS has been used as an excellent indicator of strength and overall health. It is considered a measure of the upper limbs and total body functionality (Montes-Ibarra et al., 2022; Roberts et al., 2011). HGS measurement is verified through the static force exerted

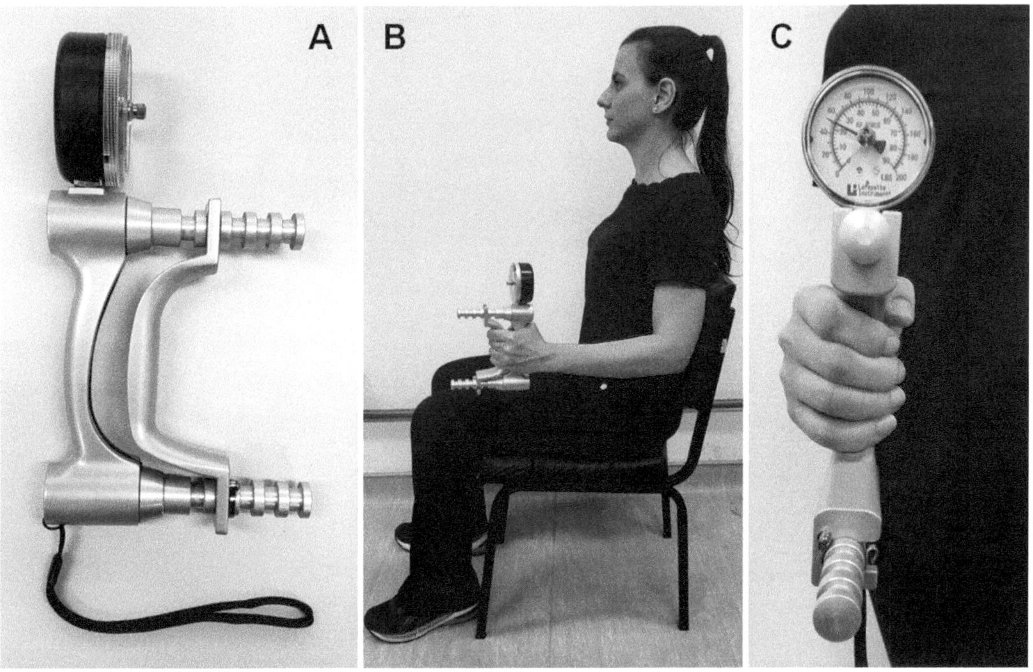

FIG. 3 Measurement of handgrip strength. This figure shows a hydraulic handgrip dynamometer (A), patient positioning (B), and handgrip measurement (C).

when the hand holds the dynamometer (Roberts et al., 2011). Reduced HGS values are associated with decreased physical and functional capacity (Roberts et al., 2011). A hydraulic handgrip dynamometer (Fig. 3A) is the most used device in scientific research and real clinical practice (Roberts et al., 2011), with accuracy and precision (Fess, 1981). It is considered a gold standard method for measuring strength and general health (Roberts et al., 2011) in kilogram-force (kgf) (Fess, 1981).

Providing some instructions regarding HGS measurement using a hydraulic dynamometer, the following step-by-step protocol is recommended (Roberts et al., 2011) (Fig. 3B): (1) Sit the patient with the back leaning against the chair and the feet flat on the floor; (2) Keep the arm supported and the wrist in a neutral position; (3) The evaluator should demonstrate how the test should be done, indicating the score recorded by HGS; (4) Position your hand in the form of a footprint, the thumb will be on one side, and the other four fingers on the other side. First, start to the right hand and then the left; (5) Guide the participant to squeeze as much as possible until the pointer stops rising; (6) Pointer indicates the maximum force in kgf (Fig. 3C); (7) Take three measurements on each side, alternately; and (8) The best measurement of six is used for statistical analysis (Roberts et al., 2011).

The European Working Group on Sarcopenia in Older People (Cruz-Jentoft et al., 2019) established the cut-off points for sarcopenia as HGS $<$ 27 kgf and $<$16 kgf for men and women, respectively. The following equation has been proposed for normative values stratified by sex and age, for both dominant and nondominant hands (Wang et al., 2018):

Male
Dominant: $-29.959 - 3.095\text{E}-^{05} \times (\text{age}^3) + 38.719 \times (\text{height}) + 0.113 \times (\text{weight})$
Nondominant: $-30.474 - 2.937\text{E}-^{05} \times (\text{age}^3) + 38.594 \times (\text{height}) + 0.099 \times (\text{weight})$
Female
Dominant: $-22.717 - 1.920\text{E}-^{05} \times (\text{age}^3) + 30.360 \times (\text{height}) + 0.048 \times (\text{weight})$
Nondominant: $-21.292 - 1.776\text{E}-^{05} \times (\text{age}^3) + 28.995 \times (\text{height}) + 0.040 \times (\text{weight})$

Example: Age in years, height in meters, and weight in kilograms. *E* is the scientific notation. $E-^{05}$ indicates 10–5 = 0.00001. For example, the predicted grip strength of the dominant hand for a 42-year-old man with a height of 1.7 m and weight of 67.1 kg is $-29.959-3.095 \times 10-5 \times (423) + 38.719 \times (1.7) + 0.113 \times (67.1) = 41.2$ kgf.

As previously mentioned, MIP and MEP measurements are useful for assessing RMS (Neder et al., 1999), which was impaired for several months after COVID-19. MIP and MEP are associated with health status, physical fitness,

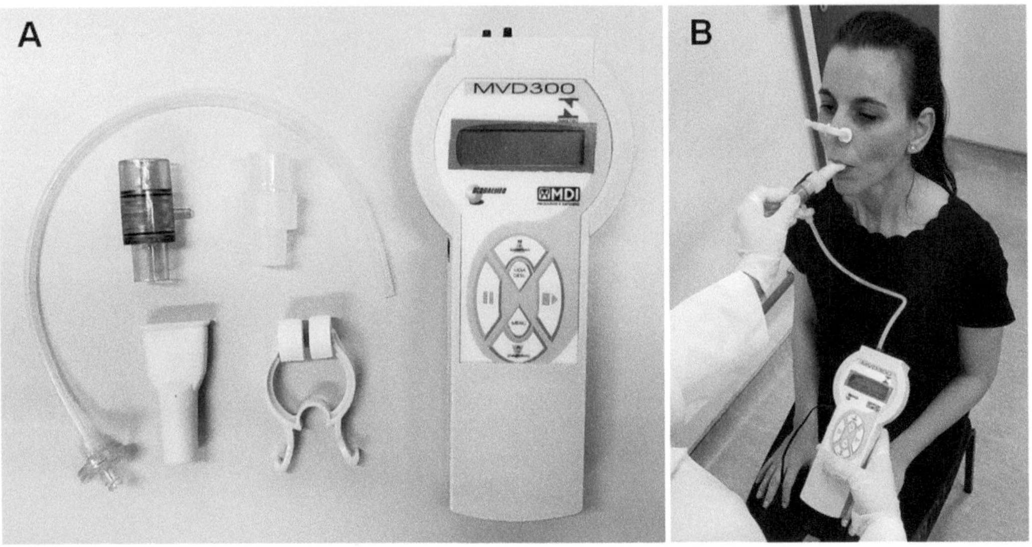

FIG. 4 Measurement of respiratory muscle strength. This figure shows a digital manovacuometer device and accessories (A), and the measurement of maximal respiratory pressures (B).

morbidity/mortality, and the diagnosis of neuromuscular and pulmonary disorders (Arora & Rochester, 1982). Their measurements are obtained by using an analog or digital manovacuometer device (Fig. 4A), which tends to be simple, cheap, and easy to acquire.

A step-by-step guide for measuring MIP and MEP according to Neder et al. (1999) by using a manovacuometer is the following: (1) Carry out biosafety procedures, separate materials (manovacuometer, note card, chair with adjustable backrest, nose clip, mouthpiece, etc.), and sit the patient with the trunk aligned at 90° with the thighs (Fig. 4B); (2) Explain the procedure to the patient; (3) Put on the nose clip to seal the exhalation/inspiration of air through the patient's nose; (4) To measure MIP, instruct the patient to exhale all the air (close to the residual volume), occlude the device's air outlet, and then ask the patient to perform a maximum inspiratory effort until the measurement pointer stabilizes; (5) To measure the MEP, instruct the patient to inhale all the air (close to total lung capacity), occlude the device's air outlet and then request the maximum expiratory effort until the measurement pointer stabilizes; (6) Perform MIP and MEP maneuvers (steps 4 and 5) at least three times and, at most, five times, if there was a variation greater than 10% between the values obtained; (7) Use at least 30 s between each MIP and MEP maneuver, or more time according to the patient's tolerance; and (8) The highest MIP and MEP values in centimeters of water (cmH_2O) are considered valid for analysis (Neder et al., 1999). The following equations have been proposed for normative values stratified by age and gender (Black & Hyatt, 1969):

Male from 20 to 80 years:	
MIP (cmH_2O) $= 143 - (0.55 \times \text{age})$	**MEP (cmH_2O)** $= 268 - (1.03 \times \text{age})$
Female from 20 to 86 years:	
MIP (cmH_2O) $= 104 - (0.51 \times \text{age})$	**MEP (cmH_2O)** $= 170 - (0.53 \times \text{age})$

Rehabilitative strategies on muscle strength in COVID-19 patients

Exercise-based rehabilitation is an effective strategy for improving health and prognosis in most respiratory diseases. However, there are not many randomized and controlled studies addressing the effects of physical exercise on muscle strength and function after COVID-19. Table 1 summarizes seven randomized controlled trials assessing the effects of exercise rehabilitation on muscle and other health outcomes in patients recovering from COVID-19. Briefly, six studies were conducted with outpatients (onset from 1 week after hospital discharge to long-COVID-19 patients with symptoms lasting more than 12 weeks) (del Corral et al., 2023; Jimeno-Almazán et al., 2023, 2022; Llurda-Almuzara et al., 2022; McNarry et al., 2022; Teixeira do Amaral et al., 2022), and one study assessed inpatients (post-COVID-19 ward) (Corna et al., 2022). Exercise rehabilitation interventions included respiratory muscle training (del Corral et al., 2023;

TABLE 1 Summary of randomized control trial involving rehabilitation strategies on muscle strength in COVID-19.

Authors/ year	Setting/Sample size Participants groups/age	Time point	Muscle group	Intervention	Duration, frequency and intensity	Effect of intervention
Llurda-Almuzara et al., 2022	Outpatient, *n*=70 IG=35, CG=35 IG: 49.5±13.7y CG: 55.1±20.9y	1st week after discharge; 1-month follow-up	**Peripheral muscle strength** – HGS (dynamometer) – LLS (5CRT)	**IG: Online exercise program** Upper/lower-extremity strength Balance exercise (single/double-legged jumps) double-legged squats Muscle stretching, walking, and breathing exercises	8-week; days-a-week, 24 sessions, 3 sets of 10–12 rep Intensity: 6 and 7 on Borg scale (0–10) Balance: progression doing exercises with eyes closed	(↑) **HGS**: IG experienced greater improvements (large effect sizes) than CG (↑) **5CRT:** IG experienced greater improvements (moderate to large effect sizes) in all *sec*ondary outcomes than those CG
Jimeno-Almazán et al., 2023	Outpatient, *n*=39 IG=19, CG=20 CG: 46±9.5y IG: 44.6±9.9y	12 weeks from onset of symptoms	**Peripheral muscle strength** – HGS (dynamometer) – LLS (5STS) – 3″ isometric knee extension test (force sensor)	**IG:** Supervised multicomponent exercise program resistance training (RT)+moderate intensity variable (MIVT)+light intensity continuous training (LICT)	**IG**: 8 weeks; 3 days-a-week and 2 days of RT (50% 1RM); 3 sets of 8 rep, 4 exercises [squat, bench press, deadlift, and bench pull])+MIVT: 4–6 × 3–5 min at 70–80% HRR/2–3 min at 55–65% HRR, and 1 day of LICT: 30–60 min, 65–70% HRR). **CG**: aerobic exercise (20–30 min), 5 days a week at an intensity that allows breathless speech. Exercises in 3 weekly sessions (3 × 10 repetitions of the 7 exercises)	(→) **HGS (Kgf)** (→) **Leg extension** (N): No differences in HG and leg extension (dominant) within groups by time or intervention. (↑) **5STS (sec):** Significant changes in the IG at post-test compared to CG
Teixeira do Amaral et al., 2022	Outpatient, n=32 IG=12, CG=20 IG: 51.9±10.2y CG: 53.3±11.6y	30–45 d after hospital discharge	**Respiratory muscle strength** – MIP and MEP (manovacuometer) **Peripheral muscle strength** – HGS (dynamometer) – LLS (5STS)	**IG: Telesupervised home-based exercise training** Exercise sessions–resistance and aerobics. Mobility, stretching, relaxing exercises **RT**: Nine multi/single-joint exercises (bodyweight squat, push-up on the wall, bodyweight lunge, one-arm row, deadlift, side lateral raise or shoulder	12-wk; resistance (3 weekly on alternate days) and aerobic (5 times a week) exercises; 6–20 rating of perceived exertion (RPE); **RT**: 1 set of 10–15 reps at week 1; 2 sets of 10–15 reps at weeks 2–3; 3 sets of 10–15 reps at weeks 4 to 6; and 3 sets of 15–20 reps at weeks 7–12. Intensity at 14–17 points of RPE scale	(→) **HGS** (Kgf): both groups increased HGS (IG: 4.5±1.3 kgf; CG: 4.6±1.0 kgf (→) **5STS (seg):** No differences during follow-up, and no groups difference **RMS**: only IG increased (↑) **MIP** (24.7±7.1 mH_2O) and (↑) **MEP** (20.3±5.8 cm H_2O)

Continued

TABLE 1 Summary of randomized control trial involving rehabilitation strategies on muscle strength in COVID-19—cont'd

Authors/ year	Setting/Sample size Participants groups/age	Time point	Muscle group	Intervention	Duration, frequency and intensity	Effect of intervention
				press, elbow flexion, calf raise, and abdominal crunch in chair)	**AT**: walking and/or cycling, 5 times a week, 10–15 min at week; 20 min at weeks 3–4; 30 min at weeks 5–12	
McNarry et al., 2022	Outpatient, $n = 148$ CG = 37, IG = 111 CG: 46.1 ± 12.7y IG: 46.8 ± 12.0y	Postacute COVID-19 (9.0 ± 4.2 months)	**Respiratory muscle strength** – MIP	**IG:** unsupervised IMT sessions **CG:** usual care	8 weeks; 3 unsupervised IMT sessions per week, on nonconsecutive days, training subsequently requiring >80% SMIP **IMT**: 6 blocks of 6 inspirations, rest periods progressively decreasing from 40 to 10 s, session durations of 20 min; inspirations prior to failure, defined as not achieving 80% SMIP on 3 breaths	**(↑) MIP** (cmH_2O): increased postintervention in the IG only
del Corral et al., 2023	Outpatient, $n = 88$ IMT = 22, IMTsham = 22, RMT = 22 RMTsham = 22 IMT: 48.9 ± 8.3y IMTsham: 45.3 ± 12.8y RMT: 46.5 ± 9.6y RMTsham: 45 ± 10.2y	Long-term post-COVID-19 (at least 3 months after the COVID-19 diagnosis confirmed)	**Respiratory muscle strength** – MIP (digital mouth pressure meter) – MEP (digital mouth pressure meter) – IME (powerbreath) **Peripheral muscle strength** – LLS (1mSTS) – HGS (dynamometer)	Home-based RMS program–threshold pressure device. Sham group: device without resistance (0 cmH_2O) IMT: only inspiratory RMT: expiratory and inspiratory training	8-week; 40 min/day, split into two 20-min sessions (morning/afternoon); 6 times per week **IMT or RMT** (real or sham): load individually tailored and increased according to the same distribution schedule for both inspiratory and expiratory muscle training	**(↑) 1mSTS** (number): IMT and RMT: large increase in lower limb muscle strength compared with sham **(→) HGS** (Kg): no difference between-group. HGS increased significantly in all groups **(↑) MIP (cmH_2O) and IME (sec):** IMT and RMT showed large increase in IMT and endurance compared with sham **(↑) MEP (cmH_2O): RMT** large increase compared with others All groups improved MIP and MEP after intervention

Corna et al., 2022	**Inpatient,** $n=32$ CG = 16, IG = 16 IG: 70.6 ± 12.2y CG: 70.5 ± 9.2y	At the time of admission in the post-COVID-19 ward	**Peripheral muscle strength** – Muscle torque of Knee extensors (handheld dynamometer) – HGS (hand dynamometer) – LLS (30sSTS)	IG and CG: standard inpatient rehabilitation program: mobilization, upper and lower limb strengthening, balance exercise, walking training, and respiratory muscle training IG: plus aerobic training	standard inpatient rehabilitation program: 10 sessions (50 min/day, 5 days/week) **AT**: arm crank ergometer for 30 min/day, 5 days/week, for a total of 10 sessions; 5 min warm-up, 20 min of training, and 5 min cool-down: fixed frequency of 60 rpm.; intensity of RPE between 11 and 14 of the Borg Scale, or within 55–85% of HRmax.	**(→) Muscle torque of knee extension** (Nm/Kg): improved in both groups, with no difference **(→) HGS (Kg)**: increased only in the IG, no difference between groups. **(→) 30sSTS (n):** improved in both groups with no difference between them
Jimeno-Almazán et al., 2022	**Outpatient,** $n=80$ CT = 20, IMT = 17 CT+IMT = 23, CG = 20 CT: 44.3 ± 7.6y RM: 43.0 ± 6.9y CT + IMT: 44.1 ± 7.4y CG: 47.8 ± 7.6y	COVID-19 patients, symptomatic lasting >12 weeks from the onset of symptoms, and not hospitalized	**Peripheral muscle strength** – ULS (bench press 1RM) – LLS (half squat 1 RM) maximal isometric—HGS (dynamometer)	CT: concurrent training program IMT: inspiratory muscle training program CT + IMT: CT+IMT CG: control (WHO guideline)	8-week; **CT:** 8 weeks; 3 days-a-week and 2 days of RT (50% 1RM); 3 sets of 8 rep, 4 exercises [squat, bench press, deadlift, and bench pull]) + MIVT: 4–6 × 3–5 min at 70–80% HRR/2–3 min at 55–65% HRR, and 1 day of LICT: 30–60 min, 65–70% HRR) IMT**:** 1 set of 30 rep (62.5% ± 4.6% of the MIP, warm-up set, twice a day, every day; resistance should be increased every 2 weeks by turning the load adjustment clockwise ¼ to 1 full turn, pending on tolerance to maintain a 12–15 RPE on modified Borg scale	**(↑) Half Squat 1RM**: improved for both groups who accomplished the multicomponent exercise training (CT and CT + IMT), no changes for IMT or CG groups. **(→) Bench press 1RM** **(→) HGS (Kgf):** no inter- or intragroup interactions were found for the dominant hand grip strength.

n: number; IG: intervention group; CG: control group; HGS: handgrip strength; ULS: upper limb strength; LLS: lower limb strength; 5CRT: 5-chair raise test; 5STS: 5-sit to stand test; RT: resistance training; 1RM: 1-repetition maximum; MIVT: moderate intensity variable training; HRR: heart rate reserve; LICT: light intensity continuous training; MIP: maximal inspiratory pressure; MEP: maximal expiratory pressure; RPE: rating of perceived exertion; AT: aerobic training; RMS: respiratory muscle strength; IMT: inspiratory muscle training; SMIP: sustained maximal inspiratory pressure; RMT: respiratory muscle training; IME: inspiratory muscle endurance; 1mSTS: 1-min sit to stand test; 30sSTS: 30-s sit to stand test; HRmax: heart rate maximal; CT: concurrent training; ↑: improvement of IG post-intervention compared to CG; →: no difference between IG and CG post-intervention; ↓: worsening of GI post-intervention compared to CG.

Jimeno-Almazán et al., 2022; McNarry et al., 2022), combined resistance and aerobic training (Jimeno-Almazán et al., 2022; Teixeira do Amaral et al., 2022), and multicomponent exercise intervention (Corna et al., 2022; Jimeno-Almazán et al., 2023; Llurda-Almuzara et al., 2022). The intervention duration ranged from 2 weeks to 12 months, and exercise intensity was widely heterogeneous among studies. All studies assessing RMS showed that exercise rehabilitation was superior to control intervention to improve MIP and/or MEP (del Corral et al., 2023; McNarry et al., 2022; Teixeira do Amaral et al., 2022). Among six studies that assessed HGS, only one showed that exercise rehabilitation was more effective than control intervention for improving this variable (Llurda-Almuzara et al., 2022). However, five studies showed a superior effect of exercise rehabilitation over control for improving several other muscle strength and/or function outcomes (del Corral et al., 2023; Jimeno-Almazán et al., 2023, 2022; Llurda-Almuzara et al., 2022; McNarry et al., 2022). Finally, only one study showed no differences between exercise intervention and standard care on any peripheral muscle strength/function parameter assessed (Corna et al., 2022). Although future studies are needed, the results of previous studies suggest that exercise rehabilitation is an effective intervention to counteract the deleterious effects of COVID-19 on muscle strength and function.

Clinical case report

Patient I. F. B., female, 51 years old, height 1.57 m, body weight 95 kg, hypertension, diabetes, preserved functional capacity (independent for walking and activities of daily living), with no history of smoking, alcoholism, or other respiratory, neurological or cardiovascular diseases. She was admitted to the emergency hospital department reporting fever, dyspnea, myalgia, and fatigue with onset of symptoms 9 days ago. Swab test for COVID-19 was positive. Chest computerized tomography revealed a bilateral lung ground-glass pattern with 50% involvement. On day 3 of hospitalization, the patient presented with refractory hypoxemia and respiratory deterioration, and the ICU team proceeded with endotracheal intubation and IMV. Patient remained mechanically ventilated for 11 days. She also received treatment with vasoactive drugs, antibiotics, corticosteroids, sedatives, and neuromuscular blockers; she was kept in a prone position. Extubation from mechanical ventilation was successfully performed. Patient was discharged after 25 days of hospitalization (day 34 after symptoms onset) (Fig. 4). In the physiotherapist predischarge evaluation, the patient presented normal vital signs. She was breathing in room air, body weight 82 kg, with impaired functional status and walking a short distance with assistance of two people. Assessments of respiratory and peripheral muscle strength showed the following results: MIP = 40 cmH_2O, MEP = 70 cmH_2O, and HGS = 12 kgf (dominant hand). Based on assessments, patient was referred to the post-COVID-19 rehabilitation outpatient clinic (Fig. 5).

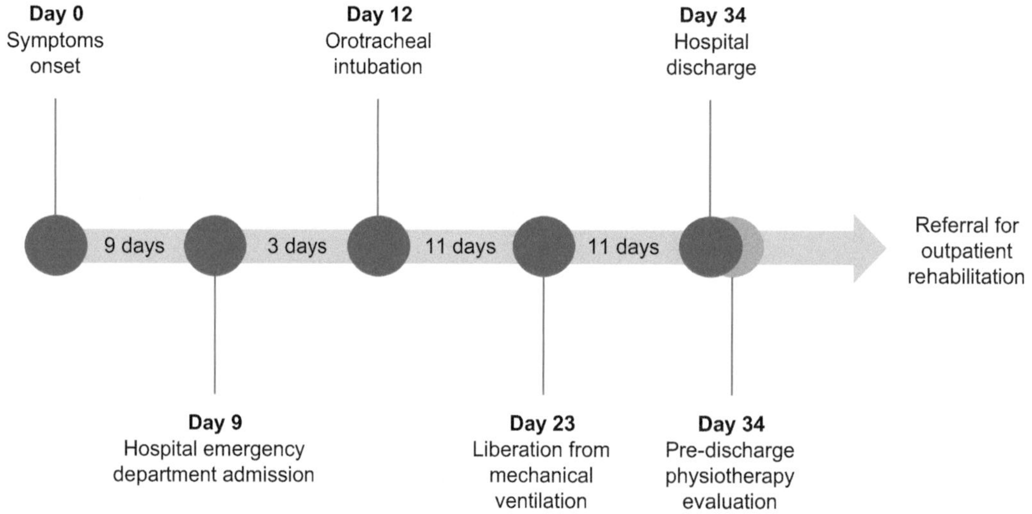

FIG. 5 Timeline of clinical case. This figure shows the timeline of key events related to COVID-19 and muscle strength assessment in the patient described in the clinical case.

Questions and answers

1. **What possible factors and mechanisms are related to muscle weakness during hospitalization for COVID-19?**
Answer: Potential factors and mechanisms of muscle weakness mainly include viral infiltration of muscle tissue, systemic inflammatory condition associated with COVID-19, hypoxemia, malnutrition, immobility, use of medications such as corticosteroids, sedatives, and neuromuscular blockers.
2. **What are the predicted values of respiratory muscle strength and peripheral muscle strength for I.F.B. patient and how is the discharge muscle impairment?**
Answer: MIP and MEP predicted values are obtained using equations proposed by Black and Hyatt (1969), according to the patient's gender (female) and age (51 years). Both inspiratory and expiratory muscle strength are reduced = 51% and 49% of predicted, respectively.

- **$MIP_{predicted}$ (cmH_2O)** = 104 – (0.51 × age) = 104 – (0.51 × 51) = **78 cmH_2O**
- **MIP $_{(\%\ of\ predicted)}$** = (MIP × 100)/MIPpredicted = (40 × 100)/78 = **51%**

- **$MEP_{predicted}$ (cmH_2O)** = 170 – (0.53 × age) = 170 – (0.53 × 51) = **143 cmH_2O**
- **MEP $_{(\%\ of\ predicted)}$** = (MEP × 100)/MEPpredicted = (70 × 100)/143 = **49%**

As shown below, HGS can be calculated using the equation proposed by Wang et al. (2018), according to the patient's gender (female), age (51 years), height (1.57 m), and weight (82 kg). Based on this, patient grip strength is reduced = 46% of the predicted value. In addition, absolute value (12) is below 16 kg, which is considered the cut-off point for sarcopenia in women.

$HGS_{predicted}$ (kgf) = (−22.717) – 1.920 × 0.00001 × (age^3) + 30.360 × (height) + 0.048 × (weight)
$HGS_{predicted}$ (kgf) = (−22.717) – 1.920 × 0.00001 × (51^3) + 30.360 × (1.57) + 0.048 × (82) = **26 kgf**
HGS (% of predicted) = (HGS × 100)/$HGS_{predicted}$ = (12 × 100)/26 = 46%

3. **What are the possible consequences of this muscle weakness and rehabilitation options for this patient after discharge?**
Answer: Overall muscle weakness is associated with exercise intolerance, reduced physical performance, impaired functional status, decreased health-related quality of life, and increased risk for disability and death. Rehabilitation strategy should include a comprehensive exercise-based rehabilitation program, including individualized and progressive therapeutic exercises. Given that this patient has respiratory and peripheral muscle weakness, with possible reduced exercise capacity, the training program should include aerobic, resistance, and inspiratory muscle training.

Conclusion

This chapter provided an overview of COVID-19-related muscle dysfunction. Peripheral and respiratory muscle dysfunction are common findings in these patients, causing exercise intolerance, impaired functional status, and decreased health-related quality of life. Handgrip strength and respiratory pressures are methods to evaluate muscle status. An exercise rehabilitation strategy, including respiratory muscle training, resistance and aerobic training, and multicomponent exercise intervention, has been demonstrated to counteract the deleterious effects of COVID-19 on muscle strength.

Summary points

- Muscle dysfunction including peripheral and respiratory muscles has been reported as an important consequence of COVID-19.
- Direct viral and immune-mediated damage, immobilization, sedation, intubation, acquired weakness in the intensive care unit, and previous comorbidities are factors associated with COVID-19-related skeletal muscle impairment.
- Debilitating musculoskeletal symptoms in patients affected by COVID-19 include myalgia, muscle weakness, and fatigue.

- Clinical consequences of muscle dysfunction include exercise intolerance, reduced physical performance, impaired functional status, decreased health-related quality of life, and increased risk for disability and death.
- Handgrip strength and maximal respiratory pressures are common measurements to assess muscle strength in COVID-19 patients.
- Exercise rehabilitation has demonstrated a superior effect over control for improving peripheral and respiratory muscle strength and/or function outcomes.

References

Agergaard, J., Leth, S., Pedersen, T. H., Harbo, T., Blicher, J. U., Karlsson, P., … Tankisi, H. (2021). Myopathic changes in patients with long-term fatigue after COVID-19. *Clinical Neurophysiology*, *132*(8), 1974–1981.

Ali, A. M., & Kunugi, H. (2021). Screening for sarcopenia (physical frailty) in the COVID-19 era. *International Journal of Endocrinology*, *2021*, 5563960. https://doi.org/10.1155/2021/5563960.

Amin, B. J. H., Kakamad, F. H., Ahmed, G. S., Ahmed, S. F., Abdulla, B. A., Mikael, T. M., … Hussein, D. A. (2022). Post COVID-19 pulmonary fibrosis; a meta-analysis study. *Annals of Medicine and Surgery*, 103590.

Arora, N. S., & Rochester, D. F. (1982). Respiratory muscle strength and maximal voluntary ventilation in undernourished patients. *American Review of Respiratory Disease*, *126*(1), 5–8.

Aschman, T., Schneider, J., Greuel, S., Meinhardt, J., Streit, S., Goebel, H. H., … Stenzel, W. (2021). Association between SARS-CoV-2 infection and immune-mediated myopathy in patients who have died. *JAMA Neurology*, *78*(8), 948–960.

Azzolino, D., & Cesari, M. (2022). Fatigue in the COVID-19 pandemic. *Lancet Healthy Longevity*, *3*(3), e128–e129.

Berentschot, J. C., Heijenbrok-Kal, M. H., Bek, L. M., Huijts, S. M., van Bommel, J., van Genderen, M. E., … Willems, E. G. (2022). Physical recovery across care pathways up to 12 months after hospitalization for COVID-19: A multicenter prospective cohort study (CO-FLOW). *The Lancet Regional Health-Europe*, *22*, 100485.

Black, L. F., & Hyatt, R. E. (1969). Maximal respiratory pressures: Normal values and relationship to age and sex. *American Review of Respiratory Disease*, *99*, 696–702.

Carfì, A., Bernabei, R., Landi, F., & Gemelli. (2020). Against COVID-19 post-acute care study group. Persistent symptoms in patients after acute COVID-19. *JAMA*, *324*(6), 603–605.

Corna, S., Giardini, M., Godi, M., Bellotti, L., & Arcolin, I. (2022). Effects of aerobic training in patients with subacute COVID-19: A randomized controlled feasibility trial. *International Journal of Environmental Research and Public Health*, *19*(24), 16383.

Cruz-Jentoft, A. J., Bahat, G., Bauer, J., Boirie, Y., Bruyère, O., Cederholm, T., Zamboni, M., … Writing Group for the European Working Group on Sarcopenia in Older People 2 (EWGSOP2), and the Extended Group for EWGSOP2. (2019). Sarcopenia: Revised European consensus on definition and diagnosis. *Age Ageing*, *48*(1), 16–31.

de Sousa, K. C. A., Gardel, D. G., & Lopes, A. J. (2022). *Postural balance and its association with functionality and quality of life in non-hospitalized patients with post-acute COVID-19 syndrome* (pp. 1–8). Physiotherapy Research International.

del Corral, T., Fabero-Garrido, R., Plaza-Manzano, G., Fernández-de-Las-Peñas, C., Navarro-Santana, M., & López-de-Uralde-Villanueva, I. (2023). Home-based respiratory muscle training on quality of life and exercise tolerance in long-term post-COVID-19: Randomized controlled trial. *Annals of Physical and Rehabilitation Medicine*, *66*(1), 101709.

dos Santos, P. K., Sigoli, E., Bragança, L. J. G., & Cornachione, A. S. (2022). The musculoskeletal involvement after mild to moderate COVID-19 infection. *Frontiers in Physiology*, *13*, 813924.

Evans, R. A., McAuley, H., Harrison, E. M., Shikotra, A., Singapuri, A., Sereno, M., … Cairns, P. (2021). Physical, cognitive, and mental health impacts of COVID-19 after hospitalisation (PHOSP-COVID): A UK multicentre, prospective cohort study. *The Lancet Respiratory Medicine*, *9*(11), 1275–1287.

Fernández-de-Las-Peñas, C., de-la-Llave-Rincón, A. I., Ortega-Santiago, R., Ambite-Quesada, S., Gómez-Mayordomo, V., Cuadrado, M. L., … Arendt-Nielsen, L. (2022). Prevalence and risk factors of musculoskeletal pain symptoms as long-term post-COVID sequelae in hospitalized COVID-19 survivors: A multicenter study. *Pain*, *163*(9), e989–e996.

Fess, E. (1981). American Society of Hand Therapists: Clinical assessment recommendations. *American Society of Hand Therapists*, 1–13.

González-Islas, D., Sánchez-Moreno, C., Orea-Tejeda, A., Hernández-López, S., Salgado-Fernández, F., Keirns-Davis, C., … Castorena-Maldonado, A. (2022). Body composition and risk factors associated with sarcopenia in post-COVID patients after moderate or severe COVID-19 infections. *BMC Pulmonary Medicine*, *22*(1), 1–8.

Gosker, H. R., Wouters, E. F., van der Vusse, G. J., & Schols, A. M. (2000). Skeletal muscle dysfunction in chronic obstructive pulmonary disease and chronic heart failure: Underlying mechanisms and therapy perspectives. *The American Journal of Clinical Nutrition*, *71*(5), 1033–1047.

Hejbøl, E. K., Harbo, T., Agergaard, J., Madsen, L. B., Pedersen, T. H., Østergaard, L. J., … Tankisi, H. (2022). Myopathy as a cause of fatigue in long-term post-COVID-19 symptoms: Evidence of skeletal muscle histopathology. *European Journal of Neurology*, *29*(9), 2832–2841.

Hennigs, J. K., Huwe, M., Hennigs, A., Oqueka, T., Simon, M., Harbaum, L., … Klose, H. (2022). Respiratory muscle dysfunction in long-COVID patients. *Infection*, *50*(5), 1391–1397.

Huang, C., Huang, L., Wang, Y., Li, X., Ren, L., Gu, X., … Cao, B. (2021). 6-month consequences of COVID-19 in patients discharged from hospital: A cohort study. *The Lancet*, *397*(10270), 220–232.

Huang, Y., Tan, C., Wu, J., Chen, M., Wang, Z., Luo, L., … Liu, J. (2020). Impact of coronavirus disease 2019 on pulmonary function in early convalescence phase. *Respiratory Research*, *21*, 1–10.

Jimeno-Almazán, A., Buendía-Romero, Á., Martínez-Cava, A., Franco-López, F., Sánchez-Alcaraz, B. J., Courel-Ibáñez, J., & Pallarés, J. G. (2023). Effects of a concurrent training, respiratory muscle exercise, and self-management recommendations on recovery from post-COVID-19 conditions: The RECOVE trial. *Journal of Applied Physiology*, *134*(1), 95–104.

Jimeno-Almazán, A., Franco-López, F., Buendía-Romero, Á., Martínez-Cava, A., Sánchez-Agar, J. A., Sánchez-Alcaraz Martínez, B. J., … Pallarés, J. G. (2022). Rehabilitation for post-COVID-19 condition through a supervised exercise intervention: A randomized controlled trial. *Scandinavian Journal of Medicine & Science in Sports*, *32*(12), 1791–1801.

Latronico, N., Peli, E., Calza, S., Rodella, F., Novelli, M. P., Cella, A., … Piva, S. (2022). Physical, cognitive and mental health outcomes in 1-year survivors of COVID-19-associated ARDS. *Thorax*, *77*(3), 300–303.

Levy, D., Giannini, M., Oulehri, W., Riou, M., Marcot, C., Pizzimenti, M., … Meyer, A. (2022). Long term follow-up of sarcopenia and malnutrition after hospitalization for COVID-19 in conventional or intensive care units. *Nutrients*, *14*(4), 912.

Llurda-Almuzara, L., Rodríguez-Sanz, J., López-de-Celis, C., Aiguadé-Aiguadé, R., Arán-Jové, R., Labata-Lezaun, N., … Pérez-Bellmunt, A. (2022). Effects of adding an online exercise program on physical function in individuals hospitalized by COVID-19: A randomized controlled trial. *International Journal of Environmental Research and Public Health*, *19*(24), 16619.

Mancuzo, E. V., Marinho, C. C., Machado-Coelho, G. L. L., Batista, A. P., Oliveira, J. F., Andrade, B. H., … Augusto, V. M. (2021). Lung function of patients hospitalized with COVID-19 at 45 days after hospital discharge: first report of a prospective multicenter study in Brazil. *Journal Brasileiro de Pneumologia*, *47*(6), e20210162.

McNarry, M. A., Berg, R. M., Shelley, J., Hudson, J., Saynor, Z. L., Duckers, J., … Mackintosh, K. A. (2022). Inspiratory muscle training enhances recovery post-COVID-19: A randomised controlled trial. *European Respiratory Journal*, *60*(4), 2103101.

Mills, G., Briggs-Price, S., Houchen-Wolloff, L., Daynes, E., & Singh, S. (2022). The prevalence and location of musculoskeletal pain following COVID-19. *Musculoskeletal Care*. https://doi.org/10.1002/msc.1657.

Montani, D., Savale, L., Noel, N., Meyrignac, O., Colle, R., Gasnier, M., … Monnet, X. (2022). Post-acute COVID-19 syndrome. *European Respiratory Review*, *31*(163), 210185.

Montes-Ibarra, M., Oliveira, C. L., Orsso, C. E., Landi, F., Marzetti, E., & Prado, C. M. (2022). The impact of long COVID-19 on muscle health. *Clinics in Geriatric Medicine*, *38*(3), 545–557.

Neder, J. A., Andreoni, S., Lerario, M. C., & Nery, L. E. (1999). Reference values for lung function tests: II. Maximal respiratory pressures and voluntary ventilation. *Brazilian Journal of Medical and Biological Research*, *32*(6), 719–727.

Noujaim, P. J., Jolly, D., Coutureau, C., & Kanagaratnam, L. (2022). Fatigue and quality-of-life in the year following SARS-Cov2 infection. *BMC Infectious Diseases*, *22*(1), 541.

Núñez-Seisdedos, M. N., Valcárcel-Linares, D., Gómez-González, M. T., Lázaro-Navas, I., López-González, L., Pecos-Martín, D., & Rodríguez-Costa, I. (2022). Inspiratory muscle strength and function in mechanically ventilated COVID-19 survivors 3 and 6 months after ICU discharge. *ERJ Open Research*, *1*(9). 00329–02022.

Paneroni, M., Simonelli, C., Saleri, M., Bertacchini, L., Venturelli, M., Troosters, T., … Vitacca, M. (2021). Muscle strength and physical performance in patients without previous disabilities recovering from COVID-19 pneumonia. *American Journal of Physical Medicine & Rehabilitation*, *100*(2), 105–109.

Piotrowicz, K., Ryś, M., Perera, I., Gryglewska, B., Fedyk-Łukasik, M., Michel, J. P., … Gąsowski, J. (2022). Factors associated with mortality in hospitalised, non-severe, older COVID-19 patients–the role of sarcopenia and frailty assessment. *BMC Geriatrics*, *22*(1), 1–12.

Piquet, V., Luczak, C., Seiler, F., Monaury, J., Martini, A., Ward, A. B., … Bayle, N. (2021). Do patients with COVID-19 benefit from rehabilitation? Functional outcomes of the first 100 patients in a COVID-19 rehabilitation unit. *Archives of Physical Medicine and Rehabilitation*, *102*(6), 1067–1074.

Pucci, G., D'Abbondanza, M., Curcio, R., Alcidi, R., Campanella, T., Chiatti, L., … Vaudo, G. (2022). Handgrip strength is associated with adverse outcomes in patients hospitalized for COVID-19-associated pneumonia. *Internal and Emergency Medicine*, *17*(7), 1997–2004.

Roberts, H. C., Denison, H. J., Martin, H. J., Patel, H. P., Syddall, H., Cooper, C., & Sayer, A. A. (2011). A review of the measurement of grip strength in clinical and epidemiological studies: Towards a standardised approach. *Age and Ageing*, *40*(4), 423–429.

Salari, N., Khodayari, Y., Hosseinian-Far, A., Zarei, H., Rasoulpoor, S., Akbari, H., & Mohammadi, M. (2022). Global prevalence of chronic fatigue syndrome among long COVID-19 patients: A systematic review and meta-analysis. *BioPsychoSocial Medicine*, *16*(1), 21.

Severin, R., Arena, R., Lavie, C. J., Bond, S., & Phillips, S. A. (2020). Respiratory muscle performance screening for infectious disease management following COVID-19: A highly pressurized situation. *The American Journal of Medicine*, *133*(9), 1025–1032.

Severin, R., Franz, C. K., Farr, E., Meirelles, C., Arena, R., Phillips, S. A., … Faghy, M. (2022). The effects of COVID-19 on respiratory muscle performance: Making the case for respiratory muscle testing and training. *European Respiratory Review*, *31*(166), 220006.

Shi, Z., De Vries, H. J., Vlaar, A. P., Van Der Hoeven, J., Boon, R. A., Heunks, L. M., … Dutch COVID-19 Diaphragm Investigators. (2021). Diaphragm pathology in critically ill patients with COVID-19 and postmortem findings from 3 medical centers. *JAMA Internal Medicine*, *181*(1), 122–124.

Silva, C. C., Bichara, C. N. C., Carneiro, F. R. O., Palacios, V. R. D. C. M., Berg, A. V. S. V. D., Quaresma, J. A. S., & Magno Falcao, L. F. (2022). Muscle dysfunction in the long coronavirus disease 2019 syndrome: Pathogenesis and clinical approach. *Reviews in Medical Virology*, *32*(6), e2355.

Soares, M. N., Eggelbusch, M., Naddaf, E., Gerrits, K. H., van der Schaaf, M., van den Borst, B., … Wüst, R. C. (2022). Skeletal muscle alterations in patients with acute Covid-19 and post-acute sequelae of Covid-19. *Journal of Cachexia, Sarcopenia and Muscle*, *13*(1), 11–22.

Suh, J., Mukerji, S. S., Collens, S. I., Padera, R. F., Pinkus, G. S., Amato, A. A., & Solomon, I. H. (2021). Skeletal muscle and peripheral nerve histopathology in COVID-19. *Neurology*, *97*(8), e849–e858.

Teixeira do Amaral, V., Viana, A. A., Heubel, A. D., Linares, S. N., Martinelli, B., Witzler, P. H. C., Orikassa de Oliveira, G. Y., Zanini, G. D. S., Borghi Silva, A., Mendes, R. G., & Ciolac, E. G. (2022). Cardiovascular, respiratory, and functional effects of home-based exercise training after COVID-19 hospitalization. *Medicine and Science in Sports and Exercise*, *54*(11), 1795–1803.

Torres-Castro, R., Vasconcello-Castillo, L., Alsina-Restoy, X., Solís-Navarro, L., Burgos, F., Puppo, H., & Vilaró, J. (2021). Respiratory function in patients post-infection by COVID-19: A systematic review and meta-analysis. *Pulmonology*, *27*(4), 328–337.

van Gassel, R. J., Bels, J., Remij, L., van Bussel, B. C., Posthuma, R., Gietema, H. A., … van de Poll, M. C. (2021). Functional outcomes and their association with physical performance in mechanically ventilated coronavirus disease 2019 survivors at 3 months following hospital discharge: A cohort study. *Critical Care Medicine*, *49*(10), 1726.

Viana, A. A., Heubel, A. D., Teixeira do Amaral, V., Linares, S. N., de Oliveira, G. Y. O., Martinelli, B., … Ciolac, E. G. (2022). Can previous levels of physical activity affect risk factors for cardiorespiratory diseases and functional capacity after COVID-19 hospitalization? A prospective cohort study. *BioMed Research International*, 7854303.

Wang, Y. C., Bohannon, R. W., Li, X., Sindhu, B., & Kapellusch, J. (2018). Hand-grip strength: Normative reference values and equations for individuals 18 to 85 years of age residing in the United States. *Journal of Orthopaedic & Sports Physical Therapy*, *48*(9), 685–693.

Xu, Y., Xu, J. W., You, P., Wang, B. L., Liu, C., Chien, C. W., & Tung, T. H. (2022). Prevalence of sarcopenia in patients with COVID-19: A systematic review and meta-analysis. *Frontiers in Nutrition*, *9*.

Section G

Case studies with mini review

Chapter 50

Case study: Oral mucosal lesions in patients with COVID-19

Juliana Amorim dos Santos[a], Rainier Luiz Carvalho da Silva[b], and Eliete Neves Silva Guerra[a]
[a]*Laboratory of Oral Histopathology, Health Science Faculty, University of Brasília, Campus Darcy Ribeiro—Asa Norte, Brasília, Brazil,*
[b]*CMF Odontologia Hospitalar, Rede Santa, Brasília, Brazil*

Abbreviations

ACE2	angiotensin-converting enzyme 2
aPDT	antimicrobial photodynamic therapy
COVID-19	coronavirus disease 2019
RT-qPCR	quantitative reverse transcription polymerase chain reaction
SARS-CoV-2	respiratory syndrome coronavirus 2

Introduction

The coronavirus disease 2019 (COVID-19) is a disease caused by severe acute respiratory syndrome coronavirus 2 (SARS-CoV-2), a virus of the coronavirus family. This presents a clinical picture that varies between asymptomatic infections (80%) and severe respiratory conditions (20%), of which approximately 5% can need ventilatory support (WHO 2020). Although COVID-19 is most well-known for causing substantial respiratory pathology, it can also result in several extrapulmonary manifestations (Gupta et al., 2020). These conditions include thrombotic complications, myocardial dysfunction, arrhythmia, acute coronary syndromes, acute kidney injury, gastrointestinal symptoms, hepatocellular injury, hyperglycemia and ketosis, neurologic illnesses, ocular symptoms, and dermatologic complications. Given that angiotensin-converting enzyme 2 (ACE2), the entry receptor for the causative coronavirus SARS-CoV-2, is expressed in multiple extrapulmonary tissues, direct viral tissue damage is a plausible mechanism of injury. In addition, endothelial damage and thromboinflammation, dysregulation of immune responses, and maladaptation of ACE2-related pathways might all contribute to these extrapulmonary manifestations of COVID-19 (Gupta et al., 2020).

Oral mucosal manifestations have been reported in patients with COVID-19. The miscellaneous clinical aspects of oral mucosal lesions (e.g., aphthous-like ulcers, herpes-like lesions, fungal infections, glossitis, and parotitis) contribute to the main challenge in defining the etiology of these oral signs (Amorim Dos Santos et al., 2020; Chaux-Bodard et al., 2020; de Maria et al., 2020; Lechien et al., 2020; Martín Carreras-Presas et al., 2021; Putra et al., 2020). It is still a question as to whether oral lesions in COVID-19 patients are due to a primary effect of coronavirus with local replication or opportunistic coinfections from other pathogens.

Due to the heterogeneity of oral lesions in COVID-19 patients, there is no standard treatment approach, and the alternatives vary from topical to systemic antifungals, antibiotics, antivirals, and corticosteroids, depending also on the patients' severity condition. Thus, this case study aims to quickly review the subject and present four additional cases of oral mucosal lesions in patients with COVID-19 treated with a topical antimicrobial photodynamic therapy and noninvasive adjuvant therapy.

Oral mucosal lesions in COVID-19: Mini review of the literature

Oral mucosal lesions in patients with COVID-19 presented multiple clinical aspects, including white and erythematous plaques, irregular ulcers, small blisters, petechiae, and desquamative gingivitis. Tongue, palate, lips, gingiva, and buccal

Management, Body Systems, and Case Studies in COVID-19. https://doi.org/10.1016/B978-0-443-18703-2.00019-0
Copyright © 2024 Elsevier Inc. All rights are reserved, including those for text and data mining, AI training, and similar technologies.

mucosa were affected. In mild cases, oral mucosal lesions developed before or at the same time as the initial respiratory symptoms; however, in those who required medication and hospitalization, the lesions developed approximately 7–24 days after onset of symptoms (Amorim Dos Santos et al., 2021a). According to the update of the living systematic review by Amorim dos Santos and colleagues (Amorim Dos Santos et al., 2021b), oral mucosal lesions were described as a case report or case series in at least 308 patients in both hospitalized and nonhospitalized conditions. In addition, 2491 patients with COVID-19 were analyzed regarding this outcome in five cross-sectional studies (Abubakr et al., 2021; Askin et al., 2020; Katz & Yue, 2021; Mina et al., 2021; Nuno-Gonzalez et al., 2021). From these patients, oral mucosal manifestations appeared in 512, representing a relative frequency of 20.5%. Although oral lesions were greatly described as aphthous-like ulcers, other descriptions such as herpes-like lesions, candidiasis, glossitis/depapillation/geographic tongue, parotitis, and angular cheilitis were also noticed. The affected anatomic area was variable between reported patients. Unspecified oral mucosal sites (130 cases) and tongue (77) were the most frequent ones, followed by unspecified lips (32), palate (24), labial commissure (19), and lower lip (15). Although oral lesions might appear earlier in some cases, most studies indicated their appearance after the onset of COVID-19 symptoms.

In addition to lesions of the oral mucosa, the overall prevalence of taste disorders was 38%. Hypogeusia, dysgeusia, and ageusia were also evaluated by a metaanalysis, and the pooled prevalence was 34% for hypogeusia, 33% for dysgeusia, and 26% for ageusia. Taste disorders were associated with a positive COVID-19 test, with mild/moderate severity of COVID-19 and with female patients with COVID-19. Furthermore, xerostomia demonstrated a prevalence of 43% in patients with COVID-19. Therefore, the reanalysis of current evidence suggests the triad xerostomia, taste dysfunction, and oral mucosal lesions as common manifestations in patients with COVID-19 (Amorim Dos Santos et al., 2021b). However, these outcomes are under discussion, and more studies will be necessary to confirm their association with direct SARS-CoV-2 infection in the oral cavity. Also, we highlight the importance of including dentists in the intensive care unit multiprofessional team to improve oral health, reduce coinfections, and contribute to evidence-based public health in critical patients with COVID-19.

Case study: Oral mucosal lesions in patients with COVID-19

This case study aims to present four additional cases of oral lesions in patients with COVID-19 treated with a topical antimicrobial and noninvasive adjuvant therapy. The patients were admitted to the Alvorada Brasilia Hospital and Orthopedic Hospital and Specialized Medicine (Brasilia, Brazil) between March 9–19, 2021. Three patients were male and one female, with at least two comorbidities and a median age of 61. All were diagnosed based on the quantitative reverse transcription polymerase chain reaction (RT-PCR) test and had severe diseases with mechanical ventilation demand. Oral manifestations were identified during hospitalization with a median appearance of 19 days (Table 1). The lesions were diagnosed as herpes-like in the lips and oral mucosa; thus, patients were immediately started on intravenous acyclovir or topical penciclovir. Antimicrobial photodynamic therapy (aPDT–methylene blue 0.01%, low-laser 4 J, 100 mW, 660 nm) was performed as an adjuvant concomitant to oral hygiene maintenance. Rapid recovery was observed in three cases (Fig. 1). Although there is still a question about whether oral lesions in COVID-19 patients are due to a primary effect of coronavirus with local replication or opportunistic coinfections from other pathogens related also to the immunity impairment, aPDT may be an

TABLE 1 Patient characteristics ($n = 4$ cases).

	Case 1	Case 2	Case 3	Case 4
Sex	Male	Female	Male	Male
Age	59	73	63	57
Comorbidities	Hypertension, diabetes mellitus, obesity	Obesity, arrhythmia, hypothyroidism	Arrhythmia, depression	Dyslipidemia, backache
Initial signs and symptoms	Acute respiratory distress	Cough; dyspnea	Acute respiratory distress	Tachypnea; fever; acute respiratory distress
COVID-19 severity	Very severe (<60%) Mechanical ventilation	Severe (40%) Mechanical ventilation	Severe (>50%) Mechanical ventilation	Very severe (>50%) Mechanical ventilation

TABLE 1 Patient characteristics ($n=4$ cases)—cont'd

	Case 1	Case 2	Case 3	Case 4
COVID-19 medication	Methylprednisolone; Enoxaparin; Ceftriaxone + tazocin + meropenem + vancomycin + ertapenem + amikacin	Dexamethasone; Enoxaparin; Levofloxacin + tazocin + daptomycin	Methylprednisolone; Enoxaparin; Ceftriaxone	Methylprednisolone; Enoxaparin; Ceftriaxone + tazocin + meropenem + vancomycin + ertapenem + amikacin
Admission date	March 17, 2021	March 19, 2021	March 17, 2021	March 9, 2021
Oral mucosal lesions appearance	April 8, 2021	April 13, 2021	April 1, 2021	March 25, 2021
Oral treatment	Two sessions of oral hygiene and aPDT + intravenous acyclovir	Four sessions of oral hygiene and aPDT + intravenous acyclovir	Four sessions of oral hygiene and aPDT + topical penciclovir	Five sessions of oral hygiene and aPDT + intravenous acyclovir
Discharge/ death	April 11, 2021 (death)	NR	April 23, 2021 (Discharge)	April 1, 2021 (death)

Abbreviations: aPDT, antimicrobial photodynamic therapy; NR, not reported.

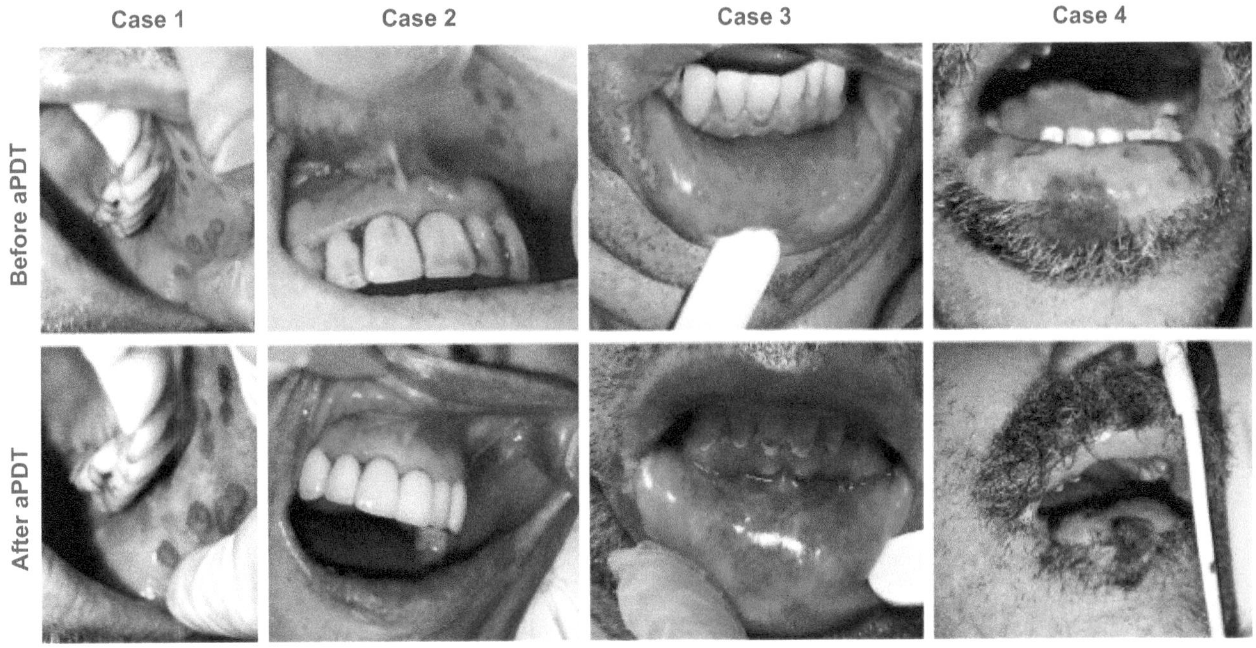

FIG. 1 Oral mucosal lesions in four patients with COVID-19. (This figure shows patient lesions before and after treatment with aPDT (antimicrobial photodynamic therapy).)

interesting alternative. aPDT is a topical and noninvasive therapy that provides antimicrobial effects, inhibiting bacterial, fungal, and viral microorganisms; this proposes applications for both suggested aetiologies. However, we refer attention to further studies accessing aPDT effectiveness and evaluating the etiopathogenesis of the oral mucosal lesions using molecular approaches in the context of COVID-19 patients. These concerns also reinforce the role of dental professionals as part of the multiprofessional team during a COVID-19 outbreak.

Summary points

- The number of studies reporting oral mucosal letions in patients with COVID-19 has increased.
- The clinical aspects of oral mucosal lesions presented in case reports and case series suggest coinfections and immunity impairment.
- Aphthous-like ulcers and herpes-like lesions were the most common oral mucosal lesions.
- We suggest studies with molecular approaches to confirm SARS-CoV-2 infection in the oral cavity to evaluate the etiopathogenesis of the oral mucosal lesions in COVID-19 patients.
- The current evidence suggests xerostomia, taste dysfunction, and oral mucosal lesions as common manifestations in patients with COVID-19.
- The case studies highlight the importance of including dentists in the intensive care unit multiprofessional team to improve oral health in COVID-19 patients.

References

Abubakr, N., Salem, Z. A., & Kamel, A. H. M. (2021). Oral manifestations in mild-tomoderate cases of COVID-19 viral infection in the adult population. *Dental and Medical Problems*, *58*(1), 7–15.

Amorim Dos Santos, J., Normando, A. G. C., Carvalho da Silva, R. L., Acevedo, A. C., De Luca Canto, G., Sugaya, N., Santos-Silva, A. R., & Guerra, E. N. S. (2021a). Oral manifestations in patients with COVID-19: A living systematic review. *Journal of Dental Research*, *100*(2), 141–154.

Amorim Dos Santos, J., Normando, A. G. C., Carvalho da Silva, R. L., Acevedo, A. C., De Luca Canto, G., Sugaya, N., Santos-Silva, A. R., & Guerra, E. N. S. (2021b). Oral manifestations in patients with COVID-19: A 6-month update. *Journal of Dental Research*, *100*(12), 1321–1329.

Amorim Dos Santos, J., Normando, A. G. C., Carvalho da Silva, R. L., De Paula, R. M., Cembranel, A. C., Santos-Silva, A. R., & Guerra, E. N. S. (2020). Oral mucosal lesions in a COVID-19 patient: New signs or secondary manifestations? *International Journal of Infectious Diseases*, *97*, 326–328.

Askin, O., Altunkalem, R. N., Altinisik, D. D., Uzuncakmak, T. K., Tursen, U., & Kutlubay, Z. (2020). Cutaneous manifestations in hospitalized patients diagnosed as COVID-19. *Dermatologic Therapy*, *33*(6), e13896.

Chaux-Bodard, A. G., Deneuve, S., & Desoutter, A. (2020). Oral manifestation of Covid-19 as an inaugural symptom? *Journal of Oral Medicine and Oral Surgery*, *26*(2), 1. https://doi.org/10.1051/mbcb/2020011.

de Maria, A., Varese, P., Dentone, C., Barisione, E., & Bassetti, M. (2020). High prevalence of olfactory and taste disorder during SARS-CoV-2 infection in outpatients. *Journal of Medical Virology*, *92*(11), 2310–2311.

Gupta, A., Madhavan, M. V., Sehgal, K., Nair, N., Mahajan, S., Sehrawat, T. S., Bikdeli, B., Ahluwalia, N., Ausiello, J. C., Wan, E. Y., et al. (2020). Extrapulmonary manifestations of COVID-19. *Nature Medicine*, *26*(7), 1017–1032.

Katz, J., & Yue, S. (2021). Increased odds ratio for COVID-19 in patients with recurrent aphthous stomatitis. *Journal of Oral Pathology & Medicine*, *50*(1), 114–117.

Lechien, J. R., Chiesa-Estomba, C. M., De Siati, D. R., Horoi, M., Le Bon, S. D., Rodriguez, A., Dequanter, D., Blecic, S., El Afia, F., Distinguin, L., et al. (2020). Olfactory and gustatory dysfunctions as a clinical presentation of mild-to-moderate forms of the coronavirus disease (COVID-19): A multicenter European study. *European Archives of Oto-Rhino-Laryngology*, *277*(8), 2251–2261.

Martín Carreras-Presas, C., Amaro Sánchez, J., López-Sánchez, A. F., Jané-Salas, E., & Somacarrera Pérez, M. L. (2021). Oral vesiculobullous lesions associated with SARS-CoV-2 infection. *Oral Diseases*, *27*(Suppl. 3), 710–712.

Mina, F. B., Billah, M., Karmakar, S., Das, S., Rahman, M. S., Hasan, M. F., & Acharjee, U. K. (2021). An online observational study assessing clinical characteristics and impacts of the COVID-19 pandemic on mental health: A perspective study from Bangladesh. *Zeitschrift fur Gesundheitswissenschaften*, *7*, 1–9.

Nuno-Gonzalez, A., Martin-Carrillo, P., Magaletsky, K., Martin Rios, M. D., Herranz Mañas, C., Artigas Almazan, J., García Casasola, G., Perez Castro, E., Gallego Arenas, A., Mayor Ibarguren, A., et al. (2021). Prevalence of mucocutaneous manifestations in 666 patients with COVID-19 in a field hospital in Spain: Oral and palmoplantar findings. *The British Journal of Dermatology*, *184*(1), 184–185. https://doi.org/10.1111/bjd.19564.

Putra, B. E., Adiarto, S., Dewayanti, S. R., & Juzar, D. A. (2020). Viral exanthem with "spins and needles sensation" on extremities of a COVID-19 patient: A self-reported case from an Indonesian medical frontliner. *International Journal of Infectious Diseases*, *96*, 355–358.

WHO, World Health Organization. (2020). *Health emergency dashboard WHO (COVID-19) Homepage*. https://covid19.who.int/.

Chapter 51

Case study: COVID-19 pneumonia presented with cavitary lesions

Bahadır M. Berktaş
Department of Pulmonology, University of Health Sciences, Atatürk Sanatorium Training and Research Hospital, Ankara, Turkey

Abbreviations

ACE2	angiotensin converting enzyme 2
c-ANCA	cytoplasmic antineutrophilic autoantibody
COVID-19	coronavirus disease 2019
CRP	C-reactive protein
CT	computed tomography
CTA	computed tomography angiography
dsDNA	double stranded deoxyribonucleic acid
ICU	intensive care unit
LMWH	low-molecular-weight heparin
MDR	multidrug resistant
MRI	magnetic resonance imaging
MRSA	methicillin-resistant *Staphylococcus aureus*
RT-PCR	reverse transcription polymerase chain reaction
SARS-CoV-2	severe acute respiratory syndrome coronavirus 2

Introduction

A novel coronavirus called SARS-CoV-2 was determined to be associated with pneumonia (Zhu et al., 2020) in December 2019. Hamming et al. (2004) showed that the surface expression of ACE2 is most abundant on the pulmonary type II alveolar epithelial cells. Apparently for this reason, the lungs are the most affected organ of a SARS-COV-2 infected person, particularly in severe COVID-19 patients.

One-fifth of infected persons remained asymptomatic after contact with SARS-CoV-2 (Kim et al., 2020). Symptoms of COVID-19 begin 3–14 days after exposure to SARS-COV-2, depending on a person's immunological status (Acter et al., 2020). While fever, cough, and fatigue are common in mild COVID-19 cases, symptoms of more severe cases that present with pneumonia are breathing difficulty, persistent chest pain or pressure, confusion, and cyanosis. A dry cough is common but sputum production may also be seen (Grant et al., 2020). Approximately 20% of SARS-COV-2 infected persons develop COVID-19 pneumonia (Camporota et al., 2022).

The infection primarily affects the respiratory system. Bilateral multilobar patchy or diffuse airspace opacities with a peripheral or posterior distribution are seen typically on chest X-ray examinations in patients with COVID-19 pneumonia (Ai et al., 2020; Salehi et al., 2020). COVID-19 pneumonia is a gradually progressing disease. Bernheim et al. (2020) found that the CT was normal in 56% of early patients. CT findings were more frequently observed after the beginning of symptoms. Although lung patterns on CT scans can considerably differ among patients with COVID-19 pneumonia, typical findings are pulmonary consolidations, ground-glass opacities, crazy paving patterns, and interlobular septal and pleural thickening. Lymphadenopathy, pleural or pericardial effusion, cavitation, and pneumothorax are rare but may be noticed with COVID-19 progression (Salehi et al., 2020).

Cavitary lung lesions secondary to COVID-19 pneumonia are rare and usually have appeared as case reports in the literature (Afrazi et al., 2021; Kurys-Denis et al., 2021; Selvaraj & Dapaah-Afriyie, 2020). Their prevalence was reported as 1.1%–11% in retrospective studies according to the severity of COVID-19 in patients in the study populations (Zarifian et al., 2021; Y. Zheng et al., 2021; Zoumot et al., 2021).

Management, Body Systems, and Case Studies in COVID-19. https://doi.org/10.1016/B978-0-443-18703-2.00040-2
Copyright © 2024 Elsevier Inc. All rights are reserved, including those for text and data mining, AI training, and similar technologies.

Recognizing unusual radiological presentations and complications is important in patients with COVID-19 pneumonia for prompt and tailored treatment to prevent irreversible lung damage. Therefore, two COVID-19 pneumonia cases complicated with cavitary lesions are presented herein.

Case one

A 74-year-old man was admitted to our hospital with shortness of breath, fatigue, loss of appetite, sore throat, and cough with sputum production for 4 days. His medical history was notable for chronic obstructive pulmonary disease, hypertension, and coronary artery disease. He was diagnosed with COVID-19 by RT-PCR with a nasopharyngeal swab. Upon presentation, he was tachypneic and hypoxic but bradycardic while a few coarse rales were heard bilaterally in his physical examination. The initial chest X-ray demonstrated bilateral air space consolidation opacities peripherally (Fig. 1A).

He was treated with enoxaparin for antithrombotic prophylaxis related to COVID-19 as well as dexamethasone 6 mg/day and cefepime for probable bacterial coinfection. However, his oxygen requirement increased and his inflammatory markers elevated 6 days after hospitalization. A chest X-ray showed that the pneumonia had progressively worsened and a new cavitary consolidation appeared on the right upper zone (Fig. 1B). A CTA scan of the chest demonstrated increased patchy reticular and ground-glass opacities and a new cavity of 3 cm in the right upper lobe without thrombosis in the pulmonary arteries (Fig. 1C and D).

The aetiologies for cavitary lung lesions including infectious, neoplastic, and autoimmune were evaluated. Rapid development in a few days was not consistent with neoplasms or mycobacterial or fungal infections. Three consecutive microscopic sputum examinations for *M. Tuberculosis* were negative and the serum galactomannan was found to be a low level (0.3 mg/L). Cultures of respiratory specimens showed no bacterial growth. A serological evaluation showed negative antinuclear antibodies, anti-dsDNA, and c-ANCA. It was concluded that the cavity was caused by COVID-19

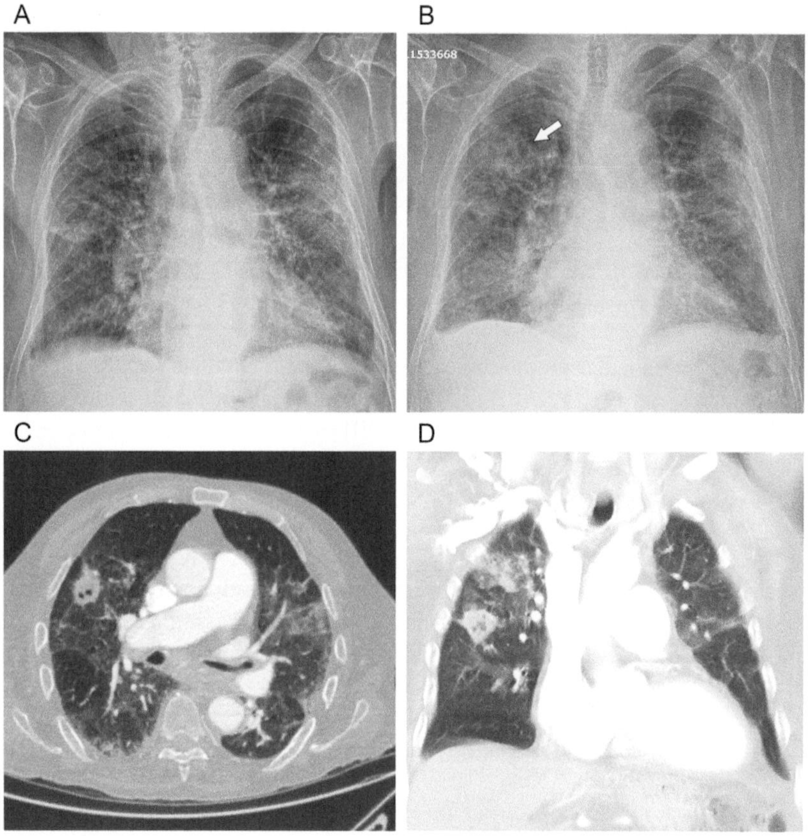

FIG. 1 Radiological images of Case One. (A) Bilateral, peripherally located air space consolidation opacities at admission. (B) Air space consolidations increased and a new cavitary consolidation (*arrow*) appeared on right upper zone on day 6. (C, D) Appearance of cavity and consolidations on transverse and coronal sections of CTA scan of the chest.

viral pneumonia in this patient. After the patient was transferred to the ICU, a third-degree atrioventricular block was diagnosed for his hypotension and arrhythmia. Unfortunately, he died just before cardiac pacemaker implantation.

Case two

A 58-year-old woman diagnosed with COVID-19 by RT-PCR with a nasopharyngeal swab a week ago was presented to another hospital with complaints of increased shortness of breath and fatigue. She had never smoked and her medical history was significant for type 2 diabetes mellitus and a peptic ulcer. Profound lymphocytopenia, hypokalemia, and hypocalcemia were found in her initial blood tests. Her serum ferritin, C-reactive protein, and D-dimer levels were high. A CT scan of her chest done at admission revealed bilateral, multifocal, patchy ground-glass opacities but no apparent cavitary lesion (Fig. 2A).

She was transferred to the ICU due to her serious hypoxia and high oxygen support needs. She treated with favipiravir, dexamethasone (6mg/day), cefepime (1g twice a day), LMWH (enoxaparin 60mg twice a day), and high-flow nasal oxygen at 60L/min initially. She was intubated and placed on mechanical ventilatory support for 12days due to persistent hypoxemia. The patient was transferred to the general COVID-19 ward after a 20-day stay in the ICU. Her RT-PCR for SARS-CoV-2 was negative after 34days.

CT angiography scans of the chest were done owing to her increased oxygen support requirement and elevated D-dimer levels. She was transferred to the ICU again on day 45. Two new thick-walled cavities within consolidation in the right lower lobe and thrombus in the right inferior pulmonary artery were detected in scans (Fig. 2B). Her RT-PCR for SARS-CoV-2 was positive again. Microscopic examinations for *M. tuberculosis* were negative, the serum galactomannan level was low, and the serology was negative for echinococcosis. Serological tests for antinuclear

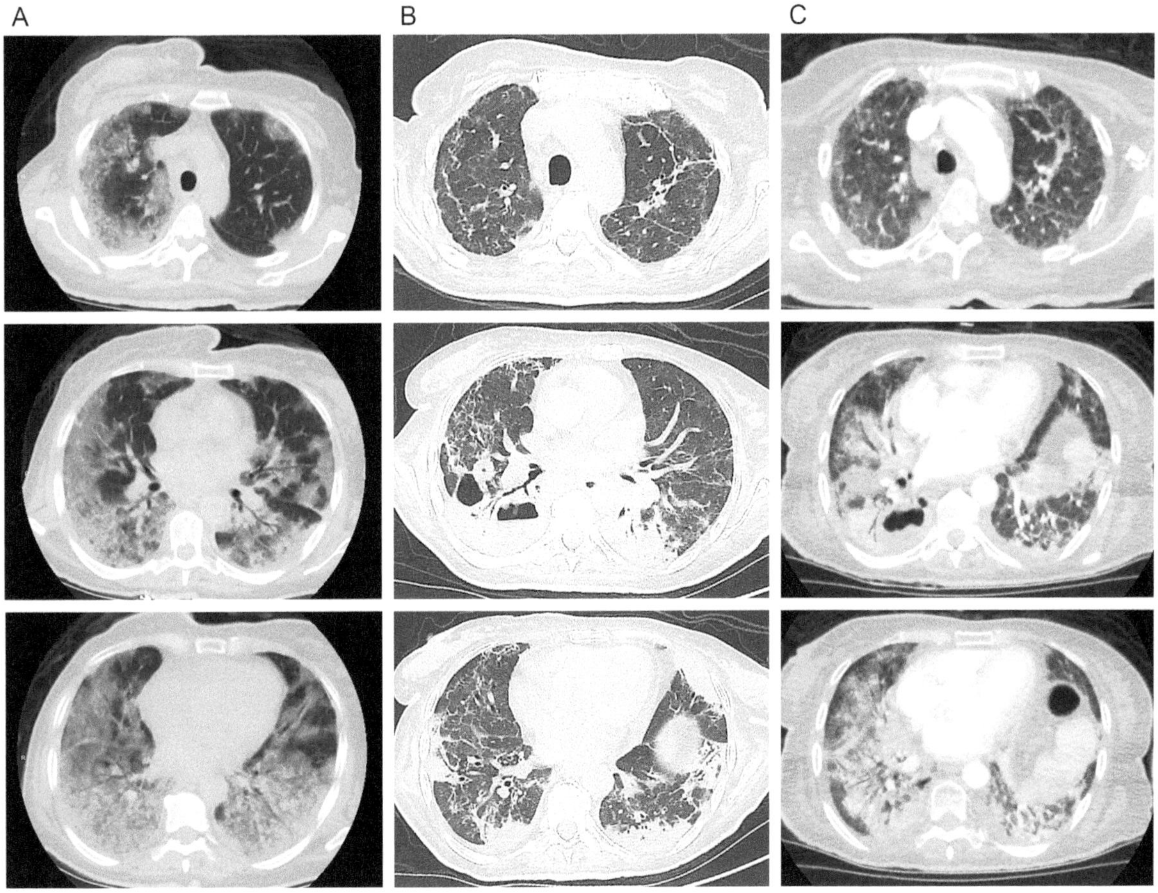

FIG. 2 Thoracic CT images of Case Two. (A) Bilateral, multifocal, patchy ground-glass opacities but no apparent cavitary lesion were seen on the chest CT at admission. (B) Two new thick-walled cavities within consolidation in the right lower lobe and thrombus in the right inferior pulmonary artery were detected on day 45 of the disease. (C) Cavitary lesions persevered and air space consolidations increased simultaneously with the deteriorating clinical status of the patient.

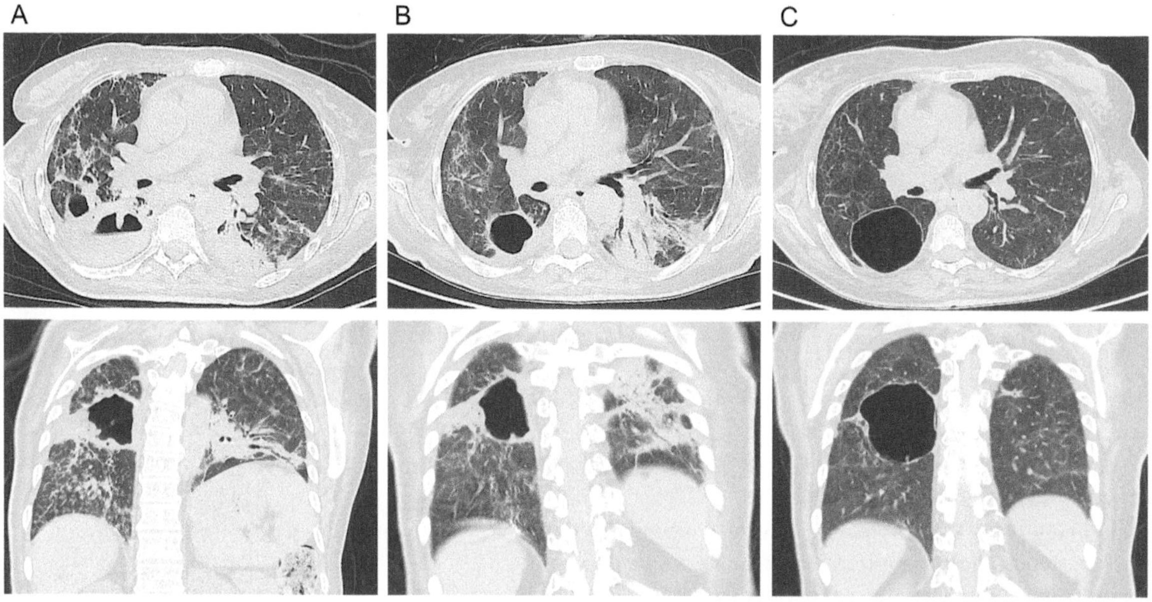

FIG. 3 Changes of cavitary lesions in thoracic CT images of Case Two. (A) Cavitary lesions continued but air space consolidations diminished simultaneously with improved clinical status on day 72. (B) Air space consolidations around cavity decreased and two cavities combined into one large, thinner-walled cavity on day 98. (C) The cavity transformed to a thinner-walled but greater-sized pneumatocele after 2 months.

antibodies and anti-dsDNA were negative. Serum CRP and procalcitonin levels increased and multidrug-resistant Klebsiella pneumonia and MRSA were isolated from cultures. Cavitary lesions persisted and air space consolidations were increased on a chest CT (Fig. 2C). Meropenem and amikacin combination therapy was started according to antimicrobial susceptibility test results. Her clinical status and oxygen support need improved while inflammatory parameters decreased with this treatment. Her RT-PCR for SARS-CoV-2 was negative on day 57.

The patient was transferred to the pulmonary ward of our hospital for pulmonary cavities and respiratory deficiency on day 72. A chest CT revealed that cavitary lesions continued but the air space consolidations were diminished (Fig. 3A). She could not walk due to diminished muscle strength in the lower extremities. After cranial and vertebral MRIs, sarcopenia caused by the long ICU stay was found to be the cause for muscle weakness. The clinical status and oxygen support need of the patient deteriorated and a fever appeared 2 weeks after hospitalization. Serum CRP and procalcitonin levels increased again. Aspergillosis DNA was found to be negative in the serum. Microscopic examinations for *M. tuberculosis* were negative. *Candida albicans*, *Pseudomonas aeruginosa*, Stenotrophomonas maltofilia, MDR Klebsiella, and MRSA were isolated from cultures.

She was treated with combined antimicrobial therapy according to the resistance patterns of microorganisms. The fever discontinued while her symptoms and inflammatory marker levels decreased in a week. Her respiratory condition improved after these treatments and she was discharged with home oxygen support. Consolidations around the cavity decreased and two cavities combined into one large, thinner-walled cavity, as revealed by a chest CT of the patient at discharge on day 98 (Fig. 3B).

The patient was reassessed for clinical status and radiological findings 2 months after discharge. Her exercise capacity improved and her oxygen dependency decreased. A CT evaluation of the chest showed that the cavity transformed to a thinner-walled but greater-sized pneumatocele (Fig. 3C). Pneumatocele often resolves spontaneously. Therefore, a conservative approach with careful follow-up was planned.

Management of cavitary lesions in COVID-19 pneumonia

A pulmonary cavity is defined as an air-filled space, perceived as lucency, within consolidation, a mass, or a nodule in the lung (Hansell et al., 2008). Cavities may present with variable wall thickness and sometimes contain a fluid level. Pulmonary cavities are typically observed as a consequence of bacterial, parasitic, or fungal infections; neoplasms; pulmonary embolisms; or autoimmune diseases (Zoumot et al., 2021).

Cavities and cystic lung changes commonly result from pulmonary coinfections and superinfections caused by bacteria, fungus, and tuberculosis in severe COVID-19 patients, particularly those treated with antibiotics, glucocorticoids, and mechanical ventilators for a long time (Chiang et al., 2004). S. Zheng et al. (2020) showed that severe pulmonary complications correlated with higher viral loads and a longer duration of COVID-19 infection.

The prevalence of coinfections and superinfections was reported as 7%–24% for COVID-19 patients in metaanalyses (Lansbury et al., 2020; Musuuza et al., 2021). The Most frequently isolated bacteria were Mycoplasma pneumonia, *P. aeruginosa*, and *Haemophilus influenzae* while the prevailing fungi were Aspergillus fumigatus and *C. albicans*; the most observed viruses as coinfections were respiratory syncytial virus and influenza A (Rawson et al., 2020). *Klebsiella pneumoniae*, *Escherichia coli*, Pseudomonas, Enterobacter and rarely Acinetobacter species were isolated as causes of late superinfections and cavities (Naranje et al., 2021). COVID-19 viral pneumonia can cause cavitary lung lesions without other aetiologies (Afrazi et al., 2021; Selvaraj & Dapaah-Afriyie, 2020; Kurys-Denis et al., 2021), as reported in the first case herein.

The repairing process of COVID-19 pneumonia is poorly understood. Many theories have been suggested for the pathophysiology of cavities in COVID-19 pneumonia. Yao et al. (2020) demonstrated by electron microscope that coronavirus particles were found in bronchial mucosa. Exfoliated bronchial epithelia and inflammatory mediators can be responsible for mucous formation. COVID-19 viral pneumonia or coinfection/superinfection-related sustained mucous in bronchioles, diffuse alveolar damage, and parenchymal necrosis may cause the development of bronchiectasis, which eventually transforms into cavitation via airway obstruction (Yao et al., 2020; Ammar et al., 2021). COVID-19-related microvascular thrombosis in pulmonary capillaries might also contribute to lung infarcts undergoing liquefaction and postobstructive dilatation via ischemic and inflammatory changes in the parenchyma of the lungs (Deshpande, 2020; Kruse et al., 2021; Ackermann et al., 2020).

The entire picture of COVID-19 pneumonia-related pulmonary pathology and postacute sequelae remains still largely unknown. Although cavitation is unusual among plenty pulmonary complications of COVID-19, cavities are associated with high morbidity and mortality (Afrazi et al., 2021). Clinical and radiological alterations in patients with COVID-19 pneumonia should be closely monitored during the hospital stay and the postdischarge follow-up period to detect complicated courses of infection such as cavity formation. Timely intervention with identification of the underlying pathology, management of the coinfection and superinfections, cautious fluid management, and supporting other affected vital organs can ensure minimal morbidity and mortality.

Summary points

- COVID-19 pneumonia is a gradually progressing disease. Clinical and radiological alterations in patients with COVID-19 pneumonia should be closely monitored.
- While cavitation is fairly unusual among plenty pulmonary complications of COVID-19, it is associated with high morbidity and mortality.
- A pulmonary cavity is defined as an air-filled space, perceived as lucency, within consolidation, a mass, or a nodule in the lung.
- Cavities commonly result from pulmonary coinfections and superinfections caused by bacteria, fungus, and tuberculosis in severe COVID-19 patients.
- COVID-19 viral pneumonia can cause cavitary lung lesions without other aetiologies.
- If COVID-19 pneumonia is complicated by cavity formation, timely intervention with identification of the underlying pathology as well as the management of coinfections and superinfections can ensure minimal morbidity and mortality.

References

Ackermann, M., Verleden, S. E., Kuehnel, M., Haverich, A., Welte, T., Laenger, F., Vanstapel, A., Werlein, C., Stark, H., Tzankov, A., Li, W. W., Li, V. W., Mentzer, S. J., & Jonigk, D. (2020). Pulmonary vascular endothelialitis, thrombosis, and angiogenesis in Covid-19. *The New England Journal of Medicine*, *383*, 120–128.

Acter, T., Uddin, N., Das, J., Akhter, A., Choudhury, T. R., & Kim, S. (2020). Evolution of severe acute respiratory syndrome coronavirus 2 (SARS-CoV-2) as coronavirus disease 2019 (COVID-19) pandemic: A global health emergency. *Science of the Total Environment*, *730*, 138996.

Afrazi, A., Garcia-Rodriguez, S., Maloney, J. D., & Morgan, C. T. (2021). Cavitary lung lesions and pneumothorax in a healthy patient with active coronavirus-19 (COVID-19) viral pneumonia. *Interactive Cardiovascular and Thoracic Surgery*, *32*, 150–152.

Ai, T., Yang, Z., Hou, H., Zhan, C., Chen, C., Lv, W., Tao, Q., Sun, Z., & Xia, L. (2020). Correlation of chest CT and RT-PCR testing for coronavirus disease 2019 (COVID-19) in China: A report of 1014 cases. *Radiology*, *296*, E32–E40.

Ammar, A., Drapé, J. L., & Revel, M. P. (2021). Lung cavitation in COVID-19 pneumonia. *Diagnostic and Interventional Imaging*, *2021*(102), 117–118.

Bernheim, A., Mei, X., Huang, M., Yang, Y., Fayad, Z. A., Zhang, N., Diao, K., Lin, B., Zhu, X., Li, K., Li, S., Shan, H., Jacobi, A., & Chung, M. (2020). Chest CT findings in coronavirus Disease-19 (COVID-19): Relationship to duration of infection. *Radiology*, *295*, 200463.

Camporota, L., Cronin, J. N., Busana, M., Gattinoni, L., & Formenti, F. (2022). Pathophysiology of coronavirus-19 disease acute lung injury. *Current Opinion in Critical Care*, *28*, 9–16.

Chiang, C. H., Shih, J. F., Su, W. J., & Perng, R. P. (2004). Eight-month prospective study of 14 patients with hospital-acquired severe acute respiratory syndrome. *Mayo Clinic Proceedings*, *79*, 1372–1379.

Deshpande, C. (2020). Thromboembolic findings in COVID-19 autopsies: Pulmonary thrombosis or embolism? *Annals of Internal Medicine*, *173*, 394–395.

Grant, M. C., Geoghegan, L., Arbyn, M., Mohammed, Z., McGuinness, L., Clarke, E. L., & Wade, R. G. (2020). The prevalence of symptoms in 24,410 adults infected by the novel coronavirus (SARS-CoV-2; COVID-19): A systematic review and meta-analysis of 148 studies from 9 countries. *PLoS One*, *15*, e0234765.

Hamming, I., Timens, W., Bulthuis, M. L., Lely, A. T., Navis, G., & van Goor, H. (2004). Tissue distribution of ACE2 protein, the functional receptor for SARS coronavirus. A first step in understanding SARS pathogenesis. *Journal of Pathology*, *203*, 631–637.

Hansell, D. M., Bankier, A. A., MacMahon, H., McLoud, T. C., Müller, N. L., & Remy, J. (2008). Fleischner society: Glossary of terms for thoracic imaging. *Radiology*, *246*, 697–722.

Kim, G. U., Kim, M. J., Ra, S. H., Lee, J., Bae, S., Jung, J., & Kim, S. H. (2020). Clinical characteristics of asymptomatic and symptomatic patients with mild COVID-19. *Clinical Microbiology and Infection*, *26*, 948.e1–948.e3.

Kruse, J. M., Zickler, D., Lüdemann, W. M., Piper, S. K., Gotthardt, I., Ihlow, J., Greuel, S., Horst, D., Kahl, A., Eckardt, K. U., & Elezkurtaj, S. (2021). Evidence for a thromboembolic pathogenesis of lung cavitations in severely ill COVID-19 patients. *Scientific Reports*, *11*, 16039.

Kurys-Denis, E., Grzywa-Celińska, A., & Celiński, R. (2021). Lung cavitation as a consequence of coronavirus-19 pneumonia. *European Review for Medical and Pharmacological Sciences*, *25*, 5936–5941.

Lansbury, L., Lim, B., Baskaran, V., & Lim, W. S. (2020). Co-infections in people with COVID-19: A systematic review and meta-analysis. *The Journal of Infection*, *81*, 266–275.

Musuuza, J. S., Watson, L., Parmasad, V., Putman-Buehler, N., Christensen, L., & Safdar, N. (2021). Prevalence and outcomes of co-infection and superinfection with SARS-CoV-2 and other pathogens: A systematic review and meta-analysis. *PLoS One*, *16*, e0251170.

Naranje, P., Bhalla, A. S., Jana, M., Garg, M., Nair, A. D., Singh, S. K., & Banday, I. (2021). Imaging of pulmonary superinfections and co-infections in COVID-19. *Current Problems in Diagnostic Radiology*, *51*, 768–778.

Rawson, T. M., Moore, L. S. P., Zhu, N., Ranganathan, N., Skolimowska, K., Gilchrist, M., Satta, G., Cooke, G., & Holmes, A. (2020). Bacterial and fungal coinfection in individuals with coronavirus: A rapid review to support COVID-19 antimicrobial prescribing. *Clinical Infectious Diseases*, *71*, 2459–2468.

Salehi, S., Abedi, A., Balakrishnan, S., & Gholamrezanezhad, A. (2020). Coronavirus disease 2019 (COVID-19): A systematic review of imaging findings in 919 patients. *American Journal of Roentgenology*, *215*, 87–93.

Selvaraj, V., & Dapaah-Afriyie, K. (2020). Lung cavitation due to COVID-19 pneumonia. *BMJ Case Reports*, *13*, e237245.

Yao, X. H., Li, T. Y., He, Z. C., Ping, Y. F., Liu, H. W., Yu, S. C., Mou, H. M., Wang, L. H., Zhang, H. R., Fu, W. J., Luo, T., Liu, F., Guo, Q. N., Chen, C., Xiao, H. L., Guo, H. T., Lin, S., Xiang, D. F., Shi, Y., … Bian, X. W. (2020). A pathological report of three COVID-19 cases by minimal invasive autopsies. *Zhonghua Bing Li Xue Za Zhi*, *49*, 411–417.

Zarifian, A., Ghasemi, N. M., Akhavan, R. A., Rahimzadeh, O. R., Abbasi, B., & Sadeghi, R. (2021). Chest CT findings of coronavirus disease 2019 (COVID-19): A comprehensive meta-analysis of 9907 confirmed patients. *Clinical Imaging*, *70*, 101–110.

Zheng, S., Fan, J., Yu, F., Feng, B., Lou, B., Zou, Q., Xie, G., Lin, S., Wang, R., Yang, X., Chen, W., Wang, Q., Zhang, D., Liu, Y., Gong, R., Ma, Z., Lu, S., Xiao, Y., Gu, Y., … Liang, T. (2020). Viral load dynamics and disease severity in patients infected with SARS-CoV-2 in Zhejiang province, China, January-march 2020: Retrospective cohort study. *BMJ*, *369*, m1443.

Zheng, Y., Wang, L., & Ben, S. (2021). Meta-analysis of chest CT features of patients with COVID-19 pneumonia. *Journal of Medical Virology*, *93*, 241–249.

Zhu, N., Zhang, D., Wang, W., Li, X., Yang, B., Song, J., Zhao, X., Huang, B., Shi, W., Lu, R., Niu, P., Zhan, F., Ma, X., Wang, D., Xu, W., Wu, G., Gao, G. F., Tan, W., & China Novel Coronavirus Investigating and Research Team. (2020). A novel coronavirus from patients with pneumonia in China, 2019. *The New England Journal of Medicine*, *382*, 727–733.

Zoumot, Z., Bonilla, M. F., Wahla, A. S., Shafiq, I., Uzbeck, M., El-Lababidi, R. M., Hamed, F., Abuzakouk, M., & ElKaiisi, M. (2021). Pulmonary cavitation: An under-recognized late complication of severe COVID-19 lung disease. *BMC Pulmonary Medicine*, *21*, 24.

Chapter 52

Case study: Optic neuritis in SARS-CoV-2 infection

Md Moshiur Rahman
Neurosurgery Department, Holy Family Red Crescent Medical College, Dhaka, Bangladesh

Abbreviation

ARDS	acute respiratory distress syndrome
COVID-19	Coronavirus 2019
CT	computer tomography
MS	multiple sclerosis
NMOSD	neuromyelitis optica spectrum disorders
ON	optic neuritis
PCR	polymerase chain reaction
SARS-CoV-2	severe acute respiratory syndrome coronavirus 2

Introduction

Coronavirus 2019 (COVID-19), a global pandemic, infected the entire world beginning in December 2019. SARS-CoV-2 is the virus linked to COVID-19 infections; the infections are primarily pulmonary, manifesting with severe respiratory complications such as acute respiratory distress syndrome (ARDS). However, extrapulmonary manifestations associated with SARS-CoV-2 have been reported in the published literature (Jiang et al., 2020). Several neuroophthalmic manifestations of COVID-19 infection have been reported. However, isolated optic neuritis in humans with COVID-19 has been reported infrequently. Optic neuritis, if it occurred, was usually part of a demyelinating syndrome.

Literature

Optic neuritis is usually unilateral, with an acute painful eye or just periorbital pain, and it manifests itself within a week (de la Cruz & Kupersmith, 2006). Multiple sclerosis (MS) or neuromyelitis Optica spectrum disorders (NMOSD) have been reported following COVID-19 vaccination (Badrawi et al., 2021; Chen et al., 2021; Ismail & Salama, 2022; Ozgen Kenangil et al., 2021; Pagenkopf & Südmeyer, 2021; Román et al., 2021), and optic neuritis (ON) is also a known complication (Khayat-Khoei et al., 2022; Yildiz Tasci et al., 2022). Although idiopathic ON following COVID-19 vaccination has been reported (Elnahry et al., 2021; Nagaratnam et al., 2022; Roy et al., 2022; Wang et al., 2022), the visual prognoses are generally favorable. Several neurological and ophthalmic manifestations of coronavirus disease 2019 (COVID-19) infection have been reported (Nasiri et al., 2021; Nepal et al., 2020). Dry eye, foreign body sensation, conjunctivitis, and keratoconjunctivitis are all common ophthalmic manifestations (Sen et al., 2021). COVID-19 neurological manifestations ranged from common indications such as headaches, hyposmia, and hypogeusia to rare indications such as ischemic and hemorrhagic stroke, myelitis, encephalomyelitis, peripheral neuropathy, and rhabdomyolysis (Nasiri et al., 2021). COVID-19 infection has been linked to optic neuritis, retinitis, uveitis, and vasculitis in animals (Hu et al., 2022). Human ophthalmic manifestations, on the other hand, are typically mild (Hu et al., 2022; Sen et al., 2021). A few cases of COVID-19 infection in humans have been linked to optic neuritis (Gold & Galetta, 2021). The majority of these cases occurred during the COVID-19 infection recovery phase or as part of a demyelination disease.

Management, Body Systems, and Case Studies in COVID-19. https://doi.org/10.1016/B978-0-443-18703-2.00031-1
Copyright © 2024 Elsevier Inc. All rights are reserved, including those for text and data mining, AI training, and similar technologies.

The case study

A 33-year-old female patient presented to the emergency room with bilateral subacute vision loss, particularly in the left eye. Her condition began a few days before presentation with a bitemporal tension headache of moderate severity that was more noticeable on the left side. She also had left eye pain that was aggravated by eye movement. She denied experiencing any diplopia, squinting, other facial pain, bulbar symptoms, or cranial nerve dysfunction. The patient had a fever, loss of smell and taste, and a sore throat 1 week before her vision loss. A computer tomography (CT) scan of the chest and a polymerase chain reaction (PCR) test for COVID-19 were both positive. Upon reviewing her past medical history, she revealed that despite 10 years of marriage, she had not been able to conceive, but her infertility had not been investigated. She stated that she had no other symptoms of hormonal dysfunction. She has no notable drug history, no relevant family history, and no notable psychosocial history.

Physical examination revealed that she was of average weight and height, with no distinguishing features. Bitemporal upper quadrantanopia was discovered during a visual field examination using the perimeter. There were no demyelinating lesions in the brain parenchyma. Except for prolactinemia, a complete hormonal profile was dull. Aquaporin 4 and anti-MOG antibodies were both negative. Pituitary macroadenoma with optic neuritis was diagnosed in the patient. Because the patient tested positive for COVID-19 infection, treatment was postponed until she recovered completely. The optic neuritis improved significantly, with full recovery achieved 3 weeks after the corticosteroid course began. An endonasal transsphenoidal approach was used to remove the pituitary macroadenoma. A senior consultant in neurosurgery performed the procedure. One month after surgery, the prolactin level returned to normal (22 ng/mL), and the visual field defect resolved. The patient's visual and endocrine complaints improved significantly, and she resumed her normal daily activities.

Optic neuritis following COVID-19 vaccination

Optic neuritis, a primarily clinical entity characterized by blurred vision, loss of color vision, and painful eye movement, is a rare but serious side effect of vaccination. Vaccinations have helped to eradicate many infectious diseases throughout history, including smallpox, poliomyelitis, and measles. Seizures and encephalopathy (Karussis & Petrou, 2014) are not uncommon after vaccination, with the earliest reports dating back to the late 19th century, following the development of neuroparalytic syndrome after Pasteur's rabies immunization (Miravalle et al., 2010). The rapid development and availability of vaccination for relief from the COVID-19 pandemic, ahead of an anticipated timetable, was an unprecedented and monumental accomplishment, leaving the prospect and question of long-term safety open and ambiguous. A metaanalysis of five randomized, double-blind, placebo-controlled trials of COVID-19 vaccine candidates found that all the reported local and systemic adverse events were mild to moderate and transient (Yuan et al., 2020). As a result, reporting an unfavorable outcome after vaccination is critical to establishing a safety threshold for widespread public use. Ophthalmic manifestations have appeared in a variety of forms following COVID-19, with neuroophthalmic manifestations being uncommon. Reporting cases of adverse reactions that occur after vaccination is a difficult but necessary task to continue developing and researching vaccines that are safe and effective for public use.

Atypical unilateral optic neuritis following COVID-19 vaccination

COVID-19 is still widespread around the world, with the development of vaccines against the causative virus severe acute respiratory syndrome coronavirus 2 (SARS-CoV-2) moving quickly and in high demand. Several vaccines have been approved for emergency use in the prevention of COVID-19, though the vaccines' critical adverse events have not been thoroughly investigated. Multiple sclerosis (MS) or neuromyelitis optica spectrum disorders (NMOSD) have been reported following COVID-19 vaccination (Ismail & Salama, 2022; Pagenkopf & Südmeyer, 2021), and optic neuritis (ON) is also a known complication (Nagaratnam et al., 2022). Although idiopathic ON has been reported following COVID-19 vaccination (Garcia-Estrada et al., 2022), the visual prognoses are generally favorable.

Conclusion

Optic neuritis, either unilateral or bilateral, is one of the possible outcomes of COVID-19 infection, to the best of our knowledge. Neuroophthalmic manifestations are uncommon in COVID-19 infection and can occur during the active disease or recovery phases. Optic neuritis is one such uncommon symptom. Keeping the ongoing pandemic in mind, we must be vigilant in identifying the neuroophthalmic features of COVID-19 infection to avoid irreversible vision loss. We recommend that any patient who experiences a sudden loss of vision or unusual headaches see a neurology specialist as soon as possible to rule out any postviral complications.

Summary points

- Optic neuritis is a rare complication of SARS-CoV-2 infection.
- Headaches, anosmia, and dysgeusia are the most common neurologic symptoms in COVID-19.
- Optic neuritis in the presence of pituitary macroadenoma following COVID-19 infection is a dilemma in determining whether the visual symptoms are due to the tumor or COVID-19 infection.

References

Badrawi, N., Kumar, N., & Albastaki, U. (2021). Post COVID-19 vaccination neuromyelitis optica spectrum disorder: Case report & MRI findings. *Radiology Case Reports*, *16*, 3864–3867.

Chen, S., Fan, X. R., He, S., Zhang, J. W., & Li, S. J. (2021). Watch out for neuromyelitis optica spectrum disorder after inactivated virus vaccination for COVID-19. *Neurological Sciences*, *42*, 3537–3539.

de la Cruz, J., & Kupersmith, M. J. (2006). Clinical profile of simultaneous bilateral optic neuritis in adults. *The British Journal of Ophthalmology*, *90*(5), 551–554.

Elnahry, A. G., Asal, Z. B., Shaikh, N., Dennett, K., Abd Elmohsen, M. N., Elnahry, G. A., Shehab, A., Vytopil, M., Ghaffari, L., Athappilly, G. K., et al. (2021). Optic neuropathy after COVID-19 vaccination: A report of two cases. *The International Journal of Neuroscience*, *1–7*.

Garcia-Estrada, C., Gomez-Figueroa, E., Alban, L., & Arias-Cardenas, A. (2022). Optic neuritis after COVID-19 vaccine application. *Clinical and Experimental Neuroimmunology*, *13*, 72–74.

Gold, D. M., & Galetta, S. L. (2021). Neuro-ophthalmologic complications of coronavirus disease 2019 (COVID-19). *Neuroscience Letters*, *742*.

Hu, K., Patel, J., Swiston, C., & Patel, B. C. (2022). Ophthalmic manifestations of coronavirus (COVID-19). In *StatPearls*. Treasure Island, FL: StatPearls Publishing.

Ismail, I. I., & Salama, S. (2022). A systematic review of cases of CNS demyelination following COVID-19 vaccination. *Journal of Neuroimmunology*, *362*, 577765.

Jiang, F., Deng, L., Zhang, L., Cai, Y., Cheung, C. W., & Xia, Z. (2020). Review of the clinical characteristics of coronavirus disease 2019 (COVID-19). *Journal of General Internal Medicine*, *35*(5), 1545–1549.

Karussis, D., & Petrou, P. (2014). The spectrum of post-vaccination inflammatory CNS demyelinating syndromes. *Autoimmunity Reviews*, *13*, 215–224.

Khayat-Khoei, M., Bhattacharyya, S., Katz, J., Harrison, D., Tauhid, S., Bruso, P., Houtchens, M. K., Edwards, K. R., & Bakshi, R. (2022). COVID-19 mRNA vaccination leading to CNS inflammation: A case series. *Journal of Neurology*, *269*, 1093–1106.

Miravalle, A., Biller, J., Schnitzler, E., & Bonwit, A. (2010). Neurological complications following vaccinations. *Neurological Research*, *32*, 285–292.

Nagaratnam, S. A., Ferdi, A. C., Leaney, J., Lee, R. L. K., Hwang, Y. T., & Heard, R. (2022). Acute disseminated encephalomyelitis with bilateral optic neuritis following ChAdOx1 COVID-19 vaccination. *BMC Neurology*, *22*, 54.

Nasiri, N., Sharifi, H., Bazrafshan, A., Noori, A., Karamouzian, M., & Sharifi, A. (2021). Ocular manifestations of COVID-19: a systematic review and meta-analysis. *Journal of Ophthalmic & Vision Research*.

Nepal, G., Rehrig, J. H., Shrestha, G. S., et al. (2020). Neurological manifestations of COVID-19: A systematic review. *Critical Care*, *24*, 421. https://doi.org/10.1186/s13054-020-03121-z.

Ozgen Kenangil, G., Ari, B. C., Guler, C., & Demir, M. K. (2021). Acute disseminated encephalomyelitis like presentation after an inactivated coronavirus vaccine. *Acta Neurologica Belgica*, *121*, 1089–1091.

Pagenkopf, C., & Südmeyer, M. (2021). A case of longitudinally extensive transverse myelitis following vaccination against Covid-19. *Journal of Neuroimmunology*, *358*, 577606.

Román, G. C., Gracia, F., Torres, A., Palacios, A., Gracia, K., & Harris, D. (2021). Acute transverse myelitis (ATM): Clinical review of 43 patients with COVID-19-associated ATM and 3 post-vaccination ATM serious adverse events with the ChAdOx1 nCoV-19 vaccine (AZD1222). *Frontiers in Immunology*, *12*, 653786.

Roy, M., Chandra, A., Roy, S., & Shrotriya, C. (2022). Optic neuritis following COVID-19 vaccination: Coincidence or side-effect?—A case series. *Indian Journal of Ophthalmology*, *70*, 679–683.

Sen, M., Honavar, S. G., Sharma, N., & Sachdev, M. S. (2021). COVID-19 and eye: A review of ophthalmic manifestations of COVID-19. *Indian Journal of Ophthalmology*, *69*(3), 488e509.

Wang, J., Huang, S., Yu, Z., Zhang, S., Hou, G., & Xu, S. (2022). Unilateral optic neuritis after vaccination against the coronavirus disease: Two case reports. *Documenta Ophthalmologica*, *145*, 65–70.

Yildiz Tasci, Y., Nalcacoglu, P., Gumusyayla, S., Vural, G., Toklu, Y., & Yesilirmak, N. (2022). Aquaporin-4 protein antibody-associated optic neuritis related to neuroendocrine tumor after receiving an inactive COVID-19 vaccine. *Indian Journal of Ophthalmology*, *70*, 1828–1831.

Yuan, P., Ai, P., Liu, Y., Ai, Z., Wang, Y., Cao, W., et al. (2020). Safety, tolerability, and immunogenicity of COVID-19 vaccines: A systematic review and meta-analysis. *medRxiv*, *11*, 20224998. https://doi.org/10.1101/2020.11.03.20224998.

Section H

Resources

Chapter 53

Recommended resources relevant to the body systems involvement and management of COVID-19

Rajkumar Rajendram[a,b], Daniel Gyamfi[c], Vinood B. Patel[d], and Victor R. Preedy[e,f]

[a]*Department of Medicine, King Abdulaziz Medical City, King Abdullah International Medical Research Center, Ministry of National Guard-Health Affairs, Riyadh, Saudi Arabia,* [b]*Department of Medical Education, College of Medicine, King Saud bin Abdulaziz University for Health Sciences, Riyadh, Saudi Arabia,* [c]*The Doctors Laboratory Ltd, London, United Kingdom,* [d]*Centre for Nutraceuticals, School of Life Sciences, University of Westminster, London, United Kingdom,* [e]*Faculty of Life Sciences and Medicine, King's College London, London, United Kingdom,* [f]*Department of Clinical Biochemistry, King's College Hospital, London, United Kingdom*

Abbreviations

ARDS	acute respiratory distress syndrome
CARDS ARDS	induced by COVID-19
COVID-19	coronavirus disease 2019
SARS-CoV-2	severe acute respiratory syndrome coronavirus 2

Introduction

In 2020, the novel coronavirus called severe acute respiratory syndrome coronavirus 2 (SARS-CoV-2) caused a global pandemic. Many patients infected with SARS-CoV-2 are asymptomatic or have few symptoms (Lei et al., 2020; Rajendram, Gyamfi, et al., 2024a). This subset of patients has a good prognosis (Lei et al., 2020). The syndrome caused by symptomatic infection with the highly contagious SARS-CoV-2 is called coronavirus disease 2019 (COVID-19) (Rajendram, Gyamfi, et al., 2024a).

Select cases require hospitalization, although most cases are mild. Symptomatic patients usually show the manifestations of a common cold (Lei et al., 2020; Rajendram, Gyamfi, et al., 2024a) or allergic rhinitis. Yet, the possible manifestations of COVID-19 are legion. Despite several large multicenter trials, few effective treatments for COVID-19 have been identified.

Common clinical manifestations of mild COVID-19 include dry cough, sore throat, fatigue, fever, headache, loss of taste and smell, malaise, myalgia, nausea, diarrhea, and vomiting (Lei et al., 2020; Magro et al., 2023). These symptoms sometimes progress to bronchiolitis and pneumonia in susceptible individuals (Lei et al., 2020). In select cases, COVID-19 can progress further, becoming a severe condition with hypoxemia and dyspnoea, and rapidly develop into acute respiratory distress syndrome (ARDS) (Cheng et al., 2020).

The specific form of ARDS induced by COVID-19 has been labeled CARDS (Rajendram, 2020). This is a multifaceted syndrome that may be further complicated by pneumothorax (Rajendram & Hussain, 2021) and intracardiac shunt (Rajendram, 2020; Rajendram, Hussain, et al., 2020; Rajendram, Hussain, et al., 2021; Rajendram, Kharal, et al., 2020a, 2020b; Rajendram, Kharal, & Puri, 2021).

Infection with SARS-CoV-2 can involve almost every organ system, resulting in a high mortality rate and multiorgan failure (Gupta et al., 2020). While most physicians focus on the respiratory manifestations of COVID-19, it is important to recognize psychiatric and behavioral features such as depression and sleep disturbances (Paunescu et al., 2023). All they are treatable when identified, these effects of COVID-19 are underrecognized and significantly impair the quality of life.

Since the start of the pandemic, the body systems involvement and management of COVID-19 have been actively investigated and covered in the thematic series of books edited by Rajendram, Patel, and Preedy (2024a, 2024b, 2024c, 2024d). In this chapter, we present tables containing resources as recommended by active clinical and nonclinical practitioners or

Management, Body Systems, and Case Studies in COVID-19. https://doi.org/10.1016/B978-0-443-18703-2.00002-5
Copyright © 2024 Elsevier Inc. All rights are reserved, including those for text and data mining, AI training, and similar technologies.

researchers. Recommended resources relating to COVID-19 are also found in Rajendram, Gyamfi, et al. (2024a, 2024b, 2024c). We acknowledge the experts who contributed to these resources.

Resources

We adopted the same format of tables used in previous compilations of resources (for example, see Rajendram et al., 2016a, 2016b, 2017). Thus, Tables 1–5 list the most up-to-date information on the regulatory bodies (Table 1), professional societies (Table 2), books (Table 3), emerging technologies (Table 4), and other resources of interest (Table 5) that are relevant

TABLE 1 Regulatory bodies or organizations relevant to the body systems involvement and management of SARS-CoV-2 or COVID-19.

Regulatory body or organization	Web address
American Board of Surgery	https://www.absurgery.org/
American College of Rheumatology	https://www.rheumatology.org/
Centers for Disease Control and Prevention	https://www.cdc.gov/
Cleveland Clinic	https://my.clevelandclinic.org/
Coronavirus Resource Center: Johns Hopkins University and Medicine	https://coronavirus.jhu.edu/
COVID-19 Global Rheumatology Alliance	https://rheum-covid.org/
COVID-19 Information Platform: Turkish Ministry of Health	https://covid19.saglik.gov.tr
COVID.gov: Department of Health and Human Services, United States	https://www.covid.gov/
European Centre for Disease Prevention and Control	https://www.ecdc.europa.eu/
European Commission Public Health	https://health.ec.europa.eu/index_en
European Medicines Agency	https://www.ema.europa.eu
International Agency for Research on Cancer, World Health Organization	https://www.iarc.who.int/
Italian Ministry of Health	https://www.salute.gov.it/portale/home.html
Mayo Clinic	https://www.mayoclinic.org/
National Foundation for Infectious Diseases	https://www.nfid.org/
National Health Service	https://www.nhs.uk/
National Health Surveillance Agency	https://www.gov.br/anvisa/pt-br
National Institute for Health and Care Excellence	https://www.nice.org.uk/
National Institutes of Health	https://www.nih.gov/
Pan American Health Organization	https://www.paho.org/en
Portuguese Directorate-General for Health	https://www.dgs.pt/
Public Health England: GOV.UK	https://www.gov.uk/government/organisations/public-health-england
Robert Koch Institute	https://www.rki.de/EN/Home/homepage_node.html
Therapeutic Goods Administration	https://www.tga.gov.au/
US Food and Drug Administration (FDA)	https://www.fda.gov/
United Nations Office on Drugs and Crime	https://www.unodc.org/unodc/index.html

TABLE 1 **Regulatory bodies or organizations relevant to the body systems involvement and management of SARS-CoV-2 or COVID-19—cont'd**

Regulatory body or organization	Web address
United States Preventive Services	https://www.uspreventiveservicestaskforce.org/uspstf/
WHO Coronavirus (COVID-19) Dashboard	https://covid19.who.int/
World Health Organization	https://www.who.int/

This table lists the regulatory bodies and organizations relevant to the body systems involvement and management of COVID-19. The links were accurate at the time of press but may change. In these cases, the use of "search" tabs should be explored at the parent address or site. In some cases, links direct the reader to pages related to COVID-19 within parent sites. Some societies and organizations have a preference for shortened terms, such as acronyms and abbreviations. See also Table 2.

TABLE 2 **Professional societies relevant to the body systems involvement and management of SARS-CoV-2 or COVID-19.**

Society name	Web address
American Association for the Surgery of Trauma	https://www.aast.org/
American Cancer Society	https://www.cancer.org/
American College of Radiology	https://www.acr.org/
American Society for Microbiology	https://asm.org/
American Society of Hematology	https://www.hematology.org/
American Thoracic Society	https://www.thoracic.org/
Asia Pacific League of Associations for Rheumatology	https://www.aplar.org/
Associazione Nazionale Medici Cardiologi Ospedalieri	https://www.anmco.it/home
Brazilian Society of Immunizations	https://sbim.org.br/
Brazilian Society of Infectology	https://infectologia.org.br/
British Society for Rheumatology	https://www.rheumatology.org.uk/
Canadian Association of Radiologists	https://car.ca/
Canadian Society of Breast Imaging	https://csbi.ca/
European Alliance of Associations for Rheumatology	https://www.eular.org/
European Respiratory Society	https://www.ersnet.org/
European Society of Breast Imaging	https://www.eusobi.org/
European Society of Cardiology	https://www.escardio.org/
European Society of Clinical Microbiology and Infectious Diseases	https://www.escmid.org/
European Society of Radiology	https://www.myesr.org/
German Society for Rheumatology	https://dgrh.de/
Infectious Diseases Society of America	https://www.idsociety.org/
Intensive Care Portuguese Association	https://www.spci.pt/
International Society for Infectious Diseases	https://isid.org/
International Society of Radiology	https://www.isradiology.org/
International Society on Thrombosis and Hemostasis	https://www.isth.org/
International Union Against Tuberculosis and Lung Disease	https://theunion.org/
Italian Respiratory Society	https://www.sipirs.it/cms/

Continued

TABLE 2 Professional societies relevant to the body systems involvement and management of SARS-CoV-2 or COVID-19—cont'd

Society name	Web address
Italian Society of General Medicine	https://www.simg.it/
Italian Society of Infectious and Tropical Diseases	https://www.simit.org/
Portuguese Medical Association	https://ordemdosmedicos.pt
Portuguese Society of Pulmonology	https://www.sppneumologia.pt/
Society of Breast Imaging	https://www.sbi-online.org/
Thrombosis and Hemostasis Society of Australia and New Zealand	https://www.thanz.org.au/
Turkish League Against Rheumatism	https://trasd.org.tr/eng
Turkish Respiratory Society	https://www.solunum.org.tr/eng/
Turkish Thoracic Society	https://www.toraks.org.tr

This table lists the professional societies relevant to the body systems involvement and management of COVID-19. The links were accurate at the time of press but may change. In these cases, the use of "search" tabs should be explored at the parent address or site. In some cases, links direct the reader to pages related to COVID-19 within parent sites. Some societies and organizations have a preference for shortened terms, such as acronyms and abbreviations. See also Table 1.

TABLE 3 Books relevant to the body systems involvement and management of SARS-CoV-2 or COVID-19.

Book title	Authors or editors	Publisher	Year of publication
Advanced Biosensors for Virus Detection Smart Diagnostics to Combat SARS-CoV-2	Khan R, Parihar A	Elsevier	2022
Computational Approaches for Novel Therapeutic and Diagnostic Designing to Mitigate SARS-CoV2 Infection	Parihar A, Khan R, Kumar A, Kaushik A, Gohel H	Elsevier	2022
Coronavirus (COVID-19) Outbreaks, Environment and Human Behavior International Case Studies	Akhtar R	Springer	2021
Coronavirus Disease—COVID-19	Rezaei N	Springer	2021
Coronavirus Disease 2019 (COVID-19): Epidemiology, Pathogenesis, Diagnosis, and Therapeutics	Saxena SK	Springer	2020
Coronavirus Disease from Origin to Outbreak (1st Edition)	Qureshi A, Saeed O, Syed U	Elsevier	2021
Coronavirus Drug Discovery	Egbuna C	Elsevier	2022
COVID-19	Ovallath S	Nova Science Publishers	2020
COVID-19 The Essentials of Prevention and Treatment	Qu JM, Cao B, Chen RC	Elsevier	2020
COVID-19 and SARS-CoV-2 The Science and Clinical Application of Conventional and Complementary Treatments	Goswami S, Dey C	CRC Press	2022
COVID-19 by Cases: A Pandemic Review	Rajaram SS, Longo A, Burak N	Nova Science Publishers	2021
COVID-19: A Critical Care Textbook	Carter C, Notter J	Elsevier	2021
COVID-19: Current Challenges and Future Perspectives	Kumar A	Bentham books	2021

TABLE 3 Books relevant to the body systems involvement and management of SARS-CoV-2 or COVID-19—cont'd

Book title	Authors or editors	Publisher	Year of publication
COVID-19: Epidemiology, Biochemistry, and Diagnostics	Niaz K, Nisar MF	Bentham books	2021
COVID-19: From Chaos to Cure: The biology behind the fight against the novel coronavirus	Yuzuki D	Silent Valley Press	2021
How to Prevent the Next Pandemic	Bill Gates	Knopf Doubleday Publishing Group	2022
SARS-CoV-2 Origin and COVID-19 Pandemic Across the Globe	Kumar V	InTechOpen	2021
Testing and Contact Tracing for COVID-19	Batista EM	Nova Science Publishers	2020
Textbook of SARS-CoV-2 and COVID-19 Epidemiology, Etiopathogenesis, Immunology, Clinical Manifestations, Treatment, Complications, and Preventive Measures	Mani S, Weitkamp JH	Elsevier	2022
The COVID-19 Pandemic: The Deadly Coronavirus Outbreak	Dhole M, Koley TK	Taylor and Francis Group	2020
The COVID-19 Pandemic: A Global High-Tech Challenge at the Interface of Science, Politics, and Illusions	Rose K	Elsevier	2022
Tourism Safety and Security Just After COVID-19	Korstanje ME	Nova Science Publishers	2022
Vaccinology and Methods in Vaccine Research	Oli AN, Esimone C, Anagu LO, Ashfield R	Academic Press	2022

This table lists the books relevant to the body systems involvement and management of COVID-19.

TABLE 4 Emerging techniques, instruments, and analytical platforms or devices relevant to the body systems involvement and management of SARS-CoV-2 or COVID-19.

Organization or company name	Web address
COVID Data Tracker: Centers for Disease Control and Prevention	https://covid.cdc.gov/covid-data-tracker/#datatracker-home
Lazzaro Spallanzani Italian Experimental Institute	https://www.istitutospallanzani.it/en/
Mammoth Biosciences: DETECTR (CRISPR-based diagnostic system)	https://mammoth.bio/diagnostics/
Meridian Bioscience: Revogene SARS-CoV-2 real-time RT-PCR device	https://www.meridianbioscience.com/diagnostics/disease-areas/respiratory/coronavirus/
OpenSAFELY: University of Oxford	https://www.opensafely.org/
Oxford Nanopore Technologies: MinION Sequencer (nanopore sequencing technology)	https://nanoporetech.com/products/minion
ProteOn XPR36 surface plasmon resonance sensor: Bio-Rad Laboratories	https://www.bio-rad.com/en-ng/product/proteon-xpr36-protein-interaction-array-system?ID=ea380548-08ca-4b4e-896b-87e5580ac411
RainDance Technologies: RainDrop Digital PCR System	https://www.selectscience.net/products/raindrop-digital-pcr-system/?prodID=116161

Continued

TABLE 4 Emerging techniques, instruments, and analytical platforms or devices relevant to the body systems involvement and management of SARS-CoV-2 or COVID-19—cont'd

Organization or company name	Web address
Roche Portugal	https://www.corporate.roche.pt
SCIEX: TripleTOF 6600+ Mass Spectrometer	https://sciex.com/landing-pages/tripletof6600plus
Seegene	https://www.seegene.com/
Sequencing Platforms: Illumina	https://www.illumina.com/systems/sequencing-platforms.html
Sherlock Biosciences: Sherlock CRISPR SARS-CoV-2 Kit	https://sherlock.bio/wp-content/uploads/2020/06/SherlockBrochure-FP_FINAL_June-2020.pdf
Thermo Fisher Scientific: QuantStudio 3D Digital PCR System	https://www.thermofisher.com/ng/en/home/life-science/pcr/real-time-pcr/real-time-pcr-instruments/quantstudio-systems/models/quantstudio-3d-digital.html
Thermo Fisher-Applied Biosystems	https://www.thermofisher.com/pt/en/home/brands/applied-biosystems.html

This table lists some emerging technologies relevant to the body systems involvement and management of COVID-19. The links were accurate at the time of press but may change. In these cases, the use of "search" tabs should be explored at the parent address or site. In some cases, links direct the reader to pages related to COVID-19 within parent sites.

TABLE 5 Other resources of interest or relevance for healthcare professionals or patients relevant to the body systems involvement and management of SARS-CoV-2 or COVID-19.

Name of resource or Organization	Web address
British Library	https://www.bl.uk/
Characteristics of COVID-19 patients: Istituto Superiore di Sanità EpiCentro, Italy	https://www.epicentro.iss.it/en/coronavirus/sars-cov-2-analysis-of-deaths
Clinical management of COVID-19—Living Guideline: World Health Organization	https://apps.who.int/iris/bitstream/handle/10665/362783/WHO-2019-nCoV-Clinical-2022.2-eng.pdf
Coronavirus Disease 2019 (COVID-19): Treatment Guidelines	https://www.covid19treatmentguidelines.nih.gov/
Coronavirus Disease: World Health Organization	https://www.who.int/emergencies/diseases/novel-coronavirus-2019
Coronavirus Free Access Collection: Cambridge University Press	https://www.cambridge.org/core/browse-subjects/medicine/coronavirus-free-access-collection
Coronavirus: Italian Society of General Medicine	https://www.simg.it/category/coronavirus/
COVID-19 advice and services: National Health Service	https://www.nhs.uk/conditions/coronavirus-covid-19/
COVID-19 Guidelines and Resources: American Board of Surgery	https://www.absurgery.org/default.jsp?covid19_resources
COVID-19 in Journal of Thrombosis and Hemostasis	https://onlinelibrary.wiley.com/journal/15387836/covid19
COVID-19 Information: Portuguese Medical Association	https://ordemdosmedicos.pt/covid-19/
COVID-19 Integrated Surveillance Data: Istituto Superiore di Sanità EpiCentro, Italy	https://www.epicentro.iss.it/en/coronavirus/sars-cov-2-dashboard
COVID-19 Overview (The Basics): Wolters Kluwer Health	https://www.uptodate.com/contents/covid-19-overview-the-basics
COVID-19 Recommendations: Intensive Care Portuguese Association	https://www.spci.pt/covid-19
COVID-19 Repository: European Alliance of Associations for Rheumatology	https://www.eular.org/rheumatic-musculoskeletal-diseases-and-covid-19-repository-for-clinicians

TABLE 5 Other resources of interest or relevance for healthcare professionals or patients relevant to the body systems involvement and management of SARS-CoV-2 or COVID-19—cont'd

Name of resource or Organization	Web address
COVID-19 Resource Center: The Lancet	https://www.thelancet.com/coronavirus
COVID-19 Resources: American Society of Hematology	https://www.hematology.org/covid-19
COVID-19 Resources: British Society for Rheumatology	https://www.rheumatology.org.uk/covid-19
COVID-19 Response: United Nations	https://www.un.org/en/coronavirus
COVID-19 Surveillance: Istituto Superiore di Sanità EpiCentro, Italy	https://www.epicentro.iss.it/en/coronavirus/sars-cov-2-surveillance-long-term
COVID-19 Vaccine Tracker and Landscape: World Health Organization	https://www.who.int/publications/m/item/draft-landscape-of-covid-19-candidate-vaccines
COVID-19: Portuguese Society of Pulmonology	https://www.sppneumologia.pt/covid-19
COVID-19: Centers for Disease Control and Prevention	https://www.cdc.gov/coronavirus/2019-ncov/index.html
COVID-19: European Centre for Disease Prevention and Control	https://www.ecdc.europa.eu/en/covid-19
COVID-19: European Medicines Agency	https://www.ema.europa.eu/en/human-regulatory/overview/public-health-threats/coronavirus-disease-covid-19
COVID-19: Food and Drug Administration	https://www.fda.gov/emergency-preparedness-and-response/counterterrorism-and-emerging-threats/coronavirus-disease-2019-covid-19
COVID-19: Istituto Superiore di Sanità EpiCentro—Epidemiology for public health, Italy	https://www.iss.it/en/coronavirus
COVID-19: Italian Ministry of Health	https://www.salute.gov.it/portale/nuovocoronavirus/homeNuovoCoronavirus.jsp
COVID-19: Massachusetts General Hospital	https://www.massgeneral.org/education/curve
COVID-19: United Nations Office on Drugs and Crime	https://www.unodc.org/unodc/en/covid-19.html
Education: International Society for Infectious Disease	https://isid.org/education/
Epidemiology for Public Health: EpiCentro—Istituto Superiore di Sanità	https://www.epicentro.iss.it/en/coronavirus/sars-cov-2-integrated-surveillance-data
PMC COVID-19 Collection: National Library of Medicine	https://www.ncbi.nlm.nih.gov/pmc/about/covid-19/
Public Health Surveillance for COVID-19: WHO	https://www.who.int/publications/i/item/WHO-2019-nCoV-SurveillanceGuidance-2022.2
SARS-CoV-2 Sequence Resources: National Library of Medicine	https://www.ncbi.nlm.nih.gov/sars-cov-2/#resources_section_1
Treatment and Management of Patients with COVID-19: IDSA Guidelines	https://www.idsociety.org/practice-guideline/covid-19-guideline-treatment-and-management/
Vaccines for COVID-19: Centers for Disease Control and Prevention	https://www.cdc.gov/coronavirus/2019-ncov/vaccines/index.html
Variants of the Virus: Centers for Disease Control and Prevention	https://www.cdc.gov/coronavirus/2019-ncov/variants/index.html
Wellcome Collection	https://wellcomecollection.org/collections
World Health Organization—Europe	https://www.who.int/europe/health-topics

This table lists some other resources relevant to the body systems involvement and management of COVID-19. The links were accurate at the time of press but may change. In these cases, the use of "search" tabs should be explored at the parent address or site. In some cases, links direct the reader to pages related to COVID-19 within parent sites.

to an evidence-based approach to the body systems involvement and management of COVID-19. Some organizations are listed in more than one table, as they occasionally fulfill several distinct roles (Rajendram et al., 2016a, 2016b, 2017).

Other resources

Other resources relevant to COVID-19 include Alzaid et al. (2015) and Rajendram et al. (Rajendram et al., 2016a, 2016b, 2017; Rajendram, Gyamfi, et al., 2024a, 2024b, 2024c).

The resources listed in these tables are included to provide general information only. This does not constitute any recommendation or endorsement of the activities of these sites, facilities, or other resources listed in this chapter by the authors or editors of this book.

Summary points

- The features of COVID-19 are usually mild and self-limiting.
- Some cases require hospitalization.
- SARS-CoV-2 may affect every body system.
- Despite several large multicenter trials, few effective treatments for COVID-19 have been identified.

This chapter lists resources relevant to the body systems involvement and management of COVID-19.

Acknowledgements

We thank the following authors for their contributions to the development of this resource. We apologize if some of the suggested material was not included in this chapter or has been moved to different sections.

Afonso Nogueira, Marta
Allievi, Sara
Amorim dos Santos, Juliana
Andrade, Maria Margarida
Ataman, Şebnem
Berktas, Bahadir M
Carvalho da Silva, Rainier
D'Angelo, Tommaso
Dey, Subo
Ducceschi, Valentino
Eker, Caglar
Elebesunu, Emmanuel
Favaloro, Emmanuel
Guerra, Eliete
Kahveci, Abdulvahap
Lazo-Langner, Alejandro
Machado, Bruna
Porres-Aguilar, Mateo
Rahman, Md Moshiur
Suki, Béla
Tari, Daniele Ugo
Zhou, James

References

Alzaid, F., Rajendram, R., Patel, V. B., & Preedy, V. R. (2015). Expanding the knowledge base in diet, nutrition and critical care. Electronic and published resources. In R. Rajendram, V. R. Preedy, & V. B. Patel (Eds.), *Diet and nutrition in critical care*. Germany: Springer.

Cheng, Y., Luo, R., Wang, K., Zhang, M., Wang, Z., Dong, L., Li, J., Yao, Y., Ge, S., & Xu, G. (2020). Kidney disease is associated with in-hospital death of patients with COVID-19. *Kidney International*, *97*(5), 829–838.

Gupta, A., Madhavan, M. V., Sehgal, K., Nair, N., Mahajan, S., Sehrawat, T. S., Bikdeli, B., Ahluwalia, N., Ausiello, J. C., Wan, E. Y., Freedberg, D. E., Kirtane, A. J., Parikh, S. A., Maurer, M. S., Nordvig, A. S., Accili, D., Bathon, J. M., Mohan, S., Bauer, K. A., … Landry, D. W. (2020). Extrapulmonary manifestations of COVID-19. *Nature Medicine*, *26*(7), 1017–1032.

Lei, F., Liu, Y. M., Zhou, F., Qin, J. J., Zhang, P., Zhu, L., Zhang, X. J., Cai, J., Lin, L., Ouyang, S., Wang, X., Yang, C., Cheng, X., Liu, W., Li, H., Xie, J., Wu, B., Luo, H., Xiao, F., … Yuan, Y. (2020). Longitudinal association between markers of liver injury and mortality in COVID-19 in China. *Hepatology (Baltimore, Md)*, *72*(2), 389–398.

Magro, P., Degli Antoni, M., Formenti, B., Viola, F., Castelli, F., Amadasi, S., & Quiros-Roldan, E. (2023). Characteristics of the population with mild COVID-19 symptoms eligible for early treatment attended in a single center in Northern Italy. *Journal of Infection and Public Health*, *16*(1), 104–106.

Paunescu, R. L., Miclutia, I. V., Verisezan, O. R., & Crecan-Suciu, B. D. (2023). Acute and long-term psychiatric symptoms associated with COVID-19. *Biomedical Reports*, *18*(1). (no pagination), Article Number: 4.

Rajendram, R. (2020). Building the house of CARDS by phenotyping on the fly. *The European Respiratory Journal*, *56*(2), 2002429.

Rajendram, R., Gyamfi, D., Patel, V. B., & Preedy, V. R. (2024a). Recommended resources for the features. In R. Rajendram, V. B. Patel, & V. R. Preedy (Eds.), *Transmission and detection of coronavirus disease 2019 (COVID-19)*. USA: Elsevier.

Rajendram, R., Gyamfi, D., Patel, V. B., & Preedy, V. R. (2024b). Recommended resources for international and life course aspects of COVID-19. In R. Rajendram, V. B. Patel, & V. R. Preedy (Eds.), *Management, body systems and case studies in COVID-19*. USA: Elsevier.

Rajendram, R., Gyamfi, D., Patel, V. B., & Preedy, V. R. (2024c). Recommended resources linking neuroscience and behaviour in COVID-19. In R. Rajendram, V. B. Patel, & V. R. Preedy (Eds.), *Management, linking neuroscience and behaviour in COVID-19*. USA: Elsevier.

Rajendram, R., & Hussain, A. (2021). Severe COVID-19 pneumonia complicated by cardiomyopathy and a small anterior pneumothorax. *BMJ Case Reports*, *14*(9), e245900. https://doi.org/10.1136/bcr-2021-245900.

Rajendram, R., Hussain, A., Mahmood, N., & Kharal, M. (2020). Feasibility of using a handheld ultrasound device to detect and characterize shunt and deep vein thrombosis in patients with COVID-19: An observational study. *The Ultrasound Journal*, *12*(1), 49.

Rajendram, R., Hussain, A., Mahmood, N., & Via, G. (2021). Dynamic right-to-left interatrial shunt may complicate severe COVID-19. *BMJ Case Reports*, *14*(10), e245301. https://doi.org/10.1136/bcr-2021-245301.

Rajendram, R., Kharal, G. A., Mahmood, N., Puri, R., & Kharal, M. (2020a). Intensive care medicine rethinking the respiratory paradigm of COVID-19: A 'hole' in the argument. *Intensive Care Medicine*, *46*, 1496–1497. https://doi.org/10.1007/s00134-020-06102-6.

Rajendram, R., Kharal, G. A., Mahmood, N., Puri, R., & Kharal, M. (2020b). Rethinking the respiratory paradigm of COVID-19: A 'hole' in the argument. *Intensive Care Medicine*, *46*(7), 1496–1497. https://doi.org/10.1007/s00134-020-06102-6.

Rajendram, R., Kharal, G. A., & Puri, R. (2021). COVID-19 may be exacerbated by right-to-left interatrial shunt. *The Annals of Thoracic Surgery*, *111*(1), 376.

Rajendram, R., Patel, V. B., & Preedy, V. R. (2016a). Recommended resources on general aspects of biomarkers. In *Biomarkers general aspects*. Germany: Springer.

Rajendram, R., Patel, V. B., & Preedy, V. R. (2016b). Recommended resources on biomarkers in cardiovascular disease. In V. B. Patel, & V. R. Preedy (Eds.), *Biomarkers in cardiovascular disease*. Germany: Springer.

Rajendram, R., Patel, V. B., & Preedy, V. R. (2017). Recommended resources on biomarkers in kidney disease. In V. B. Patel, & V. R. Preedy (Eds.), *Biomarkers in kidney disease*. Germany: Springer.

Rajendram, R., Patel, V. B., & Preedy, V. R. (Eds.). (2024a). *Features, transmission, detection and case studies in COVID-19*. USA: Elsevier.

Rajendram, R., Patel, V. B., & Preedy, V. R. (Eds.). (2024b). *Management, body systems and case studies in COVID-19*. USA: Elsevier.

Rajendram, R., Patel, V. B., & Preedy, V. R. (Eds.). (2024c). *International and life course aspects of COVID-19*. USA: Elsevier.

Rajendram, R., Patel, V. B., & Preedy, V. R. (Eds.). (2024d). *Linking Neuroscience and behaviour in COVID-19*. USA: Elsevier.

Index

Note: Page numbers followed by *f* indicate figures and *t* indicate tables.

W

Z